Advances in Intelligent Systems and Computing

Volume 1412

AF892917

Series Editor

Janusz Kacprzyk, Systems Research Institute, Polish Academy of Sciences, Warsaw, Poland

Advisory Editors

Nikhil R. Pal, Indian Statistical Institute, Kolkata, India

Rafael Bello Perez, Faculty of Mathematics, Physics and Computing, Universidad Central de Las Villas, Santa Clara, Cuba

Emilio S. Corchado, University of Salamanca, Salamanca, Spain

Hani Hagras, School of Computer Science and Electronic Engineering, University of Essex, Colchester, UK

László T. Kóczy, Department of Automation, Széchenyi István University, Gyor, Hungary

Vladik Kreinovich, Department of Computer Science, University of Texas at El Paso, El Paso, TX, USA

Chin-Teng Lin, Department of Electrical Engineering, National Chiao Tung University, Hsinchu, Taiwan

Jie Lu, Faculty of Engineering and Information Technology, University of Technology Sydney, Sydney, NSW, Australia

Patricia Melin, Graduate Program of Computer Science, Tijuana Institute of Technology, Tijuana, Mexico

Nadia Nedjah, Department of Electronics Engineering, University of Rio de Janeiro, Rio de Janeiro, Brazil

Ngoc Thanh Nguyen, Faculty of Computer Science and Management, Wrocław University of Technology, Wrocław, Poland

Jun Wang, Department of Mechanical and Automation Engineering, The Chinese University of Hong Kong, Shatin, Hong Kong

The series "Advances in Intelligent Systems and Computing" contains publications on theory, applications, and design methods of Intelligent Systems and Intelligent Computing. Virtually all disciplines such as engineering, natural sciences, computer and information science, ICT, economics, business, e-commerce, environment, healthcare, life science are covered. The list of topics spans all the areas of modern intelligent systems and computing such as: computational intelligence, soft computing including neural networks, fuzzy systems, evolutionary computing and the fusion of these paradigms, social intelligence, ambient intelligence, computational neuroscience, artificial life, virtual worlds and society, cognitive science and systems, Perception and Vision, DNA and immune based systems, self-organizing and adaptive systems, e-Learning and teaching, human-centered and human-centric computing, recommender systems, intelligent control, robotics and mechatronics including human-machine teaming, knowledge-based paradigms, learning paradigms, machine ethics, intelligent data analysis, knowledge management, intelligent agents, intelligent decision making and support, intelligent network security, trust management, interactive entertainment, Web intelligence and multimedia.

The publications within "Advances in Intelligent Systems and Computing" are primarily proceedings of important conferences, symposia and congresses. They cover significant recent developments in the field, both of a foundational and applicable character. An important characteristic feature of the series is the short publication time and world-wide distribution. This permits a rapid and broad dissemination of research results.

Indexed by DBLP, INSPEC, WTI Frankfurt eG, zbMATH, Japanese Science and Technology Agency (JST).

All books published in the series are submitted for consideration in Web of Science.

For proposals from Asia please contact Aninda Bose (aninda.bose@springer.com).

More information about this series at https://link.springer.com/bookseries/11156

Debasis Giri · Kim-Kwang Raymond Choo ·
Saminathan Ponnusamy · Weizhi Meng ·
Sedat Akleylek · Santi Prasad Maity
Editors

Proceedings of the Seventh International Conference on Mathematics and Computing

ICMC 2021

Springer

Editors
Debasis Giri
Department of Information Technology
Maulana Abul Kalam Azad University of Technology
Haringhata, India

Saminathan Ponnusamy
Department of Mathematics
Indian Institute of Technology Madras
Chennai, India

Sedat Akleylek
Department of Computer Engineering
Ondokuz Mayis University
Atakum, Turkey

Kim-Kwang Raymond Choo
The University of Texas at San Antonio
San Antonio, TX, USA

Weizhi Meng
Technical University of Denmark
Kongens Lyngby, Denmark

Santi Prasad Maity
Department of Information Technology
Indian Institute of Engineering Science and Technology
West Bengal, India

ISSN 2194-5357 ISSN 2194-5365 (electronic)
Advances in Intelligent Systems and Computing
ISBN 978-981-16-6889-0 ISBN 978-981-16-6890-6 (eBook)
https://doi.org/10.1007/978-981-16-6890-6

© The Editor(s) (if applicable) and The Author(s), under exclusive license to Springer Nature Singapore Pte Ltd. 2022
This work is subject to copyright. All rights are solely and exclusively licensed by the Publisher, whether the whole or part of the material is concerned, specifically the rights of translation, reprinting, reuse of illustrations, recitation, broadcasting, reproduction on microfilms or in any other physical way, and transmission or information storage and retrieval, electronic adaptation, computer software, or by similar or dissimilar methodology now known or hereafter developed.
The use of general descriptive names, registered names, trademarks, service marks, etc. in this publication does not imply, even in the absence of a specific statement, that such names are exempt from the relevant protective laws and regulations and therefore free for general use.
The publisher, the authors and the editors are safe to assume that the advice and information in this book are believed to be true and accurate at the date of publication. Neither the publisher nor the authors or the editors give a warranty, expressed or implied, with respect to the material contained herein or for any errors or omissions that may have been made. The publisher remains neutral with regard to jurisdictional claims in published maps and institutional affiliations.

This Springer imprint is published by the registered company Springer Nature Singapore Pte Ltd.
The registered company address is: 152 Beach Road, #21-01/04 Gateway East, Singapore 189721, Singapore

Committee

Chief Patron

Parthasarathi Chakrabarti, Director, Indian Institute of Engineering Science and Technology (IIEST) Shibpur

Patron

Arindam Biswas, Dean—International Relations & Alumni Affairs, IIEST Shibpur
Subhasis Bhaumik, Dean—Research & Consultancy, IIEST Shibpur
Sudip Kumar Roy, Dean—Academic, IIEST Shibpur

General Co-chairs

P. K. Saxena, Scientific Consultant—Cyber Security & Former Director, SAG, DRDO
P. D. Srivastava, Head of Mathematics Department, Indian Institute of Technology Bhilai
Sukanta Das, Department of Information Technology, IIEST Shibpur

Programme Co-chairs

Debasis Giri, Department of Information Technology, Maulana Abul Kalam Azad University of Technology
Kim-Kwang Raymond Choo, The University of Texas at San Antonio, USA

S. Ponnusamy, Department of Mathematics, IIT Madras
Weizhi Meng, Technical University of Denmark, Kongens Lyngby, Denmark
Sedat Akleylek, Ondokuz Mayıs University, Samsun, Turkey
Santi Prasad Maity, Department of Information Technology, IIEST Shibpur

Organizing Chair

Shyamalendu Kandar, Department of Information Technology, IIEST Shibpur

Organizing Co-chairs

Ruchira Naskar, Department of Information Technology, IIEST Shibpur
Sarita Ojha, Department of Mathematics, IIEST Shibpur

Organizing Committee

Chandan Giri, Department of Information Technology, IIEST Shibpur
Debasis Mitra, Department of Electronics and Telecommunication Engineering, IIEST Shibpur
Hafizur Rahaman, Department of Information Technology, IIEST Shibpur
Indrajit Bannerjee, Department of Information Technology, IIEST Shibpur
Prasun Ghosal, Department of Information Technology, IIEST Shibpur
Surajit Kumar Roy, Department of Information Technology, IIEST Shibpur
Tuhina Samanta, Department of Information Technology, IIEST Shibpur

Technical Program Committee

Committee for Computing

Abul Hasnat, Government College of Engineering And Textile Technology, Berhampore, India
Akram Zeki, International Islamic University, Malaysia
Aleena Swetapadma, Kalinga Institute of Industrial Technology, India
Alfredo Rial, University of Luxembourg, Luxembourg
Amit Phadikar, MCKV Institute of Engineering, India
Amitava Nag, Central Institute of Technology, Kokrajhar, India

Anirban Sarkar, National Institute of Technology Durgapur, India
Aniruddha Nag, Computer Society of India, India
Ankita Pramanik, Indian Institute of Engineering Science and Technology Shibpur, India
Anurag Singh, National Institute of Technology Delhi, India
Anwesha Mukherjee, West Bengal University of Technology, India
Archana Balmik, National Institute of Technology Rourkela, India
Arif Ahmed, University of Norway, Tromsø, Norway
Arindam Biswas, Indian Institute of Engineering Science and Technology Shibpur, India
Ashok Kumar Das, International Institute of Information Technology Hyderabad, India
Ashok Kumar Mehta, National Institute of Technology Jamshedpur, India
Asish Bera, Edge Hill University, United Kingdom
Asit Das, Indian Institute of Engineering Science and Technology Shibpur, India
Avishek Adhikari, Presidency University Kolkata, India
Axel Küpper, Technische Universität Berlin, Germany
Baisakhi Chakraborty, National Institute of Technology Durgapur, India
Bibhas Chandra Dhara, Jadavpur University, India
Bidyut Kumar Patra, National Institute of Technology Rourkela, India
Biswapati Jana, Vidyasagar University, India
Bivas Mitra, Indian Institute of Technology Kharagpur, India
Bok-Min Goi, Universiti Tunku Abdul Rahman, Malaysia
Bubu Bhuyan, North-Eastern Hill University, Shillong, India
Cem Taşkın, Trakya University, Turkey
Chanchal Patra, Maulana Abul Kalam Azad University of Technology, India
Chandan Giri, Indian Institute of Engineering Science and Technology Shibpur, India
Chandra Prakash, National Institute of Technology Delhi, India
Chandrasekhar Azad, National Institute of Technology Jamshedpur, India
Chinmoy Ghorai, Siksha 'O' Anusandhan University, India
Christina Boura, Université de Versailles Saint-Quentin-en-Yvelines, France
Chunpeng Ge, Nanjing University of Aeronautics and Astronautics, China
D P Dogra, Indian Institute of Technology Bhubaneswar, India
Danish Ali Khan, National Institute of Technology Jamshedpur, India
Debajyoty Banik, Kalinga Institute of Industrial Technology, India
Debashis De, West Bengal University of Technology, India
Debashis Nandi, National Institute of Technology Durgapur, India
Debasis Giri, Maulana Abul Kalam Azad University of Technology, India
Debbrota paul chowdhury, National Institute of Technology Rourkela, India
Debdatta Kandar, North-Eastern Hill University, Shillong, India
Deepak Singh Tomar, Maulana Azad National Institute of Technology, India
Dhananjoy Dey, Defence Research and Development Organisation, Delhi, India
Diganta Saha, Jadavpur University, India
Dilip Kumar, National Institute of Technology Jamshedpur, India
Dilip Kumar Yadav, National Institute of Technology Jamshedpur, India

Dipak Jana, Haldia Institute of Technology, India
Dipanwita Roy Chowdhury, Indian Intitute of Technology Kharagpur, India
Durmus Ozkan Sahin, Ondokuz Mayıs University, Turkey
Ecir Uğur Küçüksille, Süleyman Demirel University, Turkey
Emel Aşıcı, Karadeniz Technical University, Turkey
Engin Şahin, Çanakkale Onsekiz Mart University, Turkey
Erdal Kılıç, Ondokuz Mayıs University, Turkey
Erdinç Uzun, Tekirdağ Namık Kemal Üniversitesi, Turkey
Ergin Bayram, Ondokuz Mayıs University, Turkey
Erkan Özhan, Tekirdağ Namık Kemal University
Fagen Li, University of Electronic Science and Technology of China, China
Fahreddin Abdullayev, Kyrgyz-Turkey Manas University (Mersin University), Kyrgyzstan
Fatih Ertam, Fırat University, Turkey
Francisco Rodríguez Henríquez, CINVESTAV-IPN, Mexico
Gerardo Pelosi, Politecnico di Milano, Italy
Girraj Kumar Verma, Amity School of Engineering and Technology, Amity University, India
Gitanjali Chandwani Manocha, Thapar Institute of Engineering and Technology, Patiala, India
Han, Jinguang, Queen's University, Belfast, United Kingdom
He Debiao, Wuhan University, China
Hirak Maity, College of Engineering and Management, Kolaghat, India
Hiranmoy Roy, RCC Institute of Information Technology, India
İhsan Yilmaz, Çanakkale Onsekiz Mart University, Turkey
Ilker Ozcelik, Recep Tayyip Erdoğan University, Turkey
Imon Mukherjee, Indian Institute of Information Technology Kalyani, India
Indivar Gupta, Defence Research and Development Organisation, Delhi, India
Ismail Öztel, Sakarya University, Turkey
Jamimamul Bakas, National Institute of Technology Rourkela, India
Jaydeb Bhaumik, Jadavpur University, India
Jaydeep Howlader, National Institute of Technology Durgapur, India
Jayeeta Chakraborty, National Institute of Technology Rourkela, India
Jeevithashree D. V., Indian Institute of Science Bangalore, India
Jianting Ning, National University of Singapore, Singapore
Jorge Sá Silva, Universidade de Coimbra, Portugal
Kamal Kumar, National Institute of Technology Uttarakhand, India
Kamalika Bhattacharjee, National Institute of Technology Trichy, India
Kartick Chandra Mondal, Jadavpur University
Kemal Akyol, Kastamonu Üniversitesi, Turkey
Kim-Kwang Raymond Choo, The University of Texas at San Antonio, United States
Korhan Cengiz, Trakya University, Turkey
Kouichi Sakurai, Kyushu University, Japan
Koushlendra Kumar Singh, National Institute of Technology Jamshedpur, India
Krishan Kumar, National Institute of Technology Uttarakhand, India

Lalit Awasthi, National Institute of Technology Hamirpur, India
Liming Fang, Nanjing University of Aeronautics and Astronautics, China
Malay Kule, Indian Institute of Engineering Science and Technology Shibpur, India
Manasi Gyanchandani, Maulana Azad National Institute of Technology, India
Maroti Deshmukh, National Institute of Technology Uttarakhand, India
Marwa Qaraqe, Hamad Bin Khalifa University, Qatar
Maurantonio Caprolu, Hamad Bin Khalifa University, Qatar
Md Azharuddin, Aliah University, India
Mehmet Zübeyir Ünlü, İzmir Institute of Technology, Turkey
Mousumi Dutt, St Thomas' College of Engineering and Technology, India
Muharrem Tolga Sakallı, Trakya University, Turkey
Munmun Bhattacharya, Jadavpur University, India
Nandadulal Jana, National Institute of Technology Durgapur, India
Narasimha Shashidhar, Sam Houston State University, United States
Nashreen Nesa, Aliah University, India
Neetesh Saxena, Cardiff University, United Kindom
Neminath Hubballi, Indian Institute of Technology Indore, India
Niladri Puhan, Indian Institute of Technology Bhubaneswar, India
Nimisha Ghosh, Siksha 'O' Anusandhan University, India
Niranjan R. Chougala, Biluru Gurubasav Mahaswamiji Institute of Technology, Mudhol, India
Niyati Baliyan, Indian Institute of Technology Roorkee, India
Oishila Bandyopadhyay, Indian Institute of Engineering Science and Technology Shibpur, India
P. K. Saxena, Defence Research and Development Organisation, Delhi, India
Pawan Kumar Singh, Jadavpur University, India
Peng Xu Huazhong, University of Science and Technology, China
Prasenjit Chanak, Indian Institute of Technology (BHU), India
Prasun Ghosal, Indian Institute of Engineering Science and Technology Shibpur, India
Pratyay Kuila, National Institute of Technology Sikkim, India
Pravati Swain, National Institute of Technology Goa, India
Priyajit Biswas, Indian Institute of Engineering Science and Technology Shibpur, India
Rafał Scherer, Częstochowa University of Technology, Poland
Rama Syamala Sreepada, National Institute of Technology Rourkela, India
Ranbir Sanasam, Indian Institute of Technology Guwahati, India
Ranjan Mondal, Indian Statistical Institute Kolkata, India
Ranjit Ghosal, St Thomas' College of Engineering and Technology, India
Ratnadip Dey, Jadavpur University, India
Ravi Kanth Asv, Vellore Institute of Technology, India
Rik Chattodadhyay, Indian Institute of Engineering Science and Technology Shibpur, India
Ruchira Naskar, Indian Institute of Engineering Science and Technology Shibpur, India

S. Umer, Aliah University, India
Sabyasachi Dutta, University of Calgary, Canada
Sabyasachi Mukherjee, Indian Institute of Engineering Science and Technology Shibpur, India
Safique Ahmad, Indian Institute of Technology Indore, India
Saibal Pal, Defence Research and Development Organisation, Delhi, India
Saikat Chakraborty, National Institute of Technology Rourkela, India
Sameerudaya Venkata, National Institute of Technology Rourkela, India
Saminathan Ponnusamy, Indian Institute of Technology Madras, India
Samit Biswas, Indian Institute of Engineering Science and Technology Shibpur, India
Samrat Kumar Dey, Dhaka International University, Bangladesh
Sanjay Kumar, National Institute of Technology Jamshedpur, India
Sanjoy Saha, Jadavpur University, India
Sanjukta Bhowmick, University of Nebraska, Omaha, United States
Sankar Kumar Mridha, Indian Institute of Engineering Science and Technology Shibpur, India
Sankhayan Choudhury, University of Calcutta, India
Santi Prasad Maity, Indian Institute of Engineering Science and Technology Shibpur, India
Santoshinee Mohapatra, National Institute of Technology Rourkela, India
Saravanan Chandran, National Institute of Technology Durgapur, India
Saru Kumari, Chaudhary Charan Singh University Meerut, India
Sedat Akleylek, Ondokuz Mayıs University, Turkey
Sezer Ulukaya, Boğaziçi University, Turkey
Shehzad Ashraf Chaudhry, International Islamic University, Islamabad, Pakistan
Sherali Zeadally, University of Kentucky, United States
Shreeya Swagatika Sahoo, National Institute of Technology Rourkela, India
Shyamalendu Kandar, Indian Institute of Engineering Science and Technology Shibpur, India
S. K. Hafizul Islam, Indian Institute of Information Technology Kalyani, India
Sokratis Katsikas, Open University of Cyprus, Greece
SomeNath Bera, Kalinga Institute of Industrial Technology, India
Soubhik Chakraborty, Birla Institute of Technology Mesra, India
Soumya Sen, University of Calcutta, India
Sourav Mandal, Xavier University Bhubaneswar, India
Souvik Roy, Indian Institute of Engineering Science and Technology Shibpur, India
Souvik Sengupta, Aliah University, India
Sri Khetwat Saritha, Maulana Azad National Institute of Technology, India
Srinivasu Bodapati, Indian Institute of Technology Mandi, India
Stephan Neumann, Landesbank Saar (SaarLB), Germany
Subhankar Chatterjee, Indian Institute of Technology Delhi, India
Subhash Bhalla, University of Aizu in Aizuwakamatsu, Japan
Sudakshina Dutta, Indian Institute of Technology Goa, India
Sudeshna Sil Kar, Case Western Reserve University, Ohio, United States
Sujata Pal, Indian Institute of Technology Kharagpur, India

Sujit Kumar Das, Jadavpur University, India
Sukanta Das, Indian Institute of Engineering Science and Technology Shibpur, India
Sumanta Hati, Indian Institute of Engineering Science and Technology Shibpur, India
Sumanta Ray, Aliah University, India
Sumanto Dutta, National Institute of Technology Rourkela, India
Sumit Adak, Indian Institute of Engineering Science and Technology Shibpur, India
Sumit Hazra, National Institute of Technology Rourkela, India
Sumit Pal, Indian Institute of Technology Kharagpur, India
Sunirmal Khatua, University of Calcutta, India
Surajeet Ghosh, Indian Institute of Engineering Science and Technology Shibpur, India
Tanmoy Chakraborty, Indian Institute of Technology Delhi, India
Tanmoy Maitra, Kalinga Institute of Industrial Technology, India
Tapas Bhadra, Aliah University, India
Tapasi Bhattacharyya, Techno India University, India
Tarun Biswas, National Institute of Technology Sikkim, India
Taymaz Farshi, Ayvansaray University, Turkey
Thoudam Doren Singh, National Institute of Technology Silchar, India
Tuhina Samanta, Indian Institute of Engineering Science and Technology Shibpur, India
Turgay Yalcin, Samsun University, Turkey
Vijay Bhaskar Semwal, Maulana Azad National Institute of Technology, India
Vinay Kumar, National Institute of Technology Jamshedpur, India
Vipindev Adat, National Institute of Technology Kurukhetra, India
Vishal Saraswat, Robert Bosch Engineering and Business Solutions, India
Vivekananda Bhat K, Manipal Institute of Technology, India
Weizhi Meng, Technical Universtiy of Denmark, Denmark
Wenjuan Li, City University of Hong Kong, Hong Kong
Yannis Stamatiou, University of Patras, Greece
Yogendra Narain Singh, APJ Abdul Kalam Technical University, India
Zinnet Duygu Akşehir, Ondokuz Mayıs University, Turkey
Zulfukar Saygi, TOBB University of Economics and Technology, Turkey

Committee for Mathematics:

A. K. Dhar, Indian Institute of Engineering Science and Technology Shibpur, India
Abhijit Barman, National Institute of Technology Silchar, India
Abhishek Banerjee, Israel Institute of Technology, Israel
Ali Akbar Shaikh, Burdwan University, India
Ameeya Nayak, Indian Institute of Technology Roorkee, India
Amit Maji, Indian Institute of Technology Roorkee, India
Amit Verma, Indian Institute of Technology Patna, India
Angsuman Das, Presidency University, India

Anuradha Banerjee, Indian Institute of Technology (BHU), India
Ashish Awasthi, National Institute of Technology Calicut, India
Asoke Kumar Bhunia, Burdwan University, India
Atanu manna, Indian Institute of Carpet Technology Bhadohi, India
Atasi Deb Ray, University of Calcutta, India
Basudeb Ghosh, Jadavpur University, India
Bhagwat Ram, Indian Institute of Technology (BHU), India
Binayak S Choudhury, Indian Institute of Engineering Science and Technology Shibpur, India
Binod Chandra Tripathy, Tripura University, India
Bipan Hazarika, Gauhati University, India
Biswajit Mitra, Burdwan University, India
Debjani Chakraborty, Indian Institute of Technology Kharagpur, India
Deepmala, Indian Statistical Institute Kolkata, India
Dipak Jana, Haldia Institute of Technology, India
Dipesh Barman, Indian Institute of Engineering Science and Technology Shibpur, India
Dung Hoang Duong, University of Wollongong, Australia
Fahreddin Abdullayev, Kyrgyz-Turkish Manas University, Kyrgyzstan
Falguni Roy, National Institute of Technology Surathkal, India
Farida Parvez Barbhuiya, BITS Pilani, Hyderabad, India
Fidan Nuriyeva, Ege University, Turkey
Fouzul Atik, SRM University, India
Ganesh Ghorai, Vidyasagar University, India
Gopal Chandra Shit, Jadavpur University, India
Guru Prasad Samanta, Indian Institute of Engineering Science and Technology Shibpur, India
Hakan Kutucu, Karabak University, Turkey
Hari Shankar Prasad, National Institute of Technology Jamshedpur, India
Heinrich Begehr, Free University Berlin, Germany
Izhar Uddin, Jamia Milia Islamia University, India
Jayalakshmi Karamsi, Jawaharlal Nehru Technological University Anantapur, India
Jessica Pereira, Goa University, India
Konstantin Volkov, Kingston University, Volkov
Krishnendu Barman, Vidyasagar University, India
Kuntal Banerjee, Presidency University, India
Lakshmi Kanta Patra, Indian Institute of Technology Ranchi, India
Madan Mohan Panja, Visva-Bharati University, India
Madhumangal Pal, Vidyasagar University, India
Manotosh Kumbhakar, Texas A & M University, India
Mantu Saha, Burdwan University, India
Masaya Yasuda, Kyushu University, Japan
Md Firoz Ali, National Institute of Technology Durgapur, India
Md Nasim Akhtar, Presidency University, India
Mikheil Rukhaia, Institute of Applied Mathematics, Tbilisi State University, Georgia

Moumita Mandal, Indian Institute of Technology Jodhpur, India
Nabakumar Jana, Indian Institute of Technology (ISM) Dhanbad, India
Naren Bag, DWI—Leibnitz Institute of Interactive Materials, RWTH Aachen University, Germany
Nipen Saikia, Rajiv Gandhi University of Health Sciences, India
Önder Türk, Gebze Technical University, Turkey
Osmita Dey, Techno India University, India
P. D. Srivastava, Indian Institute of Technology Bhilai, India
Partha Sarathi Majee, Indian Institute of Technology Kharagpur, India
Paulo Mateus, Instituto Superior Tecnico, Portugal
Pawan Kumar Mishra, Indian Institute of Technology Bhilai, India
Payel Mandal, Techno India University, India
Prabir Kumar Kundu, Jadavpur University, India
Prabir Panja, Haldia Institute of Technology, India
Prakash Goswami, Indian Institute of Technology (BHU), India
Prasanta Malik, Burdwan University, India
Pratap Chandra Ray, Bethune College, India
Pratulananda Das, Jadavpur University, India
Praveen Kumar Gupta, National Institute of Technology Silchar, India
Pritha Das, Indian Institute of Engineering Science and Technology Shibpur, India
Priyanka Roy, Indian Institute of Technology Kharagpur, India
Purushottam Narain Agarwal, Indian Institute of Technology Roorkee, India
Raju Roy, Gauhati University, India
Ranjit Mehtari, National Institute of Technology Rourkella
Rifat Çolak, Firat University, Turkey
Rudrajit Sadhu, Sreegopal Banerjee College, India
Safique Ahmad, Indian Institute of Technology Indore, India
Saifur Rahman, Rajiv Gandhi University of Health Sciences, India
Sajal Mukhopadhyay, National Institute of Technology Durgapur, India
Samarjit Kar, National Institute of Technology Rourkela, India
Samiran Ghosh, University of Calcutta, India
Sanat Kumar Majumdar, Indian Institute of Engineering Science and Technology Shibpur, India
Sanjib Kumar Pal, Kazi Nazrul University, India
Sankar Kumar Roy, Vidyasagar University, India
Santanu Manna, Indian Institute of Technology Indore, India
Sarita Ojha, Indian Institute of Engineering Science and Technology Shibpur, India
Satya Bagchi, National Institute of Technology Durgapur, India
Saumya Bajpai, Indian Institute of Technology Goa, India
Shanmugam Dhinakaran, Indian Institute of Technology Indore, India
Shubhabrata Das, Presidency University, India
Shubhankar Saha, Sir Gurudas Mahavidyalaya, India
Shyamal Kumar Hui, Burdwan University, India
Shyamal Kumar Mondal, Vidyasagar University, India
Smriti Nath, Indian Institute of Technology Kharagpur, India

Snehasis Kundu, National Institute of Technology Jamshedpur, India
Somesh Kumar, Indian Institute of Technology Kharagpur, India
Soovoojeet Jana, Ramsaday College, India
Sourav Gupta, Indian Institute of Technology Kharagpur, India
Sourav Pradhan, National Institute of Technology Nagpur, India
Sree Vivek S, School of Civil Engineering, SASTRA Deemed University, Thanjavur, India
Sriparna Chattopadhyay, National Institute of Foundry and Forge Technology Ranchi, India
Subhra Bhattacharya, Presidency University, India
Suchandan Kayal, National Institute of Technology Rourkela, India
Sujit Kumar Sardar, Jadavpur University, India
Sukhendu Kar, Jadavpur University, India
Suman Rakshit, Mankar College, India
Sushil Kumar Bhuiya, Krishnath College, India
Suvra Kanti Chakraborty, Sir Gurudas Mahavidyalaya, India
Svetla Petkova-Nikova, ESAT-COSIC, KU Leuven, Belgium
Swadesh Kumar Sahoo, Indian Institute of Technology Indore, India
Vikas Kumar, Indian Institute of Technology Mandi, India
Vilem Novak, University of Ostrava, Czechia
Vishnu Pratap Singh, Visvesvaraya National Institute of Technology Nagpur, India

Preface

In the last two decades, scientific computing has become an important contributor to all scientific research programs. It is particularly important for the solution of research problems that are unsolvable by traditional theory and experimental approaches, hazardous to study in the laboratory, or time-consuming or expensive to be solved by traditional means. With mathematical modeling and computational algorithms, many more problems from the realm of science, commerce as well as other walks of life can be solved efficiently. The International Conference on Mathematics and Computing (ICMC) is such a premier forum for the presentation of new advances and research results in the fields of Cryptography, Network Security, Cybersecurity, Internet of Things, Due & Edge computing, Mathematics, Statistics and Scientific Computing. The conference will bring together leading academic scientists, experts from industry, and researchers in their domains of expertise from around the world. Earlier ICMC was organized in 2013, 2015, 2017, 2018, 2019, and 2020.

The 7th ICMC 2021 was aimed to bring together both novice and experienced scientists with developers, to meet new colleagues, collect new ideas, and establish new cooperation between research groups and provide a platform for researchers from academic and industry to present their original work and exchange ideas, information, techniques, and applications in the field of Computational Applied Mathematics, including, but not limited to the broad topics of Operations Research, Soft Computing, Cryptology, Network Security, Cybersecurity, Internet of Things, Due & Edge computing, Image Processing, Pure and Applied Mathematics, and other emerging areas of research.

7th ICMC 2021 was organized by the Department of Information Technology, Indian Institute of Engineering Science and Technology, Shibpur, India, and ICMC 2021 was honored to have Ramanujan Mathematical Society (RMS) and Cryptology Research Society India (CRSI) as its associates in organizing the conference. This conference received 366 papers from different parts of India as well as world. This book of abstracts contains 81 full papers which fall under mathematics and computing tracks. We would like to thank the directors, deans, and officials of IIEST Shibpur, general co-chairs, organizing committee members, volunteers, sponsors, attendees, and authors who attended and made this conference a success. ICMC 2021 proceedings will be published in the Springer series: Advances in Intelligent Systems and Computing.

Haringhata, India
San Antonio, USA
Chennai, India
Kongens Lyngby, Denmark
Samsun, Turkey
West Bengal, India

Debasis Giri
Kim-Kwang Raymond Choo
Saminathan Ponnusamy
Weizhi Meng
Sedat Akleylek
Santi Prasad Maity

Message from General Co-Chairs

It gives me a great pleasure in welcoming you to ICMC 2021, the 7th edition of the premier annual conference on Mathematics and Computing. This year, ICMC was held in the Department of Information Technology, Indian Institute of Engineering Science and Technology, Shibpur, India, in association with Ramanujan Mathematical Society (RMS) and Cryptology Research Society India (CRSI), where advanced technology infrastructure and collection of creative talents are gathered, making it an excellent position to develop advanced Computing. ICMC has been the most impactful conference on all aspects of Mathematics and Computing. Papers published from this impactful conference represent the hard work of many outstanding researchers from around the world. We are very delighted to report that ICMC remains at the forefront of computer networks. This year, the main conference embodied a set of 81 papers out of 366 presentations, selected through careful and rigorous peer review by TPC members as the best of the best submissions, and organized around 2 tracks and 13 keynote sessions. It has been our honor to have the most prominent scholars, as our conference invitees, Prof. R. Balasubramanian (Institute of Mathematical Sciences Chennai, India); Prof. Bimal Roy (Indian Statistical Institute Kolkata, India); Prof. Bhavani Thuraisingham (The University of Texas at Dallas, USA); Prof. Carmit Hazay (Bar-Ilan University, Israel); Prof. Aikaterini Mitrokotsa (Chalmers University of Technology, Sweden); Prof. Svetla Petkova-Nikova (ESAT-COSIC, KU Leuven, Belgium); Prof. S. Ponnusamy (Indian Institute of Technology Madras, India); Prof. Sandeep K Shukla (Indian Institute of Technology Kanpur, India); Prof. Sugata Gangopadhyay (Indian Institute of Technology Roorkee, India); Prof. Francisco Rodríguez Henríquez (CINVESTAV-IPN, Mexico City, Mexico); Prof. Ville Vuorinen (Aalto University, Finland); Prof. Mohamed M S Nasser (Qatar University, Qatar); and Prof. Harri Hakula (Aalto University, Finland).

First, we would like to thank to all program members and organizing committee members who have done an outstanding job in carrying out the paper review tasks. In particular, we would like to express our appreciation to Dr. Debasis Giri, Dr. Shyamalendu Kandar, and his team members who have spent great efforts in developing the review system. We would also like to thank Honourable Director of Indian Institute of Engineering Science and Technology, Shibpur, India, for his support and for providing Excellent Venue. Last but not the least, we thank our patrons and sponsors, especially the Ramanujan Mathematical Society, India, for their warm support. Finally, we thank you all conference participants for making ICMC a success.

P. K. Saxena
Scientific Consultant—Cyber
Security & Former Director, SAG,
DRDO

P. D. Srivastava
Indian Institute of Technology Bhilai
Sejbahar, India

Sukanta Das
IIEST Shibpur
Howrah, India

Message from Program Co-Chairs

It was a great pleasure for us to organize the 7th International Conference on Mathematics and Computing 2021 held from March 02 to 05, 2021, at the Indian Institute of Engineering Science and Technology, Shibpur, India. Our main goal was to provide an opportunity to the participants to learn about contemporary research in Mathematics and Computing, and exchange ideas among themselves and with experts present in the conference as keynote speakers.

Thirteen speakers from India and abroad delivered their talks and some of them acted as session chairs. After an initial call for papers, 366 papers were presented at the conference. All submitted papers were sent to external referees and, after refereeing, 81 papers were recommended for publication for the conference proceedings that will be published by Springer series: Advances in Intelligent Systems and Computing.

ICMC 2021 has come up as an international platform to deliver and share novel knowledge in various fields on applied mathematics and computing of interest.

We are grateful to the chief patron, patron, general co-chairs, program co-chairs, organizing chair, speakers, participants, referees, organizers, sponsors, and funding agencies for their support and help without which it would have been impossible to organize this conference. We owe our gratitude to the volunteers who worked behind the scene tirelessly in taking care of the details in making this conference a great success.

Haringhata, India — Debasis Giri
San Antonio, USA — Kim-Kwang Raymond Choo
Chennai, India — Saminathan Ponnusamy
Kongens Lyngby, Denmark — Weizhi Meng
Samsun, Turkey — Sedat Akleylek
West Bengal, India — Santi Prasad Maity

Contents

A Blind Signature Scheme Based on Bilinear Pairings

Mohan Chintamani and Laba Sa

Abstract Bilinear pairings have been used to design protocols such as identity-based encryption, one-round three-party key agreement, and aggregate signatures. Several authors have proposed many secret sharing schemes using elliptic curves and bilinear pairings. In 2004, Lee introduced bilinear self-pairing. Blind signature schemes have been the most significant obligation since the beginning of electronic money. In this article, we propose a blind signature scheme based on elliptic curves and a slightly modified Lee's self-pairing. We provide an explicit example for the illustration purpose. We also show that the signature scheme avoids well-known attacks by M. Tian, Y. Zhu, and Z. Chen related to the blind signature scheme proposed by K. Chakraborty and J. Mehta. The security aspects of the proposed blind signature scheme have been discussed in this article.

Keywords Elliptic curves · Bilinear pairings · Blind signature · ECDLP

1 Introduction

The blind signature scheme has been the most significant responsibility since the beginning of the era of electronic money (e-money). It is an interdependent agreement which includes two parties, a Bank and a client. The scheme permits a client to urge a signature on a message from the Bank without disclosing the context of the message. The message–signature pair received by the client is factually uncorrelated to the view acquired by the Bank during the execution of the agreement. Chaum [4] in 1983 first proposed the blind signature scheme. Later, many blind digital signature

The Author was supported in part by the Council of Scientific and Industrial Research, India, under award letter 09/414(1146)/2017-EMR-I.

M. Chintamani · L. Sa (✉)
School of Mathematics and Statistics, University of Hyderabad, Hyderabad 500046, India

© The Author(s), under exclusive license to Springer Nature Singapore Pte Ltd. 2022

D. Giri et al. (eds.), *Proceedings of the Seventh International Conference on Mathematics and Computing*, Advances in Intelligent Systems and Computing 1412,
https://doi.org/10.1007/978-981-16-6890-6_1

schemes were proposed (see [3, 10–12, 16]). The application of the blind signature scheme is used in securing electronic payment systems, electronic voting systems, etc.

The blind digital signature scheme should satisfy the following properties:
Blindness: It permits a user to urge a signature on a given message, while not revealing the context of the message to the signer.
Untraceability: When the user has disclosed the signature to the general public, the signer isn't able to trace the signature-message.
Unforgeability: Solely, the signer will generate a legitimate signature. This property is foremost very important and should satisfy all signature schemes.
Unlinkability: Nobody can obtain a link between a legitimate blind signature and a protocol view except the author of the message.

For detailed security aspects of the blind signature scheme, we refer to [6]. An unauthorised person can verify the blind signature publicly and meet the necessities of security-oriented agreements that have an imbalance between the author of the message and the signer.

Elliptic curve discrete logarithm problem (ECDLP) is to find an integer a such that $Q = aP$ for given two points P and Q on an elliptic curve E over a finite field with $Q \in \langle P \rangle$. We assume that ECDLP is hard to solve for the security purpose of the blind signature scheme.

Hash functions are frequently a snag for security aspects. A hash function h satisfies the *pre-image resistant* property (for a given hash value m, it is computationally hard to find a message M such that $h(M) = m$) and *collision-resistant* property (it is computationally hard to find two distinct messages M_1 and M_2 such that $h(M_1) = h(M_2)$). For more details about hash functions, see [13].

In this article, we study a stamped blind signature scheme proposed by Chakraborty and Mehta [3]. It was observed by Tian, Zhu, and Chen [15] that Chakraborty–Mehta's blind signature scheme is insecure. They gave two simple but powerful attacks. In this article, we propose the modified/analogue version of the Chakraborty–Mehta scheme [3] using self-pairing, which gives more security and avoids the attacks given by Tian, Zhu, and Chen [15].

The self-pairing was first proposed by Lee [7] in 2004. We note that Lee's self-pairing is anti-symmetric. Several secret sharing schemes were introduced based on self-pairing (see [1, 8]). In this article, a slight modification of Lee's pairing is considered.

2 Preliminaries

Consider the normal form of elliptic curve $E : y^2 = x^3 + ax + b$ over $\mathbb{F}_q$ where $\mathbb{F}_q$ is the finite field of characteristic $p > 3$, p prime, and $a, b \in \mathbb{F}_q$ with the discriminant non-zero modulo p. The group $E(\mathbb{F}_q)$ is the set of points on the elliptic curve and the point at infinity $\mathcal{O}$ is the identity element of the group.

Torsion group: Let $\overline{\mathbb{F}}_q$ be the algebraic closure of $\mathbb{F}_q$. The group $E(\overline{\mathbb{F}}_q)$ is called torsion group, that is, every point P on the curve E has finite order. Let n be a prime divisor of $\#E(\mathbb{F}_q)$. The set of n-torsion points of E, denoted by $E[n]$, is defined as $E[n] = \{P \in E(\overline{\mathbb{F}}_q) : nP = \mathcal{O}\}$. If $\gcd(n, q) = 1$ then it is well known that $E[n] \cong \mathbb{Z}_n \times \mathbb{Z}_n$ (see [2, 5, 14]). The group $E[n]$ is generated by two elements, say S and T.

We consider an explicit example of the elliptic curve $E : y^2 = x^3 + 4x + 2$ over $\mathbb{F}_5$. Then $E(\mathbb{F}_5) = \{\mathcal{O}, (3, 1), (3, 4)\}$ and $\#E(\mathbb{F}_5) = 3$. Now $\gcd(3, 5) = 1$ and $3 \mid 5^2 - 1$, i.e., the embedding degree of $E(\mathbb{F}_5)$ with respect to 3 is 2. Let $\mathbb{F}_{5^2} = \mathbb{F}_5[\gamma]/(\gamma^2 + \gamma + 1)$. Then $E(\mathbb{F}_{5^2})$ and the 3-torsion points $E[3]$ are as follows:

$$\begin{aligned} E(\mathbb{F}_{5^2}) = \{&\mathcal{O}, (3, 1), (3, 4), (0, 1 + 2\gamma), (0, 4 + 3\gamma), (1, 1 + 2\gamma), (1, 4 + 3\gamma),\\ &(2, 3 + \gamma), (2, 2 + 4\gamma), (4, 1 + 2\gamma), (4, 4 + 3\gamma), (3 + 4\gamma, 1),\\ &(3 + 4\gamma, 4), (4 + \gamma, 1), (4 + \gamma, 4), (2\gamma, 3 + 2\gamma), (2\gamma, 2 + 3\gamma),\\ &(4\gamma, 2\gamma), (4\gamma, 3\gamma), (1 + \gamma, 2 + 2\gamma), (1 + \gamma, 3 + 3\gamma),\\ &(2 + \gamma, 3 + \gamma), (2 + \gamma, 2 + 4\gamma), (1 + 4\gamma, 3 + \gamma), (1 + 4\gamma, 2 + 4\gamma),\\ &(3 + 3\gamma, 4 + 2\gamma), (3 + 3\gamma, 1 + 3\gamma)\} \end{aligned}$$

and

$E[3] = \{\mathcal{O}, (3, 1), (3, 4), (1, 1 + 2\gamma), (1, 4 + 3\gamma), (1 + \gamma, 2 + 2\gamma), (1 + \gamma, 3 + 3\gamma), (4\gamma, 2\gamma), (4\gamma, 3\gamma)\}$.

Let $S = (3, 1)$ and $T = (4\gamma, 2\gamma)$. Then every point of $E[3]$ can be written as a linear combination of S and T. For example, $0S + 0T = \mathcal{O}$, $0S + 1T = (4\gamma, 2\gamma)$, $0S + 2T = (4\gamma, 3\gamma)$, $1S + 0T = (3, 1)$, $1S + 1T = (1, 4 + 3\gamma)$, $1S + 2T = (1 + \gamma, 3 + 3\gamma)$, $2S + 0T = (3, 4)$, $2S + 1T = (1 + \gamma, 2 + 2\gamma)$, $2S + 2T = (1, 1 + 2\gamma)$. Hence, we observe that $E[3] \cong \mathbb{Z}_3 \times \mathbb{Z}_3$.

A *bilinear pairing* is a map $e : G_1 \times G_2 \longrightarrow G_T$ where G_1, G_2 are additive cyclic groups and G_T is multiplicative cyclic group, all of the same prime order, satisfying some linearity properties. In the present article, we consider a modified Lee's pairing which is defined in the next section.

3 Self-Pairings

Let E be an elliptic curve over finite field $\mathbb{F}_q$ with characteristics not equal to 2 and 3. Let S, T be a pair of n-torsion points, generating $E[n]$, where n is a prime. Consider two points $P_1 = a_1S + b_1T$ and $P_2 = a_2S + b_2T$ in $E[n]$, where $a_1, a_2, b_1, b_2 \in \mathbb{Z}_n$. For some fixed $\alpha, \beta \in \mathbb{Z}_n$, we define self-pairing map

$$e_{\alpha,\beta} : E[n] \times E[n] \longrightarrow E[n]$$

as $e_{\alpha,\beta}(P_1, P_2) = (a_1a_2 - b_1b_2)(\alpha S + \beta T)$. The trivial case ($\alpha = 0 = \beta$) has been excluded.

The following theorem list some of the properties that $e_{\alpha,\beta}$ satisfies:

Theorem 1 *For all $P_1, P_2, P_3 \in E[n]$ and the point at infinity $\mathcal{O}$,*

1. *Bilinearity: $e_{\alpha,\beta}(P_1 + P_2, P_3) = e_{\alpha,\beta}(P_1, P_3) + e_{\alpha,\beta}(P_2, P_3)$ and $e_{\alpha,\beta}(P_1, P_2 + P_3) = e_{\alpha,\beta}(P_1, P_2) + e_{\alpha,\beta}(P_1, P_3)$.*
2. *Non-degeneracy: If $e_{\alpha,\beta}(P_1, P_2) = \mathcal{O}$ for all $P_2 \in E[n]$, then $P_1 = \mathcal{O}$.*
3. *Symmetry: $e_{\alpha,\beta}(P_1, P_2) = e_{\alpha,\beta}(P_2, P_1)$ for all $P_1, P_2 \in E[n]$.*

The proof of the theorem is very similar to proposition 3.1 in [7]. For the completeness of the paper, we provide the proof below.

1. For all $P_1, P_2, P_3 \in E[n]$ with $P_3 = a_3S + b_3T$, we have

$$\begin{aligned} e_{\alpha,\beta}(P_1 + P_2, P_3) &= e_{\alpha,\beta}((a_1 + a_2)S + (b_1 + b_2)T, a_3S + b_3T) \\ &= ((a_1 + a_2)a_3 - (b_1 + b_2)b_3)(\alpha S + \beta T) \\ &= (a_1a_3 - b_1b_3)(\alpha S + \beta T) + (a_2a_3 - b_2b_3)(\alpha S + \beta T) \\ &= e_{\alpha,\beta}(P_1, P_3) + e_{\alpha,\beta}(P_2, P_3). \end{aligned}$$

   Similarly $e_{\alpha,\beta}(P_1, P_2 + P_3) = e_{\alpha,\beta}(P_1, P_2) + e_{\alpha,\beta}(P_1, P_3)$.
2. For any $P_2 = a_2S + b_2T \in E[n]$, we have

$$e_{\alpha,\beta}(P_1, P_2) = (a_1a_2 - b_1b_2)(\alpha S + \beta T) = \mathcal{O}.$$

   Thus, $a_1a_2 - b_1b_2 = 0$. By choosing P_2 with $a_2 \neq 0$ and $b_2 = 0$, we obtain $a_1 = 0$. Similarly, by choosing P_2 with $a_2 = 0$ and $b_2 \neq 0$, we obtain $b_1 = 0$. Hence $P_1 = a_1S + b_1T = \mathcal{O}$.
3. For all $P_1, P_2 \in E[n]$, $e_{\alpha,\beta}(P_1, P_2) = (a_1a_2 - b_1b_2)(\alpha S + \beta T) = e_{\alpha,\beta}(P_2, P_1)$.

Additionally, the self-pairing map satisfies the following properties:
For all $P_1, P_2 \in E[n]$, we have

1. $e_{\alpha,\beta}(P_1, P_1) = \mathcal{O}$ if $a_1 = b_1$.
2. $e_{\alpha,\beta}(P_1, \mathcal{O}) = e_{\alpha,\beta}(\mathcal{O}, P_2) = \mathcal{O}$.
3. $e_{\alpha,\beta}(aP_1, bP_2) = ab\, e_{\alpha,\beta}(P_1, P_2)$ for all $a, b \in \mathbb{Z}$.

For an illustration, we consider the elliptic curve E, 3-torsion points $E[3]$, S and T as in above example. Choose $a_1 = 2, b_1 = 2, a_2 = 2, b_2 = 1$ in $\mathbb{Z}_3$ and fixed $\alpha = 1$, $\beta = 2$ in $\mathbb{Z}_3$. Then $e_{\alpha,\beta}(P_1, P_2) = e_{1,2}(2S + 2T, 2S + T) = (2 \times 2 - 2 \times 1)(1S + 2T) = 2(1S + 2T) = 2S + T = (1 + \gamma, 2 + 2\gamma)$.

4 Proposed Blind Signature Scheme

The proposed scheme involves three parties, namely Author **A**, Signer **S**, and Verifier **V**. The scheme consists of two protocols called *signing* and *verification*. The signing protocol is carried out by both **A** and **S**, while the verification protocol is checked out by **V**. **A** wants to obtain a signature on a message from the signer without revealing the context of the message. This includes blinding the message with the goal that the **S** cannot read the message. Simultaneously, **A** needs to ensure that **S** is the assigned beneficiary of the blinded message. It is accomplished by twofold blinding (i.e., putting 2 locks) of the message. One is put by **S**, and he is the one in particular who can open it. This guarantees that he is the main individual who is accepting a blinded message from **A**. Another lock is put by **A** to make sure that **S** cannot read the actual message.

Setting Up Domain Parameters
Firstly, the signer sets up the domain parameters for the scheme.

- **S** chooses an elliptic curve E over $\mathbb{F}_q$, where $q = p^r$, and p is large prime.
- **S** chooses a random generating pair (S, T) in $E[n]$, where n is a large prime divisor of $\#E(\mathbb{F}_q)$ so that ECDLP in $E(\mathbb{F}_q)$ is hard, and some fixed $\alpha, \beta \in \mathbb{Z}_n$ for which the pairing $e_{\alpha,\beta}$ can be determined.
- Finally, **S** makes $\{E, q, n, \alpha, \beta, S, T, \alpha S + \beta T\}$ as public.

It additionally utilises a collision-resistant cryptographic hash function $h : \{0, 1\}^* \longrightarrow \mathbb{Z}_n^\times$.

Signing Protocol

- The signer **S** chooses two secret keys $a, b \in \mathbb{Z}_n$ such that gcd(a, b)=1 and makes $P = aS + bT \in E[n]$ public.
- Author **A** calculates $h(M) = m$ where M is the actual message and m is the hash value.
- Then **A** computes $G = mP = maS + mbT$ and sends G to the signer for signing.
- The signer chooses $H = cS + dT \in E[n]$ such that $ac - bd = 1$.

Remark 1 The author wants to send $m(\alpha S + \beta T)$ for signing to the signer. The value $m(\alpha S + \beta T)$ can be computed from G by only the signer, the person who can find $c, d \in \mathbb{Z}_n$ such that $ac - bd = 1$.

Signing Algorithm

1. The signer **S** receives G and calculates $A = e_{\alpha,\beta}(G, H) = m(ac - bd)(\alpha S + \beta T) = m(\alpha S + \beta T)$, where $H = cS + dT$.
2. **S** generates signature parameter z (say stamp), and calculates $h(z)$.
3. **S** selects a random integer $x \in [1, n - 1]$ such that $h(z)x \neq 1$.
4. **S** calculates a point B on elliptic curve as follows:
   $B = m(\alpha S + \beta T) + e_{\alpha,\beta}(xS + T, h(z)S + T)$ and set $J = h(z)xH$.

5. **S** sends the generated signature parameters (B, J, z) to the verifier for verification.

Verification Protocol
The verifier **V** computes $B - e_{\alpha,\beta}(J, P) + (\alpha S + \beta T)$ and observes the validity of the following equation:

$$B - e_{\alpha,\beta}(J, P) + (\alpha S + \beta T) = A.$$

The verifier **V** accepts the signature if the above expression holds. The correctness of the signature is verified as

$$\begin{aligned} B - e_{\alpha,\beta}(J, P) + (\alpha S + \beta T) &= m(\alpha S + \beta T) + (h(z)x - 1)(\alpha S + \beta T) \\ &\quad -h(z)x(ac - bd)(\alpha S + \beta T) + (\alpha S + \beta T) \\ &= m(\alpha S + \beta T) = A. \end{aligned}$$

5 Security Aspects of the Proposed Scheme

In this section, we analyse the security aspects of blindness and non-forgeability of the proposed scheme.

Blindness from Signer's point of view: The author **A** sends $G = mP = maS + mbT$ to the signer **S**. Then **S** can compute $m(\alpha S + \beta T)$ from G using his secret keys $a,\ b$ and finding $c,\ d$ such that $ac - bd = 1$. It is hard to find m from $G = mP$ as it is equivalent to solving ECDLP for large prime divisor n. Hence **S** cannot see the message.

Blindness from Adversary's point of view: An adversary can find P and mP. Finding m from mP is an instance of solving ECDLP. If the adversary performs a total break of the system, then he can find a, b, and $A = m(\alpha S + \beta T)$, but finding m from A is again an instance of solving ECDLP. Hence the message is blinded to the adversary.

Non-Forgeability of the signer's secret keys: The probability of guessing signer's keys a and b from P is negligible. So for an adversary, it is practically infeasible to guess a random signature.

1. The probability of getting two integers a and b such that $\gcd(a, b) = 1$ is $\prod_{p \le n,\ p\ prime} (1 - 1/p^2) \approx 1/\zeta(2)$ where ζ denotes the Riemann zeta function.
2. The probability of getting x such that $h(z)x \neq 1$ is $1/(n-1)$.

We have the following observations.

Theorem 2 *Let (B_1, J_1) be the signature corresponding to the message m_1 and stamp z_1. To find a random message $m_2(\neq m_1)$ for an adversary that satisfies (B_1, J_1) for a chosen stamp $z_2(\neq z_1)$ is difficult.*

Proof Suppose the adversary chooses a stamp z_2 and tries to find a m_2 satisfying the signature (B_1, J_1). The equation $m_1(\alpha S + \beta T) + e_{\alpha,\beta}(xS + T, h(z_1)S + T) = m_2(\alpha S + \beta T) + e_{\alpha,\beta}(xS + T, h(z_2)S + T)$ implies that $m_1(\alpha S + \beta T) + (h(z_1)x - 1)(\alpha S + \beta T) = m_2(\alpha S + \beta T) + (h(z_2)x - 1)(\alpha S + \beta T)$ and the equation $h(z_1)xH = h(z_2)xH$ implies that $h(z_1)x(cS + dT) = h(z_2)x(cS + dT)$ which gives $h(z_1) = h(z_2)$. Then $m_1(\alpha S + \beta T) = m_2(\alpha S + \beta T)$. Since the hash function h is collision-resistant, $h(z_1) = h(z_2)$ is not possible. In addition, $m_1(\alpha S + \beta T) = m_2(\alpha S + \beta T)$, hence $m_1 = m_2 \pmod{n}$.

Theorem 3 *Let (B_1, J_1) be the signature corresponding to the message m_1 and stamp z_1. To find a random stamp $z_2 (\neq z_1)$ for an adversary that satisfies (B_1, J_1) for a chosen message $m_2 (\neq m_1)$ is difficult.*

Proof Suppose the adversary chooses a message m_2 and wants to find a stamp z_2 which satisfies the signature (B_1, J_1). Then the equation $m_1(\alpha S + \beta T) + e_{\alpha,\beta}(xS + T, h(z_1)S + T) = m_2(\alpha S + \beta T) + e_{\alpha,\beta}(xS + T, h(z_2)S + T)$ implies that $m_1(\alpha S + \beta T) + (h(z_1)x - 1)(\alpha S + \beta T) = m_2(\alpha S + \beta T) + (h(z_2)x - 1)(\alpha S + \beta T)$ and the equation $h(z_1)xH = h(z_2)xH$ implies that $h(z_1)x(cS + dT) = h(z_2)x(cS + dT)$. Second equation gives $h(z_1) = h(z_2)$, which is not possible by collision-resistant property of the hash function h.

We will see the following possible attacks, which fail for the proposed scheme.

Attack 1. Suppose that an attacker **X** wants to get the signer's secret keys a and b, then he performs the following:

X queries a blind signature on message M. Then the signer can cipher and output a signature (B, J, z) as a response. After receiving the signature (B, J, z), **X** will try to find c and d from J. But finding c and d from J is equivalent to solving an ECDLP. Without the knowledge of c and d, **X** cannot get a and b. Hence, it is impractical to find the signer's secret keys.

Attack 2. Suppose an attacker **X** wants to generate a valid signature for a message M_1, then he can perform the following steps:

1. **X** generate a stamp z_1.
2. **X** computes $h(z_1)$ and $m_1 = h(M_1)$.
3. **X** chooses an integer $x_1 \in [1, n - 1]$ such that $h(z_1)x_1 \neq 1$.
4. Then **X** computes $B_1 = m_1(\alpha S + \beta T) + e_{\alpha,\beta}(x_1 S + T, h(z_1)S + T)$.

Now **X** cannot set $J_1 = h(z_1)x_1 H$ as H is unknown to **X**. Suppose **X** selects a random $J_2 = (h(z_1)x_1 - 1)(c_1 S + d_1 T)$ and send (B_1, J_2, z) to a verifier for verification. Then the verifier computes

$$\begin{aligned} B_1 - e_{\alpha,\beta}(J_2, P) + (\alpha S + \beta T) &= m_1(\alpha S + \beta T) + (h(z_1)x_1 - 1)(\alpha S + \beta T) \\ &\quad - h(z_1)x_1(ac_1 - bd_1)(\alpha S + \beta T) + (\alpha S + \beta T) \\ &= m_1(\alpha S + \beta T), \end{aligned}$$

provided that $(ac_1 - bd_1) = 1$, which is impractical. Hence this is resistant to the attack.

6 Conclusion

We slightly modified Lee's bilinear self-pairing. The proposed blind signature scheme is based on an elliptic curve and self-pairing, which is analogous to Chakraborty–Mehta's blind signature scheme. Our proposed scheme avoids the attacks given by Tian, Zhu, and Chen [15], which proved that Chakraborty–Mehta's scheme is insecure. We have also discussed the security aspects of our proposed scheme.

References

1. Binu VP, Sreekumar A (2015) Threshold multi secret sharing using elliptic curve and pairing. Int J Inf Process 9(4):100–112
2. Blake I, Seroussi G, Smart N (1999) Elliptic curves in cryptography. Cambridge University Press
3. Chakraborty K, Mehta J (2012) A stamped blind signature scheme based on elliptic curve discrete logarithm problem. Int J Netw Sec 14(6):316–319
4. Chaum D (1983) Blind signatures for untraceable payments. In: Advances in cryptology - crypto '82, vol 10. Springer, pp 199–203
5. Enge A (1999) Elliptic curves and their applications to cryptography: an introduction. Kluwer Academic Publishers
6. Goldwasser S, Micali S, Rivest R (1988) A digital signature scheme secure against adaptative chosen-message attacks. SIAM J Comput 17(2):281–308 April
7. Lee H-S (2004) A self-pairing map and its applications to cryptography. Appl Math Comput 151:671–678
8. Liu D, Huang D, Luo P, Dai Y (2008) New schemes for sharing points on an elliptic curve. Comput Math Appl 56:1556–1561
9. Menezes A (2005) An introduction to pairing-based cryptography, Lecture Notes. http://www.math.uwaterloo.ca/~ajmeneze/publications/pairings.pdf
10. Mohamed M (2011) A stamped hidden-signature scheme utilizing the elliptic curve discrete logarithm problem,. Int J Netw Secur 13(1):49–57
11. Nikolay Moldovyan A (2011) Blind signature protocols from digital signature standards,. Int J Netw Secur 13(1):22–30
12. Pointcheval D, Stern J (1996) Provably secure blind signature schemes. In: Asiacrypt '96. LNCS 1163. Springer, Berlin, pp 252–265
13. Preneel B (1999) The state of cryptographic hash functions, Lecture notes in computer science, vol 1561, pp 158–182. (Lectures on Data Security)
14. Silverman JH (1986) The Arithmetic of elliptic curves. Graduate texts in mathematics. Springer, New York, p. 106
15. Tian M, Zhu Y, Chen Z (2014) Two simple attacks on a blind signature scheme. Int J Netw Secur 16(6):498–500
16. Zhao ZM (2008) ID-based weak blind signature from bilinear pairings. Int J Netw Secur 7(2):265–268

Security Analysis via Algebraic Attack Against "A New Encryption Scheme for Multivariate Quadratic System"

Yasuhiko Ikematsu and Shuhei Nakamura

Abstract A Gröbner basis algorithm computes a good basis for an ideal of a polynomial ring and appears in various situations of cryptography. In particular, it has been used in the security analysis of multivariate public key cryptography (MPKC), and has been studied for a long time; however, it is far from a complete understanding. We consider the algebraic attack using a Gröbner basis algorithm for a new multivariate encryption scheme proposed by Jiahui Chen et al. at Theoretical Computer Science 2020. Their idea to construct a new scheme was to use the minus and plus modifiers to prevent known attacks, such as linearization attacks. Moreover, they discussed having a resistance to the algebraic attack using a Gröbner basis algorithm. However, in our experiments, the algebraic attack breaks their claimed 80- and 128-bit security parameters in reasonable times. It is necessary to understand whether their scheme can avoid such an attack by introducing a slight modification. In this paper, we theoretically describe why the algebraic attack breaks their scheme and give a precise complexity of the algebraic attack. As a result, we demonstrate that the algebraic attack can break the claimed 80- and 128-bit security parameters in the complexities of approximately 25 and 32 bits, respectively. Moreover, based on our complexity estimation of the algebraic attack, we conclude that the Chen et al. scheme is not practical.

Keywords Multivariate public key cryptography · Post-quantum cryptography · Encryption schemes · Gröbner basis algorithm

Y. Ikematsu (✉)
Institute of Mathematics for Industry, Kyushu University, 744 Motooka, Nishiku, Fukuoka 819-0395, Japan
e-mail: ikematsu@imi.kyushu-u.ac.jp

S. Nakamura
Department of Liberal Arts and Basic Sciences, Nihon University, Chiyoda City, Japan
e-mail: nakamura.shuhei@nihon-u.ac.jp

© The Author(s), under exclusive license to Springer Nature Singapore Pte Ltd. 2022

D. Giri et al. (eds.), *Proceedings of the Seventh International Conference on Mathematics and Computing*, Advances in Intelligent Systems and Computing 1412,
https://doi.org/10.1007/978-981-16-6890-6_2

1 Introduction

Multivariate public key cryptography (MPKC) [9] is public key cryptography using multivariate polynomials over finite fields and is considered as one of the main candidates for post-quantum cryptography (PQC) [2]. It has been studied for approximately three decades since the MI scheme [18], which is the first multivariate encryption scheme proposed in 1988. An important ingredient of MPKC is the *central map* G. Namely, $G = (g_1, \ldots, g_m)$ is a map from $\mathbb{K}^n$ to $\mathbb{K}^m$ consisting of m multivariate quadratic polynomials $g_1, \ldots, g_m$ in n variables $x_1, \ldots, x_n$ over a finite field $\mathbb{K}$ and has the property that the equation $G(x) = y$ for any $y \in \mathbb{K}^m$ can be solved easily. To hide the structure of G, we randomly choose two affine maps $U : \mathbb{K}^n \to \mathbb{K}^n$ and $T : \mathbb{K}^m \to \mathbb{K}^m$ and the public key is given by the quadratic polynomial map $P = T \circ G \circ U : \mathbb{K}^n \to \mathbb{K}^m$. The security of MPKC is based on the *MQ-Problem*, which asks for a solution to the system $P(x) = y$ of quadratic equations for a ciphertext $y \in \mathbb{K}^m$. From the fact that the MQ-problem is proven to be an NP-hard problem even for quadratic polynomials over the finite field $\mathbb{F}_2$ [15] and there exist no polynomial-time quantum algorithms at the moment, MPKC is widely considered to have the potential to resist quantum computer attacks.

In the area of digital signatures, there exist practical multivariate signature schemes such as UOV [17] and Rainbow [10]. In fact, they appear in the third round of NIST PQC standardization [19] as Rainbow [11]. Here, Rainbow is an efficient variant of the UOV scheme. However, it is considered to be difficult to construct a secure and efficient multivariate encryption scheme. In fact, the MI encryption scheme [18] was broken by Patarin in 1995 [20] using the linearization attack. Subsequently, Patarin proposed a multivariate encryption scheme extending the MI scheme, called the HFE scheme [21]. However, it was found that the HFE scheme has a serious trade-off between efficiency and security. As a result, the original HFE scheme is not practical and requires some modifiers such as HFERP [16].

In 2020, Jiahui Chen et al. proposed a new multivariate encryption scheme [7] at Theoretical Computer Science. Their original central map $G = (g_1, \ldots, g_m)$ used in [7] is constructed so that it can easily obtain any preimage of G by reducing a system of linear equations. From such a property, any equation $G(x) = y$ can be easily solved. Because the original central map G is vulnerable to the linearization attack, the authors applied the minus and plus modifiers to the original central map G and constructed the public key P. By doing so, the paper [7] claimed that the scheme can avoid the linearization attack.

In this paper, we analyze the algebraic attack using a Gröbner basis algorithm for Chen et al.'s scheme. In our experiments, the algebraic attack using a Gröbner basis algorithm breaks their claimed 80- and 128-bit parameters in a few seconds. However, it is not clear why the Gröbner basis algorithm breaks their scheme. It is necessary to understand whether their scheme can avoid such an attack by introducing a slight modification, and we theoretically have to describe why the algebraic attack breaks their scheme. Thus, we consider a precise complexity estimation of the algebraic attack in this paper. In general, the complexity of the attack using a

Gröbner basis algorithm is difficult to analyze in MPKC. Therefore, it is important for future development of MPKC to analyze the Gröbner basis algorithm against such a scheme in detail.

In our analysis, we focus on the property that the original central map G can easily obtain any preimage of G by reducing a system of linear equations. To be precise, we can recover many degree-one polynomials from the vector space spanned by the public key P even if minus and plus modifiers are applied. By substituting such degree-one polynomials into the equation $P(x) = y$ with a ciphertext y, we reduce the original MQ-problem $P(x) = y$ with m quadratic equations and n variables to a smaller MQ-problem $P'(x') = y'$. Thus, the complexity to solve the original MQ-problem $P(x) = y$ by a Gröbner basis algorithm is the same as that of the smaller MQ-problem $P'(x') = y'$, and we can obtain a precise complexity estimation of the algebraic attack using a Gröbner basis algorithm. In fact, we show that the algebraic attack can break the claimed 80- and 128-bit security parameters in the complexities of approximately 25 and 32 bits, respectively. Moreover, based on our complexity estimation of the algebraic attack, we conclude that the Chen et al. scheme is not practical.

Our paper is organized as follows. In Sect. 2, we briefly recall the general construction of multivariate encryption schemes, Chen et al. encryption scheme and the algebraic attack. In Sect. 3, we propose our attack and give experimental results. In Sect. 4, we describe the security analysis for the algebraic attack in detail. Finally, we conclude our paper in Sect. 5.

2 Preliminaries

In this section, we describe the general construction of multivariate encryption schemes, Chen et al. encryption scheme [7] and the algebraic attack.

2.1 *Multivariate Public Key Cryptography (MPKC)*

Let $\mathbb{K}$ be a finite field with q elements, and let n and m be positive integers. The public key of a multivariate public key cryptosystem consists of m multivariate quadratic polynomials $P = (p_1, \ldots, p_m)$ in n variables $x_1, \ldots, x_n$ over the finite field $\mathbb{K}$. Each polynomial $p_k(x_1, \ldots, x_n)$ $(1 \le k \le m)$ is in the form of

$$p_k(x_1, \ldots, x_n) = \sum_{i=1}^{n}\sum_{j=1}^{n} a_{ij}x_ix_j + \sum_{i=1}^{n} b_ix_i + c, \tag{1}$$

where $a_{ij}, b_i, c \in \mathbb{K}$. The security of multivariate public key schemes is based on the *MQ-Problem*, which asks for a solution of a given system of multivariate quadratic polynomials over the finite field $\mathbb{K}$. In fact, the MQ-Problem is proven to be an NP-hard problem even for quadratic polynomials over the binary field $\mathbb{F}_2$ [15].

We introduce the general construction of multivariate encryption schemes. First, the most important ingredient is a *central map* $G : \mathbb{K}^n \to \mathbb{K}^m$, which is an easily invertible quadratic map. Namely, we can compute a solution of $G(x) = y$ for any element $y \in \mathbb{K}^m$ with low complexity. Second, to hide the structure of the central map G, we randomly choose two invertible affine (or linear) maps $T : \mathbb{K}^m \to \mathbb{K}^m$ and $U : \mathbb{K}^n \to \mathbb{K}^n$. Then the *public key* is given by the composite

$$P = T \circ G \circ U : \mathbb{K}^n \to \mathbb{K}^m$$

and the *secret key* is given by $\{T, G, U\}$.

Encryption: For a plaintext $\alpha \in \mathbb{K}^n$, the ciphertext is given by $y = P(\alpha) \in \mathbb{K}^m$.

Decryption: For a given ciphertext $y \in \mathbb{K}^m$, one computes $\beta = T^{-1}(y)$, $\gamma = G^{-1}(\beta)$ and $y' = U^{-1}(\gamma)$. Then y' is the plaintext of the ciphertext y.

2.2 Chen et al.'s Encryption Scheme

In this subsection, we briefly describe the construction of the Chen et al. encryption scheme [7]. Let n be a positive integer. Assume that the characteristic of $\mathbb{K}$ is not 2. For an $(n + 1) \times n$ matrix $D = (d_{i,j})_{i,j} \in \mathbb{K}^{(n+1)\times n}$, we define $n + 1$ quadratic polynomials $g_{D,1}, \dots, g_{D,n+1}$ in n variables $x_1, \dots, x_n$ as follows:

$$\begin{aligned} g_{D,1} &= (x_1 - d_{1,1})^2 + (x_2 - d_{1,2})^2 + \cdots + (x_n - d_{1,n})^2, \\ &\vdots \\ g_{D,n+1} &= (x_1 - d_{n+1,1})^2 + (x_2 - d_{n+1,2})^2 + \cdots + (x_n - d_{n+1,n})^2. \end{aligned}$$

We recall that the polynomial map $G_D = (g_{D,1}, \dots, g_{D,n+1}) : \mathbb{K}^n \to \mathbb{K}^{n+1}$ is an easily invertible quadratic map. For an element $y = (y_1, \dots, y_{n+1}) \in \mathbb{K}^{n+1}$, we would like to solve the system of equations

$$g_{D,1}(x_1, \dots, x_n) = y_1, \cdots, g_{D,n+1}(x_1, \dots, x_n) = y_{n+1}. \tag{2}$$

It is clear that for any $1 \le i \le n$, we have

$$\begin{aligned} & y_{i+1} - y_i \\ &= g_{D,i+1}(x_1, \dots, x_n) - g_{D,i}(x_1, \dots, x_n) \\ &= 2(d_{i,1} - d_{i+1,1})x_1 + \cdots + 2(d_{i,n} - d_{i+1,n})x_n + \sum_{j=1}^{n} (d_{i+1,j}^2 - d_{i,j}^2). \end{aligned}$$

From this, we have the linear equations

$$\begin{pmatrix} y_2 - y_1 \\ y_3 - y_2 \\ \vdots \\ y_{n+1} - y_n \end{pmatrix} = D' \begin{pmatrix} x_1 \\ x_2 \\ \vdots \\ x_n \end{pmatrix} + \begin{pmatrix} \sum_{j=1}^{n} (d_{2,j}^2 - d_{1,j}^2) \\ \sum_{j=1}^{n} (d_{3,j}^2 - d_{2,j}^2) \\ \vdots \\ \sum_{j=1}^{n} (d_{n+1,j}^2 - d_{n,j}^2) \end{pmatrix}, \tag{3}$$

where $D' := (2d_{i,j} - 2d_{i+1,j})_{i,j} \in \mathbb{K}^{n \times n}$. Thus, if D' is invertible, then we can obtain the solution to (2) by solving the system of linear equations (3).

Now, we explain the key-generation of the Chen et al. encryption scheme [7]. Let a and s be two positive integers and set $m := n + 1 - a + s$. First, randomly choose a matrix $D = (d_{i,j})_{i,j} \in \mathbb{K}^{(n+1) \times n}$ such that D' is invertible. Define the polynomials $G_D = (g_{D,1}, \dots, g_{D,n+1})$ as above. Next, randomly choose quadratic polynomials $h_1, \dots, h_s$ in n variables $x_1, \dots, x_n$ over $\mathbb{K}$. Then the central map is given by

$$G = (g_{D,1}, \dots, g_{D,n+1-a}, h_1, \dots, h_s) : \mathbb{K}^n \to \mathbb{K}^m.$$

Next, randomly choose invertible affine maps $U : \mathbb{K}^n \to \mathbb{K}^n$ and $T : \mathbb{K}^m \to \mathbb{K}^m$. Then, the public key is the polynomial map $P := T \circ G \circ U$, and the secret key consists of $G_D = (g_{D,1}, \dots, g_{D,n+1})$, U and T. Note that the central map G is applied to the minus and plus modifiers, as $g_{D,n+1-a+1}, \dots, g_{D,n+1}$ are removed and some quadratic polynomials $h_1, \dots, h_s$ are added.

Encryption: For a plaintext $\alpha \in \mathbb{K}^n$, the ciphertext is given by $y = P(\alpha) \in \mathbb{K}^m$ by substituting into the public key P.

Decryption: For a given ciphertext $y \in \mathbb{K}^m$, first compute $\beta = (\beta_1, \dots, \beta_m) := T^{-1}(y)$. Second, for an element $\beta' = (\beta'_1, \dots, \beta'_a) \in \mathbb{K}^a$, solve the system of quadratic equations

$$\begin{aligned} g_{D,1}(x_1, \dots, x_n) &= \beta_1, \\ &\vdots \\ g_{D,n+1-a}(x_1, \dots, x_n) &= \beta_{n+1-a}, \\ g_{D,n+1-a+1}(x_1, \dots, x_n) &= \beta'_1, \\ &\vdots \\ g_{D,n+1}(x_1, \dots, x_n) &= \beta'_a, \end{aligned}$$

as explained above. Let $\gamma = (\gamma_1, \dots, \gamma_n)$ be the solution. If we have

$$h_1(\gamma) = \beta_{n+1-a+1}, \dots, h_s(\gamma) = \beta_m,$$

then $y' = U^{-1}(\gamma)$ is the plaintext of y; otherwise, re-choose another $\beta' = (\beta'_1, \ldots, \beta'_a) \in \mathbb{K}^a$.

2.3 *Algebraic Attack and Previous Analysis Against Chen et al. Scheme*

In this subsection, we first explain the algebraic attack. Subsequently, we describe how the authors in [7] decided the concrete parameters.

For a public key $P = (p_1, \ldots, p_m)$ and a ciphertext $y = (y_1, \ldots, y_m)$, the *algebraic attack* finds the plaintext by solving the system of equations $p_1(x) = y_1, \ldots, p_m(x) = y_m$. In the HFE challenge presented by Patarin to cryptanalysis for HFE and its minus variant [20], HFE challenge 1 was solved by this attack in [14]. The algebraic attack uses Gröbner basis algorithms such as XL [26], F_4 [12] and F_5 [13] which compute a Gröbner basis of the ideal $\langle p_1 - y_1, \ldots, p_m - y_m \rangle$. The complexity of the Gröbner basis algorithm dominates that of the algebraic attack and is estimated by

$$\mathcal{O}\left(\binom{n + D_{reg}}{D_{reg}}^{\omega}\right),$$

where n is the number of variables, $D_{reg} \in \mathbb{N}$ is the degree of regularity of the input system $p_1(x) = y_1, \ldots, p_m(x) = y_m$ and $2 < \omega \leq 3$ is a linear algebra constant. If $\{p_1, \ldots, p_m\}$ is semi-regular [3], then the degree of regularity D_{reg} is given by the degree of the first term whose coefficient in the following power series is non-positive (see [1]):

$$\frac{\prod_{i=1}^{m}(1 - t^{\deg p_i})}{(1 - t)^n}.$$

Chen et al. show in [7] that the degree of regularity against their scheme without the minus and plus modifications is low; however, such modifiers can increase the degree of regularity. In fact, they showed in some experiments that the larger the parameter a is, the more the degree of regularity increases under the condition $a = s$. Here, a is the number of removed polynomials, and s is the number of added polynomials. Then, in order to select an 80-bit security parameter for $q = 3$ and $n = 59$, they concluded that the degree of regularity needs at least 8, and it is accomplished by $a, s \geq 8$. They performed experiments under the condition $a = s$; in such a condition, the decryption failure rate is very high; $q^{-s+a-1} = q^{-1}$. Thus, a is required to be moderately small compared with s. As a result, they selected two 80- and 128-bit security parameters (A) and (B) in [7], as shown in Table 1. Here, we mention that they did not perform experiments under the condition $a < s$.

3 Revisiting the Algebraic Attack Against Chen et al. Scheme

In this section, we revisit the algebraic attack against the Chen et al. scheme. We first perform some experiments for the algebraic attack and confirm that the algebraic attack breaks 80- and 128-bit security parameters of their scheme in reasonable times. Second, we observe the property of the public key to see why the algebraic attack works. Finally, we apply the property to the algebraic attack and reconstruct the algebraic attack to make it theoretically easy to handle.

3.1 Experiments for Algebraic Attack

In this subsection, we discuss the experiments conducted in this study and show that the algebraic attack using a Gröbner basis algorithm breaks the 80- and 128-bit security parameters of their scheme in reasonable times.

Our experimental results are the average timing of 1000 experiments for each parameter in Table 1. All experiments were performed on a 3.2GHz Intel Core i7 CPU with Magma V2.24-4 [5]. The Gröbner basis algorithm we used was the F4 algorithm [12] with the graded reverse lexicographic monomial order, whose Magma command is "**GroebnerBasis**". In addition, to measure the memory, we used the Magma command "**GetMaximumMemoryUsage()**".

For example, Table 2 shows that the 80-bit parameter (A) is broken in 0.35 s with 32.1 megabytes. However, because it is not clear why the Gröbner basis algorithm breaks their parameters, we must theoretically describe it.

Table 1 Selected parameters of Chen et al. scheme in [7] at 80- and 128-bit security levels. Here, q is the cardinality of the finite field $\mathbb{K}$

	Asserted security level	(q, n, a, s, m)	Public key (KB)	Secret key (KB)
(A)	80-bit	$(3, 59, 10, 25, 75)$	134	53.9
(B)	128-bit	$(3, 83, 12, 27, 99)$	345	108

Table 2 Experimental results for the algebraic attack using the Gröbner basis algorithm F4 at two selected 80- and 128-bit parameters (A) and (B) of Chen et al.'s scheme in [7]

	Asserted security level	(q, n, a, s, m)	Experiments	
			Time (s)	Memory (MB)
(A)	80-bit	$(3, 59, 10, 25, 75)$	0.35	32.1
(B)	128-bit	$(3, 83, 12, 27, 99)$	1.07	81.6

3.2 A Key Property of the Public Key P

Here, we observe a property of the public key P in Sect. 2.2.

The polynomial $g_{D,i+1} - g_{D,i}$ is denoted by l_i for each $1 \leq i \leq n - a$. This l_i is a polynomial of degree one from the definition of $g_{D,i}$. Moreover, the set $\{l_1, \ldots, l_{n-a}\}$ is linearly independent, as D' is invertible. Denote by $\text{Span}_{\mathbb{K}} G$ the subspace in $\mathbb{K}[x_1, \ldots, x_n]$ generated by the polynomials

$$\{g_{D,1}, \ldots, g_{D,n+1-a}, h_1, \ldots, h_s\}$$

in the central map G over the finite field $\mathbb{K}$. We can easily show the following:

Lemma 1 *$\text{Span}_{\mathbb{K}} G$ is generated by*

$$\{l_1, \ldots, l_{n-a}, g_{D,1}, h_1, \ldots, h_s\}.$$

Therefore, $\text{Span}_{\mathbb{K}} G$ has a linearly independent set of $n - a$ degree-one polynomials and $s + 1$ quadratic polynomials.

Let $\text{Span}_{\mathbb{K}} P$ be the subspace in $\mathbb{K}[x_1, \ldots, x_n]$ generated by the polynomials $P = \{p_1, \ldots, p_m\}$ over $\mathbb{K}$. It is clear that $\text{Span}_{\mathbb{K}} P = \text{Span}_{\mathbb{K}} G \circ U$ because $P = T \circ G \circ U$ and T is invertible. By Lemma 1, $\text{Span}_{\mathbb{K}} G \circ U$ is generated by the set

$$\{l_1 \circ U, \ldots, l_{n-a} \circ U, g_{D,1} \circ U, h_1 \circ U, \ldots, h_s \circ U\},$$

which is linearly independent since U is invertible. As a result, we have the following theorem.

Theorem 1 *$\text{Span}_{\mathbb{K}} P$ has a linearly independent set of $n - a$ degree-one polynomials and $s + 1$ quadratic polynomials.*

Therefore, one can generate a linearly independent set of $n - a$ degree-one polynomials from only the public key P.

3.3 Reduction to a Smaller MQ-Problem

In this subsection, by using the result in Sect. 3.2, we see that solving the MQ-problem $P(x) = y$ is essentially equivalent to solving a smaller MQ-problem.

Let $y = (y_1, \ldots, y_m) \in \mathbb{K}^m$ be a ciphertext. We would like to solve the system of m quadratic equations

$$p_1(x_1, \ldots, x_n) = y_1, \ldots, p_m(x_1, \ldots, x_n) = y_m \tag{4}$$

in n variables $x_1, \ldots, x_n$ without the secret key $\{G_C, U, T\}$.

Step 1: We solve the system of linear equations in m variables $z_1, \ldots, z_m$;

$$\sum_{i=1}^{m} z_i \cdot \mathrm{Quad}(p_i) = 0, \tag{5}$$

where $\mathrm{Quad}(p_i)$ stands for the quadratic part of p_i. Here, the system $\sum_{i=1}^{m} z_i \cdot \mathrm{Quad}(p_i) = 0$ implies the linear equations arising from the relation that each coefficient of $\sum_{i=1}^{m} z_i \cdot \mathrm{Quad}(p_i)$ vanishes.

Step 2: If $a_1, \ldots, a_m$ is a solution of (5), then $\sum_{i=1}^{m} a_i p_i$ is a degree-one polynomial. By Theorem 1, the space of such degree-one polynomials is of dimension $n - a$. Thus, the solution space of this system (5) is of dimension $n - a$. Choose a basis $\mathbf{z}^{(1)}, \ldots, \mathbf{z}^{(n-a)} \in \mathbb{K}^m$ of the solution space. Then we can obtain $n - a$ linearly independent degree-one polynomials

$$\begin{aligned} r_1(x_1, \ldots, x_n) &:= \mathbf{z}_1^{(1)} p_1 + \cdots + \mathbf{z}_m^{(1)} p_m, \\ &\vdots \\ r_{n-a}(x_1, \ldots, x_n) &:= \mathbf{z}_1^{(n-a)} p_1 + \cdots + \mathbf{z}_m^{(n-a)} p_m, \end{aligned}$$

where $\mathbf{z}^{(k)} = (\mathbf{z}_1^{(k)}, \ldots, \mathbf{z}_m^{(k)})$ $(1 \leq k \leq n - a)$. Thus, $\mathrm{Span}_{\mathbb{K}} P$ is generated by $r_1, \ldots, r_{n-a}$ and other $s + 1$ quadratic polynomials.

Step 3: From (4), we have the linear equations

$$\begin{aligned} r_1(x_1, \ldots, x_n) &= \mathbf{z}_1^{(1)} \beta_1 + \cdots + \mathbf{z}_m^{(1)} \beta_m, \\ &\vdots \\ r_{n-a}(x_1, \ldots, x_n) &= \mathbf{z}_1^{(n-a)} \beta_1 + \cdots + \mathbf{z}_m^{(n-a)} \beta_m. \end{aligned} \tag{6}$$

By substituting the linear equation (6) into (4), we have a small system of $m - (n - a) = s + 1$ quadratic equations in $n - (n - a) = a$ variables.

As a result, the MQ-problem $P(x) = y$, the system of m quadratic equations in n variables, associated with Chen et al.'s scheme, can be reduced to the smaller MQ-problem with $s + 1$ quadratic equations in a variables. Algorithm 1 shows the attack stated in this subsection.

Algorithm 1: Reconstructed algebraic attack in 3.3

Input : finite field $\mathbb{K}$, public key $P = (p_1, \dots, p_m)$, and ciphertext $y = (y_1, \dots, y_m) \in \mathbb{K}^m$,

Output: the plaintext $\alpha \in \mathbb{K}^n$ of the ciphertext y, namely, $P(\alpha) = y$.

Solve the system of linear equations $\sum_{i=1}^{m} z_i \cdot \mathrm{Quad}(p_i) = 0$;
$\mathbf{z}^{(1)}, \dots, \mathbf{z}^{(n-a)} \in \mathbb{K}^m \leftarrow$ a basis of the solution space;
$r_i \leftarrow \mathbf{z}_i^{(1)} p_1 + \cdots + \mathbf{z}_i^{(m)} p_m$ for $1 \le i \le n - a$;
$R_i \leftarrow \mathbf{z}_i^{(1)} y_1 + \cdots + \mathbf{z}_i^{(m)} y_m$ for $1 \le i \le n - a$;
Substitute $r_i - R_i = 0$ $(1 \le i \le n - a)$ into $P(x) - y = 0$ and solve it by a Gröbner basis algorithm ;

4 Security Analysis for Chen et al. Scheme

In this section, we describe the security analysis for the algebraic attack in detail. In Sect. 4.1, we provide the precise complexity estimation for the reconstructed algebraic attack in Sect. 3.3 and that for the original algebraic attack in Sect. 2.3. In Sect. 4.2, using the complexity estimation, we compute the actual security level for the selected parameters (A) and (B). Moreover, we consider a parameter satisfying 80-bit security against the algebraic attack and discuss its efficiency.

4.1 *Theoretical Description of the Algebraic Attack and Complexity Estimation*

As stated in Sect. 3.3, the reconstructed algebraic attack solves the MQ-problem with $s + 1$ quadratic equations in a variables after some linear operations.

The complexity of the reconstructed algebraic attack is given as follows. The complexity of Step 1 is that of solving the linear system (5) with size m. Thus, it is $\mathcal{O}(m^{\omega})$, where $2 < \omega \le 3$ is a linear algebra constant. We can ignore the complexity of Step 2 because we must define the degree-one polynomials r_i. In Step 3, its complexity is dominated by solving a system of $(s + 1)$ quadratic equations in a variables. We use a Gröbner basis algorithm to solve the system. Under the assumption that the system is semi-regular [3], the complexity of Step 3 is

$$\mathcal{O}\left(\binom{a + D_{reg}}{D_{reg}}^{\omega}\right) \tag{7}$$

at most. Note that the degree of regularity D_{reg} is the degree of the first term whose coefficient in the following power series is non-positive:

$$\frac{(1 - t^2)^{s+1}}{(1 - t)^a}.$$

As a result, the complexity of the reconstructed algebraic attack is

$$\mathcal{O}(m^{\omega}) + \mathcal{O}\left(\binom{a + D_{reg}}{D_{reg}}^{\omega}\right). \tag{8}$$

Note that the linear operations in the reconstructed algebraic attack are contained in the computation process of a Gröbner basis algorithm of the original algebraic attack in Sect. 2.3. Thus, the degree of regularity for the original algebraic attack is the same as that of the reconstructed algebraic attack because such linear operations do not affect the degree of regularity. Therefore, the complexity of solving the MQ-problem $P(x) = y$ by the original algebraic attack is given by

$$\mathcal{O}\left(\binom{a + D_{reg}}{D_{reg}}^{\omega}\right). \tag{9}$$

4.2 *The Actual Security Level for the Selected* 80- *and* 128-*bit Security Parameters (A) and (B) in Table 1*

The selected parameters (A) (resp. (B)) are $(q, n, a, s, m) = (3, 59, 10, 25, 75)$ (resp. $(3, 83, 12, 27, 99)$). Using the formula (9), the complexities of the attack against parameters (A) and (B) are $\binom{10+3}{3}^3 = 2^{24.5}$ and $\binom{12+4}{4}^3 = 2^{32.49}$, respectively. Here we took $\omega = 3$ as a linear algebra constant. Table 2 shows the complexity for the original algebraic attack in Sect. 2.3.

Finally, we consider a parameter (q, n, a, s, m) satisfying 80-bit security against the algebraic attack under $q = 3, n = 59$. Moreover, we set $s = a + 15$, similarly to what Chen et al. took, from the perspective of the decryption failure rate. Then, we can take

$$(q, n, a, s, m) = (3, 59, 32, 47, 75)$$

as the parameter with the smallest a. However, as seen in [7, Sect. 10], the decryption complexity is given by $\mathcal{O}(q^a n^3)$. This implies that the parameter $(q, n, a, s, m) = (3, 59, 32, 47, 75)$ has a very slow decryption. As a result, based on our complexity estimation of the algebraic attack, we can confirm that Chen et al. scheme is not practical (Table 3).

5 Conclusion

In this paper, we analyzed the algebraic attack using a Gröbner basis algorithm for the Chen et al. scheme. We showed that the vector space spanned by the quadratic polynomials in the public key P has many degree-one polynomials. By applying such

Table 3 Complexities for the algebraic attack using the Gröbner basis algorithm at two selected 80- and 128-bit parameters (A) and (B) of Chen et al. scheme in [7]

	Asserted security level	(q, n, a, s, m)	Actual security level
			Complexity (bits)
(A)	80-bit	(3, 59, 10, 25, 75)	24.50
(B)	128-bit	(3, 83, 12, 27, 99)	32.49

degree-one polynomials to the equation $P(x) = y$ for a ciphertext y, we reduced the MQ-problem $P(x) = y$ to a smaller MQ-problem with $s + 1$ quadratic equations in a variables. As a result, we obtained a precise complexity estimation for the algebraic attack against the Chen et al. scheme. Our estimation shows that the claimed 80- and 128-bit security parameters in [7] are broken in the complexity of approximately 25 and 32 bits, respectively. Moreover, by discussing a parameter satisfying 80-bit security against the algebraic attack, we concluded that the Chen et al. scheme is not practical.

In the Chen et al. scheme, the minus and plus modifiers were used in order to strengthen the security against the linearization attack; however, they did not succeed in resisting the algebraic attack. As a future work, we would like to study other modifiers that can resist the algebraic attack.

Acknowledgements This work was supported by JST CREST Grant Number JPMJCR14D6, JSPS KAKENHI Grant Number JP19K20266, JP20K19802 and 2019 IMI Joint Use Research Program Short-term Joint Research "New Development of Constructing Next-Generation Cryptography via Unified Approaches of Mathematics Theory, Computation and Cryptology".

References

1. Bardet B, Faugère JC, Salvy B, Yang BY (2005) Asymptotic behavior of the index of regularity of quadratic semi-regular polynomial systems. In: 8th international symposium on effective methods in algebraic geometry (MEGA), pp 1–14
2. Bernstein DJ, Buchmann J, Dahmen E (eds) Post-quantum cryptography. Springer
3. Bettale L, Faugère JC, Perret L (2013) Cryptanalysis of HFE, multi-HFE and variants for odd and even characteristic. Des Codes Cryptogr 69:1–52
4. Beullens W, Preneel B, Szepieniec A, Vercauteren F (2020) LUOV, Technical report, National Institute of Standards and Technology, Post-quantum cryptography, https://csrc.nist.gov/Projects/Post-Quantum-Cryptography/Round-2-submissions Accessed 2020-5-10
5. Bosma W, Cannon J, Playoust C (1997) The Magma algebra system. I. The user language. J. Symbolic Comput 24:235–265
6. Casanova A, Faugére JC, Macario-Rat G, Patarin J, Perret L, Ryckeghem J (2020) GeMSS, Technical report, National Institute of Standards and Technology, Post-quantum cryptography. https://csrc.nist.gov/Projects/Post-Quantum-Cryptography/Round-2-submissions. Accessed 2020-5-10
7. Chen J, Ling J, Ning J, Lau TSC, Wang Y (2020) A new encryption scheme for multivariate quadratic systems. Theor Comput Sci 809:372–383

8. Cartor R, Smith-Tone D (2018) EFLASH: a new multivariate encryption scheme. In: SAC 2018, LNCS, vol 11349. Springer, pp 281–299
9. Ding J, Gower JE Schmidt DS (2006) Multivariate public key cryptosystems. Springer
10. Ding J, Schmidt DS (2005) Rainbow, a new multivariate polynomial signature scheme. In: ACNS 2005, LNCS, vol 3531. Springer, pp 164–175
11. Ding J, Chen MS, Petzoldt A, Schmidt DS, Yang BY (2020) Rainbow, Technical report, National Institute of Standards and Technology, *Post-quantum cryptography*. https://csrc.nist.gov/Projects/Post-Quantum-Cryptography/Round-2-submissions. Accessed 2020-5-10
12. Faugère JC (1999) A new efficient algorithm for computing Gröbner bases (F4). J Pure Appl Algebra 139:61–88
13. Faugère JC (2002) A new efficient algorithm for computing Gröbner Bases without reduction to zero (F5). ISSAC 2002:75–83
14. Faugère JC, Joux A (2003) Algebraic cryptanalysis of hidden field equations (HFE) using Groöbner bases. In: Crypto 2003, LNCS, vol 2729. Springer, pp 44–60
15. Garey MR, Johnson DS (1979) Computers and intractability: a guide to the theory of NP-completeness. W.H Freeman, Company
16. Ikematsu Y, Perlner R, Smith-Tone D, Takagi T, Vates J (2018) HFERP - a new multivariate encryption scheme. In: PQCrypto 2018, LNCS, vol 10786. Springer, pp 396–416
17. Kipnis A, Patarin L, Goubin L (1999) Unbalanced oil and vinegar schemes. In: EUROCRYPT 1999, LNCS, vol 1592. Springer, pp 206–222
18. Matsumoto T, Imai H (1988) Public quadratic polynomial-tuples for efficient signature-verification and message-encryption. In: EUROCRYPT 1988, LNCS, vol 330. Springer, pp 419–453
19. National Institute of Standards and Technology (2020) Post-quantum cryptography, Round 2 submission. https://csrc.nist.gov/Projects/Post-Quantum-Cryptography/Round-2-Submissions. Accessed 2020-5-10
20. Patarin J (1995) Cryptanalysis of the Matsumoto and Imai public key scheme of Eurocrypt 88. In:CRYPTO 1995, LNCS, vol 963, Springer, pp 248–261
21. Patarin J (1996) Hidden field equations (HFE) and isomorphisms of polynomials (IP): two new families of asymmetric algorithms. In: EUROCRYPT, LNCS, vol 1070. Springer, pp 33–48 (1996)
22. Patarin J, Courtois NT, Goubin L (2001) QUARTZ, 128-bit long digital signatures. In: CT-RSA 2001, LNCS, vol 2020. Springer, pp 282–297
23. Petzoldt A, Chen MS, Yang BY, Tao C, Ding J (2015) Design principles for HFEv- based signature schemes. In: ASIACRYPT 2015, LNCS, vol 9452. Springer, pp 311–334
24. Szepieniec A, Ding J, Preneel B (2016) Extension field cancellation: a new central trapdoor for multivariate quadratic systems. In: PQCrypto 2016, LNCS, vol 9606. Springer, pp 182–196
25. Tao C, Diene A, Tang S, Ding J (2013) Simple matrix scheme for encryption. In: PQCrypto 2013, LNCS, vol 7932. Springer, pp 231–242
26. Yang B-Y, Chen J-M (2004) All in the XL family: theory and practice. In: ICISC 2004, LNCS, vol 3506. Springer, pp 67–86

CArrency: An Energy-Efficient Proof-of-Work Scheme for Crypto-Currencies

Souvik Sur, Abhijit Das, and Dipanwita Roy Chowdhury

Abstract Proof-of-work (PoW) schemes have been an essential ingredient of crypto-currencies. These schemes enable crypto-currencies to operate as decentralized systems, and attract people to volunteer to host (mine) such services in the interest of incentives. Quite a few crypto-currencies use hash-based PoW. For example, the PoW of Bitcoin uses the SHA-256 hash function. Although SHA-256 provides 128-bit security normally considered sufficient in the domain of PoW schemes, it demands a huge amount of computational power, specifically in hardware. This paper proposes a new PoW scheme CArrency based on linear hybrid cellular automata (LHCA). The primary achievement of CArrency is a sizable reduction in the power requirements compared to SHA-256, while maintaining 128-bit security. CArrency does it by reducing the slice count required in its implementation on FPGAs. LHCAs serve this purpose effectively as any n-cell LHCA can be accommodated in a depth-two combinational circuit using $\mathcal{O}(n)$ slices in FPGAs. However, we stress that the low requirement of slices (so the power) does not dilute the practical hardness of CArrency as a PoW. Like all other PoWs, in reality the time required to solve any instance of CArrency can always be controlled by the parameter difficulty level. The hash-based PoWs rely on the one-wayness and collision resistance of hash functions in order to establish their security. On the contrary, CArrency is proven to be secure exploiting the inherent randomness of LHCAs. Further, we compare CArrency and SHA-256 in terms of their practical implementations on FPGA platforms.

Keywords Crypto-currencies · Proof-of-work · SHA-256 · Cellular automata · Discrete logarithm problem · FPGA Implementation

---

S. Sur (✉) · A. Das · D. R. Chowdhury
Department of Computer Science and Engineering, Indian Institute of Technology Kharagpur, Kharagpur, West Bengal, India
e-mail: souviksur@iitkgp.ac.in

A. Das
e-mail: abhij@cse.iitkgp.ernet.in

D. R. Chowdhury
e-mail: drc@cse.iitkgp.ernet.in

© The Author(s), under exclusive license to Springer Nature Singapore Pte Ltd. 2022
D. Giri et al. (eds.), *Proceedings of the Seventh International Conference on Mathematics and Computing*, Advances in Intelligent Systems and Computing 1412,
https://doi.org/10.1007/978-981-16-6890-6_3

1 Introduction

Crypto-currencies, introduced in the mid-1990s, deal with exchange of digital money, protected by cryptographic primitives [7]. For example, quite a few crypto-currencies including Bitcoin [14] use hash functions. Traditionally, a class of mathematical puzzles known as *proof-of-work* (PoW) is used to protect the accountability of money in such systems.

Although mostly used in crypto-currencies nowadays, the notion of PoW scheme is introduced as the "pricing functions" to prevent spam mails [8]. Rivest et al. come up with a trusted authority-based time-lock puzzle in [16]. Later, Back proposes a hash-based PoW puzzle for electronic cash in [1], which is adopted in Bitcoin too. An alternative class of PoW schemes uses the concept of memory-hard functions like Scrypt used in Litecoin [12]. The basic idea of this type of functions is to spend hardware resources (like memory) instead of computational power. We focus on reducing the energy, so we do not deal with memory-hard functions in this paper.

The term proof-of-work is coined by Jackobsson et al. [10] as "This is a protocol in which a prover demonstrates to a verifier that she has expended a certain level of computational effort in a specified interval of time." Usually, a PoW scheme consists of two phases: a challenge phase and a verification phase. The challenge phase asks the prover to solve a computationally intensive problem rigorously (not necessarily exhaustively). The verification phase checks the correctness of the solution. A PoW scheme requires that the expected effort to generate the solution is much greater than the effort to verify the solution. Moreover, there is a security constraint that no one should be able to find out the solution without spending the expected effort. An example follows.

Let $h : \{0, 1\}^* \rightarrow \{0, 1\}^{\Delta}$ be a one-way function (usually a hash function) available to both the prover and the verifier. The prover chooses an arbitrary input string $m \in \{0, 1\}^*$. Now, the prover is asked to find an l-bit number ℓ such that the application of h on the concatenation $\ell \mid\mid m$ produces an output with at least k leading zero bits, that is, $h(\ell \mid\mid m) = \overbrace{00\ldots0}^{k\ 0\text{'s}}\{0, 1\}^{(\Delta-k)}$. Since, h is a one-way function, the prover is left with only the option of trying each of the l-bit numbers as ℓ in order to find a desired output. This may lead to 2^l trials in the worst case, or $\Theta(2^k)$ trials on an average if the hash outputs are assumed random. On the other hand, when the prover comes up with an ℓ and the output of h, the verifier can verify the correctness with just a single call of h. Bitcoin employs this mechanism to allow a miner to mine a block. In case of Bitcoin, h is the SHA-256 hash algorithm [15], m is the concatenation of transactions $\langle T_1 \mid\mid T_2 \mid\mid \cdots \mid\mid T_{\gamma} \rangle$, l is 32, the number ℓ is typically known as the nonce, and k is known as the difficulty level.[1] The difficulty of finding a desired output increases exponentially with k. The Bitcoin protocol adjusts k periodically such that it takes at least 10 minutes to mine a block on an average.

---

[1] In the actual Bitcoin protocol, the difficulty level is defined in a slightly different but essentially the same way.

Another key ingredient of crypto-currencies is blockchain used in order to achieve permissionless consensus in such decentralized systems. Garay et al. establish the mathematical modeling of the Bitcoin backbone protocol in [9]. They identify two basic properties of a blockchain, common prefix and chain quality. Kiayias et al. [11] introduce a fast protocol for simplified payment verification having sublinear complexity. These improvements on blockchain consensus at the protocol level are also applicable to the proposed PoW CArrency. In particular, for any crypto-currency, we attempt to establish CArrency as a stand by PoW leaving its consensus mechanism untouched. For example, even if CArrency replaces the SHA-256-based PoW in Bitcoin, its blockchain will still follow the Nakamoto consensus [14]. The basic construction of the blockchain will remain exactly same except that the miners need to solve an instance of CArrency PoW instead of a SHA-256-based PoW to mine a block.

1.1 Our Contributions

The work reported in this paper is motivated by the possibilities of constructing a PoW scheme which offers the same security level as SHA-256 but requires less power than SHA-256. In hardware platforms, power consumption is directly proportional to the number of components in a circuit [2]. These components are slices[2] for FPGAs and elementary units like transistors or gates for ASICs. We propose a PoW scheme CArrency based on linear cellular automata. These automata fit in the current context, because they can be realized in FPGAs using only an array of slices associated with a depth-two combinational circuit. We focus on reducing the slice count so as to minimize the power consumption. Although PoW schemes are assumed to be resource-hungry, this indifference to the necessity of power does not assert CArrency as an "easier" PoW. Being a part of the real-world applications like crypto-currencies, the minimum wall-clock time required to solve a PoW matters more than its requisites of resources. Specifically for any PoW, this wall-clock time can be realized as a function of the parameter difficulty level that tunes the puzzles to ensure the expected lower bound on time (See Table 1).

We use an n-bit linear hybrid cellular automaton (LHCA). The input transactions are clubbed together to a transaction block which then passes through a transaction-to-state map to produce an n-bit value. This value is used to initialize the state of the LHCA. The LHCA is then iterated until it reaches a state containing k leading zeros, where k is the difficulty level. Under the assumption that the initializing state is a random n-bit vector (it may be a hash output), we expect $\Theta(2^k)$ state updates of the LHCA before a state with k leading zeros is reached. The count τ of update steps is supplied by the miner as the output of the challenge phase.

---

[2] We assume that a slice consists of a look-up table and a flip-flop.

Verification in our PoW scheme is a bit involved. The verifier gets the same initial state of the LHCA by a hash calculation on the transaction block. The verifier then checks whether τ updates of the LHCA from this initial state lets the LHCA reach a state with k leading zero bits. Unlike the search phase used during mining, the verifier can multiply the initial state vector of the LHCA by M^{τ}, where M is the $n \times n$ transition matrix of the (linear) LHCA. This can be done in $\mathcal{O}(n^3 k)$ time because $\tau = \Theta(2^k)$. With some precomputation, this reduces to $\mathcal{O}(n^2 k)$.

We show that the security of CArrency stands on the inherent randomness of an LHCA. Specifically, we prove that an LHCA produces a sequence of random states (i.e., bit-vectors) during its evolution (See Theorem 1). Thus, starting from an initial state (enumerated randomly) $\Theta(2^k)$ transitions are required find a state having k-leading zeros. In Theorem 5, we show that $n \geq (\lambda + k_{max} + \log_2 \lambda)$ suffices to enforce any malicious prover $\mathcal{A}$ to solve an instance of this puzzle with non-negligible probability to the security parameter $\lambda = 128$, where $k < k_{max}$ is the upper bound for the parameter difficulty level k. This bound for n can not be violated even if we allow $\mathcal{A}$ to determine the step-count τ from a given pair of initial state and the state having k-leading zeros in constant time.

We report our experimental comparisons between Bitcoin and CArrency PoW schemes. Given the widespread use of hardware platforms for Bitcoin mining, we carry out several FPGA implementations. Our experiments demonstrate that CArrency mining outperforms SHA-256-based mining by a massive margin of slice counts and so of power consumption.

1.2 Organization of the Paper

This paper is organized as follows. In Sect. 2, we present a succinct review of cellular automata, and prove some results for use in later sections. We propose our PoW scheme CArrency in Sect. 3. Section 4 establishes the security of the proposed scheme exploiting the inherent randomness of LHCAs. Section 5 provides the detailed design architecture of CArrency and its implementation on FPGA platforms. We report the results from our both software and hardware implementations in Sect. 6. Finally, Sect. 7 concludes the paper after highlighting some open problems in this context.

2 Background in Cellular Automata

A *cellular automaton* (CA) is a collection of cells on a grid of specified shape, that evolves through a number of discrete time steps according to a set of rules based on the states of neighboring cells [6]. For our purpose, we use a one-dimensional binary cellular automaton. We let n denote the number of cells in the CA. The state of the CA at time t is denoted by an n-dimensional bit vector $s(t)$, and the i-th bit of the vector $s(t)$ by b_i^t. The cells are numbered $i = 0, 1, 2, \ldots, n-1$ from right to left.

The state of the CA evolves using a set of predefined rules. We use the following two rules.

$$\text{Rule 90} : b_i^{t+1} = b_{i-1}^t \oplus b_{i+1}^t. \qquad \text{Rule 150} : b_i^{t+1} = b_{i-1}^t \oplus b_i^t \oplus b_{i+1}^t.$$

For the boundary values $i = 0$ and $i = n - 1$, the bits b_{-1}^t and b_n^t are always assumed to be 0. An n-dimensional rule vector specifies which rule is to be applied on which cell. Let r_i be the i-th bit of the rule vector. If $r_i = 0$, Rule 90 applies to the i-th cell of the CA. If $r_i = 1$, Rule 150 applies to the i-th cell of the CA. We can combine these two cases in a single equation (where product means the logical AND operation): $b_i^{t+1} = b_{i-1}^t \oplus b_i^t r_i \oplus b_{i+1}^t$. Here, all bits r_i are assumed to be constant (independent of t). These CAs are called linear hybrid CAs (LHCAs), because both Rule 90 and Rule 150 are linear, and a mixture of these rules is used. The state transition of an 8-bit LHCA under a given rule vector is illustrated in Fig. 1.

If each cell of an LHCA contains the zero bit, the CA continues to stay in that state. If the remaining $2^n - 1$ non-zero states occur in a cycle, the LHCA is called a maximum-length LHCA. A rule vector that defines a maximum-length LHCA is called a maximum-length rule vector. To design a maximum-length LHCA of n cells, we use a primitive polynomial of degree n over $\mathbb{F}_2$, and apply a deterministic algorithm [5] to generate a rule vector for an n-cell maximum-length LHCA from this primitive polynomial. Figure 2 shows a 4-cell LHCA with the rule vector 0101 generated from the primitive polynomial $x^4 + x + 1$.

An LHCA can be characterized by an $n \times n$ characteristic or state-transition matrix M over $\mathbb{F}_2$, where

$$M_{ij} \begin{cases} 1, & \text{if } b_i^{t+1} \text{ depends on } b_j^t, \\ 0, & \text{otherwise.} \end{cases}$$

The τ-th state $s(t + \tau)$ starting from the state $s(t)$ can be obtained as

$$s(t + \tau) = M^\tau s(t) \tag{1}$$

Given a rule vector, the characteristic matrix M can be efficiently constructed by placing the rule vector along the principle diagonal of M, filling the super and the sub-diagonals with 1s, and leaving all other entries as 0s.

$s(t)$ at time t	0	1	0	1	1	1	0	0
Rule vector	1	0	1	0	1	1	0	0
$s(t+1)$ at time $(t+1)$	1	0	0	1	1	0	1	0

Fig. 1 State change of a linear CA with rule vector ⟨150, 90, 150, 90, 150, 150, 90, 90⟩

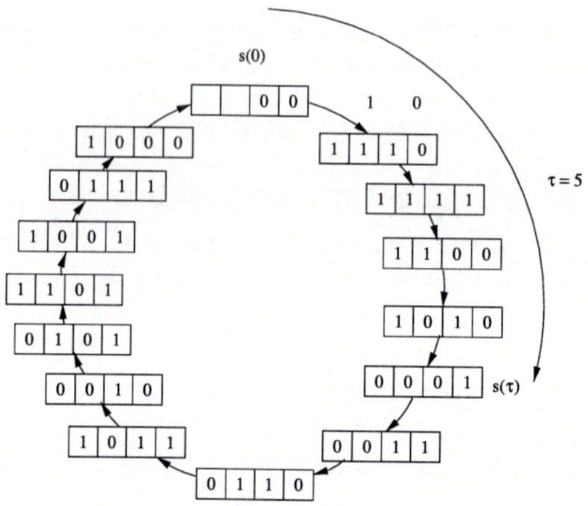

Fig. 2 A schematic view of the PoW scheme on a 4-cell LHCA

Theorem 1 *Let $\mathcal{L}$ be an n-cell LHCA governed by Rule 90 and Rule 150 only. If $\mathcal{L}$ is initialized by a random bit sequence (like the output of a well-designed hash function), then at any time t, the bits $b_i^{(t)}$ of $\mathcal{L}$ remain independent of one another with $Pr[b_i^t = 0] = Pr[b_i^t = 1] = \frac{1}{2}$ for all i.*

Proof We proceed by induction on t. For $t = 0$, the result follows from the initialization hypothesis. For the inductive step, suppose that the bits of $\mathcal{L}$ are unbiased and independent at some time $t \geq 0$. It suffices to show that

$$Pr[b_i^{t+1} = x \mid b_j^{t+1} = y] = Pr[b_i^{t+1} = x] = \frac{1}{2} \tag{2}$$

for all i, j with $i \neq j$, and for all $x, y \in \{0, 1\}$. If Rule 90 is used at the i-th cell, we have $b_i^{t+1} = b_{i-1}^t \oplus b_{i+1}^t$, and we say $support_i = \{i - 1, i + 1\}$. If Rule 150 is used at the i-th cell, we have $b_i^{t+1} = b_{i-1}^t \oplus b_i^t \oplus b_{i+1}^t$, and we say $support_i = \{i - 1, i, i + 1\}$. If $|i - j| > 2$, then $support_i$ and $support_j$ are disjoint, and the bits b_i^{t+1} and b_j^{t+1} are independent. So assume that $j = i + 1$. If the i-th bit in the rule vector is r_i, we have $b_i^{t+1} = b_{i-1}^t \oplus r_i b_i^t \oplus b_{i+1}^t$, and $b_j^{t+1} = b_{i+1}^{t+1} = b_i^t \oplus r_{i+1} b_{i+1}^t \oplus b_{i+2}^t$. The bit b_i^{t+1} depends on b_{i-1}^t but not on b_{i+2}^t. Moreover, the bit b_j^{t+1} depends on b_{i+2}^t but not on b_{i-1}^t. By induction hypothesis, b_{i-1}^t and b_{i+2}^t are independent, so Eqn. (2) holds. The cases $j = i - 1, i + 2, i - 2$ can be analogously handled. □

Theorem 2 *If $\mathcal{L}$ is a null-boundary LHCA, and the rule vector applies Rule 150 to both the boundary cells, then for all $t \geq 0$ and for all i, j, the bits b_i^{t+1} are independent of all the bits b_j^t.*

Proof Let us handle only the boundary cases. If $r_0 = 0$, then $b_0^{t+1} = b_1^t$, whereas if $r_{n-1} = 0$, then $b_{n-1}^{t+1} = b_{n-2}^t$. Copied bits cannot be independent. This forces us to take $r_0 = r_{n-1} = 1$. □

There are $\phi(2^n - 1)/n$ primitive polynomials of degree n, where $\phi()$ is Euler's totient function. So for any n, a primitive polynomial satisfying the condition of Theorem 2 can be easily found.

Theorem 1 implies that if the initialization of the LHCA is random, then the bits remain independent of one another at all times t. Theorem 2 implies that there is no leakage of information from one state to the next state. These observations are vital for our PoW scheme. Note however that in a maximum-length LHCA, the state repeats after $2^n - 1$ transitions. But $2^n - 1$ is a huge number. What is important in the current context is that there should not be information leakage after a small number of transitions.

The next theorem apparently tends to imply some pattern leakage. However, we argue in Sect. 3 that this does not make the task of the miner any easier.

Theorem 3 *Let $\mathcal{L}$ be a null-boundary CA. If its state $s(t)$ at some time t has exactly e leading zero bits, then the next state $s(t+1)$ contains exactly $e-1$ leading zero bits irrespective of the rule vector. Furthermore, for $k \ll n$, if $s(t)$ is a randomly chosen state of $\mathcal{L}$ with $k-1$ leading zero bits, then the previous state $s(t-1)$ has k leading zero bits with probability* $1/2$.

Proof The first statement is obvious. For proving the second statement, we note that there are exactly $2^{n-k+1} - 1$ states of $\mathcal{L}$ with $k-1$ leading zero bits, and exactly $2^{n-k} - 1$ states of $\mathcal{L}$ with k leading zero bits. Finally, for $k \ll n$, $(2^{n-k} - 1)/(2^{n-k+1} - 1)$ is very close to $1/2$. □

3 CArrency: A New Proof-of-work Scheme

Let $\mathcal{L}$ be a maximum-length null-boundary n-cell LHCA and $\mathcal{S}$ is the set of all states of $\mathcal{L}$. We take $\lambda = 128$ as the bit security level, $k_{max} \in \mathbb{Z}^+$ is the upper bound on the difficulty level k and $\delta = \log_2 \lambda$. We determine the size of $\mathcal{L}$ as $n \geq \lambda + k_{max} + \delta$. A miner uses a compression map $H : \{0, 1\}^* \to \mathcal{S}$ (a good hash function) to map a set of transactions $\{T_1, T_2, \ldots, T_\gamma\}$ to an n-bit vectors (basically strings). The rule vector applies Rule 150 to both the boundary cells of $\mathcal{L}$. The phases of our scheme are now described.

3.1 The Challenge Phase

The challenge phase starts with the output of $H(\cdot)$ applied on an arbitrary concatenation of the transactions: $\mathcal{T} = \langle T_1 || T_2 || \cdots || T_\gamma \rangle$. In practice, the protocol may append some additional data to the block (like time-stamps, Merkle roots, headers). The output is an n-bit string. If it is 0^n (only zero bits), then another concatenation of the transactions is tried—this case is however extremely improbable. The n bits of the (non-zero) output are used to initialize the state of $\mathcal{L}$, that is, we use $H : \mathcal{T} \longmapsto s(0)$. After this, $\mathcal{L}$ is allowed to evolve through its consecutive states. Let $s(\tau)$ be the state of $\mathcal{L}$ reached after τ clock pulses starting from $s(0)$. The challenge is to find and report one τ such that $s(\tau)$ has k leading zeros (see Fig. 2). Here, k is the difficulty level predetermined before the run of the protocol (and updated periodically if needed by the protocol). As explained shortly, we impose a restriction on τ that $0 \leq \tau < 2^{k+\delta}$, where $\delta = \log_2 \lambda$ is a small integer, again predetermined in the protocol.

By Theorem 1, the CA evolves randomly. After $\mathcal{O}(2^k)$ updates of the CA, we expect to find a τ meeting the requirement of the challenge phase. The intended way in which a miner should solve the challenge is by iterating through the updates of $\mathcal{L}$ until a desired τ is found. Section 4 shows that the existence of shortcuts is unlikely. Each state update of the CA requires computing the next value of the n cells. Depending on whether Rule 90 or Rule 150 is used for a cell, the update for that cell requires one or two 2-input XOR calculations. For τ transitions, the number of XOR operations is in the range $[n\tau, 2n\tau]$. The expected effort of the miner is $\mathcal{O}(n2^k)$. This running time does not depend on δ. The use of δ is necessitated by security reasons (see Sect. 4).

The Transaction-to-state Map The hash function H maps transaction blocks to n-bit states. A variable-output-length hash function can be used to this effect. A pertinent candidate is Keccak [3]. We can anyway use a hash function H of smaller output length. We use the transactions in $\mathcal{T}$ to generate a set of different inputs to H (for example, $\mathcal{T}, \mathcal{T}+1, \mathcal{T}+2, \ldots$). The outputs of H are concatenated until (at least) n bits are available. Another option is to use disjoint subcollections of $\mathcal{T}$ for producing different inputs to H. Under the assumption that H produces a random output for each input, the n-bit initial state obtained from the concatenation of the outputs from H can be assumed to be a random n-bit string. We highlight that H is applied only once on $\mathcal{T}$ to get $s(0)$, and is not a part of the search step in the challenge phase. Consequently, we treat the computation of $H(\mathcal{T})$ as a precomputation step.

Choice of the Bit Positions We have chosen the k most significant bit positions in the challenge phase. There are two ways of interpreting this. So far, we have worked with the interpretation that the miner is supposed to output a time τ at which the state $s(\tau)$ has *at least* k leading zero bits. Another interpretation is to demand a time τ at which $s(\tau)$ has *exactly* k leading zero bits. We now argue that the second problem is no more difficult than the first. Suppose that the miner computes a state $s(\tau)$ with exactly $k+\epsilon$ leading zero bits. If $\epsilon = 0$, the miner is done. If $\epsilon > 0$, then by Theorem 3, the state $s(\tau+\epsilon)$ has exactly k leading zero bits. The additional overhead associated with the determination of ϵ is negligible compared to the search phase.

Theorem 3 has another implication in the current context. Suppose that the miner solves the $(k-1)$-bit challenge, that is, finds a time τ at which $s(\tau)$ has $k-1$ leading zero bits. The miner can easily compute $s(\tau-1) = M^{-1}s(\tau)$. The theorem implies that $s(\tau-1)$ is a solution to the k-bit challenge with probability $1/2$. It therefore follows that the expected effort of a miner to solve the k-bit challenge is twice the expected effort to solve the $(k-1)$-bit challenge. This is precisely what the presumed difficulty of our mining scheme should satisfy.

Instead of using the k most significant bits, the protocol may predetermine a fixed set of k bit positions in the CA. These positions must remain constant in all invocations of the protocol. Since there are $n \gg k$ bit positions in the CA, one can always fix the k bit positions so that they are well separated from each other. In that case, the apparent pattern leakage mentioned in Theorem 3 vanishes. However, the argument given in the last paragraph highlights that the apparent pattern leakage does not help the miner in any statistical sense.

All the three variants for the choice of bit positions turn out to be essentially the same from a computational point of view. Switching from one variant to another calls for inconsequential changes in either implementations or their complexities. In view of this, we will deal only with the at-least-k-leading-bit-positions variant throughout the rest of this paper.

3.2 The Verification Phase

The miner publishes τ along with the transaction block $\mathcal{T}$. If needed, the miner also supplies additional information on how to use H so that any verifier can apply this function to arrive at the same n-bit string needed to initialize $\mathcal{L}$ to the same state $s(0)$ as used during mining. Let M be the characteristic (or state-transition) matrix of $\mathcal{L}$. So long as the CA and the rule vector remains fixed, M remains a constant matrix, and can be precomputed before any invocation of the protocol. The verifier uses the equation $s(\tau) = M^{\tau}s(0)$ in order to obtain the state $s(\tau)$. Verification succeeds if and only if the k leading bits of $s(\tau)$ are all zero. It is assumed that $0 \leq \tau < 2^{k+\delta}$. If this condition is not satisfied, the verifier rejects the miner's claim without doing any verification.

M^{τ} can be computed by a repeated square-and-multiply exponentiation algorithm. Given the constraint on τ and the fact that δ is a small constant, the exponentiation loop requires $\mathcal{O}(k)$ iterations, each involving squaring of an $n \times n$ matrix followed by a conditional multiplication of two matrices of the same size. Finally, the verifier computes the matrix–vector product $M^{\tau}s(0)$. For the standard matrix-multiplication algorithm, the running time is $\mathcal{O}(n^3k) = \mathcal{O}(n^3 \log \tau)$.

Matrix exponentiation is therefore much costlier than one CA update, but for the expected values of τ, which is $\mathcal{O}(2^k)$, the verification phase actually consumes much less time than the challenge phase (which runs in time exponential in k). This criterion has been one of the prime requirements of every PoW schemes, and is satisfied by our proposed scheme.

Since M is a constant matrix, we can use the idea of fixed-base exponentiation [13] in order to reduce the verification time by a factor of n. We precompute the matrices M^{2^i} for $i = 0, 1, 2, \ldots, k + \delta - 1$. If $\tau = 2^{i_1} + 2^{i_2} + \cdots + 2^{i_\kappa}$, we compute $s(\tau) = (M^{2^{i_1}}(M^{2^{i_2}}(\cdots(M^{2^{i_\kappa}}s(0))\cdots))$ using $\kappa \leq k + \delta$ matrix–vector multiplications and verification finishes in $\mathcal{O}(n^2k) = \mathcal{O}(n^2 \log \tau)$ time.

When $\tau = \mathcal{O}(n)$ (like full n-bit), the verification time becomes $\mathcal{O}(n^3)$. On the contrary, for any conceivable application with miners having moderate to heavy computing resources, we would not take $k < 64$ for mining to finish in feasible time (like ten minutes). Since δ is small ($= 7$), we have $k + \delta < n$. So restricting the value of τ speeds up verification considerably. We shortly show that this is additionally a security requirement.

4 Security Analysis of CArrency

We first formalize the notion of insecurity in the context of PoW schemes. We need to establish that a miner has no obvious easier option other than computing $\mathcal{O}(2^k)$ state transitions of $\mathcal{L}$ to mine a block. Our argument starts with the assumption that a miner can easily (that is, in time $\ll \mathcal{O}(2^k)$) produce a transaction block $\mathcal{T}$ (with $s(0) = H(\mathcal{T})$) and a τ such that $s(\tau) = M^{\tau}s(0)$ has k leading zero bits. We then show that this capability allows the miner to easily solve a computational problem which is purportedly difficult. This contradiction proves the absence of shortcuts during mining.

Let us call a miner capable of bypassing the intended protocol an attacker $\mathcal{A}$ to the scheme. We first note that $\mathcal{A}$ needs to publish a transaction block $\mathcal{T}$ along with τ as an output of the challenge phase. There is no restriction on $s(0)$ and $s(\tau)$ (except for the k zero leading bits of $s(\tau)$).

Let us first see the role played by the initial compression function H realized by a hash function. Since $s(0)$ can be arbitrary, the attacker $\mathcal{A}$ may start with an arbitrary $s(\tau)$ with k leading zero bits. The attacker may then try one or more allowed values of τ, and compute $s(0) = (M^{-1})^{\tau}s(\tau)$. Since M comes from a maximum-length CA, it is invertible. The concocted $s(0)$ must satisfy $s(0) = H(\mathcal{T})$ for some transaction block $\mathcal{T}$. If $\mathcal{A}$ can mount this attack in feasible time, he can actually invert the function H, violating the assumption of one-wayness or first-preimage resistance of hash functions.

This analysis reveals further that the attacker has limited control over $s(0)$. A transaction block typically consists of hundreds of individual transactions. If all permutations of these transactions are allowed, then the total number of inputs to H is huge. In any case, $\mathcal{A}$ always enjoys the choice of including/excluding pending transactions in/from $\mathcal{T}$. In short, the number of choices that $\mathcal{A}$ has over $\mathcal{T}$ is essentially limitless. But in the small window of time given for mining, $\mathcal{A}$ can generate and explore only a small number of inputs $\mathcal{T}$ of H. Consequently, the attacker $\mathcal{A}$ has to work with only a limited choice of the initial state $s(0)$.

Consequently, the attack mechanism of $\mathcal{A}$ must focus on $(\tau, s(\tau))$ pairs with k leading zero bits in $s(\tau)$. The intended protocol finds $s(\tau)$ and τ simultaneously. The attacker bypasses the intended protocol, and finds a τ with $s(\tau)$ having the desired pattern. We formulate the problem of obtaining τ from $s(0)$ and $s(\tau)$ as the *CA Transition Problem* (*CAT*). In the present context, the following three variants of this problem are relevant. We start by defining the computational problem relevant for our PoW scheme.

4.1 CAT: The CA Transition Problem

We formulate the problem of obtaining τ from $s(0)$ and $s(\tau)$ as the *CA Transition problem* (CAT).

Definition 1 *(CA Transition problem CAT).* Given an initial state $s(0)$ of the LHCA $\mathcal{L}$, find a time τ with $|\tau| < k + \delta$, at which the state $s(\tau)$ has k leading zero bits.

We could force the prover to find the smallest τ, for which $s(\tau)$ has k-leading zeros. But there is apparently no easily verifiable certificate that the verifier can use to confirm the "smallest"-ness of the claimed τ. We now establish that choosing τ from the restricted range $[0, 2^{k+\delta})$ should not let solving the puzzle in the challenge phase fail too often.

Theorem 4 *The probability that there is no state $s(\tau)$ with k leading zero bits for $0 \le \tau < 2^{k+\delta}$ is bounded by $e^{-2^{\delta}}$.*

Proof Suppose $\mathbb{L}$ be the set of all the states that have at least k-leading zeros. By the randomness results of Theorems 1 and 2, we can assume that $\Pr[s(t) \in \mathbb{L}] = \frac{1}{2^k}$ at any time t.[3] Therefore, the probability that no state $s(t)$ for $t \in [0, 2^{k+\delta})$ is in $\mathbb{L}$ is $(1 - 1/2^k)^{2^{k+\delta}} \le (e^{-1/2^k})^{2^{k+\delta}} = e^{-2^{\delta}}$. □

Since $\delta = \log_2 \lambda$, this failure probability becomes $e^{-\lambda}$ which is a negligible function of the security parameter λ. For $\delta = 7$, this probability is $\le 1/2^{184}$, whereas the expected number of τ in the restricted range, at which $s(\tau)$ has k leading zero bits is 128.

The security analysis of our PoW scheme is based on the randomness of the output of the hash function H and the random evolution of the LHCA $\mathcal{L}$. We show that even if there is a preprocessing algorithm $\mathcal{A}_0$ that enables the adversary $\mathcal{A}_1$ to determine τ from $s(0)$ and $s(\tau)$ in $\mathcal{O}(1)$ time, our scheme stands secure. We may assume that $s(\tau) \in \mathbb{L}$ (or even $s(\tau) \in \mathcal{S}$). The preprocessing algorithm works before the challenge x is made available.

Theorem 5 *Consider any adversary $\mathcal{A} := (\mathcal{A}_0, \mathcal{A}_1)$ is given access to an oracle Γ that determines τ from $s(0)$ and $s(\tau)$ in $\mathcal{O}(1)$ time. Even if $\mathcal{A}_0$ preprocesses the input $\mathcal{T}$ before $\mathcal{A}_1$ starts, $\mathcal{A}_1$ needs $\mathcal{O}(n2^k)$ time to find an $s(\tau) \in \mathbb{L}$ starting from $s(0) = H(\mathcal{T})$.*

Proof Upon the arrival of a challenge $\mathcal{T}$, $s(0) = H(\mathcal{T})$ is first computed. Since H is a hash function, $s(0)$ is a random state of the LHCA $\mathcal{L}$, and the adversary has

[3] The exact probability is $(2^{n-k} - 1)/(2^n - 1) \approx 1/2^k$, since $n \gg k$.

no control over this initial state. There are exactly $2^{n-k} - 1$ non-zero states of the LHCA in $\mathbb{L}$. Since the LHCA evolves randomly, the expected $\mathcal{O}(2^{\delta})$ states $s(\tau)$ with $\tau \in [0, 2^{k+\delta})$ are randomly distributed in $\mathbb{L}$. Any random choice of $s(\tau) \in \mathbb{L}$ leads to a τ in the desired range with probability $2^{\delta}/(2^{n-k} - 1) \approx 1/2^{n-k-\delta}$. Since we have chosen $n \geq \lambda + k_{max} + \delta$, this probability is $\leq 1/2^{\lambda}$. Consequently, even if the adversary is capable of determining τ from $s(0)$ and $s(\tau)$ in $\mathcal{O}(1)$ time with the help of some hypothetical oracle Γ (thanks to the preprocessing algorithm), (s)he requires an expected number of $\mathcal{O}(2^k)$ transitions. Assuming the oracle reads $s(0)$ and $s(\tau)$ and writes τ in $\mathcal{O}(n)$ time, the expected running time of $T = \mathcal{O}(n2^k)$ are still needed. □

5 Hardware Design

Since mining is the resource-hungry phase for any crypto-currency, it is relevant to implement this phase in hardware. For the sake of completeness, we implement the verification phase too in hardware.

5.1 Challenge Module

In a typical FPGA, flip-flops storing the state of the CA need $n = 199$ slices. To accommodate Rule 150, we assume that each LUT has at least three inputs. At the difficulty level k, a counter needs to count up to $2^{k+\delta} - 1$. So a $(k_{max} + \delta)$-bit counter suffices this purpose. Since $k_{max} + \delta = 64 + 7 = 71$ in practical applications, we need a 71-bit counter with 71 slices.

To check if $s(\tau) \in \mathbb{L}$, we need a σ-ary tree, where σ is the number of inputs to an LUT. The inputs of the leaf LUTs in the tree are connected to the Q' outputs of the k_{max} flip-flops at the most significant end. At the difficulty level k_{max}, the number of LUTs needed is therefore $\sum_{i=1}^{\log_\sigma k_{max}} \left\lceil \frac{k_{max}}{\sigma^i} \right\rceil \leq \left\lceil \frac{k_{max}-1}{\sigma-1} \right\rceil + \left\lceil \log_\sigma k_{max} \right\rceil$ To sum up, the total resource consumption (in number of slices) by the challenge module is bounded by $n + (k_{max} + \delta) + (\left\lceil \frac{k_{max}-1}{\sigma-1} \right\rceil + \left\lceil \log_\sigma k_{max} \right\rceil)$. As n, this quantity is linear in terms of the security parameter λ. For example, for $\lambda = 128$ and $\sigma = 3$, it evaluates to 306.

5.2 Verification Module

We use the fixed-base exponentiation strategy as discussed in Sect. 3. Since M is fixed in all runs of the protocol, all the precomputed matrices M^{2^i} (for $i = 0, 1, 2, \ldots, k + \delta - 1$) can be supplied one by one to the verifier module. So let us concentrate on a single matrix–vector multiplication.

We feed the rows of M^{2^i} one by one to the circuit. Each such row can be multiplied with the state vector in a single clock pulse, and the entire multiplication requires n clock pulses. Since the computation of $s(\tau)$ requires at most $\log \tau$ matrix–vector multiplications, verification finishes in $n \log \tau$ clock pulses. The resource requirement reduces to that for storing a row of M^{2^i}, the input state vector, and the output state vector (the result of the matrix–vector multiplication). A σ-ary tree is additionally needed to XOR the n bits in each row-vector product. For $\sigma = 3$, the number of nodes in the tree is bounded by $\lceil (n-1)/(\sigma-1) \rceil + \lceil \log_\sigma n \rceil$. To sum up, the total slice requirement of the verification circuit is $3n + \lceil (n-1)/(\sigma-1) \rceil + \lceil \log_\sigma n \rceil$. For $\lambda = 128$ it becomes 701.

6 Experimental Results

Let us begin with the performance of our software implementations. A comparison between CArrency and the existing Bitcoin SHA-256 PoW scheme is shown in Table 1 for different choices of n. The trial stops as soon as the first desired τ is located. Each row corresponds to a value of k, and reports the average time over 10,000 random trials.

In Table 1 for both the schemes, the time required in the challenge phases are nearly equal, however the values for k are different. It suggests that CArrency should use higher values of k than Bitcoin, in order to achieve the same block-generation time (in this case ≈6 s). For practical message lengths (1 Mbyte), CArrency verification is already faster than Bitcoin verification provided that we do not include the precomputation time and the time for applying the initial block-to-state map H in CArrency verification. However, the difference between the verification time for minimal message length (512-bits) is in the order of a few microseconds. A hash function typically runs in time proportional to the message length. CArrency mining, on the other hand, is not dependent on the message length (except for the initial application of H).

Table 1 also shows that the precomputation times in CArrency are quite larger than verification times. Since precomputation is done only once for fixed k_{max}, this is not a serious concern, and the benefits of precomputation are established. In all

Table 1 Timing of various phases of CArrency and Bitcoin PoW schemes to provide $\lambda = 128$-bit security

CArrency	Length of CA (n)	Difficulty level (k)	Message length	Phases		
				Challenge	Precompute	Verification
	199	29	Any	6.64 s	0.048 s	72 μs
Bitcoin	N.A	18	512-bits	6 s	N.A	4 μsec.
			1 Mbyte	≈ 3.72 h		8.2 ms

Table 2 Comparison between hardware resource consumption of CArrency and Bitcoin

Scheme	FPGA	# Available slices	# Used slices	# Cycles	Freq. (MHz)	Power (Volt)[†]
Bitcoin	Terasic Altera DE2-115 [4]	1,14,480	90,000	1*	150	12
CArrency mining	iCEstick LP4K	3758	302	1	275	1.2
CArrency verification			700	$199 \log \tau$		

*For one 512-bit message block
[†]From FPGA specifications

cases, the time for verification is much smaller than the time for the corresponding challenge phase. This is a basic requirement of any PoW scheme.

Table 2 reports a comparative study of CArrency and Bitcoin on hardware platforms. Bitcoin can use the same circuit for both mining and verification because both involve SHA-256 computations only. Hardware-level SHA-256 implementations achieve various levels of trade-off between hardware size and number of clock pulses (for each SHA-256 computation). The table shows a standard SHA-256 implementation which aim at reducing the time. CArrency uses two different circuits, one for mining and the other for verification. However, due to their small dimensions both can be accommodated on the smallest (to the best of our knowledge) FPGA board that we have chosen. The slice counts come from synthesized hardware and are slightly smaller than the estimates given earlier because of optimization by the synthesizer. The CArrency mining circuit runs in only one clock pulse for each search step (CA update), and the slice requirement so the power is significantly smaller than that needed by time-optimized SHA-256. The number of clock pulses required by each search during Bitcoin mining is equal to the number of blocks in the message (in the given implementation). On the contrary, each search in CArrency mining always takes one clock pulse.

CArrency verification is much slower than Bitcoin verification in FPGAs. However, each verification takes less than 52 ms. This is arguably a practical and tolerable compromise, particularly given that the combined slice requirement for mining and verification is quite smaller than a highly time-optimized hardware implementation of SHA-256.

In none of our implementations, we have made any effort to realize the initial block-to-state function H. We assume that this step is a part of initialization (collection of transactions and bundling them) having time-complexity linear in the length of $\mathcal{T}$, but independent of k.

7 Conclusion and Open Problems

This paper presents an idea of replacing hash-based computations in Bitcoin-like crypto-currency protocols by the state-update function of linear hybrid cellular automata. We theoretically and experimentally show that mining in our PoW scheme is orders of magnitude more energy-efficient than Bitcoin mining. This efficiency comes due to the migration from hash functions to cellular automata in the domain of PoW schemes.

A question pertaining to our design remain less than fully satisfactorily answered. Although we have been able to show that the efficiency of the preprocessing algorithm $\mathcal{A}_0$ does not affect the security of CArrency, it would be interesting to explore how efficient it can be.

References

1. Back A (1997) http://www.cypherspace.org/hashcash
2. Benini L, Hodgson R, Siegel P (1998) System-level power estimation and optimization. In: Chandrakasan, A., Kiaei, S. (eds.) Proceedings of the 1998 International Symposium on Low Power Electronics and Design, 1998, Monterey, California, USA, August 10-12, 1998. pp 173–178. ACM. https://doi.org/10.1145/280756.280881
3. Bertoni G, Daemen J, Peeters M, Assche GV (2013) Keccak. In: Johansson T, Nguyen PQ (eds) Advances in cryptology - EUROCRYPT 2013, 32nd annual international conference on the theory and applications of cryptographic techniques, Athens, Greece, May 26-30, 2013. Proceedings. Lecture notes in computer science, vol 7881. Springer, pp 313–314. https://doi.org/10.1007/978-3-642-38348-9_19
4. Bitcoin Community (2011) Open source FPGA bitcoin miner — Bitcoin Wiki. https://en.bitcoin.it/wiki/Open_Source_FPGA_Bitcoin_Miner. Accessed 27 Oct 2021
5. Cattell K, Muzio J (1996) Synthesis of one-dimensional linear hybrid cellular automata. IEEE Trans. CAD Integr Circuits Syst 15:325–335. https://doi.org/10.1109/43.489103
6. Chaudhuri,PP, Chowdhury DR, Nandi S, Chattopadhyay S (1997) Additive cellular automata: theory and applications, vol 1. Wiley
7. Chohan U (2017) Cryptocurrencies: A brief thematic review. SSRN Electron J (2017). https://doi.org/10.2139/ssrn.3024330
8. Dwork C, Naor M (1992) Pricing via processing or combatting junk mail. In: Brickell EF (ed) Advances in cryptology - CRYPTO '92, 12th annual international cryptology conference, Santa Barbara, California, USA, August 16-20, 1992, Proceedings. Lecture notes in computer science, vol 740. Springer, pp 139–147. https://doi.org/10.1007/3-540-48071-4_10
9. Garay JA, Kiayias A, Leonardos N (2015) The bitcoin backbone protocol: analysis and applications. In: Oswald E, Fischlin M (eds) Advances in cryptology - EUROCRYPT 2015 - 34th annual international conference on the theory and applications of cryptographic techniques, Sofia, Bulgaria, April 26-30, 2015, proceedings, Part II. Lecture notes in computer science, vol 9057. Springer, pp 281–310. https://doi.org/10.1007/978-3-662-46803-6_10
10. Jakobsson M, Juels A (1999) Proofs of work and bread pudding protocols (extended abstract). Springer, US, Boston, MA, pp 258–272. https://doi.org/10.1007/978-0-387-35568-9_18
11. Kiayias A, Lamprou N, Stouka A (2016) Proofs of proofs of work with sublinear complexity. In: Clark J, Meiklejohn S, Ryan PYA, Wallach DS, Brenner M, Rohloff K (eds.) Financial cryptography and data security - FC 2016 international workshops, BITCOIN, VOTING, and WAHC, Christ Church, Barbados, February 26, 2016, Revised Selected Papers. Lecture notes

in computer science, vol 9604. Springer, pp 61–78. https://doi.org/10.1007/978-3-662-53357-4_5
12. Lee C (2011) Litecoin foundation. https://litecoin-foundation.org/
13. Menezes A, van Oorschot PC, Vanstone SA (1996). Handbook of applied cryptography. https://doi.org/10.1201/9781439821916
14. Nakamoto S (2008) Bitcoin: A peer-to-peer electronic cash system
15. National Institute of Standards and Technology (2002) FIPS PUB 180-2: Secure Hash Standard. http://www.itl.nist.gov/fipspubs/fip180-2.html, supersedes FIPS PUB 180 1995 Apr 17
16. Rivest RL, Shamir A, Wagner DA (1996) Time-lock puzzles and timed-release crypto. Technical report, MIT, USA

A Study on Randomness Used in Signature Generation of UOV

Yasuhiko Ikematsu

Abstract Multivariate public key cryptography is a leading candidate for post-quantum cryptography (PQC). In 1999, Kipnis et al. proposed a multivariate signature scheme called unbalanced oil and vinegar (UOV). It has not been broken for two decades and is thus an important scheme in PQC. In particular, Rainbow, an efficient variant of UOV, was selected as one of the finalists for National Institute of Standards and Technology (NIST) PQC standardization. The signature generation of UOV is performed by substituting a random value (randomness) into quadratic equations and solving an associated linear system. In this paper, we study the randomness used in the signature generation of UOV. In particular, we consider the case in which the range of randomness is restricted and analyze the security against such a case.

Keywords Post-quantum cryptography · Multivariate public key cryptosystems · UOV signature scheme

1 Introduction

Post-quantum cryptography (PQC) [2] is to study cryptosystems that can potentially resist quantum computer attacks. This study started with discovering Shor's polynomial-time quantum algorithms [18] to factor large integers and to solve discrete logarithms. To standardize such cryptosystems, National Institute of Standards and Technology (NIST) announced a PQC standardization project [15] in 2016. This project has now advanced to the third round [16] and PQC has been actively researched.

Multivariate public key cryptosystems (MPKC) [6] have been studied since the 1980s, and are based on the multivariate quadratic polynomials (MQ) problem, which is the problem of finding a solution to a system of quadratic equations over a finite

Y. Ikematsu (✉)
Institute of Mathematics for Industry, Kyushu University, Fukuoka, Japan
e-mail: ikematsu@imi.kyushu-u.ac.jp

© The Author(s), under exclusive license to Springer Nature Singapore Pte Ltd. 2022

D. Giri et al. (eds.), *Proceedings of the Seventh International Conference on Mathematics and Computing*, Advances in Intelligent Systems and Computing 1412,
https://doi.org/10.1007/978-981-16-6890-6_4

field. (See [20].) MQ problem is NP-complete [11] and solving it is believed to be difficult even if we use a quantum computer. Thus, there is a possibility of constructing multivariate schemes that are resistant to quantum computer attacks. Therefore, MPKC is considered one of the main candidates for PQC.

Unbalanced oil and vinegar (UOV) [13] is a multivariate signature scheme proposed by Kipnis et al. in 1999. Its secret key consists of a linear (or affine) map and a system of quadratic polynomials over a finite field $\mathbb{F}_q$. Its secret quadratic polynomials have the special property of becoming linear polynomials through the substitution of a random value (randomness) into certain variables. Using this property, signature generation is performed by substituting a random value into secret quadratic polynomials and solving the associated linear equations. To our knowledge, no critical attack has been reported for the UOV scheme; hence, it is considered as a secure signature scheme. In particular, Rainbow [5, 7], the multi-layer variant of the UOV scheme, was selected as a finalist of the NIST PQC standardization [16]. The UOV scheme is important in MPKC, and further studies are desired.

In this paper, we study the randomness used in the signature generation of the UOV scheme. The signature generation of the UOV scheme is obtained by choosing a random value $\mathbf{c} \in \mathbb{F}_q^v$ with length $v \in \mathbb{N}$ and solving an associated linear system $\mathbf{x}_o \cdot A_\mathbf{c} = b$ in o-variables $\mathbf{x}_o$, where $o \in \mathbb{N}$. Here, this o-by-o matrix $A_\mathbf{c}$ is computed from only the random value $\mathbf{c}$. Now, we consider restricting the range of random values $\mathbf{c}$ to be a subset $\{\mathbf{c}^{(1)}, \mathbf{c}^{(2)}, \ldots, \mathbf{c}^{(r)}\} \subset \mathbb{F}_q^v$. There are at least two situations in which such a restriction appears. First, if we fix $\{\mathbf{c}^{(1)}, \mathbf{c}^{(2)}, \ldots, \mathbf{c}^{(r)}\}$ during key generation in advance, we can precompute the inverse $A_{\mathbf{c}^{(i)}}^{-1}$; hence, a signature is generated by just computing b and the product $b \cdot A_{\mathbf{c}^{(i)}}^{-1}$ and without computing the inverse $A_{\mathbf{c}^{(i)}}^{-1}$. This way makes the signature generation of the UOV scheme efficient and reduces the secret key size. Second, it may appear in an improper implementation of the UOV scheme. Note that similar restrictions for Rainbow [5, 7] were proposed in [21] and the ELSA scheme [17] as $r = 1$. They were proposed as efficient variants. Regarding the ELSA scheme, in [12] it was proven that the signature space of the ELSA scheme has an algebraic structure unlike that of the original Rainbow scheme, and was broken using a chosen message attack (see [12] for more detail).

In either situation, we need to study the security against randomness used in the signature generation of the UOV scheme. In this paper, we show that the restriction described earlier is vulnerable. We can see that the obtained signature space has an algebraic structure, which is different from that of the ELSA scheme. Taking differences of various pairs of two signatures, a secret key of the UOV scheme is recovered under a random message attack. As a result, we conclude that this type of restriction is vulnerable to a random message attack. Moreover, this result alerts us against the restriction of the range of random values to make the UOV scheme efficient. Note that because our attack utilizes an algebraic structure caused by the restriction, the original UOV scheme does not have such a vulnerability. Namely, the original UOV scheme is still secure.

This paper is organized as follows. In Sect. 2, we recall the unbalanced oil and vinegar (UOV) scheme and the main attacks against it. In Sect. 3, we describe the signature generation of the UOV scheme in detail and discuss the restriction of random values. In Sect. 4, we investigate a vulnerability of such a restriction by proposing a random message attack. Finally, we conclude our paper in Sect. 5.

2 Unbalanced Oil and Vinegar (UOV) Signature Scheme

In this section, we explain a general construction of multivariate signature schemes, and the UOV signature scheme [13]. Moreover, we recall the main attacks against the UOV scheme.

2.1 *Multivariate Signature Schemes*

Let $\mathbb{F}_q$ be a finite field with q elements and let n and o be two positive integers. For o quadratic polynomials $f_1, \dots, f_o$ in n variables $x_1, \dots, x_n$, we can consider $\mathcal{F} = (f_1, \dots, f_o)$ as the quadratic polynomial map

$$\mathcal{F} : \mathbb{F}_q^n \ni \mathbf{x} \mapsto (f_1(\mathbf{x}), \dots, f_o(\mathbf{x})) \in \mathbb{F}_q^o.$$

In general, to construct a multivariate signature scheme, an easily invertible quadratic map $\mathcal{F} : \mathbb{F}_q^n \to \mathbb{F}_q^o$ is used. Namely, $\mathcal{F}$ is a map such that a solution $\mathbf{x}$ to $\mathcal{F}(\mathbf{x}) = \mathbf{y}$ can be computed easily. Once such an easily invertible quadratic map is given, a signature scheme is constructed as follows. Randomly choose two invertible linear (or affine) maps $\mathcal{T} : \mathbb{F}_q^o \to \mathbb{F}_q^o$ and $\mathcal{S} : \mathbb{F}_q^n \to \mathbb{F}_q^n$ in order to hide the structure of $\mathcal{F}$. Compute the composite

$$\mathcal{P} := \mathcal{T} \circ \mathcal{F} \circ \mathcal{S} : \mathbb{F}_q^n \to \mathbb{F}_q^o,$$

which also becomes a quadratic map. Then the public key is $\mathcal{P} = \{p_1, \dots, p_o\} : \mathbb{F}_q^n \to \mathbb{F}_q^o$, and the secret key consists of $\{\mathcal{F}, \mathcal{T}^{-1}, \mathcal{S}^{-1}\}$. The signature generation and verification are performed as follows. For a message $\mathbf{m} \in \mathbb{F}_q^o$, compute $\mathbf{m}' = \mathcal{T}^{-1}(\mathbf{m}), \mathbf{m}'' = \mathcal{F}^{-1}(\mathbf{m}')$ and $\mathbf{s} = \mathcal{S}^{-1}(\mathbf{m}'')$. Here, since we took $\mathcal{F}$ as an easily invertible map, a solution $\mathbf{x} = \mathbf{m}''$ to $\mathcal{F}(\mathbf{x}) = \mathbf{m}'$ can be computed easily. This $\mathbf{s} \in \mathbb{F}_q^n$ is a signature of the message $\mathbf{m}$ and the verification is performed by checking whether $\mathcal{P}(\mathbf{s}) = \mathbf{m}$ or not.

2.2 *Construction of the UOV Scheme*

Here, we briefly recall the construction of the UOV signature scheme [13].

Let v and o be two integers such that $v > o > 0$ and set $n := v + o$. We use two variable sets $\mathbf{x}_v = (x_1, \ldots, x_v)$, and $\mathbf{x}_o = (x_{v+1}, \ldots, x_n)$, and put $\mathbf{x} = (\mathbf{x}_v, \mathbf{x}_o)$. We call the first variables $\mathbf{x}_v$, the *vinegar variables* and the second variables $\mathbf{x}_o$ the *oil variables*.

Key Generation: Randomly choose o quadratic polynomials in the variables $\mathbf{x}$ in the following form:

$$\begin{aligned} f_1(\mathbf{x}) = f_1(\mathbf{x}_v, \mathbf{x}_o) &= \sum_{i,j=1}^{v} a_{i,j}^{(1)} x_i x_j + \sum_{i=1}^{v} \sum_{j=v+1}^{n} a_{i,j}^{(1)} x_i x_j, \\ &\vdots \\ f_o(\mathbf{x}) = f_o(\mathbf{x}_v, \mathbf{x}_o) &= \sum_{i,j=1}^{v} a_{i,j}^{(o)} x_i x_j + \sum_{i=1}^{v} \sum_{j=v+1}^{n} a_{i,j}^{(o)} x_i x_j. \end{aligned} \tag{1}$$

Here, each coefficient $a_{i,j}^{(k)}$ is randomly chosen from the finite field $\mathbb{F}_q$. Then, the set $\mathcal{F} = (f_1, \ldots, f_o)$ is called a *central map* of the UOV scheme. Once we randomly choose a linear invertible map $\mathcal{S} : \mathbb{F}_q^n \to \mathbb{F}_q^n$, the public key is given by the composite $\mathcal{P} := \mathcal{F} \circ \mathcal{S} = \{p_1, \ldots, p_o\}$, which is a set of o quadratic polynomials. Moreover, the secret key is $\{\mathcal{F}, \mathcal{S}^{-1}\}$. We note that a linear invertible map $\mathcal{T} : \mathbb{F}_q^o \to \mathbb{F}_q^o$ is not necessary for the secret key, since it does not affect the security of the UOV scheme.

Signature Generation: Given a message $\mathbf{m} = (m_1, \ldots, m_o) \in \mathbb{F}_q^o$, a signature $\mathbf{s}$ is generated as follows. First, randomly choose an element $\mathbf{c} = (c_1, \ldots, c_v) \in \mathbb{F}_q^v$. Second, we can easily obtain a solution $\mathbf{d} \in \mathbb{F}_q^o$ to the equations

$$f_1(\mathbf{c}, \mathbf{x}_o) = m_1, \ldots, f_o(\mathbf{c}, \mathbf{x}_o) = m_o,$$

since they are o linear equations in oil variables $\mathbf{x}_o$ from the form of (1). If there is no solution, we choose another element $\mathbf{c}$. (We will describe how to solve such a linear system in Sect. 3 in detail.) Finally, we compute $\mathbf{s} = \mathcal{S}^{-1}(\mathbf{c}, \mathbf{d}) \in \mathbb{F}_q^n$, which is a solution to $\mathcal{P}(\mathbf{x}) = \mathbf{m}$. This $\mathbf{s} \in \mathbb{F}_q^n$ is a signature of the message $\mathbf{m}$.

Verification: It is performed by checking whether $\mathcal{P}(\mathbf{s}) = \mathbf{m}$.

2.3 Main Attacks Against the UOV Scheme

In this subsection, we recall three attacks against the UOV scheme.

Direct attack: This attack directly solves the system of quadratic equations $\mathcal{P}(\mathbf{x}) = \mathbf{m}$. To solve such a system, Gröbner basis algorithms such as F_4 [9], F_5 [10] and XL [19] are used. Because the system $\mathcal{P}(\mathbf{x}) = \mathbf{m}$ consists of o quadratic equations in n variables and $n = v + o$, it can be reduced to a system of o quadratic equations in o variables by fixing $v = n - o$ variables in $\mathbf{x}$. In general, such a reduced system is considered to be semi-regular (see [1] for the definition). Then, the complexity of solving a system of o quadratic equations in o variables using the XL Wiedemann algorithm with the hybrid approach [3] is given by

$$\min_k q^k \cdot 3 \binom{o - k + D_{reg}}{D_{reg}}^2 \binom{o - k}{2}.$$

Here, $0 \leq k \leq o$ is the number of fixed variables in the hybrid approach, and D_{reg} is given by the smallest integer d for which the coefficient of t^d in the function $\dfrac{(1-t^2)^o}{(1-t)^{o-k}}$ is non-positive.

UOV attack [14]: This attack utilizes the special form of $f_1, \ldots, f_o$ in (1). For simplicity, we assume that the characteristic of $\mathbb{F}_q$ is odd. Because each polynomial f_i has only quadratic terms, it is written using an n-by-n symmetric matrix as follows:

$$f_i(\mathbf{x}) = \begin{pmatrix} \mathbf{x}_v & \mathbf{x}_o \end{pmatrix} \cdot \begin{pmatrix} *_{v,v} & *_{v,o} \\ *_{o,v} & 0_{o,o} \end{pmatrix} \cdot \begin{pmatrix} {}^t\mathbf{x}_v \\ {}^t\mathbf{x}_o \end{pmatrix}.$$

Here, $*_{v,v}$ means a non-zero v-by-v matrix, and $0_{o,o}$ is the zero o-by-o matrix. We denote this symmetric n-by-n matrix by F_i. This F_i is called the *representation matrix* of f_i. We choose an n-by-n matrix S such that $\mathcal{S}(\mathbf{x}) = \mathbf{x} \cdot S$ for $\mathbf{x} \in \mathbb{F}_q^n$. If we set $P_i := S \cdot F_i \cdot {}^tS$, then P_i is the representation matrix of p_i since $p_i = f_i \circ \mathcal{S}$. Let $\{e_1, \ldots, e_n\}$ be a standard basis of $\mathbb{F}_q^n$; i.e., $e_1 = (1, 0, \ldots, 0)$ and so on. We set the vinegar space V and the oil space O as follows:

$$V := \mathrm{Span}\{e_1, \ldots, e_v\}, \quad O := \mathrm{Span}\{e_{v+1}, \ldots, e_n\}.$$

These are subspaces of $\mathbb{F}_q^n$. The UOV attack finds the twisted oil space

$$O \cdot S^{-1} := \mathrm{Span}\{e_{v+1}S^{-1}, \ldots, e_n S^{-1}\} \tag{2}$$

by using XY^{-1} for various two invertible matrices $X, Y \in \mathrm{Span}\{P_1, \ldots, P_o\}$. Then an attacker can find an invertible matrix S' such that $O \cdot S^{-1}S' = O$, which is an equivalent secret key. Namely, the attacker can forge a signature for any message using S'. The complexity of this attack is given by $q^{v-o} \cdot o^4$.

Reconciliation attack [8]: This attack solves the system of quadratic equations $p_1(\mathbf{x}) = 0, \ldots, p_o(\mathbf{x}) = 0$. By using a feature of the public key $p_1, \ldots, p_o$, this system can be reduced to a system of o quadratic equations in v variables. However, this attack is harder than the direct attack in general.

3 Description of Signature Generation of the UOV Scheme

Here, we explain the signature generation of the UOV scheme in more detail. We also consider restricting the range of random values to a subset of $\mathbb{F}^v$.

3.1 *How to Solve the Linear System Associated with Signature Generation*

We explain the signature generation in Sect. 2.2 in more detail. Given a message $\mathbf{m} = (m_1, \ldots, m_o) \in \mathbb{F}_q^o$, we choose a random value $\mathbf{c} = (c_1, \ldots, c_v) \in \mathbb{F}_q^v$ and substitute it into the vinegar variables $\mathbf{x}_v$ in $\mathcal{F} = (f_1, \ldots, f_o)$:

$$\begin{aligned} f_1(\mathbf{c}, \mathbf{x}_o) &= \sum_{i,j=1}^{v} a_{i,j}^{(1)} c_i c_j + \sum_{i=1}^{v} \sum_{j=v+1}^{n} a_{i,j}^{(1)} c_i x_j, \\ &\vdots \\ f_o(\mathbf{c}, \mathbf{x}_o) &= \sum_{i,j=1}^{v} a_{i,j}^{(o)} c_i c_j + \sum_{i=1}^{v} \sum_{j=v+1}^{n} a_{i,j}^{(o)} c_i x_j. \end{aligned} \tag{3}$$

Thus, we obtain o linear equations in o variables $\mathbf{x}_o = (x_{v+1}, \ldots, x_n)$:

$$f_1(\mathbf{c}, \mathbf{x}_o) = m_1, \ldots, f_o(\mathbf{c}, \mathbf{x}_o) = m_o. \tag{4}$$

We define the matrix $A_\mathbf{c} \in \mathbb{F}^{o \times o}$ and the vector $\mathbf{b}_\mathbf{c} \in \mathbb{F}_q^o$ by

$$A_\mathbf{c} := \begin{pmatrix} \sum_{i=1}^{v} a_{i,v+1}^{(1)} c_i & \cdots & \sum_{i=1}^{v} a_{i,v+1}^{(o)} c_i \\ \vdots & \ddots & \vdots \\ \sum_{i=1}^{v} a_{i,n}^{(1)} c_i & \cdots & \sum_{i=1}^{v} a_{i,n}^{(o)} c_i \end{pmatrix}, \tag{5}$$

$$\mathbf{b}_\mathbf{c} := \left(\sum_{i,j=1}^{v} a_{i,j}^{(1)} c_i c_j, \ldots, \sum_{i,j=1}^{v} a_{i,j}^{(o)} c_i c_j \right). \tag{6}$$

Then, the above system (4) is expressed as

$$\mathbf{x}_o \cdot A_{\mathbf{c}} = \mathbf{m} - \mathbf{b}_{\mathbf{c}}.$$

Computing the inverse $A_{\mathbf{c}}^{-1}$, we obtain $\mathbf{d} = (\mathbf{m} - \mathbf{b}_{\mathbf{c}}) \cdot A_{\mathbf{c}}^{-1} \in \mathbb{F}_q^o$ as a solution to (4). Then, $(\mathbf{c}, \mathbf{d}) \in \mathbb{F}_q^n$ is a solution to $\mathcal{F}(\mathbf{x}) = \mathbf{m}$. Moreover, computing $\mathbf{s} = (\mathbf{c}, \mathbf{d}) \cdot S^{-1}$, we obtain a signature $\mathbf{s}$ of the message $\mathbf{m}$. Here, S is the n-by-n matrix such that $\mathcal{S}(\mathbf{x}) = \mathbf{x} \cdot S$. Because the signature $\mathbf{s}$ depends on the choice of the random value $\mathbf{x}_v = \mathbf{c}$, we call it a signature for a random value $\mathbf{x}_v = \mathbf{c}$.

Note that the dominant part in this signature generation is the computation of the inverse of $A_{\mathbf{c}}$. Thus, if we fix the random value $\mathbf{c}$ during the key generation, a signature is generated by just computing $\mathbf{b}_{\mathbf{c}}$ and the product $(\mathbf{m} - \mathbf{b}_{\mathbf{c}}) \cdot A_{\mathbf{c}}^{-1}$ without computing the inverse $A_{\mathbf{c}}^{-1}$. This makes the signature generation of the UOV scheme efficient.

3.2 *The Restriction of Randomness*

We consider that the range of random values (randomness) is restricted to a subset $\{\mathbf{c}^{(1)}, \mathbf{c}^{(2)}, \ldots, \mathbf{c}^{(r)}\} \subset \mathbb{F}_q^v$. In this case, a random value $\mathbf{c} = \mathbf{c}^{(i)}$ is randomly chosen and the signature $\mathbf{s}$ is obtained as follows:

$$\mathbf{d} = (\mathbf{m} - \mathbf{b}_{\mathbf{c}^{(i)}}) \cdot A_{\mathbf{c}^{(i)}}^{-1}, \quad \mathbf{s} = (\mathbf{c}^{(i)}, \mathbf{d}) \cdot S^{-1}.$$

We can easily enlarge this range using scalar multiplication. Given a message $\mathbf{m}$ and a random value $\mathbf{c}^{(i)}$, we randomly choose a non-zero element $z \in \mathbb{F}_q^\times$ and compute $\mathbf{d} = z^{-1} \cdot (\mathbf{m} - z^2 \mathbf{b}_{\mathbf{c}^{(i)}}) \cdot A_{\mathbf{c}^{(i)}}^{-1}$ and $\mathbf{s} = (z\mathbf{c}^{(i)}, \mathbf{d}) \cdot S^{-1}$. Since $A_{z\mathbf{c}^{(i)}}^{-1} = z^{-1} \cdot A_{\mathbf{c}^{(i)}}^{-1}$ and $\mathbf{b}_{z\mathbf{c}^{(i)}} = z^2 \cdot \mathbf{b}_{\mathbf{c}^{(i)}}$, this $\mathbf{s}$ is also a signature of the message $\mathbf{m}$ for the random value $\mathbf{x}_v = z\mathbf{c}^{(i)}$. Then, it is clear that the range of random values $\mathbf{c}$ is the subset

$$\mathcal{C} := \left\{ z\mathbf{c}^{(1)}, z\mathbf{c}^{(2)}, \ldots, z\mathbf{c}^{(r)} \;\middle|\; z \in \mathbb{F}_q^\times \right\}.$$

4 Analysis of Randomness Used in Signature Generation of UOV Scheme

In this section, we analyze the security against randomness used in the signature generation of the UOV scheme. In particular, we consider a random message attack.

4.1 Preliminaries

We present a lemma to describe a random message attack in Sect. 4.2. In the attack, we consider recovering the twisted oil space $O \cdot S^{-1}$ in (2) from signatures. To do so, we need the following observation.

It is clear that for any element $\mathbf{w}$ in the oil space O, we have

$$f_1(\mathbf{w}) = 0, \ldots, f_o(\mathbf{w}) = 0,$$

which follows from the special form of each f_i in (1). Thus, it is also clear that for any element $\mathbf{w} \in O \cdot S^{-1}$, we have

$$p_1(\mathbf{w}) = 0, \ldots, p_o(\mathbf{w}) = 0,$$

since $p_i = f_i \circ S$.

To find elements $\mathbf{w} \in O \cdot S^{-1}$, we use signatures as follows.

Lemma 1 *Let $\mathbf{s}_1, \mathbf{s}_2 \in \mathbb{F}_q^n$ be two signatures for the UOV scheme. If $\mathbf{s}_1, \mathbf{s}_2$ are generated using the same random value $\mathbf{c} \in C$, that is, written as $\mathbf{s}_1 = (\mathbf{c}, \mathbf{d}_1) \cdot S^{-1}$ and $\mathbf{s}_2 = (\mathbf{c}, \mathbf{d}_2) \cdot S^{-1}$, then the difference $\mathbf{s}_1 - \mathbf{s}_2$ is in the twisted oil space $O \cdot S^{-1}$, and we have*

$$p_1(\mathbf{s}_1 - \mathbf{s}_2) = 0, \ldots, p_o(\mathbf{s}_1 - \mathbf{s}_2) = 0.$$

Proof This follows from the following computation:

$$\mathbf{s}_1 - \mathbf{s}_2 = (0, \mathbf{d}_1 - \mathbf{d}_2) \cdot S^{-1} \in O \cdot S^{-1}.$$

The second claim has already been stated above. □

Using this lemma, we consider a random message attack against the UOV scheme.

4.2 A Random Message Attack Against the UOV Scheme

We describe a random message attack for the case that random values $\mathbf{c}$ are chosen from

$$C = \left\{ z\mathbf{c}^{(1)}, z\mathbf{c}^{(2)}, \ldots, z\mathbf{c}^{(r)} \;\middle|\; z \in \mathbb{F}_q^{\times} \right\}.$$

First, we use $N := r + 2o$ signatures in our random message attack. Thus, N messages $\mathbf{m}_1, \ldots, \mathbf{m}_N$ are randomly chosen before key generation. After a designer generates a secret key $\{\mathcal{F}, S^{-1}\}$ and the public key $\mathcal{P} := \mathcal{F} \circ S$ of the UOV schemes, an attacker obtains signatures $\mathbf{s}_1, \ldots, \mathbf{s}_N$ for the messages $\mathbf{m}_1, \ldots, \mathbf{m}_N$.

Next, the attacker chooses a pair of signatures $\mathbf{s}$ and $\mathbf{s}'$ from $\{\mathbf{s}_1, \ldots, \mathbf{s}_N\}$. For $z \in \mathbb{F}_q^\times$, the attacker computes $\mathbf{w}_z = \mathbf{s} - z \cdot \mathbf{s}'$ and $p_1(\mathbf{w}_z), \ldots, p_o(\mathbf{w}_z)$. Assume that we can write $\mathbf{s}$ and $\mathbf{s}'$ as

$$\mathbf{s} = (z_1\mathbf{c}^{(i)}, \mathbf{d}_1) \cdot S^{-1}, \ \ \mathbf{s}' = (z_2\mathbf{c}^{(j)}, \mathbf{d}_2) \cdot S^{-1}.$$

If $i = j$ and $z = z_1/z_2$, then $\mathbf{w}_z = \mathbf{s} - z \cdot \mathbf{s}'$ is in the twisted oil space $O \cdot S^{-1}$ and $p_1(\mathbf{w}_z) = 0, \ldots, p_o(\mathbf{w}_z) = 0$ from Lemma 1. If the attacker obtains $p_1(\mathbf{w}_z) = 0, \ldots, p_o(\mathbf{w}_z) = 0$, then we can consider $\mathbf{w}_z \in O \cdot S^{-1}$. The attacker stores this $\mathbf{w}_z \in \mathbb{F}_q^n$. If there exists no such $z \in \mathbb{F}_q^\times$, the attacker chooses another pair of two signatures.

Since the attacker has $N = r + 2o$ signatures, he can store $2o$ such $\mathbf{w}_z$ at least. Let W be the subspace generated by the stored $\mathbf{w}_z$. Since W is a subspace of the o-dimensional space $O \cdot S^{-1}$ and generated by $2o$ vectors, W is equal to $O \cdot S^{-1}$ with an overwhelming probability. Therefore, the attacker can recover $O \cdot S^{-1}$.

Next, we choose an n-by-n invertible matrix S' such that $O \cdot S^{-1}S' = O$. Since $S^{-1}S'$ maps O to O, the quadratic polynomials $\{p'_1, \ldots, p'_o\} = \mathcal{P} \circ \mathcal{S}'$ are written as in the same form in (1), where $\mathcal{S}' : \mathbb{F}_q^n \to \mathbb{F}_q^n$ is a linear map such that $\mathcal{S}'(\mathbf{x}) = \mathbf{x} \cdot S'$. Thus, once such an $\mathcal{S}'$ is obtained, the attacker can forge any signature using the method of signature generation of UOV stated in Sect. 2.2. We call such $\mathcal{S}'$ an *equivalent secret key*.

Algorithm 1 describes our attack in detail.

Algorithm 1 Random message attack in 4.2

Require: The public key $\mathcal{P}$ and $N = r + 2o$ signatures $\mathbf{s}_1, \ldots, \mathbf{s}_N$.
Ensure: An equivalent secret key S'.

1: $W = \{0\}$;zero-subspace in $\mathbb{F}_q^n$
2: **while** $\dim_{\mathbb{F}_q} W \neq o$ **do**
3: Choose two signatures $\mathbf{s}_i$ and $\mathbf{s}_j$ $(1 \le i \ne j \le N)$
4: **for** $z \in \mathbb{F}_q^\times$ **do**
5: $\mathbf{w}_z = \mathbf{s}_i - z \cdot \mathbf{s}_j$
6: **if** $\mathcal{P}(\mathbf{w}_z) = 0$ **then**
7: $W = W + \mathbb{F}_q \cdot \mathbf{w}_z$ and return Step 2
8: **end if**
9: **end for**
10: **end while**
11: Choose a basis $\{\mathbf{w}_1, \ldots, \mathbf{w}_o\}$ of W
12: Choose elements $\mathbf{w}'_1, \ldots, \mathbf{w}'_v$ that generate a complementary subspace of W in $\mathbb{F}_q^n$
13: $S'' = \begin{pmatrix} \mathbf{w}_1 \\ \vdots \\ \mathbf{w}_o \\ \mathbf{w}'_1 \\ \vdots \\ \mathbf{w}'_v \end{pmatrix}$; n-by-n matrix
14: $S' = S''^{-1}$; an equivalent secret key

We estimate the complexity of our attack. The dominant part is from Steps 2–10 in Algorithm 1; i.e., the iteration of checking whether $\mathbf{w}_z = \mathbf{s}_i - z \cdot \mathbf{s}_j$ satisfies $\mathcal{P}(\mathbf{w}_z) = 0$ for each $\mathbf{s}_i, \mathbf{s}_j$ and $z \in \mathbb{F}_q^\times$. This is repeated $(q-1)\binom{r+2o}{2} = \mathcal{O}(q(r+2o)^2)$ times at most, where q is the cardinality of $\mathbb{F}_q$. Here, the complexity to compute $\mathcal{P}(\mathbf{w}_z)$ is $\mathcal{O}(n^2 o)$. Therefore, the complexity of Algorithm 1 is given by

$$\mathcal{O}(q(r+2o)^2 n^2 o).$$

If the number r of random values is a polynomial in n, then the complexity of our attack is polynomial time. In contrast, if the number r is an exponential in n, i.e., $q^{\mathcal{O}(n)}$, then the complexity is exponential. In particular, if we take $\mathcal{C} = \mathbb{F}^v$ as the original UOV scheme, then this attack is not practical.

4.3 *Experimental Results with Small r for the Random Message Attack in Sect. 4.2*

In this subsection, we confirm the random message attack in Sect. 4.2 by performing some experiments with a small r. Based on three main attacks stated in Sect. 2.3, we prepare two 128-bit security parameters (I) and (II) of the UOV scheme as follows:

$$\text{(I)}\ (q, v, o, n) = (4, 142, 71, 213)$$
$$\text{(II)}\ (q, v, o, n) = (31, 94, 47, 141)$$

All the experiments in this subsection were performed on a CPU with 2.3 GHz Intel Core i5 and Magma V2.24-5 [4]. Tables 1 and 2 present the experimental results of the random message attack in Sect. 4.2 for parameters (I) and (II), respectively. Here, we took $r = o, 2o, 4o$ as the number of $\mathbf{c}^{(i)}$ in

$$\mathcal{C} = \left\{ z\mathbf{c}^{(1)}, z\mathbf{c}^{(2)}, \ldots, z\mathbf{c}^{(r)} \,\middle|\, z \in \mathbb{F}_q^\times \right\}.$$

The number $N = r + 2o$ is that of the signatures required in our random message attack. In Tables 1 and 2, we describe the average times of 100 experiments to recover equivalent secret keys $\mathcal{S}'$ by Algorithm 1. These experimental results indicate that our attack is valid and is efficient for recovering an equivalent secret key if the number r of $\{\mathbf{c}^{(1)}, \mathbf{c}^{(2)}, \ldots, \mathbf{c}^{(r)}\}$ is small.

Table 1 Experimental result of our random message attack in 4.2 against the parameter (I). The number $N = r + 2o$ is that of the signatures required in our random message attack

The number r of random values $\mathbf{c}$	The number N of signatures	Experiment times
$o = 71$	213	10.0 s
$2o = 142$	284	20.2 s
$4o = 284$	426	44.1 s

Table 2 Experimental result of our random message attack in 4.2 against the parameter (II). The number $N = r + 2o$ is that of the signatures required in our random message attack

The number r of random values $\mathbf{c}$	The number N of signatures	Experiment times
$o = 47$	141	56.0 s
$2o = 94$	188	116.2 s
$4o = 188$	282	236.7 s

5 Conclusion

The UOV scheme is an important signature scheme in MPKC; further studies on the UOV scheme are required. In this paper, we studied the randomness used in the signature generation of the UOV scheme. We considered restricting the range of random values (randomness) to a subset C in $\mathbb{F}_q^v$. There are at least two situations in which such a restriction appears. The first one is for obtaining efficiency, and the second one is in an improper implementation of the UOV scheme. In this paper, we proposed a random message attack against the UOV scheme. Under the random message attack, by taking the differences of various signature pairs, an attacker can recover an equivalent key and thus can forge any signature. Our experiments showed that this attack is efficient if the subset C is small (i.e., polynomial size). As a result, we conclude that this type of restriction is vulnerable to a random message attack. Moreover, this result alerts us against the restriction of the range of random values to make the UOV scheme efficient. Note that since our attack utilizes an algebraic structure coming from the restriction, the original UOV scheme does not have such a vulnerability. As a future work, we would like to study the randomness of the Rainbow signature scheme.

Acknowledgements This work was supported by JST CREST Grant Number JPMJCR14D6 and JSPS KAKENHI Grant Number JP19K20266.

References

1. Bardet M, Faugère JC, Salvy B, Yang BY (2005) Asymptotic behavior of the index of regularity of quadratic semi-regular polynomial systems. In: 8th international symposium on effective methods in algebraic geometry (MEGA), pp 1–14
2. Bernstein DJ, Buchmann J, Dahmen E (eds) (2009) Post-quantum cryptography. Springer, Berlin
3. Bettale L, Faugère JC, Perret L (2009) Hybrid approach for solving multivariate systems over finite fields. J Math Cryptol 3:177–197
4. Bosma W, Cannon J, Playoust C (1997) The Magma algebra system. I. The user language. J Symb Comput 24:235–265
5. Ding J, Chen MS, Petzoldt A, Schmidt DS, Yang BY Rainbow, technical report, national institute of standards and technology, post-quantum cryptography. https://csrc.nist.gov/Projects/Post-Quantum-Cryptography/Round-2-submissions
6. Ding J, Gower JE, Schmidt DS (2006) Multivariate public key cryptosystems. Springer, Berlin
7. Ding J, Schmidt DS (2005) Rainbow, a new multivariate polynomial signature scheme, ACNS 2005, LNCS, vol 3531. Springer, pp 164–175
8. Ding J, Yang BY, Chen CHO, Che MS, Cheng CM (2008) New differential-algebraic attacks and reparametrization of Rainbow. ACNS 2008, LNCS, vol 5037. Springer, pp 242–257
9. Faugère JC (1999) A new efficient algorithm for computing Gröbner bases (F4). J Pure Appl Algebra 139:61–88
10. Faugère JC (2002) A new efficient algorithm for computing Gröbner Bases without reduction to zero (F5). ISSAC 2002:75–83
11. Garey MR, Johnson DS (1979) Computers and intractability: a guide to the theory of NP-Completeness. W.H. Freeman and Company
12. Hashimoto Y, Ikematsu Y, Takagi T (2018) Chosen message attack on multivariate signature ELSA at Asiacrypt 2017, IWSEC 2018. LNCS, vol 11049. Springer, pp 3–18
13. Kipnis A, Patarin L, Goubin L (1999) Unbalanced oil and vinegar schemes, EUROCRYPT 1999. LNCS, vol 1592. Springer, pp 206–222
14. Kipnis A, Shamir A (1998) Cryptanalysis of the oil and vinegar signature scheme. CRYPTO 1998. LNCS, vol 1462. Springer, pp 257–266
15. National institute of standards and technology, post-quantum cryptography standardization. https://csrc.nist.gov/projects/post-quantum-cryptography
16. National institute of standards and technology, post-quantum cryptography, round 3 submission. https://csrc.nist.gov/Projects/post-quantum-cryptography/round-3-submissions
17. Shim KA, Park CM, Koo N (2017) An existential unforgeable signature scheme based on multivariate quadratic equations. Asiacrypt 2017. LNCS, vol 10624. Springer, pp 37–64
18. Shor P (1997) Polynomial-time algorithms for prime factorization and discrete logarithms on a quantum computer. SIAM J Comput 26(5):1484–1509
19. Yang BY, Chen JM (2004) All in the XL family: theory and practice, ICISC 2004. LNCS, vol 3506. Springer, pp 67–86
20. Yasuda T, Dahan X, Huang Y-J, Takagi T, Sakurai K, MQ challenge: hardness evaluation of solving multivariate quadratic problems, IACR e-Print. https://eprint.iacr.org/2015/275.pdf
21. Zambonin G, Bittencourt M, Custódio R (2019) Handling vinegar variables to shorten rainbow key pairs, AFRICACRYPT 2019. LNCS, vol 11627. Springer, pp 391–408

On the Inadequacy of Multimodal Maps for Cryptographic Hash Applications with Phase-Space Modulation

Imad El Hanouti, Hakim El Fadili, Fawad Ahmed, and Jawad Ahmad

Abstract In this paper, we outline a specific problem concerning the use of multimodal chaotic maps in the core compression-construction of cryptographic hash functions. In cryptographic hash applications, the security level is measured according to their collision resistance property. This paper analyzes the strength of this property, by showing how it could be affected in the case where message units are modulated directly into the chaotic phase space of a multimodal map. Mathematical properties of multimodal maps are discussed and used to prove the risk of collision existence in such hash schemes. Four examples of widely used chaotic-maps in hash constructions are given and analyzed. Finally, we demonstrate the risk of collision existence using a practical example inherited from a previously proposed hash function.

Keywords Cryptographic hash function · Chaos-based hash function · Multimodal map · Collision analysis

1 Introduction

In information security, cryptographic hash function is a special kind of one-way function with some cryptographic properties. It produces a fixed-length digest from a variable-length message. The main cryptographic properties of a hash function are its one-wayness and its resistance to collisions [12]. In recent years, many cryptographic hash functions based on chaos theory have been proposed [8, 11, 15, 18]. The main

I. El Hanouti (✉) · H. El Fadili
LIPI Laboratory, Sidi Mohamed Ben Abdellah University, Fez, Morocco
e-mail: imad.elhanouti@usmba.ac.ma

F. Ahmed
Department of Electrical Engineering, HITEC University Taxila, Rawalpindi 47080, Punjab, Pakistan

J. Ahmad
School of Computing, Edinburgh Napier University, Edinburgh EH10 5DT, UK

© The Author(s), under exclusive license to Springer Nature Singapore Pte Ltd. 2022

D. Giri et al. (eds.), *Proceedings of the Seventh International Conference on Mathematics and Computing*, Advances in Intelligent Systems and Computing 1412,
https://doi.org/10.1007/978-981-16-6890-6_5

motivation behind this is to try using some chaotic models which may reduce the design complexity while keeping the same level of security. Moreover, the disclosure of some widely used hash functions (e.g., MD5, SHA-1 and RIPEMD-128) [3, 5] has laid to open the door in front of both academia and enterprises to provide new proposals based on new research directives such as neural networks, chaos theory and parallel computing [8, 11, 16, 18].

Regardless of the theoretical foundation behind the design of a cryptographic hash function, three main properties have to be met in order to consider the hash function as secure [12]:

(i) Pre-image resistance (or one-wayness): this property turns out to make it computationally infeasible to find any message that hashes to a given digest.
(ii) Second pre-image resistance (or weak collision resistance): giving a specific message, it is computationally infeasible to find any second message that hashes to the same hash value as the first one.
(iii) Collision resistance (or strong collision resistance): it is computationally infeasible to construct two distinct messages such that they hash to the same digest.

Any failure against one of these requirements makes the hash scheme cryptographically vulnerable and insecure.

Chaos-based hash functions are new models of hashing schemes that use chaotic functions as the main component in their compression function. Chaotic functions provide some good and essential requirements for hash functions like the one-wayness and highly sensitive dependence to initial conditions. Generally, the majority of proposed schemes try to modulate message information (i.e., message units) into the chaotic phase space or into the parameter space of the core chaotic map [4]. Nevertheless, so many chaos-based hash functions have been cryptanalyzed and disclosed in some subsequent works [7, 17].

Multimodal maps have been seemed also to be widely used in the construction of compression functions for hash schemes (one can refer, for example, to works presented in [8, 9, 11, 16]). Multimodal maps, discussed in this paper, are maps defined on a real interval I to itself, where I has a partition into finitely many subintervals on which the map is strictly monotone [13, 14]. In this paper, we will analyze one of the main cryptographic properties of cryptographic hash functions, that is, the collision resistance. This analysis will be carried out on one type of chaos-based hash function that uses message information modulation into the phase space of multimodal chaotic maps. And we will show that under these conditions, a potential risk of collision may exist. We demonstrate our results about the risk of collision using a practical example of such schemes inherited from the literature.

The rest of this paper is organized as follows: In Sect. 2, we will introduce some basic mathematical background related to our scope of study. In Sect. 3, we will analyze in depth the risk of collision existence in chaos-based hash schemes that use multimodal maps, and we will focus on weak collision analysis. In Sect. 4, we will give a practical demonstration based on a slightly modified scheme inherited from a real proposition, and we will show the critical risk of such vulnerability when dealing with binary digital images as an example. Main results and some recommendations are discussed briefly in Sect. 5. Finally, Sect. 6 concludes the paper.

2 Mathematical Introduction of the Scope Under Study

An m-dimensional discrete dynamical system (or map) is defined by the function f that updates the state $x = (x_1, x_2, \ldots, x_m) \in I$, where $I \subset \mathbb{R}^m$ is the state space such that $\mathbb{R}^m$ is an m-dimensional vector space, under the control parameter $p = (p_1, p_2, \ldots, p_d) \in M$, where $M \subset \mathbb{R}^d$ is the set of parameters, according to the following rule:

$$\begin{aligned} f : I \times M &\to I \\ (x, p) &\mapsto f(x, p) \end{aligned} \tag{1}$$

Thus, the map that maps a state $x_n \in I$ to the state $x_{n+1} \in I$ under the fixed control parameter p could be written as the difference equation:

$$x_{n+1} = f_p(x_n), n \in \mathbb{N} \tag{2}$$

In this paper, we will consider only one family of maps that is called *multimodal maps*.

Definition 1 For a one-dimensional map (i.e., $m = 1$), let $I \subset \mathbb{R}$; a multimodal map is defined by the function f that is continuous and piecewise monotone on I. That is, if f is strictly increasing on the subinterval $I_n \subset I$, then it is strictly decreasing on I_{n+1}.

Unimodal maps [2], and piecewise maps [13], all belong to this family. We denote this family of maps by $\mathcal{F}$.

We give four examples of maps that belong to the set $\mathcal{F}$ and are used in some chaos-based hash functions. These maps are

(a) *The logistic map*: The evolution rule of this map is shown in Fig. 1a and defined according to the following equation:

$$x_{n+1} = px_n(1 - x_n) \tag{3}$$

with $I = [0, 1], x_0 \in I$ and $M = (0, 4]$. The parameter space that leads to chaotic behavior is within the range [3.57, 4] excluding some non-chaotic windows. This map is used, for example, in hash functions proposed in [1, 16, 17].

(b) *The symmetric (full) tent map*: the evolution rule of this map is shown in Fig. 1b and given by

$$x_{n+1} = \begin{cases} \frac{x_n}{p}, & \text{if } 0 \le x_n < p \\ \frac{1-x_n}{1-p}, & \text{if } p \le x_n < 1 \end{cases} \tag{4}$$

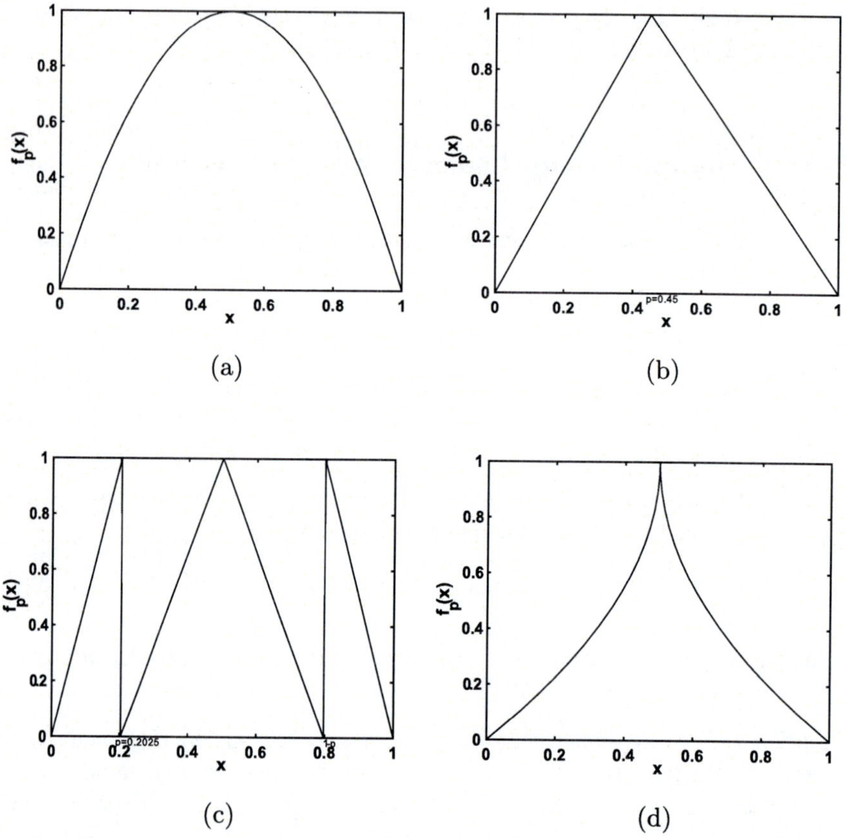

Fig. 1 Evolution rule of the four maps: **a** The logistic-map's evolution rule for $p = 4.0$. **b** The full tent map's evolution rule for $p = 0.45$. **c** The piecewise tent map's evolution rule for $p = 0.2025$. **d** Ref. [6] map's evolution rule for $p = 1$

with $M = (0, 1)$, $I = [0, 1]$ and $x_0 \in I$. This map is used, for example, in hash functions proposed in [8, 18].

(c) *The piecewise tent map*: This map is a generalization form of the full tent map. The evolution rule of this map is shown in Fig. 1c and defined as follows:

$$x_{n+1} = \begin{cases} \frac{x_n}{p}, & \text{if } 0 \leq x_n < p \\ \frac{x_n - p}{0.5 - p}, & \text{if } p \leq x_n < 0.5 \\ \frac{1 - x_n - p}{0.5 - p}, & \text{if } 0.5 \leq x_n < 1 - p \\ \frac{1 - x_n}{p}, & \text{if } 1 - p \leq x_n < 1 \end{cases} \tag{5}$$

with $M = (0, \frac{1}{2})$, $I = [0, 1]$ and $x_0 \in I$. This map is used, for example, in hash functions proposed in [8, 10].

(d) *The map used in* [6]: The evolution rule of this map is shown in Fig. 1d and defined as

$$x_{n+1} = \begin{cases} \sqrt{|2x_n - 1|}, & \text{if } p = 0 \\ 1 - \sqrt{|2x_n - 1|}, & \text{if } p = 1 \end{cases} \tag{6}$$

with $M = \{0, 1\}$, $I = [0, 1]$ and $x_0 \in I$.

3 Collision Analysis of Hash Functions Based on Multimodal Maps

According to Definition 1, we have the following lemma:

Lemma 1 *Let $f \in \mathcal{F}$ be a multimodal map and $p \in M$ be a fixed parameter value, then, there exists, a subset $U \subseteq I$, such that for all $y \in U$, there exist at least two distinct values $x_1, x_2 \in I$ such that $f_p(x_1) = f_p(x_2) = y$.*

Proof Let $I = [a, b]$, and $f \in \mathcal{F}$ strictly monotone (i.e., strictly increasing) on the interval $I_1 = [a, c] \subset I$, and strictly monotone (i.e., strictly decreasing) on the interval $I_2 = (c, b] \subset I$. Here p is a fixed parameter value.

Since f_p is continuous on I, then $f_p([a, c]) = [f_p(a), f_p(c)] = U_1 \subset I$ and $f_p((c, b]) = [f_p(b), f_p(c)) = U_2 \subset I$.

Let $U = U_1 \cap U_2$ then $U \neq \emptyset$ since $f_p(a) < f_p(c)$ and $f_p(b) < f_p(c)$. We have (by the fact of continuity and strict monotony of f_p) $\forall y \in U, \exists! x_1 \in [a, c] : f_p(x_1) = y$ and $\forall y \in U, \exists! x_2 \in (c, b] : f_p(x_2) = y$.

Thus, for any given value $y \in U$, there exist two distinct values x_1 and x_2 from the interval I (i.e., $x_1 \in [a, c]$, and $x_2 \in (c, b]$) such that $f_p(x_1) = f_p(x_2) = y$. □

Consider our four previous examples in Eqs. (3),(4),(5) and (6). According to the proof above, for every map of them we have

(a) For the logistic map, we have $U = \big[0, f_p(1/2)\big)$ and $f_p(x_1) = f_p(x_2 = 1 - x_1)$ with $x_1 \neq x_2$.
(b) For the full tent map, we have $U = \big[0, f_p(p)\big)$ and $f_p(x_1) = f_p(x_2)$ where $x_1 \in [0, p)$ and $x_2 = 1 - \frac{x_1(1-p)}{p} \in [p, 1]$.
(c) For the piecewise skew tent map, we have $U = \big[0, f_p(1/2)\big)$ and $f_p(x_1) = f_p(x_2) = f_p(x_3) = f_p(x_4)$ where $x_1 \in [0, p), x_2 = p + \frac{x_1(0.5-p)}{p} \in [p, \frac{1}{2}), x_3 = 1 - p - \frac{x_1(0.5-p)}{p} \in [\frac{1}{2}, 1 - p)$ and $x_4 = 1 - x_1 \in [1 - p, 1]$.
(d) For the map used in [6], we have $U = [0, f_p(1/2))$ and $f_p(x_1) = f_p(x_2 = 1 - x_1)$ with $x_1 \neq x_2$.

In other words, logistic map, full tent map and the map used in [6] are subjects of an easy construction of a second pre-image collision attack, while for the piecewise skew tent map, we can easily construct up to four other pre-images.

According to [4], three dominant modulation methods used in chaos-based hash functions exist: the message units are modulated into the chaotic map by influencing the phase space, parameter space or iteration times. For our analysis, we consider only the case where the message information is modulated into the phase space such as in works proposed in [1, 6, 16]. Since the compression function is the main component in every hash function, therefore, if a multimodal chaotic map f is used in the core compression function of the hash scheme, then often attacking the compression function leads to the total disclosure of the hash function.

According to Lemma 1, for any given message unit (after original message pre-processing and expansion) $m \in U$, there exists at least one other message entity $m' \in U$ such that $f_p(m) = f_p(m')$. Hence, we can theoretically construct at least one other message that does not affect the behavior of the chaotic evolution of the map. This message m' could lead to the construction of a second pre-image collision. The compression function in this particular case will not be sensitive to message changes and this could be represented as follows:

$$F\left(f_p(m), K\right) = F\left(f_p(m'), K\right) \quad (7)$$

where F is the compression function and K is the input state to the compression round.

To demonstrate the risk of collision existence in schemes that are based on multimodal chaotic maps, we adopt without loss of generality, a model of cryptographic hash function inherited from the design proposed in [16] that uses 2D coupled map lattices based on the logistic map. Then we conduct a collision analysis on it in the next section to show the vulnerability of using multimodal maps in cryptographic hash construction.

4 Collision Analysis of a Cryptographic Hash Function

In this section, we present a practical demonstration to show the risk of collision existence in cryptographic hash schemes that uses message information modulation into the chaotic phase space of a multimodal map. We rely on a real model proposed in [16] to build a slightly modified version, and we conduct a weak collision (i.e., a second pre-image) analysis on it. We briefly describe the model proposed in [16] in the following steps:

Step 1. Take an n-bit length input message M and pad it with zero bits in order to make it divisible into R blocks of 128 bits. Then arrange blocks m_i to construct a new message M' according to the following manner:

$$IM = M||pad = m_1||m_2||m_3||\dots||m_R \tag{8}$$

$$M' = m_1||m_2||m_3||\dots||m_R||m_R||m_{R-1}||\dots||m_1 \tag{9}$$

where $||$ denotes the concatenation of two bit streams.

Step 2. Map each message byte b of M' to its corresponding float value c in a new message sequence (containing $2R$ blocks of float numbers) $C = C_1C_2 \dots C_R C_R C_{R-1} \dots C_1$ according to a linear transformation:

$$c = s.b + r \tag{10}$$

where $s = 1/256$ and $0 \leq r \leq 1/256$.

Step 3. Construct the starting lattice $L_0 = \{x_{i,j}\}$, where $1 \leq i \leq 4$ is the row index of the lattice and $1 \leq j \leq 8$ is its column index, as follows:

$$\begin{cases} x_{1,j} = C_0(j) \\ x_{2,j} = C_1(j) \\ x_{3,j} = C_0(j+8) \\ x_{4,j} = C_1(j+8) \end{cases} \tag{11}$$

where $C_1(j)$ is the jth element of the block C_1 and C_0 is the block resulting from the application of the linear transformation in Eq. (10) to the following initial fixed 16 bytes IV =(01,23,45,67,89,AB,CD,EF,FE,DC,BA,98,76, 54,32,10).

Step 4. Iterate the current lattice L_n (L_0 if no block has been processed yet) for $k = 45$ times to get a final round lattice L_{n+45} according to the following coupled map lattice, i.e., 2D CML:

$$x_{i,j}^{n+1} = f_\mu(x_{i,j}^n) + f_\epsilon(x_{i+1,j}^n) + f_\epsilon(x_{i,j+1}^n) \tag{12}$$

where f is the logistic map as depicted in Eq (3), $\mu = 3.8$ and $\epsilon = 0.1$ is the coupling constant. Row and column indexes are summed modulo 4 and 8, respectively, $x_{i,j}^n$ are elements of the lattice L_n and $x_{i,j}^{n+1}$ are elements of the lattice L_{n+1} that results from iterating the lattice L_n once.

Step 5. If there are blocks in C that are not yet processed, then update the last lattice according to the following rule and go to 4:

$$\begin{cases} x_{2,j} = C_l(j) \\ x_{4,j} = C_l(j+8) \end{cases} \tag{13}$$

where l is the order of the current block.

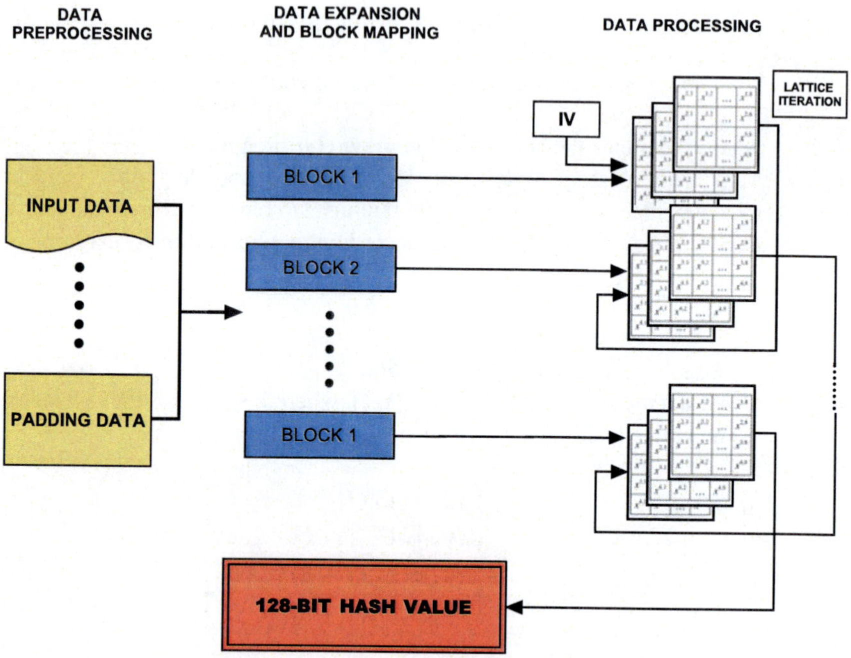

Fig. 2 General flowchart of the hash algorithm proposed in [16]

Step 6. After processing all blocks in C, we extract a 128-bit hash value H from the last lattice by extracting 8-bits (from $9th$ to $16th$) from the float point representation of some elements in $\{x_{i,j}\}$. These elements are taken in the raster order such that $i = 1, 3$ and $j = 1, 2 \ldots 8$.

For a detailed description of the scheme and its parameters, one can refer to [16]. Figure (2) shows the general flowchart of the process of the hash function using the described process

Our version is very slightly changed form of the original one. Considering Eq. (10), in [16], they take $b = 0.8/256$. In this paper, we take $b = 0.5/256$ to make our demonstration more concise and practical.

We know that the map in Eq. (12) is a multimodal map since the logistic map Eq. (3) is a multimodal one. From Lemma 1 and from example (a in the previous section, we know that

$$f_p(x_{i,j}) = f_p(1 - x_{i,j}) \tag{14}$$

for every parameter value p.

Thus, for every element $x_{i,j}$ in the lattice L_n, we can construct another lattice $\hat{L_n}$ such that we can modify any desired element $x_{i,j}$ by $(1 - x_{i,j})$ without affecting the resulting iterated lattice L_{n+1}.

For this reason, for each character (i.e., byte) b, we can easily construct a different one $\hat{b}$ that gives the same hash value. This attack is a second pre-image attack (or weak collision attack). The relationship between b and $\hat{b}$ is given by

$$\frac{\hat{b}+0.5}{256} = 1 - \frac{(b+0.5)}{256} \tag{15}$$

that is,

$$\hat{b} = 255 - b = \bar{b} \tag{16}$$

where $\bar{b}$ is the complement of b in $\mathbb{Z}/256\mathbb{Z}$. To demonstrate our attack with real examples, we take an arbitrary two input message-texts:

T_1 ="The parameter μ is equal to 0.5"
T_2 ="HE IS THE LEADER OF OUR TRAINED STAFF"

with their hash values, respectively, as follows:

$H(T_1)$ =EDECC3E0B7EFD7BBD1E6EDD3EFE1E3C1
$H(T_2)$ =BAD9E6ECEEEECFDECFEAE7EAE6BDCBE9

Then we flip bits of the character 'μ' in the message T_1 that yields the character 'J' in the message $\hat{T}_1$, and we change the word 'trained' in the message T_2 to the word 'raine' in the message $\hat{T}_2$.Then we hash the two changed messages:

$\hat{T}_1$ ="The parameter J is equal to 0.5"
$\hat{T}_2$ ="HE IS THE LEADER OF OUR «RAINE» STAFF"

their hash values are re-calculated, and remain unchanged:

$H(\hat{T}_1)$ =EDECC3E0B7EFD7BBD1E6EDD3EFE1E3C1
$H(\hat{T}_2)$ =BAD9E6ECEEEECFDECFEAE7EAE6BDCBE9

That is, $H(T_1) = H(\hat{T}_1)$ and $H(T_2) = H(\hat{T}_2)$. The hash values remain unchanged which demonstrates the second pre-image collision existence in this hash function.

In general, for every message with l byte length (after padding), we can easily construct up to 2^l distinct messages with the same length l such that all of them hash to the same value H. To demonstrate the gravity of this result, we consider a set of digital images that are encoded on 8-bits/pixel and that consist of only two colors, black and white (i.e., binary images, scanned books ...). We denote this set by $\mathcal{M}$.

Property 1 *Let $\mathcal{M}_{w,h} \subset \mathcal{M}$ be the set of digital images that belongs to $\mathcal{M}$ and that they have the same size (width, height). Let $M_0 \in \mathcal{M}_{w,h}$ be the all blank image, i.e., all bits are set to 1, of size (w, h). Then, $\forall M \in \mathcal{M}_{w,h} : H(M) = H(M_0)$*

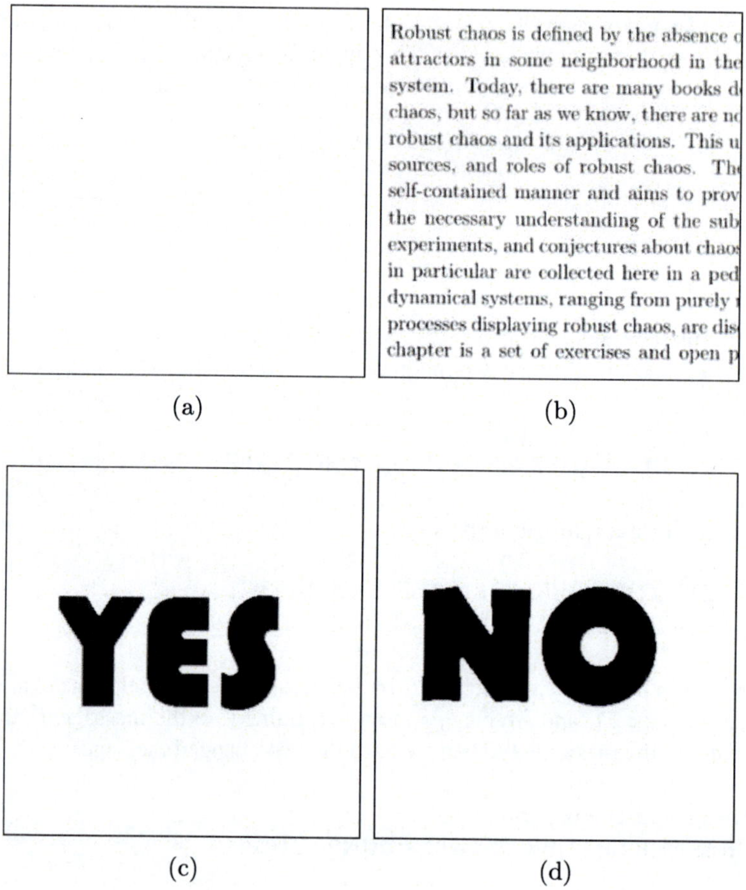

Fig. 3 The four images used in our collision demonstration: **a** A blank image. **b** A snippet of text from [19]. **c** YES image. **d** NO image

Proof Trivially prove from the fact that every pixel in M is either black or white (0 or 255). Since ($255 = \bar{0}$), then according to Eq. (16) all images of the set $\mathcal{M}_{w,h}$ will have the same hash value. □

To demonstrate our last result with real examples, we consider the four images belonging to the set $\mathcal{M}_{300,300}$ shown in Fig. 3a–d. We calculate their hash values, respectively, $H(m_1)$, $H(m_2)$, $H(m_3)$ and $H(m_4)$:

$H(m_1)$ =E3EEE9EAD6EDEFEFCBEFEECCEDBFD7EA
$H(m_2)$ =E3EEE9EAD6EDEFEFCBEFEECCEDBFD7EA
$H(m_3)$ =E3EEE9EAD6EDEFEFCBEFEECCEDBFD7EA
$H(m_4)$ =E3EEE9EAD6EDEFEFCBEFEECCEDBFD7EA

We have $H(m_1) = H(m_2) = H(m_3) = H(m_4)$, which confirms our results about the hash collision vulnerability.

5 Discussion

Chaotic maps exhibit good characteristics and respond to a lot of requirements demanded by information security applications. However, the choice of the model should be well justified and well-studied before direct implementation in cryptographic applications. In the case of cryptographic hash applications, we do not recommend the direct use of multimodal maps in the core compression function. But they can be used as secondary components to extract the required entropy used in compression functions. The use of higher dimensional maps and multi-parameter ones along with parameter modulation could make it extremely difficult to construct message collisions in hash schemes compared to simple multimodal mappings. This last judgment should be studied and proved in further works. However, chaos-based models could offer more benefits to the domain of information security, but every choice must be well analyzed before being implemented in real schemes. We recommend the adoption of deep security analysis over newly proposed schemes to enlarge the field of security results and to enhance the robustness of further proposals.

6 Conclusion

In this paper, we analyzed multimodal maps and their use in cryptographic hash functions. We have shown that under the direct modulation of message information into the chaotic phase space, there exists a high risk of weak collision. To show the gravity of our result, we did a demonstration on a very slightly modified version inherited from a real construction proposed in the literature. We easily constructed some second pre-images to affirm the inadequacy and the weakness of using a multimodal map in the compression function. Those layouts are rarely analyzed; thereafter much other analysis (namely cryptanalysis) must be achieved. Hence, we do not recommend the use of multimodal maps in the core compression function of cryptographic hash schemes. This result should be considered in further works, and the selection of certain chaotic maps in hash function constructions should be studied more in depth.

References

1. Ahmad M, Khurana S, Singh S, AlSharari HD (2017) A simple secure hash function scheme using multiple chaotic maps. 3D Research. https://doi.org/10.1007/s13319-017-0123-1
2. Arroyo D, Amigó J, Li S, Alvarez G (2010) On the inadequacy of unimodal maps for cryptographic applications. First publ. In: RECSI 2010 : IX [i.e. XI] Reunión Española sobre Criptología y Seguridad de la Información, Tarragona 7-10 de septiembre 2010 ; coordinado por Josep Domingo Ferrer ... Tarragona : Publicacions URV, 2010, pp 37–42 (01 2010)
3. Biham E, Chen R, Joux A (2015) Cryptanalysis of sha-0 and reduced sha-1. J Cryptol 28(1):110–160
4. Kocarev L, Lian S (2011) Chaos-based cryptography: theory, algorithms and applications, vol 354. Springer Science & Business Media
5. Landelle F, Peyrin T (2016) Cryptanalysis of full ripemd-128. J Cryptol 29(4):927–951. https://doi.org/10.1007/s00145-015-9213-5
6. Li HD, Feng DG (2003) Composite nonlinear discrete chaotic dynamical systems and keyed hash functions. Jisuanji Xuebao/Chinese J Comput
7. Li P, Zhang M, Shi J (2012) The design problems of one kind of chaotic one-way hash function. In: 2012 fourth international conference on multimedia information networking and security. IEEE, pp 592–594
8. Li Y, Ge G (2019) Cryptographic and parallel hash function based on cross coupled map lattices suitable for multimedia communication security. Multimed Tools Appl 78(13):17973–17994. https://doi.org/10.1007/s11042-018-7122-y
9. Li Y, Xiao D, Deng S (2012) Secure hash function based on chaotic tent map with changeable parameter. High Technol Lett. https://doi.org/10.3772/j.issn.1006-6748.2012.01.002
10. Li Y, Xiao D, Deng S, Han Q, Zhou G (2011) Parallel Hash function construction based on chaotic maps with changeable parameters. Neural Comput Appl https://doi.org/10.1007/s00521-011-0543-4
11. Liu GJ, Shan L, Dai YW, Sun JS, Wang ZQ (2006) One-way Hash function based on chaotic neural network. Wuli Xuebao/Acta Physica Sinica. https://doi.org/10.7498/aps.55.5688
12. Paar C, Jan P (2009) Understanding cryptography : a textbook for students and practitioners. Springer, Berlin, London. https://search.library.wisc.edu/catalog/9910092139102121
13. Sahnoune A, Berkani D (2018) On the correlation of chaotic signals generated by multimodal skew tent map. Signal Image Video Proc. https://doi.org/10.1007/s11760-018-1279-8
14. Van Strien S, Vargas E (2004) Real bounds, ergodicity and negative schwarzian for multimodal maps. J Am Math Soc. https://doi.org/10.1090/S0894-0347-04-00463-1
15. Wang S, Hu G (2007) Hash function based on chaotic map lattices. Chaos. https://doi.org/10.1063/1.2735812
16. Wang Y, Liao X, Xiao D, Wong KW (2008) One-way hash function construction based on 2D coupled map lattices. Inf Sci. https://doi.org/10.1016/j.ins.2007.10.008
17. Xiao D, Liao X, Wong KW (2006) Improving the security of a dynamic look-up table based chaotic cryptosystem. In: IEEE transactions on circuits and systems II: express briefs. https://doi.org/10.1109/TCSII.2006.875336
18. Yi X (2005) Hash function based on chaotic tent maps. In IEEE transactions on circuits and systems II: express briefs. https://doi.org/10.1109/TCSII.2005.848992
19. Zeraoulia E, Sprott JC (2011) Robust chaos and its applications. World Sci. https://doi.org/10.1142/8296, https://www.worldscientific.com/doi/abs/10.1142/8296

Image Encryption with Parallel Evolution of 2-D Cellular Automata

Shramona Chakraborty, Soma Kundu, and Dipanwita Roy Chowdhury

Abstract The paper introduces an image encryption scheme using two-dimensional Cellular Automata (CA). The random evolution of CA along with masking makes the scheme secure as well as very well suited for software encryption. Interestingly, the model has achieved a noticeable speedup using a window-based parallel implementation of CA. Security analysis has been performed and the results show that the scheme with high diffusion and confusion is strong against correlation and differential attacks.

Keywords Cellular automata · Image encryption · Symmetric-key cipher

1 Introduction

In recent times, the requirement of the multimedia application is surging with the progress of digital communication. Multimedia includes text, image, audio, video, and animation. Among them, image plays a crucial role, and assuring its security has become pertinent in the storage and communication system. Image encryption is a process of encoding image with the help of some encryption algorithm in such a way that unauthorized users do not get access to it. Various well-known methods are the Chaos-based method [1], SCAN-based method [2], Tree structure method [3], etc. The conventional methods for image encryption are RGB color shuffling, bit manipulation, chaotic mapping method, etc. Image encryption technique using elementary Cellular Automata (CA) is also very popular [4]. In cryptography, the first approach of using CA has been proposed by Wolfram Stephen [5] which is a stream cipher constructed on a simple one-dimensional (1-D) CA. The simple and

S. Chakraborty (✉) · S. Kundu · D. R. Chowdhury
Crypto Research Lab, Department of Computer Science and Engineering, Indian Institute of Technology, Kharagpur 721 302, India
e-mail: shramo.cse@iitkgp.ac.in

D. R. Chowdhury
e-mail: drc@cse.iitkgp.ac.in

© The Author(s), under exclusive license to Springer Nature Singapore Pte Ltd. 2022
D. Giri et al. (eds.), *Proceedings of the Seventh International Conference on Mathematics and Computing*, Advances in Intelligent Systems and Computing 1412,
https://doi.org/10.1007/978-981-16-6890-6_6

regular structure of CA along with good pseudo-random evolution properties eases its utility. For image encryption, the pseudo-random numbers generated by CA are used to encrypt each pixel of the image by XOR or substitution operation [6–8]. Alvarez et al. [9] proposed one image encryption scheme based on 1-D reversible CA. Here, the next state of a cell has been set on by both current and previous states of the cell's neighbors. The cipher image has been outlined by the identical color palette as the plain image but the size is double the size of the plain image. Afterwards, Maleki et al. [10] modified the model at the cost of the image quality where the Least Significant Bit (LSB) of the image is not recoverable, hence the cryptosystem is lossy. Yu et al. [11] introduced an idea of encrypting an image based on symmetrical-couple toggle 1-D CA. The method discards the image quality loss and cipher image enlargement but there is no interaction between pixels as each pixel has been encrypted independently. Hence, the algorithm does not depict global sensitivity property.

The above-mentioned researches introduce a new area of work to explore. An image can be represented by a two-dimensional (2-D) matrix and this feature has been exploited by using 2-D CA to encrypt and decrypt images. Both the structures fit exactly with each other making the computation easier. In 2-D CA, the cells are positioned in a 2-D grid and it is the generalization of 1-D CA. Chaudhuri et al. [12] proved that 2-D CA generates better random patterns than 1-D CA. Wang et al. [13] introduced an image encryption algorithm using 2-D partitioned CA. Chen et al. [14] have also presented an image encryption system using 2-D progressive CA. Recently, Choi et al. [15] have introduced an image encryption algorithm using programmable complemented maximum length CA. Satyabrata et al. [16] have proposed an image encryption algorithm using 2-D CA which requires three different keys for a single RGB image channel. In this work, a novel image encryption algorithm using elementary 2-D CA has been proposed. This is a symmetric-key block cipher. The proposed system follows the general structure called iterated product cipher [17], which constitutes a continual application of substitution. This kind of framework is used in many modern encryption methods including Advanced Encryption Standard, Data Encryption Standard, etc. The proposed model is secured from brute-force, cipher text, differential, and statistical attacks and has achieved significant performance enhancement utilizing parallel computing. A number of comparisons in security analysis are made with existing recent algorithms which involves both CA and non-CA methodologies.

The organization of rest of the paper is as follows: The background of the 2-D CA is delivered in Sect. 2. In Sect. 3 the proposed method of image encryption is presented. Section 4 gives the security analysis and outcomes. Lastly, the conclusion has been drawn in Sect. 5.

2 Fundamentals of Two Dimensional (2-D) Cellular Automata (CA)

2-D CA consists of a regular grid of cells. A cell is a machine or object which remains in a particular state at a particular time. The cell can take values from a finite set called alphabet S. Neighborhood is a finite-order subset that denotes all the neighboring nodes corresponding to a particular cell. CA consists of 4 tuples (d, N, S, f), where d defines the dimension, N is the neighborhood, S is the set of possible states, and f is the transition function. Here, a single cell can choose two types of neighborhoods. In von Neumann neighborhood (Fig. 1a), a cell has 4-neighbor cells (top, bottom, left, right), whereas in Moore neighborhood (Fig. 1b) a cell has 8-neighbor cells (left, top, right, down, top-left, bottom-left, top-right, bottom-right). In this paper, a single cell's 5-neighbors are considered where the particular cell and its von Neumann neighborhood are taken. The combination has $2^5 = 32$ possible configurations of states.

Rule of 2-D CA: The CA structure can be observed as a discrete lattice [4] of cells where each cell stays in either state 0 or 1. A 5-neighborhood 2-D CA can be depicted as the following notations:
i, j: Row and column of a single cell in the 2-D array
t: Time step
$q_{i,j}(t)$: stores the state of an (i, j)th cell after tth time step.
$q_{i,j}(t+1)$: stores the state of an (i, j)th cell after $(t+1)$th time step.
The expression for the state transition function of a 5-neighborhood cell is

$$q_{i,j}(t+1) = f(q_{i,j}(t), q_{i-1,j}(t), q_{i,j-1}(t), q_{i+1,j}(t), q_{i,j+1}(t))$$

where the f is a local transition function known as a rule of the CA. Horizontal dependency constitutes the states: $q_{i,j-1}(t)$, $q_{i,j}(t)$, and $q_{i,j+1}(t)$, whereas vertical dependency constitutes the states: $q_{i-1,j}(t)$ and $q_{i+1,j}(t)$. It can be observed that these neighborhood configurations have $2^5 = 32$ distinct arrangements possible, hence, 32 bits are required to represent the transition rule of 2-D CA which is impractical to manage.

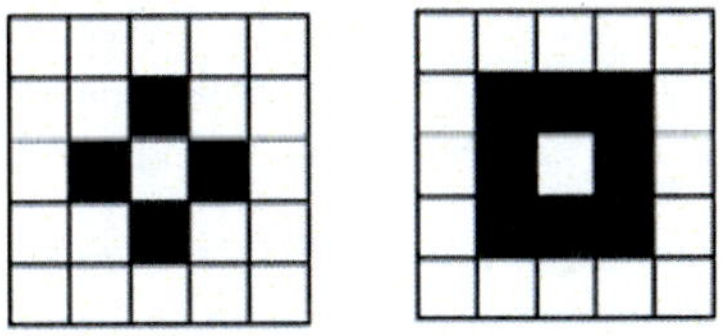

Fig. 1 Types of neighborhood

(a) von Neumann **(b)** Moore

2.1 Linear Algebraic Characterization of CA

A N-cell 1-D CA is denoted by an $N \times N$ characterization matrix over GF(2). The construction of the matrix is as follows:

$$T_{ij} = \begin{cases} 1, \text{if the current state of jth cell depends on the next state of ith cell} \\ 0, \text{otherwise} \end{cases} \tag{1}$$

The next state of the CA with the corresponding rule is achieved by matrix multiplication of T from Eq. 1 with the current state of the CA [12]. To characterize a 2-D RVN (Restricted Vertical Neighborhood where either top neighbor or bottom neighbor from vertical dependency is considered) CA [12], a $M^2 \times N^2$ sized characterization matrix is used. Here, all the 5-neighborhoods are considered by different matrices of size $M \times N$. The 2-D CA matrix of size $M^2 \times N^2$ is represented by T as shown in Eq. 2 where each of the matrices S_i, B_i, and C_i are of size $M \times N$. S_i, B_i, and C_i are horizontal, bottom, and top dependencies of row i, respectively. After calculating dependencies, the resultant state is calculated by bit wise XOR operations. The 2-D CA is defined as uniform CA if every S_is are identical. Otherwise, it is called a hybrid CA.

$$T = \begin{bmatrix} S_1 & B_1 & & \\ C_2 & S_2 & B_2 & \\ & & & \\ & & & B_{M-1} \\ & & & S_M \end{bmatrix} \tag{2}$$

Group CA: If a CA forms a cyclic group under some transformation operation T, then for all $f(x)$ where f is a transition function, there should be an integer M such that Eq. 3 is satisfied [12].

$$T^M = I \tag{3}$$

$$f_M(x) = T^M \cdot f(x)$$

$$det(T) = 1$$

The CA which has such a property is called Group CA. Every state of such type of CA lies in some cycle. Various 2-D Group CA properties referred from [12] are applied in the proposed image encryption algorithm and are described below. Rule matrices are restricted according to the properties of Group CA.

Theorem 1 *A 2-D RVN CA is a group CA iff each row of the CA, examined as 1-D, neglecting the bottom and top dependencies also constitutes a group CA.*

Theorem 2 *For a $M \times N$ group CA, the cycle lengths will be the factors of $lcm(r, k)$, where r is the smallest integer for which $S^r = I$. Here, I is the identity matrix, S is the horizontal dependency matrix, and $k = 2^{\lceil log_2 M \rceil}$.*

Theorem 3 *For any group RVN CA, the maximum possible cycle length is* $2^{\lceil log_2 M \rceil}(2^N - 1)$.

3 Image Encryption Using 2-D CA

The proposed algorithm is illustrated on color images where the RGB channels are treated as three separate 2-D matrices. Each of the channel runs parallelly and merges together to produce the final encrypted image. Every single pixel of an image is of by 8 bits. The prime objective of the image encryption scheme is to replace the pixel values of the plain image with some unpredictable values. The functionalities of the total algorithm is represented in Fig. 2.

It can be seen in Fig. 2 that the transposer function is applied as a pre- and post-computation for encryption and decryption, respectively. The backbone of the algorithm is 2-D CA evolution followed by masking. To compute the 2-D CA evolution, a group rule is selected which satisfies Theorem 1 and is represented as 2-D rule matrix which is chosen as the **key**. The evolution and masking procedures are then iterated N_C number of times, where "N_C" is the number of cycles used for encryption. Subsequently, for decryption, $N_T - N_C$ number of cycles is used. Here, "N_T"

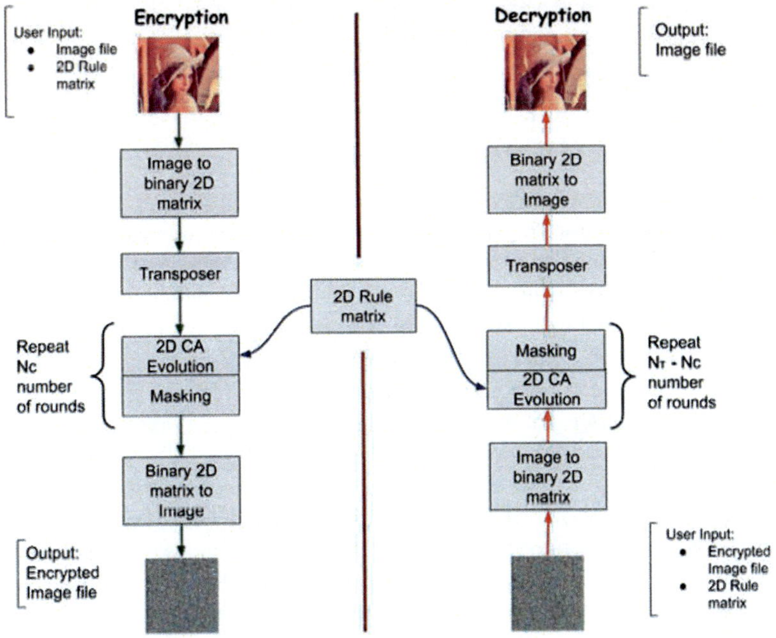

Fig. 2 Block diagram of Image encryption and decryption

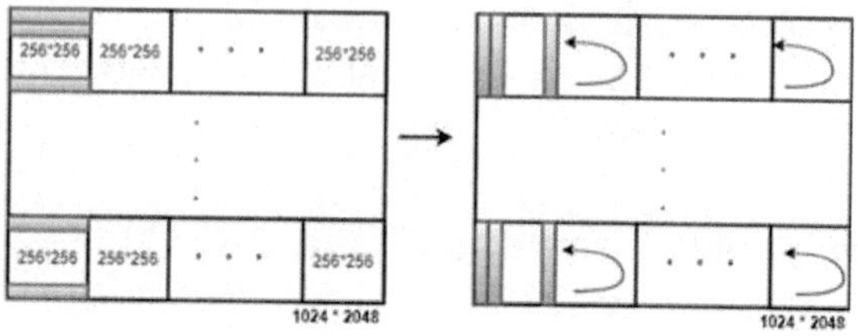

Fig. 3 Transposer example

is the cycle length chosen from the group rule for encryption and decryption. The output obtained after N_C and $N_T - N_C$ number of cycles are converted into image format to get the encrypted and decrypted images, respectively.

3.1 Transposer

Transposer function is used to jumble the effect of edges in an image. For an image of dimension $M \times P$, the binary matrix of that image is of size $M \times N$ (where $N = 8 \times P$, as each pixel value ranges from 0 to 255). The binary matrix $B_{M \times N}$ is divided into $(M \times N)/(x \times x)$ sub-matrices, where $x \leq M/2$ and $x \mid M$. After that, the transpose matrix operation is applied parallelly to each sub-matrix and placed in their respective position. An example is shown in Fig. 3, here a binary matrix of size 1024×2048 is divided into 32 sub-matrices where x is 256.

3.2 Encryption and Decryption

Encryption and decryption consist of two parts: 2-D CA Evolution and Masking. A 2-D rule matrix is chosen of size $n \times n$. Here, n is the window size which can be varied. The computation using different window sizes is discussed later. The binary matrix $B_{M \times N}$ is divided into $(M \times N)/(n \times n)$ different sub-matrices $b_{n \times n}$ of size $(n \times n)$ each. If the size of the B matrix is not divisible by the size of the rule matrix, reduction in size is done. The rows and columns sizes are reduced till they are multiple of n, remaining parts are left as it is.

2-D CA Evolution: At a time, a 2-D rule matrix and one sub-matrix are taken. For each value b_{ij} of the sub-matrix where $i \in [\,0, n-1]$ and $j \in [\,0, n-1]$, the next state of that value is generated by capturing the 5-neighborhoods of the sub-matrix. Now according to the 2-D rule matrix, XOR operation is applied. Here, null-boundary CA is considered. The procedure is defined in Algorithm 1.

Algorithm 1: 2-D CA Evolution

```
Input: 2-D sub-matrix b and 2-D rule matrix R of size n × n
Output: Next state of 2-D sub-matrix b
Function Evolution(b,R):
    NextVal ← 0
    for i ← 0 to n do
        for j ← 0 to n do
            Left ← j − 1, Self ← j, Right ← j + 1          // Horizontal dependencies
            Top ← i − 1, Bottom ← i + 1                    // Vertical dependencies
            if R_{i,Self}==1 then
                NextVal = NextVal ⊕ b[i][Self]             // Checking for Self-position
            if R_{i,Left}==1 then
                NextVal = NextVal ⊕ b[i][Left]             // Checking for Left-position
            ⋮
            /* Same way, check for rest of the positions                  */
            b[i][Self] ← NextVal
    return b
```

Masking: At a time, an initial matrix of size $n \times n$ and one sub-matrix are taken. The initial matrix is a logical matrix and each value is XORed with the respective sub-matrix's value. Masking is applied repeatedly by shifting one bit of both row and column of the matrix $B_{M \times N}$ to mix the effect of the boundary.

Both the procedures can run parallelly as each cell gets updated simultaneously and independently. The parallel computation results in speeding up the time needed to obtain the solution and the advantage of using it is described in Sect. 4.4. The encryption operation runs for "N_C" number of times and the decryption operation runs for "$N_T - N_C$" number of times. Here, N_C is taken as a parameter for encryption and decryption methods and N_T is the total number of cycles which is calculated as $N_T = lcm(r, k)$, where r and k are described in Theorem 2. Both N_C and N_T depends on the chosen 2-D rule matrix. The entire encryption algorithm is described in Algorithm 2.

Encryption Using Window Size Variation: Results are observed using 1000 varieties of images of different sizes with various rule matrices and a different number of iteration steps. Four random results are shown in Fig. 4 using rule matrices $R1_{5\times5}$ and $R2_{6\times6}$. The maximum cycle length N_T for both R1 and R2 is 120.

Algorithm 2: Image Encryption Algorithm using 2-D CA

```
Input: Image file, Key = 2-D rule matrix R
Output: Encrypted image file
Function Transposer(B):
    for each sub-matrix b_{i,j} in B do
        Transpose b_{i,j}
    return B

Function EncryptionAlgorithm(B, R, N_C):
    k ← 1
    while k ≤ N_C do
        for each sub-matrix b_{i,j} in B do
            Evolution (b, R)
            Mask row-wise b_{i,j}
            Mask column-wise b_{i,j}
        k ←k + 1
    return B

Function Main:
    /* Convert the image file into 2-D binary matrix B. Use N_C
       as the number of iterations for encryption                */
    Transposer(B)
    EncryptionAlgorithm(B, R, N_C)
    Write the encrypted image file as output from the resultant binary matrix B
    return 0
```

$$R1 = \begin{bmatrix} 16 & 17 & 21 & 5 & 20 \\ 24 & 25 & 29 & 5 & 20 \\ 24 & 25 & 29 & 13 & 20 \\ 24 & 25 & 29 & 5 & 28 \\ 16 & 25 & 29 & 5 & 28 \end{bmatrix} \qquad R2 = \begin{bmatrix} 17 & 5 & 17 & 21 & 5 & 20 \\ 25 & 13 & 25 & 29 & 5 & 28 \\ 25 & 13 & 25 & 29 & 13 & 28 \\ 25 & 5 & 25 & 21 & 13 & 28 \\ 25 & 5 & 17 & 29 & 13 & 28 \\ 25 & 13 & 25 & 29 & 13 & 28 \end{bmatrix}$$

The number of evolution steps for encryption (N_C) for R1 and R2 is 89. The remaining 31 number of evolution steps are used with the encrypted image (shown in Fig. 4) to regain the image.

4 Security Analysis

The performance of the encryption method is measured by its ability to withstand all kinds of security attacks like the brute-force attack, cipher text attack, statistical attack, differential attack, etc.

(a) Image-1 (lena.jpeg)
(b) Encrypted Image-1 using Rule $R1_{(5\times5)}$
(c) Encrypted Image-1 using Rule $R2_{(6\times6)}$
(d) Decrypted Image-1 (lena.jpeg)
(e) Image-2 (Fruit.jpeg)
(f) Encrypted Image-2 using Rule $R1_{(5\times5)}$
(g) Encrypted Image-2 using Rule $R2_{(6\times6)}$
(h) Decrypted Image-2 (Fruit.jpeg)

Fig. 4 Original, encrypted, and decrypted images

4.1 Key-Space Analysis

The von Neumann neighborhood (Fig. 1a) is considered in our algorithm, so each entry of the rule matrix can have $2^5 = 32$ number of possible values. Thus, the possible number of 2-D rule matrices of size $n \times n$ for a 2-D RVN CA (5-neighborhood) are 32^{n^2}. Considering $n = 6$, the possible number of rule matrices are $32^{6^2} = 2^{180}$ which is computationally infeasible for brute-force attack as in Modern Cryptography, 2^{128} is termed as secure [17].

4.2 Statistical Analysis

Statistical analysis is a powerful method of cryptanalysis of different ciphers [4]. Confusion and diffusion of ciphers are treated as a quantitative measure of security against statistical attacks.

Confusion property: Confusion property increases ambiguity and conceals the relation between encrypted output and the key. To define the confusion property, different categories of images of various sizes are taken for histogram calculation. An image histogram is a type of histogram that calculates pixel intensity value. A few examples

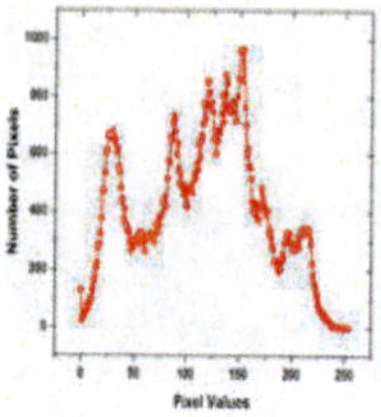

(a) Histogram of lena image-4a

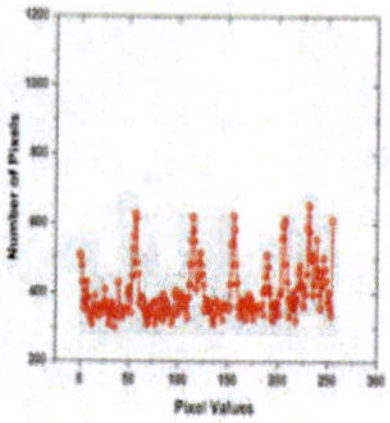

(b) Histogram of encrypted image-4b using $R1_{(5\times5)}$

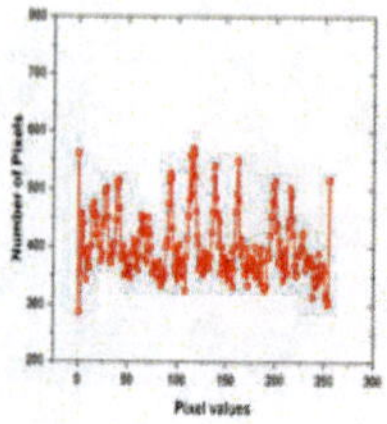

(c) Histogram of encrypted image-4c using $R2_{(6\times6)}$

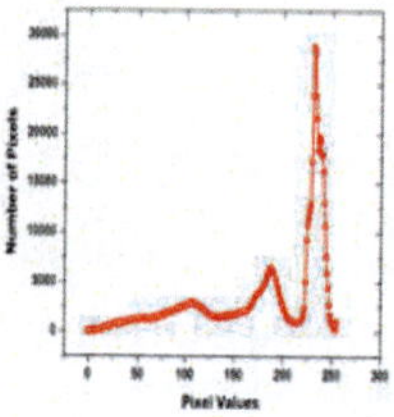

(d) Histogram of fruit image-4e

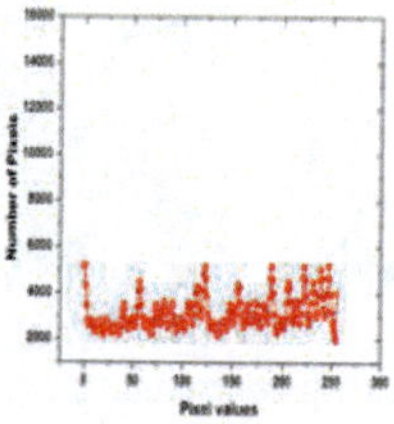

(e) Histogram of encrypted image-4f using $R1_{(5\times5)}$

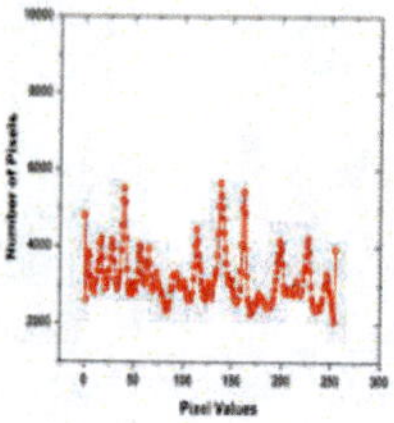

(f) Histogram of encrypted image-4g using $R2_{(6\times6)}$

Fig. 5 Histogram analysis of original and encrypted images

are shown in Fig. 5 where it can be observed that the histograms of cipher images are comparatively different and uniform than the histograms of the corresponding plain images. This uniform pattern increases the uncertainty between pixel values.

Diffusion property: The diffusion property is basically hiding the interrelation between the plain and cipher image. It is determined by the sensitivity of the proposed model. The sensitivity is measured by checking propagation of difference in pixel values in the cipher image by converting a single bit in the plain image. An example is shown in Fig. 6. Here, lena's image (Fig. 4a) is considered as plain image (P_k), and cipher image's (C_k) transition is checked for each round, where r denotes rounds and

$k = 1$ for original value

$k = 2$ for disputed value and

$$Diff^r = \begin{cases} 1, \; if \; C^r_{k=1} = C^r_{k=2} \\ 0, \; if \; C^r_{k=1} \neq C^r_{k=2} \end{cases} \tag{4}$$

In Fig. 6, for $k = 2$ in P_k, the only different bit is the middle bit of the plain image. The corresponding cipher image differs with a few more points after one round. Further, the difference between the two cipher images has increased with an increase in

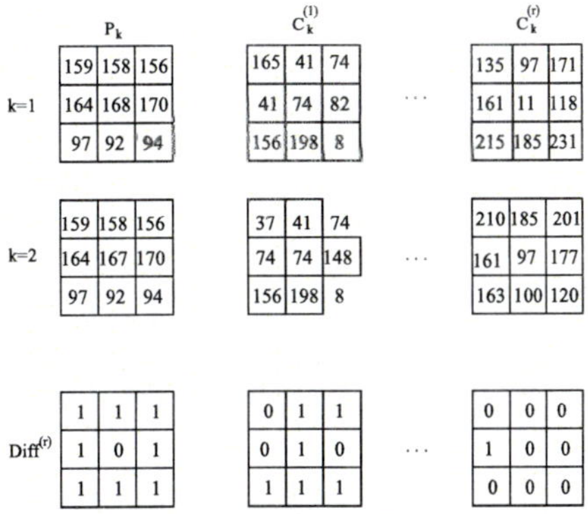

Fig. 6 Difference in pixels of the original image vs the cipher images (Fig. 4a)

number of rounds. After 100 experiments, it is found that at round number N_c, the maximum cipher image pixel difference has occurred. Hence, N_c is considered as well suited for an encryption parameter. The obtained result is comparable with the previous work of Yu et al. [13].

Correlation of Two Adjacent Pixels: Correlation attack is the most threatening for image encryption where the high correlation among the pixels of plain and encrypted images are exploited to attack the system. So, to increase the security, correlation between the cipher images pixels must be minimized. For finding a correlation between two adjacent pixels, 1000 random pairs of adjacent pixels are taken from each image. Correlations among different angles are observed. The calculation is shown in Eq. 5:

$$E(x) = \frac{1}{n}\sum_{i=1}^{n} x_i \tag{5}$$

$$Covariance(x_1, x_2) = \frac{1}{n}\sum_{i=1}^{n}(x_{1i} - E(x_1)) \times (x_{2i} - E(x_2))$$

$$Var(x) = \frac{1}{n}\sum_{i=1}^{n}(x_i - E(x))^2$$

$$Correlation(x_1, x_2) = \frac{Covariance(x_1, x_2)}{\sqrt{Var(x_1)}\sqrt{Var(x_2)}}$$

where x_1, x_2 represents the adjacent pixels each with dimension n, and $E(x)$ and $Var(x)$ are the mean and variance of vector x, respectively. The correlation along different directions are shown in Fig. 7 and the total result along all the direction are depicted in correlation Tables 1 and 2 using the two images shown in Fig. 4a and e. It is observed that for two adjacent pixels, the average correlation in horizontal,

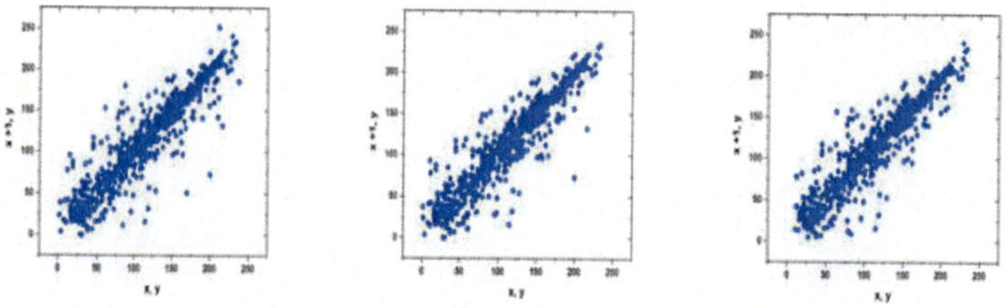

(a) Horizontal Correlation of image-4a

(b) Vertical Correlation of image-4a

(c) Diagonal Correlation of image-4a

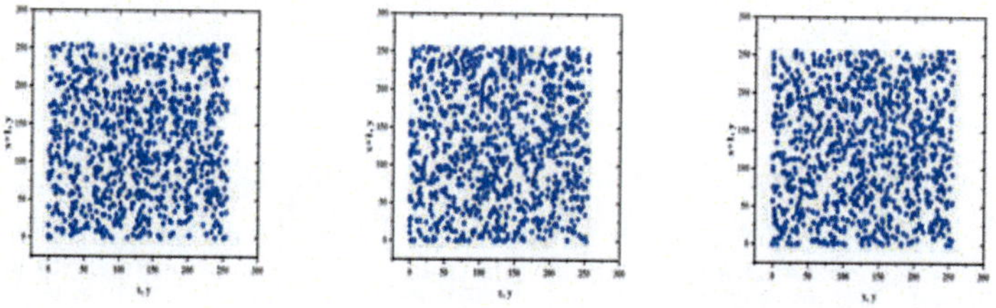

(d) Horizontal Correlation of encrypted image-4c using Rule $R2_{(6\times6)}$

(e) Vertical Correlation of encrypted image-4c using Rule $R2_{(6\times6)}$

(f) Diagonal Correlation of encrypted image-4c using Rule $R2_{(6\times6)}$

Fig. 7 Correlation of two adjacent pixels in horizontal, vertical, and diagonal directions

Table 1 Correlation coefficients of the image: lena (Fig. 4a)

Type of adjacency	Plain image: (Fig. 4a)	Cipher image- $R1_{(5\times5)}$	Cipher image- $R2_{(6\times6)}$
Horizontal	0.9321	0.0493	0.0185
Vertical	0.9700	−0.0456	−0.0405
Diagonal	0.9172	0.0048	−0.016

Table 2 Correlation coefficients of the image: fruit (Fig. 4e)

Type of adjacency	Plain image (Fig. 4e)	Cipher image- $R1_{(5\times5)}$	Cipher image- $R2_{(6\times6)}$
Horizontal	0.9817	0.0211	0.0419
Vertical	0.9834	−0.0382	−0.0167
Diagonal	0.9709	−0.0456	−0.0734

vertical and diagonal directions are near to 0 in the cipher image, whereas it is close to 1 in the plain image. The performance of proposed model is compared with other models and has been shown in Table 3. All these results show that the algorithm resists statistical attacks.

Table 3 Comparison of Correlation coefficients of the image: (Fig. 4a) with other methods

Scheme	Proposed	Ref. [15]	Ref. [16]	Ref. [18]	Ref. [19]	Ref. [20]	Ref. [21]	Ref. [22]
Horizontal	0.0185	0.0006	0.0053	−0.0011	−0.0285	0.0027	−0.0038	−0.0245
Vertical	−0.0405	0.0015	0.0078	0.0014	0.0014	0.0005	−0.0026	−0.0226
Diagonal	−0.0161	−0.0009	0.0042	0.0005	0.0013	−0.0045	0.0017	−0.0193

4.3 Differential Attack

An adversary may produce a small change in image or may change one pixel of the image and can observe the change in result to find the connection between the cipher and plain image. Thus, if one minor change causes a notable change in the encrypted image, then it is said to be secure from the differential attack. Two popular measurements UACI (Unified average change intensity) and NPCR (Number of Pixel Change Rate) have been used to test the impact of one pixel change in an image.

NPCR: Let P and C represent the plain and encrypted images of dimension $M \times N$, respectively. P_{ij} and C_{ij} are the pixel values of the image located in the (i, j) position. $B(P, C)$ in Eq. 6 stands for NPCR [1, 23].

$$B(P, C) = \frac{1}{MN} \sum_{i=1}^{M} \sum_{j=1}^{N} \delta(i, j) \tag{6}$$

where

$$\delta(i, j) = \begin{cases} 1, if\, P_{ij} = C_{ij} \\ 0, if\, P_{ij} \neq C_{ij} \end{cases}$$

UACI: Let $G(P, C)$ in Eq. 7 stands for UACI between two images [1, 23].

$$G(P, C) = \frac{1}{MN} \sum_{i=1}^{M-1} \sum_{j=1}^{N-1} \mid P(i, j) - C(i, j) \mid \tag{7}$$

For ideal image, NPCR should be within 0.4% and the UACI variation between the two images should be less than 0.5 [23]. Lena's image (Fig. 4a) is taken as a test image with five different groups of keys to encrypt and evaluate. As a result, five cipher images are generated. Table 4 gives the result of the analysis where the variation of differences between the plain and cipher image, under different key conditions, are very uniform. The difference between the minimum and maximum UACI change value is not more than 0.5. The value of NPCR also lies within 0.4%.

Table 4 Scrambling evaluation parameter

Plain image	Cipher image				
Lena (Fig. 4a)	Lena1	Lena2	Lena3	Lena4	Lena5
B(P, C)	0.0036	0.0039	0.0036	0.0036	0.0038
G(P, C)	76.5917	76.1926	76.2167	76.2368	76.2782

Table 5 Comparison of NPCR and UACI of image: (Fig. 4a) with other methods

Scheme	Proposed	Ref. [15]	Ref. [16]	Ref. [18]	Ref. [19]	Ref. [20]	Ref. [21]	Ref. [22]
NPCR (%)	99.6075	99.5909	99.5268	99.6063	99.6200	99.6173	99.6552	99.6094
UACI (%)	33.4633	33.5341	32.7136	33.4621	33.4600	29.5664	33.4846	33.4635

On the other hand, [18] stated the NPCR and UACI equations in a slightly different way from Eqs. 6 and 7. Two cipher images $C1$ and $C2$ are considered for calculation, where $C1$ and $C2$ are cipher image before and after single pixel of plain image is changed. For NPCR, the difference is taken as 0 if $C1 = C2$, otherwise 1. And similarly for UACI, the difference is taken as $\mid (C1 - C2) \mid$. Next, the expectation of the NPCR and UACI are calculated as mentioned in Eq. 8 which are described in [24]

$$NPCR_E = (1 - 2^{-n}) \times 100\% \tag{8}$$

$$UACI_E = \frac{1}{MN} \sum_{i,j} \frac{\mid (C1 - C2) \mid}{2^n - 1} \times 100\%$$

where n denotes the digit. The ideal values for NPCR_E and UACI_E are 99.6094% and 33.4635%, respectively, for 8 bits digits. It can be observed in Table 5 that the values are close to the ideal values, which specifies that for one pixel, the rate of influence is very high. A comparison is made with other methods and is shown in Table 5.

4.4 Speed Analysis

Along with good security factors, a good encryption scheme needs to retain real-time requirements. For computation, Linux-Ubuntu operating system is used with a frequency of 2.80 GHz and memory 8 GB. To check the efficiency of the algorithm, different sizes of images are chosen. Apart from serial, parallel implementation is done using Open Multi-Processing (OpenMP). To check the encryption speed, each

Table 6 Speed analysis unit: seconds

Image	Serial implementation	Parallel implementation
Lena (128 × 128)	0.5797	0.5138
Lena (256 × 256)	0.9850	0.8608
Lena (320 × 320)	1.2902	0.9339
Lena (512 × 512)	2.9950	1.6847
Lena (1024 × 1024)	7.0318	5.9672
Lena (4096 × 4096)	45.9306	32.8653

experiment is conducted 100 times, then the average of the encrypted time is taken and is shown in Table 6. It can be observed that the parallel implementation has resulted in a speedup of at least 1.40 times than serial implementation and ensures lossless decryption. It is analyzed that CA evolution step consumes 90% of the entire encryption procedure. Hence, the computing efficiency in CA evolution is the core developmental area.

5 Conclusion

The paper presents a new image encryption algorithm with the parallel evolution of 2-D CA. Efficient usage of masking has improved boundary scattering. The simple and lightweight feature of 2-D CA makes the algorithm's run-time very fast. The software encryption shows great unpredictable pixel values generation, key-space, speedup using parallel evolution, confusion, and diffusion properties. The simulation experiment indicates that the proposed method exhibits good cryptographic properties. 2-D CA withstand substantial randomness test and the key-space is sufficiently large to be cracked by brute-force method. The technique is also efficient for rule matrices with a small cycle length. All the results and comparisons express that the proposed algorithm is appropriate and fits well for digital image encryption.

References

1. Chen G, Mao Y, Chui CK (2004) A symmetric image encryption scheme based on 3d chaotic cat maps. Chaos, Solitons, Fractals 21(3):749–761
2. Maniccam SS, Bourbakis NG (1999) Scan based lossless image compression and encryption. In: Proceedings 1999 international conference on information intelligence and systems (Cat. No.PR00446), pp 490–499
3. Li X, Knipe J, Cheng H (1997) Image compression and encryption using tree structures1this research is supported in part by motorola wireless data group and the canadian natural sciences and engineering research council under grant ogp9198.1. Pattern Recogn Lett 18(11):1253–1259

4. Jin J (2009) Image encryption method based on elementary cellular automata. IEEE southeastcon 2009:345–349
5. Wolfram S (1986) Cryptography with cellular automata. In: Williams HC (ed) Advances in cryptology – CRYPTO '85 proceedings. Springer, Berlin, pp 429–432
6. Xuelong Z, Qianmu L, Manwu X, Fengyu L (2005) A symmetric cryptography based on extended cellular automata. In: 2005 IEEE international conference on systems, man and cybernetics, vol 1, pp 499–503
7. Chen R-J, Lai J-L (2007) Image security system using recursive cellular automata substitution. Pattern Recogn 40(5):1621–1631
8. Perales JCM, de Cómputo ES (2015) Color image encryption by cellular automata
9. Alvarez G, Hernandez A, Hernandez L, Martin A, Rodriguez G (2005) A new graphic cryptosystem based on one-dimensional memory cellular automata. In: Proceedings 39th annual 2005 international carnahan conference on security technology, pp 200–203
10. Maleki F, Mohades A, Hashemi SM, Shiri ME (2008) An image encryption system by cellular automata with memory. In: 2008 third international conference on availability, reliability and security, pp 1266–1271
11. Yu L, Li Y, Xia X (2008) Image encryption algorithm based on self-adaptive symmetrical-coupled toggle cellular automata. In: 2008 Congress on image and signal processing, vol 3, pp 32–36
12. Chaudhuri PP, Chowdhury DR, Nandi S, Chattopadhyay S (1997) Additive cellular automata: theory and applications, vol 43. Wiley, New York
13. Wang Y, Zhao Y, Zhou Q, Lin Z (2018) Image encryption using partitioned cellular automata. Neurocomputing 275:1318–1332
14. Chen R-J, Chen Y-H, Chen C-S, Lai J-L (2006) Image encryption/decryption system using 2-d cellular automata. In: 2006 IEEE international symposium on consumer electronics, pp 1–6
15. Choi US, Cho SJ, Kim JG, Kang SW, Kim HD (2020) Color image encryption based on programmable complemented maximum length cellular automata and generalized 3-d chaotic cat map. Multimedia Tools Appl 79:22825–22842
16. Roy S, Shrivastava M, Pandey CV, Nayak SK, Rawat U (2020) Ievca: an efficient image encryption technique for iot applications using 2-d von-neumann cellular automata. Multimedia Tools Appl 1–39
17. Brassard G et al (1988) Modern cryptology: a tutorial, vol 325. Springer, Berlin
18. Xiao S, Yu ZJ, Deng YS (2020) Design and analysis of a novel chaos-based image encryption algorithm via switch control mechanism. Security and communication networks 2020
19. Chai X, Gan Z, Yuan K, Chen Y, Liu X (2019) A novel image encryption scheme based on dna sequence operations and chaotic systems. Neural Comput Appl 31(1):219–237
20. Kulsoom A, Xiao D, Abbas SA et al (2016) An efficient and noise resistive selective image encryption scheme for gray images based on chaotic maps and dna complementary rules. Multimedia Tools Appl 75(1):1–23
21. Pak C, Huang L (2017) A new color image encryption using combination of the 1d chaotic map. Signal Proc 138:129–137
22. Zhang Y, Chen A, Tang Y, Dang J, Wang G (2020) Plaintext-related image encryption algorithm based on perceptron-like network. Inf Sci 526:180–202
23. Zhu K, Lin Z, Ding Y (2019) A new RSA image encryption algorithm based on singular value decomposition. Int J Pattern Recogn Artif Intell 33(01):1954002
24. Luo Y, Cao L, Qiu S, Lin H, Harkin J, Liu J (2016) A chaotic map-control-based and the plain image-related cryptosystem. Nonlinear Dyn 83(4):2293–2310

A New Identity-Based Multivariate Signature Scheme

Satyam Omar, Sahadeo Padhye, and Dhananjoy Dey

Abstract Authentication of public keys in asymmetric cryptography is an important objective to fulfil. Identity-based schemes are an efficient solution for this scenario. There are various identity-based schemes that are based on either factoring problem or DLP problem, but these problems can be solved in polynomial time using quantum computers. Multivariate cryptography is one of the main alternatives for the construction of post-quantum digital signature schemes. Digital signature construction needs modest computations in multivariate public key cryptography, so these schemes are considered quite efficient. In this paper, we present a new identity-based signature scheme in which the signature size and user key size are relatively small. The design of our proposed signature scheme is based on MQDSS construction.

Keywords Multivariate public key cryptography · MQ problem · Identity-based signature scheme

1 Introduction

A mechanism is required to authenticate the public keys in public key cryptography. Generally, certificates signed by a certification authority are used to authenticate the public keys. The certificate contains the identity and public key of the person, along with the signature of certification authority on the person's identity and public key. Due to the amount of the information contained by the certificate, often its size

S. Omar · S. Padhye (✉)
Department of Mathematics, Motilal Nehru National Institute of Technology, Prayagraj 211004, India
e-mail: sahadeo@mnnit.ac.in

S. Omar
e-mail: satyamomar@mnnit.ac.in

D. Dey
Department of Mathematics, Indian Institute of Information Technology, Lucknow 226 002, India
e-mail: ddey@iiitl.ac.in

© The Author(s), under exclusive license to Springer Nature Singapore Pte Ltd. 2022
D. Giri et al. (eds.), *Proceedings of the Seventh International Conference on Mathematics and Computing*, Advances in Intelligent Systems and Computing 1412,
https://doi.org/10.1007/978-981-16-6890-6_7

becomes very large. The distribution of this large certificate is also a cumbersome problem. The identity-based approach is an excellent option to denunciate these problems. Moreover, the identity-based approach solves the problem of the large size of public keys also.

The identity-based approach was first suggested by Shamir as an alternative of certificates in 1984 [16]. In this perspective, the identity of the person, such as e-mail address, identification number issued by the government, is used to derive his public key with the help of a publicly known hash function. There is a vital role of a central trusted authority (*TA*) who has his master public key and master secret key. With the help of these keys, *TA* performs the algorithm *Extract* and computes the private key of the person using the public key of the person. So, we can say that *TA* provides a private key rather than a certificate for the authentication of the public key. Hence, the significant advantage of using an identity-based signature scheme is the elimination of the need for certificates. The only problem in identity-based schemes is that *TA* knows the private keys of all the persons authenticated by him, which may not be the case in using certificates.

Several identity-based schemes have been constructed [2, 6, 21] in conventional public key cryptography, i.e., schemes based on either factoring or DLP problem, but the schemes based on these problems will not be secure any more after applying quantum algorithms of Peter Shor [18]. Multivariate Public Key Cryptography (MPKC) is a promising branch of post-quantum cryptography that produces small-size signatures compared to all other post-quantum techniques. Moreover, due to the modest computations on a finite field, these schemes are very efficient to implement on low-cost devices such as smart cards and RFID chips [1, 3]. Most of the early signature schemes in MPKC, such as Oil and Vinegar [5], SFLASH [4], TTS [20], were broken. Some of the secure and efficient MPKC's are UOV [12], Rainbow [7], MQDSS [5], on which we can design identity-based signature schemes.

In 2013, Shen et al. [19] proposed an identity-based signature scheme first time in MPKC, which was based on UOV signature scheme [12]. Then, Le van Luyen proposed an improved model of the identity-based signature scheme based on Rainbow signature scheme [13] in 2019. In this paper, we propose another approach to design an identity-based signature scheme with the help of MQDSS [5] construction. The security of this scheme is based on the MQ problem, which is proven NP-hard even over the field $GF(2)$. The size of the designed signature and private keys of the signer are relatively small in comparison to the schemes of Shen et al. [19] and Le van Luyen [13]. We have shown the comparison of our scheme with these schemes in 128-bit security level.

MQDSS [5] is constructed with the help of a 5-pass variant of the identification scheme of Sakumoto et al. [17], which uses zero-knowledge proof to identify the prover without leaking any information about the secret key of the prover. Hence, MQDSS signature [5] also does not reveal any private information about the signer. Our construction is based on MQDSS [5] which is existentially unforgeable under adaptive chosen message attack (EU-CMA) in the random oracles, provided the commitment functions are computationally hiding and computationally binding. Rest of the paper is organized as follows.

Table 1 Notation table

Notation	Stands for
λ	Security parameter
TA	Trusted authority
$\mathbb{E}$	Finite field
μ, ν	Affine transformations
MSK	Master secret key
MPK	Master public key
ID	Identity
U	User
τ	User's public key
s	User's secret key
$D, \bar{D}$	Quadratic system of equations
$\bar{H}, H, H_0, H_1, H_2$	Hash functions
Com_0, Com_1	Commitment functions
l, v, k	positive integers
ω	Number of rounds
M	Message
MD	Message digest
$\bar{G}$	Polar form of $\bar{D}$
$\|$	Concatenation
α, ch	Challenge
$Sign$	Signature
u, v, w	Elements of fields
$\mathbf{A}$	Adversary

We introduce notations, the background of the identity-based signature scheme, and the necessary tools to design our scheme in Sect. 2. After that, we discuss the general construction of multivariate digital signatures in Sect. 3, and then we propose our identity-based multivariate signature scheme along with security proof in Sect. 4. Then, we compare our scheme with the existing multivariate identity-based signature schemes in Sect. 5. In Sect. 6, we show the efficiency of our signature scheme. Finally, we conclude the paper in Sect. 7 (Table 1).

2 Preliminaries

In this section, we discuss a few basic concepts required to explain our proposed scheme. The following notations are used in this paper.

2.1 Identity-Based Signature Scheme

An identity-based signature scheme consists of four polynomial time algorithms, namely *Setup*, *Extract*, *Signature generation*, and *Signature verification*.
Setup is a randomized algorithm executed by *TA*, which takes input (1^λ), where λ is the security parameter, and outputs master keys (master public key (MPK) and master secret key (MSK)) and common parameters. A publicly known hash function (H) is used whose input is a binary string of arbitrary length, and output length is according as the desired security. The user (U) has an identity $(ID_U) \in \{0, 1\}^*$ and his public key (PK_U) is $H(ID_U)$.
Extract is a randomized algorithm executed by *TA*, which takes public key (PK_U) of the user and master keys as input and produces secret key (SK_U) for the user U.
Signature Generation is also a randomized algorithm executed by the signer, which takes user's secret key (SK_U), public key (PK_U), message (m), and master public key (MPK) as input, and outputs an identity-based signature $(Sign)$.
Signature Verification is a deterministic algorithm executed by the verifier, which takes message (m), signature $(Sign)$, master public key (MPK), and identity of the signer (ID_U) as input, and outputs 1 or 0 according as the acceptance or rejection of the signature.

The standard security notion for an identity-based signature scheme is existential unforgeability against adaptive chosen message attack (EU-CMA).

2.2 Existential Unforgeability Against Adaptive Chosen Message Attack

Suppose a PPT adversary has only the public key of the signer and a signing oracle, to which he can query dynamically. If the probability of constructing a valid signature on at least one document by this adversary is negligible in λ, we say that the signature scheme is EU-CMA secured.

In general, we talk about universal forgery (weaker security notion) instead of existential forgery (strongest security notion) in MPKC signature schemes, where the challenger (signer) wants that the adversary should be able to attack on any message (any arbitrary message or given by the signer) to successfully forge the signature scheme.

2.3 Commitment Scheme

A commitment scheme [11] is a cryptographic tool that is used to commit a chosen value or statement while keeping it hidden from others and able to reveal the committed value later. There are two essential properties satisfied by a commitment scheme, viz. hiding and binding.

Hiding property allows the committing party to hide the committed value after sending commitment on it and before revealing the committed value himself. This property ensures the unforgeability from the verifier's side.

Binding property prevents the committing party to change the committed value after computing commitment on it. This property ensures honesty from the side of the committing party.

The commitment scheme must be computationally hiding and computationally binding for its use in zero-knowledge protocols or signature schemes. Computationally hiding means that it should be computationally difficult for a verifier to obtain the committed value in polynomial time before the reveal phase, and computationally binding means that it should be computationally difficult for the committing party to obtain the collision on commitment function in polynomial time.

2.4 *Identification Scheme*

The Identification scheme [5] is an interactive protocol which allows a prover to convince a verifier of its identity, i.e., the prover convinces the verifier that he (prover) is having a secret without leaking any information about the secret. This protocol uses zero-knowledge proof to accomplish this purpose. The two most important properties that a secure Identification scheme must satisfy are soundness and honest-verifier zero-knowledge. These properties indicate the fulfilment of the properties required in the definition of the identification scheme.

2.5 *Fiat–Shamir Transformation*

Fiat and Shamir [8] presented the idea of transforming a $(2n + 1)$-pass identification scheme into a signature scheme by introducing appropriate use of hash functions. Their concept was implemented on a factoring problem-based scheme. The resulting signature was just collecting all the information sent by the prover in the identification scheme associated with the message. The security notions satisfied by the identification scheme are also satisfied by the resulting transformed signature scheme. The core of this transformation is the number of repetitions of the identification scheme to obtain the desired security level. Here, we have used the same idea to construct an identity-based signature scheme with the help of Sakumoto et al.'s identification scheme [17].

3 Multivariate Public Key Cryptography

Multivariate public key cryptography is a branch of post-quantum cryptography which is mostly used for the construction of digital signatures. The security of these schemes is basically based on two problems, viz., IP-problem and MQ problem. The security of our proposed scheme depends on the MQ problem.

3.1 MQ Problem

Let p is a prime number and r is an integer and $\mathbb{E} = GF(p^r)$ be a finite field of order p^r. Then, the quadratic system of equations

$$\begin{aligned} d_1(z_1, z_2, \ldots, z_l) &= \eta_1 \\ d_2(z_1, z_2, \ldots, z_l) &= \eta_2 \\ &\vdots \\ d_v(z_1, z_2, \ldots, z_l) &= \eta_v \end{aligned}$$

where $d_r \in \mathbb{E}[z_1, z_2, \ldots, z_l]$, $\eta_r \in \mathbb{E}$, for all $r = 1, 2, \ldots, v$.
Problem: To find a vector $(s_1, s_2, \ldots, s_l) \in \mathbb{E}^l$ satisfying $d_r(s_1, s_2, \ldots, s_l) = \eta_r$; for all r with $1 \leq r \leq v$. This problem is called Multivariate Quadratic polynomial (MQ) Problem, which is NP-hard [10] even over the field $GF(2)$. The security of our identity-based signature scheme is based on MQ Problem.

3.2 Sakumoto Et Al.'s Identification Scheme

Sakumoto et al. [17] proposed two variants of the multivariate identification scheme as a 3-pass variant and a 5-pass variant. The knowledge error in the 3-pass scheme is $\frac{2}{3}$ while the knowledge error in the 5-pass scheme is $\frac{1}{2}$, so the 5-pass variant will take fewer rounds for the desired level of security. The size of the transcript was also large in the case of the 3-pass variant. Hence, a 5-pass variant was used to design MQDSS signature scheme [5], which is just Fiat-Shamir transformation [8] of the 5-pass variant of Sakumoto et al.'s identification scheme [17].

In an identification scheme, there is a prover P who tries to convince a verifier V of its identity. The prover has secret key (s) and corresponding public key (τ) such that $\bar{D}(s) = \tau$, where $\bar{D}$ is the quadratic system of equation which is difficult to invert. The polar form of quadratic function [9] $\bar{D}$ is $\bar{G}(x, y) = \bar{D}(x + y) - \bar{D}(x) - \bar{D}(y)$, where $\bar{D}$ is the quadratic system of equations having constant term zero and $\bar{G}$ be a bilinear function. Here, P wants to convince V that P has the secret key s without revealing any information about s. At first, P breaks s into two parts as $s = u_0 + u_1$, and hence, $\tau = \bar{D}(s) = \bar{D}(u_0) + \bar{D}(u_1) + \bar{G}(u_0, u_1)$. To make the polar form $\bar{G}$ free from one of u_0 or u_1, u_0 and $\bar{D}(u_0)$ is again broken as $\alpha u_0 = v_0 + v_1$ and

$\alpha\bar{D}(u_0) = w_0 + w_1$. Now, we will discuss the way in which 5-pass Sakumoto et al.'s identification scheme [17] works.

I-pass: P chooses randomly $u_0, v_0 \in_R \mathbb{E}^l$, $w_0 \in_R \mathbb{E}^v$. Now, P computes $u_1 = s - u_0$ and the commitments $c_0 = Com(u_0, v_0, w_0)$, $c_1 = Com(u_1, G(v_0, u_1) + w_0)$ and sends c_0, c_1 to V.

II-pass: V chooses randomly $\alpha \in_R \mathbb{E}$ and sends this first challenge to P.

III-pass: P computes $v_1 = \alpha u_0 - v_0$, $w_1 = \alpha\bar{D}(u_0) - w_0$. Now P sends this first response rsp_1 to V.

IV-pass: V chooses second challenge ch randomly from $\{0, 1\}$ and sends it to P.

V-pass: According as ch is 0 or 1, P sends the second response $rsp_2 = u_0$ or $rsp_2 = u_1$ to V.
After getting rsp_2, V checks the validity of commitments with the help of rsp_1 and rsp_2 as if $ch = 0$, check $c_0 \stackrel{?}{=} Com(u_0, \alpha u_0 - v_1, \alpha\bar{D}(u_0) - w_1)$, and if $ch = 1$, check $c_1 \stackrel{?}{=} Com(u_1, \alpha(\tau - \bar{D}(u_1)) - \bar{G}(v_1, u_1) - w_1)$.

Theorem 1 *The 5-pass Sakumoto et al.'s identification scheme [17] is zero-knowledge and sound with soundness error* $\frac{1}{2} + \frac{1}{2|\mathbb{E}|}$ *if underlying commitment scheme is computationally hiding and computationally binding, where* $|\mathbb{E}|$ *is the cardinality of the underlying finite field.*

3.3 Rainbow Signature Scheme

Rainbow Signature scheme [7] is the most efficient digital signature scheme in MPKC. It is a multi-layer UOV signature scheme [12]. UOV is one of the unbreakable digital signature schemes in MPKC, but the signature size and scheme's efficiency are not much appreciable for secured parameters. In 2005, Jintai Ding proposed a multi-layer construction of UOV [12], which possesses a smaller signature and key size as well as good efficiency. The central map in Rainbow construction is the collection of different oil and vinegar polynomials, due to which it is easily invertible. After hiding this with two suitable invertible affine transformations, the resulting quadratic system of equations is difficult to invert.

4 Proposed Signature Scheme

We propose a new identity-based signature scheme based on MQDSS [5] construction. The steps of the algorithms *Setup*, *Extract*, *Signature Generation*, *Signature Verification* are as follows:

Setup: The security parameter $\lambda \in \mathbb{N}$ is used to produce the private key and public key for *TA*. Let l, v and k be the positive integers and $\mathbb{E} = \mathbb{F}_{31}$ be the finite field. The private key of *TA* is (μ, D, ν) called master secret key (*MSK*) and the public key of *TA* is $\mu \circ D \circ \nu = \bar{D}$ called master public key (*MPK*), where $\mu : \mathbb{E}^v \rightarrow \mathbb{E}^v$ and $\nu : \mathbb{E}^l \rightarrow \mathbb{E}^l$ are invertible affine transformations and $D : \mathbb{E}^l \rightarrow \mathbb{E}^v$, $\bar{D} : \mathbb{E}^l \rightarrow \mathbb{E}^v$ are quadratic system of equations. It means $D, \bar{D}$ are quadratic system of equations with l variables and v equations. These master keys will be same as in Rainbow signature construction [7] for small key size and better efficiency of *Extract* algorithm. Let ω be the number of rounds of Sakumoto et al.'s identification [17] to obtain λ security level.

The size of *MPK* is $\frac{v \cdot (l+1) \cdot (l+2)}{2}$ *field elements*, and the size of *MSK* is

$$v \cdot (v+1) + l \cdot (l+1) + \sum_{j=1}^{2} o_j \left(\frac{\hat{v}_j \cdot (\hat{v}_j + 1)}{2} + \hat{v}_j \cdot o_j + \hat{v}_{j+1} + 1 \right) \textit{field elements};$$

where o_j, v_j be the oil and vinegar variables in the Rainbow construction [7].

Let $\bar{H} : \{0,1\}^* \rightarrow \mathbb{E}^v$, $H : \mathbb{E}^l \times \{0,1\}^* \rightarrow \mathbb{E}^l$, $H_0 : \{0,1\}^{2k\omega} \rightarrow \{0,1\}^k$, $H_1 : \mathbb{E}^l \times \{0,1\}^k \rightarrow \mathbb{E}^\omega$, $H_2 : \mathbb{E}^l \times \{0,1\}^k \times \mathbb{E}^\omega \times \mathbb{E}^{(l+v)\omega} \rightarrow \{0,1\}^\omega$ be the collision-resistant hash functions and $Com_0 : \mathbb{E}^l \times \mathbb{E}^l \times \mathbb{E}^v \rightarrow \{0,1\}^k$, $Com_1 : \mathbb{E}^l \times \mathbb{E}^v \rightarrow \{0,1\}^k$ be the computationally hiding and computationally binding commitment functions. Hash functions and commitment functions are publically known.

Extract: After identifying the user U having identity *ID*, *TA* computes public key (τ) of the user as $\tau = \bar{H}(ID)$. Now, *TA* computes private key (s) of U with the help of *MSK* as $s = \bar{D}^{-1}(\tau) = (\mu \circ D \circ \nu)^{-1}(\tau) = \nu^{-1} \circ D^{-1} \circ \mu^{-1}(\tau)$.

Signature Generation: Let $M \in \{0,1\}^*$ be the message to sign. Then the signer proceeds as follows:

1. Computes message dependent random value $R = H(s \parallel M)$.
2. Computes randomized message digest $MD = H(R \parallel M)$.
3. Randomly chooses $u_{0_1}, u_{0_2}, \ldots, u_{0_\omega}, v_{0_1}, v_{0_2}, \ldots, v_{0_\omega} \in \mathbb{E}^l$, $w_{0_1}, \ldots, w_{0_\omega} \in \mathbb{E}^v$ and computes $c_{0_j} = Com_0(u_{0_j}, v_{0_j}, w_{0_j})$, $c_{1_j} = Com_1(u_{1_j}, \bar{G}(v_{0_j}, u_{1_j}) + w_{0_j})$ for all rounds $j = 1, 2, \ldots \omega$ as in identification scheme [17].
4. Computes $Sign_0 = H_0(c_{0_1} \parallel c_{1_1} \parallel c_{0_2} \parallel c_{1_2} \ldots \parallel c_{0_\omega} \parallel c_{1_\omega})$.
5. Produces the challenges α_j as the components of $h_1 = H_1(MD, Sign_0)$ and computes $v_{1_j} = \alpha_j u_{0_j} - v_{0_j}$ and $w_{1_j} = \alpha_j \bar{D}(u_{0_j}) - w_{0_j}$; for all $j = 1, 2, \ldots \omega$.
6. Computes $Sign_1 = (v_{1_1} \parallel w_{1_1} \parallel v_{1_2} \parallel w_{1_2} \ldots \parallel v_{1_\omega} \parallel w_{1_\omega})$.
7. Produces the final challenges $ch_1, ch_2, \ldots, ch_\omega$ as the digits of $h_2 = H_2(MD, Sign_0, h_1, Sign_1)$ and computes $Sign_2 = (u_{ch_1}, u_{ch_2}, \ldots, u_{ch_\omega}, c_{1-ch_1}, c_{1-ch_2}, \ldots, c_{1-ch_\omega})$.
8. Sends $Sign = (R, Sign_0, Sign_1, Sign_2)$ as the signature on the document M along with his identity *ID*.

Signature Verification: After getting $Sign$ on M and *ID*, the verifier proceeds as follows:

1. Computes public key of the signer as $\tau = \bar{H}(ID)$.
2. Computes randomized message digest $MD = H(R \parallel M)$.

3. Computes $h_1 = H_1(MD, Sign_0)$ and parses h_1 to obtain $\alpha_1, \alpha_2, \ldots, \alpha_\omega$.
4. Computes $h_2 = H_2(MD, Sign_0, h_1, Sign_1)$ and parses h_2 to obtain $ch_1, ch_2, \ldots, ch_\omega$.
5. Parses $Sign_1$ and $Sign_2$ to obtain $v_{1_1}, w_{1_1}, v_{1_2}, w_{1_2}, \ldots, v_{1_\omega}, w_{1_\omega}$ and $u_{ch_1}, u_{ch_2}, \ldots, u_{ch_\omega}$.
6. Half of the commitments are obtained by $Sign_2$ and other half are obtained according as the value of $ch_1, ch_2, \ldots, ch_\omega$ as

If $ch_j = 0$ $\quad c_{0_j} = Com_0(u_0, \alpha u_0 - v_1, \alpha\bar{D}(u_0) - w_1)$. If $ch_j = 1$ $\quad c_{1_j} = Com_1(u_1, \alpha(\tau - \bar{D}(u_1)) - \bar{G}(v_1, u_1) - w_1)$.

7. Computes $Sign_0' = H_0(c_{0_1} \parallel c_{1_1} \parallel c_{0_2} \parallel c_{1_2} \parallel \ldots \parallel c_{0_\omega} \parallel c_{1_\omega})$.
8. Checks the validity of the equation $Sign_0' \stackrel{?}{=} Sign_0$. If equality holds then accepts the signature and outputs 1, otherwise rejects the signature and outputs 0.

Correctness: From the properties of bilinear function $\bar{G}$, we can check the validity of the equality

$$\bar{G}(v_{0_j}, u_{1_j}) + w_{0_j} = \alpha(\tau_j - \bar{D}(u_{1_j})) - \bar{G}(v_{1_j}, u_{1_j}) - w_{1_j};\ \text{for all } j = 1, 2, \ldots, \omega.$$

This equality helps at the time of computing commitment functions in *Signature Verification*, which is the verifying step of the scheme.

Unforgeability: According to the standard security notions, the signature scheme must be EU-CMA secure. The following theorem shows EU-CMA security of our scheme.

Theorem 2 *The proposed scheme is EU-CMA secure in random oracle model, if*

- *MQ Problem is intractable, even in search version.*
- *All the used hash functions $\bar{H}$, H, H_0, H_1, H_2 must be modeled as random oracles.*
- *The commitment functions Com_0, Com_1 must be computationally hiding and computationally binding, and the error probabilities should be negligible functions of security parameter λ.*

Proof Let us assume that the adversary **A** has signing oracle and he gets signature $\bar{Sign} = (\bar{R}, \bar{Sign}_0, \bar{Sign}_1, \bar{Sign}_2)$ on any message $\bar{M}$. We have to prove the unforgeability of our scheme, so on the contrary, let **A** has computed the signature $Sign' = (R', Sign_0', Sign_1', Sign_2')$ on any message M'. **A** can not get secret key s of the signer from $\bar{R}$ and $\bar{Sign}_0$ due to pre-image resistance property of H and computational hiding property of Com_0 and Com_1. So, let **A** would have used another element $s' \in \mathbb{E}^l$ as secret key and proceeded as the signer proceeds in *Signature Generation*. **A** generates the signature $Sign' = (R', Sign_0', Sign_1', Sign_2')$ but according to *Signature Verification*, the following equation must be satisfied:

$$\bar{G}(v'_{0_j}, u'_{1_j}) + w'_{0_j} = \alpha(\tau - \bar{D}(u'_{1_j})) - \bar{G}(v'_{1_j}, u'_{1_j}) - w'_{1_j};\ \forall j = 1, 2, \ldots \omega.$$

So, using $s' = u'_{0_j} + u'_{1_j}$, we will get

$$\begin{aligned}
\bar{G}(v'_{0_j}, u'_{1_j}) + w'_{0_j} + w'_{1_j} &= \alpha'(\tau - \bar{D}(u'_{1_j})) - \bar{G}(v'_{1_j}, u'_{1_j}) \\
\bar{G}(v'_{0_j}, u'_{1_j}) + \bar{G}(v'_{1_j}, u'_{1_j}) + \alpha'\bar{D}(u'_{0_j}) &= \alpha'(\tau - \bar{D}(u'_{1_j})) \\
\bar{G}(v'_{0_j} + v'_{1_j}, u'_{1_j}) + \alpha'\bar{D}(u'_{0_j}) &= \alpha'(\tau - \bar{D}(u'_{1_j})) \\
\bar{G}(\alpha' u'_{0_j}, u'_{1_j}) + \alpha'\bar{D}(u'_{0_j}) &= \alpha'(\tau - \bar{D}(u'_{1_j})) \\
\alpha'\bar{G}(u'_{0_j}, u'_{1_j}) + \alpha'\bar{D}(u'_{0_j}) &= \alpha'(\tau - \bar{D}(u'_{1_j})) \\
\bar{G}(u'_{0_j}, u'_{1_j}) + \bar{D}(u'_{0_j}) &= \tau - \bar{D}(u'_{1_j}) \\
\bar{D}(u'_{0_j} + u'_{1_j}) - \bar{D}(u'_{0_j}) - \bar{D}(u'_{1_j}) + \bar{D}(u'_{0_j}) &= \tau - \bar{D}(u'_{1_j}) \\
\bar{D}(s') &= \tau
\end{aligned}$$

It shows that **A** has solved the MQ problem which has been assumed to be intractable. Thus, our identity-based signature scheme is EU-CMA secure.

5 Comparison

There are two identity-based schemes [13, 19] already in MPKC, which are based on UOV [12] and Rainbow [7] respectively. Rainbow-based scheme [13] has shown that it is better than UOV-based scheme [19] in terms of signature size and key sizes in 128-bit security level. In [13], Le van Luyen has shown that the signature size reduction is nearly 50% and the key size reduction is nearly 65% when using the field $\mathbb{E} = \mathbb{F}_{31}$. So, we will compare our scheme to Rainbow-based scheme [13] to show our betterments in 128-bit security level. The parameters of our scheme are according to the updated parameters in the second round of NIST standardization of MQDSS [5]. We have compared the sizes of signature, master public keys (*MPK*), master secret key (*MSK*), and user secret key(*USK*). Our identity-based signature is $Sign = (R, Sign_0, Sign_1, Sign_2)$, where the sizes[1] of the components of $Sign$ are as follows:

size of $R = 5 \cdot l$ bits
size of $Sign_0 = k$ bits
size of $Sign_1 = 5 \cdot l \cdot \omega + 5 \cdot v \cdot \omega$ bits
size of $Sign_2 = 5 \cdot l \cdot \omega + k \cdot \omega$ bits

Thus, the total signature size in our scheme is $5 \cdot l + (1 + \omega) \cdot k + 5 \cdot \omega \cdot (2 \cdot l + v)$ bits, where $k = 256$ is recommended in [5]. In our scheme, $(31, 64, 376)$ stands for underlying field $\mathbb{E} = \mathbb{F}_{31}, l = 64$ (l is distributed in oil and vinegar variables as in Rainbow construction, i.e., $(23, 21, 20), l = 64, v = l - 23 = 41$), and $\omega = 376$ rounds of identification scheme [17] to obtain 128-bit security level.

In Table 2, it is clear that the sizes of signature, master public key, and master secret key are reduced up to $83\%, 63\%, and 64\%$, respectively, and the size of user secret key is nearly negligible in our scheme. Hence, in terms of signature size and key sizes, our scheme is much better than the existing identity-based schemes [13] in MPKC.

[1] The size of each element of $\mathbb{F}_{31}$ is 5 bits.

Table 2 Comparison table

Size(KB)	IBS-Rainbow (31, 36, 28, 28) [13]	Our scheme (31, 64, 376)
Signature	304.3	50.60
MPK	148.3	53.68
MSK	103.7	37.40
USK	330	0.039

Table 3 Performance analysis

Algorithm	Intel x64–86	Intel x64–86 AVX2
Extract	0.76	0.69
Sign. Gen.	33.34	4.10
Sign. Ver.	24.18	2.79

6 Performance Analysis

We show the performance of our identity-based signature scheme on instruction sets Intel x64-86 and Intel x64-86 AVX2. In [5], the performance of signature scheme is shown in terms of cycles and it is stated that 3502311 cycles are measured in 1 milli seconds (ms). According to updated version (2.1) of MQDSS[5], on Intel x64-86, the number of cycles in *Extract*, signing and verification algorithms are 2670612, 116772480, 84685222 respectively, while on Intel x64-86 AVX2, the number of cycles in *Extract*, signing and verification algorithms are 2406494, 14355450, 9806182, respectively. According to time consumption on both the instruction sets, we can compare our scheme as follows:

In Table 3, the time taken in key generation, signature generation, and signature verification is shown in ms. From Table 3, it is clear that our signature scheme is working as per our expectations using AVX2 instructions.

7 Conclusion

In this paper, we have proposed a new identity-based multivariate signature scheme in which signature and key size are relatively small in comparison to existing identity-based signature schemes in MPKC. Due to multivariate construction, our scheme is a good alternative in the post-quantum perspective. Our scheme is EU-CMA secure because of MQDSS [5]. This scheme is simply the Fiat-Shamir transformation [8] of Sakumoto et al.'s identification scheme [17], therefore, having a simple design, small key and signature size, and flexible parameters, our scheme is the better suite to use as an identity-based signature scheme in MPKC.

References

1. Bogdanov A, Eisenbarth T, Rupp A, Wolf C (2008) Time-area optimized public-key engines: MQ-cryptosystems as replacement for elliptic curves? CHES 2008, LNCS, vol 5154. Springer, pp 45–61
2. Boneh D, Franklin M (2001) Identity-based encryption from the weil pairing. In: J Kilian (ed) Advances in cryptology CRYPTO 2001. CRYPTO 2001. LNCS, vol 2139. Springer, Berlin, Heidelberg, pp 213–229
3. Chen AIT, Chen MS, Chen TR, Cheng CM, Ding J, Kuo EL, Lee FY, Yang BY (2009) SSE implementation of multivariate PKCs on modern x86cpus. CHES 2009, LNCS, vol 5747. Springer, pp 33–48
4. Courtois NT, Goubin L, Patarin J (2003) SFLASHv3, a fast asymmetric signature scheme. IACR Cryptology ePrint archive, report 2003/211. Citeseer
5. Chen MS, Hlsing A, Rijneveld J, Samardjiska S, Schwabe P (2016) From 5-pass MQ-based identification to MQ-based signatures. In: Advances in cryptology—ASIACRYPT 2016—22nd international conference on the theory and application of cryptology and information security. LNCS, vol 10032. Springer, pp 135–165
6. Cocks C (2001) An identity based encryption scheme based on quadratic residues. In: Honary B (ed) Cryptography and coding. Cryptography and coding 2001. LNCS, vol 2260. Springer, Berlin, Heidelberg, pp 360–363
7. Ding J, Schmidt DS (2005) Rainbow, a new multivariate polynomial signature scheme. ACNS 2005, LNCS, vol 3531. Springer, pp 164–175
8. Fiat A, Shamir A (1987) How to prove yourself: practical solutions to identification and signature problems. In: Odlyzko AM (ed) Advances in cryptology CRYPTO 86. CRYPTO 1986. LNCS, vol 263. Springer, Berlin, Heidelberg, pp 186–194
9. Fouque PA, Granboulan L, Stern J (2005) Differential cryptanalysis for multivariate schemes. In: Cramer R (ed) Advances in cryptology EUROCRYPT 2005. EUROCRYPT 2005. LNCS, vol 3494. Springer, Berlin, Heidelberg, pp 341–353
10. Garey Michael R, Johnson David S (1991) Computers and intractability, a Guide to the theory of NP-completeness. W.H. Freeman
11. Goldreich O (2001) Foundations of cryptography, vol. 1, Basic tools. Cambridge University Press
12. Kipnis A, Patarin L, Goubin L (1999) Unbalanced oil and vinegar schemes. EUROCRYPT 1999, LNCS, vol 1592. Springer, pp 206–222
13. Luyen LV (2019) An improved identity-based multivariate signature scheme based on rainbow. Cryptography 2019, vol 3. MDPI
14. Patarin J (1997) The oil and vinegar signature scheme. In: Dagstuhl workshop on cryptography
15. Petzoldt A, Chen MS, Yang BY, Tao C, Ding J (2015) Design Principles for HFEv-based Signature schemes. ASIACRYPT 2015—Part 1, LNCS, vol 9452. Springer, pp 311–334
16. Shamir A (1985). Identity-based cryptosystems and signature schemes. In: Blakley GR, Chaum D (eds) Advances in cryptology. CRYPTO 1984. LNCS, vol 196. Springer, Berlin, Heidelberg, pp 47–53
17. Sakumoto K, Shirai T, Hiwatari H (2011) Public-key identification schemes based on multivariate quadratic polynomials. In: Rogaway P (ed) Advances in cryptology CRYPTO 2011. CRYPTO 2011, LNCS, vol 6841. Springer, Berlin, Heidelberg, pp 706–723
18. Shor P (1997) Polynomial-time algorithms for prime factorization and discrete logarithms on a quantum computer. SIAM J Comput 26(5):1484–1509
19. Shen W, Tang S, Xu L (2013) IBUOV, a provably secure identity-based UOV signature scheme. In: Proceedings of the 2013 IEEE 16th international conference on computational science and engineering, CSE 2013. Sydney, Australia, pp 388–395. 35 Dec 2013

20. Yang BY, Chen JM, Chen YH (2004) TTS: high-speed signatures on a low-cost smart card. In: Joye M, Quisquater JJ (eds) Cryptographic hardware and embedded systems—CHES 2004. CHES 2004. Lecture notes in computer science, vol 3156. Springer, Berlin, Heidelberg, pp 371–385
21. Zhang F, Liu S, Kim K (2002) ID-based one round authenticated tripartite key agreement protocol with pairings. Cryptology ePrint archive, report 2002/122. Citeseer

Encrypted Operator Design with Domain Aware Practical Performance Improvement

Ayan Ghosh, Abhishek Raj, and Ayantika Chatterjee

Abstract Cloud has established itself as an essential part of the new-age computation. However, using a public platform like cloud for data storage and computing, raises severe data security concerns. Storing and uploading data in encrypted form to the cloud may conform data confidentiality. However, data encrypted with traditional encryption schemes do not support processing on encrypted data. Hence, this diminishes the computing possibility in the cloud domain, and each time data need to be taken back and forth. Homomorphic Encryption (HE) is a possible solution to this problem, supporting direct processing of encrypted data. However, realizing any algorithm in the encrypted domain is not straightforward and requires algorithms to be represented in circuit-based form which is a challenging task considering underlying existing processors are unencrypted. Further, performance issues of HE techniques pose a real bottleneck to practical applications. In this work, we focus on a domain aware approach to realize homomorphic computations in practical time which are otherwise too slow or infeasible to implement in encrypted domain. In this work we discussed exponent, logarithm, and sigmoid operators; highlighted why exponent computation on encrypted data with traditional multiple multiplications incur huge performance overhead and logarithm, sigmoid computations on encrypted values are impossible considering existing underlying unencrypted processors. To the best of our knowledge, this work is first in literature to handle these operators fully in the encrypted domain without any cloud domain intermediate decryption.

Keywords Security · Cloud · Fully homomorphic encryption · Encrypted computing

---

A. Ghosh (✉) · A. Raj · A. Chatterjee
Indian Institute Of Technology, Kharagpur, India
e-mail: ABHISHEK.RAJ@iitkgp.ac.in

© The Author(s), under exclusive license to Springer Nature Singapore Pte Ltd. 2022

D. Giri et al. (eds.), *Proceedings of the Seventh International Conference on Mathematics and Computing*, Advances in Intelligent Systems and Computing 1412,
https://doi.org/10.1007/978-981-16-6890-6_8

1 Introduction

Homomorphic Encryption (HE) techniques have enabled data privacy measures to be applied while outsourcing data to untrusted public domain storage. The main advantage of HE is to enable data processing while data is encrypted. This arrangement is particularly ground-breaking in domains like cloud computing and big data processing. Maintaining cloud security while being able to process data is a significant region of research these days. Privacy of information can be kept by imposing traditional encryption before storing it in the cloud. This procedure guarantees security however introduces additional difficulties to process such encrypted information on cloud. To get the required information, encrypted data have to send back to customer side to process as a decrypted plain text. This prompts a few security issues as the ciphertext is continually passed through the communication channel and this leads to possible exposure to trespassers. HE techniques have acknowledged and solved this problem to some extent.

Gentry's work [1] first effectively performed arithmetic operations (addition, multiplication, subtraction, etc.) on encrypted data without revealing or access to secret keys, and the scheme was named Fully Homomorphic Encryption (FHE). The generic idea behind Fully Homomorphic Encryption is similar to this: Take a set of plain texts $m_1, \ldots, m_t$ and corresponding ciphertexts $c_1, \ldots, c_t$ under some FHE scheme with predefined keys. For some computable function f over a set of plain texts, the FHE scheme allows us to perform some polynomial-time execution to generate another ciphertext that corresponds to $f(m_1, \ldots, m_t)$ under the same encryption keys. Theoretically, this solution is promising. However, the performance of Homomorphically Encrypted data processing is still a bottleneck. While working with real-world algorithms or even operators, application of FHE theory is not straightforward since our existing underlying processors can not perform loop execution or any other decision making in the encrypted domain [2]. Considering this limitation, most of the traditional operators need to be revisited to design their FHE supported counterpart [3].

In this work, we focus on implementing a few domain aware encrypted operators that will in turn be used to design an end-to-end homomorphic framework. The operators design conforms the data security as per FHE standard while making computation on encrypted data feasible and performance practical for most of the applications. By domain aware we mean, the inputs of the operators are within a fixed range and the range is known beforehand. We have shown in this work, how domain aware exponent function (e^x where x can be any number), sigmoid function and logarithm function can be computed homomorphically within a practical time requirement. Exponent calculation has many uses in data processing: from bi-variate Gaussian function in population growth calculation [4] to sigmoid, softmax activation functions in neural networks [5]. Logarithm is a reverse operation of exponent that has many applications like entropy calculation in information theory and probability [6]. It is to be noted most of the popular data processing libraries include exponent and logarithm functions as part of their distribution packages due to frequent uses of these functions. The traditional method of e^x calculation may use repetitive multiplication, which is not practical to use over the encrypted domain. Logarithm and sigmoid are non-linear functions that can't be implemented directly over encrypted data. Previous works used approximation polynomials for most of these non-linear operators. Furthermore, Homomorphic computation can only take place when algorithms are converted to circuit-based representation, which is not a trivial task for complex operations like an exponent, logarithm, or sigmoid. In this work, we focus on designing such complex operators in the encrypted domain with an underlying homomorphic library. We have chosen Fully homomorphic encryption as underlying homomorphic scheme to evaluate the worst-case performance. However, our proposed implementation method is generic enough to be applied to any framework supported by leveled or somewhat homomorphic encryption scheme [7].

The rest of the paper is organized as follows: Sect. 2 describes a chronological description of related works. Section 3 explains our proposed method of encrypted operators like exponent, logarithm, and sigmoid calculation in the domain of cloud computing. In Sect. 4, we explain the performance analysis of experimental details, and Sect. 5 concludes the paper with future work to be done.

2 Related Works

Realization of any traditional algorithms in encrypted domain should be done in their circuit-based representation as all FHE libraries like TFHE [8], SEAL [9] provide mostly encrypted bitwise addition and multiplication modules along with bitwise encrypted multiplexer circuits. Hence, encrypted operator design is a serious challenge, and that is not much explored in literature. Encrypted operators are mostly important for encrypted database design as discussed in case of CryptDB [10], ZeroDB [11], Mylar, Arx, and Seabed [12], etc. However, these databases are prone to either security problems or computing limitations. In ZeroDB, all data processing activity occurs at the client-side, diminishing the purpose of cloud computing. On the other hand, CryptDB achieves encrypted processing by executing SQL queries on encrypted data in the different forms of encryption schemes, which requires an intermediate change of encryptions and leads to different attacks. Further, complex queries are not executed in these encryption schemes. Microsoft SQL Server and the Microsoft Azure SQL database [13] with "Always Encrypted" (AE) feature are capable of encrypting and decrypting data with deterministic encryption scheme within client applications or application servers. However, analysis shows [14] databases with deterministic encryption are susceptible to possible data leakage. In the case of other encrypted databases, schemes are prone to several vulnerabilities due to the property-preserving property [14].

Few other schemes in literature are either somewhat homomorphic and allow limited homomorphic operations [15], tensor-based FHE solution for special blend arithmetic operations [16] or suitable for approximate numbers [17]. Another scheme supports limited Computing on Masked Data (CMD) to improve data veracity [18] in the case of Big data. Even FHE-based secure cloud analytics solutions [19, 20], only use encrypted addition and squaring operators. Hence, it is important to focus on efficient design of a few important operators like exponential operator or exponent [21], logarithm, sigmoid in encrypted domain, and this is the motivation of our work.

3 Designs of Encrypted Operators

It is not difficult to observe that a function's nonlinearity makes it more complicated and hard to compute. The direct approach may need several fundamental arithmetic operations like multiplications and divisions that are, unfortunately, still not practical to implement on encrypted data. If we know that a particular application may use a fixed range of variables for a complex function, then our work shows that how a precomputed method can speed up calculations on encrypted data. We have chosen the exponent function to show how the precomputed approach can improve performance as a whole. Further, we discuss how similar techniques can be applied to implement encrypted logarithm and sigmoid implementation.

3.1 Exponent Operator Computation

Traditionally, to compute the exponential function (e^x), there may be two ways, either to perform repetitive multiplication $e^x = e \times e \times \ldots e$ or to exponent calculation through series expansion as $e^x = 1 + \frac{x}{1!} + \frac{x^2}{2!} + \frac{x^3}{3!} + \ldots$. In both cases, it requires significant computationally expensive operations like repetitive encrypted multiplications. Moreover, each multiplication leads to the doubling of output bits. In this way, there is a growth in terms of bit sizes for the next level multiplication. It is difficult to incorporate such an increase of bit sizes in each step when encrypted multiplication module need to be designed in their circuit-based representation to incorporate underlying FHE libraries for homomorphic domain inputs. For these challenges, we propose a precomputation based method to compute e^x efficiently and within a constant time when x varies within a fixed range based on the application. We have analyzed a few practical applications and datasets from different domains and observed that different applications require different ranges of e^x computation, which we can prefix from existing data patterns and perform some precomputations. Based on this, our proposed technique includes the following steps:

1. Select a few intermediate short ranges and selective values within the overall range of exponent computation requirement. For example, let us consider the exponent function values required for a particular application ranges from 0 to 100. Within this range, we choose the following short ranges (Range[i]) and selective values (Val[i]). According to our proposed precomputation and storage-

based method, to avoid costly multiplications, final encrypted exponent values will be computed performing some encrypted comparisons with the exponents of short-range boundaries and selective values.

2. For that, we store the encrypted values of the exponents of short-range boundaries and selective values. For each of the ranges, upper boundaries of Range[i] is stored in an array R[i] as shown in Table 1, where $R[i] = Enc(i)$, with respective encrypted exponents for future comparisons. As selective values, we choose 0, 1,..., 9 and store Enc(Val[i]) versus $Enc(e^{Val[i]})$ as shown in Table 1 (ii).
3. These precomputed R[i] and Val[i] values will be used to compute any exponent e^x, where x is within the predetermined range. This exponent computation is to be done with a Giant step- Baby step method, which will be detailed next with an example.

Algorithm 1: Algorithm for Giant Steps

Result: $Enc(e^{lowerIntRange})$, Lower boundary of intermediate range: lowerIntRange
initialization;
for $i \leftarrow 0$ **to** $MaxRCount - 1$ **by** 1 **do**
  if *R[i] < x < R[i+1]* **then**
    lowerIntRange=R[i];
  else
    continue;
  end
  Output $Enc(e^{lowerIntRange})$.
end

Let us consider, we are interested to compute e^{17} in encrypted domain. That implies value ($x = 17$) and respective exponent ($Enc(e^{17})$) both are encrypted with some randomized FHE scheme. According to Algorithm 1, each R[i] value (that is $Enc(0), Enc(1), \ldots, Enc(100)$ in our example) will be compared with $Enc(17)$ to find the $R[i]$ for which $R[i] < x < R[i + 1]$. Thus, the lower boundary of intermediate range (lowerIntRange) $= Enc(10)$ and Algorithm 1 will output $Enc(e^{lowerIntRange})$, that is $Enc(e^{10})$. This long range detection for the exponent value is termed as Giant Step. Next, through baby step, we need to compute the difference between $Enc(x)$ and $Enc(lowerIntRange)$ with FHE subtraction module, that is $Enc(17) - Enc(10)$ in our example. Now, with encrypted comparison $Enc(e^{x-lowerIntRange})$ need to be found from Table 1. Finally, $Enc(e^x) = Enc(e^{10}) \times Enc(e^{17-10})$ gives the final exponent value. In the subsequent sections, we explain how to realize this Giant Step-Baby step method on encrypted data with underlying FHE gates.

Giant Step In general, by the giant step we try to identify range of x, such that $R[i] < x < R[i + 1]$, where i ranges from 0 to MaxRCount -1, where MaxRCount is number of rows in precomputed table. As explained in Algorithm 1, we iterate over each rows of R to identify $R[i]$ for which $R[i] < x < R[i + 1]$ and finally obtain

Table 1 (i) $Enc(e^{R[i]})$ precomputations, (ii) $Enc(e^{Val[i]})$ precomputations

Boundaries R[i]	$Enc(e^{R[i]})$
$Enc(0)$	$Enc(e^0)$
$Enc(10)$	$Enc(e^{10})$
$Enc(20)$	$Enc(e^{20})$
$Enc(30)$	$Enc(e^{30})$
$Enc(40)$	$Enc(e^{40})$
$Enc(50)$	$Enc(e^{50})$
$Enc(60)$	$Enc(e^{60})$
$Enc(70)$	$Enc(e^{70})$
$Enc(80)$	$Enc(e^{80})$
$Enc(90)$	$Enc(e^{90})$
$Enc(100)$	$Enc(e^{100})$

Sel. Values Val[i]	$Enc(e^{Val[i]})$
$Enc(0)$	$Enc(e^0)$
$Enc(1)$	$Enc(e^1)$
$Enc(2)$	$Enc(e^2)$
$Enc(3)$	$Enc(e^3)$
$Enc(4)$	$Enc(e^4)$
$Enc(5)$	$Enc(e^5)$
$Enc(6)$	$Enc(e^6)$
$Enc(7)$	$Enc(e^7)$
$Enc(8)$	$Enc(e^8)$
$Enc(9)$	$Enc(e^9)$

the value $e^{R[i]}$. However, this is not so straightforward when x and $R[i]$ values are encrypted and general bitwise matching or array search will not work. Figure 1 shows how Algorithm 1 can be realized by underlying basic FHE circuits.

As already mentioned, required exponents of specific values and range boundaries need to be precomputed in tables like Table 1. As shown in Fig. 1, rows from Table 1 will be selected and each $R[i]$ and $R[i+1]$ values will be compared with x. Since all the values are encrypted, a general bitwise comparison will not work, and the comparison is to be done with encrypted comparators [3]. The comparison results from both the comparators will be either $Enc(0)$ or $Enc(1)$, and these two encrypted bit values will be multiplied with the encrypted bitwise AND (FHE_AND) module of the supporting library. Output of this encrypted bitwise multiplication will act as the select the respective $E(e^{R[i]})$ from the Table 1. Encrypted selection of particular $E(e^{R[i]})$ is to be done by encrypted decision making with the help of encrypted multiplexer [3]. Once the output of FHE_AND module is Enc(1), that indicates both the conditions $R[i] < x$ and $x < R[i+1]$ have been satisfied. This is the only case when both of the encrypted comparators will give $Enc(1)$, and thus the output of the FHE_AND will be $Enc(1)$. Otherwise, if any of the comparator outputs is false, FHE_AND will give $Enc(0)$. For each row iteration, when both the comparator conditions are satisfied for a particular $R[i]$ value, the FHE_Mux outputs $E(e^{R[i]})$. Otherwise, the MUX outputs whatever value is in the buffer. The buffer initially contains encrypted zero; once the $R[i]$ is selected for MUX output, it will be stored in the buffer and held for the next iterations.

In our example, to calculate 17 at first, the buffer is set to contain Enc(0). Next, we will start comparing the encrypted exponent (Enc(17) in this case) with $R[i]$ values-starting from 0, the comparator output will be zero. So, the MUX will select its 0 position input which is $Enc(0)$. After the comparison with 10, the Buffer_1 will

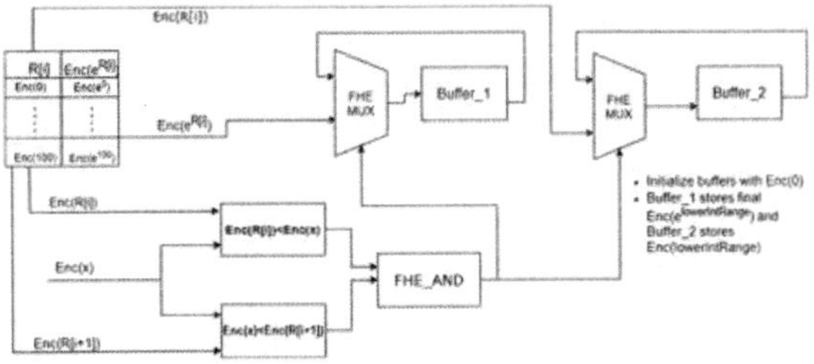

Fig. 1 Giant Step Execution

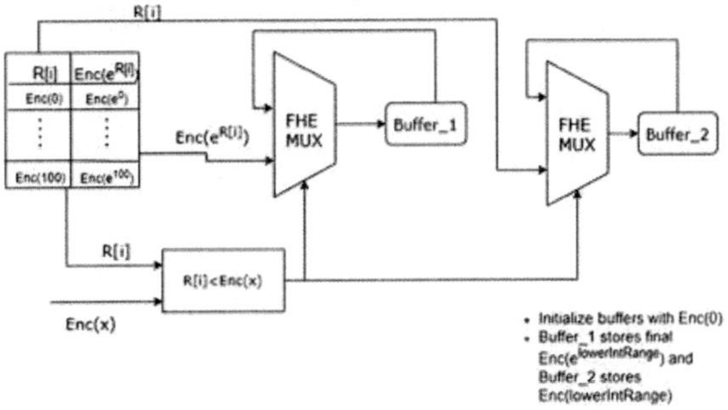

Fig. 2 Giant Step Modified with less Comparison

contain $Enc(e^{10})$. Then for the next iterations, again selection line will be 0, and the buffer value of $Enc(e^{10})$ will be held. Similarly, for Buffer_2, ultimately, $Enc(10)$ will be held.

The direct implementation of Algorithm 1 can be further improved with just one comparator module needed. The improvement is shown in Fig. 2. Once the required lowerIntRange is found, the technique to hold the value in the respective buffers is also modified. Here till $R[i]$ is smaller than the exponent, the comparator output will be Enc(1), and this will be fed as selection line to FHE_MUX. Till this, $Enc(e^{R[i]})$ will be selected for MUX output and stored in buffer_1 and $R[i]$ will be stored in Buffer_2. Once the $R[i]$ is greater than the exponent, then the buffer value will be selected on line 0 which contains the $Enc(e^{R[i-1]})$ value in Buffer_1 and $R[i-1]$ in the Buffer_2. For all next iteration, $R[i]$ will be greater than the exponent, so the buffer value will be held, and after all iterations, the final Buffer_1 output will be the $Enc(e^{lowerIntRange})$ and Buffer_2 output will be Enc(lowerIntRange), which is $R[i]$ just below the exponent.

Algorithm 2: Algorithm for Baby Step

Result: The encrypted exponent: expVal
initialization;
$digitval = x - lowerIntRange$ $EncMatch(expVal1, babyRange, digitval)$
$EncMatch(expVal2, giantRange, lowerIntRange)$ $expVal = expVal1 * expval2$

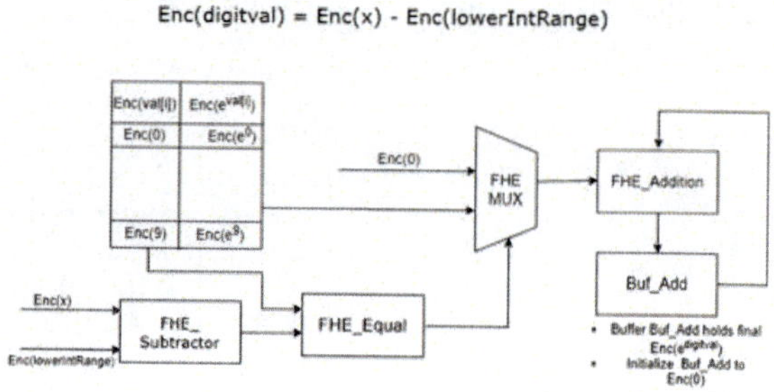

Fig. 3 Baby Step Execution

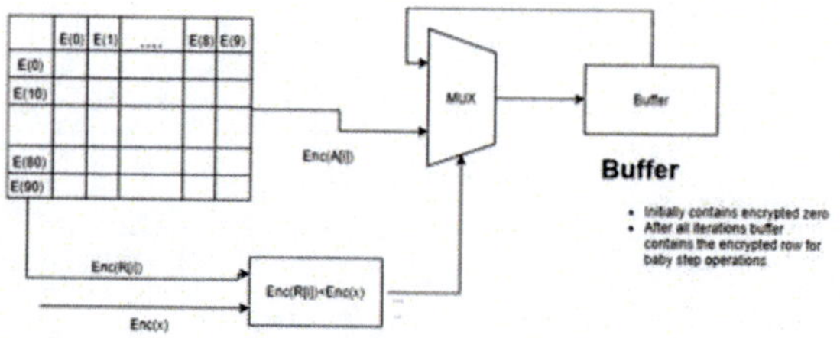

Fig. 4 Logarithm Giant step

Baby Step Now, we subtract the result ($Enc(lowerIntrange)$) from exponent ($Enc(x)$) to get encrypted single digit value. We shall check for the equality with encrypted match function for this subtraction result and encrypted digit values (Fig. 3) as mentioned in Algorithm 2. Baby steps start from $Enc(0)$ and is increased by $Enc(1)$ and check with single digit exponent. With this after seventh iteration match is found for $Enc(17)$, so we store $Enc(7)$ as result of baby step. Now, the calculation of exponent 17 will be done by exponent 10 and exponent 7 values. The only part that remains is to multiply those $Enc(e^{10})$ and $Enc(e^{7})$ values to get value of $Enc(e^{17})$. The baby step can also be modified to use the optimized giant step trap technique to eliminate the summation of all the FHE_MUX results.

We need to realize three main suboperations to complete the full implementation:

Encrypted Comparator: This module takes two encrypted numbers $Enc(a)$ and $Enc(b)$ and produces an output of Enc(1) if $b > a$ and Enc(0) otherwise. We used the already developed subtraction module [2], and checked the sign bit of the subtraction result for our implementation. If and only if $b > a$ the Most Significant Bit (MSB) of encrypted subtraction $a - b$ will be negative and that will yield an $Enc(1)$ as a result.

Encrypted Match Function: The baby step execution (Fig. 3) is based on a match function over a list of data. For our implementation the baby step finds out the encrypted exponent value ($Enc(e^{y})$) of an input digit $Enc(y)$). For this, we need a list of all encrypted digit and corresponding encrypted exponent values. When y matches with $R[i]$, then only the FHE_Equal module generates an $Enc(1)$, which will be fed to the FHE_MUX module as a selection input. The encrypted MUX selects $Enc(e^{R[i]})$ when selection line is Enc(1) otherwise $Enc(0)$. The MUX output is then added with a buffer value (initialized with $Enc(0)$) and stored into the buffer. So, after the full iteration of the list, the buffer contains the sum of $Enc(e^{R[i]})$ and $Enc(0)$, which is the desired result. FHE_Equal module is designed with an encrypted comparator module; when both $a > b$ and $b > a$ is false, then only FHE_Equal results an $Enc(1)$ (detailed discussion is in [22]).

Normalization: For any practical application, input data has to be in some predefined range. However, the raw data may not be confined to that range only; hence a normalization process is necessary. For our implementation, there is an extra constraint for normalization as existing FHE libraries support only integer input. For exponent computation, we need to store initial $Enc(e^{R[i]})$ and to process on them, which is not possible if we consider $e^{R[i]}$ values in floating-point representation. Hence, for example, if we have any $e^{R[i]}$ as m_1-bit before the decimal and m_2-bit after the decimal, we consider up to say k-bits after the decimal and multiply the number with 10^k for normalization. Thus, we need a total of $m_1 + k$-bit to store the normalized exponent values for processing.

Now, for example, if $x > 15$ then e^x can surpass 32 bit variable size (as e is approximately 2.718), so if we intend to use 32 bit-sized variables, then we need to normalize the e^x values to that range. As for our implementation, the x can lie between any range, so suitable normalized range needs to be considered.

3.2 *Logarithm Operator Computation*

Similar to the exponent operator, logarithm function table can be precomputed to be stored in the cloud. Again the domain of datapoints determines the precomputed values. Once the log table is precomputed and stored in the cloud, then the log function computations turn to be a search into the table and finding a match entry problem. Similar to exponent, giant steps can be used for 10th digit comparison, and then baby steps are used to reach the desired entry. The only difference is here that unlike exponent computation here whole table should be precomputed and stored (as log operation doesn't support multiplicative addition like exponent). The precomputed table for logarithm is like a matrix where each rows have 10 values, for example, first row contains $log(0)$ to $log(9)$ values for a specific base. Hence, the giant step is a bit different here as shown in Fig. 4, unlike exponent here giant step result is an encrypted row than just single encrypted value. After the giant step is over, the buffer contains the single encrypted row that will be used in the baby step, the same as Fig. 3. $Enc(R[i])$ indicates a full row in the buffer result for, e.g., if $Enc(R[i][0]) < Enc(x)$ and $R[i][0] = 80$ then the buffer contains values like this:

E(80)	E(81)	E(82)	...	E(89)

Total 10 encrypted comparisons are required in giant step and then 10 more encrypted comparisons for baby step. Hence, total 20 encrypted comparisons are required to determine a log value within a range of $Enc(0)$ to $Enc(100)$. The rest of the operation is quite similar to the exponent operator computation which is discussed in detail in the previous subsection.

Fig. 5 Sigmoid Calculation Circuit diagram

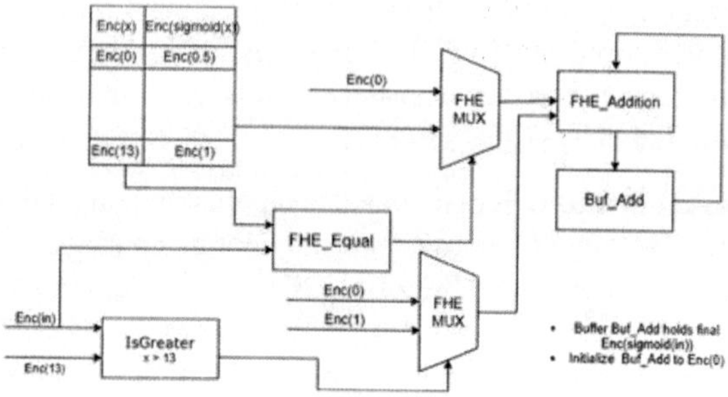

3.3 Sigmoid Function Computation

Our mentioned scheme can be promising to compute e^x avoiding a huge number of encrypted multiplications. However, it is important to note that sometimes algorithms translation to encrypted domain needs something more and not the only line by line translation of intermediate operators; a closer look may create an opportunity to reduce overall operational cost. For example, the sigmoid activation function in Artificial Neural Network(ANN) also includes exponent operation in the form [5]:

$$f(x) = \frac{1}{1 + e^{-x}}$$

To make Machine Learning as a Service (MLaaS) in encrypted domain practical, it is essential to realize such activation functions in encrypted domain efficiently. It also may be noted that sigmoid function is used not only as an activation function in ANN, but a variety of other applications had been using it for different logistic purposes. Other uses of the sigmoid function are waveshaper transfer functions in audio signal processing, Hill equation, and Hill-Langmuir equation in biochemistry and pharmacology where the requirement of encrypted data processing can be prominent. It also may be noted that sigmoid function is used not only as an activation function in ANN, but a variety of other applications had been using it for different logistic purposes. Other uses of the sigmoid function are waveshaper transfer functions in audio signal processing, Hill equation, and Hill-Langmuir equation in biochemistry and pharmacology where the requirement of encrypted data processing can be prominent. However, a step-by-step realization of sigmoid function requires encrypted exponent computation followed by encrypted division implementation, which leads to costly timing requirement and can make the realization impractical.

input(x)	sigmoid(x)
0	0.5
1	0.73106
2	0.8808
3	0.95257
4	0.98201
5	0.99331
6	0.99753
7	0.99909
8	0.99966
9	0.99988
10	0.99995
11	0.99998
12	0.99999
13	1

In literature, logistic sigmoid functions are polynomially approximated to be used in the encrypted domain. Early works [23] in this direction used Taylor approximation. Taylor approximation suffers from significant errors as it is a local approximation, not specialized for input variable intervals. As a remedy, minimax approximation [24] was proposed based on Remez algorithm [25]. In the financial and medical domain, such approximations are implemented for Logistic regression over encrypted data for fully homomorphic encryption [26, 27]. However, our proposed method is different from these approaches as we did not use any approximating polynomial.

It is straightforward for the sigmoid function to use the precomputed method as the range of input variable(x) is quite less. Sigmoid function achieves its maximum value at $x = 13$. Also, the normalization here works pretty well according to the precision required in the application. Thus if any encrypted value Enc(input) is greater than Enc(13), the respective sigmoid value will always be Enc(1). Otherwise, it will follow precomputed table. This same logic can be realized with a much simpler circuit-based representation, as shown in Fig. 5. It is important to note that this way of sigmoid computation does not require costly encrypted multiplication, and sigmoid value can be computed in constant time for any range of x, even with negative integers if proper normalization technique is applied.

4 Results

We analyzed standard applications in medical (Pima Indians diabetes data set [28], Wisconsin breast cancer database [29], Habermans survival data [30], Cardiotocography data set [31], Dermatology data [32]) and financial (Stock market prediction [33], bank failures prediction [34], credit card fraud detection [35]) domain, which shows the required exponent ranges after normalization are approximately 0 to 25. Our per-

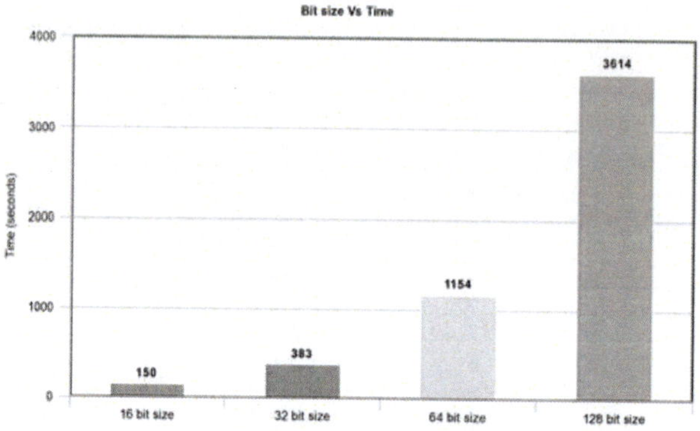

Fig. 6 Processing Time versus bit size comparison for exponent operator

Table 2 Processing time change varying bit size in single processor

Bit size	Processing time
16 bit	150 sec
32 bit	383 sec
64 bit	1154 sec
128 bit	3614 sec

formance results in Table 2 gives an idea about the timing requirements for exponent calculations in case of such applications. However, our proposed scheme is scalable enough for larger or smaller ranges with changed range boundaries. According to the study of different datasets, for most of the practical Machine Learning applications 32 to 64 bit size precision for exponent computation will suffice.

Our proposed scheme has been implemented on Intel i5-7200U CPU @ 2.50GHz x 4 with 8GB memory and Ubuntu 18.04.4 LTS 64-bit platform using TFHE homomorphic library [8]. Table 2 shows the execution time with varying bit sizes, and Fig. 6 shows how the execution time increases with the increase of bit sizes for exponent computation. Here bit size refers to the size of x in case of exponent operator e^x, and accordingly, the values are normalized as discussed in Sect. 3.1. As a comparison, the graph shows that doubling the bit size affects the execution time to be increased approximately thrice in our case. However, it has to be noted that for exponent or power computation with repetitive bitwise multiplication, each n-bit multiplication requires $O(n^2)$ timing, and upon doubling the bit size from n to $2n$, the timing requirement increases $O(4n^2)$ order. In comparison to that, our proposed scheme with precomputation performs the exponent or power computation within a constant time for within a fixed range.

Using domain knowledge once precomputation is done, it is seen that different encrypted functions can be computed in almost a similar time. From Table 3 it can be seen for 16bit data Exponent, Logarithm, and Sigmoid functions can be computed in

Table 3 Processing time of different 16 bit encrypted operators

Different operators	Processing time
Exponent	150 sec
Logarithm	198 sec
Sigmoid	184 sec

Table 4 Processing time improvement using OpenMP parallelization

Parallelization	Processing time
Sigmoid (single thread)	184 sec
Sigmoid (multi thread)	83 sec

between 150 and 200 sec. As we are now equipped with cloud computing, hardware resources are now can be on a rent basis. For these advantages, High Performance Computing (HPC) is mainly shifted to cloud-based resources. With flexible pricing, even small enterprises are now capable of using the advantages of cloud computing. As we have already addressed that homomorphic computations are time-consuming and performance is a serious bottleneck for practical applications, a parallel computation paradigm can be helpful in boosting performance. For multi-core processing OpenMP is a standard [36] programming technique to assign thread-based operations to available processing cores. We have used openMP in our framework and have seen more than 2 times improvement with a quadcore processor. With cloud computing, hundreds or even thousands of cores can be deployed as per requirement. We have explored OpenMP implementation using ubuntu OpenMP library in a quadcore system and found out that performance improvement is more than twice. Table 4 shows how sigmoid function processing time improved after OpenMP multi-threaded multi-core shared resource implementation.

5 Conclusion

It has to be noted that design of homomorphic operators for end to end encrypted framework is still in the nascent stage. According to the present literature, all encrypted operators are optimized to some extent either in terms of limited precision or polynomially approximated functions and implemented over limited homomorphic encryption schemes to improve performance. Theoretically, any operation is claimed to be possible by FHE, but in practice, operators that are computationally expensive need a close revisit to design their encrypted counterparts. In this work, we highlighted the importance of domain knowledge in realizing the encrypted operators. If the range values for a particular operation are available to the cloud, using a precomputation method, complex operators can be transformed into a search and compare problem. The security is kept intact as the cloud doesn't learn anything and data is kept always encrypted throughout the computation. The result is also encrypted which is sent back to the user to decrypt. In our work, we demonstrated the concept with the exponent operator (e^x) in the encrypted domain, and proposed a precomputation-based technique to improve the computation time. We further show how this technique computes large exponents within a fixed range in constant time, which otherwise requires a considerable number of encrypted multiplication operations. The method is scalable enough to be used with other complex functions in a fixed range of values, as shown for logarithm and sigmoid function. In this work, we also explored parallel execution possibilities to improve overall performance. In our future work, we plan to map sufficient operators to perform certain tasks for a particular domain (medical/financial) and propose an end-to-end framework to utilize domain knowledge in favor of performance boost.

References

1. Van Dijk M, Gentry C, Halevi S, Vaikuntanathan V (2010) Fully homomorphic encryption over the integers. In: Annual international conference on the theory and applications of cryptographic techniques. Springer, pp 24–43
2. Chatterjee A, Sengupta I (2020) Sorting of fully homomorphic encrypted cloud data: can partitioning be effective? IEEE Trans Serv Comput 13(3):545–558
3. Chatterjee Ayantika, Sengupta Indranil (2015) Translating algorithms to handle fully homomorphic encrypted data on the cloud. IEEE Trans Cloud Comput 6(1):287–300
4. Law R, Murrell DJ, Dieckmann U (2003) Population growth in space and time: spatial logistic equations. Ecology 84(1):252–262
5. Nwankpa C, Ijomah W, Gachagan A, Marshall S (2018) Activation functions: comparison of trends in practice and research for deep learning. arXiv:1811.03378
6. Liang J, Shi Z, Li D, Wierman MJ (2006) Information entropy, rough entropy and knowledge granulation in incomplete information systems. Int J Gen Syst 35(6):641–654
7. Acar A, Aksu H, Selcuk Uluagac A, Conti M (2018) A survey on homomorphic encryption schemes: theory and implementation. ACM Comput Surv (CSUR) 51(4):1–35
8. Chillotti Ilaria, Gama Nicolas, Georgieva Mariya, Izabachène Malika (2020) Tfhe: fast fully homomorphic encryption over the torus. J Cryptol 33(1):34–91
9. Chen H, Laine K, Player R (2017) Simple encrypted arithmetic library-seal v2. 1. In: International Conference on Financial Cryptography and Data Security. Springer, pp 3–18
10. Ada Popa R, Redfield CMS, Zeldovich N, Balakrishnan H (2011) Cryptdb: protecting confidentiality with encrypted query processing. In: Proceedings of the twenty-third ACM symposium on operating systems principles, pp 85–100
11. Egorov M, Wilkison M (2016) Zerodb white paper. arXiv:1602.07168
12. Papadimitriou A, Bhagwan R, Chandran N, Ramjee R, Haeberlen A, Singh H, Modi A, Badrinarayanan S (2016) Big data analytics over encrypted datasets with seabed. In: 12th {USENIX} symposium on operating systems design and implementation ({OSDI} 16), pp 587–602
13. Robb D (2020) Always encrypted: Database security product overview and analysis. https://www.esecurityplanet.com/products/always-encrypted-database-security.html. Accessed: 2020-03-03
14. Grubbs P, Ristenpart T, Shmatikov T (2017) Why your encrypted database is not secure. In: Proceedings of the 16th workshop on hot topics in operating systems, pp 162–168
15. Pisa PS, Abdalla M, Duarte OCMB (2012) Somewhat homomorphic encryption scheme for arithmetic operations on large integers. In: 2012 global information infrastructure and networking symposium (GIIS), pp 1–8

16. Gai K, Qiu M (2018) Blend arithmetic operations on tensor-based fully homomorphic encryption over real numbers. IEEE Trans Ind Inform 14(8):3590–3598
17. Cheon JH, Kim A, Kim M, Song Y (2017) Homomorphic encryption for arithmetic of approximate numbers. In: Takagi T, Peyrin T (eds) Advances in cryptology – ASIACRYPT 2017. Springer International Publishing, Cham, pp 409–437
18. Kepner J, Gadepally V, Michaleas P, Schear N, Varia M, Yerukhimovich A, Cunningham RK (2014) Computing on masked data: a high performance method for improving big data veracity. In: 2014 IEEE high performance extreme computing conference (HPEC), pp 1–6
19. Kumarage H, Khalil I, Alabdulatif A, Tari Z, Yi X (2016) Secure data analytics for cloud-integrated internet of things applications. IEEE Cloud Comput 3(2):46–56
20. Gilad-Bachrach R, Dowlin N, Laine K, Lauter K, Naehrig M, Wernsing J (2016) Cryptonets: applying neural networks to encrypted data with high throughput and accuracy. In: International conference on machine learning, pp 201–210
21. Microsoft (2020) Arithmetic operators. https://docs.microsoft.com/en-us/sql/mdx/arithmetic-operators?view=sql-server-ver15. Accessed: 2020-01-30
22. Chatterjee A, Kaushal M, Sengupta I (2013) Accelerating sorting of fully homomorphic encrypted data. In: International conference on cryptology in India. Springer, pp 262–273
23. Bos JW, Lauter K, Naehrig M (2014) Private predictive analysis on encrypted medical data. J Biomed Inform 50:234–243
24. Fraser W (1965) A survey of methods of computing minimax and near-minimax polynomial approximations for functions of a single independent variable. J ACM (JACM) 12(3):295–314
25. Remes Eugene (1934) Sur le calcul effectif des polynomes d'approximation de tchebichef. CR Acad. Sci. Paris 199:337–340
26. Cheon JH, Jeong J, Lee J, Lee K (2017) Privacy-preserving computations of predictive medical models with minimax approximation and non-adjacent form. In: International conference on financial cryptography and data security. Springer, pp 53–74
27. Chen Hao, Gilad-Bachrach Ran, Han Kyoohyung, Huang Zhicong, Jalali Amir, Laine Kim, Lauter Kristin (2018) Logistic regression over encrypted data from fully homomorphic encryption. BMC Med Genomics 11(4):81
28. Kayaer K, Yildirim T, et al (2003) Medical diagnosis on pima indian diabetes using general regression neural networks. In: Proceedings of the international conference on artificial neural networks and neural information processing (ICANN/ICONIP), vol 181, pp 184
29. Mangasarian OL, Nick Street W, Wolberg WH (1995) Breast cancer diagnosis and prognosis via linear programming. Oper Res 43(4):570–577
30. Landwehr JM, Pregibon D, Shoemaker AC (1984) Graphical methods for assessing logistic regression models. J Am Stat Assoc 79(385):61–71
31. Borowik G (2014) Data mining approach for decision and classification systems using logic synthesis algorithms. In: Advanced methods and applications in computational intelligence. Springer, pp 3–23
32. Übeyli ED, Doğdu E (2010) Automatic detection of erythemato-squamous diseases using k-means clustering. J Med Syst 34(2):179–184
33. Vui CS, Soon GK, On CK, Alfred R, Anthony P (2013) A review of stock market prediction with artificial neural network (ann). In: 2013 IEEE international conference on control system, computing and engineering. IEEE, pp 477–482
34. Nur Ozkan-Gunay E, Ozkan M (2007) Prediction of bank failures in emerging financial markets: an ann approach. J Risk Financ
35. Sahin Y, Duman E (2011) Detecting credit card fraud by ann and logistic regression. In: 2011 international symposium on innovations in intelligent systems and applications. IEEE, pp 315–319
36. Dagum Leonardo, Menon Ramesh (1998) Openmp: an industry standard api for shared-memory programming. IEEE Comput Sci Eng 5(1):46–55

A Comprehensive Review of Fraudulent Email Detection Models

Hitesh Narayan Soneji, Aniruddh Sajith Soman, Aniruddh Vyas, and Shubha Puthran

Abstract Emails have become an integral part of our lives; be it personal or professional, emails are used everywhere. Most websites require the customers to enter their email ID, increasing the risk of exposure to spammers who attack them by sending spam messages. To overcome this problem, researchers have come up with different techniques to classify emails as legitimate or spam mails. The conducted literature survey includes 32 relevant papers out of the 47 selected papers. The paper is based on a comprehensive six research question-based approach which has been implemented to find the Machine Learning (ML) and Deep Learning (DL) techniques used, datasets used, pre-processing methods implemented, spam types, different evaluation metrics used and the strengths and weaknesses of the models in the literature. The domains analyzed in ML and DL are Naïve Bayes (NB), Support Vector Machines (SVM), K-Nearest Neighbor (K-NN), Random Forest, Neural Networks (NN), etc. This review will help the researchers in identifying the present and the future context of research in the different approaches for the classification of spam emails.

Keywords Fraud · Email · Machine learning · Deep learning

1 Introduction

Electronic mail or email is a method of exchanging messages between people. It first saw usage in the 1960s, but gained popularity in the 1970s. Back then, the person communicating via email needed to be online to receive an email, but that is no longer the case; one can receive emails while offline as well. Emails are one of, if not the most common methods of communication and as such, they are used by everyone, from the common man to the CEOs of MNCs. Like any other form of communication, emails are prone to attacks from hackers and phishers. The major email protocols

H. N. Soneji (✉) · A. S. Soman · A. Vyas · S. Puthran
Computer Engineering Department, MPSTME, NMIMS University, Mumbai, India

S. Puthran
e-mail: Shubha.Puthran@nmims.edu

© The Author(s), under exclusive license to Springer Nature Singapore Pte Ltd. 2022
D. Giri et al. (eds.), *Proceedings of the Seventh International Conference on Mathematics and Computing*, Advances in Intelligent Systems and Computing 1412,
https://doi.org/10.1007/978-981-16-6890-6_9

in use, SMTP and POP, are easily accessible to everyone, making them vulnerable to attacks. Email fraud is the intentional deception made for personal gain or to damage another individual through email. Email fraud started happening around the time email became a mainstream method of communication. There are various types of email fraud; some of the most prominent ones are Spoofing, Phishing for Data and Bogus Offers. The general public is not aware of such methods of fraud and often falls prey to them. As per a recent statistic in [1], more than a hundred million phishing/fake emails related to Coronavirus were sent to people. The spammer sends an email claiming to collect donations for relief efforts in relation to COVID-19, and obtain the credit card credentials, bank details and money from good Samaritans. These are a small subset of all spam emails, relating only to COVID-19. Considering all spam emails related to banking, marketing, advertising, anti-virus spam, etc. the number would be much higher.

Fraudulent Emails are of many types, some of the most common ones are 1. Spam—the emails sent by spammers to sell fake products or commit fraud; 2. Phishing—these emails are used to trick unsuspecting users into divulging confidential information such as usernames, passwords and bank account details. They often mimic official emails from companies such as PayPal, eBay and Citibank; 3. CEO Fraud—this fraud involves sending emails pretending to be the CEO of the company and asking for money transfers from the finance department. This type of fraud is becoming more commonplace; 4. Spoofing—spoofing emails trick users into visiting malicious sites or downloading software which then steal the users' sensitive information.

Distinguishing between legitimate and fraudulent emails is not an easy task for the layman. With the increase in reach of malicious emails and poor spam filters, the common person often falls prey to scams and ends up losing money or their personal data. Cybercrime via emails has become a common problem. According to [2], even big company executives fall victim to such problems and lose millions of dollars in these scams. The necessity of a smart email spam filter has become the urge of society, with people of all ages getting caught in scams. Many spam filters become obsolete as soon as the spammers find a way to bypass such filters.

This paper reviews the various techniques used and presented for email spam detection over the years. There are many ways to combat email fraud, one being using an email spam filter. Such a filter will simply remove any and all spam emails. First, the email will have to be collected. Then the feature of the email which is to be analyzed must be extracted (body, sender, URLs). Once it has been extracted, ML techniques are used to classify the emails as spam or legitimate. It is checked whether certain keywords are present which are often seen in spam emails. If the email is detected as spam, it will automatically delete the email or store it in a separate folder. This paper reviews 32 papers which deal with email fraud detection. The different techniques in each paper have been studied, and a detailed study showing the different methods used in each paper, their strengths and weaknesses, the evaluations metrics used, etc. has been presented.

The remaining paper is structured as follows. Section 2 is the methodology used. Section 3 contains the analysis of the research questions. Finally, Sect. 4 is the discussion and conclusion.

2 Methodology

This paper's methodology has been proposed with the help of some Research Questions which will help to understand the different approaches taken across the literature, identify the datasets involved, the advantages and disadvantages, etc. The questions listed below will help analyze the different techniques and implementations used in Email Fraud Detection which are present in the literature.

2.1 *Research Questions*

1. Types of Spam and how they are used by Spammers.
2. Which data pre-processing techniques have been implemented for preparing the data?
3. Which ML and DL techniques have been used for the classification of the emails?
4. What datasets have been utilized across the literature?
5. What Evaluation metrics have been most commonly used in the literature?
6. What are some of the strengths and shortcomings of the different techniques used?

3 Analysis of the Research Questions

3.1 *RQ1: Types of Spam and How They Are Used by Spammers*

The first spam emails were documented in the mid-1990s. Sending fraudulent or phishing emails started around the time emails were gaining popularity; it also urged the development of methods to filter out such emails for the users. As the number of spam classifiers built increased, the scammers evolved their methods to deceive the classifiers and reach the users. Below, the major types of spam emails and how they are used are listed.

Spoofing—Spoofing is a method where emails are sent impersonating someone else, and trying to convince the victim of being someone else to steal information.

Phishing for Data—These emails purport to be from an existing company or relative, or someone the victim already has a business or personal relationship with. One example can be an email from the victim's bank asking to confirm their details, which then collects their data and they are robbed.

Bogus Offers—Emails to purchase goods or services can be categorized as bogus offers. In this type of scam, the sender tries to send some irresistible offer, to which the victims fall prey and invest money, only to find out later that the offer was fake and they have been robbed.

As per statistics in [3], 45% of all emails are spam, email spam costs businesses around 20.5 billion dollars a year. These statistics show the need for better spam filters in the market, which can classify spam emails more efficiently.

3.2 RQ2: Which Data Pre-processing Techniques Have Been Implemented for Preparing the Data?

Pre-processing data is one fundamental step to convert the data in a suitable format to feed the machine. There are various pre-processing techniques used in the literature, out of which with this question we would address the most commonly used techniques across the literature. The various pre-processing techniques are as follows.

Stop Word Removal: Stop Word Removal is one of the most basic pre-processing techniques. Stop Words (we, us, are, if, of, etc., a, an, the, ...) are words which are repeated very often and can create problems during classification due to the high frequency of such words available. With the help of Stop Word Removal, these words are removed from the text.

Word Tokenization: Word Tokenization is performed to convert the given text into a list of individual words from the text. A 2D array-like list is formed where each row contains a new list consisting of words from the corresponding text. Word Tokenization can be performed easily with the help of the Natural Language Toolkit (NLTK) module of Python.

Term Frequency Inverse Document Frequency (tf-idf): tf-idf is a statistical measure that tells how relevant a word is to a document in a collection of documents. tf-idf is calculated by multiplying 2 measures: 1. The number of times a word appears in a document and 2. The inverse document frequency of the word across a set of documents.

The term frequency is calculated by a simple raw count of instances of a word that appear in that document. Inverse document frequency means how rare or common a word is in the document set; it is calculated by taking the number of documents, and dividing it by the number of documents that contain the word, and taking the log function of it. When the inverse document frequency value approaches zero, it means that the word is very common and if it approaches 1 then the word is rare.

After multiplying these 2 values, we get our tf-idf score. This is then fed to the ML or DL model, which greatly improves its accuracy compared to just using word count.

Word2Vec (Word Embeddings): Word2Vec is a method to obtain word embeddings. Word2Vec is a two-layer model which processes the given text by vectorizing the words. The input is a body of text and the output is a set of vectors. Word2Vec is used to group words of similar meaning together in a vector space. Word2Vec can be achieved with two methods—(1) Bag of Words and (2) Skip Gram.

The bag-of-words model is used to predict a single word with the help of a feature text. The skip gram model is used when from the features of a word the context of the word needs to be predicted which is also called the company of the word.

For example, consider the following sentence 'Today is sunny and nice'. In the bag-of-words method, with the help of the word's features {today, is, nice}, the word itself {sunny} is predicted. In the skip gram method, with the help of the words {sunny}, its company {today is nice} is predicted. The larger the text corpus, the better can be the similarity between words in the vector space obtained. There are many pre-trained Word2Vec models available online which can be used for creating word embeddings for the desired dataset. In [3], the authors have used the Google Word2Vec model which consists of more than 3 million words and vector size of 300. With the help of such pre-trained Word2Vec models, the making of word embeddings becomes easy and more accurate.

Tf-idf and Word2Vec are pre-processing techniques which are generally applied in the later stages of pre-processing, when the data needs to be fed to the model for predictions or classification. Either of the two methods is implemented in the pre-processing stage. By applying these methods, the overall accuracy of the classification increases, as can be seen in many papers in the literature.

Feature Correlation Techniques—Correlation techniques are applied for feature selection, which is used to reduce the number of features and select only the relevant features from the data. This helps in reducing the training time, data dimensionality and data size. In the paper [14], the authors have used a correlation technique to select the useful features from the pre-processed data.

Some of the correlation techniques studied across the literature are

Cramer's V—Cramer's V is a measure of association between two variables, giving a value between 0 and +1. It is based on the Chi-Square test of Association. It is a statistical method and like any other statistical method, it has both a null and an alternative hypothesis. This method is used to find any significant relationship between the two variables. 0 correlation value indicates that the two variables are not correlated and that the variables can be discarded for the next steps, while a value of +1 indicates that the variables are highly correlated.

Pearson's Correlation—Pearson's correlation coefficient is a measure of the strength of the association between the two variables. Pearson's correlation values range from −1 to +1, where −1 indicates that there is a negative correlation between the variables and as one increases the other decreases, whereas +1 indicates that it's a positive correlation and as one value increases the other also increases. A 0 value indicates that there is no correlation between the two variables.

Information Gain—Information Gain measures the reduction in entropy by splitting a dataset according to a given value of a random variable. A larger Information gain value indicates lower entropy, and a lower information gain indicates higher entropy. Information gain is the technique used in ID3 algorithms for tree branch selection. Information gain values also range from 0 to +1, where 0 indicates no correlation and +1 indicates strong correlation.

3.3 RQ3: Which ML and DL Techniques Have Been Used for Classification of the Emails?

With the help of this fundamental question, the various Machine Learning (ML) and Deep Learning (DL) techniques used in the literature have been identified. As shown in Table 1, ten different classifiers have been identified from the literature. The following were the techniques identified:

Support Vector Machine (SVM)—SVM uses the concept of hyperplane to categorize the input. Support Vectors are created by finding the nearest neighbor in each category to the new input. The area enclosed between the 2 support vectors is known as the hyperplane (positive hyperplane and negative hyperplane) and the input is categorized in whichever hyperplane part it lies. Reference [4] shows the comparison between two types of SVM models—Extreme Learning Machines (ELM) and SVM, which shows SVM to be performing better than ELM. In [5], the authors have used an SVM model for classification which is incremental; it retrains itself after certain intervals of time.

Table 1 Distribution of ML and DL models used across the literature

Classifier used	Count
NB	14
SVM	12
DT	11
K-NN	6
ANN	5
CNN	5
RF	5
LR	4
LSTM	3
C-Means	1

Decision Trees (DT)—DT is one of the oldest methods of Machine Learning Classification algorithms. Trees are built using leaves, where each leaf is a category having 2 decisions YES and NO. The input is categorized by comparing it with the leaf and is assigned to the category having the maximum similarities. DT (ID3) works on the principle of Entropy and Information Gain. The construction of a DT is all about finding the attribute that gives the highest information gain. Information gain is calculated from the entropy. Entropy is derived from information theory. The higher the entropy value, the more is the information content. The other types of DT model used in the literature is J48. In [6], the authors have used the J48 type of DT, which gives an accuracy of about 55% in the classification.

Random Forest (RF)—It works on the same principles as Decision trees. A big decision tree is divided into small decision trees to create a forest, and N random trees are selected out of this forest and the analysis is done on these N trees. The final output will be the majority of the results given by N trees. In [7], the authors have proposed an RF model along with polynomial kernel as a filter, and along the Linguistic-analysis tool used as pre-processing. In [8], the authors have compared different ML models, and RF gave the best accuracy among all the models, though it takes more time to train the model compared to others. In [9], the authors have used RF as the classifier because the model can handle many features easily, and they have used a huge dataset collected over a period of 2.5 months; they got an accuracy of around 75%.

K-Nearest Neighbor (K-NN)—In this classification technique as the name suggests, the nearest K neighbors to a new input which is to be categorized are checked. The new input is then classified into the category where the maximum number of its neighbors belong. The distance between the input and the neighbors is calculated using the Euclidean distance. In [10], the authors have used a two-step model, where in the first stage they have used K-NN to expand the training set and assign labels to the expanded features, and in the second stage, using the expanded training set obtained with the help of K-NN, they classified the emails using Long Short-Term Memory (LSTM). Reference [11] uses the principle of persuasion, which counts whether the word from the feature appears or not. Such 25 features are selected. Different ML models were used, from which K-NN gave the best accuracy of 97.2% among all the classifiers used.

Logistic Regression (LR)—Logistic Regression Classifier algorithm is derived from linear regression. Logistic Regression is appropriate when the result of the analysis is binary. Probability is used to predict the value of the dependent variable by using the sigmoid function. In [12], the authors have used a hybrid DT + LR model along with false negative threshold. First, along the false negative threshold, LR is used to reduce the noisy data, then its output is fed to DT, which increases the overall accuracy of the model.

Naïve Bayes (NB)—Naïve Bayes is a well-known Classifier algorithm which works on the Bayes theorem. It works by calculating the probability value of each feature category, and with the help of these values predicts the category of the new input.

NB works on the basis of the below equation:

$$P(Y|Z) = \frac{P(Z|Y)P(Y)}{P(Z)} \quad (1)$$

where P(Y|Z) is the probability of event Y occurring given that event Z exists. P(Z|Y) shows the probability of event Z occurring given that event Y exists. P(Y) is the probability of event Y occurring and P(Z) denotes the probability of event Z occurring. The other classifier based on Bayes theorem is Multinomial Naïve Bayes. In [13], the authors have used NB for email classification using Particle Swarm Optimization (PSO). Correlation-based feature selection is used for selecting the features, and a dataset containing 1000 emails is used. After feature selection, the NB model is used to classify the emails and then PSO is used to further optimize the results. This resulted in 7.5% better accuracy. In [14], an intelligent text modification is used which converts the given symbols in the text to possible letters, and then spell check is used on it to verify the integrity of the word on which NB is used to classify the emails; this approach gave an accuracy of 62%.

Fuzzy C-Means—This model works by assigning a class to each data label which corresponds to a cluster on the basis of the distance between the data point and the cluster center. In [16], the authors have used the Fuzzy C-Means model for classification; it considers the membership value of an individual element in the set; a threshold value is set which corresponds to the maximum value for the membership matrix, then the model is trained and finally filtering is performed.

Artificial Neural Network (ANN)—ANN works on the idea of simulating a lot of interconnected cells, and by reducing the number of cells as each layer passes, finding the required output. This output can be a binary classification, image segmentation, image classification, etc. ANN is a type of Deep Learning Model. The type of ANN in the literature is Multi-Layer Perceptron (MLP). In [17], ANN has been used. Word embeddings have been used to feed the model and ten-fold-cross validation is used for training, testing and validation. The authors have used two different datasets to collect the emails, and classification has been performed on five pre-defined features like number of URLs, HTML email, etc. The model gave a high accuracy of 92.2%. In [18], the authors have compared various ML and DL models for email fraud classification, and used the SpamAssassin dataset for training and testing; among all the models ANN gave the best F-measure, recall and precision which was followed by SVM.

Convolutional Neural Networks (CNN)—CNN has 3 types of layers, input layer, hidden layer and output layer; there is no limit on the number of hidden layers. CNN works on the same principle as ANN of simulating the multiple interconnected cells. The major difference between the 2 models is that CNN has only the last layer as fully connected, whereas in ANN all the layers are fully connected. The other type of CNN specified in the literature is Recurrent Convolutional Neural Networks (RCNN).

Long Short-Term Memory (LSTM)—LSTM is a type of Recurrent Neural Network (RNN) used in deep learning applications. LSTM unlike RNN has feedback connections, which helps the model to handle single data points and an entire sequence of data. It is more used in applications like handwriting recognition, speech recognition, etc.

Table 1 illustrates how many times the different ML and DL models have been used in the literature. It is quite imminent that NB and SVM have been used the maximum, while the least popular technique used is the Fuzzy C-Means algorithm.

3.4 RQ4: What Datasets Have Been Utilized Across the Literature?

Across the literature, there are nine public datasets; namely, UCI, SpamBase, PhishTank, SpamAssassin, Ling Spam, SpamCorpus, PhishCorpus, Jose Nazario and Enron are part of the public datasets. and there are 13 self-created datasets which are distributed among Universities, Private and Others which are not open-source.

Figure 1 shows the distribution of these datasets across the literature.

University datasets contain emails from different universities of the authors in the literature. Private datasets contain email data from different corporations and some private emails of the authors. In Others, the source of the dataset has not been mentioned by the authors.

SpamAssassin is the most widely used dataset in the literature, followed by Enron. These datasets are easily accessible on the Web and contain a large amount of emails.

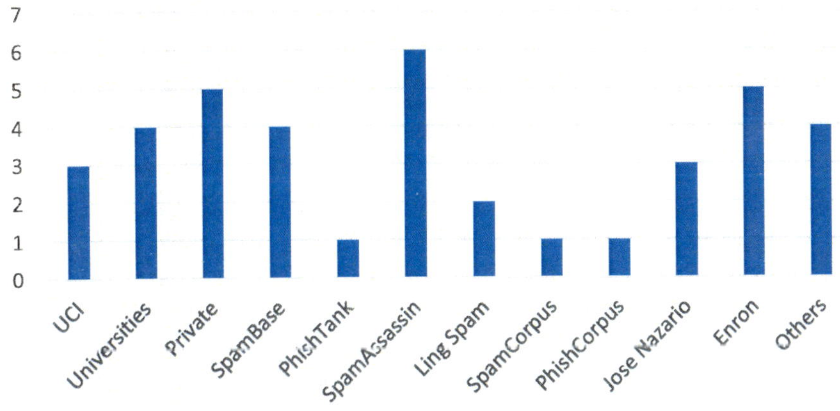

Fig. 1 Datasets used in the literature

3.5 RQ5: What Evaluation Metrics Have Been Most Commonly Used in the Literature?

Evaluation Metrics are a way of showing how the machine learning or deep learning model is performing. Accuracy cannot be used as the sole metric to evaluate a model as it doesn't always give the full picture of the model's performance. There are various other evaluation metrics in the literature which are discussed in this question.

The most basic metric used for evaluation of a classification model is True Positive (TP), True Negative (TN), False Positive (FP) and False Negative (FP), which are defined as follows:

- **TP**: When a True value is correctly predicted as True by the model.
- **TN**: When a False value is correctly predicted as False by the model.
- **FP**: When a False value is incorrectly predicted as True by the model.
- **FN**: When a True value is incorrectly predicted as False by the model.

These values can be illustrated with the help of a confusion matrix. Figure 2 illustrates a confusion matrix, which is used to easily visualize TP, TN, FN and FP. The above metrics are most commonly used for evaluating the performance of classification models which have been mentioned in the literature.

The following metrics are also calculated which help us further analyze the results obtained.

Accuracy: Equation (2) shows the formula to calculate accuracy. Accuracy tells us how often the model makes correct predictions. It is calculated by dividing the total number of correct predictions by the total number of predictions.

$$\text{Accuracy} = \frac{TP + TN}{TP + TP + FP + FN} \tag{2}$$

Recall: Recall is the ratio of correctly predicted True values to the total number of True values. Essentially, it tells us how often a True value is predicted as True. The result of recall varies between zero and one, with zero meaning there were no

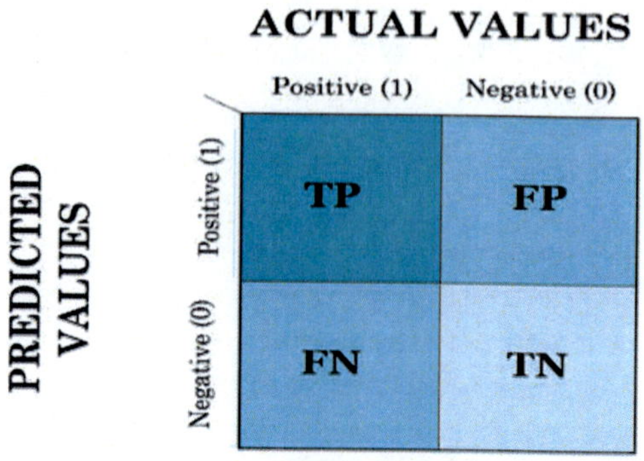

Fig. 2 Confusion matrix illustration

correctly predicted true values, and one meaning all the true values were correctly predicted. Equation (3) shows the formula to calculate recall.

$$Recall = \frac{TP}{FN + TP} \tag{3}$$

Precision: Precision is the fraction of relevant instances among the retrieved instances. Equation (4) shows the formula to calculate precision.

$$\text{Precision} = \frac{TP}{TP + FP} \tag{4}$$

During the evaluation phase of a model, having both high precision and high recall is not possible. There is a precision-recall trade-off, meaning that although maximizing both would be ideal, increasing one reduces the other, so an equilibrium needs to be maintained. For some datasets, deciding between having high precision or high recall depends upon the application of the dataset. Most of the times, however, an equilibrium needs to maintained, where F1-score is used to calculate the precision-recall situation.

F1-score: Equation (5) shows the formula to calculate F1-score. F1-score is a metric that combines recall and precision by taking their harmonic mean.

$$\text{F1} - \text{score} = 2 * \frac{Recall * Precision}{Recall + Precision} \tag{5}$$

Specificity: Specificity is the ratio of correctly predicted False values to the total number of False values. Equation (6) shows the formula to calculate specificity.

$$\text{Specificity} = \frac{TN}{TN + FP} \tag{6}$$

False Positive Rate (FPR): It is the ratio between false positive values to the total actual negative values. Equation (7) shows the formula to calculate accuracy.

$$\text{FPR} = \frac{FP}{FP + TN} \tag{7}$$

Accuracy cannot be taken as the sole metric to evaluate a model. Considering an example where a dataset contains majority labels as 0 and there are very few labels as 1, and the model predicts the majority class correctly but there are many inconsistencies in the evaluation of the minority class, the model would still have a high accuracy, but it would not reflect the shortcomings of the model.

3.6 RQ6: What Are Some of the Strengths and Shortcomings of the Different Techniques Used?

Table 2 illustrates the strengths and weaknesses of the models studied across the literature.

4 Conclusion

The aim of this literature survey was to study all the different models used in email spam detection. Each of these models uses different ML and DL algorithms to classify the mails. 32 papers have been studied and 6 questions were formulated which helped us differentiate each model described in these papers.

RQ1: Spam emails are sent with the intent of causing fraud. The major type of spam emails is bogus offers, phishing for data and spoofing. It causes loss of money as well as personal information.

RQ2: We found the pre-processing techniques across the literature and illustrated them. Word2Vec and tf-idf are the commonly used techniques in the literature, and they also help to increase the overall accuracy of the model. Data pre-processing is the most fundamental step for preparing the data and finally explained feature correlation techniques for variable selection.

RQ3: The different ML and DL techniques were illustrated and explained; in the literature, we saw that NB was the most widely used ML technique which was used 14 times, followed by SVM and DT, while in DL models CNN and ANN both have a frequency of 5. These models were used with different pre-processing models, filters, optimizers, etc. and DL models had the most promising results.

RQ4: Research question 3 illustrates the different datasets used in the literature, from which 9 are public and 13 are self-created datasets. Out of those, SpamAssassin and Enron datasets were most widely used.

RQ5: The different Result Analysis metrics used in the literature were summarized in this question. Only accuracy cannot be considered as the sole metric for analyzing the performance of a model and more parameters should be considered for properly evaluating the model.

RQ6: The Strengths and Weaknesses of all the papers in the literature were analyzed. We found that the THEMIS model based on the RCNN model produced the highest accuracy of 99.84% among all the ML and DL techniques, while among the ML techniques RF gave the best accuracy. In the literature, many papers had just proposed models and not implemented them, so their analysis could not be carried out thoroughly.

Table 2 Strength and weaknesses of the models used in the literature

Methods used	Strengths	Weaknesses
Model named THEMIS based on RCNN [4]	THEMIS model gives a high accuracy of 99.848%. It has a recall of 99% and precision of 99.664%. It performs better than CNN and LSTM. It has a low FPR which reduces the risk of filtering out the legitimate email to a greater extent	THEMIS cannot detect phishing emails with no email header and only an email body
SVM and ELM [5]	The proposed SVM spam detector yielded the best results. The difference in their accuracies is low but the difference in the speeds is very high. The results of these two machines were much better than any other classifier that was used on the same dataset	ELMs essentially make use of Neural Networks In general, Neural Networks require more time to process and large datasets for better accuracy
Incremental SVM [6]	The retraining in this model helps to keep the model up to date and is ready to train it with new spam words	The retraining could make the model take more time to run in the longer period
Clustering, J48, Naive Bayes, SVM, ANN, Decision tree [7]	Naïve Bayes technique gives faster results and better accuracy over other techniques except SVM, which has a better accuracy	SVM gives better accuracy than Naïve Bayes but takes more time to build a model
Polynomial filter + RF [8]	Out of all algorithms used, Random Forest gave the highest accuracy and shows promising results	Adding a filter to reduce duplicate emails
LR, DT, NB, K-NN and SVM [9]	Out of the algorithms used, RF outperforms other algorithms	K-NN takes much more time to train the model but gives the same result as RF. RF takes less time to train
RF [10]	(a) RF can handle a high number of features (b) RF can handle a mixture of binary, numeric and categorical features (c) RF can estimate which features are more important than others	RF in general requires a huge amount of time to train as it generates lots of trees and makes a decision on the majority of votes

(continued)

Table 2 (continued)

Methods used	Strengths	Weaknesses
K-NN for data expansion + LSTM for classification [11]	LSTM captures the induced statements in the message better. This method performs better than the existing phishing email detection method as it improves accuracy, reduces false negative rate and false positive rates	In general, LSTM models require a high memory bandwidth to train and need more time as well
K-NN, DT and NB algorithms [12]	The experimental results show high values of TPR and precision, and the FPR is controlled within an acceptable range when a suitable classifier is used	The model will need to be updated regularly to keep it up to date with the changes made by phishers
Hybrid DT + LR with FN Threshold [13]	The proposed method shows high accuracy in detecting spam email compared to other methods, and it outperforms prior research. The proposed method has less complexity	Feature Selection is required as the Email dataset has many features
NB + Particle Swarm Optimization [14]	The proposed method produces much better results than what Naïve Bayes would show on the same dataset. The PSO helps to optimize the Naive Bayes parameters which helps to improve the accuracy	Naïve Bayes can be swapped out and be replaced by any other classifiers which would help to improve the accuracy even more
Multinomial NB [15]	The algorithm consistently reduced the false positive rate	Speed and efficiency of the model need to be improved
Fuzzy C-Means [16]	Only paper to propose a different model like Fuzzy C-Means. Gives best result for overlapped dataset and comparatively better than k-means algorithm	-
Feature extraction using NLP and ANN for classification [17]	The accuracy with which the mails were detected is 92.2% ANN gives high accuracy for the given dataset	ANN model in general requires a large amount of data for training; high accuracies cannot be achieved with small datasets

(continued)

Furthermore, one of the findings was that there is a lack of datasets built on new data; all the datasets used were old. Hence, to tackle this issue effectively and at par with the current techniques of spammers to bypass current systems, updated datasets are required in this field. Finally, the implementation of traditional ML techniques

Table 2 (continued)

Methods used	Strengths	Weaknesses
ANN, Naïve Bayes, SVM, Decision tree and Random Forest [18]	Neural network model provides a better result than other classifiers on the given dataset Neural network model worked as the dataset is complex with high dimensionality, and missing data	Interpretability of neural network is low compared to the other models. Due to correlation between variables all other algorithms except neural network, SVM and C5.0 did not perform well Tested on only one dataset
J48, SVM and Naïve Bayes [19]	The accuracy of the model is good irrespective of which classifier is applied to it Frequency-based features attain high accuracy	This model deals with only the body of the mail. It does not consider the header of the mail
NB, LR DT, K-NN, RF and SVM [20]	SVM gives the highest accuracy when used with Doc2Vec for mails with and without headers Doc2Vec models performed better compared to tf-idf models in both header and without header	Huge variations in experimentation were implemented but only a small dataset was used; using different datasets would give a better comparison and result analysis
A CNN-based model called CNNPD [21]	This model achieves high accuracy, precision, recall and F-measure	This model currently uses small datasets. Results with large datasets are not known
Genetic Programming (GP) + RF [22]	GP is an evolutionary model that evolves classification models that are represented as trees. It is adaptable and its generated models are interpretable SMOTE oversampling along with GP gives a better overall performance of accuracy	GP when used by itself shows competitive results but it's not enough, especially the accuracy, recall for spam and g-mean
Data Insertion + RF classifier [23]	This method is effective in increasing the number of phishing samples without changing the malicious attributes. It also solves the problem of spatial bias during model training and reduces the difference in statistical characteristics between benign and malicious samples	The constructed samples are missing some text features or samples and there's a certain randomization in the data inserted in the sample which changes the sample's natural language processing result

(continued)

Table 2 (continued)

Methods used	Strengths	Weaknesses
NB [24]	Because of its strong independent assumption conditions, the size of parameters needed for model prediction is reduced which greatly saves memory consumption and computing time	The paper does not implement the model, which does not help to understand how the model performed on datasets
K-nearest community classifier is used [25]	This model works well when all features are considered and the Local Community Detection Based on Dependency is used without closure	Not as good as other classifiers like Bayesian or Vector Space
NB, SVM, DT, LSTM and ANN [26]	DL models outperformed the ML models in classification	Other choices for data pre-processing other than one-hot encoding could have been used
Proposed a classification model using M [27]	The proposed model is a self-learning system that is customizable to each user and based on their dataset. The accuracy will only increase depending on the size of the dataset	It is a proposed model
Cuckoo-search SVM (CS-SVM) [28]	CS-SVM performs better compared to the traditional SVM model	The model cannot work on a distributed system
NB and SVM used for classification [29]	SVM and NB as per the paper provide better accuracies compared to other models in their literature	This is just a proposed model. There are no experimental results
Proposed a header-based system [30]	The paper suggests that only looking at the email body is not enough and hence header-based approach can be used to get potential features	The authors have not mentioned any results for the model
Classification based on Geographical Location of sender [31]	SVM was used and the accuracy achieved was 98.94%. The only paper in the literature survey using the geographical location of sender	Phishers could possibly evade this method by posing as small banks. Another way they could evade is by compromising a bank's system by sending phishing messages or hosting a fake site

(continued)

Table 2 (continued)

Methods used	Strengths	Weaknesses
URL-based feature extraction + NB classification [32]	It yielded a probabilistic phishing detection framework that can quickly adapt to new attacks with reasonably good true positive rates and close to zero false positive rates. The system exploits the lexical and host-based features of URLs	The model was tested only on emails containing URLs
URL-based feature extraction and classification [33]	Lexical and textual analysis of email helps to effectively classify email. FPR rate is reduced	The authors have proposed the model but no experimental results were listed
Weighted SVM [34]	Improvised WSVM exhibits lower misclassification rate than SVM. In Improvised WSVM, when weights are produced by KFCM algorithm, the result is more promising	System performance can still be improved in terms of accuracy and precision. There may be inconsistency in case of larger number of iterations compared with the average number of data

has become very common, and newer techniques like hybrid models and more use of DL techniques need to be implemented.

References

1. Tidy J (17 April 2020) Google blocking 18m coronavirus scam emails every day. https://www.bbc.com/news/technology-52319093
2. Keyworth M, Wall M (8 Jan 2016) The 'bogus boss' email scam costing firms millions. https://www.bbc.com/news/business-35250678
3. Budanović N (11 Dec 2019) 20 spam statistics that shed light on the dark side of your inbox. https://dataprot.net/statistics/spam-statistics
4. Fang Y, Zhang C, Huang C, Liu L, Yang Y (2019) Phishing email detection using improved RCNN model with multilevel vectors and attention mechanism. 2019 IEEE Access 7, 56329–56340
5. Olatunji SO (2017) Extreme learning machines and support vector machines models for email spam detection. In: 2017 IEEE 30th Canadian conference on electrical and computer engineering (CCECE). Windsor, ON, pp 1–6
6. Tseng C, Chen M (2009) Incremental SVM model for spam detection on dynamic email social networks. In: 2009 international conference on computational science and engineering. Vancouver, BC, pp 128–135
7. Vyas T, Prajapati P, Gadhwal S (2015) A survey and evaluation of supervised machine learning techniques for spam e-mail filtering. In: 2015 IEEE international conference on electrical, computer and communication technologies (ICECCT). Coimbatore, pp 1–7
8. More S, Kulkarni SA (2013) Data mining with machine learning applied for email deception. In: 2013 international conference on optical imaging sensor and security (ICOSS). Coimbatore, pp 1–4

9. Nandhini S, Marseline DJKS (2020) Performance evaluation of machine learning algorithms for email spam detection. In: 2020 international conference on emerging trends in information technology and engineering (ic-ETITE). Vellore, India, pp 1–4
10. Deshmukh P, Shelar M, Kulkarni N (2014) Detecting of targeted malicious email. In: 2014 IEEE global conference on wireless computing & networking (GCWCN). Lonavala, pp 199–202
11. Li Q, Cheng M, Wang J, Sun B (2020) LSTM based phishing detection for big email data. IEEE Trans Big Data 2020:1–1
12. Li X, Zhang D, Wu B (2020) Detection method of phishing email based on persuasion principle. In: 2020 IEEE 4th information technology, networking, electronic and automation control conference (ITNEC). Chongqing, China, pp 571–574
13. Wijaya A, Bisri A (2016) Hybrid decision tree and logistic regression classifier for email spam detection. In: 2016 8th international conference on information technology and electrical engineering (ICITEE). Yogyakarta, pp 1–4
14. Agarwal K, Kumar T (2018) Email spam detection using integrated approach of naïve bayes and particle swarm optimization. In: 2018 second international conference on intelligent computing and control systems (ICICCS). Madurai, India, pp 685–690
15. Peng W, Huang L, Jia J, Ingram E (2018) Enhancing the naive bayes spam filter through intelligent text modification detection. In: 2018 17th IEEE international conference on trust, security and privacy in computing and communications/12th IEEE international conference on big data science and engineering (TrustCom/BigDataSE). New York, NY, pp 849–854
16. Singh AK, Bhushan S, Vij S (2019) Filtering spam messages and mails using fuzzy C means algorithm. In: 2019 4th international conference on internet of things: smart innovation and usages (IoT-SIU). Ghaziabad, India, pp 1–5
17. Moradpoor N, Clavie B, Buchanan B (2017) Employing machine learning techniques for detection and classification of phishing emails. In: 2017 computing conference. London, pp 149–156
18. Panigrahi PK (2012) A comparative study of supervised machine learning techniques for spam e-mail filtering. In: 2012 fourth international conference on computational intelligence and communication networks. Mathura, pp 506–512
19. Nizamani S, Memon N, Glasdam M, Nguyen DD (2014) Detection of fraudulent emails by employing advanced feature abundance, pp 169–174
20. Unnithan NA, Harikrishnan NB, Vinayakumar R, Soman KP, Sundarakrishna S (2018) Detecting phishing E-mail using machine learning techniques CEN-secureNLP. In: 2018 proceedings of the 1st anti-phishing shared task pilot 4th ACM IWSPA Co-Located 8th ACM conference data application security privacy (CODASPY), pp 51–54
21. Alotaibi R, Al-Turaiki I, Alakeel F (2020) Mitigating email phishing attacks using convolutional neural networks. In: 2020 3rd international conference on computer applications & information security (ICCAIS). Riyadh, Saudi Arabia, pp 1–6
22. Habib M, Faris H, Hassonah MA, Alqatawna J, Sheta AF, Al-Zoubi AM (2019) Automatic email spam detection using genetic programming with SMOTE. In: 2018 fifth HCT information technology trends (ITT). Dubai, United Arab Emirates, pp 185–190
23. Yu G, Fan W, Huang W, An J (2020) An explainable method of phishing emails generation and its application in machine learning. In: 2020 IEEE 4th information technology, networking, electronic and automation control conference (ITNEC). Chongqing, China, pp 1279–1283
24. Chen R, Zhang C, Guo J, Wang X (2019) Application of naive bayesian algorithms in E-mail classification. In: 2019 Chinese automation congress (CAC). Hangzhou, China, pp 3933–3938
25. Prilepok M, Kudelka M (2015) Spam detection based on nearest community classifier. In: 2015 international conference on intelligent networking and collaborative systems. Taipei, pp 354–359
26. Bagui S, Nandi D, Bagui S, White RJ (2019) Classifying phishing email using machine learning and deep learning. In: 2019 international conference on cyber security and protection of digital services (cyber security). Oxford, United Kingdom, pp 1–2
27. Alurkar AA, et al (2017) A proposed data science approach for email spam classification using machine learning techniques. In: 2017 internet of things business models, users, and networks. Copenhagen, pp 1–5

28. Niu W, Zhang X, Yang G, Ma Z, Zhuo Z (2017) Phishing emails detection using CS-SVM. In: 2017 IEEE international symposium on parallel and distributed processing with applications and 2017 IEEE international conference on ubiquitous computing and communications (ISPA/IUCC). Guangzhou, pp 1054–1059
29. Zaid A, Alqatawna J, Huneiti A (2016) A proposed model for malicious spam detection in email systems of educational institutes. In: 2016 cybersecurity and cyberforensics conference (CCC). Amman, pp 60–64
30. Bin S, Razak A, Bin Mohamad AF (2013) Identification of spam email based on information from email header. In: 2013 13th international conference on intellient systems design and applications. Bangi, pp 347–353
31. Sanchez F, Duan Z (2012) A sender-centric approach to detecting phishing emails. In:2012 international conference on cyber security. Washington, DC, pp 32–39
32. Azeez NA, Oluwatosin A (2016) CyberProtector: identifying compromised URLs in electronic mails with bayesian classification. In: 2016 international conference on computational science and computational intelligence (CSCI). Las Vegas, NV, pp 959–965
33. Hajgude J, Ragha L (2012) Phish mail guard: phishing mail detection technique by using textual and URL analysis. In: 2012 world congress on information and communication technologies. Trivandrum, pp 297–302
34. Vishagini V, Rajan AK (2018) An improved spam detection method with weighted support vector machine. In: 2018 international conference on data science and engineering (ICDSE). Kochi, pp 1–5

DDoS Detection in Software-Defined Network Using Entropy Method

Ashidha Anil, T. A. Rufzal, and Vipindev Adat Vasudevan

Abstract In the internet era, Software-defined Network (SDN) is the most alluring technology, which decouples the control plane and data plane in the network. The decoupling approach in networking has tremendous advantages compared to the traditional one. Although SDN provides several benefits by decoupling the control plane from the data plane, there is a contradictory relationship between SDN and Distributed Denial of Service (DDoS) attacks. Contradiction comes because the centralized SDN controller can detect and mitigate DDoS easily although centralization makes it vulnerable to these attacks. The DDoS attack is a Denial of Service (DoS) attack utilizing multiple distributed attack sources and can be easily detected using entropy method. Every network in the system has an entropy and an increase in randomness causes an increase in entropy. In this paper, detection and mitigation of DDoS attacks have been tackled by the entropy method which calculates the entropy variations at the destination IP address and the threshold. If the entropy values drop below a threshold value, we are blocking the certain port in the switch and bringing the port down. The proposed method detects the DDoS attack packets early and the solution is lightweight.

Keywords Software-defined network · Entropy · Denial of service · Distributed DoS · Threshold

1 Introduction

Software-defined Networking (SDN) is an emerging network architecture where network control is decoupled from forwarding and is directly programmable as defined by the Open Networking Foundation Definition [1]. This approach in networking

A. Anil (✉)
Tata Consultancy Services, Mumbai, India

T. A. Rufzal
Accuver Calibration Laboratory Co LLC, Dubai, UAE

V. Adat Vasudevan
University of Vigo, Vigo, Spain

© The Author(s), under exclusive license to Springer Nature Singapore Pte Ltd. 2022

D. Giri et al. (eds.), *Proceedings of the Seventh International Conference on Mathematics and Computing*, Advances in Intelligent Systems and Computing 1412,
https://doi.org/10.1007/978-981-16-6890-6_10

decouples the control plane and data plane in the network and allows programmers to design innovative network functions/protocols in an easy, efficient, and flexible way. Although SDN brings many benefits in the era of networking, there are certain challenges also that need to be tackled to deploy it properly.

In this research work, we are concentrating on one of the leading security threats among several security issues in SDNs; Distributed Denial of Service (DDoS) attacks. A network attack that sends large number of packets with an intention to either block the service or reduce the performance of a network service is defined as DDoS attack. It uses different sources that are distributed, to interrupt the service of a target network. The flooding-based attack [2] is an example of DDoS attack, in which a victim network is attacked by forwarding large number of packets to saturate the available bandwidth. The whole purpose of DDoS attack is to target a victim from different locations utilizing a substantial amount of distributed sources. A large number of such attacks have been reported in the recent past, especially with the advancements in the Internet of Things (IoT) and connected devices [3]. Normally botnets are used to set up DDoS attacks as a huge collection of available compromised hosts (also referred as zombies). Classification of attacks [4] are given below.

1.1 Reflection-Based Flooding Attacks

- Smurf Attack: In Smurf Attack [5], Internet Control Message Protocol (ICMP) packets are used to create network congestion by broadcasting these packets with the victim's forged source IP. Broadcasting can be done using an IP broadcast address and it creates a huge amount of illegitimate traffic on the network.
- Fraggle Attack: Fraggle attack [6] has a similarity to Smurf attack, but it uses illegitimate UDP packets to create network congestion instead of ICMP packets.

1.2 Protocol Exploitation Flooding Attacks

- SYN Flooding attack: Attackers utilize the TCP three-way handshake protocol and sends the packet with the SYN bit set off in TCP SYN flood attack [7]. The victim responds back to the source address with the packet where the SYN-ACK bit is set. But the victim never get any response from the attacker, because either the source address of the packet is spoofed or the attacker is purposefully unresponsive. TCP receive queue of the victim will be flooded with these forged packets and denies new legitimate TCP connections from the clients.
- UDP Fragmentation attack: Attackers forward huge UDP packets with a size of more than 1500 bytes in order to consume more bandwidth with lesser packets in UDP Fragmentation attack [6]. These fake packets are designed to look like the size of the packets are larger than the network's maximum transmission unit. But only parts of the packets are transmitted and can not be reassembled by the

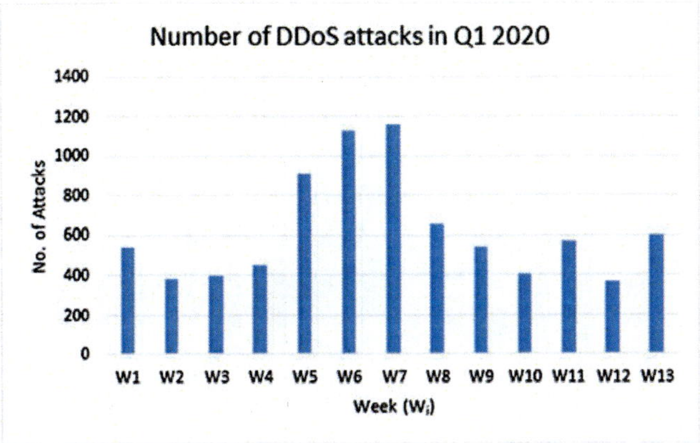

Fig. 1 Number of DDoS attacks in 2020 Quarter 1

victim's resources. Without knowing these, the server will try to reassemble the packets and lead to the server's unavailability to the legitimate packets.

1.3 *Reflection- and Amplification-Based Flooding Attacks*

- DNS amplification flooding attack: The botnets generate DNS queries with small size and direct toward the victim. These queries make a large volume of network traffic which has forged source IP address in domain name system (DNS) amplification flooding attack [8].
- NTP amplification flooding attack: Network Time Protocol (NTP) amplification attack [9] has a similarity to DNS amplification attack, but it utilizes NTP servers instead of DNS servers.

In the time of the coronavirus pandemic, the internet traffic and presence has increased tremendously which also has lead to more DDoS attacks. Quarterly reports from Kaspersky Lab [10] show that there is a clear increase in the quality and quantity of DDoS attacks during the pandemic period. Figure 1 illustrates that the number of attacks have peaked and reached almost 1200 per week in the first quarter of 2020 while Fig. 2 shows that the number of attacks have reached 1400 per week in the second quarter. These figures show the urgency of addressing DDoS attacks and we have proposed a solution which can effectively tackle the same in SDN networks.

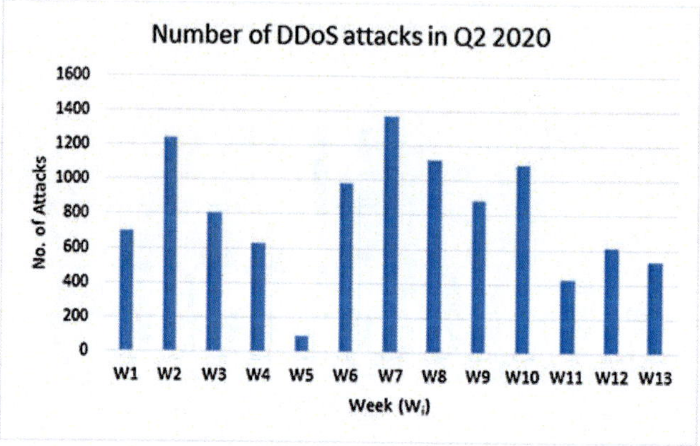

Fig. 2 Number of DDoS attacks in 2020 Quarter 2

2 DDoS Attack Detection Techniques in SDN

Some of the popular SDN based DDoS attack detection techniques are using Entropy, Machine learning, Traffic pattern analysis, Connection rate, and SNORT and Open-Flow [11] integration.

2.1 *Entropy*

Entropy-based method used to detect anomalous network activities by utilizing network feature distributions. Source IP address, destination IP address, and port numbers are the various probability distribution features of network [12] that are used to calculate the entropy. The presence of anomalies can be identified using threshold values which are predefined based on the obtained entropy values.

2.2 *Machine Learning*

Machine learning techniques categorize the SDN-based DDOS attacks as benign and malicious based on the network features and it is widely used in wired and wireless networks. Support Vector Machine (SVM), Decision tree, Fuzzy Logic, Artificial Neural Network (ANN), Evolutionary algorithms, k-means clustering algorithms, and Naive Bayes are the commonly used techniques [13]. The method extracts and analyzes the traffic information from switches, repeats the collection process, and stores the data for training and testing using different algorithms. It has self-learning capability and intelligence even though the effectiveness of these techniques depend on the training data set.

2.3 *Traffic Pattern Analysis*

This technique is extensively used for detection of bots (infected machines) in botnet attacks because of the correlated or similar traffic pattern among the bots. These behavioral patterns are the same because of the commands which are sent to the machines in the same botnet. Typical botnet attacks include DDoS attacks, and fraudulent activities such as spam, identity theft, phishing, and information exfiltration. Consumer IoT devices like smart TV and CCTV cameras are easy to compromise [14] and also susceptible to DDoS attacks [15]. These attacks can be identified using features like common destination, common connection time, and common platform features.

2.4 *Connection Rate*

Connection success ratio over overall connections and number of connections established in a particular window of time are the two types of connection rate [16]-based anomaly detection techniques. The first technique considers that a benign host has more successful connection than a malicious host and sets a threshold successful ratio to identify the infected hosts. The second technique considers that the malicious host will try to establish a large number of connections in a short time compared to benign hosts and is not likely to repeat the connection. This will also use a threshold value for established connections to detect the anomaly and to identify the infected hosts.

2.5 *SNORT and OpenFlow Integration*

These techniques use a combination of the Intrusion detection systems (for example, SNORT) and OpenFlow to detect attacks and reconfigure the network dynamically. An intrusion detection system monitors the traffic to detect malicious activities. OpenFlow switches are then reconfigured dynamically based on the identified attacks in real time. SNORT is an attack detection method which monitors the network traffic and analyzes each packet for any anomalies to detect attacks like DOS, CGI, and buffer overflow attacks. Snortflow [17]integrates SNORT and OpenFlow for a better anomaly detection process particularly in cloud-based systems which utilizes capabilities of both.

3 Entropy-Based Approaches for DDoS Attack Detection

DDoS attack detection attempts are always popular among industry and academia but addressing the same in SDN using entropy is rare. In this section, we are analyzing related works of entropy-based DDoS detection approaches.

Kubra Kalkan et al. proposed a Joint Entropy-Based Security Scheme (JESS) [18] which detect and mitigate DDoS in SDN environment using joint entropy method. The method has three stages, which starts with Nominal stage and will generate information such as nominal profile and joint entropy values following Preparatory stage utilizes these values to detect attack, and get activated when bandwidth is more than it's nominal value. Once the second stage is completed, the third stage is initialized which is Activation mitigation. It will drop the attack packet while protecting the normal packets. Success rate is 80% for an unfamiliar attack type and 70% for a mixed attack.

Jie Zhang et al. introduced Advance Entropy-Based Scheme (AEB) [19] to detect DDoS in a traditional network. This method categorizes variable rate DDoS attacks and treat separately to find optimum threshold and for accurate DDoS detection. Performance of AEB is similar to the normal entropy-based method for high-rate DDoS attacks, but has better performance only in case of low-rate DDoS detection.

Seyed Mohammad Mousavi et al. proposed an entropy method [20] to detect DDoS attack which utilizes the centralized control of SDN. Entropy variations are obtained from the destinations IP addresses and this method could detect attacks in the first 500 packets. In our proposed method, attack can be detected as early as in the first 250–500 packets of the randomly generated traffic.

4 Proposed Entropy Method

In this paper, entropy is the method which is used to detect DDoS attacks in SDN. Entropy is the measure of disorder in thermodynamics and it was proposed by Rudolf Clausius in the 1850s and later adopted to information theory by Shannon and defined as a measure of the uncertainty associated with a random variable [21]. Every network in the system has entropy, the higher the randomness the higher is the entropy and vice versa. Entropy method can do the effective analysis even though the important information about the distribution of the feature which is analyzed is lost. However, it's not possible to detect anomalies which do not affect randomness (Fig. 3).

As in the flowchart, when the packets are received from the host, controller calculate the entropy using the Eq. (3). For entropy calculation, we need to consider the data set $\mathcal{W}$ as in (1), called window which consists of events (X_i). The data set we are referring to is a packet header and there will be a continuous stream of incoming data and it needs to be counted. The probability of X_i in the window can be calculated using the Eq. (2). Finally, the entropy H can be obtained by summing up the probabilities of all events in the same window.

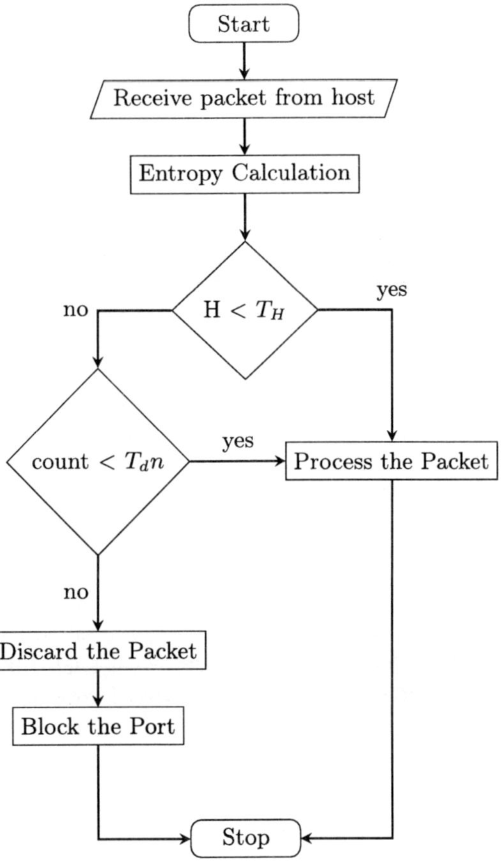

Fig. 3 Flowchart for proposed method

In the proposed method, detection can only be possible after threshold calculation, because only the threshold can differentiate the illegitimate and legitimate packets. From the experiments conducted, the attack entropy values are only in the range of zero to one and the threshold can be set as 1. Once the threshold is set, if the entropy value (H) is less than the threshold entropy value (T_H) and count of packets is less than threshold of duplicate value ($T_d n$), which is a carefully chosen limit for duplicate packet, it can be inferred that an attack is in progress and need to protect the controller. So, we need to discard the packet and bring the port down before it exhausts the controller.

$$\mathcal{W} = \{X_1, X_2, X_3, \ldots X_n\} \quad (1)$$

$$P_i = \frac{X_i}{n} \quad (2)$$

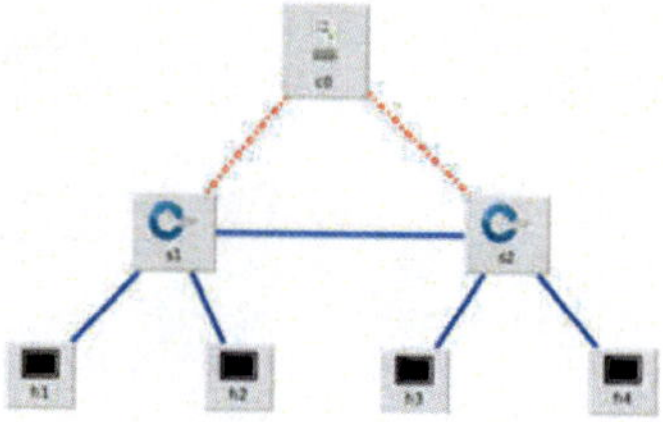

Fig. 4 Simplified example of mininet topology

$$\text{Entropy}, \mathcal{H} = \sum_{i=1}^{n} P_i \log P_i \tag{3}$$

5 Simulation

This challenging network needs an efficient emulator to implement OpenFlow network on laptops. Mininet is a tool suggested by most of the users and it is simple and open source. With the help of remote controllers, we can create our own topology for the network efficiently. It can deploy large networks with limited resources of virtual machine or a single computer. Also offers ease of use, accuracy of performance, and scalability. With the help of remote controllers, we can create our topology for the network efficiently. Different types of SDN controllers include NOX, POX, Ryu, Beacon, and Trema. Here, we choose POX as the remote controller. We need to implement the routing in SDN with the help of the POX controller. It is easy to do the traffic analysis and updating of rules because it is feasible to change dynamically in the SDN. POX, an open-source OpenFlow/SDN Controller rooted on python is chosen for its ability for faster development and prototyping of new network application. Mininet Virtual Machine version includes a pre-installed of POX controller. Using the POX controller, dumb OpenFlow devices can be converted into hub, switch, load balancer, and firewall devices. The POX controller facilitates a simple way to run OpenFlow/SDN experiments. POX can pass multiple and different parameters in accordance to real or experimental topologies, thus allowing user to run experiments on mininet emulator as well as on hardware implementations, and testbeds (Figs. 4 and 5).

In this paper, two experiments are conducted to study the efficiency of the implemented method. It can be tested with any scenarios, but considering these two experiments give us the benefit of analysis easiness as well as the optimal size for networks. If the experiment uses too small networks with less number of switches such as 3 or 4, performance of proposed method on larger networks will not be able to assume properly. If it uses more number of switches, then the analysis of performance will be complex. So the authors have considered these problems and conducted two experiments with 9 and 10 switches. Experiment 1 has incorporated 9 switches and 64

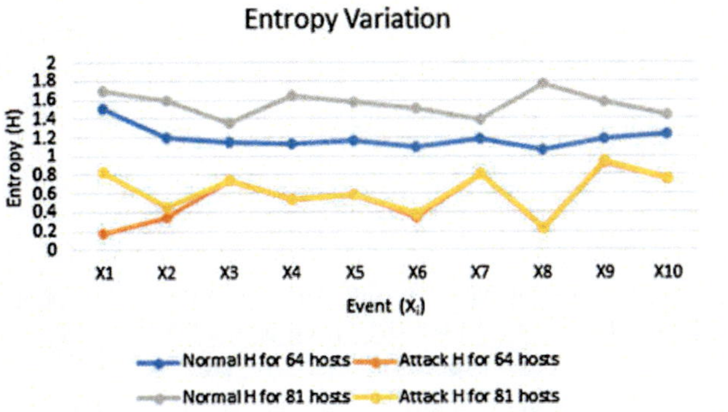

Fig. 5 Normal and attack entropy values of networks from experiments

Table 1 Entropy obtained from network with 64 hosts

H_{N1}	H_{A1}	H_{N2}	H_{A2}
1.50	0.183	1.70	0.830
1.20	0.354	1.60	0.453
1.15	0.743	1.35	0.743
1.14	0.541	1.64	0.541
1.17	0.592	1.57	0.593
1.10	0.346	1.50	0.376
1.19	0.812	1.39	0.812
1.06	0.221	1.76	0.221
1.18	0.931	1.58	0.941
1.24	0.750	1.44	0.750

hosts for a tree topology with a depth of 2 and fan-out [2] of 8. Experiment 2 has incorporated 10 switches and 81 hosts for a tree topology with the depth of 2 and fan-out of 9. For each experiment, a virtual network was created with a remote POX controller. A default IP 127.0.0.1 and port 6633 was allocated to the controller. Also switches, hosts, and links need to be added for the simulation of virtual networks. Packet transfer in the network was executed using xterms in mininet and the flow details of the hosts were obtained from the receiver host.

From the analysis of the graph, it is observed that the Normal Entropy value of the SDN network is always greater than 1. Least normal entropy values observed from experiment 1 with 64 hosts and experiment 2 with 81 hosts are 1.06 and 1.35, respectively. In case of attack entropy values, experiment 1 shows the highest value as 0.931, while experiment 2 has the highest value of 0.941. Table 1 listed the normal entropy values of network with 64 hosts (H_{N1}), normal entropy values of network with 81 hosts (H_{N2}), attack entropy values of network with 64 hosts (H_{A1}), and attack entropy values of network with 81 hosts (H_{A2}). These entropy values from

SDN with window size 10 made clear that the optimal threshold value should be 1 for the detection of proposed method.

6 Conclusions

Since the beginning of 2020, because of the COVID-19 pandemic, people are working online or being online more than they used to be. This attracted more attacks compared to the previous years especially the DDoS attacks. Now it's the peak time to detect and mitigate the same and the solution should not bring any burden to the system. In this time, popular SDN needs the solution for DDoS more than the traditional networks. Even though there are new techniques to do the detection of DDoS for SDN, we choose entropy method because of the easiness and it can do the effective analysis even though the necessary information about the distribution of the feature which is analyzed is forgotten.

Concluding the results of our studies, it is observed that early detection is possible, the attack can be detected as early as in the first 250–500 packets of the randomly generated traffic. We have tested different scenarios and found that the success rate is 96%. This success rate is obtained when the attack packets were 25% of the total traffic. Implemented code has provided less CPU load in both attack and normal scenarios and minimal code addition which was our first priority.

References

1. O. N. Fundation (2012) Software-defined networking: the new norm for networks. ONF White Pap 2(2–6):11
2. Jensen M, Gruschka N, Luttenberger N (2008) The impact of flooding attacks on network-based services. In: 2008 third international conference on availability, reliability and security. IEEE, pp 509–513
3. Adat V, Gupta B (2018) Security in internet of things: issues, challenges, taxonomy, and architecture. Telecommun Syst 67(3):423–441
4. Bawany NZ, Shamsi JA, Salah K (2017) Ddos attack detection and mitigation using sdn: methods, practices, and solutions. Arab J Sci Eng 42(2):425–441
5. Kumar S (2007) Smurf-based distributed denial of service (ddos) attack amplification in internet. In: Second international conference on internet monitoring and protection (ICIMP 2007). IEEE, p 25
6. Dong S, Abbas K, Jain R (2019) A survey on distributed denial of service (ddos) attacks in sdn and cloud computing environments. IEEE Access 7:80 813-80 828
7. Geetha K, Sreenath N (2014) Syn flooding attack–identification and analysis. In: International conference on information communication and embedded systems (ICICES2014). IEEE, pp 1–7
8. Saharan S, Gupta V (2019) Prevention and mitigation of dns based ddos attacks in sdn environment. In: 2019 11th international conference on communication systems & networks (COMSNETS). IEEE, pp 571–573
9. Rudman L, Irwin B (2005) Characterization and analysis of ntp amplification based ddos attacks. In: 2015 information security for South Africa (ISSA). IEEE, pp 1–5

10. Senn M (2009) The most basic entry for a website consists of the author name(s), page title, website title, web address, and date accessed (accessed February 3, 2014). [Online]. http://securelist.com/
11. Braun W, Menth M (2014) Software-defined networking using openflow: protocols, applications and architectural design choices. Future Internet 6(2):302–336
12. Sharma S, Sahu SK, Jena SK (2015) On selection of attributes for entropy based detection of ddos. In: 2015 international conference on advances in computing, communications and informatics (ICACCI). IEEE, pp 1096–1100
13. Cervantes J, Garcia-Lamont F, Rodriguez-Mazahua L, Lopez A (2020) A comprehensive survey on support vector machine classification: applications, challenges and trends. Neurocomputing 408:189–215
14. Adat V, Dahiya A, Gupta B (2018) Economic incentive based solution against distributed denial of service attacks for iot customers. In: 2018 ieee international conference on consumer electronics (ICCE). IEEE, pp 1–5
15. Adat V, Gupta B (2017) A ddos attack mitigation framework for internet of things. In: 2017 international conference on communication and signal processing (ICCSP). IEEE, pp 2036–2041
16. Mehdi SA, Khalid J, Khayam SA (2011) Revisiting traffic anomaly detection using software defined networking. In: International workshop on recent advances in intrusion detection. Springer, pp 161–180
17. Xing T, Huang D, Xu L, Chung C-J, Khatkar P (2013) Snortflow: a openflow-based intrusion prevention system in cloud environment. In 2013 second GENI research and educational experiment workshop. IEEE, pp 89–92
18. Kalkan K, Altay L, Gür G, Alagöz F (2018) Jess: joint entropy-based ddos defense scheme in sdn. IEEE J Sel Areas Commun 36(10):2358–2372
19. Zhang J, Qin Z, Ou L, Jiang P, Liu J, Liu AX (2010) An advanced entropy-based ddos detection scheme. In: 2010 international conference on information, networking and automation (ICINA), vol 2. IEEE, pp V2–67
20. Mousavi SM, St-Hilaire M (2015) Early detection of ddos attacks against sdn controllers. In: 2015 international conference on computing, networking and communications (ICNC). IEEE, pp 77–81
21. Bereziński P, Szpyrka M, Jasiul B, Mazur M (2015) Network anomaly detection using parameterized entropy. In: IFIP international conference on computer information systems and industrial management. Springer, pp 465–478

New Records of Pre-image Search of Reduced SHA-1 Using SAT Solvers

Emanuele Bellini, Alessandro De Piccoli, Rusydi Makarim, Sergio Polese, Lorenzo Riva, and Andrea Visconti

Abstract Although the first SHA-1 collision attack has been carried out in 2017, the cryptographic hash function seems still robust against pre-image attacks. The aim of this work is to measure how many rounds of SHA-1 can be inverted in practice using SAT solvers. To do so, we have first modeled the problem of finding a pre-image of SHA-1 as a system of Boolean formulas, explicitly describing the procedure to obtain such model. Then, we configured the model based on different combinations of number of rounds and number of free bits in our target pre-image. Finally, to find a solution of the model, we used a SAT solver on our server, testing several combinations of restart policies and polarity modes. We analyze and report the number and positions of the pre-image bits that can be fixed to influence the ability of the SAT solver to find the remaining free bits of the pre-image in a shorter time. In particular, we execute partial pre-image attacks on 64-, 80- 96-, 112- and 128-bit messages, outperforming the current state-of-the-art records.

Keywords SHA-1 · Pre-image · SAT solver

E. Bellini · R. Makarim
Cryptography Research Centre, Technology Innovation Institute, Abu Dhabi, UAE
e-mail: emanuele.bellini@tii.ae

R. Makarim
e-mail: rusydi.makarim@tii.ae

A. De Piccoli · S. Polese · L. Riva · A. Visconti (✉)
Computer Science Department, Università degli Studi di Milano, Milan, Italy
e-mail: andrea.visconti@unimi.it

A. De Piccoli
e-mail: alessandro.depiccoli@unimi.it

S. Polese
e-mail: sergio.polese@unimi.it

L. Riva
e-mail: lorenzo.riva3@studenti.unimi.it

© The Author(s), under exclusive license to Springer Nature Singapore Pte Ltd. 2022
D. Giri et al. (eds.), *Proceedings of the Seventh International Conference on Mathematics and Computing*, Advances in Intelligent Systems and Computing 1412,
https://doi.org/10.1007/978-981-16-6890-6_11

1 Introduction

SHA-1 (Secure Hash Algorithm 1) is a cryptographic hash function which takes as input a bit string of arbitrary length and produces an output message digest of 160-bit. The algorithm was designed by the United States National Security Agency, and is a U.S. Federal Information Processing Standard, described in FIPS 180-4 Secure Hash Standard [1]. According to the NIST document SP 800-17 [2], SHA-1 hash function presents *collision resistance* strength of less than 80 bits, *a second pre-image resistance* strength of 105–160 bits, and *a pre-image resistance* strength of 160 bits. The collision resistance strength of a hash function is expected to be half the length of its hash value. This is, however, not the case for SHA-1 due to the cryptanalytic results in [3], which demonstrated that SHA-1 has a collision resistance strength that is considerably less than 80 bits. The second pre-image resistance strengths of the approved hash functions depend not only on the function itself, but also on the sizes of the messages that the hash functions process [4]. Information on determining the actual second pre-image resistance strengths of the approved hash functions for different message lengths is provided in Appendix A of [2]. At the time that SP 800-17 [2] was written, there had been no known short cuts for finding pre-images of the hash values generated by SHA-1. In spite of the fact that collisions have been found on SHA-1 and because of its pre-image resistance, the adoption of SHA-1 is still widespread and allowed in certain applications by some of the most important standards. For example, NIST [2] suggests that SHA-1 can still be used in certain *randomized hashing for Digital Signatures*. Furthermore, the use of SHA-1 is allowed in the construction of the widely adopted message authentication code HMAC-SHA-1 (see, e.g., [2, 5]).

In this work, we analyze what is the difficulty of finding pre-images of a round-reduced version of SHA-1 by modeling the problem as a satisfiability problem and then using a SAT solver to find solutions of those instances. We show what is the maximum number of rounds that an attacker can hope to invert in a reasonable amount of time. In our testing activities, we set new records on the maximum number of rounds for which a partial pre-image can be found. By using SAT solvers, we invert up to 31 rounds with 64 input bits free, where previous works were stuck at 27 rounds with 40 input bits free [6].

The paper is organized as follows. In Sect. 2 we summarize the most successful results on SHA-1 pre-image attacks. In Sects. 3 and 4 we introduce SHA-1 and SAT solvers, respectively. We describe our SAT model and its simplifications in Sect. 5. Finally, experimental results are shown in Sect. 6.

2 Related Works

Since 2004, a vast literature studied the collision resistance of SHA-1 and of its predecessor SHA-0, see, e.g., [7, 8] and their references, that resulted in the 2017 discovery of the first collision of the full hash by Stevens et al. [8] in a computation that required $2^{63.1}$ SHA-1 applications.

SHA-1 is also subject to pre-image attacks. To date, none of these attacks to the full SHA-1 algorithm have been published in the literature. Anyway, many applications of hash functions rely on the pre-image resistance and breaking this kind of resistance also allows to find collisions. Theoretical pre-image attack history on SHA-1 starts in 2008 with the work of De Cannière and Rechberger [9] (attack complexity of 2^{157} for 45-round-reduced SHA-1). In [10], the authors exploited a meet-in-the-middle attack to present a pre-image attack on 48 out of 80 rounds with complexity of 2^{159}. Their work was improved by Knellwolf and Khovratovich, thanks to a differential view on the MiTM approach ($2^{158.7}$ for 57-round-reduced SHA-1), and finally by Espitan et al. in [11] thanks to higher order differential (62-round two-block pre-image with padding with complexity $2^{159.3}$).

In all the above-mentioned approaches, the purpose of the authors is to show that the analyzed hash function is not ideal, but the computational cost of the attacks is really high and practically out of reach.

One less explored approach in studying the security of a hash function is to determine which is the maximum number of rounds for which a pre-image of the hash can be found "in practice". Of course, this expression mostly depends on the computational power the attacker has at his disposal. This second approach is very common in public key cryptography, where, for example, challenges are proposed to break the largest possible parameters of a hard mathematical problem such as among many others, factorization [12] or decoding [13]. The most recent work we could find in this direction is from Nejati et al. in 2017 [14]. Here, the problem of finding a pre-image of a reduced version of SHA-1 is modeled into a SAT instance, and then different SAT solvers are used to solve the instance. The authors claim that, with this method, pre-images for more than 23 steps cannot be constructed in a reasonable amount of time even with the latest techniques and hardware [15–17]. Moreover, they also assert 27 round is their best result for a partial pre-image attack when leaving only 40 free bits of the input message.

3 A Brief Introduction to SHA-1

SHA-1 has been designed following the so called *Merkle-Damgård construction* [18]. This is a method of building collision-resistant cryptographic hash functions from collision resistant one-way compression functions. SHA-1 algorithm can be described in two stages: preprocessing and hash computation. Preprocessing involves padding an input and parsing the padded input into 512-bit blocks, each of which is divided into 16 words W_i, $i = 0, \ldots, 15$. The hash computation is performed for each block. It generates a message schedule W_i, $i = 16, \ldots, 79$ and uses, along with functions F_i, its constants and word operations to iteratively transform the fixed initial hash value, 160 bits in 5 words (A_0, B_0, C_0, D_0, E_0), through the repeated use of the round function shown in Fig. 1. Then, a modular

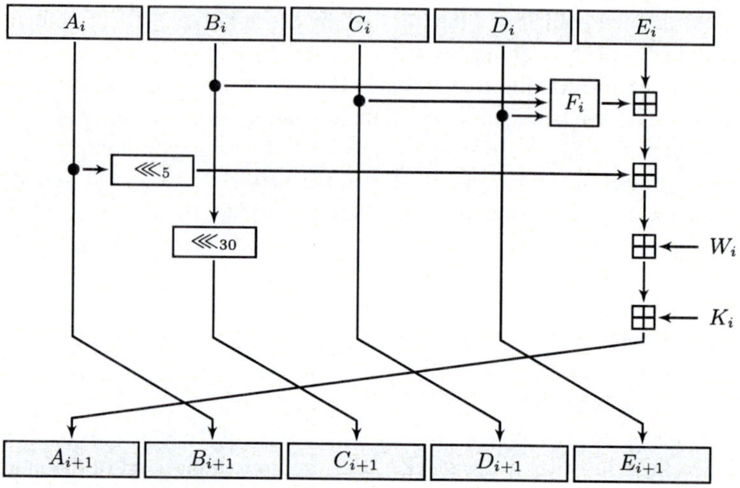

Fig. 1 SHA-1: round function

addition is performed between the obtained hash value and the initial state. The result is the initial hash value for the next block. After running out all blocks, we get the final hash value called *digest*. We remark that (a) F_i is a nonlinear function for $i = 0, \ldots, 19$, and for $i = 40, \ldots, 59$—F_i is equal to $(B \wedge C) \vee ((\neg B) \wedge D)$, and $(B \wedge C) \vee (B \wedge D) \vee (C \wedge D)$, respectively—and a linear function for $i = 20, \ldots, 39$ and $i = 60, \ldots, 79$—i.e., $F_i = B \oplus C \oplus D$; (b) the message schedule structure is given by $W_i = (W_{i-3} \oplus W_{i-8} \oplus W_{i-14} \oplus W_{i-16}) \lll_1$, for $i = 16, \ldots, 79$. As noted in [19], in specific cases the message schedule structure can be cleverly rewritten exploiting many optimizations.

4 SAT Solvers

Boolean satisfiability, usually referred to as SAT, is the problem of checking if a propositional logic formula can ever evaluate to true. In practice, one is not only interested in this decision ("satisfiable/unsatisfiable") problem, but also in finding an actual satisfying assignment if there exists one. All practical satisfiability algorithms are known as SAT solvers and do produce such an assignment if it exists. Every SAT problem can be easily reduced to formulae in CNF (Conjunctive Normal Form), a restricted form of propositional logic formulae, consisting of only conjunction of clauses, which are disjunctions of literals. In order to solve a CNF, modern SAT solvers usually adopt two approaches, namely the DPLL algorithm and the CDCL (Conflict-Driven Clause Learning) algorithm [20]. Also the efficiency heavily relies on various features that have been developed and tested over the last decade. Among others, restart policies have been an object of many of these researches. The randomized restarts allow clause learning algorithms to arbitrarily stop the search and restart their branching process from decision level zero. Most of the current SAT solvers employ aggressive restart strategies such as *geom* [21], *glue* [22], and *luby* [23].

5 Modeling SHA-1 as Satisfiability Problem

SHA-1 can be represented as a function $S : \{0, 1\}^{512} \times \{0, 1\}^{160} \rightarrow \{0, 1\}^{160}$ with $S(M, IV) = C$, where M is a message block, IV is the initialization of the state, and C is the hash output corresponding to M and IV. Theoretically, it is possible to explicitly express each hash output bit as a Boolean expression depending on M and IV, and to gather 160 such expressions in a system representing the full hash function. Although 160 equations are sufficient to do that, the number of literals of each expression is exponential in the number of bits in M and IV—it is exponential even if we applied heuristics [24]—making this approach not feasible in practice. A better solution is to model each component of the cipher, by expressing the output bit of each component as an expression of its input bits. In what follows, we detail how to construct the SAT model that we use to find pre-images of SHA-1.

Proposition 1 *Given two* 32 *bits words x and y, the system of Boolean formulas that represents the modular addition modulo 2^{32} between x and y and returning the word z is*

$$\begin{cases} c_0 = x_0 \wedge y_0 & \\ c_i = (x_i \wedge y_i) \vee (x_i \wedge c_{i-1}) \vee (y_i \wedge c_{i-1}) & i = 1, \dots, 30 \\ z_0 = x_0 \oplus y_0 & \\ z_i = x_i \oplus y_i \oplus c_{i-1} & i = 1, \dots, 31. \end{cases}$$

The index i denotes the bit number where the MSB is indexed by 31.

a Input message M is divided in sixteen 32-bit words $W_0, \dots, W_{15}$. Every bit of M is assigned to a variable, resulting in **512** variables for a one-block message.

b Other sixty-four words $W_{16}, \dots, W_{79}$ are computed through the message expansion. For every 32-bit word we need 32 equations. The total number of equations required is $64 \cdot 32 = \mathbf{2048}$. **2048** variables are introduced.

c 160 variables are added to represent the first state, that is A_0, B_0, C_0, D_0, E_0. The IVs H_0, H_1, H_2, H_3, H_4 are assigned to this state. To do so, we need $5 \cdot 32 = \mathbf{160}$ equations and $5 \cdot 32 \cdot 2 = \mathbf{320}$ new variables.

d The SHA-1 step function is obtained as follows. Notice that the most important operation is the addition modulo 2^{32}. We split the computation of *TEMP*, a 32-bit word, in four additions modulo 2^{32}.

$$\begin{cases} t_{t-1} = K_{(t-1)//20} \boxplus E_{t-1} \\ u_{t-1} = t_{t-1} \boxplus F_{t-1}(B_{t-1}, C_{t-1}, D_{t-1}) \\ v_{t-1} = u_{t-1} \boxplus A_{t-1} \\ TEMP_{t-1} = v_{t-1} \boxplus W_{t-1} \\ E_t = D_{t-1} \\ D_t = C_{t-1} \\ C_t = B_{t-1} \lll 30 \\ B_t = A_{t-1} \\ A_t = TEMP_{t-1} \end{cases} \tag{1}$$

Using Proposition 1, four additions modulo 2^{32} are represented by $4 \cdot 63 = 252$ equations. Moreover, we have five variables reassignments, namely, $5 \cdot 32 = 160$ equations, hence $252 + 160 = 412$ equations per round, that means a total of $412 \cdot 80 = \mathbf{32960}$ equations. For each addition modulo 2^{32} we need an extra variable for the carry bit, therefore $(16 + 4) \cdot 32 = 640$ variables per round are required, that means a total of $640 \cdot 80 = 51200$ variables. Notice that such number includes the 160 variables for each bit of A_0, B_0, C_0, D_0, and E_0, hence it can be reduced to $51200 - 160 = \mathbf{51040}$.

e We rewrite the final step: $Z_0 = H_0 \boxplus A$, $Z_1 = H_1 \boxplus B$, $Z_2 = H_2 \boxplus C$, $Z_3 = H_3 \boxplus D$, $Z_4 = H_4 \boxplus E$. Again, using Proposition 1, these five additions modulo 2^{32} require $5 \cdot 63 = \mathbf{315}$ equations. The variables used in these modular additions come from the initial state and from the last round of step **d** respectively. Therefore, we only need Z for each word and each bit, namely $5 \cdot 32 = \mathbf{160}$ variables.

To sum up, our model is represented by a system of **35483** equations and **54080** variables that can be simplified. In particular, as described in [25, 26] we can substitute constant, input values and remove equations with different names but same meaning. Notice that if we have $x = 0$ and $x + y + z = 0$, after substituting the value of x in the second equation, we get $y = z$ and we can substitute the occurrences of y. Therefore, this strategy is repeated iteratively until no more substitution is possible. In order to speed up the execution of our model we tried to understand the relationship between variables that appear in our CNF. We suppose that the more a variable is connected to other variables, the more important it will be because its truth value will affect many others. In our specific case, the number of connections of a variable i is defined by the number of variables that appear in the same clause of i. Because SAT solvers usually work with a top-down approach, we rearrange the CNF by placing the clauses that contain the most connected variables at the top. Doing so, the SAT solver will immediately assign a value to these variables, providing an answer to our satisfiability problem in a shorter time. Looking closely at Table 1, it is possible to measure this speed up.

Table 1 SHA1 pre-image attack before and after the optimizations, 440 bits free

Number of rounds	After simplification (s)	After reorganization clauses (s)
19	0.24	0.20
20	0.681	0.61
21	403.334	335.93

6 Experimental Results

We implemented a Python tool called *Shanatomy* [27], to help us analyze the Boolean representation of SHA-1. The tool can be used to generate the Boolean representation of SHA-1 variants operating on words with any number of bits. Moreover, it translates the Boolean system into DIMACS standard, commonly used by SAT solvers. Handling the extension of the SAT solver, the tool can also preserve XOR operations (which usually is not part of the DIMACS standard). The advantages of keeping XORs are obvious: the solver has much less equations since a XOR with n inputs needs 2^{n-1} equations to be represented.

In order to test the efficiency of a pre-image attack to reduced versions of SHA-1, we choose CryptoMiniSat5 [28]. In the plethora of SAT solvers, this is not a random choice. Indeed, we experimentally observed that MiniSat [29], MathSat [30], and Yices [31] seems to perform better than CryptoMiniSat up to round 20, but they take more time than CryptoMiniSat to compute round 21 and above. We run our code on a server equipped with (CPU) 2 x Intel ® Xeon ® Gold 6258R @2.70 GHz, (Memory) 768 GB 2933 MHz, (OS) Ubuntu 18.04.5 LTS and on a cluster equipped with (CPU) 128 x IBM POWER9 AC922 @3.1 GHz, (Memory) 256 GB 2666 MHz, (OS) Red Hat Enterprise Linux Server 7.6. The SAT solver adopted is CryptoMiniSat v5.8.0. In our testing activities we run our code using all combinations between four polarity modes (auto, false, random, true) and three restart policies (geom, glue, luby), enabling the XOR extension of CryptoMiniSat. In order to run many SAT instances and compare their running time, we set two upper thresholds. Such thresholds are set on the basis of the average execution times of solved instances measured on our server. More precisely, we set 10 s as maximum execution time for round 20 and 180 s for round 21. For each polarity mode (four), each restart policy (three) and each hardware configuration (four), we executed 100 runs for a total of $4 \times 3 \times 4 \times 100 = 4800$ execution times collected. Average values are reported in Table 2. Notice that we highlighted in bold the number of pre-images found (out of 100), and we used the word "`n.o.t.`" to indicate the number of threads used. Due to space constraint we omit data collected on our cluster where we have not found significant new results but only better timings. Experimental results (see Table 2) show that (a) although there isn't a combination of "polarity mode–restart policy" which always performs better than others, in our testing activities, on average, we achieved better results with "auto", "random" (polarity mode), and "geom" (restart policy). Notice that we received evidence of these results measuring the execution time; (b) sometimes we

Table 2 Server: Average times (secs) to compute a pre-image, 440 bits free

Polar	Restart	n.o.t.	Round 20	Round 21	Polar	Restart	n.o.t.	Round 20	Round 21
Auto	Geom	1	96/2.640	16/167.619	Rnd	Geom	1	96/2.488	4/175.615
		2	99/1.660	29/157.426			2	100/1.541	7/174.211
		4	100/1.138	36/149.510			4	100/0.921	23/158.744
		8	100/0.926	43/146.887			8	100/0.744	31/148.996
	Glue	1	97/2.662	23/163.576		Glue	1	69/5.914	2/178.810
		2	99/2.074	19/166.040			2	100/1.840	10/171.939
		4	100/1.421	16/167.765			4	100/1.425	11/168.737
		8	100/1.224	19/167.341			8	100/1.181	5/176.104
	Luby	1	94/3.389	26/160.880		Luby	1	90/3.523	15/175.700
		2	100/1.632	23/162.603			2	100/1.799	11/171.910
		4	100/1.437	23/163.843			4	100/1.126	12/172.044
		8	100/1.036	35/151.600			8	100/0.777	27/153.750
False	Geom	1	95/2.558	17/171.267	True	Geom	1	96/2.460	27/163.243
		2	99/1.837	11/173.127			2	100/1.430	21/167.822
		4	100/1.357	20/165.835			4	100/1.267	29/161.265
		8	100/0.940	29/156.152			8	100/0.974	35/151.634
	Glue	1	91/4.062	12/173.967		Glue	1	87/4.566	21/170.374
		2	100/1.801	18/167.625			2	100/1.956	21/168.125
		4	100/1.544	9/176.643			4	100/1.613	22/168.964
		8	100/1.210	13/172.139			8	100/1.274	24/168.082
	Luby	1	97/3.304	8/174.264		Luby	1	91/3.698	25/165.653
		2	100/1.686	14/168.858			2	100/1.741	19/169.599
		4	100/1.237	18/167.466			4	100/1.374	35/153.619
		8	100/1.044	25/163.203			8	100/0.961	43/152.214

had indirect evidence of the performance improvement by measuring the number of pre-images found. In this case, also the combination "true–luby", round 21, is noteworthy; (c) a greater number of threads allows to decrease the execution time, but this does not happen in a linear way. This behavior can be observed analyzing the computational time spent to execute, for example, round 21, [auto, geom, 1-8] and [rnd, geom, 1-8]. Interestingly, the difficulty of finding a pre-image on round 22, 440 input bits free is sensibly higher. Indeed, our server was able to find only 3 pre-images (out of 100) in less than one hour using 32 threads. Therefore, to find pre-images for a larger number of rounds, we reduce the number of free input bits as near as possible to 160—recall that the complexity of a brute-force pre-image attack is 2^{160}. Firstly, we run our code leaving two consecutive 32-bit words W_i free, i.e., 64 bits free. Collected data are shown in Table 3 and they represent an average value of 30 measurements. In particular, this table shows that leaving 64 bits free at the beginning of our input data, for example, W_6, W_7, does not have the same effect of leaving them at the

Table 3 SAT based attack with two consecutive W_i unknown words

	Rounds						
Unknown words W_i	21	22	23	24	25	26	27
6&7	4.16 s						
7&8	0.45 s	1 h 51 m 19.23 s					
8&9	1 m 17.31 s	1 m 40.51 s					
9&10	1.17 s	2 m 30.03 s	1 m 57.31 s				
10&11	0.02 s	2.40 s	2 m 31.21 s	2 m 22.43 s			
11&12	0.02 s	0.02 s	2.10 s	13.83 s	2 m 29.93 s		
12&13	0.01 s	0.08 s	3.75 s	50.42 s	56.13 s	2 h 12 m 39.69 s	
13&14	0.01 s	0.02 s	0.19 s	14.41 s	17.33 s	6 m 43.31 s	
14&15	×	0.01 s	0.02 s	0.33 s	13.46 s	13.83 s	7 m 7.29 s

end, namely, W_{14}, W_{15}. The measured gap is noteworthy. In one case we talk about hours while in the other of hundredths of a second. Secondly, we adopt a slightly different approach. Due to its structure—we are performing XOR operations on a ring of integers modulo 2, where $W_i = (W_{i-3} \oplus W_{i-8} \oplus W_{i-14} \oplus W_{i-16}) \lll_1$, for $i = 16, \ldots, 79$—it is possible to unfold the entire message expansion of SHA-1 and rewrite such expansion as a set of any sixteen consecutive words, which do not have to be necessarily the first sixteen. When we try to break round r, 64 bits free, fixing words from W_{r-14} to W_{r-1} is particularly effective. In fact, guessing the value of the two words W_{r-15} and W_{r-16} we are able to get the 16 words of the input message. Notice that, exploiting the equation of the message schedule we can obtain relations between some input words, e.g., $W_0 \oplus W_2$, also without guessing W_{r-15} and W_{r-16}. Doing so, the computational time spent by CryptoMiniSat to execute a pre-image attack gives better results than those obtained fixing the input message words in other ways (e.g., first 14 words). In addition to 64 bits free, we also executed an attack leaving 80, 96, 112, and 128 bits free. Gathered data are reported in Table 4 using the bold notation to represent average values on 30 runs. Notice that, due time consuming, 25th round 112 bits free represents an average value on 16 runs.

Table 4 Computational time (secs) spent to execute a partial SHA-1 pre-image attack

	Equivalent free bits				
Rounds	64	80	96	112	128
20	**<0.01**	**<0.01**	**<0.01**	**0.01**	**0.03**
21	**<0.01**	**<0.01**	**0.01**	**0.02**	**5.94**
22	**<0.01**	**<0.01**	**0.22**	**6.84**	**2679.73**
23	**<0.01**	**0.01**	**0.17**	**5.32**	6266.63
24	**0.01**	**0.02**	**13.92**	**550.90**	1766488.38
25	**0.03**	**2.21**	**1727.79**	**106508.62**	
26	**1.37**	**89.17**	172552.81	528294.05	
27	**4.98**	**137.46**			
28	**131.02**	4717.71			
29	**4588.24**	593366.71			
30	9924.96				
31	16255.33				

7 Conclusions

In this paper we showed how to model the problem of finding SHA-1 pre-images into a satisfiability problem and how to mount a practical pre-image attack in different frameworks. In particular, leaving more than 160 bits free, we compared different restart policies and polarity modes concluding that there isn't a specific combination that always performs better than others, but on average we obtained better result with "auto", "random" (polarity mode), and "geom" (restart policy). Then, leaving 64, 80, 96, 112, and 128 bits free, we improved the state-of-art records executing a partial pre-image attack on 31, 29, 26, 26, 24 rounds, respectively. To do so, we carefully studied the structure of SHA-1 in order to understand which variables should be fixed in the system of equations. In particular, we analyzed the message expansion and show that choosing an appropriate set of variables W_i, we can considerably reduce the solving time of our CNF. Finally, we provided the entire Python source code [27] to reproduce our experiments.

Acknowledgements UniMi's researchers thank (a) TII that supported this research and provided insight and expertise; (b) the CINECA award under the ISCRA initiative, for the availability of high performance computing resources and support.

References

1. Dang Q (2015) FIPS PUB 180–4, Secure Hash Standard (SHS). US Department of Commerce, NIST

2. Dang Q (2012) NIST SP 800–107 Rev 1, Recommendation for applications using approved hash algorithms. US Department of Commerce, NIST
3. Wang X, Yin YL, Yu H (2005) Finding collisions in the full SHA-1. In: Proceedings of EUROCRYPT 2005. Springer
4. Kelsey J, Schneier B (2005) Second preimages on n-bit hash functions for much less than 2^n work. In: Proceedings of EUROCRYPT 2005. Springer
5. Bellare M, Canetti R, Krawczyk H (1996) Keying hash functions for message authentication. In: Proceedings of CRYPTO 1996. Springer
6. Nejati S, Liang JH, Gebotys C, Czarnecki K, Ganesh V (2017) Adaptive restart and CEGAR-based solver for inverting cryptographic hash functions. In: Proceedings of VSTTE 2017. Springer
7. Leurent G, Peyrin T (2019) From collisions to chosen-prefix collisions application to full SHA-1. In: Proceedings of EUROCRYPT 2019. Springer
8. Stevens M, Bursztein E, Karpman P, Albertini A, Markov Y (2017) The first collision for full SHA-1. In: Proceedings of CRYPTO2017. Springer
9. De Cannière C, Rechberger C (2008) Preimages for reduced SHA-0 and SHA-1. In: Proceedings of CRYPTO 2008. Springer
10. Aoki K, Sasaki Y (2009) Meet-in-the-middle preimage attacks against reduced SHA-0 and SHA-1. In: Proceedings of CRYPTO 2009. Springer
11. Espitau T, Fouque PA, Karpman P (2015) Higher-order differential meet-in-the-middle preimage attacks on SHA-1 and BLAKE. In: Proceedings of CRYPTO 2015. Springer
12. The New RSA Factoring Challenge. http://tiny.cc/osbbtz, Last accessed Dec 2020
13. Decodingchallenge.org. http://decodingchallenge.org, Last accessed Dec 2020
14. Ganesh V (2017) Adaptive restart and cegar-based solver for inverting cryptographic hash functions. In: Proceedings of VSTTE 2017. Springer
15. Legendre F, Dequen G, Krajecki M (2012) Encoding hash functions as a sat problem. In: 24th international conference on tools with artificial intelligence. IEEE
16. Legendre F, Dequen G, Krajecki M (2014) Logical reasoning to detect weaknesses about sha-1 and md4/5. IACR Cryptol. ePrint Arch. 2014:239
17. Nossum V (2012) SAT-based preimage attacks on SHA-1. Master's thesis, University of Oslo
18. Merkle R (1979) Secrecy, authentication, and public key systems. PhD Thesis, Stanford University
19. Visconti A, Gorla F (2020) Exploiting an HMAC-SHA-1 optimization to speed up PBKDF2. IEEE Trans Dependable Secur Comput 17(4)
20. Van Harmelen F, Lifschitz V, Porter B (2007) Handbook of knowledge representation. Elsevier Science
21. Walsh T (1999) Search in a small world. In: Proceedings of IJCAI 1999. Morgan Kaufmann Publishers Inc
22. Gomes CP, Selman B, Kautz H (1998) Boosting combinatorial search through randomization. AAAI/IAAI 98
23. Luby M, Sinclair A, Zuckerman D (1993) Optimal speedup of las vegas algorithms. Inf Process Lett 47(4)
24. Visconti A, Schiavo CV, Peralta R (2018) Improved upper bounds for the expected circuit complexity of dense systems of linear equations over GF(2). Inf Process Lett 137
25. Iuorio AF, Visconti A (2019) Understanding optimizations and measuring performances of PBKDF2. In: Proceedings of WIDECOM 2019. Springer
26. Steube J (2013) Optimising computation of hash-algorithms as an attacker. http://hashcat.net/events/p13/js-ocohaaaa.pdf
27. Shanatomy. https://github.com/Crypto-TII/shanatomy, Last accessed Jan 2021
28. Soos M, Nohl K, Castelluccia C (2009) Extending SAT solvers to cryptographic problems. In: Proceedings of SAT 2009. Springer
29. Eén N, Sörensson N (2004) An extensible sat-solver. In: Proceedings of SAT 2004. Springer
30. Cimatti A, Griggio A, Schaafsma B, Sebastiani R (2013) The MathSAT5 SMT solver. In: Proceedings of TACAS 2013. Springer
31. Dutertre B (2014) Yices 2.2. In: Proceedings of CAV 2014. Springer

OpenPGP Email Forwarding via Diverted Elliptic Curve Diffie-Hellman Key Exchanges

Francisco Vial-Prado and Aron Wussler

Abstract An offline OpenPGP user might want to forward part or all of their email messages to third parties. Given that messages are encrypted, this requires transforming them into ciphertexts decryptable by the intended forwarded parties, while maintaining confidentiality and authentication. It is shown in recent lines of work that this can be achieved by means of proxy re-encryption schemes, however, while encrypted email forwarding is the most mentioned application of proxy re-encryption, it has not been implemented in the OpenPGP context, to the best of our knowledge. In this paper, we adapt the seminal technique introduced by Blaze, Bleumer, and Strauss in EUROCRYPT'98, allowing a Mail Transfer Agent to transform and forward OpenPGP messages without access to decryption keys or plaintexts. We also provide implementation details and a security analysis.

Keywords Proxy re-encryption · Forwarding · Elliptic curve cryptography · OpenPGP

1 Introduction

Proxy re-encryption is the process of transforming a ciphertext so that it can be decrypted by a different party than was originally intended . This transformation is carried out without access to decryption secrets, plaintexts, or interactive communication with secret key holders. First defined and designed by M. Blaze, G. Bleumer, and M. Strauss [8], some forms of proxy re-encryption have already been applied to ElGamal PGP-encrypted mailing lists [10] in the context of encrypted email. In fact, the most highlighted application scenario is email redirection (see, e.g., [3, 4, 6]),

F. Vial-Prado
Fortanix, Bengaluru, India
e-mail: aron@wussler.it

A. Wussler (✉)
Proton Technologies AG, Geneva, Switzerland
e-mail: francisco@vialprado.com

© The Author(s), under exclusive license to Springer Nature Singapore Pte Ltd. 2022

D. Giri et al. (eds.), *Proceedings of the Seventh International Conference on Mathematics and Computing*, Advances in Intelligent Systems and Computing 1412,
https://doi.org/10.1007/978-981-16-6890-6_12

however, the current OpenPGP protocol specification does not include support for this functionality and, to the best of our knowledge, it has not been implemented in well-known encrypted email service providers.

A straightforward way to achieve delivery of forwarded encrypted email is to simply transfer private keys, but there are several drawbacks to this practice. On the one hand, if keys are transferred to forwarded parties, an attacker controlling them would gain decryption rights to unforwarded emails, which is not the case in unencrypted forwarding. On the other hand, private keys could be given to a trusted *Mail Transfer Agent* (MTA) so that they can decrypt and re-encrypt to forwarded parties, but this would contradict the trust model of end-to-end encrypted email services. The technique we describe aligns with these trust models, proposing to distribute trust among forwarded parties and the MTA. Security is provided as long as there is no collusion involving the MTA, i.e., we consider that the MTA that takes care of the forwarding is a semi-trusted proxy that is not able to decrypt.

Using two widely used OpenPGP implementations [1, 2], we verify the correctness of the technique in the case where the MTA is also a *Mail Delivery Agent* (MDA), allowing automatic forwarding between addresses entrusted to this MDA. However, in order to enable interoperability with other agents, a certain *key fingerprint* consideration needs to be ensured by those services, requiring the modification of a specific OpenPGP packet to indicate forwarding support.

In this paper, we adopt the technique presented in [8] based on diverted Diffie-Hellman key exchanges, which allows to emulate proxy re-encryption abilities in the context of symmetrically encrypted communications following current implementations of the OpenPGP protocol [11] and involving no additional trust hypotheses, other than those expected naturally of any email forwarding. We provide a security analysis and a simple simulation-based proof of the semi-honest security of the forwarding protocol.

1.1 Forwarding PGP Messages

The scenario we address is the following: Bob wants to allow Charles to decrypt an email that was originally encrypted to Bob's public key without having access to Bob's private key or any online interaction. Naturally, MTAs should not have the ability to read the contents of such messages. We achieve this by proposing a protocol that requires one-time communications between Bob, Charles, and a trusted MTA: Bob generates two specific secret elements (a regular secret key, and a *proxy factor*), securely transfers one to Charles, and the other to the trusted MTA.

With the proxy factor, the MTA gains the ability to transform any PGP message encrypted to Bob's public key into another PGP message that can be decrypted with the newly generated private key, which is now held by Charles. At the same time, the MTA cannot decrypt the message, nor transform it to another public key. The ciphertext transformation is also efficient; upon participating in ECDH key

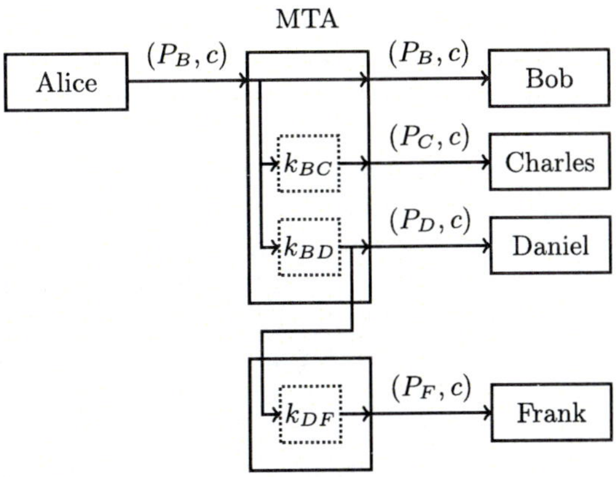

Fig. 1 Forwarding scheme—Alice sends a deferred ECDH ciphertext (P_B, c) to Bob, which the authorized MTA transforms into (P_C, c), (P_D, c), and (P_F, c) using the proxy factors

exchanges, proxies need to store one random field element and two 16-byte key identifiers per forwarding, and compute a single scalar multiplication on the elliptic curve per forwarded ciphertext (Fig. 1).

1.2 A BBS-like Transformation in OpenPGP

The ciphertext transformation technique we describe here is an instance of the ElGamal-based proxy re-encryption scheme presented in the seminal paper by Blaze, Bleumer, and Strauss [8], and as such, is *bidirectional*, meaning that if a proxy can transform ciphertexts from Alice to Bob, they also have the inverse ability. While this property may be undesirable in most contexts, this is not an issue in this OpenPGP application. If Bob forwards encrypted an email to Charles with this technique, Charles's forwarded private key is used exclusively to decrypt Bob's forwarded messages and he has no incentive to receive messages encrypted to the corresponding forwarded public key, since they could be potentially forwarded to Bob.

As indicated in [5], these schemes are also *transitive*, meaning that if Alice forwards an email to Bob and Bob forwards it to Charles, nothing prevents proxies from forwarding Alice's emails directly to Charles and not to Bob. More generally, any malicious forwarding MTA could compose re-encryptions and selectively deliver messages to some recipients, ignoring the rest. While transitiveness appears to be unavoidable here, such an MTA would be driving against its deliverability incentives and may be detected by end parties. Transitiveness is also a property of manual PGP email forwarding; it is generally not assumed that the original sender or any intermediate forwarder is informed about further forwardings.

The reason why this scheme was chosen over more modern re-encryption schemes is that an important objective is to maintain backward compatibility with incoming messages. Furthermore, the sender is often oblivious to automatic mail forwarding happening on the recipient's side, so the chosen scheme must be transparent from the sender's perspective. The BBS technique provides a fast and secure scheme that is compatible with this constraint.

An undesirable property that is partially mitigated by the use of the OpenPGP protocol, also pointed out in [5], is the fact that any collusion between a forwarded party and the proxy results in the knowledge of the forwarding party's private key. Fortunately, OpenPGP's structure prevents impersonation attacks: If Charles above steals Bob's private key with the help of a rogue MTA, he could not sign messages in the name of Bob or authenticate as Bob. The forwarding key is derived from an encryption-only sub-key and does not serve for signing or authentication purposes, because it will not be recognized as a signing or authenticating key by any compliant OpenPGP implementation. Nevertheless, in this case, Charles and the rogue MTA can decrypt every message encrypted to Bob they can obtain, even old or unforwarded. As a mitigation against collusion attacks, we advice to use short-lived keys, avoid the use of filters, generate new keys upon forwarding setup, and finally to deprecate these keys when discontinuing the forwarding. In other words, we assume that either no collusion with MTAs is possible, or that forwarded parties are trusted with unforwarded messages.

2 The Forwarding Scheme

While we describe the forwarding technique in terms of large subgroups of elliptic curves (given the well-known implementation advantages, e.g., [7]), the same technique and security proof can be applied to any large subgroup where the Discrete Logarithm and Diffie-Hellman problems are hard, and subgroup membership testing is efficient. In the remaining sections, let E be an elliptic curve defined over a finite field $\mathbb{F}$, and $G \in E$ be a generator of a large subgroup of E of prime order n.

2.1 *Asynchronous ECDH Exchanges*

In existing implementations of OpenPGP, asynchronous Elliptic Curve Diffie-Hellman exchanges are performed to address encrypted email messages (see Fig. 2), and work as follows. Let $d_B \in \mathbb{Z}_n$ be Bob's private key and $Q_B := d_B G \in E$ be Bob's public key. Alice uses Q_B to generate an ephemeral challenge P and a shared secret $S := d_A d_B G$ that is known only to the two of them, as described in Fig. 2.

Our purpose is to show that a selected proxy transmitting a message encrypted to Bob can grant access to the Alice-Bob shared secret S defined above to selected third parties, without the proxy knowing S or leaking information about the forwarding to Alice or other involved proxies.

Alice		Bob
Retrieves Q_B	$\xleftarrow{Q_B}$	Long-term key pair (Q_B, d_B)
Secretly picks ephemeral d_A		
Computes $P = d_A G$, $S = d_A Q_B = d_A d_B G$		
Encrypts $c = \mathrm{Enc}_S(m)$	$\xrightarrow{P, c}$	Computes $S = d_B P = d_A d_B G$ Decrypts $m = \mathrm{Dec}_S(c)$

Fig. 2 Alice sends an encrypted message to Bob with the Deferred ECDH exchange. P: Ephemeral exchange value; S: Shared secret; m: Plaintext; (P, c): Ciphertext

2.2 *Diverting the Secret*

Let us now describe the BBS-like proxy transformation. Suppose Bob, whose long-term key pair is (d_B, Q_B), wants to forward incoming messages sent with the protocol defined in 2.1 to Charles. First, Bob generates a new key pair (d_C, Q_C) for Charles and computes the proxy factor

$$k_{BC} := d_B / d_C \mod n$$

for the proxy.

Note that it is not possible to guess d_B from the sole knowledge of d_C or k_{BC}, since the mapping of $\mathbb{F}_n$ onto itself given by $\phi_y : x \mapsto x y^{-1}$ is bijective for every $y \in \mathbb{F}_n^*$ and both d_B and d_C are sampled uniformly from a subset of $\mathbb{F}_n$ (in other words, they are indistinguishable from random elements of a subset of $\mathbb{F}_n$, depending on the chosen curve). This means that unless the MTA and the forwarded party collude, they cannot access the secret d_B.

Now, given a plaintext m, an ephemeral DH shared secret between Alice and Bob $S = d_A d_B G$ (also, $S = d_A Q_B$), and a ciphertext (P_B, c) where $c = \mathrm{Enc}_S(m)$ encrypts m, the MTA's objective is to transfer the knowledge of S to Charles, such that he can decrypt c. To achieve this, upon receival of (P_B, c), the MTA first verifies that P_B is not in a small subgroup of the curve, and then computes $P_C := k_{BC} P_B$. Why is verifying P_B necessary? Note that $k_{BC} = d_B / d_C$ is not necessarily a private key of the scheme. In particular, when interpreted as an integer, k_{BC} may not be a multiple of h, the cofactor of the curve, and therefore the above computation is vulnerable to small subgroup attacks. Namely, if P_B belongs to a small subgroup of the curve, then P_C is not necessarily 0, as it would be $d_i P_B$ for any private key d_i. An implementation must abort if $h P_B \neq 0$ (for Curve25519 and Curve448, h is 8 and 4, respectively). According to [12], "*a large number of existing implementations do not [check the all-zero output]*", but in this case, this is mandatory in order to avoid leakage of information about the proxy factors.

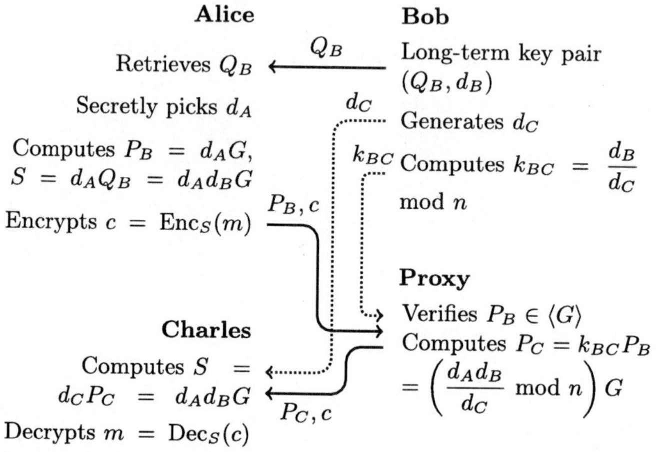

Fig. 3 Forwarded ECDH exchange. Dotted exchanges are done over an existing secure channel. P_B, P_C: Ephemeral exchange values; m: Plaintext; c: Ciphertext

Once P_B is verified, $P_C = k_{BC} P_B$ is computed, and the MTA transfers (P_C, c) to Charles, who in turn computes $d_C P_C = d_C d_C^{-1} d_B P_B = d_A d_B G = S$, allowing him to decrypt.

In other words, the following equation is computed by involved parties:

$$
\begin{aligned}
S &= d_A d_B G && \text{(Computed by Alice)}\\
&= d_B d_A G && \text{(Computed by Bob)}\\
&= d_B d_C^{-1} d_C d_A G && \text{(Computed by Charles).}
\end{aligned}
$$

This procedure is described in Fig. 3. Note that Bob is able to set more than one forwarding address by generating several valid private keys d_i and corresponding proxy factors. Given that the forwarding MTA could be selectively forwarding mail to different users, e.g., using filters, or that the user might want to interrupt the forwardings at different times, it is compulsory to use unique d_i values for every different forwarded party.

The scheme is transitive. In fact, Charles can also forward emails further by repeating the same procedure. Namely, he could generate a new key pair (Q_F, d_F) that he shares with Frank, and compute k_{CF} to share with the MTA.

2.3 *Transformation Proxy as a Multiplication Oracle*

In the procedure described above, Charles could gain access to an arbitrary stream of forwarded ciphertexts by sending messages to Bob using the same protocol, and use the forwarded message to obtain information. In this case, the proxy acts as

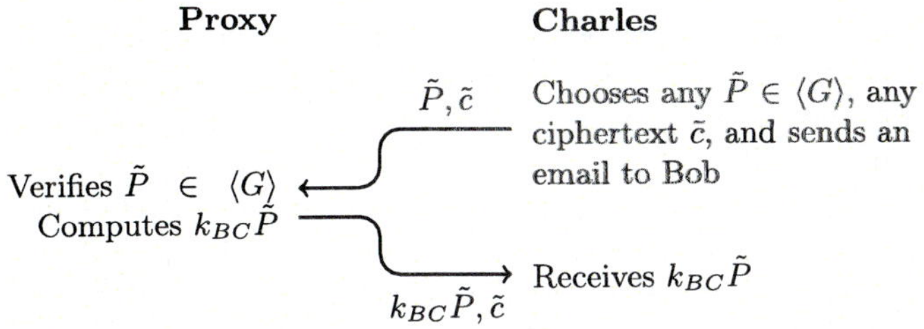

Fig. 4 The proxy acts as a scalar multiplication oracle when receiving messages from Charles

a multiplication oracle by the secret factor k_{BC}: Charles can choose any $\tilde{P} \in E$ and any valid ciphertext $\tilde{c}$, and submit $(\tilde{P}, \tilde{c})$ to the proxy as a message to Bob. The proxy verifies and transforms the ciphertext, returning $(k_{BC}\tilde{P}, \tilde{c})$. Note that guessing the secret factor is exactly solving ECDLP, thus Charles is not able to obtain any information about k_{BC}, or ultimately d_B, since $k_{BC}\tilde{P}$ is indistinguishable from random. This is described in Fig. 4.

We assume that malicious parties have complete freedom in submitting encrypted messages to Bob, but note that this activity may be detected by Bob or other forwarded parties (since they also receive these messages).

3 OpenPGP Implementation

In this section, we describe some considerations regarding OpenPGP implementations for parties willing to achieve the encrypted forwarding procedure described above.

3.1 Client Tasks

Assume Bob holds the following long-term keys in order to ensure deferred ECDH exchanges:

- An EdDSA long-term primary key, signing-only;
- An ECDH sub-key (Q_B, d_B), encryption-only.

Setting up the Forwarding In order to allow forwarding to Charles, Bob generates another key using the same curve and parameters as his existing OpenPGP key, with both the EdDSA and ECDH parts. Let (Q_C, d_C) be the parameters of the encryption-only ECDH part that Bob transfers securely to Charles. Finally, Bob computes $k_{BC} = d_B/d_C \bmod n$ and sends it to the MTA.

Fingerprint Selection for KDF As mandated in the OpenPGP specification [11], Sect. 13.5, a key-derivation function is called in order to obtain the decryption key for a given message; in particular, the original recipient fingerprint is needed as an input to this KDF. Therefore, if a message was originally encrypted to Bob and forwarded to Charles, the decrypting implementation needs to use Bob's fingerprint when decrypting a forwarded message, instead of Charles' fingerprint. In other words, Bob must specify that the fingerprint associated with Q_B must be used when decrypting instead of Q_C in the key-derivation function; otherwise, a fingerprint mismatch will not allow decryption, since Q_B was used to originally encrypt this particular message.

This feature prevents tampering with the recipient, but since we need Charles to decrypt the ciphertext, we propose to change the decryption by altering the field containing the KDF parameters in the algorithm-specific part for ECDH keys. By adding the original fingerprint and specifying a new version 2, whose field is already defined and "reserved for future extensions", we emulate Bob's fingerprint in Charles' decryption procedure. This would not reduce the tampering protection's effectiveness, since this information is embedded in Charles' key, and at the same time backward compatibility for the sender. Concretely, we propose to alter *variable-length field containing KDF parameters* defined in [11], Sect. 13.5, as follows:

- *(Unchanged)* a one-octet size of the following fields; values `0` and `0xff` are reserved for future extensions;
- *(Upgraded)* a one-octet value `02`, indicating forwarding support;
- *(Unchanged)* a one-octet hash function ID used with the KD;
- *(Unchanged)* a one-octet algorithm ID for the symmetric algorithm used to wrap the symmetric key for message encryption;
- *(Added)* one-octet of flags with value `0x01`, indicating to expect a key fingerprint;
- *(Added)* a 20-octet fingerprint to be used in the KDF function; for version 5 keys, the 20 leftmost bytes of the fingerprint.

The forwarded parties' OpenPGP implementation will use this value in the key wrapping instead of the original fingerprint in order to obtain the right session key. Note that with this key pair, only forwarded ciphertexts can be decrypted, and it does not allow anyone to decrypt messages encrypted to Q_B that were not transformed by the MTA.

3.2 Server Tasks

The MTA acts as the re-encryption proxy; it safely stores the factors k_{ij} along with the key IDs to replace them in the ciphertext metadata. When a matching incoming email arrives, it alters the asymmetric key packet corresponding to the correct key ID.

An incoming message from Alice to Bob has a *Public-Key Encrypted Session Key Packets* that wraps a symmetric key to decrypt the data packet. This packet contains

- the curve OID, identifying the correct field;
- the ephemeral value P_B;

- the key ID of Bob's public key;
- the encrypted session key.

The first step is to parse the ephemeral exchange value P_B and verify that it belongs to the subgroup generated by G and not other small subgroups of the curve, as mentioned in Sect. 2.2. This can be achieved by generating any private key s and checking $sP_B \stackrel{?}{=} 0$, or equivalently, checking that $hP_B \stackrel{?}{=} 0$, where h is the cofactor of the curve ($h = 8$ for Curve25519). If P_B does not belong to the large subgroup, the MTA must refuse to process this ciphertext, as the transformation would leak information about the proxy factors.

Once P_B is verified, the MTA replaces the ephemeral value of the above packet with $k_C P_B$. Also, it replaces the key ID with Charles' key ID to ensure that his OpenPGP implementation will accept the packet for decryption (in OpenPGP terminology, the MTA eventually *un-armors* the message, replaces both values, and *armors* the result).

The MTA can also implement filters based on the unencrypted fields, e.g., sender and recipient addresses, or headers. These, of course, rely purely on trust, e.g., a misbehaving MTA could forward every mail to Charles.

3.3 Implementation Details

Using the widely adopted GopenPGP [1] and OpenPGP.js [2], we emulated the forwarding and verified decryption correctness using those implementation APIs. Note that setting up the forwarding is essentially generating two private scalars, and ciphertext transformation is essentially one scalar multiplication on the curve, therefore, there is a negligible extra cost when supporting this feature.

Due to its generic design, forwarding can be implemented with all curves supported by the OpenPGP standard. Our curve of choice was Curve25519 [7, 12], defined in the finite field of $2^{255} - 19$ elements, and whose base point $G : x = 9$ generates a large subgroup of prime order

$$n = 2^{252} + 27742317777372353535851937790883648493.$$

Secret keys are sampled randomly from $2^{254} + 8\{0, 1, 2, \ldots, 2^{251} - 1\}$, and proxy factors are interpreted in $\mathbb{F}_n$ where n is the prime number displayed above. The design properties of Curve25519 make it stand at a comfortable security level against all known attacks. It is worth noting that scalar multiplication is implemented in constant time, since double and adding use the same formulæ. This consideration is important since any forwarded party can submit ciphertexts for transformation and measure the time of the MTA's reaction, with the objective of learning about proxy factors. Note also that some modern OpenPGP implementations are not implemented in constant time for other curves.

The verification of small subgroup points in this particular curve is simply checking if $8P$ is 0, in which case the proxy must refuse to transform the ciphertext.

4 Security Analysis

This section describes how the forwarding protocol is secure against eavesdroppers, semi-honest and malicious adversaries, except for collusions between the proxy and any forwarded party (since they can trivially recover Bob's secret).

A simulation-based proof using the standard techniques from [13] is provided, and works as follows. We first define an *ideal functionality* $\mathcal{F}$ associated with the forwarding protocol Π. Given a set of participants of Π, we describe their *views* and construct *simulators* that produce random corresponding views. Finally, we show that the simulated views and the output of $\mathcal{F}$ (i.e., the *ideal world*) are computationally indistinguishable from the execution of the protocol and its output (the *real world*).

This proof can be found in Appendix 6. We provide here an overall analysis of the security provided by the forwarding protocol.

4.1 Threat Model

The threat model we consider is an expanded version of the deferred Diffie-Hellman exchange:

- Bob, the original receiver, is always honest: he follows the protocol and samples from the correct distributions.
- External eavesdroppers do not participate but may collect all values exchanged in the protocol except for private keys and proxy factors (i.e., they know all elements in paths defined by bold arrows in Fig. 3).
- F, the set of forwarded parties, may contain a subset of colluded parties that may also send messages to Bob and eavesdrop the protocol.
- Alice, the original sender, may collude with forwarded parties.
- T, the transformation proxy, could misbehave and/or collude with other parties.

We show that the only collusion that succeeds in attacking is when T colludes with a forwarded party. This is expected, since they have multiplicative shares of Bob's secret d_B. To ease notation, for each forwarded party $F_i \in F$, let $k_i := k_{BF_i}$ be the proxy factor held by T, $d_i := d_{F_i}$ the secret scalar held by F_i (generated by Bob), and $P_i := P_{F_i}$ the transformed shares (generated by the proxy).

4.2 Semi-honest Parties

We discuss the semantic security of the execution of the protocol, establishing that no party can extract information about the plaintext or secrets from the elements collected throughout the execution, assuming that other parties are honest (this includes ciphertexts and messages received during the protocol).

External Eavesdroppers Let us first discuss security against an adversary that is not participating in the protocol nor controlling any party, but who intercepts all communications except for private scalars and proxy factors (these communications are denoted by dashed paths in Fig. 3). Additionally, they could send messages to Bob and eavesdrop the forwarding as described in Sect. 2.3. Namely, they could choose any ciphertext $\tilde{c}$ and a point $\tilde{X}$ of the large subgroup and eavesdrop $k_i\tilde{X}$. Such a party holds

$$\begin{array}{ll} Q_B = d_B G & \text{Bob's long-term public key,} \\ P_B = d_A G & \text{Alice's DH share to Bob,} \\ \{P_{F_i} = d_A k_i G : i = 1, \ldots, |F|\} & \text{set of transformed shares,} \\ \{k_i\tilde{X} : i = 1, \ldots, |F|\} & \text{for chosen } \tilde{X} \in E, \end{array}$$

and is interested in extracting $d_A d_B G$ (the session secret), any proxy factor $k_i = d_B d_i^{-1}$, or any named scalar. Since proxy factors and secret scalars were generated honestly, note that this party holds values that are indistinguishable from uniformly random elements of the large subgroup. Obtaining any named scalar solves instances of the ECDL problem, and producing the session secret $d_A d_B G$ solves the computational ECDH problem. This proves that the protocol is semantically secure against passive eavesdroppers.

Transformation Proxy The proxy collects the following elements throughout the protocol

$$\begin{array}{ll} Q_B = d_B G & \text{Bob's long-term public key,} \\ P_B = d_A G & \text{Alice's DH share to Bob,} \\ \{k_i : i = 1, \ldots, |F|\} & \text{proxy factors,} \\ \tilde{X}_1, \tilde{X}_2, \ldots & \text{shares submitted by other parties (2.3).} \end{array}$$

It is clear that all these elements are uniformly random, provided by honest parties. Also, again by the ECDH assumption, the proxy alone cannot produce the session secret $d_A d_B G$. Naturally, if the proxy and Alice collude (and Alice is not a forwarded party), they can produce the session secret but cannot extract Bob's secret d_B. In addition, the proxy cannot compute d_B from the list of proxy factors: recall that $k_i = d_B d_i^{-1} \bmod n$, but since these integers are interpreted in $\mathbb{F}_n$, there is no notion of GCD (namely, for every $x, y, z \in \mathbb{F}_n$ there exist u, v such that $x = uz$, $y = vz$).

Colluded Forwarded Parties Any forwarded party F_i that also submits a stream of points $\tilde{X}_1, \tilde{X}_2, \ldots$ (as in Sect. 2.3) collects

$Q_B = d_B G$	Bob's long-term public key,
d_i	the decryption share,
$P_i = d_A d_B d_i^{-1} G$	the transformed DH share,
$k_i \tilde{X}_1, k_i \tilde{X}_2, \ldots$	transformed DH shares of submitted messages.

Assume further that this party eavesdrops $P_B = d_A G$ from the communication between Alice and the proxy (or, equivalently, it colludes with Alice). Note that multiplying all DH shares by d_i, this party holds exactly a DH triplet $(d_A G, d_B G, d_A d_B G)$ and a set of ECDL samples $(\tilde{X}, d_B \tilde{X})$ for chosen X. A collusion between forwarded parties will only extend the list of ECDL samples, thus, security follows from both the ECDL and ECDH assumptions.

4.3 Malicious Parties

Malicious Proxy Consider a rogue transformation proxy. It follows from the ECDH and ECDL assumptions that a malicious proxy cannot extract secrets d_A, d_B, d_{F_i}, $d_A d_B$, or $d_A d_B G$ from its view only. However, there are other possible misbehaviors of a malicious proxy: restrict the forwarding to selected parties, alter proxy factors (sabotaging ciphertexts) or forward messages even if instructed not to. We consider these actions as unavoidable for the forwarding functionality, probably detectable in the context of MTAs, and not part of our trust assumptions.

Malicious Forwarded Parties Note that forwarded parties do not control any value in the protocol, with the exception described in Sect. 2.3 where they send messages to Bob themselves and wait for the proxy's reaction. In this case, they only learn samples of the form $k_i X$ for chosen $X \in E$ and proxy factors k_i (these are uniform elements of the large subgroup). Since this is not a deviation from the protocol, we include these samples in the semi-honest security proof, therefore, security against malicious forwarded parties follows.

Collusion Between the Proxy and Forwarded Parties Any forwarded party that colludes with the proxy can recover Bob's key:

$$d_i k_i = d_i d_i^{-1} d_B = d_B \mod n.$$

We point out that while recovering this private key may allow decrypting other messages, it does not allow to impersonate Bob and generate valid signatures, since compliant OpenPGP implementations consider this key as encryption-only.

5 Conclusions

In this article, we adapted the DH diverting techniques presented in [8] to provide the forwarding functionality to an encrypted email service compatible with the OpenPGP protocol. This allows encrypted emails to be securely forwarded with small modifications to an MTA server and the forwarded parties' clients. In this context, the scheme we propose

- is **transitive**: Bob can forward his encrypted messages to Charles, Charles can forward them to Daniel, and so forth;
- is **non-interactive**: Bob's keys are sufficient to derive the proxy transformation factors without any further exchange;
- is **transparent**: Transformed messages are indistinguishable from regular PGP messages;
- and **distributes trust**: We proved that the only way of gaining access to unforwarded emails is for at least one forwarded party to collude with the MTA. In addition, we showed that any set of forwarded parties cannot gain access to the private key without collusion with the proxy, since they confront several independent instances of ECDL and/or ECDH problems.

We verified the correctness of the forwarded decryption using two well-known OpenPGP implementations. While this can already be implemented to allow forwarding within users of the same mail provider, we described a concrete proposal to the OpenPGP specification that would imply forwarding compatibility between different implementations.

Acknowledgements We thank anonymous reviewers for their helpful comments. We also thank Ilya Chesnokov, Eduardo Conde, and Daniel Huigens for extensive discussions, helpful Comments, and important insights intn ECC and OpenPGP. We would also like to thank Ben Caller for his help with proofreading the paper.
Aron Wussler thanks Daniel Kahn Gillmor for introducing him to proxy re-encryption schemes for PGP mailing lists. Also, he wishes to acknowledge the help provided by Prof. Gerhard Dorfer, who provided valuable help with algebra.

6 Simulation-Based Proof

Using the simulation techniques from [9, 13], we present a security analysis of the forwarding protocol. This proof addresses the security against semi-honest adversaries, i.e., parties that correctly follow the protocol and sample from the correct distributions, but use all available information to steal secrets. Also, a set of semi-honest adversaries can collude.

The idea of simulation-based proofs is to emulate any set of colluded participants by a random simulator residing in an ideal world where a trusted party exists and the protocol is secure by definition. The rationale behind this proof is that if it was possible to extract secrets from the views of colluded parties, then it would be trivial to distinguish them from the simulator (just select the view that allows extracting secrets). Conversely, if one cannot distinguish between the views, colluded parties cannot extract secrets. Therefore, the objective is to show that the set of elements held by the semi-honest parties (their *view* of the protocol) are indistinguishable from the elements held by the simulator.

6.1 Security Definitions

Ideal Functionalities A *functionality* is a process that maps tuples of inputs to tuples of outputs of a protocol Π, one for each party involved. More precisely, for a fixed set $P = \{P_1, \ldots, P_k\}$ of k parties participating in the protocol, functionalities are k-ary functions $\mathcal{F} : (\{0, 1\}^*)^k \to (\{0, 1\}^*)^k$ mapping inputs to outputs of Π. We write $\mathcal{F} = (f_1, \ldots, f_k)$ where each f_i is a k-ary function that outputs a string. In addition, if $\mathcal{F}$ computes the desired outcome by means of a trusted party in an ideal world that can communicate over perfectly secure channels with all participants, we say that $\mathcal{F}$ is an *ideal functionality*.

Views Given an execution of a protocol Π on inputs $\mathcal{X} = (x_1, \ldots, x_k)$, the *view* of party P_i consists in all elements accessible to P_i throughout the protocol:

$$\text{view}_{P_i}(\mathcal{X}) = (x_i, r_i, m_1^i, \ldots, m_j^i),$$

where x_i is P_i's input, r_i is the content of its internal random tape used to sample elements, and m_j^i is the jth message it received. Given a set of colluded parties, their joint view is defined as the tuple consisting in the concatenation of their views. Also, let $\text{output}_{P_i}(\mathcal{X})$ be the elements held by party P_i identified as the output of the protocol (note that the inputs or outputs may be empty for some parties).

Simulators Let Π be a protocol with inputs $\mathcal{X} = (x_1, \ldots, x_k)$. A simulator is a PPT algorithm that, given an input x_i corresponding to a party of the protocol, produces a tuple $\text{sim}_i(x_i)$ similar to the view of this party. A *joint simulator* takes a set of inputs and produces a tuple similar to the concatenation of the corresponding parties' views.

Simulation-Based Proof Following [13], our notion of security is based on emulating ideal functionalities defined by the forwarding protocol. This means that, given a protocol Π with inputs $\mathcal{X}$, an ideal functionality $\mathcal{F}$ computing the output of $\Pi(\mathcal{X})$, and a set of semi-honest colluded parties, we construct a simulator that takes the inputs of these parties and produces a random joint view. We then show that these views along with the output of the protocol (i.e., the *real world*) and the simulator along with the ideal functionality result of these parties (i.e., the *ideal world*) are computationally indistinguishable. For instance, when simulating party i, we show that

$$\left(\text{view}_{P_i}(\mathcal{X}), \text{output}_{P_i}(\mathcal{X})\right) \simeq \left(\text{sim}_i(x_i), f_i(\mathcal{X})\right)$$

and similarly for joint views of colluded parties, achieving the proof.

6.2 Forwarding Ideal Functionality

Consider parties Alice, Bob, T (the proxy), and Charlie (the forwarded party). As before, let χ be the distribution of $\mathbb{Z}$ that samples private keys uniformly from a subset of $\mathbb{Z}$, according to the security requirements of E (for instance, in Curve25519 [7], private keys are random samples of the form $2^{254} + 8m$ for some $m < 2^{252}$). Assume Alice sends a message (P_B, c) to Bob, and Bob's public key is $Q_B = d_B G$.

A basic forwarding ideal functionality could be given by

$$\big((P_B, c), (Q_B, k_{BC}, d_C), \cdot, \cdot\big) \mapsto \big(\cdot, \cdot, \cdot, (P_C, c)\big)$$

such that $d_C P_C = d_B P_B$. This works since Charles receives d_C at some point in the protocol, and can decrypt c given $S = d_C P_C$. However, note that this functionality does not take into account multiple forwarded parties, nor the fact that those parties can also submit messages to Bob (as in Sect. 2.3). Recall also that, for each forwarded party $F_i \in F$, we note by $k_i := k_{BF_i}$ the proxy factor held by T and $d_i := d_{F_i}$ the secret scalar held by F_i.

Definition 1 Consider parties Alice, Bob, T (the proxy), and $F = \{F_1, \ldots, F_m\}$ (the forwarded parties). Let

$$\mathcal{X} := (\mathcal{X}_A, \mathcal{X}_B, \mathcal{X}_T, \mathcal{X}_{F_1}, \ldots, \mathcal{X}_{F_m})$$

be the input of the protocol where

$$\begin{array}{ll} \mathcal{X}_A := P_B, c & \text{Alice's message,} \\ \mathcal{X}_B := Q_B, (k_i, d_i)_{i=1}^m & \text{Bob's public key, proxy} \\ & \text{factors, and secret shares,} \\ \mathcal{X}_T := \cdot & \\ \mathcal{X}_{F_1} := (X_{1j})_{j=1}^{n_1} & F_1 \text{ sends } n_1 \text{ messages to Bob as in 2.3,} \\ \vdots & \\ \mathcal{X}_{F_m} := (X_{mj})_{j=1}^{n_m} & F_m \text{ sends } n_m \text{ messages to Bob as in 2.3.} \end{array}$$

The forwarding functionality is $\mathcal{F} = (\mathcal{F}_A, \mathcal{F}_B, \mathcal{F}_T, \mathcal{F}_{F_1}, \ldots, \mathcal{F}_{F_m})$ where

$$\begin{array}{l} \mathcal{F}_A : \mathcal{X} \mapsto \cdot \\ \mathcal{F}_B : \mathcal{X} \mapsto c, d_B, P_B, ((X_{ij})_{j=1}^{n_i})_{i=1}^m \\ \mathcal{F}_T : \mathcal{X} \mapsto \cdot \\ \mathcal{F}_{F_1} : \mathcal{X} \mapsto c, d_1, P_1, ((k_1 X_{ij})_{j=1}^{n_i})_{i=1}^m \\ \vdots \\ \mathcal{F}_{F_m} : \mathcal{X} \mapsto c, d_m, P_m, ((k_m X_{ij})_{j=1}^{n_i})_{i=1}^m \end{array}$$

such that $d_1 P_1 = d_2 P_2 = \cdots = d_m P_m = d_B P_B$.

Note that for every message X_{ij} sent as in Sect. 2.3, forwarded parties also receive an encryption c_{ij} of some message. Without loss of generality, we omit these encryptions from the ideal functionality, since these are trivial to simulate and provide no information to attackers (as they are simply transmitted unchanged throughout the protocol).

More precisely, following Sect. 2.3, forwarded parties pick any message $\tilde{m}$ and a secret key $\tilde{d} \leftarrow \chi$, set $\tilde{S} = \tilde{d}Q_B$ where Q_B is Bob's public key, and let $X_{ij} = \tilde{d}G$, $c_{ij} = \mathrm{Enc}_{\tilde{S}}(m)$. Instead, without loss of generality, we let forwarded parties freely choose X_{ij} as an input to the protocol.

6.3 *Security Against Colluded, Eavesdropper Semi-honest Parties*

Simulating the Semi-honest Proxy According to the protocol described in Sect. 2.2, we have the following view of the proxy throughout the protocol:

$$\mathrm{view}_T(\mathcal{X}) = \big(\cdot, c, P_B, k_1, \ldots, k_m, (X_{ij})_{ij}\big)$$

where each k_i was provided by Bob (for the sake of notation, $(X_{ij})_{ij}$ consists in all points chosen by forwarded parties, in Definition 1). Indeed, for each i, Bob sampled $d_i \leftarrow \chi$, $k_i := d_B/d_i \bmod n$, and sent k_i to the proxy. Now, consider the simulator that samples $y \leftarrow \chi$, $x_i \leftarrow \chi$, and $z_i \leftarrow \chi$ for $i = 1, \ldots, m$, a tuple of random points $(\tilde{X}_{ij})_{ij}$ and sets

$$\mathrm{sim}_T(\mathcal{X}_T) := \big(\cdot, c, yG, x_1 z_1^{-1}, \ldots, x_m z_m^{-1}, (\tilde{X}_{ij})_{ij}\big).$$

Recall that (i) there is no input or output for T in this protocol, and also (ii) all other parties behave honestly in this case (in particular, X_{ij} are uniformly random points of the curve). Given these facts, it is straightforward to see that $\mathrm{view}_T(\mathcal{X})$ and $\mathrm{sim}_T(\mathcal{X}_T)$ are computationally indistinguishable.

Simulating Forwarded Parties As described in Sect. 2.3, each forwarded party F_i computes the session secret $S = d_i P_i$ from the output, and also has a stream of pairs of the form $(X_{ij}, k_i X_{ij}) \in E^2$ for chosen X_{ij} (this is the result of sending messages encrypted to Bob and parsing the forwarded ciphertexts).

$$(\mathrm{view}_{F_i}(\mathcal{X}), \mathrm{output}_{F_i}(\mathcal{X})) = \big((X_{ij})_{j=1}^{n_i}; c, d_i, P_i, (k_i X_{ij})_{j=1}^{n_i}\big)$$

. Note that, since there are no intermediate values computed by forwarded parties, a simulator that can access the input and output of a forwarded party can simulate it trivially. In this case, we have simply

$$\big(\mathrm{sim}_{F_i}(\mathcal{X}_{F_i}), \mathcal{F}_i(\mathcal{X})\big) \equiv \big(\mathrm{view}_{F_i}(\mathcal{X}), \mathrm{output}_{F_i}(\mathcal{X})\big).$$

Naturally, the joint view of colluded forwarded parties is also trivially simulated by the joint simulators.

Simulating Colluded, Eavesdropper Forwarded Parties Additionally, let us assume further that this party eavesdropped the share $P_B = d_A G$ from the communication between Alice and the proxy, and recall that they also have the public key $Q_B = d_B G$. Such party has the following view and output:

$$(\text{view}_{F_i}(\mathcal{X}), \text{output}_{F_i}(\mathcal{X})) = \left((X_{ij})_{j=1}^{n_i}; d_A G, d_B G; c, d_i, P_i, (k_i X_{ij})_{j=1}^{n_i}\right)$$

. Consider the simulator that samples $x, y \leftarrow \chi$ and sets

$$\left(\text{sim}_{F_i}(\mathcal{X}_{F_i}), \mathcal{F}_i(\mathcal{X})\right) = \left((X_{ij})_{j=1}^{n_i}; xG, yG; c, d_i, P_i, (k_i X_{ij})_{j=1}^{n_i}\right).$$

The only distinct elements are $d_A G, d_B G$ and xG, yG. Note that, since $d_i P_i = d_A d_B G$, a DH triplet $(d_A G, d_B G, d_A d_B G)$ can be composed in the view. The simulator, on the other hand, can compose the tuple $(xG, yG, d_A d_B G)$. However, since $d_A d_B G$ is uniformly random, this tuple is indistinguishable from a proper DH triplet (xG, yG, xyG) by the ECDH assumption. It follows that the view and the simulator are computationally indistinguishable. It is straightforward to extend the simulator and address the case where multiple semi-honest forwarded parties collude: The joint view and output will only have more independent ECDL samples of the form $(X, k_i X)$ and parties can compose the same ECDH triplet (also, note that the additional samples d_i, P_i of the form are also accessible by the simulator, and the same argument holds).

Putting all these cases together, it follows that the protocol securely computes $\mathcal{F}$ in the presence of semi-honest, colluded adversaries.

References

1. Gopenpgp. https://gopenpgp.org/
2. Openpgp.js. https://openpgpjs.org/
3. Aono Y, Boyen X, Phong LT, Wang L (2013) Key-private proxy re-encryption under LWE. In: Proceedings of the 14th international conference on progress in cryptology INDOCRYPT 2013—vol 8250. Springer-Verlag, Berlin, Heidelberg, pp 1–18. https://doi.org/10.1007/978-3-319-03515-4_1
4. Ateniese G, Benson K, Hohenberger S (2009) Key-private proxy re-encryption. In: Fischlin M (ed) Topics in cryptology—CT-RSA 2009. Springer, Berlin Heidelberg, Berlin, Heidelberg, pp 279–294
5. Ateniese G, Fu K, Green M, Hohenberger S (2005) Improved proxy re-encryption schemes with applications to secure distributed storage. vol 2005, issue 01
6. Ateniese G, Fu K, Green M, Hohenberger S (2006) Improved proxy re-encryption schemes with applications to secure distributed storage. ACM Trans Inf Syst Secur 9(1):1–30. https://doi.org/10.1145/1127345.1127346 10.1145/1127345.1127346
7. Bernstein DJ (2006) Curve25519: new diffie-hellman speed records. In: In public key cryptography (PKC), Springer LNCS 3958, p 2006

8. Blaze M, Bleumer G, Strauss M (1998) Divertible protocols and atomic proxy cryptography. In: Advances in cryptology—EUROCRYPT '98, international conference on the theory and application of cryptographic techniques, Espoo, Finland, May 31–June 4, Proceeding. Lecture notes in computer science, vol 1403. Springer, pp 1–18. https://doi.org/10.1007/BFb0054122
9. Canetti R (2006) Security and composition of cryptographic protocols: a tutorial (part i). SIGACT News 37(3):67–92. https://doi.org/10.1145/1165555.1165570 10.1145/1165555.1165570
10. Khurana H, Heo J, Pant M (2006) From proxy encryption primitives to a deployable secure-mailing-list solution. In: Ning P, Qing S, Li N (eds) Inf. Commun. Secur. Springer, Berlin Heidelberg, Berlin, Heidelberg, pp 260–281
11. Koch W, Carlson B, Tse R, Atkins D, Gillmor D (2020) Openpgp message format draft-ietf-openpgp-rfc4880bis-09. RFC 4880bis, RFC Editor. https://tools.ietf.org/html/draft-ietf-openpgp-rfc4880bis-09
12. Langley A, Hamburg M, Turner S (2016) Elliptic curves for security. RFC 7748, RFC Editor
13. Lindell Y (2017) How to simulate it—a tutorial on the simulation proof technique. Springer International Publishing, Cham, pp 277–346. https://doi.org/10.1007/978-3-319-57048-8_6

A Survey on Machine Learning-Based Ransomware Detection

Nanda Rani, Sunita Vikrant Dhavale, Amarjit Singh, and Atul Mehra

Abstract Ransomware is a program used by an attacker or hacker, that locks or encrypts target files or data. The user or the owner of data cannot access these without the explicit assistance of the attacker. After locking or encrypting, the attacker demands ransom generally in the form of cryptocurrencies to permit user to regain access to the locked data. However, there is no guarantee that the user can access seized data again even after that ransom has been paid. Researchers have proposed various tools and techniques to protect and fight against ransomware. Existing tools and methods are not sufficient to detect ransomware early because several new ransomware variants are being introduced every day. Machine learning techniques are used efficiently in various applications like ransomware detection, spam detection, text classification, pattern recognition, etc. Further, deep learning, a subfield of machine learning, eliminates the burden of re-engineering the features for the new types of malware or network attacks that may arise. In this paper, several machine learning-based detection techniques against ransomware are reviewed.

Keywords Machine learning · Ransomware detection · Malware · Malicious software · Cybersecurity

N. Rani (✉) · S. V. Dhavale
Defence Institute of Advanced Technology, Pune, India
e-mail: nanda_mcse19@diat.ac.in

S. V. Dhavale
e-mail: sunitadhavale@diat.ac.in

A. Singh
Persistent Systems, Pune, India
e-mail: amarjit_singh1@persistent.com

A. Mehra
National Critical Information Infrastructure Protection Centre, Mumbai, India
e-mail: atulmehra@nciipc.gov.in

© The Author(s), under exclusive license to Springer Nature Singapore Pte Ltd. 2022

D. Giri et al. (eds.), *Proceedings of the Seventh International Conference on Mathematics and Computing*, Advances in Intelligent Systems and Computing 1412,
https://doi.org/10.1007/978-981-16-6890-6_13

1 Introduction

In recent months, numerous attacks have taken place globally because of the current pandemic situation and dependence on work from home. This situation increases the vulnerability of networks and the IT infrastructure across the government as well as private sectors globally. The attackers are using malware or malicious programs called Ransomware to target the Information Infrastructure and infect the targets either by exploiting vulnerabilities or social engineering. Although, Malware has several classes like viruses, worms, rootkits, spyware, trojans, ransomware, etc. Ransomware is scareware that encrypts all essential files of users and demands a ransom to access those files.

Ransomware is a widespread malware that encrypts the victim's data mutely and makes it unavailable to the user. They can only decrypt data after the victim pays the demanded ransom. Generally, the demanded ransom is in digital currency, and the anonymous nature of digital currency enables attackers to escape prosecution. It also gives reason to increase in the number of ransomware attacks [1]. Ransomware is of two types: Crypto ransomware and Locker ransomware [2]. Crypto ransomware encrypts files on the system and makes them inaccessible. Crypto ransomware is also known as File-Locker ransomware. Locker ransomware does not tamper with data instead prevents users from using their system either by showing a popup that never closes or locks their desktop. It is also known as Screen-Locker ransomware [3]. In the current scenario, ransomware has given rise to more dangerous variants such as WannaCry [4], Cryptowall [5], Locky [6], GandCrab [7], CryptoLocker [8], TeslaCrypt [9].

The year 2016 is remarkable in the history of malware because of the WannaCry ransomware attack. It is the most successful ransomware attack and distinguished as the "Year of ransomware" [10]. Wannacry ransomware exploited a Weakness named EternalBlue present in the Server Message Block (SMB) protocol of Windows frameworks. It infected thousands of users worldwide. For encrypted user data, The attackers demanded bitcoin in exchange for a decryption key [11]. Ransomware steals the victim system's essential functions, denies the user access to those function, and lastly demands payment to regain access. According to the threat research report of Exambeam,[1] ransomware follows a common six-stage attack flow kill-chain shown in Fig. 1 [12]:

- *Distribution Campaign:* Attacker mutely tricks the victim into downloading dropper code, which initiates infection. The attackers use techniques like social engineering, email Phishing, and many more.
- *Malicious code injection:* In this stage, malicious code download happens and ransomware introduces on the target's machine.
- *Malicious Payload Staging:* Ransomware inserts itself into the system and set up persistency.

---

[1] https://www.exabeam.com/wp-content/uploads/2017/07/Exabeam_Ransomware_Threat_Report_Final.pdf.

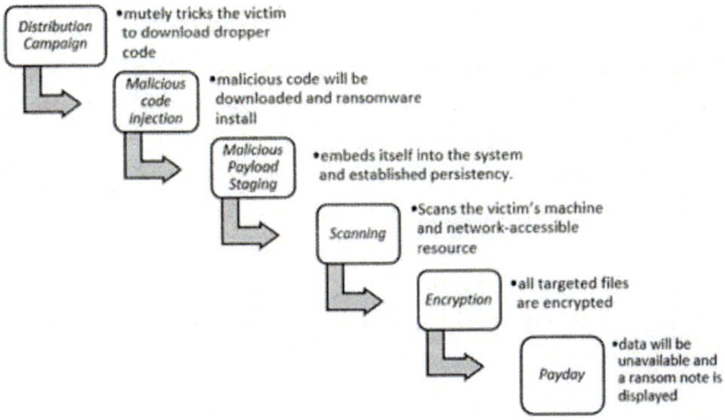

Fig. 1 Attack Flow Kill Chain of Ransomware

- *Scanning:* Scans the target machine and network-reachable resource for encryption.
- *Encryption:* Encryption starts and all targeted documents are encrypted.
- *Payday:* Victim's data being unavailable to them and a payment note is shown to the targeted system's screen demanding payment.

Most of the ransomware uses symmetric encryption to encrypt the victim's files. However, some ransomware also uses hybrid symmetric and asymmetric encryption techniques. This technique involves an asymmetric key to encrypt the targeted file and an asymmetric public key to encrypt the symmetric key [2].

The organization of the paper is as follows. Description of various machine learning algorithms is presented in Sect. 2. Ransomware datasets are given in Sect. 3. Several ML-based methods and techniques for ransomware detection are discussed in Sect. 4. Detailed discussion and efficacy of these techniques are explained in Sect. 5 and Sect. 6 gives the conclusion.

2 Description of the Machine Learning Algorithms

Ransomware has different characteristics from other types of malware. Ransomware generally performs a large number of file operations for encryption in a short period. They are much similar to benign file characteristics. The signature-based detection mechanism is not suitable to protect system data from risky unknown ransomware. The development of artificial intelligence researchers started shifting towards a detection mechanism based on combining machine learning and deep learning. We have several detection techniques, which are using many machine learning algorithms like RF (Random Forest), SVM (Support Vector Machine), NB (Naïve Bayes), KNN (K-Nearest Neighbours), SGD (Stochastic Gradient Descent), and LF (Logistic Regression).

Random Forest is an ensemble learning method for regression and classification analysis and gives output using numerous decision trees. SGD focussed on minimizing losses between predicted and real values by updating parameters. NB is for classification analysis based on the Bayesian theorem. SVM is broadly used for pattern recognition and creates a classification model to determine which category a new sample belongs. KNN is mostly used for regression and classification in pattern recognition, in which the model calculates K-nearest elements and decides which class a new sample belongs to. Usage of these algorithms has boosted the detection mechanism efficiency and enhanced the detection mechanism to detect unknown ransomware by observing processes or file systems' activity patterns.

There are various evaluation metrics on which we decide the efficiency of any machine learning model. The value of these metrics is present in the confusion matrix of the Ml model. Those parameters are the following:

1. *True Positive (TP):* Values that are actually positive and predicted positive.
2. *False Positive (FP):* Values that are actually negative but predicted to positive.
3. *False Negative (FN):* Values that are actually positive but predicted to negative.
4. *True Negative (TN):* Values that are actually negative and predicted to negative.
5. *Accuracy:* Ration of Correctly classified to the total number of test samples.

$$Accuracy = \frac{TP + TN}{TotalTestSamples} \tag{1}$$

6. *Precision:* Ratio of true positive to Actual positive.

$$Precision = \frac{TP}{TP + FP} \tag{2}$$

7. *Recall:* Ratio of True positive to Predicted Results.

$$Recall = \frac{TP}{YP + FN} \tag{3}$$

There are some challenges to implement a machine learning detection system. One of the most common limitations is the training cost. To train large datasets with an extensive feature set, we may face high training costs to train our detection model. To reduce this cost, we may adopt dimensionality reduction. Dimensionality reduction reduces the cost of training. One more biggest concern of the ML solution is Adversarial machine learning (AML). AML means the attacker uses tactics to bypass the ML-based detector. Attackers are sometimes as skilled as security researchers. They are regularly working to develop malware variants to confuse detectors. By creating a hybrid classifier, these problems can be solved, i.e., hybrid feature selection (static and dynamic both utilized), and multiple classification algorithms can be trained (ensemble malware detector).

3 Description of the Ransomware Research Datasets

For ransomware detection, we have many datasets shared by several researchers, and also one can create customized datasets based on the need of the model. One can collect ransomware samples from VirusShare[2] and benign samples from Softonic. Some of the popular datasets are the following:

- This dataset is prepared by Sgandurra et al. [14], which includes 582 samples of ransomware and 942 goodware. They used the Cuckoo sandbox for analyzing this dataset consists of various goodware and ransomware. Goodware samples include generic utilities for Windows, drivers, browsers, file utilities, multimedia tools, network utilities, emulators, and virtual machine monitors [14]. Ransomware samples belong to different families, and it identifies them with the codes shown in Table 1. For each family, they downloaded all the samples found on VirusShare. They also downloaded and analyzed a dataset of 942 benign applications (the goodware set). They selected these from the top list of the most popular software from a software aggregator website.[3] To ensure that the downloaded goodware samples did not contain suspicious components inside their payload, checked with the results of VirusTotal.[4] This dataset contains 1524 records and 30970 features and detailed information available at [15]. The different sets of features are identified with the codes shown in Table 2.
- This dataset [17] contains 7,222 bitcoin seed addresses related to 67 families and identified by applying the expansion procedure described in Paquet-Clouston et al. [16]. The open-source GraphSense Cryptocurrency Analytics platform (v.0.3.1) has implemented the extraction procedure. The complete expanded dataset available at zenodo.org
- The ISOT Research lab has given a dataset that consists of over 420 GB of ransomware and benign program execution traces [18]. The ransomware samples were obtained from VirusTotal under an academic license and got several samples from anti-malware companies. The dataset consists of 669 ransomware samples representing most of the widespread ransomware families and variants. In addition to the ransomware samples, the dataset includes data from 103 benign applications representing the most popular Windows user's most popular software applications. The total size of the dataset on disk is 428 GB. Cuckoo sandbox is used for analysis and used Windows 7 (64 bit) to run samples and collect behaviour data.

We have numerous tools available to collect features from ransomware samples either by static analysis or dynamic analysis. We have various tools like IDA Pro, PEid, PeView, PsFile, SSDeep, and many more for static analysis [41]. IDA Pro Can automatically analyze executable files and disassemble them with the most appropriate configuration [42]. PEid is used to check whether obfuscation of malware. PeView gives complete headers information of Windows executable files like code,

[2] https://virusshare.com/.

[3] https://software.informer.com/software/.

[4] https://www.virustotal.com/gui/.

Table 1 Family Names ID

Family Name	ID
GoodWare	0
Critroni	1
Cryptlocker	2
CryptoWall	3
KOLLAH	4
Kovter	5
Locker	6
MATSNU	7
PGPCODER	8
Reveton	9
TeslaCrypt	10
Trojan-Ransom	11

Table 2 Set of Features ID

ID	Description
API	API invocations
DROP	Extensions of the dropped files
REG	Registry key operations
FILES	File operations
FILE_EXT	Extension of the files involved in file operations
DIR	File directory operations
STR	Embedded strings

data, and the file's metadata. PsFile tool displays information of opened files on the system, which helps detect whether the system is operated remotely. SSDeep calculates the fuzzy hash value of executables, which is used to verify variants of malware.

We have various tools like process Explorer, Wireshark, ProcMon, Sandboxes, Regshot, FakeDNS, and many more for Dynamic analysis [41]. Process Explorer is a task manager and system monitor tool that gives detailed information about the system's running processes. Wireshark is a network traffic analyzer, provides various features like data stream, port address, and URLs for further investigation. ProcMon captures activities done by running processes, i.e., calling every system calls of registry modification, memory and file system operations, and network activities. Regshot saves the state of the Windows registry before and after the malware sample's execution. FakeDNS is used to give the response to DNS queries and provides a virtual networking service to capture the actual behaviour of malware. We have various sandboxes to perform automatic analysis like Cuckoo, Panda, Limon, Parsa, etc. These sandboxes are integrated with various analysis tools to provide the facilities to extract different malware features.

4 ML-based Ransomware Detection Methods and Techniques

Combating with ransomware is not at all an easy task for several reasons. The operation of ransomware is similar to benign software and acts without anyone getting to know about them. Hence, detection of ransomware in zero-day attacks is much needed this time. The main goal is to prevent damage to the system by ransomware, detect zero-day (previously unseen) ransomware and be as unnoticeable as possible.

There are different methods and techniques to detect ransomware [19–21]. Static analysis- based methods disassemble source code without executing it. They suffer from high false positive rates and fail to detect obfuscated ransomware. New variants are coming up frequently and attackers are also using different packing techniques to mutate their codes. Researchers are moving towards dynamic behaviour analysis-based methods that monitor the interactions of the executed code in a virtual environment for fighting these issues. However, these detection methods are slower and require considerable memory resources. The machine learning approach is best suited for the analysis of the behaviour of any process or application. Some highly conventional methodologies of ML-based detection techniques are the following:

- Al-rimy et al. [22] proposed a method that recognizes a zero-day attack by implementing behavioural detection and anomaly detection. They used a group of SVM classifier for the behavioural model. The results of this classifier are integrated using the majority voting method. One class SVM is trained for the anomaly detection model and this model is trained on benign programs. The results of these two classifiers are associated with using OR logic.
- RWGuard [23], proposed by Mehnaz et al., detect file locker in real-time using three monitoring techniques—process monitoring, file change similarity, and decoy monitoring. They have chosen a threshold value for every change, and if changes crossed that threshold value sent for future assessments. Here changes belong to file changes like entropy, file size change, similarity, and file type change.
- Continella et al. [24] introduced ShieldFS detects at runtime and uses a supervised classifier trained with IRP (I/O Request Packet) data. They collected these data from both infected and clean machine. The classifier uses two models: System-centric: considering IRP information originating from a single process. Process-centric: consider each process individually and train on data coming from individual processes. The proposed solution uses an array of such classifiers, each one trained at a different level of file accessed percentage.
- Alhawi et al. [25] introduced NetConverse, a machine learning method based on a Decision Tree classifier, by observing ransomware network traffic. After observation, this algorithm kept track of the characteristic of each packet originated by the infected system.
- Harikrishnan et al. [26] proposed GURLS, a supervised learning method rooted in the Regularized Least Square Algorithm. They completed the training on the frequency of API calls and strings used as features—the best performance achieved

with the Radial basis function kernel. However, the Linear kernel results in false-positive, i.e., benign class misclassified as ransomware.

- To detect crypto-ransomware, a static approach to classify malicious and benign executables files proposed in [27]. They used the SMO [5] (Sequential Minimal Optimization) classifier of the WEKA tool and selected features as density histograms of opcodes extracted from those executables.
- DRTHIS [28] proposed by Homayoun et al., classify executables as benign and malicious and also map ransomware to its family. DL techniques like LSTM (Long Short-Term Memory) and CNN (Convolutional Neural Network), along with the SoftMax algorithm, are used for classification.
- DNAact-Ran [2] proposed by Khan et al., is a digital DNA sequencing engine for detecting and classification ransomware. DNAact-Ran selects key features using MOGWO (Multi-Objective Grey Wolf Optimizer) and BCS (Binary Cuckoo Search) algorithm and utilizes the k-mer frequency vector. They used DNA sequencing design constraints to generate DNA sequences for selected features. It characterizes ransomware based on their "digital genotypes and inferred digital phenotypes" to recognize their malicious function. This innovation uses a set of "genome rules" and utilized an active learning algorithm to categorize software programs and data as either goodware or ransomware.
- Sgandurra et al. [14] proposed a tool named EldeRan, an ML approach for dynamically analyzing ransomware. This tool monitors the running application's set of registry-related and system actions in a sandbox environment. It extracts features using a mutual information algorithm for further classification using a Regularized Logistic Regression classifier.
- Azmoodeh et al. [29] proposed a detection technique based on the concept of power consumption patterns. This detection technique applies to specifically android IoT devices. When $k = 1$, The KNN (K Nearest Neighbour) classifier gives maximum accuracy.
- Poudyal et al. [30] proposed a framework based on reverse engineering for detection. This framework does multi-level analysis like function calls, assembly codes, libraries, raw binaries, and interpreting the malware code segment's purpose. They formed a feature database by the output of reverse-engineered and pre-processed binary samples. Next, passed to the machine learning model for training and testing. They used eight machine learning classifiers in this research and their detection varied between 76% and 97%. The highest detection rate is 97.95%, achieved by the Random Forest.
- Almashhadani et al. [31] extracted features from network traffic in their detection method. They used TCP (Transmission Control Protocol), NBNS (NetBIOS Name Service), DNS (Domain Name System), HTTP (Hyper Text Transfer Protocol) traffic, and extracted 20 features for analysis, then proposed a detection module based on two different levels: flow level and packet level. They used several classification algorithms like Random Tree, LibSVM, Random Forest, Bayes

[5] An algorithm which solves quadratic programming problem arises in SVMs.

Net to build a packet-based and flow-based classifier. The algorithm which gives the highest accuracy is used for each classifier accordingly.

- Karimi et al. [32] focussed on detecting android mobiles and IoT (Internet Of Things). They detected ransomware by transforming a sequence of executable instructions into a grayscale image and then used LDA (Latent Dirichlet Allocation) for image reduction and classification methods.
- Shaukat and Ribeiro [33] proposed a layered defense system to protect against ransomware based on a hybrid method (the combination of dynamic and static approach). They used the Strong Trap layer for early detection, and Machine learning layer for zero-day intrusion and File Backup layer for protecting user data. They tested results using Random Forest, Support Vector Machines, Gradient Tree Boosting, and Artificial Neural Networks.
- Moussaileb et al. [34] detected ransomware behaviour by analyzing file system traversal action performed. As most ransomware starts encryption from the hard disk's root, they can detect it using decoy folders. They used various machine learning techniques for ransomware classification, such as KNN, decision tree, Random Forest.
- Alam et al. [35] proposed a two-step unsupervised detection tool RAPPER (Ransomware Prevention via PERformance counter). They observed hardware performance counters present in modern processors to recognize activities of encryption based on CPU events. RAPPER issues an alarm if observed any suspicious process activity. This two-step framework uses ANN and Fast Fourier Transformation for being the highest accurate, fast, and reliable. They didn't provide any ransomware examples in the training phase and their framework identifies anomaly that deviates from learned behaviour.
- Maiorca et al. [36] proposed R-PackDroid for the android operating system. They used system API packages as crucial information to discriminate between benign, ransomware, or generic malware by Random Forest classifier.
- Hasan and Rahman [37] proposed RansHunt, a detection framework based on Support Vector Machine (SVM). They extracted more than 60 thousand features from dynamic and static operations such as network actions, file system operations, API calls, process operations, registry operations. Chen et al. [38] observed a flow graph of API calls used by the suspected process and extracted 70,000 features for analysis. For dimensionality reduction, they used correlation techniques and compared results using SVM, Simple Logistic algorithm, Random Forest, and Naïve Bayes.
- Vinayakumar et al. [39] evaluate deep and shallow networks for ransomware classification and detection. After evaluation, they found that the deep network has performed well than the traditional shallow network. They proposed a Multi-Layer perceptron (MLP) architecture for detection using API calls.

5 Detailed Discussion

Malwares are analyzed based on dynamic and static features [40]. In static analysis, researchers analyze static features of executable files like memory compactness and different sections. For dynamic analysis, researchers used dynamic features like registry, DLLs, APIs, and various researchers applied machine learning to these dynamic feature combinations [40]. Detection tools based on signature matching are still widely used by many anti-virus software. If the new signature doesn't include in the database, then new variants can be undetected by those tools. Attackers are exploring new variants of ransomware frequently today. So, signature-based detection cannot detect unseen ransomware. Researchers evaluated each possible way to apply machine learning for ransomware detection. Researchers proposed two different ways to detect ransomware using machine learning: Host-based and Network-based.

5.1 Host-Based Detection

Researchers applied machine learning techniques to monitor process behaviour, file system activity, API calls, Registry Key Operations, energy consumption patterns, and many more for host-based detection methods.

Researchers focussed on process anomaly detection to enhance their detection rate. The technique discussed in the paper uses Windows API calls [22, 36, 39], I/O request Packets (IRP) logs, File system operations, set of operation performed per file extension, directories operations, dropped files, registry key operation, strings [14, 22–24, 26] for detection. For file system activity detection, researchers recorded folder listing, Files written, Files Renamed, files read, write entropy, file type coverage [24]. The various researchers recorded IRP open, IRP write, IRP create for IRP logs [23]. Chen et al. [38] generated API CFG (Call Flow graph) by API sequence extracted from the API monitor tool. They converted CFG, into a feature vector for further processing.

Baldwin et al. [27] focussed on the real CPU instruction (opcode), not the data that is to be processed (operand). For dataset creation, They used IDAPro with an instruction counter and extracted features like count, density, rank, opcode, and the total number of opcodes.

Houmayoun et al. [28] record all executed events within the initial ten seconds when users launch any application and transform the captures sequence to detection purposes. Azoomdeh et al. [29] proposed detection methods for IoT networks by observing the energy consumption pattern of different processes to detect ransomware. They used PowerTutor to observe and sample power consumption of all running processes in an interval of 500 ms.

Alam et al. [35] used Hardware Performance Counters (HPCs) to observe hardware-related activities to detect malicious programs dynamically. They model the normal system behaviour instead of ransomware behaviour. They used LSTM-based encoder to detect suspicious behaviour of the system.

Some researchers have proposed a hybrid method of dynamic and static analysis [33, 37]. Hasan and Rahaman [37] combined static features like Function length frequency (FLF), Printable String Information (PSI) and dynamic features like Process Operations, API function calls, File system operations, registry key operations, and network activities.

Khan et al. [2] proposed a unique solution for ransomware detection. They proposed digital DNA sequencing, i.e., a pseudo DNA illustration of a digital artefact denoted as a sequence of DNA characters (A, C, G and T) [2]. The sample binary features are encoded with corresponding biological DNA. They trained the generated training dataset using an active learning classifier.

5.2 *Network-Based Detection*

Attackers commonly prefer symmetric key encryption due to its speed and efficiency. However, some attackers used an asymmetric key or hybrid symmetric and asymmetric encryption methodologies. In all the cases, after installing the malicious program to an infected machine, it needs to connect with C&C (Control & Command) server either to send a symmetric key or to request a public key. Then only that malicious program can starts file encryption. It is also possible to detect ransomware at the network level during their connection and conversation with the C&C server.

Alhawi et al. [25] collected network traffic samples and compared benign application characteristics using several classification techniques. They considered network traffic features like destination and source IP address, protocol type, source and destination port number, the total number of bytes and packets per conversation, and many more [25].

Almashhadani et al. [31] has done a behaviour analysis of crypto-ransomware network activities. To set up the connection with the C&C server, generally, ransomware uses an encrypted list of hardcoded IP addresses to establish a TCP connection with a remote attacker network. Also, they can find an attacker server by running the DGA (Domain Generation Algorithm). Alternatively, they use NBNS (NetBIOS Name Service) protocol, which finds any previously infected system on the local network connected with the C&C server previously. Using machine learning, they leverage the statistics of NBNS, DNS, and HTTP protocol from collected PCAP files [31].

In the network-based analysis, there is a presumption that malware will contact their C&C server for purposes, such as to get an encryption key. But this method may fail if malware didn't contact to C&C server or used any local key. Also, malware may use proxy servers while contacting C&C servers, and then the network-based analysis may not be feasible to detect.

A comparison between all explained methods shown in Table 3 shows how successful methods can detect ransomware using machine learning.

Table 3 Performance comparison of various machine learning-based ransomware detection

Proposed solution	Concept	Base	TPR (%)	FPR (%)	Advantage	Limitation
Al-rimy et al. [22]	Behavioural detection as well as anomaly detection	Host-based	99	2.4	Detect the novel attack, maintaining a low false alarms rate	Reliance on the fixed time-based threshold to define the early phase of attacks
Mehnaz et al. [23]	Process monitoring, file change similarity, and decoy monitoring	Host-based	100	0.9	Detect file locker in real-time	Overhead and cannot recover files encrypted by a custom-written cryptographic library
Continella et al. [24]	Uses supervised classifier trained with IRP	Host-based	97.7	0.038	High accuracy even on unseen variants	Susceptibility to Targeted Evasion
Alhawi et al. [25]	Based on Decision Tree classifier	Network based	97.1	1–6	Good accuracy	No recovery of affected files
Harikrishnan et al. [26]	Based on Regularized Least Square Algorithm	Host-based	87.7	7.5	Best performance achieved with Radial basis function kernel	With Linear kernel resulted in false positive
Baldwin et al. [27]	Static analysis-based approach for executables files	Host-based	95	0–5	High precision (>95%) for the 2-class model	Focuses Only on Crypto-Ransomware
Homayoun et al. [28]	Uses Deep Learning Technique like LSTM and CNN	Host-based	97.2	2.7	Can detect unseen ransomware	Only for Locker Ransomware
Khan et al. [2]	Based on Digital DNA sequencing	Host-based	87.9		Early alerting system	Accuracy needs to be improved
Sgandurra et al. [14]	ML approach for dynamically analysis and classification	Host-based	93.3	1.6	Early detection of new variants	Not suitable for silent ransomware
Azmoodeh et al. [29]	Based on power consumption patterns	Host-based	95.65	–	Unique local fingerprint of ransomware's energy utilization	Only focus on energy consumption

(continued)

Table 3 (continued)

Proposed solution	Concept	Base	TPR (%)	FPR (%)	Advantage	Limitation
Poudyal et al. [30]	Combination of reverse engineering and ML	Host-based	97.95	2.5	High Accuracy with Random Forest	No information for unknown ransomware
Almashha-dani et al. [31]	Behavioural analysis of crypto-ransomware network activities	Network-based	97	2	Strong foundation for the evolution of next generation network-based malware detection system	No early detection
Karimi et al. [32]	Apply LDA to a grayscale image of transformed executables instruction	Host-based	97.5	-	Detect pre-seen samples	Considered only at locker and koler ransomware
Shaukat and Ribeiro [33]	Layered defense system based on a combination of the dynamic and static approach	Host-based	98.25	0.56	Early and zero-day detection	System overhead
Moussaileb et al. [34]	Analyzing file system traversal action performed	Host-based	99.35	<1	Highest accuracy	Raises false positives
Alam et al. [35]	Based on ANN and Fourier transformation	Host-based	100	Almost 0	Good accuracy	Cannot recover from infected files
Maiorca et al. [36]	Based on information Extracted from system API packages	Host-based	97	1	Detect novel ransomware by only using previously released training samples	Not tested against obfuscated application
Hasan and Rahman [37]	Combination of static and dynamic analysis	Host-based	97	3	Able to detect worm behaviour of ransomware	
Chen et al. [38]	Based on Dynamic API Calls Flow Graph	Host-based	97.6	1.2	High Accuracy	Implemented on small dataset
Vinayakumar et al. [39]	API calls passed to 3-layer MLP network	Host-based	98	<1	Tested against both crypto and locker ransomware	Failed to learn the complete patterns to differentiate samples of Cryptolocker and CryptoWall

6 Conclusion

There are several ways to detect ransomware, such as signature matching, hashing, entropy analysis, and many more. Nowadays, machine learning-based solutions are gaining more popularity for ransomware detection because of their efficiency and robustness. No method guarantees 100% foolproof and zero-day attack detection. Also, none of the proposed solutions are guaranteed to detect every family of ransomware. Most of them show success in detecting some types of ransomware and fail for other types of ransomware. A good backup strategy can save our data from being unavailable to us, such as using immutable backup or offline backups. The machine learning model's efficiency depends upon how many optimized features are chosen for training. Here, feature engineering plays a vital role in creating an efficient detection system. In ransomware programs, the uncommonly executed/used things can be used to detect anomalies like different string patterns in ransomware binaries, function calls related to cryptography algorithms/file access substantially modified/deleted data/files, accessing large volume of different directories, etc. Techniques such as PCA (Principal Component Analysis), MI (Mutual Information), and N-gram can be used to decide which features correctly differentiate between benign and ransomware behaviour. Further, multi-layered DL models having the inherent capability to handle large datasets with automatic feature engineering can overcome ML limitations. Semi-supervised solutions can be preferred to tackle the limited availability of labelled datasets in ransomware detection.

Acknowledgements The authors would like to thank National Critical Infrastructure Protection Centre, Mumbai for supporting this research work.

References

1. Sultan H, Khalique A, Alam SI, Tanweer S (2018) A survey on ransomware: evolution, growth and impact. Int J Adv Res Comput Scie 9(2)
2. Khan F, Ncube C, Ramasamy LK, Kadry S, Nam Y (2020) A digital DNA sequencing engine for ransomware detection using machine learning. IEEE Access 8:119710–119719. https://doi.org/10.1109/ACCESS.2020.3003785
3. Sahay SK et al (2020) A survey on ransomware detection techniques. In: SKM 2019, CCIS 1186. Springer Nature Singapore Pte Ltd, pp 55–68
4. Mohurle S, Patil M (2017) A brief study of WannaCry threat: ransomware attack 2017. Int J Adv Res Comput Sci 8(5):1938–1940
5. The current state of ransomware (2020) CryptoWall. https://news.sophos.com/en-us/2015/12/17/the-current-state-of-ransomware-cryptowall/. Accessed 27 Aug 2020
6. Look into locky ransomware (2020) https://blog.malwarebytes.com/threat-analysis/2016/03/look-into-locky. Accessed 02 Aug 2020
7. Evolution of GandCrab Ransomware (2020) https://www.acronis.com/en-in/articles/gandcrab/. Accessed 10 Sept 2020
8. CryptoLocker (2020) Everything You Need to Know. https://www.varonis.com/blog/cryptolocker/. Accessed 23 Sept 2020

9. TeslaCrypt ransomware (2020) https://www.kaspersky.co.in/resource-center/threats/teslacrypt. Accessed 20 Nov 2020
10. 2016 (2020) Year of the ransomware attacks. http://techgenix.com/2016-ransomware/. Accessed 16 Jul 2020
11. Mimoso M (2017) Leaked NSA exploit spreading ransomware world wide. https://threatpost.com/leaked-nsa-exploitspreading-ransomware-worldwide/125654. Accessed 03 Sep 2020
12. The Anatomy of a Ransomware Attack (2016) The threat research report. https://www.exabeam.com/wp-content/uploads/2017/07/Exabeam_Ransomware_Threat_Report_Final.pdf. Accessed 20 sept 2020
13. Proofpoint (2017) 2017 Q3 threat report. https://www.proofpoint.com/sites/default/files/pfpt-us-tr-q317-threat-report_1.pdf
14. Sgandurra D, Muñoz-González L, Mohsen R, Lupu EC (2016) Automated dynamic analysis of ransomware: Bene_ts, limitations, and use for detection. arXiv:1609.03020
15. Ransomware Dataset RISSP group (2020). http://rissgroup.org/ransomware-dataset/. Accessed 10 Nov 2020
16. Paquet-Clouston M, Haslhofer B, Dupont B (2019) Ransomware payments in the Bitcoin ecosystem. J Cybersec. 5(1):tyz003. https://doi.org/10.1093/cybsec/tyz003
17. Ransomware bitcoin datasets (2020). https://github.com/behas/ransomware-dataset. Accessed 01 Nov 2020
18. ISOT Ransomware dataset (2020) https://www.uvic.ca/engineering/ece/isot/datasets/botnet-ransomware/index.php. Accessed 25 Nov 2020
19. Scaife N, Carter H, Traynor P, Butler KR (2016) Cryptolock (and drop it): stopping ransomware attacks on user data. In: 2016 IEEE 36th international conference on distributed computing systems (ICDCS), pp 303–312
20. Kharaz A, Arshad S, Mulliner C, Robertson W, Kirda E (2016) UNVEIL: a largescale, automated approach to detecting ransomware. In: 25th USENIX security symposium (USENIX Security 2016), Austin, TX, pp 757–772. USENIX Association
21. Gomez-Hernandez JA, 'Alvarez-Gonzaalez L, Garcıa-Teodoro P, (2018) R-Locker: thwarting ransomware action through a honeyfile-based approach. Comput Sec 73:389–398
22. Al-rimy BAS, Maarof MA, Prasetyo YA, Shaid SZM (2018) Ariffin, AFM: Zero-day aware decision fusion-based model for crypto-ransomware early detection. Int J Integr Eng 10(6):82–88
23. Mehnaz S, Mudgerikar A, Bertino E (2018) RWGuard: a real-time detection system against cryptographic ransomware. In: Bailey M, Holz T, Stamatogiannakis M, Ioannidis S (eds) RAID 2018. LNCS, vol 11050. Springer, Cham, pp 114–136. https://doi.org/10.1007/978-3-030-00470-5_6
24. Continella A et al (2016) ShieldFS: a self-healing, ransomware-aware filesystem. In: Proceedings of the 32nd annual conference on computer security applications, ACSAC 2016. ACM, New York, pp 336–347
25. Alhawi OMK, Baldwin J, Dehghantanha A (2018) Leveraging machine learning techniques for windows ransomware network traffic detection. In: Dehghantanha A, Conti M, Dargahi T (eds) Cyber threat intelligence. AIS, vol 70. Springer, Cham, pp 93–106. https://doi.org/10.1007/978-3-319-73951-9_5
26. Harikrishnan NB, Soman KP (2018) Detecting ransomware using GURLS. In: 2018 second international conference on advances in electronics, computers and communications (ICAECC). IEEE
27. Baldwin J, Dehghantanha A (2018) Leveraging support vector machine for opcode density-based detection of crypto-ransomware. In: Dehghantanha A, Conti M, Dargahi T (eds) Cyber threat intelligence. AIS, vol 70. Springer, Cham, pp 107–136. https://doi.org/10.1007/978-3-319-73951-9_6
28. Homayoun S et al (2019) DRTHIS: deep ransomware threat hunting and intelligence system at the fog layer. Futur Gener Comput Syst 90:94–104
29. Azmoodeh A, Dehghantanha A, Conti M, Choo K-KR (2017) Detecting cryptoransomware in IoT networks based on energy consumption footprint. J Ambient Intell Hum Comput 9(4):1141–1152. https://doi.org/10.1007/s12652-017-0558-5

30. Poudyal S, Subedi KP, Dasgupta D (2018) A framework for analyzing ransomware using machine learning. In: 2018 IEEE symposium series on computational intelligence (SSCI). https://doi.org/10.1109/SSCI.2018.8628743
31. Almashhadani AO, Kaiiali M, Sezer S, O'Kane P (2019) A multiclassifier network-based crypto ransomware detection system: a case study of locky ransomware. In: IEEE access, vol 7 (2019), pp 47 053–47 067
32. Karimi A, Moattar MH (2017) Android ransomware detection using reduced opcode sequence and image similarity. In: 2017 7th international conference on computer and knowledge engineering (ICCKE). IEEE, pp 229–234
33. Shaukat SK, Ribeiro VJ (2018) Ransomwall: a layered defense system against cryptographic ransomware attacks using machine learning. In: 2018 10th international conference on communication systems & networks (COMSNETS). IEEE, pp 356–363
34. Moussaileb R, Bouget B, Palisse A, Le Bouder H, Cuppens N, Lanet J-L (2018) Ransomware's early mitigation mechanisms. In: Proceedings of the 13th international conference on availability, reliability and security. ACM, p. 2
35. Alam M, Bhattacharya S, Mukhopadhyay D, Chattopadhyay A (2018) RAPPER: ransomware prevention via performance counters CoRR, vol. abs/1802.03909. arXiv:1802.03909
36. Maiorca D, Mercaldo F, Giacinto G, Visaggio CA, Martinelli F (2017) R-PackDroid: API package-based characterization and detection of mobile ransomware. In: Proceedings of the symposium on applied computing. ACM, pp 1718–1723
37. Hasan MM, Rahman MM (2017) Ranshunt: a support vector machines-based ransomware analysis framework with integrated feature set. In: 2017 20th international conference of computer and information technology (ICCIT), pp 1–7
38. Chen Z-G, Kang H-S, Yin S-N, Kim S-R (2017) Automatic ransomware detection and analysis based on dynamic API calls flow graph. In: Proceedings of the international conference on research in adaptive and convergent systems. ACM, pp 196–201
39. Vigneswaran R, Vinayakumar R, Soman KP, Poornachandran P (2018) Evaluating shallow and deep neural networks for network intrusion detection systems in cyber security, pp 1–6. https://doi.org/10.1109/ICCCNT.2018.8494096
40. Ijaz M, Durad H, Ismail M (2019) Static and dynamic malware analysis using machine learning, pp 687–691. https://doi.org/10.1109/IBCAST.2019.8667136
41. Jagsir S, Singh J (2020) A, survey on machine learning-based malware detection in executable files. J Syst Arch 101861. ISSN 1383-7621. https://doi.org/10.1016/j.sysarc.2020.101861
42. Zhang D, Zhang Z, Jiang B, Tse TH (2018) The impact of lightweight disassembler on malware detection: an empirical study. In: (2018) IEEE 42nd annual computer software and applications conference (COMPSAC). Tokyo 2018, pp 620–629. https://doi.org/10.1109/COMPSAC.2018.00094

Variant of Niederreiter-Based Signature Scheme for Blockchain Technology

Feng-Wei Lee, Denis C. K. Wong, Bok-Min Goi, and Wai-Kong Lee

Abstract The well-known Niederreiter encryption scheme is an improved version of the classical McEliece encryption scheme in terms of key size and encryption efficiency, but the conversion of the scheme to a usable signature scheme is a very challenging process. Moreover, the current blockchain architecture still relies on hard problems that cannot resist quantum algorithms attacks. In this paper, we construct a Niederreiter-based signature scheme with a comprehensive security analysis that is suitable to be used together with the blockchain technology and resistance to Shor's algorithm for better underlying performance and security of the blockchain architecture.

Keywords Niederreiter scheme · Syndrome decoding · Post quantum

1 Introduction

Cryptography is a field of study that focused on securing communications between different parties by adapting various cryptographic primitives and is extremely important to ensure the security of data sent through a channel with the presence of an adversary. By constructing and analyzing various cryptographic primitives such as encryption scheme, signature scheme, and signcryption scheme, it helped to prevent the public to read the private messages that sent through the channel or to verify the authenticity of messages. Since the initiation of the field of quantum computers, many quantum algorithms such as Shor's algorithm have been formulated to tackle

Supported by Universiti Tunku Abdul Rahman Research Fund (UTARRF), project number: IPSR/RMC/UTARRF/2019-C2/D01, vote number: 6200/D13.

F.-W. Lee (✉) · D. C. K. Wong · B.-M. Goi
Lee Kong Chian Faculty of Engineering and Science, Universiti Tunku Abdul Rahman, Jalan Sungai Long, Bandar Sungai Long, 43000 Kajang, Selangor, Malaysia
e-mail: lihong652008@1utar.my

W.-K. Lee
Gachon University, Seongnam, South Korea

© The Author(s), under exclusive license to Springer Nature Singapore Pte Ltd. 2022
D. Giri et al. (eds.), *Proceedings of the Seventh International Conference on Mathematics and Computing*, Advances in Intelligent Systems and Computing 1412,
https://doi.org/10.1007/978-981-16-6890-6_14

the classical hard problems that are used for constructing most of the conventional cryptographic primitives, which aimed to secure information and communications in real life. Therefore, the construction of more quantum resistance cryptographic primitives are important to ensure the security level of communications remained even the post-quantum era begins. The current post-quantum cryptography can be divided into several approaches which includes hash-based cryptography, code-based cryptography, and lattice-based cryptography. In this paper, we concentrated on the study of code-based cryptography. One of the most famous code-based encryption scheme was the McEliece encryption scheme [1] that already exists for more than 40 years. It is one of the candidate for post-quantum encryption scheme as it has not yet been broken by any known algorithm and currently still show resistance to Shor's algorithm [2]. Since the proposal of the McEliece encryption scheme, there are many researchers that have been working on code-based cryptographic primitives by modifying and improving the security issues of McEliece scheme. In 1986, Niederreiter proposed an encryption scheme that was actually a variation of the McEliece encryption scheme [3]. The scheme proposed by Niederreiter uses the parity check matrix instead of a generator matrix that was originally used in the McEliece encryption scheme. Niederreiter encryption scheme is an improved version of McEliece encryption scheme with a smaller key size and an encryption process that is about ten times faster. From the encryption schemes proposed by McEliece and Niederreiter, many researchers start to construct signature schemes based on these encryption schemes. One of the well-known construction was the Courtois–Finiasz–Sendrier (CFS) signature scheme [4]. An improved version of CFS scheme, namely, modified CFS signature scheme (mCFS) was then proposed in [5]. Since Niederreiter encryption scheme is also a code-based encryption scheme and our proposed scheme is based on Niederreiter's work, thus from now onward, we called our constructed scheme as the Niederreiter-based signature scheme. During the announcement of post-quantum cryptography standardization project by NIST, several code-based cryptographic primitives was submitted. Some code-based signature schemes—pqsigRM [6], RaCoSS [7], and RankSign [8] was submitted during the first round of NIST standardization project. However, none of them moved on to the second round. Only seven code-based encryption schemes that are able to move forward to round two of the standardization project which including HQC [9] and ROLLO [10].

Cryptography is one of the main component that forms the architecture of blockchain technology. In short, blockchain is a chain of blocks that was secured with cryptography primitives which first described in 1991 and the first conceptualized blockchain was only proposed in 2008 by Satoshi Nakamoto [11]. Current blockchain technology rely on cryptographic primitives such as Elliptic Curve Digital Signature Algorithm (ECDSA), which are not able to resist attacks of quantum algorithms [12]. In this paper, we investigate how to apply our proposed Niederreiter-based signature scheme to the blockchain technology. This proposed scheme can be adapted in blockchain applications such as smart contract-based decentralized voting system with slight modifications.

Paper contributions. In this paper, we construct a Niederreiter-based signature scheme with detailed security analysis. The proposed signature scheme is applied to blockchain architecture with further modification to form a code-based signature scheme for a single blockchain transaction.

Paper organization. This paper consists of five sections. Section 2 is the preliminaries section and some related works are discussed. Section 3 is our proposed Niederreiter-based signature scheme with its security proof. Section 4 shows how to apply our proposed signature scheme to existing blockchain technology. Finally, we conclude in Sect. 5.

2 Preliminaries

In this section, we state some preliminary results which will be needed later. We start by giving a brief introduction to error correcting code which is the main ingredient of a code-based cryptographic primitive. Throughout this paper, all codes are linear codes which is defined as a subspace of a predetermined vector space over a finite field $\mathbf{F}_q$, where q is a prime power. Since all subspaces have bases, we can randomly construct one basis $\mathcal{B}$ for an arbitrary linear code C. By using $\mathcal{B}$, we can construct two important matrices which are used for the encoding and the decoding processes as discussed below.

Generator Matrix. A generator matrix, G_C of C is a $k \times n$ matrix where its rows form $\mathcal{B}$. The encoding process $Encode$ is defined by using G_C as follows: $Encode : \mathbf{F}_q^k \to \mathbf{F}_q^n, \quad Encode(m) = mG_C = v$, where m is called the message word and v is the corresponding codeword which will be transmitted through an insecure channel.

Parity Check Matrix. A parity check matrix H_C of C is a $n \times n - k$ matrix where its columns form a basis of the code $C^\perp$ which is known as the dual code of C. Thus, the syndrome decoding process $Decode$ of C can be defined by using H_C as follows: $Decode : \mathbf{F}_q^{n-k} \to \mathbf{F}_q^n, \quad Decode(wH_C) = \hat{v}$, where w is the received word, $\hat{v}$ is the estimated codeword and $wH_C = s$ usually is called the syndrome.

Since the basis of $C^\perp$ and the basis of C form the matrices H_C and G_C, respectively, we can obtain H_C from G_C or vice versa by following the well-known method known as the t-method.

t-Method

1. **Step 1:** Reduce G_C to its row echelon form G'.
2. **Step 2:** Permute the columns of G' by using a permutation p_f to obtain a $G'' = (I_k \ A)$ in standard form. Form $H' = \begin{pmatrix} -A \\ I_{n-k} \end{pmatrix}$.

3. **Step 3:** Permute the rows of H' by using permutation p_f to obtain H_C.

Note that the code C can be defined by using G_C as $\{v = mG_C \in \mathbf{F}_q^n \mid m \in \mathbf{F}_q^k\}$ with three important parameters n, k, and d, where n is the length of the code, k is the dimension of the code and the minimum distance $d = min\{d(x, y) \mid x, y \in C, x \neq y\}$. We said that C is a t-error correcting code if it can correct all errors of weight less than or equal to t and there exists at least one error of weight $t + 1$ that cannot be corrected. As our proposed scheme is the Niederreiter-based signature scheme and so the scheme is based on the hardness of solving the following syndrome decoding problem [3].

Syndrome Decoding Problem. Given a syndrome $s \in \mathbf{F}_q^{n-k}$ together with the parity check matrix H_C of a random linear code C, find $w \in \mathbf{F}_q^n$ such that $s = wH_C$.

The underlying security of the blockchain architecture make use of various cryptographic primitives, one of the most important is the signature scheme. The standard model of a signature scheme is given in the following definition.

Definition 1 A signature scheme is a 3-tuple $(Gen, Sign, Ver)$ defined as follows:

1. Gen is the key generation phase upon input a security parameter n, output a pair of public and private keys (pk, sk), that is, $Gen(n) \leftarrow (pk, sk)$.
2. $Sign$ is the signing phase upon input a message m, sign the message m with sk to obtain a signed message with signature σ and we write $Sign(m, sk) \leftarrow \sigma$.
3. Ver is the verification phase upon input a signed message (m, σ) and verify its validity using the public key pk. Output 1 if the signature is valid else return 0.

   Symbolically, $Ver(m, \sigma, pk) := \begin{cases} 1, & \text{valid,} \\ 0, & \text{invalid.} \end{cases}$

The current blockchain architecture that are implemented was highly depends on cryptographic primitives which are not resist the attack of quantum algorithms. Therefore, a blockchain with quantum resistance cryptographic primitives is much concerned. Blockchain technology that based on post-quantum signature scheme should satisfy several properties: consists of quantum resistance signature scheme and technology of blockchain, resist classical security attacks, resist quantum algorithm attacks, and preserves the linkability and traceability property of blockchain's signature scheme [13].

3 Niederreiter-Based Signature Scheme

We constructed a signature scheme that based on Niederreiter's encryption scheme. First, the private keys used in the decryption process of Niederreiter's encryption scheme which includes a decoding algorithm, permutation matrix, invertible matrix, and parity check matrix are used to form the signing process of the signature scheme with an addition of syndrome formation proposed and hash process. Public key used in Niederreiter's encryption is now used for signed message verification by computing and comparing the syndromes with the hash value. The details of the scheme is shown as follows:

Scheme 1 Variant of Niederreiter-based signature scheme

Gen

Consider a binary t-error correcting linear code C of length n and dimension k with an efficient decoding algorithm $Decode$ and a $k \times n$ generator matrix G of C. Compute a $n \times (n-k)$ parity check matrix H of C and construct matrices S, P, H', where S is an invertible $(n-k) \times (n-k)$ matrix, P is a $n \times n$ matrix and H' is formed by computing $H' = PHS$. Define the encoding function as $Encode : \mathbf{F}_q^k \rightarrow \mathbf{F}_q^n$, $Encode(m) = mG = v \in C$ and the decoding function as $Decode : \mathbf{F}_q^{n-k} \rightarrow \mathbf{F}_q^n$, $Decode(uH) = u$. Finally, consider a public cryptographic hash function $h : \{0, 1\}^* \rightarrow \mathbf{F}_q^n$. The private keys are $S, H, P, Decode$ and the public key is H'.

Sign

Given a message $x \in \{0, 1\}^*$, we first compute $x' = h(x)$. Then, find decodable $y \in \mathbf{F}_2^{n-k}$ such that $y = x'H$. Compute $z = Decode(yS)P$ and output the signed message (x, z).

Ver

Given a signed message (x, z). We first compute $x' = h(x)$, and then compute $w = x'H'$ and $w' = zH'$. The signed message is valid if $w = w'$ and $wt(z) \leq t$.

From the signing step, we see that $x \in 0, 1*$ and $z \in \mathbf{F}_q^n$, thus size of x is q^α and size of z is q^n and the size of signed message is $(q^{\alpha+n})$. To show the correctness of signature scheme constructed, given the private keys $S, H, P, Decode$, public key H', and public hash functions h, we see that

$$\begin{aligned} zH' &= Decode(yS)PH' \\ &= Decode(x'HS)PH' \\ &= Decode(x'P^{-1}H')PH' \\ &= x'P^{-1}PH' \\ &= x'H' \end{aligned} \tag{1}$$

Scheme 1 is based on syndrome decoding problem as its hard problem. Therefore, solving the syndrome decoding problem will then break the scheme easily. Syndrome decoding problem states that given a syndrome s and a parity check matrix H, find a w such that $s = wH$. From Scheme 1, since H' is publicly known, x' is message hashed using a public hash function and hence we can compute a w such that $w = x'H'$, and thus if anyone that can solve the syndrome decoding problem can easily break the scheme by finding a random $z \neq x'$ such that $w = zH'$ provided w and H' are known. Next, we proof the following theorem.

Theorem 1 *Scheme 1 is existential unforgeable under chosen message attack provided the syndrome decoding problem is hard to solve.*

Proof Suppose the message is $x \in \{0, 1\}^*$, we use the hash oracle to compute $x' = h(x) \in \mathbf{F}_q^n$. Since $\mathbf{F}_q^n = C \bigcup C^\perp$, where C is the linear code used in the proposed scheme and $C^{\perp}$ is its corresponding dual code of C. If $x' = h(x) \in C$, then by using the public parity check matrix H', we see that $x'H' = \mathbf{0}$. Next, we choose another codeword $x'' \neq x' \in C$, then we also have $x''H' = \mathbf{0}$ and hence $x'H' = x''H'$. On the other hand, if $x' = h(x) \in C^{\perp}$, then by using the t-method and taking transpose to H' to obtain G_C which is the generator matrix of C. Hence, $G_C x'^T = \mathbf{0}$. We find another codeword $x'' \neq x' \in C^{\perp}$, then $G_C x''^T = \mathbf{0}$ and so $G_C x'^T = G_C x''^T$. Therefore, we are done.

Finally, if $x' = h(x) \in C$, we need to find an arbitrary $x'' \in \mathbf{F}_q^n$ such that $x'H' = x''H'$. The proof proceed as follows:

Suppose there is an adversary $\mathcal{A}$ that can break Scheme 1, consider the experiment where a challenger $\mathcal{C}$ is given the system parameters and return the public key H' and the private keys (H, P, S). H' is given to $\mathcal{A}$ and (H, P, S) was kept secretly by $\mathcal{C}$. Next, $\mathcal{A}$ with messages, $x_1, \ x_2, \ \ldots \ , \ x_r$, queries the challenger for signature. The challenger that has access to the sign oracle returns the corresponding signature to A as follows:

$$Q = \{(x_1, z_1), \ (x_2, z_2), \ \ldots \ , \ (x_r, z_r)\} \ , \tag{2}$$

where Q is the set of pairs of message and signature. Since the hash oracle was publicly known, the adversary queries the hash oracle to obtain $X' = \{h(x_1) = x_1', \ h(x_2) = x_2', \ \ldots \ , \ h(x_r) = x_r'\}$. The adversary then uses his key H' to compute $Y' = \{y_1 = x_1'H', \ y_2 = x_2'H', \ \ldots \ , \ y_r = x_r'H'\}$. Next, the adversary randomly picks a P' and a S' and tries to decode to obtain

$$\begin{aligned} Z' = \{z_1 = Decode(y_1 S')P', \ z_2 = Decode(y_2 S')P', \\ \ldots , z_r = Decode(y_r S')P'\} \ . \end{aligned} \tag{3}$$

The adversary forms the following table to store all information he/she queried and computed:

Next, to forge a valid signed message, the adversary first computes $\sum_{i=1}^{r} y_i = (\sum_{i=1}^{r} x_i')H'$. Suppose $y^* = \sum_{i=1}^{r} x_i'$. Then, as the syndrome decoding process is a linear transformation, the adversary computes

$$\begin{aligned} z^* &= Decode(y^* H'S')P' \\ &= Decode((\textstyle\sum_{i=1}^{r} x_i')H'S')P' \\ &= \textstyle\sum_{i=1}^{r} Decode(x_i'H'S')P' \\ &= \textstyle\sum_{i=1}^{r} z_i \ . \end{aligned} \tag{4}$$

Thus, randomly picks a x^* and queries the hash oracle for a successive number of time until he/she obtains $h(x^*) = \sum_{i=1}^{r} x_i' = y^*$. Together with $z^* = Decode(y^* H'S')P'$, the adversary successfully forges a signed message (x^*, z^*). If (x^*, z^*) is in the list in Table 1, the adversary repeats the process with less number of y_i. If (x^*, z^*) is not in the table, computes $h(x^*) = x'$, then computes $x'H'$, and z^*H'. If $x'H' = z^*H'$, then the adversary successfully forges a signed message. It turns out that solving $x'H' = z^*H'$ is equivalent to solve the syndrome decoding problem.

Table 1 Informations obtained by adversary

Q	X'	Y'	Z'
(x_1, z_1)	$h(x_1) = x'_1$	$y_1 = x'_1 H'$	$z_1 = Decode(y_1 S')P'$
(x_2, z_2)	$h(x_2) = x'_2$	$y_2 = x'_2 H'$	$z_2 = Decode(y_2 S')P'$
.	.	.	.
.	.	.	.
(x_r, z_r)	$h(x_r) = x'_r$	$y_r = x'_r H'$	$z_r = Decode(y_r S')P'$

Furthermore, there are two additional cases to be considered: Case 1—the hashed messages returned, $x'_i \in X'$ are not corresponded to the messages, x_i and Case 2—both hashed messages, $x'_i \in X'$ and signatures, $z_i \in Z'$ returned are not corresponded to the messages, x_i and syndromes computed using the hashed messages, $y_i \in Y'$. We perform a more thorough analysis as follows:

Case 1. If the hash oracle returned a set of hashed messages, X' of messages in X, where

$$X' = \{x'_1,\ x'_2,\ \dots\ ,\ x'_r\}\ , \tag{5}$$

but x'_i is not correspond to x_i, for $x'_i \in X'$, $x_i \in Q$, $i = 1, 2, \dots, r$, then the adversary uses his key H' to compute $Y' = \{y_i = x'_i H' \mid i = 1, 2,\ \dots\ , r\}$. Next, the adversary randomly picks a P' and S' and tries to decode to get

$$Z' = \{z'_i = Decode(y_i S')P' \mid i = 1, 2,\ \dots\ , r\}\ , \tag{6}$$

where $\exists z_j \in Q$ and $z'_i \in Z'$ such that $z'_i = z_j\ ,\ for\ i, j \in \{1, 2, \dots, r\}$ and i may not be equal to j.

If $z'_i \in Z'$ are unique, then forms the set $V = \{(x'_i, z'_i, x_j, z_j) | i, j \in \{1, 2, \dots, r\}\}$, where x'_i that correspond to z'_i is the hashed message of $x_j \in Q$ such that $z'_i = z_j$, where $z_j \in Q$. By using the set of unique signatures $z'_i \in Z'$ and their corresponding hashed messages $x'_i \in X'$ and the corresponding $y_i \in Y'$, the signed message can be forged by using the method stated before.

For X' and Z' as defined in (5) and (6), if $z'_i \in Z'$ are not unique, which means one $z'_i \in Z'$ correspond to at least two $x'_i \in X'$. Adversary performs the following:

a. For each z'_i that correspond to l messages, constructs a set $L_i = \{x'_k | k = 1, 2, 3, \dots, l\}$ where all x'_k are hashed messages that correspond to z'_i.
b. For each z'_i with its corresponding L_i, randomly picks a $x'_k \in L_i$ to form a tuple of hashed message with corresponding signature (x'_k, z'_i).
c. Gathers all tuples formed to get a set of unique hashed messages with their corresponding signatures.

d. Repeats using different combinations of hashed messages in different L_i.
e. Output: Multiple sets of unique hashed messages with their corresponding signatures. Lastly, adversary collects all these sets as multiset T.

Next, adversary constructs multiple sets of messages using all combinations of messages $x_1, x_2, \ldots, x_r$ and gathers all these sets as R. Adversary then queries the hash oracle to get their corresponding sets of hashed messages which collected as R'. Compares each set of hashed messages in T with sets of hashed messages in R', adversary obtains the corresponding set of messages in R for every set in T.

Thus, the adversary proceeds to forge the new signed message using the sets in T with their corresponding set of messages in R by following same argument as shown in the beginning of the proof.

Case 2. If the hash oracle returned a set of hashed messages, X' of messages in X, where $X' = \{x'_1, x'_2, \ldots, x'_r\}$, but x'_i is not correspond to x_i, $\forall x'_i \in X', x_i \in Q, i = 1, 2, \ldots, r$, the adversary uses his key H' to compute $Y' = \{y_i = x'_i H' \mid i = 1, 2, \ldots, r\}$. Then, by using Y', the adversary queries and gets a set of signature, Z' where

$$Z' = \{z'_1, z'_2, \ldots, z'_r\} , \tag{7}$$

but z'_i is not correspond to y_i, $\forall z'_i \in Z', y_i \in Y', i = 1, 2, \ldots, r$.

For Q and Z' as defined in (2) and (7), by comparing with the set of messages and signatures in Q using those unique signatures in Z', adversary performs the following:

a. For each unique signature $z'_i \in Z'$, find its corresponding messages $x_i \in Q$. Forms a tuple of message and signature (x_i, z'_i) where the signature is unique. Gather all tuples as set ζ.
b. By using messages in ζ, queries the hash oracle to get set of corresponding unique hash messages θ.
c. Output: ζ and θ.

Adversary forges the signed message using method shown in the beginning of the proof by using ζ and θ.

If signatures are not unique, adversary performs the following:

a. For each non unique signature z'_i, forms a tuple (x_i, z'_i) by randomly picks one of its corresponding message $x_i \in Q$.
b. Gathers all tuples to form a set of unique messages with their corresponding signatures as β_i.
c. Repeats with different combinations of message for different signature. Collects all β_i as a multiset β.
d. For each $\beta_i \in \beta$, queries the hash oracle for the corresponding set of hashed messages.
e. Output: Sets of unique messages with signatures and their corresponding hashed messages.

The adversary then forges the signed message using the method shown in the beginning of the proof with these set of messages and unique signatures. □

Let ϵ' be the probability and t_s' be the time takes to solve the syndrome decoding problem, we assume that the syndrome decoding problem is (t_s', ϵ')-secure. Let ϵ be the probability and t_s be the time takes to break the scheme, q_{sig} be the number of signing queries, q_h be the number of hash queries and $T(n, k)$ be the time to construct an error pattern.

Theorem 2 *Scheme 1 is $(t_s, \epsilon, q_{sig}, q_h)$-secure, where*

$$\epsilon \leq \frac{2^n \epsilon' q_h}{2^n + q_{sig} q_h((2^n)^{-1} \sum_{k=1}^{t} \binom{n}{t} - 1) - \sum_{k=1}^{t} \binom{n}{t}}$$

and

$$t_s = t_s' - q_{sig}(q_h + q_{sig})\mathcal{O}(2^n) - q_{sig} T_s(n, k).$$

Proof Assume an adversary which $(t, \epsilon, q_{sig}, q_h)$—breaks Scheme 1 that is existential unforgeable under chosen message attack, we construct a challenger that (t_s', ϵ')-solves the syndrome decoding problem. From the proof of Theorem 1, we see that in order to forge a signed message, we need to find a decodable syndrome $y \in \mathbf{F}_2^{n-k}$ in order to form a signature $z \in \mathbf{F}_2^n$ such that $wt(z) \leq t$. Thus, the probability $Pr(error)$ for adversary to form z successfully in $\mathbf{F}_2^n$ under the scheme is

$$Pr(error) \geq \left(1 - \frac{q_{sig} q_h}{2^n}\right)\left(1 - \frac{\sum_{k=1}^{t} \binom{n}{t}}{2^n}\right)\epsilon. \quad (8)$$

Since Scheme 1 is existential unforgeable under chosen message attack, then the message queried must not be requested for a signature before. Then the probability ϵ' of solving the syndrome decoding problem by using Scheme 1 subroutine is

$$\epsilon' \geq \frac{1}{q_h}(1 - \frac{q_{sig} q_h}{2^n})(1 - \frac{\sum_{k=1}^{t} \binom{n}{t}}{2^n})\epsilon. \quad (9)$$

Thus, by expanding Eq. (9), the probability ϵ of breaking Scheme 1 is

$$\epsilon \leq \frac{2^n \epsilon' q_h}{2^n + q_{sig} q_h ((2^n)^{-1} \sum_{k=1}^{t} \binom{n}{t} - 1) - \sum_{k=1}^{t} \binom{n}{t}}. \tag{10}$$

Since the adversary can queries the hash oracle for q_h times and sign oracle for q_{sig} times, time to solve the decoding instance is t_s', then the time t of breaking the scheme is

$$t_s = t_s' - q_{sig}(q_h + q_{sig})\mathcal{O}(2^n) - q_{sig} T_s(n, k). \tag{11}$$

4 Code-Based Signature Scheme on Blockchain

Cryptography is one of the main component that build up the blockchain architecture. However, the current blockchain technology still relies on cryptographic primitives that are not quantum resistance such as the Elliptic Curve Digital Signature Algorithm (ECDSA). To align with the development of physical quantum computers with ability to run quantum algorithms, research and development of post-quantum cryptographic primitives for blockchain technology is important. In this section, we applied our proposed quantum resistance signature scheme to blockchain technology with further modification. As mentioned before, for the technology of blockchain, pairs of public and private keys are used to make transaction. In a public blockchain, to store these keys that are needed for transaction inside a blockchain, a blockchain wallet is required. The blockchain wallet stored the user's private key. Before performing a transaction, a public key is generated by using the private key with cryptographic primitives and several hashing is done to generate an address from the corresponding public key. The address then uniquely used for each transaction. To apply our signature scheme on blockchain, we first need a user identity, ID. Before performing transaction, we need to input a security parameter to obtain the corresponding public parameters, p and a master key, ms that stored in a wallet for public blockchain. By using the user's identity, master key and public parameters, we generate pairs of user's public and private keys for transaction by using our proposed signature scheme. The private key aimed to sign a transaction, while public key used to verify the corresponding transaction. An address is formed by using the public key with several hashing. The details are shown in the following scheme:

Scheme 2 Proposed signature scheme for Blockchain

Setup

Input a randomly generated security parameter n. Select a hash function $h_0 : \{0, 1\}^* \rightarrow \mathbf{F}_q^n$, that maps binary string of arbitrary length to a n—dimensional vector space over $\mathbf{F}_q$. Select another hash function $h_1 : \{0, 1\}^* \rightarrow \{0, 1\}^*$, that maps a binary string of arbitrary length to binary string of arbitrary length.

Using n, randomly form a $k \times n$ generator matrix which is the master key, ms where ms is a generator matrix of a t-error correcting (n, k) -binary linear code C_1. The public parameters are h_0, h_1, n, k and q.

Identity Extraction (IDExt)

Given the user identity, ID, compute $h_0(ID) = d$. By using the master key, ms which is a generator matrix of C_1, randomly choose $k - 1$ codewords in C_1 together with d to form another $k \times n$ generator matrix, G which is a generator matrix of another t-error correcting (n, k)-binary linear code C, where d is one of the row of G. The $k - 1$ codewords in C_1 and d must be a linearly independent set and hence span C as well. Compute the corresponding $n \times (n - k)$ parity check matrix H of G by using t-method. G act as the master private key of the user.

Key Generation (Gen)

By using the public parameter n, k, and q, construct matrices S, P, H' where S is an invertible $(n - k) \times (n - k)$ matrix, P is a $n \times n$ permutation matrix and $H' = PHS$. Define the encoding function, $Encode : \mathbf{F}_q^k \rightarrow \mathbf{F}_q^n$, where $Encode(m) = mG = v \in C$ and the decoding function, $Decode : \mathbf{F}_q^{n-k} \rightarrow \mathbf{F}_q^n$, $Decode(uH) = u \in \mathbf{F}_q^n$.

Private Keys: $S, H, P, Decode$. **Public Key:** H'.

Message Construction (MsgCon)

To sign a transaction in a blockchain, first take the previous transaction details, $prevtx$ and public key of receiver for the next transaction pk_b, form the message by computing $h_1(prevtx||pk_b) = x$.

Signing (Sign)

Given a message $x \in \{0, 1\}^*$, compute $x' = h_0(x)$. Find decodable y, where y is a vector of length $n - k$ such that $y = x'H$, then compute $z = Decode(yS)P$, where z is an error vector. **Output:** (x, z).

Verification

Given a signed message (x, z), compute $x' = h_0(x)$, then compute $w = x'H'$ and $w' = zH'$. Signature is valid if $w = w'$ and $wt(z) \leq t$.

Since the signature scheme that applied on blockchain was modified from the signature scheme that we constructed in the previous section, the security model and correctness of the blockchain's signature scheme followed the same as previous security model with some additional steps. The adversary able to forge the signature for the transaction if given pairs of message with the corresponding valid signature. In the blockchain, the public key was publicly known, while the private keys were generated from the security parameters, user's identity and randomly generated master key which kept secretly. Therefore, if an adversary wants to forge a signature, he needs to guess the master key which generate the private key for signing. The adversary queries to get the signature using the identity of his choice and tries to guess the master key to form private key in order to forge the signature. After this additional step, the security model remained the same as previous signature scheme. When the signature is valid and the new transaction satisfy unspent transaction output

(UTXO), this transaction will be included in a new block of the blockchain. Since d is the hashed identity that obtained from user's identity, then every d that we formed must be unique. Therefore, to ensure the uniqueness of d, we first need to ensure that the identity of the user is unique. In other word, restriction of d can be done through restriction of identity that the user provided. To ensure the uniqueness of the identity of the user, the identity of the user can either be formed from: (i) combination of user's email, username and password, (ii) unique identity number provided to the user upon joining the blockchain, and (iii) combination of user's password or username together with unique token created for user upon joining the blockchain while joining through external link.

The hashed identity, d can only be recovered by using the public key H' together with the public parameters that is known. To recover d, we use H' and the public parameters to construct the following algorithm $dRec$:

a. Input: H' and public parameters.
b. Randomly construct a permutation matrix P of size n and an invertible matrix S of size $n - k$.
c. Compute $H = P^{-1}H'S^{-1}$ and $G = (t - method(H^T))^T$.
d. Output: Rows of G.

The hashed identity, d is one of the row of G. Since the way of forming S and P is $q^{(n-k)^2}$ and q^{n^2}, respectively, thus the possibility of hitting both successful guess

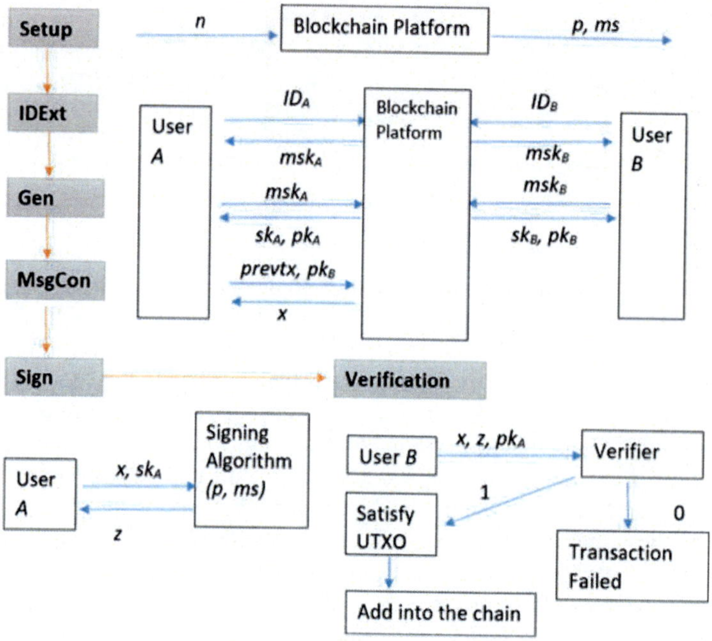

Fig. 1 Transaction Between Parties in Blockchain Architecture

for S and P is $q^{2n^2-2nk+k^2}$ to recover d. The hash function is a one way function, thus it is infeasible to recover the user's identity. The figure (Fig. 1) simplified the connections and communications between parties using the proposed scheme inside a blockchain architecture.

5 Conclusions

In conclusion, since digital signature schemes has been used as the underlying security architecture for blockchain technology, the security level of signature scheme used and its future development potential is much concerned. Therefore, to align with the research and development of quantum algorithms, we constructed a quantum resistance Niederreiter-based signature scheme with security model. The signature scheme is secure, provided the syndrome decoding problem is hard to solve. Besides, we applied the signature scheme we constructed on blockchain by adding an identity extraction phase for further generation of unique private and public keys for users in the chain to improve the performance of current blockchain technology that relies on cryptographic primitives that are not quantum resistance.

References

1. McEliece RJ (1978) A public-key cryptosystem based on algebraic coding theory. DSN Progress Report, Jet Propulsion Laboratory, California Institute of Technology, Pasadena, CA
2. Shor PW (1994) Algorithms for quantum computation: discrete logarithms and factoring. In: Proceedings 35th annual symposium on foundations of computer science. IEEE Computer Society Press. https://doi.org/10.1109/SFCS.1994.365700
3. Niederreiter H (1986) Knapsack-type crytosystems and algebraic coding theory. Probl Contr Inf Theory 15(2):157–166
4. Courtois N, Finiasz M, Sendrier N (2001) How to achieve a McEliece-based digital signature scheme. In: Boyd C (ed) Advances in cryptology - ASIACRYPT 2001, (ASIACRYPT 2001), vol 2248. Lecture Notes in Computer Science. Springer, Berlin
5. Dallot L (2007) Towards a concrete security proof of courtois, finiasz and sendrier signature scheme in research in cryptology, Second Western European Workshop, WEWoRC 2007. Bochum, Germany, Revised Selected Papers, pp 65–77
6. Lee W, Kim Y-S, Lee Y-W, No J-S (2017) Post quantum signature scheme based on modified Reed-Muller code : pqsigRM, NIST Post-quantum cryptography standardization project
7. Roy PS, Xu R, Fukushima K, Kiyomoto S, Morozov K, Takagi T (2017) RaCoSS. NIST post-quantum cryptography standardization project, Random code-based signature scheme
8. Nicolas A, Philippe G, Hauteville A, Ruatta O, Zémor G (2017) RankSign - a signature proposal for the NIST's call, NIST post-quantum cryptography standardization project
9. Melchor CA, Aragon N, Bettaieb S, Bidoux L, Blazy O, Deneuville J-C, Gaborit P, Persichetti E, Zémor G, HQC: hamming quasi-cyclic, NIST post-quantum cryptography standardization project
10. Melchor CA, Aragon N, Bardet M, Bettaieb S, Bidoux L, Blaze O, Deneuville J-C, Gaborit P, Hauteville A, Otmani A, Ruatta O, Tillich JP, Zémor G (2017) ROLLO - Rank-ouroboros. LAKE and LOCKER, NIST post-quantum cryptography standardization project

11. Nakamoto S (2019) Bitcoin: a peer-to-peer electronic cash system. Cryptography Mailing list. https://metzdowd.com. Accessed 25 May 2019
12. Johnson D, Menezes A, Vanstone SA (2001) The elliptic curve digital signature scheme. Int J Inf Secur. https://doi.org/10.1007/s102070100002
13. Yulong G, Xiubo C, Ying S, Xinxin N, Yixian Y (2018) A secure cryptocurrency scheme based on post-quantum blockchain. IEEE Access 1–1. https://doi.org/10.1109/ACCESS.2018.2827203

Chaos-Based Uncompressed Frame Level Video Encryption

Jagannath Sethi, Jaydeb Bhaumik, and Ananda S. Chowdhury

Abstract In recent years, video encryption based on chaotic maps have emerged as an important research problem. Properties like high randomness, ergodicity, sensitivity to controlling parameters, and initial states make chaotic map-based video encryption more popular than the conventional encryption techniques. In this paper, we propose a compression-independent method for video encryption at frame level where a new confusion technique, based on sort index permutation, is introduced. During shuffling, a new sequence is generated for every video frame under consideration. Generated pseudorandom chaotic sequence depends on initial values, controlling parameters, and the video frames. We have used a logistic-sine-coupling map (LSCM) for generating highly pseudorandom sequence. Block-level diffusion has been employed for scrambling the frames. The proposed framework is more robust against known and chosen-plaintext attacks as we have incorporated the size and sum of all pixel values for generating random chaotic sequence. Experimental results clearly demonstrate that our method is effective in terms of security, efficiency, and robustness compared to some of the existing approaches.

Keywords Chaotic map · Video encryption · Sort index permutation · Block-level diffusion

J. Sethi (✉) · J. Bhaumik · A. S. Chowdhury
Department of Electronics and Telecommunication Engineering, Jadavpur University, Kolkata 700032, India
e-mail: jaydeb.bhaumik@jadavpuruniversity.in

A. S. Chowdhury
e-mail: as.chowdhury@jadavpuruniversity.in

© The Author(s), under exclusive license to Springer Nature Singapore Pte Ltd. 2022

D. Giri et al. (eds.), *Proceedings of the Seventh International Conference on Mathematics and Computing*, Advances in Intelligent Systems and Computing 1412,
https://doi.org/10.1007/978-981-16-6890-6_15

1 Introduction

Chaos-based multimedia security is an important research topic for protecting images and videos from unauthorized access and violation of security. Protection of data during the transmission and storage requires confidentiality and integrity. Several applications of video in the field of medical, video conferencing, military imaging system need robust security for storing and transmitting the video data. Existing encryption methods like advanced encryption standard (AES), data encryption standard (DES), and international data encryption algorithm (IDEA) are not suitable for encrypting digital video in real-time applications [1]. Chaotic systems are mostly used for video encryption because of properties like complex chaotic behavior, sensitivity to initial values and controlling parameters, ergodicity and unpredictability. Our compression-independent video encryption method can be used for military and defence applications where important video is stored for a long time and attacker has enough time for brute-force attack. This is because our proposed method is more secure as it has larger key space.

Since a joint compression and encryption algorithm has to support various issues like format compliance, codec compliance, and so on, it is not suitable for a wide variety of applications [2, 3]. One solution for this is to use compression-independent video encryption. Several analyses indicate that for permutation use of only cipher is not a secure strategy as this is vulnerable to known and chosen-plaintext attacks [4]. So, both confusion and diffusion operations are needed to achieve higher degree of security. In the proposed scheme, we have used confusion and block-level diffusion for encrypting the uncompressed video.

The proposed video encryption framework consists of three main steps. In the first step, chaotic sequence are generated by logistic-sine-coupling map (LSCM) [5]. In the second step, individual frames are shuffled by sort index-based permutation. A new chaotic sequence is generated using previous shuffled frame. In the third and final step, we perform block-level diffusion of shuffled frames to get the encrypted video frames. The main contributions of our proposed compression-independent video encryption algorithm are as follows:

(1) We introduce a sort index-based permutation for the confusion step. Every frame is shuffled by a separate chaotic sequence. Chaotic sequences are updated by the previous shuffled frame and in particular, for shuffling each frame, its previous shuffled frame and the already generated chaotic sequence are used.

(2) We develop a block-level diffusion in which an unique sequence is generated for each block. Previous ciphered block is used for updating the initial values and controlling parameter while generating the sequence for each block.

The rest of this paper is organized as follows. In Sect. 2, we discuss the related works. In Sect. 3, we describe our proposed framework, pseudorandom sequence generation, confusion, and diffusion process. In Sect. 4, we present our experimental results with detailed analysis. Finally, the paper is concluded in Sect. 5 with an outline for directions of future research.

2 Related Work

Video encryption methods can be classified into two approaches, (a) compression-dependent approach (b) compression-independent approach. Compression-dependent encryption does not yield a better trade-off between efficiency and security [6]. For highly sensitive video applications compression-independent methods are preferred for encrypting the video. Many compression-independent video encryption methods for uncompressed video have been discussed in [6–9]. Zhang et al. [10] presented a layered cellular automata-based surveillance videos encryption strategy where only regions of interest are encrypted. Further, there are some methods [11–13] which performed selective video encryption for increasing the speed of the encryption process. However, selective encryption methods cannot provide high security. In [14], the authors presented scan-based permutation and substitution method for encrypting the video. In [15], a probabilistic technique was proposed for encrypting the keyframes by using logistic adjusted sine map [16]. This method is energy friendly which supports IoT systems. The authors in [5] proposed a new chaotic map to achieve complex chaotic behavior. In their method they first coupled Sine and Logistic maps and applied a sine function to the coupled result. In our work we use [5] for generating the chaotic sequence. Two methods are proposed by Valli et al. in their paper [6]. The first work is based on a 12-dimensional chaotic map and the second is based on the Ikeda delay differential equation. The authors claim that their frame-by-frame video encryption scheme supports real-time application. Socek et al. in [3] proposed correlation preserving video encryption by using permutation. A compression-independent chaotic video encryption scheme (CVES) was proposed by Li et al. [17] which depends on multiple chaotic systems which encrypt the video frame by frame. In [7], the authors presented a puzzle-based compression-independent video encryption algorithm which supports real-time application. Kezia et al. [8] encrypt video data frame by frame using high dimensional Lorenz chaotic system. But only changing the positions is not sufficient to provide higher security. Using Chebyshev maps a public key encryption (PKE) of images and videos was proposed by Ganesan et al. [18]. In [19], authors presented a three-level security for encrypting video using combined Logistic and Tent chaotic maps. In their scheme, they have used block-level confusion and diffusion. Two compression-independent methods have been proposed in [20] for encrypting video using four chaotic maps. Any change in an original video frame can only affect the corresponding encrypted frame as most of the above-mentioned methods used frame-by-frame encryption. Also any highly secure video encryption method based on higher dimensional chaotic map needs many parameters in generation of chaotic sequence. But the needs of many parameters correspond to larger key sizes that make key management more difficult. So our aim is to propose a robust as well as efficient encryption method for uncompressed video communication. In our proposed method, an inter frame dependent shuffling technique is used. So any change in one original frame can affect all the subsequent encrypted frames in the video which makes our method more robust. Our method is secure and it is better because a lesser number of secret parameters have to be exchanged securely between sender and receiver before sending the intended video confidentially.

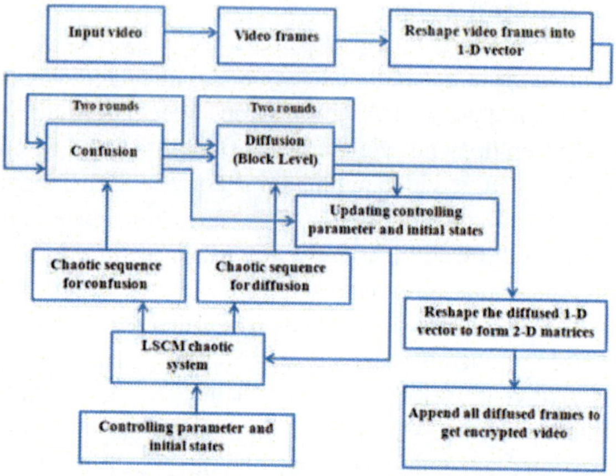

Fig. 1 Flow chart of the proposed video encryption technique

3 Proposed Framework

This section narrates our proposed method in detail. The section contains three parts. In the first part, we discuss the generation of pseudorandom chaotic sequence. We then present our sort index-based confusion method where we show how to shuffle the pixels positions in different frames. Finally, we discuss the technique of block-level diffusion of confused frames. The proposed framework is shown in Fig. 1. The encryption method is completed by two rounds of confusion followed by two rounds of block-level diffusion.

3.1 Pseudorandom Sequence Generation

A 2D logistic-sine-coupling map (LSCM) [5] is used for generating sequence. We select this map as it is very sensitive to initial states, controlling parameter and it holds complex chaotic behavior with high efficiency. It has more chaotic and hyperchaotic range which produces more random sequences. To make a relation between generated sequence and original frame, we have computed the sum of all pixel values of the original frame. We have done the randomness test for generated chaotic sequence using Diehard statistical suite [21] and its result has shown in Table 1. For randomness test we have considered of $x = 0.5684$, $y = 0.4278$, $\theta = 0.9121$. In our test, a sequence of 16777216 bytes are generated by using Algorithm-1 and then analyzed by the Diehard statistical test. The mathematical equation of the LSCM is as follows:

$$\begin{aligned} x_{i+1} &= \sin(\pi(4\theta x_i(1 - x_i) + (1 - \theta)\sin(\pi y_i))) \\ y_{i+1} &= \sin(\pi(4\theta y_i(1 - y_i) + (1 - \theta)\sin(\pi x_i))), \end{aligned} \tag{1}$$

where θ is the controlling parameter and $(x, y, \theta) \in [0, 1]$.

Table 1 Randomness test using diehard suite

Statistical test	P value	Result
Birthday spacings	0.135756	Pass
Overlapping 5-permutation	0.589286	Pass
Binary rank for 31 × 31 matrices	0.613440	Pass
Binary rank for 32 × 32 matrices	0.378504	Pass
Binary rank for 6 × 8 matrices	0.080413	Pass
Bitstream	0.46525	Pass
Overlapping-Pairs-Sparse-Occupancy	0.582900	Pass
Overlapping-Quadruples-Sparse-Occupancy	0.518500	Pass
DNA	0.456200	Pass
Count-the-1's on a stream of bytes	0.435546	Pass
Count-the-1's for specific bytes	0.123514	Pass
Parking lot	0.581103	Pass
Minimum distance	0.350740	Pass
3D spheres	0.278296	Pass
Squeeze	0.214191	Pass
Overlapping sums	0.798832	Pass
Runs	0.907553	Pass
Craps	0.536751	Pass

Algorithm 1 Generation of pseudorandom chaotic sequence

Input: (F, x_0, y_0, θ), where F is input video frame of size [1,3MN]
Output: (Chaoticseq)
Size of each frame $\leftarrow [M, N, 3]$
Length of Chaotic sequence $\leftarrow [1, 3MN]$
$S = \sum_{i=1}^{3MN} F(i)$
if $S = 0$
$t \leftarrow 0$
else
$t = \frac{\sqrt{S}+5}{(S+255)}$
end
$x_1 = x_0 + t, y_1 = y_0 + t, \theta = \theta + t;$
for i = 1 to 3MN **do**
$x_{i+1} = \sin(\pi(4\theta x_i(1 - x_i) + (1 - \theta)\sin(\pi y_i)))$
$y_{i+1} = \sin(\pi(4\theta y_i(1 - y_i) + (1 - \theta)\sin(\pi x_i)))$
$Chaoticseq(i) = floor(10^{10} \times (x_{i+1} + y_{i+1}))$ mod 256
end for

3.2 Sort Index-Based Shuffling

In our proposed method, we have selected the initial states and controlling parameters $(x_0, y_0, \theta_0, m_1, n_1, \theta_1)$ as secret key for encrypting the video. At first each color video frame of size [M, N, 3] is arranged into an one-dimensional array [1, 3 MN]. Secondly, a chaotic sequence is generated using Algorithm-1. The sequence is represented by D1 as follows:

$$D1 = Chaoticseq(\mathbf{0}, x_0, y_0, \theta_0), \tag{2}$$

where $\mathbf{0}$ is a null vector of length same as an individual video frame. First frame is shuffled by the sort index of chaotic sequence D1. A new chaotic sequence is generated using Algorithm-1 which is denoted by D2 as follows:

$$D2 = Chaoticseq(\text{Previous confused frame}, x_0, y_0, \theta_0). \tag{3}$$

Then D2 is modified to get another sequence denoted by D3 as follows:

$$D3 = ((\text{Previous confused frame} + \text{D2}) \mod 256). \tag{4}$$

Finally, the remaining frames are shuffled by the sort index of D3 using Eqs. 3 and 4. Second round of confusion for all frames are completed by Eqs. 2, 3 and 4 taking secret parameter as (m_1, n_1, θ_1).

3.3 Block-Level Diffusion

Let CF, C be the shuffled frame array and scrambled frame array after confusion and diffusion process, respectively. Block-level diffusion is used for scrambling the frames. Each shuffled frame array is divided into P numbers of blocks having block length (L). The value of L is such that total numbers of pixels in a frame (3 MN) are divisible by L. A new chaotic sequence is generated for each block using Algorithm-1 by replacing the length of sequence from 3MN to L. The block-level diffusion process for each video frame arrays in first round are illustrated as

$$C_{i(Block)} = \begin{cases} CF_{i(Block)} \oplus CF_{(LastBlock)} \oplus S_1, & \text{for 1st Block} \\ C_{i-1(Block)} \oplus CF_{i(Block)} \oplus S_2, & \text{for other Blocks,} \end{cases} \tag{5}$$

where S_1 and S_2 are generated by Eqs. 6 and 7, respectively, using Algorithm-1 and are given by

$$S_1 = Chaoticseq(\mathbf{0}, x_0, y_0, \theta_0) \tag{6}$$

$$S_2 = Chaoticseq(\text{Previous diffused block}, x_0, y_0, \theta_0). \tag{7}$$

Second round of diffusion for all frame arrays are completed by Eqs. 5–7 considering secret parameter as (m_1, n_1, θ_1). Finally, each diffused array is reshaped into 2D RGB matrices and all diffused 2D frames are appended to form an encrypted video. Two rounds of inverse block diffusion followed by two rounds of inverse confusion operations successively decrypt an encrypted video.

Algorithm 2 Encryption of video in frame level

Input: $(F, P, x_0, y_0, \theta_0, m_1, n_1, \theta_1)$, where F is input video frame of size [1,3MN] and P is the number of blocks in a frame whose length is L.

Output: (Encrypted video)

1: **for** i=1 to Last frame **do**
2: if $i = 1$
3: $D1 = Chaoticseq(0, x_0, y_0, \theta_0)$
4: $D1^{'} = sort(D1)$
5: $CF_{(i,:)} = F_{(i,:)}(D1')$
6: else
7: $D2 = Chaoticseq(CF_{(i-1,:)}, x_0, y_0, \theta_0)$
8: $D3 = (CF_{(i-1,:)} + D2) \mod 256$
9: $D3^{'} = sort(D3)$
10: $CF_{(i,:)} = F_{(i,:)}(D3')$.
11: end if
12: **end for**
13: For second round of shuffling is completed by repeating step.1 to step.12 taking secret parameter as (m_1, n_1, θ_1)
14: **for** i=1 to Last Frame **do**
15: **for** j=1 to Last block **do**
16: if $j = 1$
17: $S1 = Chaoticseq(0, x_0, y_0, \theta_0)$
18: $C_{(i,(j-1)*L+1:j*L)} = CF_{(i,(P-1)*L+1:P*L)} \oplus S1 \oplus CF_{(i,(j-1)*L+1:j*L)}$
19: else
20: $S2 = Chaoticseq(C_{(i,(j-2)*L+1:(j-1)*L}, x_0, y_0, \theta_0)$
21: $C_{(i,(j-1)*L+1:j*L)} = C_{(i,(j-2)*L+1:(j-1)*L)} \oplus S2 \oplus CF_{(i,(j-1)*L+1:j*L)}$.
22: end if
23: **end for**
24: **end for**

3.4 *Time Complexity*

Let each frame consists of r numbers of pixels for each color component in a video, where r $= MN$ and M, N are the number of rows and columns in a frame, respectively. Assume that video consists of α number of frames. The proposed encryption scheme consists of three steps. Firstly separate chaotic sequence is generated using LSCM map for different frames. So time complexity of this step is $O(r)$. Secondly each frame is shuffled by sort index-based permutation. So this step is $O(r^2)$. Thirdly, each block in shuffled frame is scrambled by bitwise xor with chaotic sequence and previous scrambled block. Let P be number of blocks found when dividing r by length of each block (L). A unique chaotic sequence is generated for each block using previous scrambled block with $O(L)$ time. Time complexity for scrambling, P number of blocks is of $O(PL)$. So, the overall complexity of each encrypted frame is $(O(r) + O(r^2) + O(PL)) \approx O(r^2)$ (as $P, L << r^2$). For α number of frames in a video the overall complexity of our algorithm is $O(\alpha r^2)$.

4 Experimental Results

We have applied our proposed method to a set of standard videos as used in [6] which include Rhinos.avi ($240 \times 320 \times 3$), Train.avi ($192 \times 352 \times 3$), Viptrain.avi ($240 \times 360 \times 3$), and Flamingo.avi ($192 \times 352 \times 3$). All experiments are carried out on a desktop PC with Intel(R) Core (TM) i5-9330H CPU having 2.40 GHz clock speed and 8 GB of memory (RAM). For implementation, we have used MATLAB 2020a. We have experimentally set $x_0 = 0.5684$, $y_0 = 0.4278$, $\theta_0 = 0.9121$, $m_1 = 0.3255$, $n_1 = 0.5258$, $\theta_1 = 0.9091$ as secret key and block length as 256 for block-level diffusion. In this section, we compare our experimental results with three existing compression-independent video encryption methods [6, 19].

4.1 Correlation Analysis

Correlation coefficient between two pixels p and q, denoted by $CC_{p,q}$ is given by

$$CC_{pq} = \frac{cov(p, q)}{\sqrt{D(p) \times D(q)}}, \tag{8}$$

where $cov(p, q)$ is the covariance between p and q, $E(p)$, $E(q)$ and $D(p)$, $D(q)$ are the expectation and variance of variable p and q respectively. The value of CC_{pq} is close to 1 for original video frames and is near to 0 for encrypted frames. In Table 2, we have compared average correlation coefficient of the videos before and after encryption of two adjacent pixels along horizontal, vertical, and diagonal directions with other existing methods. In our method the encrypted frames have correlation coefficients nearly equal to zero. We have randomly selected 3000 pixels in input video frames and their respective adjacent pixels in the horizontal, vertical, and diagonal directions for finding correlation coefficient.

4.2 Entropy Analysis

Entropy is a parameter to measure the randomness. Higher entropy indicates higher uncertainty. Entropy of encrypted frames is nearer to 8 which indicates that the

Table 2 Mean correlation coefficient

Methods	Video files	CC before encryption			CC after encryption		
		Horizontal	Vertical	Diagonal	Horizontal	Vertical	Diagonal
12D Map [6]	Rhinos	0.9866	0.9765	0.9673	0.0196	0.0192	0.0138
	Train	0.937	0.9088	0.8735	0.0159	0.0195	0.0191
	Flamingo	0.9753	0.9697	0.948	0.0155	0.0146	0.0171
	Viptrain	0.9517	0.9497	0.9029	0.0203	0.0175	0.0224
Ikeda DDE [6]	Rhinos	0.9866	0.9765	0.9673	0.0181	0.0140	0.0107
	Train	0.937	0.9088	0.8735	0.0154	0.0105	0.0121
	Flamingo	0.9753	0.9697	0.948	0.0150	0.014	0.0148
	Viptrain	0.9517	0.9497	0.9029	0.0176	0.0147	0.0158
[19]	Rhinos	0.9866	0.9765	0.9673	0.0181	0.0140	0.0107
	Train	0.937	0.9088	0.8735	0.0154	0.0105	0.0121
	Flamingo	0.9753	0.9697	0.948	0.0150	0.014	0.0148
	Viptrain	0.9517	0.9497	0.9029	0.0176	0.0147	0.0158
8D Map [6]	Rhinos	0.9866	0.9765	0.9673	0.0254	0.0184	0.0135
	Train	0.937	0.9088	0.8735	0.0186	0.0179	0.0155
	Flamingo	0.9753	0.9697	0.948	0.0227	0.015	0.0159
	Viptrain	0.9517	0.9497	0.9029	0.0192	0.0133	0.0165
Our proposed method	Rhinos	0.9839	0.9741	0.9735	0.0040	0.0080	0.0127
	Train	0.9400	0.9287	0.8822	0.0063	0.0154	0.0126
	Flamingo	0.9783	0.9767	0.9568	0.0074	0.0061	0.0094
	Viptrain	0.9505	0.9566	0.9197	0.0020	0.0048	0.0114

encrypted frames are highly random. We have also tested our method for all pixels of values 0 and 255 where it gives complete random like encrypted frames. The entropy of a video frame can be found as

$$H(k) = -\sum_{i=1}^{255} P(k_i)\log_2 P(k_i), \tag{9}$$

where H(k) is the entropy, $P(k_i)$ indicates the probability of symbol k_i. We have shown the entropy values of original video frame (4th) and encrypted video frame (4th) in Table 3.

Table 3 Entropy test

Method	Video files	Entropy of original frame	Entropy of encrypted frame
Our proposed method	Rhinos	7.0769	7.9992
	Train	7.2074	7.9998
	Flamingo	6.3881	7.9994
	Viptrain	7.4370	7.9993
	Frame having all value 0	0	7.9993
	Frame having all value 255	0	7.9992

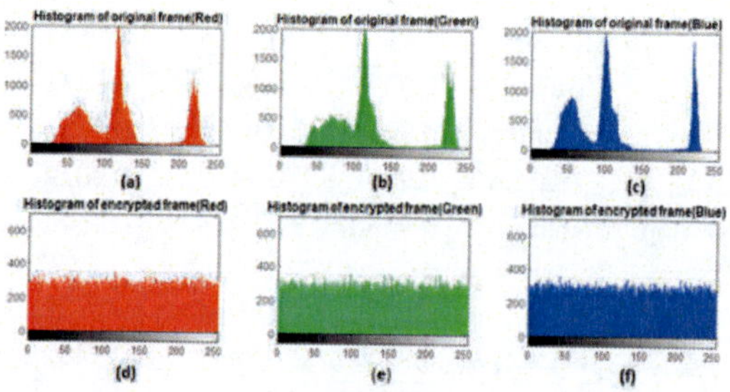

Fig. 2 Histogram of Red, Green, Blue components: **a–c** Original video frame; **d–f** Encrypted frame

4.3 Histogram Analysis

Histogram of original video frame and its encrypted version with individual color (R, G, B) components for the 4th frame of the Rhinos video has shown in Fig. 2. The histograms of the original video frame and its encrypted frame are different. The histogram of latter frame is very uniform. So our proposed encryption methods can defend the statistical attacks.

4.4 NPCR and UACI Analysis

The number of changing pixel rate (NPCR) and the unified averaged changed intensity (UACI) values have to be greater than ideal values of 99.5693 and 33.4635, respectively, to defend differential attack. The expressions for NPCR and UACI are as follows:

$$NPCR(C_1, C_2) = \sum_{i,j} \frac{W(i, j)}{M \times N} \times 100\% \tag{10}$$

$$UACI(C_1, C_2) = \sum_{i,j} \frac{|C_1(i, j) - C_2(i, j)|}{255 \times M \times N} \times 100\%, \tag{11}$$

Table 4 Comparison of NPCR and UACI values of encrypted video frames

Methods	Rhinos video		Train video		Flamingo video		Viptrain video	
	NPCR	UACI	NPCR	UACI	NPCR	UACI	NPCR	UACI
12D Map [6]	99.44	33.485	99.36	33.499	99.37	33.4985	99.34	33.4596
Ikeda DDE [6]	99.95	33.5914	99.94	33.5065	99.94	33.5886	99.95	33.599
[19]	99.51	33.54	99.61	33.50	99.52	33.52	99.58	33.62
8D Map [6]	99.32	33.44	99.36	33.46	99.36	33.4077	99.28	33.4427
Our proposed method	99.64	33.5021	99.62	33.4918	99.62	33.5088	99.63	33.4611

Table 5 Comparison of mean encryption time in seconds

Methods	Video files			
	Rhinos	Train	Flamingo	Viptrain
12D Map [6]	1.7964	1.6493	0.8402	2.3137
Ikeda [6]	1.2147	0.9908	1.1124	1.3574
[19]	0.5403	0.5178	0.5982	0.4723
8D Map [6]	0.9124	0.8631	0.8062	1.0588
Our proposed method	0.7946	0.8548	0.8538	0.9890

where C_1, C_2 are two encrypted frames generated from same video frame and secret key but changing a single pixel value. $M \times N$ is the size of video frames and $W(i, j) = 1$ if $C_1 \neq C_2$ and is 0 otherwise. We have used [22] for testing purpose to show that our scheme can resist differential attack. Table 4 shows the results for NPCR and UACI. We have tested for all cases like changing one bit pixel value in original video frame, changing initial conditions and controlling parameters. Our method gives NPCR and UACI values above the threshold value for all the cases thereby yielding high level of security against differential attacks.

4.5 Mean Encryption Time

Table 5 shows a comparison of mean encryption time of our method with some other existing methods [6, 19]. Our method takes lesser time for encryption compared to [6]. When block length increases our method takes lesser time for encryption. But there is a trade-off between block length and mean encryption time.

Table 6 Comparison of key space

Methods	Keyspace
12D Map [6]	3072×2^{128}
Ikeda DDE [6]	49152×2^{128}
[8]	2^{128}
[19]	2^{212}
8D Map [6]	8×2^{128}
Our proposed method	256×2^{256}

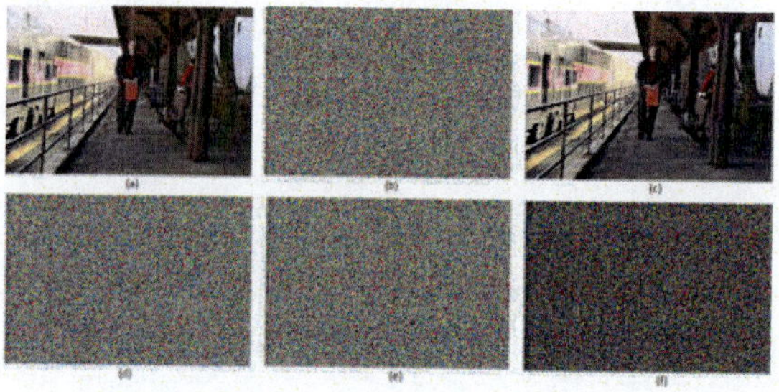

Fig. 3 **a** Original frame No.4 of Viptrain video **b** encrypted frame with key 0.5684, 0.4278, 0.9121, 0.3255, 0.5258, 0.9091 **c** decrypted frame with key 0.5684, 0.4278, 0.9121, 0.3255, 0.5258, 0.9091 **d** decrypted frame with key $0.5684 + 10^{-15}$, 0.4278, 0.9121, 0.3255, 0.5258, 0.9091 **e** decrypted frame with key 0.5684, $0.4278 + 10^{-15}$, 0.9121, 0.3255, 0.5258, 0.9091 **f** difference frame between **d** and **e**

4.6 Analysis of Secret Key

Generally 2^{128} is the standard key space required to resist the exhaustive attack in any encryption method. Our proposed security algorithm consists of six key parameters $x_0, y_0, \theta_0, m_1, n_1, \theta_1$. Precision of 10^{-15} is considered for each parameter. So the secret key space is $10^{15*6} = 10^{90}$. In our case key space is $10^{90} \approx 256 \times 2^{256}$. So it provides sufficient resistance against brute-force attack. In Table 6 we have compared key space of our method with some recent methods. Any minor change in key, decryption process cannot produce original video frames. So decryption process of proposed methods is robust against differential attack. In Figs. 3 and 4, it has been shown that how any minor change in secret key can make the decryption impossible for our proposed method. Decryption will be possible only when the sender's secret key matches with receiver's secret key.

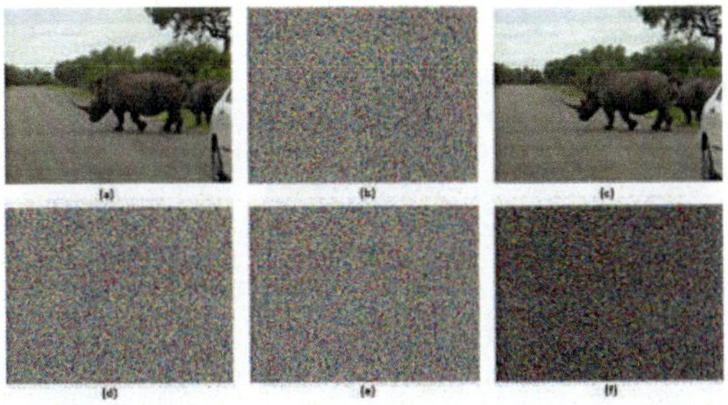

Fig. 4 **a** Original frame No.4 of Rhinos video **b** encrypted frame with key 0.5684, 0.4278, 0.9121, 0.3255, 0.5258, 0.9091 **c** decrypted frame with key 0.5684, 0.4278, 0.9121, 0.3255, 0.5258, 0.9091 **d** decrypted frame with key $0.5684 + 10^{-15}$, 0.4278, 0.9121, 0.3255, 0.5258, 0.9091 **e** decrypted frame with key 0.5684, $0.4278 + 10^{-15}$, 0.9121, 0.3255, 0.5258, 0.9091 **f** difference frame between **d** and **e**

4.7 Robustness Against Noise

The salt and pepper noise with density 0.05, Gaussian noise with 0 mean and 0.03 variance and speckle noise with density 0.05 are applied to encrypted video frames. As the video after decryption with exact key is noisy, the receiver can detect that someone has tampered the information in transmission. The peak signal-to-noise ratio (PSNR) for encrypted video has to be less for better security. Mean PSNR for first fifteen encrypted frames is shown in Fig. 5. The PSNR value for decrypted video should be high so that it can resist different channel noise. We have shown decrypted video frame after addition of different noises in encrypted video frames in Fig. 6. The PSNR is computed as

$$PSNR_{dB} = 10 \times \log_{10} \frac{255^2}{MSE}, \tag{12}$$

where MSE is the mean squared error given as

$$MSE = \sum_{i=0}^{M-1} \sum_{j=0}^{N-1} \frac{|I(i,j) - C(i,j)|^2}{M \times N}, \tag{13}$$

where I is original frame and C is encrypted frame.

4.8 Security Analysis

To make an encryption method computationally secure it has to satisfy two conditions. Firstly, to break the cipher its cost has to be more than the real value of encrypted data. Secondly, time to break the encryption method should be more than

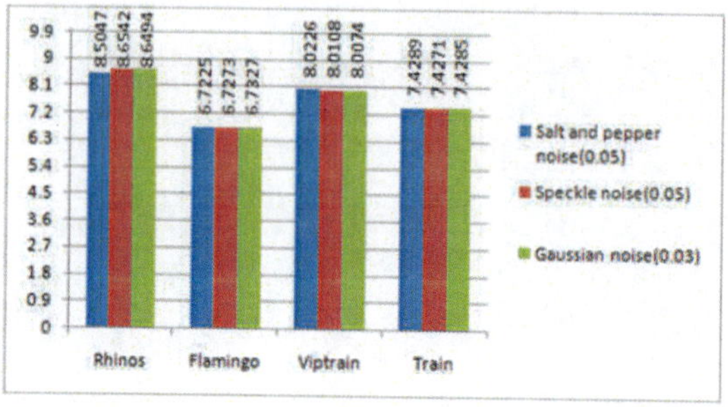

Fig. 5 Mean PSNR of encrypted video

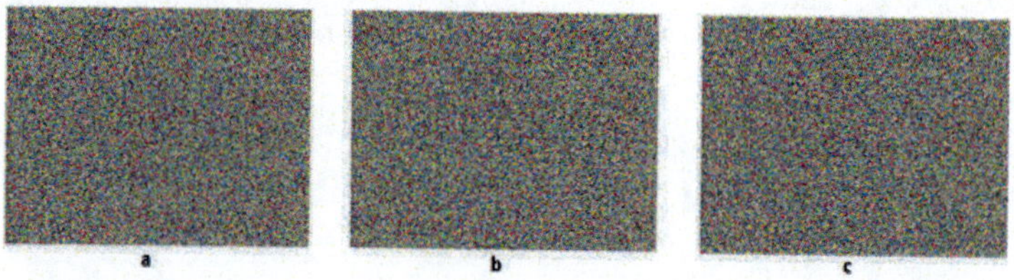

Fig. 6 Decrypted video frame under: **a** salt and pepper noise **b** speckle noise **c** Gaussian noise

functional lifetime of data. Our proposed method satisfies both conditions. The security analysis has been done by considering known and chosen-plaintext attacks.

In case of known-plaintext attack, the attacker has access to plaintext, ciphertext, and encryption algorithm. The adversary has to construct different security keys for different frames. Security key for each frame needs the summation of all the pixels present in previous shuffled frame. Keyspace for our proposed method is 10^{90}. So it needs 10^{90} combination of secure key to break the encryption method. The security key is changed for every iteration, so it is very difficult to extract the security key. So our proposed method is robust against known-plaintext attack.

In case of chosen-plaintext attack, it is assumed that the adversary knows one pair of original and encrypted frame as (I1, C1). Also attacker can choose another frame I2 of same size as original which differs from I1 by 1 in the first row only. Let T1 and T2 are the chaotic sequences for shuffling I1 and I2. T1 and T2 are different as the summation of pixels in I1 and I2 are different so as their sort indices. Hence I1 and I2 are shuffled differently. During scrambling a unique sequence is generated for each block which is dependent on the summation of previous scrambled block. So the final scrambled frame for I1 and I2 are different. We considered previous shuffled frame in chaotic sequence generation and shuffling the current frame. Similarly we considered previous scrambled block for chaotic sequence generation and scrambling current block which gives higher security and resistance to chosen-plaintext attack.

Table 7 Performance comparison of different methods

Reference	Process used	Type of encryption	Security	Application
12D [6]	Confusion, Diffusion	Complete	Moderate	Wireless communication
Ikeda DDE [6]	Confusion, Diffusion	Complete	High	Video storage and communication
[8]	Confusion, Diffusion	Complete	Moderate	Video communication
[18]	Random Modulation	Complete	Low	Wireless communication
[13]	Confusion, Diffusion	Complete	High	Video storage
Proposed method	Confusion, Diffusion	Complete	High	Video storage and communication

4.9 External Comparisons

Our proposed scheme encrypts the complete video instead of partial or selective encryption. Performance of different video encryption schemes is summarized and compared in Table 7. Experimental results are compared with three methods given in [3, 6]. Our results show better performance in terms of average correlation coefficient after encryption, key space, mean encryption time, entropy, and PSNR of encrypted video. So proposed method is efficient and can be used in real-time video security applications. The NPCR and UACI values of our method are higher as compared to 12D map [3, 6]. In their first method (12D chaotic map) [6], twelve initial conditions were used. So it is difficult to communicate all the initial values over the channel if sufficient bandwidth is not available. But in our case we have used six initial states which require lesser bandwidth as compared to 12D chaotic map method. Histogram of encrypted video frames looks more uniform. By changing a pixel value in any position of any frame by 1 bit our propose method affects several pixels in all frames. So it provides higher NPCR and UACI. Higher security is needed when a video is stored for long time as there is sufficient time for an attacker. Our proposed method is highly sensitive to any minor change in original video frames, initial states, and controlling parameter. So our method is robust, efficient and can provide high security. Our proposed method gives higher entropy which indicates that encrypted video frames are unpredictable.

5 Conclusion

In this paper, we proposed a compression-independent video encryption method at frame level. For each frame, a new chaotic sequence is generated by employing previous frame and already generated chaotic sequence. Secure sorting permutation method is used for shuffling the video frames. Similarly, an unique chaotic sequence is generated for every block in the frame and block-level diffusion method is used for scrambling the video frames. Experimental comparisons using various measures clearly establish the superiority of the proposed method over some of the existing approaches. We have also analyzed the security of our proposed method with regard to brute-force attack, sensitivity to key for decryption, PSNR of encrypted video, entropy analysis, differential attacks, known and chosen-plaintext attack analysis. The comprehensive analysis clearly establishes that our proposed video encryption method is highly secure. In future, we will be exploring compression-dependent video encryption methods.

References

1. Lian S, Sun J, Wang Z, Dai Y (2004) A fast video encryption scheme based-on chaos. In: ICARCV, vol 1, pp 126–131
2. Liu F, Koenig H (2010) A survey of video encryption algorithms. Comput Secur 29(1):3–15
3. Socek D, Magliveras S, Ćulibrk D (2007) Digital video encryption algorithms based on correlation-preserving permutations. EURASIP J Info Secur 052965. https://doi.org/10.1155/2007/52965
4. Li S, Li C, Chen BN, Lo K-T (2008) A general quantitative cryptanalysis of permutation-only multimedia ciphers against plaintext attacks. Signal Process Image Commun 23:212–223
5. Hua Z, Jin F, Xu B, Huang H (2018) 2D logistic-sine-coupling map for image encryption. Signal Process 149:148–161
6. Valli D, Ganesan K (2017) Chaos based video encryption using maps and Ikeda time delay system. Eur Phys J Plus 132:542. https://doi.org/10.1140/epjp/i2017-11819-7
7. Liu F, Koenig H (2005) Puzzle - A Novel Video Encryption Algorithm. In: Dittmann J, Katzenbeisser S, Uhl A (eds) CMS 2005, vol 3677. LNCS. Springer, Berlin, Heidelberg, pp 88–97
8. Kezia H, Sudha GF (2008) Encryption of digital video based on Lorenz chaotic system. In: Proceedings of the 16th international conference on advanced computing and communications, pp 40–45
9. Hole RN, Kolhekar M (2018) Robust encryption of uncompressed videos with a selective frame scheme. In: 3rd international conference for convergence in technology (I2CT), Pune, pp 1–7
10. Zhang X, Seo S, Wang C (2018) A lightweight encryption method for privacy protection in surveillance videos. IEEE Access 6:18074–18087. https://doi.org/10.1109/ACCESS.2018.2820724, https://doi.org/10.1155/2009/429581
11. Cheng H, Li X (2000) Partial encryption of compressed images and videos. IEEE Trans Signal Process 48(8):2439–2451
12. Zeng W, Lei S (2003) Efficient frequency domain selective scrambling of digital video. IEEE Trans Multimed 5(1):118–129
13. Raju CN, Umadevi G, Srinathan K, Jawahar CV (2008) Fast and secure real-time video encryption. In: Sixth indian conference on computer vision, graphics and image processing, Bhubaneswar, pp 257–264. https://doi.org/10.1109/ICVGIP.2008.100
14. Maniccam SS, Bourbakis NG (2004) Image and video encryption using SCAN patterns. Pattern Recognit 37(4):725–737

15. Muhammad K, Hamza R, Ahmad J, Lloret J, Wang H, Baik SW (2018) Secure surveillance framework for IoT systems using probabilistic image encryption. IEEE Trans Ind Inform 14(8):3679–3689
16. Hua Z, Zhou Y (2016) Image encryption using 2d logistic-adjusted-sine map. Inf Sci 339:237–253
17. Li SJ, Zheng X, Mou X, Cai Y (2002) Chaotic encryption scheme for real-time digital video. In: Proceedings of SPIE real-time imaging VI, San Jose, CA, USA, vol 4666, pp 149–160
18. Ganesan K, Singh I, Narain M (2008) Public key encryption of images and videos in real time using Chebyshev maps. In: Fifth international conference on computer graphics, imaging and visualisation. CGIV
19. Ranjithkumar R, Ganeshkumar D, Suresh A, Manigandan K (2019) A new one round video encryption scheme based on 1D chaotic maps. In: 5th international conference on advanced computing and communication systems (ICACCS), Coimbatore, India, pp 439–444
20. Elkamchouchi H, Salama WM, Abouelseoud Y (2020) New video encryption schemes based on chaotic maps. IET Image Process 14(2):397–406
21. Marsaglia G, Diehard statistical tests. http://stat.fsu.edu/pub/diehard/
22. Wu Y, Noonan JP, Agaian S (2011) NPCR and UACI randomness tests for image encryption. Cyber J Multidiscip J Sci Technol, J Sel Areas Telecommun (JSAT) 2:31–38

Comparison of Algorithms for Style Transfer of Images Using Texture Synthesis

Vishakh Rao, K. U. Hemanth, and Bhawana Rudra

Abstract Style transfer is a process of transferring a style from one image to an other image, obtaining a new image which is an artistic mixture of the two. Recent work on this problem involved adopting convolutional neural-networks (CNN). This ignited a renewed interest in this field, due to the very impressive results obtained. Gaty's approach utilizes this method. There exists an alternative path toward handling the style transfer task via generalization of texture synthesis algorithms. Kwatra's approach utilizes this method. We shall implement style transfer using Kwatra's and Gaty's approach and compare their performance. Based on the results obtained, after using 2 metrics, we shall find out which works the best.

Keywords Style transfer · Texture synthesis · Convolutional Neural Network (CNN) · Segmentation

1 Introduction

Style transfer is a method of moving a style from one image (Style-Image) to another (Content-Image). The goal is to synthesize a new image that reflects a creative mixture of content and style. To illustrate this function, Fig. 1 presents an example of a style transfer that we perform by the application of the two algorithms to pairs of content and style. This transfer function is not well-defined at its heart, and as such, it may be interpreted and approached in several different ways.

V. Rao · K. U. Hemanth · B. Rudra (✉)
National Institute of Technology Karnataka, Surathkal, Mangalore, India
e-mail: bhawanarudra@nitk.edu.in

© The Author(s), under exclusive license to Springer Nature Singapore Pte Ltd. 2022

D. Giri et al. (eds.), *Proceedings of the Seventh International Conference on Mathematics and Computing*, Advances in Intelligent Systems and Computing 1412,
https://doi.org/10.1007/978-981-16-6890-6_16

Fig. 1 Example of style transfer

2 Motivation

Photo editing software such as B612, Retrica, Snapchat filters, and Instagram, is the inspiration behind Artistic Style Transfer via Texture Synthesis. Recent success of Convolutionary Neural Networks has generated renewed interest in the subject. Moreover, with the emergence of new applications such as Snapchat and Instagram, which create open avenues for art technology for the general public, the problem has become even more relevant to the scientific community. Subsequent research was undertaken to capture the nature of the machine-learning process, and to differentiate between style and content in terms of pure image processing, with the goal to reduce overhead computational learning and run time [2]. Our goal was to try to improve the artistic quality of style transfer and to compare the images produced by the two algorithms.

3 Literature Review

3.1 Related Work

One style transfer strategy is to build off the related field of texture synthesis. Texture synthesis attempts to create a realistic picture of the material from a reference material. When it comes to the task of style conversion, we can think of the theme as the texture we try to imitate. Kwatra et al. [3] proposed one efficient algorithm for texture synthesis, which is based on minimizing an energy function. This technique of global optimisation is similar to other methods that use region growing.

Elad and Milanfar [4] applied the Kwatra et al. ideas to the transition of types. Patches from a style image can be robustly and efficiently mapped to regions in the material image by first deriving an analogous objective function for the energy minimization, then performing the optimization in stages, producing satisfying artistic results. An alternative approach to achieving style transfer is through the study of image components, as Zhang et al. [5] have shown. This method decomposes the style and content images into the draft, paint, and edge components, then modifies the paint and edge components of the content image to match those of the style image, and eventually reconstructs the final image from the new content image components.

There are drawbacks in Gatys's portrait-style transfer algorithm. In [6], a method where the image style can be captured while retaining as much as possible of the portrait structure is suggested. This [6] specifies (1) Generate background mask for the Content-Image to distinguish the foreground from the background by semantic segmentation; (2) Apply fine-grained style transfer to the background, and simultaneously transfer coarse-grained style to the entire image. Thus, by preserving certain salient features of the content image, the style image shall be applied [7].

3.2 *Outcome of Literature Survey*

Transfer of image style is commonly used worldwide so that a better algorithm is required. Significant focus on portraits is provided in real life. Thus, it is found that the approach of Gatys and Kwatra was not up to the mark so [6] specifies a method for overcoming this problem.

3.3 *Problem Statement*

Performing Style Transfer of images using two different methods and comparing their performance using structural similarity index (SSIM) and Peak Signal to Noise Ratio (PSNR).

3.4 *Objectives*

1. Transferring of Style from one image to another image using Kwatra's approach.
2. Transferring of Style from one image to another image using modified Gatys' approach.
3. Comparison between texture synthesis of Kwatra's and Gatys' Method using SSIM and PSNR.

4 Methodology

For our study, we shall not be using any dataset. We have handpicked several content images and style images.

4.1 For Style Transfer Using Kwatra's Approach

This algorithm draws inspiration from Elad and Milanfar's recent work [4], which draws several stages of optimization from an energy minimization perspective. We adapt these stages, while making our own implementation possible.

Terminology Terminology for sequence of energy functions from the research of Elad [4] in the optimization of Kwatra's texture synthesis algorithm [3] for style transfer application. Elad finds an algorithm that minimizes this cost function to generate a weight mask image that blends style and material. Our approach is building on that concept.

$$\frac{1}{c} \sum_{(i,j)\in\Omega_{L,n}} \min_{(k,l)} \left\| R^n_{ij} X - Q^n_{kl} S \right\|^r_2 + \|X - C\|^2_W + \lambda\{X\} \tag{1}$$

Here:
X represents the estimated image
C represents the content image
S represents the style image
R^n_{ij} represents extraction of the i,j-th patch of size nxn
Q^n_{kl} represents extraction of the k,j-th patch of size nxn
W represents a weight mask
L represents the working scale
r & c represent regularization/normalization factors
λ represents image prior statistics

Initialization Before running the algorithm, some preprocessing and initialization is performed on the input images shown in Fig. 2.

For this job, we presume (and ensure) as a zero stage that all images input are 400×400 pixels, with 3 RGB color channels. This algorithm can, however, be generalized to images of variable dimensions and properties, even though this is not done.

The reason for adding noise (as shown in Fig. 3) is to make more venturous guesses possible with the matching phase of patch. In fact, failure to do so would

Fig. 2 Input content image and input style image

result in repeated patches (from the patchmatching step), particularly in areas of high uniformity (which are areas that contain similar patches by definition) as shown in Fig. 4.

Finally, we create a picture that's "hallucinated". This picture is intended to solely reflect the theme, with little to no material. It is created with a few minor alterations by running the mentioned style transfer algorithm on a significantly blurred content image (Gaussian kernel, ÿ = 100). Most importantly, we are missing the fusion style and fusion steps with material. Such steps do not apply because the fusion style depends on and uses the hallucination image; and while creating a hallucination, we do not care about preserving the material. This image will be used later during the fusion-style phase in the algorithm.

Style Fusion We have applied a weighted average between the projected image and a hallucinated image in each iteration to increase the variation in the style transferred while still remaining true to the material. The weighting of the two images is adjustable as needed. We picked fairly conservative values for the projected performance and hallucinated images of 75% and 25% for our implementation. An example of Style Fusion and Hallucinated Image is shown in Fig. 5.

Patch Matching The aim of patch matching is to extract areas of the style image that mimic areas of the content image. To do so, we first pick a patch in the content picture for which we want to find a style-match. In the style image, we perform a nearest-neighbor (NN) search: we iterate through patches in the style image and pick

Fig. 3 Example of strong additive Gaussian noise and example of color transfer from "Starry Night" to "house" [8]

Fig. 4 Example of patch matching on uniform white image with no noise and example of patch matching on uniform image with added noise [8]

Fig. 5 Example of style fusion applied to "Starry Night" style image and "house" content image, noise initialization not shown for clarity Left: Hallucinated image Right: Resulting image [8]

the patch that has the lowest L2-norm regarding the patch of content image. The NN search can be formulated as the following problem of optimization.

$$\{k^*, l^*\} = \underset{\text{NN search}}{\text{Argmin}_{(k,l)} \left\| R^n_{ij} X - Q^n_{kl} S \right\|_2} \tag{2}$$

This method is intense computationally, and is structured in two ways. First, we note that considering every possible patch of each image (style and content), the quality of the output image does not improve over skipping any patches. The explanation is that consecutive patches are identical to each other, so to contribute useful information about the style of that area, only one patch from a certain area is needed. Therefore, we have opted to add a regular grid of overlapping patches from which to select our material and design patches. The measurement reduction is equal to the 1 + square, however, several pixels are skipped; in our case, we obtain an average reduction of 99.4% . Since we were able to cover the patch space by a wide margin without losing fidelity significantly, we wanted to include the three other major rotations of each patch as well. Doing this gives the matching algorithm a more complex set of style patches to search from which, in certain cases, the selected style patch's minimum L2-norm may be reduced [9].

The matching of patches with key component analysis (PCA) was also optimized. By projecting every possible patch of the image style into a smaller space before computing the standard, the computational effort of minimization (with the trade-off of using more memory) is greatly reduced. The basis (and dimension) of this projection space is chosen from the patented vectors whose energy contribution is 95 per cent of the original total energy of the patch space. The remaining own-vectors do not store vital information from each other in differentiating type patches, so they are not considered. By performing PCA, we can reduce the size of the original patch space by 85% on average [10]. We add noise to each measured metric associated with a patch after calculating the L2-norm of each style patch to the content patch in order to allow the matching algorithm to make venturous guesses during run-time. The Gaussian noise we introduced had a normal 10% deviation of what should have been the minimum L2-norm. The noise level was designed to allow us to select a style patch from a similar set of L2-norms. Finally, in the style picture, we report the original patch corresponding to the selected projected patch as being most similar to the given content patch I_j based on the L2-norm metric.

$$z_{ij} = Q^n_{k^*,l^*} S \tag{3}$$

Most Similar Content Patch

Style Synthesis We are now trying to use the patches in design, found using NN, to estimate the image performance. Essentially, each I j patch in the output image will roughly suit the correct NN style patch, zij. This can be described as the following problem of optimization:

$$\tilde{X} = \underset{X}{\text{Argmin}} \sum_{(i,j)\in\Omega} \left\| R^n_{ij} X - z_{ij} \right\|_2^r \tag{4}$$

Style Synthesis optimization

which can be approximated directly using the iterative re-weighted minimum squares (IRLS), as stated by Kwatra [3].

Content Fusion We use a pixel by pixel weighted average of the actual image and the projected style image to ensure the original content is retained. In areas with more "content" the weight on the object picture would be heavier. The weight will promote the estimated picture in areas where there is less "material". The weight mask of pixel by pixel is determined by performing a series of segmentation algorithms on the image of material. First, because the content is usually in the foreground of the image, we assign a small weight to the context of the image content, and a larger weight to the foreground. The separation of the foreground and background is accomplished using active segmentation based on contour. Next, since the edges of the content image hold the most details in an image, the edges of the content image are assigned a greater weight. Those edges were removed from the Gaussian operation's Laplacian zero crossings. Eventually, we add a Gaussian smoothing to the mask to make the transitions between high content areas and lower style areas smoother [11]. Figure 6 displays one example mask. Equation 5 is used to apply the merge of content and to update the calculation.

$$[H]\hat{X} = (W + I)^{-1}(\tilde{X} + WC) \tag{5}$$

Figure 7 shows the style synthesis output before content fusion and after content fusion using above mask.

Color Transfer You can select the color palette of the final output image to be what you want. For example, we could either retain the original content image color palette, or take the colors from the image design. Because the colors used in the picture style can be viewed as part of the theme, we opted for the latter approach. This often results in more visually appealing images, as the color palette of the design

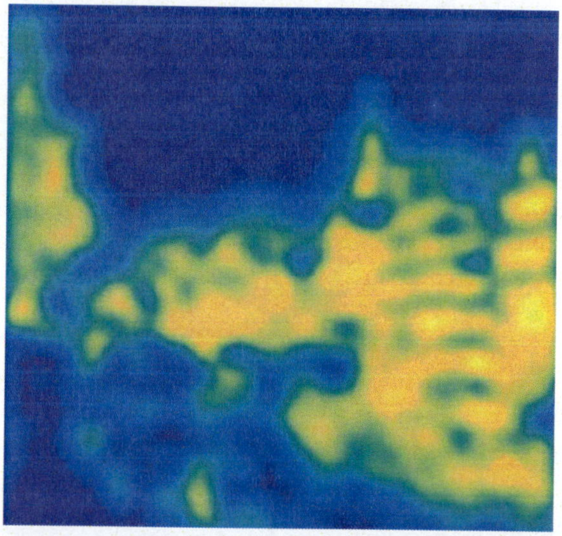

Fig. 6 Example mask generated from "house" image [8]

Fig. 7 The style synthesis output before content fusion and after content fusion using above mask [8]

picture is usually more varied. We shall use the MATLAB function, imhistmatch, in our implementation.

Denoising The ultimate step of the innermost loop is to restore our heavily processed image to a general image's prior statistics. Yes, in the output picture we see objects of the grid for patch selection and edges for content fusion. To minimize these effects, we must choose a filter that smoothens weak edges and preserves strong ones-a bilateral filter, precisely the Domain Transfer Filter [12].

Scale Space The above technique is extended to several resolution scales in order to achieve a more complex and reliable model transference. The larger the scale (i.e., lower resolution) we use for a given patch size, the greater would be the features extracted. As described in [4], it is possible to combine the different scales into a single optimisation problem. They sweep sequentially over the sizes, but for convenience, from coarsest to finest.

Patch Size Another way we can control the scale of the transition of style is by selecting various patch sizes. This has the added advantage of smoothing the resulting patch grid artifacts as smaller patches are built to lie across the larger boundaries. In addition, we increase the variation in the distribution of patch choices by selecting various patch sizes and increase the refinement of the information in the design transition.

In general, differentiating between scale space and patch size should provide the algorithm with a diverse array of structures from different scales and proportions from which to compare patch match.

4.2 *For Style Transfer Using Portrait Aware Artistic Style Approach*

The algorithm [6] is a two-stage method: Portrait segmentation and portrait-aware styling transfer. In the portrait segmentation stage, the algorithm separates the portrait domain from the background, and obtains a masked image. Apply a fine-grained style transfer on the background and a coarse-grained style transfer on the whole image simultaneously. Given a content image x_c and style image x_s, the goal is to synthesize an image $\hat{x}$ which simultaneously shares the content representation of x We vector its CNN representation in layer l for any input image x.

Style transfer aims at producing a synthetic picture $\hat{x}$ [1], which simultaneously shares the content representation of and the style representation of x_s. To this end, we initialize the image using white noise $\hat{x}$, which is of the same size as x_c. Meanwhile, we apply the masked image y [11] which is an image map of values in {0, 1} to label the background domain of $\hat{x}$. This background domain is defined as $\hat{x}' = \hat{x} \circ y$ where $\circ$ denotes element-wise multiplication. Together with the Content-image x_c and the Style-image x_s, these four images ($\hat{x}$, x_c, x_s, y) are fed into the CNN, as shown in Fig. 8.

Then loss function

$$E_{total} = \alpha E_c + \beta E_s + \gamma E_{s'} \tag{6}$$

where

$$E_c = \frac{1}{N_l D_l(x_c)} \Sigma (F_l(\hat{x}) - F_l(x_c))^2_{i,j} \tag{7}$$

and

$$E^l_{s'} = \frac{1}{4N_l^2} \Sigma (G_l(\hat{x}) - G_l(x_c))^2_{i,j} \tag{8}$$

where G is the Gram matrix

$$G_l(x) = \frac{1}{D_l(x)} F_l(x)^T F_l(x) \tag{9}$$

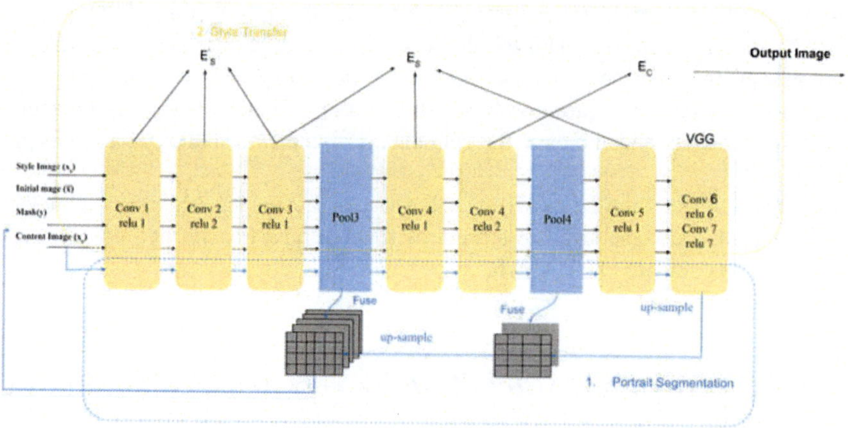

Fig. 8 CNN being Used

The Style image and content image is fed to the 19-layer version of VGG-Net, which is pre-trained on the ImageNet data set and has a classification error rate of just 7% [13]. This is used extensively in object detection. We can use the layers of the VGG-Network to reduce the number of parameters and the time of convergence. We need the VGG network to detect objects and to apply the style accordingly.

The lower layers of CNN tends to catch local styles, while the upper layers will result in a smoother and more consistent visual experience for reflecting styles. Inspired by these observations, we pick some lower layers for E_s to achieve fine-grained context style transfer, and add upper layers to the entire picture to capture the Style-image global shape. In this way, the upper layers for E_s can not only help handle the local color-chunks problem effectively, but also make the synthetic image style more compatible with the overall design image. Thus, these two loss terms can guide the background and the whole image to simultaneously transfer style with brushes of varying strengths and sizes [6].

The Style Transfer function is a loss optimization problem where the aim is to generate a new T image, the content is close to the C image type, and the style is close to the S image style. Therefore, the loss for the problem can be defined as the sum of loss between the style features of style and the target image and the content features of content and target image [14], and the loss between mask features and content image.

The original idea of Gatys [1] suggests only one style loss function to guide style transfer, the modified approach style transfers the portrait and background separately by using different style loss functions for the portrait domain and background, which requires portrait segmentation to separate the foreground from the background. In the modified approach, style loss is split into two parts: the E_s focusing on the entire image and $E_{s'}$ focusing on the background [6].

4.3 Evaluation Metrics

The mechanism of human visual perception is highly capable of distinguishing structural details from a scene and thus identifying the discrepancies between a reference and a sample scene information extracted. Hence, on tasks that require differentiating between a sample and a reference image, a metric that replicates this behavior would perform better. The Structural Similarity Index (SSIM), is driven by the need to capture structural losses in the picture. The Structural Similarity Index (SSIM) is a perceptual metric that quantifies processing-induced deterioration in image quality, such as data compression or data transmission losses. It is a total reference metric, requiring two images from the same image capture—a reference image and a processed image. Consequently, an SSIM calculation can provide a reasonable approximation of the perceived picture quality.

Peak signal-to-noise ratio, abbreviated as PSNR, is the ratio that affects the fidelity of its representation between the maximum possible power of a signal and the power of corrupting noise. PSNR is usually expressed on the logarithmic decibel scale. It is a calculation of the consistency between the compressed image and the real image [15]. These two metrics have been used in Python.

5 Results and Analysis

The algorithms were run on a set of 13 images and the resulting output was noted. The style images and the content images were resized to 512×512 to perform our tests.

5.1 Tests Performed on Algorithm Based on Kwatra's Approach

After testing the algorithm with various parameters, we have found the following set of parameters to give impressive results without compromising the computational time and efficiency:

Set of Patch Sizes: 36×36, 22×22, 13×13, 9×9.
Set of Scales: 0, 1, 2.
Number of update iterations per patch size: 3
Threshold and Sigma values for segmentation: 0.5 and 20, respectively.

The sizes of patches were designed to give us a wide spectrum of image style structures while not compromising the usage of speed and memory. In general, as patch decreases, the duration of the calculation time will increase as there will be more matches on NN to consider. In addition,with more patches, we can remove grid artifacts from the choice our patch shape. These artifacts are reduced because each consecutive patch size was designed so that a patch would exist above the previous patch's grid line. A scale of L implies that the images are resized by a factor of 1/L.

5.2 Tests Performed on Modified Gatys' Approach

Experimental Environment: The modified Gatys approach was experimented on—i7 (6cores), 8 GB RAM, NVidia GTX 1650 4 GB, and it was performed on Ubuntu 20.04 LTS operating system. We have chosen Python 3.7 as our programming environment. TensorFLow, Scikit-learn were the various packages that were used to help us in defining the CNNs.

Modified Approach After performing various tests on the Modified Gatys' approach, we have found

$$Content\ weight\ \alpha = 6.0$$

$$Style\ weight\ \beta = 600$$

$$Mask\ weight\ \gamma = 2500$$

$$total\ variation\ weight\ = 200$$

These were the optimal weights to obtain the required image.

We pass the content images through the Block2_Conv2 layer, the style images through Block3_Conv2 layers and mask images through Block1_Conv1, block2_Conv2, Block3_Conv1 layer. The images are passed through these layers in the VGG19 network which was trained on imagenet dataset and then Content loss, Style loss and Masked loss is computed. These layers were found to work better after experimenting on these layers. We run them for 10 epochs.

Experimentation The original paper suggests to use

$$Content\ weight\ \alpha = 5.0$$

$$Style\ weight\ \beta = 200$$

$$Mask\ weight\ \gamma = 20000$$

$$total\ variation\ weight\ = 200$$

The total variation weight was left to us to choose so we chose 200.

We pass the content images through the Block4_Conv2 layer, the style images through Block3_Conv1, Block4_Conv2, Block5_Conv1 layers and mask images through Block1_Conv1, block2_Conv2, Block3_Conv1 layer. The images are passed through these layers in the VGG19/16 network which was trained on imagenet dataset and then Content loss, Style loss and Masked loss is computed. For our analysis, we set the iterations for 200 where [6] had set them to 1000.

The two methods were tested with "House.jpg" and "starry_night.jpg" with 2 sets of values for number of epochs they are 10 and 200 (Table 1).

In the above experiment our modified approach stopped at 113th iteration because there was no change in the loss value and the original approach described by the [6] stopped at 152nd iteration as there was no change in loss value.

The style content is too much in the image produced by the original approach [6], therefore, its SSIM value is very low. Whereas our modified approach is too smooth and has less brushstrokes. In Fig. 9 shown above, the left image is obtained from our modified approach and the right image is obtained from the original approach [6]. We can clearly see that our modified approach has the style content that is not too much and the structure is also not sacrificed, hence this yields a higher SSIM value. Whereas in the original approach [6], there are way too many brushstrokes which sacrifices the structural integrity of the image produced.

Table 1 Values obtained after running them for 200 iterations

	Original approach	Our approach
Time taken (s)	1978	833
Original loss value	4.531831e12	1.039431e13
Final SSIM value	0.334	0.770
Final loss value	1.556327e12	1.153559e12

Fig. 9 The left image—Our approach; the right image—the original approach [6]

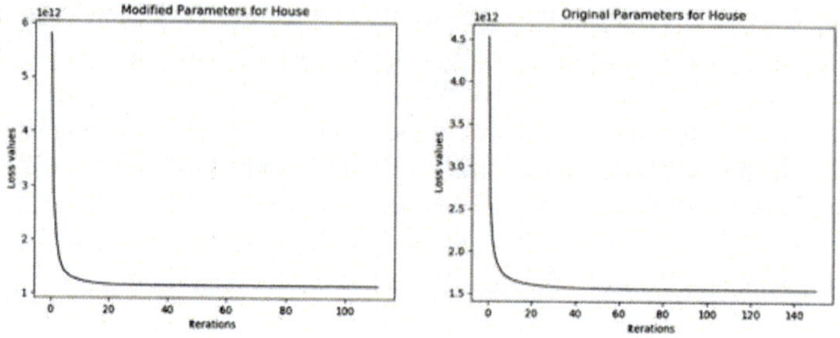

Fig. 10 Graphs of loss versus iterations

Table 2 Results for original approach [6] and our Modified approach. 10 iterations

	Original approach	Modified approach
Time taken (s)	134	77
Original loss value	4.531831e12	1.039431e13
Final SSIM value	0.286	0.706
Final loss value	1.6918441e12	1.2716841e12

Thus from our experimental analysis (as shown in Fig. 10), we chose to use our own weights and after various trial error experiments, we concluded that passing the images through the set of layers that was defined above yielded better and faster results (Table 2).

In Fig. 11, clearly the left image has the right amount of style content infused in it, whereas the right image has too much of style content infused in it. The left image has maintained its structure as close to the original image. Whereas in the right image too much of style content has sacrificed the structure of the image. Hence, the right image has lower SSIM value when compared to the left image.

Figure 12 shows the graphs of loss versus iterations for our approach (left image) and the original approach (right image), it shows the convergence of loss values.

5.3 *Final Output*

In Fig. 13, from left to right—The content Image, The style Image, Output from Kwatra's Approach, Output from the Modified Gatys' approach. In these sets of outputs the images produced by Kwatra's seem distorted and the images obtained from modified Gatys' approach yield better images (Tables 3 and 4).

Lesser the PSNR score Better the image quality and higher the SSIM score better is the structural quality of the picture.

In Fig. 14, from left to right—The content Image, The style Image, Output from Kwatra's Approach, Output from the Modified Gatys' approach. Again the images

Fig. 11 Number of iterations = 10

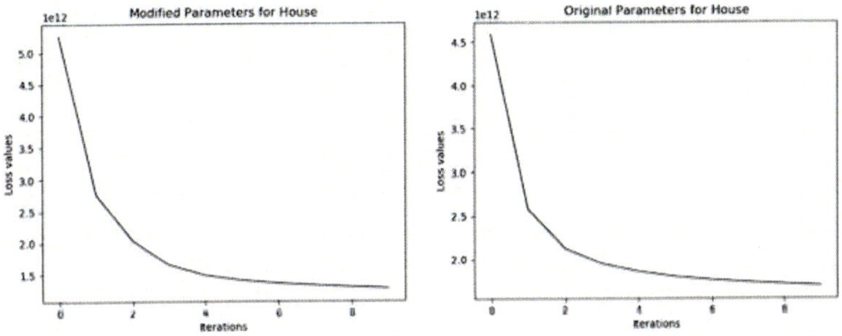

Fig. 12 Graphs of loss versus iterations

Fig. 13 The obtained outputs. Set-1

Table 3 The name of the content image, the style image and the obtained SSIM value

Content image	Style image	Kwatra's algorithm	Modified gatys approach
Angelina	Abstract	0.622	0.7399
Angelina	Ocean2	0.712	0.522
Didi	Scream	0.608	0.569
Emilia	Ocean2	0.357	0.490
Emilia	Fire	0.222	0.556
Emilia	Rooster	0.507	0.619
Friends	Starry_night	0.601	0.557
Girl	Style2	0.578	0.679
House	Starry_night	0.639	0.706
Nat	Pixels	0.641	0.349
Nat	Ocn	0.673	0.519
Keanu	Fire	0.222	0.454
Keanu	Ocn	0.557	0.544

Table 4 Table representing the name of the content image, the style image, and the obtained PSNR value

Content image	Style image	Kwatra's algorithm	Modified gatys approach
Angelina	Abstract	16.674	14.19
Angelina	Ocean2	20.96	14.69
Didi	Scream	12.71	11.81
Emilia	Ocean2	7.15	10.62
Emilia	Fire	7.61	9.56
Emilia	Rooster	12.04	10.47
Friends	Starry_night	11.40	9.93
Girl	Style2	19.19	13.37
House	Starry_night	18.24	15.68
Nat	Pixels	15.73	14.44
Nat	Ocn	13.73	13.36
Keanu	Fire	7.61	10.33
Keanu	Ocn	11.83	11.40

Fig. 14 The obtained outputs. Set-2

produced by Kwatra's algorithm yielded distorted images and modified Gatys' approach yielded images with too much style content in them which resulted in poor SSIM values.

6 Conclusion and Future Work

We were successful in implementing both the algorithms and comparing the results given by both of them. We were also successful in getting better results through our modifications. The obtained SSIM values seem like they vary but the PSNR scores indicate Modified Gatys Approach is better. If we examine the resultant images individually, the images produced by Modified Gatys approach seems more pleasant to look at. Moreover, portraits give better results by Modified Gatys Approach.

Future work involves exploring and comparing other algorithms and also implement and compare the algorithms for style transfer of videos.

Conflict of Interest Statement

On behalf of all authors, the corresponding author states that there is no conflict of interest.

Acknowledgements *We have used images from various web sources for experimental purposes only.*

References

1. Gatys LA, Ecker AS, Bethge M (2016) Image style transfer using convolutional neural networks. In: IEEE CVPR, Las Vegas, USA, pp 2414–2423
2. Jing Y, Yang Y, Feng Z, Ye J, Yu Y, Song M, Neural style transfer: a review. IEEE Trans Visual Comput Graph
3. Gastal ESL, Oliveira MM (2011) Domain transform for edge-aware image and video processing. ToG 30(4):169
4. Elad M, Milanfar P (2017) Style-transfer via texture-synthesis. IEEE Trans Image Proc 26(5):2338–2351
5. Zhang Wei, Cao Chen, Chen Shifeng, Liu Jianzhuang, Tang Xiaoou (2013) Style transfer via image component analysis. IEEE Trans Multimedia 15(7):1594–1601
6. Xing Y, Li J, Dai T, Tang Q, Niu L, Xia S-T (2018) Portrait-aware artistic style transfer. In: IEEE ICIP, Athens, Greece, pp 2117–2121
7. Liu Y et al (2019) Image neural style transfer with preserving the salient regions. IEEE Access 7:40027–40037
8. Wang E, Tan N (2017) Artistic style transfer. Digital image processing. Stanford University, pp 1–7
9. Xiao C, Liu M, Yongwei N, Dong Z (2011) Fast exact nearest patch matching for patch-based image editing and processing. IEEE TVCG 17(8)
10. Barnes C, Schechtman E, Goldman D, Finkelstein E (2010) The generalized PatchMatch correspondence algorithm. ECCV, pp 29–43
11. Handa A, Garg P, Khare V (2014) Masked neural style transfer using convolutional neural networks. In: 2018 international conference on recent innovations in electrical, electronics & communication engineering (ICRIEECE), Bhubaneswar, India, pp 2099–2104
12. Kwatra V, Essa I, Bobick A, Kwatra N (2005) Texture optimization for example-based synthesis. ACM ToG 24(3):795–802. https://www.djangoproject.com
13. Daru P, Gada S, Chheda M, Raut P (2017) Neural style transfer to design drapes. In: 2017 IEEE international conference on computational intelligence and computing research (ICCIC), Coimbatore, pp 1–6
14. Jeong T, Mandal A (2018) Flexible selecting of style to content ratio in neural style transfer. In: 2018 17th IEEE international conference on machine learning and applications (ICMLA), Orlando, FL, pp 264–269
15. Horé A, Ziou D (2010) Image quality metrics: PSNR vs. SSIM. In: 2010 20th international conference on pattern recognition, Istanbul, pp 2366–2369

Ear Localization Based on YOLOv3 Detector

Huy Nguyen-Quoc and Vinh Truong Hoang

Abstract Nowadays, biometric factors have been received lots of attention, especially in human ear. A large number of ear-based applications are proposed due to the valuable shape of ears in identification. However, there is still a lack of efficient and fast ear detection approaches. In this paper, we present an approach based on YOLOv3 framework for ear detection. A wild and large-scale facial image dataset is built for evaluating the proposed method.

Keywords Ear detection · YOLOv3 · Ear localization

1 Introduction

Person identification is known to have an important place in our daily life. With the development of computer vision, automatic identification is now mostly based on biometric factors, such as fingerprint, voice, gait, palm-print, face, and iris. This is a major research topic in computer vision with various approaches introduced in the last decade. For example, Ranjan et al. proposed a fast and precise face identification system [14]. Wang and Kumar [19] enhance the accuracy of iris recognition using dilated residual features. Umer et al. fused iris and periocular features to identify person [18]. Petterson et al. [12] applied wavelet decomposition and Gabor filter for iris identification.

Among the proposed biometric patterns, the human ear has also been received a lot of attention from researchers these days. According to [13], the human ear structure is not randomly conducted but still relative to the cell segmentation. Therefore, each human ears have a unique shape which is as distinguishable as other proposed biometric patterns, and yet this information is twice due to each normal person has

H. Nguyen-Quoc · V. T. Hoang (✉)
Faculty of Information Technology, Ho Chi Minh City Open University, Ho Chi Minh, Vietnam
e-mail: vinh.th@ou.edu.vn

H. Nguyen-Quoc
e-mail: 1751010049huy@ou.edu.vn

© The Author(s), under exclusive license to Springer Nature Singapore Pte Ltd. 2022
D. Giri et al. (eds.), *Proceedings of the Seventh International Conference on Mathematics and Computing*, Advances in Intelligent Systems and Computing 1412,
https://doi.org/10.1007/978-981-16-6890-6_17

two ears. Furthermore, ear images are easy to achieve by cropping from portraits, full-body images, or videos. From this discovery, many identification systems via human ear have been proposed. Specifically, in 1963, Zimberoff Manuel illustrates the first ear identification system through a photo. After that, a countless number of approaches have been proposed to enhance the performance of ear identification. A basic ear identification system can be simply divided into two stages, including detection and recognition. The detection stage is where human ears are found in images or videos. The obtained ears are then processed by machine to predict whom these ears belong to in the recognition stage.

Nowadays, various ear recognition methods have been introduced and give promising results. For instance, Alshazly et al. applied many sets of deep learning models and transfer learning for ear recognition [1]. Hassaballah et al. introduce a comparative study of ear recognition based on local binary patterns [7]. Emervsivc et al. presented a CNN-based ear recognition pipeline [5]. Alshazly et al. proposed a deep CNN for unconstrained ear recognition [2]. Ganapathi applied geometric statistics-based descriptor for 3D ear recognition [6].

However, the lack of efficient ear detection methods is still remaining. Islam et al. proposed a fast and automatic ear detection system based on cascaded AdaBoost [8]. Cintas et al. proposed a CNN model which automatically localize ear from occluded profile face images [4]. Kamboj et al. introduced CED-Net, which is an ear detector for unconstrained images [10]. Ganapathi applied ensemble-based CNN model for unconstrained ear detection [6]. Additionally, YOLO is mostly used in many detection tasks due to its fast execution time which is an advantage for real-time applications. For example, Hendry and Chen [3], Jamtsho et al. [9] applied YOLO for license plate localization. Li et al. detect driver manual distraction via hand and ear using YOLO [11]. Wu et al. used pruning-based YOLOv4 for apple flowers detection [20]. In this paper, we proposed a new approach of ear detection based on YOLOv3.

In order to evaluate the proposed method, we collected 10,000 facial images of more than 100 Asian celebrities for experimental data. The remaining of this paper is organized as follows. Section 2 shortly introduces YOLO. Experimental data and results are then described in Sect. 3. Finally, conclusion and future works are discussed in Sect. 4.

2 Proposed Approach

Inspired by the human visual system, Redmon et al. present a significant object detector, named You Only Look Once (YOLO) [15]. Most previous detectors repurposed classifiers for detection. Unlike them, YOLO defines detection as a regression problem which makes this method outperforms the other one in execution time. To compare to the performance of state-of-the-art detectors, such as Faster R-FCN, SSD, and RetinaNet, YOLO still struggles to precisely localize objects. However, the accuracy of YOLO has been incrementally improving in the latter versions.

YOLO is not complexly designed, the input image is first divided into an $S \times S$ grid. If the central part of an object locates in a grid cell, it means that object is detected by that grid cell. Each grid cell returns B bounding boxes which contains 5 predictions: x, y, w, h and confidence score. Where, x, y are coordinate of the center point of the bounding box, w, h are width and height of the bounding box. The confidence score represents the confidence of the model for detecting each box which equal the intersection over union (IOU) between the predicted box and the ground truth. The YOLOv1 architecture is mostly based on GoogLeNet model for image classification which contains 24 convolutional layers and 2 fully connected layers. Instead of using inception modules, the author simply applies 1×1 reduction layers and 3×3 convolutional layers. Moreover, YOLOv2 added batch normalization for on all of the convolutional layers and get more than 2% improvement in mAP [16]. YOLOv2 also replaces fully connected layers with anchor boxes to predict bounding boxes. Furthermore, YOLOv3 applied a new feature extraction network and replaced softmax in multi-class classification to enhance the performance [17]. In this paper, we apply the latest version of YOLO which is YOLOv3 for the best performance.

3 Experiments

3.1 Dataset Description

In order to evaluate the proposed approach, we collected facial images of Asian celebrities from social network to create a database with 10,000 unconstrained images. For that, 3,000 images are feed to YOLOv3 for training, 2,000 images are used for validation and the rest is testing set. Figure 1 illustrates several samples of our dataset. It is a challenging task since some images have only one ear side.

Fig. 1 Illustration of our facial dataset

Fig. 2 Illustration of several testing images

Fig. 3 Illustration of some fail cases

3.2 Experimental Result

We employ a YOLOv3 detector with 100 epochs and batch size equals to 8. Other parameters are set to default. The achieved performance is significant with 95.98% of validation accuracy and 96.97% of testing accuracy. Figures 2 and 3 illustrate several testing images and fail cases.

The reason for achieved fail cases is mostly due to the occluded ears. Other conditions, such as rotation, blur, and illumination, may also affect the detection. Therefore, a larger number of training images are required for better performance.

4 Conclusion

In this paper, we proposed a new ear detection approach which is YOLOv3. The performance is significantly achieved with 96.97% of testing accuracy. With this promising experimental result, we believe YOLOv3 is now one of the most robust detectors for ear detection task. Moreover, a fast execution time makes our approach more suitable for real-time applications. For the future works, the learning database is enlarged for better performance. Several algorithms are applied to get rid of the geometric shape of ears. A complete and fast real-time ear detector is presented based on YOLOv3.

References

1. Alshazly H, Linse C, Barth E, Martinetz T (2019) Ensembles of deep learning models and transfer learning for ear recognition. Sensors 19(19):4139
2. Alshazly H, Linse C, Barth E, Martinetz T (2020) Deep convolutional neural networks for unconstrained ear recognition. IEEE Access 8:170295–170310
3. Chen RC et al (2019) Automatic license plate recognition via sliding-window darknet-yolo deep learning. Image Vis Comput 87:47–56
4. Cintas C, Delrieux C, Navarro P, Quinto-Sanchez M, Pazos B, Gonzalez-Jose R (2019) Automatic ear detection and segmentation over partially occluded profile face images. J Comput Sci & Technol 19
5. Emeršič Ž, Križaj J, Štruc V, Peer P (2019) Deep ear recognition pipeline. In: Recent advances in computer vision. Springer, pp 333–362
6. Ganapathi II, Ali SS, Prakash S (2020) Geometric statistics-based descriptor for 3d ear recognition. Vis Comput 36(1):161–173
7. Hassaballah M, Alshazly HA, Ali AA (2019) Ear recognition using local binary patterns: a comparative experimental study. Expert Syst Appl 118:182–200
8. Islam SM, Bennamoun M, Davies R (2008) Fast and fully automatic ear detection using cascaded adaboost. In: 2008 IEEE workshop on applications of computer vision. IEEE, pp 1–6
9. Jamtsho Y, Riyamongkol P, Waranusast R (2020) Real-time bhutanese license plate localization using yolo. ICT Express 6(2):121–124
10. Kamboj A, Rani R, Nigam A, Jha RR (2020) Ced-net: context-aware ear detection network for unconstrained images. Pattern Anal Appl 1–22
11. Li L, Zhong B, Hutmacher C Jr, Liang Y, Horrey WJ, Xu X (2020) Detection of driver manual distraction via image-based hand and ear recognition. Acc Anal & Prev 137:105432
12. Petterson HN, Rehnholm J, Vikström S, Åslund M, Åstrand E, Tomasic I (2020) Iris identification using wavelet decomposition and gabor filter. In: 2020 43rd international convention on information, communication and electronic technology (MIPRO). IEEE, pp 265–270
13. Pflug A, Busch C (2012) Ear biometrics: a survey of detection, feature extraction and recognition methods. IET Biometrics 1(2):114–129
14. Ranjan R, Bansal A, Zheng J, Xu H, Gleason J, Lu B, Nanduri A, Chen JC, Castillo CD, Chellappa R (2019) A fast and accurate system for face detection, identification, and verification. IEEE Trans Biometr Behav Iden Sci 1(2):82–96
15. Redmon J, Divvala S, Girshick R, Farhadi A (2016) You only look once: unified, real-time object detection. In: Proceedings of the IEEE conference on computer vision and pattern recognition, pp 779–788
16. Redmon J, Farhadi A (2017) Yolo9000: better, faster, stronger. In: Proceedings of the IEEE conference on computer vision and pattern recognition, pp 7263–7271
17. Redmon J, Farhadi A (2018) Yolov3: an incremental improvement. arXiv:1804.02767
18. Umer S, Sardar A, Dhara BC, Rout RK, Pandey HM (2020) Person identification using fusion of iris and periocular deep features. Neural Netw 122:407–419

19. Wang K, Kumar A (2019) Toward more accurate iris recognition using dilated residual features. IEEE Trans Inf Forensics Secur 14(12):3233–3245
20. Wu D, Lv S, Jiang M, Song H (2020) Using channel pruning-based yolo v4 deep learning algorithm for the real-time and accurate detection of apple flowers in natural environments. Comput Electron Agric 178:105742

A Novel Reversible Data Hiding in Encrypted Images by Controlled Swapping of Adjacent Pixels

V. M. Manikandan

Abstract Reversible data hiding (RDH) got much attention in the past few years due to its applications in various areas. In this research, we have considered the challenging task of performing RDH in encrypted images. The security of the image encryption process is mainly evaluated using two types of analysis: one is the entropy measure and the other is histogram analysis. The RDH schemes in encrypted images always result in some alterations in the entropy measure and some changes in the histogram of the encrypted images. In this paper, we propose a novel RDH scheme that will be capable to generate an encrypted image with hidden data without any change in the entropy and in the histogram. The new algorithm is a block-wise RDH process in which the data hider can hide one single bit of additional data in an encrypted image block having size $B \times B$ pixels. The pixels of a selected block will be divided into two groups: *white pixels* and *black pixels*. The classification of pixels is done based on the checkerboard pattern. The block will remain unchanged to hide a bit value 0, and all the adjacent white pixels and black pixels will be swapped to hide a bit value 1. To extract the hidden bits at the time of image restoration, the closeness between the pixel intensity value is analyzed. The experimental results show that the new RDH schemes provide a good embedding rate without affecting the bit error rate.

Keywords Reversible data hiding · Image encryption · Image decryption · Pixel swapping · Secret message transmission · Medical image transmission

1 Introduction

The data hiding schemes are widely used for the last 30 years in which a digital cover medium is used as a concealer to hide some secret information for secure message communication [2]. The hidden information may be a secret message that

V. M. Manikandan (✉)
SRM University AP, Andhra Pradesh, India
e-mail: manikandan.v@srmap.edu.in
URL: https://srmap.edu.in/

© The Author(s), under exclusive license to Springer Nature Singapore Pte Ltd. 2022

D. Giri et al. (eds.), *Proceedings of the Seventh International Conference on Mathematics and Computing*, Advances in Intelligent Systems and Computing 1412,
https://doi.org/10.1007/978-981-16-6890-6_18

the sender wants to transmit or it may be a watermark for ensuring data authentication or copyright protection. When we are using data hiding schemes for secret message communication, the scheme will be termed as *steganography* and when we are using the data hiding scheme for data authentication or copyright protection, the scheme will be known as *watermarking*. The secret message that we are hiding in the cover medium during watermarking is termed as *watermark* and it will be some unique information for both the sender and the receiver.

Reversible data hiding (RDH) is a widely studied topic in the domain of data hiding [3]. In a RDH scheme, the data hider can conceal a message with the help of a cover medium. Later, the receiver can take out the hidden message while restoring the original cover image. The RDH schemes can be exploited to authenticate the medical images in the telemedicine domain, the diagnosis report can be treated as a message and it can be hidden in the medical image during transmission, the cloud service provider can use the RDH schemes to store metadata in digital contents.

The design of RDH scheme for encrypted images is a challenging research problem. The RDH in encrypted images considers an encrypted image as the input and the data hider does not have any information about the original content of the image. Since the pixels are randomly distributed in the encrypted image without any correlation between adjacent pixels, the RDH schemes suitable for the natural image will not be suitable for the RDH in encrypted images. A few RDH schemes are available for hiding secret messages in encrypted images [1, 4–7, 9–11], but the embedding rate of the existing schemes comparatively low for the usage of the same in real-life applications. This motivated us to work on this domain to propose a new RDH scheme in encrypted images.

The RDH in encrypted images considers an encrypted image as the input and the data hider does not have any information about the original content of the image. Since the pixels are randomly distributed in the encrypted image without any correlation between adjacent pixels, the RDH schemes suitable for the natural image will not be suitable for the RDH in encrypted images. A few RDH schemes are available for hiding secret messages in encrypted images, but the embedding rate of the existing schemes is comparatively low for the usage of the same in real-life applications. This motivated us to work on this domain to propose a new RDH scheme in encrypted images.

1.1 Challenges in the Existing Scheme

The existing RDH scheme in encrypted images alters the pixel values and results in a modified encrypted image with a hidden message. The alterations of the pixels in the encrypted images will affect the security of the image encryption process. In this paper, we introduce a new RDH scheme in the encrypted image which will ensure that the entropy and histogram of the original image will remain the same in the final encrypted image which contains the hidden message. The readers may note that entropy and histogram analysis is two common parameters that are widely used to analyze the security of the image encryption process. From an encrypted grayscale image (8-bit representation), we suppose to get an entropy measure close to 8 and the histogram of the encrypted image may look flat (all pixel values are equally distributed). In the proposed scheme, we have used a controlled pixel swapping approach during data hiding and a smoothness measure is used to identify the correctly recovered image block at the time of data extraction and image recovery phase.

2 Proposed Scheme

The new RDH algorithm in the encrypted image is discussed in this section.

Algorithm 1: Controlled pixel swapping algorithm for RDH in encrypted images

Input: The encrypted image E, the message D (sequence of bits). Let us assume that the size of E is $R \times C$ pixels.

Output: An image in encrypted form, say F which is embedded with message bits. The size of the F will be same as E and the entropy and histogram of E and F will be same

Divide the encrypted image E into non-overlapping blocks of $B \times B$ pixels. Without loss of generality, let us assume that $R\%B = 0$ and $C\%B = 0$.

Let us assume the length of D is same as the total number of $B \times B$ sized non-overlapping blocks in the image. The number such blocks N is equal to $\lfloor \frac{R}{B} \rfloor \times \lfloor \frac{C}{B} \rfloor$

Generate a checkerboard pattern P having size of $B \times B$ size in which 0 represents *black* and 1 represents *white*.

Initialize a two-dimensional array E' of zeros having size $R \times C$ to keep the final encrypted image with hidden message.

$p = 1; \quad q = 1; \quad K = 1$

while $((p \leq (R - B - 1))$ **do**

  while $(q \leq (C - B - 1))$ **do**

    $T = E^B_{p,q}$; /* $E^B_{p,q}$ is a block of size $B \times B$ pixels from the encrypted image E in such a way that (p, q) will be left top index at E. */

    if $(D^K == 1)$ **then**

      Swap all the adjacent black pixels and *white* pixels by accessing the pixels in column-wise linear order to get modified image block T.

    $F^B_{p,q} = T$; /* The image block T having size $B \times B$ (may be modified or not modified based on the secret message bit) will be stored at the respective position in F */

    $K = K + 1$;

    $q = q + 1$;

  $p = p + 1$;

Return the final encrypted image F where the secret message D is embedded.

Algorithm 2: Controlled pixel swapping algorithm for RDH in encrypted images

Input: The encrypted image with hidden message $F$and the decryption key Y. Let us assume that the size of F is $R \times C$

Output: The decrypted and recovered image G of size $R \times C$ pixels and the extracted secret message D

Generate the pseudo-random integer matrix M of size $R \times C$ using decryption key Y.

Perform the bit-wise XOR operation between $M_{x,y}$ and $F_{x,y}$ to get a partially decrypted and recovered image I where $x = 1, 2, \ldots, R$ and $y = 1, 2, \ldots, C$.

Generate a checkerboard pattern P of size $R \times C$ where 0 represents *black* and 1 represents *white*.

Access all the pixels in the encrypted image F and swap all the adjacent *black* pixels and *white* pixels by considering the checkerboard pattern P to get a new version of F called J'.

Perform the bit-wise XOR operation between $M_{x,y}$ and $J'_{x,y}$ to get a partially decrypted and recovered image J where $x = 1, 2, \ldots, R$ and $y = 1, 2, \ldots, C$.

Divide the partially recovered images I and J into non-overlapping blocks in such a way that each block will have a size of $B \times B$ pixels. The total number of blocks N=$\lfloor \frac{R}{B} \rfloor \times \lfloor \frac{C}{B} \rfloor$

Initialize a one-dimensional binary array D of size N to keep the extracted secret message bits.

$p = 1; \quad q = 1; \quad K = 1$

while (($p \leq (R - B - 1)$) **do**

    while ($q \leq (C - B - 1)$) **do**

        $T^0 = I^B_{p,q}$;

        $T^1 = J^B_{p,q}$;

        Find the smoothness measure M^0 and M^1 from the image blocks T^0 and T^1, respectively, using Algorithm 3.

        if $M^0 < M^1$ **then**

            $R^B_{p,q} = T^0$;

            $D^K = 0$

        else

            $R^B_{p,q} = T^0$;

            $D^K = 1$

        $K = K + 1$;

        $p = q + 1$;

    $p = p + 1$;

Return the recovered image R and the extracted message sequence.

The scheme proposed in this manuscript is a RDH scheme in encrypted images. Since our major research contribution in the area of RDH, we have used a basic stream cipher-based image encryption to generate encrypted images for the experimental study. In a stream cipher-based image encryption, an encrypted image E will be generated from the original image I through the bit-wise XOR between a matrix M which is generated using a pseudo-random generator. This process is defined below:

$$E_{x,y} = I_{x,y} \oplus M_{x,y}, \tag{1}$$

Algorithm 3: Computation of smoothness measure from the given image block

Input: One small image block L that contains $B \times B$ pixels.
Output: The measure of smoothness M from the image block L.

$p = 2; q = 2; M = 0$
while $(p \leq B - 1)$ **do**
  while $(q \leq B - 1)$ **do**
    $C = L_{p,q};$
    Find all the 8-neighborhood pixels of C and keep it in a list S. Let us assume, $S'' = [P_1, P_2, P_3, P_4, P_5, P_6, P_7, P_8]$ where P_i represents i^{th} 8-neighborhood pixel.
    Find the a list of absolute differences between P_i and C, where $i = 1, 2, \ldots, 8$ and keep it in S.
    Sort the list S to get S'.
    Find the sum of the first five values Q from S'.
    $M = M + Q$;
    $p = p + 1$;
  $q = q + 1;$
Return M.

where $x = 1, 2, \ldots, R$ and $y = 1, 2, \ldots C$ when the original image having $R \times C$ pixels. The generate the pseudo-random matrix M we shall use any secure pseudo-random generator like RC4 algorithm. The sender and receiver should be agreed with the same key value and the pseudo-random generator will regenerate the same pseudo-random integer matrix M when we feed the same key value. So the image decryption is quite easy whenever it is required. The decrypted image can be generated in the following way:

$$I_{x,y} = E_{x,y} \oplus M_{x,y}, \tag{2}$$

where $x = 1, 2, \ldots, R$ and $y = 1, 2, \ldots C$ when the encrypted image size is $R \times C$ pixels.

2.1 Proposed RDH Scheme in Encrypted Images

The proposed RDH scheme will take an encrypted image E as the input and let us assume that it consists of $R \times C$ pixels. During the RDH process, the image E will be divided into the non-overlapping blocks. Each block will have a size of $B \times B$ pixels. As per the new scheme, in each block, one-bit from the message can be embedded.

Let us assume that we have considered a block T consists of $B \times B$ pixels from E at a time. There are two possibilities during data hiding:

- Case 1: Need to embed message bit 0 in T
- Case 2: Need to embed message bit 1 in T

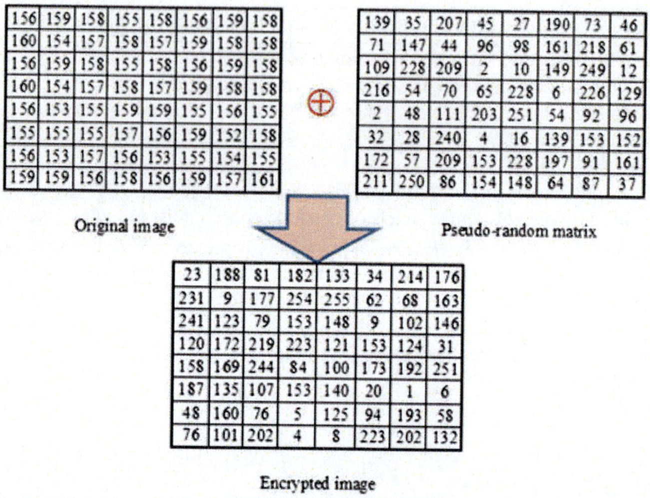

Fig. 1 A sample 8 × 8 image block and its encryption to get the encrypted image block

Fig. 2 Classification of pixels into *black* and *white*

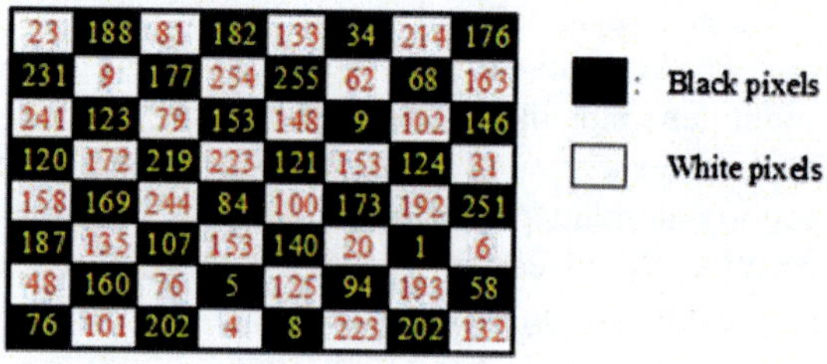

Case 1: Need to Embed a Bit 0 in T In case 1, no need to do any kind of changes in the image block T. Just keep T in the corresponding position of the final encrypted image F.

Case 2: Need to Embed a Bit 1 in T To embed message bit 1 into the image block T, we have introduced a pixel swapping scheme. The *white* pixels and the *black* pixels will be considered during swapping operation. Note that the pixel grouping is done based on the checkerboard pattern. For embedding a bit value 1 in the block, the pixels will be accessed in the column-wise linear order and the adjacent black and white pixels will be swapped.

For better understanding, an illustrative example is discussed here. Consider an encrypted image block obtained from image encryption as shown in Fig. 1.

Let us assume the encrypted image block obtained in Fig. 1 is considered during RDH to embed a bit value 1. The first step is classify the pixels in the image into two categories *black pixels* and *white pixels*. The classification of pixels is schematically shown in Fig. 2.

188	23	182	81	34	133	176	214
9	231	254	177	62	255	163	68
123	241	153	79	9	148	146	102
172	120	223	219	153	121	31	124
169	158	84	244	173	100	251	192
135	187	153	107	20	140	6	1
160	48	5	76	94	125	58	193
101	76	4	202	223	8	132	202

Fig. 3 The image block obtained after swapping *black* and *white* pixels

Since we need to embed a secret message bit 1 in the block shown in Fig. 2, the adjacent *black* and *white* pixels will be swapped. The image block that we will get after swapping the adjacent *black* and *white* is shown in Fig. 3.

The block that we have obtained after swapping the adjacent *black* and *white* will be the final encrypted image block with a hidden message. This block will be placed in the same location to generate the final encrypted image F. The pixels in the encrypted image and the pixels in the final encrypted image after data hiding remain the same (only positions of the pixels are changed). Since no change in the pixel values during the data hiding process, the histogram of the encrypted image and the final encrypted after data hiding will be the same.

A statistical measure called Entropy is commonly used to analyze the randomness in the pixel distribution of an image. A high entropy measure indicates that the image pixels are having high amount of randomness. The entropy (E) measure from an image is defined as follows:

$$E = -\sum_{K=0}^{255} P(K).\log_2 P(K). \tag{3}$$

Since the proposed data hiding scheme does not change any pixel values, the entropy from the image that we have considered for data hiding and the final encrypted image block with a hidden message will remain the same. The entropy computation does not consider the position and during our proposed RDH we didn't modify any pixel values.

2.2 Data Extraction and Image Recovery

During extraction of concealed message and the restoration of the original is carried out by analyzing two different versions of the image blocks. The two versions of the received image will be considered.

- Version 1: Decrypt the received image F by using M (pseudo-random matrix generated using decryption key) to get the first version of the image, say I.
- Version 2: Swap all the adjacent *black* and *white* pixels to get an image J'. Then decrypt J' using the pseudo-random integer matrix M to get the second version of the image, say J.

55	52	121	124	57	59	249	248
78	116	210	209	92	94	121	121
22	21	72	77	3	1	107	106
116	78	153	154	125	127	253	253
171	174	59	63	86	82	167	160
167	167	105	111	4	7	159	153
12	9	212	213	186	184	97	96
182	182	82	80	75	72	211	239

⇩

7901

156	159	158	155	158	156	159	158
160	154	157	158	157	159	158	158
156	159	158	155	158	156	159	158
160	154	157	158	157	159	158	158
156	153	155	159	159	155	156	155
155	155	155	157	156	159	152	158
156	153	157	156	153	155	154	155
159	159	156	158	156	159	157	161

⇩

222

Fig. 4 The smoothness measure from two versions of the image blocks during data extraction and image recovery

The images I and J will be partitioned into blocks of size $B \times B$ pixels. Then each of these blocks will be considered in a predefined order. Let us assume that T^0 is a block from I and T^1 is the corresponding block in J. It should be noted that any one of the image blocks (T^0 or T^1) will be correctly recovered block. A measure is introduced to find the closeness between the pixels in T^0 and T^1, and whichever the block is having a low measure (as per Algorithm 3), it indicates that in that block the pixels are very close. If the block T^0 gives a low smoothness measure then T^0 will be the recovered block and we will extract 0 from T^0. If T^1 gives a low smoothness measure then T^1 will be correctly recovered block and we will extract the bit value 1 from this block.

For better comprehension, the two versions of the block that will be generated during data extraction and image recovery is given in Fig. 4. The smoothness measure obtained from those two blocks (as per Algorithm 3) is also mentioned in Fig. 4. It can be seen that the correctly recovered block gives a very low smoothness measure as compared to the other block.

3 Experimental Study and Result Analysis

The algorithms discussed in this manuscript are implemented using Matlab 2020a for the result analysis. One of the most commonly used image data set called USC-SIPI is accessed during the experimental study [8]. The USC-SIPI image dataset is managed by the University of Southern California and it consists of four different classes of images. The four categories of images are: *Miscellaneous, Sequences, Textures* and *Aerials*. The 8-bit grayscale versions of the images are taken as the input during our study. All images are resized into 512×512 pixels. A bit sequence generated using pseudo-random generator is used as the message. The length of the secret message is assumed as the total number of blocks in the image while performing the data hiding process.

The efficiency of the proposed RDH schemes is evaluated using the following efficiency parameters.

Table 1 Number of images recovered correctly versus Block size used

Image category	Total No. of images	4 × 4	8 × 8	16 × 16	32 × 32	64 × 64
Misc.	39	3	39	39	39	39
Aerial	38	3	38	38	38	38
Sequences	69	4	69	69	69	69
Textures	58	5	65	65	65	65

3.1 Analysis of Embedding Rate and Bit Error Rate

The embedding rate of a RDH scheme is the number of bits that we can embed per pixel. The embedding rate is denoted by *bits per pixels* (bpp) and a RDH scheme with a high embedding rate is preferred for practical applications. In the proposed scheme the embedding rate is directly proportional to the block size that we are using. If the block size is $B \times B$ pixels then the embedding rate will be $\frac{1}{B^2}$ bpp, so we prefer to use small block sizes to improve the embedding rate. But in a RDH scheme, we are also concerned about the recovery of the original image. From the experimental study, it is observed that using small block sizes may lead to errors in image recovery. So a minimum block size is mandatory to recover the image and to extract the hidden message correctly.

To identify the optimum block size, we have carried out an experimental study on all four classes of images by varying the block sizes. We have considered the block sizes such as $2^2, 2^3, 2^4, 2^5$, and 2^6 and identified the image recovery capability for each category of the images. The number of images recovered from each category of images while using different block sizes is given in Table 1. It can be seen that except for the *Texture* images all other image classes are recovered perfectly while using a block size of 8×8 pixels or more. So we can say that the embedding rate for all the images in the classes such as *Misc., Aerial,* and *Sequence* will be $\frac{1}{8^2} = 0.0156$ bpp. For the *Texture* image class the embedding rate is $\frac{1}{16^2} = 0.0039$ bpp.

The bit error rate is 0 for all the images *Misc., Aerial,* and *Sequence* while using block size 8×8 pixels which means all the images are recovered correctly.

The comparison of embedding rate for the four well-known images from the proposed scheme and a few existing schemes is given in Table 2.

3.2 Analysis of Entropy and Histogram

As we mentioned earlier, one of the novelties of the proposed scheme is that the entropy and the histogram of the encrypted image before data hiding and after data hiding will remain the same. The average entropy measure obtained from original image ($E1$), the entropy measure from the encrypted image ($E2$), and the entropy

Table 2 Comparison of embedding rate for the well-known images

Scheme	Airplane	Peppers	Boat	Baboon
Scheme in [9]	0.0039	0.0015	0.0039	0.0009
Scheme in [5]	0.0039	0.0039	0.0039	0.0009
Scheme in [10]	0.0300	0.0300	0.0300	0.0100
Scheme in [4]	0.0020	0.0020	0.0020	0.0020
Scheme in [7]	0.0040	0.0040	0.0040	0.0040
Scheme in [11]	0.0080	0.0080	0.0080	0.0080
Scheme in [6]	0.0080	0.0080	0.0080	0.0080
Scheme in [1]	0.0039	0.0039	0.0039	0.0039
Proposed Scheme	0.0156	0.0156	0.0156	0.0156

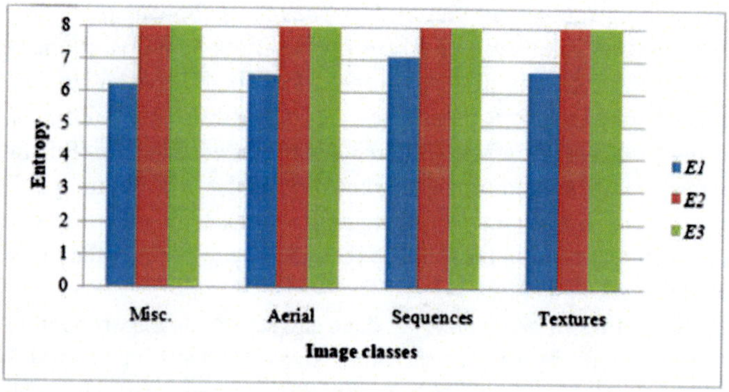

Fig. 5 Comparison of average entropy measure

measure from the final encrypted image with a hidden message ($E3$) are shown in Fig. 5.

From Fig. 5, it can be observed that the entropy measure from the encrypted image and the entropy measure from the final encrypted image with the hidden message are exactly same (7.99).

Another parameter that is used to evaluate the efficiency of the image encryption process is histogram analysis. An image encryption process suppose to generate an encrypted image with flat histogram (all pixel values are uniformly distributed). Our claim in the proposed scheme is the data hiding process will not affect the shape of the histogram since we are not modifying the pixel values during data hiding rather than we are doing a controlled pixel swapping operation. The histogram of original *Peppers* image, the histogram of encrypted *Peppers* image, and the histogram of final image obtained after data hiding process are given Figs. 6, 7, and 8, respectively.

From the histogram figures, it can be observed that the histogram of encrypted image and the histogram obtained after data hiding remains same.

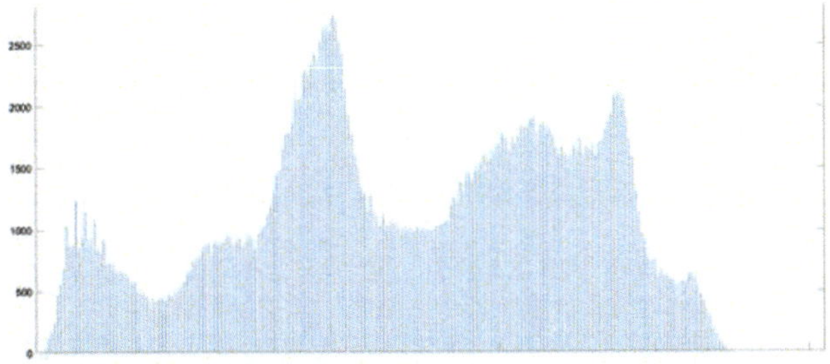

Fig. 6 Histogram of *Peppers* image

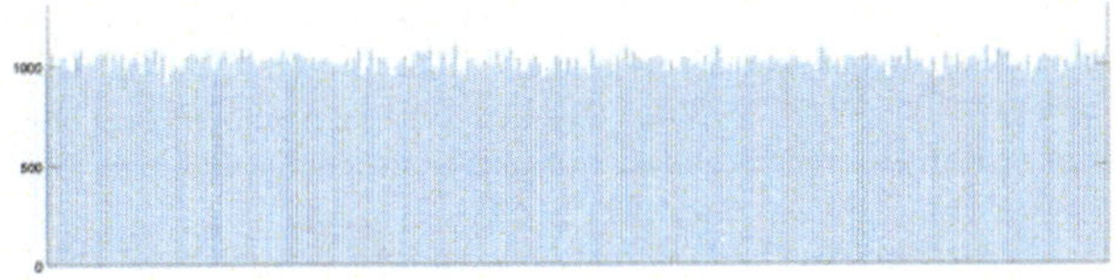

Fig. 7 Histogram of encrypted *Peppers* image

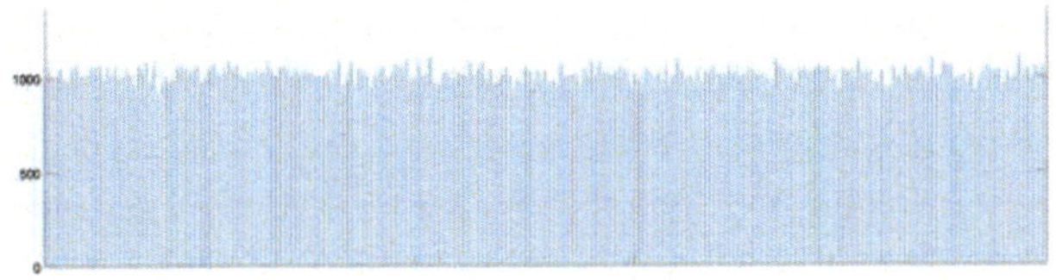

Fig. 8 Histogram of final image obtained after data hiding

4 Conclusion

A new RDH scheme in encrypted images that will not alter the entropy and histogram of the encrypted images after the data hiding process is introduced in this manuscript. The data hiding process is carried out by classifying the encrypted image pixels into two classes called *black* pixels and *white* pixels. The encrypted image blocks where we need to embed a secret message bit 1 only will undergo a controlled pixel swapping process between adjacent *black* pixels and *white* pixels. The experimental study shows that the proposed scheme is capable to successfully recover all the images from *Misc., Aerial,* and *Sequences* categories while using a block size of 8×8 pixels, which yields to an effective embedding rate of 0.0156 bits per pixels. The comparative study with the well-known existing schemes shows that the proposed scheme is having a very good embedding rate and it can be used for practical applications such as medical image transmission or cloud computing for additional data hiding purposes.

References

1. Agrawal S, Kumar M (2017) Mean value based reversible data hiding in encrypted images. Optik-Int J Light Electron Optics 130(2017):922–934
2. Bender W, Gruhl D, Morimoto N, Lu A (1996) Techniques for data hiding. IBM Syst J 35(3.4):313–336
3. Celik MU, Sharma G, Tekalp AM, Saber E (2002) Reversible data hiding. In: Proceedings. International conference on image processing, vol 2. IEEE, pp II–II
4. Chen YC, Shiu CW, Horng G (2014) Encrypted signal-based reversible data hiding with public key cryptosystem. J Vis Commun Image Represent 25(5):1164–1170
5. Hong W, Chen TS, Wu HY (2012) An improved reversible data hiding in encrypted images using side match. IEEE Signal Proc Lett 19(4):199–202
6. Li M, Li Y (2017) Histogram shifting in encrypted images with public key cryptosystem for reversible data hiding. Signal Proc 130:190–196
7. Shiu CW, Chen YC, Hong W (2015) Encrypted image-based reversible data hiding with public key cryptography from difference expansion. Signal Proc: Image Commun 39:226–233
8. USC: image database (2009). Accessed 3 Feb 2014
9. Zhang X (2011) Reversible data hiding in encrypted image. IEEE Signal Proc Lett 18(4):255–258
10. Zhang X (2011) Separable reversible data hiding in encrypted image. IEEE Trans Inf Forens Secur 7(2):826–832
11. Zhang X, Long J, Wang Z, Cheng H (2015) Lossless and reversible data hiding in encrypted images with public-key cryptography. IEEE Trans Circuits Syst Video Technol 26(9):1622–1631

Color Image Contrast Enhancement Through Hybrid PCA-Linear Image Fusion

Shahid Aziz, Vinit Kumar, and Syed Shahnawazuddin

Abstract A simplistic and an efficient approach toward color image contrast enhancement is proposed in this paper. The proposed approach effectively combines linear and principal component analysis (PCA) based fusion methods for contrast enhancement. In the proposed technique, we first obtain three different enhanced images using three pre-existing established techniques. Next, two of the enhanced images are combined using the PCA-based fusion method. The image resulting from first level of fusion is then combined with the third image using linear fusion technique to secure the final enhanced image. Since two different fusion techniques are employed, therefore, the proposed method is referred to as *hybrid image fusion technique*. Again, as we are employing three different images to secure the final enhanced image, there are 3 possible ways in which hybrid image fusion can be implemented. All the possible combinations of hybrid image fusion are compared with the existing techniques qualitatively as well as quantitatively in this paper. The proposed approach is observed to be better than the contemporary image enhancement methods.

Keywords Image enhancement · Image fusion · Linear image fusion · PCA-based image fusion

1 Introduction

In the present era, smartphones with inbuilt high precision cameras are ubiquitously present. This has made the art of clicking good quality pictures trivial even for novice people who have no prior training in the field of photography. But, due to

S. Aziz · V. Kumar (✉) · S. Shahnawazuddin
National Institute of Technology Patna, Patna, India
e-mail: vinitk.pg19.ec@nitp.ac.in

S. Aziz
e-mail: shahida.phd20.ec@nitp.ac.in

S. Shahnawazuddin
e-mail: s.syed@nitp.ac.in

© The Author(s), under exclusive license to Springer Nature Singapore Pte Ltd. 2022
D. Giri et al. (eds.), *Proceedings of the Seventh International Conference on Mathematics and Computing*, Advances in Intelligent Systems and Computing 1412,
https://doi.org/10.1007/978-981-16-6890-6_19

lack of proper know-how, some kind of deficiency in the image capturing device itself, and adverse environmental conditions (such as overcast conditions, lack of proper sunlight, and artificial lighting arrangement), contrast of the captured image is found to be below par. The aforementioned vagaries have led to the growth of image enhancement techniques. Image enhancement is the process of improving the visual quality of an input image using some predefined algorithms [3]. There are multifarious techniques for enhancing the visual appearance and/ or the quantitative parameters of an image. It has been observed that an algorithm which produces desirable results for a particular quantitative parameter may not yield a desirable value for another quantitative parameter. Hence, in order to obtain the desired values for all the quantitative parameters, image fusion has come up as a powerful alternative [4].

Image fusion is the process of combining two different images to form a new image depicting enhanced information and details of the image. Needless to say, the sole purpose of image fusion is to construct images that are more suitable, pertinent, and intelligible for human comprehension. Two of the popular existing image fusion techniques explored in this paper are linear fusion and principal component analysis (PCA) based image fusion.

Normally, a single-level image fusion is carried out for contrast enhancement. In order to further enhance the quality of an image, in this paper, we have studied the role of two-level hybrid image fusion. In the proposed approach, first we obtain enhanced images using three different existing image enhancement techniques. Next, single-level PCA-based image fusion is carried out using any of the two enhanced images. Finally, the third enhanced image is combined with the output of the earlier performed single-level PCA-based fusion using the linear technique. The obvious advantage of incorporating two-level hybrid image fusion is that it culminates in better quantitative parameters as compared to those obtained using only single-level image fusion. Simultaneously, even the perceptual quality of the image obtained through two-level fusion is improved remarkably. The same has been experimentally established on a large number of low-contrast color images. As mentioned earlier, images enhanced by the three different existing enhancement techniques are utilized to perform hybrid fusion. Consequently, three different combinations of hybrid fusion are possible. All the three possible combinations are studied and experimentally compared in this work.

The rest of the paper is organized as follows. In Sect. 2, a brief overview of the explored existing image enhancement techniques is presented. The proposed two-level hybrid image fusion method is discussed in Sect. 3. The experimental results substantiating the efficacy of the proposed approach are given in Sect. 4. Finally, the paper is concluded in Sect. 5.

2 Review of Explored Image Enhancement Techniques

In this section, we have shed some light on the existing image enhancement techniques. It is worthwhile to note that the explored existing approaches are reported to be very efficient for the task of color image enhancement.

2.1 Adaptive Gamma Correction with Weighting Distribution

Gamma-correction-based image enhancement technique is reported to be quite effective. However, considering a constant value of Gamma and applying the same to all pixels may result in over or under enhancement for some of the pixels. To overcome this shortcoming, using an adaptive value of gamma as per requirement is proposed in [10]. Employing adaptive Gamma correction leads to superior results.

2.2 Adaptive Contrast Enhancement Using Modified Histogram Equalization

Adaptive contrast enhancement using modified histogram equalization, abbreviated as ACHME [8], divides the histogram into four parts based on the median brightness. This method consists of six steps. Those are finding average difference matrices, clipping process, new gray point array distribution, histogram equalization, enhancement rate controlling, and adjusting the level of enhancement.

2.3 Weighted Adaptive Approximation Histogram Equalization

Weighted adaptive approximation histogram equalization (WAHE) [2] is a global contrast enhancement method. Here histogram equalization is performed using a weighted histogram approximation approach. The steps involved in implementing WAHE are adjustable histogram equalization, histogram smoothing, weighted histogram approximation, and black and white stretching.

3 Proposed Two-Level Hybrid Image Fusion Technique

In this section, the proposed two-level hybrid image fusion method is described. The section begins with an overview of two of the existing single-level image fusion techniques, i.e., linear and PCA-based fusion. This is followed by the description of the proposed approach.

3.1 Overview of the Explored Image Fusion Approaches

Principal component analysis which is abbreviated as PCA [7] involves a mathematical procedure that transforms a number of correlated variables into uncorrelated variables called principal components. The algorithm for PCA-based image fusion is given next.

Let the input images which are to be fused be I_1 and I_2. First, the input images are arranged in column vectors. The resulting matrix, say Z, is of dimension $2 \times n$ where n is the length of each image vector. The empirical mean M_e is then calculated for each of the columns which is of dimension 1×2. Next, the mean vector M_e is subtracted from each column of the matrix Z. The resulting matrix X is found to be of dimension $2 \times n$. This is followed by the task of determining the covariance matrix C of X. The eigen-vectors V and the eigen-values D of C are calculated next. Both V and D are of dimensions 2×2. Let the matrices D and V be as follows:

$$D = \begin{bmatrix} d_{11} & d_{12} \\ d_{21} & d_{22} \end{bmatrix}$$

$$V = \begin{bmatrix} v_{11} & v_{12} \\ v_{21} & v_{22} \end{bmatrix}.$$

Now, if $d_{11} \geq d_{22}$, then

$$P_1 = \left(\frac{v_{11}}{v_{11} + v_{21}} \right) \tag{1}$$

$$P_2 = \left(\frac{v_{21}}{v_{11} + v_{21}} \right). \tag{2}$$

However, if $d_{11} < d_{22}$, then

$$P_1 = \left(\frac{v_{12}}{v_{12} + v_{22}} \right) \tag{3}$$

$$P_2 = \left(\frac{v_{22}}{v_{12} + v_{22}} \right). \tag{4}$$

Finally, the fused image I_f is obtained using the following equation:

$$I_f = P_1 I_1 + P_2 I_2 \tag{5}$$

such that $P_1 + P_2 = 1$.

In linear image fusion [5], on the other hand, two images are combined linearly with the help of a weighting parameter α. Let there be two input images I_1 and I_2 obtained using different enhancement techniques. The fused output image I_f using linear image fusion is then given by

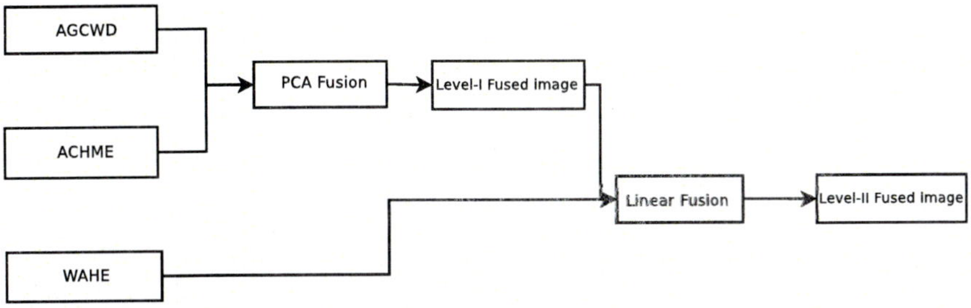

Fig. 1 Block diagram for the Combination-I of the proposed two-level hybrid image fusion. In this case, first the images obtained using AGCWD and ACHME are combined through the PCA-based fusion technique. The Level-I fused image is then combined with the image obtained using WAHE through linear fusion resulting in the final output, i.e., the Level-II image

$$I_f = \alpha I_1 + (1 - \alpha) I_2, \tag{6}$$

where α is the weighting parameter. If $\alpha = 0$, then $I_f = I_2$. Similarly, if $\alpha = 1$, then $I_f = I_1$. The value of α is generally obtained using a grid search [5].

3.2 *Proposed Technique*

As already discussed earlier, three images obtained using the existing enhancement techniques are considered in this case. In the proposed hybrid image fusion approach, the three input images are fused at two levels in the following manner:

- First, Using PCA-based fusion technique, any of the two input images are fused;
- Next, the output image obtained from the first level is combined with the remaining third image using linear fusion method.

Since, two different fusion techniques are employed, the proposed approach is referred to as *hybrid image fusion.*

As a consequence of considering three different images, the total number of possible two-level fusion combinations is three. The block diagrams describing those three combinations are shown in Figs. 1, 2, and 3, respectively.

4 Results and Discussion

The CSIQ database [1] was used in this work for qualitative and quantitative analyses presented in this section. A total of 10 images were considered. The presented quantitative parameters were averaged over those images. The experimental evaluations were performed using MATLAB version 2018A, running on a desktop with

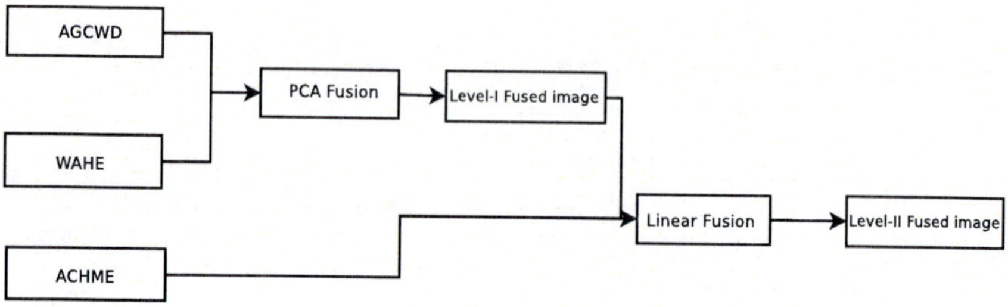

Fig. 2 Block diagram for the Combination-II of the proposed two-level hybrid image fusion. In this case, first the images obtained using AGCWD and WAHE are combined through the PCA-based fusion technique. The Level-I fused image is then combined with the image obtained using ACHME through linear fusion resulting in the final output, i.e., the Level-II image

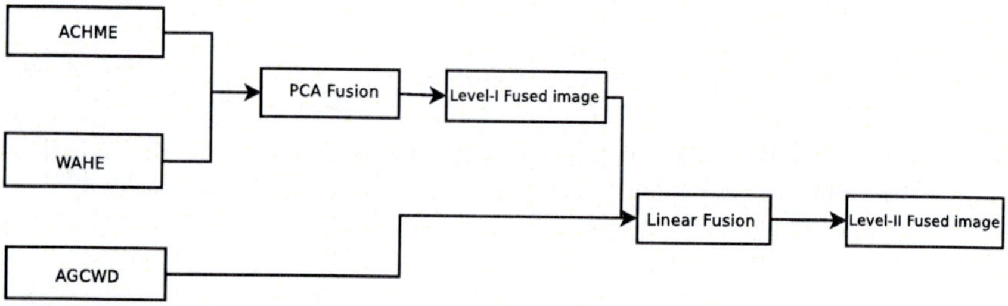

Fig. 3 Block diagram for the Combination-III of the proposed two-level hybrid image fusion. In this case, first the images obtained using ACHME and WAHE are combined through the PCA-based fusion technique. The Level-I fused image is then combined with the image obtained using AGCWD through linear fusion resulting in the final output, i.e., the Level-II image

8GB RAM. As mentioned earlier, the optimal value for the fusion parameter α was obtained through a grid search over the values ranging from 0 to 1. The quantitative parameters used for performance evaluation are entropy [10], absolute mean brightness error (AMBE) [6], peak signal to noise ratio (PSNR), and structural similarity index measurement (SSIM) [9]. The PSNR values reported in this paper are in decibels (dB).

The average values of the quantitative parameters for the different combinations of the proposed approach are given in Table 1. The quantitative parameter values are enlisted for the existing techniques as well. As evident from Table 1, fusion of images results in better values for all the considered quantitative parameters. Furthermore, the results for the Combination-II of the proposed hybrid fusion approach are better than those of the other techniques. It is worth highlighting here that the value of AMBE falls by a huge margin when compared to those of the existing approaches on applying hybrid fusion. At the same time, PSNR also increases by a large margin. Further to that, the change in SSIM is also significantly high. Moreover, entropy values are also better when compared to those for the existing techniques. Thus, we can claim that the proposed technique is superior to the explored existing approaches.

Table 1 Quantitative parameter values for existing techniques as well as the proposed hybrid two-level fusion approach. The values are enlisted for all the three possible combinations of the proposed method

Combinations	Quantitative Parameters	Existing techniques					Proposed
		Original	AGCWD	ACHME	WAHE	Level-1 Fusion	Level-2 Fusion
Combination-I	Entropy	5.7873	6.9022	6.5480	6.8345	7.0217	6.9992
	AMBE		50.7722	24.5226	7.2320	15.3556	5.8303
	PSNR		13.4894	17.6201	19.5329	18.4519	19.5722
	SSIM		0.7371	0.7110	0.7672	0.7776	0.7904
Combination-II	Entropy	5.7873	6.9022	6.5480	6.8345	7.0350	6.9258
	AMBE		50.7722	24.5226	7.2320	23.6781	7.2030
	PSNR		13.4894	17.6201	19.5329	17.2558	19.8103
	SSIM		0.7371	0.7110	0.7672	0.7874	0.7930
Combination-III	Entropy	5.7873	6.9022	6.5480	6.8345	6.7981	7.0103
	AMBE		50.7722	24.5226	7.2320	15.7682	16.2818
	PSNR		13.4894	17.6201	19.5329	18.6685	18.5170
	SSIM		0.7371	0.7110	0.7672	0.7488	0.7974

Fig. 4 Images obtained using the existing techniques as well as the proposed Combination-I for the *cactus* image

Apart from the quantitative analysis, the visual appearance analysis of the output image is also equally important. Hence, we have shown some images obtained using our proposed method and compared them with those obtained by employing the existing techniques. The set of four images obtained in the case of Combination-I are shown in Figs. 4, 5, 6 and 7, respectively. Similarly, the images obtained in the case of Combination-II are shown in Figs. 8, 9, 10, and 11, respectively, Finally, those for the case of Combination-III are shown in Figs. 12, 13, 14, and 15, respectively. In all the cases, its clearly evident that the image obtained by the proposed hybrid fusion technique is perceptually better.

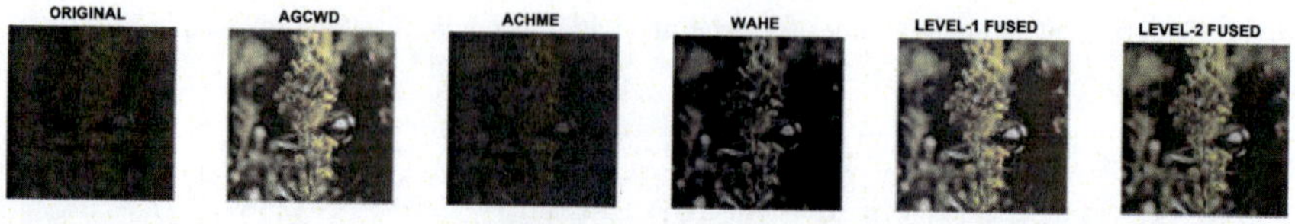

Fig. 5 Images obtained using the existing techniques as well as the proposed Combination-I for the *butter-flower* image

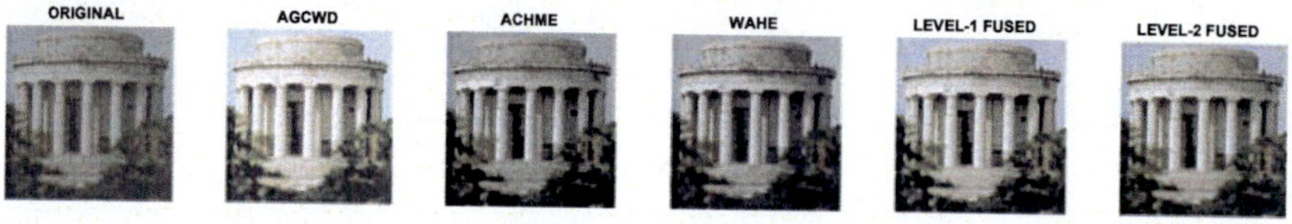

Fig. 6 Images obtained using the existing techniques as well as the proposed Combination-I for the *monument* image

Fig. 7 Images obtained using the existing techniques as well as the proposed Combination-I for the *woman* image

Fig. 8 Images obtained using the existing techniques as well as the proposed Combination-II for the *cactus* image

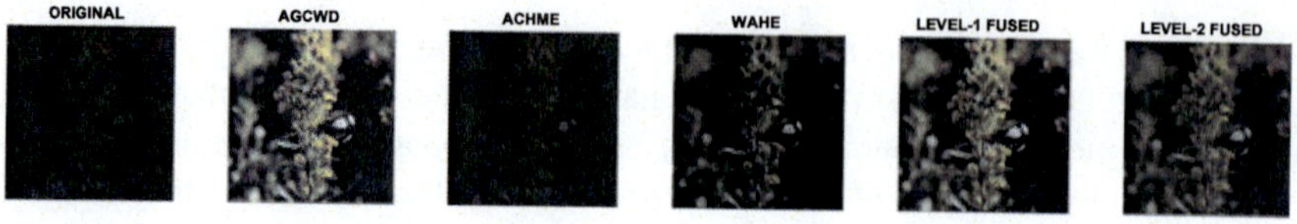

Fig. 9 Images obtained using the existing techniques as well as the proposed Combination-II for the *butter-flower* image

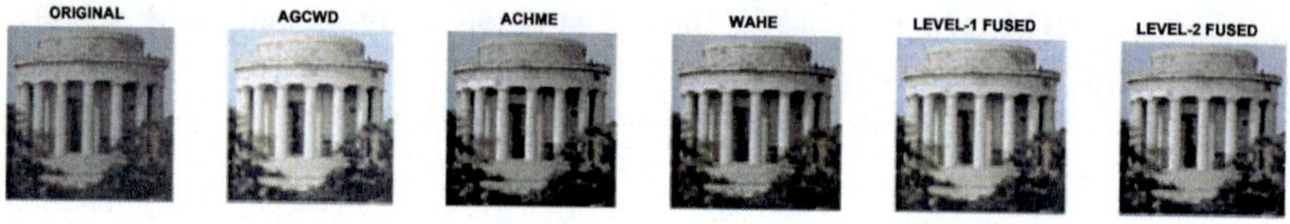

Fig. 10 Images obtained using the existing techniques as well as the proposed Combination-II for the *monument* image

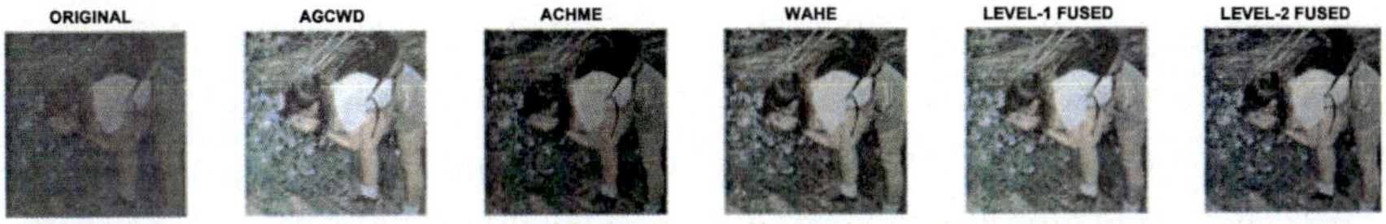

Fig. 11 Images obtained using the existing techniques as well as the proposed Combination-II for the *woman* image

Fig. 12 Images obtained using the existing techniques as well as the proposed Combination-III for the *cactus* image

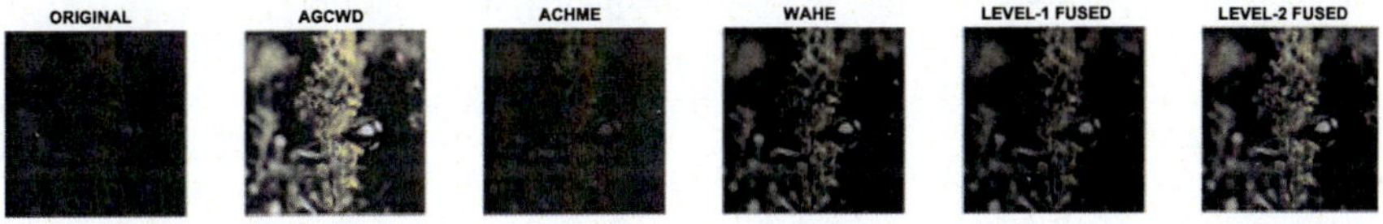

Fig. 13 Images obtained using the existing techniques as well as the proposed Combination-III for the *butter-flower* image

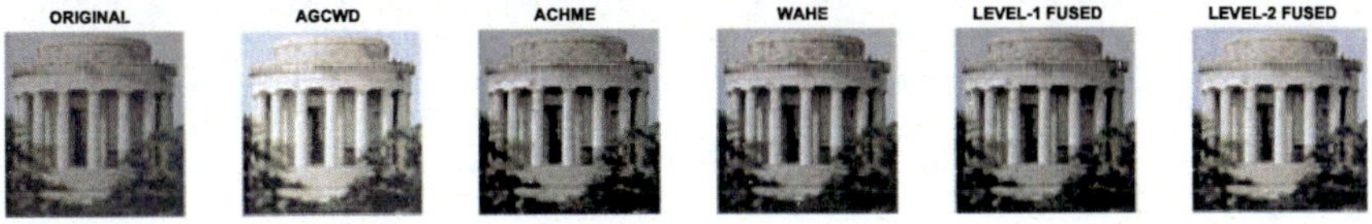

Fig. 14 Images obtained using the existing techniques as well as the proposed Combination-III for the *monument* image

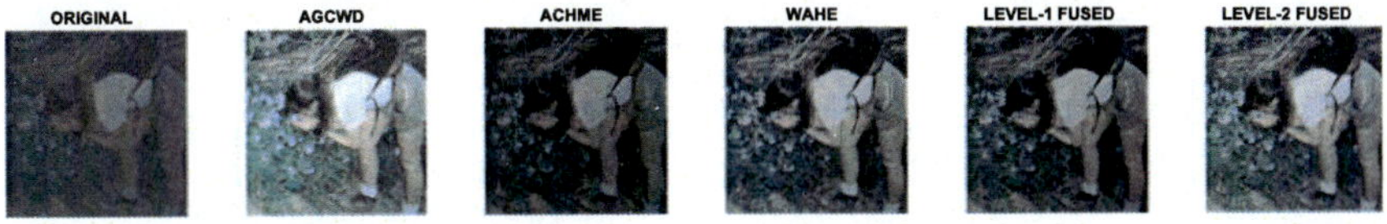

Fig. 15 Images obtained using the existing techniques as well as the proposed Combination-III for the *woman* image

5 Conclusion

In this study, a simple and efficient technique for color image contrast enhancement has been proposed. The said approach effectively combines linear and principal component analysis-based fusion methods for contrast enhancement. In the presented method, we first obtain three different enhanced images using three pre-existing established techniques. Two of those images were then combined using PCA-based fusion method. The resultant output image was then combined with the remaining third image using linear fusion method. This two-level fusion gives the final enhanced output image. Since three different images were employed to obtain the final enhanced one, there were 3 possible ways in which hybrid image fusion could be performed. All the possible combinations of the proposed hybrid image fusion have been compared with the existing techniques qualitatively as well as quantitatively in this paper. By observing the presented results, we can eventually conclude that the proposed technique enhances the input image quality substantially.

References

1. CSIQ: http://vision.okstate.edu/index.php
2. Arici T, Dikbas S, Altunbasak Y (2009) A histogram modification framework and its application for image contrast enhancement. Image Process, IEEE Trans Image Process 18(10):1921–1935
3. Gonzalez RC, Woods RE (2008) Digital image processing. Prentice Hall
4. Goshtasby A, Nikolov SG (2007) Image fusion: advances in the state of the art. Inf Fusion 8:114–118
5. Hsieh CH, Chen BC, Lin CM, Zhao Q (2010) Detail aware contrast enhancement with linear image fusion 126(11)
6. Huang Z, Zhang T, Li Q, Fang H (2016) Adaptive gamma correction based on cumulative histogram for enhancing near-infrared images. Infrared Phys Technol 79(11)
7. Naidu DV, Raol J (2008) Pixel-level image fusion using wavelets and principal component analysis. Def Sci J 58(05)
8. Santhi K, Banu R (2015) Adaptive contrast enhancement using modified histogram equalization. Optik (05)
9. Wang Z, Bovik A, Sheikh H, Simoncelli E (2004) Image quality assessment: from error visibility to structural similarity. Image Process, IEEE Trans Image Process 13(05):600–612
10. Yeh WC, Cheng FC, Chiu YS (2012) Efficient contrast enhancement using adaptive gamma correction with weighting distribution. IEEE Trans Image Process 22(10)

A New Class of Quantum Codes Associate with a Class of Maps

Debashis Bhowmik, Dipendu Maity, and Ashish Kumar Upadhyay

Abstract If the cyclic sequence of faces for all the vertices in a map are of the same types, then the map is a semi-equivelar map. In particular, a semi-equivelar is equivelar if the faces are the same type. Topological quantum codes are being introduced as an alternative quantum code. The homological quantum code is a subclass of topological quantum code. We produce a technique to construct homological quantum codes associate with semi-equivelar maps. We also present homological quantum codes associate with maps on the surfaces with Euler characteristic -1, -2. Furthermore, we will present fifteen classes of homological quantum codes associate with covering maps of the class of maps on the surface with Euler characteristic -1 and -2.

Keywords Topological quantum code · Homological quantum code · Semi-equivelar maps

1 Introduction

Quantum error correction codes (QECs) have been developed to protect quantum information from decoherence and quantum noise. Shor [13] first introduced QECs in 1995, and in 1998, Calderbank et al. [7] proposed a systematic way to create QECs

Dr. Dipendu Maity is supported by NBHM, DAE (No. 02011/9/2021-NBHM(R.P.)/R&D-II/9101) and Dr. Ashish Kumar Upadhyay is supported by SERB-DST, India.

D. Bhowmik (✉) · A. K. Upadhyay
Department of Mathematics, Indian Institute of Technology Patna, Patna 801 106, India
e-mail: debashis.pma15@iitp.ac.in

A. K. Upadhyay
e-mail: upadhyay@iitp.ac.in

D. Maity
Department of Sciences and Mathematics, Indian Institute of Information Technology Guwahati, Bongora, Assam 781 015, India
e-mail: dipendu@iiitg.ac.in

© The Author(s), under exclusive license to Springer Nature Singapore Pte Ltd. 2022

D. Giri et al. (eds.), *Proceedings of the Seventh International Conference on Mathematics and Computing*, Advances in Intelligent Systems and Computing 1412,
https://doi.org/10.1007/978-981-16-6890-6_20

from classical error-correcting codes. Topological quantum computation is an analog type of quantum computation with the benefit of being inherently fault-tolerant because of the physical system's topological properties. Kitaev [11] implemented topological quantum code (TQC), a subclass of stabilizer code, which constitutes a class of quantum error-correcting codes that are constructed based on linear code structure. Such codes are related to tessellations of a surface. In this article, a surface will mean a connected compact 2-manifold without boundary. A cellular embedding of a simple finite graph on a surface is called a map. Semi-equivelar maps are a generalization of Archimedean solids to the surfaces other than the 2-sphere. If G denotes an embedded graph on a surface S, then the vertices and edges of the graph G are called vertices and edges of the map, and closure of the connected components of $S \setminus G$ are called faces. Let $\mathcal{K}$ denote a map, and $V(\mathcal{K})$ denote the set of vertices of $\mathcal{K}$. For $x \in V(\mathcal{K})$, the faces containing x form a cycle (called the *face cycle* at x) C_x in the dual graph of $\mathcal{K}$. So, C_x can be written as $A_1 - A_2 - \cdots - A_l - A_1$, where A_j is a path consisting of n_j many a_j-gons and $a_j \neq a_{j+1}$ for $1 \leq j \leq l$. Addition in the suffix is modulo l. A map $\mathcal{K}$ is said to be *semi-equivelar map* if for any $x, y \in V(\mathcal{K})$, C_x, and C_y are of same type. We say that the semi-equivelar map is a map of type $[a_1^{n_1}, \ldots, a_l^{n_l}]$. In particular, if the type of a map is $[p^q]$, then it is called *equivelar map*. We suggest [8] and [2] to the reader to find out more on this topic. We define dth cover map of $\mathcal{K}$ as follows. Let C be a non-separable cycle of length k in $\mathcal{K}$. Cut $\mathcal{K}$ along C, then we have a map $\mathcal{K}_*$ with two boundary cycles as $A(p_1, p_2, \ldots, p_k)$ and $B(q_1, q_2, \ldots, q_k)$, where identifying A with B by the map $p_i \longmapsto q_i$ $(i = 1, 2, \ldots, k)$ produce the cycle C, and hence $\mathcal{K}$. Consider two copies of $\mathcal{K}_*$, denoted as $\mathcal{M}_1$ and $\mathcal{M}_2$. Thus, $\mathcal{M}_1$ and $\mathcal{M}_2$ have boundary cycles $A_1(p_1^1, p_2^1, \ldots, p_k^1)$, $B_1(q_1^1, q_2^1, \ldots, q_k^1)$ and $A_2(p_1^2, p_2^2, \ldots, p_k^2)$, $B_2(q_1^2, q_2^2, \ldots, q_k^2)$, respectively. Identify $A_1(p_1^1, p_2^1, \ldots, p_k^1)$ with $B_2(q_1^2, q_2^2, \ldots, q_k^2)$ by the map $p_i^1 \longmapsto q_i^2$, and $B_1(q_1^1, q_2^1, \ldots, q_k^1)$ with $A_2(p_1^2, p_2^2, \ldots, p_k^2)$ by the map $q_i^1 \longmapsto p_i^2$, we get a map on the surface of Euler characteristic $2\chi(\mathcal{K})$ with $2 \times |V(\mathcal{K})|$ number of vertices. Again, consider d copies of $\mathcal{K}_*$ as $\mathcal{M}_j$ with cycles A_j, B_j for $j = 1, 2, \ldots d$ $(d \geq 1)$. As before, identify B_i with A_{i+1} for $1 \leq i \leq d-1$ and B_d with A_1, we get a map of same type, which is dth cover of $\mathcal{K}$. We denote this map as $\mathcal{K}^d$. Clearly, $d \times |V(\mathcal{K})|$ is the number of vertices of $\mathcal{K}^d$ with Euler characteristic $d\chi(\mathcal{K})$.

Homological Quantum Code is a subclass of TQCs. Bombin and Martin-Delgado [6] introduce it. They also presented some homological quantum code on the surface of the arbitrary genus. The natural parameters of these codes are $[[n, k, d_{min}]]$ with code length n having k encoded qubits, and minimum distance d_{min}. This article attempts to present some new classes of homological quantum code associate with a class of the equivelar and semi-equivelar maps. These codes are constructed by two parity-check matrices H_X and H_Z such that $H_X H_Z^T = 0$, where entries of H_X, H_Z are from $\mathbb{Z}_2$. These codes can systematically correct all errors up to $\lfloor \frac{d_{min}-1}{2} \rfloor$ qubits.

This paper aims to construct a technique to calculate the homological quantum code associate with semi-equivelar maps. We also present homological quantum codes for the semi-equivelar maps on the surface with Euler characteristic -1 and -2. Furthermore, we build 15 new classes of homological quantum codes, namely $[[40d, 2+d, 4]]$, $[[84d, 2+d, 4]]$, $[[63d, 2+d, 4]]$, $[[126d, 2+d, 4]]$, $[[60d, 2+d, 4]]$, $[[72d, 2+d, 4]]$, $[[48d, 2+d, 4]]$, $[[36d, 2+d, 3]]$, $[[30d, 2+d, 4]]$, $[[36d, 2+d, 3]]$, $[[42d, 2(1+d), 3]]$, $[[36d, 2(1+d), 3]]$, $[[30d, 2(1+d), 3]]$, $[[36d, 2(1+d), 4]]$, $[[36d, 2(1+d), 3]]$ with $d \geq 1$. In this list, the first ten classes are associated with the dth cover map of the semi-equivelar maps on the surface with Euler characteristic -1, and the remaining codes are associated with the dth cover map of the semi-equivelar maps on the surface with Euler characteristic -2. For more on the other classes, we refer the reader to [3].

This paper is organized in the following manner. In Sect. 2, we present an overview of Homological Quantum code for general complex, and homological quantum code for an embedded map on the surface of genus g. In Sect. 3, we give an example to calculate codes associated with equivelar maps and semi-equivelar maps. We produce a list of codes associate with equivelar and semi-equivelar maps on the surface with Euler characteristic -1, -2 along with its covering maps in Sect. 4 and conclude this paper in Sect. 5.

2 Homological Quantum Codes Associate with Maps

Proposition 1 ([1, 12]) *Let C_X and C_Z be two classical binary linear code of length n with parity-check matrices H_X and H_Z, respectively. If $H_X H_Z^T = 0$ then the stabilizer code with binary check matrix*

$$A = \begin{bmatrix} H_X & 0 \\ 0 & H_Z \end{bmatrix}$$

is a $[[n, k, d_{min}]]$ quantum code where n is the codeword length, $k = n - \dim C_X^{\perp} - \dim C_Z^{\perp}$, $d_{min} = min\{wt(x) : x \in (C_Z \setminus C_X^{\perp}) \cup (C_X \setminus C_Z^{\perp})\}$.

Let $(C_\bullet, \partial_\bullet)$ be a chain complex associated with a complex K

$$\cdots \to C_{i+1} \xrightarrow{\partial_{i+1}} C_i \xrightarrow{\partial_i} C_{i-1} \to \cdots. \quad (1)$$

Then, $H_i(K) = \frac{Ker(\partial_i)}{Im(\partial_{i+1})}$ is called the ith Homology of K, [10]. We denote $H_i(K)$ by $H_i(K, \mathbb{Z}_2)$ and $H_i(K, \mathbb{Z})$ when the chain complex is over $\mathbb{Z}_2$ and $\mathbb{Z}$, respectively.

Let $H_X = [\partial_i]_{p\times q}$ and $H_Z = [\partial_{i+1}]^T_{r\times q}$ where $p = \dim C_{i-1}$, $q = \dim C_i$, $r = \dim C_{i+1}$ and $[\partial_j]$ is a matrix corresponding to linear map ∂_j with respect to a basis of C_j and C_{j-1}. Clearly, $H_X H_Z^T = [\partial_i \circ \partial_{i+1}] = 0$. So, by Proposition 1, we can construct a $[[n, k, d_{min}]]$ quantum codes, where $n = \dim C_i$, $k = n - \dim C_X^{\perp} - \dim C_Z^{\perp} = \dim(H_i)$, and $d_{min} = min\{wt(x) : x \in (C_X \setminus C_Z^{\perp}) \cup (C_Z \setminus C_X^{\perp})\}$. Note that $H_i = \frac{Ker(\partial_i)}{Im(\partial_{i+1})} = \frac{Ker(H_X)}{Im(H_Z^T)} = C_X \setminus C_Z^{\perp}$, and by dualizing (1), we can obtain the ith Cohomology $H^i = \frac{Ker(H_Z)}{Im(H_X^T)} = C_Z \setminus C_X^{\perp}$. For more information on this, see [1, 12, 15].

Let M be a surface of genus g, and $\mathcal{K}$ be a map embedded on M with f_0, f_1 and f_2 as vertex, edge and face set, respectively. Let $\mathcal{V}$, $\mathcal{E}$, $\mathcal{F}$ be $\mathbb{Z}_2$-vector spaces with basis f_0, f_1, and f_2, respectively. We can now define a chain complex

$$\{0\} \rightarrow \mathcal{F} \overset{\partial_2}{\rightarrow} \mathcal{E} \overset{\partial_1}{\rightarrow} \mathcal{V} \rightarrow \{0\}$$

with $\partial_1 \circ \partial_2 = 0$. Hence, $H_X H_Z^T = [\partial_1 \circ \partial_2] = 0$, where $H_X = [\partial_1]$, i.e., vertex–edge incident matrix of $\mathcal{K}$, and $H_Z = [\partial_2]^T$, i.e., face–edge incident matrix of $\mathcal{K}$. So, $[[n, k, d_{min}]]$ quantum code is obtained with $n = |f_1|$, $k = \dim(H_1(\mathcal{K}, \mathbb{Z}_2))$, and $d_{min} = min\{wt(x) : x \in (\mathcal{C}_X \setminus \mathcal{C}_Z^{\perp}) \cup (\mathcal{C}_Z \setminus \mathcal{C}_X^{\perp})\}$ i.e. length of the shortest non-contractible cycle of $\mathcal{K}$ or $\mathcal{K}^*$, where $\mathcal{K}^*$ is the dual of $\mathcal{K}$. Now if $H_2(\mathcal{K}, \mathbb{Z}) \cong \mathbb{Z}$, then $H_1(\mathcal{K}, \mathbb{Z}_2) \cong H_1(M, \mathbb{Z}_2) \cong (\mathbb{Z}_2)^{2g}$, and if $H_2(\mathcal{K}, \mathbb{Z}) \cong \{0\}$ then $H_1(\mathcal{K}, \mathbb{Z}_2) \cong H_1(M, \mathbb{Z}_2) \cong (\mathbb{Z}_2)^g$. Therefore, for $H_2(\mathcal{K}, \mathbb{Z}) \cong \mathbb{Z}, k = 2g$, and for $H_2(\mathcal{K}, \mathbb{Z}) \cong \{0\}$, $k = g$. But $H_2(\mathcal{K}, \mathbb{Z}) \cong \mathbb{Z}$ implies $2g = 2 - \chi(\mathcal{K})$, and $H_2(\mathcal{K}, \mathbb{Z}) \cong \{0\}$ implies $g = 2 - \chi(\mathcal{K})$. Therefore, for any $H_2(\mathcal{K}, \mathbb{Z})$, $k = 2 - \chi(\mathcal{K})$, where $\chi(\mathcal{K})$ denotes Euler characteristic of $\mathcal{K}$.

3 Codes Associate with *d*th Cover Maps

We know from [9] that equivelar maps of type $[p^q]$ exist for $p = 3, q = 7$ on double torus with 12 vertices, namely$\mathcal{N}_i$, $i = 1, 2, 3, 4, 5, 6$. For every $\mathcal{N}_i$, it is obvious to envision that each of these maps have a non-contractible cycle of length three, therefore, for each $\mathcal{N}_i$, $d_{min} = 3$. Now, H_X and H_Z corresponding to $\mathcal{N}_1$ is given by Eqs. (2) and (3), respectively. It is clear that $H_X H_Z^T = 0$ over $\mathbb{Z}_2$. Thus, we can get a code corresponding to this. Observe that in this case $n = |EG(\mathcal{N}_i)| = 42$, $k = 2 - \chi(\mathcal{N}_i) = 4$, and therefore $[[42, 4, 3]]$ code is obtained. Now, for dth cover ($d \geq 1$) of each map $\mathcal{N}_i$ produce a map $\mathcal{N}_i^d$ on the orientable surface with $12d$ vertices, $42d$ edges, and Euler characteristic $-2d$. Therefore, $k = 2 - \chi(\mathcal{N}_i) = 2 \times (1 + d)$, and hence, produce a class of codes with parameter $[[42d, 2(1 + d), 3]]$ associate with the map $\mathcal{N}_i^d$.

In [2, 4, 14, 16], it is shown that there are at least 17 semi-equivelar maps with Euler characteristic -1, but all the map type is listed for which map exist. These maps are denoted by $\mathcal{K}_i$ for $i = 1, 2, \ldots, 17$. In $\mathcal{K}_3$, $|f_0| = 20$, $|f_1| = 40$ and $|f_2| = 19$. So, the corresponding vertex–edge and face–edge incidence matrices H_X and H_Z are of size 20×40 and 19×40, respectively. They are given by Eqs. (4) and (5).

Now, we see that $H_X H_Z^T = 0$ over $\mathbb{Z}_2$. Therefore, we get a $[[n, k, d_{min}]]$ code with $n = |f_1| = 40$, $k = 2 - \chi(\mathcal{K}_3) = 3$, and $d_{min} = 4$, the length of the shortest non-contractible cycle in $\mathcal{K}_3$ and $\mathcal{K}_3^*$, *i.e.*, a $[[40, 3, 4]]$ code is obtained. Similarly, codes related to maps $\mathcal{K}_i, i = 1, 2, \ldots, 17$ can obtain, and are listed in Table 1.

As in the case of equivelar maps on double torus, dth ($d \geq 1$) covering of semi-equivelar maps $\mathcal{K}_i$, listed in Table 1, produce a map $\mathcal{K}_i^d$ with Euler characteristic $-d$, and therefore for every $\mathcal{K}_i^d$, $k = 2 + d$. In $\mathcal{K}_i^d$, $i = 1, 2, 3,$

Table 1 Table of quantum codes associate with semi-equivelar maps on the surface of $\chi = -1$

Map Type	Maps	$n = \lvert f_1 \rvert$	$k = 2 - \chi$	d_{min}	$[[n, k, d_{min}]]$
$[4^3, 5^1]$	$\mathcal{K}_1, \mathcal{K}_2, \mathcal{K}_3$	40	3	4	$[[40, 3, 4]]$
$[3^1, 4^1, 7^1, 4^1]$	$\mathcal{K}_4$	84	3	4	$[[84, 3, 4]]$
$[6^2, 7^1]$	$\mathcal{K}_5$	63	3	4	$[[63, 3, 4]]$
$[4^1, 6^1, 14^1]$	$\mathcal{K}_6$	126	3	4	$[[126, 3, 4]]$
$[[4^1, 8^1, 10^1]]$	$\mathcal{K}_7$	60	3	4	$[[60, 3, 4]]$
$[4^1, 6^1, 16^1]$	$\mathcal{K}_8, \mathcal{K}_9$	72	3	4	$[[72, 3, 4]]$
$[3^1, 4^1, 8^1, 4^1]$	$\mathcal{K}_{10}, \mathcal{K}_{11}$	48	3	4	$[[48, 3, 4]]$
$[6^2, 8^1]$	$\mathcal{K}_{12}, \mathcal{K}_{13}$	36	3	3	$[[36, 3, 3]]$
$[3^1, 4^1, 3^1, 4^2]$	$\mathcal{K}_{14}$	30	3	4	$[[30, 3, 4]]$
$[3^5, 4^1]$	$\mathcal{K}_{15}, \mathcal{K}_{16}, \mathcal{K}_{17}$	36	3	3	$[[36, 3, 3]]$

$(|f_0|, |f_1|) = (20d, 40d)$. Similarly, $(|f_0|, |f_1|) = (42d, 84d)$ for $\mathcal{K}_4^d$; $(|f_0|, |f_1|) = (42d, 63d)$ for $\mathcal{K}_5^d$; $(|f_0|, |f_1|) = (84d, 126d)$ for $\mathcal{K}_6^d$; $(|f_0|, |f_1|) = (40d, 60d)$ for $\mathcal{K}_7^d$; $(|f_0|, |f_1|) = (48d, 72d)$ for $\mathcal{K}_i^d$, $i = 8, 9$; $(|f_0|, |f_1|) = (24d, 48d)$ for $\mathcal{K}_i^d$, $i = 10, 11$; $(|f_0|, |f_1|) = (24d, 36d)$ for $\mathcal{K}_i^d$, $i = 12, 13$; $(|f_0|, |f_1|) = (12d, 30d)$ for $\mathcal{K}_{14}^d$, and $(|f_0|, |f_1|) = (12d, 36d)$ for $\mathcal{K}_i^d$, $i = 15, 16, 17$. Therefore, we obtained the code parameters $[[n, k, d_{min}]] = [[40d, 2 + d, 4]]$ associated with the map $\mathcal{K}_i^d$ for $i = 1, 2, 3$; $[[n, k, d_{min}]] = [[84d, 2 + d, 4]]$ associated with the map $\mathcal{K}_4^d$; $[[n, k, d_{min}]] = [[63d, 2 + d, 4]]$ associated with the map $\mathcal{K}_5^d$; $[[n, k, d_{min}]] = [[126d, 2 + d, 4]]$ associated with the map $\mathcal{K}_6^d$; $[[n, k, d_{min}]] = [[60d, 2 + d, 4]]$ associated with the map $\mathcal{K}_7^d$; $[[n, k, d_{min}]] = [[72d, 2 + d, 4]]$ associated with the map $\mathcal{K}_i^d$ for $i = 8, 9$; $[[n, k, d_{min}]] = [[48d, 2 + d, 4]]$ associated with the map $\mathcal{K}_i^d$ for $i = 10, 11$; $[[n, k, d_{min}]] = [[36d, 2 + d, 3]]$ associated with the map $\mathcal{K}_i^d$ for $i = 12, 13$; $[[n, k, d_{min}]] = [[30d, 2 + d, 4]]$ associated with the map $\mathcal{K}_{14}^d$ and $[[n, k, d_{min}]] = [[36d, 2 + d, 3]]$ associated with the map $\mathcal{K}_i^d$ for $i = 15, 16, 17$. The above details are presented in compact form in Table 2.

$$H_X = \begin{pmatrix}
1\,1\,1\,1\,1\,1\,1\,0\,0\,0\,0\,0\,0\,0\,0\,0\,0\,0\,0\,0\,0\,0\,0\,0\,0\,0\,0\,0\,0\,0\,0\,0\,0\,0\,0\,0\,0\,0\,0\,0\,0\,0 \\
1\,0\,0\,0\,0\,0\,0\,1\,1\,1\,1\,1\,1\,0\,0\,0\,0\,0\,0\,0\,0\,0\,0\,0\,0\,0\,0\,0\,0\,0\,0\,0\,0\,0\,0\,0\,0\,0\,0\,0\,0\,0 \\
0\,1\,0\,0\,0\,0\,0\,0\,0\,0\,0\,1\,0\,1\,1\,1\,1\,1\,0\,0\,0\,0\,0\,0\,0\,0\,0\,0\,0\,0\,0\,0\,0\,0\,0\,0\,0\,0\,0\,0\,0\,0 \\
0\,0\,0\,0\,0\,0\,0\,0\,0\,0\,0\,0\,1\,0\,0\,0\,0\,1\,1\,1\,1\,1\,1\,0\,0\,0\,0\,0\,0\,0\,0\,0\,0\,0\,0\,0\,0\,0\,0\,0\,0\,0 \\
0\,0\,1\,0\,0\,0\,0\,0\,0\,0\,0\,0\,0\,0\,0\,0\,1\,1\,1\,0\,0\,0\,0\,1\,1\,1\,1\,0\,0\,0\,0\,0\,0\,0\,0\,0\,0\,0\,0\,0\,0\,0 \\
0\,0\,0\,0\,0\,0\,0\,0\,0\,1\,0\,0\,0\,0\,0\,0\,0\,0\,0\,0\,0\,0\,1\,0\,0\,0\,1\,1\,1\,1\,1\,0\,0\,0\,0\,0\,0\,0\,0\,0\,0\,0 \\
0\,0\,0\,1\,0\,0\,0\,0\,0\,0\,0\,0\,0\,1\,0\,0\,0\,0\,0\,0\,0\,0\,0\,0\,0\,1\,0\,1\,0\,0\,0\,1\,1\,1\,0\,0\,0\,0\,0\,0\,0\,0 \\
0\,0\,0\,0\,0\,0\,0\,1\,0\,0\,0\,0\,0\,0\,0\,0\,0\,0\,0\,1\,0\,0\,0\,0\,0\,0\,0\,0\,1\,0\,0\,0\,1\,0\,1\,1\,1\,0\,0\,0\,0\,0 \\
0\,0\,0\,0\,1\,0\,0\,0\,0\,0\,0\,0\,0\,0\,1\,0\,0\,0\,0\,0\,0\,0\,0\,1\,0\,0\,0\,0\,0\,0\,0\,0\,0\,1\,0\,0\,1\,1\,1\,0\,0\,0 \\
0\,0\,0\,0\,0\,0\,0\,0\,1\,0\,0\,0\,0\,0\,0\,0\,0\,0\,0\,0\,0\,1\,0\,0\,0\,0\,0\,0\,0\,0\,1\,0\,0\,0\,1\,0\,0\,1\,0\,1\,1\,0 \\
0\,0\,0\,0\,0\,1\,0\,0\,0\,0\,0\,0\,0\,0\,0\,1\,0\,0\,0\,0\,0\,0\,0\,0\,1\,0\,0\,0\,0\,0\,0\,1\,0\,0\,0\,0\,0\,0\,1\,1\,0\,1 \\
0\,0\,0\,0\,0\,0\,1\,0\,0\,0\,1\,0\,0\,0\,0\,0\,0\,0\,0\,0\,1\,0\,0\,0\,0\,0\,0\,0\,0\,1\,0\,0\,0\,0\,0\,1\,0\,0\,0\,0\,1\,1
\end{pmatrix}_{12\times 42} \tag{2}$$

Table 2 Code table associate with covering map of semi-equivelar maps on the surface of $\chi = -1$

Map type	$n = \|f_1\|$	k	d_{min}	$[[n, k, d_{min}]]$	$\frac{k}{n}$ as $n \to \infty$
$[4^3, 5^1]$	$40d$	$2+d$	4	$[[40d, 2+d, 4]]$	$\frac{1}{40}$
$[3^1, 4^1, 7^1, 4^1]$	$84d$	$2+d$	4	$[[84d, 2+d, 4]]$	$\frac{1}{84}$
$[6^2, 7^1]$	$63d$	$2+d$	4	$[[63d, 2+d, 4]]$	$\frac{1}{63}$
$[4^1, 6^1, 14^1]$	$126d$	$2+d$	4	$[[126d, 2+d, 4]]$	$\frac{1}{126}$
$[4^1, 8^1, 10^1]$	$60d$	$2+d$	4	$[[60d, 2+d, 4]]$	$\frac{1}{60}$
$[[4^1, 6^1, 16^1]]$	$72d$	$2+d$	4	$[[72d, 2+d, 4]]$	$\frac{1}{72}$
$[3^1, 4^1, 8^1, 4^1]$	$48d$	$2+d$	4	$[[48d, 2+d, 4]]$	$\frac{1}{48}$
$[6^2, 8]$	$36d$	$2+d$	3	$[[36d, 2+d, 3]]$	$\frac{1}{36}$
$[3^1, 4^1, 3^1, 4^2]$	$30d$	$2+d$	4	$[[30d, 2+d, 4]]$	$\frac{1}{30}$
$[3^5, 4^1]$	$36d$	$2+d$	3	$[[36d, 2+d, 3]]$	$\frac{1}{36}$

$$H_Z = \begin{pmatrix}
000000000000000000000000000000000000000111 \\
000000000000000000001100000000000000000010 \\
000000000001100001000000000000000000000000 \\
000000010000100000010000000000000000000000 \\
000000000000000000011000000000000001000000 \\
000000000000000000000000000011000001000000 \\
000000000000000000000000000000001100100000 \\
000000000000000000000000000110001000000000 \\
000000000000000000000000011100000000000000 \\
000000000000000000000000110000010000000000 \\
000000000000000000000000000000000000011100 \\
000000000000001100000000000000000000001000 \\
000000000000010100000000000000010000000000 \\
010100000000010000000000000000000000000000 \\
000110000000000000000000000000000001000000 \\
001010000000000000000001000000000000000000 \\
000000000000001010000001000000000000000000 \\
000000000000000011100000000000000000000000 \\
001001000000000000000000100000000000000000 \\
000001100000000000000000000000000000000001 \\
100000100010000000000000000000000000000000 \\
110000000010000000000000000000000000000000 \\
000000001010000000000000000001000000000000 \\
000000001100000000000000000000100000000000 \\
000000011000000000000000000000000010000000 \\
000000000000000000000000000000000010110000 \\
000000000000000000000110000000100000000000 \\
000000000000000000100010001000000000000000
\end{pmatrix}_{28 \times 42} \quad (3)$$

$$H_X = \begin{pmatrix}
1111000000000000000000000000000000000000\\
1000111000000000000000000000000000000000\\
0000100111000000000000000000000000000000\\
0100000100110000000000000000000000000000\\
0000000000101110000000000000000000000000\\
0010000000001001100000000000000000000000\\
0000000000000001011100000000000000000000\\
0001000000000000010011000000000000000000\\
0000000000000000000010111000000000000000\\
0000000001000000000000100110000000000000\\
0000001000000100000000000001100000000000\\
0000000010000000100000000001010000000000\\
0000000000000000000000000100101100000000\\
0000000000000010000001000000000010000001\\
0000000000000000000000010000001011000000\\
0000000001000000000000000010000000110000\\
0000000000000000000000001000000000011100\\
0000000000000000000000000000010100100010\\
0000000000010000001000000000000000001001\\
0000000000000000000100000000000001000110
\end{pmatrix}; \tag{4}$$

$$H_Z = \begin{pmatrix}
1100100100000000000000000000000000000000\\
1001010000000000000010100000000000000000\\
0110000000101000000000000000000000000000\\
0011000000000001010000000000000000000000\\
0000101010000000000000000001000000000000\\
0000011000000000000000000100100000000000\\
0000000101010000000000000000000000011000\\
0000000011000000000000000000010000100000\\
0000000000110010000010111000000000000001\\
0000000000001100100000000001000000000000\\
0000000000000110000000000000101010000000\\
0000000000000001100100000000010000000010\\
0000000000000000011001000000000000000001\\
0000000000000000001100000000000000001100\\
0000000000000000000011010000000010000000\\
0000000000000000000000011000000001000100\\
0000000000000000000000101010000000010000\\
0000000000000000000000000110000100100000\\
0000000000000000000000000000001101000010
\end{pmatrix} \tag{5}$$

In [5], Bhowmik and Upadhyay classified all semi-equivelar maps on the surface of Euler characteristic -2. In particular, they have presented that there are 11 semi-equivelar maps for the vertices up to 12. Similar to the above, we can also calculate the homological quantum code associate with these maps and its d-cover. Therefore, the codes associated with the semi-equivelar maps for vertices up to 12 is listed in Table 3, and the codes associated with d-cover maps of the map listed in Table 3 are presented in Table 4.

4 Table of Quantum Codes

See Tables 1, 2, 3 and 4.

Table 3 Table of codes associate to SEMs on the surface with $\chi = -1$

Map type	Maps	$n = \|EG(\mathcal{K})\|$	$k = 2 - \chi(\mathcal{K})$	d_{min}	$[[n, k, d_{min}]]$
$[3^4, 4^2]$	$KO_{1[3^4,4^2]}$, $KO_{2[3^4,4^2]}$, $KNO_{[3^4,4^2]}$	36	4	3	$[[36, 4, 3]]$
$[3^1, 4^4]$	$KNO_{[3^1,4^4]}$, $KNO_{2[3^1,4^4]}$	30	4	3	$[[30, 4, 3]]$
	$KO_{[3^1,4^4]}$	30	4	4	$[[36, 4, 4]]$
$[3^3, 4^1, 3^1, 4^1]$	$KO_{1[3^3,4^1,3^1,4^1]}$, $KO_{2[3^3,4^1,3^1,4^1]}$, $KNO_{1[3^3,4^1,3^1,4^1]}$, $KNO_{2[3^3,4^1,3^1,4^1]}$, $KNO_{3[3^3,4^1,3^1,4^1]}$	36	4	3	$[[36, 4, 3]]$

Table 4 Table of codes associate to dth cover SEMs with $\chi = -1$

Map type	Maps ($\mathcal{K}$)	$n = \|EG(\mathcal{K})\|$	$k = 2 - \chi(\mathcal{K})$	d_{min}	$[[n, k, d_{min}]]$
$[3^7]$	$\mathcal{N}_i^d, i = 1, 2, \ldots, 6$	$42d$	$2(1 + d)$	3	$[[42d, 2(1 + d), 3]]$
$[3^4, 4^2]$	$KO^d_{1[3^4,4^2]}$, $KO^d_{2[3^4,4^2]}$, $KNO^d_{[3^4,4^2]}$	$36d$	$2(1 + d)$	3	$[[36d, 2(1 + d), 3]]$
$[3^1, 4^4]$	$KNO^d_{[3^1,4^4]}$, $KNO^d_{2[3^1,4^4]}$	$30d$	$2(1 + d)$	3	$[[30d, 2(1 + d), 3]]$
	$KO^d_{[3^1,4^4]}$	$30d$	$2(1 + d)$	4	$[[36d, 2(1 + d), 4]]$
$[3^3, 4^1, 3^1, 4^1]$	$KO^d_{1[3^3,4^1,3^1,4^1]}$, $KO^d_{2[3^3,4^1,3^1,4^1]}$, $KNO^d_{1[3^3,4^1,3^1,4^1]}$, $KNO^d_{2[3^3,4^1,3^1,4^1]}$, $KNO^d_{3[3^3,4^1,3^1,4^1]}$	$36d$	$2(1 + d)$	3	$[[36d, 2(1 + d), 3]]$

5 Conclusion

This paper has presented a technique to construct a homological quantum code associated with semi-equivelar maps. Using this technique, we constructed codes associated with semi-equivelar maps on the surface with Euler characteristic -1 and -2. We also have constructed 15 new classes of homological quantum codes associated with dth cover of equivelar and semi-equivelar maps existed on the surface with Euler characteristic $-1, -2$. This work can be extended to other classes of maps on higher genus surfaces.

References

1. Albuquerque CD, Palazzo R, Silva EB (2009) Topological quantum codes on compact surfaces with genus $g \geq 2$. J Math Phys 50:023513
2. Bhowmik D, Maity D, Upadhyay AK, Yadav BP (2020) Semi-equivelar maps on the surface of Euler genus 3. http://arxiv.org/abs/2002.06367
3. Bhowmik D, Maity D, Yadav BP, Upadhyay AK (2020) New classes of quantum codes associated with surface maps. https://arxiv.org/abs/2007.01684
4. Bhowmik D, Upadhyay AK (2019) A Classification of Semi-equivelar maps on the surface of Euler characteristic -1. Indian J Pure Appl Math (accepted)
5. Bhowmik D, Upadhyay AK (2020) Some Semi-equivelar Maps of Euler Characteristics -2. Natl Acad Sci Lett. https://doi.org/10.1007/s40009-020-01026-7
6. Bombin H, Martin-Delgado MA (2007) Homological error correction: cassical and quantum codes. J Math Phys 48:052105
7. Calderbank AR, Rains E, Shor PW, Sloane N (1998) Quantum error correction via codes over $GF(4)$. IEEE Trans Inf Theory 44:1369–1387
8. Datta B, Maity D (2018) Semiequivelar maps on the torus and the Klein bottle are Archimedean. Discret Math 341(12):329–3309
9. Datta B, Upadhyay AK (2006) Degree-regular triangulations of the double-torus. Forum Math 18:1011–1025
10. Hatcher A (2002) Algebraic topology. Cambridge University Press
11. Kitaev AY (2003) Fault-tolerant quantum computation by anyons. Ann Phys 303:2–30
12. Naghipour A (2019) New classes of quantum codes on closed orientable surfaces. Cryptogr Commun 11:999–1008
13. Shor PW (1995) Scheme for reducing decoherence in quantum memory. Phys Rev A 2:2493–2496
14. Tiwari AK, Upadhyay AK (2017) Semi-equivelar maps on the surface of Euler characteristic -1. Note Math 37:91–102
15. Tillich JP, Zémor G (2009) Quantum LDPC codes with positive rate and minimum distance proportional to $n^{1/2}$. In: Information Theory, 2009. ISIT 2009. IEEE International Symposium, pp 799–803. IEEE
16. Upadhyay AK, Tiwari AK, Maity D (2014) Semi-equivelar maps. Beiträge Algebra Geom 55:229–242

Big Five Personality Prediction from Handwritten Character Features and Word 'of' Using Multi-label Classification

Salankara Mukherjee, Ishita De Ghosh, and Debatree Mukherjee

Abstract Handwriting is an expressive behavior and communicates one's personality characteristics. In this work, we try to build a correlation between personality traits and handwriting features employing machine learning techniques. Quantitative analysis is done for correlating handwritten features and personality traits. Two datasets are created for the work, personality trait value dataset and handwriting feature value dataset. The Big Five personality assessment model along with BFI questionnaire are used to collect personality data. It is observed that in handwriting samples of a person, there is a consistency of certain features in respect of graphometric measurements which are both repeatable and reliable. We utilize the measurements related to four lowercase letters of English handwritten script, namely 'a', 'g', 'n' and 't', and the features of the word 'of' as well. Personality trait prediction from handwriting features is formulated as a multi-label classification problem. Each of the five traits corresponds to a class in the classifier. Both non-binary and binary label values are examined by using three multi-label classifiers, Classifier Chain (CC), Binary Relevance (BR) and Label-Power set (LP). We employ KSTAR, K-nearest neighbor and Multi-layer Perceptron (MLP) as base classifiers. CC-KSTAR (94.1% with non-binary label values) and CC/LP-KSTAR (97.6% with binary label values) provide the best result considering train-test (67:33) as test option. CC-KNN (ED) (98.1% with non-binary label values) displays the best result considering cross validation as a test option.

Keywords Personality analysis · Big Five model · Handwriting feature analysis · Multi-label classification

S. Mukherjee (✉)
Department of Computer Science, West Bengal State University, Barasat, Kolkata 700126, India

I. D. Ghosh
Department of Computer Science, Barrackpore Rastraguru Surendranath College, Kolkata 700120, India

D. Mukherjee
Department of Psychology, Barrackpore Rastraguru Surendranath College, Kolkata 700120, India

© The Author(s), under exclusive license to Springer Nature Singapore Pte Ltd. 2022

D. Giri et al. (eds.), *Proceedings of the Seventh International Conference on Mathematics and Computing*, Advances in Intelligent Systems and Computing 1412,
https://doi.org/10.1007/978-981-16-6890-6_21

1 Introduction

The personality of an individual is defined as a set of characteristic features related to behavior, cognition and emotional patterns evolving from biological factors as well as socio-cultural environment and experiences [1]. Personality psychology deals with interesting questions like what is personality and why a person is unique as an individual [2]. A common way to understand personality is by assessing personality traits or factors. The Five-factor model (FFM), commonly known as Big Five model, is a widely accepted personality assessment model for adult human beings [3, 4]. According to the model, personality can be understood considerably by assessing the five factors, viz., ***Openness to experience, Conscientiousness, Extraversion, Agreeableness*** and ***Neuroticism***, known collectively as *OCEAN*.

Similar to personality, the handwriting of an individual is shaped by biological factors as well as the environment and experiences. A person possesses a particular handwriting because handwriting is produced by distinctive neuromuscular movements triggered by the brain [5]. Since handwriting and personality both are influenced by genetic and memetic (environmental) factors, many studies have been done to correlate the two [6–8]. For example, in Graphology, psychological states, temperament and behavior are interpreted from handwriting by trained graphologists. There are some recent works on graphology-based personality or behavior analysis using automated techniques [9, 10]. But the use of graphology for identifying personality traits is debatable because it lacks a standard, and thrives on human insight and observation power [11].

Present work proposes a method to predict Big Five personality traits from handwritten text features without using graphology and employing machine learning techniques. For our work, two datasets are created by collecting data from a number of persons. The first one is a dataset of personality trait scores calculated from self-reports using the Big Five Inventory (BFI questionnaire) collected from a number of subjects. The second one is a dataset of handwriting feature values extracted from English handwritten text samples collected from the same set of subjects. Machine learning tools are used then to find out the inter-relationship between personality scores and feature values. Prediction of personality traits based on the Big Five model is treated as a multi-label classification problem, because each of these traits corresponds to a class. Results obtained with multi-label classifiers like Classifier Chains (CC) (for non-binary binary label values) and Classifier Chains (CC), Binary Relevance (BR) and Label Power-set (LP) (for binary label values) are presented here. Along with the multi-label classifiers, three base classifiers, namely KSTAR, K-nearest neighbor (KNN) and Multi-layer Perceptron (MLP) are used. Considering train-test (67:33) as a testing option, CC-KSTAR (94.1% with non-binary label values) and CC/LP-KSTAR (97.6% with binary label values) with a combination of features provide the best result and when cross validation as the testing option is considered, CC-KNN (ED) with word 'of' feature (98.1% with non-binary label values) displays the best result.

The contribution of the proposed work is given now. It establishes a correlation between personality factors and handwriting attributes in a quantitative manner. It employs machine learning tools for multi-label classification to predict Big Five personality traits from handwriting features. Literature survey shows that problems employing multi-label classifiers consider in general binary label values. The proposed method successfully employs non-binary label values (ranging from 1 to 10) to be used in Classifier Chain as a multi-label classifier. The dataset built by us is probably the second dataset after 'Solar flair' employing such non-binary label values to be used in the MEKA software (version 1.9.0) [12]. Comparison with state of the art of the work shows a reasonably high score. Finally, the work is done without the intervention of graphology. The paper is organized as follows. A review on related work is given in Sect. 2; basic concepts like the Big Five model, types of classification used and evaluation metrics are discussed in Sect. 3; the proposed work is presented in Sect. 4; experimental results are given in Sect. 5; and finally concluding remarks are given in Sect. 6.

2 Previous Work

Proposed work deals with three different aspects, viz. (i) feature extraction from handwritten text documents, (ii) personality prediction based on handwriting and (iii) multi-label classification. This section presents a brief review of the fields. Feature extraction at the macro and micro levels is being utilized in many applications, viz. handwriting recognition, writer identification, individuality detection and personality prediction. Srihari et al. have done feature analysis at the macro as well as micro levels for establishing the individuality of handwriting in [13]. Pervouchine et al. have analyzed structural features of characters 'd', 'y', 'f' and graphemes like 'th', for writer identification purpose in [14]. Srihari et al. have worked with a very common word 'and' for determining individuality among children's handwriting in [15]. Rachmawanto et al. used character-level features like eccentricity, roundness, perimeter and area for handwriting recognition purposes in [16]. Sharma et al. have used capital letters for writer identification purpose in [17]. For the same purpose, Mukherjee et al. have analyzed discriminating features of line, word, character-based features and their combinations in [18].

Coll et al. have explored connections between handwriting features and personality characteristics in a particular hiring scenario in [19]. Prasad et al. have explored allograph combinations in handwritings to assess the behavior of writers by the segmentation method and SVM classifier in [20]. Besides automated handwriting analysis, manual psychological analysis is done for each subject. Two-thirds of the data is used for training and the rest one-third is used for testing. The method gives results with 90.3% of accuracy. Champa et al. have used a rule-based classifier to assess personality traits of writers using baseline, pen pressure, slant of writing, letter 't' and the lower loop of the letter 'y' [21]. Each feature was evaluated by a different technique. A writer-independent handwriting analysis system to predict the personality traits of a writer is suggested in [22]. It is implemented by linear regression techniques using features like slant of baseline, pen pressure, slant of letters and size of writing. Joshi et al. have used a method to predict personality traits from handwriting features by using a machine learning approach in [23]. It is implemented by the handwriting features like baseline, margin, slant of the words and height of t-bar. Personality prediction based on handwriting and signature recognition using artificial neural network and the multi-structure algorithm is proposed by Tarare et al. in [24]. Manimala et al. have proposed computer-assisted handwriting analysis by considering eight features such as size and slant of words, space between words, connectivity within a word, pressure, margin, baseline, loop of 'e' and distance between i-dot and i-stem in [25]. Experiments have been conducted for 80 samples and the accuracy obtained is 93.77%. In every work mentioned above, experimental design is not stated clearly and evaluation is done manually by trained graphologists. Garoot et al. have performed a survey on computerized analysis of handwriting in the light of graphology in [26]. It provides information about various techniques used for the automatic assessment of personality. To determine personality from offline handwriting, Bhade et al. have classified the photos of handwriting based on only three features, left margin, right margin and word spacing in [27]. Extracted features are mapped with personality using the rule-based technique. Lemos et al. have explored personality prediction based on handwriting analysis based on line-based features using Convolutional Neural Network (CNN) [28]. The last two works have been done recently, but still they use macro features only, though micro features are more expressive in terms of individuality and hence personality. Ghosh et al. proposed an automatic system to analyze English lowercase letters except 'g', 'j', 'n' and 'z' to identify human behavior [10]. To evaluate the system, subjects are requested to check the system response based on his or her handwriting by choosing agree or disagree options. The accuracy of the system is 86.70%.

Fallah et al. have used Minnesota Multiphasic Personality Inventory (MMPI) for training their system in [29]. Hidden Markov Model (HMM) is employed to classify writer-specific attributes, whereas a neural network (NN)-based approach is used to classify attributes which are not writer-related. The system achieves more than 70% of accuracy. An exploratory study is conducted by Chen et al. using handwriting information and seven personality dimensions (using QZPS questionnaire [30]) in [31]. The authors have used Support Vector Machine (SVM), AdaBoost and K-Nearest Neighbor classifiers for each of the seven personality dimensions in order to analyze a unique set of handwriting features. The results are promising with accuracy ranging from 62.5% to 83.9%. Mihai et al. have used the Myers–Briggs personality detection model and handwriting features: baseline, word slant, pen-pressure, letter 't' and letter 'f' in [9, 32]. They have used handwriting feature values in binary coded form, called handwriting map (HM), which are used in a three-layer architecture based on neural networks to predict Big Five traits in [33]. It is shown that predefined texts in the training stage give better results offering 84.4% accuracy for intra-subject tests and 80.5% accuracy for inter-subject tests.

Personality prediction methods described above are based on handwriting data. There are some recent works in which Lima et al. have tried to predict Big Five personality based on social media data collected from platforms like Twitter and Facebook [34]. The system achieves approximately 83% accuracy for prediction from Twitter messages. Relevant work which inspires us to use multi-label classification is the following. Comparison of different problem transformation methods for multi-label classification using the MEKA system is given by Modi et al. in [35]. The multi-label classification method to classify Tamil phonemes is described by Pushpa et al. in [36]. A comprehensive review of the currently available multi-label learning software like MEKA, MULAN, CLUS, etc. are provided in [37] by Charte. A brief discussion on basic concepts is given in the next section.

3 Preliminary Concepts

Basic concepts on the Big Five personality model and multi-label classification are discussed one by one now. Evaluation metrics used in this work are discussed after that.

3.1 Big Five Personality Model

The Big Five personality model or the Five Factor model (FFM) provides a general consensus on the taxonomy of personality traits, derived from the analysis of natural language terms people use to describe themselves and others [38, 39]. It integrates diverse systems of personality descriptions in a common framework instead of replacing them [40]. It is a well-explored model across multiple cultures and in diverse populations [2]. The five factors are described in brief now [34, 41].

- Openness: Openness is related to one's intellect and ability to create and accept novel ideas and things. People high on this dimension are in general imaginative, witty, original and artistic, whereas people low on this dimension have in general a preference for familiarity and conventionality.
- Conscientiousness: It is also called lack of impulsivity. Conscientious people are in general cautious, dependable, organized and responsible; they are also oriented toward results and achievements. Impulsive people tend to be careless, disorderly and not dependable.
- Extraversion: People high on this dimension are in general energetic, enthusiastic, dominant, sociable and talkative; they express positive emotions easily, and enjoy other people's company. Introvert people tend to be timid, shy, retiring and submissive.
- Agreeableness: Agreeable people are in general friendly, cooperative, trusting and warm. People low on this dimension are in general cold, quarrelsome and unkind.

- Neuroticism: Neuroticism is related to people who lack control over emotions and tend to experience negative emotions such as anger and anxiety easily; they are also more vulnerable to depression. People low on this dimension are in general calm and contented.

3.2 Multi-label Classification

Personality prediction based on the Big Five model is considered as a *multi-label classification (MLC)* problem because an individual is evaluated for the five traits each of which corresponds to a class. In MLC, the training set is composed of instances each of which is associated with a set of class labels, and the task is to predict the label-set of an unknown instance by analyzing training instances with known label-sets. Now multi-label learning is formally defined [42].

Let T be a training set with N instances $E_i, i = 1, \ldots, N$, where E_i is represented by M-dimensional feature vector $x_i = (x_{i1}, x_{i2}, \ldots, x_{iM})$. Let E_i be associated with a subset of class labels $Y_i \subseteq L$, where $L = \{y_j : j = 1, \ldots, q\}$ is the set of q possible labels. So $E_i = (x_i; Y_i)$, where $i = 1, \ldots, N$. The representation is shown in Fig. 1. The task of a multi-label learning algorithm is to generate a classifier H, which predicts Y, the subset of labels associated with an unlabeled instance $E = (x, ?)$.

Existing methods for MLC fall into two main categories, (i) problem transformation and (ii) algorithm adaptation. Categorization is done according to the way multi-label tasks are mapped to single-label tasks. *Problem transformation (PT)* maps a multi-label learning task into one or more single-label learning tasks. If the task is mapped into more than one single-label task, then it is treated as several independent classification problems, one for each label [42]. Binary Relevance (BR), Label Power-set (LP) and Classifier Chains (CC), which are an improvement of BR, are popular PT algorithms used in our work [36, 43]. All the three are discussed below. Moreover, KSTAR, KNN (ED) and MLP are used as base classifiers in our work. We choose these traditional classification methods because their performance is good in comparison to other methods.

- **Binary Relevance (BR):** In BR, a multi-label problem with q labels is converted into q single-label problems with binary label values [36]. Each problem is solved independently to get the final result. The concept is illustrated in Fig. 2a [44].

Fig. 1 Multi-label data

Instances	Feature vector X				Label Y
E_1	x_{11}	x_{12}	....	x_{1M}	Y_1
E_2	x_{21}	x_{22}	....	x_{2M}	Y_2
•	•	•	....	•	•
•	•	•	....	•	•
E_N	x_{N1}	x_{N2}	....	x_{NM}	Y_N

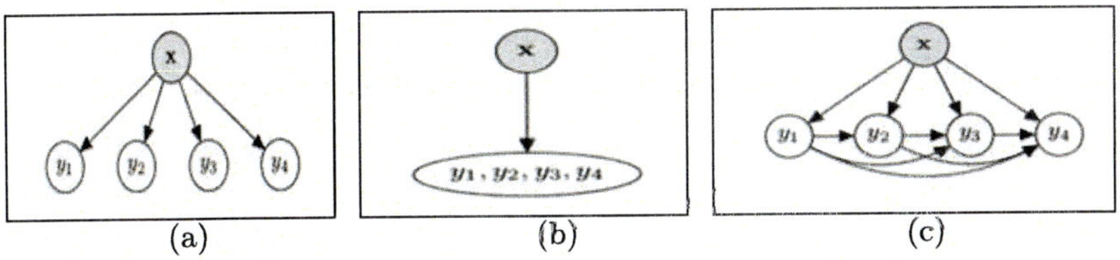

Fig. 2 Problem transformation methods: **a** Binary Relevance (BR), **b** Label Power-set (LP), **c** Classifier Chain (CC)

- **Label Power-set (LP):** In LP, the relationship between labels is considered, which is not so in BR. LP generates a new class for every combination of labels and then solves the problem using multiclass classification approaches. Its drawback is that an exponential number of classes may be generated, leading to very few instances in some classes. The concept is illustrated in Fig. 2b [44].
- **Classifier Chain (CC):** Classifier chains is based on BR, but it overcomes the disadvantages of BR and achieves higher performance. It also provides low time complexity. Target variables $(y_1, y_2, y_3, \ldots, y_q)$ are not fully independent. The features $(x_1, x_2, x_3, \ldots, x_m)$ are initially used to predict y_1. Next $(x_1, x_2, x_3, \ldots, x_m, y_1)$ is used to predict y_2.
  At the qth step, $(x_1, x_2, x_3, \ldots, x_m, y_1, \ldots, y_{q-1})$ predicts y_q. The ordering in which the labels are predicted can be specified by the user and can greatly influence the results. The method CC is illustrated in Fig. 2c [44].

3.3 Evaluation Metrics

Evaluation metrics, specially designed for MLC, are given now [36, 42]. Let N be the number of instances in a multi-label dataset and, L be the set of all labels defined for the dataset. For the ith instance, let Y_i be the set of predicted labels, and Z_i be the set of true labels, $Y_i, Z_i \subseteq L$.

Accuracy is the average ratio of correctly predicted labels to the total number of true or false labels for all instances. Greater accuracy means better classifier.

$$Accuracy = \frac{1}{N} \sum_{i=1}^{N} \frac{|Y_i \cap Z_i|}{|Y_i \cup Z_i|}$$

Hamming loss is the average ratio of incorrectly predicted labels to the total number of labels for all instances. For Hamming Loss, smaller value means better performance.

$$Hamming\_Loss = \frac{1}{N} \sum_{i=1}^{N} \frac{|Y_i \Delta Z_i|}{|L|}$$

where Δ is the symmetric difference between Y_i and Z_i.

Hamming score is simply the difference of Hamming loss from unity. It is also the average value of the labels in a non-binary multi-label classification.

Exact_match is the number of instances for which the predicted set of labels is equal to the true set of labels, divided by the total number of instances. This is a strict measure as it requires the predicted set of labels to be an exact match of the true set of labels.

$$Exact\_match = \frac{1}{N}\sum_{i=1}^{N} I\left(Y_i = Z_i\right)$$

where $I\left(True\right) = 1$ and $I\left(False\right) = 0$.

Average precision is the proportion of labels ranked ahead of a certain relevant label. The goal is to establish how many positions have to be traversed until this label is found. The best performance is reached when average precision is equal to 1. The bigger the value of average precision is, better the performance.

$$Avg\_precision =$$

$$\frac{1}{N}\sum_{i=1}^{N}\frac{1}{Y_i}\sum_{y \in Y_i}\frac{|\{y\prime \in Y_i : rank(x_i, y\prime) \leq rank(x_i, y)\}|}{rank(x_i, y)}$$

Proposed work is described now.

4 Proposed Work

As stated already, the objective of the proposed work is to establish a correlation between handwriting and personality by empirical methods. The flow of our work is now given in an algorithmic way.

Proposed Algorithm:

- **Step I:** Preprocess input images using the following steps.
  1. RGB to grayscale conversion
  2. Noise removal from grayscale image using median filter
  3. Grayscale to binary conversion
- **Step II:** Separate words and characters from preprocessed images.
  1. Apply procedure 'Word separation' to separate the words.
  2. Apply procedure 'Character separation' to separate the characters.
  3. Select words 'of' from the list of separated words manually.
  4. Select characters 'a', 'g', 'n' and 't' manually from list of separated characters.
- **Step III:** Calculate Big Five trait (OCEAN) scores for each subject using the method described in Sect. 4.1.

- **Step IV:** Measure handwriting feature values for each subject using the methods described in Sect. 4.2.
- **Step V:** Create datasets for (i) OCEAN score values, (ii) handwriting feature values.
- **Step VI:** Train and test multi-label classifiers using the datasets. 33% of the dataset is used as testing purpose, and the rest is used for training purpose.
- **Step VII:** Evaluate performance parameters for predicting personality values.

Procedure: Word Separation

- **Step 1:** Copy preprocessed image I into a temporary matrix T.
- **Step 2:** Apply morphological closing operation on T with *rectangular* structuring element on temporary image matrix.
- **Step 3:** Determine *Bounding box* of each closed object in the image.
- **Step 4:** Extract words by mapping *corner-coordinates of bounding boxes* found in T onto I.

Procedure: Character Separation

- **Step 1:** Apply *connected component separation* method on each extracted words.
- **Step 2:** If characters within an extracted word are not connected, obtain individual characters from the word as a result of the above step.
- **Step 3:** If characters within an extracted word are connected, obtain individual characters manually from the word.

Only data structure required for the algorithm is two-dimensional arrays, and time complexity is $O(M \times N)$, where $M \times N$ is the size of the input images. Schema diagram in Fig. 3 portrays the flow of our work pictorially. Creation of dataset is a necessary pre-requisite of the work, since no dataset is available. It is described now.

4.1 Dataset Creation

Two types of data are collected for our work, viz. handwritten text samples and personality self-study reports. The subjects under study are Bengali-speaking individuals from middle-class families residing in West Bengal, India. There are 50 subjects in total, of which, 27 are male and 23 are female. All of them are aged between 25 and 70, and are educated, the minimum qualification being a graduation degree. Before providing data, they fill up the consent form approved by West Bengal State University Ethical Committee. The subjects are requested to provide a self-study report on five traits together known as OCEAN, based on a 44-item Big Five inventory (BFI). They have to fill up the BFI response form on a five-point scale (1–5) as given now: 1 for *disagree strongly,* 2 for *disagree a little,* 3 for *neither agree nor disagree,* 4 for *agree a little,* 5 for *agree strongly.* For each subject, OCEAN score values are calculated from the report by using BFI scoring formula [40]. Resultant values are floating-point values in the range of 1–5. For our work, the values are

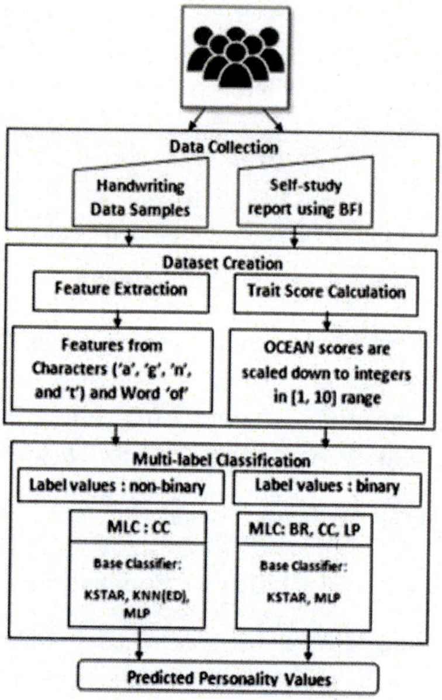

Fig. 3 Schema diagram of proposed work

scaled in the range of 1–10. These *non-binary values*, are then used as trait label values to create the dataset.

In order to collect handwriting data, the subjects are requested to write three pages of English text on A4-size sheets with black or blue ball-point pens. Acquisition of handwriting data is done using a mobile scanner app. To extract features from the scanned images, a total number $600(=50 \times 3 \times 4)$ of the word 'of' (considering four occurrences per set per sample) and $2400(=50 \times 3 \times 4 \times 4)$ characters (considering average four samples of each type of four letters from each set) are segmented and analyzed. Feature extraction procedures are discussed in detail now.

4.2 *Handwriting Feature Extraction*

Feature selection and extraction are important parts of handwriting analysis because quantitative values are obtained through the features. In this work, we pay attention to individuality analysis, because personality has an important relation with individuality. Features [45] like the design of allographs, their structures, dimensions, starting and ending strokes of letters and structures of some special words are useful features for individuality analysis. Now, we describe the features used in this work, viz. four characters *'a'*, *'g'*, *'n'* and *'t'* and a popularly used word *'of'*.

Character-Based Features: Character-based features are more suitable than global or macro features for personality prediction because they better express the individuality of writing [18]. Some common strokes in English lowercase letters are

Table 1 List of character features

Character	Features
Small 'a'	Orientation
	Ellipticity
Small 'g' lower stem loop	Lower stem loop closed or not
	Horizontal width within the loop
	Height to width ratio of the loop area
	Nature of loop
	Lower stem slant angle
Small 'n'	Eccentricity of the hump
	Hump shape Arcade or Angular
Small 't'	t-bar connected with stem or not
	Nature of 't' w.r.t. upper and lower parts of stem
	't' stem slant
	't' bar position

Fig. 4 Different writing styles of small 'a'

Doughnut, Stem loop, Hump and *Stick*. In general, a writer follows the same stroke patterns for similarly constructed letters [46]. In this work, we deal with four lower-case letters, namely *'a', 'g', 'n'* and *'t'*, which contain the above four strokes. Each of the letters and their various combinations offer a variety of writing styles to explore individuality. The list of character-based features is given in Table 1. For feature analysis, the letters are segmented at first. For each letter, a region of interest (ROI) is selected [47]. ROI for each letter is listed below [18].

- Doughnut-like shape in letter 'a'.
- Lower stem loop pattern in letter 'g'.
- Hump-like pattern in letter 'n'.
- Stick-like pattern in letter 't'.

Features of Character 'a': Different writing styles for letter 'a' are displayed in Fig. 4. To examine the doughnut-shaped pattern (conic portion) present in letter 'a', we consider two features, *orientation* and *ellipticity* [18]. The angle between the horizontal axis and the major axis of the conic section is called orientation. The value is calculated in degrees, ranging from $-90°$ to $+90°$. To find out whether the doughnut-shaped part of 'a' is circular or elliptical, its similarity or dissimilarity with a circle is examined. A measure called ellipticity is used for that. It is calculated by the *angle* θ between the major and the minor axis of an elliptic shape.

Fig. 5 Characteristics of letter ‘a’: **a** Positive orientation, almost circle **b** Negative orientation, elliptic in nature **c** measurement of Ellipticity

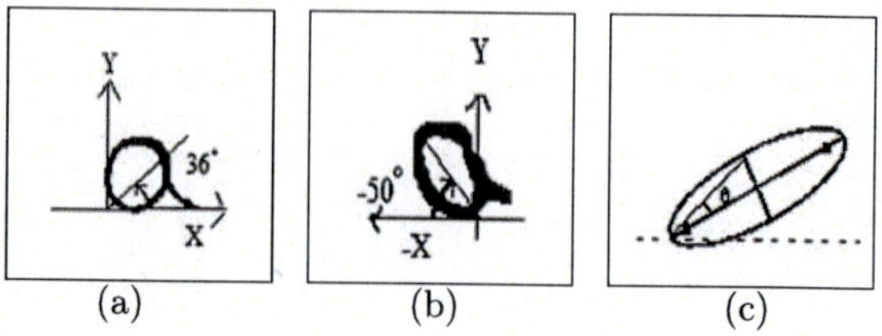

$$\theta = \tan^{-1} \frac{minor\_axis}{major\_axis}$$

When θ is 45°, the shape is a perfect circle. When θ is between 40° and 50°, the shape is almost a circle. When θ is less than 40° or more than 50°, the shape is elliptical in nature. Figure 5 illustrates the concepts of orientation and ellipticity.

Features of Character ‘g’: In this work, we consider only the features related to the lower stem loop of the letter ‘g’. Different writing styles of ‘g’ stem loop are depicted in Fig. 6. Five characteristics of the lower stem of ‘g’ are analyzed. They are (i) whether it forms a loop or not, (ii) nature of the loop, (iii) width of the loop, (iv) height to width ratio of the loop and (v) slant angle of the ‘g’ stem. The algorithms for extracting the above-mentioned features are as follows:

Procedure: Detection of ‘g’ stem loop

- **Step 1:** Thin the segmented ‘g’ stem part to one pixel.
- **Step 2:** Check the 8 neighbors of each thinned pixel to count the number of neighbors.
- **Step 3:** Check for the neighbor count.

  if neighbor count for each pixel<=2 **then**
    no closed loop is formed
  else
    closed loop is formed
  end if

Procedure: Identification of ‘g’ stem loop nature

- **Step 1:** Check for stem loop dimension or pattern.

  if ‘g’ stem has a closed loop **then**
    (i) Calculate the horizontal width within the loop.

Fig. 6 Different writing styles of small letter ‘g’

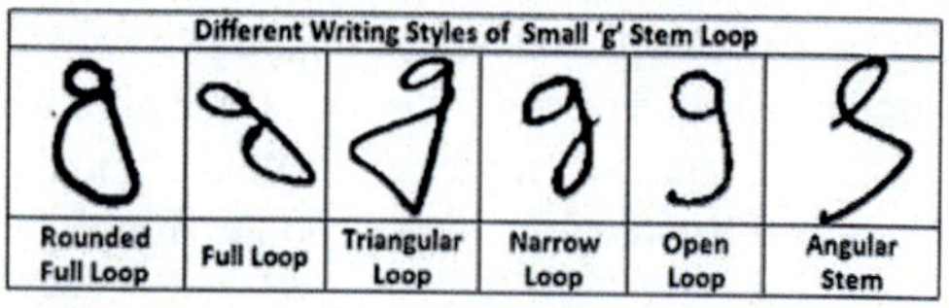

(ii) Calculate the horizontal height within the loop.
(iii) Calculate the height to width ratio to determine the loop pattern depending on some threshold.
else
Calculate only the width inside the curvature part of open stem loop.
end if

Procedure: Calculation of slant angle of 'g' stem
Slant angle of 'g' stem is calculated using the word-slant calculation method mentioned in [48].

- **Step 1:** Perform shear transformation for some range of angles on the 'g' stem image matrix.
- **Step 2:** Find the vertical projection profile.
- **Step 3:** The maximum vertical projection profile value is determined to find the slant angle.

Features of character 'n': For small letter 'n', we consider two features, (i) eccentricity and (ii) nature of hump-like structure [18]. *Eccentricity (EC)* is 1 or unity for a perfect parabola and its diminishing value indicates deviation from a parabolic shape.

$$Eccentricity = \sqrt{1 - \frac{(Minor\_axix)^2}{(Major\_axix)^2}}$$

Here, EC is calculated using the formula given above and its illustration with respect to letter 'n' is given in Fig. 7. Different hump-like structures of the letter 'n' are displayed in Fig. 8. To determine whether the nature of the hump-like structure is arcade, angular or flat arcade, we can use the following algorithm.

Procedure: Detection of hump type of letter 'n'

- **Step 1:** Thin the segmented arched part or hump part of letter 'n' to one pixel.
- **Step 2:** Consecutive black pixels at each row are calculated from top to bottom manner.
- **Step 3:** Count consecutive black pixels.
  if Count $<$ threshold th **then**
  The hump shape is angular.
  else if Count $>$ $th + 3$ **then**
  The hump shape is flat arcade.
  else
  The hump shape is normal arcade.
  end if

Features of character 't': Five characteristics of the letter small letter 't' are examined. They are (i) connectivity, (ii) position of t-bar on the stem, (iii) presence of

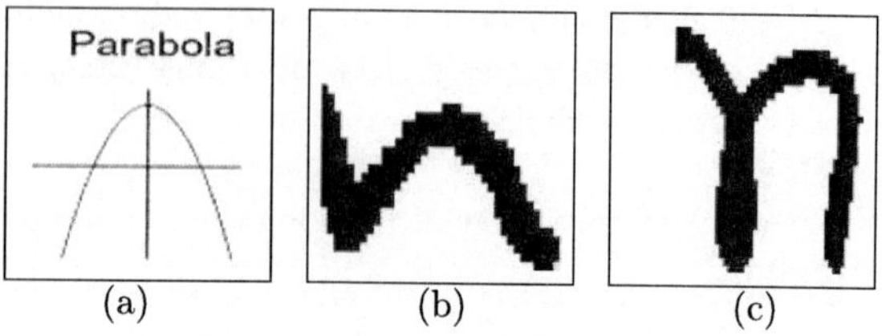

Fig. 7 Illustration of EC, specifically for letter 'n': **a** For a parabola EC = 1, **b** Letter 'n' with EC = 0.885, **c** Letter 'n' with EC = 0.580

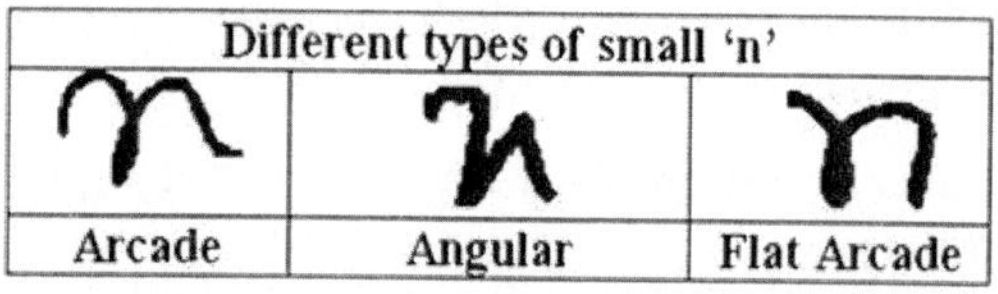

Fig. 8 Different writing styles of small letter 'n'

loop in t-stem, (iv) nature of right end of t-stem and (v) stem slant. The procedures for determining these features of letter 't' are described as follows.

Procedure: Detection of connectedness of 't'-bar and 't'-stem

- **Step 1:** Apply *connected component separation* method on image matrix of letter 't'.
- **Step 2:** Component count operation.

  if no. of component==1 **then**
    Letter 't' is connected.
  else
    Letter 't' stem and t-bar is not connected.
  end if

Procedure: 't' bar position detection

- **Step 1:** Calculate number of consecutive black pixels in each row from top to bottom manner.
- **Step 2:** bar_pos = row number with highest count of consecutive black pixels.
- **Step 3:** th = threshold value

  if bar_pos < th **then**
    't' bar position is above.
  else if bar_pos > th **then**
    't' bar position is below.
  else
    't' bar position is middle.
  end if

Procedure: Detection of upper stem loop of 't'

- **Step 1:** Apply one pixel thinning at the segmented upper stem part of 't'.
- **Step 2:** Check the 8 neighbors of each thinned pixel to count the number of neighbors.

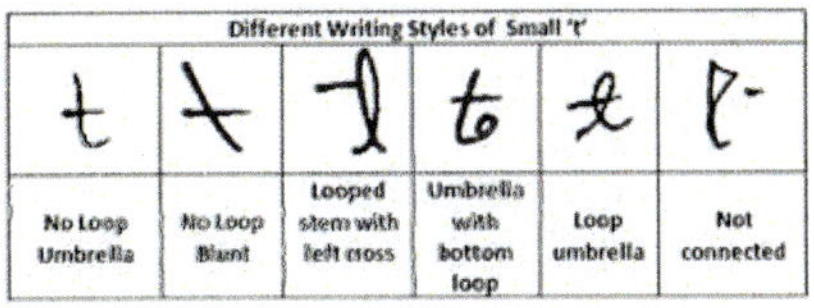

Fig. 9 Different writing styles of small 't'

- **Step 3:** Check for the neighbor count.

  if neighbor count for each pixel<=2 **then**
    no upper stem closed loop is present
  else
    upper stem closed loop is formed.
  end if

Procedure: Identification of right-end pattern of 't'-stem

- **Step 1:** Trace row-wise (5–10 rows) from the bottom to the top of the image matrix.
- **Step 2:** Check alternating presence of consecutive black pixels, then consecutive white pixels, and then consecutive black pixels again.

  if pattern is maintained **then**
    't' has an umbrella handle shaped curved end
  else
    't' has an blunt end.
  end if

Procedure: Calculation of slant angle of 't'-stem Slant angle of 't' stem is calculated similar to the 'g' slant calculation method mentioned in [48].

- **Step 1:** Perform shear transformation for some range of angles on the 't' image matrix.
- **Step 2:** Find the vertical projection profile.
- **Step 3:** The value of maximum vertical projection profile value is determined to find the slant angle.

Examples of small letter 't' is given in Fig. 9.

Features of Word 'of': Now we analyze features of the word 'of'. Previously, the authors obtained encouraging performances in the field of writer identification and individuality study using the word 'and' and grapheme 'th' [14, 15]. Now we explore characteristics of the word 'of' to make a correlation between handwriting features and personality. The choice of the word 'of' is based on several considerations:

- The word 'of' is a very popular word as well as a preposition which occurs frequently in handwriting samples.
- It is observed manually that the construction of the word 'of' is unique for most of the individuals. Experimental results also support that.
- The presence of the small letter 'f' in the word 'of' is very useful because the letter 'f' is a popular alphabet in the English script which covers all the three zones of writing.

Type of 'of'	Features	Type of 'of'	Features
	'o' in upper zone; word slant: extreme left; 'f' nature: Simple no loop blunt.		'o' in middle zone; word slant: vertical 'f' nature: fluid double stroke, 'f' bar absent.
	'o' in upper middle zone; word slant: vertical; 'f' nature: Small upper loop umbrella.		'o' in upper middle zone; word slant: moderate right; 'f' nature: lower arc stroke back.
	'o' in lower zone; word slant: vertical; 'f' nature: Simple no loop blunt.		'o' in middle zone with eyelet word slant: moderate right; 'f' nature: Upper narrow lower full loop
	'o' in middle zone; word slant: vertical; 'f' nature: Big lower loop		'o' in middle zone with eyelet word slant: moderate left; 'f' nature: balanced.

(a)

SL_no	Features of the word 'of'	
1.	Boundary values	Height
		Width
		Height to width ratio
2.	Position of 'o' w.r.t 'f'	
3.	Slant angle of word 'of'	
4.	'o' and 'f' connected or not	
5.	Gap between 'o' and 'f'	
6.	No. of loops present in word 'of'	
7.	Nature of 'o'	
8.	Nature of 'f'	
9.	'f' bar position	

(b)

Fig. 10 **a** Different writing styles of word 'of'. **b** List of features of word 'of'

- We are able to obtain the relative position of the middle zone letter 'o' w.r.t. the three-zone character letter 'f'. Most of the individuals maintain this relative positioning of 'o' and 'f' unconsciously.

List of features is depicted in Fig. 10a and different writing styles of the word 'of' are displayed in Fig. 10b.

Extraction procedures of these features are given in detail now.

Procedure: Boundary value calculation

- **Step 1:** Determine the bounding box of the image matrix.
- **Step 2:** height = bottom_row − top_row, width = right_col − left_col
- **Step 3:** h_w_ratio = height/width

Procedure: Determination of position of 'o' w.r.t. 'f'

- **Step 1:** Divide the image into three zones.
- **Step 2:** up_r = up_zone_width − mid_zone_width low_r = low_zone_width − mid_zone_width
- **Step 3:** Position identification of 'o' w.r.t. 'f'

  if up_r > low_r **then**
    'o' is at the lower zone w.r.t 'f'
  else if up_r < low_r **then**
    'o' is at the upper zone w.r.t 'f'.
  else
    'o' is in almost middle zone w.r.t 'f'.
  end if

Procedure: Slant angle calculation of word 'of'
Slant angle of word 'of' is calculated by the method given in [48].

- **Step 1:** Perform shear transformation for some range of angles on the image matrix.
- **Step 2:** Find the vertical projection profile.

- **Step 3:** The value of maximum vertical projection profile value is determined to found the slant angle.

Procedure: 'o' and 'f' connected or not

- **Step 1:** Apply *connected component separation* method on image matrix of letter 't'.
- **Step 2:** Component count operation.

  if no. of component==1 **then**
  'o' and 'f' are connected.
  else
  'o' and 'f' are not connected.
  end if

Procedure: Calculation of gap between 'o' and 'f'

- **Step 1:** Identify the right-most black pixel position (row and column) of letter 'o'.
- **Step 2:** Traverse column-wise until another black pixel appears in that particular row.
- **Step 3:** The number of columns traversed represent the distance between 'o' and 'f'.

Procedure: Position of 'f' bar

- **Step 1:** Calculate number of consecutive black pixels in each row from top to bottom manner.
- **Step 2:** bar_pos = row number with highest count of consecutive black pixels.
- **Step 3:** th = threshold value

  if bar_pos < th **then**
  'f' bar position is above.
  else if bar_pos > th **then**
  'f' bar position is below.
  else
  'f' bar position is middle.
  end if

Loops in character stokes are mostly circular or elliptical. By identifying the number of circular shapes in the image, we will count the **number of loops present** in the word 'of'.

Nature of 'o' is to determine whether there is an eyelet present within the 'o' structure or not. To identify this, we traverse row-wise within the circular area of 'o', if we get consecutive black pixels in one or two rows sequentially then an eyelet is present within 'o'.

Nature of 'f' is determined by identifying the upper part loop, lower part loop, presence of both upper and lower part loop and absence of loop structure. To identify whether there is any closed loop present in the upper or lower part of 'f', the specific part, after segmentation, is thinned to one-pixel width at first. Then for each pixel in the thinned portion, the number of neighbors is counted. If each pixel has at most two neighbors, then it is concluded that no closed loop is formed, otherwise it is concluded that a closed loop is formed.

5 Experimental Results and Discussion

The algorithms described above are implemented using MATLAB version R2014a (32bit) in a machine with Intel Core-i5 processor, 4 GB RAM and 64-bit operating system. For the classification process, we have used MEKA (version 1.9.0) for Multi-label Classification (MLC) [49].

Prediction results using classifier chain as multi-label classifier and non-binary label values, for various feature combinations of four characters and the word 'of' are shown in Table 2. Multi-label classification is used at the first stage of the classification process, because these MLC algorithms will convert the multi-label problem into several single-label problems and in the next stage different base classifiers will operate on those single-label problems.

CC is used as multi-label classifier and it is important to note that only CC can handle non-binary values as label values but similar attempts with BR and LP classifiers have failed. We have used two types of learners, i.e. Lazy and Eager. Base classifiers KSTAR and KNN (with Euclidean Distance (ED)) are an example of lazy learner and MLP is an example of eager learner. Accuracy values at each of the five labels, corresponding Hamming score, i.e. average value of the accuracy score of five labels, Hamming Loss and Exact match values are presented here. We use two different test options; first, the whole dataset is divided into train set and test set in a 2:1 ratio, and the next is Cross Validation with 10 folds.

Some observations from Table 2 are discussed now.

- KSTAR as base classifier in multi-label classification shows the best accuracy results *94.1%* among all base classifiers for the feature combinations (Character+word 'of') using Train-Test (67:33) option and non-binary label values.
- In the same scenario, MLP and KNN (ED) as base classifiers show comparatively lower accuracy values w.r.t. KSTAR.
- KNN (ED) as base classifier shows the best accuracy *92.9%* among all the base classifiers when the features of the word 'of' is only considered. But when the character feature is considered solely, KSTAR represents the best result, *91.8%* (for Train-Test (67:33) option).
- Using cross validation as test option and KNN (ED) as base classifier, the word 'of' shows the best accuracy *98.1%*.
- In the case of a 10-fold cross-validation test option, we can compare the three feature extraction cases with MLP as the base classifier. A combination of character features and the word 'of' features performs best, *96.4%* w.r.t. the other two cases.

Table 2 Prediction results using classifier chain as multi-label classifier and non-binary values (1, 2, 3, 4, 5, 6, 7, 8, 9) of label

Handwritten features	Test options	Types of base classifiers	Label wise accuracy					Hamming score	Exact match	Hamming loss
			O	C	E	A	N			
Word 'of'	Train: Test (67:33)	KSTAR	0.824	0.804	0.804	0.863	0.824	0.824	0.784	0.176
		KNN (ED)	0.961	0.902	0.902	0.980	0.902	0.929	0.902	0.070
		MLP	0.922	0.784	0.745	0.824	0.725	0.80	0.627	0.200
	Cross validation	KSTAR	0.935	0.909	0.916	0.948	0.909	0.923	0.896	0.077
		KNN (ED)	1.00	0.968	0.968	1.00	0.968	**0.981**	0.968	0.019
		MLP	0.955	0.870	0.896	0.883	0.805	0.882	0.701	0.118
Character features	Train: Test (67:33)	KSTAR	0.941	0.902	0.922	0.922	0.902	0.918	0.902	0.082
		KNN (ED)	0.882	0.863	0.941	0.922	0.902	0.902	0.863	0.098
		MLP	0.686	0.804	0.784	0.804	0.745	0.765	0.529	0.235
	Cross validation	KSTAR	0.974	0.974	0.994	0.981	0.974	0.974	0.974	0.026
		KNN (ED)	0.974	0.961	0.948	0.942	0.948	0.955	0.942	0.045
		MLP	0.857	0.877	0.864	0.851	0.812	0.852	0.701	0.148
Combination (Word 'of' + Character Features)	Train: Test (67:33)	KSTAR	0.941	0.941	0.941	0.941	0.941	**0.941**	0.941	0.059
		KNN (ED)	0.941	0.922	0.941	0.922	0.941	0.933	0.922	0.067
		MLP	0.882	0.902	0.902	0.941	0.902	0.906	0.804	0.094
	Cross validation	MLP	0.974	0.955	0.981	0.987	0.922	0.964	0.864	0.036

From the above points, we can conclude that individually the word 'of' features and character features both are very strong choices for feature selection in the personality prediction process; and we have also combined both of them to observe the results, whether it is encouraging or not. So more minute inspection of small or discriminating elements of handwriting leads to more intra-person similarity.

As our experiment with non-binary label values fails for BR and LP, only CC can handle this successfully. So to explore the results with other multi-label classifiers, we have tried to convert our non-binary label values to binary format. For this conversion, we count the frequency of each value (1–10) for each label. The label value with the highest frequency in each label is now treated as the threshold value. We denote the label values as '0' which are less than the threshold value and as '1' which are greater than the threshold value. Multi-label classifiers BR, CC and LP along with base classifiers KSTAR and MLP are used in this process. Only the Train-Test percentage split (67:33) test option is used in this case. Label-wise accuracy, Hamming Score, Exact Match, Accuracy and Average Precision values are displayed in Table 3.

Observations from Table 3 are now discussed in the following section:

- It is interesting to note that when the word 'of' features are treated individually, MLP as base classifier performs better than KSTAR as a base classifier in all the three multi-label classifications. MLP with LP combination shows the highest accuracy *94.1%* in the case of word 'of'.
- In case of only character features, KSTAR along with three MLC performs better than MLP. Both CC and LP with base classifier KSTAR provide the highest accuracy *97.6%*.
- Finally, when we perform the classifications with the combination of character features and the word 'of' features, we get the best results compared to the previous two cases. CC and LP along with base classifier KSTAR show the highest performance, *97.6%*. MLP with all the MLC also performs better than the previous cases.

A line chart in Fig. 11a and a bar chart in Fig. 11b are displayed to portray the overall scenario of this experiment with non-binary and binary label values, respectively.

Comparison with other work: In Table 4, we have shown the comparison of our work with some state-of-the-art work. Fallah et al. [29] and Chen et al. [31] have proposed two different methods on automated personality prediction using Arabic and Chines scripts and achieved accuracy 78.9% and between 62.5 and 83.9%, respectively. In [33], the authors propose a three-layer architecture based on binary data and neural networks that aims to determine the Big Five personality traits of an individual by analyzing offline handwriting. Their approach also offers 84.4% accuracy for intra-subject tests and 80.5% accuracy for inter-subject tests. Ghosh et al. proposed an automatic system to analyze English lowercase letters to identify human behavior using SVM and CNN classifiers and acquire 86.70% accuracy [10]. Our proposed work has framed a method of Big Five personality prediction based on handwriting features and multi-label classification. Datasets used for all work mentioned in

Table 3 Prediction results for multi-label classification using binary data as label values

Handwritten features	Types of base classifiers	Label wise accuracy					Hamming score	Exact match	Accuracy	Average precision
		O	C	E	A	N				
Word 'of'	KSTAR (BR)	0.941	0.863	0.922	0.882	0.902	0.902	0.765	0.892	0.808
	KSTAR (CC)	0.941	0.863	0.922	0.863	0.941	0.906	0.765	0.892	0.682
	KSTAR (LP)	0.941	0.882	0.922	0.882	0.941	0.914	0.784	0.902	0.682
	MLP (BR)	0.961	0.902	0.961	0.863	0.941	0.925	0.784	0.892	0.774
	MLP (CC)	0.961	0.941	0.922	0.922	0.922	0.933	0.784	0.916	0.701
	MLP (LP)	0.941	0.961	0.922	0.941	0.941	0.941	0.843	0.934	0.682
Character features	KSTAR (BR)	0.941	0.980	1.00	0.922	0.941	0.957	0.863	0.949	0.755
	KSTAR (CC, LP)	0.961	0.980	1.00	0.980	0.961	0.976	0.941	0.975	0.657
	MLP (BR)	0.882	0.902	0.941	0.902	0.843	0.894	0.608	0.862	0.793
	MLP (CC)	0.863	0.863	0.941	0.882	0.804	0.871	0.608	0.839	0.695
	MLP (LP)	0.941	0.902	1.00	0.980	0.882	0.941	0.843	0.937	0.686
Combination (Word 'of' + Character Features)	KSTAR (BR)	1.00	0.941	1.00	0.922	1.00	0.973	0.922	0.970	0.706
	KSTAR (CC, LP)	1.00	0.941	1.00	0.941	1.00	**0.976**	0.941	**0.976**	0.656
	MLP (BR)	0.941	0.941	0.961	0.922	0.941	0.941	0.784	0.907	0.754
	MLP (CC)	0.941	0.922	0.961	0.922	0.922	0.925	0.765	0.899	0.689
	MLP (LP)	0.902	0.980	0.980	0.941	0.902	**0.949**	0.863	**0.938**	0.683

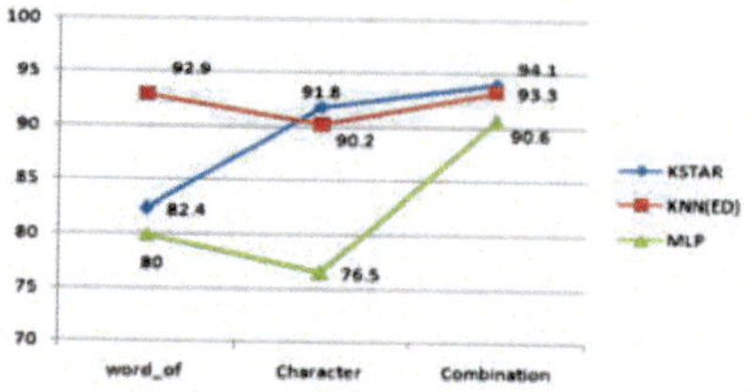

(a)

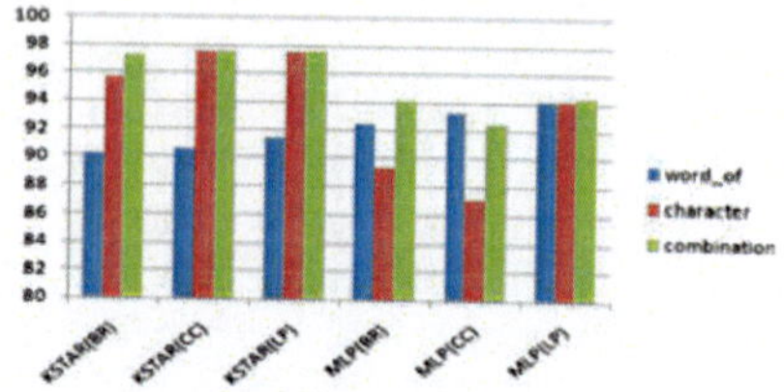

(b)

Fig. 11 **a** Comparison of performances of KSTAR, KNN (ED) and MLP as base classifiers with three feature extraction cases and non-binary label values, **b** Comparison of BR, CC and LP as MLC with two base classifiers and three feature extraction cases with binary label values

Table 4 Comparison of proposed method with state of the art

Work	Year	Method	Personality model	Dataset	Prediction accuracy	
					Binary label value	Non-Binary label value
Fallah and Khotanlou [29]	2016	Hidden Markov Models and neural networks	MMPI	70 writers	78.9%	–
Chen and Lin [31]	2017	Support vector machines, K-nearest neighbors, and AdaBoost classifiers	QZPS	56 writers	62.5–83.9%	–
Gavrilescu and Vizireanu [33]	2018	Feed-forward neural networks; Binary label values	Big five	128 writers	84.4% (intra-subject), 80.5% (inter-subject)	–
Ghosh et al. [10]	2020	SVM and CNN	Graphology based	30 writers	86.70%	–
Proposed work	2020	MLC (BR, CC, LP) Base classifier (KSTAR, KNN (ED), MLP)	Big five	50 writers	97.6%	94.1% (Train-Test), 98.1% (Cross validation)

Table 4 are private. All the previous work mentioned here used binary data as label values; we have used both binary and non-binary data as label values for classification and achieved satisfactory performance in both cases, 97.6% for binary data and 94.1% for non-binary data.

6 Conclusion

Big Five personality prediction based on handwriting features is explored using multi-label classification. Character features, word 'of' features and their combinations are analyzed to explore the usefulness of a person's individuality in the prediction process. To ignore the debatable role of Graphology, we utilize the personality trait scores collected through the BFI questionnaire. Multi-label classification with non-binary label values is used successfully for the correlation. We also explore the performances of BR, CC and LP in the case of binary label values. Using non-binary label values, CC with KSTAR provides an accuracy of 94.1% and using binary label values CC and LP with KSTAR gives accuracy 97.6%. Hence, correlation between handwritten features and personality traits is established quantitatively. Results extracted from a combination of features conclude that both character and word 'of' play the key role in the accuracy determination phenomenon. Our future plan is to extend the work by increasing the dataset by collecting data from many more subjects and increasing the number of feature values from handwriting samples. Critical analysis of more variety of features may result in better prediction.

Acknowledgements We express our gratitude to Prof. Kaushik Roy of West Bengal State University, Barasat, India, for guiding us to get the ethical clearance for our work.

References

1. Hall CS, Lindzey G, Campbell JB (2020) Theories of personality. Wiley, Berlin
2. Friedman HS, Schustack MW (2009) Personality: classic theories and modern research. Pearson Education, New Delhi
3. Rothmann S, Elize C (2003) The big five personality dimensions and job performance. South Afr J Ind Psychol 29:68–74
4. Corr PJ, Matthews G (2009) The Cambridge handbook of personality psychology. Cambridge University Press, Cambridge
5. Plamondon R (2010) Neuromuscular studies of handwriting generation and representation. In: International conference on frontiers in handwriting recognition (ICFHR), Kolkata, India
6. Galbraith D, Wilson W (1964) Reliability of the graphoanalytic approach to handwriting analysis. Perceptual Motor Skills 19:615–618
7. Sheikholeslami G, Srihari SN, Govindaraju V (1996) Computer aided graphology. In: International workshop on frontiers in handwriting recognition, England, pp 457–460
8. Kedar S, Nair V, Kulkarni S (2015) Personality identification through handwriting analysis: a review. J Adv Res Comput Sci Softw Eng 5(1)
9. Gavrilescu M (2017) 3-layer architecture for determining the personality type from handwriting analysis by combining neural networks and support vector machines. Electr Eng Comp Sci 79:135–152
10. Ghosh S, Shivakumara P, Roy P, Pal U, Lu T (2020) Graphology based handwritten character analysis for human behavior identification. CAAI Trans Intell Technol
11. Tett RP, Palmer CA (1997) The validity of handwriting elements in relation to self-report personality trait measures. Personal Ind Diff 22:11–18
12. Meka:a multi-label extension to weka (2018). https://sourceforge.net/projects/meka/files/Datasets. Accessed 9 Oct 2020

13. Srihari SN, Cha SH, Arora H, Lee S (2002) Individuality of handwriting. J Foren Sci 47(4)
14. Pervouchine V, Leedham G (2007) Extraction and analysis of forensic document examiner features used for writer identification. Pattern Recogn 40(03):1004–1013
15. Srihari SN, Meng L, Hanson L (2016) Development of individuality in children's handwriting. J Forens Sci 61(5):1292–1300
16. Rachmawanto EH, Anarqi GR, Setiadi DRIM, Sari CA (2018) Handwriting recognition using eccentricity and metric feature extraction based on k-nearest neighbors. In: 2018 international seminar on application for technology of information and communication, pp 411–416
17. Sharma MK, Chanderiya V (2020) Writer identification using graphemes. Indian Academy of Sciences
18. Mukherjee S, Ghosh ID (2020) Writer identification based on writing individuality and combination of features. In: 2nd international conference on applied signal processing
19. Coll R, Fornes A, Llados J (2009) Graphological analysis of handwritten text documents for human resources recruitment. In: 10th international conference on document analysis and recognition, pp 1081–1085
20. Prasad S, Singh VK, Sapre A (2010) Handwriting analysis based on segmentation method for prediction of human personality using support vector machine. Int J Comput Appl 08(12):0975–8887
21. Champa HN, AnandaKumar KR (2010) Automated human behavior prediction through handwriting analysis. In: 2010 first international conference on integrated intelligent computing, pp 160–165
22. Rahiman A, Varghese D, Kumar M (2013) Handwritten analysis based individualistic traits prediction. Int J Image Proc 07(2):209–218
23. Joshi P, Agarwal A, Dhavale A, Suryavanshi R (2015) Handwriting analysis for detection of personality traits using machine learning approach. Int J Comput Appl 130(15)
24. Tarare PR, Jadhav D (2015) Study of personality prediction based on handwriting and signature recognition using multiple artificial neural network and multi-structure algorithm. Int J Pure Appl Res Eng Technol 3(9):681–693
25. Manimala S, Gopal M, Gokhale PG, Chandrashekar S (2016) Automated handwriting analysis for human behavior prediction. Int J Comput Sci Eng 5(5):7–14
26. Garoot AH, Safar M, Suen CY (2017) A comprehensive survey on handwriting and computerized graphology. In: 14th IAPR international conference on document analysis and recognition (ICDAR), vol 01, pp 621–626
27. Bhade V, Baraskar T (2018) A model for determining personality by analyzing off-line handwriting. Springer, Singapore
28. Lemos N, Shah K, Rade R, Shah D (2018) Personality prediction based on handwriting using machine learning. In: International conference on computational techniques, electronics and mechanical systems (CTEMS), pp 110–113
29. Fallah B, Khotanlou H (2016) Identify human personality parameters based on handwriting using neural network. In: Artificial intelligence and robotics, pp 120–126
30. Dengfeng W, Hong C (2003) Processes and preliminary results in the construction of the chinese personality scale (qzps). J Chinese Psychol Acta Psychol Sinica 35:127–136
31. Chen Z, Lin T (2017) Automatic personality identification using writing behaviours: an exploratory study. Behav Inf Technol 1–8
32. Gavrilescu M (2015) Study on determining the myers-briggs personality type based on individual's handwriting
33. Gavrilescu M, Vizireanu N (2018) Predicting the big five personality traits from handwriting. EURASIP J Image Video Proc 57:1–17
34. Lima ACE, de Castro LN (2014) A multi-label, semi-supervised classification approach applied to personality prediction in social media. Neural Netw 58:122–130
35. Modi H, Panchal M (2012) Experimental comparison of different problem transformation methods for multi-label classification using meka. Int J Comput Appl 59(15):0975–8887
36. Pushpa M, Karpagavalli S (2017) Multi-label classification: problem transformation methods in tamil phoneme classification. In: 7th international conference on advances in computing and communications, Cochin, pp 572–579

37. Charte F (2020) A comprehensive and didactic review on multilabel learning software tools. IEEE Access 8:50330–50354
38. Goldberg LR (2006) An alternative "description of personality": the big-five factor structure. J Personal Soc Psychol 59(6):1216–1229
39. Costa P, McCrae R (1992) Normal personality assessment in clinical practice: the neo personality inventory. Psychol Ass 4:5–13
40. John O, Naumann LP, Soto CJ (2008) Handbook of personality: Theory and research, 3rd edn. Guilford Press, University of California at Berkeley
41. Goldberg L, Johnson J, Eber HW, Hogan R, Ashton MC, Cloninger R, Gough HG (2006) The international personality item pool and the future of public-domain personality measures. J Res Personal 40:84–96
42. Cherman EA, Monard MC, Metz J (2011) Multi-label problem transformation methods: a case study. CLEI Electron J 14(1)
43. Multi-label classification with scikit-multilearn (2018). https://xang1234.github.io/multi-label. Accessed 08 Oct 2020
44. Multi-label classification (2015). https://users.ics.aalto.fi/jesse/. Accessed Jan 2021
45. Srihari SN, Cha SH, Arora H, Lee S (2001) Individuality of handwriting: a validation study. In: 6th international conference on document analysis and recognition
46. Skills for action. https://skillsforaction.com. Accessed 11 oct 2020
47. Ratha NK, Govindaraju V (2008) Advances in biometrics sensors, algorithms and systems. Springer, London
48. Mahanta LB, Deka A (2013) Skew and slant angles of handwritten signature. Int J Innov Res Comput Commun Eng 1(9)
49. Read J, Reutemann P, Pfahringer B, Holmes G (2016) Meka: a multi-label/multi-target extension to weka. J Mach Learn Res 17:1–5

A Comparative Study of Pre-trained Gene Embeddings for COVID-19 mRNA Vaccine Degradation Prediction

U. Vamsi Krishna, B. Premjith, and K. P. Soman

Abstract Messenger Ribonucleic acid (mRNA) vaccine faces a challenge of structural instability, due to which the production of vaccine becomes a big challenge. The sequence information of the mRNA vaccine can provide possible degradation sites. Recently, Deep learning areas like Natural Language Processing have shown great promise in understanding these sequences. An appropriate sequence to vector representation is necessary to apply deep learning methods effectively. In this paper, pre-trained dna2vec, rna2vec, and lshvec gene embeddings are compared to identify the best vector representation for predicting the amount of degradation given the mRNA vaccine sequences. The comparison shows that dna2vec embedding performs best.

Keywords COVID-19 vaccine · mRNA sequencing · Vaccine degradation · Gene sequencing · Gene embedding

1 Introduction

In the given circumstances, the requirement of a quick and effective vaccine is necessary. mRNA vaccine has shown great promise for future vaccine production [12]. The development of these vaccines is simple and faster [3]. They have shown excellent results for a few types of cancer [13]. The major drawback of the mRNA vaccine is its structural instability [13]. An mRNA sequence expires in two days under highly controlled temperature conditions. The liquid-based vaccines can be stored for a week at room temperatures [23], but the effectiveness is still needed to be tested. Storing vaccines without damage becomes challenging which makes the distribution of vaccines almost near to impossible. Understanding the structural stability becomes the prime

U. V. Krishna · B. Premjith (✉) · K. P. Soman
Center for Computational Engineering and Networking (CEN), Amrita School of Engineering, Coimbatore, Amrita Vishwa Vidyapeetham, India
e-mail: b_premjith@cb.amrita.edu

K. P. Soman
e-mail: kp_soman@amrita.edu

© The Author(s), under exclusive license to Springer Nature Singapore Pte Ltd. 2022
D. Giri et al. (eds.), *Proceedings of the Seventh International Conference on Mathematics and Computing*, Advances in Intelligent Systems and Computing 1412,
https://doi.org/10.1007/978-981-16-6890-6_22

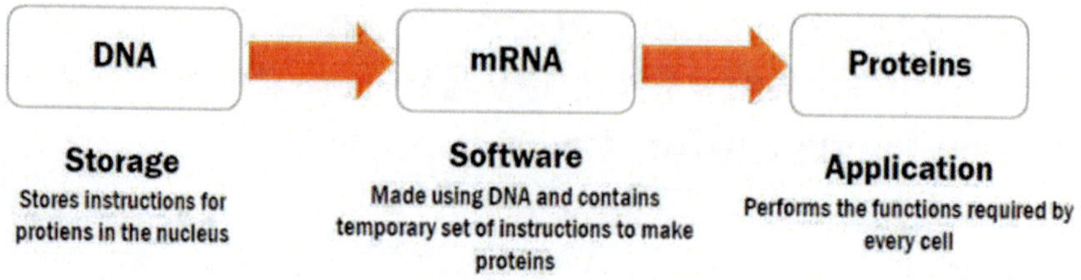

Fig. 1 The block diagram of the working of mRNA

goal for effective vaccine production. Reasons for the instability lies in the working of the mRNA. mRNA are temporary instructions made by DNA to produce particular proteins. The temporary nature of mRNA makes it very unstable and requires extreme temperature conditions to keep it stable. Figure 1 illustrated a computer science terminology to get a better intuition of how mRNA works. Deoxyribonucleic acid (DNA) acts as a storage of instructions whereas mRNA acts as a temporary software between the protein production and DNA. Protein production is like the application layer where the functions are performed based on the instruction provided by the DNA. DNA contains four nucleotides each of which is represented by a single letter: either A, T, C, or G [22]. The basic difference between DNA and RNA is the nucleotide 'T' which is replaced with nucleotide 'U'. The word analogy for DNA sequences is codon. Each codon contains only three nucleotides which implies that each word contains only three characters. Four characters with a word length of three making a total possibility of $(4 \times 4 \times 4)$64 codons or words [8]. The sequential and structural information of the mRNA can determine the degradation of the vaccine [13]. Recently, Stanford University released an mRNA vaccine dataset on Kaggle. Further information on the dataset is provided in Sect. 3. This dataset allows data scientists and machine learning enthusiasts to contribute to curing the global crises.

Recently, a lot of Deep learning methods are incorporated into mRNA sequencing and have shown promising results in interpreting the sequences. Natural Language Processing (NLP) has shown the capabilities in interpreting the human language appropriately. NLP uses word embedding for representing the words into vectors. These vectors serve as input to the machine learning and deep learning models to perform the necessary task. The basic embeddings are classified into local vector representation and distributed vector representations. Count-based methods like Bag-of-Words and Term Frequency-Inverse Document Frequency (TF-IDF) are used for local vector representation [16]. Word2vec and fastText are examples for distributed vector representations, which use either Skip-gram or Continuous Bag-of-Words (CBOW) for training the models [7, 17]. Embeddings used for converting DNA and RNA sequences into vectors are called Gene Embeddings [2]. Selecting an apt gene embedding becomes essential for retrieving the contextual information present in the sequences.

2 Literature Review

K-mers one-hot encoding is a common representation in DNA sequencing [18]. K-mers are used to convert the sequences into subsequences similar to words in a sentence and one-hot encoding is used to convert the k-mers into vectors. The major problem of one-hot encoding is that the distance between the k-mers vectors remains the same. Therefore, a distributed vector representation, known as dna2vec, was introduced [10]. Instead of using one-hot encoding, this method uses a shallow two-layer neural network to get the vector representation of the k-mers. The experimental results show that the addition of vectors is similar to nucleotide concatenation. The dna2vec has also been used in mRNA sequencing task [9] for predicting protein binding sites using one-dimensional Convolutional Neural Networks (CNNs). rna2vec is a gene embedding used for converting RNA sequences into vectors [11]. The rna2vec embedding with CNN architecture was used in the methylation prediction problem. The problem faced by most of the gene embeddings like dna2vec is that they cant use k-mers length more than 8 due to the computational cost. A gene embedding with local sensitivity hashing is presented in [20] known as lshvec to solve this computational problem. The embedding uses a fastText model and discards the sub-word feature to reduce the computational cost. This way, any number of k-mers can be used. Moreover, hashing helps in improving the searching in the lookup table. Long Short-Term Memory (LSTM) is a common model used in sequential modeling [5, 14]. Bidirectional LSTM (Bi-LSTM) is used to extract the forward and backward context of the sequences [4, 15]. CNN with Bi-LSTM is a commonly used architecture in DNA and RNA sequencing [6]. Bi-LSTM paired with CNN is effective compared to other sequential models [19].

3 Dataset Description

The data is taken from the Kaggle competition dataset titled 'Open Vaccine COVID-19 vaccine degradation prediction' by Stanford University. The task of the competition is to predict the degradation rates at various locations along the RNA sequence. The dataset contains the sequence, structure, and predicted loop information of 2100 mRNA vaccine sequences with corresponding degradation rates. The task is to predict five degradation indices. The five degradation indices are as follows: One is reactivity, and the other four indices are the degree of fragility in four different conditions Degree of fragility at 10 PH value, degree of fragility at 10 PH value with magnesium, degree of fragility at 50 C temperature, and degree of fragility at 50 C temperature with magnesium. More details about the data can be found here [21]. The sequence information is used to compare the gene embeddings.

Fig. 2 The block diagram of the experimental setup for the comparative study

Fig. 3 CNN Bi-LSTM architecture diagram for rna2vec embedding

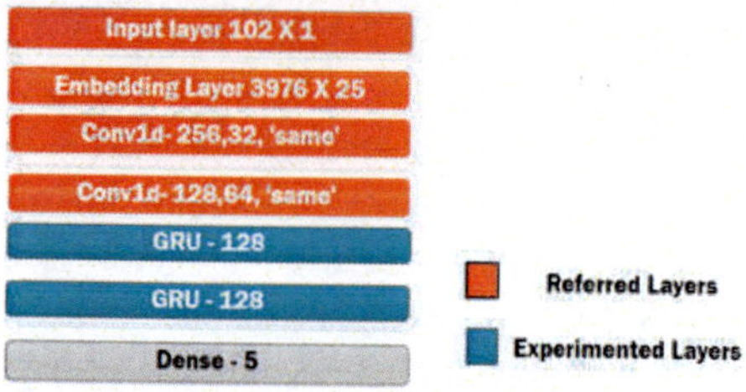

4 Experimental Setup

The most common architectures used are Bi-LSTM with CNN and without CNN, as mentioned in Sect. 2. In this work, a hybrid Bidirectional Gated Recurrent Unit (Bi-GRU) with CNN and Bi-GRU models are used for comparing the performance. GRUs are similar to LSTMs but they have a fewer number of parameters [1]. The train, validation, and test split is 70, 10, and 20, respectively. Initially, the input sequence is converted into k-mers. The k-mers are converted into vectors using pre-trained gene embeddings. It's a regression task where the model has to predict continuous values given the sequence. The metric used to evaluate the predicted values is Mean Column Root Mean Square Error (MCRMSE). Equation 1 shows the formulation of MCRMSE. MCRMSE is also used as the loss function for training the models. The vectors provided by the embeddings serve as input to the deep learning model. The model predicts the values based on the input sequence. Figure 2 shows the block diagram of the experimental setup for the comparative study. The architecture for a few embeddings is different depending on the way they are used. Vectors are directly used as input to the neural network in dna2vec and lshvec, whereas the rna2vec considers the vectors for initializing the embedding layer of the network. Hence, dna2vec and lshvec architectures do not contain an embedding layer, whereas the ran2vec uses it as the initial layer. Figures 3 and 4 show the architectures used for rna2vec and dna2vec, respectively. The pre-trained models are trained on different k-mers sizes. The dna2vec is trained on k-mer length from 3 to 8. The rna2vec and lshvec used 4-mers and 9-mers for training the embeddings, respectively.

$$\frac{1}{N}\sum_{j=1}^{N}\sqrt{\frac{1}{n}\sum_{i=1}^{n}\left(y_{ij}-\hat{y_{ij}}\right)^2} \tag{1}$$

where N is the number of target indices and n is the toatl number of sequences. y and $\hat{y}$ are actual and predicted values, respectively.

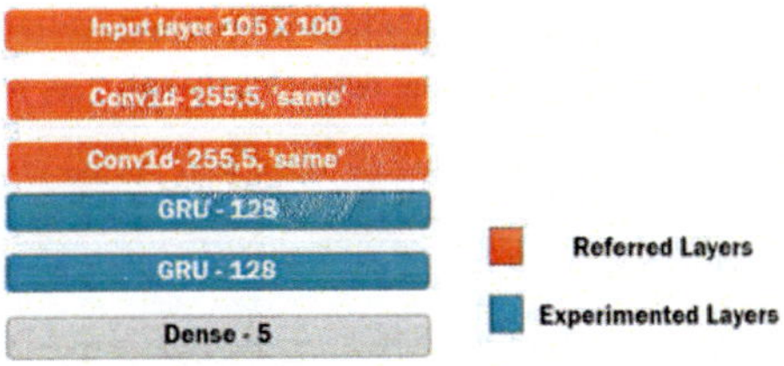

Fig. 4 CNN Bi-LSTM architecture diagram for dna2vec embedding

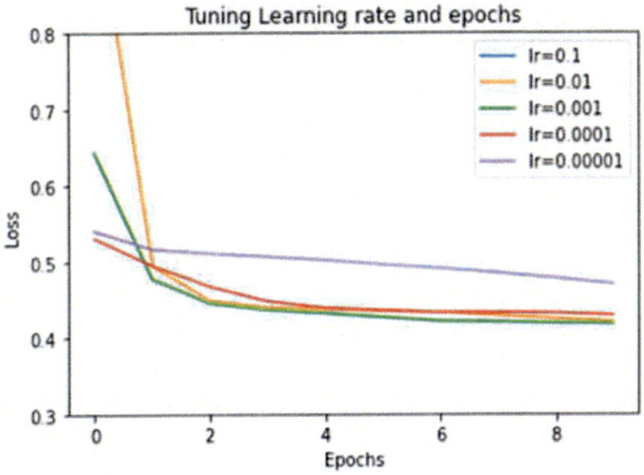

Fig. 5 Tuning epochs and learning rate based on training loss for CNN-BiGRU architecture with rna2vec embedding

5 Hyperparameter Tuning

The main challenge deep learning models face is reproducibility. Hyperparameter tuning improves the model's quality and reproducibility. The epoch and learning rate are tuned using training MCRMSE loss. All the other parameters are tuned using test MCRMSE loss. Figure 5 shows a graphical illustration of how loss varies with respect to the epochs while training the CNN-BiGRU architecture using rna2vec embedding. A learning rate of 0.001 gives the least MCRMSE loss. The loss saturates around eight epochs. Similar to BiGRU architecture and for all the embeddings, the learning rate and epochs are tuned based on training loss and testing loss. Batch-size of 8, 16, 32, 64, and 128 were used for training both Bi-GRU and CNN-BiGRU architecture. All the batch sizes performed similarly, but batch-size 32 gave improved loss for both the models. Different GRU models with 2–10 GRU layers containing 16–256 hidden layer neurons were taken for experimentation, out of which two layers and 128 hidden layer size gave the least MCRMSE loss for all the embeddings. CNN layers are referred from the existing architectures used in [9, 11]. Figures 3 and 4 show which layers are referred to and experimented with. The fine-tuned hyperparameters which gave the least MCRMSE are shown in Tables 1 and 2 for Bi-GRU and CNN-BiGRU architectures, respectively.

Table 1 Hyperparameters for BiGRU architecture using rna2vec

Embeddings	Learning rate	No. of epochs	Batch-size	Dropout	GRU layers
One-hot encoding	0.001	68	32	0.2	2
dna2vec	0.001	100	32	0.2	2
rna2vec	0.001	12	32	0.1	2
lsh2vec	0.001	78	32	0.3	2

Table 2 Hyperparameters for CNN-BiGRU architecture using rna2vec

Embeddings	Learning rate	No. of epochs	Batch-size	Dropout	CNN layers	GRU layers	CNN filter size	No. of filters
One-hot encoding	0.0001	52	32	0.2	1	2	256	32
dna2vec	0.001	100	32	0.2	2	2	255, 255	5, 5
rna2vec	0.001	8	32	0.1	2	2	256, 128	32, 64
lsh2vec	0.001	80	32	0.3	2	2	255, 255	5, 5

6 Results and Discussions

Evaluation of the embeddings is done based on the MCRMSE loss. Table 3 shows the MCRMSE loss for embeddings with different architectures. One-hot encoding is done to get a basic understanding of the efficiency of the distributed gene embeddings. The loss values obtained with one-hot encoding as the feature by the Bi-GRU and CNN-Bi-GRU models are 0.39 and 0.36, respectively. rna2vec gives a maximum loss of 0.42 and 0.40. The lsh2vec measures 0.33 and 0.29 for both the architectures. rna2vec performed poorly on the vaccine sequences. dna2vec achieved a loss of 0.26 and 0.24 for both the architectures. dna2vec performance was better than other models because it has a linear representation and also since it is capable of providing vector representation for variable-length k-mer. Such uncomplicated representation can be very helpful in inferring the contextual information present in the sequence. dna2vec,one-hot encoding, and lshvec used the k-mer length of 3, 3, and 9, respectively, which are multiple of 3. The three embeddings performed better than rna2vec which is trained on 4-mers. This could be because each codon contains only 3 characters as mentioned in Sect. 1. This might indicate that the embeddings with a multiple of 3 as k-mer size are capturing codon information present in the sequence.

Table 3 MCRMSE obtained by the embeddings for Bi-BRU with and without CNN

Architecture	Embeddings			
	One-hot encoding	dna2vec	rna2vec	lsh2vec
BiGRU	0.39	0.26	0.42	0.33
CNN-BiGRU	0.36	0.24	0.4	0.29

7 Conclusion and Future Scope

The comparative study shows that among the mentioned gene embedding, dna2vec performed the best. The gene embeddings which are trained on human genome sequences, i.e. dna2vec and lshvec performed better than rna2vec embedding and one-hot encoding. Results also indicate that the k-mer length of 3 or multiple of 3 gives better performance for the particular prediction task. The CNN-BiGRU architecture performed better than Bi-GRU indicating that the sequences contain frequency variations that are captured by convolutional layers. To improve the results, different deep learning architectures with dna2vec embedding can be experimented with. Mentioned gene embeddings are trained on different k-mer sizes. Experimenting with k-mer size can improve the results. The lshvec embeddings are capable of representing higher length k-mers. Experimenting with large size k-mers is possible. Pre-trained embeddings were compared to get a preliminary idea of how the existing embeddings perform for the particular prediction task. Training the embeddings using the input sequences can improve results and can also give a better insight into the contextual information present in the sequences.

References

1. Cho K, Van Merriënboer B, Gulcehre C, Bahdanau D, Bougares F, Schwenk H, Bengio Y (2014) Learning phrase representations using RNN encoder-decoder for statistical machine translation 06
2. Choy CT, Wong CH, Chan SL (2019) Embedding of genes using cancer gene expression data: biological relevance and potential application on biomarker discovery. Front Genet 9:682
3. Rachlin MWE (2017) mrna vaccines: disruptive innovation in vaccination. Moderna 17:05
4. Graves A, Schmidhuber J (2005) Framewise phoneme classification with bidirectional LSTM and other neural network architectures. Neural Netw 18(5–6):602–610
5. Hochreiter S, Schmidhuber J (1997) Long short-term memory. Neural Comput 9(8):1735–1780
6. Hu S, Ma R, Wang H (2019) An improved deep learning method for predicting dna-binding proteins based on contextual features in amino acid sequences. PLOS ONE 14:1–21, 11
7. Joulin A, Grave E, Bojanowski P, Douze M, Jégou H, Mikolov T (2016) Fasttext.zip: compressing text classification models. arXiv:1612.03651
8. Alexey V Lobanov, Anton A Turanov, Dolph L Hatfield, and Vadim N Gladyshev. Dual functions of codons in the genetic code. *Critical reviews in biochemistry and molecular biology*, 45(4):257–265, 2010

9. Mostavi M, Salekin S, Huang Y. Deep-2'-o-me: Predicting 2'-o-methylation sites by convolutional neural networks. In: 2018 40th annual international conference of the IEEE engineering in medicine and biology society (EMBC), pp 2394–2397
10. Ng P (2017) dna2vec: Consistent vector representations of variable-length k-mers 01
11. Pan X, Shen H-B (2018) Learning distributed representations of RNA sequences and its application for predicting RNA-protein binding sites with a convolutional neural network. Neurocomputing 305:51–58
12. Pardi N, Hogan M, Porter F, Weissman D (2018) mRNA vaccines—a new era in vaccinology. Nat Rev Drug Discov 17:01
13. Pardi N, Hogan MJ, Weissman D (2020) Recent advances in mRNA vaccine technology. Curr Opin Immunol 65:14–20
14. Premjith B, Soman KP, Kumar MA (2018) A deep learning approach for malayalam morphological analysis at character level. Procedia Comput Sci 132:47–54
15. Premjith B, Soman KP, Poornachandran P (2018) A deep learning based part-of-speech (POS) tagger for sanskrit language by embedding character level features. In: Proceedings of the 10th annual meeting of the forum for information retrieval evaluation, pp 56–60. ACM
16. Ramos J (2003) Using tf-idf to determine word relevance in document queries, 01
17. Rehurek R (2011) Scalability of semantic analysis in natural language processing
18. Rizzo R, Fiannaca A, La Rosa M, Urso A (2016) A deep learning approach to DNA sequence classification 9874:129–140, 07
19. Sasidhar TT, Premjith B, Soman KP (2020) Emotion detection in hinglish (hindi+ english) code-mixed social media text. Procedia Comput Sci 171:1346–1352
20. Shi L, Chen B (2019) A vector representation of DNA sequences using locality sensitive hashing. BioRxiv
21. Stanford University (2016) Openvaccine: Covid-19 mrna vaccine degradation prediction. https://www.kaggle.com/c/stanford-covid-vaccine/data
22. Watson JD, Crick FHC (1953) Molecular structure of nucleic acids: a structure for deoxyribose nucleic acid. Nature 171(4356):737–738
23. Zhang N-N, Li X-F, Deng Y-Q, Zhao H, Huang Y-J, Yang G, Huang W-J, Gao P, Zhou C, Zhang R-R, Guo Y, Sun S-H, Fan H, Shu-Long Z, Chen Q, He Q, Cao T-S, Huang X-Y, Qiu H-Y, Nie J-H, Jiang Y, Yan H-Y, Ye Q, Zhong X, Xue X-L, Zha Z-Y, Zhou D, Yang X, Wang Y-C, Ying B, Qin C-F (2020) A thermostable mRNA vaccine against covid-19. Cell 182(5):1271-1283.e16

MACI: Malicious API Call Identifier Model to Secure the Host Platform

Ayushi Acharya, Hari Prasad, Vinod Kumar, Ishu Gupta, and Ashutosh Kumar Singh

Abstract The system of mobile agents is a widely used model for distributed computations and distributed applications. Mobile agents are assuming critical jobs as they are able to ship their state starting with one platform then onto the next, with their information unblemished, and are equipped for performing fittingly in the new platform's conditions. Albeit mobile agents are used extensively, they can transport with them obscure malign codes leading to the agent-hosting platform's disruption. To detect any malicious addition to the code being carried, we propose a model named Malicious API Call Identifier (MACI) model which uses the API function call strings that can mirror the functional qualities of a program and can be used to find malicious codes. Certain particular malicious functions (present in API call sequences) are commonly included in these malicious codes which the agent carries. The MACI model uses the n-gram analysis generally used on text/words for NLP tasks to find inherent patterns/coordination among the function calls present in each API call sequence. The model tests four classifiers based on five evaluation metrics, some with ensemble techniques at optimum parameters with tenfold cross-validation. The model applies a suitable feature reduction technique and achieves up to 98.31% accuracy and a 2.89% miss rate to outperform the other state-of-the-art works done in detecting malign APIs.

Keywords N-grams · Ensemble techniques · Vectorization

1 Introduction

In the field of heterogeneous networking and distributed computing, mobile agents are exceptionally known and are used extensively. The mobile agent is an assembly of executable codes that immigrate starting with one hosting platform onto the next for its execution. They perform different jobs to accomplish their objectives. Host

A. Acharya · H. Prasad · V. Kumar · I. Gupta (✉) · A. K. Singh
National Institute of Technology, Kurukshetra, India

© The Author(s), under exclusive license to Springer Nature Singapore Pte Ltd. 2022

D. Giri et al. (eds.), *Proceedings of the Seventh International Conference on Mathematics and Computing*, Advances in Intelligent Systems and Computing 1412,
https://doi.org/10.1007/978-981-16-6890-6_23

platform is a machine or a framework that gives mobile agents the asset or information needed for the execution of the code they carry.

Though mobile agents are valuable, they can cause incredible harm to the hosting platform due to being impacted by a malware attack (like a virus) during their migration. The major concern is the agent migrating through unknown hosting platforms, at the same time catching information, sharing information with them, and calculating the results. Thus, the mobile agent system has serious security risks. The agent loses control of its owner whenever it starts its journey towards the destination platform. Also, agents can be made to leak the data they carry [1–5].

Three introductions seen as critical for the agent's working are *Host Platform faced with an agent*: It refers to the masking of the intention, reliability, and accessibility by an incoming agent for misuse. The agent does so by changing the information it carries and mishandling the host's assets since it was privileged by the host to have access to the host's network. Since the mobile agent is vastly used and it works in all types of distributed heterogeneous networks (mostly over the cloud), the risk of an incoming malign agent disrupting the host is very high and may cause various breaches, misuse, and leakage of the data residing in the host computer or the data traveling through cloud [6–9]. Therefore, a proper data leakage prevention mechanism is also needed. *Agent faced with the Host Platform:* The moment the agent initiates the movement from a node to another in the network, it is mandated to give the information about the data it is carrying and the status at present, making it ineffective in terms of data privacy. There also stands a chance of interference by the intermediate hosting platforms that it encounters in the path to the final node [10–14]. These platforms/hosts may modify the code carried by the agent to have a result favoring them. *Agent faced with an agent:* Other agents may try to modify the metadata or the variables. They may try to get the gist about the communication happening or the data being shared without actually modifying it [15–18].

The major concern is a malicious agent (containing malicious API calls) pitched against an unaware platform. The paper addresses this challenge with MACI: The proposed model to detect incoming code as benign or malicious to keep our hosting platform safe and secure. In this static approach, the code is not let to run in the host's environment, rather first the model checks the API call sequences for malicious behavior/patterns among different function calls. The effort is to try and model an optimum classifier based on proper evaluation metrics using the n-gram approach to find behavioral patterns among function calls.

2 Related Work

Budong [19] summed up the potential malign characteristics with their viability to get detected. The AI approaches dependent on the static and dynamic characteristic investigation are concentrated on the benefits and restrictions. At last, potential arrangements are proposed and novel malware recognition is advanced. Yuntao [20] advanced a novel attribute determination strategy for hybrid "H-gram" with joint cross-entropy. In a virtual setting, the examination of the dynamic nature and attributes of local API was done. 4 Classifiers: AdaBoost, Random Forest, ID3, and Ensemble Bagging were tested with tenfold cross-validation on the WEKA platform. Sanchit [21] adopted a unique strategy to use the API call traces (from the Windows OS) to sort out the malicious behavior of the applications. More than 500 huge APIs were utilized and planned to be categorized into 26 various classes. Any malicious application then has its intent/behavior captured through the API call grouping in form of a unique impression. These impressions were then compared to one another for assessing API call grouping's closeness among testing and training models.

Pallavi [22] classified the mobile agents into two (binary) categories: malign and benign using various classifiers like NB, SVM, K-NN, and Decision Tree on the API call traces (from the Windows OS). The assistance of n-gram was taken to snare the features. Various features corresponding to three-grams till eight-grams were taken. Forty occurrences of every malignant and benign sample were prepared, utilized with five-fold cross-validation for a reduction in the problem of class-imbalance while taking training and test samples. Classifiers were tested with the various boundary condition settings with the goal that ideal results are captured. The WEKA tool was utilized. Yuxin et al. [23] used a generative neural network model having multiple layers as an encoder to diminish the components of the information vectors. As the raw data to be learned into information, the n-gram vectors were sent to the network. Through setting up the unlabeled data, they achieved preciseness beating that of the K-Nearest Neighbor, Support Vector Machines, and Decision Tree computations.

Zhi [24] proposed distinguishing the static Portable Executable (PE) Windows records through opcode assessment by calculating the occurrence of each opcode. K-Nearest Neighbors (K-NN) and back-spread neural framework (BP) calculations were used alongside 5-overlay cross-validation. To diminish the components of the eigenvectors, their model used a viable mathematical strategy called Power Iteration to include extraction. Yi-Ting et al. [25] proposed a depiction of malware with an embedder, an encoder, with a channel called a filter. The model had two top tier structures BERT and Sent2Vec to get relations between API calls. The encoder contained two neural layers. The neural structure was utilized to complete malware profiling. The system catches 62 warily picked Windows API calls and arranges them into five distinct classes. Overall they tested more than 24,000 samples and achieved more than 97.5% accuracy in determining them as malign using neural networks. Yen-Hung Frank [26] examined the working of a few algorithms. Ten machine inclining algorithms were utilized to label nine malware families. Sensitivity, F-measure, and AOC-ROC curve area were the parameters for evaluation.

3 Malicious API Call Identifier Model

As malicious coders often insert harmful codes and sometimes use devices to compose their code, the n-gram analysis generally used on text/words for NLP tasks [27] is being used to take advantage in the proposed model: MACI. The n-gram technique distinguishes the specific attributes of collections of devices, code implanters, compilers, and programming conditions. Likewise, features and impressions that are expressed for malicious code makers, malign coding approaches, and functioning of malign code can be easily grabbed by the n-gram technique. In the n-gram features, the formed impressions are evident and unique. Therefore, it is difficult for pernicious coders to deliberately create threatening code that stunts the n-gram examination when all access to rules is there. In the MACI model, the mobile agent encounters a review with the help of different channels and procedures for categorizing. This is to check for harmful code at whatever point an agent goes to the host platform. Figure 1 describes the framework of our model for identifying malicious API calls made to the host by an incoming agent. For the cases in which the mobile agent is viewed as malignant, that agent is eliminated and isn't allowed to execute in order to not let it hurt the host platform. Then, the following agent is tested and if isn't viewed as dangerous, it is allowed to execute in the system. Then, the framework checks other incoming mobile agents arranged by their arrival-time.

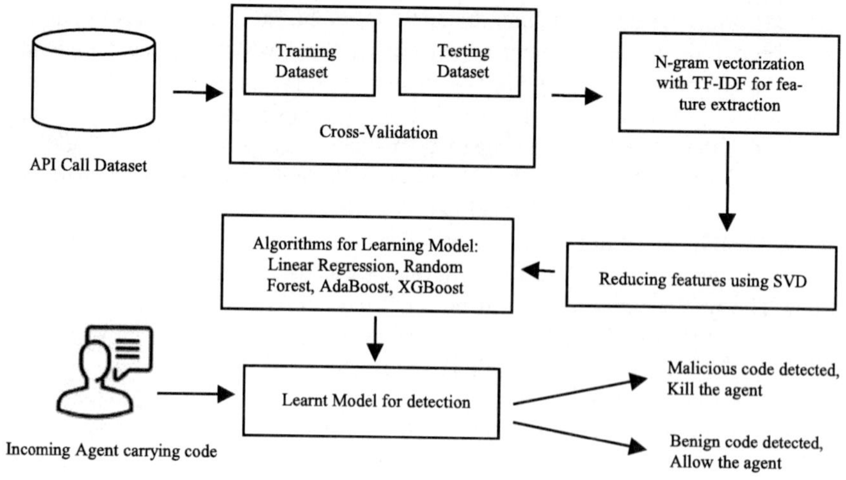

Fig. 1 MACI framework

3.1 Methodology

The model uses the Windows API call sequence corpora consisting of malicious and benign API call sequences/trace files. The dataset is then split into testing and training sets and 4 classifiers are applied (each tested for 3-g to 8-g). Each call sequence is then vectorized using n-gram by TF-IDF vectorization. Since the number of features thus produced by n-gram vectorization are very, so the MACI model makes use of the Singular Value Decomposition (SVD) feature reduction technique, applying it to work on the sparse matrix produced by TF-IDF vectorization and reducing the no. of extracted features to suitably low (for faster detection). Model is then trained and unknown API calls are tested with various metrics like accuracy, miss rate being recorded. The model makes the different classifiers: Random Forest, AdaBoost, XGBoost, and Logistic Regression to go through tenfold cross-validation to lessen the bias and get the best-averaged outputs. Evaluation metrics are compared for each classifier and the one with the best results at a lower n-value is declared the best classifier. The model considers a lower n-value to keep real-time detection possible since a greater n-value has thousands of features to choose from thus taking a considerable amount of time for classification/detection of malicious or benign API traces.

3.2 Extracting API Call Sequences

For each sample API call sequence, the model extracts the occurrence count related to every function call as the feature. The extraction procedure is explained and shown in Fig. 2. Like when any 3 sample API call sequences are taken: sample1 = {API_1, API _2, API _3, API_2, API_3}, sample2 = {API_3, API_6, API_4, API_3}, sample3 = {API_5, API_4, API_4, API_4, API_2}. An array "A" stores the API function calls

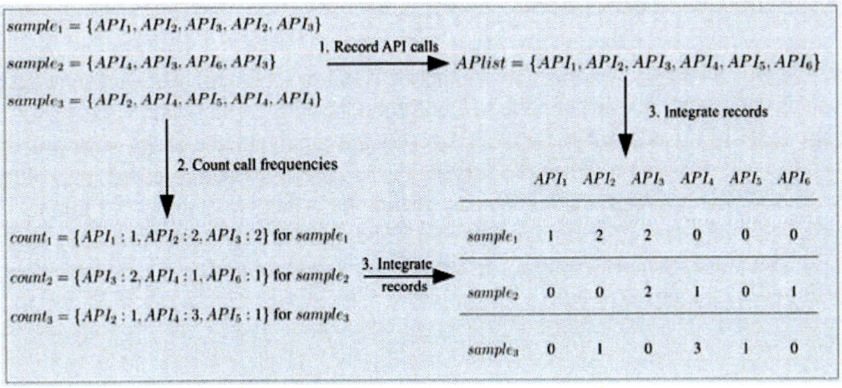

Fig. 2 API call feature extraction

Fig. 3 Confusion matrix

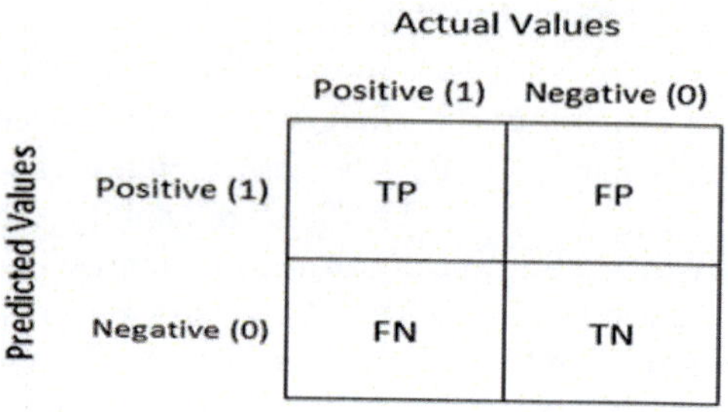

that have been called by all (malicious/benign) programs, i.e., A = {API 1, API 2, API 3, API 4, API 5, API 6}. The frequency of each called API corresponding to the samples is counted. Then, the model cross-refers the array "A" and API function call occurrences to build a new table for classification.

3.3 Evaluation Parameters

Classification models are decided for the degree of rightness or inaccuracy in their results with the assistance of the matrix appeared in Fig. 3. It helps with computing the scores for assessment metrics by using instances like True-Negative (TN), True-Positive (TP), False-Negative (FN), and False-Positive (FP). Further, Table 1 contains the necessary evaluation parameters and acceptable ranges with a description of the evaluation metrics.

4 Performance Evaluation and Analysis

4.1 Benchmarks

The model is tested against the state-of-the-art Windows API call sequence corpora consisting of malicious and benign API call sequences/trace files. The widely used CSDMC2010 [28] and APIMDS [29] datasets are amalgamated along with other call traces collected from various sources. These API call traces are then verified from the VFXHeavens website for being malicious or benign samples. In totality, the model works upon 600 malicious and 150 benign API call traces, forming the dataset. The dataset has a column for the API call trace sequence and another for labeling them as malign or not.

Table 1 Performance evaluation metrics

Parameters	Definition	Formula	Expectation
Recall/ Sensitivity	Potential to correctly classify true instances as true only. Also called the measure of Sensitivity	TP/(TP + FN)	Highest (0–1)
Precision	The proportion of instances labeled as true that are indeed true	TP/(TP + FP)	Highest (0–1)
Miss Rate	The proportion of all the actually true instances that are falsely labeled. Opposite of Recall	FN/(TP + FN)	Lowest (0–1)
ROC-AUC	Degree of separability. Assesses the potential of the model while differentiating classes from one another	NA	0.85 to 1
Accuracy	Potential of the classifier to categorize instances as they are for the whole dataset's classification	(TP + TN)/(TP + FP + TN + FN)	Highest (0–100)
F-Score	Exact, symphonious averaging of Precision and Recall. Also known as F-metric	(2* Recall *Precision)/(Recall + Precision)	Highest (0–1)

4.2 Results and Observations

The model achieves up to 98.31% accuracy in detecting the API calls having malicious function call patterns. The mapping of accuracy and miss rate versus "n" value is shown in Figs. 4 and 5. The mapping of precision, F-score, and AUC-ROC versus "n" value is shown in Figs. 6, 7, and 8, respectively. At n-value 3, the XGBoost classifier gives the greatest accuracy, the least miss rate (i.e., the number of API call

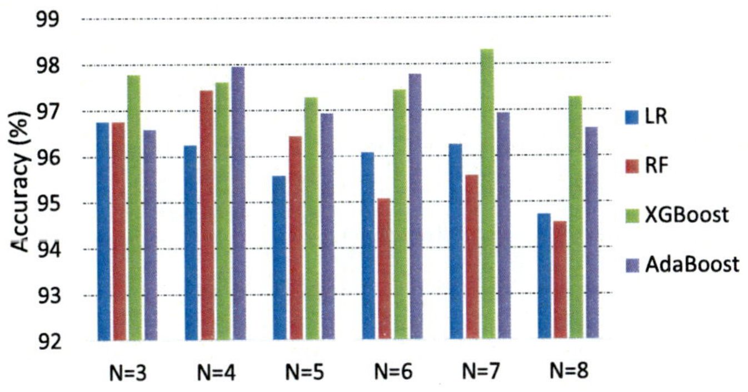

Fig. 4 Accuracy (average) w.r.t "N"—gram sized features

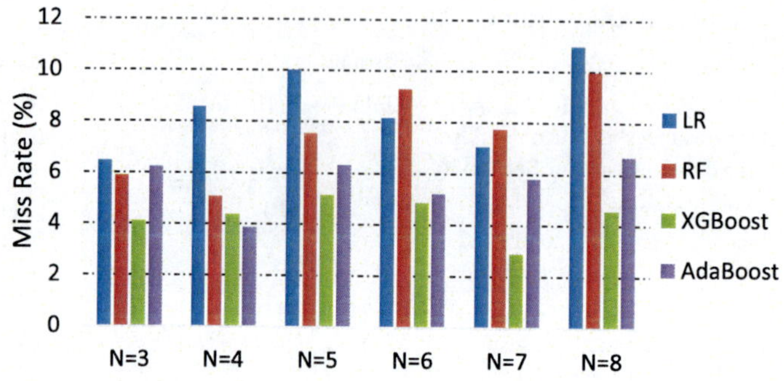

Fig. 5 Miss rate (average) w.r.t "N"—gram sized features

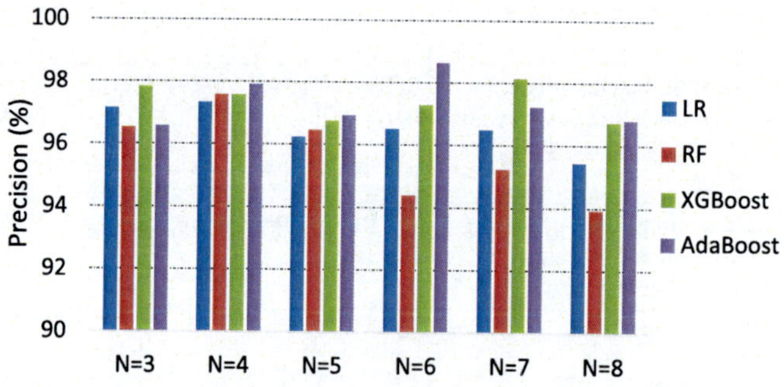

Fig. 6 Precision (average) w.r.t "N"—gram sized features

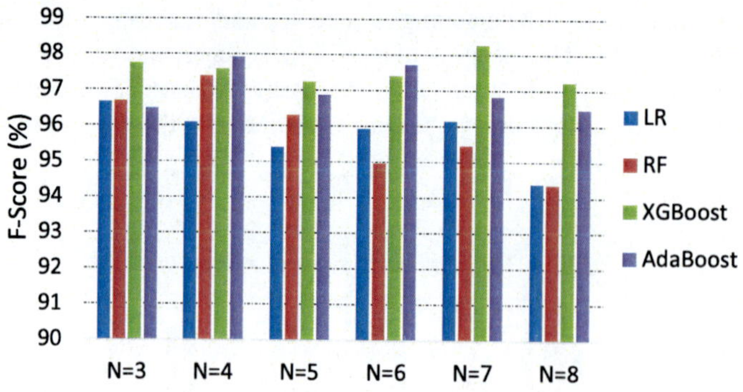

Fig. 7 F-Score (average) w.r.t "N"—gram sized features

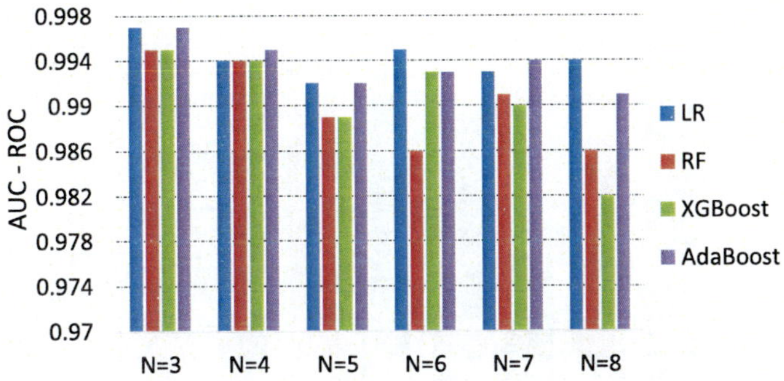

Fig. 8 AUC-ROC (average) w.r.t "N"—gram sized features

sequences actually malicious but not detected by our model), and also the greatest F-score. For n-value 4, the AdaBoost classifier gives the greatest accuracy, precision while also maintaining the least miss rate. For n-value 5, the XGBoost classifier and for n-value 6, the AdaBoost classifier does the best in terms of parameter scores. For n-values 7 and 8, the XGBoost classifier provides the highest accuracy, precision, and F-score while also maintaining the least miss rate. Overall, it can be seen that for "n" values 3, 5, 7, and 8, the XGBoost classifier performs the best and for the "n" values 4 and 6, the AdaBoost classifier does the best by scoring well in the 5 chosen parameters. In the MACI model, the AdaBoost classifier at 4-g produces a very good detection accuracy of 97.96% but the XGBoost classifier at 7-g produces an even better accuracy of 98.31%. The trade-off here is between accuracy and detection speed as more no. of features means more computation time is required.

As it can be seen from the graphs plotted that the XGBoost classifier consistently outperforms the other classifiers in terms of accuracy and F-score. It also has the least miss rate overall while scoring well in precision and AUC-ROC score also. The Logistic Regression classifier churns out the best AUC-ROC score while the AdaBoost classifier has the best precision score.

4.3 Comparative Analysis with State-of-Art Works

The comparative analysis of the MACI model based on the accuracy achieved by other related works (for malicious API identification) is shown in Fig. 9. The comparison of precision and miss rates achieved through related works done in the field of identification of malign and benign APIs for various purposes are displayed in Figs. 10 and 11.

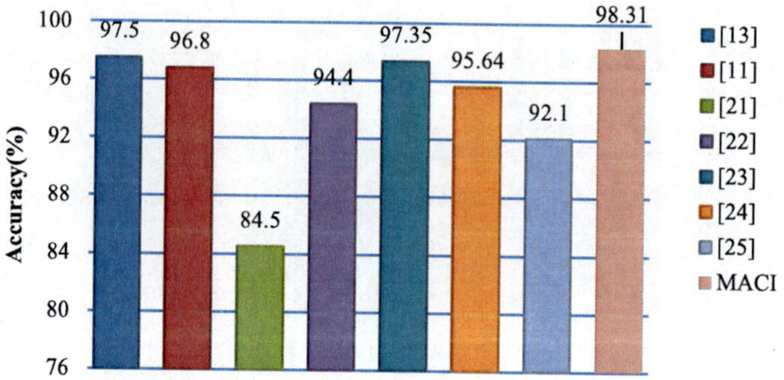

Fig. 9 The accuracy achieved over the state-of-the-art works

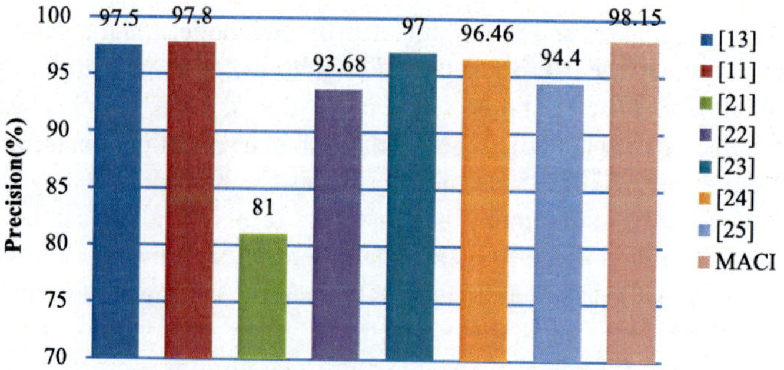

Fig. 10 The precision achieved over the state-of-the-art works

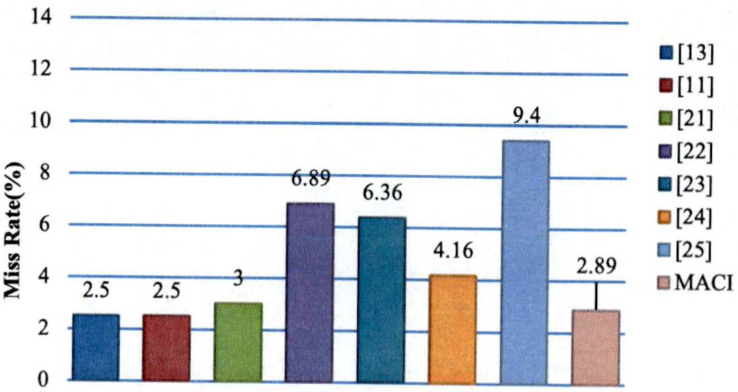

Fig. 11 Miss rate achieved over the state-of-the-art works

5 Conclusion and Future Scope

To detect malicious API calls, the correlation among the function calls needs to be studied well. Our approach to making use of the n-gram feature space for detecting malicious codes proves to be very effective. N-grams provided a good view of the relationship among function calls in the API call strings, which is evident from the results achieved. The MACI model achieves the highest accuracy of 98.31% and the least miss rate of 2.89% for n-value 7. The XGBoost classifier continuously provides the best detection accuracy with the lowest low miss rates for various "n"-gram feature spaces and is pronounced as the best classifier. Compared to the other similar works done, the MACI model uses more advanced ensemble and feature reduction techniques, thus securing an improvement over the accuracy and precision achieved by around 3% and 2%, respectively. While the miss rate achieved by MACI is also among the lowest in comparison. Therefore, the MACI model's results impose superiority over the other state-of-the-art works done. The proposed model's feature extraction method also works well and is very scalable as any number of required features can be extracted. More effort is required to check greater "n"-gram features above the "n" value 9 (i.e., 9-g), and at the same time reducing these enormous numbers of features associated with those higher "n"-gram values. The model uses the SVD feature reduction technique to reduce the number of features to suitably low for faster detection. The future scope will be in the direction of making use of higher "n"-valued features. High-quality feature reduction techniques need to be applied to make the detection in real time by reducing the features to just the essential ones while using higher "n"-gram features.

References

1. Singh AK, Gupta I (2020) Online ınformation leaker ıdentification scheme for secure data sharing. Multimed Tools Appl 79(41):31165–31182
2. Gupta I, Singh AK (2020) SELI: statistical evaluation based leaker ıdentification stochastic scheme for secure data sharing. IET Commun 14(20):3607–3618
3. Kaur K, Gupta I, Singh AK (2017) A comparative study of the approach provided for preventing the data leakage. Int J Netw Secur Appl (IJNSA) 9(5):21–33
4. Kaur K, Gupta I, Singh AK (2018) Data leakage prevention: email protection via gateway. J Phys: Conf Ser 933(1). IOP Publishing
5. Gupta I, Singh AK (2019) A confidentiality preserving data leaker detection model for secure sharing of cloud data using ıntegrated techniques. In: Seventh ınternational conference on smart computing and communication systems (ICSCC), Sarawak, Malaysia. IEEE, pp 1–5
6. Gupta I, Singh AK (2020) A framework for malicious agent detection in cloud computing environment. Int J Adv Sci Technol (IJAST) 135, 49–62
7. Gupta I, Gupta R, Singh AK, Buyya R (2020) MLPAM: a machine learning and probabilistic analysis based model for preserving security and privacy in cloud environment. IEEE Syst J
8. Gupta I, Singh AK (2020) An integrated approach for data leaker detection in cloud environment. J Inf Sci Eng 36(5):993–1005
9. Gupta I, Singh N, Singh AK (2019) Layer-based privacy and security architecture for cloud data sharing. J Commun Softw Syst (JCOMSS) 15(2):173–185

10. Ali M, Shiaeles S, Bendiab G, Ghita B (2020) MALGRA: machine learning and N-Gram malware feature extraction and detection system. Electronics 9(11). MDPI
11. Han W, Xue J, Wang Y, Huang L, Kong X, Mao L (2019) MalDAE: detecting and explaining malware based on correlation and fusion of static and dynamic characteristics. Comput Secur 83, 208–233
12. Ma X, Guo S, Bai W, Chen J, Xia S, Pan Z (2019) An API semantics-aware malware detection method based on deep learning. Secur Commun Netw 1–9
13. Hardy W, Chen L, Hou S, Ye Y, Li X (2016) DL4MD: a deep learning framework for intelligent malware detection. In: International conference on data mining DMIN'16
14. Liu C, Zhang Z, Wang S (2016) An android malware detection approach using bayesian inference. In: IEEE international conference on computer and information technology (CIT), Nadi. IEEE, pp 476–483
15. Gupta I, Singh AK (2020) GUIM-SMD: guilty user identification model using summation matrix based distribution. IET Inf Secur 14(6):773–782
16. Gupta I, Singh AK (2019) Dynamic threshold based ınformation leaker ıdentification scheme. Inf Process Lett 147, 69–73
17. Gupta I, Singh AK (2018) A probabilistic approach for guilty agent detection using bigraph after distribution of sample data. Procedia Comput Sci 125, 662–668
18. Gupta I, Singh AK (2017) A probability based model for data leakage detection using bigraph. In: 7th ınternational conference on communication and network security (ICCNS). Tokyo, Japan. ACM, pp 1–5
19. Xu B, Li Y, Yu X (2020) Malware detection based on static and dynamic features analysis. In: Machine learning for cyber security ML4CS 2020. Lecture notes in computer science, vol 12486. Springer, Cham
20. Zhao Y, Bo B, Feng Y, Xu C, Yu B (2019) A feature extraction method of hybrid gram for malicious behavior based on machine learning. Secur Commun Netw, Hindawi, 2674684:1–2674684:8
21. Gupta S, Sharma H, Kaur S (2016) Malware characterization using windows API call sequences. In: Carlet C, Hasan M, Saraswat V (eds) Security, privacy, and applied cryptography engineering SPACE 2016. Lecture notes in computer science, vol 10076. Springer, Cham, pp 271–280
22. Bagga P, Hans R, Sharma V (2017) N-grams based supervised machine learning model for mobile agent platform protection against unknown malicious mobile agents. Int J Interact Multimed Artif Intell 4(6):33–39
23. Yuxin D, Siyi Z (2019) Malware detection based on deep learning algorithms. Neural Comput Appl 31(2):461–472
24. Sun Z, Rao Z, Chen J, Xu R, He D, Yang H, Liu J (2019) An opcode sequences analysis method for unknown malware detection. In: ICGDA 2019, Proceedings of the 2nd international conference on geoinformatics and data analysis. ACM, New York, pp 15–19
25. Huang Y, Ting-Yi C, Sun YS, Chen YC (2019) Learning malware representation based on execution sequences. arXiv:abs/1912.07250
26. Hu Y, Ali A, Hsieh C, Williams A (2019) Machine learning techniques for classifying malicious API calls and N-Grams in kaggle data-set. SoutheastCon, 1–8
27. Ahmed H, Traore I, Saad S (2018) Detecting opinion spams and fake news using text classification. Secur Priv 1(1). Wiley
28. CSDMC2010 Dataset: https://www.azsecure-data.org/other-data.html
29. APIMDS Dataset: http://ocslab.hksecurity.net/apimds-dataset

Multivariate Deep Learning Model with Ensemble Pruning for Time Series Forecasting

Mohit Kosuri, Cherry Tandu, Sobhan Sarkar, and J. Maiti

Abstract To predict future events using historical data, Time Series Forecasting (TSF) should be used to get precise and accurate predictions. It has been a challenging issue to deal with the errors and value loss while predicting the future; hence, a dynamic error correction is proposed to overcome the errors. Additionally, it is important to find out a fast optimization technique to avoid this difficulty. Therefore, it is proposed in this study to use an improved stacking-based ensemble pruning method, namely Genetic Algorithm (GA)-II to produce high accuracy and strong stability in time series forecasting. A meta predictor known as Kernel Ridge Regression (KRR) is proposed for stacking ensemble models for its improved forecasting performance. The main goal of this study is to attain reliable and precise time-series forecasting. In the process of extracting various types of data features, particular types of Deep Neural networks are effective. Therefore, these types of models combine and increase the use of Deep Learning and Ensemble Learning techniques. It is better to use different Deep Neural Networks as Deep Learning models and use boosting and stacking techniques as neural networks take more time by using these types of methods, and the results would be better with low calculations. In time-series data, the value changes dynamically which may increase or decrease the accuracy of the prediction, so to overcome this type of problem, some error correction methods like Dynamic Error Correction (DEC) and a technique like Non-Dominated Sorting Genetic Algorithm (NSGA) and Multi-Populated Non-Dominated Sorting Genetic Algorithm-II (GA-II) to get optimal solutions in terms of accuracy are used.

M. Kosuri
Department of Mechanical Engineering, IIT Kharagpur, Kharagpur 721302, India

C. Tandu
Department of Ocean Engineering and Naval Architecture, IIT Kharagpur, Kharagpur 721302, India

S. Sarkar (✉)
Business School, University of Edinburgh, 29 Buccleuch Pl, EH8 9JS Edinburgh, UK
e-mail: sobhan.sarkar@ed.ac.uk

J. Maiti
Department of Industrial and Systems Engineering, IIT Kharagpur, Kharagpur 721302, India

© The Author(s), under exclusive license to Springer Nature Singapore Pte Ltd. 2022

D. Giri et al. (eds.), *Proceedings of the Seventh International Conference on Mathematics and Computing*, Advances in Intelligent Systems and Computing 1412,
https://doi.org/10.1007/978-981-16-6890-6_24

Keywords Time series forecasting · Multivariate analysis · Deep learning model · Ensemble pruning

1 Introduction

In time series forecasting, it is difficult for an individual deep learning model to maintain robust forecasting performance. To achieve the accurate and robust forecasting performance, a novel ensemble deep learning model is proposed to obtain accurate and stable time series forecasting which is limited to univariate data and cannot be directly scaled to multi-dimensional data. When a univariate data is given to the models, it does not satisfy the need when more than one feature is affecting the forecasting results. It does not check which features are affecting the accuracy result before giving the input data to the models. To overcome this problem, a multivariate ensemble model is proposed for forecasting with more than one input feature. Multivariate ensemble deep learning model with dynamic error correction and multi-objective ensemble pruning for time series forecasting to maintain and increase the forecasting accuracy is given to the models as input. Construction of DNN models as basic predictor with multivariate dataset along with stacking and boosting ensemble techniques makes the result better with low calculation. To increase the forecasting accuracy, an ensemble pruning technique known as Multi-Population Non-Dominated Sorting Genetic Algorithm-II (NSGA-II) is used to get an optimal solution.

Deep Learning (DL) is one of the powerful Machine Learning (ML) approaches [22]. There are different ML models available, such as support vector machine [18, 28, 29], artificial neural network [24, 26], decision tree [23, 27], clustering [20, 31], rule generation models [19, 21], and convolutional neural network [14, 15]. These models are comparatively more powerful than their traditional counterpart [6, 11, 30]. The DL models have the potential to obtain better features from the data by transforming original data to a more precise level by combining simple non-linear modules layer by layer. In the process of extracting various types of features, certain types of deep neural networks are effective. Many DL models have been developed, including Multilayer Perceptron (MLP) [32] also called as "vanilla" neural network and Convolutional Neural Network (CNN) [5]. Therefore, it is better not to rely on only one predictor as every model has its complexity, and with this consideration of their advantages, it is suggested to use more than one DL model and combine them using improved ensemble techniques to make the prediction more accurate. Keeping this in mind, the following two basic ensemble models are considered: Multilayer Perceptron (MLP) [32] and Convolutional Neural Network (CNN) [5]. In MLP, the layers are fully connected starting from the input layer and to the hidden layers [32]. Finally, the desired output is achieved from this model. CNN has a grid-like structure used for local weight sharing. Multiple types of convolution kernels are used to receive the local data from the region for accessing the different types of feature information and then convoluted to extract more effective features. CNN is composed of a (1 $\times$ 3) 1-D convolution kernel with 1-D maximum pooling and flatten network parameters.

After achieving the forecasting result from the construction of basic ensemble models (MLP and CNN), the forecasted output is then introduced to the boosting [34] method called AdaBoost for dynamically modifying the weight distribution obtained from the basic ensemble models. In boosting, the two main factors are Loss FFunction (LF) and Weight Updating Function (WUF) [33]. While finding the effectiveness for forecasting, the second function is used to obtain the LF and WUF for more accurate and precise forecasting results. The obtained results from boosting [3] may contain error values. Therefore, to rectify these errors, a Dynamical Error Correction (DEC) method is proposed for future forecasting results. In DEC, it uses the past three period data to predict a fourth period new data value, and this new fourth value is then replaced with the actual fourth data value.

The genetic algorithm is used for pruning the outputs from boosting the basic predictors for the best combination to increase the stability of the predictors. Then, the best combination from the genetic algorithms is used for stacking with Kernel Ridge Regression (KRR) as it is a meta predictor to strengthen the basic predictors. The target of this article is to propose a strong and stable time series forecasting using multivariate time series data. DL models like MLP [32] and CNN [5] have high bias and variance. The major advancement of using DL models is to sort all the major and minor errors by transforming the original multivariate data from a small level of depiction to a very improved and sophisticated level of representation by joining models layer by layer to extract the best output feature from the data.

The remainder of the paper is explained as follows: Sect. 2 presents the literature review on time series forecasting and deep learning model. Section 3 illustrated the methodology in detail. The experimental analyses with dataset information, evaluation metrics, and complexity analysis are discussed in Sect. 4. Experimental results are discussed in Sect. 5. Finally, the key findings, limitations of the study, and scopes for future works are discussed briefly in Sect. 6.

2 Literature Review

An ensemble learning model proposed by Qiu et al. [16] increases the precision and reliability of DL models having the same network architecture for training different training parameters. A non-linear ensemble deep learning [1] model has been proposed by Chen et al. for Time Series Forecasting to increase the precision in DL models. These studies mainly focused on different network architectures for improving the accuracy of the basic predictors. Zhou et al. proposed a boosting [34] algorithm known as AdaBoost algorithm for Time Series Forecasting (TSF). An extended AdaBoost algorithm [7] for an ensemble model is proposed by [12]. Based on the proposal given by Barrow and Crone [10], an ensemble model increases the precision when applied with different types of DL models for TSF. Liu et al. [32] introduced a technique to improve the precision by dynamically updating the weights to increase the TSF accuracy for DL models. Researchers have come up with an innovative idea of combining the two, three, or more number of models to achieve very high accuracy. One of the proposed models by Sarkar et al. is the stacking method [25]. Many researchers have used this method for many different DL and ML models to get higher accuracy. A unique DL ensemble model is constructed to employ Multi-population NSGA-II [2] for precise accuracy. The main advantages of three various Multi-Objective Optimal Evolutionary Algorithms (MOOEA) [4] for calculating and selecting the best accuracy from given DL models are clearly explained by Peimankar et al. For MOOEA [8], the most popular technique is Ensemble Pruning (EP) as its key purpose is to improve the forecasting results in terms of precision and reliability and to maintain the performance speed during running. Therefore, in this work, GA-II [9, 17] is used for EP to gain more precision and to maintain reliability and stability during TSF. From the review of literature, the following research issues are identified:

(i) When univariate data is given to the models, it does not satisfy the need when more than one feature is affecting the forecasting results.
(ii) It does not scale TSF from one variable to multi-variable data.
(iii) It does not check which features are affecting the accuracy result before giving the input data to the models.

Based on the research issues discussed above, the present study contributes to the body of literature as follows:

(i) A multivariate ensemble model is proposed for predicting the forecasting with more than one input feature.
(ii) This model maintains and increases the forecasting accuracy when more than one feature is given to the models as an input.
(iii) Construction of DNN models as basic predictor with multivariate dataset along with stacking and boosting ensemble techniques makes the result better with minimum calculations.

3 Methodology

The selected deep learning and ensemble learning models for construction in this work for TSF are described below.

(i) *Multilayer Perceptron (MLP):* MLP is proposed in this work as it is a classical type of ANN. They are composed of neurons in each layer. Initially, the data is given to the first layer, i.e., input; then, the prediction calculations are done in the interior layer, i.e., hidden layer. The interior layers can be added or removed according to the requirement and for better precision of results. Finally, the predictions are repatriated to the last layer, i.e., the output layer. MLPs [32] are very easy to use and can easily map the inputs and outputs received from various types of datasets, including TSF datasets.

(ii) *Convolutional Neural Network (CNN):* CNN captures the weights in the form of the 1-D array by collecting all the local weights from the given input data to increase the reliability based on unique feature extraction. Overfitting of the data is being avoided because of the fully connectedness of this network. Hierarchical patterns are identified and taken advantage of to find complex patterns using smaller and simpler patterns. Multiple layers have been constructed in the architecture of CNN [5] which include input layers, output layers, and multiple series of hidden layers. Activation and pooling are the parts of hidden layers.

(iii) *Extensive Deep Neural Network (Boosting):* A boosting DL algorithm is introduced to improve the precision of the basic DL model. For the introduced model, one of the important steps is to calculate the Forecasting Effectiveness (FE). During the calculation of the second-order forecasting, FE metric is used to optimize the LF and WUF for each sample, which can be expressed as Eq. (1):

$$m^k = \sum_{n=1}^{N} Q_n A_n^k \tag{1}$$

where m indicates forecasting effectiveness, N denotes the total number of time points, Q_n indicates the discrete probability distribution, and A_N implies the forecasting accuracy. A_N can be expressed as $A_n = 1 - |\epsilon_n|$, where

$$\epsilon_n = \begin{cases} -1, & (y_n - \hat{y}_n)/y_n < -1 \\ (y_n - \hat{y}_n)/y_n, & -1 \leqslant (y_n - \hat{y}_n)/y_n \leqslant 1 \\ 1, & (y_n - \hat{y}_n)/y_n > 1 \end{cases} \tag{2}$$

In Eq. (2), y_n and $\hat{y}_n$ indicate actual value of point n and forecasted value, respectively. In each DL model, the most important feature information is obtained based on weight distribution from the actual size. Even though it requires more time for computation, storing the memory to train and save data, it has very unique performance and offers a higher forecasting accuracy.

(iv) *Dynamic Error Correction (DEC):* Each basic DL model can more effectively reduce the forecasting error by learning the new Dynamic Error Correction (DEC) method. It is employed to reduce the forecasting error of the model. The time point of the present forecasting is updated using the previous time point forecasting errors. Hence, it tracks the error sequence before the particular time point and can dynamically reduce the forecasting error. The exponential of $-d$ helps us to see the error dynamically which means as time increases, the error decreases. It is important to analyze and correct error sequences of DL models before boosting. DEC uses the previous few time points to achieve a new time point and automatically updates the present forecasting error at the present time point following Eq. (3):

$$\overline{y'}_{d+1} = \overline{y}_{d+1} + \sum_{d=1}^{k} (y_d - \bar{y}_d) * \frac{(y_d - \bar{y}_d)}{MAXAE} * exp^{(-d)} \tag{3}$$

where $\overline{y'}_{d+1}$ = corrected forecasting value, $exp^{(-d)}$ = time delay factor, $k = 3$ (by trail and error method), and $MAXAE$ = Maximum absolute error.

(v) *Genetic Algorithm-II:* GA-II is an algorithm mostly performed to rectify the numeric and categorical data by optimization. Multi-population NSGA-2 [17] is built to further explore the optimization strategies. The steps involved in this algorithm are given below.

(i) *Initializing the required algorithm parameters:* The following parameters of the algorithms are initialized: g= number of iterations, $Maxgen$ = Highest iteration value, Np = size of the population, Cl = Length of the chromosome, Cr = Rate of crossover, Mr = Rate of mutation, P= Total population, P_g= First half population, and P_g^0= Second half population.

(ii) *Initializing population:* It is done based on a binary code scheme to convert each gene to chromosomes of the population by encoding. Each gene of the population can only be coded with 1 or 0. A reverse population technique is used to generate half of the population randomly and the other half is generated by encoding in a reverse way from the first generated population. Binary coding is carried out to fill each chromosome of the population with 1 or 0. $gene(i, j) = rand(0, 1)$ where $gene(i, j)$ represents the jth gene of ith chromosome.

(iii) *Intra-crossing:* In the intra-crossing process, crossover (Cr) and mutation (Mr) operations are done. The probability of crossover and mutation are taken as 0.75 and 0.25, respectively. Crossover and mutation techniques are performed for both halves of the population to obtain the respective offspring. Crossover operation equation is expressed in Eq. (4):

$$gene_{i,j}^{g} = gene_{i,j}^{g} + (-1)^{gene_{i,j}^{g}} * \left| gene_{i,j}^{g} - gene_{r1,j}^{g} \right| \tag{4}$$

where $gene_{i,j}^{g}$ represents the jth of the ith chromosome at the gth generation. $R1$ is a random value between $(1, Np)$ except i (current chromosome number).

(iv) *Cross population inter-crossing:* Cross population inter-crossing operation is done populations OPg and OPg' (offspring of both halves of the population). Inter-crossing operation on some universes of the population OPg and OPg' avoids the problem of not achieving global optima by allowing information dissemination to happen between the universes of the population hence used to update the populations allowing dissemination to happen. Inter-crossing operation is expressed in Eqs. (5) and (6):

$$gene_{i,j1:j1+r2}^{g}(OP) = gene_{i,j1:j1+r2}^{g}(OP') \tag{5}$$

$$gene^{g}_{i,j2:j2+r3}(OP^{'}) = gene^{g}_{i,j2:j2+r3}(OP) \tag{6}$$

where $r2$ and $r3$ are randomly generated integers in a range of 1 to 3. $j1$ and $j2$ are gene index values.

(v) *Optimization:* In this step, the optimal values of Bias and Variance are computed. Calculating respective objective function values of obtained offspring. Combining offspring populations with their respective parent population. Non-dominated universes are arranged by their objective values and a new population is generated from the Cr distance value of the sorted non-dominated universes.

(vi) *Termination:* The process is iterated until the max generation is reached. Increment the iteration number g by 1. Continue (iii) to (v) steps on new generations until the $Maxgen$ is reached.

4 Experiment Analysis

The best way to check whether the proposed model yields a good result or not is to check the stability and accuracy. The proposed method is applied on five datasets, i.e., Hourly dataset with PM2.5 concentration data obtained from five cities, namely "Beijing", "Shanghai", "Guangzhou", "Chengdu", and "Shenyang". Three EM are used to calculate the loss in five baseline models performance. For all the five datasets, the same parameters are used to calculate the generalized performance, where the length of the time step is taken as equal to ten and the remaining settings are described in Table 1 and data split is described in Table 2.

Table 1 Parameters used in constructing model

Model	Experiment settings	Value
MLP/CNN	Rate of learning	0.01
	Epochs	200
	Size of the batch	32
	Activator function	"relu"
K.R.R	Alpha	10
GA-II	Rate of crossover	0.75
	Ratc of mutation	0.25
	Rate of inter-crossing	0.85
	Total population	50
	Total iterations taken	30

Table 2 Summary of total train and test split of $PM_{2.5}$ Concentration for five datasets

Dataset	Train split	Test split	Validation split	Total size
Beijing	1960	613	489	3062
Chengdu	1960	613	489	3062
Guangzhou	1960	613	489	3062
Shanghai	1960	613	489	3062
Shenyang	1944	608	486	3038

4.1 Variables and Data Types

A great source of multivariate time series data is available in the UCI machine learning repository. Hourly dataset with $PM_{2.5}$ concentration data in "Beijing", "Shanghai", "Guangzhou", "Chengdu", and "Shenyang" cities. The time period is taken between Jan 1, 2010 to Dec 31, 2015. In the meteorological data, there are 86 attributes (integer and real) with the existence of missing values within it, which are denoted as "NA".

4.2 Evaluation Metric

For evaluating the forecasting models' performance of the time series, there are three statistical metrics that are widely used. These are absolute mean error (MAE), square means error (MSE), and percentage error of the absolute symmetric mean (SMAPE) [13] whose mathematical expressions are shown in Eqs. (7), (8), and (9), respectively. Five concentration levels of $PM_{2.5}$ taken from various regions are tested by forecasting. Finally, the results of loss performance in every DL model are described in Table 3.

$$MAE = \frac{1}{D}\sum_{d=1}^{D}|y_d - \bar{y}_d| \tag{7}$$

$$MSE = \frac{1}{D}\sum_{d=1}^{D}(y_d - \bar{y}_d)^2 \tag{8}$$

$$SMAPE = \frac{1}{D}\sum_{d=1}^{D}\frac{|y_d - \bar{y}_d|}{(|y_d| + |\bar{y}_d|)^{/2}} \tag{9}$$

where y_d = prediction value, $\bar{y}_d$ = Actual value, and D = Total number of data points.

Table 3 Total evaluation metric comparison for all the five datasets

Dataset	Model	MAE	MSE	SMAPE
Beijing	MLP_BOOSTING_BEST	14.514	348.657	0.204
	CNN_BOOSTING_BEST	10.346	166..819	0.189
	BOOSTING_KRR	10.316	188.388	0.383
	BOOSTING_KRR_DEC	11.093	187.570	0.353
	GA-II_BOOSTING_KRR_DEC	9.401	153.463	0.343
Chengdu	MLP_BOOSTING_BEST	14.199	315.570	0.293
	CNN_BOOSTING_BEST	12.419	238.299	0.253
	BOOSTING_KRR	9.577	132.604	0.287
	BOOSTING_KRR_DEC	7.593	85.001	0.209
	GA-II_BOOSTING_KRR_DEC	5.462	59.450	0.122
Guangzhou	MLP_BOOSTING_BEST	8.394	141.669	0.203
	CNN_BOOSTING_BEST	5.618	59.328	0.152
	BOOSTING_KRR	11.155	201.731	0.430
	BOOSTING_KRR_DEC	7.629	85.697	0.306
	GA-II_BOOSTING_KRR_DEC	6.108	53.969	0.245
Shanghai	MLP_BOOSTING_BEST	14.517	430.574	0.194
	CNN_BOOSTING_BEST	8.132	120.811	1.249
	BOOSTING_KRR	22.529	761.312	0.542
	BOOSTING_KRR_DEC	19.063	487.541	0.495
	GA-II_BOOSTING_KRR_DEC	17.357	384.885	0.452
Shenyang	MLP_BOOSTING_BEST	02.595	288.241	0.340
	CNN_BOOSTING_BEST	6.963	87.459	0.221
	BOOSTING_KRR	13.868	229.866	0.483
	BOOSTING_KRR_DEC	7.5109	100.394	0.337
	GA-II_BOOSTING_KRR_DEC	6.585	45.812	0.245

4.3 *Complexity and Scalability Analysis*

It is important to practically check the proposed model. $PM_{2.5}$ dataset is taken for computing several methods used in this study and other base models. All these methods and models are done in Python-3 GPU with 12.78 GB RAM and 68.40 GB disk. Each and every model running time is reported in Table 4.

Table 4 Running times for different models for five datasets

Cities	Dataset	Time(s)
Beijing	MLP_BOOSTING_BEST	46.823011
	CNN_BOOSTING_BEST	58.086994
	BOOSTING_KRR	58.487484
	BOOSTING_KRR_DEC	0.160580
	GA-II_BOOSTING_KRR_DEC	812.627046
Chengdu	MLP_BOOSTING_BEST	72.939346
	CNN_BOOSTING_BEST	90.367785
	BOOSTING_KRR	91.0681288
	BOOSTING_KRR_DEC	0.226677
	GA-II_BOOSTING_KRR_DEC	794.549375
Guangzhou	MLP_BOOSTING_BEST	71.639781
	CNN_BOOSTING_BEST	89.648378
	BOOSTING_KRR	90.313273
	BOOSTING_KRR_DEC	0.255202
	GA-II_BOOSTING_KRR_DEC	797.649196
Shanghai	MLP_BOOSTING_BEST	45.409239
	CNN_BOOSTING_BEST	59.052728
	BOOSTING_KRR	59.503863
	BOOSTING_KRR_DEC	0.161241
	GA-II_BOOSTING_KRR_DEC	466.944375
Shenyang	MLP_BOOSTING_BEST	44.2410666
	CNN_BOOSTING_BEST	57.631364
	BOOSTING_KRR	58.075797
	BOOSTING_KRR_DEC	0.216350
	GA-II_BOOSTING_KRR_DEC	472.369035

5 Results and Discussions

Multivariate time series data analysis is eventually more challenging as compared to univariate data analysis. A great source of multivariate time series data is available as open source in the UCI machine learning repository. PM2.5 concentration data (hourly data) collected from Beijing, Shanghai, Guangzhou, Chengdu, and Shenyang are used in this study. The time period is considered between Jan 1, 2010 to Dec 31, 2015. The datasets consist of missing data which are denoted as NA. The results obtained are discussed for different models for the Beijing dataset as follows: Using the Beijing dataset, the comparison graphs between the results of basic DL models generated by the boosting model are shown in Fig. 1a. The white region in the heat map refers to the maximum correlation magnitude and collinearity for the DL models between the features. The black region in the heat map refers to the least correlation magnitude and collinearity between the features. From Table 3, the forecasting

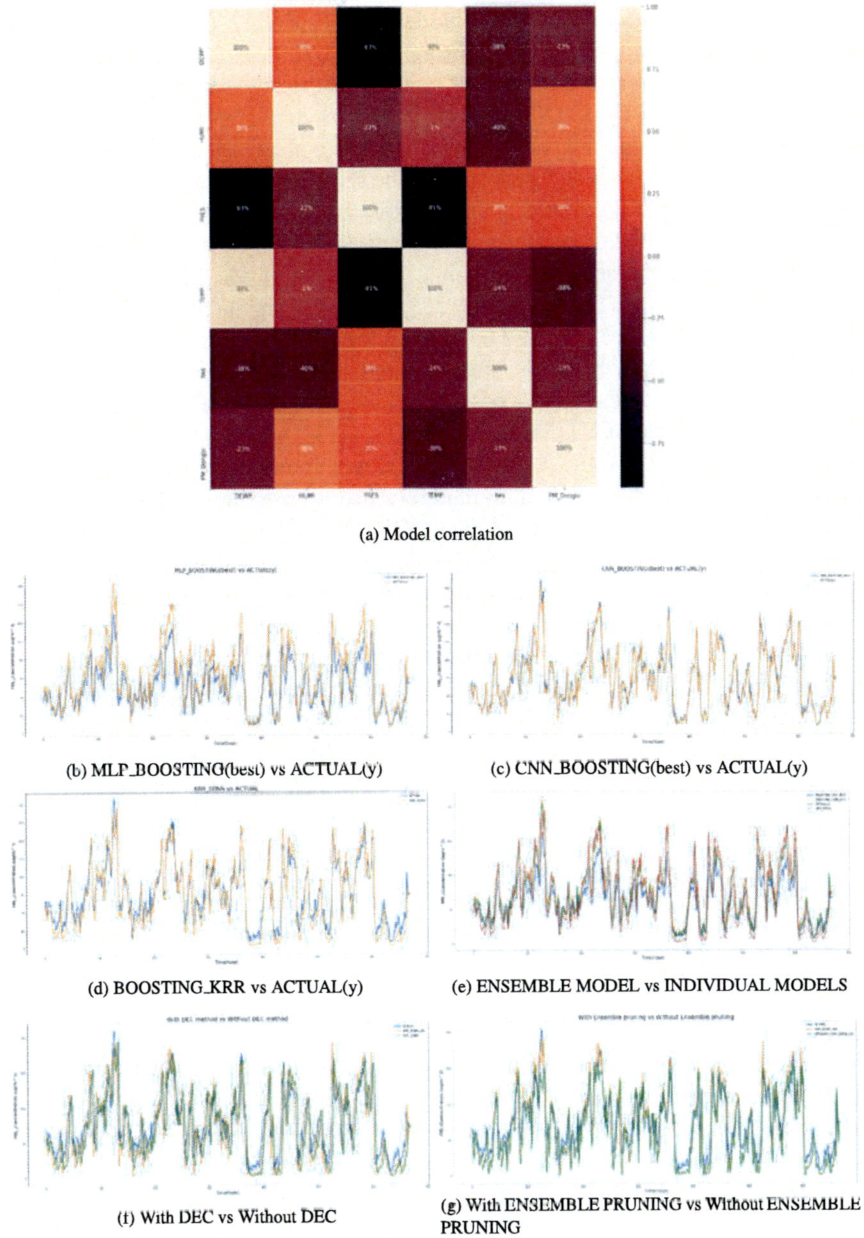

(a) Model correlation

(b) MLP_BOOSTING(best) vs ACTUAL(y)

(c) CNN_BOOSTING(best) vs ACTUAL(y)

(d) BOOSTING_KRR vs ACTUAL(y)

(e) ENSEMBLE MODEL vs INDIVIDUAL MODELS

(f) With DEC vs Without DEC

(g) With ENSEMBLE PRUNING vs Without ENSEMBLE PRUNING

Fig. 1 Experimental forecasting results on BEIJING dataset performances of the two individual DL models are observed to exhibit some differences in different time series datasets. From Table 3, the MAE value for the CNN model after boosting is found to be 10.095 which is the best when compared with the MLP model after BOOSTING. Therefore, this result evidently states that combining different individual DL models improves the ability of the proposed model for handling different datasets. The results of the stacking model are obtained using KRR (BOOSTING_KRR); BOOSTING_KRR is observed to generate the best output as KRR can deal with the influence of multi-collinearity within the basic predictors in the ensemble model. For example, for the Beijing dataset in Table 3, the SMAPE values of BOOSTING_ KRR is 0.383. By comparing the BOOSTING_KRR model with the BOOSTING_KRR_DEC model, the implemented DEC method and the performance can be observed. After employing the DEC method to each DL model, the performance has improved. For instance, for dataset Beijing in Table 3, the SMAPE values of BOOSTING_KRR_DEC is 0.353, which is better when compared with BOOSTING_KRR. The resulting curves of the ensemble model after DEC are closer to the actual value with minimum forecasting error. For the dataset Shanghai, in Table 3, the MSE value of GA-II_BOOSTING_KRR_DEC is 384.88 when compared with BOOSTING_KRR_DEC which is 487.541 and BOOSTING_KRR which is 761.312. The result indicates that the GA-II_BOOSTING_KRR_DEC is the best as compared to other ensemble models. In addition, when the graphs are compared with the BOOSTING_KRR_DEC without ensemble pruning, the resulting curve of GA-II_BOOSTING_DEC_KRR is closer to the actual curve. Therefore, the proposed GA-II [17] exhibits better reliability when compared with the other models. Likewise, the performance of the models on datasets Chengdu, Guangzhou, Shanghai, and Shenyang is compared based on the evaluation metric scores.

6 Conclusion

Time series forecasting plays a very important role in decision making. There are different types of forecasting models available, including ARIMA, CNN, and LSTM. However, the basic forecasting algorithm fails to provide efficient results in terms of accuracy and stability. To overcome this issue, we proposed a novel ensemble pruning method that combines MLP and CNN algorithms. We have used Boosting and Stacking ensemble techniques to strengthen the forecasting model. We have also used a Multi-population Non-dominated sorting genetic algorithm to improve the stability of the forecasting model. We proposed dynamic error correction to reduce the forecasting error of the model. Instead of implementing stacking directly, using a genetic algorithm before stacking is found to be the best for pruning, showing better results by increasing the stability of the model. The DEC method reduces the forecasting error efficiently.

There are a few limitations in this study. One of the limitations is during the time series forecasting, the proposed model cannot handle the high dimensionality in the dataset. Additionally, our proposed method demands higher computational time and memory resources. As a future scope, multi-objective optimizers can be used to further enhance the efficiency and stability of the model.

References

1. Chen S, Billings S, Grant P (1990) Non-linear system identification using neural networks. Int J Control 51(6):1191–1214
2. Deb K, Pratap A, Agarwal S, Meyarivan T (2002) A fast and elitist multiobjective genetic algorithm: Nsga-ii. IEEE Trans Evol Comput 6(2):182–197
3. Drucker H (1997) Improving regressors using boosting techniques. In: ICML. vol 97, pp 107–115
4. Dutta S, Das KN (2019) A survey on pareto-based eas to solve multi-objective optimization problems. In: Soft computing for problem solving. Springer, pp 807–820
5. Ferreira A, Giraldi G (2017) Convolutional neural network approaches to granite tiles classification. Expert Syst Appl 84:1–11
6. Gautam S, Maiti J, Syamsundar A, Sarkar S (2017) Segmented point process models for work system safety analysis. Saf Sci 95:15–27
7. Hastie T, Rosset S, Zhu J, Zou H (2009) Multi-class adaboost. Stat Interface 2(3):349–360
8. Jiménez F, Sánchez G, García JM, Sciavicco G, Miralles L (2017) Multi-objective evolutionary feature selection for online sales forecasting. Neurocomputing 234:75–92
9. Knowles JD, Corne DW (2000) Approximating the nondominated front using the pareto archived evolution strategy. Evol Comput 8(2):149–172
10. Kourentzes N, Barrow DK, Crone SF (2014) Neural network ensemble operators for time series forecasting. Expert Syst Appl 41(9):4235–4244
11. Krishna OB, Maiti J, Ray PK, Samanta B, Mandal S, Sarkar S (2015) Measurement and modeling of job stress of electric overhead traveling crane operators. Saf Health Work 6(4):279–288
12. Liu RY, Parelius JM, Singh K et al (1999) Multivariate analysis by data depth: descriptive statistics, graphics and inference,(with discussion and a rejoinder by liu and singh). Ann Stat 27(3):783–858
13. Panigrahi S, Behera HS (2017) A hybrid ets-ann model for time series forecasting. Eng Appl Artif Intell 66:49–59
14. Pramanik A, Djeddi C, Sarkar S, Maiti J and others (2020) Region proposal and object detection using Hog-based cnn feature map. In: International conference on data analytics for business and industry 2020, Way Towards a Sustainable Economy (ICDABI), IEEE, pp 1–5
15. Pramanik A, Pal SK, Maiti J, Mitra P (2021) Granulated rcnn and multi-class deep sort for multi-object detection and tracking. IEEE Trans Emerg Top Comput Intell (2021)
16. Qiu X, Zhang L, Ren Y, Suganthan PN, Amaratunga G (2014) Ensemble deep learning for regression and time series forecasting. In: 2014 IEEE symposium on computational intelligence in ensemble learning (CIEL). IEEE, pp 1–6
17. Raymer ML, Punch WF, Goodman ED, Kuhn LA, Jain AK (2000) Dimensionality reduction using genetic algorithms. IEEE Trans Evol Comput 4(2):164–171
18. Sarkar S, Verma A, Maiti J (2018) Prediction of occupational incidents using proactive and reactive data: A data mining approach. In: Industrial safety management- 21st century perspective of Asia. Springer, Singapore, pp 65–79
19. Sarkar S, Baidya S, Maiti J (2017) Application of rough set theory in accident analysis at work: A case study. In: ICRCICN 2017. pp 245–250 (2017)

20. Sarkar S, Ejaz N, Maiti J (2018) Application of hybrid clustering technique for pattern extraction of accident at work: A case study of a steel industry. In: 2018 4th international conference on recent advances in information technology (RAIT), IIT Dhanbad. IEEE, pp 1–6
21. Sarkar S, Lohani A, Maiti J (2017) Genetic algorithm-based association rule mining approach towards rule generation of occupational accidents. In: Communications in computer and information science, vol 776. Springer, Singapore, pp 517–530
22. Sarkar S, Maiti J (2020) Machine learning in occupational accident analysis: A review using science mapping approach with citation network analysis. Saf Sci 131:104900
23. Sarkar S, Patel A, Madaan S, Maiti J (2017) Prediction of occupational accidents using decision tree approach. In: INDICON 2017. IEEE, pp. 1–6
24. Sarkar S, Pateshwari V, Maiti J (2017) Predictive model for incident occurrences in steel plant in India. In: ICCCNT 2017, pp 1–5
25. Sarkar S, Pramanik A, Khatedi N, Balu A, Maiti J (2020) Gsel: A genetic stacking-based ensemble learning approach for incident classification. In: Proceedings of ICETIT 2019. Springer, pp 719–730
26. Sarkar S, Pramanik A, Maiti J, Reniers G (2020) Predicting and analyzing injury severity: a machine learning-based approach using class-imbalanced proactive and reactive data. Saf Sci 125:104616
27. Sarkar S, Raj R, Sammangi V, Maiti J, Pratihar D (2019) An optimization-based decision tree approach for predicting slip-trip-fall accidents at work. Saf Sci 118:57–69
28. Sarkar S, Sammangi V, Raj R, Maiti J, Mitra P (2019) Application of optimized machine learning techniques for prediction of occupational accidents. Comput Oper Res 106:210–224
29. Sarkar S, Vinay S, Pateshwari V, Maiti J (2017) Study of optimized svm for incident prediction of a steel plant in India. In: INDICON 2017. IEEE, pp 1–6
30. Singh K, Raj N, Sahu S, Behera R, Sarkar S, Maiti J (2015) Modelling safety of gantry crane operations using petri nets. Int J Injury Control Saf Promot 1–12
31. Verma A, Chatterjee S, Sarkar S, Maiti J (2018) Data-driven mapping between proactive and reactive measures of occupational safety performance. In: Industrial safety management- 21st century perspective of Asia. Springer, Singapore, pp 53–63
32. Wang L, Teo KK, Lin Z (2001) Predicting time series with wavelet packet neural networks. In: IJCNN'01. international joint conference on neural networks. Proceedings (Cat. No. 01CH37222). vol 3. IEEE, pp 1593–1597
33. Zhang S, Chen Y, Zhang W, Feng R, A novel ensemble deep learning model with dynamic error correction and multi-objective ensemble pruning for time series forecasting. Inf Sci 544:427–445
34. Zhi R, Zhou C, Li T (2019) Dynamic facial feature learning by deep evolutionary neural networks. In: Cyberspace data and intelligence, and cyber-living, syndrome, and health. Springer, pp 312–327

Deep Photo Classification Based on Geometrical Shape of Principal Object Presents in Photographs via VGG16 DCNN

Soma Debnath, Molla Saroyar Hossain, and Suvamoy Changder

Abstract In this paper, we propose a framework using a deep convolutional neural network (DCNN) to classify the photographs automatically based on different geometrical shapes of principal object exist in a photograph. Classification of photographs based on by identifying the shape of main object is challenging when a photograph contains multiple objects of different shapes. For this purpose, we choose only those images to train and test which contain either only single object as it's principal object or multiple objects of same geometrical shape, because our system is capable to identify only one type of shape of object in a photograph and classify the photograph based on the shape of main object. Shapes are very important element in photography composition. It has different meaning in different shapes and creates different feelings in human perception. Photograph containing a principal object with a geometrical shape is very visually appealing and achieve aesthetically high score in different photography competitions. Related study reveals that shape detection of objects using DCNN is very rare. Our model is a multiclass classifier which can classify the photographs into three different categories like photographs containing main object of circle, square and triangle shapes. Instead of using manually designed features the model automatically extracts features using a pre-trained DCNN VGG16. The proposed system has been executed on the ground truth dataset of 1700 images and achieve very satisfactory result. Another contribution of this work is to produce an image dataset on different geometrical shape of objects called Shape-Detection dataset which will be useful for future work in this domain.

Keywords Classification · Deep neural network · Geometrical shape · Photo aesthetics · Photographic composition

S. Debnath (✉) · M. S. Hossain · S. Changder
Department of CSE, National Institute of Technology Durgapur, Durgapur, India
e-mail: sd.16ca1102@phd.nitdgp.ac.in

S. Changder
e-mail: suvamoy.changder@cse.nitdgp.ac.in

© The Author(s), under exclusive license to Springer Nature Singapore Pte Ltd. 2022

D. Giri et al. (eds.), *Proceedings of the Seventh International Conference on Mathematics and Computing*, Advances in Intelligent Systems and Computing 1412,
https://doi.org/10.1007/978-981-16-6890-6_25

1 Introduction

In the current era, along with all the technological enhancements, digital media technology is also being enhanced. Common people are very attracted to digital media and camera technology. They are becoming addicted to use such technologies in their daily life. This is one of the reasons why all the companies are focusing more on research to enhance camera technology and so photography also becomes a very popular hobby of people. Photography is an art of capturing lights through digital sensor devices like cameras. It's a psychology. A photograph always portrays the photographer's psychological perception about the phenomenon captured in the device. In order to click a good photograph, the photographer has to keep in mind a number of factors, among which the rules of composition are very important factor. Geometrical shape of object is such a composition in photography that attracts people because of its presence. The repetition of a particular shape creates a pattern in photography which is visually interesting.

Different shapes convey different meaning in photography like circle conveys a sense of completeness or fullness. It indicates unity [1]. Viewer's first focus goes on an object of circular shape in photographs. A circular object as the primary subject plays a dominant role in photography. Similarly, triangle forms a group connecting three points in a photograph. It portrays a feeling of stability and aggression. Square formats in a photograph form a sense of balance and stability. The presence of all such geometrical shapes in photographs carry a good visual feeling. Due to this in various photography competitions, it has been seen that the photographs containing a principal object with a specific geometrical shape gain few extra points in terms of aesthetic score.

Most of research has done on aesthetic score assessment of photographs instead of focusing on various rules of composition in photography. The classification of digital photographs on the basis of the presence of geometrical shapes of principal object is an essential aspect in determining the aesthetic appeal of a digital photograph. So, in this paper, we focus to detect various geometrical shapes of main object exist in photograph automatically and classify the photographs based on the presence of shapes. When the size of the dataset is small then train a complete CNN from scratch is very difficult. That's why in this paper we present a DCNN-based classifier using transfer learning approach which can apply over a small dataset. The proposed framework fine-tune the pre-trained VGG16 DCNN on our proposed Shape-Detection database. Figure 1 shows the photographs from our database containing geometrical shapes.

The proposed model extracts feature automatically to detect shapes of main object presence in a photograph and build a DCNN-based classifier to classify the photographs into three different categories such as photographs containing circular object, square object and triangular object. The proposed classifier would be more useful for other works like identifying geometrical shapes of main object in photographs can add a good amount of score in aesthetic evaluation process and the kind of photographs in which the principal object has a specific geometrical shape and we want to search for such images automatically from any random database

Fig. 1 Photographs from our dataset Shape-Detection of three different categories Circle, Square and Triangle

without manual labelling. To build a classifier we use a transfer learning technique based on fine-tuning the pre-trained VGG16 convolutional neural network (CNN) which can classify the photographs into three categories. The model has been examined on the ground truth dataset Shape-Detection which is specially designed for this purpose and available online.

2 Literature Review

In this section, we briefly discuss about the automatic shape detection techniques in digital photographs. Our objective of this research is to detect regular geometrical shape of principal object in digital photographs automatically that it can be used as an important feature to photo aesthetic assessment task. Most of the research have done on this domain using digital image processing and machine learning techniques. In [2], Abbadi et al. proposed an architecture to detect different shapes and recognize them automatically in a digital image. They used contrast-limited adaptive histogram equalization (CLAHE) technique and Gaussian filter to detect shape and recognize them automatically.

Using optimal edge of shapes Moon et al. [3] proposed a two-dimensional shape detection technique. Minimizing noise power and mean squared error derived an optimal step edge operator in one dimension. They detect shape using pixel defined global contour detection. In the application of vehicle detection in aerial images, track contour in video images and features detection from human faces this method is used. In [4], Gomez et al. used a machine learning approach to detect shape automatically. First they find out the contour of the shapes and employed noise removal Gaussian filter. To recognize the different shapes, they used binary classification tree model.

Debnath et al. [5] proposed a framework to detect different geometrical shapes in digital photographs using machine learning approach. They applied boot strap aggregating function of Random Forest classifier to classify the photographs into four different categories based on four different geometrical shapes. In [6], Zakaria et al. proposed a method of shape recognition using the intensity values of image, and they applied Otsu's method to get threshold values. Singh et al. [7] proposed a method to detect shape from the real life tactile images. Such images contain real life objects. They used linear support vector machine (SVM) classifier to classify the objects based on their shapes. In [8], ayyildiz et al. proposed a predictive model to identify geometric shapes of different objects using image processing techniques and an artificial neural network (ANN) modelling by Backpropagation technique. Spasojevic et al. [9] proposed a geometrical shape classifier using multilayer neural network. Their model used the number of 30 images for training and testing data on cubic, pyramid and cylindrical shapes. In [10], a multilayer perceptron model (MLP) identify geometrical shapes with real-time neural networks.

3 Ground Truth Dataset (Shape-Detection)

The size of dataset and the labelling of images play a very vital role in any research work. For the proposed work we create own ground truth dataset from the popular photography websites https://www.dpchallenge.com/ and https://www.flickr.com/. We collect 1700 images of three different categories like photographs containing circular principal object, square principal object and triangular principal object manually from both the sites, out of them 1300 images used for training purpose, 200 images for validation and remaining 200 images for testing purpose. The principal object can be oval or sphere in shape in the category of photograhs with circular principal object. Simmilarly, the shape of main object can be rectangle or cube and pyramid in the categories of photographs with square principal object and triangular principal object, respectively. The dataset is publicly available at https://www.somadebnath.com/post/shape-detection-dataset. We introduce a new small size database to detect shape of principal object in photographs automatically. The database is named as Shape-Detection dataset. Our intention to form such dataset is for the research work on shape composition in photography, because the existence of such dataset on any specific composition is very rare.

4 Proposed Method

This section describes the proposed classification framework using a pre-trained DCNN VGG16 [11]. Our ground truth dataset Shape-Detection is very small in size and insufficient, due to this reason it's not possible to train whole cnn from scratch. So, we fine-tune the pre-trained VGG16 neural network on the dataset Shape-Detection. We use transfer learning scheme to fine-tune the pre-trained VGG16 network with its base weight ImageNet. Transfer learning is such a technique where deep neural network learns some basic features in its initial layer from the images and transferred those features to next layer.

4.1 Fine-Tuning the Pre-Trained VGG16 Network

Our ground truth dataset Shape-Detection is very different in naturewise compare to base dataset ImageNet, which was used to train VGG16 CNN. So, we chose to fine-tune the network weights of pre-trained VGG16 CNN on our dataset Shape-Detection. The framework of proposed method is described in Fig. 2. It is the overall process of fine-tuning VGG16 network. Here, we used 3-way softmax function to classify the photographs into three categories. Softmax is a generalized logistic function or exponential function. It takes a vector of Z for K real numbers. The real number set $\mathbb{R}$ take as input to this function. It can be written as:

$$\sigma : \mathbb{R}^k \rightarrow \mathbb{R}^k$$

$$\sigma(\overrightarrow{Z})_i = \frac{e^{Z_i}}{\sum_{j=1}^{k} e^{Z_j}} \text{fori} = 1, 2, \ldots, \text{kand} Z(z_1, z_2, \ldots, z_k) \epsilon \mathbb{R}^k \quad (1)$$

Initially, we fine-tune the full VGG16 network with the base weights of ImageNet and next we fine-tune half network by freezing the initial 9 layers. We retrain the remaining layers of VGG16 CNN on the ground truth dataset Shape-Detection.

Our network is fine-tuned on VGG16 model but to get comparatively good result it is tested on other CNN like 'InceptionV3' and 'Resnet'. But, we can observe from Table 1 that fine-tuning of VGG16 full neural network gives better result as compared to others.

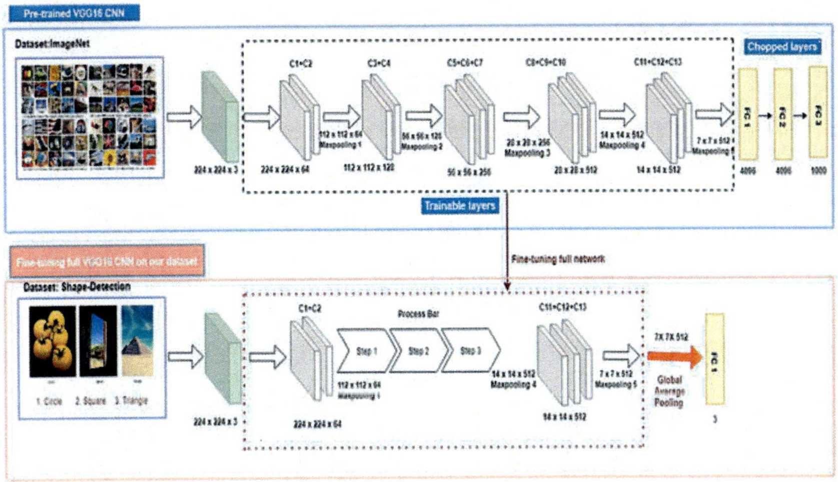

Fig. 2 Overall process of fine-tuning full VGG16 neural network

Table 1 Comparison of accuracies of diferent CNN

DCNN	Training	Validation	Test
VGG16_Full_Network	0.9664	0.875	0.875
VGG16_Half_Network	0.97	0.895	0.855
ResNet50	0.6362	0.635	0.555
InceptionV3	0.9641	0.84	0.82

4.1.1 Fine-Tuning the Pre-trained Full VGG16 Network

In this section, we fine-tune the full VGG16 neural network with initial weights of pre-trained neural network. Here, ImageNet is used as weights of VGG16 model. We made the last three dense layers of the VGG16 false and retrain upto last maxpooling layer of VGG16 CNN. Figure 3 shows the architecture of chopped VGG16 CNN, and Fig. 4 shows the building CNN model of loaded network. In the time of compilation 'adam', 'categorical_crossentropy' are used as optimizer and loss function, respectively. Suppose we have two classes X and Y. Where X ϵ {c, 1-c} and $Y\varepsilon\{c^{\wedge}, 1 - c^{\wedge}\}$.. Here c is the class label.

The cross-entropy measure between X and Y is

$$F(X, Y) = -\sum_i X_i \log Y_i = -c \log c - (1 - c) \log(1 - c)$$
$$= -(c \log c + (1 - c) \log(1 - c)) \tag{2}$$

We keep the learning rate very small (0.00001) to overcome the ovefitting problem and number of epochs are 5.

Image Preprocessing:

The image size is normalized to 224*224 pixels and rescaled in range of 255. The image data set of 3 classes were divided into training set, validation set and test set in 76, 12 and 12%. In order to minimize over-fitting 'ImageDataGenerator' is used to expand the data set by random horizontal flip zooming operations on training set image. After this the images of training and validation set is prepared in batch size of 32 and 'categorical' as class mode for feeding into network.

Data feeding and Training of Model:

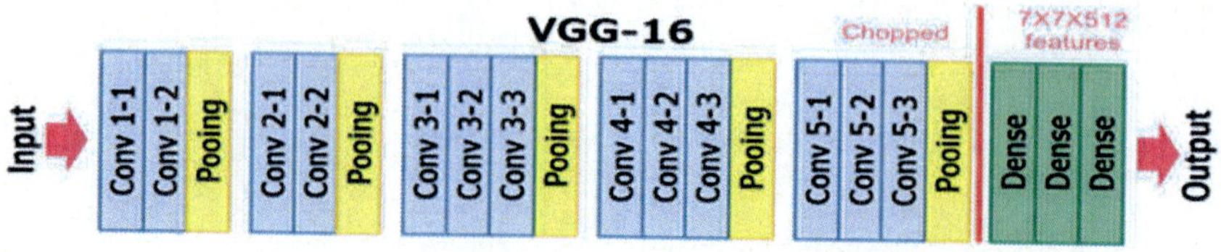

Fig. 3 The architecture of chopped last three dense layers of VGG16 CNN

```
[INFO] loading network...
WARNING:tensorflow:From
/home/soma/.local/lib/python2.7/site-packages/keras/backend/tensorflow_backend.p
y:4070: The name tf.nn.max_pool is deprecated. Please use tf.nn.max_pool2d
instead.

Model: "vgg16"
_________________________________________________________________
Layer (type)                 Output Shape              Param #
=================================================================
input_1 (InputLayer)         (None, None, None, 3)     0
_________________________________________________________________
block1_conv1 (Conv2D)        (None, None, None, 64)    1792
_________________________________________________________________
block1_conv2 (Conv2D)        (None, None, None, 64)    36928
_________________________________________________________________
block1_pool (MaxPooling2D)   (None, None, None, 64)    0
_________________________________________________________________
block2_conv1 (Conv2D)        (None, None, None, 128)   73856
_________________________________________________________________
block2_conv2 (Conv2D)        (None, None, None, 128)   147584
_________________________________________________________________
block2_pool (MaxPooling2D)   (None, None, None, 128)   0
_________________________________________________________________
block3_conv1 (Conv2D)        (None, None, None, 256)   295168
_________________________________________________________________
block3_conv2 (Conv2D)        (None, None, None, 256)   590080
_________________________________________________________________
block3_conv3 (Conv2D)        (None, None, None, 256)   590080
_________________________________________________________________
block3_pool (MaxPooling2D)   (None, None, None, 256)   0
_________________________________________________________________
block4_conv1 (Conv2D)        (None, None, None, 512)   1180160
_________________________________________________________________
block4_conv2 (Conv2D)        (None, None, None, 512)   2359808
_________________________________________________________________
block4_conv3 (Conv2D)        (None, None, None, 512)   2359808
_________________________________________________________________
block4_pool (MaxPooling2D)   (None, None, None, 512)   0
_________________________________________________________________
block5_conv1 (Conv2D)        (None, None, None, 512)   2359808
_________________________________________________________________
block5_conv2 (Conv2D)        (None, None, None, 512)   2359808
_________________________________________________________________
block5_conv3 (Conv2D)        (None, None, None, 512)   2359808
_________________________________________________________________
block5_pool (MaxPooling2D)   (None, None, None, 512)   0
=================================================================
Total params: 14,714,688
Trainable params: 14,714,688
Non-trainable params: 0
```

Fig. 4 Building CNN model of loaded VGG16

Images are fitted in the model using fit.generator function, training dataset feed for training and validation set for vaildate the model with epoches of 5. After training in return we get accuracy and loss of training and validation set. Model accuracy and cross entropy loss function is displayed in Fig. 5. From the plot we can observe that after 5th epoch the training accuracy is 95%, validation accuracy is 86% and test accuracy is 85%.

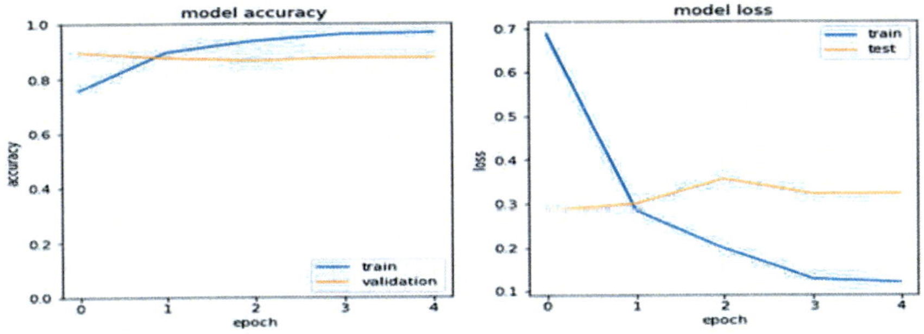

Fig. 5 Model accuracy and cross-entropy loss function for fine-tuned full VGG16 network

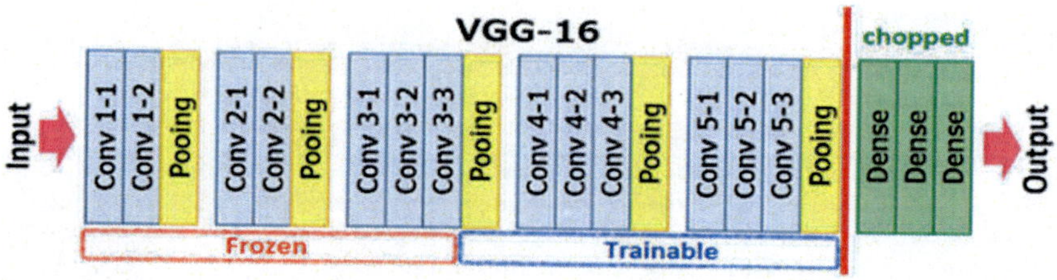

Fig. 6 The framework of fine-tuned half VGG16 network

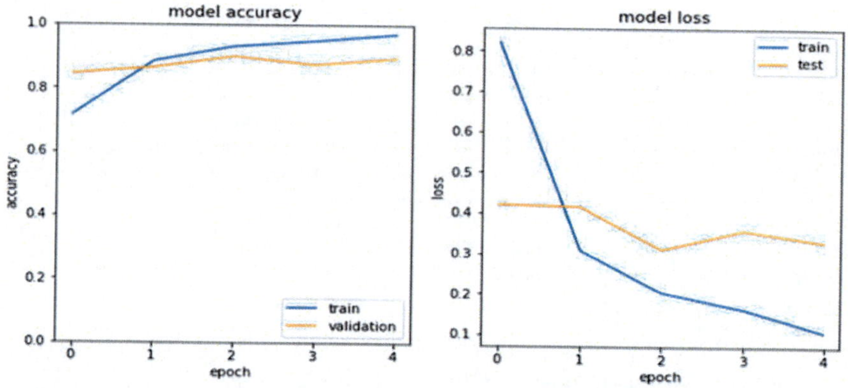

Fig. 7 The accuracy and cross-entropy loss for fine-tuned half VGG16 model

4.1.2 Fine-Tuning the Pre-trained Half VGG16 Network

In this section fine-tuning half VGG16 network by freezing the initial 9 layers of VGG16 CNN and retrain the rest of the layers. Figure 6 shows the architecture of half fine-tuned VGG16 model.

Training of half VGG16 model:

We retrain the half VGG16 model from 10th layer to last maxpooling layer. We keep the learning rate 0.00001, batch size 32 and number of epochs 5. After retraining the model we get the training accuracy 97% and validation accuracy 89%. Figure 7 shows the accuracy and cross-entropy loss for fine-tuned half VGG16 model.

4.2 Experimental Result

Since automatic shape detection of principal object present in photographs using DCNN technique is very rare as per literature study, so we are unable to compare our method with other existing DCNN techniques over shape detection. Hence, to get comparative result, we compare four different CNNs by fine-tuning the neural networks. Table 1 shows the comparison of accuracies of four different methods.

From Table 1, we can observe that by fine-tuning full VGG16 neural network we get best result in automatic shape detection of main object and classification of photographs based on the shape of object. Figure 8 and Fig. 9 show the accuracy and cross-entropy loss for ResNet50 and InceptionV3, respectively. Table 2 shows the comparison of accuracies of different shape detection approaches and we can see our proposed method gives outstanding result.

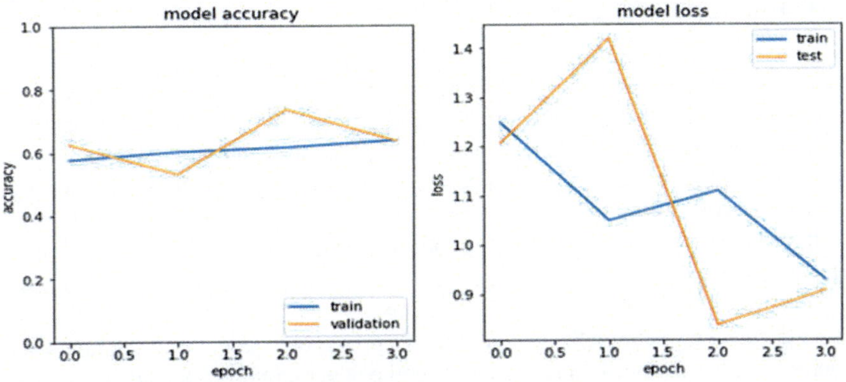

Fig. 8 Accuracy and cross-entropy loss for ResNet50

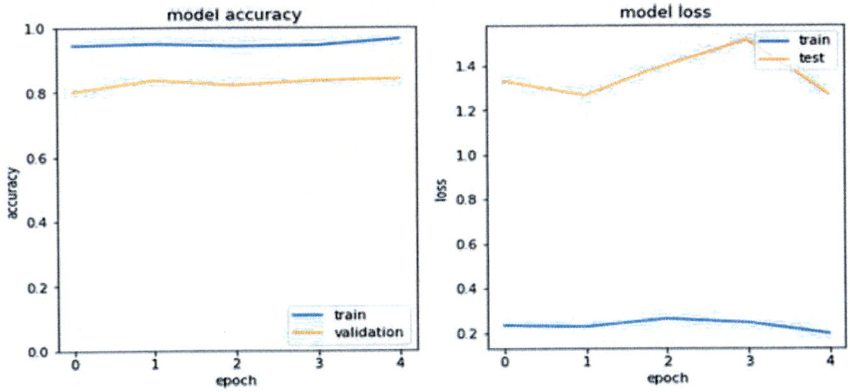

Fig. 9 Accuracy and cross-entropy loss for InceptionV3

Table 2 Comparison of accuracies of diferent shape detection methods

Authors	Used method	Dataset	Result (*Accuracy/RMSE*)
Our Proposed method	VGG16 DCNN	Own (1700)	88%
ayyildiz et al. [8]	ANN-Backpropagation	Unknown	RMSE: 0.0011(Training) 0.0045(Test)
Spasojevic et al. [9]	Multilayer NN	Own(30)	NA
Villatoro et al. [10]	ANN-MLP	Unknown	84%

5 Conclusion and Future Work

In this paper, we proposed a DCNN framework to detect the geometical shape of main object present in digital photographs automatically and classify the photographs based on the shape. A multiclass classifier is build up by fine-tuning the pre-trained VGG16 neural network with its basic weights of ImageNet. We just fine-tune the learned network weights on our ground truth dataset and achieve the satisfactory result. In future, we will concentrate on increasing the detection accuracy as well as expanding the Shape-Detection dataset for better training and testing. As a purpose of geometrical shape detection, we can develop an interesting camera application for assisting amateur photographers by finding the principal object in a photograph with specific geometrical shape in real time to get the better composition and we can add this compositional feature to existing researches on evaluation of aesthetics.

References

1. Suler J (2013) Photographic psychology: image and psyche. http://truecenterpublishing.com/photopsy/psylines.htm/
2. Abbadi N, Saadi L (2013) Automatic detection and recognize different shapes in an image. IJCSI 10(6, 1):162–166
3. Moon H, Chellappa R, Rosenfeld A (2002) Optimal edge-based shape detection. IEEE Trans Image Process 11(11)
4. Gomez S (2011) Shape recognition using machine learning. In: 6th Colombian computing congress
5. Debnath S, Changder S (2018) Automatic detection of regular geometrical shapes in photograph using machine learning approach. In: 2018 tenth international conference on advanced computing (ICoAC). Chennai, India, pp 1–6. https://doi.org/10.1109/ICoAC44903.2018.8939083
6. Firdaus ZM, Seng CH, Azmin SS (2012) Object shape recognition in image for machine vision application. Int J Comput Theory Eng 4(1):76–80
7. Singh G, Jati A, Khasnobish A, Bhattacharyya S, Konar A, Tibarewala DN, Nagar AK (2012) Object shape recognition from tactile images using regional descriptors. In: Fourth world congress on nature and biologically inspired computing (NaBIC), pp 53–58
8. Ayyildiz M, Cetinkaya K (2017) Predictive modeling of geometric shapes of different objects using image processing and an artificial neural network. Proc Inst Mech Eng, Part E: J Process Mech Eng 231(6):1206–1216

9. Spasojevic SS, Susic MZ, Durovic ZM (2012) Recognition and classification of geometric shapes using neural networks. In: 11th symposium on neural network applications in electrical engineering. IEEE, pp 71–76
10. Castillejos-Villatoro E, Nangusé-Vazquez G, Medina-Santiago A, Velazquez-Martinez JA (2016) Identification of geometric shapes with real-time neural networks. Int J Adv Netw Appl 8(1):2973
11. Simonyan K, Zisserman A (2014) Very deep convolutional networks for large-scale image recognition. arXiv:1409.1556

Multitask Deep Learning Model with Efficient Encoding Layer and Enhanced Parallel Convolution Block

Anupam Biswas, Angshuman Bora, Debashish Malakar, Subham Chakraborty, and Suman Bera

Abstract In recent years, multitask learning has turned out to be of great success in various applications. Though single task model training has promised great results throughout these years, it ignores valuable information which might help in estimating better learning parameters. By virtue of learning-related tasks, multitask learning has been able to generalize the models even better. In this paper, we try to enhance the feature mapping of multitask deep learning models by sharing features among related tasks and inductive transfer learning. We have explored the learning of task relationship among various tasks for acquiring more benefits from multitask learning. The proposed enhanced model is compared with two state-of-the-art multitask deep learning models, namely Hyperface and FaceNet. The results show better performance of the proposed model in predicting age, gender and ethnicity on the UTK face dataset.

Keywords Neural networks · Multitask deep learning · Deep learning · Recurrent neural networks · Convolution neural networks

A. Biswas (✉) · A. Bora · D. Malakar · S. Chakraborty · S. Bera
Department of Computer Science and Engineering, National Institute of Technology Silchar, Silchar-788010, Assam, India
e-mail: anupam@cse.nits.ac.in

A. Bora
e-mail: Angshuman_ug@cse.nits.ac.in

D. Malakar
e-mail: debashish_ug@cse.nits.ac.in

S. Chakraborty
e-mail: subham_ug@cse.nits.ac.in

S. Bera
e-mail: suman_ug@cse.nits.ac.in

© The Author(s), under exclusive license to Springer Nature Singapore Pte Ltd. 2022
D. Giri et al. (eds.), *Proceedings of the Seventh International Conference on Mathematics and Computing*, Advances in Intelligent Systems and Computing 1412,
https://doi.org/10.1007/978-981-16-6890-6_26

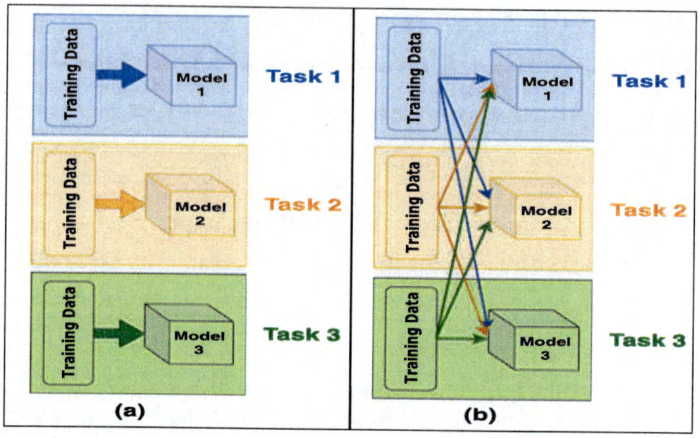

Fig. 1 Single task Llarning verses multitasking learning

1 Introduction

Deep learning models are in general designed and trained to learn a single task [2]. Though complicated and extensive, the tasks are sometimes decomposed into relatively small and individualistic sub-problems, which are then learned separately. Finally, towards the end, all are joined to get the final solution [17]. Models are trained to accomplish the solution by focusing on a single task overlooking data that may lead to improving the solution further. On the other hand, in multitask deep learning models, multiple tasks are learned together to obtain solutions to different related problems. In particular, by distributing information among related tasks, the model is better equipped to solve different related problems. Figure 1 shows the key difference between single task learning and multitask learning.

Several multitask deep learning models have been designed to solve relevant problems of different application domains (detailed in Sect. 2). Though multitask learning improves accuracy over single tasks, jointly learning different tasks comes with its own set of challenges. Determining the overlapping tasks is critical to the success of the model. In deep learning models designed for image-driven tasks, the most important part is the upper layer of convolution that performs image encoding. If the information from the input image is not extracted and encoded properly, the model might fail to perform well. In doing so, there are many models that can perform this task efficiently, but it results in many layers between input and output, which in turn increases the depth of the model and hence unintentionally leads to high computation cost. In this paper, an efficient multitask deep learning has been proposed. The proposed model overcomes the above-mentioned issues and improves the performance in the following way.

- An efficient image encoding layer using multiple blocks of sequential convolution layers in parallel is proposed. This input image encoder is designed to perform feature extraction efficiently from an input image.
- The dense layers of the model are designed in a way such that the model can still perform well, even if there is weak independence between the tasks. This is done by allowing the model to share the learnt features using hard parameter sharing.
- This proposed model architecture is able to remove the post-processing part after the calculation of the output which is required by many other existing multitask models.

The rest of the paper is organized as follows. Section 2 briefly explains the various works that are relevant to multitask learning. Section 3 presents the architecture of the proposed multitask deep learning and elaborates different blocks of the model. Section 4 details the experimental setup including datasets, baseline models and evaluation measures. Section 5 discusses the performance of the proposed model and performs a comparative analysis of the results with other state-of-the-art models. Section 6 concludes highlighting the performance of the proposed model.

2 Related Work

Multitask learning is a sub-domain of machine learning whic h focuses on processing multiple (more than a single) tasks at the same time, by taking into account the likeness between the various tasks. The neural networks used today try to solve a single task at once be it regression, classification, predicting structured objects, etc. All these tasks try to perform a single task and in the end only a single type of attribute is expected.

Let an image 'I' consist of an object with $A = (a_1, a_2, a_3, a_4, ..., a_n)$ number of attributes. The task is to design a model which determines these attributes present in the image. One of the approaches is to design a model using a multi-class classification neural network. But due to the presence of variations in the attribute classes, it will hinder the performance of the model.

The use of MTL is an efficient approach to tackle this problem domain. The objects present in an image have got multiple attributes and some of the attributes are correlated with each other of different objects of the same class. The features need to be shared with all other sub-tasks to advance the learning and the accuracy of the model. This can be done using feature joint learning in MTL [9]. Hard parameter and soft parameter sharing are the two MTL models which can be used for training the model.

The Convolutional Neural Network architecture comprises a number of layers which involves a lot of activation maps. This results in the increase in the dimensionality of the hyper-features [1]. The hyper-features need to be linked in such a way that they retrieve the features which are similar to the multiple tasks in an efficient manner. One of the basic approaches to address this kind of problem is the use of feature fusion which basically combines feature vectors in order to get a single feature vector. As CNN has the capability to evaluate complex functions, therefore, it is able to perform a fusion of the hyperparameters.

The approaches for computer vision in multitask learning are often approached by sharing the convolutional neural network layers. After that, the task-specific work is learnt by the model using fully connected layers. The paper [10] proposed a methodology to enhance certain models by using Deep Relationship Networks, where along with the shared layers and specific layers for each task of MTL, a matrix was placed prior to the fully connected layers, which allowed the deep learning model to learn various relationships among the tasks. This method is comparable to some Bayesian models. But, there is a drawback here. This proposed methodology still depends on a pre-defined structure to share different parameters. The structure is normally suitable for already studied different problems of computer vision, but is proven to be prone to error for novel tasks.

Multitasking has been able to flourish in the image domain. Its benefits can be found in semantic scene parsing, pose estimation, dynamic MR image construction, diagnostic medical imaging, etc. The majority of the semantic scene parsing models use supervised learning which are quite dependent on densely labelled data and do not promise good precision. The amount of data presently available for this field is also less which hinders performance measures of the prevailing deep learning models and other computer vision solutions. In the Semantic analysing scene, the main challenge in the fully textured RGB images is the highly variable content it contains in it.

Damien Fourure et al. [5] introduced a gradient reversal strategy for domain adaptation, where the target data from multiple datasets and from various tasks is utilized using a selective loss function, and strong correlations among tasks are captured using an auto-context procedure.

Yu and Lane [15] proposed a multitasking deep learning model which performs two image processing tasks: image segmentation and classification. The accuracy of the segmentation task doubled at the pixel level while using the multitasking model instead of a single task image processing model. The accuracy of the classification task is also enhanced by 2% while using the multitasking model compared to a single task model.

Pim Moeskops [11] trained a multitasking ConvNet for segmentation of six tissues in brain MR images. The CNN learns to identify the anatomical structures of the brain, the imaging modality and tissue classes. The model learns the diverse segmentation without the use of any task-specific training.

Multitasking Dynamic Contrast-Enhanced MR [13] imaging is used for the quantitative evaluation of neovascular architecture of the carotid artery walls. A strong correlation has been found between images which induced the images to be linearly dependent. Fully supervised deep neural networks normally require a large quantity of pixel-level labels that are quite costly to generate manually. Rihuan Ke [6] put forward a deep learning multitask model using a recursive approximation of tasks to reduce the manual cost of pixel-level labels. They divided the segmentation problem into a series of sub-problems which are defined recurrently based on the improving level of performance in approximation. The majority of the images used in training were only labelled by coarse partial masks, and the model learns these partial masks statistics in the training phase and these regions are expanded towards the boundaries of the object backed by the information learned in a data-driven manner.

Multitasking has also been used for real-time image enhancement [7]. A bilateral guided upsampling approach has been proposed. The overall model is on the basis of an encoder-decoder architecture in which the image processing shares all the encoding and decoding components. This multitasking model achieves higher performance compared to deep learning-based joint upsampling methods.

Multitask learning has also found applications in local seismic image processing [14]. There are three basic image processing tasks related to seismic images—detection of fault in structures, estimation of the seismic normal vector to evaluate local orientation in structures and removing noise by smoothing. These processing tasks are connected to each other as they are based on analysis of the same seismic structural features. However, traditional techniques used for seismic image processing consider these tasks as independent and hence face considerable challenges.

The idea of pooling information from multiple datasets using multitasking can be used for image classification in MNIST-like image datasets. The information sharing across hidden layers gives the potential of better learning and hence increased performance. The medical image processing domain suffers from the collection of large datasets with clean annotations due to the little availability of experts required for data labelling. Le et al. [8] put forward a multitask deep learning model for segmentation and classification of images in cancer diagnosis in mammography which combines pixel-level segmentation with global-level classification annotations. The model is based on Fully Convolutional Networks that enable better feature sharing and fast prediction.

Multitasking is also used for CT image-based analysis for Covid-19 [3]. Instead of performing the classification and segmentation independently, they modelled an architecture composed of two decoders and an encoder for reconstruction and segmentation and a multi-layer perceptron is used for classification purposes.

3 Proposed Multitask Model

The architecture of the proposed multitask model is presented in Fig. 2. The main components of the model are as follows:

1. Input Layer;
2. Image Encoding Layer with Parallel Blocks of Convolution;
3. Dense layers for desired classification.

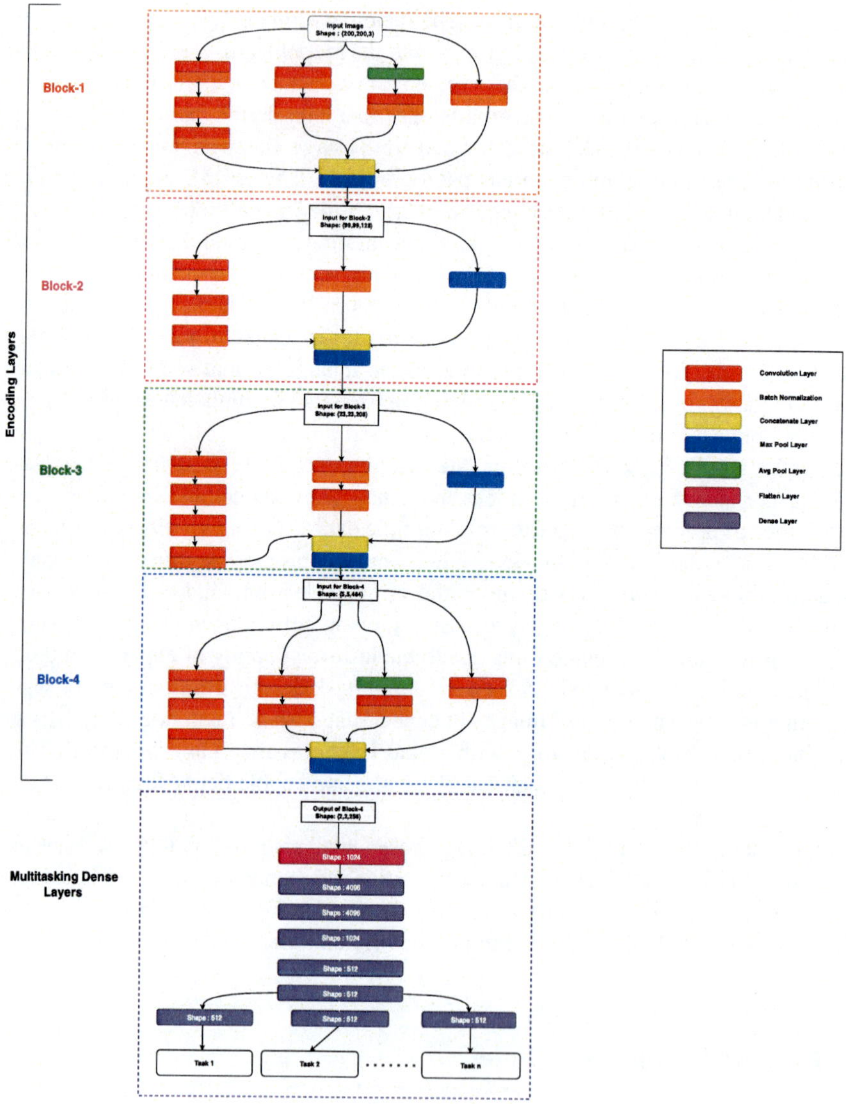

Fig. 2 Architecture of proposed multitask deep learning model

3.1 Input Layer

For this proposed multitask model, we have fixed the input image shape $(200 \times 200 \times 3)$. That is, the input image is a 200×200 pixel size colour image.

3.2 Image Encoding Layer with Parallel Blocks of Convolution

This part of the proposed model consists of four blocks of convolution layers that have multiple parallel layers in each of the blocks.

Block 1 This block consists of 4 sequences of convolution layers arranged in parallel. The first sequence consists of three alternate layers of convolution and batch normalization layers with 32, 48 and 48 filters in the convolutional layers. Consider the max pool window size to be (2×2) if not mentioned otherwise. The second sequence consists of two alternate layers of convolution and batch normalization layers with 24 and 32 filters in the convolutional layers. The third sequence consists of 1 average polling and one alternate layer of convolution and batch normalization layers with 16 filters in the convolutional layers. The fourth sequence consists of one layer of convolution and batch normalization layers with 32 filters in the convolutional layers. The output from each of the sequences is concatenated and a max pool layer is applied with poll window size 2. This is the output of block 1 with a shape $(99 \times 99 \times 128)$.

Block 2 The output from block-1 is passed through block 2. This block consists of three parallel sequences of convolution layers. The first sequence contains two alternate layers of convolution and batch normalization layers with 32, 48 and 48 number of filters in the convolutional layers. The second sequence in this block contains one alternate layer of convolution and batch normalization layers with 32 filters in the convolutional layer. The third sequence is a max pool layer. The output of these three sequences is then concatenated and passed through a max pool layer. The output of block 2 is of shape $(23 \times 23 \times 208)$.

Block 3 The output from block-2 is passed through block 3. This block consists of three parallel sequences of convolution layers. The first sequence contains four alternate layers of convolution and batch normalization layers with 96 filters in each of the convolutional layers. The second sequence in this block contains two alternate layers of convolution and batch normalization layers with 96 and 160 filters in the convolutional layer. The third sequence is a max pool layer. The output of these three sequences is then concatenated and passed through a max pool layer. The output of block 2 is of shape $(5 \times 5 \times 464)$.

Block 4 This block consists of 4 sequences of convolution layers arranged in parallel. The first sequence consists of three alternate layers of convolution and batch normalization layers with 64, 96 and 96 filters in the convolutional layers. The second sequence consists of two alternate layers of convolution and batch normalization layers with 48 and 64 filters in the convolutional layers. The third sequence consists of 1 average polling and one alternate layer of convolution and batch normalization layers with 32 filters in the convolutional layers. The fourth sequence consists of one layer of convolution and batch normalization layers with 64 number of filters in the convolutional layers. The output from each of the sequences is concatenated and a max pool layer is applied with poll window size 2. This is the output of block 1 with a shape $(2 \times 2 \times 256)$.

3.3 Dense Layers for Desired Classification

Following block 4 of encoding layers, the output is flattened and passed through a series of dense layers. Here, for the purpose of multitask learning, the concept of hard parameter sharing is being used as it reduces the chances of overfitting. The flattened layer is being passed to a dense layer with an input shape of 1024 and output shape of 4096, which is followed by a dense layer of input shape 4096 and output shape of 4096, passed through another three dense layers with input and output shapes as (4096, 1024), (1024, 512) and (512, 512), respectively. These layers are basically the feature sharing layers of hard parameter sharing.

Now, for task-specific layers consider there are a total of n tasks for any dataset under consideration. For each of the tasks, a dense layer is initialized with the shape of 512. Corresponding to those layers, n output layers are also initialized with their respective required output shape. Therefore, for each of the tasks, there is a dense layer of 512 nodes followed by an output layer with the required output shape. Each of those output sequences of layers is connected with the last dense layer of size 512 mentioned in the paragraph above.

4 Experimental Setup

The model is trained with the usage of graphics processing units (GPUs) in order to advance the training process. The system setup for training the model is as follows:

GPU: NVIDIA Quadro RTX 5000.
Software: CUDA Toolkit 9.0, CuDNN 7.0.5, Python 3.7+, TensorFlow GPU v2.0.
System type: 64-bit Operating System, x64-based processor.
Installed Memory (RAM): 128 GB.

4.1 Baseline Multitask Models

4.1.1 Model Architectures

The model architecture of the implemented multitask model on the UTK face dataset is as follows:

- **Hyperface** This model is proposed by Ranjan et al. [12]. The model network comprises 5 convolutional layers. The network is initialized with the weights of the R-CNN Face network. They fused the max1, conv3 and pool5 layers which belong to AlexNet using an external network. Conv1a and conv3a are added to the pool1 and conv3 layers of the AlexNet to acquire a feature map of dimension $6 \times 6 \times 256$ at the output. The output of these layers is concatenated to the conv5 of the AlexNet architecture. A 1×1 convolution layer is added to reduce the dimensionality of the feature map. Since we have three classification tasks, we have split the network into 3 parts unlike the model proposed by the authors.
- **FaceNet** The next model is based on the work proposed by Das et al. [4]. The model is mainly based on the ResNet V1 architecture. The network comprises a cluster of input layers and the ResNet Deep neural architecture along with L2 normalization. They have added three convolution layers one for each of the tasks, i.e. age, gender and race classification and a softmax layer for label prediction in individual tasks. The Softmax layer is followed by another layer of L2 normalization and dropout layer.

4.2 Datasets

In the proposed model, the UTK Face dataset [16] is used for training and obtaining results. It is a large-scale face dataset containing almost 24,000 face images with a very wide range of ages (it spans from 0 to 116 years old). The facial features can be seen properly, though it contains some contrasts in resolutions, lighting, posture, etc. Here, each image is nicely aligned and contains only a single picture of a face. The labelling of each image is done on four parts joined together in a single name, age (0–116), gender (male or female), race (White, Black, Asian, Indian and Others) and finally date and time. The dataset is hugely popular due to its applications, particularly in age prediction, face detection, etc.

In designing the dense layers for the dataset, the network is split into three parts for determining the three attributes, namely age, gender and ethnicity as detailed in [16]. The output layers are preceded with a dense layer of size 512 nodes that are connected to the last common dense layer of size 512 in the hard parameter sharing part of the proposed model.

4.3 Evaluation Measures

Along with the proposed multitask model, two other models, namely Hyperface and FaceNet were selected for comparison analysis. For the evaluation of the model, the dataset was divided into a 80 : 20 train and test ratio. The train data was used to train all the models mentioned above. The epoch count was set for each model to be 50. Each batch consisted of 64 images. There were 234 steps involved per epoch of training. Four evaluation measures considered are as follows:

1. Execution time per epoch;
2. Average time taken per step in an epoch;
3. Loss;
4. Accuracy.

5 Result Analysis and Discussion

The dataset was divided into the 80 : 20 ratio of train and test data. After that, the model was trained. The accuracy and loss in determining the age, gender and ethnicity can be seen in Fig. 3. For comparison of results, two models, namely **FaceNet** and **Hyperface** were taken. The comparison results can be seen below.

5.1 Execution Time During Training Phase

Time taken during the training phase of the proposed multitask model and the other 2 models is shown in Fig. 4a and in Fig. 4b. Figure 4a shows the total execution time taken by each of the models in each epoch. Figure 4b shows the average time taken by the models per step in an epoch. It can be seen that both the values are almost constant. Hence, the execution time for all the model is almost the same. Also,

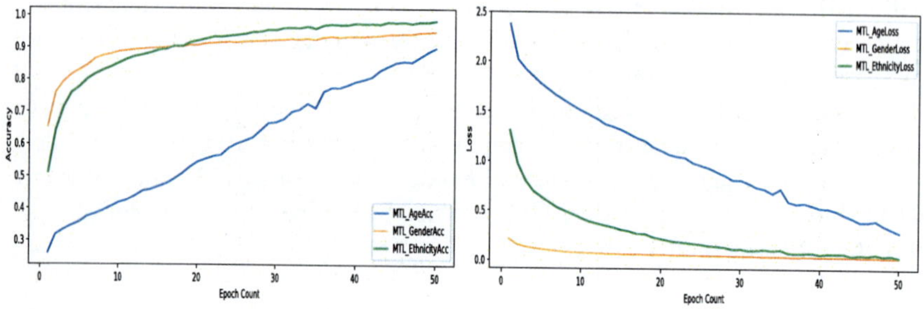

Fig. 3 Accuracy and loss of proposed model in determining age, gender and ethnicity

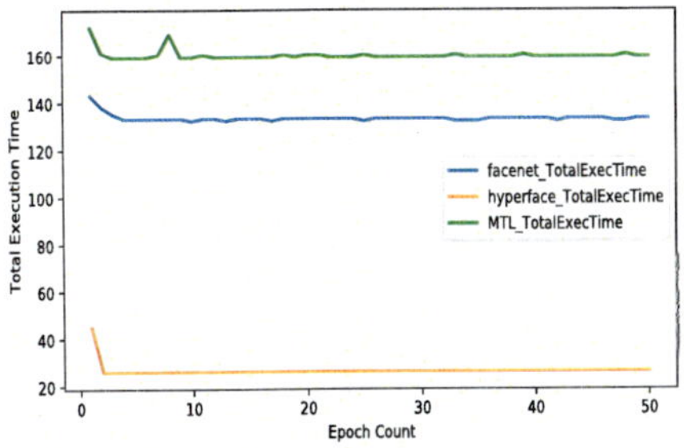

(a) Total Execution Time

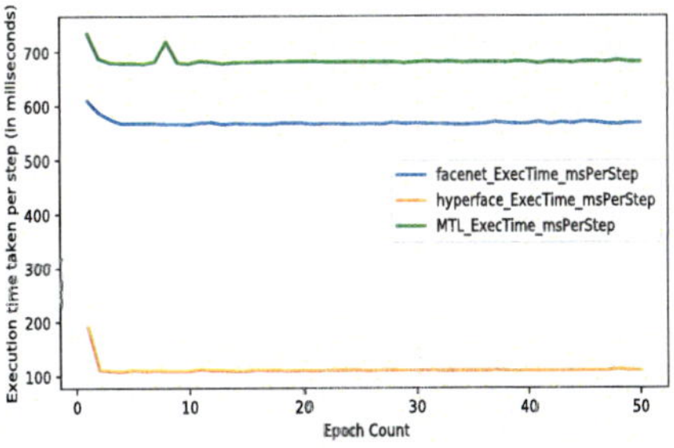

(b) Per Step Execution Time

Fig. 4 Comparison of total execution time of proposed multitask, Hyperface and FaceNet model

it can be seen that the order of execution time required shows that our proposed model architecture takes the most amount of time to run a batch of data. However, comparing the accuracy of the proposed model as compared to the other models under consideration, it can be said that the time consumption is reasonable.

5.2 *Loss Comparison During Training Phase*

This section shows the comparison between the loss of the proposed multitask model with Hyperface and FaceNet models. Figure 5a shows the total amount of loss each of the models above faced during its training. Figure 5b–d shows the loss in age, gender and ethnicity while training. In all the cases, the proposed multitask model outperforms all the other existing models considered for comparison. Not only the proposed model remains at the bottom of the loss curve, but it also does it efficiently. This implies that the encoding layer used for image encoding is effective very much. Due to this, there is a very small amount of information that is lost during the encoding part.

5.3 *Accuracy Comparison During Training Phase*

The accuracy value while training each model is presented in Fig. 6. As can be seen in Fig. 6a, among all the three models the age estimation accuracy in training on the same dataset is increasing steadily as well as continuously. This implies that the model was able to learn the age feature very well. Since age estimation is a tricky task, therefore, age was considered for some overlapping classes. The proposed multitask model shows exceptional accuracy in training.

The accuracy in gender estimation gradually increases and reaches a steady state as can be seen in Fig. 6b. The curve starts converging after the first 10 epochs. This

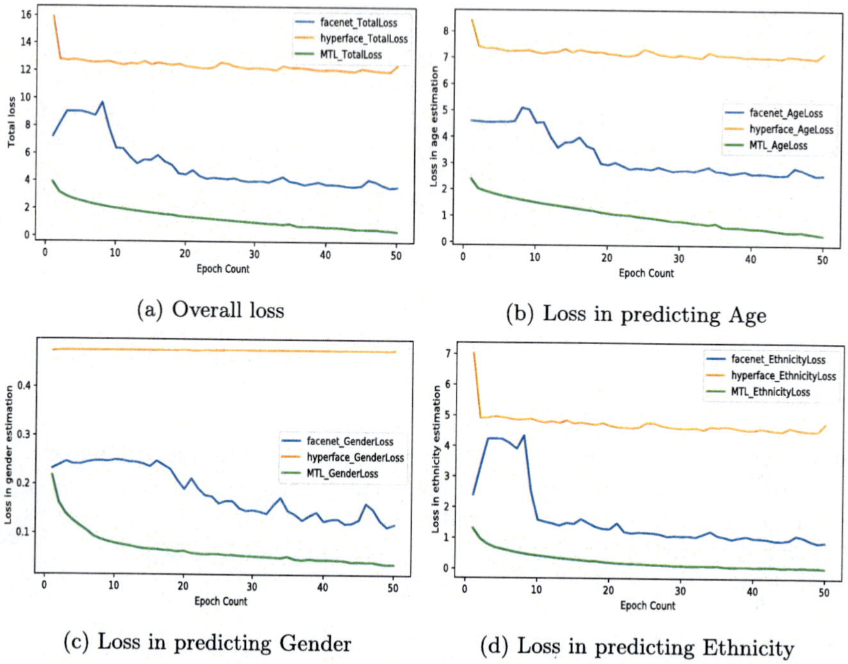

(a) Overall loss

(b) Loss in predicting Age

(c) Loss in predicting Gender

(d) Loss in predicting Ethnicity

Fig. 5 Comparison of loss of proposed model, Hyperface and FaceNet models

implies that the model was able to efficiently extract information from the input images. This learning curve for the proposed multitask model is significantly ahead of the learning curve generated by other models considered for comparison.

The same can be seen in Fig. 6c in calculating the ethnicity accuracy. The learning curve in ethnicity estimation is seen to start converging after the first 30 epoch. Not only the curve attains a stable state, but it also surpasses the learning curves for other models by a significant margin. This indicates that feature learning was very efficient.

5.4 *Classification Accuracy with Test Data*

After successful training of the model, the performance of proposed multitask model, Hyperface and FaceNet models was compared with their test data results. The result comparison table can be seen in Table 1.

As can be clearly seen in the table, the training accuracy for the proposed model is significantly higher than the other two models. The performance of the trained model even holds more accuracy for the test data. The proposed multitask model performs better than the Hyperface model and FaceNet model. This shows that the proposed image encoding layer for the multitask model is very efficient in feature extraction.

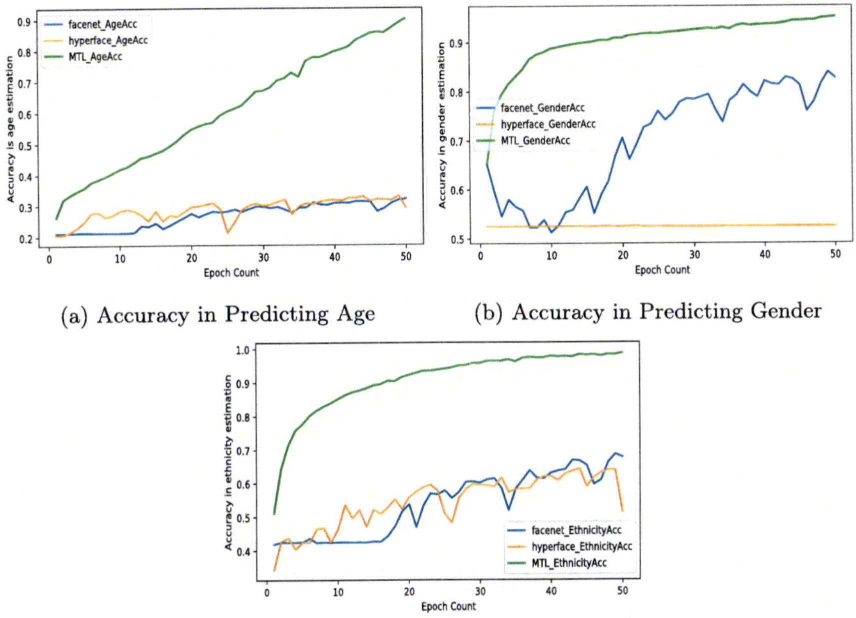

(a) Accuracy in Predicting Age

(b) Accuracy in Predicting Gender

(c) Accuracy in Predicting Ethnicity

Fig. 6 Accuracy comparison of proposed model with Hyperface and FaceNet

Table 1 Model performance on test data

	Model name	Age accuracy (%)	Gender accuracy (%)	Ethnicity accuracy (%)
Train data	Hyperface	30	53	55
	FaceNet	31	83	68
	Proposed model	83	94	96
Test data	Hyperface	30	52	54
	FaceNet	31	79	67
	Proposed model	35	88	78

6 Conclusion

The main challenge in multitask learning is the understanding of what to share between the hidden layers. Sharing unrelated tasks might lead to overfitting of the model. We have proposed a multitask deep learning model to solve this issue. Our proposed model has shown significant advancement on the UTK face dataset compared to two state-of-the-art multitasking models—Hyperface [12] and FaceNet [4]. On the training dataset, the model has shown an accuracy of 83% compared to 30% and 31% of the other models. Similar results have been reported in the test data too.

These results are evident for the potential of the proposed model to generalize for other datasets, which has outperformed the two existing models.

Acknowledgements This work is supported by the Science and Engineering Board (SERB), Department of Science and Technology (DST) of the Government of India under Grant No. ECR/2018/000204 and Grant No. EEQ/2019/000657.

References

1. Agarwal A, Triggs W (2008) Multilevel image coding with hyperfeatures. Int J Comput Vis 78 (06 2008). https://doi.org/10.1007/s11263-007-0072-x
2. Alom MZ, Taha TM, Yakopcic C, Westberg S, Sidike P, Nasrin MS, Hasan M, Van Essen BC, Awwal AA, Asari VK (2019) A state-of-the-art survey on deep learning theory and architectures. Electronics 8(3):292
3. Amyar A, Modzelewski R, Ruan S (2020) Multi-task deep learning based ct imaging analysis for covid-19: Classification and segmentation. medRxiv
4. Das A, Dantcheva A, Bremond F (2018) Mitigating bias in gender, age and ethnicity classification: a multi-task convolution neural network approach. In: Proceedings of the European conference on computer vision (ECCV). pp 0
5. Fourure D, Emonet R, Fromont E, Muselet D, Neverova N, Trémeau A, Wolf C (2017) Multitask, multi-domain learning: application to semantic segmentation and pose regression. Neurocomputing 251:68–80
6. Ke R, Bugeau A, Papadakis N, Kirkland M, Schuetz P (2020) Schönlieb. Multi-task deep learning for image segmentation using recursive approximation tasks, C.B
7. Kong K, Lee J, Song W, Kang M, Kwon K, Kim SG (2019) Multitask bilateral learning for real-time image enhancement. J Soc Inf Displ 27. https://doi.org/10.1002/jsid.791
8. Le TLT, Thome N, Bernard S, Bismuth V, Patoureaux F (2019) Multitask classification and segmentation for cancer diagnosis in mammography. arXiv:1909.05397
9. Li Y, Tian X, Liu T, Tao D (2015) Multi-task model and feature joint learning. In: Twenty-fourth international joint conference on artificial intelligence
10. Long M, Cao Z, Wang J, Yu PS (2015) Learning multiple tasks with multilinear relationship networks. arXiv:1506.02117
11. Moeskops P, Wolterink JM, van der Velden BH, Gilhuijs KG, Leiner T, Viergever MA, Išgum I (2016) Deep learning for multi-task medical image segmentation in multiple modalities. In: International conference on medical image computing and computer-assisted intervention. Springer, pp 478–486
12. Ranjan R, Patel VM, Chellappa R (2017) Hyperface: a deep multi-task learning framework for face detection, landmark localization, pose estimation, and gender recognition. IEEE Trans Pattern Anal Mach Intell 41(1):121–135
13. Wang N, Christodoulou AG, Xie Y, Wang Z, Deng Z, Zhou B, Lee S, Fan Z, Chang H, Yu W, Li D (April 2019) Quantitative 3d dynamic contrast-enhanced (dce) mr imaging of carotid vessel wall by fast t1 mapping using multitasking. Mag Resonan Med 81(4):2302–2314
14. Wu X, Liang L, Shi Y, Geng Z, Fomel S (2019) Multitask learning for local seismic image processing: fault detection, structure-oriented smoothing with edge-preserving, and seismic normal estimation by using a single convolutional neural network. Geophys J Int 219(3):2097–2109
15. Yu B, Lane I (2014) Multi-task deep learning for image understanding. In: 2014 6th international conference of soft computing and pattern recognition (SoCPaR). IEEE, pp 37–42
16. Zhang, Zhifei SY, Qi H (2017) Age progression/regression by conditional adversarial autoencoder. In: IEEE conference on computer vision and pattern recognition (CVPR). IEEE
17. Zhang Y, Yang Q (2017) A survey on multi-task learning. arXiv:1707.08114

Genetic Algorithm-Based Deep Learning Models: A Design Perspective

Debojyoti Sarkar, Nilima Mishra, and Anupam Biswas

Abstract Deep learning models are immensely used for numerous problems in different domains and have proven to be superior over feature-based machine learning techniques. However, the success of any deep learning model is dependent on several factors like the tuning of appropriate different hyper-parameters, neural network architectures, optimizers, etc. To learn neural network weights, gradient-based optimizers such as stochastic gradient descent, min-batch gradient descent and the Adam optimizer are widely used. However, the architecture of neural networks and hyper-parameters has to be tuned manually for better performance of the model. Evolutionary deep learning models are gaining attention in recent years to overcome the manual tuning of hyper-parameters and the network architecture. This paper presents a design perspective of recently developed Genetic Algorithm-driven evolutionary deep learning models along with the inherent challenges.

Keywords Deep learning · Evolutionary algorithm · Evolutionary deep learning · Genetic algorithm · Neural network

1 Introduction

Deep learning is the subset of machine learning that emphasizes learning by example [16]. With the evolution of Artificial Intelligence and Machine Learning, strategies to mimic the human mind in data classification and analysis pave the way to deep

D. Sarkar (✉) · A. Biswas
Department of Computer Science and Engineering, National Institute of Technology Silchar, Silchar 788010, Assam, India
e-mail: debojyoti_rs@cse.nits.ac.in

A. Biswas
e-mail: anupam@cse.nits.ac.in

N. Mishra
Department of Electronics and Telecommunication Engineering, International Institute of Information Technology, Bhubaneswar, Odisha, India
e-mail: b217024@iiit-bh.ac.in

© The Author(s), under exclusive license to Springer Nature Singapore Pte Ltd. 2022
D. Giri et al. (eds.), *Proceedings of the Seventh International Conference on Mathematics and Computing*, Advances in Intelligent Systems and Computing 1412,
https://doi.org/10.1007/978-981-16-6890-6_27

learning. The name deep learning indicates the use of several layers in an architecture to process and manage data. In these techniques, we make the system to learn from experience, understanding the hierarchy of concepts involved. The complex concepts are then trained into the model by building them out of the simpler ones. Thus, it indicates how the concepts are built on top of each other.

The flow of ideas comes from the very basic observation of what happens in the mind in response to the surrounding. When we identify it, the question of why it happens comes around. This leads to the development of the learning models, where we train the model with a set of data, to identify and classify. Thereafter, we analyse what to do while training the model. When the model is trained, the outputs obtained are matched with the expected ones. This leads to a deviation of the output from the expected value. Algorithms are then implemented to reduce the error by optimizing the operation through the layers.

Deep learning models prove to be highly effective while dealing with large sets of labelled data [15]. They work with better computing power and can be used in handling huge varieties of data. Deep learning is implemented through a wide range of architectures in terms of Multi-layered perceptrons and neural networks (NN). The core functionality of deep learning lies in its neural network architecture that deals with training and learning of the model and finally identifying the input. Challenges faced are mostly related to architectural design and adaptation of the connection weights [20]. An effective network design asks for a well-trained human effort and a lot of time. For someone who is interested only in utilizing the neural networks for a certain task, unaware of its entire working mechanism, it gets tedious to dig into the details of the architecture design. Most of the ways to address the network design or weight adaptation issues use gradient-based algorithms. The gradient-based algorithms, though quite fast, manifest their age-old drawbacks such as local optima stagnation and lack of convergence [10]. Moreover, the gradient-based methods demand continuous, differentiable, uni-modal and noise-free search spaces for execution. This sets a hard limitation to the efficacy of gradient-based methods while applied in neural network training or learning. Evolutionary Algorithms (EAs) are population-based, non-deterministic search algorithms that deal with optimal solution generations [8]. The fundamental strategy behind EAs comes from Charles Darwin's Theory of Evolution. EAs have two fundamental constructs: exploration and exploitation. Exploration refers to a global search of potential solutions while exploitation is an intensified local search. A group of individuals, each representing a potential solution, form a population which is engaged in finding the best individual (or solution) in a certain solution space. The individuals perform exploration and exploitation in a balanced manner to finally converge to an optimal solution. The non-deterministic nature of EAs ensures a random but dynamically evolving population [5]. Genetic algorithm (GA) is the most common EA.

GA can easily work on discontinuous search space. A proper balance maintained in the exploration and exploitation processes tackles the problem of local optima stagnation. Therefore, they can be safely applied to multi-modal problem spaces. As the GA is meta-heuristic, stochastic algorithms, devoid of any gradient descent methods, they can be applied to non-differential landscapes as well [21]. GA is applied in deep learning at two different stages:

(a) Finding the most efficient network architecture for the NNs.
(b) Assigning the best suited weights to the connections between nodes of NNs.

The NN architecture search and weight adaptation problems in deep learning can be formulated as a problem of minimization of an error function, where the error represents the quality of the network.

The rest of the paper is organized as follows: Sect. 3 describes GA-based deep learning which consists of network automation and parameter tuning, and Sect. 4 discusses the shortcomings of GA and the challenges faced while using GA in deep learning problems. We then conclude in Sects. 5, and 6 contains the acknowledgement.

2 Evolutionary Deep Learning Models

2.1 Generic Framework of Evolutionary Algorithms

Evolutionary algorithms operate on the basis of two fundamental constructs—exploration or diversification and exploitation or intensification. The algorithm starts with initializing a random set or a population of potential solutions known as individuals. The individuals in the population now undergo some evaluation known as the fitness evaluation. This stage of computation is about attaining a proper balance between exploration and exploitation. Too much exploitation will end up with us having a local optimum, known as local optima stagnation, while on the other hand, an excessive exploration is time-consuming and might never lead us to a solution, i.e., the convergence might never occur. Here, the individuals collaborate to improve their individual solutions. This is done by digging deep in a search landscape and at the same time communicating with the other individuals in the population to have an idea of the direction of the search in order to reach the global optimal solution. After a certain number of repeated searches, it becomes obvious that some individuals (or solutions) in the population can be no good a solution. Here is where these individuals undergo a selection process so as to establish a set of individuals, which are potential solutions, to create a new improved population.

Stochastic methods of computation [4] are largely dependent on randomness, therefore, in order to bring some randomness, some variations are made to the new population by tuning some parameters. This mechanism varies from algorithm to algorithm. For instance, in Particle Swarm Optimization (PSO), the weights and constants related to the social and the cognitive component, and the velocity or the direction of search triggered by the 'inertia' are modified [12]. In Ant Colony Optimization (ACO), the modifications are involved in pheromone trail levels [7] while in GA, the variations are achieved through crossover and mutation [24]. The minor

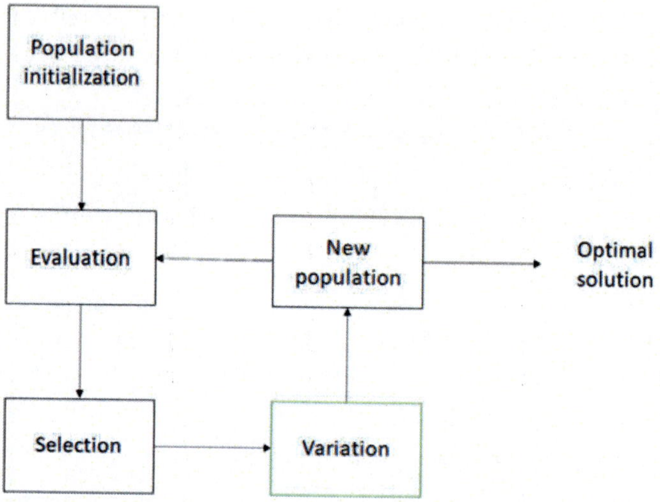

Fig. 1 Generic framework of EA

variations in the intermediate stages help keep intact the previous solution quality while improving the present solution quality in the new population. The new population once again undergoes the fitness evaluation and this process continues until the iteration threshold is exhausted. The final population is expected to be composed of the best quality individuals, in other words, the final set of solutions contains the best possible solutions. The best individual from this final refined population is now considered to be the optimal solution (Fig. 1).

2.2 *Evolutionary Deep Learning Model*

Evolutionary algorithms are applied in deep learning to address two major problems: optimal network architecture search and nodal weight tuning in neural networks.

The neural networks work on a complex architecture inspired by the human nervous system. The information is fed to the receiver (or input) node(s) which then goes through several levels of intermediate hidden nodes to finally provide a result at the output node(s). The combination of input, output and hidden nodes is called the neural network architecture. The architecture maps a set of inputs to a corresponding output. The key is to search for an architecture among innumerable architectures, well suited for a given problem. In evolutionary computation, all the possible architectures are considered to be the search space of the problem. A finite group of randomly selected potentially optimal architectures then undergo fitness evaluations through repeated training of the neural network over repeated iterations. The fitness value is normally calculated based on either of the two factors in a single objective

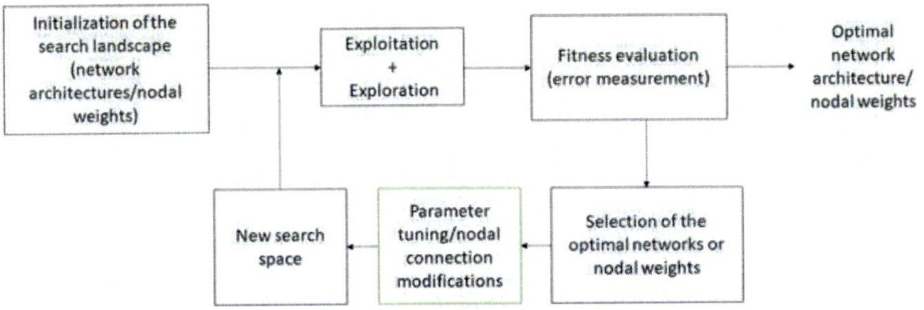

Fig. 2 Evolutionary deep learning model

evolutionary algorithm: sum squared error (SSE) and complexity of the architecture [18, 27]. SSE depicts the error in the prediction of a neural network, hence, the lesser the SSE, the better the network. However, in a multi-objective evolutionary strategy, both SSE and complexity might be considered. The network architectures passing the fitness test go through some variations to create a new search space, which upholds the quality of the previous search space and at the same time, is much refined in a sense that it gives a way of finding more optimal networks. The new search space is again evaluated and this process continues for a number of iterations.

The nodes in a neural network, also called perceptrons, are the primary calculating components. They are responsible for receiving information and feeding information to the next nodes. Perceptrons accept and output binary values, and use weighted sum to determine whether the output should be 0 or 1. The challenge here is to find a suitable weight for each node in the network. In traditional neural networks, the weights are adjusted to their optimal value by a series of repetitive runs with varying weights, biases and inputs, to minimize the loss function, known as training [9]. The loss function is basically the difference between the predicted value and the observed value. With a fewer number of inputs it is an easy task to do but while dealing with a larger input set, for example, a set of image pixels, where there are thousands of inputs, the training becomes a tedious task. The traditional neural network training is carried out by a gradient-based backpropagation algorithm [11]. A gradient-based learning requires the loss function to be continuous and differentiable. Also, backpropagation proves to be much slower in the case of networks with a large number of nodes and sometimes fails to converge to the required training error when the number of nodes is less.

In EA-based weight adaptation, the training process is formulated as the evolution of the nodal weights in the environment determined the network's architecture, with an objective of minimizing the loss function [27]. The initial search space or population for the problem contains a random set of probable weights for the nodes. The weights, represented by individuals in the population, evolve with each iteration of the GA. The value of the loss function is regarded as the measurement for fitness evaluation. Lesser the value, the fitter the individuals. The final iteration sees through an optimal set of weights for the nodes in the network (Fig. 2).

3 GA-Based Deep Learning

Genetic Algorithms are used in deep learning especially for the optimization of neural networks. GAs are invoked for solving two important neural network problems—searching for optimal network architectures and parameter tuning [14].

3.1 Automation of Network Design

3.1.1 Need for Evolutionary Network Adaptation

Designing neural networks is hard for humans. Even a small network with fewer nodes can behave in a mystifying way, defying comprehension. Designing a large, multi-layer, non-linear neural network consumes a lot of time and requires keen supervision. It is very difficult to understand the primary principles governing information processing in such networks and even more difficult to exploit in order to come up with any new useful network architectures [14, 18, 27]. Therefore, most of the commonly used neural networks employ the basic network architecture schemes such as layered feedforward design [23] or simple recurrent schemes [13]. Those seeking optimal designs are thus cast into uncharted darkness.

3.1.2 Use of GA for Automation

The problem of network design can be reduced to a search problem for an architecture which performs well for a given problem. A few challenges faced in network architecture design are enumerated as follows:

- The search space is extensive as there are uncountable possible architectures for a problem.
- Since the network modification is done by a discrete change of nodes or nodal connections, therefore, the search space is non-differentiable.
- The discontinuous performance of the varying network design makes the overall search space discontinuous.
- The mapping between network design and network performance is indirect, epistatic and dependent on the initial random selection of network structures, making the space complex and noisy.
- Structurally similar networks may show varying performances, making the search space deceptive.
- Also, structurally dissimilar networks show similar performances depicting the multi-modality of the space.

GA is sought as an automation technique as it can address discontinuous, non-differentiable, noisy, deceptive and multi-modal search spaces [27].

3.1.3 The GA-Based Network Automation Algorithm

The backbone of the GA-based network search algorithm lies in two adaptive processes: 1. Genetic search through the network architecture space; 2. Backpropagation to calculate the fitness.

3.1.4 Generation of Population

The network is represented by a connectivity constraint matrix C of the order $N \times (N+1)$, where N represents the number of nodes in the architecture. The first N columns represent the constraints on the connection between the N nodes and the $(N+1)$th column represents the constraints for the threshold biases for each node. Every element in C, $c_{i,j}$ is a subset of S, a connectivity constraint set, and indicates the nature of connection from node j to node i or the bias to node i if $j = N + 1$. Summing up, the columns and rows depict the fan-out and fan-in connections in the network, respectively [14]. The elements contained in S tell us about the connection weights and it could be anything among the following:

- Fixed weight;
- Learnable weight (L);
- Learnable but restricted to positive values only (L+);
- Learnable but restricted to negative values only (L-);
- Learnable but restricted to negative values only (L-);
- 0 or no weight.

For simplicity, we could choose only 0 and L implementation. A network with 5 nodes and its matrix representation as well as the genotype (individual) are shown in Fig. 3. We observe that the genotype or a chromosome is nothing but a bit string formed by the concatenation of all the row entries in C. Thus, the initial population with a distinct number of chromosomes, each representing a unique network architecture, is ready to be fed into the algorithm.

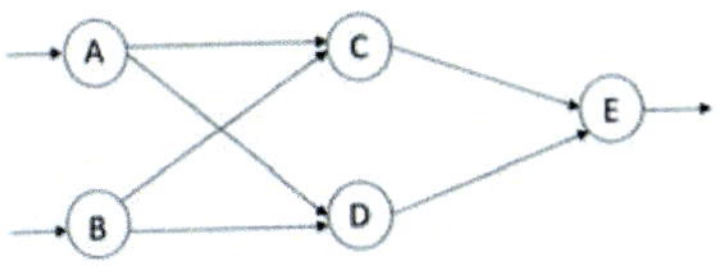

000000 000000 LLOOOL LLOOOL OOLLOL

	A	B	C	D	E	bias
A	0	0	0	0	0	0
B	0	0	0	0	0	0
C	L	L	0	0	0	L
D	L	L	0	0	0	L
E	0	0	L	L	0	L

Fig. 3 The network, its corresponding constraint matrix (C) and the $N \times (N+1)$ bit string representation

3.1.5 Genetic Operations

Crossover operation is performed in the N-bit string generated from each row [22]. A random position is selected and the bit information is swapped. Each row of matrix C constitutes an independent unit of information processing as they deal with the fan-in connections only. Thus, the N-bit string possesses only the receptive information, based on which the computation in the nodes depends. Crossover is performed by swapping because it is simple for a bit string chromosome. Mutation is done by flipping a random bit (from 0 to 1 and vice versa) in a position among all the N*(N+1) positions in the C matrix [14, 17]. A mutation performed on a C matrix transforms it into a whole new C matrix representing a new network architecture. Therefore, mutation alone has the capacity of exploring the entire search space of all the possible architectures.

3.1.6 Fitness Evaluation

Fitness is evaluated by the following steps:

1. Instance of an architecture is created.
2. Connections are established between nodes where L (learnable) is present in the C matrix with no connections where it is 0.
3. L connections are initialized with random weights.
4. The initialized architectures are trained for a certain number of epochs using the Backpropagation learning rule.
5. Total sum squared error (TSSE) of each network's performance at the last epoch finally determines the networks' fitness [27].
6. The lower the TSSE, the better the learning.

3.1.7 Problems for Experimentation

- XOR—it's a simple 2-input 1-output system where two binary inputs map to a single binary output as $00->0$, $01->1$, $10->1$ and $11->0$. On repeated learning, it is found that a 5-nodal architecture gives an optimal TSSE.
- Four-Quadrant problem or 2-dimensional XOR—it's a complex XOR as the inputs can be any real number from 0 to 1. The x-axis and y-axis represent two inputs. The lower left and upper right inputs map to 0 and the lower right and upper left inputs map to 1.
- Pattern copying—a simple transformation where the inputs are directly copied to the output.

3.2 Parameter Tuning

In a neural network architecture, be it a feedforward or a recurring architecture, there is a constant need for the assignment of suitable nodal weights [1]. Finding a set of optimal values for the weights in order to increase the accuracy of the network and the system as a whole is a very frequent and challenging task. The process involves weight adaptation through learning. Learning can be processed in three different ways—supervised, unsupervised and reinforced [19]. The classical methods involve weight training through gradient descent-based algorithms such as Backpropagation and conjugate gradient algorithm [11]. The gradient-based algorithms suffer from two major problems—local minima stagnation of the error function and inability to converge in the case of multi-modal functions. GA can be swiftly invoked in the weight training or learning process to overcome the aforementioned shortcomings. The problem of parameter tuning or weight adjustment can be projected as minimization of the error function in the neural network [27]. The genetic algorithm applied for parameter tuning is a two-step approach:

1. Representation of the connection weights.
2. Operational methodology of the genetic operators.

3.2.1 Connection Weight Representation

There are mainly two ways of representing the connection weights of a neural network so as to fit into the genetic algorithm framework—binary representation and real number representation [25]. In binary representation, integral weights are represented in bit strings. An independent nodal structure is considered to be comprised of only the fan-in connections and therefore, the input weights, represented in bits, are kept together, and finally all the bit strings are concatenated to form the final string. A bit conversion of a network is shown in Fig. 4a. The real number representation, however, allows fractional weights. The weights for all the inputs to a node are written in decimals as shown in Fig. 4b.

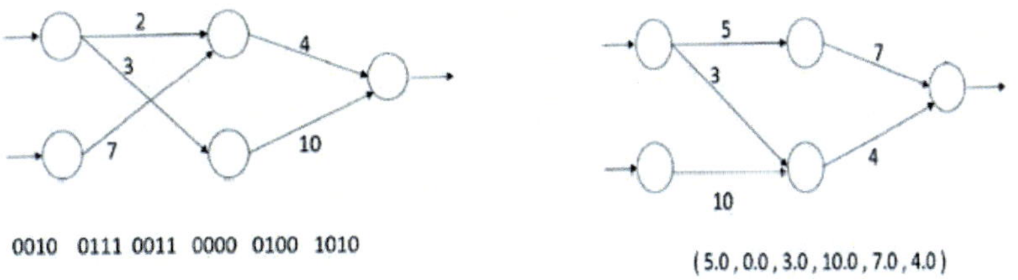

(a) Binary representation (b) Real number representation

Fig. 4 Network architecture representation schemes

3.2.2 Genetic Operations

The bit string or the real number array is arranged in such a way that the information about the input connection weights is placed together in the structure. This protects the integrity of the nodal weights against the crossover operations applied throughout the generations of evolution. The different types of genetic crossovers such as order crossover, random crossover, partial crossover and greedy crossover when applied on the chromosome, flip the information of only independent nodes. The Mutation is done in single or multiple stages by flipping a bit, in the case of bit strings, at a certain index or a few indices, in the chromosome. Mutation in a real number representation can be achieved by a probabilistic up- or down-scaling of the real numbers.

3.2.3 Fitness Evaluation

Fitness evaluation is solely based on the SSE measure. SSE is basically the difference in performance between the actual output and the desired output [26]. The difference is, however, subtracted from a standard value for normalization [27].

4 Discussions

4.1 Inefficacy of GA

Genetic algorithm, as promising as it may look, has certain limitations [3, 6]. We have cited a few as follows:

(a) A simultaneous evolution of the network architecture and adaptation of the connection weights is difficult even using GA.
(b) The fitness evaluation considering the performance error and the network complexity converts the system into a multi-objective optimization problem, which is tedious to perform.
(c) GA fails to converge when the network size is small.
(d) With a large number of problem-specific evaluation algorithms available, the absence of a rigid evaluation scheme of optimization algorithms causes a dilemma in its application, hence, compromising its efficiency. However, there are, at present, plenty of evaluation strategies available [2].

4.2 Challenges in Application of GA in Deep Learning

(a) It is a difficult task to represent the network architecture that is to be fed to the GA computational framework.

(b) The GA operations such as crossover and mutation should be carefully used. More often than not, they ruin a good population.
(c) Gradient descent techniques, such as BP, employed in the learning stage might lead to a local minimum error, hence, deceiving the fitness evaluation process.

5 Conclusion

Evolutionary algorithms have made noticeable improvements in deep learning pertaining to neural networks. The search for an optimal network architecture design and the adaptation of nodal weights are the two broad problem areas in neural networks where significant improvements could be brought in by EAs through the minimization of error or loss function. We have seen how the genetic algorithm eases its way into the optimization process of NNs and makes way for a tremendous amount of efficiency and accuracy.

Acknowledgements This work is supported by the Science and Engineering Board (SERB), Department of Science and Technology (DST) of the Government of India under Grant No. ECR/2018/000204 and Grant No. EEQ/2019/000657.

References

1. Abraham A (2005) Artificial neural networks. In: Handbook of measuring system design
2. Biswas A, Biswas B (2017) Analyzing evolutionary optimization and community detection algorithms using regression line dominance. In: Information sciences, vol 396, pp 185–201
3. Blanco A, Delgado M, Pegalajar MC (2001) A real-coded genetic algorithm for training recurrent neural networks. Neural Netw 14(1):93–105
4. Guus C, Boender E, Edwin Romeijn H (1995) Stochastic methods. In: Handbook of global optimization. Springer, pp 829–869
5. Carlos A, Coello C, Lamont GB, Van Veldhuizen DA et al (2007) Evolutionary algorithms for solving multi-objective problems, vol 5. Springer
6. Ding S, Su C, Yu J (2011) An optimizing BP neural network algorithm based on genetic algorithm. Artif Intell Rev 36(2):153–162
7. Dorigo M, Birattari M, Stutzle T (2006) Ant colony optimization. IEEE Comput Intell Mag 1(4):28–39
8. Fortin F-A et al (2012) DEAP: evolutionary algorithms made easy. J Mach Learn Res 13(1):2171–2175
9. Gaier A, Ha D (2019) Weight agnostic neural networks. In: Advances in neural information processing systems, pp 5364–5378
10. Jin W et al (2000) The improvements of BP neural network learning algorithm. In: WCC 2000-ICSP 2000. 2000 5th international conference on signal processing proceedings. 16th world computer congress 2000, vol 3. IEEE, pp 1647–1649
11. Johansson EM, Dowla FU, Goodman DM (1991) Backpropagation learning for multilayer feed-forward neural networks using the conjugate gradient method. Int J Neural Syst 2(04):291–301
12. Kennedy J, Eberhart R (1995) Particle swarm optimization. In: Proceedings of ICNN'95-international conference on neural networks, vol 4. IEEE, pp 1942–1948
13. Medsker LR, Jain LC (2001) Recurrent neural networks. Des Appl 5

14. Miller GF, Todd PM, Hegde SU (1989) Designing neural networks using genetic algorithms. In: ICGA, vol 89, pp 379–384
15. Najafabadi MM et al (2015) Deep learning applications and challenges in big data analytics. J Big Data 2(1):1
16. O'Shea T, Hoydis J (2017) An introduction to deep learning for the physical layer. IEEE Trans Cognit Commun Netw 3(4):563–575
17. Palmes PP, Hayasaka T, Usui S (2005) Mutation-based genetic neural network. IEEE Trans Neural Netw 16(3):587–600
18. Sagar GVR, Venkata Chalam S (2011) Evolutionary algorithm for connection weights in artificial neural networks
19. Schmidhuber J (2015) Deep learning in neural networks: an overview. Neural Netw 61:85–117
20. Shin R, Packer C, Song D (2018) Differentiable neural net-work architecture search
21. Sivanandam SN, Deepa SN (2008) Genetic algorithms. In: Introduction to genetic algorithms. Springer, pp 15–37
22. Spears WM, Anand V (1991) A study of crossover operators in genetic programming. In: International symposium on methodologies for intelligent systems. Springer, pp 409–418
23. Svozil D, Kvasnicka V, Pospichal J (1997) Introduction to multi-layer feed-forward neural networks. Chem Intell Lab Syst 39(1):43–62
24. Whitley D (1994) A genetic algorithm tutorial. Stat Comput 4(2):65–85
25. Whitley D, Starkweather T, Bogart C (1990) Genetic algorithms and neural networks: Optimizing connections and connectivity. Parallel Comput 14(3):347–361
26. Widrow B, Lehr MA (1990) 30 years of adaptive neural net- works: perceptron, madaline, and backpropagation. Proc IEEE 78(9):1415–1442
27. Yao. X (1999) Evolving artificial neural networks. Proc IEEE 87(9):1423–1447

Automated System for Face-Mask Detection Using Convolutional Neural Network

Neha Sengar, Akriti Singh, Saumya Yadav, and Malay Kishore Dutta

Abstract Coronavirus Disease 2019 (COVID-19) pandemic is affecting the health of the global population severely. It is one of the deadliest diseases in history and has severely affected all the countries. The only way to prevent the spread of corona is to cover faces and follow social distancing norms until a vaccine is developed. The face mask is effective in blocking the droplets that contain the COVID-19 virus. Hence, it is necessary to wear a face mask as a precautionary measure against it. In the proposed work, the face mask detection model is generated using an optimized neural network architecture for performing the classification task (mask or no mask). For training and model assessment, a dataset of 8695 images has been taken from four different sources. The model achieves a validation accuracy of 99.52%.

Keywords NASNet-mobile · Face mask · COVID-19 · Convolutional neural network · Transfer learning

1 Introduction

COVID-19 is caused by the beta coronavirus named severe acute respiratory syndrome coronavirus 2 (SARS-CoV-2) which originated in Wuhan, China [1]. Its symptoms are mild symptoms like dry cough, fever and difficulty in breathing. Patients with mild symptoms can be treated easily. It can be fatal for patients who develop complications like pneumonia and acute respiratory distress syndrome (ARDS). Earlier it was believed that it would remain confined to China [2]. But it has spread in almost all countries. It has infected nearly 26 million people and caused about 8.6 lakh deaths worldwide. It is transmitted through the tiny droplets that are emitted during coughing, sneezing and speaking. It can also be transmitted through aerosol which remains suspended in the air for a long time. It can also infect people if they touch their faces after coming in contact with the contaminated surfaces.

N. Sengar (✉) · A. Singh · S. Yadav · M. K. Dutta
Centre for Advanced Studies, Dr. A.P.J. Abdul Kalam Technical University, Lucknow, U.P, India

© The Author(s), under exclusive license to Springer Nature Singapore Pte Ltd. 2022
D. Giri et al. (eds.), *Proceedings of the Seventh International Conference on Mathematics and Computing*, Advances in Intelligent Systems and Computing 1412,
https://doi.org/10.1007/978-981-16-6890-6_28

Spreading of coronavirus could be prevented using a surgical face mask by the people [3]. WHO has advised people to wear face masks if they have breathing problems or if they take care of someone having breathing problems [4]. Besides, most of the public service providers don't serve the customers if they do not wear the face masks, shopkeepers don't allow anyone to enter without a face mask [5]. Wearing face masks contains the transmission of COVID-19, by blocking the emission of respiratory droplets and saliva from the infected patients [6]. Thus, it has become necessary to develop face mask detection methods that give real-time results with maximum accuracy.

Two-component methods for face mask detection is proposed by Loey et al. in [7]. The first component uses ResNet50 for feature extraction, and the second component consists of traditional machine learning methods for training, validation and testing. Madhura Inamdar et al. [8] proposed Facemasknet which is a deep learning architecture. It achieved an accuracy of 98.6%. Z. Mingjie Jiang et al. [9] propose a face mask detector which comprises a feature pyramid network to combine high-level semantic information with several feature maps, and a context attention module for performing the mask detection. Prateek Khandelwal et al. [10] proposed a social distancing measurement algorithm and a face mask detection method. MobileNetV2 was used for the development of face mask detection methods. Jian Xiao et al. [11] proposed a face mask detection model based on VGG-19. It achieved a precision of 97.62% and a recall of 96.31%.

The main contribution of the proposed work is to make an artificial intelligence-based system that automatically identifies the face mask in real time. For this purpose, a dataset of 8695 images is taken which include 4281 images with mask and 4414 images without a mask. Collected images are augmented and trained on state-of-the-art memory-efficient convolutional neural network (CNN) architecture NASnet-Mobile model for performing the classification task. The rest of the paper is organized as follows. Section II elaborates on the materials and methodology and section III consists of the results. Section IV concludes the paper.

2 Materials and Methodology

2.1 Dataset

The dataset used in this paper consists of 4281 images with mask and 4414 images without mask. These images were collected from four different sources. Dataset graph is shown in Fig. 1. Dataset1 (DS1) is from the open-source repository [12] which has 690 images with masks and 686 images without a mask. Dataset2 (DS2) contains 694 images with masks and 543 images without a mask from an open repository [13]. Dataset3 (DS3) which is an open repository [14] containing 503 images with mask and 503 without face mask images. Dataset4 (DS4) is manually

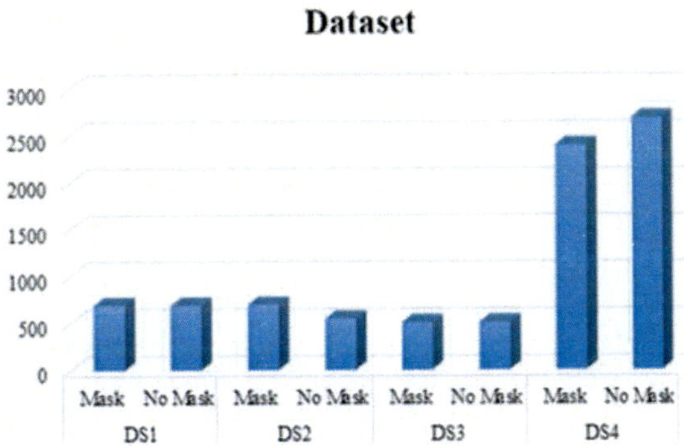

Fig. 1 Face mask dataset distribution graph

collected using smartphones and digital cameras which contain 2394 images with masks and 2682 images without masks.

2.2 *Data Labelling and Augmentation*

All the images were labelled into two categories, with masks and without masks. One hot encoding was performed to convert the categorical labels into the numeric labels. The DataImageGenerator was used for data augmentation. It performs in-place data augmentation which transforms batch of images and replaces them with the transformed images. It applied various transformations like rotation, zoom, width shift, height shift, shearing and horizontal flip so that different variations of the images are used at every epoch. The final dataset is divided for training and validation in the ratio of 3:1, respectively.

2.3 *The Deep Learning Model: NASNet-Mobile*

The model used for face mask detection consists of two stages. The first stage uses the NASNet-Mobile model, and the second stage uses a fully connected layer. It takes the output of the NASNet-Mobile model as an input for fully connected layer. The conceptual schematic diagram of the proposed work is given in Fig. 2. Once the model is trained with the dataset it is utilized in public places and hot spot areas through CCTV surveillance.

NASNet-Mobile Model: Architecture The NASNet was developed using the concept of neural architecture search (NAS) [15]. It uses a regularization approach called Scheduled DropPath which enhances the generalization of the model. Images size of 224 × 224 pixels are given as an input to train in NASNet-Mobile. The

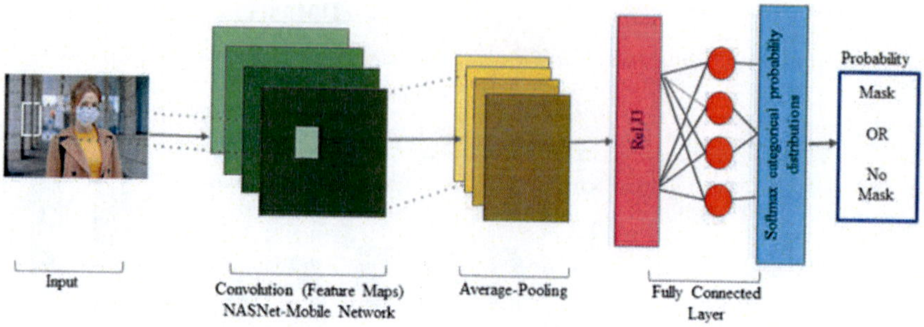

Fig. 2 Architecture of the proposed model

motive behind taking the NASNet-Mobile model is that it generates a very small-sized training model. It can be installed easily on hardware devices with less memory consumption.

In this proposed work, NASNet-Mobile architecture has been used which has 53, 26,716 parameters. The two-stage proposed work has a total 4,405,270 parameters in which 4,368,532 number of parameters are defined as trainable parameters and 36,738 are non-trainable parameters. The smallest unit of the NASNet model is the block and a group of blocks which is called cell. These blocks combine the present and the previous input to a single output feature map. The cells are not predefined and are searched with the help of a search method employing reinforcement learning. The initial convolutional filters are used for scaling and are known as normal cells and reduction cells. The normal cells return feature maps of the same dimension and the reduction cells reduce the dimensions of the feature maps by a factor of two. The controller recurrent neural network (RNN) searches the structure of normal and redundant cells. Convolution layers are interleaved with average pooling layers that is used to decrease the spatial size of the convolved feature matrix. It looks for a larger area of the input face mask images matrix and takes aggregate information. Rectified linear unit (ReLU) is used as an activation function for the hidden layers to increase non-linearity without effecting its receptive field. Keras framework with tensorflow back-end has been used for experimental setup to train the model. Last layer is fully connected (FC) layer which is a dense layer. It is the final learning phase of CNN architecture performing classification tasks. Generated model is fine-tuned to improve the performance of the model [16]. For the proposed methodology, transfer learning is used in which CNN is trained for performing one task whose generated weights are used for performing other related task. Pre-trained weights are utilized for generating the mask detection model which is trained on ImageNe [17] dataset and has capability to categorize images into 1000 different classes. The dense layers are trained from their randomly initialized weights. FC layer in the architecture is modified with the total number of classes, i.e. two classes mask and no mask.

2.4 Real-Time Video-Based Detection

Real-time video-based mask detection is performed with the help of a pre-trained model using the Caffe [18] (Convolutional architecture for fast feature embedding) model weights. It is an open-source deep learning framework made with expression, modularity and speed in mind. It has found applications in a wide range of areas like object classification, object detection and learning semantic features. The NasNet-Mobile is trained to detect masks on the faces in the video using caffe model. Algorithm 1 is used for the real-time face mask detection in a video.

Algorithm 1: Face Mask Detection from input video

Input: Video Stream;
Output: Bounding boxes detecting faces and labels;
Step1: Save the captured frame, f;
Step2: Pre-process f
Resize the frame in dimension 400*400;
Step3: Load the trained deep learning face detection model;
Step4: Image f is processed with the trained model
detections = detectObjectsFromImage(input image = f);
Step5: Detect class label in the frame and compute the probability that they are wearing masks;
Step6: Determining the class label and the color of the bounding box;
if mask > withoutMask: then
label='mask' color='green'
else
label='no mask' color='red'
end

These two trained models are working together to detect the people wearing face masks from the video stream.

3 Results

All the methods and experiments were performed on an Intel Xeon processor running at 2.50 GHz using 4.0 GB of RAM and Windows 10 pro. The dataset was collected from four different sources and split into a ratio of 3:1 for training and testing the model. The batch size is taken as 32 for training the model.

Adam optimizer was used for updating the weights of the network during the training. The learning rate is initialized to a value 0.001 to achieve an accurate loss. For training the model as a binary classifier, binary cross-entropy loss is used as a loss function. The model is trained for 95 epochs. It achieves a validation accuracy of 99.52% with 0.94 training loss. Accuracy and loss curves for the proposed model are presented in Fig. 3.

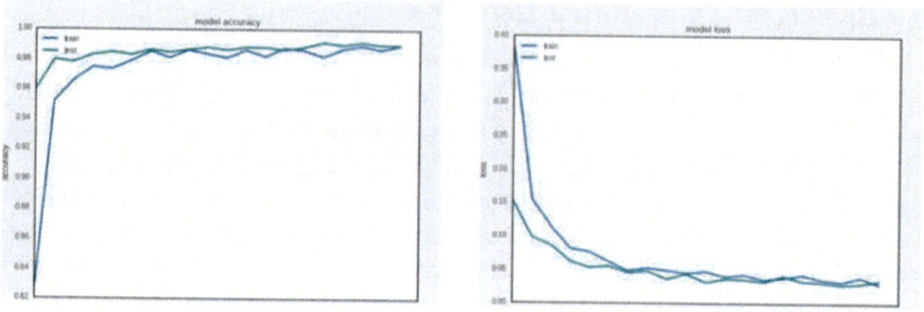

Fig. 3 Accuracy curve and loss curve

To authenticate the model, testing is performed on a test dataset which contains 72 images with masks and 33 images of a person without a mask. The confusion matrix of the obtained results is shown in Fig. 4.

Table 1 represents the result of Precision, Recall, accuracy and F1-score in real-time where precision, recall, accuracy and F1-score are calculated as

$$Precision = \frac{TruePositive}{TruePositive + FalsePositive} \tag{1}$$

$$Recall = \frac{TruePositive}{TruePositive + FalseNegative} \tag{2}$$

$$Accuracy = \frac{TruePositive + TrueNegative}{Positive + Negative} \tag{3}$$

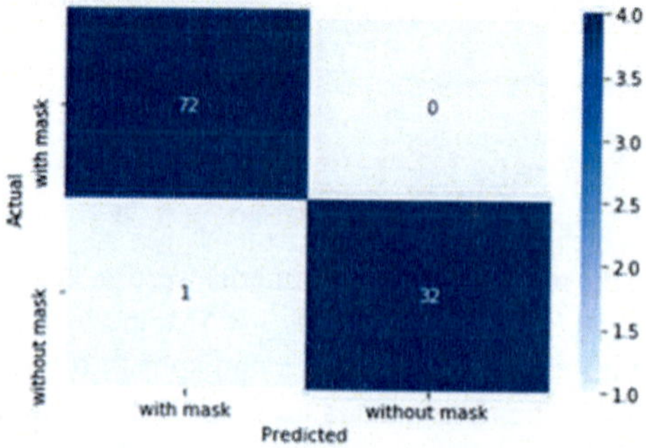

Fig. 4 Confusion Matrix of the proposed work

Table 1 Precision, Recall, F1-score and Accuracy obtained on the test dataset

Category	Test samples	Precision (%)	Recall (%)	F1-score (%)	Accuracy (%)
With mask	72	98.63	100	99.31	99.05
Without mask	33	100	96.97	98.46	99.05

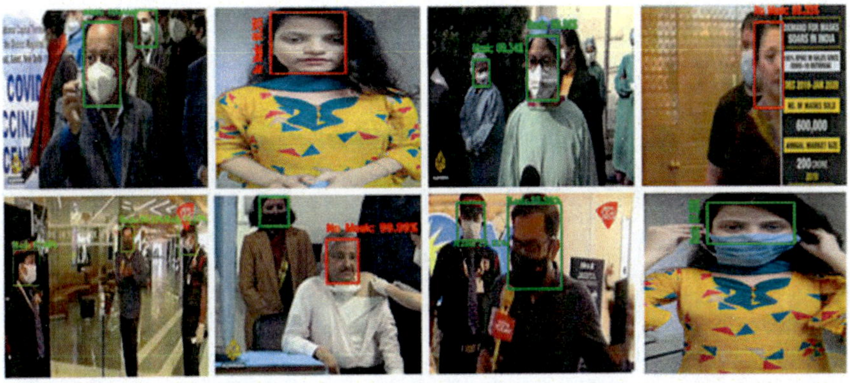

Fig. 5 Results of the real time video-based face mask detection model

$$F1Score = 2 \times \left(\frac{Precision \times Recall}{Precision + Recall}\right) \tag{4}$$

The result of detected and classified images of face mask by the proposed model on a newly generated video of 23 s and another video of 105 s uploaded by WION [19] is shown in Fig. 5. The proposed model performs the mask detection for all the faces present in the captured frame.

4 Conclusion and Future Work

COVID-19 has infected the global population and affected the economy of all countries. The only way to prevent its spread is to ensure that everyone wears face masks and follows the social distancing rules. This has led to a lot of research in artificial intelligence to develop tools and techniques to check its spread among the people. In this paper, a face mask detection algorithm using NASNet-Mobile architecture has been proposed to ensure that everybody wears a face mask. In the proposed work, a face mask detection model is generated to perform real-time mask detection. Datasets are collected from different sources and trained with a fine-tuned NASNet model. It used a pre-trained model using the weights of the Caffe model to detect faces and NASNet for detecting face masks on those faces. The performance of the developed CNN model is giving the validation accuracy of 99.52% and the test accuracy for the new dataset is achieved as 99.05%. Since in the current scenario identification of face mask is the most important task, the experimental results of the proposed model support the mask identification with very high accuracy.

In the future, this work can be extended to include more features like social distance measurement for monitoring whether people are maintaining a safe distance from each other or not. It will ensure better safety against COVID-19.

References

1. Sohrabi C, Alsafi Z, O'Neill N, Khan M, Kerwan A (2020) World health orga- nization declares global emergency: a review of the 2019 novel coronavirus (COVID-19). Int J Surg 76, 71–76.https://doi.org/10.1016/j.ijsu.2020.02.034
2. Ozili PK, Arun T (27 Mar 2020) Spillover of COVID-19: impact on the global economy. Available at SSRN: https://ssrn.com/abstract=3562570 or https://doi.org/10.2139/ssrn.3562570
3. NHL Leung DKW Chu EYC Shiu 2020 Respiratory virus shedding in exhaled breath and efficacy of face masks Nat Med 26 676 680 https://doi.org/10.1038/s41591-020-0843-2
4. Feng S, Shen C, Xia N, Song W, Fan M, Cowling BJ (2020) Rational use of face masks in the COVID-19 pandemic. Lancet Respir Med (20)30134-X: S2213–2600. https://doi.org/10.1016/S2213-2600(20)30134-X
5. Y Fang Y Nie M Penny 2020 Transmission dynamics of the COVID-19 outbreak and effectiveness of government interventions: a data- driven analysis J Med Virol 92 https://doi.org/10.1002/jmv.25750
6. VCC Cheng SC Wong VWM Chuang SYC So 2020 The role of community-wide wear- ing of face mask for control of coronavirus disease 2019 (COVID-19) epidemic due to SARS-CoV-2 J Infect 81 1 107 114 https://doi.org/10.1016/j.jinf.2020.04.024
7. Loey M, Manogaran G, Taha MHN, Khalifa NEM (2020) A hybrid deep trans- fer learning model with machine learning methods for face mask detection in the era of the COVID-19 pandemic. Measurement 167. https://doi.org/10.1016/j.measurement.2020.108288
8. Inamdar M, Mehendale N (29 July 2020) Real-time face mask identification using facemasknet deep learning network. Available at SSRN: https://ssrn.com/abstract=3663305 or https://doi.org/10.2139/ssrn.3663305
9. Jiang M, Fan X (2020) RetinaMask: a face mask detector. https://arxiv.org/abs/2005.03950
10. Khandelwal P, Khandelwal A, Agarwal S (2020) Using computer vision to enhance safety of workforce in manufacturing in a post COVID world. https://arxiv.org/abs/2005.05287
11. Xiao J, et al. Application of a novel and improved VGG-19 network in the detection of workers wearing masks. J Phys: Conf Ser 1518, 01204
12. PrajnaBhandary face mask detection dataset https://www.kaggle.com/chandrasekarank/prajna-bhandary-face-mask-detection-dataset. Accessed 17 Aug 2020
13. Dataset for mask detection. https://www.kaggle.com/harry418/dataset-for-mask-detection. Accessed 17 Aug 2020
14. COVID Face Mask Detection Dataset. https://www.kaggle.com/prithwirajmitra/covid-face-mask-detection-dataset/data. Accessed 17 Aug 2020
15. Radhika K, Devika K, Aswathi T, Sreevidya P, Sowmya V, Soman KP (2020) Performance analysis of NASNet on unconstrained ear recognition. In: Rout M, Rout J, Das H (eds) Nature inspired computing for data science. Studies in computational intelligence, vol 871. Springer, Cham. https://doi.org/10.1007/978-3-030-33820-6
16. N Tajbakhsh May 2016 Convolutional neural networks for medical image analysis: full training or fine tuning? IEEE Trans Med Imaging 35 5 1299 1312 https://doi.org/10.1109/TMI.2016.2535302
17. K rizhevsky A, Sutskever I, Hinton G (3–6 Dec 2012) ImageNet classification with deep convolutional neural networks. In: Proceedings of the 25th international conference on neural information processing systems. Lake Tahoe, NV, USA, pp 1097–1105
18. Jia Y, Shelhamer E, Donahue J, Karayev S, Long J, Girshick R, Guadarrama S, Darrell T (2014) Caffe: convolutional architecture for fast feature embedding. arXiv:1408.5093
19. India faces spike in demand of face masks, sanitizers amid Coronavirus outbreak around the world. https://www.youtube.com/watch?v=sNZopgds-Y8. Accessed 7 Aug 2020

Using Protein Interactome Similarity to Improve *Random Walk with Restart Model* for Drug Repurposing

I. T. Anjusha, N. Saleena, and K. Abdul Nazeer

Abstract Development of new drugs has the limitations of high cost, high time requirement and low success rate. If an existing drug can be used to treat a drug-seeking disease, we can reduce these limitations. The process of using existing drugs to treat new diseases is called drug repurposing. Since the drugs are already approved for human use, the success rate becomes high. In recent years, numerous random walk models have been proposed on the disease-drug heterogeneous network and become a popular drug repurposing approach. The performance of random walk-based approach depends on the network similarity measures used to build the heterogeneous network. In this paper, we improve the network similarity measures by integrating the similarity between the disease and drug-specific protein interactomes in human Protein-Protein Interaction network. We then run a random walk with restart algorithm over the modified network to predict disease-drug relations. Our experiments reveal that performance of random walk model has improved after integrating protein interactome similarity.

Keywords Drug repurposing · Network-based · Random walk · Protein-protein interaction network

1 Introduction

Tremendous biological knowledge and the availability of modern tools helped improve the precision of identifying the disease-causing factors in each individual separately. Based on the difference in disease-causing factors in individual patients,

I. T. Anjusha (✉) · N. Saleena · K. Abdul Nazeer
National Institute of Technology Calicut, Calicut, India
e-mail: anjusha_p180127cs@nitc.ac.in

N. Saleena
e-mail: saleena@nitc.ac.in

K. Abdul Nazeer
e-mail: nazeer@nitc.ac.in

© The Author(s), under exclusive license to Springer Nature Singapore Pte Ltd. 2022
D. Giri et al. (eds.), *Proceedings of the Seventh International Conference on Mathematics and Computing*, Advances in Intelligent Systems and Computing 1412,
https://doi.org/10.1007/978-981-16-6890-6_29

we can divide wide disease classes into smaller subtypes. The treatment will be more safe and effective if we can develop personalized drugs for each of these disease subtypes.

Development of new drugs has the limitations of high cost, high time, and low success rate. If an existing drug can be used to treat these disease subtypes, we can circumvent the limitations. The process of using existing drugs to treat new diseases is called *drug repurposing*. Since the drugs are already approved for human use, it's success rate becomes high. For example, the drug *Raloxifene* originally developed to treat *osteoporosis* is found to be efficient to treat *breast cancer*.

Computational drug repurposing approaches can be broadly classified into *network model*, *text mining*, *machine learning*, *molecular docking*, and *signature matching* based approaches. Among them, network-based approaches are found to be one of the most promising ones. In network-based approaches, a disease-drug heterogeneous network is constructed by combining disease-disease similarity network, drug-drug similarity network and known disease-drug interaction networks. Then some computational approaches are applied to this network to predict novel disease-drug relationships.

Radom walk models [2–4] are efficiently used in network-based drug repurposing approaches in which, the relevance of a given disease-drug relationship is computed based on the steady probability that a random walker will reach the given disease(or drug) node starting from the given drug(or disease) node. The underlying assumption of this model is that similar diseases can be treated by similar drugs. Some recent studies [5–7] suggest that, for a drug to be used against a specific disease, its associated targets and the disease-specific genes should be nearby in the Protein-Protein Interaction(PPI) network.

The performance of random walk-based approach depends on the network similarity measures used to build the heterogeneous network. In this paper, we improve the network similarity measures by integrating protein interactome similarity with the traditional similarity measures used to build the disease-drug heterogeneous network. Then, a random walk model is applied to this modified network to predict disease-drug relations.

The rest of this paper is structured as follows. In Sect. 2, related works are presented. Section 3 discusses the proposed method. Experimental results are discussed in Sect. 4. By highlighting the main contributions and providing some future directions, Sect. 5 concludes the paper.

2 Related Work

This section reviews the important approaches related to the proposed method.

Xing Chen et al. [1] implemented the Random Walk with Restart(RWR) algorithm on a drug-target heterogeneous network and predicted drug-target relations. On the basis of [1], the following approaches are proposed on the disease-drug heterogeneous network and predicted disease-drug associations. Huimin Luo et al. [2]

implemented random walks on the disease similarity network and drug similarity network separately and, the average of the prediction scores assigned by the two walks is taken as the final prediction score. Hui Liu et al. [3] presented a two-pass random walk with restart on the disease-drug heterogeneous network(TP-NRWRH) algorithm, where two random walks are implemented on the disease-drug heterogeneous network. For each disease-drug pair, one random walk starts from the disease node and its known drugs and predicts the probability of reaching objective drug node and the second random walk starts from the drug node and its known diseases and predicts the probability of reaching objective disease node. Mean of the prediction scores retrieved by the two walks is taken as the final prediction score.

In these two approaches, the disease-drug heterogeneous network is built by combining the disease-disease similarity network, drug-drug similarity network, and known disease-drug relation network. The disease-disease phenotype similarity matrix and the disease-disease similarity matrix based on the known disease-drug associations are combined to form the disease-disease similarity matrix. Similarly, the drug-drug chemical similarity matrix and drug-drug similarity matrix based on the known disease-drug relations are combined to form the drug-drug similarity matrix. Huimin Luo et al. [4] combined six subnetworks, namely drug-drug network, disease-disease network, target-target network, known disease-drug network, known drug-target network, and known disease-target network to build a heterogeneous network and then, implemented the random walk on this improved network to predict disease-drug relations. Here, drug-drug network, disease-disease network, target-target network are built based on chemical structure similarity, phenotype similarity, and protein sequence similarity, respectively.

Some recent studies [5–7] suggests that, for a drug to be effective against a specific disease, its associated targets and the disease-specific genes should be nearby in the human Protein-Protein Interaction(PPI) network. Feixiong cheng et al. [5] computed the network-based proximity measure as the average shortest path length between the drug targets and the nearest disease genes in the human PPI network and used this proximity measure to predict novel disease-drug relations. Yadi Zhou et al. [6] identified human proteins functionally associated with the coronavirus infection. Then, the subnetwork corresponding to these identified proteins is extracted from the human PPI network. The known drug-target(protein) network is mapped into the Human coronavirus (HCoV)-specific subnetwork to search for druggable targets. Network proximity measure is used to quantify the relationship between HCoV-specific subnetwork and drug targets. They prioritize 16 potential anti-HCoV repurposable drugs and that are validated by gene set enrichment analysis. Giulia Fiscon et al. [7] modified the system by performing a cluster analysis to detect groups of drugs and diseases. If a drug and disease belongs to the same cluster, corresponding cluster quality score is integrated to the similarity measure. YuHang Zhang et al. [8] propose a random walk model to identify COVID-19 infection-related human genes. Human proteins that have the same Gene Ontology functions of COVID-19 are picked up as seed nodes. Then, implemented the random walk algorithm on the protein-protein interaction network to predict COVID-19 infection-related human genes.

3 Proposed Method

In this section, we discuss the proposed method for drug repurposing. We integrated protein interactome similarity with the traditional similarity measures used to build the disease-drug heterogeneous network. Then, the performance of RWR [1] and TP-NRWRH [3] algorithms before and after integrating the protein interactome similarity are compared. We used the PPI network used in [5] for calculating protein interactome similarity. We only considered those interactions with p-value<0.05. Network proximity measure between each pair of protein interactome is computed using the "shortest distance" based method [5]. The proposed method is given in Algorithm 1.

Algorithm 1 Proposed Algorithm for Drug Repurposing

Input: η: importance of starting network(drug/disease network) of random walk.
λ: probability of moving from drug network to disease network or vice versa.
γ: restart probability of random walk.
Output: Predicted disease-drug interaction matrix.
Steps:
Construct drug-drug chemical structure similarity matrix, S^c_{RR}.
Construct drug-drug protein interactome similarity matrix, S^i_{RR}.
Construct the matrix, S^{ci}_{RR} by combining S^c_{RR} and S^i_{RR} as follows:

$$S^{ci}_{RR}[i, j] = max(S^c_{RR}[i, j], S^i_{RR}[i, j])$$

Construct drug-drug similarity matrix, S^d_{RR} based on known disease-drug relations.
Construct integrated drug-drug similarity matrix, S_{RR} as follows:

$$S_{RR} = 1 - (1 - S^{ci}_{RR})(1 - S^d_{RR})$$

Construct disease-disease phenotype similarity matrix, S^p_{DD}.
Construct disease-disease protein interactome similarity matrix, S^i_{DD}.
Construct the matrix, S^{pi}_{DD} by combining S^p_{DD} and S^i_{DD} as follows:

$$S^{pi}_{DD}[i, j] = max(S^p_{DD}[i, j], S^i_{DD}[i, j])$$

Construct disease-disease similarity matrix, S^d_{DD} based on known disease-drug relations.
Construct integrated disease-disease similarity matrix, S_{DD} as follows:

$$S_{DD} = 1 - (1 - S^{pi}_{DD})(1 - S^d_{DD})$$

Construct disease-drug adjacency matrix, S^a_{RD}.
Construct disease-drug protein interactome similarity matrix, S^i_{RD}.
Apply relevant RWR model with parameters η, λ and γ on the improved network to predict disease-drug relations.

The process of drug repurposing consists of two phases, as follows:

1. Disease-drug heterogeneous network creation.
2. Applying relevant random walk model on the heterogeneous network.

3.1 Disease-Drug Heterogeneous Network Creation

The heterogeneous network consists of three subnetworks, namely, disease-disease similarity network, drug-drug similarity network, and disease-drug interaction network. Networks are represented by matrices.

3.1.1 Construct Integrated Disease-Disease Similarity Matrix, S_{DD}

The integrated disease-disease similarity matrix is constructed by combining disease-disease phenotype similarity, protein interactome similarity, and the similarity based on known disease-drug relations.

- Construct disease-disease phenotype similarity matrix, S^p_{DD}: Similarity between each pair of diseases in the dataset are computed based on the mesh terms present in the medical description of the diseases by using MimMiner [11].
- Construct disease-disease protein interactome similarity matrix, S^i_{DD}: In order to compute protein interactome similarity, we first identified the human genes functionally associated with each disease in the dataset by searching Online Mendelian Inheritance in Man (OMIM) database [12]. Disease-specific protein interactome is the subnetwork corresponding to these identified genes in the human PPI network (Fig. 1).

  Distance between the protein interactomes corresponding to disease d_i and disease d_j is computed using (1) [6]:

$$d\_avg_{D_i D_j} = \frac{\sum\limits_{p_1 \in D_i} \min\limits_{p_2 \in D_j} d(p_1, p_2) + \sum\limits_{p_2 \in D_j} \min\limits_{p_1 \in D_i} d(p_1, p_2)}{||D_i|| + ||D_j||} \quad (1)$$

  where
  D_i: Set of genes related to disease d_i.
  D_j: Set of genes related to disease d_j.
  $d(p_1, p_2)$: Shortest distance between genes p_1 and p_2 in the human PPI network.
  Protein interactome similarity between disease d_i and disease d_j is computed using (2):

$$S^i_{DD}[i, j] = \frac{1}{1 + d\_avg_{D_i D_j}} \quad (2)$$

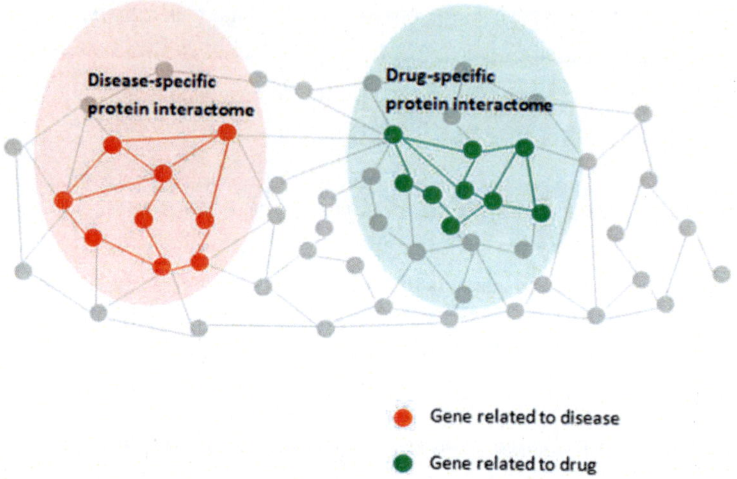

Fig. 1 Disease-specific and drug-specific protein interactome in the human PPI network

- Construct the matrix, S^{pi}_{DD} by combining S^{p}_{DD} and S^{i}_{DD} as given in (3):

$$S^{pi}_{DD}[i, j] = max(S^{p}_{DD}[i, j], S^{i}_{DD}[i, j]) \tag{3}$$

- Construct disease-disease similarity matrix, S^{d}_{DD} based on the known disease-drug relations using bipartite network projection method [3, 10].
- Construct integrated disease-disease similarity matrix, S_{DD} as given in (4):

$$S_{DD} = 1 - (1 - S^{pi}_{DD})(1 - S^{d}_{DD}) \tag{4}$$

3.1.2 Construct Integrated Drug-Drug Similarity Matrix, S_{RR}

The integrated drug-drug similarity matrix is constructed by combining drug-drug chemical structure similarity, protein interactome similarity, and the similarity based on the known disease-drug relations.

- Construct drug-drug chemical structure similarity matrix, S^{c}_{RR}: Chemical structure of drugs in the dataset are downloaded from DrugBank [9]. The similarity of two drugs is calculated as the ratio of the number of chemical fingerprints common in the two drugs to the total number of chemical fingerprints present in these drugs.
- Construct drug-drug protein interactome similarity matrix, S^{i}_{RR}: In order to compute protein interactome similarity, we first identified the interacting genes of each drug from Drugbank. Drug-specific protein interactome is the subnetwork corresponding to these identified genes in the human PPI network (Fig. 1).

Distance between the protein interactomes corresponding to drug r_i and drug r_j is computed using (5):

$$d\_avg_{R_i R_j} = \frac{\sum\limits_{p_1 \in R_i} \min\limits_{p_2 \in R_j} d(p_1, p_2) + \sum\limits_{p_2 \in R_j} \min\limits_{p_1 \in R_i} d(p_1, p_2)}{||R_i|| + ||R_j||} \quad (5)$$

where
R_i: Set of genes related to drug r_i.
R_j: Set of genes related to drug r_j.
$d(p_1, p_2)$: Shortest distance between genes p_1 and p_2 in the human PPI network.
Protein interactome similarity between drug r_i and drug r_j is computed using (6):

$$S^i_{RR}[i, j] = \frac{1}{1 + d\_avg_{R_i R_j}} \quad (6)$$

- Construct the matrix, S^{ci}_{RR} by combining S^c_{RR} and S^i_{RR} as given in (7):

$$S^{ci}_{RR}[i, j] = max(S^c_{RR}[i, j], S^i_{RR}[i, j]) \quad (7)$$

- Construct drug-drug similarity matrix, S^d_{RR} based on the known disease-drug relations using bipartite network projection method [3, 10].
- Construct integrated drug-drug similarity matrix, S_{RR} as given in (8):

$$S_{RR} = 1 - (1 - S^{ci}_{RR})(1 - S^d_{RR}) \quad (8)$$

3.1.3 Construct Disease-Drug Adjacency Matrix, S^a_{RD} and Disease-Drug Protein Interactome Similarity Matrix, S^i_{RD}

Construct disease-drug adjacency matrix, S^a_{RD} and disease-drug protein interactome similarity matrix, S^i_{RD} as follows:

- Construct disease-drug adjacency matrix, S^a_{RD}: disease-drug adjacency matrix is constructed using (9):

$$S^a_{RD}[i, j] = \begin{cases} 1 & \text{if there exist a known relation} \\ & \text{between drug } r_i \text{ and disease } d_j. \\ 0 & \text{otherwise} \end{cases} \quad (9)$$

- Construct disease-drug protein interactome similarity matrix, S^i_{RD}: In order to compute protein interactome similarity, we first identified the interacting genes of each drug from Drugbank and the genes functionally associated with each disease from OMIM database. Distance between the protein interactomes corresponding to drug r_i and diseases d_j is computed using (10):

$$d\_avg_{R_i D_j} = \frac{\sum\limits_{p_1 \in R_i} \min\limits_{p_2 \in D_j} d(p_1, p_2) + \sum\limits_{p_2 \in D_j} \min\limits_{p_1 \in R_i} d(p_1, p_2)}{||R_i|| + ||D_j||} \quad (10)$$

where,
R_i: Set of genes related to drug r_i.
D_j: Set of genes related to disease d_j.
$d(p_1, p_2)$: Shortest distance between genes p_1 and p_2 in the human PPI network.
Protein interactome similarity between drug r_i and disease d_j is computed as given in (11):

$$S^i_{RD}[i, j] = \frac{1}{1 + d\_avg_{R_i D_j}} \quad (11)$$

3.2 *Applying Relevant Random Walk Model on the Heterogeneous Network*

Here, first we build the transition matrix, M and initialize probability vectors, P_D and P_R for the random walk starting from the disease network and the drug network, respectively. Then, we run the random walk algorithm using these parameters.

3.2.1 Decide the Transition Matrix, *M*

Let λ be the probability of moving from drug network to disease network or vice versa. The transition matrix, representing the probability of transition between one node to other node in the network is defined as in (12):

$$\text{Transition matrix, } M = \begin{bmatrix} M_{RR} & M_{RD} \\ M_{DR} & M_{DD} \end{bmatrix} \quad (12)$$

where

- M_{RR} is the matrix representing the transition probability between two drug nodes and is defined as in (13):

$$M_{RR}[i, j] = \begin{cases} \frac{S_{RR}[i,j]}{\sum\limits_k S_{RR}[i,k]} & \text{if} \sum\limits_k S^a_{RD}[i, k] = 0 \\ \frac{(1-\lambda) S_{RR}[i,j]}{\sum\limits_k S_{RR}[i,k]} & \text{otherwise} \end{cases} \quad (13)$$

- M_{DD} is the matrix representing the transition probability between two disease nodes and is given as in (14):

$$M_{DD}[i,j] = \begin{cases} \frac{S_{DD}[i,j]}{\sum\limits_k S_{DD}[i,k]} & \text{if} \sum\limits_k S^a_{RD}[k,i] = 0 \\ \frac{(1-\lambda)S_{DD}[i,j]}{\sum\limits_k S_{DD}[i,k]} & \text{otherwise} \end{cases} \quad (14)$$

- M_{RD} is the matrix representing the probability of transition from a drug node to a disease node and is given as in (15):

$$M_{RD}[i,j] = \begin{cases} \frac{\lambda S^a_{RD}[i,j]}{\sum\limits_k S^a_{RD}[i,k]} & \text{if} \sum\limits_k S^a_{RD}[i,k] \neq 0 \\ 0 & \text{otherwise} \end{cases} \quad (15)$$

- M_{DR} is the matrix representing the probability of transition from a disease node to a drug node and is defined as in (16):

$$M_{DR}[i,j] = \begin{cases} \frac{\lambda S^a_{RD}[j,i]}{\sum\limits_k S^a_{RD}[k,i]} & \text{if} \sum\limits_k S^a_{RD}[k,i] \neq 0 \\ 0 & \text{otherwise} \end{cases} \quad (16)$$

3.2.2 Initialize Probability Vectors, P_{D_0} and P_{R_0}

For each disease-drug interaction in the dataset, we initialize two probability vectors: P_{D_0} and P_{R_0} for the random walk starting from the disease node and the drug node, respectively. The probability vector contains the prediction score assigned to each interaction in the network after each iteration of RWR algorithm.

- Initialize the probability vector, P_{D_0} for the random walk starting from disease node: The initial probability vector, for the random walk starting from disease node is defined as in (17):

$$P_{D_0} = \begin{bmatrix} (1-\eta)P_{D\_R_0} \\ \eta P_{D\_D_0} \end{bmatrix} \quad (17)$$

where
η: $\eta \in [0, 1]$, importance of disease network in the heterogeneous network.
$P_{D\_R_0}$: The initial probability of the drug network for the random walk starting from disease node.
$P_{D\_D_0}$: The initial probability of the disease network for the random walk starting from disease node.
For each disease-drug association, the probability of the drug network, $P_{D\_R_0}$ is initialized in such a way that all drugs which are already related to the disease are assigned equal probabilities, with the sum equal to 1. The probability of the disease network, $P_{D\_D_0}$ is initialized such that probability 1 is assigned to the disease present in the disease-drug relation, and probability 0 is assigned to other diseases.

– Initialize the probability vector, P_{R_0}, for the random walk starting from drug node: The initial probability vector, for the random walk starting from drug node is defined as in (18):

$$P_{R_0} = \begin{bmatrix} \eta P_{R\_R_0} \\ (1-\eta) P_{R\_D_0} \end{bmatrix} \tag{18}$$

where
η: $\eta \in [0, 1]$, importance of drug network in the heterogeneous network.
$P_{R\_R_0}$: The initial probability of the drug network for the random walk starting from drug node.
$P_{R\_D_0}$: The initial probability of the disease network for the random walk starting from drug node.
For each disease-drug association, the probability of the drug network, $P_{R\_R_0}$ is initialized in such a way that probability 1 is assigned to the drug present in the disease-drug relation and probability 0 is assigned to other drugs. The probability of the disease network, $P_{R\_D_0}$ is initialized in such a way that all diseases which are already related to the drug are assigned equal probabilities, with the sum equal to 1.

3.2.3 Run Random Walk Algorithm on the Modified Heterogeneous Network

The RWR algorithm is applied to the modified network to predict disease-drug associations. Equations (19) and (20) correspond to the random walk starting from drug node and disease node, respectively. Equation (19) and\or (20) can be computed iteratively until the walker reaches a steady state. After each iteration, the random walker can restart the walk with probability γ or proceed the walk with probability $(1-\gamma)$.

$$P_{R_{t+1}} = (1-\gamma) M^T P_{R_t} + \gamma P_{R_0} \tag{19}$$

$$P_{D_{t+1}} = (1-\gamma) M^T P_{D_t} + \gamma P_{D_0} \tag{20}$$

We then, rank the disease-drug associations in the decreasing order of the final prediction score.

4 Experiments and Results

The experiments conducted to determine the efficiency of the proposed system are discussed in this section, and the results are recorded.

4.1 Experimental Setup

The performance evaluation is carried out on the gold standard dataset constructed by Gottlieb et al. [13] and Cdataset constructed by Huimin Luo et al. [2]. The gold standard dataset contains 1933 known disease-drug relations involving 313 diseases and 593 drugs and the Cdataset contains 2532 known disease-drug relations involving 409 diseases and 663 drugs. We run RWR [1] and TP-NRWRH [3] algorithms on a disease-drug heterogeneous network before and after integrating the protein interactome similarity. We compared the result by conducting 10-fold cross-validations. That is, we divided the known disease-drug relations in the dataset into ten subsets with almost same size. After that, we run the random walk algorithm ten times by taking each of the ten subsets exactly once as the test set and the remaining nine subsets as training sets. We modified the test set for the proposed system such that it's elements are replaced with corresponding elements in the protein interactome similarity matrix. The results from all iterations are then combined to get the final result. We implemented the algorithms in R-programming language using RStudio IDE.

4.2 Evaluation Measures

We evaluated the proposed system based on Area Under the Curve (AUC) and number of correctly predicted known disease-drug relations for different top rank thresholds.

4.2.1 Area Under the Curve(AUC)

We plotted Receiver Operating Characteristic(ROC) curve, representing the true-positive rate (sensitivity) versus false-positive rate (1-specificity) at different cutoffs. Then, AUC is computed for each ROC curve. Higher AUC value represents a higher performance of the system.

4.2.2 Number of Correctly Predicted Known Disease-Drug Relations

Here, we ranked the disease-drug relations corresponding to each drug based on the prediction score. We considered a known disease-drug relation in the dataset as correctly predicted, if its prediction score assigned by the algorithm is greater than the score corresponding to the rank threshold. Since top-ranked results are more important in practical applications, we determined the numbers of correctly retrieved known disease-drug relations with respect to different top-rank thresholds. We considered the top 1,10, 20, 50, and 100 ranks as thresholds for the evaluation. We represented the final result as a graph.

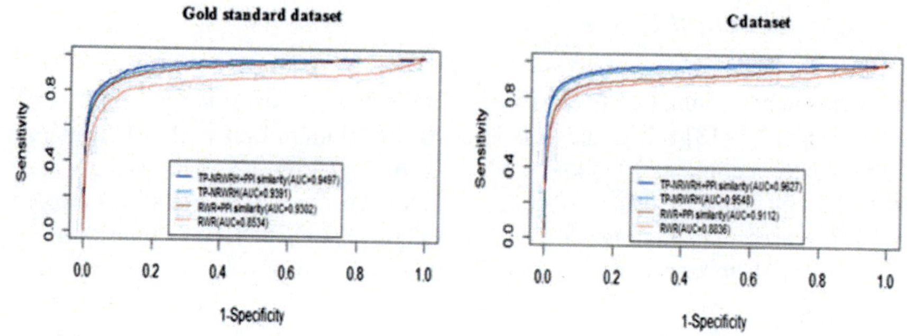

Fig. 2 ROC curves and AUC values

4.3 Evaluation Results

The ROC curves of RWR and TP-NRWRH before and after integrating protein interactome similarity are shown in Fig. 2. In gold standard dataset, the AUC values of RWR and TP-NRWRH models are incremented by 8.99% and 1.12%, respectively, after integrating protein interactome similarity. In Cdataset, the AUC values of RWR and TP-NRWRH models are incremented by 3.12% and 0.827%, respectively, after integrating protein interactome similarity.

Figure 3 shows the number of correctly predicted known disease-drug relations by RWR and TP-NRWRH applied on the heterogeneous network before and after integrating protein interactome similarity. In gold standard dataset, RWR and TP-

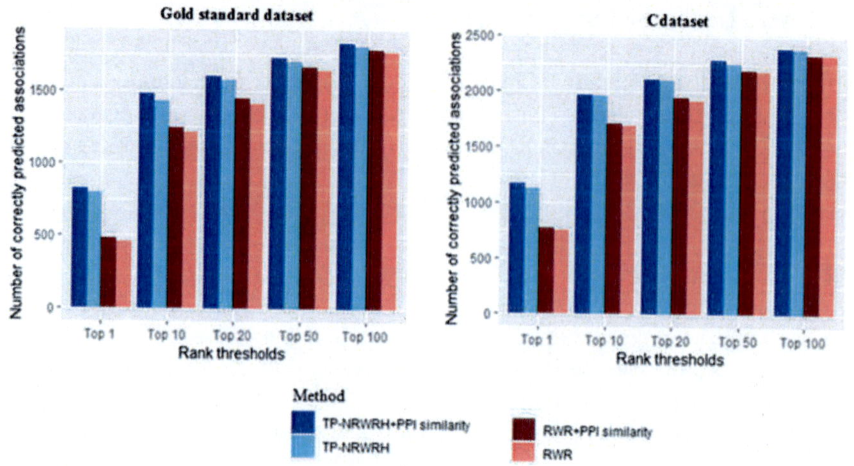

Fig. 3 The number of correctly predicted disease-drug associations with respect to different top-ranked thresholds

NRWRH models retrieved 26 and 33 more correct predictions, after integrating protein interactome similarity. In Cdataset, RWR and TP-NRWRH models retrieved 15 and 24 more correct predictions, after integrating protein interactome similarity.

5 Conclusion and Future Work

Drug repurposing is the process of using existing drugs to treat new diseases. In this paper, we presented a network-based approach for drug repurposing by integrating protein interactome similarity measures to a random walk with restart model. Our experiments reveal that the random walk model has higher performance after integrating the protein interactome similarity.

Some future works to pursue include:

- Using text mining approaches to identify novel drug-gene, disease-gene, and disease-drug associations from biomedical literature.
- Using more biological knowledge to enhance the similarity measures.
- Extending the two-layer disease-drug network to the multi-layer network by including component networks connecting drug and disease networks.
- Specializing this approach to predict potential drugs for new diseases.

We hope these improvements will lead to a system that can predict disease-drug associations more accurately.

References

1. Chen X, Liu M, Yan G (2012) Drug-target interaction prediction by random walk on the heterogeneous network. In: Molecular bioSystems
2. Luo H, Wang J, Li M, Luo J, Peng X, Wu F-X, Pan Y (2016) Drug repositioning based on comprehensive similarity measures and Bi-Random walk algorithm. In: Bioinformatics
3. Liu H, Song Y, Guan J, Luo L, Zhuang Z (2016) Inferring new indications for approved drugs via random walk on drug-disease heterogeneous networks. In: BMC Bioinformatics
4. Luo H, Wang J, Li M, Luo J, Ni P, Zhao K, Wu F-X, Pan Y (2019) Computational drug repositioning with random walk on a heterogeneous network. In: IEEE/ACM transactions on computational biology and bioinformatics
5. Cheng F, Lu W, Liu C, Fang J, Hou Y, Handy DE et al (2019) A genome-wide positioning systems network algorithm for in silico drug repurposing. Nat Commun 10:3476
6. Zhou Y, Hou Y, Shen J, Huang Y, Martin W, Cheng F (2020) Network-based drug repurposing for novel coronavirus 2019-nCoV/SARS-CoV-2. Cell Discov
7. Fiscon G, Conte F, Farina L, Paci P (2020) SAveRUNNER: a network-based algorithm for drug repurposing and its application to COVID-19
8. Zhang Y, Zeng T, Chen L, Ding S, Huang T, Cai Y-D (2020) Identification of COVID-19 infection-related human genes based on a random walk model in a virus-human protein interaction network. BioMed Res Int
9. Wishart DS, Knox C, Guo AC, Cheng D, Shrivastava S, Tzur D, Gautam B, Hassanali M (2008) DrugBank: a knowledgebase for drugs, drug actions and drug targets. Nucl Acids Res

10. Zhou T, Ren J, Medo M, Zhang YC (2007) Bipartite network projection and personal recommendation. Phys Rev E Stat Nonlin Soft Matter Phys
11. van Driel M, Bruggeman J, Vriend G, Brunner H, Leunissen J (2006) A text-mining analysis of the human phenome. Eur J Hum Genet
12. Hamosh A, Scott AF, Amberger JS, Bocchini CA, McKusick VA (2002) Online Mendelian Inheritance in Man (OMIM), a knowledgebase of human genes and genetic disorders. Nucl Acids Res 30(1):52–55
13. Gottlieb A, Stein GY, Ruppin E, Sharan R (2011) Predict: a method for inferring novel drug indications with application to personalized medicine. Mol Syst Biol

COVID-19 Detection on Chest X-Ray and CT Scan Images Using Multi-image Augmented Deep Learning Model

Kiran Purohit, Abhishek Kesarwani, Dakshina Ranjan Kisku, and Mamata Dalui

Abstract COVID-19 is a deadly and highly infectious pneumonia type disease. RT-PCR is a proven testing methodology for the detection of coronavirus infection in spite of having a lengthy testing time. Sometimes, it gives false-positive results more than the desired rates. To support the conventional RT-PCR methodology or testing independently without RC-PCR methodology for correct clinical diagnosis, COVID-19 testing can be acquired with images of X-Ray and CT Scan of a person. This image-based analysis will make a radical change in the detection of coronavirus in the human body with negligible false-negative and false-positive results. For the detection of COVID-19 in CT Scan and X-Ray images of coronavirus suspected individuals, this paper uses a multi-image augmented Convolutional Neural Network (CNN). For training the CNN model, multi-image augmentation utilizes discontinuity information acquired from the edged images to increase the meaningful examples. With this method, the proposed model exhibits a higher classification accuracy of around 98.97% for X-Ray and 95.38% for CT Scan images. Using multi-image augmentation, X-Ray images achieve a specificity of 98.88% and a sensitivity of 99.07% whereas a specificity of 95.98% and sensitivity of 94.78% are achieved in CT Scan images. The experimental results are also compared with VGG-16 and ResNet-50 models. The evaluation has been performed on publicly available databases comprising chest images of both X-Ray and CT Scan.

Keywords Coronavirus · CNN · Image augmentation · X-Ray · CT Scan

K. Purohit (✉) · A. Kesarwani · D. Ranjan Kisku · M. Dalui
Department of Computer Science and Engineering, National Institute of Technology, Durgapur, India
e-mail: drkisku@cse.nitdgp.ac.in

M. Dalui
e-mail: mamata.dalui@cse.nitdgp.ac.in

© The Author(s), under exclusive license to Springer Nature Singapore Pte Ltd. 2022
D. Giri et al. (eds.), *Proceedings of the Seventh International Conference on Mathematics and Computing*, Advances in Intelligent Systems and Computing 1412,
https://doi.org/10.1007/978-981-16-6890-6_30

1 Introduction

COVID-19 or Coronavirus is a deadly disease which so far has infected millions of people and deaths are increasing day by day. Due to the deadly infectious nature of coronavirus, it is spreading rapidly among people who are exposed to COVID-19 infected individuals.

Due to unknown causes of pneumonia type infection and the ability to generate new strain by mutation, it is almost impossible to have a cure in the form of vaccine or medicine for COVID-19 affected individuals. In the affected countries, a standard diagnostic method called (RT-PCR) reverse transcription polymerase chain reaction is adopted for detection of viral nucleic acid as coronavirus infection in COVID-19 suspected individuals. The test takes four to six hours or even a whole day to give the results. The time for the generation of the result is more compared to the time of the spread of coronavirus and sometimes, it gives false-negative and false-positive results. The false-negative rate for SARS-CoV-2 RT-PCR testing is profoundly factor: most noteworthy in the initial 5 days after exposure (up to 67%), and least on day 8 after exposure (21%). In view of this investigation, SARS-CoV-2 RT-PCR false-negative rate is very high, even at its least on day 8 post-exposure, or 3 days after symptoms. One out of five individuals tested is COVID-19 negative. RT-PCR tests showed 5% false-positive rates. Therefore, for testing COVID-19 rapidly and in a more efficient way, chest images of COVID-19 suspected individuals of CT Scan and X-Ray could be an answer. Moreover, time taken by RT-PCR test, shortage of test kits, and false-positive errors make it inefficient.

In contrast, CT Scan and X-Ray images are widely accepted traditional forms of diagnosing individuals for a number of diseases which is a common practice adopted by radiologists and medics in healthcare. The CT Scan and X-Ray technologies are used for several decades since their inception in medical diagnosis. In many highly affected areas or regions, it is difficult to provide a sufficient number of RT-PCR test kits for testing COVID-19 infection for thousands of corona suspected people. Therefore, to address this issue, COVID detection can be made from chest CT Scan and X-Ray images of corona suspected individuals who are suffering from COVID-19 symptoms.

The previous investigations uncover that contaminated individuals display different visual qualities in Chest X-Ray pictures, as appeared in Fig. 1. In non-ICU patients, these qualities regularly incorporate multifocal, inconsistent reticular opacities and reciprocal ground glass opacities, while thick aspiratory unions in ICU patients. Nonetheless, the manual investigation of these visual attributes on Chest X-Ray pictures requires domain expertise and is challenging. Also, the exponential increment in the quantity of contaminated patients makes it hard to finish the finding in time, prompting high bleakness and mortality.

For CT SCAN, imaging discoveries have mostly included single or numerous injuries, sketchy or segmental GGOs, and reticular markings that essentially followed peribronchovascular and subpleural disseminations. Interlobular septal thickening

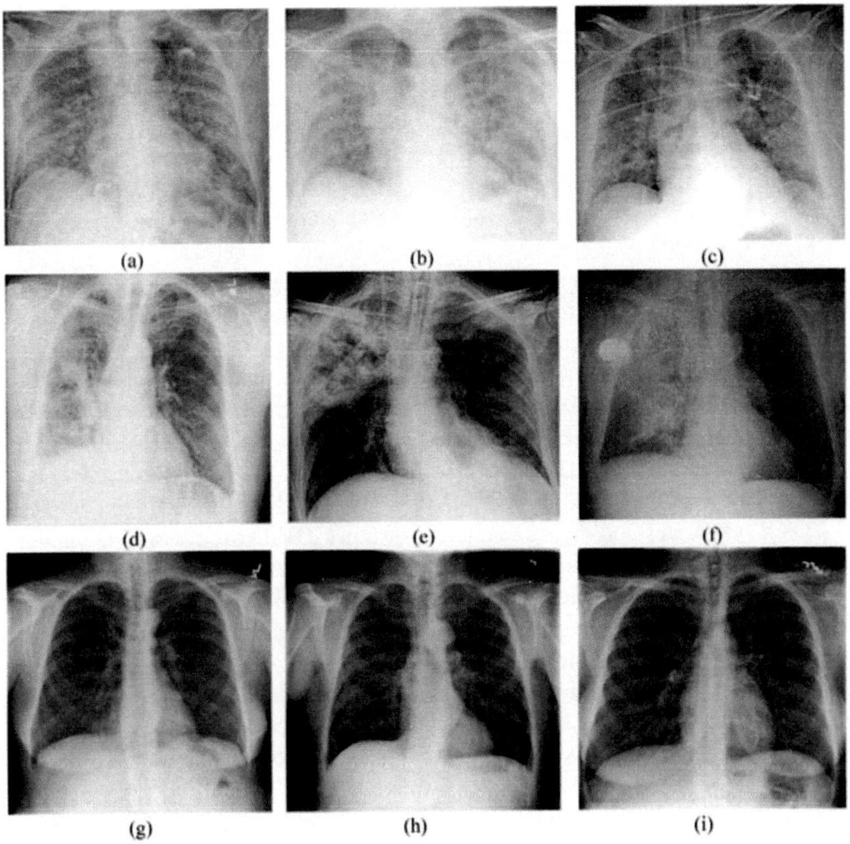

Fig. 1 **a–c** Chest X-Ray images infected by COVID-19, **d–f** Chest X-Ray images infected by Pneumonia, and **g–i** X-Ray images of normal persons [3]

may be available, and pleural emission and extended mediastinal lymph hubs were infrequently observed.

To deal with the issues related to the RT-PCR testing kit, we have come up with a solution by developing an AI-based multi-image augmentation method for the detection of COVID-19 infection in corona suspected persons. With this integrated framework, both chest CT Scan and X-Ray images could be tested for virus detection. This application makes use of multiple representations having sharp discontinuity information of same images of CT Scan and X-Ray, produced through first- and second-order derivative operators, which are mixed up with visible band images of X-Ray and CT Scan for training the deep model. This deep model can learn the underlying pattern of COVID-19 infected images of CT Scan and X-Ray in a more effective way from representative images as well as original images of the same person used for training. Moreover, with a simple configuration of the CNN model, this work performs well for a range of COVID-19 infected chest images of CT Scan and X-Ray.

The main objective of using the deep learning model [21] is to achieve higher accuracy of classification with chest images of CT Scan and X-Ray by separating the COVID-19 cases from non-COVID-19 cases. It is well known that someone needs a huge number of example images of both COVID-19 and non-COVID-19 individuals to train a deep model and to make the model learn about the patterns more effectively. To achieve this target, a number of representative images are generated using sharpening filters technique driven by first- and second-order derivative operators [9] and then these discontinuity information of the images are mixed up with the original CT Scan and X-Ray images separately, and further, these large number of multi-image augmentation are used for training the CNN-based deep model. The databases of images of CT Scan [26] and X-Ray [5] are publicly available in GitHub repository for experimental purpose. Both these datasets contain chest images of COVID-19 and non-COVID-19 individuals. The X-Ray database contains 67 COVID examples and the same number of non-COVID examples whereas the CT Scan database contains 345 COVID examples and the same number of non-COVID examples. To conduct the experiment, images are downsampled to 50×50 dimension from their original size. The random subsampling or holdout method is adopted for testing the model efficacy. In the holdout method, the whole dataset containing COVID positive and negative samples is divided into a number of ratios. It has been observed that when the number of training examples is increased, the model exhibits higher accuracy. Moreover, this result exhibits more consistency while layers are being changed in CNN-based deep model. To evaluate the framework in a robust and effective way, a number of metrics used are classification accuracy, F1-score, recall, precision, confusion matrix, loss, and area under curve (AUC). The values of these metrics have been determined on different ratios of training and test samples considering a number of layers in the deep model. The model correctly classifies the chest CT Scan and X-Ray images of COVID-19 cases from non-COVID-19 cases. Contributions are summarized as follows:

- Hybrid edge detection operator (novel operator) is used with other edge detectors to obtain the augmented images of CT Scan and X-Ray.
- The experiment used both images of CT Scan and X-Ray to evaluate the performance of the proposed methodology in a more exhaustive way.
- A fair comparison of the proposed methodology with the existing methods is presented to prove the efficacy for detecting COVID-19.
- The proposed methodology is evaluated with two powerful deep networks such as ResNet-50 and VGG-16 on the same set of databases of CT Scan and X-Ray.

The paper is organized as follows. Section 2 proposes the methodology to detect COVID-19 infections in chest images of CT Scan and X-Ray. Experimental results, analysis, and comparison among various deep learning models are presented in Sect. 3. In the last section, concluding remarks are made.

2 Proposed Work

For COVID-19 detection in chest images of CT Scan and X-Ray, the proposed methodology uses a number of sharpening filters such as Roberts, Prewitt, Sobel, Laplacian, Canny, Scharr, and a novel filter Hybrid.

2.1 *Hybrid Filter Generation*

For strengthening the process of detection of sharp discontinuity on images of CT Scan and X-Ray, a novel edge detector is used, which is called Hybrid filter [12]. It uses both Canny [9] and Sobel [9] detector to normalize the noise content as well as provides the high-frequency spatial information. This combination makes the filter helpful for image segmentation and edge detection. The image texture properties are also increased. Here, texton image [10] is used for unique texture which is generated from derivative of Gaussian [25]. Then the texton gradient image is determined by the paired disk masks. The paired disk masks are the pair of half-disk binary images. To get the texton gradient image, the pair of mask is convolved with the texton image, and the distance is calculated between two convolved images with the use of Chi-square metric [13]. The output is obtained by the combination of Sobel and Canny edge detection operator and texton gradient image as shown in Eq. 1.

$$O = Tg \times (w1 \times Canny\_edged\_image + w2 \times Sobel\_edged\_image) \quad (1)$$

where O is the image output, Tg is the image of texton gradient, and $w1$ and $w2$ are the weights ($w1 + w2 = 1$). The design process of the Hybrid filter is shown in Fig. 2.

2.2 *Multi-image Representation*

Sharpening of an image [9] expands the contrast between dark and bright areas to draw out the features. The sharpening method uses a high-pass kernel to an image. Sharpening is only inverse to blurring. We decrease the edge content in case of blurring, and in sharpening, we increment the edge content. For noisy images, firstly smoothing is done and then a sharpening filter is applied.

The majority of the information about the shape of an image is enclosed in the edges. Firstly, edges are distinguished in an image by the use of first- and second-order derivative operators [9], and afterward by enhancing those edges, image sharpness will increment and the image will become more clear. Consequently, COVID-19

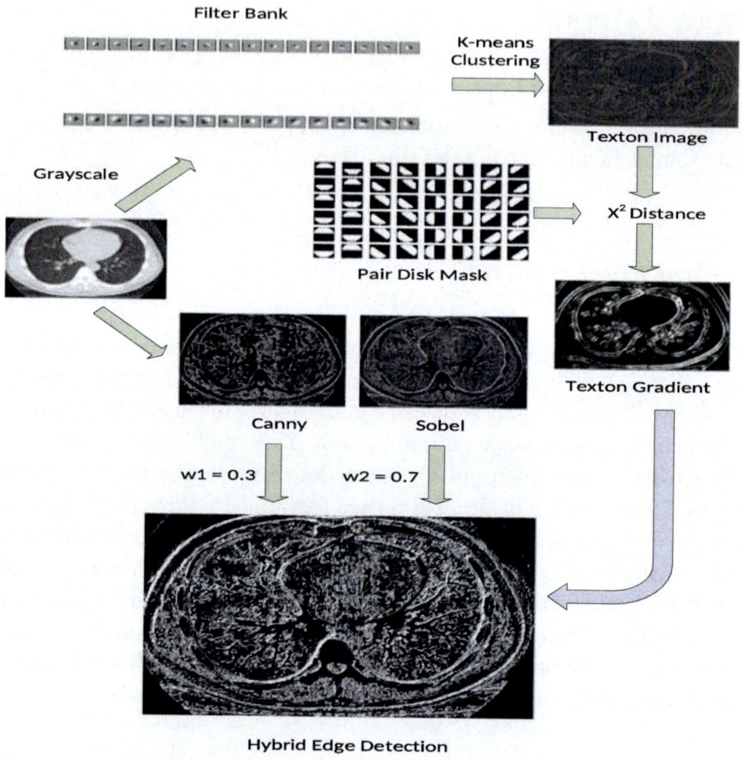

Fig. 2 Flow chart of Hybrid Edge Detector

detection on chest images of CT Scan and X-Ray will be more precise and accurate if we detect on the edged image.

The CT Scan [26] and X-Ray [5] databases contain a smaller number of images which may not be useful for training the CNN model. Moreover, with a small number of examples, desired classification accuracy may not be achieved. So, to resolve this issue, we apply multi-image augmentation by increasing the number of examples as well as the diversity of available characteristics with images of CT Scan and X-Ray. For multi-image augmentation, the input is converted into a grayscale image and then histogram equalization is done to correct the contrast of the input grayscale image. To achieve better image representation with discontinuity information, a number of first- and second-order edge detectors such as Roberts, Prewitt, Laplacian, Sobel, Canny, Scharr, and newly developed Hybrid are applied. Then, the results of the edge detection operator are appended to our dataset.

Edge detection operators perform a crucial job in separating low-level features or discovering information of the lungs. A sharp discontinuity change over the boundaries of the gray levels is called an edge. In chest images of CT Scan and X-Ray, edges represent lung boundaries, which occurs due to the change in the gray levels

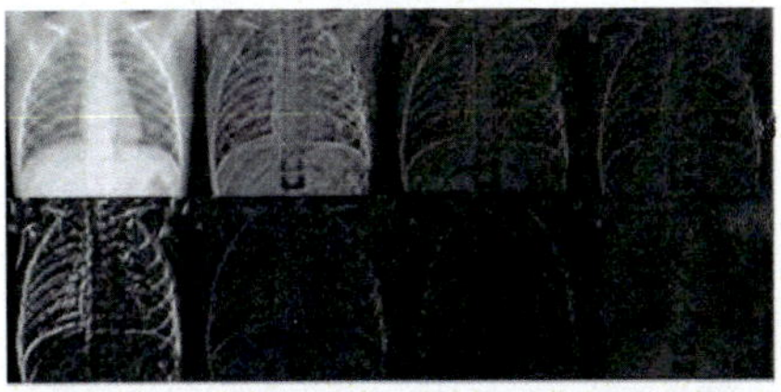

(a) Chest image of X Ray and its Multiple Representations of a COVID-19 Non-Infected Individual.

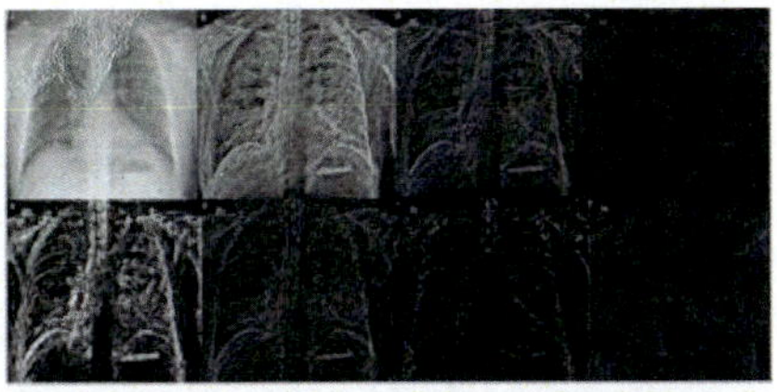

(b) Chest image of X Ray and its Multiple Representations of a COVID-19 Infected Individual.

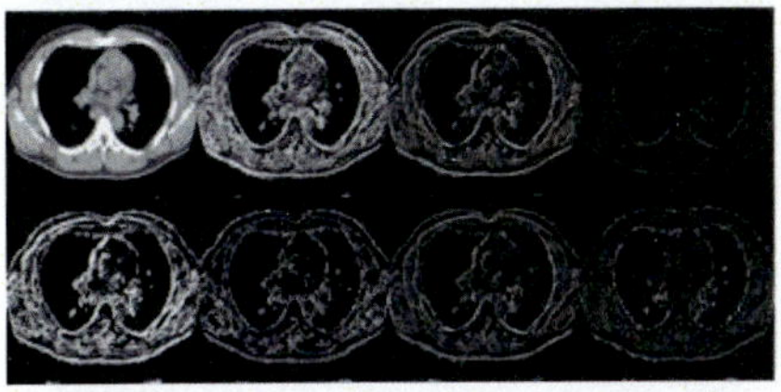

(c) Chest CT scan and its Multiple Representations of a COVID-19 Non-Infected Individual.

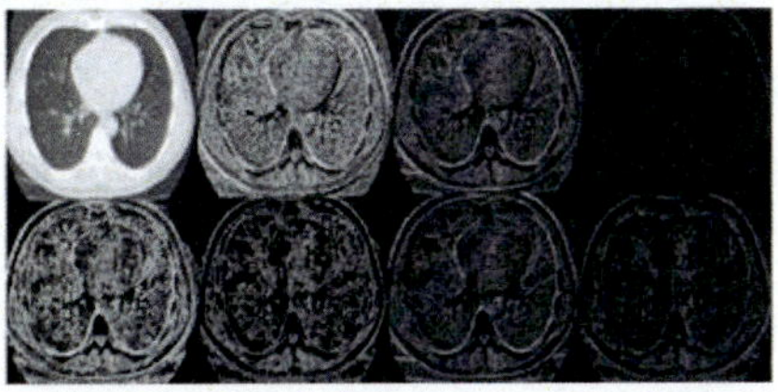

(d) Chest CT scan and its Multiple Representations of a COVID-19 Infected Individual.

Fig. 3 Multiple representations of chest images of X-Ray and CT Scan

at these lung boundaries. Edges are detected to remove basic and smaller details, increasing the speed and reducing the complexity without losing useful data. Here, significant data is retained and non-essential data is separated out. We get more exact and accurate outcomes if the processing is performed on these edged images. Thus, COVID-19 detection on chest images of CT Scan and X-Ray is found to be more precise if detected on the edged image.

Figure 3a shows a chest X-Ray (top left) and its multiple representations obtained by applying sharpening filters, of a person not infected by COVID-19 whereas Fig. 3b shows a chest X-Ray and its multiple representations obtained by applying sharpening filters, of COVID-19 infected person. Figure 3c shows a chest CT Scan image and its multiple representations obtained by applying sharpening filters, of a person not infected by COVID-19 whereas Fig. 3d shows a chest CT Scan image and its multiple representations obtained by applying sharpening filters, of COVID-19 infected person.

2.3 *Training and Classification of CNN-Based Deep Learning Model*

To perform training and classification with a multi-image augmented CNN model, the basic architecture of LeNet [6] model is exploited. It is used to predict COVID

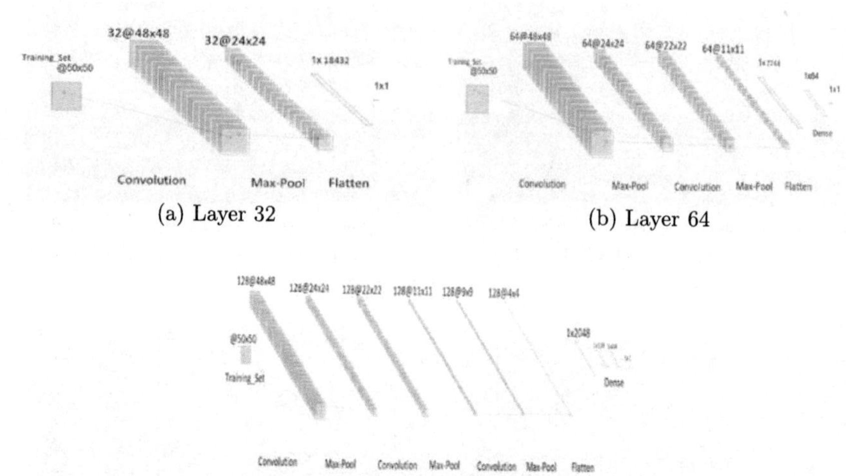

(a) Layer 32

(b) Layer 64

(c) Layer 128

Fig. 4 Convolution neural network of layer size 32, 64, and 128

and non-COVID examples from images of X-Ray and CT Scan. The deep CNN model uses three layers such as convolutional, pooling, and fully connected layers. Two activation functions viz. RELU and sigmoid are used. RELU is used after the convolutional layer and the sigmoid function is used for the classification of test images into COVID and non-COVID classes. Figure 4 shows the deep model with a number of parameters. In the training stage, a binary cross entropy-based loss function and the standard Stochastic Gradient Descent (SGD) optimizer with 32 as batch size are used. The learning rate is set as 0.01, which is linearly decayed and the maximum epoch is set to 30. To conduct the experiment, images are downsampled to 50×50 dimension from their original size. The random subsampling or holdout method is adopted to test model efficacy. In the holdout method, the whole dataset containing COVID positive and negative samples is divided into a number of ratios like 90:10, 80:20, 70:30, and 60:40 as training and testing samples. In order to alleviate the overfitting of the model, multi-image augmentation is used for training the model using sharpening filters. This augmentation generates a large number of representative images carrying discontinuity information.

Steps for detection of coronavirus infection in the images of CT Scan and X-Ray of suspected individuals:

Step 1: Accept the colored input images from the dataset.

Step 2: Convert the image into grayscale.

Step 3: Downsample the images to 50×50 dimension from their original size.

Step 4: In CNN-based deep learning model, we choose $layer\_sizes = [32, 64, 128]$, $dense\_layers = [0, 1, 2]$, and $conv\_layers = [1, 2, 3]$.

Step 5: Convolution with a 3×3 filter size is applied.

Step 6: Activation function RELU is used after convolutional layer.

Step 7: Then Max Pooling is applied with 2×2 filter size.
Step 8: Go to Step 5 if $conv\_layer - 1 > 0$.
Step 9: Flatten the matrix.
Step 10: Activation function Sigmoid is used for classification of test image into COVID and non-COVID classes.
Step 11: In training stage, the standard binary cross entropy loss, first-order SGD optimizer with 32 as batch size, and maximum epochs 30 are used.
Step 12: The random subsampling or holdout method is adopted. Whole dataset is divided in ratios like 90:10, 80:20, 70:30, and 60:40 as training and testing samples.
Step 13: Various metrics are evaluated such as F1 Score, recall, precision, accuracy, confusion matrix, loss, and area under ROC curve (AUC).
Step 14: Once the model is trained, one can easily predict whether the individual is affected by Coronavirus infection or not.

3 Evaluation

3.1 Experimental Protocol and Metrics

The proposed augmented deep model, VGG-16 and ResNet-50, uses Python and Keras package with TensorFlow [7] with Nvidia Geforce GTX 1050 GPU, 16 GB RAM, and Intel i5 Processor. To evaluate the framework in a robust and effective way, a number of metrics such as the area under curve (AUC), precision, F1-score, recall, classification accuracy, confusion matrix, and loss have been used. The values of these metrics have been determined on different ratios of training and test samples considering a number of layers in the deep model. The model correctly classifies the chest images CT Scan and X-Ray of COVID-19 cases from non-COVID-19 cases. The evaluation protocol makes use of the holdout method where the whole dataset containing COVID positive and negative samples are divided into a number of ratios. During testing, the layers are changed in the augmented-based deep model to realize the consistency of the model.

3.2 Experimental Results

It has been observed that when there is an increase in training examples, the model exhibits higher classification accuracy. For the traditional train and test ratio of 70:30, the classification accuracy is obtained around 98.97 and 95.38% for images of X-Ray and CT Scan, respectively. Examples are randomly split into training and testing sets for evaluation. The experiment is conducted twice for every split and it is noted that the accuracy of the model comes similar on both evaluations. F1 scores remain slightly similar and model loss (%) changes as the layer size is increased while setting the layer size to different values. For less training samples, a smaller layer size gives good results whereas if the training samples are increased, then more layers are required for better accuracy. The proposed deep augmented model is also compared with VGG-16 [19] and ResNet-50 [1] architecture.

Visual Geometry Group (VGG), a research group from Oxford University in the United Kingdom, proposed VGG networks. There are two commonly used networks, i.e., VGG-19 and VGG-16, here "19" represents 19 deep layers consisting of 3 Fully Connected (FC) layers and 16 Convolutional Layers (CL) whereas the "16" means 3 FC layers and 13 CL. The size of the input image in this network is $224 \times 224 \times 3$. VGG network uses an ImageNet pre-trained model which cannot improve the accuracy of COVID-19 screening as it contains a new set of images with labels. "Residual Network" also called ResNet, contains the width and depth of the neural network. These two core factors decide the complexity of the neural network. As the depth of the neural network increases, the training error increases. So, the residual network is used to solve this problem and increase the network performance (precision and accuracy) as compared to the traditional neural model. The commonly used residual networks are ResNet 18, 34 (two deep layers) and ResNet-50, 101, 152 (three deep layers). ResNet does not perform well in our case due to less amount of training examples.

The proposed deep augmented method achieves the sensitivity of 99.07% when the ratio of train and test samples is 70:30 and specificity of 98.88% when the ratio of train and test samples is 70:30 and layer size is 32 on the X-Ray dataset that contains 536 COVID-19 images and 536 non-COVID-19 images including augmented images. CT Scan dataset contains 2760 COVID-19 images and 2760 non-COVID-19 images individuals including augmented images. The sensitivity of 94.78% and specificity of 95.98% are achieved when the ratio of train and test samples is 70:30 and layer size is 64. Tables 1 and 2 show the result for original and augmented images, respectively.

The model is correctly able to classify chest images of X-Ray and CT Scan of COVID-19 examples from non-COVID-19 examples. The confusion matrix is shown in Fig. 5 for the standard ratio of train and test samples i.e., 70:30 for images of CT Scan and X-Ray with multi-image augmentation and without multi-image augmentation on three different architectures as proposed. ROC curves for images of CT Scan and X-Ray exhibiting higher accuracy when the ratio of train and test is 70:30 are shown in Fig. 7.

The ROC curve determined on X-Ray images with 64 layers is not shown because it seems like an overfitting having AUC as 1. Table 3 shows COVID-19 screening performance of models having higher accuracy with different train-test values. Figure 6 shows the optimal accuracy comparison between original and augmented techniques on both datasets of CT Scan and X-Ray. It is found that in the X-Ray dataset, the accuracy slightly differs in both original and augmented techniques due to less number of examples. Whereas, in the CT Scan dataset, the augmented technique performed with a number of representative images is found better than the non-augmented technique with original images. Proposed and existing models classification accuracies are shown in Table 4. The existing models either used CT Scan images or X-Ray

Table 1 COVID-19 screening performance in original images with CT scan and X-Ray datasets

	CT Scan images					X-Ray images				
	Proposed			ResNet	VGG	Proposed			ResNet	VGG
Train-Test	Layers					Layers				
60–40%	**32**	**64**	**128**	**50**	**16**	**32**	**64**	**128**	**50**	**16**
Accuracy (%)	88.7	90.87	87.68	88.7	78.7	98.51	99.25	98.51	50	98.51
Loss (%)	31.45	36.13	65.75	70.9	45.73	6.49	2.59	13.16	23.89	6.25
AUC (%)	94.5	95.2	93.5	88.7	78.7	99.9	100	98.5	50	98.5
Precision	0.86	0.9	0.89	0.91	0.72	0.97	1	0.97	0.5	0.97
Sensitivity (%)	92.17	92.17	85.51	85.5	93.62	100	98.51	100	1	100
Specificity (%)	85.22	89.57	89.86	91.88	63.76	97.01	100	97.01	0	97.01
F1 Score	0.89	0.91	0.87	0.88	0.81	0.99	0.99	0.99	0.67	0.99
70–30%	**32**	**64**	**128**	**50**	**16**	**32**	**64**	**128**	**50**	**16**
Accuracy (%)	89.57	92.03	90.29	78.84	77.54	99.25	97.1	100	50	98.51
Loss (%)	29.79	26.81	50.84	15.6	45	5.21	12.09	2.22	55.6	6.19
AUC (%)	94.8	95.9	95.7	78.8	77.5	100	99.8	100	50	98.5
Precision	0.91	0.93	0.87	0.71	0.76	0.99	1	1	0.5	0.97
Sensitivity (%)	87.25	91.01	95.07	97.39	81.15	100	94.03	100	88.05	100
Specificity (%)	91.88	93.04	85.51	60.28	73.91	98.51	100	100	11.94	97.01
F1 Score	0.89	0.92	0.91	0.82	0.78	0.99	0.97	1	0.64	0.99
80–20%	**32**	**64**	**128**	**50**	**16**	**32**	**64**	**128**	**50**	**16**
Accuracy (%)	94.06	95.22	94.35	93.48	78.7	99.25	97.01	99.25	51.49	98.51
Loss (%)	18.93	17.16	41.09	28.3	42.97	3.84	6.27	12.05	10.57	5.02
AUC (%)	98.1	98	96.8	93.5	78.7	100	100	98.6	51.5	98.5
Precision	0.93	0.95	0.94	0.93	0.72	0.99	1	0.99	1	0.97
Sensitivity (%)	95.07	95.65	94.78	94.49	93.04	100	94.03	100	2.98	100
Specificity (%)	93.04	94.78	93.91	92.46	64.34	98.51	100	98.51	1	97.01
F1 Score	0.94	0.95	0.94	0.94	0.81	0.99	0.97	0.99	0.06	0.99
90–10%	**32**	**64**	**128**	**50**	**16**	**32**	**64**	**128**	**50**	**16**
Accuracy (%)	94.93	**98.26**	98.26	94.49	83.33	99.25	**100**	99.25	50	99.25
Loss (%)	20.65	**9.47**	11.68	31.83	40.26	3.21	**1.08**	2.2	68.33	3.95
AUC (%)	98.4	**99.5**	99	0.91	83.3	100	**100**	100	50	99.3
Precision	0.97	0.97	**0.98**	0.91	0.79	0.99	**1**	0.99	0.5	0.99
Sensitivity (%)	92.46	**99.42**	98.55	98.26	91.01	100	**100**	100	1	100
Specificity (%)	97.39	97.1	**97.97**	90.72	75.65	98.51	**100**	98.51	0	98.5
F1 Score	0.95	**0.98**	0.98	0.95	0.85	0.99	**1**	0.99	0.67	0.99

Table 2 COVID-19 screening performance in augmented images with CT scan and X-Ray datasets

	CT scan images					X-Ray images				
	Proposed			ResNet	VGG	Proposed			ResNet	VGG
Train-Test	Layers					Layers				
60–40%	**32**	**64**	**128**	**50**	**16**	**32**	**64**	**128**	**50**	**16**
Accuracy (%)	93.46	92.54	91.59	85.53	85.07	98.88	98.79	98.04	93.66	90.39
Loss (%)	22.82	48.46	65.33	59.9	37.15	8.92	10.38	12.12	21.54	21.54
AUC (%)	97.2	96.5	96	85.5	92	99.7	99.4	99.4	93.7	90.4
Precision	0.92	0.93	0.91	0.86	0.88	0.99	0.99	0.98	0.95	0.99
Sensitivity (%)	94.67	91.99	91.74	84.31	81.15	99.06	98.7	97.95	91.97	81.52
Specificity (%)	92.25	93.08	91.45	86.75	88.99	98.7	98.88	98.13	95.33	99.25
F1 Score	0.93	0.93	0.92	0.85	0.84	0.99	0.99	0.98	0.94	0.89
70–30%	**32**	**64**	**128**	**50**	**16**	**32**	**64**	**128**	**50**	**16**
Accuracy (%)	95.05	95.38	92.46	89.02	88.99	98.97	99.44	98.69	90.76	92.26
Loss (%)	19.44	26.97	37.33	52.16	35.49	7.4	**1.91**	8.05	48.19	17.81
AUC (%)	97.8	98.4	97.2	89	93.4	99.7	100	99.6	90.8	92.3
Precision	0.96	0.96	0.95	0.9	0.9	0.99	1	0.99	0.99	0.99
Sensitivity (%)	93.64	94.78	89.38	87.71	87.83	99.07	99.07	98.33	82.27	85.82
Specificity (%)	96.48	95.98	95.54	90.32	90.14	98.88	99.81	99.06	99.25	98.69
F1 Score	0.95	0.95	0.92	0.89	0.89	0.99	0.99	0.99	0.9	0.92
80–20%	**32**	**64**	**128**	**50**	**16**	**32**	**64**	**128**	**50**	**16**
Accuracy (%)	96.47	96.99	95.45	92.75	91.16	99.07	98.88	99.16	96.18	93.1
Loss (%)	13.79	16.39	21.58	31.42	26.25	7.63	5.18	4.87	20.58	17.35
AUC (%)	98.6	99	98.2	92.7	97.1	99.5	99.7	99.8	96.2	93.1
Precision	0.96	0.97	0.95	0.94	0.96	0.99	0.99	0.99	0.99	0.97
Sensitivity (%)	96.73	97.39	95.83	90.83	86.38	99.25	99.25	99.44	92.91	88.99
Specificity (%)	96.2	96.6	95.07	94.65	95.94	98.88	98.52	98.89	99.44	97.2
F1 Score	0.96	0.97	0.95	0.93	0.91	0.99	0.99	0.99	0.96	0.93
90–10%	**32**	**64**	**128**	**50**	**16**	**32**	**64**	**128**	**50**	**16**
Accuracy (%)	98.5	**98.41**	96.9	96.09	92.61	**100**	99.63	99.63	93.38	93.84
Loss (%)	**7.27**	8.75	13.84	14.8	22.28	3.66	2.57	2.5	24.7	16.41
AUC (%)	99.4	**99.6**	99	96.1	97.8	**100**	99.9	100	93.4	93.8
Precision	0.98	**0.98**	0.96	0.95	0.9	**1**	1	1	0.89	0.98
Sensitivity (%)	**98.69**	98.37	97.5	96.88	96.23	**100**	99.63	99.44	99.44	89.92
Specificity (%)	98.29	**98.44**	96.3	95.3	88.99	**100**	99.63	99.81	87.31	97.76
F1 Score	**0.98**	0.97	0.97	0.96	0.93	**1**	1	1	0.94	0.94

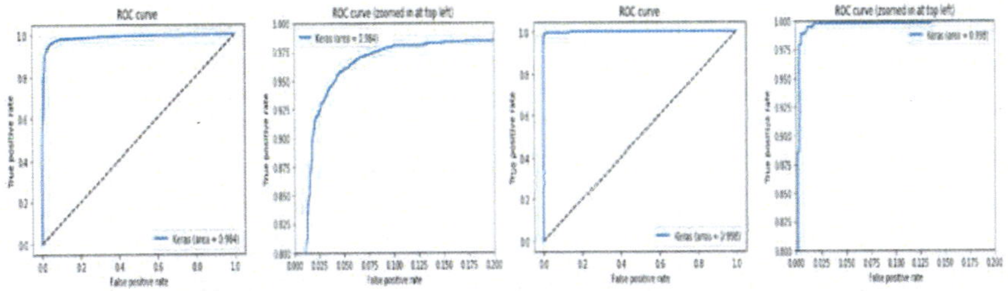

(a) CT Scan images with multi-image augmentation having 64 layers when train and test ratio is 70:30 (accuracy=95.38%)

(b) X-Ray images with multi-image augmentation having 32 layers when train and test ratio is 70:30 (accuracy=98.97%)

Fig. 5 ROC curves

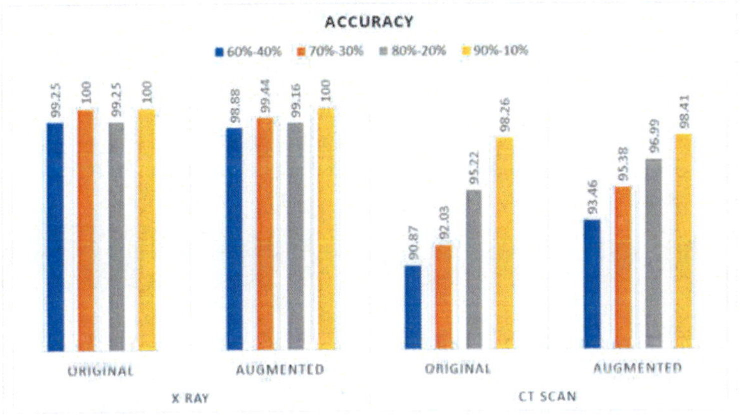

Fig. 6 COVID-19 screening accuracy

images for evaluation. Whereas, the proposed augmented deep model has used both images of CT Scan and X-Ray with a quite large number of samples.

3.3 Comparison

Authors in [11] made a COVIDX-Net for detection of COVID-19 in images of X-Ray. An accuracy of 90% is obtained with 25 COVID-19 (+ve) and 25 normal examples. Authors in [22] developed COVID-Net using deep network to detect COVID-19. It achieved 92.4% accuracy determined on 53 COVID-19 (+ve) and 8066 normal X-Ray examples. Ghoshal et al. [8] used the CNN model on 25 COVID-19 (+ve) examples and obtained an accuracy of 92.9%. Authors in [2] used transfer learning on 224 COVID-19, 700 pneumonia, and 504 normal examples. It obtained 98.75% accuracy in two-class problem and 93.48% accuracy in the three-class problem. Murugan and

Table 3 COVID-19 screening performance of outperforming model with respect to accuracy having different train and test values

	Original		Augmented		Original		Augmented	
	CT scan	X-Ray	CT scan	X-Ray	CT scan	X-Ray	CT scan	X-Ray
Train-Test	**60–40%**				**70–30%**			
Layers	**64**	**64**	**32**	**32**	**64**	**128**	**64**	**64**
Accuracy (%)	90.87	99.25	93.46	98.88	92.03	100	95.38	99.44
Loss (%)	36.13	2.59	22.82	8.92	26.81	2.22	26.97	1.91
AUC (%)	95.2	100	97.2	99.7	95.9	100	98.4	100
Precision	0.9	1	0.92	0.99	0.93	1	0.96	1
Sensitivity (%)	92.17	98.51	94.67	99.06	91.01	100	94.78	99.07
Specificity (%)	89.57	100	92.25	98.7	93.04	100	95.98	99.81
F1 Score	0.91	0.99	0.93	0.99	0.92	1	0.95	0.99
	Original		Augmented		Original		Augmented	
	CT Scan	**X-Ray**	**CT Scan**	**X-Ray**	**CT Scan**	**X-Ray**	**CT Scan**	**X-Ray**
Train-Test	**80–20%**				**90–10%**			
Layers	**64**	**32**	**64**	**128**	**64**	**64**	**64**	**32**
Accuracy (%)	95.22	99.25	96.99	99.16	98.26	100	98.41	100
Loss (%)	17.16	3.84	16.39	4.87	9.47	1.08	8.75	3.66
AUC (%)	98	100	99	99.8	99.5	100	99.6	100
Precision	0.95	0.99	0.97	0.99	0.97	1	0.98	1
Sensitivity (%)	95.65	100	97.39	99.44	99.42	100	98.37	100
Specificity (%)	94.78	98.51	96.6	98.89	97.1	100	98.44	100
F1 Score	0.95	0.99	0.97	0.99	0.98	1	0.97	1

Goel [14] used the E-DiCoNet model on three different classes and achieved an accuracy of 94.07%. Authors in [18] made use of CNN to obtain the features of the image and classified it using Support Vector Machine (SVM). They attained an accuracy of 95.38% using SVM and ResNet-50. Authors in [15] applied three different deep models which are Inception-ResNetV2, InceptionV3, and ResNet-50. They achieved 98% accuracy using 50 COVID-19 positive images of X-Ray and 50 normal examples. Tulin et al. [16] applied DarkCovidNet, a deep model to detect COVID-19. They used 1125 images which include 125 COVID-19 positive, 500 (pneumonia), and 500 without any findings to train the model and achieved accuracies of 98.08% for two classes and 87.02% for three classes.

Wang et al. [23] achieved 82.9% accuracy by applying a deep model called modified Inception using CT Scan images of 195 COVID-19 (+ve) and 258 normal examples. Ying et al. [20] obtained 86% accuracy using 777 COVID-19 (+ve) and 708 normal CT Scan examples using DRE-Net, a deep pre-trained ResNet-50 model. Authors in [24] obtained 86.7% accuracy for detection of COVID-19 by applying ResNet coupled with 175 Healthy, 219 COVID-19 positive, and 224 pneumonia CT

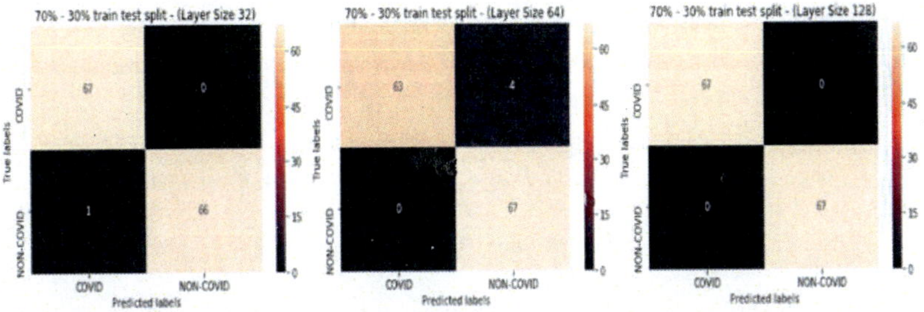

(a) X-Ray images without multi-image augmentation

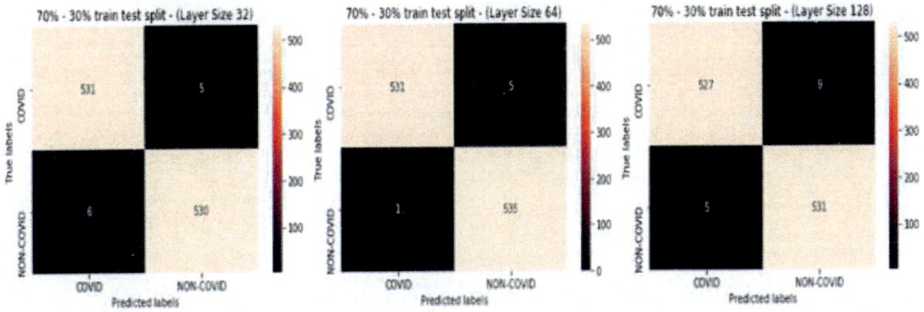

(b) X-Ray images with multi-image augmentation

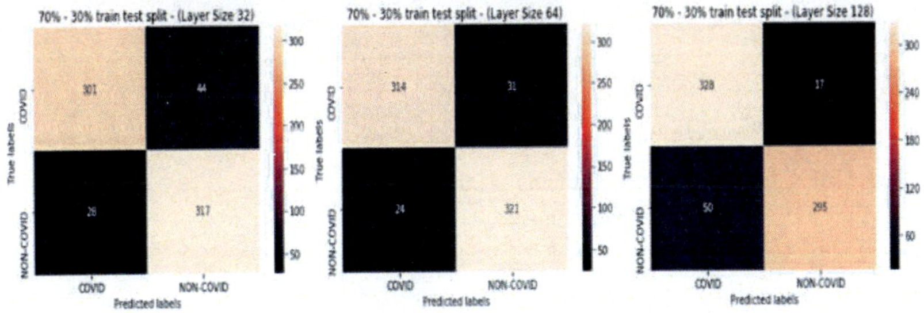

(c) CT Scan images without multi-image augmentation

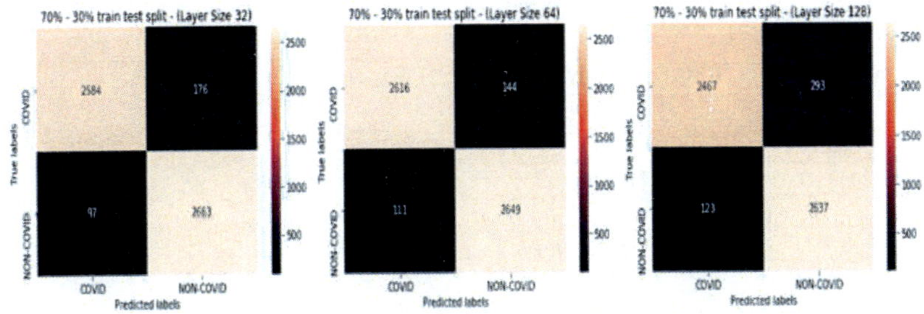

(d) CT Scan images with multi-image augmentation

Fig. 7 Confusion Matrix of our model when train and test ratio is 70:30

Table 4 Comparison among various COVID-19 techniques based on deep learning

Study	Images	Subjects	Methodology	Accuracy (%)
Hemdan et al. [11]	X-Ray	Covid 19(+ve) – 25	CovidX Network	90
		Normal(25)		
Wang and Wong [22]	X-Ray	Covid 19(+ve) – 53	Covid Network	92.4
		Healthy(8066)		
		Covid 19(-ve) – 5526		
Ghoshal et al. [8]	X-Ray	Covid 19(+ve) – 25	CNN	92.9
		Others - #Not available		
Ioannis et al. [2]	X-Ray	Covid 19(+ve) – 224	VGG-19 Network	93.48
		Healthy(504)		
		Pneumonia(700)		
Murugan & Goel [14]	X-Ray	Covid 19(+ve) – 900	E-DiCoNet	94.07
		Healthy(900)		
		Pneumonia(900)		
Sethy et al. [18]	X-Ray	Covid 19(+ve) – 25	ResNet-50 and SVM	95.38
		Covid 19(-ve) – 25		
Narin et al. [15]	X-Ray	Covid 19(+ve) – 50	ResNet-50 and Deep CNN	98
		Covid 19(-ve) – 50		
Tulin et al. [16]	X-Ray	No Findings – 500	DarkCovidNet	98.08
		Covid 19(+ve) – 125		
		Pneumonia(500)		
Wang et al. [23]	CT Scan	Covid 19(+ve) – 195	M-Inception	82.9
		Covid 19(-ve) – 258		
Ying et al. [20]	CT Scan	Healthy(708)	DRE-Net	86
		Covid 19(+ve) – 777		
Xu et al. [24]	CT Scan	Healthy(175)	Location Attention + ResNet	86.7
		Covid 19(+ve) – 219		
		Viral pneumonia(224)		

(continued)

Table 4 (continued)

Study	Images	Subjects	Methodology	Accuracy (%)
Zheng et al. [27]	CT Scan	Covid 19(+ve) – 313	3D Deep Network + UNet	90.8
		Covid 19(-ve) – 229		
Pathak et al. [17]	CT Scan	Covid 19(+ve) – 413	Fine-tuned ResNet32	93.01
		Normal(439)		
Chen et al. [4]	CT Scan	Covid 19(+ve) – 51	UNet++ Network	95.2
		Others(55)		
Proposed method	X-ray	Covid 19(+ve) – 536	Multi-image augmentation + CNN	99.44
		Covid 19(-ve) – 536		
	CT Scan	Covid 19(+ve) – 2760		95.38
		Covid 19(-ve) – 2760		

Scan images. Authors in [27] introduced a 3D deep CNN for COVID-19 prediction and obtained 90.8% accuracy using 313 COVID-19 positive and 229 COVID-19 negative CT Scan examples. Authors in [17] made use of fine-tuned ResNet32 on CT Scan images having two classes and achieved an accuracy of 93.01%. Chen et al. [4] introduced a UNet++ Network for COVID-19 detection and achieved 95.2% accuracy using 51 COVID-19 (+ve) and 55 COVID-19 (-ve) CT Scan examples.

Due to less number of X-Ray or CT Scan, examples that are used to train the deep learning models often exhibit the classification accuracies not up to the mark. Whereas, the proposed augmented deep model uses a large number of images of CT Scan and X-Ray for training. This multi-image augmentation has driven the CNN to exhibit higher classification accuracies while a basic deep learning architecture is used. Moreover, the proposed model is compared with VGG-16 and ResNet-50 on the same set of images of CT Scan and X-Ray. The proposed augmented model outperforms ResNet-50 and VGG-16 as well as the existing models too.

4 Conclusion and Future Work

This paper has proposed an augmented CNN to detect COVID-19 on chest images of CT Scan and X-Ray and classify from non-COVID-19 cases. The proposed model need not require to extract the feature manually, it is automated with end-to-end structure. Most of these previous studies have fewer examples for training the deep models. In contrast, the proposed model has used a multi-image augmentation technique driven by first- and second-order derivative edge operators and this augmentation generates a number of representative edged images. This augmentation technique provides a sufficient number of examples for training the model, making it robust. While CNN is trained with these augmented images, the classification accuracies of 99.44% for X-Ray images and 95.38% for CT Scan images are obtained. The experimental results are found to be highly convincing and emerged as a useful application for COVID-19 testing on chest images of CT Scan and X-Ray of corona suspected individuals. The subtle responses of various abnormalities like tuberculosis, pneumonia, influenza, and so forth confound the classifier, restricting the performance of the system. Future works may include the detection of multiple conditions such as pneumonia, bronchitis, and tuberculosis along with COVID-19 of suspected individuals having the respiratory illness.

References

1. Akiba T, Suzuki S, Fukuda K (2017) Extremely large minibatch SGD: training resnet-50 on imagenet in 15 min. arXiv:1711.04325
2. Apostolopoulos ID, Mpesiana TA (2020) Covid-19: automatic detection from x-ray images utilizing transfer learning with convolutional neural networks. Phys Eng Sci Med 1
3. Chandra TB, Verma K, Singh BK, Jain D, Netam SS (2020) Coronavirus disease (covid-19) detection in chest x-ray images using majority voting based classifier ensemble. Expert Syst Appl 165:113909
4. Chen J, Wu L, Zhang J, Zhang L, Gong D, Zhao Y, Hu S, Wang Y, Hu X, Zheng B et al (2020) Deep learning-based model for detecting 2019 novel coronavirus pneumonia on high-resolution computed tomography: a prospective study. MedRxiv
5. Cohen JP, Morrison P, Dao L (2020) Covid-19 image data collection. arXiv:2003.11597, https://github.com/ieee8023/covid-chestxray-dataset
6. El-Sawy A, Hazem EB, Loey M (2016) Cnn for handwritten arabic digits recognition based on lenet-5. In: International conference on advanced intelligent systems and informatics. Springer, pp 566–575
7. Géron A (2019) Hands-on machine learning with Scikit-Learn, Keras, and TensorFlow: Concepts, tools, and techniques to build intelligent systems. O'Reilly Media
8. Ghoshal B, Tucker A (2020) Estimating uncertainty and interpretability in deep learning for coronavirus (covid-19) detection. arXiv:2003.10769
9. Gonzalez RC, Woods RE, Eddins SL (2004) Digital image processing using MATLAB. Pearson Education India
10. He C, Li S, Liao Z, Liao M (2013) Texture classification of polsar data based on sparse coding of wavelet polarization textons. IEEE Trans Geosc Remote Sens 51(8):4576–4590
11. Hemdan EED, Shouman MA, Karar ME (2020) Covidx-net: a framework of deep learning classifiers to diagnose covid-19 in x-ray images. arXiv:2003.11055
12. Kesarwani A, Purohit K, Dalui M, Kisku DR (2020) Measuring the degree of suitability of edge detection operators prior to an application. In: 2020 IEEE applied signal processing conference (ASPCON), pp 128–133. https://doi.org/10.1109/ASPCON49795.2020.9276678
13. Liu X, Wang D (2002) A spectral histogram model for texton modeling and texture discrimination. Vis Res 42(23):2617–2634
14. Murugan R, Goel T (2021) E-diconet: extreme learning machine based classifier for diagnosis of covid-19 using deep convolutional network. J Ambient Int Humaniz Comput 1–12

15. Narin A, Kaya C, Pamuk Z (2020) Automatic detection of coronavirus disease (covid-19) using x-ray images and deep convolutional neural networks. arXiv:2003.10849
16. Ozturk T, Talo M, Yildirim EA, Baloglu UB, Yildirim O, Acharya UR (2020) Automated detection of covid-19 cases using deep neural networks with x-ray images. Comput Biol Med 103792 (2020)
17. Pathak Y, Shukla PK, Tiwari A, Stalin S, Singh S (2020) Deep transfer learning based classification model for covid-19 disease. IRBM
18. Sethy PK, Behera SK (2020) Detection of coronavirus disease (covid-19) based on deep features. Preprints 2020030300:2020
19. Simonyan K, Zisserman A (2014) Very deep convolutional networks for large-scale image recognition. arXiv:1409.1556
20. Song Y, Zheng S, Li L, Zhang X, Zhang X, Huang Z, Chen J, Zhao H, Jie Y, Wang R et al (2020) Deep learning enables accurate diagnosis of novel coronavirus (covid-19) with CT images. MedRxiv
21. Szegedy C, Liu W, Jia Y, Sermanet P, Reed S, Anguelov D, Erhan D, Vanhoucke V, Rabinovich A (2015) Going deeper with convolutions. In: Proceedings of the IEEE conference on computer vision and pattern recognition, pp 1–9
22. Wang L, Wong A (2020) Covid-net: a tailored deep convolutional neural network design for detection of covid-19 cases from chest x-ray images. arXiv:2003.09871
23. Wang S, Kang B, Ma J, Zeng X, Xiao M, Guo J, Cai M, Yang J, Li Y, Meng X et al (2020) A deep learning algorithm using CT images to screen for corona virus disease (covid-19). MedRxiv
24. Xu X, Jiang X, Ma C, Du P, Li X, Lv S, Yu L, Ni Q, Chen Y, Su J et al (2020) A deep learning system to screen novel coronavirus disease 2019 pneumonia. Engineering
25. Zhang B, Zhang L, Zhang L, Karray F (2010) Retinal vessel extraction by matched filter with first-order derivative of gaussian. Comput Biol Med 40(4):438–445
26. Zhao J, Zhang Y, He X, Xie P (2020) Covid-ct-dataset: a CT scan dataset about covid-19. arXiv:2003.13865
27. Zheng C, Deng X, Fu Q, Zhou Q, Feng J, Ma H, Liu W, Wang X (2020) Deep learning-based detection for covid-19 from chest CT using weak label. MedRxiv

Identification of Relevant Quantities in Arithmetic Word Problems Using Siamese Neural Network

Sourav Mandal, Arif Ahmed Sekh, and Sudip Kumar Naskar

Abstract Presence of irrelevant sentences containing irrelevant quantities in Math Word Problems (MWPs) throws an immense challenge in formulating the final equation(s). Failure in identification of the relevant quantities with respect to the question asked in an MWP reduces the overall system performance. This paper demonstrates a novel deep learning-based approach using Siamese neural network to classify the relevant and irrelevant sentences towards identification of the relevant quantities in arithmetic MWPs. The proposed relevance classifier produced an accuracy of 91.0% on a dataset variant (RQDV) prepared from a *Combined* dataset of two standard datasets consisting of arithmetic word problems. Our relevance classifier may benefit any existing arithmetic MWP solvers in improving their system performance, and we test this by performing a case study on a prototype system in which the integration of our relevance classifier improved the accuracy of the MWP solver from 68.0 to 81.0% on the *Combined* dataset.

Keywords Relevant quantity identification · Solving arithmetic word problems · Siamese neural network application

1 Introduction

Solving Math Word Problems (MWPs) of diverse types is a challenging research problem in *Information Extraction* (IE) and *Natural Language Understanding* (NLU), which require world knowledge, linguistic knowledge, and knowledge of mathematical concepts [1]. One of the primary challenges in solving MWPs is to

S. Mandal (✉) · A. A. Sekh
XIM University, Bhubaneswar, OR, India
e-mail: sourav.mandal@ieee.org

A. A. Sekh
UiT The Arctic University of Norway, Tromsø, Norway

S. K. Naskar
Jadavpur University, Kolkata, Kolkata, India

© The Author(s), under exclusive license to Springer Nature Singapore Pte Ltd. 2022

D. Giri et al. (eds.), *Proceedings of the Seventh International Conference on Mathematics and Computing*, Advances in Intelligent Systems and Computing 1412,
https://doi.org/10.1007/978-981-16-6890-6_31

identify relevant sentences in an MWP for extracting relevant quantities to form the final equation(s) (cf. example given below), which requires overall natural language understanding of the word problems. A student who is lacking such knowledge may not answer correctly, and a tutor in real has to explain why and how a quantity is relevant or irrelevant in the global context of the question asked.

Word problem (WP): Tim had 7 quarters and 9 nickels in his bank. His dad gave him 3 nickels and 5 pennies. How many nickels does Tim have now ?
Simplified word problem (SWP): Tim had 7 quarters. **Tim had 9 nickels in his bank. Tim's dad gave Tim 3 nickels**. Tim's dad gave 5 pennies. How many nickels does Tim have now ?
Word problem with relevant sentences (RWP): Tim had 9 nickels in his bank. Tim's dad gave Tim 3 nickels. How many nickels does Tim have now ?
Relevant Quantities: 9 and 3; **Irrelevant Quantities:** 7 and 5; **Equation**: 9+3=**12** (Answer)

Word Problem Formulation: An aritmetic Math word problem, WP is a set of word problem descriptors as sentences, $WP = \{S_1, S_2, ..., S_q\}$, and contains n ($s_{i=1}^n$) number of sentences including a single question sentence (S_q). A word problem WP is preprocessed (cf. Sect. 3) to create a **Simplified word problem** (SWP) by splitting complex and compound sentences into a set of 'nil' (no quantity) or single quantity sentences ($s_{i=1}^m$) such as $SWP = \{s_1, s_2, ..., s_m\}$.

Quantity, $Q = \{q_1, q_2, ..., q_n\}$, is a set of numeric values present in the word problem, WP, and may contain irrelevant quantities.

A **Relevant quantity**, $RQ \in Q$ is a quantity which is a cue (semantically close match) with the question (S_q) asked. A quantity which is not relevant to solving the question is an **Irrelevant quantity**, IRQ. **Word problem with relevant sentences**, RWP ($\in SWP$), is the rephrased word problem excluding the irrelevant sentences (i.e., containing irrelevant quantities).

In this work, our major contribution is in using a Siamese (aka twin) neural network [2] with a binary classifier to identify the relevant sentences to rephrase WP as RWP so that we can easily extract RQs from Q in a WP. Additionally, we perform a case study and develop a prototype system with a BiLSTM-based deep neural network to predict the desired operation $op \in \{+, -, *, /\}$ to finally fit with the extracted relevant quantities into the desired single equation as in Eq. 1 to generate the final answer (cf. Fig. 2) of an input WP.

$$Single\ EQ = \{RQ_1\ op\ RQ_2, ..., op\ RQ_p\} \quad (1)$$

The following section gives an idea about some related research works. Section 3 describes the dataset we used and the variant we prepared for this work. We explain our proposed system in Sect. 4 in identifying relevant quantities. Section 5 describes system prototype, experimental results, system performance, and success and failure scenarios followed by conclusion in Sect. 6.

2 Related Work

This research of automatically solving word problems attracts AI and machine learning researchers since a long time as this research problem is very close to human intelligence and learning. Reference [3–5] since the year 1960s [6–9]. Some remarkable systems in solving word problems attempted in earlier days are STUDENT [7], WORDPRO [8], ARITHPRO [9], ROBUST [10] (attempted first time to solve word problems with irrelevant quantities). [10] introduced a methodology to solve MWPs in *multi-step* (multiple equation). The system is also capable to identify *extraneous* or irrelevant information in the context of answer generation using propositional logic. Later, [11] proposed a method for identifying *irrelevant sentences* in the MWPs using classification techniques.

In recent years, we find some research works (and systems) in solving arithmetic word problems, such as KAZB [12], ARIS [13], ALGES [14], 'Roy and Roth' [15–17], SigmaDolphin [18], Tag-based [19, 20], Formulation-based [21], and OOMWPS [22]. Some other deep neural network-based approaches have been attempted very recently by [23–28]. However, in literature, we do not find enough study in identifying relevant quantities with the state-of-the-art word problem solvers, since this task has an immense effect on the final accuracies of such systems. Some relevant quantity identification procedures as available with the existing works are given precisely in the list below.

- **Rule-based approaches:** matched entities, attributes, variables, adjectives, verb categories, etc., among sentences (especially with question sentence) [13, 14, 20, 22, 28, 29]. [13] used an entity-attribute relation with matching keywords to match key attributes with the relevant quantities and selects them only for equation formation. [22, 29] adopted a interesting rule-base approach to identify relevant quantities within word problems. They represent an word problem in object-oriented paradigm which can represent all key attributes as an separate entities like owner, item, item_attribute, etc., associated with the quantities. After such representation they are able to identify the relevant quantities using a set theory based approach.
- **Classification-based approaches**: built relevance classifiers based on features such as unit (variable) features, related NP features, etc. [15], attributes of variables, cross-attribute relations, etc. [21]. [15] trained a binary relevance classifier using binary SVM (relevant or not) to identify relevant quantities. [21] used cross-attribute relations between question sentence and the other sentences in an MWP to identify relevant quantities.
- **Deep neural network-based approaches**: An instance of the relevant quantity labeled as 'True', otherwise 'False' to train the dataset [25] with the capabilities

of mapping the relevant numbers in problem to equation template learning has been found in [23, 24]. They actually proposed an LSTM network (RNN-based) for this purpose.

In recent years, we have few more large datasets containing different types of irrelevant information. It has been observed that the performance of different systems are largely depend on how the irrelevant information (and/or quantities) are handled. Therefore, it throws additional challenges to the research of solving word problems.

3 Dataset

We use two standard arithmetic (*single equation* and *single operation*) datasets—*AddSub* [13] and *SingleOp* [15] from an arithmetic word problem repository MAWPS [30]. These 2 datasets together contain a total of 917 problems among which there are 159 word problems with irrelevant quantities. We preprocess (rule-based) the *Combined* dataset into a collection of simplified word problems (SWP) with 'nil' (no quantity) or single quantity sentences by eliminating conjunctions and substituting coreferences if any. We restructured the original MWPs in the *Combined* dataset and prepared a new variant (**RQDV**) of the dataset to make the dataset suitable for the relevant quantity identification (RQI) task. The datasets are available publicly for view.[1]

For a word problem WP in the *Combined* dataset, for each simple sentence $s_i \in SWP$ ($s_{i=1}^m$) and $S_q \in WP$, we build a sentence-question input pair (for our Siamese-based network) such as $\{s_i, S_q\}$ with the binary classifier 0 (relevant) or 1 (irrelevant) to denote whether s_i is relevant or not with respect to solving $\{S_q\}$. Therefore, for example, if a WP in the *Combined* dataset has 3 simple sentences s_i ($s_{i=1}^3$) excluding the S_q, we have 3 such $\{s_i, S_q\}$ pairs in RQDV. The data in RQDV is directly used in our Siamese-based network (cf. Fig. 1b).

4 Sentence Relevancy Classification Using Siamese Deep Neural Network with Binary Classifier

We carried out sentence relevance classification using a Siamese network-based deep recurrent neural network (LSTM) (cf. Fig. 1). Siamese (twin) neural network is successfully used in finding sentence similarities like in Quora to identify and retrieve similar questions asked before by some other users. We are motivated by this concept and implement the same in this work to identify similar sentences with respect to question sentence in a word problem. Our method employs a binary classification (relevant and irrelevant) to select relevant sentences from the word problem. Firstly,

[1] https://sites.google.com/view/datasets-for-relevant-quantity/home.

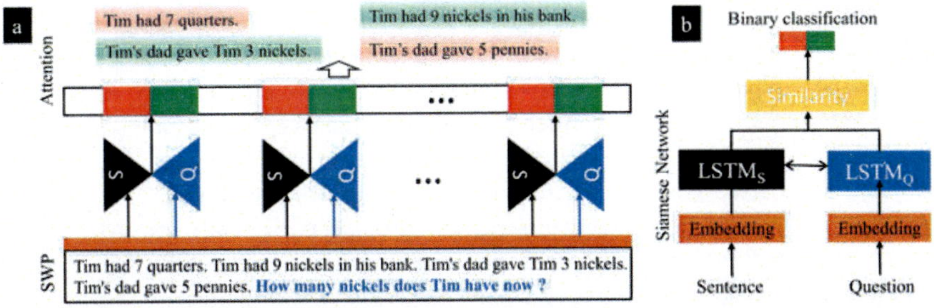

Fig. 1 **a** Siamese network with binary classifier for relevant sentence identification. **b** Structure of siamese network for single sentence-question pair input

the complex sentences are converted into simple sentences which are then represented as a vector using pre-trained word vectors (Word2Vec) [31]. Our proposed network consists of two parallel LSTM cells, sentence LSTM ($LSTM_s$) and question LSTM ($LSTM_q$). This LSTM pair shares the same weight during training, i.e., $weight(LSTM_s) == weight(LSTM_q)$). We design our network using the state-of-the-art LSTM method reported in [32]. One LSTM cell consists of a input gate (i_t) and a forget gate (f_t); a memory unit (c_t); and a output gate (o_t). Intermediate weights ($W_i, W_o, W_c, W_f, Z_i, Z_o, Z_c, Z_f$) and bias ($b_i, b_o, b_c, b_f$) are used as the backbone of the LSTM where, W_i and b_i are the weight and bias given to the input gate i at timestamp t. The functionalities of the LSTM units are defined as below.

$$\begin{aligned} i_t &- sigmoid(W_i x_t + Z_i h_{t-1} + b_i) \\ f_t &= sigmoid(W_f x_t + Z_f h_{t-1} + b_f) \\ \tilde{c}_t &= tanh(W_c x_t + Z_c h_{t-1} + b_c) \\ c_t &= i_t \odot c_t + f_t \odot c_{t-1} \\ o_t &= sigmoid(W_o x_t + Z_c h_{t-1} + b_o) \\ h_t &= o_t \odot tanh(c_t) \end{aligned} \quad (2)$$

The final layers of the LSTMs are used as the input to a similarity layer as given in Fig. 1. First, each of the variable length sentences paired with the question sentence ($\{s_i, S_q\}$) are input (from RQDV) to the hidden layers of the LSTMs (cf. Fig. 1a). Fig. 1b represents a single such input as sentence-question pair. The hidden representation of sentence-question pair ($h^{(s)}_{T_s}, h^{(q)}_{T_q}$), $s = \{1, ..., m\}$ (m is the number of simple sentences excluding question sentence in the SWP) is taken as the input to the similarity identification layer. This procedure is repeated for estimating the similarities of each sentence-question pairs in SWP. The relevancy between such a pair is basically regarded as the sentence similarity which is defined as in Eq. 3.

$$S(h^{(s)}_{T_s}, h^{(q)}_{T_q}) = exp(-||(h^{(s)}_{T_s} - h^{(q)}_{T_q}||_1) \quad (3)$$

The similarity layer uses Manhattan distance (l_1 norm) and it forces to learn the semantic similarity between the question and the sentences. Finally, a binary softmax layer is used to classify each sentence as either relevant or irrelevant (with respect to the question sentence) and we further use this outcome as the binary classification to filter out the irrelevant sentences from a SWP to prepare RWP.

For our proposed Siamese-based network, we used 50-dimensional LSTM hidden units to learn from manually labeled sentence-question pairs. We chose word2vec word embedding of dimension 300 with maximum sequence length of 20. We worked with the training batch size of 16, learning rate 0.001, *Adam* optimizer, and binary cross-entropy loss. We employed conventional early stopping and learning rate reduction on the validation loss. The loss (L) is minimized over the sentence-question pair as given below, where y_i is the ground truth and $\hat{y}_i$ is predicted class.

$$L = -\frac{1}{N}\sum_{i=1}^{N} y_i . \log(\hat{y}_i) + (1 - y_i). \log(1 - \hat{y}_i) \tag{4}$$

5 System Prototype (Case study) and Results

Our relevant identification module can be added with existing systems to enhance their present performances. We performed a case study and developed a word problem solver prototype of our own to test the importance of identification of such relevant sentences and quantities. Figure 2 shows a implementation of a word problem solver prototype system consisting of 3 major modules (circled in blue) which is inspired by the work of [29]. Input to the system is WP which is then prepossessed (cf. Sect. 3) to convert it into SWP for further 'operation prediction' and 'relevant quantities identification' (RQI). The 'operation prediction' module shown in Fig. 2 can be of any type (or methodology) as available in actual in the literature with state-of-the-art systems. As a case study we reused a BiLSTM-based operation predication method (similar to [29]) and get the test accuracy 81.4% on the *Combined* dataset. The RQI module first predicts the relevant sentences from an SWP to prepare RWP, and

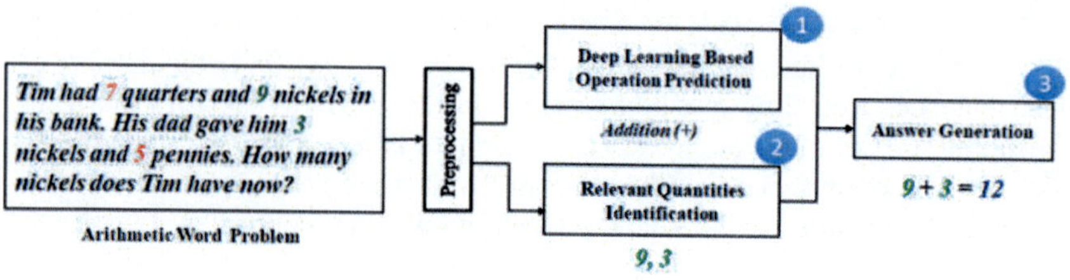

Fig. 2 System prototype of a word problem solver

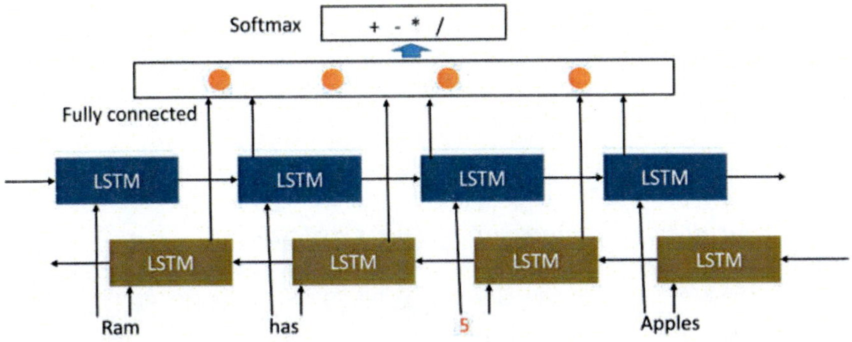

Fig. 3 Operation classification using BiLSTM [29]

further extracts RQs from RWP to form the equation as given in Eq. 1 with the predicted operation to generate the final answer of a WP.

In this work, we used *bi-directional* LSTM memory blocks (see Fig. 3) to classify the four basic operations (+, -, *, and /) (similar to [29]). We basically did 'word problem to operation' ($WP\,2\,Op$) classification.

We follow standard 80:20 split of the **RQDV** dataset into train, and test sets, with further random 20% split of the train set into validation set. We train the network using batch size of 16, and learning rate 0.001. We employ conventional early stopping and learning rate reduction on the validation loss.

5.1 *Experimental Results*

We followed standard 80:20 split of the RQDV dataset as well as *Combined* dataset into train and test sets, with further 20% random split of the train set into validation set. We obtained the test accuracy of 91.0% and validation accuracy of 92.1% for the RQI task on the RQDV dataset. Figure 4a, b shows the training loss and accuracy, and confusion matrix of the RQI task. For the MWP solver, we obtained the system

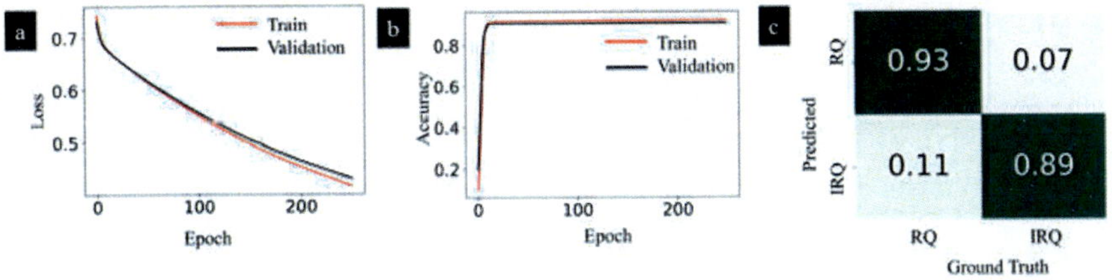

Fig. 4 **a** Training loss and accuracy over epochs of the proposed siamese network-based deep neural network for the RQI task. **b** Confusion matrix of the RQI task

accuracy of 68.0% without the RQI module and 81.0% accuracy with the RQI module on the test set, and overall 84.7% accuracy with the RQI module on the *Combined* dataset as a whole. This 84.7% accuracy is basically the percentage of word problems answered correctly by our proposed system prototype as a whole.

5.2 *Performance Comparison of Our Case Study*

We achieved test accuracy of 81.4% for operation prediction with training accuracy 99.0%, validation accuracy 98.2% on actual dataset after preprocessing (i.e., SWP form of combined dataset). We got the testing accuracy of 91% of our RQI task with the validation accuracy of 92% on **RQDV** dataset. We got the overall system (case study) accuracy of 68.0% without RQI module and 81.0% with RQI module.

Overall system performance comparison with the state-of-the-art systems is given in the Table 1. Here, we test our trained RQI and 'operation classification' models on all the word problems (total 917 problems in combined dataset) and achieve accuracy of 84.7%. Table 2 shows some random examples of correctly assigned class and failure cases.

5.3 *Success and Failure Scenarios in Identifying Relevant Quantities: Examples*

Table 2 shows the performance of our RQI module on some example word problems from the *Combined* dataset. Table 2 gives an overview on how the Siamese-based neural networks performs on the dataset.

Table 1 Comparison of the performance of our system (case study) with other state-of-the-art systems

Systems	AddSub	SingleOp	Combined
KAZB [12]	64.0	73.7	68.85
Roy & Roth [15]	72.0	73.9	72.95
MathDQN [28]	78.5	73.3	75.9
Our prototype system with RQI	–	–	84.7

Table 2 Examples of some random correct and the failure Cases (**red** color refers irrelevant and **blue** color refers relevant class)

Original problem	Simplified problem with relevant class	Remark
Benny found 66 seashells and 49 starfish on the beach He gave 52 of the seashells to Jason How many seashells does Benny now have?	Benny found 66 seashells on the beach Benny found 49 starfish on the beach Benny gave 52 of the seashells to Jason How many seashells does Benny now have?	Correct
Albert has two snakes The garden snake is 10 inches long The boa constrictor is 7 times longer than the garden snake How long is the boa constrictor?	Albert has two snakes The garden snake is 10 inches long The boa constrictor is 7 times longer than the garden snake How long is the boa constrictor?	Correct
Keith spent $6.51 on a rabbit toy $5.79 on pet food, and a cage cost him $12.51 He found a dollar bill on the ground What was the total cost of Keith 's purchases?	Keith spent $6.51 on a rabbit toy Keith spent $5.79 on pet food Keith spent a cage cost Keith $12.51 Keith found a dollar bill on the ground What was the total cost of Keith 's purchases?	Correct
Mary got fast food for lunch Mary spent $1.08 on soup and $4.80 on a salad Mary paid with a 20 dollar bill What was the total of the lunch bill?	Mary got fast food for lunch Mary spent $1.08 on soup Mary spent $4.80 on a salad Mary paid with a 20 dollar bill What was the total of the lunch bill?	Correct
Suzie found 2 worms in the yard and measured them with a ruler 1 worm was 0.8 inch long The other worm was 0.1 inch long How much longer was the longer worm?	Suzie found 2 worms in the yard Suzie measured them with a ruler 1 worm was 0.8 inch long The other worm was 0.1 inch long How much longer was the longer worm?	Wrong 1st sentence
When Jake had 1 cat, he needed to serve 0.5 can of cat food each day Now that Jake has adopted a second cat, he needs to serve a total of 0.9 can each day. How much extra food is needed to feed the second cat?	When Jake had 1 cat Jake needed to serve 0.5 can of cat food each day Now that Jake has adopted a second cat Jake needs to serve a total of 0.9 can each day How much extra food is needed to feed the second cat?	Wrong 1st sentence
There are 247 students in each school 256 students are going to the zoo They have to be divided into groups so that each teacher has one group How many students will be in each group?	There are 247 students in each school 256 students are going to the zoo They have to be divided into groups so that each teacher has one group How many students will be in each group?	Wrong 3rd sentence

6 Conclusion

In this paper, we presented a system for classifying relevant and irrelevant sentences to extract the relevant quantities with respect to the question asked in an MWP. The proposed relevance classifier significantly improved the overall performance of the MWP solver on standard datasets. Our novel Siamese network-based relevance classifier can enhance the performance of available arithmetic MWP solvers. We used entirely a new approach which have the potential to increase the system performances given in Sect. 2 by handling irrelevant information more effectively. As an extension of this work, we will explore other types of deep learning-based models and binary classification mechanisms for the RQI task.

References

1. Diane JL (2016) Natural language processing for enhancing teaching and learning. In: Proceedings of the thirtieth AAAI conference on artificial intelligence, Phoenix, Arizona, USA, 12–17 Feb 2016, pp 4170–4176
2. Bromley J, Guyon I, LeCun Y, Säckinger E, Shah R (1993) Signature verification using a siamese time delay neural network. In: 7th NIPS conference on advances in neural information processing systems, Denver, Colorado, USA, vol 6, pp 737–744
3. Mukherjee A, Garain U (2008) A review of methods for automatic understanding of natural language mathematical problems. Artif Intell Rev 29(2):93–122
4. Dongxiang Z, Lei W, Nuo X, Bing TD, Heng TS (2018) The gap of semantic parsing: a survey on automatic math word problem solvers. arXiv:abs/1808.07290
5. Sourav M, Sudip KN (2017) Solving arithmetic mathematical word problems: a review and recent advancements. In: Information technology and applied mathematics (ICITAM 2017), Haldia, WB, India, 30 Oct–1 Nov 2017, pp 95–114
6. Feigenbaum EA, Feldman J (1963) Computers and thought, vol 7. McGraw-Hill Inc, New York, USA
7. Daniel GB (1964) Natural language input for a computer problem solving system. Phd thesis, Massachusetts Institute of Technology, Cambridge, MA, USA, Sept 1964
8. Charles RF (1985) Understanding and solving arithmetic word problems: a computer simulation. Behav Res Methods Inst Comput 17(5):565–571 (1985)
9. Dellarosa D (1986) A computer simulation of children's arithmetic word-problem solving. Behav Res Methods Inst Comput 18(2):147–154
10. Bakman Y (2007) Robust understanding of word problems with extraneous information, Jan 2007. arXiv:0701393
11. Cetintas S, Si L, Xin YP, Zhang D, Park JY, Tzur R (2010) A joint probabilistic classification model of relevant and irrelevant sentences in mathematical word problems. JEDM-J Educ Data Min 2(1):83–101
12. Kushman N, Zettlemoyer L, Barzilay R, Artzi Y (2014) Learning to automatically solve algebra word problems. In: Proceedings of the 52nd Annual Meeting of the Association for Computational Linguistics (ACL 2014), Baltimore, MD, USA, 22–27 June 2014, pp 271–281
13. Hosseini MJ, Hajishirzi H, Etzioni O, Kushman N (2014) Learning to solve arithmetic word problems with verb categorization. In: Proceedings of the 2014 conference on empirical methods in natural language processing (EMNLP 2014), a meeting of SIGDAT, a special interest group of the ACL, Doha, Qatar, 25–29 Oct 2014, pp 523–533

14. Koncel-Kedziorski R, Hajishirzi H, Sabharwal A, Etzioni O, Ang SD (2015) Parsing algebraic word problems into equations. Trans Assoc Comput Linguist 3:585–597 (2015)
15. Roy S, Roth D (2015) Solving general arithmetic word problems. In *Proceedings of the 2015 Conference on Empirical Methods in Natural Language Processing (EMNLP 2015)*, Lisbon, Portugal, 17–21 Sept 2015, pp. 1743–1752
16. Roy S, Vieira T, Roth D (2015) Reasoning about quantities in natural language. Trans Assoc Comput Linguist 3:1–13 January
17. Subhro Roy and Dan Roth. Unit dependency graph and its application to arithmetic word problem solving. In *Proceedings of the Thirty-First AAAI Conference on Artificial Intelligence*, pages 3082–3088, San Francisco, California, USA, February 4-9, 2017
18. Shi S, Wang Y, Lin C-Y, Liu X, Rui Y (2015) Automatically solving number word problems by semantic parsing and reasoning. In: Proceedings of the 2015 conference on empirical methods in natural language processing (EMNLP 2015), Lisbon, Portugal, 17–21 Sept 2015
19. C-C Liang, Hsu K-Y, Huang C-T, Li C-M, Miao S-Y, Su K-Y (2016) A tag-based English Math word problem solver with understanding, reasoning and explanation. In: Proceedings of the demonstrations session, the 2016 conference of the North American chapter of the association for computational linguistics: human language technologies (NAACL HLT 2016), San Diego, CA, USA, 12–17 June 2016, pp 67–71
20. Liang C-C, Tsai S-H, Chang T-Y, Lin Y-C, Su K-Y (2016) A meaning-based English Math word problem solver with understanding, reasoning and explanation. In: Proceedings of the conference system demonstrations, 26th international conference on computational linguistics (COLING 2016), Osaka, Japan, 11–16 Dec 2016, pp 151–155
21. Mitra A, Baral C (2016) Learning to use formulas to solve simple arithmetic problems. In: Proceedings of the 54th annual meeting of the association for computational linguistics (ACL 2016), Berlin, Germany, 7–12 Aug 2016, pp 2144–2153
22. Mandal S, Naskar SK (2019) Solving arithmetic word problems by object oriented modeling and query-based information processing. Int J Artif Intell Tools 28(4):1940002:1–1940002:23 (2019)
23. Huang D, Shi S, Lin C-Y, Yin J, Ma W-Y (2016) How well do computers solve math word problems? large-scale dataset construction and evaluation. In: Proceedings of the 54th annual meeting of the association for computational linguistics, (ACL 2016), Berlin, Germany, 7–12 Aug 7-12 2016, pp 887–896
24. Wang Y, Liu X, Shi S (2017) Deep neural solver for math word problems. In: Proceedings of the 2017 conference on empirical methods in natural language processing (EMNLP 2017), Copenhagen, Denmark, 9–11 Sep 2017, pp 845–854
25. Huang D, Liu J, Lin C-Y, Yin J (2018) Neural math word problem solver with reinforcement learning. In: Proceedings of the 27th international conference on computational linguistics, COLING 2018, Santa Fe, New Mexico, USA, 20–26 Aug 2018
26. Huang D, Yao J-G, Lin C-Y, Zhou Q, Yin J (2018) Using intermediate representations to solve math word problems. In: Proceedings of the 56th annual meeting of the association for computational linguistics (ACL 2018), Melbourne, Australia, 15–20 July 2018, pp 419–428
27. Wang L, Wang Y, Cai D, Zhang D, Liu X (2018) Translating math word problem to expression tree. In: Proceedings of the 2018 conference on empirical methods in natural language processing, Brussels, Belgium, 31 Oct–4 Nov 2018, pp 1064–1069
28. Wang L, Zhang D, Gao L, Song J, Guo L, Shen HT (2018) Mathdqn: solving arithmetic word problems via deep reinforcement learning. In: Proceedings of the thirty-second AAAI conference on artificial intelligence, (AAAI-18), New Orleans, Louisiana, USA, 2–7 Feb 2018, pp 5545–5552
29. Mandal S, Ahmed Sekh A, Naskar SK (2020) Solving arithmetic word problems: a deep learning based approach. J Intell Fuzzy Syst 39(2):2521–2531
30. Koncel-Kedziorski R, Roy S, Amini A, Kushman N, Hajishirzi H (2016) MAWPS: a math word problem repository. In: Proceedings of The 2016 conference of the North American chapter of the association for computational linguistics: human language technologies (NAACL HLT 2016), San Diego, California, USA, 12–17 June 2016, pp 1152–1157

31. Mikolov T, Chen K, Corrado G, Dean J (2013) Efficient estimation of word representations in vector space. In: Bengio Y, LeCun Y (eds) 1st International conference on learning representations, ICLR 2013. Arizona, USA, May, Scottsdale, pp 2–4
32. Greff K, Srivastava RK, Koutník J, Steunebrink BR, Schmidhuber J (2017) LSTM: a search space odyssey. IEEE Trans Neural Netw Learn Syst 28(10):2222–2232 (2017)

Time Synchronization in Underwater Wireless Sensor Networks: A Survey

Amrita Datta and Mou Dasgupta

Abstract Time synchronization in underwater sensor networks (UWS-Ns) has come up as an active area of research in recent years. The drastic differences in physical and chemical characteristics of underwater environments as compared to terrestrial environments have necessitated that underwater time synchronization protocols be different from their terrestrial counterparts. The protocols reviewed in this survey range from static to dynamic, doppler assisted, and ordinary least squares-based protocols. This paper is an attempt to review ten such UWSN time synchronization protocols that are in extant literature and elaborate on the advantages and disadvantages of each of them.

Keywords UWSNs · Time synchronization

1 Introduction

Under Water Sensor Networks (UWSNs) have found widespread use owing to their applications in civilian and military domain [1, 2]. These applications include undersea exploration, habitat monitoring, target tracking, disaster prevention, military surveillance, etc. Time synchronization, which pertains to sensors in a network working on a common time scale, is a very important component of a UWSN. Without proper time synchronization the sensor devices would not be able to work in tandem with each other in an accurate manner. As data in a network flows from multiple sensor devices, the need for a correct global time becomes necessary for data merging.

Environment underwater is different from a terrestrial environment. The radio waves do not propagate well in seawater. Soundwaves that travel underwater are much slower than radio waves. As a result, time synchronization protocols that are used in terrestrial networks [3, 4] cannot directly be ported to underwater sensor networks. The nature of underwater environment is such that there is long propagation delays and water currents and waves make sensor nodes mobile [5, 6].

A. Datta (✉) · M. Dasgupta
National Institute of Technology Raipur, Raipur, India

© The Author(s), under exclusive license to Springer Nature Singapore Pte Ltd. 2022

D. Giri et al. (eds.), *Proceedings of the Seventh International Conference on Mathematics and Computing*, Advances in Intelligent Systems and Computing 1412,
https://doi.org/10.1007/978-981-16-6890-6_32

This propagation delay is because speed of sound underwater is five orders of magnitude slower than radio waves in air or vacuum [7, 8]. This propagation delay also depends on the chemical and physical characteristics of underwater environment like pressure, temperature, node mobility, salinity, etc. These are variables and their accurate estimation in real time is difficult. The mobility of sensor nodes in UWSNs is due to wind and ocean current action [9]. On an average, a sensor node can move around with currents at the rate of 3-6 km/hr [10]. In UWSNs where Autonomous Underwater Vehicles (AUV) are deployed, the movement of AUV also needs to be considered [11] as Doppler shift must be considered in such cases [12].

Underwater sensor networks also suffer many constraints like limited bandwidth and low transmission rate [13]. Such constraints compel communications to take place at low carrier frequency that have low energy and require a lot of power [12]. This type of communication takes place is a narrow bandwidth [11]. The problems and constraints in UWSN like propagation delay, limited bandwidth, ambient characteristics, etc. make time synchronization challenging, and many protocols have been proposed which this paper discusses in the subsequent sections.

The rest of the paper is arranged as follows. Section 2 gives a brief overview of the time synchronization in underwater. Section 3 explains the existing approaches of time synchronization. Section 4 comparative study and Sect. 5 concludes the paper.

2 Background of Time Synchronization

Extant time synchronization models in UWSNs make use of message exchanges between sensor nodes. Underwater environments are dynamic owing to physical and chemical characteristics that can change from time to time. Such non-deterministic characteristics result in inconsistent time stamps when sensor nodes communicate with each other. Koeptz and Schwab [14, 15] outline common sources of error which leads to unsynchronized time stamps in sensor to sensor communication-

- *Send time*: It is a highly non-deterministic parameter dependent on operating system and current load. This time is taken by the sensor nodes in constructing a message that is sent to the MAC layer.
- *Access time*: Sensor nodes may not be able to immediately access the channel and send information. Access time represents the delay they face before they can begin communicating through the channel. The current network traffic and the nature of the MAC protocol influence this least deterministic parameter. Physical layer time stamping can be used to eliminate this error.
- *Propagation time*: The time taken for the information to reach from one node to another is propagation time. This depends on the distance between two nodes which are communicating with each other. In terrestrial network, propagation time is not much of an issue as the sheer speed of radio waves mitigates any synchronization error.

- *Receive time*: This indicates the time taken to process the notification of an incoming message by the synchronizing nodes.
- *Byte Alignment Time*: This deterministic parameter arises due to any change in byte alignment at the receiver.

Most of the time synchronization protocols that have been proposed in context of terrestrial wireless sensor networks can be broadly divided into two categories—receiver-receiver and sender-receiver-based approaches. For example, Reference Broadcast Synchronization [16] falls in the former category while Light Weight Tree-Based Synchronization (LTS) [17] falls in the latter category. However, in UWSN, the approaches are categorized into static or dynamic. For example, TSHL [18] and WATER-Sync [19] falls under static approach, whereas, Mu-Sync [20] and Mc-Sync [21] falls under the dynamic approach.

3 Existing Approaches of Time Synchronization

UWSNs have received considerable attention over the last few years. Research in time synchronization in UWSNs has received lesser attention as compared to other issues pertaining to UWSNs. In this section we will discuss several time synchronization protocols in both static approach and dynamic approach categories.

3.1 Static Approaches

3.1.1 TSHL

This time synchronization protocol has been designed for UWSNs exhibiting high latency. The protocol can handle long propagation delays. At the same time energy efficiency is also maintained. The basic setup involves two phases [18]-sensor nodes estimate the clock skew in the first phase and in the second phase, message exchange takes place between the sensor nodes to estimate the offset. These messages exchange is essentially a swap of skew compensated synchronized messages (as shown in Fig. 1).

3.1.2 Water-Sync

WATER-Sync, an acronym for water-quality monitoring sensor network [19] has been proposed for shallow water bodies and for military applications. WATER-Sync is a correlation-based UWSN time synchronization protocol. This protocol seeks to integrate the time synchronization procedure with the network routing topology. The routing topology has a tree-like architecture with the surface station at the top being

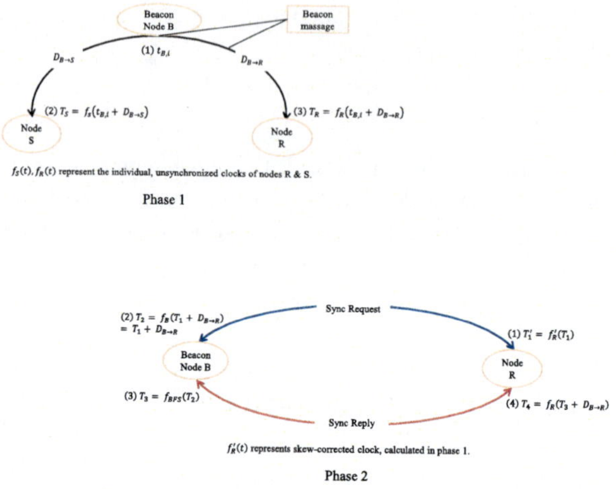

Fig. 1 TSHL time synchronization phases [18]

the root of the tree. Gradual Depth Timing (GDT) and Level 1 Skew Compensation (LSC) are the two stages proposed in this protocol. GDT makes use of the surface station (root) as the initiation of the synchronization process. The two steps use the same channel for data transmission and due to energy considerations, LSC phase is prevented from occurring frequently.

3.2 Dynamic Approaches

3.2.1 MU-Sync

Among cluster-based algorithms for time synchronization in UWSNs, Mu-Sync [20] (shown in Fig. 2) uses linear regression technique for accurate time synchronization. As Mu-Sync determines both clock skew and offset in the regression step, frequent re-synchronization is avoided. After message exchange between sensor nodes is completed and local time information is obtained, Mu-Sync runs linear regression twice. The first regression is run to determine the effect of variable propagation delay. The second regression is run to estimate the clock skew and clock offset. Since Mu-Sync is cluster-based protocol, it can be used in UWSNs that employ multi-hop architecture.

3.2.2 Mc-Sync

Mobile Counteracted Time Synchronization Approach [21] considers node mobility. Mc-Sync makes use of two reference nodes that are deployed in the direction of the

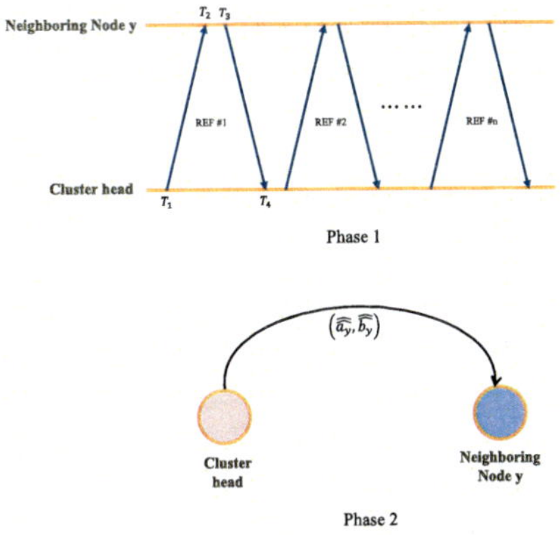

Fig. 2 MU-Sync time synchronization phases [20]

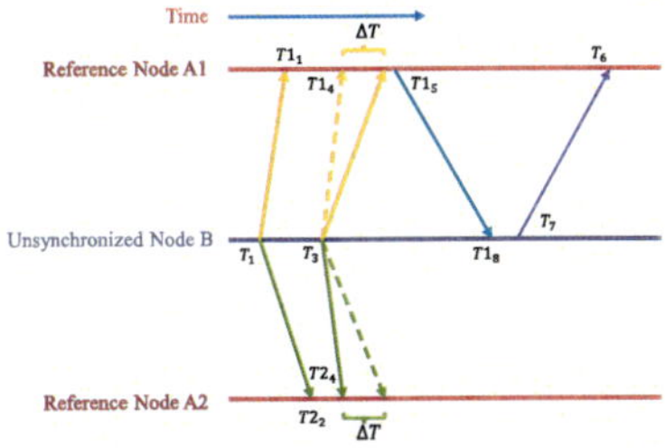

Fig. 3 Mobile counteracted time synchronization [21]

ocean current and the unsynchronized node lies on the line joining the two reference nodes (shown in Fig. 3). One of these reference nodes can move autonomously, while the other reference node, which is joined to the autonomously moving node by a light cable, moves passively with ocean current. Once deployed at the opposite ends of a target area, the reference nodes move through the sensor nodes in that area and keep on synchronizing the time.

3.2.3 E²DTS

The main focus of this paper is proposing an energy efficiency distributed time synchronization [22] algorithm in underwater acoustic node mobility networks. E^2DTS is a time synchronization protocol which is energy efficient and is based on distributed architecture. It can be used for time synchronization in conditions of node mobility. The clock skew and clock offset are estimated by considering both node mobility and propagation delay. A mobile beacon moves around the network sending reference packets (shown in Fig. 4). A neighboring node receives these packets and marks their local time. Once enough packets have been received, clock skew is calculated. Once skew synchronized, the node then estimates clock offset again by receiving messages. Since receiving messages consumes lesser energy that sending messages, E^2DTS is energy efficient as in this unidirectional message exchange protocol, half the energy gets conserved as compared to bidirectional message exchange.

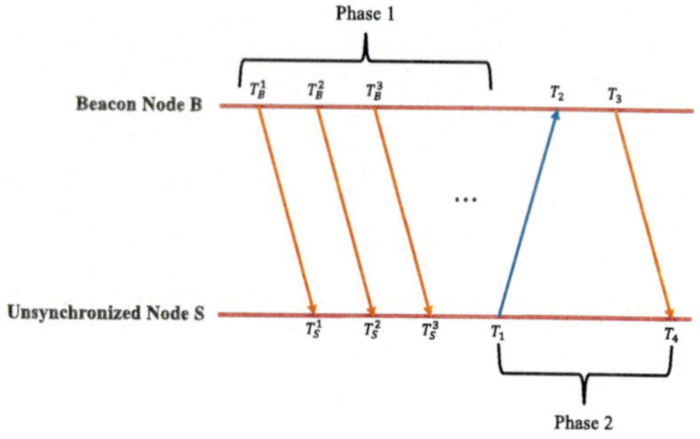

Fig. 4 E^2DTS two-phase message exchange [22]

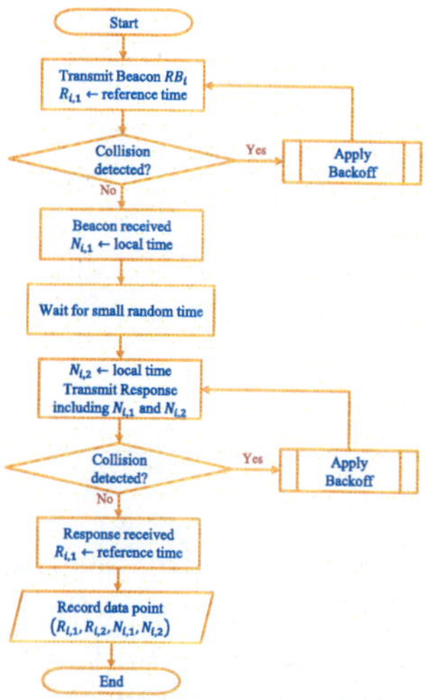

Fig. 5 RUSN time synchronization [23]

3.2.4 RSUN

Robust Synchronization for underwater networks [23] focuses on packet collision. The paper proposes that packet collision is an often-ignored aspect in USWN time synchronization protocols. When there is network traffic, packet collision leads to loss of data packets and synchronization process must wait for a random amount of time till resolution is obtained. Also, this leads to inconsistent time stamps which if not removed can lead to inaccurate time synchronization. RSUN protocol (flowchart shown in Fig. 5) works in three steps. First, the time stamps are received. Second, the time stamps are analyzed for inconsistency and outliers are removed using Cook's distance. And finally, linear regression is run to estimate clock skew and offset.

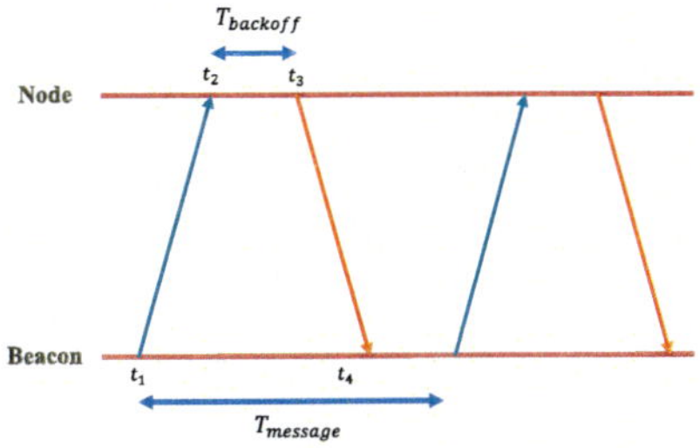

Fig. 6 D-Sync message scheme [6]

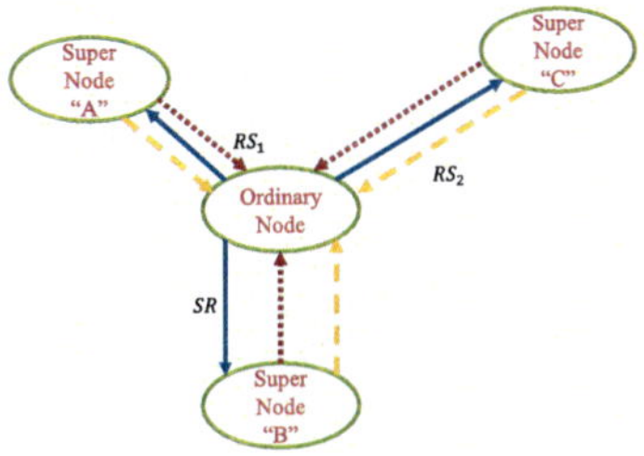

Fig. 7 Message exchange for Mobi-Sync [5]

3.2.5 D-Sync

D-Sync [6] (shown in Fig. 6) takes advantage of the Doppler Shift phenomenon to synchronize node time in UWSN. Doppler shift happens when there is a frequency shift in waves between source and observer one or both of which could be moving. Doppler Shift estimated time synchronization, however, does not consider clock skew while determining Doppler scaling factor and the accuracy is completely dependent on measurement of the shift. As such, time synchronization could be prone to errors.

3.2.6 Mobi-Sync

A first of its kind protocol that considers the spatial correlation between the nodes in a UWSN [5]. The sensor nodes underwater can move randomly due to currents and waves. Once spatial correlation in such movements is determined, long dynamic propagation delay can be estimated much more accurately. Mobi-Sync (shown in Fig. 7) protocol synchronizes time using the following three phases. In the first step of first phase, messages are exchanged where ordinary nodes send out request messages to super nodes, which then respond back. Then delay calculation, which is completed after considering spatial correlation between mobile nodes. Once, enough data points are obtained by ordinary nodes, linear regression takes place. This completes the first phase of the protocol. In the second phase, weighted least squares estimation is used to relax the assumptions that were made in the previous phase. The final phase involves calibration to improve accuracy of the estimation where a second round of ordinary least squares regression is used to estimate clock skew and offset.

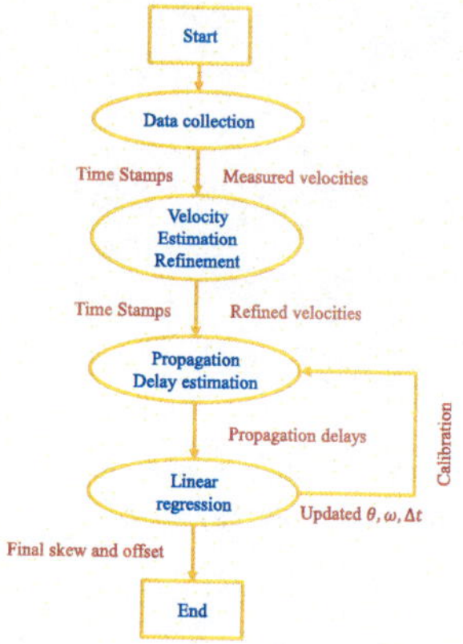

Fig. 8 DA-Sync message scheme [24]

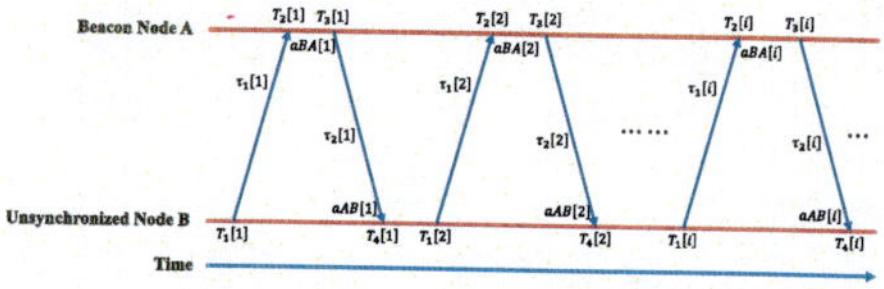

Fig. 9 DE-Sync massage exchange scheme [25]

3.2.7 DA-Sync

Doppler Assisted time synchronization for UWSN unlike other UWSN time synchronization protocols works in the MAC cross layer. Doppler Shift estimation followed by accuracy maintenance using Kalman filter is done in DA-Sync. The node mobility necessitates the protocol to use Doppler scaling factor algorithm which is further attuned using Kalman filter. In DA-Sync [24], accuracy improves a lot. The fact that time stamp information from the MAC layer is used, further reduces non-deterministic errors in time synchronization Fig. 8 depicts the DA-Sync protocol flowchart.

3.2.8 DE-Sync

This protocol (shown in Fig. 9) exhibits the following features-First, DE-Sync [25] does not measure sound speed, which depend on multiple non-deterministic factors like salinity, temperature, etc., thereby reducing the error in estimation of clock offset and clock skew. Secondly, while determine the Doppler scaling factor, clock skew is considered. This results in more accurate estimation of the scaling factor and subsequent time synchronization with a high degree of accuracy. And thirdly, rather than being contained in relative motion of nodes, is contained in the messages that are exchanged between nodes, thereby reducing computation overhead.

3.2.9 BE-Sync

BE-Sync [26] takes into account packet collisions and proposes a back-off technique using non-homogenous Poisson process to ensure time synchronization of sensor nodes.

4 Comparative Study of Time Synchronization

UWSN have found a multitude of applications and hence numerous time synchronization protocols have been proposed, each with its own unique advantages and disadvantages. Table 1 gives a ready reckoner on the advantages and disadvantages of each protocol which would help one to decide on which might best suit a scenario or application. This is a combined comparison study in brief which would help the people working in the domain to understand the nature of the protocols. This would also help them to solve major issues at root level should they arise.

Table 1 Advantages and disadvantages of reviewed time synchronization

Time-Sync for UWSNs	Advantages	Disadvantages
TSHL	– Minimal energy cost	– It cannot be handled mobile sensor nodes.
MU-Sync	– To solve sensor nodes mobility Energy Conservation	Extra error creates – Experiment time is very long
Mc-Sync	– Eliminate the problems of node mobility and improve the synchronization	– No energy efficient – It uses many two-way messages exchange
E^2DTS	– Reduce the synchronization error	– There is no mention of how energy efficiency is calculated
RUSN	– Removed the inconsistent time stamps	
D-Sync	– Measure speed with Doppler shift	– Does not consider the effect of the skew during the process of estimating Doppler scaling factor
DA-Sync	– Reduce the non-deterministic errors	– Incorporates the moving in between sensor nodes
DE-Sync	– DE-Sync is a high accuracy time synchronization	– Compared to the existing schemes, DE-Sync does not directly use the Doppler scale factor to calculate the relative speed between node
Mobi-Sync	– Calibration procedure more efficient as compare to MU-Sync – More energy efficient as compare to MU-Sync – Achieve more high accuracy as compare to TSHL and MU-Sync	– Message overhead affects the accuracy of Mobi- Sync that is more messages exchanges provides more sample data in which we need to conduct linear regression to each of these samples

5 Conclusion

The current paper has reviewed UWSN time synchronization protocols and their accompanying advantages and disadvantages. Two of the proposed protocols were static while the rest were dynamic. The ever-changing nature of aquatic environment makes dynamic protocols more desirable over static ones. Most protocols seek to balance the trade-off between energy efficiency and synchronization accuracy. Time synchronization in UWSNs is an active area of research and has seen protocols that used novel techniques that seek to improve accuracy in time synchronization.

References

1. Yifan W, Zaichen Z, Guangguo B (2010) An overview on underwater acoustic sensor networks. In: Proceedings of the 2010 17th international conference on telecommunications, Doha, pp 779–783. https://doi.org/10.1109/ICTEL.2010.5478812
2. J. Heidemann, Ye W, Wills J, Syed A, Li Y (2006) Research challenges and applications for underwater sensor networking. In: Proceedings of the IEEE wireless communications and networking conference (2006) WCNC 2006, Las Vegas, NV, pp 228–235. https://doi.org/10.1109/WCNC.2006.1683469
3. Lasassmeh SM, Conrad JM, Time synchronization in wireless sensor networks: a survey. In: Proceedings of the IEEE SoutheastCon. Concord, NC 2010, pp 242–245. https://doi.org/10.1109/SECON.2010.5453878
4. Wu Y, Chaudhari Q, Serpedin E (2011) Clock synchronization of wireless sensor networks. IEEE Signal Proc Mag 28(1):124–138. https://doi.org/10.1109/MSP.2010.938757 Jan.
5. Liu J, Zhou Z, Peng Z, Cui J, Zuba M, Fiondella L (2013) Mobi-Sync: efficient time synchronization for mobile underwater sensor networks. IEEE Trans Parallel Distrib Syst 24(2):406–416. https://doi.org/10.1109/TPDS.2012.71 Feb.
6. Lu F, Mirza D, Schurger C (2010) D-Sync: doppler-based time synchronization for mobile underwater sensor networks. In: Proceedings of the WUWNet '10: proceedings of the fifth ACM international workshop on underwater networks, Woods Hole, MA, pp 1–8, Sept 2010
7. Guo Y, Zhang Z (2014) Clock synchronization study for large scale underwater sensor networks. J Electron Inf Technol 36(3):1498–1503
8. Huang SH, Yang TC, Tsao J (2013) Tracking time-varying correlated underwater acoustic channels in the signal subspace. In: Proceedings of the Eighth ACM international conference on underwater networks and systems
9. Caruso A, Paparella F, Vieira LFM, Erol M, Gerla M (2008) The meandering current mobility model and its impact on underwater mobile sensor networks. In: Proceedings of the IEEE INFOCOM 2008—the 27th conference on computer communications, pp 221–225. https://doi.org/10.1109/INFOCOM.2008.53
10. Parrish N, Roy S, Arabshahi P (2009) Symbol by symbol Doppler rate estimation for highly mobile underwater OFDM. In: Proceedings of the fourth ACM international workshop on underwater networks, New York, NY, USA, pp 1–8
11. Qarabaqi P, Stojanovic M (2012) Small scale characterization of underwater acoustic channels. In: Proceedings of the seventh ACM international conference on underwater networks and systems
12. Zhou Z, Peng Z, Cui J, Shi Z (2011) Efficient multipath communication for time-critical applications in underwater acoustic sensor networks. IEEE/ACM Trans Netw 19(1):28–41. https://doi.org/10.1109/TNET.2010.2055886 Feb.

13. Stojanovic M (2007) On the relationship between capacity and distance in an underwater acoustic communication channel. In: Proceedings of the ACM SIGMOBILE mobile computing and communications review (MC2R), vol 11, no 4, pp 34–43
14. Kopetz H, Schwabl W (1989) Global time in distributed real-time system, Technical Report 15/89, TechnischeUniversit at Wien
15. Kopetz H, Schwabl W (1987) Clock synchronization in distributed real-time systems. IEEE Trans Comput 36(8):933–939 Aug.
16. Elson J, Girod L, Estrin D (2002) Fine-grained network time synchronization using reference broadcasts. In: Proceedings of the ACM SIGOPS operating systems review, Boston, MA, USA, pp 147–163
17. Greunen JV, Rabaey J (2003) Lightweight time synchronization for sensor networks. In: Proceedings of the 2nd ACM international conference on wireless sensor networks and applications, Sept 2003, pp 11–19
18. Syed AA, Heidemann J (2006) Time synchronization for high latency acoustic networks. In: Proceedings of the IEEE INFOCOM 2006, 25TH IEEE international conference on computer communications, Barcelona, pp 1–12. https://doi.org/10.1109/INFOCOM.2006.161
19. Hu F, Wilson S, Xiao Y (2008) Correlation-Based security in time synchronization of sensor networks. In: Proceedings of the IEEE wireless communications and networking conference, Las Vegas, NV 2008, pp 2525–2530. https://doi.org/10.1109/WCNC.2008.444
20. Chirdchoo N, Soh WS, Chua KC (2008) MU-Sync: a time synchronization protocol for underwater mobile networks. In: Proceedings of the WUWNet08, pp 35–42
21. Guo Y, Liu Y (2013) Time synchronization for mobile underwater sensor networks. J Netw 8(1):116–123
22. Li Z, Guo Z, Hong F, Hong L (2013) E2DTS: an energy efficiency distributed time synchronization algorithm for underwater acoustic mobile sensor networks. Ad Hoc Netw 11:1372–1380. https://doi.org/10.1016/j.adhoc.2011.03.008
23. Khandoker TI, Fang L, Huang D, Sreeram V (2012) A robust time synchronization approach for underwater networks with inconsistent timestamps. In: Proceedings of the international symposium on communications and information technologies (ISCIT), Gold Coast, QLD, pp 1086–1091. https://doi.org/10.1109/ISCIT.2012.6380854
24. Liu J, Wang Z, Zuba M, Peng Z, Cui J, Zhou S (2014) DA-Sync: a doppler-assisted time-synchronization scheme for mobile underwater sensor networks. IEEE Trans Mob Comput 13(3):582–595. https://doi.org/10.1109/TMC.2013.13 March
25. Zhou F, Wang Q, Nie D, Qiao G (2018) DE-Sync: a doppler-enhanced time synchronization for mobile underwater sensor networks. Sensors 18(6):1710. https://doi.org/10.3390/s18061710
26. Datta A, Dasgupta M (2020) BE-Sync: a bandwidth efficient time synchronization for underwater wireless sensor networks. In: Proceeding 3rd international conference on VLSI, communication and signal processing (VCAS 2020), 09th–11th Oct 2020 at MNNIT Allahabad, Prayagraj (UP), India

Stock Market Predictions Using FastRNN-Based Model

Konark Yadav, Milind Yadav, and Sandeep Saini

Abstract Predicting the correct values of stocks in fast fluctuating high-frequency financial data is always a challenging task. Existing state-of-the-art models are very efficient in terms of accuracy but lags in prediction speed. In this work, we aim to develop a deep-learning-based fast model for live predictions of stock values with minimum errors. The proposed model is based on fast recurrent neural networks (FastRNNs), which provides us with both of the desired features. We have considered the 1-min time interval stock data of four companies for a period of one day. The model is aimed to have a low computational complexity as well so that it can be run for live predictions as well. The model's performance is measured by root mean square error (RMSE) along with computation time. The model outperforms LSTM, CNN, and other deep learning models for live predictions of stock values.

Keywords Fast recurrent neural network (FastRNN) · Convolutional neural network (CNN) · Time series forecasting · Root mean square error (RMSE) · Stock price prediction

1 Introduction

Every economy around the globe is more and more interconnected with each other nowadays. The role of multinational companies in this dependence is very vital. Such companies are listed on the most values stock exchanges and the fluctuations in the stock values of such company cause tremors around the globe. Thus, forecasting the stock values is becoming more crucial now. Forecasting is the process of predicting

K. Yadav · S. Saini (✉)
Department of Electronics and Communication Engineering, The LNM Institute of Information Technology, Jaipur, India
e-mail: sandeep.saini@lnmiit.ac.in

M. Yadav
Department of Computer Science and Engineering, Rajasthan Technical University, Kota, RJ, India

© The Author(s), under exclusive license to Springer Nature Singapore Pte Ltd. 2022
D. Giri et al. (eds.), *Proceedings of the Seventh International Conference on Mathematics and Computing*, Advances in Intelligent Systems and Computing 1412,
https://doi.org/10.1007/978-981-16-6890-6_33

the future value of any series by considering the previous patterns or long historical data. This is also known as time series forecasting. For example, if the price of gold is increasing every year at Christmas time, then we can predict a similar trend for the current year as well and plan the purchase well in advance to avoid the high rates at Christmas time. Similarly, computational models can help us in predicting the weather for the next day, week, or months as well. With the high volume of money involved, stock market values have attracted the attention of computer scientists as well to design models and architectures for precise stock value prediction. A lot of such systems have been developed with high accuracies during the past decades as well.

Stocks are not a simple time series with only one factor affecting the outcome. These can be either univariate or multivariate. Univariate stocks are rare and they are dependent on only one factor or only one company's performance. With the emerging partnerships and dependence of every big company on its partner's stocks, the second type of stock, i.e. multivariate, is more common now. So in such cases, the prediction of exact future stock values can help a lot of investors and stakeholders. This is the motivation behind our proposed model.

The first model to predict the outcome of a time series was first proposed by Ahmad and Cook [2] in 1979. In this work, auto regressive integrated moving average (ARIMA) model was introduced which is one of the most trusted models for time series forecasting even now. This is also a reason behind the fact that a lot of conventional models are based on auto regression (AR) and moving average (MA) and exponential smoothing [20], and GARCH models [15]. The existing stock market prediction methods can be classified as follows:

1. **Fundamental analysis** is concerned with the company that underlies the stock itself. In such methods, the historical performance, as well as the credibility of the company's accounts, is evaluated to predict future stock values. These methods are stochastic in nature and have limited options to consider unprecedented events to predict the stock in such cases.
2. **Technical analysis**-based methods are not considering the working principles of the company but mostly consider the trends and historical stock values. AR, MA, ARIMA, and similar methods are popular methods in this category. These methods are more effective for short-term predictions rather than long-term predictions.
3. **Machine-learning-based (Time Series Forecasting)** methods are the state-of-the-art models for stock market predictions. These methods are based on artificial neural networks (ANNs) and genetic algorithms (GA). In recent years, recurrent neural network (RNN) has become the best ANN for time series forecasting.

We have focused on the deep-learning-based approaches in this work. Deep-learning-based model are best suited for applications like time series forecasting, text and language processing [29, 30, 35], and image processing [26]. We have studied the latest models and architectures proposed explicitly for stock market prediction. We have focused on two performance parameters, i.e. execution time and RMSE. A model can be very accurate in prediction but very slow in computing the output. Such a model cannot be applied for live stock value prediction but very useful for long-term predictions. On the other hand, a faster model with acceptable RMSE can be utilized for live predictions as well. The proposed model is based on fast-RNN architecture. This model is designed to provide faster and accurate results. The rest of the paper is organized as follows. We have provided the literature review of the existing machine-learning and deep-learning-based models for stock market prediction in Sect. 2. We have focused on a single-network-based as well as hybrid models in this section. The proposed model is described in complete detail in Sect. 3. We have explained the mathematical model as well as structural augmentations made to implement this model in the same. Section 4 consists of details of our experimental setup, datasets, and simulation results for short and long-term stock price prediction. We conclude our work in Sect. 5.

2 Literature Review

We have described that there are three major methods for stock market prediction. We have focused only on machine learning and more specifically on deep-learning-based approaches in our study. One of the earliest applications of machine learning models for stock market forecasting was from Francis et al. [31] in 2001. The authors used a very robust machine learning algorithm, i.e. support vector machine (SVM) to verify the feasibility of machine learning algorithms for financial series forecasting. The promising results motivated a lot of research groups to explore more machine learning algorithms for the same. In 2003, Egeli et al. [12] proposed a multi-layer perceptron and generalized feed-forward network-based architecture for financial forecasting. These models were trained and tested on moving averages for 5 and 10-day periods. The results outperformed conventional ARIMA-based approaches. In 2005 David et al. [13] combined conventional data mining algorithms with neural networks to predict stock values. The authors used the probabilistic neural network in their model that was having three feed-forward layers.

With the encouraging results from single-network-based models for time series forecasting, hybrid models were also explored for the purpose. In 2001, Abraham et al. [1] proposed one of the first hybrid models that were based on neuro-fuzzy logic and ANN. In 2005, Armano et al. [6] proposed the model with hybridization of a genetic classifier designed to control the activation of a feed-forward artificial neural network for performing a locally scoped forecasting activity. In 2007, Fu et al. [16] analyzed the problem based on a fuzzy cerebellar model articulation controller–Bayesian Ying Yang (FCMAC-BYY) neural network. This proposed model was motivated by the Chinese ancient Ying-Yang philosophy that states that everything in the universe can be viewed as a product of constant conflict between opposites (Ying and Yang). In 2008, Choudhary et al. [10] proposed the GA-SVM model for stock market prediction. The model is a hybrid version of the genetic algorithm and support vector machines. The system was tested on three of India's biggest companies, i.e. TCS, Infosys, and RIL that had the trading data of 1386 days from

2002 to 2008. Further, it was tested on the data of 30 companies. The hybrid model outperformed the SVM-based models in terms of RMSE. In 2009, Tsai et al. [33] also performed the stock price forecasting by using a hybrid model designed using a decision tree (DT) and artificial neural networks (ANNs). The proposed model, DT-ANN, had an accuracy of 77% and was among the most precise systems at that time.

In the last decade, the focus has shifted to deep-learning-based models. Ding et al. [11] developed an event-driven deep learning model. In this model, a variant of convolutional neural network (CNN), i.e. deep CNN was used to predict the stock values. The events are extracted from the news articles and stored as dense vectors, trained using a novel neural tensor network. This dense network was trained using the deep CNN. Chen et al. [8] used RNN boost for stock market prediction. Pawar et al. used both LSTM as well as RNN for the same [27]. Akita et al. [3] applied deep learning models such as paragraph vector and long short-term memory (LSTM), to financial time series forecasting. The model also utilized the news article data and converted those into the paragraph vector which was then fed to LSTM to predict the stock prices. This model was texted on the Tokyo Stock Exchange. Fischer et al. [14] also proposed a similar LSTM-based architecture. The model outperformed random forest (RAF), a deep neural net (DNN), and a logistic regression classifier-based (LOG) models. Convolutional neural network (CNN) has been employed by several researchers for the problem. Gudelek et al. [18] enhanced the regular CNN model to the two-dimensional model and employed it for the task. The first convolutional layer had 32 filters of size 28×28 and the second one had 64 filters of the same sizes. This improved the RMSE values. In the next section, we explain our proposed model which aims at improving the prediction speed of the current state-of-the-art models by maintaining the RMSE values as well.

3 Proposed Model

In designing a stock market price prediction model, the foremost requirement is the availability of a suitable dataset. We have considered four companies for our study and those are Facebook Inc., Uber Inc., Apple Inc., and Nike Inc. from the New York Stock Exchange (NYSE). We have obtained the stock values from Yahoo finance.[1] The dataset includes information about day stamp, time stamp, transaction id, the stock price (open and close), and volume of stock sold in each minute interval. For our model, we have used the close price for each stock. Our work also aims on creating a prototype for live prediction. We consider a working duration of 8 h and divide those hours into training and testing time. We are predicting the future prices of each minute for the next 50 min by keeping the initial 7 h 10 min data in training. The best window length was identified by calculating the root mean square error for various window sizes.

[1] https://finance.yahoo.com.

We have kept the size of the data the same for all the stocks, i.e. each stock has 430 rows and trained the model on 40 epochs. We have considered the error and computation time of each model for our study. If the loss (mean squared error) for the current epoch is less than the value obtained from the previous epoch, the weight matrix for that particular epoch is stored. After the completion of the training process, each of these models was tested on the remaining 50 values. In this process the model with the least root mean squared error (RMSE) is taken as the final model for prediction.

To compare our model with existing similar models, we have initially considered a few baseline models and then the state-of-the-art models. We have considered the following baseline models for our study.

1. **ARIMA** is a widely known model for time series forecasting that works on the concept of moving average. It is based on the trend stationarity, seasonality of the data. Here, we have to first make the input data stationary and then after finding their auto-correlation values and partial auto-correlation, we would be able to make forecasting. The only glitch with this model is that it can't be automated for all kinds of stocks as we have to set different p, d, q (number of auto regressive terms, number of non-seasonal differences needed for stationarity, number of lagged forecast errors) values for different stocks.
2. **Long short-term memory (LSTM)** is a well-established model used for time series forecasting and frequently applied for stock market price prediction [7, 17]. It is a derived RNN with forget gate functionality. LSTMs are very effective for long as well as short-term predictions.
3. **FBProphet** is another time series forecasting model introduced by Facebook Inc. [32]. This model required very less computational time in comparison to other models [25].
4. **A hybrid model of convolutional neural network and LSTM:**. This is a hybrid model designed recently and extensively used in stock price prediction. The model consists of both LSTM and CNN layers [22].

3.1 Proposed FastRNN-Based Model

FastRNN was proposed by Microsoft in 2018 [23]. The model is based on the residual network (ResNet) [19]. If we look at the structure of ResNet, then in this model, every network layer is not stacked over each other but they form a residual mapping which is always easier to optimize. The model is based on the VGG model and modified by adding residual layers. ResNet is trained on 25.5 million parameters. This makes ResNet a very heavy network as well. FastRNN exploits the learned residual connections of ResNet and provides stable training along with comparable training accuracies at a much lower computational cost. The prediction accuracies of FastRNN are 19% higher than regular RNN structures and a little less than the gated RNN.

Let X = [x_1, ..., x_T] be the input vector where $x_t \epsilon R^D$ denotes the t-th step feature vector. So, for a multi-class RNN we aim to learn a function $F : R^{DxT} 1, ..., L$ that will predict one of these L classes for a given data point X. The standard RNN architecture hidden state vectors are represented by:

$$h_t = tanh(Wx_t + Uh_{t-1} + b) \quad (1)$$

Learning U, W in the above architecture is difficult as the gradient can have an exponentially large (in T) condition number. Unitary network-based methods can solve this problem by expanding the network size and in this process, they become significantly large and the model accuracy may not remain the same.

However, FastRNN uses a simple weighted residual connection to stabilize the training by generating well-conditioned gradients. In particular, FastRNN updates the hidden state ht as follows:

$$h'_t = \sigma(Wx_t + Uh_{t-1} + b) \quad (2)$$

$$h_t = \alpha.h'_t + \beta.h_{t-1} \quad (3)$$

Here α and β are trainable weights which are parameterized by sigmoid function and limited in the range $0 < \alpha, \beta < 1$. σ is a nonlinear function such as tanh, sigmoid, or ReLU, and can vary across datasets.

FastRNN updates hidden state in a controlled manner with α, β limiting the extent to which the current feature vector xt updates the hidden state. Also, FastRNN has only two more parameters than RNN and requires fewer computations, which is a tiny fraction of the per-step computation complexity of RNN. Unlike unitary methods [5, 21, 36], FastRNN does not introduce expensive structural constraints on U and hence scales well to large datasets with standard optimization techniques. The proposed FastRNN-based model for stock market predictions is shown in Fig. 1.

The model is tested on the stock values of four companies. The experimental setup, environment and the results are elaborated in the next section.

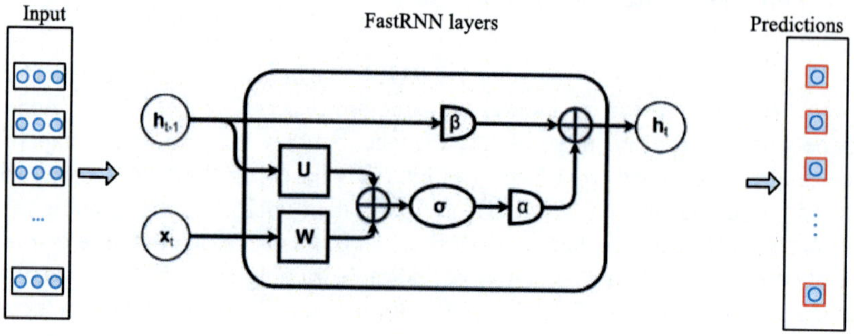

Fig. 1 FastRNN-based proposed model for faster stock values prediction

Table 1 Root mean square error (RMSE) and computation time calculated for the state-of-the-art and proposed model for Apple Inc. stock values

Model Name	RMSE value	Time (in sec)
ARIMA [4]	0.796109	1.63
BiLSTM_Attention_CNN_BiLSTM [34]	0.234644	25.72292113
CNN_LSTM_Attention_LSTM [37]	0.214821	16.00164294
FBProphet [9]	0.935556	0.659962893
LSTM [28]	0.228731	13.28157353
LSTM_Attention_CNN_BiLSTM [34]	0.263613	19.96186757
LSTM_Attention_CNN_LSTM [28]	0.27994	17.32732081
LSTM_Attention_LSTM [28]	0.299334	**19.28274226**
LSTM_CNN_BiLSTM [24]	0.23489	**16.76800251**
FastRNN (proposed)	**0.202456**	**3.337492943**

4 Experimental Results and Discussions

4.1 Experimental Setup

We have used Google Colab as our execution environment with 12 GB (adjustable) assigned RAM, with one socket, two threads per core, 13684K of L3 cache, CPU of 2250 MHz, and with no power level. We have developed our model using the Python programming language (Python 3.5). We used Keras (with TensorFlow backend) library for making our deep learning models. For making the data into a suitable format, we have used Pandas, and for dividing our data into test and train, we have used Numpy. For visualization, we have matplotlib library. The dataset and the corresponding codes are available at GitHub.[2]

4.2 Results

We have used 430 stock values for training and 70 for testing. We have tested our proposed model on the stock values of four companies, i.e. Apple, Facebook, Nike, and Uber. These proposed models are compared with nine other state-of-the-art models. We have trained the models for 40 epochs. The RMSE values and the computation time for each of the nine models along with the proposed model for four companies, i.e. Apple, Facebook, Nike, and Uber, are shown in Tables 1, 2, 3, and 4, respectively. These tables highlight the best values for each column.

[2] https://github.com/MilindYadav-97/Stock_Predictions_II.

Table 2 Root mean square error (RMSE) and computation time calculated for the state-of-the-art and proposed models for Facebook Inc. stock values

Model name	RMSE value	Time (in sec)
ARIMA [4]	0.86664567	**1.60001**
BiLSTM_Attention_CNN_BiLSTM [34]	0.26834072	26.56901288
CNN_LSTM_Attention_LSTM [37]	0.15718991	16.61885238
FBProphet [9]	1.59050836	**0.405165672**
LSTM [28]	0.18714926	14.00746107
LSTM_Attention_CNN_BiLSTM [34]	0.22215487	20.65539312
LSTM_Attention_CNN_LSTM [28]	0.24017975	17.97418046
LSTM_Attention_LSTM [28]	0.29138532	19.86988878
LSTM_CNN_BiLSTM [24]	0.19898068	17.52065849
FastRNN (proposed)	**0.15137204**	**3.447671652**

Table 3 Root mean square error (RMSE) and computation time calculated for the state-of-the-art and proposed models for Nike Inc. stock values

Model name	RMSE value	Time (in sec)
ARIMA [4]	0.465822234	**1.63001**
BiLSTM_Attention_CNN_BiLSTM [34]	0.106373725	26.56245756
CNN_LSTM_Attention_LSTM [37]	0.051930262	17.05163288
FBProphet [9]	0.550859082	**0.509964705**
LSTM [28]	0.052901	14.01486564
LSTM_Attention_CNN_BiLSTM [34]	0.062675555	20.4820962
LSTM_Attention_CNN_LSTM [28]	0.061914832	17.52905965
LSTM_Attention_LSTM [28]	0.102347907	19.73444676
LSTM_CNN_BiLSTM [24]	0.047966405	17.31208062
FastRNN	**0.039458341**	**3.807400703**

The training process is fast and could be carried out on a CPU because the data size was not very high. The values of loss functions for each of the four companies' data training are shown in the graphical form in Fig. 2.

From the above-mentioned tables, Tables 1, 2, 3 and 4, it can be observed that the proposed model is much faster than all the compared models except FBProphet and ARIMA. While, when we compare the RMSE values, this outperforms these two mentioned models as well. Thus, as an overall combination the proposed model is a better choice for live stock values predictions. There is a visualized comparative study in Fig. 2, which shows how the validation loss (mean square error) on the validation dataset is improving after each epoch for all the studied stocks. Here, Fig. 3 shows the comparative study of the actual and predicted values of the baseline

Table 4 Root mean square error (RMSE) and computation time calculated for the state-of-the-art and proposed models for Uber stock values

Model name	RMSE value	Time (in sec)
ARIMA [4]	0.163886332	**1.610950985**
BiLSTM_Attention_CNN_BiLSTM [34]	0.025170157	24.91984868
CNN_LSTM_Attention_LSTM [37]	0.023596089	15.85278392
FBProphet [9]	0.064013152	**0.489181757**
LSTM [28]	0.024660817	13.3820951
LSTM_Attention_CNN_BiLSTM [34]	0.031076263	19.3135848
LSTM_Attention_CNN_LSTM [28]	0.032733788	16.7529645
LSTM_Attention_LSTM [28]	0.027326071	18.67898536
LSTM_CNN_BiLSTM [24, 24]	0.021864702	16.29291153
FastRNN (proposed)	**0.020682922**	**3.35441184**

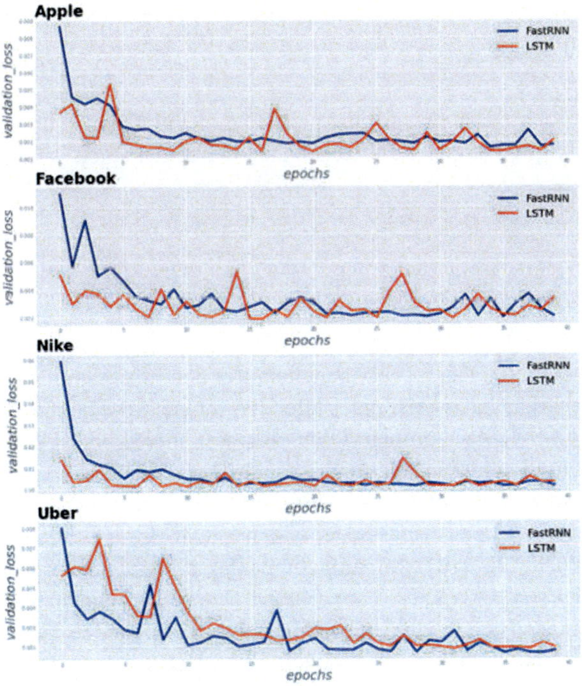

Fig. 2 Comparison plots of validation losses (mean square error) of our proposed and baseline models at each epoch for all the studied stocks

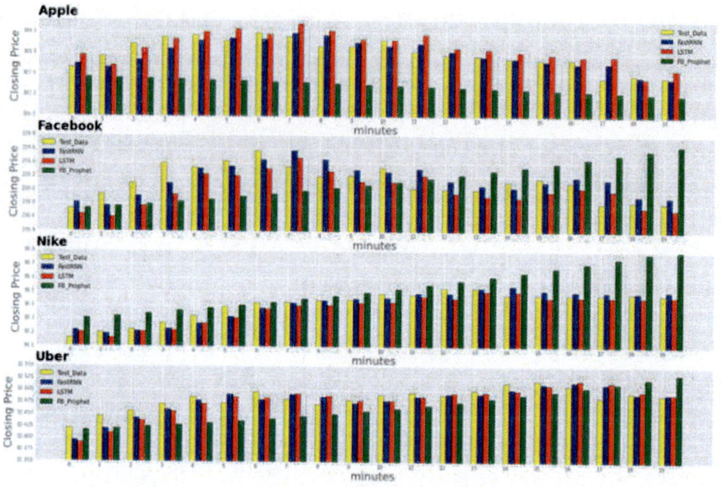

Fig. 3 Predicted stock values with proposed model and baseline models

model and our proposed models for the next 20 min and it can be observed that the proposed models' predictions were closer than the actual stock values for each of the visualized stock.

5 Conclusions

Financial data prediction can be very beneficial for companies and investors. Deep-learning-based models have been quiet effective for such predictions in the recent years. In this work, we have proposed deep learning model which consists of FastRNN. The proposed model is able to provide faster predictions when compared to the state-of-the-art models (except ARIMA and FBProphet). While improving the speed of prediction, it also provides better or at par RMSE values as well. The model is tested on the stock values of four major companies of NYSE and provided accurate predictions for these stock values. Thus, compared to FBProphet and ARIMA also, this is a better choice for a reliable model. In future, these models can be applied to other time series prediction problems as well.

References

1. Abraham A, Nath B, Mahanti PK (2001) Hybrid intelligent systems for stock market analysis. In: International conference on computational science. Springer, pp 337–345
2. Ahmed MS, Cook AR (1979) Analysis of freeway traffic time-series data by using Box-Jenkins techniques. No 722 in Urban systems operations, Transportation Research Board

3. Akita R, Yoshihara A, Matsubara T, Uehara K (2016) Deep learning for stock prediction using numerical and textual information. In: 2016 IEEE/ACIS 15th international conference on computer and information science (ICIS). IEEE, pp 1–6
4. Ariyo AA, Adewumi AO, Ayo CK (2014) Stock price prediction using the arima model. In: 2014 UKSim-AMSS 16th international conference on computer modelling and simulation. IEEE, pp 106–112
5. Arjovsky M, Shah A, Bengio Y (2016) Unitary evolution recurrent neural networks. In: International conference on machine learning, pp 1120–1128
6. Armano G, Marchesi M, Murru A (2005) A hybrid genetic-neural architecture for stock indexes forecasting. Inf Sci 170(1):3–33
7. Chen K, Zhou Y, Dai F (2015) A lstm-based method for stock returns prediction: a case study of china stock market. In: 2015 IEEE international conference on big data (big data). IEEE, pp 2823–2824
8. Chen W, Yeo CK, Lau CT, Lee BS (2018) Leveraging social media news to predict stock index movement using RNN-boost. Data Knowl Eng 118:14–24
9. Chikkakrishna NK, Hardik C, Deepika K, Sparsha N (2019) Short-term traffic prediction using sarima and fbprophet. In: 2019 IEEE 16th India council international conference (INDICON). IEEE, pp 1–4
10. Choudhry R, Garg K (2008) A hybrid machine learning system for stock market forecasting. World Acad Sci Eng Technol 39(3):315–318
11. Ding X, Zhang Y, Liu T, Duan J (2015) Deep learning for event-driven stock prediction. In: Twenty-fourth international joint conference on artificial intelligence
12. Egeli B et al (2003) Stock market prediction using artificial neural networks. Dec Sup Syst 22:171–185
13. Enke D, Thawornwong S (2005) The use of data mining and neural networks for forecasting stock market returns. Expert Syst Appl 29(4):927–940
14. Fischer T, Krauss C (2018) Deep learning with long short-term memory networks for financial market predictions. Eur J Oper Res 270(2):654–669
15. Franses PH, Van Dijk D (1996) Forecasting stock market volatility using (non-linear) garch models. J Forecast 15(3):229–235
16. Fu J, Lum KS, Nguyen MN, Shi J (2007) Stock prediction using fcmac-byy. In: International symposium on neural networks. Springer, pp 346–351
17. Gers FA, Eck D, Schmidhuber J (2002) Applying lstm to time series predictable through time-window approaches. In: Neural Nets WIRN Vietri-01. Springer, pp 193–200
18. Gudelek MU, Boluk SA, Ozbayoglu AM (2017) A deep learning based stock trading model with 2-D CNN trend detection. In: 2017 IEEE symposium series on computational intelligence (SSCI). IEEE, pp 1–8
19. He K, Zhang X, Ren S, Sun J (2016) Deep residual learning for image recognition. In: Proceedings of the IEEE conference on computer vision and pattern recognition, pp 770–778
20. Huang SJ, Shih KR (2003) Short-term load forecasting via ARMA model identification including non-gaussian process considerations. IEEE Trans Power Syst 18(2):673–679
21. Jing L, Gulcehre C, Peurifoy J, Shen Y, Tegmark M, Soljacic M, Bengio Y (2019) Gated orthogonal recurrent units: on learning to forget. Neural Comput 31(4):765–783
22. Kim T, Kim HY (2019) Forecasting stock prices with a feature fusion lstm-cnn model using different representations of the same data. PloS one 14(2):e0212320
23. Kusupati A, Singh M, Bhatia K, Kumar A, Jain P, Varma M (2018) Fastgrnn: a fast, accurate, stable and tiny kilobyte sized gated recurrent neural network. In: Advances in neural information processing systems, pp 9017–9028
24. Le T, Vo MT, Vo B, Hwang E, Rho S, Baik SW (2019) Improving electric energy consumption prediction using CNN and BI-LSTM. Appl Sci 9(20):4237
25. Mohan S, Mullapudi S, Sammeta S, Vijayvergia P, Anastasiu DC (2019) Stock price prediction using news sentiment analysis. In: 2019 IEEE fifth international conference on big data computing service and applications (BigDataService). IEEE, pp 205–208

26. Paliwal N, Vanjani P, Liu JW, Saini S, Sharma A (2019) Image processing-based intelligent robotic system for assistance of agricultural crops. Int J Soc Hum Comput 3(2):191–204
27. Pawar K, Jalem RS, Tiwari V (2019) Stock market price prediction using LSTM RNN. In: Emerging trends in expert applications and security. Springer, pp 493–503
28. Qiu J, Wang B, Zhou C (2020) Forecasting stock prices with long-short term memory neural network based on attention mechanism. PloS one 15(1):e0227222
29. Saini S, Sahula V (2020) Chapter 11—setting up a neural machine translation system for English to Indian languages. In: Sinha G, Suri JS (eds) Cognitive informatics, computer modelling, and cognitive science. Academic Press, pp 195–212. https://doi.org/10.1016/B978-0-12-819443-0.00011-8, http://www.sciencedirect.com/science/article/pii/B9780128194430000118
30. Saini S, Sahula V (2020) Cognitive architecture for natural language comprehension. Cogn Comput Syst 2(1):23–31 https://doi.org/10.1049/ccs.2019.0017, https://ietresearch.onlinelibrary.wiley.com/doi/abs/10.1049/ccs.2019.0017
31. Tay FE, Cao L (2001) Application of support vector machines in financial time series forecasting. Omega 29(4):309–317
32. Taylor SJ, Letham B (2018) Forecasting at scale. Am Stat 72(1):37–45
33. Tsai C, Wang S ()209 Stock price forecasting by hybrid machine learning techniques. In: Proceedings of the international multiconference of engineers and computer scientists, vol 1, p 60
34. Wang M, Cheng J, Zhai H (2020) Life prediction for machinery components based on CNN-bilstm network and attention model. In: 2020 IEEE 5th information technology and mechatronics engineering conference (ITOEC). IEEE, pp 851–855
35. Yadav K, Lamba A, Gupta D, Gupta A, Karmakar P, Saini S (2020) Bilingual sentiment analysis for a code-mixed Punjabi English social media text. In: 2020 5th international conference on computing, communication and security (ICCCS) (2020), pp 1–5. https://doi.org/10.1109/ICCCS49678.2020.9277309
36. Zhang J, Lei Q, Dhillon IS (2018) Stabilizing gradients for deep neural networks via efficient SVD parameterization (2018). arXiv:1803.09327
37. Zhang Y, Zheng J, Jiang Y, Huang G, Chen R (2019) A text sentiment classification modeling method based on coordinated CNN-LSTM-attention model. Chinese J Electron 28(1):120–126

Ceiling Improvement on Breast Cancer Prediction Accuracy Using Unary KNN and Binary LightGBM Stacked Ensemble Learning

Ajeet Singh, Vikas Tiwari, and Appala Naidu Tentu

Abstract The employment of data mining, granular knowledge exploration and machine learning approaches in the medical field (several of diseases in bioinformatics) are proven to be prolific as such approaches are thought of computationally empirical within the higher cognitive process of medical practitioners. As an intelligent healthcare strategy, this paper presents a fast and accurate framework for breast cancer disease classification and predictive analysis. In an experimental simulation, we obtained ceiling improvement on prediction accuracy using Unary KNN and Binary LightGBM Stacked Ensemble Learning. We have given the visualization of results and judged the obtained various statistical performance measures. Our empirical evaluation and experimental analysis reflect that this strategy with provable performance guarantees performs well in comparison with other commonly used significant state-of-the-art methods. We observed that the proposed framework can be a powerful tool for simplifying or speeding up computations, and when employed appropriately it can lead to little loss in classification quality.

Keywords Intelligent decision system · Machine learning · Healthcare · Breast cancer · Chronic disease prediction · Prognostic accuracy

1 Introduction

Various types of diseases in bioinformatics domain have been explored and they require new varieties of treatment phenomena to enhance the quality of healthcare and derive all-in-one solutions for associated well-being. The earlier forecast and location of disease cells can be useful in curing the illness in medical applications. The most common types are breast and lung cancer with 2,68,676 and 2,34,038 approximated new cases that occurred in 2019. Applying machine learning for survival prediction, i.e. predicting whether a patient will survive a given period of time

A. Singh (✉) · V. Tiwari · A. N. Tentu
C.R. Rao Advanced Institute of Mathematics, Statistics and Computer Science,
University of Hyderabad Campus, Prof. CR Rao Road, Hyderabad 500046, Telangana, India
e-mail: ajeetcs@uohyd.ac.in

© The Author(s), under exclusive license to Springer Nature Singapore Pte Ltd. 2022

D. Giri et al. (eds.), *Proceedings of the Seventh International Conference on Mathematics and Computing*, Advances in Intelligent Systems and Computing 1412,
https://doi.org/10.1007/978-981-16-6890-6_34

after diagnosis, can increase the prognostic accuracy and might ultimately lead to better-informed decision-making. Cancer is the second leading cause of death globally. The detection of cancer cells and their treatment at the beginning stages of breast cancer in women are the essence of the success of infected patient survival. In the last two decades, incredible technological developments happened for primary stage diagnosis and recurrence detection of breast cancer disease. There is a need for new and practically empirical AI-oriented research methodologies like knowledge exploration and information retrieval, which can be exploited by discovering correlations [5, 7, 10] among the massive amount of expected variables to predict the recurrence of breast cancer disease utilizing the trained model. Breast cancer is a kind of malignant tumor which develops when the cells present inside the breast tissue split and grow without the normal control on cell demise and cell dissection.

Though finding the specific cause of breast cancer remains a riddle, but by observing the recurrence patterns [13, 14] of this disease in women, some common risk factors are extracted. Generally, these factors include variables such as age, genetic risk and family chronicle. Rules for predicting the prognosis of new patients or explaining specific outcomes can be extracted by advanced data mining methods with extensive capability to support physicians in their intelligent decision-making.

1.1 Contribution Highlights

Efficient processing mechanisms to make the intelligent learning and prediction systems capable of dealing with multivariate data and effective learning technologies to find out the rules to describe the data are still of urgent need. The contributions in this paper are as follows:

- We propose a framework for breast cancer disease classification and predictive analysis. In an experimental simulation, we obtained ceiling improvement on prediction accuracy using Unary KNN and Binary LightGBM Stacked Ensemble Learning.
- We have given the visualization of results and judged the obtained various statistical performance measures. Our empirical evaluation and experimental analysis reflect that this strategy with provable performance guarantees performs well in comparison with other commonly used significant state-of-the-art methods. In this work, *R* language (v'3.4.4) and *H2O* framework (v'1.5.2) with Python (version 3.7) are exploited for simulation and statistical computing purposes.

1.2 Organization of the Paper

The remaining paper is organized as follows: Sect. 2 reviews previous studies and development in this domain. Section 3 describes some related preliminaries and definitions. Section 4 covers the framework for breast cancer disease classification and predictive analysis, detailed modeling and experimental procedure, obtained results along with the comparative performance analysis. Finally, Sect. 5 concludes the paper.

2 Literature Review

In the recent study, the knowledge discovery approach predicts the survival condition of breast cancer patients with high confidence by applying on the breast cancer dataset [1, 2]. The authors in [3] have used deep neural networks to design new models which predict the patient survival and conditional survival of colon cancer. In [4] the authors have proposed an inference algorithm which learns the model parameters from time-to-event data while handling right censoring. Bellaachia et al. [5] have used a data mining algorithm to analyze the prediction of survivability rate of breast cancer patients. Authors in [6, 7] have used three famous data mining algorithms such as decision trees, artificial neural networks and support vector machines along with logistic regression which is commonly used for statistical analysis to develop prediction models for prostate cancer survivability. Decision tree (C5) is the best predictor with 93.6% accuracy followed by artificial neural networks with 91.2% accuracy. Authors in [8, 9] have used J48 implementation of the C4.5 algorithm and a Naive Bayes classification technique to predict lung cancer survivability from a huge dataset with 15 years of patient records.

Authors in [10–12] have given optimal models to predict the survival rate of breast cancer patients. They have implemented seven common algorithms such as Logistic Regression Model, ANN, Naive Bayes, Bayes Network, Decision Tree with Naive Bayes, Decision Tree (ID3) and Decision Tree (J48) for the development of disease prediction model. Khan et al. [15] have examined the combination of accuracy and interpretability in the form of Fuzzy Logic and Decision Trees. Authors in [16] have proposed SSL (semi-supervised learning) co-training algorithm. They have considered the concept of tagging virtual labels to unlabeled patient data, that is, pseudo-labels, and treating them as if they were labeled. Authors in [17] have presented various extensions that enhance the quality of the vectors as well as the training speed. Authors in [18] have focused to design a predictive model for breast cancer patients and identify the relationships between certain predictive variables and survival. Authors in [19] have proposed a combined approach of two machine learning models, semi-supervised learning (SSL) co-training as well as Decision Tree. Authors in [20] have used a self-organizing map (SOM) and density-based spatial clustering of applications with noise (DBSCAN) to create patient cohort clusters. Authors in [21] have presented a survivability prediction model with machine learning techniques for colon cancer. The performance of the six-feature model is close to the model which is using 18 features. A new technique is given by the authors [22] for prognostic prediction, by using non-parametric statistics to utilize the available information in a neural architecture. This technique is applied to breast cancer prognosis and results in accurate model that can play a vital role in preventing unnecessary surgeries. Authors in [23] have proposed a set of new algorithms to increase the effectiveness of classification for 5-year survivability of breast cancer patients from large data set with an imbalanced property. To demonstrate the applicability of the proposed method [24] to medical diagnosis and decision-making, the authors have used the diagnosis of breast cytology. Zhang et al. [25] present an instance-based learning approach in which typical instances are selected to store as concept descriptions.

3 Preliminaries and Statistical Terminologies

This section presents some significant preliminaries and statistical terminologies.

3.1 *Information and Decision System*

An *information system* [33] is represented as pair (U, A) where,
U : denotes—a kind of non-empty finite set.
A : denotes—a non-empty finite set of features (attributes).
However, A *decision table* is a system of the form $S = (U, C \cup \{d\})$ where
C : denotes set of conditional attributes.
d : denotes the decision attribute.

3.2 *Linearly Separable Data-Aided Binary Classification*

- **Goal:** Find the hyperplane (i.e. decision boundary) linearly separating our classes. Our boundary will have equation: $w^T x + b = 0$.
- Anything above the decision boundary should have label 1. That is, x_i s.t. $w^T x + b > 0$ will have corresponding $y_i = 1$.
- Similarly, anything below the decision boundary should have label -1. That is, x_i s.t. $w^T x + b < 0$ will have corresponding $y_i = -1$.

4 Our Framework, Modeling Pipeline and Results Discussion

This section covers computational environment, simulation setup, dataset details, modeling procedure and results discussion.

4.1 Modeling and Simulation Setup

The modeling and experimental simulation setup consists of system environment as Ubuntu 16.04 LTS with 64-bit OS, 8 GB RAM, processor as Intel Core i7-860 @ 2.80 GHz clock speed, number of cores as 4 and totally 8 threads available. *R* language (v'3.4.4) and *H2O* framework (v'1.5.2) with Python version 3.7 are used for simulation and statistical computing purposes.

4.2 Dataset Detail

In this work, for experiments, Wisconsin Breast Cancer Dataset [26–28] is used. Each record in the dataset represents one breast cancer tissue sample. The adopted methodology and general framework are shown in Fig. 1. This general framework is taken into consideration as a baseline while designing Scenario I and Scenario II modeling pipelines.

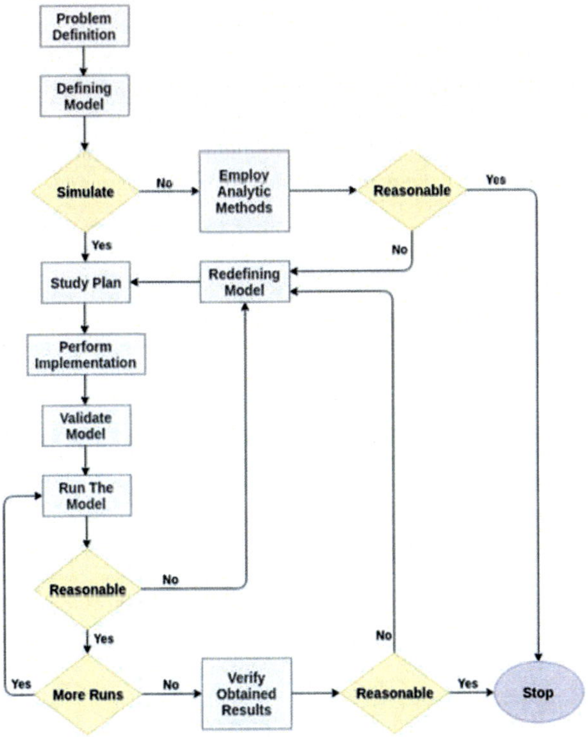

Fig. 1 Framework for adopted methodology

4.3 Scenario I: Model Pipeline and Output Results

In modeling Scenario I, first the dataset is loaded. We have splitted out the dataset, where 20% of the data is kept for validation, the remaining portion (80%) is selected for training and testing the models. The ID column vector is removed and the rest of the conditional variables' values are converted into numeric. Summary of attributes of the entire data in terms of *<Min, 1st Qu, median, mean, 3rd Qu, Max>* is given in Table 1. The observed class distribution is
(benign : malignant ⇒ 65.4 : 34.6)
Then the correlation among the input conditional variables is computed.

The statistical distribution of model variables and data visualization in the form of histograms, density plots, box plot, scatterplot matrix plots and bar plots are represented as Figs. 2, 3, 4, 5 and 6, respectively.

By analyzing the nature of the data, we applied an optimal suite of linear and nonlinear learning algorithms. The 10-fold cross-validation with four repeats strategy is utilized here. The obtained results of the statistical parameters are given in Tables 2 and 3. Corresponding graphical representation for baseline accuracy along with Kappa statistical measure is presented as Fig. 7.

Next, the transform function is applied to normalize the skewed distribution and further execution is performed on the transformed dataset keeping the same runtime parameters as previous.

Further, to optimize the model performance, SVM (with RBF kernel) parameters are tuned with final values as—sigma ← 0.15 and C ← 1. Next, KNN is tuned on its parameter k (in values range of 1–18). The final value used for the model was k = 17. The performance gain can be observed in Tables 4 and 5 and Figs. 8 and 9.

Table 1 Dataset Summary; (*Denotions:* Cl.th → Cl.thickness, C.si → Cell.size, C.sh → Cell.shape, M.ad → Marg.adhesion, E.si → Epith.c.size, B.nu → Bare.nuclei, B.cr → Bl.cromatin, N.nu → Normal.nucleoli, Mit → Mitoses)

Cl.th	C.si	C.sh	M.ad	E.si	B.nu	B.cr	N.nu	Mit
Min.: 1.000	Min.: 1.000	Min.: 1.000	Min.: 1.000	Min.: 1.000	Min.: 1.000	Min.: 1.000	Min.: 1.000	Min.: 1.000
1st Qu.: 2.000	1st Qu.: 1.000	1st Qu.: 1.000	1st Qu.: 1.000	1st Qu.: 2.000	1st Qu.: 1.000	1st Qu.: 2.000	1st Qu.: 1.000	1st Qu.: 1.000
Median: 4.000	Median: 1.000	Median: 2.000	Median: 1.000	Median: 2.000	Median: 1.000	Median: 3.000	Median: 1.000	Median: 1.000
Mean: 4.384	Mean: 3.116	Mean: 3.198	Mean: 2.875	Mean: 3.232	Mean: 3.468	Mean: 3.405	Mean: 2.877	Mean: 1.611
3rd Qu.: 6.000	3rd Qu.: 5.000	3rd Qu.: 5.000	3rd Qu.: 4.000	3rd Qu.: 4.000	3rd Qu.: 5.000	3rd Qu.: 4.250	3rd Qu.: 4.000	3rd Qu.: 1.000
Max.: 10.000	Max.: 10.000	Max.: 10.000	Max.: 10.000	Max.: 10.000	Max.: 10.000	Max.: 10.000	Max.: 10.000	Max.: 10.000

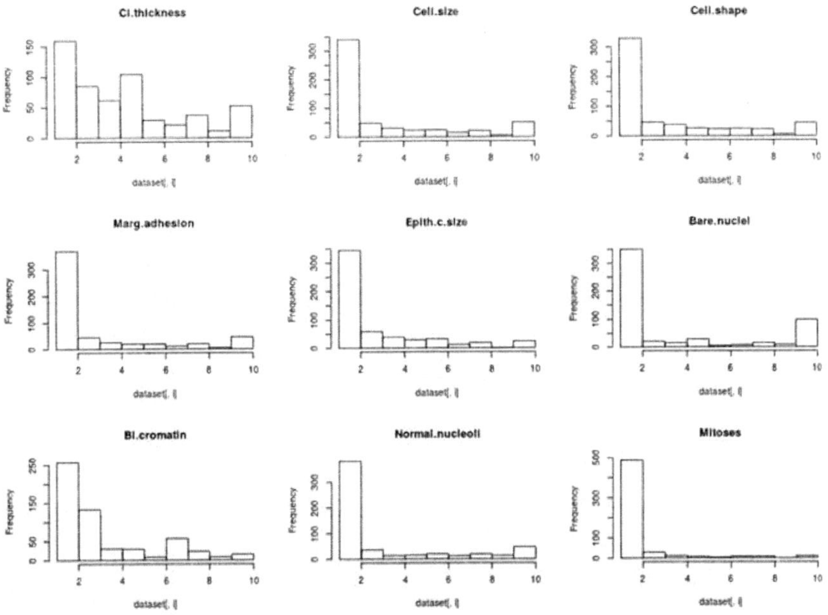

Fig. 2 Histograms representation

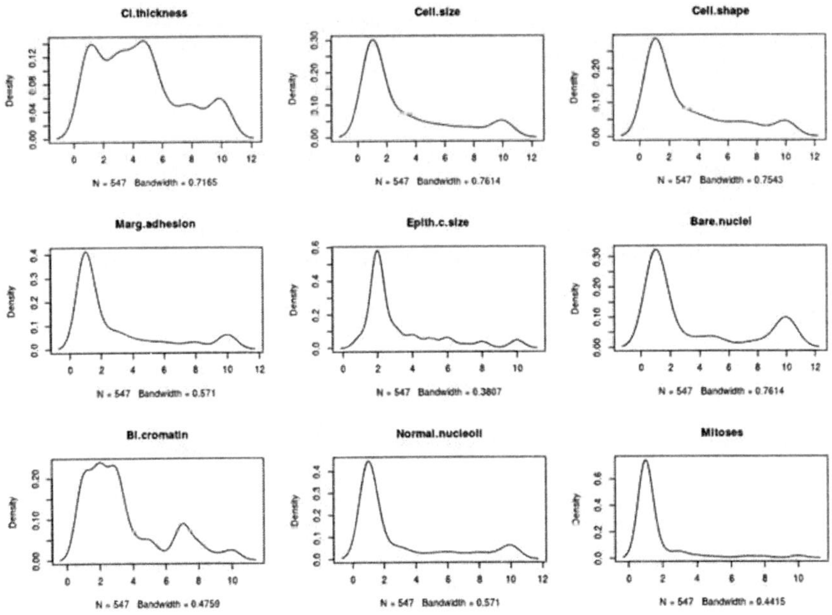

Fig. 3 Density distribution

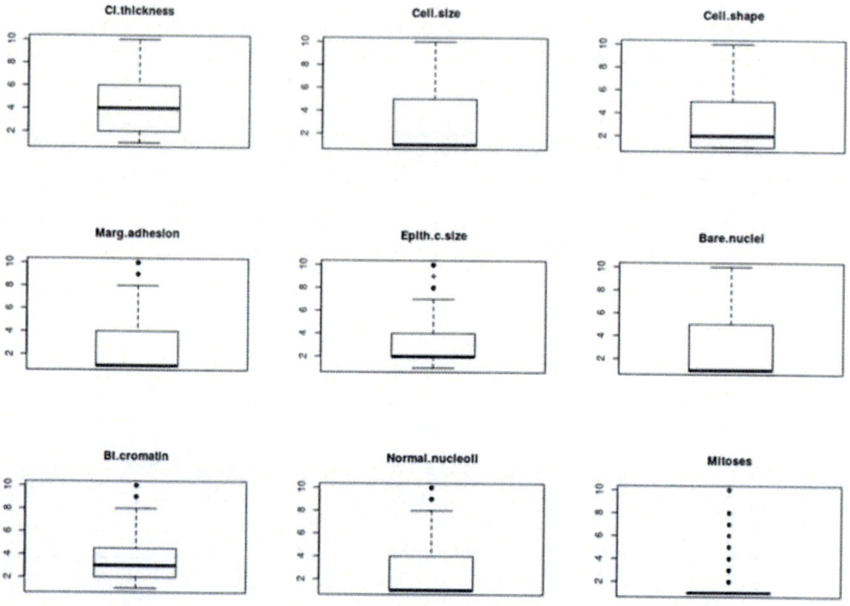

Fig. 4 Boxplot representation

Fig. 5 Scatterplots

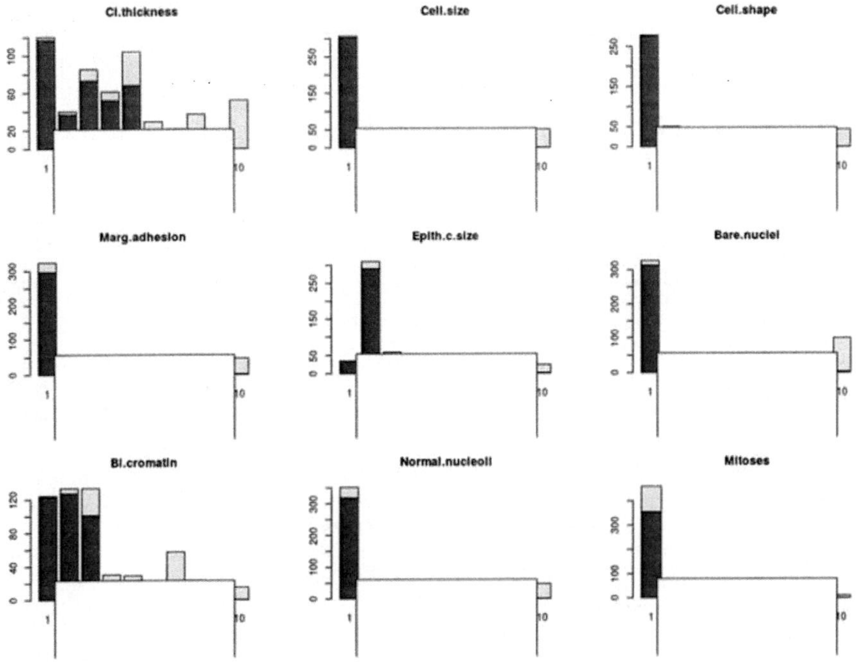

Fig. 6 Bar plots representation

Table 2 summary_resamples (object = results); Accuracy

Method	Min.	1st Qu.	Median	Mean	3rd Qu.	Max	NA's
LG	0.8909091	0.9498316	0.9636364	0.9609977	0.9814815	1	0
LDA	0.8888889	0.9454545	0.9629630	0.9585734	0.9771825	1	0
GLMNET	0.8909091	0.9498316	0.9636364	0.9640392	0.9818182	1	0
KNN	0.9090909	0.9629630	0.9728836	0.9707620	0.9818182	1	0
CART	0.8571429	0.9090909	0.9272727	0.9324311	0.9461851	1	0
NB	0.9090909	0.9446970	0.9636364	0.9621878	0.9817340	1	0
SVM	0.8909091	0.9444444	0.9629630	0.9518951	0.9641234	1	0

Table 3 summary_resamples (object = results); Kappa

Method	Min.	1st Qu.	Median	Mean	3rd Qu.	Max.	NA's
LG	0.7526237	0.8903491	0.9200813	0.9136612	0.9591931	1	0
LDA	0.7440758	0.8778682	0.9187970	0.9073541	0.9493543	1	0
GLMNET	0.7526237	0.8903491	0.9207048	0.9203519	0.9601869	1	0
KNN	0.8014440	0.9195906	0.9402224	0.9355909	0.9598811	1	0
CART	0.7083333	0.8011046	0.8464753	0.8531748	0.8859322	1	0
NB	0.7964471	0.8796434	0.9205797	0.9172069	0.9597332	1	0
SVM	0.7755102	0.8799491	0.9207048	0.8972223	0.9220518	1	0

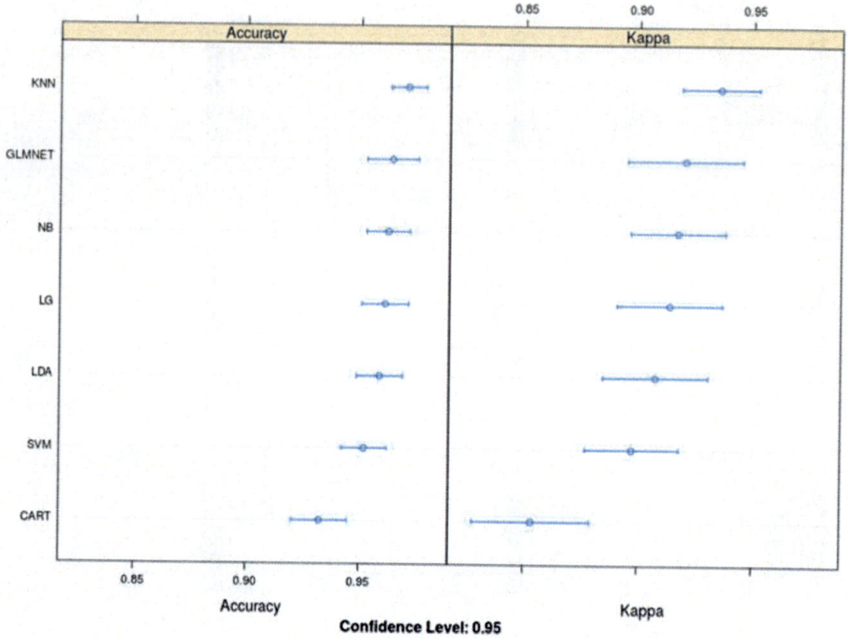

Fig. 7 Baseline accuracy and kappa representation

Table 4 Re-sampling results across tuning parameters (The final values used for the model were sigma = 0.1 and C = 1)

Sigma	C	Accuracy	Kappa
0.025	1	0.9689109	0.9329235
0.025	2	0.9683269	0.9314403
0.025	3	0.9695278	0.9341394
0.025	4	0.9701343	0.9354965
0.025	5	0.9701343	0.9354965
0.025	6	0.9689109	0.9329293
0.025	7	0.9676876	0.9302032
0.025	8	0.9676876	0.9302032
0.025	9	0.9682937	0.9315590
0.025	10	0.9670591	0.9288215
0.050	1	0.9713464	0.9382360
0.050	2	0.9725585	0.9408533
0.050	3	0.9707291	0.9368374
0.050	4	0.9701118	0.9354950
0.050	5	0.9694945	0.9341274
0.050	6	0.9694945	0.9341274

(continued)

Table 4 (continued)

Sigma	C	Accuracy	Kappa
0.050	7	0.9694945	0.9341274
0.050	8	0.9688885	0.9327716
0.050	9	0.9688997	0.9328178
0.050	10	0.9682937	0.9314271
0.100	1	0.9731646	0.9422986
0.100	2	0.9713352	0.9382492
0.100	3	0.9701231	0.9355412
0.100	4	0.9688997	0.9327829
0.100	5	0.9664751	0.9275447
0.100	6	0.9646457	0.9233957
0.100	7	0.9640396	0.9220387
0.100	8	0.9628275	0.9192922
0.100	9	0.9628387	0.9192111
0.100	10	0.9634560	0.9204856
0.150	1	0.9725585	0.9409753
0.150	2	0.9707291	0.9368926
0.150	3	0.9670924	0.9289103
0.150	4	0.9640508	0.9220156
0.150	5	0.9616154	0.9165448
0.150	6	0.9622214	0.9179356
0.150	7	0.9622326	0.9177849
0.150	8	0.9628387	0.9190402
0.150	9	0.9622214	0.9175415
0.150	10	0.9628163	0.9189235

Next, bagging and boosting ensemble methods are employed in further finetuned modeling. Bagged classification and regression trees, random forest, stochastic gradient boosting, C5.0 methods are modeled with the data. The obtained results are given in Tables 6 and 7. Performance on these ensembling methods on the data is shown as Fig. 10. Finally, after validation of the model, test and predictive analysis is performed on unseen data. We obtained ceiling improvement on disease prediction accuracy (≈99.4%) using Unary KNN with ensembling. The final simulation results consisting of various statistical performance parameters are given in Table 8.

Table 5 Accuracy was used to select the optimal model using the largest value. (The final value used for the model was k = 17)

k	Accuracy	Kappa
1	0.9500325	0.8894817
2	0.9531069	0.8967630
3	0.9628495	0.9188475
4	0.9603363	0.9128179
5	0.9688997	0.9323375
6	0.9695166	0.9339598
7	0.9719633	0.9391947
8	0.9719525	0.9391787
9	0.9713572	0.9378967
10	0.9701451	0.9351851
11	0.9713244	0.9378279
12	0.9713464	0.9378517
13	0.9719525	0.9392355
14	0.9725585	0.9405913
15	0.9719525	0.9392355
16	0.9713464	0.9378517
17	0.9725585	0.9405913
18	0.9719412	0.9392854

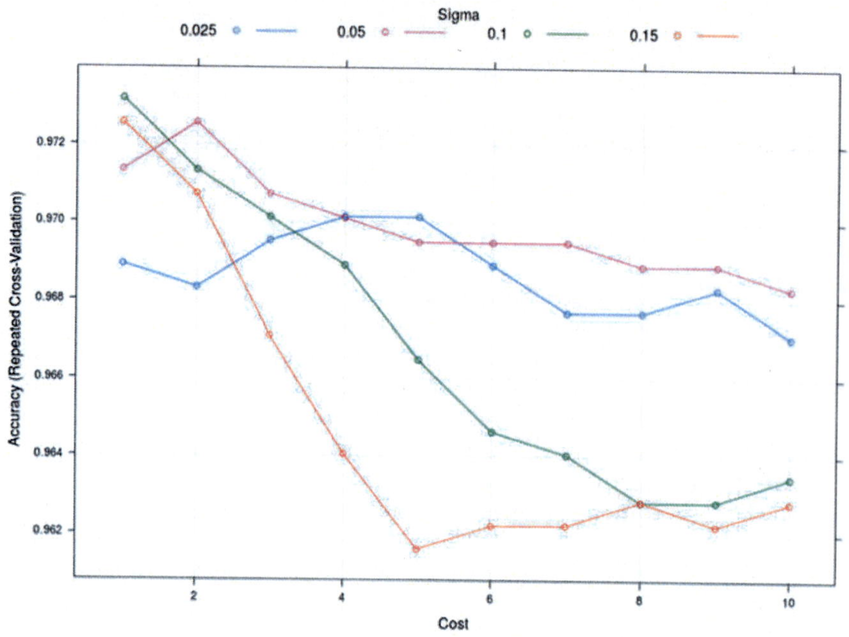

Fig. 8 Tuning $< sigma, C >$ parameters in SVM

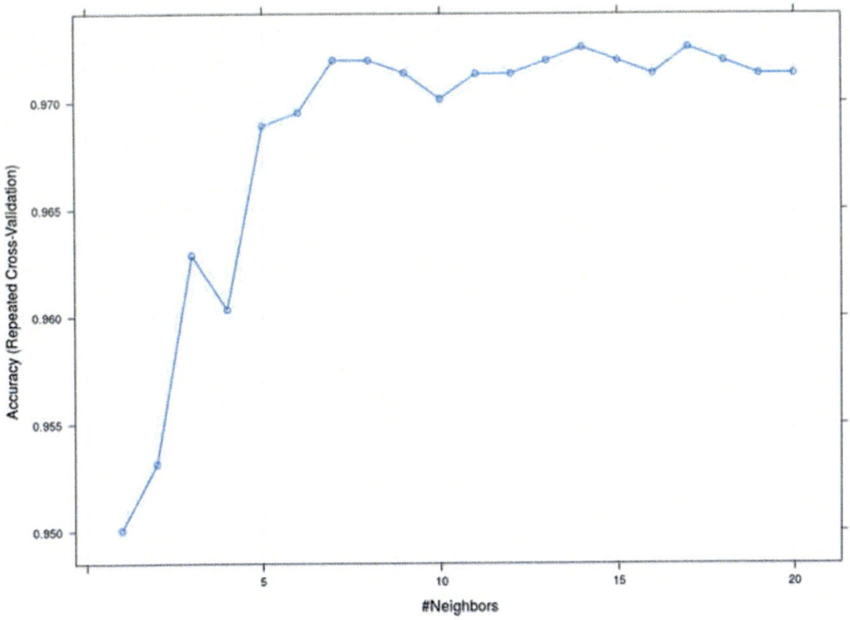

Fig. 9 Tuning < k > parameter in KNN

Table 6 Ensembling—obtained accuracy

BAG	0.8727273	0.9454545	0.9632997	0.9622439	0.9818182	1	0
RF	0.9272727	0.9629630	0.9816498	0.9731309	0.9818182	1	0
GBM	0.8727273	0.9498316	0.9636364	0.9622439	0.9818182	1	0
C50	0.8909091	0.9498316	0.9636364	0.9640284	0.9818182	1	0

Table 7 Ensembling—obtained kappa

BAG	0.7286822	0.8808664	0.9207048	0.9173235	0.9597332	1	0
RF	0.8391813	0.9207048	0.9595853	0.9415904	0.9602888	1	0
GBM	0.7076689	0.8900351	0.9207048	0.9162823	0.9601869	1	0
C50	0.7646220	0.8903491	0.9207048	0.9214000	0.9602888	1	0

4.4 *Scenario II: Model Pipeline and Output Results*

In this scenario, we have performed simulation on input Wisconsin Breast Cancer dataset using Binary LightGBM Stacked models. Dataset ($\mathcal{W}$) is splitted into $\mathcal{W}_{trn}, \mathcal{W}_{val}, \mathcal{W}_{tst}$ sets. Since the dataset consists of totally 11 variables (attribute vectors), out of which $C_1, C_2 \cdots C_{10}$ are denoted as conditional variables and last

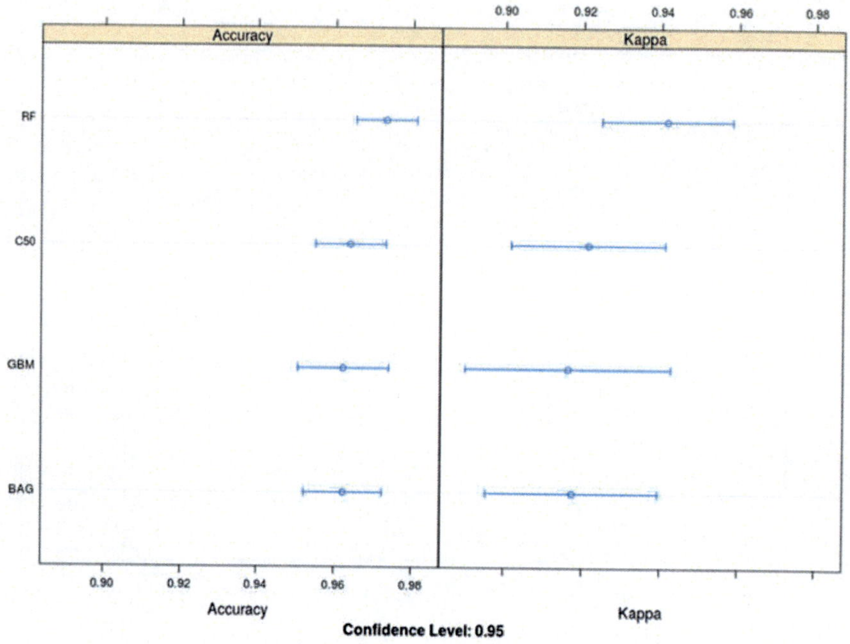

Fig. 10 Performance on ensembling methods

Table 8 Final model—statistical performance results

Parameter	Value
Accuracy	0.9926
95% CI	(0.9597, 0.9998)
No Information Rate	0.6471
P-Value [Acc > NIR]	<2e-16
Kappa	0.984
Mcnemar's Test P-Value	1
Sensitivity	0.9886
Specificity	1.0000
Pos Pred Value	1.0000
Neg Pred Value	0.9796
Prevalence	0.6471
Detection Rate	0.6397
Detection Prevalence	0.6397
Balanced Accuracy	0.9943

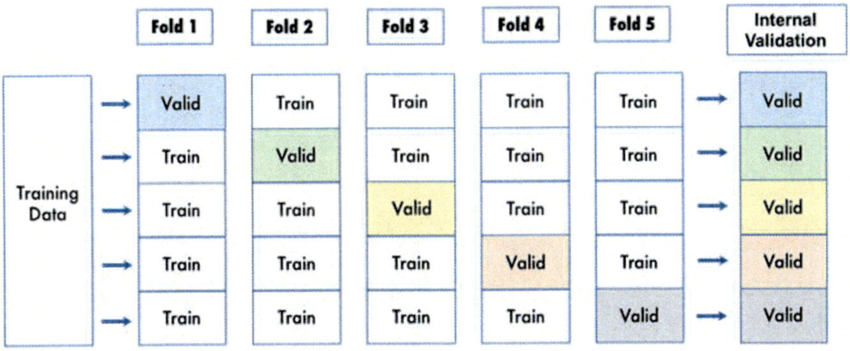

Fig. 11 Visualization: cross validation with 5 folds

one (C_{11}) is decision variable/class. The experimental pipeline consists of the following steps:

Ingest Data ▷
- Column types are detected.

Feature Preprocessing ▷
- Raw features are turned into numeric.

Model and Feature Tuning ▷
- Find the optimal parameters for XGBoost and LightGBM models by training models with different parameters.
- The best parameters are those that generate the greatest AUC on the internal validation data.
- 433 trained and scored to evaluate attributes and model parameters.

Feature Blending and Evolution ▷
- Observed the best representation of the data for the final model training by creating and evaluating 304 features over 59 iterations.
- 360 trained and scored to further evaluate engineered features.

Final Model ▷
- The final model is a stacked ensemble of binary number of LightGBM models.
- Features of these models are the best features found during the feature engineering iterations.

Create Scoring Pipeline ▷
- Created and exported the Python MOJO Scoring Pipeline.

The cross-validation process is repeated three times to ensure the validation metrics is robust. The visualization is shown in Fig. 11. Final StackedEnsemble pipeline has a learning rate (lr) $\rightarrow$ 0.04.

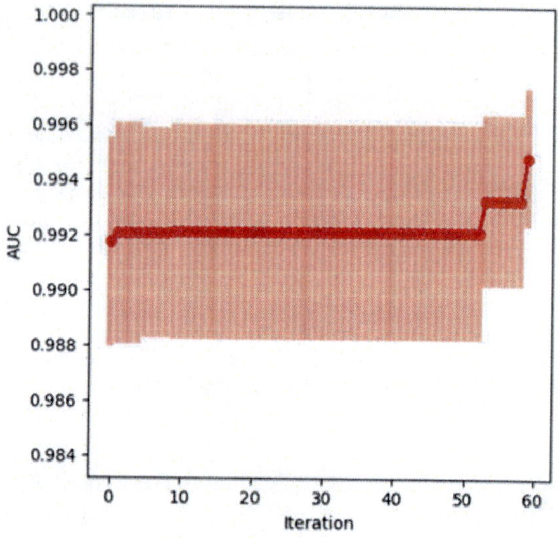

Fig. 12 Model and feature tuning stage and feature evolution stage overall impact

Table 9 Final model—statistical performance results

Performance Parameter	Value
Accuracy	0.98959
AUC	0.99486
AUCPR	0.9896
F05	0.9706
F1 Score	0.9722
F2 Score	0.9854
GINI	0.9897
LOGLOSS	0.0917
MCC (correlation coefficient)	0.9576
Simulation Time	1042 s

The feature evolution stage uses a genetic algorithm to search the large feature engineering space. The graph in Fig. 12 shows the effect that model and feature tuning stage and feature evolution stage had on the performance.

Performance of the Final Model—

The final model performance summary in terms of obtained statistical parameters, i.e. accuracy, AUC, AUCPR, F05, F1 Score, F2 Score, Gini, LOGLOSS, MCC (Matthews Correlation Coefficient) and simulation time is given in Table 9. It can be observed that we obtained AUC (area under the ROC curve) $\approx$ 0.9948. AUC provides an aggregate measure of performance across all possible classification thresholds.

Figure 13 presents receiver operating characteristic curve. Precision versus Recall curve is represented by Fig. 14.

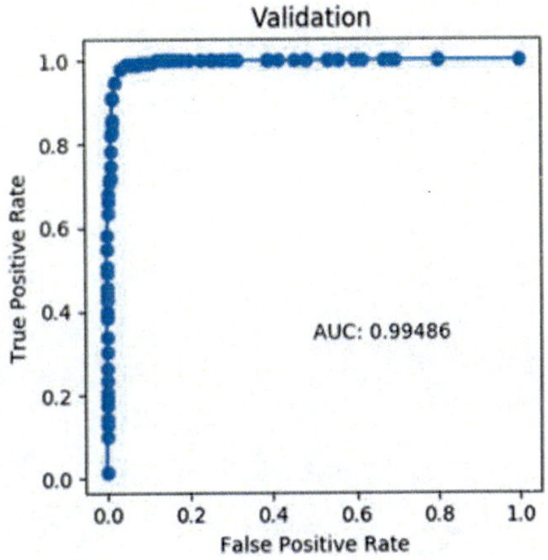

Fig. 13 ROC visualization

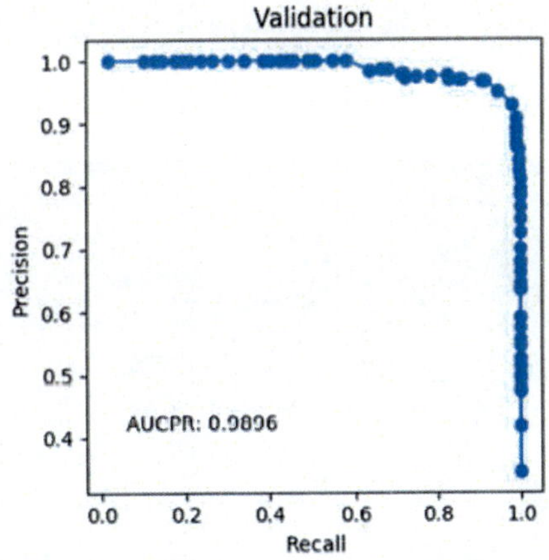

Fig. 14 Precision recall curve

It displays multivariate data in the form of a two-dimensional chart. Figures 15, 16 and 17 are the representation for {Confusion Matrix, Precision Recall Curve, ROC Curve, respectively, } @ test on 20% splitted unseen data.

4.5 *Comparative Analysis*

Table 10 presents the comparative analysis where the obtained experimental results are compared with some significant existing state-of-the-art methods on the same dataset. In experimental simulation, we obtained ceiling improvement on prediction accuracy. From comparative analysis, it is proved that our adopted model pipelines outperforms other existing approaches.

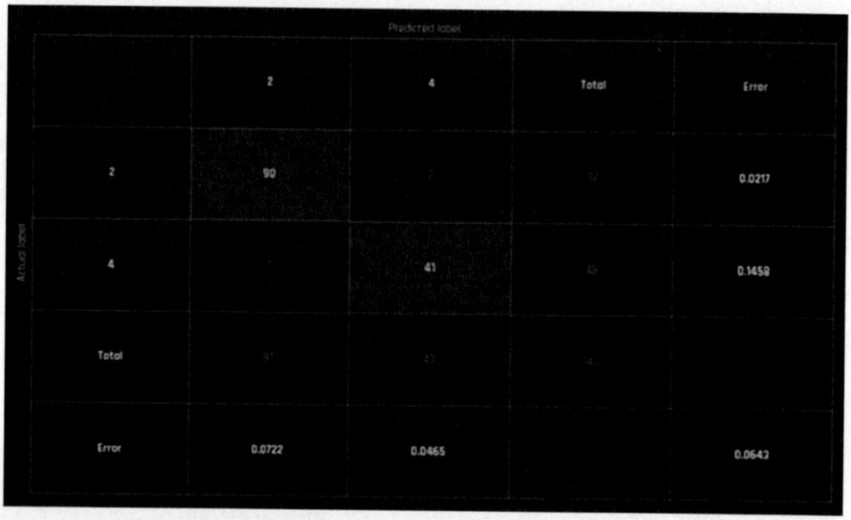

Fig. 15 Confusion matrix_test on 20%

Fig. 16 Precision recall curve_test on 20%

Fig. 17 ROC curve_test on 20%

Table 10 Comparative analysis with existing approaches

Reference	Method	Prediction accuracy (%)
Zafiropoulos et al. [29]	SVM with Gaussian RBF	89.28
Ali et al. [30]	SVM with poly kernel	88
Ali et al. [30]	Neural Networks model	92
Shajahaan et al. [31]	C4.5 algorithm	95.57
Shajahaan et al. [31]	CART	92.42
Shajahaan et al. [31]	Naive Bayes	97.42
Ivancakova et al. [32]	Random Forests	97.37
Ivancakova et al. [32]	C4.5	95.61
Proposed Model Pipeline (Scenario I)	Unary KNN with ensembling	99.4
Proposed Model Pipeline (Scenario II)	Binary LightGBM Stacked Ensemble Learning	99

5 Conclusion

The earlier forecast and location of disease cells can be useful in curing the illness in medical/healthcare applications. Predictive analytics and modeling encompasses a variety of statistical methodologies from machine learning that can analyze the present along with historical facts to make predictions about future events. We have given a framework for breast cancer disease classification and predictive analysis. In the experimental simulation, we obtained ceiling improvement on prediction accuracy using Unary KNN and Binary LightGBM Stacked Ensemble Learning.

References

1. Afshar HL, Ahmadi M, Roudbari M, Sadoughi F (2015) Prediction of breast cancer survival through knowledge discovery in databases. Global J Health Sci 7(4):392
2. Al-Allak A, Leonard R, Lewis PD (2010) The naive-bayes decision tree classifier predicting the probability of survival in breast cancer, p4-09-19
3. Al-Bahrani R, Agrawal A, Choudhary A (2017) Survivability prediction of colon cancer patients using neural networks. Health Inf J Sep 1:1460458217720395. https://doi.org/10.1177/1460458217720395
4. Alaa AM, van der Schaar M (2017) Deep multi-task gaussian processes for survival analysis with competing risks. In: Proceedings of the 30th conference on neural information processing systems (NIPS 2017)
5. Abdelghani B, Erhan G (2006) Predicting breast cancer survivability using data mining techniques 58(13):10–110
6. Collins FS, Tabak LA (2014) Nih plans to enhance reproducibility. Nature 505(7485):612
7. Dursun D (2009) Analysis of cancer data: a data mining approach. Expert Syst 26(1):100–112
8. Dursun D, Glenn W, Amit K (2005) Predicting breast cancer survivability: a comparison of three data mining methods. Artif Intell Med 34(2):113–127
9. Dimitoglou G, Adams JA, Jim CM (2012) Comparison of the c4.5 and a naive bayes classifier for the prediction of lung cancer survivability. arXiv:1206.1121
10. Dooling V, Kim A, McAneny B, Webster J (2016) Personalized prognostic models for oncology: a machine learning approach. arXiv:1606.07369
11. Elfiky A, Pany M, Parikh R, Obermeyer Z (2017) A machine learning approach to predicting short-term mortality risk in patients starting chemotherapy. bioRxiv, p 204081
12. Arihito E, Takeo S, Hiroshi T (2008) Comparison of seven algorithms to predict breast cancer survival. Int J Biomed Soft Comput Human Sci J Biomed Fuzzy Syst Ass 13(2):11–16
13. Gao P, Zhou X, Wang Z-N, Song Y-X, Tong L-L, Xu Y-Y, Yue Z-Y, Xu H-M (2012) Which is a more accurate predictor in colorectal survival analysis? Nine data mining algorithms vs. the tnm staging system. PLoS One 7(7):e42015
14. Johnson AEW, Pollard TJ, Mark RG (2017) Reproducibility in critical care: a mortality prediction case study. In: Machine learning for healthcare conference, pp 361–376
15. Khan U, Shin H, Choi JP, Kim M (2008) WFDT: weighted fuzzy decision trees for prognosis of breast cancer survivability. In: Proceedings of the 7th Australasian data mining conference, vol 87. Australian Computer Society, Inc., pp 141–152
16. Juhyeon K, Hyunjung S (2013) Breast cancer survivability prediction using labeled, unlabeled, and pseudo-labeled patient data. J Amer Med Inf Ass 20(4):613–618
17. Mikolov T, Sutskever I, Chen K, Corrado GS, Dean J (2013) Distributed representations of words and phrases and their compositionality. In: Advances in neural information processing systems, pp 3111–3119
18. Rostami SM, Ahmadzadeh M (2017) Extracting predictor variables to construct breast cancer survivability model with class imbalance problem. J AI Data Mining
19. Nam Y, Shin H (2013) A hybrid cancer prognosis system based on semi-supervised learning and decision trees. In: International conference on neural information processing. Springer, pp 640–648
20. Nagesh S, Markus H, Win KT, Yang J (2018) Breast cancer data analysis for survivability studies and prediction. Comput Methods Prog Biomed 155:199–208
21. Ana S, Tiago O, Jose N, Novais P (2016) Treating colon cancer survivability prediction as a classification problem. ADCAIJ: Adv Distri Comput Artif Intell J 5(1):37–50
22. Street WN (1998) A neural network model for prognostic prediction. In: ICML, pp 540–546
23. Kung-Jeng W, Bunjira M, Kun-Huang C, Kung-Min W (2014) A hybrid clas- sifier combining smote with pso to estimate 5-year survivability of breast cancer patients. Appl Soft Comput 20:15–24
24. Wolberg WH, Mangasarian OL (1990) Multisurface method of pattern separation for medical diagnosis applied to breast cytology. Proc Nat Acad Sci 87:9193–9196

25. Zhang J (1992) Selecting typical instances in instance-based learning. In: Proceedings of the ninth international machine learning conference. Morgan Kaufmann, Aberdeen, Scotland, pp 470–479
26. Mangasarian OL, Wolberg WH (1990) Cancer diagnosis via linear programming. SIAM News 23(5):1–18
27. Mangasarian OL, Setiono R, Wolberg WH (1990) Pattern recognition via linear programming: theory and application to medical diagnosis. In: Coleman TF, Li Y (ed) Large-scale numerical optimization. SIAM Publications, Philadelphia, pp 22–30
28. Bennett KP, Mangasarian OL (1992) Robust linear programming discrimination of two linearly inseparable sets. Optim Methods Softw 1:23–34 (Gordon & Breach Science Publishers)
29. Zafiropoulos E, Maglogiannis I, Anagnostopoulos I (2006) A support vector machine approach to breast cancer diagnosis and prognosis. Artif Intell Appl Innov 2006:500–507
30. Ali EEE, Feng WZ (2016) Breast cancer classification using support vector machine and neural network. Int J Sci Res 5(3):1–6
31. Shajahaan SS, Shanthi S, ManoChitra V (2013) Application of data mining techniques to model breast cancer data. Int J Emerg Technol Adv Eng 3(11):362–369
32. Ivancakova J, Babi F, Butka P (2018) Comparison of different machine learning methods on wisconsin dataset. In: IEEE 16th world symposium on applied machine intelligence and informatics, Slovakia, February 7–10
33. Hastie T, Tibshirani R, Friedman J (2009) The elements of statistical learning, 2nd edn. Springer, New York. https://doi.org/10.1007/978-0-387-84858-7

An Opportunistic Ensemble Learning Framework for Network Traffic Classification in IoT Environments

Sujit Bebortta and Sumanta Kumar Singh

Abstract The advancements of Internet of Things (IoT)-based platforms have considerably motivated the growth in malicious attacks to IoT devices and the network infrastructure. This has given rise to information theft and vulnerabilities arising due to the flow of anomalous traffic throughout the network causing a partial or complete interruption in normal network activities. Hence, it is crucial for network administrators and service providers to track the devices associated with a network as well as the type of traffic flowing through it. In relation to this, the present work provides an ensemble-based approach for characterizing the traffic flows pertaining to IoT systems. The proposed framework comprises of adaptive boosting (AdaBoost) algorithm and Naïve Bayes (NB) algorithm for identifying anomalous traffic flow behavior. Further, the performance of the proposed algorithm is comparatively assessed with conventional learning models. It is experimentally observed that the proposed algorithm provides a prediction accuracy of 96.316% along with a sensitivity value of 0.931, positive prediction value (PPV) of 0.992, and F-score of 0.959 for identifying anomalous traffic flows in the network. The proposed algorithm also provides a substantially high AUC-ROC value of 0.991. Hence, it can be evidently observed that the proposed approach may prove to be a befitting choice for designing intrusion detection system (IDS) and management of traffic flow behavior in large heterogeneous networks.

Keywords Internet of Things · Network traffic flow · Machine learning · Intrusion detection system (IDS) · Ensemble learning

1 Introduction

The emergence of Internet of Things (IoT)-based paradigms has leveraged uninterrupted interactions between sensory devices and humans. The various application areas of IoT include smart homes, smart cities, smart healthcare services, smart enter-

S. Bebortta (✉) · S. K. Singh
Department of Computer Science and Engineering, College of Engineering and Technology, Bhubaneswar 751003, India

© The Author(s), under exclusive license to Springer Nature Singapore Pte Ltd. 2022

D. Giri et al. (eds.), *Proceedings of the Seventh International Conference on Mathematics and Computing*, Advances in Intelligent Systems and Computing 1412,
https://doi.org/10.1007/978-981-16-6890-6_35

prises, and so on. These operations require large-scale interconnectivity for exchanging and storing information over different platforms among IoT devices and smart software applications. IoT devices may typically refer to low-power sensor/actuator-based devices which perform autonomous communications among themselves and other application servers in the network. This involves relaying data and information between large autonomous networks and the application servers. The network complexity associated with these growing IoT applications gives rise to network traffic management issues and may consequently lead to adversarial threats over the network.

In accordance with the proliferation observed in IoT-based applications in recent times, several service providers and network operators have encountered difficulty in capturing the device as well as network dynamics. Apart from this, it is also crucial to handle vulnerabilities or anomalous traffic behavior propagated from various attackers or hacked devices in a network. Therefore, it is momentous for network administrators to trace the devices in a network and the type of traffic flowing through the network (i.e., malicious or normal). Hence, most of these tasks can be accomplished by employing a robust intrusion detection system (IDS) for timely tracing and preventing attacks. Several works have proposed the use of different machine learning techniques for efficiently identifying attacks as well as devices compromised for causing the attacks [1–3]. It has been observed that most of the contemporary learning techniques have achieved promising results in training the models and efficiently identifying attacks.

In large-scale IoT-based systems, keeping track of every IoT device connected to the network is crucial to manage. It is also pertinent to manage the type (i.e., normal or malicious) and amount of traffic flow generated from these devices which contribute for impeding the network performance. Different frameworks have been suggested for handling these issues in IoT platforms which could lead to better analysis of IoT devices and the underlying network [4–9]. However, these approaches have some limitations as they may prove to be less capable in handling the dynamic behavior of IoT environments. Therefore, conventional learning algorithms may not efficiently capture all the features of these networks for efficiently differentiating between anomalous and normal traffic flow behavior. In this paper, a robust ensemble-based learning framework for the characterization of network traffic in IoT environments is proposed. Our framework combines the adaptive boosting (AdaBoost) algorithm which is a dynamic ensemble approach with Naïve Bayes algorithm to detect network traffic types from a composite collection of traffic samples. The proposed approach is experimentally observed to accomplish a high prediction accuracy of 96.316% along with a sensitivity value of 0.931, an F-score of 0.959, and a positive predictive value (PPV) of 0.992 for characterizing anomalous traffic flow behavior. The performance of the proposed algorithm is comparatively studied with conventional learning algorithms like Bayesian network (BayesNet), classification and regression tree (CART), Naïve Bayes, and random forest algorithms. The respective results pertaining to the performance of each model are provided. From our study, it is inferred that ensemble-based approach can prove to be much robust in classifying network traffic flows, specifically for IoT-based environments which possess a vast number of traffic features. Hence, these frameworks can also assist in designing autonomous intrusion detection systems (IDSs) for real-time traffic monitoring applications.

2 Background Studies

In this section, a comprehensive background of some state-of-the-art works for network traffic classification in IoT systems based on machine learning approaches is presented. Glennan et al. [10] proposed a semi-supervised fuzzy clustering approach for labeling different classes of traffic. Their proposed technique is proclaimed to handle traffic flows being generated from anonymous applications. The authors propose two basic approaches for enhancing the performance of their proposed network traffic classification framework which uses fuzzy cluster labeling and feature selection techniques. It was observed from their analysis that their proposed framework provided an accuracy of 95.10% and considerably improved the classification performance over every class. Khan et al. [11] proposed a technique based on artificial neural networks (ANNs) for the detection of DDoS amplification attacks. These attacks are beyond the scope of traditional attack detection and isolation strategies as the attack packets associated constitute authentic data. The authors compute the Lyapunov exponent to assess the complexity of the packet flows for analyzing the nature of traffic depending on the computed exponent. Further, they adopted the proposed chaotic neural network algorithm and ANN algorithm for classifying network traffic. A classification accuracy of approximately 98% was achieved for the chaotic model in classifying the network traffic data.

Perera et al. [12] proposed a comparative study for automatic classification of network traffic based on supervised machine learning models. The study accounts for managing the network QoS based on the performance of the underlying communication network. Two feature selection techniques such as correlation-based feature selection and relief attribute evaluation techniques were employed. Further, six machine learning techniques such as NB, DT, BayesNet, NB-Tree, RF, and MLP were used for classifying the network traffic. It was observed from their analysis that RF and DT provided promising results for classifying network traffic with high accuracy and lower computational complexity. Sharma and Gupta [13] proposed a framework for privacy threat analysis in android-based applications. The ease of availability of android-based applications and poor software infrastructures makes them prone to attackers. This leads to stealthy attacks and propagation of malwares into the users' devices. For performing experimental analysis, the authors considered three datasets viz., AndroZoo, M0-Droid, and Drebin for identifying malware and benign applications. The performance of the considered machine learning algorithms was evaluated for performance parameters like sensitivity, specificity, F-score, and accuracy. The model achieved the best accuracy of 92.98% and an F-score of 92.25% in recognizing the samples appended in the dataset. From their study, it was inferred that the proposed machine learning technique can be used efficiently for identifying and analyzing permissions for android-based malwares.

Rathore and Park [14] provided a framework for classifying different IoT security attacks by analyzing network traffic characteristics. These attacks are identified by considering an attack detection mechanism working over a given network at a central point capable of collecting and analyzing network data. The authors have also suggested employing a semi-supervised learning approach over supervised machine learning algorithms for detecting IoT device attacks for fog computing environments. It was proclaimed that their proposed scheme assisted in handling issues with unlabeled datasets and most QoS requirements of IoT-based devices such as lower latency, scalability, resource constraints, and ubiquity. The authors employed the NSL-KDD dataset for performing their evaluations, and an accuracy of 86.53% was achieved for their proposed approach. Kozik et al. [15] proposed the use of extreme learning machines (ELMs) model for facilitating automatic IoT device attack detection mechanism over cloud computing platform. Their framework leverages the inherent capabilities of a cloud platform for providing a deep analysis of network traffic and classification of attacks. This is, however, contrary to resource constraint edge computing platforms as such applications may require building and upgrading the detection models for training voluminous traffic data. Hence, their work facilitates the use of ELM models pre-built over the cloud for handling massive traffic data. It was observed that their approach can be efficiently used for performing deep network inspection and can further be implemented for the analysis and classification of aggregated traffic corresponding to edge nodes.

3 Proposed Framework

The approach suggested in this paper combines an ensemble learning algorithm, i.e., AdaBoost algorithm with NB algorithm, which is a statistical learning model. Figure 1 depicts the architecture of the proposed framework where the IoT network traffic data is initially collected over different IoT networks through specific network traffic capturing tools [16]. These data are further supplied to the ensemble learning module which forms the attack detection framework by combining AdaBoost and NB algorithms to train the model. After training the model with labeled instances for the test data, it is loaded into the system upon which various performance parameters are comparatively analyzed. Based upon the prediction outcome from the test data, the model is considered to develop IDS for detecting malicious network activities. The IoT controller module as well as the network administrator also play a crucial role to this end in identifying the devices responsible for generating malicious traffic and managing the number of IoT devices associated with a network. These observations act as key attributes toward security enforcement mechanisms in IoT and preserving the network policies. In the succeeding subsection, we provide a background of each approach used for developing the proposed framework viz., AdaBoost and NB algorithms. Further, the performance obtained for the proposed ensemble-based algorithm is comparatively studied with some conventional machine learning models such as BayesNet, CART, NB, and RF.

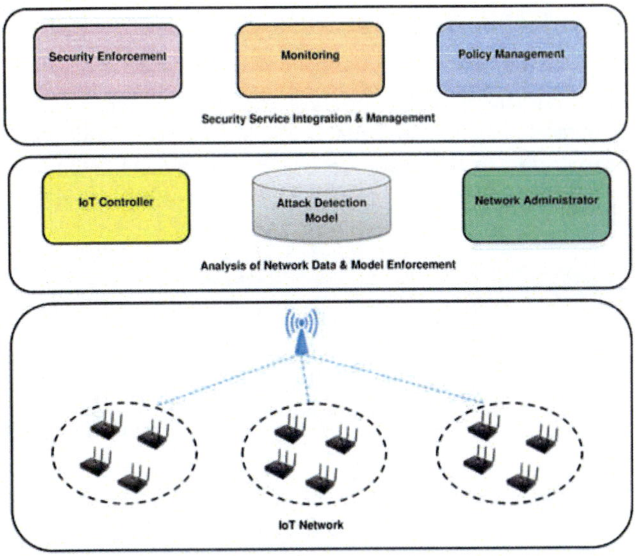

Fig. 1 Architecture of IoT-based network traffic framework

3.1 AdaBoost Algorithm

The adaptive boosting (AdaBoost) algorithm is a popularly used ensemble learning approach that is used to learn from datasets by constructing an explicit set of learners [17, 18]. It trains the learners and hence combines the joint prediction accuracy of these learning models to maximize the prediction outcome. The AdaBoost algorithm can comprise multiple base learners or a single base learner. In this paper, we employ a single base learner to train the model, i.e., the NB algorithm, which is a statistical learning technique. Here, AdaBoost algorithm considers a set of homogeneous learners pertaining to the NB algorithm. The AdaBoost algorithm further boosts the performance of these learners by generating accurate prediction outcomes.

Now considering a binary classification problem, the AdaBoost algorithm can be expressed as a composite weighted additive formulation for ith base learners as

$$F(x) = sign\left(\sum\nolimits_i \alpha_i k_i(x)\right), \tag{1}$$

where α_i represents linear composite coefficient for $k_i(x)$ base learners for x sample data, such that $i = 1, 2, \ldots, n$ and $\alpha_i > 0$.

3.2 Naïve Bayes Algorithm

The NB algorithm performs classification tasks by assuming each feature as independent for a given dataset [19]. This algorithm implements Bayes' rule for identifying class labels corresponding to a given dataset based on maximal likelihood hypothesis [20]. Hence, for any given dataset, we can obtain the probability for each attribute as

$$P\left(C_i|a_1, a_2, \ldots, a_j\right), \quad (2)$$

where C_i represents ith class variables for $i = 1, 2, \ldots, n$, and a_j represents the respective attributes corresponding to each class such that $j = 1, 2, \ldots, m$ and $a_j \in A$.

Hence, employing Bayes' rule, the conditional probability for Eq. (2) is given as follows:

$$P\left(C_i|A\right) = \frac{P\left(C_i\right) P\left(A|C_i\right)}{P\left(A\right)}. \quad (3)$$

According to [19], the joint probability function for A attributes over C_i can be obtained as

$$P\left(A|C_i\right) = \prod_{j=1}^{m} P\left(a_j|C_i\right). \quad (4)$$

Now, in order to predict the network traffic behavior for a given dataset, the maximum a posteriori (MAP) function can be computed by maximizing Eq. (4) as follows:

$$P\left(C|A\right) = \underset{i \in \{1,2,\ldots,n\}}{\arg\max}\ P\left(C_i\right) \prod_{j=1}^{m} P\left(a_j|C_i\right), \quad (5)$$

where $\prod_{j=1}^{m} P\left(a_j|C_i\right)$ is the joint probability function (as obtained in Eq. (4)) which provides the weighted multiplicative combination of attribute set A for C_i classes, respectively.

4 Results and Discussions

In this section, we report the performance metrics used for evaluating the performance of the proposed ensemble learning technique for network traffic classification in convergence with other well-known machine learning models. We then provide the experimental results obtained from our analysis in predicting both normal as well as anomalous traffic for IoT scenarios.

4.1 Performance Evaluation

In order to assess the performance of our proposed ensemble learning framework, we have computed sensitivity, positive predictive value (PPV), F-score, prediction accuracy, and ROC curves for each model. The sensitivity (or, recall) value can be represented as the ratio between true positives (t_p) and the sum of true positives and false negatives (f_n). Here, t_p signifies the amount of anomalous traffic accurately identified from the sample data, whereas f_n indicates the anomalous traffic flows which have been inaccurately detected by the learning model. The sensitivity for a given model is represented as follows:

$$Sensitivity = \frac{t_p}{(t_p + f_n)}. \tag{6}$$

Now, PPV provides accurately identified attacks in traffic flows from the entire sample space which has been detected as anomalous. It can be expressed as the ratio between t_p and $(t_p + f_p)$, and it is given as

$$PPV = \frac{t_p}{(t_p + f_p)}, \tag{7}$$

where f_p indicates false positive values.

The F-score is a crucial measure to assess the predictive competency of a model and can be obtained by considering harmonic mean of PPV and sensitivity values (as obtained in Eqs. (6) and (7), respectively). Hence, F-score for a learning model can be obtained as

$$F - score = 2 \times \left(\frac{PPV \times Sensitivity}{PPV + Sensitivity} \right). \tag{8}$$

The accuracy of a model provides the amount of correctly classified samples (i.e., normal and anomalous traffic types) from the entire sample space. Thus, the accuracy of a model (expressed in %) can be provided as follows:

$$ACC = \frac{t_p + t_n}{(t_p + t_n + f_p + f_n)} \times 100. \tag{9}$$

Further, the ROC curves indicate the predictive performance of a learning model with respect to certain thresholds. The ROC curves which are close to 1 indicate an optimally performing prediction model. The ROC curves indicate probability curves for a specific prediction model which are plot against the true positive rate (TPR) and false positive rate (FPR) [21–23]. Here, a comparison between the ROC curves pertaining to our proposed ensemble-based framework and conventionally used machine learning models viz., BayesNet, CART, NB, and RF are presented for the classification of network traffic flows. Further, the area under curve (AUC) for the respective ROC curves provides a precise summary of ROC threshold values. It can be used to distinguish between anomalous and normal classes of traffic. In general, a high AUC-ROC value is desired, where a value close to 1 indicates a better performing model.

4.2 Experimental Results

In this section, we report the results obtained by performing traffic analysis over the NSL-KDD dataset for performance metrics discussed in the above section. The dataset comprises specifically four network attack types viz., user to root (U2R), probing, DoS, and remote to local (R2L) attacks. However, the instances are categorized into normal and anomalous which facilitates a binary categorization of the dataset. The training data comprises 90% of the entire sample with 125,973 instances, whereas the testing data has 22,544 instances. The dataset has been widely used for identifying malicious traffic inputs and specifically is used for capturing advanced Internet traffic behavior [24].

In Fig. 2a, the performance of the proposed ensemble learning framework and conventional learning algorithms is presented for characterization of normal network traffic flows. The performance metrics considered here are sensitivity, PPV, and F-score. It is observed that the proposed ensemble-based algorithm, i.e., AdaBoost-NB provides the highest values for the considered performance metrics. The sensitivity value of the proposed algorithm is 0.992, the PPV value is 0.942, and F-score is 0.966. The respective performance metrics for all considered algorithms are presented in Table 1, where the optimal values are highlighted using bold fonts.

The performance of the algorithms in identifying anomalous traffic flows is presented in Fig. 2b. It is observed that the proposed ensemble-based approach outperforms the efficiency of conventionally used learning algorithms in characterizing malicious traffic behavior in the network. The proposed approach provides a remarkably high F-score of 0.959, a sensitivity value of 0.931, and a PPV of 0.992. It can also be observed that the RF algorithm also provides a fairly high PPV value of 0.971. Here, it is worth noting that the PPV value plays an essential role in identifying anomalous traffic flows as it considers the accurately detected malicious flows from the total sample which has been detected as malicious in a given sample space. Further, toward this end, the efficacy of obtaining high sensitivity value can also be realized as it provides the true detection rate of anomalous flows from the entire sample space. Table 2, provides performance metrics obtained for the proposed ensemble learning approach (i.e., AdaBoost-NB) along with conventional learning methods for characterizing anomalous traffic flow behavior.

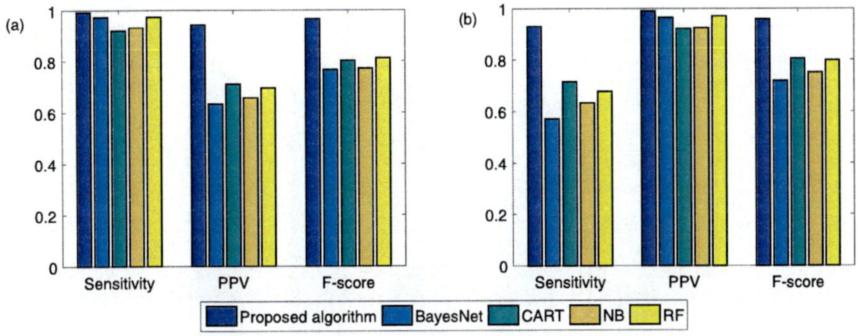

Fig. 2 Performance of learning algorithms in characterizing different traffic types **a** Normal traffic flows and **b** Abnormal traffic flows

Table 1 Performance metrics for the proposed ensemble algorithm and conventional learning algorithms for normal traffic flow

Algorithms	Sensitivity	PPV	F-score
Proposed algorithm	**0.992**	**0.942**	**0.966**
BayesNet	0.973	0.632	0.766
CART	0.919	0.711	0.801
NB	0.931	0.657	0.771
RF	0.973	0.695	0.811

Table 2 Performance metrics for the proposed ensemble algorithm and conventional learning algorithms for anomalous traffic flow

Algorithms	Sensitivity	PPV	F-score
Proposed algorithm	**0.931**	**0.992**	**0.959**
BayesNet	0.571	0.965	0.718
CART	0.715	0.921	0.805
NB	0.633	0.924	0.751
RF	0.677	0.971	0.798

The prediction accuracy for each learning algorithm is represented in Fig. 3. It is observed that the proposed ensemble-based algorithm provides the highest prediction accuracy of 96.316%. Further, the RF algorithm provides the next highest prediction accuracy of 80.452% which is much lower than the accuracy of the proposed approach for network traffic characterization. The ROC curves for all the learning models are provided in Fig. 4. The proposed algorithm obtains a steep ROC curve and its performance converges with 1, hence indicating high predictive power for the proposed ensemble-based algorithm. Table 3 provides the prediction accuracy

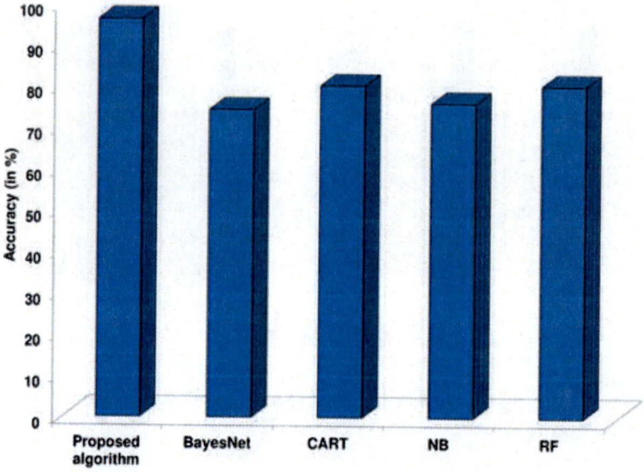

Fig. 3 Accuracies for the proposed ensemble algorithm and conventional learning algorithms

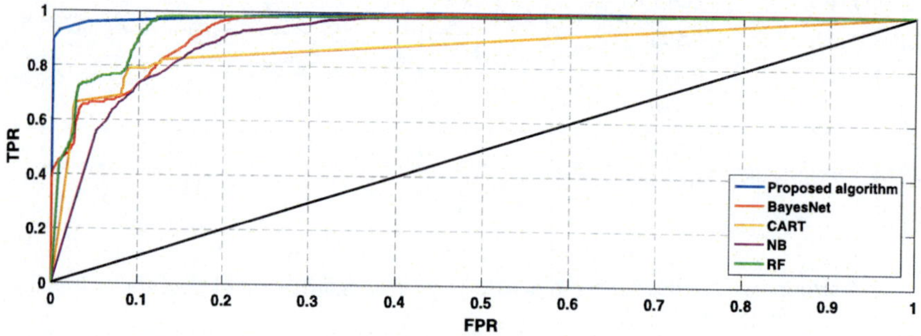

Fig. 4 ROC curves for the proposed algorithm and conventional algorithms

Table 3 Prediction accuracy and AUC-ROC for learning algorithms

	Proposed algorithm	BayesNet	CART	NB	RF
Accuracy (in %)	**96.316**	74.432	80.323	76.118	80.452
AUC-ROC	**0.991**	0.945	0.878	0.908	0.959

pertaining to each algorithm expressed in % along with the respective AUC-ROC values. It is clearly indicated from the table that the proposed algorithm proves to be much efficient in characterizing network traffic as compared to conventional learning models.

5 Conclusion and Future Challenges

In this paper, we studied the role of network traffic analysis in dynamic IoT-based systems. Pertaining to our observations, an ensemble-based learning framework combining AdaBoost and NB algorithm was suggested. The efficacy of the proposed framework was studied in convergence with some conventional learning algorithms like BayesNet, CART, NB, and RF. It was observed from our experimental results that the proposed algorithm provided a substantially high predictive performance of 96.316% along with F-score, PPV, and sensitivity values of 0.959, 0.992, and 0.931, respectively, for detecting anomalous traffic flows. Further, the AUC-ROC value for the proposed algorithm is obtained as 0.991. Thus, the proposed ensemble-based framework provides comparatively higher predictive competency in precisely identifying the traffic features.

For achieving real-time network traffic analysis, a regular and timely assessment of IoT traffic behavior is required to identify abnormalities in their intrinsic patterns of network usage. Further, a major drawback in developing IDS for IoT systems is to obtain sufficiently labeled traffic datasets. Hence, in such scenarios, the role of futuristic techniques like semi-supervised and unsupervised learning approaches comes into importance.

References

1. Salman O, Elhajj IH, Chehab A, Kayssi A (2019) A machine learning based framework for IoT device identification and abnormal traffic detection. Transactions on emerging telecommunications technologies, e3743
2. Panda S, Panda G (2020) Intelligent classification of IoT traffic in healthcare using machine learning techniques. In: 2020 6th international conference on control, automation and robotics (ICCAR). IEEE, pp 581–585
3. Singh SK, Mishra AK, Rain fall prediction using bigdata analytics
4. Yao H, Gao P, Wang J, Zhang P, Jiang C, Han Z (2019) Capsule network assisted IoT traffic classification mechanism for smart cities. IEEE Int Things J 6(5):7515–7525
5. Tahaei H, Afifi F, Asemi A, Zaki F, Anuar NB (2020) The rise of traffic classification in IoT networks: A survey. J Network Comput Appl 154:102538
6. Bebortta S, Singh AK, Mohanty S, Senapati D (2020) Characterization of range for smart home sensors using Tsallis' entropy framework. In: Advanced computing and intelligent engineering. Springer, Singapore, pp 265–276
7. Bebortta S, Panda M, Panda S (2020) Classification of pathological disorders in children using random forest algorithm. In: 2020 international conference on emerging trends in information technology and engineering (ic-ETITE). IEEE, pp 1–6
8. Bebortta S, Senapati D, Rajput NK, Singh AK, Rathi VK, Pandey HM, ... Tiwari P (2020) Evidence of power-law behavior in cognitive IoT applications. Neural Comput Appl 1–13
9. Kumar A, Lim TJ (2019) EDIMA: early detection of IoT malware network activity using machine learning techniques. In: 2019 IEEE 5th world forum on internet of things (WF-IoT). IEEE, pp 289–294
10. Glennan T, Leckie C, Erfani SM (2016) Improved classification of known and unknown network traffic flows using semi-supervised machine learning. In: Australasian conference on information security and privacy. Springer, Cham, pp 493–501

11. Khan MS, Ferens K, Kinsner W (2014) A chaotic complexity measure for cognitive machine classification of cyber-attacks on computer networks. Int J Cogn Inf Natural Intell (IJCINI) 8(3):45–69
12. Perera P, Tian YC, Fidge C, Kelly W (2017) A comparison of supervised machine learning algorithms for classification of communications network traffic. In: International conference on neural information processing. Springer, Cham, pp 445–454
13. Sharma K, Gupta BB (2019) Towards privacy risk analysis in Android applications using machine learning approaches. Int J E-Services Mobile Appl (IJESMA) 11(2):1–21
14. Rathore S, Park JH (2018) Semi-supervised learning based distributed attack detection framework for IoT. Appl Soft Comput 72:79–89
15. Kozik R, Choraś M, Ficco M, Palmieri F (2018) A scalable distributed machine learning approach for attack detection in edge computing environments. J Parallel Distr Comput 119:18–26
16. Lin H, Yan Z, Fu Y (2019) Adaptive security-related data collection with context awareness. J Network Comput Appl 126:88–103
17. Hastie T, Tibshirani R, Friedman J (2009) The elements of statistical learning: data mining, inference, and prediction. Springer Science & Business Media, Berlin
18. Zhou ZH (2012) Ensemble methods: foundations and algorithms. CRC Press, Boca Raton
19. Rish I (2001) An empirical study of the naive Bayes classifier. In: IJCAI 2001 workshop on empirical methods in artificial intelligence, vol 3, No 22, pp 41–46
20. Moustafa N, Turnbull B, Choo KKR (2018) An ensemble intrusion detection technique based on proposed statistical flow features for protecting network traffic of internet of things. IEEE Int Things J 6(3):4815–4830
21. Adekitan AI, Abolade J, Shobayo O (2019) Data mining approach for predicting the daily Internet data traffic of a smart university. J Big Data 6(1):11
22. Rahmat S, Niyaz Q, Mathur A, Sun W, Javaid AY (2019) Network traffic-based hybrid malware detection for smartphone and traditional networked systems. In: 2019 IEEE 10th annual ubiquitous computing, electronics & mobile communication conference (UEMCON). IEEE, pp 0322–0328
23. Gratian M, Bhansali D, Cukier M, Dykstra J (2019) Identifying infected users via network traffic. Comput Secur 80:306–316
24. Khonde SR, Ulagamuthalvi V (2019) Ensemble-based semi-supervised learning approach for a distributed intrusion detection system. J Cyber Secur Technol 3(3):163–188

Predicting and Analysing Pedestrian Injury Severity: A Machine Learning-Based Approach

Anjali Rao, Sobhan Sarkar, Anima Pramanik, and J. Maiti

Abstract Pedestrians have a higher risk of casualties on the road. To improve the situation, it is essentially needed to understand the factors contributing to fatal accidents and their contributions to frame better regulations. Yet there is a dearth of substantial superior indexes that classifies the severity of the pedestrian injury. Machine learning (ML) has been successfully implemented in many domains but it is not widely utilized in the development of pedestrian road safety rules and regulations. In this paper, an ML approach is presented to generate a set of rules for predicting the severity of an accident based on several factors, including the number of vehicles, speed of vehicles, and environment. Out of thirty-four variables, 20 variables are selected using the Chi-square technique. Three popular ML algorithms, namely Support Vector Machine (SVM), Random Forest (RF), and Decision Tree (DT) are used for the prediction of injury severity. From experiments, RF and DT are found to have similar accuracy scores, and hence, DT has been selected since it is simple to interpret and to generate a set of 'if-then' rules. A total of 20 safety decision rules are retrieved from the dataset, which explains the factors behind the severity of the pedestrian injury.

Keywords Pedestrian injury · Machine learning · Prediction · Hyper-parameter tuning · Safety rules

1 Introduction

A pedestrian can be defined as a person walking or running on foot for either short-distance travelling for relaxation activities. Pedestrians are more exposed to severe injuries. According to the Traffic Safety Facts released by National Highway Traf-

A. Rao · A. Pramanik · J. Maiti
Department of Industrial and Systems Engineering, IIT Kharagpur, Kharagpur 721302, India

S. Sarkar (✉)
Business School, University of Edinburgh, 29 Buccleuch Pl, Edinburgh EH8 9JS, UK
e-mail: sobhan.sarkar@ed.ac.uk

© The Author(s), under exclusive license to Springer Nature Singapore Pte Ltd. 2022

D. Giri et al. (eds.), *Proceedings of the Seventh International Conference on Mathematics and Computing*, Advances in Intelligent Systems and Computing 1412,
https://doi.org/10.1007/978-981-16-6890-6_36

fic Safety Administration,[1] pedestrian fatalities in total vehicle fatalities have been constantly growing, from 11 to 16% [7]. In any accident, the severity of an injury to a pedestrian could be accredited to various factors, including the speed of the vehicle, and environment [8]. In India, there has been increasing attention on pedestrian zones which are being established. If a non-polluting and safer pedestrian environment is ensured, more walking trips are prominent. A pedestrian injury could be reduced only if pedestrian-vehicle conflicts can be eliminated [30]. There has been intensive research observed to determine the factors influencing pedestrian injury. To reduce conflicts or casualties, it is thus very essential to have effective precautionary measures and road safety strategies.

Recognizing factors that affect pedestrian-vehicle crash and then identifying how these factors affect severity of a crash are critical in planning and implementing road safety management techniques. The studies dealing with crash and severity prediction are increasing over time [3, 4, 9, 31]. Many researchers have, therefore, realized a need to establish a predictive machine learning (ML) model which can connect possible and explanatory factors, including characteristics of the traffic, environment on road, and human behaviour with the probability and severity of the pedestrian injury. They have also ventured to model these crashes to identify factors responsible for it [5, 6, 8, 25]. ML techniques have specialty in classification tasks [17, 19, 29], which is considered to be efficient than other conventional approaches used in safety domain [26]. Training a classifier for a particular function is, from a geometric perspective, equivalent to finding the equation which best separates different classes into feature spaces in a multidimensional surface [22]. Hence, a discriminative model builds up a decision boundary between the classes. Visible by its numerous examples as exercised in a different spectrum, ML continues to be lauded as an efficient tool for prediction [10]. Even today, ML has been extensively used in prediction modeling, for example, forecasting, classification, and diagnosis. ML is also capable to map non-linear and complex phenomena into useful equations and models that, in turn, help management to make better decisions [18, 21]. ML-based methods can be deployed more efficiently to predict the severity as they have been successfully applied in different fields and disciplines.

Although there are many studies conducted in pedestrian accident analysis, predictive modeling of pedestrian injury severity using ML is rather new. Moreover, it has also been observed that the use of hyper-parameter tuning of the ML algorithms has not been addressed in earlier studies. Another important observation is related to the recommendation of accident analysis, which is often done using either normal statistical approach or qualitative approach. Recommendation using rule generation [23] or pattern extraction [13, 14] for pedestrian safety by ML model is relatively a new topic of research.

[1] https://www.nhtsa.gov/road-safety/pedestrian-safety.

1.1 Research Issues

From the above-mentioned brief discussion, the following research issues can be identified, which are as follows:

(i) Very less amount of studies are conducted so far on pedestrian severity prediction using ML approaches.
(ii) None of the approaches have presented hyper-parameter tuning of classifiers for obtaining better prediction.
(iii) Although the factors for accidents are described, analysis of factors using rule generation has not been carried out so far.

1.2 Contributions of the Work

Based on the issues mentioned above, the present study contributes to the body of literature as follows:

(i) ML-based model has been proposed to predict the pedestrian severity.
(ii) Grid search has been adopted for hyper-parameter tuning of the ML-based classifiers.
(iii) Decision tree (tuned) has been used to generate rules for analysis of the factors attributable to accidents.

Therefore, the objective of the present study is to predict the severity of pedestrian injury using ML techniques. In our study, three popular, as well as powerful ML algorithms or classifiers, namely, Decision Tree (DT), Support Vector Machine (SVM), and Random Forest (RF), have been used. Their parameters are tuned by using a grid search technique. Besides, the preprocessing of data has been discussed. Missing data is handled by using RF imputation technique. In this study, since we deal with 34 attributes initially, it is very essential to reduce the dimensionality of the dataset. The 24 attributes are finally considered to be most important according to the Chi-square technique. The decision tree is evaluated as the best one to have maximum accuracy and robustness from the analyses. Through this study, we are trying to classify the severity of an accident as fatal, serious or slight based on different responses recorded after the occurrence of the accident. A list of implementable recommendations based on rules generated can thus be very effective for the management to reduce the injury severity. Therefore, rule extraction, as well as interpretation, is a very essential task in road accident analysis and prevention.

The remainder of the paper is organized as follows. Section 2 describes the materials and methods used for this study. Section 3 provides the results and inferred rules from the classification tree and finally, Sect. 4 provides a conclusion of the study with a set of scopes for future works.

2 Materials and Methods

The road crash data used in this study are collected from in STATS19 dataset and provided by the English Department of Transport.[2] The data refers to crashes occurred in Great Britain (GB) in the past three years 2016–2018. All reported crashes occurred on a public highway (including footways), in which at least one vehicle or a vehicle in collision with a pedestrian is involved and which becomes known to the police within 30 days of its occurrence. The data on pedestrian crashes (1000 incidents) consist of 34 variables.

2.1 *Data and Data Pre-processing Stage*

In this stage, missing data is imputed by using the RF algorithm. The strategy that has been incorporated for RF missing data imputation is growing a random forest using each variable that has missing values, and then prediction of the missing values is done using the grown forest. Next, we iterate the algorithm for better results. For imputing values, RF method is preferred, as this approach can capture interactions between different independent variables and non-linearity of features more effectively, and thus can help avoid the over-fitting issue by scaling the data into higher dimensions [27]. The data initially obtained are from a large dataset. Hence, feature selection is required since: (i) it eliminates over-fitting as there is less opportunity to make decisions based on noise and (ii) training time is significantly reduced. For feature selection, Chi-Square technique has been used which enable us to identify 20 features/variables as important, and therefore, they are used for further analyses.

2.2 *Methods*

We are interested in classifying a pedestrian injury severity into three main classes: 'Fatal', 'Serious', and 'Slight'. Thereafter, we extract rules that lead to the occurrence of each class. There are three ML classifiers, namely, decision tree, random forest, and support vector machine used in this study, which are briefly discussed below. The methods used in this study have been coded with the help of both R [20] and Python.

Classification/Decision Tree (DT): In DT modeling, classification, and regression tree (CART) algorithm is used, where classification is performed through the generation of trees [2]. In the feature space, separate regions are generated. The most probable value of the dependent attribute in each of the regions is then predicted. The algorithm employs different measures of impurity, such as Gini index and Entropy

---

[2] Link: www.gov.uk.

Table 1 The optimal values of parameters used for each algorithm

Algorithms	Parameters used	Optimal values
DT	Criterion	Entropy
	max_depth	5
	min_samples_leaf in tree	30
	min_samples_split	2
RF	Criterion	Entropy
	n_estimators	20
	max_depth	15
	min_sample_leaf	10
	min_sample split	15
SVM	Kernel	RBF
	Gamma	0.01
	Regularization	0.729

for splitting. The main disadvantage comes from the higher complexity in computation. Hence, pruning is done on the tree to minimize the complexity [22]. Using Grid search, hyper-parameters of it, namely, 'maxdepth', 'minleaf', 'impurity', and 'minsamplesplit' are tuned (refer to Fig. 3a–d), and their optimal values are reported in Table 1.

Random Forest (RF): To improve stability of DT, [1] proposed the RF method. It constructs a large number of classification trees by bootstrapping and employing random feature selection. The RF selects the predicted class as the class which receives maximum votes considering all trees together [22]. Using Grid search, hyper-parameters of it, namely, 'regularization', 'kernel', and 'min_sample_split' are tuned (refer to Fig. 3j–l), and their optimal values are reported in Table 1.

Support Vector Machine (SVM): SVM, developed by [28], is a popular ML technique used for classification analysis. Theoretically, it explores less over-fitting problem and exhibits a better generalization ability. However, the main issue in building a SVM model is to adequately obtain training parameter values else poor prediction can be occurred [24]. Using Grid search, hyper-parameters of it, namely, 'n_estimators', 'impurity', 'max_depth', and 'min_sample_leaf' are tuned (refer to Fig. 3e–i), and their optimal values are reported in Table 1.

3 Results and Discussion

3.1 *Training and Validation Design and Attribute Importance*

In this study, we have used a k-fold cross-validation (CV) for model training and validation. It is one of the quickest and easiest ways to re-sample the data. The process involves choosing a value for only a single parameter called k. To determine the optimal value of k, a bias-variance trade-off is considered. Keeping the considerations in view, we are setting $k = 5$, so that the test error rates do not suffer either from high bias or from very high variance. In this study, we have used the Chi-square technique for evaluating attribute importance [24]. From Fig. 1, it is evident that 'Number of pedestrians' is the most important attribute; whereas, 'Junction details' is the least important one. The attributes, namely, 'Vehicle types', 'X1-st Road class', 'Pedestrian age' are also found important.

Modeling and Hyper-Parameter Tuning: Three popular ML algorithms, namely, Random Forest, SVM, and Decision Tree have been used to predict pedestrian injury severity by first training the models on the preprocesses dataset. All the models' Hyperparameters have been tuned for optimized results using the grid search. The experimental results are exhibited in Fig. 2.

Evaluation of Algorithms: The measures, namely accuracy, precision, recall, and F1-Score have been calculated from the confusion matrix. Two other measures have been calculated, namely: AUC and G-Mean (refer to Fig. 3a). In the domain of road

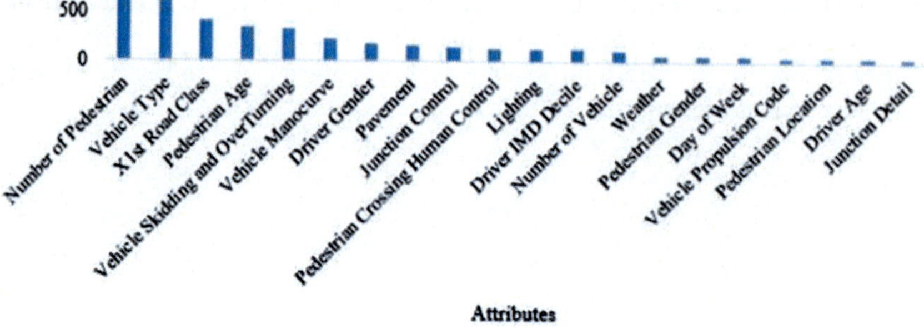

Fig. 1 Evaluation of feature/attribute importance

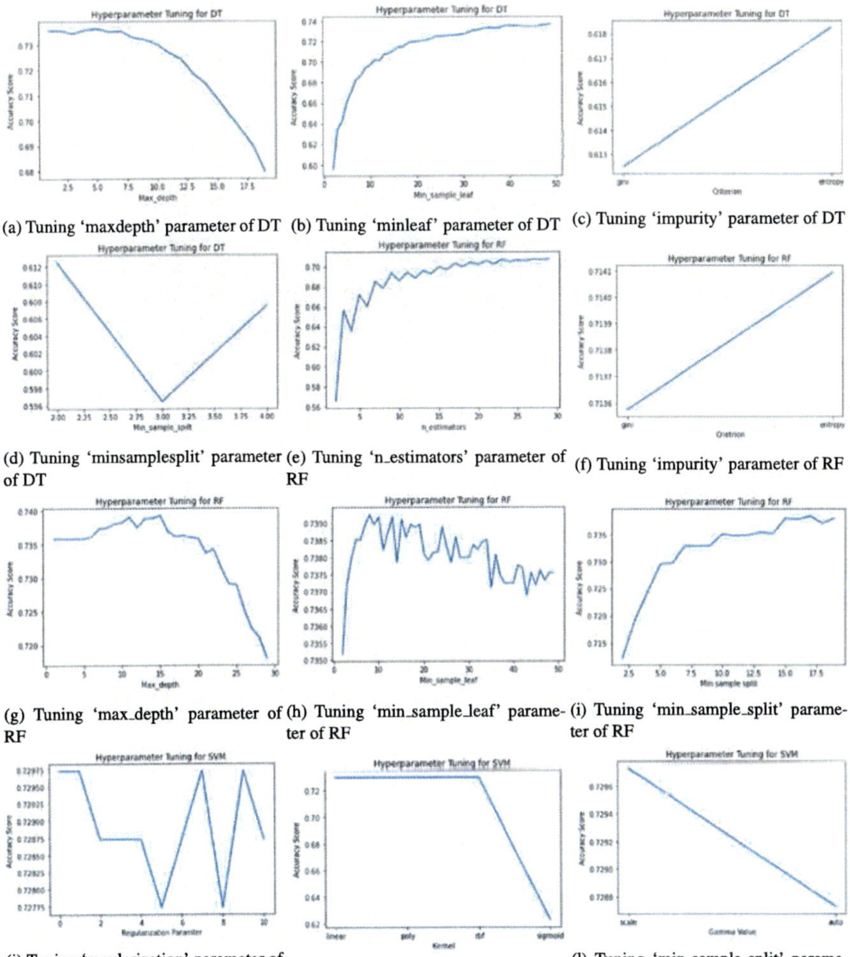

(a) Tuning 'maxdepth' parameter of DT (b) Tuning 'minleaf' parameter of DT (c) Tuning 'impurity' parameter of DT

(d) Tuning 'minsamplesplit' parameter of DT (e) Tuning 'n_estimators' parameter of RF (f) Tuning 'impurity' parameter of RF

(g) Tuning 'max_depth' parameter of RF (h) Tuning 'min_sample_leaf' parameter of RF (i) Tuning 'min_sample_split' parameter of RF

(j) Tuning 'regularization' parameter of SVM (k) Tuning 'kernel' parameter of SVM (l) Tuning 'min_sample_split' parameter of SVM

Fig. 2 Experimental results on hyper-parameter tuning of classifiers

safety, the false negative is more dangerous than false positive. Hence, recall is the best suitable measure to take into account for our analysis. From experiments, DT and RF are out-performing SVM provided with the prediction performance with a recall value of 0.98 as compared to SVM. (refer to Fig. 3a). Additionally, in Fig. 3b, the receiver operating characteristic (ROC) curves for all three classifiers are depicted.

Robustness Checking: Robustness is the capacity of an ML model to remain unaffected by small, deliberate variations in model parameters. A plot is created showing a box and whisker summarizing the distribution of accuracy for each algorithm eval-

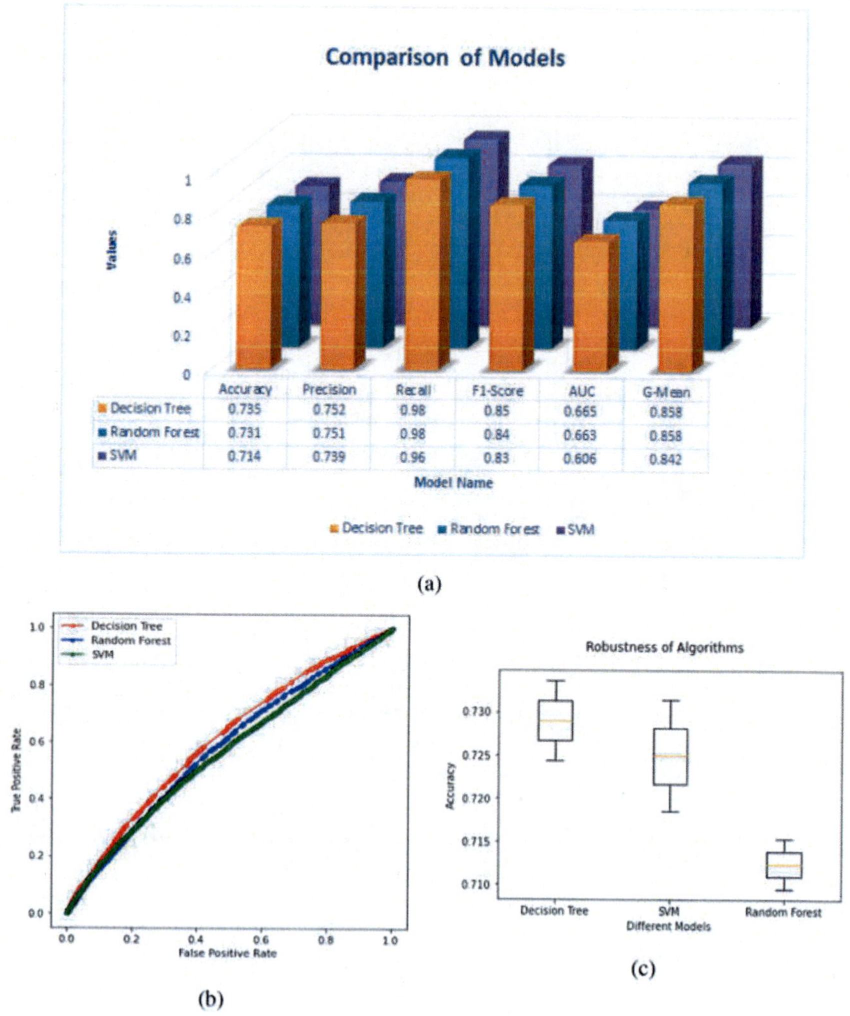

Fig. 3 **a** Comparison of three models based on different performance metrics, **b** ROC curves for all three classification algorithms, and **c** Robustness checking for all classifiers based on accuracy

uated, which is illustrated in Fig. 3c [21]. From this figure, it is found that RF is more robust than the other two algorithms.

Significance Testing: After the robustness checking, we investigate the existence of significant differences among the performances of the algorithms. A non-parametric statistical hypothesis known as Wilcoxon Signed-Rank test has been used for comparisons between algorithms at a significance level ($\alpha = 0.05$) [21, 22]. The value of statistics given in Table 2 indicates that there exists no significant difference in the prediction made by those models.

Table 2 Results of Wilcoxon Signed-Rank test with $\alpha = 0.05$

Classifier	DT	RF	SVM
DT	*		
RF	Statistics = 18.5, p = 0.072	*	
SVM	Statistics = 21.0, p = 0.059	Statistics = 20.6, p = 0.068	*

3.2 *Rules Generation and Interpretation*

The DT has been selected as the prediction model. Thereafter, safety decision rules have been extracted using it. The generation of meaningful rules to classify pedestrian injury severity is one of the major parts in our study. Each rule has some conditions depending on which either of the 'Fatal', or 'Serious' or 'Slight' is predicted. A total of 20 safety decision rules have been obtained from the DT algorithm and reported in Table 3. Some of the rules are discussed here. For example, *Rule 1* predicts severity as serious based on attributes vehicle manoeuvers, road type, no of vehicles present, and weather. It is observed when the driver was reversing on a one way street with foggy weather conditions there's a possibility of a serious accident. In another example, *Rule 5*, predicts severity as fatal based on attributes such as vehicle manoeuvers, Speed_limit, Pavement and Day of the week. It is predicted that in a case where

Table 3 Rules generated for explaining pedestrian injury severity

Rule No.	Rules	Severity category
1.	If(['VM']<= Reversing) & (['RT']<= One way street) & (['PCPF']<= Central Refuge) & (['JD']<= Not at junction) & (['VM']<= Others) & (['NV']<= >2) & (['DW']<= weekend) & (['Wea']<= Fog/Mits)	Serious
2.	If['VM']<=Turning Left/Right/U) & (['RT']<= One way street) & (['PCPF']<= Foot Bridge) & (['JD']<= More than four arms) & (['VM']>Moving off) & (['VSO']>Yes)) & (['Pav']<= Dry) & (['VJL']<= 4.5) &	Slight
3.	If(['VM']<= Turning Left/Right/U) & (['RT']<= One way street) & (['PCPF']<=Central Refuge) & (['JD']>Not at junction) & (['VSO']>Yes) & (['VTA']<= Articulated Vehicle) & (['VM']>Overtaking) (['Ar']<= Urban) & (['DW']<= Weekday)	Serious
4.	If(['VM']<= Turning Left/Right/U) & (['RT']<= Dual Carriageway) & (['PCPF']>Central Refuge) & (['VM']<= Overtaking) & (['Wea']>Fine/No High Winds) & (['VTA']<= No Tow/Articulation) & (['VM']<= Going Ahead) & (['VJL']<= Not at junction) & (['JD']<= Mini-Roundabout)	Serious
5.	If(['VM']>=Turning Left/Right/U) & (['PCPF']>Central Refuge) & (['Pav']<= Wet or damp) & (['S']<= >=50) & (['XRC.1']<= A) & (['XRC']>B) & (['DW']<= Weekend) & (['NV']>>2)	Fatal

(continued)

Table 3 (Continued)

Rule No.	Rules	Severity category
6.	If ([‘VM’]<= Turning Left/Right/U) & ([‘RT’]>Roundabout) & ([‘JD’]>Not at junction) & ([‘Pav’]<=Wet or damp) & ([‘Lt’]<= Darkness/Lightning unknown) & ([‘VSO’]<= No) & ([‘RT’]<= One way street) & ([‘VM’]<= Overtaking) & ([‘VT’]<= Bicycle)	Serious
7.	If([‘PCHC’] <= School/Crossing/ Patrol) && ([‘VJL’] <= Not at junction) && ([‘XRC’] >C) && ([‘LT’]= Daylight)	Slight
8.	If ([‘VM’]<= Reversing) & ([‘RT’]>One way street) & ([‘JD’]>Not at junction) & ([‘Pav’]<= Wet or Damp) & ([‘Lt’]<= Darkness) & ([‘VSO’]>Yes) & ([‘RT’]>Roundabout) & ([‘JC’]>Auto Traffic Signal)	Serious
9.	If([‘VM’]>Turning Left/Right/U) & ([‘PCPF’]<= Central Refuge) & ([‘JD’]<= More than four arms(not roundabout)) & ([‘JC’]<=Auto Traffic Signal) & ([‘VJL’]>Not at junction) & ([‘Pav’]<= Dry) & ([‘NV’]<=1) & ([‘Lt’]<=Darkness-Lightening Unknown)	Slight
10.	If([‘VM’]<= Turning Left/Right/U) & ([‘RT’]>One way street) & ([‘JD’]>Not at junction) & ([‘Pav’]>Wet or damp) & ([‘PCHC’]<= School crossing patrol) & ([‘VTA’]<= Articulated Vehicle) & ([‘DW’]<= Weekday) & ([‘VSO’]>No)	Serious
11.	If([‘VM’]<=Reversing) & ([‘RT’]<=One way street) & ([‘PCPF’]>=Foot Bridge/Subway) & ([‘VM’]<= Other) & ([‘Wea’]>Fine/No High wind) & ([‘VTA’]>No tow/articulation) & ([‘Pav’]<= Snowy/Frozen) & ([‘VJL’]>Not at junction)& ([‘PCPF’]<= No physical crossing facility within 50 m)	Fatal
12.	If([‘VM’]>Turning Left/Right/U) & ([‘Pav’]<= Wet or damp) & ([‘S’]<= 20) & ([‘XRC.1’]<= A) & ([‘XRC’]>B) & ([‘DW’]>Weekend) & ([‘VJL’]>Not at junction)	Serious
13.	If([‘PCHC’] <= School /Crossing/ Patrol) && ([‘VT’] <= Bicycle) && ([‘NP’] <=>2) &&([‘Lt’] <= Darkness-Light Lit) && ([‘PA’] <= 35-44)	Slight
14.	If(([‘VM’]=Going Ahead) && ([‘VT’] <= Bicycle) && ([‘NP’] <= 1) && ([‘XRC’] <= B) && ([‘Pav’]= Dry) && ([‘VTA’]= No tow/ Articulation)	Slight
15.	If([‘VM’]= Moving off) & ([‘PAV’]= dry) & ([‘XRC’]=A) & ([‘JD’]=T or Staggered Junction) & ([‘PCPF’]= Zebra)	Slight
16.	If ([‘VM’]<= ‘Going Ahead’ & ([‘RT’]<= ‘One way street’) & ([‘PCPF’]<= ‘No Physical Crossing facility within 50 m’) & ([‘JD’]<= ‘Not at junction’) & ([‘XRC’]<= ‘A’) & ([‘NV’]<= 2) & ([‘DW’]<= ‘Week day’)	Serious
17.	If([‘VM’]<= Moving off) & ([‘RT’]>One Way street) & ([‘JD’]>Not at junction) & ([‘Pav’]<= Wet or damp) & ([‘Lt’]<= Darkness/Lightening Unknown) & ([‘VSO’]>Yes) & ([‘VTA’]>Articulated vehicle) & ([‘Pav’]>Snowy/Frozen)	Fatal
18.	If([‘VM’]>Turning Left Right/U) & ([‘PCPF’]>Central Refuge) & ([‘Pav’]>Wet or damp) & ([‘VJL’]>Not at junction) & ([‘PCHC’]<=School crossing patrol) & ([‘NV’]<= >2) & ([‘XRC’]>A) & ([‘Ar’]<= Rural) & ([‘VSO’]>Yes)	Serious
19.	If(([‘VM’]=Going Ahead) && ([‘VT’] <= Bicycle) && ([‘NP’] <= 1) && ([‘XRC’] <= B) && ([‘Pav’]= Dry) && ([‘VTA’]=No tow/Articulation)	Fatal
20.	If([‘VM’]= Moving off) & ([‘PAV’]= dry) & ([‘XRC’]=A) & ([‘JD’]=T or Staggered Junction) & ([‘PCPF’]= Zebra)	Slight

a vehicle is turning left/right/u and speed limit >50 Kmph, it is a weekend with road condition as wet, then at this instance, it would be a fatal accident. Through this study, the rules that are extracted can be very useful while framing road safety regulation. The inferences that are taken out from these rules can be very effective for road safety management to help control the severity of any accident that might happen Rule-wise measures if implemented we believe can control rising cases of road accidents.

4 Conclusions

In this study, we evaluated different ML models, namely, SVM, DT, and RF for our prediction of severity of pedestrian accidents. DT and RF outperforms the SVM model on accuracy and recall scores. The hyper-parameters of all the ML models are optimized/tuned by grid search technique. Besides the application of the classifiers, data preprocessing by using RF imputation is also done. The main part of our study is to use the DT algorithm for the generation of if-then rules. It has thus been concluded from the rules that more fatal accidents have occurred when the pedestrian was alone on the street and lightning condition was bad, i.e., in dim-lit areas. Further, it is also interpreted that when the vehicle speed was greater than 50 Kmph and the vehicle was overtaking, there is a possibility of a serious accident, and there was no physical crossing in 50 m range. Hence, this rule emphasizes the importance of physical crossing for pedestrians on high-speed tracks. The meaningful rules also establish that accidents just never occur at random but a pattern there exists which they follow, which can be predicted and such factors, when identified, can be vital for framing regulations for pedestrian safety.

However, there are few limitations to the approach. We considered only three models for our study and chose one out of them for our further evaluation. Hence, our results are limited to these models. However, there could be some other model which may provide us with better prediction accuracy. In this paper, a grid search approach is used to tune the hyper-parameters of classifiers; whereas, alternatively, other stochastic algorithms can be used. Besides, only 20 features out of 34 features are considered after applying Chi-Square feature importance algorithm, there exist other well-established methods which can offer interesting results, and hence the if-then rules, so generated, may be different. With this regard, rough set theory [15] or Bayesian model [15] can also be deployed to find more road safety rules, which may be of higher importance in road safety management. In the future, a decision support system can be developed [16] or computer vision [11, 12] can be used for pedestrian road safety management. However, the approach presented in this paper can be utilized as a reference for any future work in other domain also.

References

1. Breiman L (2001) Random forests. Mach Learn 45(1):5–32
2. Breiman L, Friedman J, Stone CJ, Olshen RA (1984) Classification and regression trees. CRC Press, Boca Raton
3. Chen C, Zhang G, Qian Z, Tarefder RA, Tian Z (2016) Investigating driver injury severity patterns in rollover crashes using support vector machine models. Accid Anal & Prev 90:128–139
4. Cunto FJ, Ferreira S (2017) An analysis of the injury severity of motorcycle crashes in brazil using mixed ordered response models. J Trans Safety & Secur 9(sup1):33–46
5. Gårder PE (2004) The impact of speed and other variables on pedestrian safety in maine. Accid Anal & Prev 36(4):533–542
6. LaScala EA, Gerber D, Gruenewald PJ (2000) Demographic and environmental correlates of pedestrian injury collisions: a spatial analysis. Accid Anal & Prev 32(5):651–658
7. Liu J, Hainen A, Li X, Nie Q, Nambisan S (2019) Pedestrian injury severity in motor vehicle crashes: an integrated spatio-temporal modeling approach. Accid Anal & Prev 132:105272
8. Loo BP, Tsui K (2007) Factors affecting the likelihood of reporting road crashes resulting in medical treatment to the police. Inj Prev 13(3):186–189
9. Milton JC, Shankar VN, Mannering FL (2008) Highway accident severities and the mixed logit model: an exploratory empirical analysis. Accid Anal & Prev 40(1):260–266
10. Poh CQ, Ubeynarayana CU, Goh YM (2018) Safety leading indicators for construction sites: a machine learning approach. Autom Constr 93:375–386
11. Pramanik A, Harshvardhan, Djeddi C, Sarkar S, Maiti J (2020) Region proposal and object detection using hog-based cnn feature map. In: 2020 international conference on data analytics for business and industry: way towards a sustainable economy (ICDABI). IEEE, pp 1–5
12. Pramanik A, Sarkar S, Maiti J (2019) Oil spill detection using image processing technique: an occupational safety perspective of a steel plant. In: Emerging technologies in data mining and information security, vol 814. Springer, Singapore, pp 247–257
13. Pramanik A, Sarkar S, Maiti J (2021) A real-time video surveillance system for traffic pre-events detection. Accid Anal & Prev 154:106019
14. Pramanik A, Sarkar S, Maiti J, Mitra P (2021) Rt-gsom: rough tolerance growing self-organizing map. Inf Sci 566:19–37
15. Sarkar S, Baidya S, Maiti J (2017) Application of rough set theory in accident analysis at work: a case study. In: ICRCICN 2017, pp 245–250
16. Sarkar S, Chain M, Nayak S, Maiti J (2019) Decision support system for prediction of occupational accident: a case study from a steel plant. In: Emerging technologies in data mining and information security, vol 813. Springer, Singapore, pp 787–796
17. Sarkar S, Ejaz N, Maiti J (2018) Application of hybrid clustering technique for pattern extraction of accident at work: a case study of a steel industry. In: 2018 4th international conference on recent advances in information technology (RAIT), IIT Dhanbad. IEEE, pp 1–6
18. Sarkar S, Maiti J (2020) Machine learning in occupational accident analysis: a review using science mapping approach with citation network analysis. Saf Sci 131:104900
19. Sarkar S, Pateshwari V, Maiti J (2017) Predictive model for incident occurrences in steel plant in India. In: ICCCNT 2017, pp 1–5
20. Sarkar S, Pramanik A, Khatedi N, Maiti J (2020) An investigation of the effects of missing data handling using 'r'-packages. In: Data engineering and communication technology. Springer, Singapore, pp 275–284
21. Sarkar S, Pramanik A, Maiti J, Reniers G (2020) Predicting and analyzing injury severity: a machine learning-based approach using class-imbalanced proactive and reactive data. Saf Sci 125:104616
22. Sarkar S, Raj R, Vinay S, Maiti J, Pratihar DK (2019) An optimization-based decision tree approach for predicting slip-trip-fall accidents at work. Saf Sci 118:57–69
23. Sarkar S, Vinay S, Djeddi C, Maiti J (2021) Text mining-based association rule mining for incident analysis: a case study of a steel plant in india. Pattern Recog Artif Intell 1322:257

24. Sarkar S, Vinay S, Raj R, Maiti J, Mitra P (2019) Application of optimized machine learning techniques for prediction of occupational accidents. Comput & Oper Res 106:210–224
25. Schneider RJ, Ryznar RM, Khattak AJ (2004) An accident waiting to happen: a spatial approach to proactive pedestrian planning. Accid Anal & Prev 36(2):193–211
26. Singh K, Raj N, Sahu S, Behera R, Sarkar S, Maiti J (2015) Modelling safety of gantry crane operations using petri nets. Int J Injury Control Safety Prom 1–12
27. Tang F, Ishwaran H (2017) Random forest missing data algorithms. Stati Anal Data Min: ASA Data Sci J 10(6):363–377
28. Vapnik V (1995) The nature of statistical learning theory, vol 8. Search PubMed, p 188
29. Verma A, Chatterjee S, Sarkar S, Maiti J (2018) Data-driven mapping between proactive and reactive measures of occupational safety performance. In: Industrial safety management- 21st century perspective of Asia. Springer, Singapore, pp 53–63
30. Wong S, Sze NN, Li YC (2007) Contributory factors to traffic crashes at signalized intersections in hong kong. Accid Anal & Prev 39(6):1107–1113
31. Ye F, Lord D (2014) Comparing three commonly used crash severity models on sample size requirements: multinomial logit, ordered probit and mixed logit models. Anal Methods Accid Res 1:72–85

Fog Computing Applications with Decentralized Computing Infrastructure—Systematic Review

Deepak Kumar, Rishu, and Sangeetha Annam

Abstract With the help of sensors and other mobile devices, the Internet of Things (IoT) generates a large amount of data. It improves daily computer efficiency and provides each mobile payment model that collects data and improves the expansion and performance of end-to-end devices. Even the cloud emigrates itself to the edge, but some IoT application does not get real-time access as they needed. In real-time situations, short seconds also produce the effect when delays or jitters are not acceptable. In this article, the authors have suggested a review of a well-crafted platform; fog computing, cloud expansion that provides computing, networking, storage, networking, and overcoming existing challenges. Analysis of many applications proves that fog reacts as a medium between cloud devices and IoT devices that bring devices closer to the edge of the network. Also, the paper highlights other simultaneous opportunities and challenges with research trends or safety indicators and privacy concerns of computer fog in various locations.

Keywords Cloud computing · Fog computing · Internet of Things (IoT) · Fog network · Fogging · Edge computing

1 Introduction

Due to the enhancement of technology, lifestyles have become hectic in today's society. Users are demanding a way to get a more luxurious experience and to fortune smart devices to reform the life aspects. So it becomes essential to adequately distribute and decentralize the techniques to implement real-time monitoring systems. Then, the Internet of Things (IoT) comes into the approach to perform this kind of operation [1, 2]. There are various types of smart devices and sensors that have been developed as a result of the amazing electronic testing and miniaturization which is an important part of IoT. In addition to this, Smart devices are connected to a variety of mobile technologies such as 3G, 4G, Wi-Fi, and 5G as

D. Kumar · Rishu (✉) · S. Annam
Institute of Engineering and Technology, Chitkara University, Rajpura, PB, India

© The Author(s), under exclusive license to Springer Nature Singapore Pte Ltd. 2022

D. Giri et al. (eds.), *Proceedings of the Seventh International Conference on Mathematics and Computing*, Advances in Intelligent Systems and Computing 1412,
https://doi.org/10.1007/978-981-16-6890-6_37

well as their smart-looking protocols [3]. Besides, the data generated by these IoT devices is transferred to a remote cloud server for real-time processing, diagnosis, and detection.

As people often know that cloud computing works effectively and stores data, but they do not know that those smart devices often face enhanced storage challenges and threats, computer power, network bandwidth, and latency that delivers delayed service quality and user experience [4]. Providing timely services for applications, infrastructure, software, and a platform with flexible resources at low cost to deploy users, the cloud is considered a paradigm of computer domination and reduces the barrier of limited resources on smart devices. Therefore, cloud computing is not a common solution for all IoT applications. To overcome the aforementioned challenges between cloud and end-users, we introduced a new concept known as Fog Computing which facilitates computer-based components of the content delivery system and provides services to billions of connected devices [5]. Fogging is a distributed decentralized approach; it acts as an insightful gateway that enables the cloud to make more efficient processing, analysis, and data processing. This review article summarizes the uses and benefits of fog, along with it, its focus on the primary context of IoT applications also which are based on existing cloud models.

This paper provides a view of fog computing in different applications of the area for IoT. This paper is organized as follows: Sect. 2 summarizes the benefits and various applications of IoT cloud models. Section 3 describes the overview and architecture of fog computing along with applications-based fog system scenarios, respectively. In Sect. 4, relevant literature and various application areas of Fog computing are discussed. In Sect. 5, we discuss some future challenges that need to be eliminated in Fog computing. Lastly, Sect. 6 describes the conclusion.

2 Benefits and Applications of IoT Cloud Models

This section gives various applications on IoT cloud-based model. These IoT-enabled programs open up a new era of smart applications and a large number of smart services that are making a huge impact on people's lives and boosting economic growth. The IoT paradigm gives assurance to make consumer devices such as electronics, home appliances, medical devices, refrigerators, cameras, and sensors to be a part of the Internet world. As IoT supports analytics and integration and transfers the data generated by sensors, it envisions a new world among humans and smart devices that enhance the quality of life, easily access health services along efficiently recover the disasters. For real-time decisions and analytics, these devices gathered data from the end devices and transfer them to the cloud. According to a CISCO report, more than 50 billion devices are associated with the Internet in 2020, and it is expected that produced data will extend up to 500 Zbytes [6]. Recent Statista analysis says that the number of connected devices exceeds 75 billion in 2020, which will produce data in a huge amount and have to process rapidly and in a feasible manner. Moreover, the

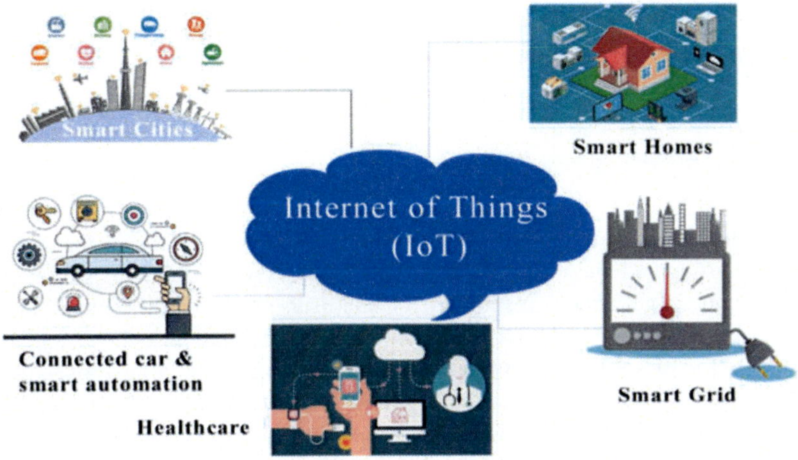

Fig. 1 Applications of IoT

data achieved with IoT-enabled devices appropriate 45% will be analyzed, processed, and gathered [7] at the network edge by 2019.

Smart cities [8], smart grids [9], healthcare [10], smart homes, smart transportation, traffic light systems, security, smart irrigation, and surveillance are a few known examples of smart IoT-enabled applications [11] as discussed in Fig. 1.

(a) With the use of Health and fitness-oriented wearable devices, it becomes possible to analyze the patient health status, send to smartphones, and easily track the symptoms, for example, heart rate, blood pressure, perspiration and glucose levels, alcohol levels, and body temperature to get an indication whether a cold or the flu is on the way or not. With the advancements of technology, IoT provides smart healthcare systems.
(b) IoT makes the home smart by providing home security and sensors to detect changes in environmental situations such as air-conditioning, temperature, and systems to be monitored. Cameras used in homes for security purposes can capture any person or intruder that is going to enter into the home and send warnings or alerts to the homeowners by the mobile apps to an enhanced security mechanism of smart homes in a different manner.
(c) Under Smart Cities, traffic light systems and transportations can be monitored by IoT to achieve smart cities.

According to data generated by IoT-based applications, the latest research shows that these applications generate 25,000 data tuples in 25,000 per second [12]. To manage large volumes of data, it is difficult for tempering. As a result of high throughput and processing, other issues appear such as network bandwidth, high latency, security, mobility, geo-distribution, location information, and privacy. In some cases, IoT-enabled applications are not accepted as less than smart transport, robot systems, smart grids [13], smart healthcare applications [14, 15], and other important low-latency applications [14] require a faster response, which is why network delays become a huge cloud limit.

3 Fog Computing

In this section, we get an overview of the fog module and its configuration for the four layers of hosting and the purpose of the fog network system. CISCO has proposed a fog common to get a quick response from end-users and bring them closer to the end. The standard Fog Computer is entirely suited to real-time or delayed applications [16]. Therefore, the fog computer provides large storage, high connectivity, and brings end-users quicker for a cloud to improve privacy, mobility, and security. Generally, fog applications either IoT devices or edge networks connected to continuous cloud-connected notes are called Fog nodes, which provide a faster response in milliseconds to seconds than real-time connections which is why computer fog is fully suited to low-latency applications [17]. According to the Open Fog association [18], Fog computing is a state-of-the-art facility that distributes computers, storage, resources, communications, and controls anywhere in a cloud-based network form to eliminate devices or objects. Fog computing is a distributed paradigm proposed by CISCO in 2014 that brings cloud end-to-end devices that reduce cloud limitations such as latency, bandwidth, mobility, and much more efficiently and works on IoT devices in milliseconds bringing a huge amount of benefits to smart devices. Fog Computing has empowered the near-surface cloud and has lots of benefits namely benefits such as Heterogeneity, instantaneous access, location allotment, and minimal stoppage [19]. CISCO has also proposed the development of Fog computing platform technology which is defined from computer hardware to component components of the system.2, discussing the structure of the fog house. Allying cloud, a cloud-based system, was also developed by IAAS, PAAS, and SAAS, respectively, on the network. On the edge of the hardware layer such as Router and Switches become Fog nodes that can be transferred to IoT-based applications and clouds openly through user-based agreements. Therefore, for connection PAAS and SAAS CISCO DSX are used (Fig. 2).

Layer 4 consists of several sensor nodes distributed to different public infrastructures to monitor the data or condition that varies from time to time. Layer 3 included the small nodes that are less power consuming and highly performable which belong to the local community and neighborhood. In layer 2, each computing node is connected to layer 3 nodes and detects obstacles and hazardous events earlier and prevents cities from any kind of damage like gas leakage, water pipe breakage, etc. Cloud data center as a top layer works on the large-scale, city-wide analysis of patterns and behaviors. In this way, this four-layered fog computing architecture provides quick response, high performance, and intelligence and detects disasters and natural hazards timely to prevent smart cities from any damage.

The main purpose of launching Fog computing is to overcome the limitations and restrictions that arise in the cloud while processing, processing, or responding to

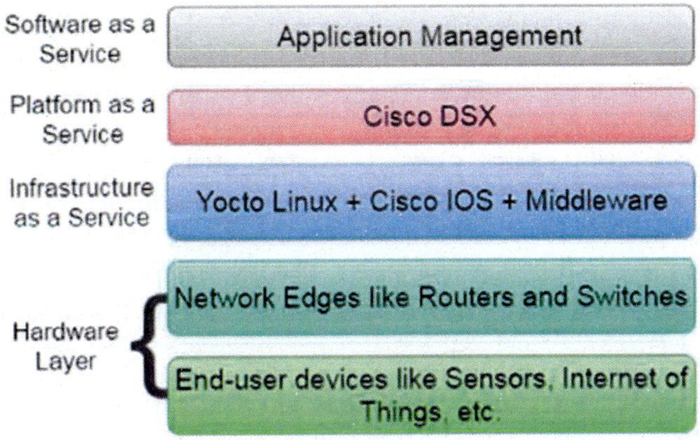

Fig. 2 According to CISCO, the structural design of the fog computing platform

network applications. The situation is introduced in [19], where a layer of fog is placed between the clouds and IoT devices to make cities smarter. Likely in smart cities, a lot of IoT-driven applications such as traffic light and management system, smart buildings, health management, parking system, sanitation, and many more that are latency based and data generated by these devices are on the huge volume that required large bandwidth. But the cloud does not fulfill these mentioned requirements, that's why fog computing comes into an approach to accomplish these issues.

4 Related Work

Generally, Fog computing is a broad area of the cloud and it deals with real-time IoT-based applications namely Healthcare, smart grids, and smart cities. This section presents literature works that were reviewed by various researchers in different applications which are discussed below.

4.1 Healthcare

If we come into the healthcare or medical field, healthcare applications provide huge data to store in the cloud instead of limited storage and computing devices in an efficient manner. Fog computing is the best method due to the following reasons: its application location awareness, low latency, high mobility, etc. for healthcare applications, produced a fairly large data volume, latency-sensitive-based application, and requires low response time. To build fog computing infrastructure, fog nodes connected jointly as connected nodes improve elasticity, scalability, and redundancy.

In the healthcare and medicine industry, developing IoT-based applications is a challenging task due to its drawbacks toward clouds. This section focuses on the existing work done or proposed by researchers and reviews Fog-driven IoT-based scenarios that overcome the limitation of IoT Cloud-based models.

Prabal Verma [20] proposed a five-layered DAL, ECL, IML, DML, CSL, fog-based smart remote patient monitoring approach. In this mechanism, the Fog node collects the patient's health data and history from the cloud, and then fog sends the necessary information to the cloud for further future actions. Ahmad et al. [21] introduced a Fog computing-based Health Fog framework in which the fog layer act as an intermediate layer between the cloud and the end devices. The main purpose of this designed framework is to enhance flexibility by integration with CSAB and resolve security issues with the use of cryptographic techniques and primitives. In paper [22], the authors presented a smart healthcare paradigm based on fog computing. In this paper, the measurement of blood is taken as Use case. But it is difficult to calculate the efficiency and effectiveness of a developed system without performing any kind of challenging task.

In [23], the related work presented by Stantchev et al. for smart healthcare infrastructure conferred the three-level fog-based architecture, from a business point of view. In this, they just modeled high-level Architecture without any experimental evaluation and real-world implementation discussion. Rahimi et al. [24] proposed a competent authorization, authentication, and secure approach, at the edge of IoT-based health systems which need some processing power. An intelligence layer that acts as an intermediator is defined between the sensing nodes and cloud.

In [25], the author presented the gateway called UT-GATE, smart e-health architecture that is IoT-based health monitoring application. In this gateway, data is collected from sensor nodes and sent to the cloud that is fog assisted through UT-GATE, where the Fog layer used in architecture is beneficial for latency as less transmission time is required and makes it energy efficient. To keep securing the stored data, they used a Python-based crypto library, wherewith a public key, all the compressed files are encrypted and in the cloud, only the data gathered service has the private key to perform the process of decrypting.

In [26], the author developed a fog computing-based ECG monitoring system. This mechanism extracted various ECG features and implemented various fog computing services such as less transmission time, location awareness, interoperability, real-time notifications, and many more. Rajinder Sandhu et al. in [27] show the proposed framework to effectively detect the patient who is infected, severely infected, and uninfected using a dataset built through Beatty et al. in 2010 and monitor the dengue virus. In this developed architecture, Fog provides a large amount of storage capacity for data related to dengue and provides a connection between IoT devices and cloud data storage effectively to prevent life from this threatening disease. In paper [28], the author proposed a tri-tired fog-based health monitoring system in which all the data are processed at the fog layer to the context of the cloud and the IoT devices included sensors. Sensors consist of wearable or non-wearable devices in the form of smartphones, watches, fitness bands, glasses, and so on, which are attached to the patients to monitor their health status.

4.2 *Smart Cities*

With the leading access to public assets such as healthcare, education, better job opportunities, and other conveniences, more and more rural areas people shifted to cities. Smart cities (SC) make the use of cities resources in an efficient way that enhances the quality, safety, and ease the life of citizens along with the good quality of infrastructure. However, rapid urbanization does not just provide the amenity, it also heightens the existing problem faced by modernized cities such as pollution, traffic congestion, quality of education, insufficient public and health budget, resources and education, sanitation, roads conditions, public safety, and emergency responses.

Reference [29] proposed a video summarization fog-enabled architecture that is based on low powers PCs or computers that worked on Raspberry Pi. Message and Queue Telemetry Transport protocol (MQTT) is used to communicate among the master nodes of each region. A small number of researchers [30] presented a fog-based SC architecture known as Fog Computing Architecture Network (FOCAN), and this is a multilevel model. Communication among the FOCAN's component is done through Primary/inter primary communications which consist of TCP/IP communication protocol and secondary communication which consists of end-to-end connections of TCP/IP, IEEE802.11/15 for wireless and optic fiber, and Cat 5 or 6 for wired connection.

Mohamed et al. [31] designed a service-oriented architecture, as a holistic approach which is very helpful for the decision process. During designing, the authors classify the numerous kinds of decisions that can be made in buildings for using the energy resources. Therefore, the decision process includes the control layer, configuration layer, and human layer work on platforms.

The authors [32] describe a multitier fog model that is based on data analytics service and is used for SC applications. Therefore, this model consists of ad-hoc (A fog) and dedicated fogs (D fog). To measure the performance of the fog computing-based model, machine learning techniques namely Logistic regression (LR) and support vector machine (SVM) are used. QoS-aware scheme is also designed in the article to perform analytical services, job and resource management schemes.

4.3 *Vehicular Fog Computing (VFC)*

The current transportation generation is not about getting from source to destination, it also seeks safety, raw, and more efficient manner. As we see or heard accidents daily in newspapers, radio, or televisions, and the reasons behind this are inadequate traffic management and congestion. Through this road safety and management are becoming challenging issues. Nowadays present vehicles act as intelligent devices appointed with multiple sensors and use mobile applications that performed multiple tasks at one time such as collecting information about traffic, navigation communication along with providing entertainment, safer, and more decisive travel [33]. The numerous applications of VFC are used increasingly such as in smart navigation, traffic lights, road condition detection, emergency warning, road hazards alerts, and entertainment, i.e. multimedia.

Some of the authors [34, 35] proposed a certificateless aggregate signcryption (CLASC) scheme along with fog-based road condition monitoring system (RCMS). RCMS consists of a Control Sensor (CS), Mobile Sensors (MSs), Roadside unit (RSU), and cloud server for storage.

(a) **CS**: Partial Private Key for registers are generated in CC as its work is to generate keys.
(b) **MSs**: These are equipped with smart devices and within vehicles to generate data and provide real-time access such as location, time, and signals, and on-road incidents.
(c) **RSU**: It is a fog node or server that sends data or a warning about road accidents or accidents in a particular area, which hears MSs to quickly process and make decisions near end-users. The biggest advantage of RSU is that it eliminates latency problems as it does calculations at the edges and only sends outputs or outputs to cover and eliminate users.
(d) **Cloud**: It is a data storage center that stores all historic data on the cloud for further utilization of data.

5 Outcomes

In this section, we go through some security challenges and threats driven by Fog Computing in future work, which need to be identified and overcome as discussed below:

(a) As security is one of the issues in IoT, to defeat this aspect, Fog computing enters into the technology world [36]. But there are so many security attacks namely as man in middle attack and denial of service attack still is an issue. To conquer this limitation, a lot of security parameters, algorithms, and devices are required to implement to enhance security mechanisms [37]. The collected or analyzed data from the sensors need to be secured at the source, processing time through the network, and at the destination location while communicating among devices and platforms. Further, instead of the original text, cipher text exhausted more disk space that affected the working operations of database and application layers.
(b) Disclosing keen information is one of the defy tasks that arise due to revealing cryptographic keys. According to recent research cache, interference attacks are the most common and prevention from these attacks is too expensive and this kind of attack only protects against a specific type of attack, and its elimination requires both software and hardware changes [38].
(c) In fog computing, huge numbers of small fog nodes are connected in a network that generates a small amount of data but when multiple large numbers of

devices are connected and huge data volume is generated, it becomes cardinally challenging to handle [39].

(d) Malware, Trojans, worms, and spyware attacks are sometimes difficult to identify in the Fog-based network so fog computing also requires lightweight, detection services based on the network to fully defend against these kinds of attacks in effective manners.

(e) Fog computing minimizes the rogue access points, packet-sniffing, and akin limitations and challenges by developing both authentication and encryption processes but due to the visibility of wireless devices, ensuring the security is difficult to implement for fog network.

(f) An insider threat issue that arises like user identity, scalability, data security is difficult to manage in a fog network.

(g) A lot of security parameters confidentiality, integrity, data backup, non-repudiation, DOS, and recovery need to be redefined in Fog computing networks.

6 Conclusion

In the cloud era, Fog computing is deliberate as one of the continuous indicators of travel that enhances the benefits of IoT healthcare systems. This paper introduces many computer functions and practical applications to fog-based systems connected to the cloud, in many scenarios, and use cases using a fog computer that offers many benefits to the industry such as grids, cities, buildings, medicines, and healthcare. Fog computing is the ambient or extent of cloud computing that brings benefits to support all IoT-enabled devices. But still, there are some security challenges that need to be overcome in fog computing that is mentioned above in the Future Challenges section, consisting of attacks such as a man-in-middle attack, authentication, integrity, and many more. Whenever fog nodes gathered data from the sensing nodes or end devices, there is no proof of identity that it is real or not. In such cases, attackers and intruders get a chance to interrupt or fabricate the data that result in loss of security aspects.

References

1. Atzori L, Iera A, Morabito G (2010) The internet of things: a survey. Comput Netw 54(15):2787–2805
2. Ning H, Liu H, Ma J, Yang LT, Huang R (2016) Cybernetics: cyber-physical–social–thinking hyperspace based science and technology. Futur Gener Comput Syst 56:504–522
3. Dastjerdi AV, Buyya R (2016) Fog computing: helping the internet of things realize its potential. Computer 49(8):112–116
4. Miorandi D, Sicari S, De Pellegrini F, Chlamtac I (2012) Internet of things: vision, applications and research challenges. Ad Hoc Netw 10(7):1497–1516
5. Luan TH, Gao L, Li Z, Xiang Y, Wei G, Sun L (2015) Fog computing: focusing on mobile users at the edge. arXiv:1502.01815

6. Evans D (2011) The internet of things: how the next evolution of the internet is changing everything. CISCO White Paper 1(2011):1–11
7. https://www.cisco.com/c/dam/en_us/solutions/trends/iot/docs/computing-overview.pdf
8. Qiu T, Zheng K, Song H, Han M, Kantarci B (2017) A local-optimization emergency scheduling scheme with self-recovery for a smart grid. IEEE Trans Industr Inf 13(6):3195–3205
9. Cao Y, Chen S, Hou P, Brown D (2015) FAST: a fog computing assisted distributed analytics system to monitor fall for stroke mitigation. In: 2015 IEEE international conference on networking, architecture and storage (NAS). IEEE, pp 2–11
10. Stantchev V, Barnawi A, Ghulam S, Schubert J, Tamm G (2015) Smart items, fog, and cloud computing as enablers of servitization in healthcare. Sens Transducers 185(2):121
11. Qiu T, Qiao R, Wu DO (2018) EABS: an event-aware backpressure scheduling scheme for emergency internet of things. IEEE Trans Mob Comput 1:72–84
12. Li J, Jin J, Yuan D, Palaniswami M, Moessner K (2015) EHOPES: data-centered fog platform for smart living. In: 2015 International, telecommunication networks and applications conference (ITNAC). IEEE, pp 308–313
13. Bilal K, Khalid O, Erbad A, Khan SU (2018) Potentials, trends, and prospects in edge technologies: fog, cloudlet, mobile edge, and micro data centers. Comput Netw 130:94–120
14. Escamilla-Ambrosio PJ, Rodríguez-Mota A, Aguirre-Anaya E, Acosta-Bermejo R, Salinas-Rosales M (2018) Distributing computing in the internet of things: cloud, fog and edge computing overview. In: NEO 2016. Springer, Cham, pp 87–115
15. Vora J, Tanwar S, Tyagi S, Kumar N, Rodrigues JJ (2017). FAAL: fog computing-based patient monitoring system for ambient assisted living. In: 2017 IEEE 19th international conference on e-health networking, applications and services (Healthcom). IEEE, pp 1–6
16. Giang NK, Kim S, Kim D, Jung M, Kastner, W (2014) Extending the EPCIS with building automation systems: a new information system for the internet of things. In 2014 eighth international conference on innovative mobile and internet services in ubiquitous computing (IMIS). IEEE, pp 364–369
17. Chiang M, Zhang T (2016) Fog and IoT: an overview of research opportunities. IEEE Internet Things J 3(6):854–864
18. Tang B., Chen Z, Hefferman G, Wei T, He H, Yang Q (2015) A hierarchical distributed fog computing architecture for big data analysis in smart cities. In: Proceedings of the ASE big data & social informatics 2015. ACM, p 28
19. Bonomi F, Milito R, Zhu J, Addepalli S (2012) Fog computing and its role in the internet of things. In: Proceedings of the first edition of the MCC workshop on mobile cloud computing. ACM, pp 13–16
20. Verma P, Sood SK (2018) Fog assisted-IoT enabled patient health monitoring in smart homes. IEEE Internet Things J
21. Singh S, Bansal A, Sandhu R, Sidhu J (2018) Fog computing and IoT based healthcare support service for dengue fever. Int J Pervasive Comput Commun 14(2):197–207
22. Paul A, Pinjari H, Hong WH, Seo HC, Rho S (2018) Fog computing-based IoT for health monitoring system. J Sens
23. Ahmad M, Amin MB, Hussain S, Kang BH, Cheong T, Lee S (2016) Health fog: a novel framework for health and wellness applications. J Supercomput 72(10):3677–3695
24. Stantchev V, Barnawi A, Ghulam S, Schubert J, Tamm G (2015) Smart items, fog and cloud computing as enablers of servitization in healthcare. Sens Transducers 185(2):121
25. Moosavi SR, Gia TN, Rahmani AM, Nigussie E, Virtanen S, Isoaho J, Tenhunen H (2015) SEA: a secure and efficient authentication and authorization architecture for IoT-based healthcare using smart gateways. Procedia Comput Sci 52, 452–459
26. Moosavi SR, Gia TN, Nigussie E, Rahmani AM, Virtanen S, Tenhunen H, Isoaho J (2015) Session resumption-based end-to-end security for healthcare internet-of-things. In 2015 IEEE international conference on computer and information technology; ubiquitous computing and communications; dependable, autonomic and secure computing; pervasive intelligence and computing (CIT/IUCC/DASC/PICOM). IEEE, pp 581–588

27. Rahmani AM, Thanigaivelan NK, Gia TN, Granados J, Negash B, Liljeberg P, Tenhunen H (2015) Smart e-health gateway: Bringing intelligence to internet-of-things based ubiquitous healthcare systems. In: Consumer communications and networking conference (CCNC), 2015 12th annual IEEE. IEEE, pp 826–834
28. McGrath MJ, Scanaill CN (2013). Body-worn, ambient, and consumer sensing for health applications. In: Sensor technologies. Apress, Berkeley, CA. pp 181–216
29. Nasir M, Muhammad K, Lloret J, Sangaiah AK, Sajjad M (2019) Fog computing enabled cost-effective distributed summarization of surveillance videos for smart cities. J Parallel Distrib Comput 126:161–170
30. Naranjo PGV, Pooranian Z, Shojafar M, Conti M, Buyya R (2018) Focan: a fog-supported smart city network architecture for management of applications in the internet of everything environments. J Parallel Distrib Comput
31. Mohamed N, Al-Jaroodi J, Jawhar I (2018) Service-oriented big data analytics for improving buildings energy management in smart cities. In: 2018 14th International Wireless Communications & Mobile Computing Conference (IWCMC). IEEE, pp 1243–1248
32. He J, Wei J, Chen K, Tang Z, Zhou Y, Zhang Y (2018) Multitier fog computing with large-scale iot data analytics for smart cities. IEEE Internet Things J 5(2):677–686
33. Karagiannis G, Altintas O, Ekici E, Heijenk G, Jarupan B, Lin K, Weil T (2011) Vehicular networking: a survey and tutorial on requirements, architectures, challenges, standards and solutions. IEEE Commun Surv Tutor 13(4):584–616
34. Basudan S, Lin X, Sankaranarayanan K (2017) A privacy-preserving vehicular crowdsensing-based road surface condition monitoring system using fog computing. IEEE Internet Things J 4(3):772–782
35. Huang C, Lu R, Choo KKR (2017) Vehicular fog computing: architecture, use case, and security and forensic challenges. IEEE Commun Mag 55(11):105–111
36. Tang C, Wei X, Zhu C, Chen W, Rodrigues JJ (2018) Towards smart parking based on fog computing. IEEE Access 6:70172–70185
37. Stojmenovic I, Wen S, Huang X, Luan H (2016) An overview of fog computing and its security issues. Concurr Comput: Pract Exp 28(10):2991–3005
38. Page D (2003) Defending against cache-based side-channel attacks. Inf Secur Tech Rep 8(1):30–44
39. Hu F, Hao Q, Bao K (2014) A survey on software-defined network and openflow: from concept to implementation. IEEE Commun Surv Tutor 16(4):2181–2206

Investigating External Audit Records to Detect Fraudulent Firms Employing Various Machine Learning Methods

Farhana Akter Sunny, Md. Imran Khan, Md. Shahriare Satu, and Mohammad Zoynul Abedin

Abstract Fraudulent firm detection is a vital issue for the development of the economy worldwide. Auditing is a review process of financial statements that controls abnormal activities in the organization. Machine learning methods provide a faster solution to identify the fraudulent firm more precisely. The objective of this work is to propose a machine learning model that can investigate, predict fraudulent cases and generate more stable classification results. Therefore, we gathered audit data from the Kaggle repository and explored its characteristics. Then, highly collinear features have been reduced from this dataset. After that, various feature transformation methods were implemented and generated transformed datasets. Later, we employed several state-of-the-art classifiers into these datasets and investigated the performance of the classifier. Logistic regression shows the highest accuracy and F-measure with a confidence interval (98.6%±0.9%) for the squared transformed, and XGBoost provides the highest area under the curve with confidence interval (98.4%±0.9%) for the box-cox transformed dataset.

Keywords Fraudulent firm · Machine learning · Feature transformation method · Classifiers

F. A. Sunny
Department of Computer Science and Engineering, Green University of Bangladesh, Dhaka, Bangladesh

Md. I. Khan
Department of Computer Science and Engineering, Gono Bishwabidyalay, Dhaka, Bangladesh

Md. S. Satu
Department of Management Information Systems, Noakhali Science and Technology University, Noakhali, Bangladesh

M. Z. Abedin (✉)
Department of Finance and Banking, Hajee Mohammad Danesh Science and Technology University, Dinajpur 5200, Bangladesh
e-mail: abedinmz@hstu.ac.bd

© The Author(s), under exclusive license to Springer Nature Singapore Pte Ltd. 2022
D. Giri et al. (eds.), *Proceedings of the Seventh International Conference on Mathematics and Computing*, Advances in Intelligent Systems and Computing 1412,
https://doi.org/10.1007/978-981-16-6890-6_38

1 Introduction

Financial fraudulent activity is a critical issue to hamper the productivity of any organization. Hence, it is a serious matter where every firm wants to reduce the fraudulent behavior from their stakeholders. Auditing is a process to identify anomalies by checking imbalance and illegal tasks in the firms [1]. Planned works are executed during auditing and provided a report about this organization. In this case, many risks had been identified which were required to evaluate the financial criteria for any organization and impart the outcomes of intrigued clients [2]. Besides, auditing statements are used as a trustworthy source of accounting information for external clients. Machine learning is provided more efficient, accurate and faster solutions to produce sustainable technologies in various applications. Like other fields, it helps to take proper decisions about financial analysis [3]. Hence, it can scrutinize different characteristics of audit data and identify irregularities of inspected firms. Moreover, there were happened a small number of works to identify fraudulent firms employing machine learning methods [4, 5]. The International Auditing and Assurance Standard Board had explored the utilization of data analytical methods in some areas [7]. Gepp et al. [6] investigated some auditing information to give outlines about the future opportunities of it using several big data analytical methods. Chen [8] proposed an optimal classification model about fraudulent financial reports that used Bayesian belief network (BBN), support vector machine (SVM), artificial neural network (ANN), decision tree with chi-squared automatic interaction detector (DT-CHAID) along with classification and regression trees (CART) where DT-CHAID showed 92.69% accuracy. Hajek and Henriques studied various classification methods for audit planning [10] where DTNB performed to select fraudulent firms and BNN performed for detecting non-fraudulent firms. Mohammadi et al. [11] implemented five classifiers into audit datasets where ANN shows the highest 75% accuracy using nine predictors.

Financial fraud is a noteworthy phenomenon for any financial organization that hinders economic development. This work aims to propose a machine learning methodology to detect fraud using different classifiers in audit data and generate more stable classification results. Firstly, we have collected an audit dataset where several records were gathered from different sectors in India. Then, various features were analyzed and eliminated from the highly correlated features in this dataset. Then, we extracted a secondary dataset by dropping out some factors based on audit risk, and several transformed datasets were produced from it. Later, we used several machine learning classifiers to investigate their outcomes and compare the performance based on the accuracy, F-measure and area under the curve (AUC) with confidence interval, respectively. This work can provide a significant contribution to optimize audit reports and get transparent information about financial organizations. Also, it increases the economical value of any financial firm.

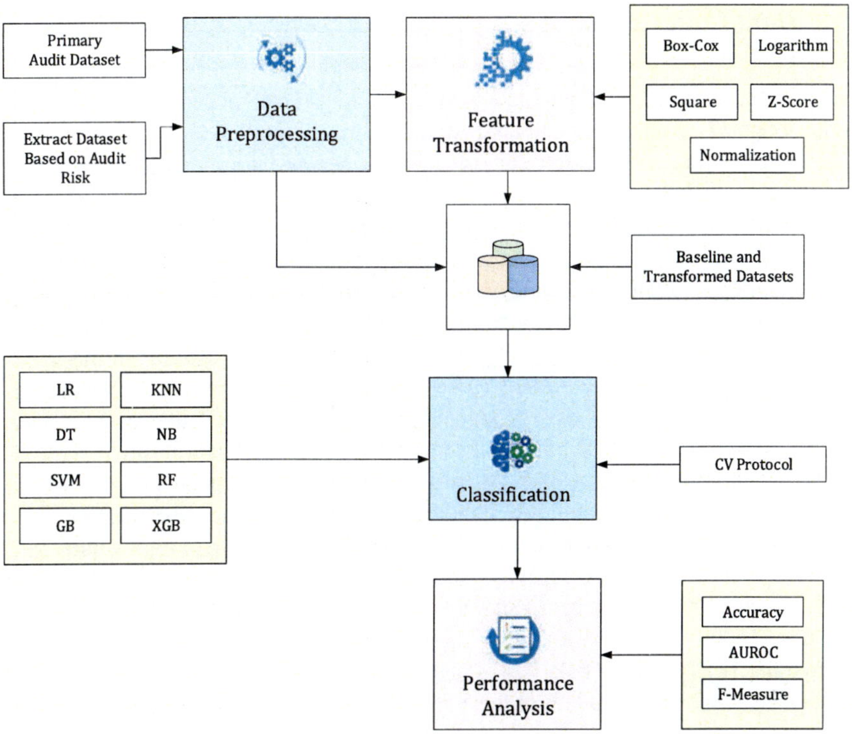

Fig. 1 Working methodology

Organization: We assembled this paper as follows: Sect. 2 presents the step-by-step analytical procedure of this work. Therefore, Sect. 3 provides the experimental result of individual classifiers and compares them with others. Finally, we summarize this work and indicate contribution with the consequence and future ramifications in Sect. 4.

2 Proposed Methodology

To investigate audit data, we dropped out the minimum number of features, and several machine learning techniques such as feature transformation and classification were implemented and generated more improved results. The proposed methodology is shown briefly in Fig. 1.

2.1 Data Collection

We collected audit dataset from the Kaggle repository (Data set Link) where various instances of 777 firms were gathered by the controller and auditor of 46 cities in India. These firms were targeted at 14 different sectors and identified fraudulent firms based on risk detection. During this procedure, many risk factors were evaluated from the on-going and past records in the firms. Some factors like profit value, loss value, parameters, reputation summary, environmental condition, past reports and the probability of their discrepancy were considered as input for determining risk factors. The greater audit risk and the higher probability are estimated as fraudulent behavior. Table 1 presents the descriptive statistics of various features in the audit dataset. Location id (F2) and money value (F13) contained more negligible missing values. Almost all variables have produced greater mean and median than the mode. So, the skewness values of individual features are indicated as a positive skewed dataset. Several features (e.g., Para A, Para B, Risk B, Total, Money Value and Risk D) contained high kurtosis and low standard deviation called heavy-tailed features. Besides, these features have provided the tendency of spreading to high values and their mean values are dispersed from standard deviation. In that case, only detection risk is provided with a constant value where it obtains symmetric distribution and deviation. However, there are usually three major risk factors considered, namely inherent risk (F23), control risk (F24) and detection risk (F25). Finally, we determine the audit risk that is the product of these risks by the following equation:

$$AR = \sum_{i=1}^{m}(P_{IN}(L_{IN})) \times \sum_{i=1}^{n}(P_{CO}(L_{CO}) \times DR\{\eta = m + n\}. \quad (1)$$

where L denotes the loss, P is the probability of loss, m and n indicate the number of risk factors.

2.2 Correlation Analysis and Feature Transformation Methods

In this work, different correlation coefficients were needed to determine and explore highly correlated features. This matrix with individual features are shown in Fig. 2 where the collinearity of all variables was compared and some of them were shown their coefficients > 80, such as F3-F5, F6 with F8 and F9, F8-F9, F10 with F11 and F12, F11-F12, F13-F15, F16-F18, F19 with F20 and F21, F22-F7, F23 with F13 and F15 as well as F26 with F6, F8 and F9. Then, we have found a dataset that showed positive skewed and heavily-tailed features. Therefore, we utilized feature transformation methods to generate secondary data and obtained stable results. There were widely used different methods such as box-cox, logarithm, normalization, z-score and square-root. The box-cox method transforms various features of the

Table 1 Descriptive statistics

S.N.	Variable Name	Description	Missing Values	Mean	Median	Mode	Variance	Skewness	Kurtosis
F1	Sector Score	Historical risk score	0	20.18	3.89	55.57	591.41	0.77	−1.34
F2	Location Id	Unique ID of the city/province	3	14.86	13.00	8.00	97.84	0.90	0.16
F3	Para A	Discrepancy of planned spending of inspection and summary report A in Rs (in crore)	0	2.45	0.88	0.00	32.25	8.51	100.09
F4	Score A	Derived from the Para A	0	0.35	0.20	0.20	0.03	0.49	−1.50
F5	Risk A	Derived from the Para A	0	1.35	0.18	0.00	11.84	8.36	97.33
F6	Para B	Discrepancy of unplanned spending of inspection and summary report B in Rs (in crore)	0	10.80	0.41	0.00	2508.37	20.54	508.22
F7	Score B	Derived from the Para B	0	0.31	0.20	0.20	0.03	0.96	−0.92
F8	Risk B	Derived from the Para B	0	6.33	0.08	0.00	904.38	20.51	507.08
F9	Total	Total discrepancy of another reports Rs (in crore)	0	13.22	1.37	0.00	2633.01	19.26	463.95
F10	Numbers	Historical discrepancy score	0	5.07	5.00	5.00	0.07	6.74	71.66
F11	Score B1	Derived from the number	0	0.22	0.20	0.20	0.01	3.56	12.20
F12	Risk C	Derived from the number	0	1.15	1.00	1.00	0.29	4.00	17.00

(continued)

Table 1 (continued)

S.N.	Variable Name	Description	Missing Values	Mean	Median	Mode	Variance	Skewness	Kurtosis
F13	Money value	Amount of money associated with misstatements in the past audits.	1	14.14	0.09	0.00	4436.43	10.53	127.21
F14	Score MV	Derived from the Money Value	0	0.29	0.20	0.20	0.03	1.30	−0.17
F15	Risk D	Derived from the Money Value	0	8.27	0.02	0.00	1597.67	10.53	127.18
F16	District loss	Historical risk score of a district in the last 10 years	0	2.51	2.00	2.00	1.51	2.23	3.37
F17	PROB	Probability of District Loss	0	0.21	0.20	0.20	0.00	6.56	47.14
F18	Risk E	Product of District Loss and Prob	0	0.52	0.40	0.40	0.08	3.01	11.55
F19	History	Average historical loss of firm in the last 10 years	0	0.10	0.00	0.00	0.28	9.28	120.88
F20	Prob	Probability of Historical Loss score	0	0.22	0.20	0.20	0.00	4.35	19.09
F21	Risk F	Product of History and prob	0	0.05	0.00	0.00	0.09	10.39	143.79
F22	Score	Deciding factor for classifying a firm	0	2.70	2.40	2.00	0.74	1.06	−0.21
F23	Inherent risk	Risk present in the transactions	0	17.68	2.21	1.40	2996.49	9.17	104.53
F24	Control risk	Undetected risk of control system	0	0.57	0.40	0.40	0.20	5.16	41.46
F25	Detection risk	Not even detected by the audit procedures	0	0.50	0.50	0.50	0.00		
F26	Audit risk	Product of Inherent, control and detection risks	0	7.17	0.56	0.28	1495.18	20.11	482.37

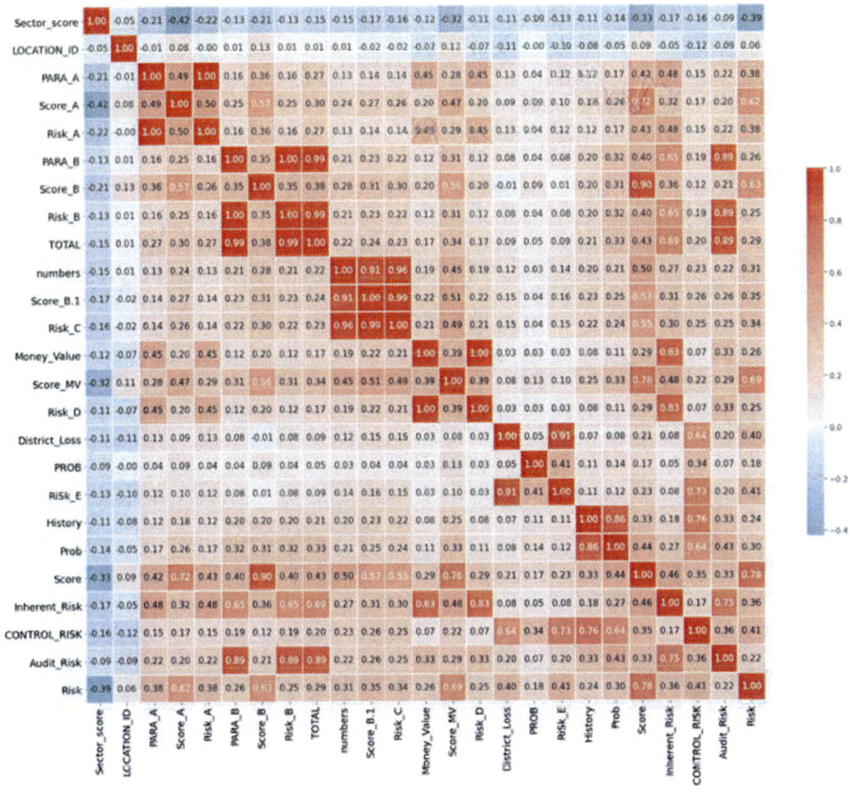

Fig. 2 Correlation matrix

dataset by applying power-law distribution [12]. In log-transformation, it determines the logarithm of the attributes from the original dataset. Normalization transforms individual attribute precisely into [0,1] range based on min–max method [13]. When any dataset is heavy-tailed distributions, this method is highly recommended. The z-score method converts all attributes into the same scale. To reduce skewness, the square-root transformation can be used.

2.3 *Classification Methods*

We identified different state-of-the-art classifiers like [14–16] logistic regression (LR), K-nearest neighbor (KNN), decision tree (DT), Naïve Bayes (NB), support vector machine (SVM), random forest (RF), and extreme gradient boosting (XGB). LR determines the class probabilities based on the boundary distance to provide precise prediction [17]. KNN is a widely used instance-based classifier that investigates given training instances based on K closest neighbors and classifies them as groups. Grounded on the Bayes theorem, NB is supposed as an independence model, where the characteristics of a certain class of totally independent from other characteristics. [18]. SVM is a well-performed classifier that isolates instances into two classes using hyperplane [19]. In this work, we used the radial basis function as a kernel to investigate audit data. RF is a randomized collection of DT where each tree provides a vote for its esteemed and identifies appropriate class from the majority of these votes [20]. Besides, XGB is a particular form of gradient boosting that utilizes progressively precise approximations and defines second-order gradients to analyze numerous records and reduce loss function more accurately.

2.4 Performance Metrics

Several evaluation metrics like accuracy, F-measure, and AUC have been used to measure the performance of an individual classifier. After this analysis, there were generated four types of instances, such as true positive (TP), false positive (FP), true positive and false negative (FN).

- **Accuracy** is used to assess how a classifier can estimate true value more appropriately. It can be measured by the ratio of the total number of correctly predicted and predicted records which is given as follows.

$$\text{Accuracy} = \frac{\text{TP} + \text{TN}}{(\text{TP} + \text{TN} + \text{FP} + \text{FN})} \tag{2}$$

- **F-measure** measures the harmonic mean of precision and recall [13]. It calculates using the following formula:

$$F - \text{Measure} = \frac{2 \times \text{precision} \times \text{recall}}{(\text{precision } + \text{recall})} = \frac{2\text{TP}}{2\text{TP} + \text{FP} + \text{FN}} \tag{3}$$

- **AUC** defines how well a positive class can be isolated from a negative class. True positive rate (TPR) is defined as sensitivity and classifiers predict the fraudulent firm by measuring this rate. On the other hand, a true negative rate is defined as specificity. ROC curve is a graph in which the trade-off between TPR and FPR is plotted.

$$TPR(Sensitivity) = \frac{\text{TP}}{\text{TP} + \text{FN}} \tag{4}$$

$$TNR(Specificity) = \frac{\text{TN}}{\text{TN} + \text{FP}} \tag{5}$$

$$AUC = \frac{\text{TPR} + \text{TNR}}{2} \tag{6}$$

Table 2 Comparison of the performance individual classifiers for audit risk prediction

Classifier	Metrics	Baseline	Box-cox	Logarithm	Normalize	Z-score	Square
LR	Accuracy	0.988±0.008	0.899±0.022	0.961±0.014	0.926±0.019	0.955±0.015	**0.986±0.009**
	F-measure	0.988±0.008	0.897±0.022	0.960±0.014	0.926±0.019	0.955±0.015	**0.986±0.009**
	AUC	0.986±0.008	0.883±0.023	0.954±0.015	0.917±0.020	0.949±0.016	**0.984±0.009**
KNN	Accuracy	0.976±0.011	0.900±0.021	**0.966±0.013**	0.957±0.015	0.962±0.014	**0.966±0.013**
	F-measure	0.976±0.011	0.899±0.021	**0.966±0.014**	0.957±0.015	0.962±0.014	**0.966±0.013**
	AUC	0.970±0.012	0.886±0.023	0.959±0.014	0.952±0.015	0.956±0.015	**0.960±0.014**
B	Accuracy	0.971±0.012	0.918±0.020	0.951±0.015	**0.971±0.012**	**0.971±0.012**	0.928±0.018
	F-measure	0.971±0.012	0.919±0.019	0.951±0.015	**0.971±0.012**	**0.971±0.012**	0.926±0.019
	AUC	0.973±0.012	0.927±0.019	0.955±0.015	**0.973±0.012**	**0.973±0.012**	0.912±0.020
SVM	Accuracy	0.987±0.008	0.897±0.022	0.968±0.014	0.930±0.018	0.941±0.017	**0.978±0.010**
	F-measure	0.987±0.008	0.895±0.022	0.958±0.014	0.929±0.018	0.941±0.017	**0.978±0.010**
	AUC	0.984±0.009	0.879±0.023	0.950±0.015	0.922±0.019	0.935±0.018	**0.976±0.011**
RF	Accuracy	0.976±0.011	0.982±0.009	0.982±0.009	**0.983±0.009**	0.980±0.010	0.979±0.010
	F-measure	0.976±0.01	0.982±0.009	0.982±0.009	**0.983±0.009**	0.980±0.010	0.979±0.010
	AUC	0.975±0.011	0.982±0.009	0.982±0.009	**0.984±0.009**	0.980±0.010	0.978±0.010
DT	Accuracy	0.975±0.011	0.978±0.011	0.978±0.011	0.982±0.009	**0.983±0.009**	0.976±0.011
	F-measure	0.974±0.011	0.978±0.011	0.978±0.011	0.982±0.009	**0.983±0.009**	0.976±0.011
	AUC	0.972±0.012	0.975±0.011	0.975±0.011	0.979±0.010	**0.981±0.010**	0.973±0.011
XGB	Accuracy	0.984±0.009	0.984±0.009	**0.984±0.009**	**0.984±0.009**	0.983±0.009	0.976±0.009
	F-measure	0.984±0.009	**0.984±0.009**	**0.984±0.009**	**0.984±0.009**	0.983±0.009	0.976±0.009
	AUC	0.984±0.009	**0.984±0.009**	0.984±0.009	0.984±0.009	0.981±0.009	0.973±0.009

3 Experiment Result

In this work, we manipulated various features to identify fraudulent firms using machine learning methods. Almost all of them were implemented using the sci-kit learn library in Python. Since most of these features were highly correlated to each other, hence we reduced the linearity by dropping out several features that were responsible to produce better results. Besides, five feature transformation methods have been implemented into the baseline dataset and several transformed datasets are generated . Then, different classifiers such as LR, KNN, NB, SVM, RF, DT, and XGB have been implemented into primary (baseline) and its transformed datasets. To evaluate individual classifiers, we used several metrics like accuracy, F-measure, and AUC with confidence interval respectively. Table 2 shows the performance of all classifiers, both the baseline and transformed datasets which are described as follows:

- LR shows the highest accuracy with confidence interval (98.6%±0.9%), F-measure with confidence interval (98.6%±0.9%) and AUC with confidence interval (98.4%±0.9%) for the squared dataset.

- Again, KNN provides the best accuracy with confidence interval (96.6%±1.3%), F-measure with confidence interval (96.6%±1.4%) for logarithm and squared dataset. Maximum AUC is measured for the squared dataset.
- NB gives the highest accuracy with confidence interval (97.1%±1.2%), F-measure with confidence interval (97.1%±1.2%) and AUC with confidence interval (97.3%±1.2%) for normalized and z-scored dataset.
- SVM generates the leading accuracy and F-measure with confidence interval (97.8%±1.0%) and AUC with confidence interval (97.6%±1.1%) for the squared dataset.
- RF shows the best accuracy with confidence interval (98.3%±0.9%), F-measure with confidence interval (98.3%±0.9%) and AUC with confidence interval (98.4%±0.9%) for the normalized dataset.
- DT produces the highest accuracy with confidence interval (98.3%±0.9%), F-measure with confidence interval (98.3%±0.9%) and AUC with confidence interval (98.1%±1.0%) for the z-scored dataset.
- XGB provides the maximum accuracy with confidence interval (98.4%±0.9%) and F-measure with confidence interval (98.4%±0.9%) for the logarithm and normalized dataset. It also provides the highest AUC with confidence interval (98.4%±0.9%) for the Box-cox transformed dataset.

The performance of these classifiers delivered better outcomes for detecting fraudulent firms. However, the only accuracy is not considered as the best metrics for binary classification. After investigating the overall outcomes, LR shows the best accuracy and F-measure for squared dataset among all classifiers. Instead, the AUC values are assisted to rank fraud and non-fraud firms where XGB presents the maximum AUC for the box-cox dataset, respectively. Fig. 3a–e shows the ROC curves of different classifiers for baseline, box-cox, squared, normalized and z-scored transformed dataset, respectively.

4 Discussion

There were a small number of works about fraudulent firms detection using machine learning. Hooda et al. [16] prepared this fraudulent firms dataset in India and proposed a decision-making framework that identified the firms more accurately. In that work, they used particle swarm optimization (PSO) to rank and select features. But the selection criteria of PSO were not explained properly. The characteristics of the baseline dataset and high correlation among features were not found. To justify this work, our primary purpose is to compare the results of various classifiers that give better accuracy by dropping out some linear features in the baseline data. Linearity reduces the variability and predictability of features. So, we removed more collinear features and converted them into suitable formats to get stable results. In addition, related works and this dataset were also analyzed using several classifiers and their

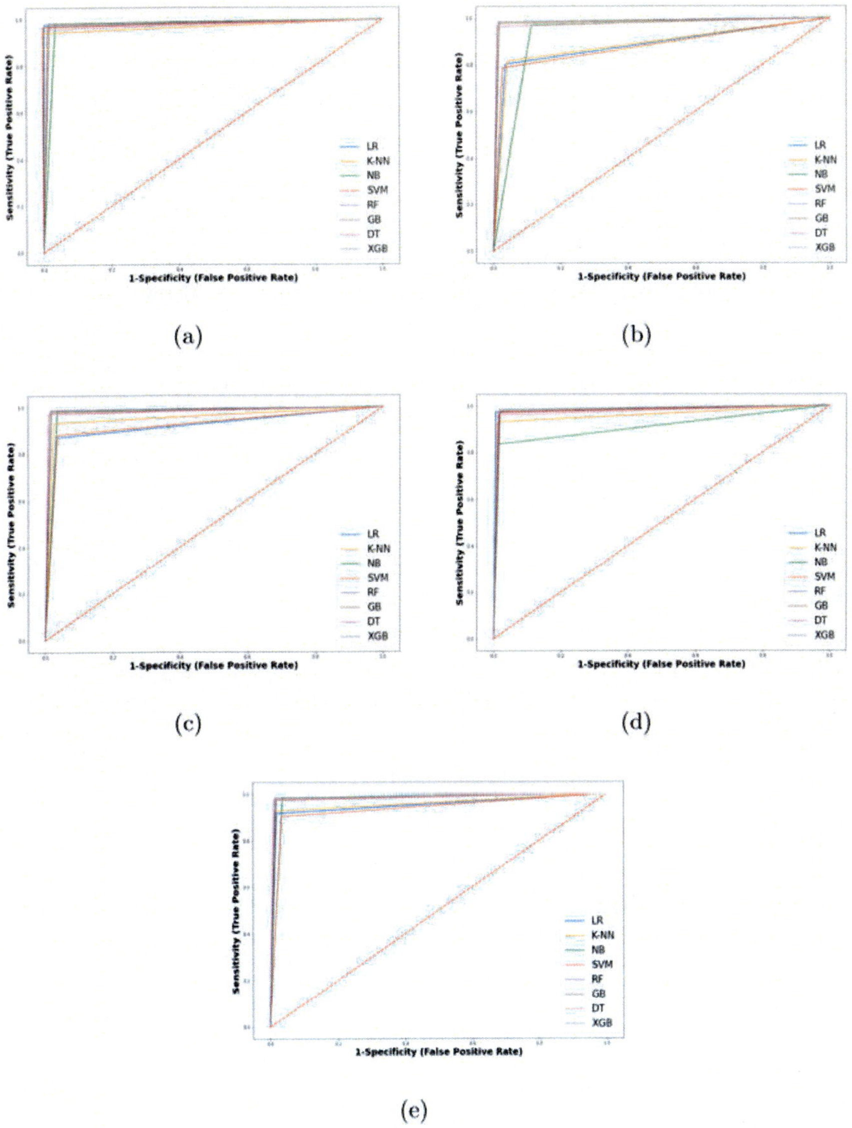

Fig. 3 ROC Curve for **a** Baseline **b** Box-cox **c** Normalized **d** Squared **e** Z-scored Dataset

performance for detecting fraudulent firms is shown . We have investigated the performance using correlation, transformation and classification techniques, where the performance of different classifiers outperforms the existing works [16]. In addition, the proposed model extracted significant factors that will be helpful to investigate fraudulent firms in India as well as worldwide.

5 Conclusion and Future Work

Auditing is an important task to identify fraudulent behavior of any financial organization. Machine learning methods are more helpful to inspect audit data. In this work, we gathered, cleaned and explored the characteristics of the dataset. Therefore, seven state-of-the-art classifiers and five feature transformation methodologies were employed to investigate audting instances and detect fraudulent firms. In this case, LR measures at most (98.6%±0.9%) accuracy and F-measure with confidence interval for the squared dataset and XGB measures maximum AUC for the box-cox dataset. From these analyses, these classifiers can predict the fraudulent behavior of firms. But, we cannot investigate multiple datasets and their findings to identify fraudulent firms which is the more stable performance in this work. In future work, we will optimize more statistical and machine learning procedures to improve this work. Again, an ensemble methodology will be proposed for audit data analysis that provides consistent outcomes for heterogeneous audit and fraudulent data.

References

1. Hakami TA, Rahmat MM, Yaacob MH, Saleh NM (2020) Fraud detection gap between auditor and fraud detection models: evidence from gulf cooperation council. Asian J Account Gov 13:1–13
2. Shajalal M, Petr H, Abedin MZ (2021) Product backorder prediction with deep neural network on imbalance data. Int J Prod Res. https://doi.org/10.1080/00207543.2021.1901153
3. Abedin MZ, Guotai Chi, Moula FE, Azad ASMS, Khan MSU (2019) Topological applications of multilayer perceptrons and support vector machine in financial decision support systems. Int J Finan Econ 24:474–507
4. Guotai Chi, Abedin MZ, Moula FE (2017) Modeling credit approval data with neural networks: an experimental investigation and optimization. J Bus Econ Manage 18(2):224–240. https://doi.org/10.3846/16111699.2017.1280844
5. Moula FE, Guotai Chi, Abedin MZ (2017) Credit default prediction modeling: an application of support vector machine. Risk Manag 19(2):158–187
6. Gepp A, Smith T, Linnenluecke MK (2018) Big data in accounting and finance: a review of influential publications and a research agenda. J Account Lit
7. IAASB (2016) Exploring the growing use of technology in the audit, with a focus on data analytics. International Auditing and Assurance Standards Board (IAASB). https://www.ifac.org/publications-resources/exploringgrowing-use-technology-audit-focus-data-analytics
8. Chen S (2016) Detection of fraudulent financial statements using the hybrid data mining approach. SpringerPlus: a SpringerOpen J
9. Sadgali I, Sael N, Benabbou F (2019) Performance of machine learning techniques in the detection of financial frauds. Proc Comput Sci Elsevier 148:45–54
10. Hajek P, Henriques R (2017) Mining corporate annual reports for intelligent detection of financial statement fraud - A comparative study of machine learning methods. Elsevier, pp 139–152
11. Mohammadi M, Yazdani S, Khanmohammadi M, Maham K (2020) Financial reporting fraud detection: an analysis of data mining algorithms. Int J Financ Manag Account Iran. **4**
12. Amina A, Shahb B, Khattakb AM, Moreirac FJL, Alid G, Rochae A, Anwara S (2018) Cross-company customer churn prediction in telecommunication: A comparison of data transformation methods. Int J Inf Manag Elsevier

13. Zhang F, Keivanloo I, Zou Y (2017) Data transformation in cross-project defect prediction. Springer, Empir Software Eng
14. Satu MS, Tasnim F, Akter T, Halder S (2018) Exploring significant heart disease factors based on semi supervised learning algorithms. In: 2018 international conference on computer, communication, chemical, material and electronic engineering (IC4ME2), Rajshahi, 2018, pp 1–4. https://doi.org/10.1109/IC4ME2.2018.8465642.
15. Ahammed K, Satu MS, Satu MS, Khan MI, Whaiduzzaman M (2020) Predicting infectious state of hepatitis C virus affected patient's applying machine learning methods. In: (2020) IEEE region 10 symposium (TENSYMP). Dhaka, Bangladesh, vol 2020, pp 1371–1374. https://doi.org/10.1109/TENSYMP50017.2020.9230464
16. Hooda N, Bawa S, Rana PS (2018) Fraudulent firm classification: a case study of an external audit. J Appl Artif Intell An Int J Taylor & Francis
17. Abedin MZ, Guotai Chi, Uddin MM, Shahriare S, Imran K, Petr H (2021) Tax default prediction using feature transformation-based machine learning. IEEE Access 9:19864–19881. https://doi.org/10.1109/ACCESS.2020.3048018
18. Li Y (2017) Multi-class sentiment classification: the experimental comparisons of feature selection and machine learning algorithms. Elsevier 477:448–465
19. Thaher T, Asghar A, Mafarja HM, Dong JS, Mirjalili S (2019) Binary harris hawks optimizer for high-dimensional, low sample size feature selection. Evolutionary machine learning techniques. Algorithms for intelligent systems. Springer, Singapore, pp 251–272
20. Tan X, Su S, Huang Z, Guo X, Zuo Z, Sun X, Li L (2019) Wireless sensor networks intrusion detection based on smote and the random forest algorithm. Sensors, vol 19
21. Petr H, Abedin MZ (2020) A profit function-maximizing inventory backorder prediction system using big data analytics. IEEE Access 8:58982–58994. https://doi.org/10.1109/ACCESS.2020.2983118
22. (1997) The problem of multicollinearity . In: Understanding regression analysis. Springer, Boston, MA. https://doi.org/10.1007/978-0-585-25657-3_37

An Enhanced Quality of Service-Based Trust Model for Cloud Computing

Mohamed Fazil Mohamed Firdhous and Rahmat Budiarto

Abstract Cloud computing has taken the computing services market by storm due to its advantages over traditional computing platforms. Cloud computing makes its resources available over the Internet, so clients can purchase them and pay only for what is used similar to other utilities. For cloud computing to become acceptable to a wider business clientele including the ones with high expectations, it is vital to maintain the QoS commitments. Hence, it is necessary to have mechanisms and tools to measure and quantify the performance of cloud service providers. In this paper, the authors propose a mathematical model that could measure multiple QoS attributes, quantify them and compute a single score called the trust score to help customers identify the most suitable service provider. The proposed enhanced QoS (eQoS) trust model was tested using simulations. The results show that the proposed mechanism performs better than the other two QoS-based trust models, namely QoS Trust and Turnaround Trust published in the literature.

Keywords Cloud computing · Quality of service · Mathematical model · QoS metrics · Trust computing

1 Introduction

Cloud computing has taken the computing services industry by storm through the revolutionary way it offers the computing resources as services. With the innovative pay as you use charging scheme, it has joined the bandwagon of electricity, water, gas and telephony as the fifth utility [4]. Customers can purchase computing resources such as infrastructure, development platform or applications as services over the

M. F. M. Firdhous (✉)
Department of Information Technology, University of Moratuwa, Moratuwa 10400, Sri Lanka
e-mail: firdhous@uom.lk

R. Budiarto
College of Computer Science and Information Technology, Albaha University, Al Baha, Saudi Arabia
e-mail: rahmat@bu.edu.sa

© The Author(s), under exclusive license to Springer Nature Singapore Pte Ltd. 2022

D. Giri et al. (eds.), *Proceedings of the Seventh International Conference on Mathematics and Computing*, Advances in Intelligent Systems and Computing 1412,
https://doi.org/10.1007/978-981-16-6890-6_39

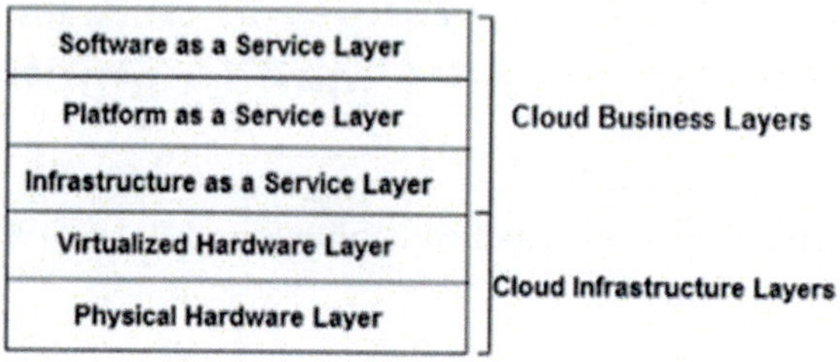

Fig. 1 Cloud computing layered architecture

Internet as they use them as if they are hosted locally and have dedicated access to them. Cloud computing services are divided into three main groups as Infrastructure as a Service (IaaS), Platform as a Service (PaaS) and Software as a Service (SaaS) [4]. These services along with the underlying infrastructure layers are shown in Fig. 1.

The infrastructure that makes the cloud services possible is divided further into physical and virtualized hardware layers. The physical hardware provides the workhorse for the computing requirements and the virtualization slices that hardware into individual computing elements to meet the requirements of the customers. The physical hardware can be made up of individual computers or computers working as a group in a network. The computing elements can be brought up and removed dynamically as well as extend and shrink on demand in parallel and independent to each other. Hosting multiple parallel systems increases the utility and efficiency of the underlying hardware resources. With cloud computing, organizations are able to outsource their computing requirements to the experts at the lowest possible cost enabling them to concentrate on their core business functions.

The economic advantage of cloud computing as opposed to in-house computing can be explained by comparing the resource allocation patterns under both schemes. Figure 2 shows the capacity utilization curve developed by Amazon Web Services

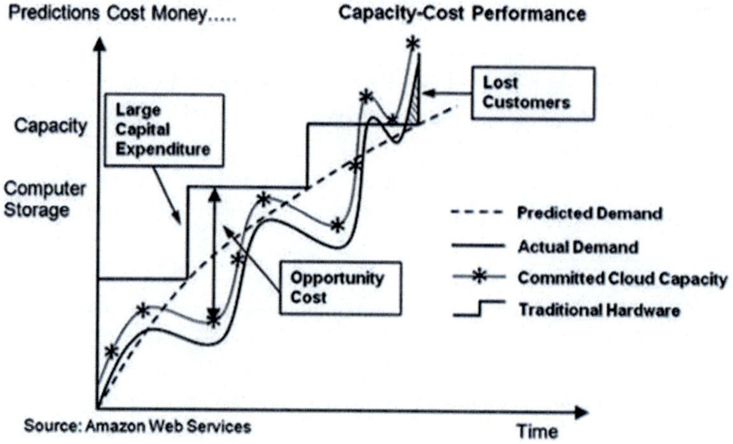

Fig. 2 Capacity versus utilization curve [2]

(AWS) for a demand versus allocation of storage under cloud computing and in-house implementation schemes [2].

Figure 2 shows that the demand pattern is going through multiple fluctuations in both the short as well as long runs. The short-run fluctuations are due to the active hours during a day as well as a week and the long-term fluctuations are generally the results of seasonal variations and growth/decline in demand patterns. From Fig. 2, it can be seen that the in-house implemented systems cannot follow the demand patterns closely as done by the cloud systems. This results in losses either due to resource idling or loss of business. Since cloud implemented systems can follow the demand patterns closely during both short- and long-term fluctuations, it is able to provide 100% return on investment. Cloud systems gain this capability closely as virtual systems can be brought up and removed on the fly on demand. As physical resources are consumed only by active systems, the resources released by the inactive users can be allocated to other users enabling the providers to increase their profits from the same hardware. As cloud computing provides enhanced profits for providers, it has attracted many to this industry increasing the competition. As multiple customers are hosted in parallel on the same hardware, the service quality will depend on the capacity of the hardware and the number of concurrent users hosted. Increasing the number of active users beyond a certain point will affect every user active at that point in time. Hence, from the customers' point of view, it is good to have some independent measure indicating whether a selected provider would be able to meet his or her requirements.

In [16], Manuel has shown that trust scores can be used as a measure to quantify the QoS performance of cloud service providers. Literature reports several trust computing mechanisms based on different principles, criteria and functions [8, 9, 11, 16, 20, 21]. Except for [16, 8], other mechanisms use only a single metric to compute the overall trust score. This is against the basic concept of trust, which is based on multiple criteria [17]. Thus, they are not suitable for computing QoS-based trust computing as they may not be able to meet performance demands based on multiple parameters. The trust model proposed by Manuel in [16] is very weak as the QoS metrics are computed using the service provider's credentials only. No continuous monitoring or updates are carried out on the trust scores. Though the mechanism proposed in [8] is a strong one with statistical validation, the mathematical derivation of the trust score is not provided. In this paper, the author attempt to rectify this shortcoming by providing a strong mathematical background to it. This paper has been divided into five main sections as follows: Sect. 1 lays the foundation by providing an introduction to the paper. Section 2 critically analyzes the related work already presented in the literature. Sections 3 and 4 present the mathematical model for trust along with the experimental setup used for testing the model with an in-depth analysis of the results. Finally, Sect. 5 concludes the paper summarizing the findings.

2 Related Work

This section presents a critical analysis of related literature on two specific areas. They are namely quality of service in cloud computing and trust computing in distributed systems.

2.1 *Quality of Service in Cloud Computing*

The business operation of any organization goes through fluctuations in the short run as well as long run due to various factors. Irrespective of the causes and nature, businesses must be able to handle these fluctuations and convert them to their advantage. Computing is one of the most important operations for any business as it provides many advantages. The most important advantages of computers to businesses include the increased productivity and accuracy of the operations, enhanced reach beyond the traditional physical or geographical boundaries and the ability to run 24*7 operations with minimal human intervention. With the aid of computers, now business executives can identify trends with almost 100% accuracy by processing exorbitant amounts of data within short times. This enables them to organize their operations proactively to take advantage of the emerging trends while minimizing threats.

As many businesses depend on computers heavily for their core business functions, the quality and dependability of resources and operations are of utmost importance to them. When the requirements are outsourced to third-party public operators, the assurance that these services would meet expectations is also of paramount importance. Hence, these services must be procured from a service provider who has a good reputation for meeting them. With time, more and more customers sign up for services. Eventually, they would all compete for the resources available in the provider's data center. After reaching the critical mass of customers for a given amount of resources, the service quality would deteriorate with every additional customer signing up, if the system is not upgraded. This will affect both the customer as well as the service provider in the long run. Hence, managing the quality of service is important for both service providers and the customers.

In [15], Liu et al. have proposed a generic QoS framework for a cloud workflow system covering all four stages. The main shortcomings of this framework are not specifically identifying any QoS parameters and also not discussing the methods for differentiating different types of clients demanding different QoS levels. On the other hand, the basis for the workflow scheduling algorithm proposed by Chen and Zhang is set-based discrete particle swarm optimization [6]. This mechanism can handle up to seven parameters specified by the users compared to the traditional optimization techniques that consider only a single workflow execution time. The downside of this mechanism is a lack of a monitoring scheme for QoS violation and punishment.

The Service Level Agreement (SLA) management framework proposed by Buyya et al. in [3] has a special emphasis on managing QoS requirements. The proposed architecture has the ability to integrate the market-based resource provisioning within virtualization technologies for flexible resource allocations to user applications. The main shortcoming of this framework is its inability to support different cloud service offerings such as IaaS, PaaS and SaaS in a combined fashion.

The optimal resource allocation model for the revenue maximization model proposed in [7] has been mathematically derived from basic principles. The model has been validated through simulations using both synthetic and traced datasets. The proposed model performs better than heuristic optimization of resources while maximizing profits. But the application of this method in real-world applications is limited as it considers only the mean response time as the QoS attribute.

The inclusion of trust into the QoS targets is a novelty in the workflow scheduling model proposed by Li et al. in [14]. The two-stage workflow model consists of two stages named a macro multi-workflow stage and a micro single workflow stage. The first stage is based on trust and the second stage classifies workflows into time-sensitive and cost-sensitive ones based on QoS demands. The QoS attributes used in this model are limited to average values of response time, bandwidth, storage, reliability and cost with no guarantee of service delivery, which can be considered as the major limitation of this model.

The importance of dynamically monitoring the QoS of virtualized services has been highlighted in [1]. The continuous monitoring of the performance of cloud services helps both service providers as well as the application providers to keep their resources at maximum efficiency. As the proposed QoS monitoring is based on SNMP, messages generated by the systems themselves may lead to erroneous conclusions resulting from misconfiguration of services or external attacks as SNMP is one of the most commonly attacked layers in computer networks.

The method proposed by Manuel in [16] for monitoring and quantifying QoS of cloud services could be used to rank service providers based on their performance. But this method is based on a very weak foundation as the QoS values are computed using service providers' credentials as opposed to their actual performance. In [11], Gholami and Arani have proposed an extension to Manuel's model as the Turnaround_Trust trust model. This model incorporates run speed along with the parameters that are already with the model proposed by Manuel. Hence, the enhancement is very minor and cannot overcome the shortcomings that are with the original model, which it is based on.

2.2 *Trust Computing in Cloud Systems*

Trust is an idea and a concept that has been studied by researchers in different fields for the purpose of understanding and quantifying certain characteristics in their field of interest [10]. In social sciences, trust is formally studied with the objective of understanding the nature and behavior of human societies. For psychologists, trust is a mental attitude and they investigate what happens in a human mind when one trusts or distrusts another. Trust has a utility value for economists. Computation models developed by computer scientists for representing different conditions and situations are generally based on all or some of these studies and findings [19]. Almost all the different distributed computing paradigms including e-commerce, peer-to-peer networks, grid computing, semantic web, web services, mobile networks and cloud computing have been using trust and reputation systems for the purpose of identifying the right peer to interact with [18]. These models are generally based on some well-known mathematical functions.

In [13], Kim et al. have proposed a trust evaluation model for guaranteeing the QoS requirements of the customers. In this model, the trust of different units within a data center has been pre-computed using historical information and used when selecting a resource for serving requests. This method helps the service provider to allocate his resources optimally without sacrificing the QoS requirements of customers. The main shortcoming of this method is that it is essentially an internal resource management tool to help service providers rather than the customers. If this tool can be implemented outside the service provider as a third-party tool, it may help customers to select the right service provider based on their requirements.

Habib et al. in [12] have presented an in-depth analysis on the importance of trust and reputation in cloud computing. Their analysis focuses more on the impediments preventing users from adopting cloud computing. In [12], they emphasize the importance of the role of trust as a mediator between service providers and customers. They strongly stress the need for incorporating QoS metrics into trust and reputation systems as these attributes play an important role in the provision of dependable services. The main outcome of this work is the identification of the essential properties and the difficulties of incorporating QoS attributes into trust and reputation systems.

3 Trust Metrics

In order to arrive at a mathematical model for trust based on QoS attributes, it is important to identify a set of attributes that a user generally considers when selecting a cloud service provider. In this article, the author considers the following attributes as important in computing the QoS of cloud systems. The selected attributes are response time, processing time, latency, availability, cost, reliability, operational integrity and data integrity. It must also be noted that the model proposed in this article is very flexible, so that it can be extended accommodating any new attribute that may be necessary to meet any specific requirement.

3.1 Performance Metrics

The definitions of some of the important performance metrics along with their mathematical formula to compute them are given below.

Availability is the percentage of times a system is up and ready to accept work from a client. A system may be unavailable for many reasons including non-availability of power, failure of another device between the system and the client, system maintenance or upgrading or overloaded by services already running on it or diverted by an attacker. Irrespective of the reason, the system is considered unavailable by the users.

Availability (A_v) can be expressed as a fraction between the amount of time it is available for service with respect to the total time as given in Formula (1) below:

$$A_v = \frac{T_A}{T_T} \quad (1)$$

where T_A is the amount of time the system is available to service a client request and T_T is the total time.

Availability can also be computed as a fraction of the no. of requests (jobs) serviced (R_S) against the total no. of requests (R_T) within a given time period. Availability (A_{vr}) in this case will be defined as follows:

$$A_{vr} = \frac{R_S}{R_T} \quad (2)$$

Response time is defined as the time elapsed between the times a customer makes the request and the system accepts it. Response time efficiency R_t is given by

$$R_t = \frac{R_p}{R_a} \quad (3)$$

where R_p is the promised response time and R_a is the actual response time.

Processing time is the total time taken for a process to complete the work starting from the submission to the delivery of the final results. Processing time efficiency P_t is given by

$$P_t = \frac{P_p}{P_a} \quad (4)$$

where P_p and P_a are the promised and the actual processing times, respectively.

Reliability, also known as the success rate, is the number of jobs successfully completed in a given time T out of the total number of jobs accepted within the same time T. Reliability (R_e) is computed by

$$R_e = \frac{C_j}{N_j} \quad (5)$$

where C_j is the number of jobs successfully completed in a given in T and N_j is the total number of jobs accepted for processing over the same time T.

Operational integrity is the assurance given by a remote system that a program submitted to it would be executed as expected making results to be authentic and reliable. Operational integrity is achieved by ensuring that the program code is not tampered with prior to or during execution by a rogue user or system. Operational integrity factor I_o is given by

$$I_o = \frac{J_o}{J_T} \quad (6)$$

where J_o is the total no. of jobs completed preserving operational integrity in a given in T and J_T is the total number of jobs processed over the same time.

Data integrity is the assurance that the data is not modified by an unauthorized person, an agent or in an unauthorized manner. Data integrity factor I_D is given by

$$I_D = \frac{D_o}{D_T} \quad (7)$$

where D_o is the number of jobs completed preserving data integrity in a given in T and D_T is the total number of jobs processed over the same time period T.

It could be seen that the attributes shown above all follow similar pattern in computing the efficiencies. Hence, the list of attributes can be extended to include any attribute not included there without much difficulty to cater to any user requirement or future demand.

3.2 *Computation of Overall Trust*

After defining the metrics, Manuel in [16] straight away creates the trust value of a resource by adding the weighted sum of attributes. Also, it is gone noted that the weights assigned are ad-hoc values that do not have any scientific explanation why such a weight is assigned to that given metric. Assigning weights directly to the metrics does not reflect the dynamic nature of cloud computing. The performance of a cloud computing system depends on many factors such as the capacity of the system, loading and the nature of the external environment including the network links, devices, etc. Hence, in this paper, the author proposes a trust computing scheme based on the posterior probability defined by Bayes' rule.

Trust based on QoS metrics can be represented by a directed graph and the associated performance table as shown in Fig. 3 and Table 1.

where $O = 1$ represents the successful outcomes and $O = 0$ represents unsuccessful outcomes. s_i and n_i represent the number of times the metric i is considered to

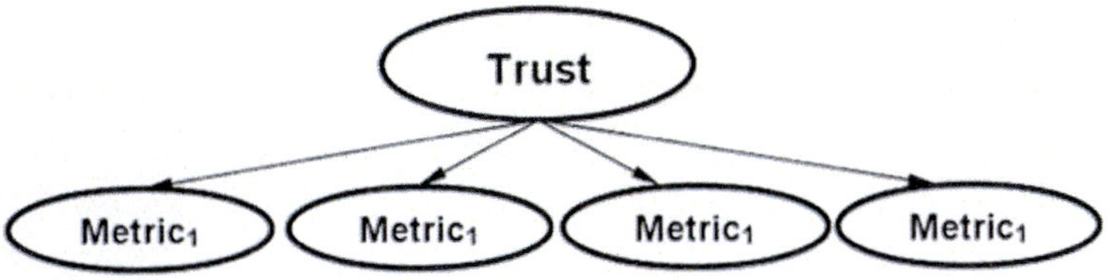

Fig. 3 Naive Bayesian network for a cloud computing node

Table 1 Performance table

Metric	No. of interactions with O = 1	No. of interactions with O = 0
$Metric_1$	s_1	n_1
$Metric_2$	s_2	n_2
$Metric_3$	s_3	n_3
$Metric_4$	s_4	n_4

have met its requirements when the overall outcome is successful and unsuccessful, respectively.

From Table 1, it is possible to compute all the required conditional and absolute probabilities. The trust of a given cloud computing node for a given set of performance metrics is given by the joint probability:

$$T = p(O \cap m_1 \cap m_2 \cdots \cap m_n) \tag{8}$$

where O and m_i represent the outcome and ith metric when all of them are met and the final outcome is considered successful. The joint probability can be represented as a conditional probability and simplified as given below using Bayes' theorem.

$$T = p(O)p(m_1, m_2 \cdots m_n|O)$$

and

$$T = p(O)p(m_1|O)p(m_2|O) \cdots p(m_n|O) \tag{9}$$

provided the metrics (m_i) are independent of each other.

Table 1 is to be continuously updated by the trust provider, whenever a client obtains the services from a given cloud computing node. Hence, it must maintain a performance table similar to Table 1 for each cloud computing node and continuously update it. Whenever a client requests the trust value of a node, it is possible to compute it using Eq. 9 based on any set of performance metrics provided by the client.

Table 2 Modified performance table

Metric	Weight	No. of interactions with $O = 1$		No. of interactions with $O = 0$	
		Original	Modified	Original	Modified
m_1	α_1	S_1	$\alpha_1 S_1$	n_1	$\alpha_1 n_1$
m_3	α_3	S_3	$\alpha_3 S_3$	n_2	$\alpha_3 n_3$
m_7	α_7	S_7	$\alpha_7 S_7$	n_3	$\alpha_7 n_7$

3.3 *Computing Trust Score with Different Priorities*

It is also possible to compute the trust value using Eq. 9 when a client has different priorities for different metrics. When such a prioritization of metrics is required, it will be given by metric weights as shown in Eq. 10.

$$\tau = \frac{\alpha_1 m_1 + \alpha_2 m_2 + \cdots + \alpha_n m_n}{\alpha_1 + \alpha_2 + \cdots + \alpha_n} \tag{10}$$

and

$$\alpha_1 + \alpha_2 + \cdots + \alpha_n = 1$$

where m_i is the ith parameter and α_i is the weight applied to it.

For example, when a client seeks a certain service with QoS requirements on metrics m_1, m_3 and m_7 with the priorities of α_1, α_3 and α_7, then the performance table would be modified as shown in Table 2.

The values in the modified columns will be used for computing the conditional probabilities and the trust score for the node.

4 Results and Discussion

The proposed model has been evaluated in a simulated environment using the CloudSim simulation tool. CloudSim provides the cloud computing researchers and developers a platform for modeling and testing their proposals before implementing them on real systems. CloudSim is capable of simulating multiple cloud systems with many data centers equipped with different types of computers and devices similar to real-world systems [5]. CloudSim has been selected as the simulation platform evaluating the proposed model due to its above-mentioned advantages. The model was subjected to extensive testing using random and consistent behavior of a predefined cloud computing unit. The simulation environment was set up as follows:

No. of data centers—1

No. of servers—1

Server Properties:
Architecture— ×86;
No. of processors—1 (Quad Core)
Memory—4 GB
Storage—1 TB
Network—1 Gbps
VM allocation policy—Simple
Method of VM scheduling—time and space shared
cost—0.

The simulation results were validated by comparing them against the performance of QoS Trust and Turnaround Trust models proposed in [16, 11], respectively, under similar conditions. Initially, the response time of the system was selected as the performance metric for evaluation. During this experiment, the required response time was fixed at 30n and the actual performance time was allowed to fluctuate around this value slightly. The system was initially allowed to settle down at the beginning as the random fluctuations allowed on the system are not to disturb the long-term behavior of the system.

Figure 4 shows the performance of the three trust models under small fluctuations in response time. From Fig. 4, it can be seen that the proposed enhanced QoS (eQoS) trust model shows a steady performance at 0.5 while the other two models fluctuate even for minor changes in performance. As the fluctuations allowed are very minor in nature, it should not be taken as the result of a permanent degradation or improvement and the trust value should not be modified. Hence, we can safely conclude that the model proposed in this work is more representative of the performance than the other two models.

Figures 5 and 6 show the behavior of the models when the performance of the system was consistently worse than the required performance of 30 s and when the system consistently performs better, respectively. From Figs. 5 and 6, it can be clearly

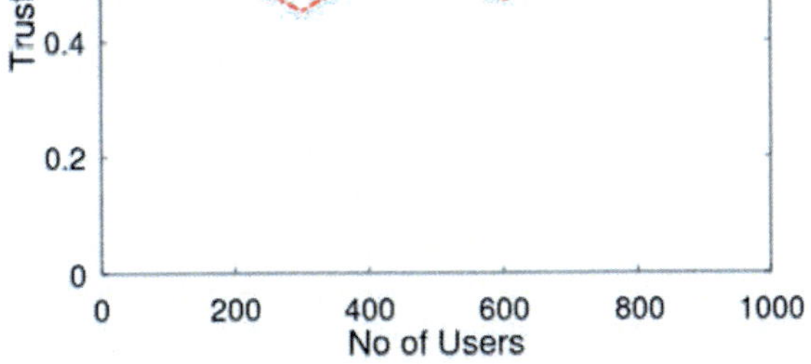

Fig. 4 Performance of the trust model under small fluctuations

Fig. 5 Performance of the trust models under consistent underperformance

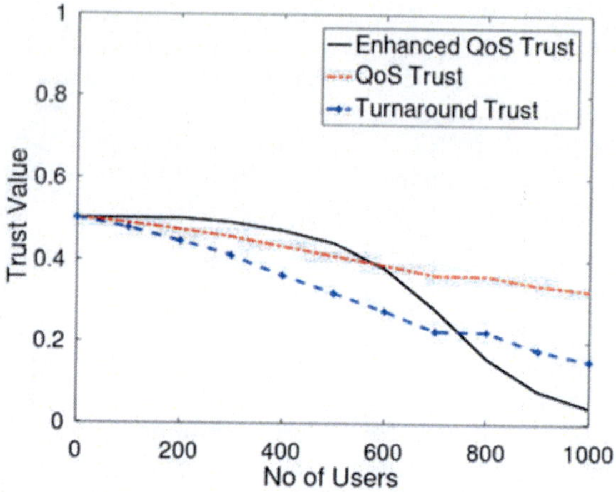

Fig. 6 Performance of the trust models under consistent performance

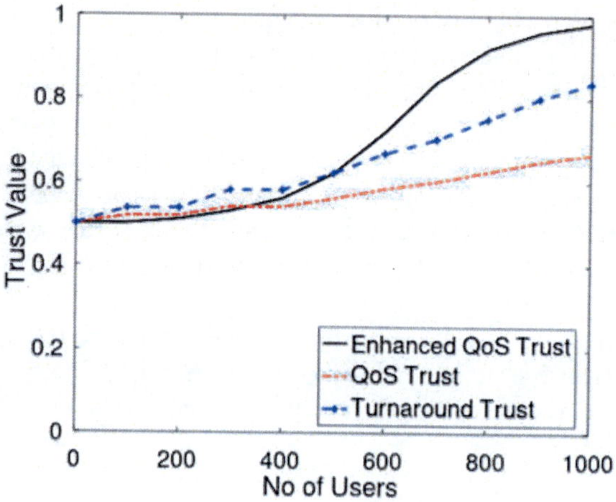

seen that the trust values computed by the proposed eQoS model are realistic as it responds to the changes as expected. Though the QoS model and the Turnaround model also either improve or diminish the trust scores, they show fluctuations in between.

Figure 7 shows the performance of the models when the cloud computing node was continuously loaded initially with relaxed demands and then with more and more stringent demands with time in such a manner that its performance slowly and steadily diminishes.

Figure 7 shows that the proposed eQoS trust model initially holds the trust value at fixed 0.5 as the performance of the node was acceptable. Then it starts declining in the middle as the system performance gets more and more deteriorated. On the other hand, the other two models show a steady decline from the beginning itself. These steady declines do not represent the behavior of the cloud system as its performance

Fig. 7 Trust scores computed under steady loading

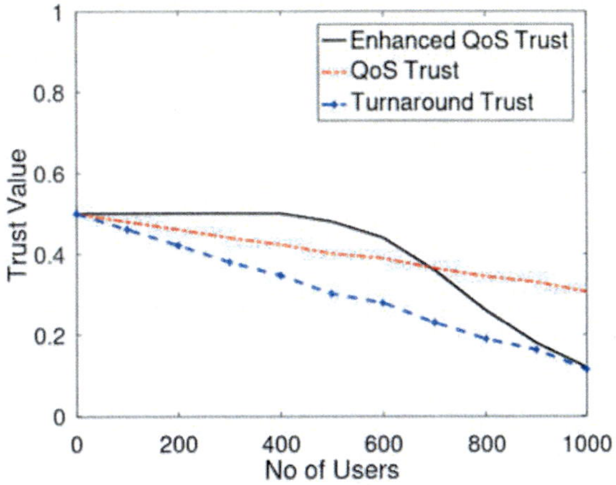

was better at the beginning and in the middle only, it becomes lower than the requirement. Hence, it can be concluded that the eQoS trust model proposed in this article computes more realistic trust scores that are more representative of the performance of the system than the trust values computed by both the QoS and the Turnaround trust models proposed in [11, 16].

5 Conclusions

This paper presented a trust computing model based on the quality of service for cloud computing systems. The proposed model is based on the Naive Bayesian network for conditional probability. The Bayesian model computes the posterior probability based on past performance. Hence ,our model is strongly grounded on an established theory as opposed to many models discussed in the related work section. The trust scores computed using our model are very realistic and strongly connected to the common expectations under the given behavior of the system. The simulation environment used for testing the model and validating the results is very realistic and very similar to the practical environment. Hence, it can be concluded that the model proposed in this paper performs better than the models discussed in the literature.

References

1. Alhamazani K, Ranjan R, Rabhi F, Wang L, Mitra K (2012) Cloud monitoring for optimizing the QoS of hosted applications. In: 4th IEEE international conference on cloud computing technology and science. Taipei, Taiwan, pp 765–770
2. AWS: Capacity versus utilization curve (2020) Retrieved May 15, 2020, from Amazon Web Services. http://www.amazon.com/economics

3. Buyya R, Garg SK, Calheiros RN (2011) \SLA-oriented resource provisioning for cloud computing: challenges, architecture, and solutions. In: International conference on cloud and service computing. Hong Kong, pp 1–10
4. Buyya R, Yeo CS, Venugopal S, Broberg J, Brandic I (2009) Cloud computing and emerging IT platforms: vision, hype and reality for delivering computing as the 5th utility. Futur Gener Comput Syst 25(6):599–616
5. Calheiros RN, Ranjan R, Beloglazov A, de Rose CAF, Buyya R (2011) CloudSim: a toolkit for modeling and simulation of cloud computing environments and evaluation of resource provisioning algorithms. Softw: Pract Exp 41(1):23–50
6. Chen WN, Zhang J (2012) A set-based discrete PSO for cloud workflow scheduling with user-defined QoS constraints. In: IEEE international conference on systems, man and cybernetics. Seoul, Korea, pp 773–778
7. Feng G, Garg S, Buyya R, Li W (2012) Revenue maximization using adaptive resource provisioning in cloud computing environments. In: 13th ACM/IEEE international conference on grid computing. Beijing, China, pp 192–200
8. Firdhous M, Ghazali O, Hassan S (2017) Robust multi-dimensional trust computing mechanism for cloud computing. J Telecommun, Electron Comput Eng (JTEC) 9(1–2):73–76
9. Firdhous M, Ghazali O, Hassan S, Harun NZ, Abas A (2011) Honey bee based trust management system for cloud computing. In: 3rd international conference on computing and informatics (ICOCI 2011). Bandung, Indonesia, pp 327–332
10. Gan Z, He J, Ding Q (2009) Trust relationship modelling in e-commerce based social network. In: International conference on computational intelligence and security. Beijing, China, pp 206–210
11. Gholami A, Arani MG (2015) A trust model based on quality of service in cloud computing environment. Int J Database Theory Appl 8(5):161–170
12. Habib SM, Hauke S, Ries S, Muhlhauser M (2012) Trust as a facilitator in cloud computing: a survey. J Cloud Comput 1(19):1–18
13. Kim H, Lee H, Kim W, Kim Y (2010) A trust evaluation model for QoS guarantee in cloud systems. Int J Grid Distrib Comput 3(1):1–10
14. Li W, Zhang Q, Wu J, Li J, Zhao H (2012) Trust-based and QoS demand clustering analysis customizable cloud workflow scheduling strategies. In: IEEE international conference on cluster computing workshops. Beijing, China, pp 111–119
15. Liu X, Yang Y, Zhang G, Li W, Cao D (2011) A generic QoS framework for cloud workflow systems. In: 9th IEEE international conference on dependable, autonomic and secure computing. Sydney, Australia, pp 713–720
16. Manuel P (2015) A trust model of cloud computing based on quality of service. Ann Oper Res 233:281–292
17. McKnight DH, Chervany NL (2001) Conceptualizing trust: a typology and e-commerce customer relationships model. In: 34th Hawaii international conference on system sciences. Maui, HI, USA
18. Momani M, Challa S (2010) Survey of trust models in different network domains. Int J Adhoc, Sens Ubiquitous Comput 1(3):1–19
19. Mui L (2002) Computational models of trust and reputation: agents, evolutionary games, and social networks. Ph.D. thesis, University of Boston, Boston, MA, USA
20. Noor TH, Sheng QZ, Yao L, Dustdar S, Ngu AHH (2016) Cloudarmor: Supporting reputation-based trust management for cloud services. IEEE Trans Parallel Distrib Syst 27(2):367–380
21. Saeed O, Shaikh RA (2018) A user-based trust model for cloud computing environment. Int J Adv Comput Sci Appl 9(3):337–346

DEA for Measuring the Performance of Higher Education Sector: A Case Study of Myanmar

Marlar Tin, Ankita Panwar, and Millie Pant

Abstract The higher education system is an important part of the social and economic progress of a country and should be monitored from time to time to identify its shortcomings and suggest measures for its growth. In this paper, we used a linear programming-based, multi-criteria decision-making technique, Data Envelopment Analysis (DEA), to assess the relative efficiency of 14 different regions and states of Myanmar over three academic years 2015, 2016, and 2017.

Keywords Data envelopment analysis · Higher education · Efficiency · Myanmar

1 Introduction

The performance assessment of the education sector plays an essential role in the socio-economic development of a nation [1]. Measuring the performance often depends on a number of criteria, input as well as output, and such problems are called multi-criteria decision-making (MCDM) problems. Literature reveals DEA to be one of the prominent methods for assessing the relative performance of educational institutions, favored by researchers from time to time. Tomkins and Green [3] used DEA to measure the overall relative efficiency of the UK university's accounting department. They used the number of staff and number of students as input and output variables, respectively, and used a total of 6 models in their studies. Results showed that the different pairs of input and output variables produced a stable efficiency score. Tyagi et al. [2] studied the performance of the 19 departments of the IIT Roorkee. They calculated the technical, pure technical, and scale efficiencies and verified

M. Tin
Computer University Dawei, Tanintharyi, Myanmar

A. Panwar (✉) · M. Pant
Applied Science and Engineering, Indian Institute of Technology (IIT) Roorkee, Roorkee, India
e-mail: apanwar@as.iitr.ac.in

M. Pant
e-mail: pant.milli@as.iitr.ac.in

© The Author(s), under exclusive license to Springer Nature Singapore Pte Ltd. 2022

D. Giri et al. (eds.), *Proceedings of the Seventh International Conference on Mathematics and Computing*, Advances in Intelligent Systems and Computing 1412,
https://doi.org/10.1007/978-981-16-6890-6_40

the peer groups for the inefficient departments. They also implemented sensitivity analysis for measuring the teaching and research performance separately. Chen and Chen [4] discussed in their paper innovation and improving the quality performance of the Taiwan universities. They used 99 Taiwan universities and divided them into five categories: research, teaching, professions, research-teaching, and education-in practice-intensive. The empirical results showed that almost 73% of universities are inefficient, so improving innovation and improving quality performance. Zhang and Shi [5] evaluated the teaching performance in Chinese universities. They used a total of 24 universities and colleges as DMUs, and to find the efficiencies and find the differences between them, Zhang et al. [6] studied the performance of the enrollment quota allocation and evaluation for the higher education institutes in China. They used a total of three input variables and three output variables.

In the present study, DEA is implemented to assess the efficiency of the higher education sector of Myanmar in terms of its different administrative divisions, details of which are given in later section.

The present study is divided into five sections including an introduction. In Sect. 2, a brief description of DEA is given, including its three important mathematical models. In Sect. 3, the case study is described. In Sect. 4, results obtained through DEA are discussed and finally, the concluding remarks are provided in Sect. 5.

2 Data Envelopment Analysis (DEA)

DEA was introduced by Charnes, Cooper, and Rhodes in 1978 [7] for measuring the relative efficiency of decision-making units when multiple inputs and outputs are to be considered [2]. It was further extended by Banker, Charnes, and Cooper in 1984 as the BCC model [8]. The relative efficiency is defined as the ratio of the weighted sum of output variables to the weighted sum of input variables [9] as given below.

$$\text{Efficiency} = \frac{\text{The weighted sum of output}}{\text{The weighted sum of input}}$$

$$\mathrm{E}_k = \frac{\mathrm{u_1 y_{j1} + u_2 y_{j2} + \ldots}}{\mathrm{v_1 x_{i1} + v_2 x_{i2} + \ldots}}$$

where $\mathrm{E_k}$ = efficiency of the kth DMU, $\mathrm{v_1}$ = weight given to output 1, $\mathrm{y_{j1}}$= amount of output 1 from unit j, $\mathrm{u_1}$= weight given to input 1, and $\mathrm{x_{i1}}$= amount of input 1 used by unit i.

Mathematical form of the CCR model

$$maximize E_k = \sum_{j=1}^{M} u_{j0} y_{jk} \tag{1}$$

Subject to

$$\sum_{i=1}^{N} v_{i0}\mathrm{x}_{ik} = 1 \tag{2}$$

$$\sum_{j=1}^{s} v_{j0}x_{j0} - \sum_{i=1}^{m} u_{i0}\mathrm{y}_{i0} \leq 0, \forall n \tag{3}$$

$$v_{j0}, u_{i0} \geq 0, \forall i, j \tag{4}$$

Mathematical form of the BCC model

$$maximize E_k = \sum_{j=1}^{M} u_{j0}y_{jk} + \mu_{o0} \tag{5}$$

Subject to

$$\sum_{i=1}^{N} v_{i0}\mathrm{x}_{ik} = 1 \tag{6}$$

$$\sum_{i=1}^{N} v_{i0}\mathrm{x}_{ik} - \sum_{i=1}^{m} u_{i0}\mathrm{y}_{i0} + \mu_{o0} \leq 0, \forall n \tag{7}$$

$v_{j0}, u_{i0} \geq 0, \forall i, j\ and\ \mu_{o0}$ is unrestricted in sign.

If the efficiency score is one, then the unit is efficient; otherwise, it is called inefficient.

Scale efficiency: The ratio of the CCR to BCC represents the value of the scale efficiency (SE) [10]. It measures the impact of scale size on the efficiency of a DMU.

$$Scale\ efficiency = \frac{E_{CCR}}{E_{BCC}} \tag{8}$$

where

E_{CCR} = CCR score of the DMU

E_{BCC} = BCC score of the DMU.

3 Case Study: Higher Education Sector in Myanmar

In Myanmar, there are 163 Art and Science Universities, Colleges, Technological Universities (TU), and Computer Universities (CU) located in different administrative divisions. The objective is to determine the relative efficiency of the different administrative divisions of Myanmar, on the basis of the performance of higher education institutions situated in these areas. Administratively Myanmar is divided into five administrative units, which are further divided into seven regions and seven states. Dataset related to the input and output variables are shown in Table 1.

Steps for applying DEA

The entire procedure of applying DEA can be done in four steps, given below:

Step 1: Decision-making units (DMUs)

The seven states (Kachin, Kayin, Mon, Kayah, Shan, Rakhine, and Chin) and seven regions (Sagaing, Ayeyarwady, Yangon, Bago, Tanintharyi, Magway, and Mandalay) of Myanmar are taken as DMUs. So overall, there are 14 DMUs.

Step 2: Input/output variables

Two input and one output variable are considered for the present study. These variables are taken after consulting different literature and are defined below:

Input Variables

a. **Higher education Institutes (HEIs)**: The total number HEIs dispersed in the different states and regions of Myanmar.
b. **Number of teachers:** The Number of teachers in the different states and regions of Myanmar. The number of teachers has been taken as a variable by a number of researchers [3, 11, 12].

Output Variables

a. **Number of students:** This variable gives information about the number of students in the different states and regions of Myanmar. The number of students has been considered as an output in [12, 13].

Step 3: Calculation of relative efficiency

Once the DMUs and the input/output variables have been finalized according to step 1 and step 2, the relative efficiency of each of the 14 DMUs is calculated using Eqs. (1) though (8).

Step 4: Analysis of results

In this step, the results obtained in the previous step are analyzed. In the present study, the obtained results are discussed in the next section.

Table 1 Data set for 14 states and regions

	2015			2016			2017		
DMUs	Students (Output)	HEIs (Inputs)	Teachers (Inputs)	Students (Output)	HEIs (Inputs)	Teachers (Inputs)	Students (Output)	HEIs (Inputs)	Teachers (Inputs)
Kachin	35,674	80	1411	34,680	82	1599	40,754	84	1582
Kayah	6647	18	382	7194	22	434	8631	24	425
Kayin	19,528	57	841	21,282	60	1197	23,998	62	1252
Chin	10,151	41	584	10,690	45	772	11,532	45	706
Sagaing	94,522	180	4038	98,897	198	4439	108,381	200	4475
Tanintharyi	20,385	53	872	24,086	58	1186	25,958	59	1278
Bago	72,208	171	3094	81,268	187	4113	92,097	187	4352
Mon	31,162	86	1586	33,974	90	1687	39,085	92	1702
Magway	71,265	124	2965	75,837	142	3127	84,348	147	3109
Mandalay	96,589	189	4772	95,833	201	4361	103,846	203	4422
Rakhine	37,171	95	1933	42,401	104	2128	47,592	115	2213
Yangon	127,342	226	4820	126,369	233	5154	131,068	235	5190
Shan	54 54,173	215	2540	58,234	221	3111	64,126	225	3001
Ayeyarwady	95,469	197	3342	106,739	231	4716	120,996	241	4862

4 Results and Discussion

4.1 Performance Assessment of the 14 DMUs

The scale efficiency score calculated through Eq. (8) of the 14 states and regions for the three academic years 2015, 2016, and 2017 are shown in Table 2. Codes R or S represent the region or state present in that division.

Assessment for the year 2015: In 2015, Magway (R9), Yangon (R12), and Ayeyarwady (R14) are evaluated to be 100% efficient. The average efficiency as calculated through the SE model, and for the year 2015, it is 0.924. The least efficient unit is Kayah (S2) with an efficiency score of 0.658 or it is only 65.8% efficient.

Assessment for the year 2016: In the academic year 2016, Yangon (R12), situated in the Lower Burma division, is efficient while the remaining 13 DMUs are inefficient. The mean value of the model is 0.939 and DMU Kayah, S2, is with an efficiency score of 0.676 is the least efficient.

Assessment for the year 2017: In the academic year 2017, out of the 14 DMUs, only the Magway (R9) region, located in central Burma, is efficient; the other 13 DUMs are inefficient. The average value of this model is 0.943. Kayah, coded as S2 and situated in the East Burma region, is the least efficient with an efficiency score of 0.749.

Table 2 Efficiency comparison of HEI for three academic year

Administrative divisions	Code	DMUs	2015	2016	2017
			SE	SE	SE
Central Burma	R9	Magway	1.000	0.989	1.000
	R10	Mandalay	0.985	0.997	0.977
East Burma	S2	Kayah	0.658	0.676	0.749
	S13	Shan	0.984	0.989	0.999
Lower Burma	R12	Yangon	1.000	1.000	0.972
	R7	Bago	0.993	0.996	0.984
	R14	Ayeyarwady	1.000	0.998	0.938
North Burma	S1	Kachin	0.943	0.941	0.962
	R5	Sagaing	0.998	0.996	0.980
South Burma	R6	Tanintharyi	0.857	0.907	0.891
	S3	Kayin	0.850	0.908	0.941
	S8	Mon	0.950	0.946	0.967
West Burma	S11	Rakhine	0.959	0.965	0.984
	S4	Chin	0.762	0.835	0.864
		Mean	0.924	0.939	0.943

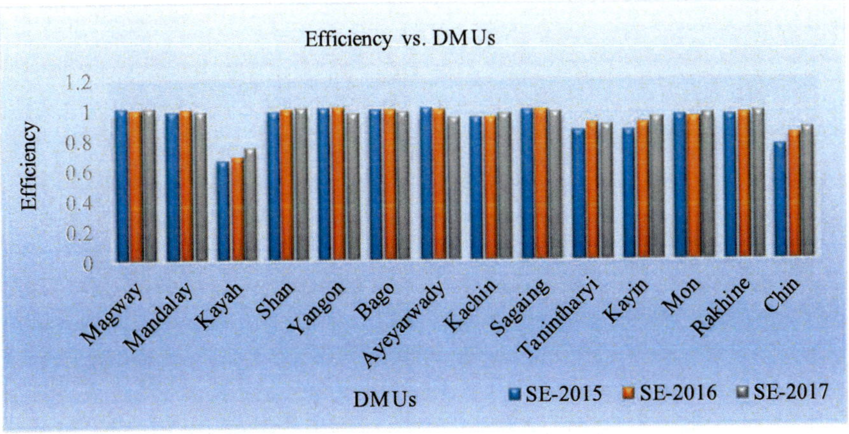

Fig. 1 Efficiency versus all the fourteen DMUs

The results are shown graphically in Fig. 1.

4.2 *Performance Assessment in Terms of Regions and States*

The SE score of seven different regions, Sagaing, Ayeyarwady, Yangon, Bago, Tanintharyi, Magway, and Mandalay, for the three consecutive academic years, 2015, 2016, and 2017, is as follows: average value of SE for the year 2016 is 0.983, which is the higher in comparison to the other two academic years. The average score for the years 2015 and 2017 are 0.976 and 0.963, respectively. For the seven states, Kachin, Kayin, Mon, Kayah, Shan, Rakhine, and Chin, the average value of the academic year 2017 is 0.923, which is higher than the other two academic years. The mean value of the other two academic years 2015 and 2016 is 0.872 and 0.894, respectively. Corresponding results are provided in Table 3 and the graphical illustrations

Table 3 Average scores of regions and states

Regions	SE-2015	SE-2016	SE-2017	States	SE-2015	SE-2016	SE-2017
Sagaing	0.998	0.996	0.980	Kachin	0.943	0.941	0.962
Ayeyarwady	1.000	0.998	0.938	Kayin	0.850	0.908	0.941
Yangon	1.000	1.000	0.972	Mon	0.950	0.946	0.967
Bago	0.993	0.996	0.984	Kayah	0.658	0.676	0.749
Tanintharyi	0.857	0.907	0.891	Shan	0.984	0.989	0.999
Magway	1.000	0.989	1.000	Rakhine	0.959	0.965	0.984
Mandalay	0.985	0.997	0.977	Chin	0.762	0.835	0.864
Average Score	0.976	0.983	0.963	**Average Score**	0.872	0.894	0.923

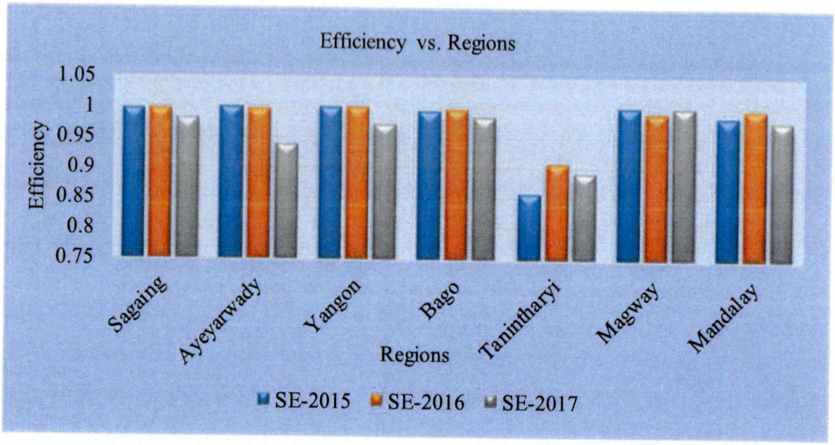

Fig. 2 Average efficiency of the regions

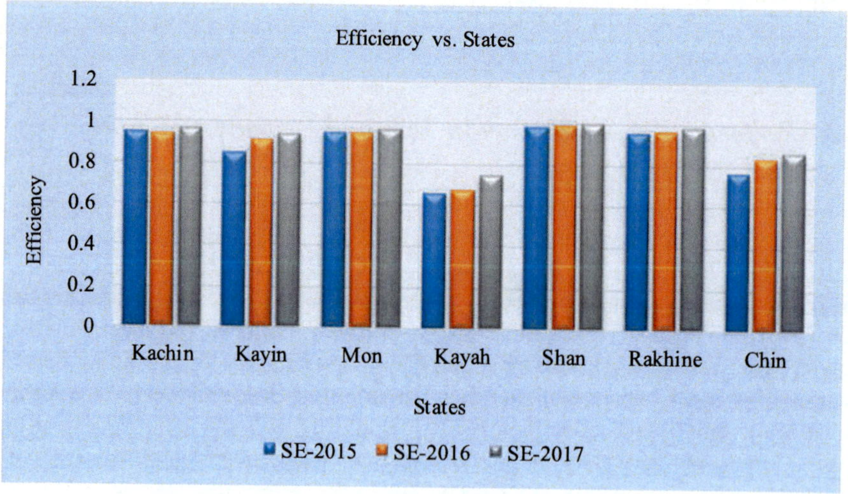

Fig. 3 Average efficiency of the states

are given in Figs. 2 and 3.

4.3 Comparison in Terms of Administrative Divisions

Lastly, the comparison is done in terms of administrative divisions, for which the results are given in Table 4. It can be seen that the highest average efficiency for the year 2017 is for the Central Burma division (=0.988), for the year 2016, the best

Table 4 Average value of administrative divisions

Administrative divisions	Average value (SE-2015)	Average value (SE-2016)	Average value (SE-2017)	Overall average
Central Burma	0.992	0.993	0.988	0.991
Lower Burma	0.997	0.998	0.964	0.986
North Burma	0.970	0.968	0.971	0.969
South Burma	0.885	0.920	0.933	0.912
West Burma	0.860	0.900	0.924	0.894
East Burma	0.821	0.832	0.874	0.842

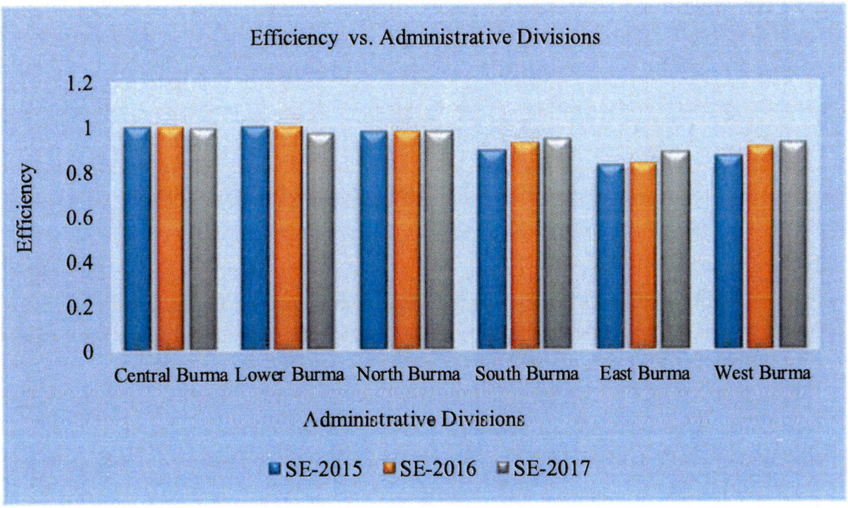

Fig. 4 Relative efficiency measurement of administrative divisions

average efficiency is calculated for the Lower Burma division (=0.998), while for the year 2015, the highest efficiency is again obtained for Lower Burma division (=0.997). However, if we look at the overall average for the three years, the best score is obtained by the Central Burma division (=0.991) (Figs. 4 and 5).

5 Conclusion

In the present study, an analysis is done for the higher education sector of Myanmar on the basis of the Higher Education Institutions located in its different administrative divisions, states, and regions. The methodology used is DEA, a linear programming-based method for calculating the relative efficiency of DMUs (HEIs in this case).

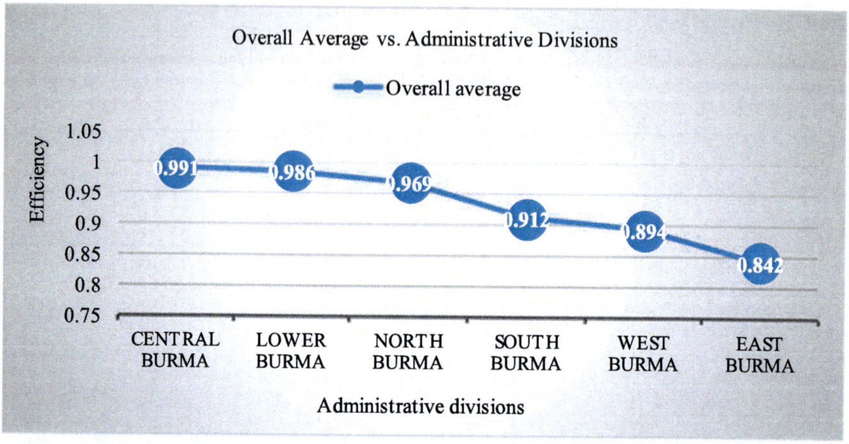

Fig. 5 Relative efficiency of the administrative divisions

Based on the results analyzed, some conclusions drawn from the present study are as follows:

a. In terms of an overall comparison of the seven states and seven regions, the performance of state Magway, denoted as R9 and located in the administrative division of Central Burma, can be said to be the best with an overall average efficiency score for three consecutive years as 0.996.
b. If the assessment of the seven states and the seven regions is done separately, then we can observe that the average performance of the regions (=0.896) is better than the average performance of the states (=0.974) for the three consecutive years.
c. Out of the seven states, the performance is shown by Shan, with an average efficiency of 0.968 for the three consecutive years. While out of the seven regions, the best performance is shown by Magway with an average efficiency score of 0.996 for the three consecutive years.
d. In terms of the administrative divisions, one can see that the performance of Central Burma, consisting of two regions (Magway and Mandalay), gave the best performance with an average value of 0.991, for three consecutive years.

The present study gives a basic idea about the status of the higher education sector of Myanmar. It may be beneficial for the Government and the policymakers for developing the future policies on the basis of these results. However, in order to get a deeper insight into the system, a larger dataset consisting of more numbers of input and output variables is needed which is the agenda of future research.

Acknowledgements The authors acknowledge with thanks the financial support provided by the project as DST-1628-DPT.

References

1. Jauhar SK, Pant M, Dutt R (2018) Performance measurement of an Indian higher education institute: a sustainable educational supply chain management perspective. Int J Syst Assur Eng Manag 9(1):180–193. https://doi.org/10.1007/s13198-016-0505-4
2. Tyagi P, Yadav SP, Singh SP (2009) Relative performance of academic departments using DEA with sensitivity analysis. Eval Program Plann 32(2):168–177. https://doi.org/10.1016/j.evalprogplan.2008.10.002
3. Tomkins C, Green R (1988) An experiment in the use of data envelopment analysis for evaluating the efficiency of uk university departments of accounting. Financ Account Manag 4(2):147–164. https://doi.org/10.1111/j.1468-0408.1988.tb00296.x
4. Chen JK, Chen IS (2011) Inno-Qual efficiency of higher education: empirical testing using data envelopment analysis. Expert Syst Appl 38(3):1823–1834. https://doi.org/10.1016/j.eswa.2010.07.111
5. Zhang X, Shi W (2019) Research about the university teaching performance evaluation under the data envelopment method. Cogn Syst Res 56:108–115. https://doi.org/10.1016/j.cogsys.2018.11.004
6. Zhang G, Wu J, Zhu Q (2020) Performance evaluation and enrollment quota allocation for higher education institutions in China. Eval Program Plann 81(May):101821. https://doi.org/10.1016/j.evalprogplan.2020.101821
7. Charnes A, Cooper WW, Rhodes E (1978) Measuring the efficiency of decision making units. Eur J Oper Res 2(6):429–444
8. Banker RD, Charnes A, Cooper WW (1984) Some models for estimating technical and scale inefficiencies in data envelopment analysis. Manage Sci 30(9):1078–1092. https://doi.org/10.1287/mnsc.30.9.1078
9. Singh N, Pant M, Goel A (2018) ANN embedded data envelopment analysis approach for measuring the efficiency of state boards in India. Int J Syst Assur Eng Manag 9(5):1092–1106. https://doi.org/10.1007/s13198-018-0743-8
10. Mohamed Shahwan T, Mohammed Hassan Y (2013) Efficiency analysis of UAE banks using data envelopment analysis. J Econ Adm Sci 29(1):4–20. https://doi.org/10.1108/10264111311319204
11. Forget BG (1979) Molecular genetics of human hemoglobin synthesis. Ann Intern Med 91(4):605–616. https://doi.org/10.7326/0003-4819-91-4-605
12. Avkiran NK (2001) Investigating technical and scale efficiencies of Australian universities through data envelopment analysis. Socioecon Plann Sci 35(1):57–80. https://doi.org/10.1016/S0038-0121(00)00010-0
13. Abbott M, Doucouliagos C (2003) The efficiency of Australian universities: a data envelopment analysis. Econ Educ Rev 22(1):89–97. https://doi.org/10.1016/S0272-7757(01)00068-1

Synchronizing Words for Real-Time Deterministic Pushdown Automata

Extended Abstract

Tomoyuki Yamakami and Eitatsu Mikami

Abstract Synchronizing finite(-state) automata have been widely studied in numerous fields of mathematics and computer science for theoretical and practical purposes. This exposition attempts to extend the scope of the synchronization property from one-way deterministic finite automata to one-way real-time deterministic pushdown automata. In particular, it argues on the characteristics of synchronizing words (or reset words), each of which is read by such a pushdown automaton that starts with arbitrary inner states with the empty stack and eventually reaches a unique inner state. Analogously to the Černý function for finite automata, we discuss the behaviors of the function $S(n, k, e)$ that produces the maximal word length over all the shortest synchronizing words of n-state real-time deterministic pushdown automata with size-k stack alphabets (excluding the bottom marker) and of push size e. We prove reasonably good lower bounds and, in a certain restricted setting, upper bounds of $S(n, k, e)$. We also study two additional variations of $S(n, k, e)$ based on the empty- and the same-stack models.

Keywords Synchronizing automata · Real-time pushdown automata · Synchronizing word · Stack alphabet · Lower and upper bounds

1 Background and New Challenges

1.1 Finite-State Automata and Synchronizing Words

In case of a sudden mechanical failure during an operation of any automated operational system, we wish to reset the whole system from wherever the system stops by performing a simple sequence of emergency "actions." Setting up such an emergency sequence in the process of designing any complex operational system is of great importance in practice. This leads to the concept of "synchronization" of mechanical

T. Yamakami (✉) · E. Mikami
Faculty of Engineering, University of Fukui, 3-9-1 Bunkyo, Fukui 910-8507, Japan

© The Author(s), under exclusive license to Springer Nature Singapore Pte Ltd. 2022

D. Giri et al. (eds.), *Proceedings of the Seventh International Conference on Mathematics and Computing*, Advances in Intelligent Systems and Computing 1412,
https://doi.org/10.1007/978-981-16-6890-6_41

machinery and the concept actually dates back to the 1950s. Moore [6] and Ginsburg [4] argued how suitably chosen sequences of actions (which were called *experiments*) restore the current status of machines by observing their visible outputs produced only by the chosen actions. Such a concept of synchronization has found numerous applications to robotics and biocomputing nowadays. A one-way deterministic finite(-state) automaton is called *synchronizing* (or *synchronizable*) with a *synchronizing word* (or a *reset word*) w if, no matter which inner state of the automaton to begin with, the automaton must enter exactly a unique inner state just after completely reading off w. The study of such synchronizing automata has been connected to semigroup theory, multi-valued logic, and symbolic dynamics. The past literature was partly overlooked in, e.g., surveys [8, 12].

Of all interests of ours, we are now focused on a close connection between the number of machine's inner states and the length of synchronizing words. In 1964, Černý [10] discovered that the maximal word length over all the shortest synchronizing words of n-state synchronizing automata is at least $(n-1)^2$ for any $n \geq 2$; in other words, certain words of length at least $(n-1)^2$ can reset all n-state synchronizing automata. Concerning an upper bound, he also proved a loose bound of $2^n - n - 1$. The best known upper bound today is $0.1654n^3 + O(1)$ obtained in 2019 by Shitov [9]. It was conjectured in [11] that $(n-1)^2$ is also an upper bound. This is known today as the *Černý conjecture*. Despite much efforts, this conjecture remains unproven. For our further discussion, we write $C(n)$ for the Černý function computing the maximal word length over all the shortest synchronizing words of n-state deterministic finite automata.

In 1990, Eppstein [2] showed that the problem of determining whether a given finite automaton has a synchronizing word of length at most a given number is NP-complete (via a reduction from 3SAT). Moreover, the problem of determining whether a given finite automaton has the shortest synchronizing word of a given length was shown in 2010 by Olschewski and Ummels [7] to be BH_2-complete, where BH_2[1] is the second level of the Boolean hierarchy over NP.

The finite-automata model, in general, exemplifies a simple memoryless machinery and it has found numerous real-world applications. Nevertheless, recent technological advancement makes it much easier to embed a small, restricted form of memory chips into the mainframe structure of the machinery, and therefore it has become not so far fetched to manipulate a small amount of memory at will. For instance, small autonomous robots with built-in memory chips have become more common and they could be designed to support synchronizing mechanism in order to migrate into an initial formation at any time when an operator wishes to collect all the robots at once. In reviewing such circumstances of the world, our challenge taken in this exposition is to expand the scope of the existing synchronizing automata study to a whole new direction. We wish to make one more critical step forward to deal with more complicated machinery equipped with small memory chips in hope of finding useful, real-life applications.

[1] The complexity class BH_2 was sometimes expressed as DP in the past literature and it is composed of all languages of the form $A \cap \overline{B}$ for two languages A and B in NP.

1.2 From Finite Automata to Pushdown Automata

This exposition will look into synchronizing mechanism and synchronizing words of *one-way real-time deterministic pushdown automata*[2] in place of (one-way) deterministic finite automata dealt in the past literature. In the rest of this exposition, we succinctly call such machines "pushdown automata" for readability. Such a pushdown automaton uses a well-structured memory device, known as a *stack*, whose access is, however, limited to its topmost entry and the stored order of the data items cannot be altered. It is well known that pushdown automata are significantly more powerful than finite automata in recognition power due to the use of memory storage despite its limited access. For such pushdown automata, we pay our attention to three key parameters: (1) the number of inner states in use, (2) the size of a stack alphabet, and (3) the push size (i.e., the maximal number of stack symbols that can replace in the topmost stack symbol at once).

For our pushdown automata, in a similar way as finite automata, we consider the maximal length of all the shortest synchronizing words. Let $n, e \geq 2$ and $k \geq 1$. To distinguish it from $C(n)$, we denote by $S(n, k, e)$ the maximal word length over all the shortest synchronizing words of n-state pushdown automata with size-k stack alphabets (excluding the bottom marker $\perp$, for simplicity) and of push size e. Each synchronizing pushdown automaton reaches the same end state but the final stack content may differ along every computation path of the machine.

Different from the case of finite automata, it is possible to consider two more natural variations of synchronizing pushdown automata. When, in all final configurations of a pushdown automaton M on an input w, all end states and all final stack contents are exactly the same, we call this machine M and the input w a *same-stack synchronizing pushdown automaton* and a *same-stack synchronizing word*, respectively. We write $S_{\text{same}}(n, k, e)$ if we use same-stack synchronizing pushdown automata. When we further demand the final stack to be empty (except for $\perp$), the corresponding machine and word are called an *empty-stack synchronizing pushdown automaton* and an *empty-stack synchronizing word*. We then define $S_{\text{empt}}(n, k, e)$ using such empty-stack synchronizing pushdown automata.

In the rest of this exposition, we will explore basic features of $S(n, k, e)$ and its variants $S_{\text{same}}(n, k, e)$ and $S_{\text{empt}}(n, k, e)$ through fundamental analyses of the behaviors of synchronizing pushdown automata.

All detailed proofs omitted from this exposition due to the page limit will be included in its complete version.

---

[2] We should remark that Caucal's [1] concept of "synchronization" of pushdown automata significantly differs from our concept introduced in this exposition. Our concept is a natural extension of the synchronizing finite automata discussed in Sect. 1.1 whereas Caucal's is obviously not.

2 Preparations: Basic Notions and Notation

2.1 Numbers, Sets, and Languages

Natural numbers are nonnegative integers and they form the set denoted by $\mathbb{N}$. For positive integers, we use the notation $\mathbb{N}^{+}$ and $\mathbb{N}^{++}$ for $\mathbb{N} - \{0\}$ and $\mathbb{N} - \{0, 1\}$, respectively. Given two integers m and n with $m \leq n$, $[m, n]_{\mathbb{Z}}$ denotes the *integer interval* $\{m, m+1, m+2, \ldots, n\}$. In particular, when $n \geq 1$, we abbreviate $[1, n]_{\mathbb{Z}}$ as $[n]$. For any set S, $\mathcal{P}(S)$ denotes the *power set* of S, that is, the collection of all subsets of S.

An *alphabet* is a finite nonempty set of "symbols" or "letters." A *string* (or *word*) over alphabet Σ is a finite sequence of symbols in Σ. Given a string x, $|x|$ denotes the *length* of x. A special string, ε, refers to the *empty string* of length 0. For any number $n \in \mathbb{N}$, Σ^{n} (resp., $\Sigma^{\leq n}$) denotes the set of all strings of length exactly n (resp., length at most n) and we set $\Sigma^{*} = \bigcup_{n \in \mathbb{N}} \Sigma^{n}$. Given two strings x and y, xy denotes the *concatenation* of x and y. For a string w and a language L, wL (resp., Lw) denotes the language $\{wx \mid x \in L\}$ (resp., $\{xw \mid x \in L\}$). Given two languages A and B over the same alphabet Σ (which includes two distinguished symbols, say, σ_0 and σ_1), $A \oplus B$ denotes the *disjoint union* of A and B, defined as $A \oplus B = \sigma_0 A \cup \sigma_1 B$.

2.2 Real-Time Deterministic Pushdown Automata

Throughout this exposition, we deal with special one-way deterministic pushdown automata, equipped with no initial state and no final state, which perform no ε-move (namely, a step in which a tape head reads no input symbol but may change an inner state and make a stack operation). The last requirement is often referred to as "real time" in the literature. Hereafter, we briefly give their formal definition. A *(one-way) real-time deterministic pushdown automaton* (with no fixed initial and no fixed final states) M is a quintuple $(Q, \Sigma, \Gamma, \delta, \perp)$, where Q is a finite set of inner states, Σ is an input alphabet, $\perp$ is the bottom marker indicating the bottom of the stack, Γ is a stack alphabet (not including $\perp$), and δ is a transition function mapping $Q \times \Sigma \times \Gamma_{\perp}$ to $Q \times (\Gamma^{*} \cup \Gamma^{*}\perp)$, where $\Gamma_{\perp} = \Gamma \cup \{\perp\}$. As noted in Sect. 1.2, we succinctly call such a machine M a "pushdown automaton" in the rest of this exposition, because we will not use any other variant of pushdown automata. The machine can begin with *any* inner state in Q with an empty stack except for the bottom marker $\perp$. We allow the machine to replace a topmost stack symbol with any string over Γ of length at most a fixed constant described by δ. In this exposition, we always demand that M neither removes $\perp$ nor replaces $\perp$ by any other symbol at any step. A *stack content* refers to a series of stack symbols (including $\perp$) stored inside the stack and such a stack content is expressed as $a_1 a_2 \cdots a_m$, where a_1 is the topmost symbol and a_m is the lowest symbol, which is the bottom marker $\perp$. The *push size* refers to the maximal word length over all the strings substituted for any topmost symbol of the stack in a single step, that is, $\max_{(p,q,\sigma,a)}\{|z| : \delta(q,\sigma,a) = (p,z)\}$, where the maximality is taken over all $p, q \in Q$, $\sigma \in \Sigma$, and $a \in \Gamma_{\perp}$.

Let us consider the *extended transition function* $\delta^* : Q \times \Sigma^* \times \Gamma^*\perp \to Q$ of δ. When $\delta^*(q, w, \perp) = (p, r)$, the first entry p of (p, r) is expressed by $\delta^*(q, w, \perp)_{\text{state}}$ and the second entry r is denoted by $\delta^*(q, w, \perp)_{\text{stack}}$. For our pushdown automaton M and for any input $w \in \Sigma^*$, we always treat an arbitrary inner state q in Q as a *start state* of M and $\delta^*(q, w, \perp)_{\text{state}}$ as an *end state* of M from q on w. Moreover, $\delta^*(q, w, \perp)_{\text{stack}}$ is called a *final stack content* of M obtained by M from $(q, \perp)$ on w.

2.3 Synchronizing Pushdown Automata and Synchronizing Words

Using pushdown automata explained in Sect. 2.2, we give the formal definitions of synchronizing pushdown automata and their synchronizing words.

Definition 1 We say that a pushdown automaton $M = (Q, \Sigma, \Gamma, \delta, \perp)$ is *synchronizing* if there exists a word $w \in \Sigma^*$ such that, for any pair $p, q \in Q$, $\delta^*(q, w, \perp)$ and $\delta^*(p, w, \perp)$ both exist and $\delta^*(q, w, \perp)_{\text{state}} = \delta^*(p, w, \perp)_{\text{state}}$ holds. This word w is called a *synchronizing word* (or a *reset word*) of M. We write $S(n, k, e)$ for the maximal word length over all the shortest synchronizing words of n-state synchronizing pushdown automata of push size e with input alphabets and size-k stack alphabets. The *best synchronizing pushdown automata* and the *best synchronizing words* are defined so as to witness the value $S(n, k, e)$.

As a concrete example, consider a pushdown automaton M with $Q = \{q_1, q_2, q_3\}$, $\Sigma = \{a, b, x, y, z, x', y', z'\}$, and $\Gamma = \{X\}$ shown in Fig. 1. It is not difficult to check that the word $xyzax'y'z'bx'y'z'bxyza$ is a synchronizing word of M.

It is important to remark that, in this exposition, the stack alphabet Γ *does not* contain the bottom marker $\perp$ for our convenience. We define $\phi(n, k, e)$ to be the maximal height of stack contents (including $\perp$) generated during the execution of the best synchronizing pushdown automaton on the best synchronizing word. Obviously, $\phi(n, k, e) \leq e \cdot S(n, k, e)$ holds.

For any parameters $n, e \in \mathbb{N}^{++}$ and $k \in \mathbb{N}^{+}$, if $n \leq n'$, $k \leq k'$, and $e \leq e'$, then $S(n, k, e) \leq S(n', k', e')$ holds (by appropriately expanding the best synchronizing pushdown automaton achieving $S(n, k, e)$ with the use of additional states and stack symbols). An important property of synchronizing finite automata is the *pairwise synchronization*, in which the synchronization of computations on every pair of start states leads to the synchronization of computations on all start states. Unlike such machines, unfortunately, synchronizing pushdown automata may not meet the pairwise synchronization property and this phenomenon makes our analysis of synchronizing pushdown automata even harder.

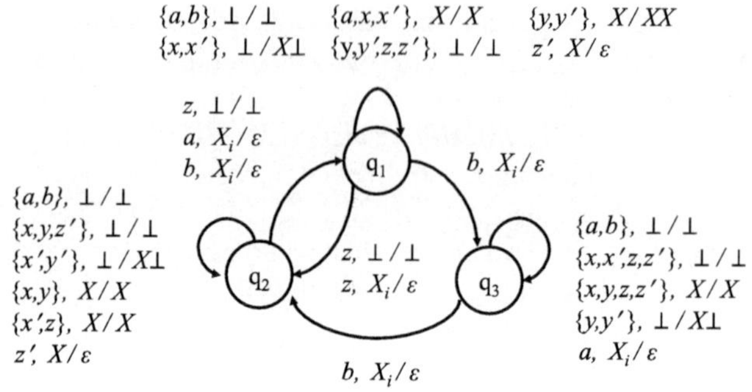

Fig. 1 The transition diagram of M. The notation, e.g., "$\{a, b\}, \perp/\perp$" indicates both "$a, \perp/\perp$" and "$b, \perp/\perp$"

In our synchronizing pushdown automata, inner states are allowed to migrate into a unique end state with an *arbitrary* stack content. It is possible for us to further demand the machines to end up with the *same* stack content when they terminate. Furthermore, we can require the machines to *empty* their stack (except for $\perp$) when they eventually stop. We next formalize those two notions, which obviously impose stronger requirements on the behaviors of $S(n, k, e)$.

Definition 2 A pushdown automaton is *same-stack synchronizing* if there exists a word w such that $\delta^*(q, w, \perp) = \delta^*(p, w, \perp)$ holds for any pair $p, q \in Q$. Similar to $S(n, k, e)$, $S_{\text{same}}(n, k, e)$ is defined using same-stack synchronizing words. If we require both $\delta^*(q, w, \perp) = \delta^*(p, w, \perp)$ and $\delta^*(q, w, \perp)_{\text{stack}} = \varepsilon$ for any pair $p, q \in Q$, then we obtain the notion of *empty-stack synchronizing*. The notation $S_{\text{empt}}(n, k, e)$ is used for empty-stack synchronizing automata.

Among the aforementioned three synchronization notions imposing various restrictions on final stack contents, it follows that $S(n, k, e) \le S_{\text{same}}(n, k, e) \le S_{\text{empt}}(n, k, e)$ for any triplet $(n, k, e) \in \mathbb{N}^{++} \times \mathbb{N}^{+} \times \mathbb{N}^{++}$. It is possible, however, to reduce the value of k down to 2 at the cost of an increase of the values of n and e, as shown below.

Lemma 3 *For any integer* $k \ge 3$, $S(n, k, e) \le (\ell + 2) \cdot S(n(2^{\ell} - 1), 2, e\ell)$, *where* $\ell = \lceil \log(k + 1) \rceil$.

Similarly, the push size e can be reduced to 2 if we increase the value of k.

Lemma 4 *Let* $n, e \in \mathbb{N}^{++}$ *and* $k \in \mathbb{N}^{+}$. *If* $k = 1$, *then* $S(n, 1, e) \le S(n, e, 2)$, *or else* $S(n, k, e) \le S(n, \frac{k(k^e-1)}{k-1}, 2)$.

For readability, hereafter, we mostly fix e to 2 and wish to estimate the value of $S(n, k, 2)$ and the values of its two variants.

To make it easier to describe the proofs of our main assertions in later sections, we want to introduce necessary terminology. The *state pool* at step t on input w refers to the set of all inner states reachable from any start state by reading the first t symbols of w. Similarly, the *stack pool* at step t on w is the set of all stack contents obtained after reading the first t symbols of w. Moreover, simply replacing "inner states" in the above definition by "configurations," we obtain the notion of *configuration pool*.

3 Properties of Synchronizing Pushdown Automata

3.1 Length of Synchronizing Words

Let us focus on the values of $S(n, k, e)$ and its variants given in Definition 2. Since pushdown automata are a natural extension of finite automata, we can obtain the following simple lower bound of $S(n, k, 2)$ in terms of the Černý function $C(n)$.

Theorem 5 *For any* $n \in \mathbb{N}^{++}$ *and* $k \in \mathbb{N}^{+}$, $S(n, k, 2) \geq k \cdot C(n)$.

Proof To show that $S(n, k, 2) \geq k \cdot C(n)$, let $m = C(n)$ and consider the best synchronizing finite automaton $M = (Q_M, \Sigma_M, \delta_M)$ and the best synchronizing word $w = w_1 w_2 \cdots w_m$ with $w_i \in \Sigma_M$ for M, where δ_M is a transition function from $Q_M \times \Sigma_M$ to Q_M. For convenience, let $Q_M = \{q_1, q_2, \ldots, q_n\}$ and $\Sigma_M = \{a_1, a_2, \ldots, a_r\}$. Note that $\{w_1, \ldots, w_m\} \subseteq \Sigma_M$. We wish to construct a pushdown automaton N as follows. We first prepare k distinct symbols $x_1, x_2, \ldots, x_k$ not in Σ_M and set $x = x_1 x_2 \cdots x_k$. The desired pushdown automaton N is of the form $(Q, \Sigma, \Gamma, \delta, \bot)$ with $Q = Q_M$, $\Sigma = \Sigma_M \cup \{x_1, x_2, \ldots, x_k\}$, $\Gamma = \{X_1, X_2, \ldots, X_k\}$, and $\delta : Q \times \Sigma \times \Gamma_\bot \to Q \times (\Gamma^* \cup \Gamma^* \bot)$. We further set $\delta(q, x_1, \bot) = (q, X_1 \bot)$ and $\delta(q, x_i, X_{i-1}) = (q, X_i)$ for any index $i \in [2, k]_{\mathbb{Z}}$. For an arbitrary symbol $a \in \{w_1, \ldots, w_m\}$, when $\delta_M(q, a) = p$, we set $\delta(q, a, X_k) = (p, \varepsilon)$. For all other cases, we automatically set $\delta(q, b, r) = (q, r)$. The word $\hat{w} = x w_1 x w_2 \cdots x w_m$ is synchronizing for N. Next, we wish to show that $\hat{w}$ is the shortest synchronizing word. Assume otherwise. Note that, in order to apply δ_M, we need to read x once and empty the stack. Moreover, by ignoring x, δ behaves exactly like δ_M. Since w is a synchronizing word of M, so is $\hat{w}$ for N. The length of the word $\hat{w}$ is exactly km $(= k \cdot C(n))$. Therefore, we conclude that $S(n, k, 2) \geq |x w_1 x w_2 \cdots x w_k| = k \cdot C(n)$. □

Since $C(n) \geq (n-1)^2$ [10], we immediately obtain the following corollary.

Corollary 6 *For any* $n \in \mathbb{N}^{++}$ *and* $k \in \mathbb{N}^{+}$, $S(n, k, 2) \geq k(n-1)^2$.

As a special case of $n = 2$, Corollary 6 yields $S(2, k, 2) \geq k$. Actually, we can obtain a much better lower bound of $S(2, k, 2)$ as shown below.

Theorem 7 *For any* $k \in \mathbb{N}^{+}$, $S(2, k, 2) \geq 3k + 1$ *and* $S_{\mathrm{empt}}(2, k, 2) \geq 4k$.

Proof Firstly, we wish to show that $S(2, k, 2) \geq 3k + 1$ for any $k \geq 1$. We split our proof into two separate cases. Let us consider the first case where k is odd.

We set $t = \frac{k+1}{2}$. Our pushdown automaton $M_k = (Q, \Sigma, \Gamma, \delta, \bot)$ is given by the transition diagram shown in Fig. 2 with $Q = \{q_0, q_1\}$, $\Sigma = \{a, b, c, d\}$, and $\Gamma = \{X_1, \ldots, X_t, Y_1, \ldots, Y_{t-1}\}$. Notice that $|\Gamma| = k$.

Let w denote the word $(abc)^k d$ of length $3k + 1$. It is not difficult to show by checking all computation paths that w is indeed a synchronizing word of M_k. This indicates that, if w is the shortest synchronizing word of M_k, then we can conclude that $S(2, k, 2) \geq |w| = 3k + 1$. In what follows, our goal is to prove that w is indeed the shortest synchronizing word of M_k. To describe a computation of M_k, we use *simplified configurations* of the form (q, γ), which indicates that M is in state q with stack content γ.

To lead to a contradiction, we assume that there exists another synchronizing word $x = x_1 x_2 \cdots x_m$ of length smaller than $|w|$ of M_k. Since $|x| = m$, the state pool at step m must become a singleton for the first time; therefore, the state pool at step $m - 1$ must be $\{q_0, q_1\}$. From this property, the last symbol x_m must be d and the following transitions (i)–(ii) must occur while reading d: (i) $\delta(q_0, d, X_t) = (q_1, \varepsilon)$ and (ii) either $\delta(q_1, d, \bot) = (q_1, \bot)$ or $\delta(q_1, d, r) = (q_1, \varepsilon)$ for a certain symbol $r \in \Gamma$. In other words, the final transitions change $(q_0, X_t\xi)$ to (q_1, ξ) in a single step, and either $(q_1, \bot)$ changes to $(q_1, \bot)$ or $(q_1, Z\gamma)$ changes to (q_1, γ) in a single step for certain strings $\xi, \gamma \in \Gamma^*$ and $Z \in \Gamma$. To push X_t into the stack during the execution of M_k, we need to read abc and push Y_{t-1} into the stack, that is, $\delta(q_1, a, Y_{t-1}) = (q_1, X_t Y_{t-1})$, $\delta(q_1, b, X_t) = (q_1, X_t X_t)$, and $\delta(q_1, c, X_t) = (q_0, \varepsilon)$. To push Y_{t-1} into the stack, however, we need to read rbc and push X_{t-1} into the stack for a certain input symbol $r \in \{a, b\}$, namely, $\delta(q_0, r, X_{t-1}) = (q_0, Y_{t-1} X_{t-1})$, $\delta(q_0, b, Y_{t-1}) = (q_0, Y_{t-1} Y_{t-1})$, and $\delta(q_0, c, Y_{t-1}) = (q_1, \varepsilon)$.

In general, for each index $i \in [2, t-1]_{\mathbb{Z}}$, (i) to push X_i into the stack, we need to read abc and push Y_{i-1} first and (ii) to push Y_i into the stack, we need to read rbc and push X_{i-1} first. In the case of pushing X_1, we need to read abc and apply $\delta(q_1, a, \bot) = (q_1, X_1 \bot)$, $\delta(q_1, b, X_1) = (q_1, X_1 X_1)$, and $\delta(q_1, c, X_1) = (q_0, \varepsilon)$. Thus, the input word x must have length at least $t|abc| + (t-1)|rbc| + |d| = 6t - 2 = 3k + 1$. This is clearly a contradiction against the choice of x.

The second case where k is even can be treated in essence in the same way as before, and thus we omit its proof in this exposition.

For the empty-stack model, let us consider the word $(abc)^k d^k$ of length $4k$. After reading $(abc)^k d$, as shown above, we obtain $(q_1, X_{t-1} \cdots Y_1 X_1)$ and $(q_1, Y_{t-1} X_{t-1} \cdots Y_1 X_1)$. By further reading d^{k-1}, we can empty the stack (except for $\bot$). Thus, this word is empty-stack synchronizing. □

Concerning an upper bound of $S(n, k, 2)$, we are hoping to obtain a concrete value that upper bounds $S(n, k, 2)$ for each fixed value of (n, k). Meanwhile, we can manage to prove the following simple bound for synchronizing pushdown automata whose stack heights are bounded from above by a fixed constant, say, τ. We say that a synchronizing pushdown automaton M has *stack height at most* τ on input w if M's stack height (including $\bot$) never exceeds τ during the execution of M on w starting from an arbitrary inner state. Given an arbitrary fixed number τ in $\mathbb{N}^{++}$, $S_\tau(n, k, e)$ denotes the maximal word length over all the shortest synchronizing words of n-

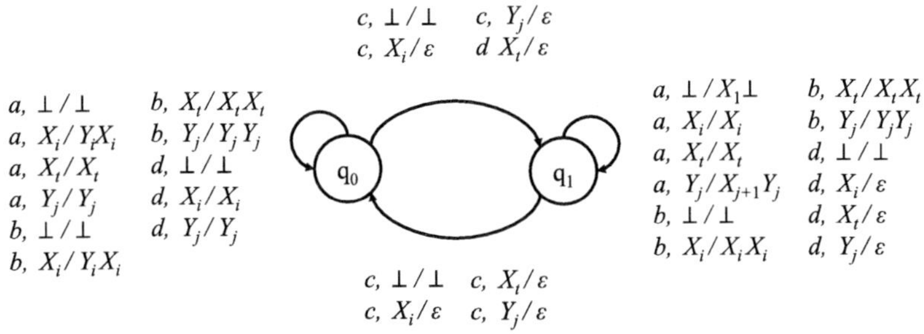

Fig. 2 The transition diagram of M_k for an odd number k with two parameters $i, j \in [t-1]$

state synchronizing pushdown automata with size-k stack alphabet and of push size e allowing stack height to be at most τ along all computation paths. Similarly, we define $S_{\text{same},\tau}(n,k,e)$ and $S_{\text{empt},\tau}(n,k,e)$, respectively, using same-stack and empty-stack synchronizing pushdown automata with stack height bounded by τ.

Proposition 8 *Let τ be any constant in $\mathbb{N}^{++}$. For any $n \in \mathbb{N}^{++}$ and $k \in \mathbb{N}^{+}$, we define $\theta_1 = \tau$ and $\theta_k = \frac{k^{\tau}-1}{k-1}$ for any $k \geq 2$. It follows that $S_{\tau}(n,k,2) \leq 2^{n\theta_k} - n(2^{\theta_k} - 1) - 1$, $S_{\text{same},\tau}(n,k,2) \leq 2^{n\theta_k} - n\theta_k - 1$, and $S_{\text{empt},\tau}(n,k,2) \leq 2^{n\theta_k} - n - 1$.*

Proof Sketch. Hereafter, we are focused on verifying that $S_{\text{same},\tau}(n,k,2) \leq 2^{n\theta_k} - n\theta_k - 1$, and we omit the other cases of $S_{\text{empt},\tau}(n,k,2)$ and $S_{\tau}(n,k,2)$. Let $m = S_{\text{same},\tau}(n,k,2)$ and consider a synchronizing pushdown automaton $M = (Q, \Sigma, \Gamma, \delta, \bot)$ whose stack height is at most τ on a synchronizing word, say, w of length m. Let $\hat{\Gamma} = \Gamma^{\leq \tau-1}\bot$. Note that, if $k = 1$, then $|\hat{\Gamma}| = \tau$ holds, or else $|\hat{\Gamma}| = \sum_{i=0}^{\tau-1} k^i = \frac{k^{\tau}-1}{k-1}$ holds. We thus obtain $|\hat{\Gamma}| = \theta_k$ for any $k \in \mathbb{N}^{+}$.

Let us consider a new one-way deterministic finite automaton $N = (\hat{Q}, \hat{\Sigma}, \hat{\delta})$ with $\hat{Q} = Q \times \hat{\Gamma}$, and $\hat{\Sigma} = \Sigma$. The new transition function $\hat{\delta} : \hat{Q} \times \hat{\Sigma} \to \hat{Q}$ works as follows. If $\delta(q, a, \bot) = (p, \beta\bot)$, then we set $\hat{\delta}((q,\bot), a) = (p, \beta\bot)$. If $\delta(q, a, X) = (p, \beta)$ with $X \in \Gamma - \{\varepsilon\}$, then $\hat{\delta}((q, X\alpha\bot), a) = (p, \beta\alpha\bot)$ for any string $\alpha \in \Gamma^*$ satisfying $|\beta\alpha\bot| \leq \tau$. Between δ and $\hat{\delta}$, it follows by induction that $\delta^*(q, w, \bot)_{\text{state}} = \hat{\delta}^*((q,\bot), w)_{\text{state}}$ for any $q \in Q$ and $w \in \Sigma^*$.

We want to define a variant of the *power(-set) automaton* of N, say, $P = (Q', \Sigma', \delta', R_0, F')$ in the following manner. Let $Q' = \mathcal{P}(\hat{Q}) - \{\emptyset\}$ and $\Sigma' = \Sigma$. Note that $|Q'| = 2^{n\theta_k} - 1$ since $|\hat{Q}| = n\theta_k$. For any $R \in Q'$, let $\delta'(R, a) = \{\delta((q,\gamma), a) \mid (q,\gamma) \in R\}$ $(\in Q')$. The initial state R_0 of P is $R_0 = \{(q,\bot) \mid q \in Q\}$. All final states in F' are of the form $\{(q,\gamma)\}$ for any $q \in Q$ and $\gamma \in \hat{\Gamma}$.

Let us consider all computation paths of N on input w from any start state $(q,\bot)$ and reaching a unique "end" configuration. Those computation paths correspond to a single unique computation path, say, ξ_w of P starting from the initial state R_0 and reaching a certain final state in F'. Since the length $|w|$ corresponds to the length of ξ_w, it suffices to estimate the value of $|\xi_w|$.

For each index $\ell \in [n]$, we define $REACH_{\ell}^{(\text{same})}$ as the set $\{R \in Q' \mid |R| = \ell\}$. It follows that ξ_w may possibly contain all elements in $\bigcup_{\ell:2\leq\ell\leq n} REACH_{\ell}^{(\text{same})}$ but it must contain only one element in $REACH_1^{(\mathrm{same})}$. Since $|REACH_1^{(\mathrm{same})}| = |\hat{Q}| = n\theta_k$, we conclude that $|\xi_w| \leq |Q'| - n\theta_k = 2^{n\theta_k} - n\theta_k - 1$. □

Since it looks quite difficult to obtain the maximum stack height of each synchronizing pushdown automaton on any given input string, Proposition 8 does not seem to help us calculate any general upper bound of $S(n, k, e)$. When $k = 1$, pushdown automata are customarily called *counterautomata*. In particular, when $(n, k, e) = (2, 1, 2)$, it is possible to determine the precise value of $S(n, k, e)$ by a rather simple argument.

Theorem 9 $S(2, 1, 2) = 4$.

Proof Sketch. The lower bound $S(2, 1, 2) \geq 4$ immediately follows from Theorem 7 simply by setting $k = 1$. To prove the theorem, it thus suffices to show that $S(2, 1, 2) \leq 4$. Toward a contradiction, let us assume that $S(2, 1, 2) \geq 5$.

Take the best synchronizing pushdown automaton $M = (Q, \Sigma, \Gamma, \delta, \perp)$ and the best synchronizing word $w = w_1 w_2 \cdots w_m$ with $m \geq 5$ and $w_i \in \Sigma$ for any $i \in [m]$. Let $Q = \{q_0, q_1\}$ and $\Gamma = \{X\}$. For simplicity, we assume that the state pool at step m becomes $\{q_1\}$. Let us recall the notion of simplified configuration given in the proof of Theorem 7. Since M is a counter automaton, we further simplify it and use the new notation (q, n) to indicate a simplified configuration in which M is in state q with stack height $n + 1$ (including $\perp$). Consider two such simplified configurations associated with two inner states q_0 and q_1. We write (n_0, n_1) to indicate that (q_0, n_0) and (q_1, n_1) are simplified configurations of M obtained simultaneously after a certain step. We conveniently call this a *combined stack configuration*. In the case, where we have two simplified configurations (q_1, n'_1) and (q_1, n'_2) for $n'_1, n'_2 \in \mathbb{N}$, we write $(*, \{n'_1, n'_2\})$, where $*$ indicates that M does not enter q_0. With this succinct notation, M is assumed to start with $(0, 0)$ and end with $(*, A)$ for a certain set $A \subseteq \mathbb{N}$ with $1 \leq |A| \leq 2$ after completing reading w. We say that an input symbol b *transforms* (n_1, n_2) to (n'_1, n'_2) (resp., $(*, A)$) if M changes combined stack configuration (n_1, n_2) to (n'_1, n'_2) (resp., $(*, A)$) by reading b.

We first claim that, after reading any one of $w_1, w_2, \ldots, w_{m-1}$, the combined stack configuration $(0, 0)$ never occurs. This is because, if $(0, 0)$ occurs after reading w_i for a certain index $i \in [m - 1]$, then the word $w_{i+1} w_{i+2} \cdots w_m$ is also synchronizing, a contradiction against the best synchronizability of w. Let us argue three separate cases (1)–(3) depending on the behaviors of M when it reads w_{m-1}.

(1) Assume that w_m transforms a combined stack configuration of the form $(n_1 + 1, n_2 + 1)$ with $n_1, n_2 \in \mathbb{N}$ to $(*, A)$. It then follows that, after reading any one of $w_1, w_2, \ldots, w_{m-2}$, no combined stack configuration of the form $(n'_1 + 1, n'_2 + 1)$ with $n'_1, n'_2 \in \mathbb{N}$ occurs because, otherwise, we obtain a shorter synchronizing word. Thus, all combined stack configurations obtained after reading any one of $w_1, w_2, \ldots, w_{m-2}$ must have the form $(0, n + 1)$ for a certain $n \in \mathbb{N}$ and they are all distinct. Since $m \geq 5$, there exist three combined stack configurations $(0, 1)$, $(0, 2)$, and $(0, 3)$. It is possible to delete the input symbol that transforms $(0, 2)$ to $(0, 3)$ even by keeping the synchronizability. This is an obvious contradiction.

(2) Assume that w_m transforms $(0, n+1)$ for a certain $n \in \mathbb{N}$ to $(*, A)$. We first claim that, after reading any one of $w_1, w_2, \ldots, w_{m-2}$, we obtain no combined stack configuration of the form $(0, n''+1)$ with $n'' \in \mathbb{N}$. Hence, any combined stack configuration must have the form either $(n''+1, 0)$ or (n_1+1, n_2+1) for $n'', n_1, n_2 \in \mathbb{N}$. In particular, w_1 transforms $(0, 0)$ to either $(1, 0)$ or $(1, 1)$. Take the smallest index $i_0 \in [m-1]$ such that (n_1+1, n_2+1) with $n_1, n_2 \in \mathbb{N}$ occurs after reading w_{i_0}, if such an i_0 actually exists. If i_0 does not exist, then we conveniently set $i_0 = 0$. We further argue three cases (a)–(c) separately.

(a) Consider the case where $i_0 = m-1$, namely, w_{m-1} transforms (n_1+1, n_2+1) to $(0, n+1)$. In this case, either $n_1 = 0$ or $n_2 = 0$ must happen. Moreover, w_1 must transform $(0, 0)$ to $(1, 0)$. Consider the first subcase of $n_1 = 0$. We run M on w_{m-1} from $(1, 0)$. Clearly, w_{m-1} can transform $(1, 0)$ to either $(0, n'+1)$ or $(1, n'+1)$ for a certain $n' \in \mathbb{N}$. If the former case happens, then the word $w_1 w_{m-1} w_m$ is synchronizing. In contrast, if the latter case happens, then $w_1 w_{m-1} w_{m-1} w_m$ is synchronizing. In the both cases, we obtain a contradiction against the choice of w. In the second subcase of $n_2 = 0$, w_{m-1} can transform $(1, 0)$ to either $(0, n'+1)$ or $(1, n'+1)$ for a certain $n' \in \mathbb{N}$. Therefore, this case can be treated similarly as the subcase of $n_1 = 0$. Due to the page limit, we omit the remaining two cases: (b) the case of $i_0 = 0$ and (c) the case of $0 < i_0 < m-1$.

(3) Assume that w_m transforms $(n+1, 0)$ for $n \in \mathbb{N}$ to $(*, A)$. This case is treated similarly to (2).

In all the discussed cases, we obtain the desired contradictions. Therefore, it follows that $S(2, 1, 2) \leq 4$, as requested. □

3.2 Complexity of Searching Synchronizing Words

Hereafter, we shift our focal point to the computational complexity of the decision problem of asking for the existence of synchronizing words of any given (not necessarily synchronizing) real-time deterministic pushdown automaton.[3] Let us recall that, throughout this exposition, real-time deterministic pushdown automata are always abbreviated as "pushdown automata." It seems difficult to check whether a given pushdown automaton is synchronizing in polynomial time even if the search area size of synchronizing words is specified. To clarify this point, we start with considering the simple decision problem, RESET-WORD, of determining whether, for a given word w and a given pushdown automaton M, w is a synchronizing word of M. We then assert that RESET-WORD belongs to the complexity class P.

Lemma 10 *There exists an algorithm that solves* RESET-WORD *in time* $O((|w| + |M|)e \log |Q||\Gamma||w|)$, *where* w *is an input word over alphabet* Σ *and a pushdown automaton* M *has an inner state set* Q *and a stack alphabet* Γ *of push size* e.

[3] After the technical report [5] of this paper, the decision problem of whether or not a given pushdown automaton is synchronizing has been shown to be unsolvable [3].

The problem LIMITED-RESET-WORD asks if, for any given length $\ell \geq 1$, expressed in unary as 1^{ℓ}, and any given pushdown automaton M, there exists a synchronizing word of length *exactly* ℓ for M. Its finite-automata version is known to be NP-complete [2]. An important point is that the length bound ℓ of synchronizing words is given as a part of each input and we need to concentrate only on all words of length exactly ℓ in order to look for synchronizing words. This leads to the following.

Proposition 11 LIMITED-RESET-WORD *is* NP-*complete.*

The complexity class BH_2 is the second level of the *Boolean hierarchy over* NP, which is included within Δ_2^{P}, the second Δ-level of the *polynomial hierarchy*. Note that every language in BH_2 has the form $A \cap \overline{B}$ for certain two languages $A, B \in \mathrm{NP}$. Concerning pushdown automata, we define MIN-RESET-WORD as the problem of determining whether, for a given number $\ell \geq 1$, expressed in unary as 1^{ℓ}, and a pushdown automaton M, the shortest synchronizing word of M, if any, has length exactly ℓ. This problem MIN-RESET-WORD turns out to be BH_2-complete.

Theorem 12 MIN-RESET-WORD *is* BH_2-*complete.*

References

1. Cauca D (2006) Synchronization of pushdown automata. In: Proceedings of DLT 2006, vol 4036. Springer, pp 120–132
2. Eppstein D (1990) Reset sequences for monotonic automata. SIAM J Comput 19:500–510
3. Fernau H, Wolf P, Yamakami T (2020) Synchronizing deterministic push-down automata can be really hard. In: Proceedings of MFCS 2020, vol 170, pp 33:1–33:15
4. Ginsburg S (1958) On the length of the smallest uniform experiment which distinguishes the terminal states of a machine. J ACM 5:266–280
5. Mikami E, Yamakami T (2020) Synchronizing pushdown automata and reset words. Comp2019-54(2020-03), Technical report of the institute of electronics, information and communication engineers (2020). (in Japanese)
6. Moore EF (1956) Gedanken-experiments on sequential machines. Autom Stud Ann Math Stud 34:129–153
7. Olschewski J, Ummels M (2010) The complexity of finding reset words in finite automata. In: Proceedings of MFCS 2010, vol 6281. Springer, pp 568–579
8. Salomaa A (2002) Synchronization of finite automata: contributions to an old problem. In: Mogensen TÆ et al (ed) The essence of computation, vol 2566. Springer, pp 37–59
9. Shitov Y (2019) An improvement to a recent upper bound for synchronizing words of finite automata. J Aut Lang Comb 24:367–373
10. Černý J (1964) Poznámka k homogénnym experimentom s konečnými automatami. Matematicko-fyzikálny Časopis Slovensk. Akad. Vied. 14:208–216 (in Slovak)
11. Černý J, Pirická A, Rosenauerová B (1971) On directable automata. Kybernetika 7:289–298
12. Volkov MV (2008) Synchronizing automata and the Černý conjecture. In: Proceedings of LATA 2008, vol 5196. Springer, pp 11–27

A 2-Competitive Largest Job on Least Loaded Machine Online Algorithm Based on the Multi List Scheduling Model

Debasis Dwibedy and Rakesh Mohanty

Abstract Design of algorithms without complete future input information is a challenging area in Computing. In online scheduling, inputs are given one by one in order and an irrevocable decision is made on each input prior to the availability of subsequent inputs. Online scheduling with identical machines finds extensive applications in emerging areas such as High Performance Computing (HPC), Cloud Computing, Intelligent Computing, and Edge Computing. Given multiple lists of independent jobs and a number of identical parallel machines as inputs, the objective of the online scheduler is to allocate a batch of incoming jobs irrevocably to the machines before the arrival of the next batch. On simultaneous submission of jobs from multiple sources, a non-trivial research challenge is to design a novel framework and an efficient constant competitive online scheduling algorithm. We address this challenge by designing a new 2-competitive deterministic online algorithm, known as *Largest Job on Least Loaded Machine (LJLLM)* based on our proposed *Multi List Scheduling (MLS)* Model.

Keywords Competitive analysis · Identical machines · Multi list · Makespan · Online scheduling · Parallel submissions

1 Introduction

In real-world scenarios, inputs are given one by one in order and an irrevocable decision on each received input is required by the algorithm prior to the availability of subsequent inputs [1]. Such algorithms are known as online algorithms [2, 3]. However, traditional offline algorithms receive the whole set of inputs at the outset and process the received inputs simultaneously. In this paper, we study the online non-preemptive scheduling of n independent jobs on m identical parallel machines, where $m \geq 2$ and $n > m$. We consider a realistic scenario, where jobs are submitted to a scheduler simultaneously by k distinct parallel sources in an online fashion

D. Dwibedy (✉) · R. Mohanty
Veer Surendra Sai University of Technology, Burla 768018, OR, India

© The Author(s), under exclusive license to Springer Nature Singapore Pte Ltd. 2022

D. Giri et al. (eds.), *Proceedings of the Seventh International Conference on Mathematics and Computing*, Advances in Intelligent Systems and Computing 1412,
https://doi.org/10.1007/978-981-16-6890-6_42

such that each source can submit at most one job at a particular time step, where $1 \leq k \leq n$. The scheduler must schedule irrevocably a batch of r available jobs at each time step prior to the availability of next batch of jobs with an objective to minimize the completion time (i.e., makespan) of the entire job schedule, where $1 \leq r \leq k$.

1.1 Motivation

Practical Motivation. In practical applications, several parallel sources can be traced such as users of multiprocessing systems, vendors ordering for products to a production company, up-links submitting packets to a router and clients requesting for services to cloud servers [4]. For instance, let us consider an *Edge Computing* facility with multiple edge nodes, that provides services to several research centers of a University. Most often, servicing requests from all centers are submitted simultaneously in an online fashion and each center requires an immediate processing of its submissions. If the centers share equal importance, then the computing facility must ensure processing of the requests in parallel to yield a minimum time of completion for each request.

Research Motivation. Overlooking the communication bandwidth related issues and focusing on the computational model, we raise the following non-trivial research challenge.

Can we design a model and an efficient constant competitive online algorithm that explicitly schedules the simultaneous arrival of jobs from multiple distinct sources with an objective to minimize the makespan of the schedule?

Scheduling of independent jobs on $m (\geq 2)$ identical machines for makespan minimization has been proved to be NP-complete by a polynomial time reduction from the well-known Partition problem [6]. One can verify combinatorially that n jobs can be scheduled on m machines in m^n possible ways. In the worst case, the probability of obtaining an optimum schedule is $\frac{1}{m^n}$. The problem becomes much harder as soon as the jobs are given one by one unlike the whole set of jobs at the outset.

The performance of an online algorithm is widely evaluated by *Competitive Analysis* method with *Competitive Ratio (CR)* [7] as a performance measure. *CR* is the ratio between the cost obtained by an online algorithm (*ONL*) and the cost incurred by the optimum offline algorithm (*OPT*). Let $C_{OPT}(J)$ and $C_{ONL}(J)$ be the costs incurred by algorithms *OPT* and *ONL*, respectively, for processing an input sequence $J : \langle J_1, J_2, \ldots, J_n \rangle$ of n jobs. Algorithm *ONL* is c-competitive, if for smallest $c\ (\geq 1)$, $C_{ONL}(J) \leq c \cdot C_{OPT}(J) + d$, for all permutations of job sequence in J, where d is a non-negative constant. Here, c is referred to as *CR*. A non-trivial challenge is to obtain a *CR* bounded by a constant and closer to 1. In the next sections, we present a brief overview of the seminal contributions related to online scheduling with parallel submissions from the literature.

1.2 Related Work

Online scheduling problem has been well studied in the literature with multitude of variants [1, 8]. However, little is known about online scheduling with parallel submissions. Here, the objective is to optimize the makespan of the job schedule while ensuring timely acceptance of simultaneous submissions to guarantee fairness in scheduling with respect to distinct sources [5]. In particular, the objectives, minimization of makespan, and fairness in scheduling have been widely studied in offline settings (for a comprehensive review of seminal contributions; please see [9]). We now present a brief overview of related works on online scheduling and important known results as follows.

Online scheduling problem has received significant research interest since the pioneering work of Graham in [10]. Graham introduced the famous *List Scheduling (LS)* model for scheduling a sequence of jobs that are given one by one from a list with an objective to minimize makespan. Here, sources of job requests are not considered explicitly. Graham proposed the $(2 - \frac{1}{m})$-competitive *List Scheduling Algorithm (LSA)*. In [11], Graham considered the offline setting of the *LS* model, where a sequence of jobs are ordered in a list in non-increasing processing time. For this setting, he proposed the algorithm *Largest Processing Time (LPT)*, which follows the order of the list to schedule jobs on the idle machines. Graham achieved a worst-case performance ratio of $\frac{4}{3} - \frac{1}{3m}$ for algorithm *LPT*. Thereafter, many improved online algorithms [12–20] have been proposed for the *LS* model. The best known upper bound on the *CR* of 1.9201 was obtained by Fleischer and Wahl in [18]. The current best lower bound of 1.88 on the *CR* was contributed by Rudin III in [19].

Recently, *resource augmentation* policies have gained significant attention for improving the performance of online scheduling algorithms. Here, an online algorithm is given some *Extra Piece of Information (EPI)* on future jobs or power in terms of additional space while processing a sequence of jobs. One of the most considered *EPI* in the literature [21–23] for obtaining improved competitive online scheduling algorithms is the *arrival order of jobs*. The best competitive ratio of 1.25 for scheduling jobs that arrive one by one in non-increasing order of their processing times was achieved by Cheng et al. [22]. Many authors [24–28] have considered a buffer of certain length $k (\geq 1)$ for keeping k incoming jobs before making a scheduling decision. Lan et al. [27] obtained the best competitive ratio of 1.5 with a buffer of size of $(\frac{3}{2}) \cdot m$ for any $m \geq 2$. Albers and Hellwig [29] proposed an $(\frac{4}{3} + \epsilon)$-competitive algorithm, where $0 < \epsilon \leq 1$. They considered extra space for the algorithm to maintain parallel virtual schedules for processing an online job sequence and finally chose the best schedule for actual assignments of the jobs.

To the best of our knowledge, limited work has been reported in the current literature [30] for online scheduling in the context of parallel submissions. Less attention has been devoted on explicit ordering of jobs with respect to distinct sources and scheduling of simultaneously arriving jobs. This naturally pave the way for our contribution in this paper.

1.3 Our Contribution

In Sect. 2, we propose the *MLS* model as a framework to theoretically analyze *online scheduling of parallel submissions*. In the *MLS* model, we consider multiple parallel sub-lists, one for each source of job requests. The *MLS* model inherently ensures fairness in scheduling by considering at each time step at most one job from each of the sources of job requests prior to a scheduling decision. In Sect. 3, we develop a deterministic online algorithm named *Largest Job on Least Loaded Machine (LJLLM)* for online scheduling over *MLS* model. At any time step, algorithm *LJLLM* receives a batch of r jobs (i.e., exactly one job from each of the sources as long as a source has an unscheduled job), where $1 \leq r \leq k$ and assigns the jobs one by one in non-increasing order of their processing times to the machine with minimum load after each assignment of a job. Here, every time a decision is made by considering r distinct sources of job requests to ensure fairness in scheduling. The objective is to balance total load among m machines so as to minimize the maximum load incurred on any of the machines. In Sect. 4, we show that any algorithm for online scheduling over *MLS* model is at least $(2 - \frac{1}{m})$-competitive. Furthermore, we prove that algorithm *LJLLM* is at most 2-competitive for larger m.

2 Our Proposed Multi List Scheduling Model

We present the basic terminologies and notations related to our work in Table 1 as mentioned below.

Our proposed *MLS* model for online scheduling of n jobs with two sub-lists and two machines is shown in Fig. 1. Let M_1, M_2 be two identical machines and L_1, L_2 be two sub-lists representing two sequences of n_1, n_2 number of jobs, which are submitted in parallel by two sources, respectively, where L_1: $\langle J_1^1, J_2^1, \ldots, J_{n_1}^1 \rangle$ and

Table 1 Basic terminologies and notations

Notation	Terminology
M_j	jth machine, $1 \leq j \leq m$
l_j	Load of M_j
L_r	rth sub-list, $1 \leq r \leq k$
J_i^r	ith job from rth sub-list
p_i^r	Processing time or size of J_i^r
c_i^r	Completion time of J_i^r
n_r	Number of jobs in L_r, $n_r \geq 1$
φ_j	Idle time of M_j
C_{max}	Makespan of a schedule

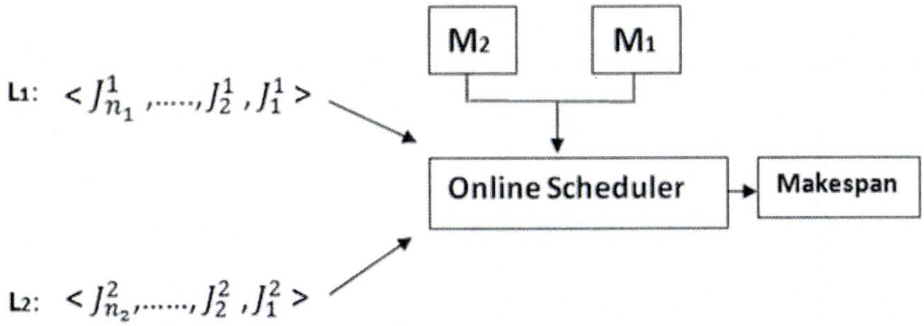

Fig. 1 MLS model with two sub-lists and two identical machines

L_2: $\langle J^2_1, J^2_2, \ldots, J^2_{n_2}\rangle$. Jobs from L_1 and L_2 are submitted one by one in parallel such that the online scheduler receives at most two jobs (i.e., one each from L_1 and L_2) as a batch at any time step t_i for $1 \leq i \leq n$ and irrevocably schedules the jobs before next batch of jobs are given at time step $t_{i+1} = t_i + \epsilon$, where ϵ (> 0) amount of time is roughly required to schedule each batch of jobs. The outcome of the scheduling (i.e., makespan) can be represented in two ways. In the first case, makespan is the earliest time by which all jobs complete their processing, i.e., $C_{max} = \max\{c^r_i | 1 \leq r \leq 2\}$. In the second case, makespan is the largest load incurred on any of the machines, i.e., $C_{max} = \max\{l_j | 1 \leq j \leq m\}$, where $l_j = \sum_r \sum_i (p^r_i)$.

We define our proposed *MLS* model by extending the concept of 2 machines and 2 sub-lists to m machines and k sub-lists as follows. Let us consider a list L of n independent jobs, which is partitioned into k sub-lists, representing k distinct sources, where $L = \bigcup_{r=1}^{k} L_r$, and $L_x \cap L_y = \phi$, where $x \neq y$ and $1 \leq x, y \leq k$. We have $|L_r| = n_r$, $1 \leq r \leq k$ and $\sum_{r=1}^{k} n_r = n$. We are given m (≥ 2) identical machines, where p^r_i of J^r_i is same in all M_j. At most, one job from each sub-list L_r is submitted at each time step t_i such that a batch of r jobs are available to the scheduler and an irrevocable scheduling decision is required before the availability of next batch of r jobs at time step t_{i+1}. The processing time of a job is known only upon its availability and a job must be executed with no preemption. The goal is to attain a minimum makespan by efficient job scheduling. A machine M_j is said to be *idle* for a time period, when it is not executing any J^r_i. Total idle time incurred by all M_j at the completion of a schedule is: $\sum_{j=1}^{m} \varphi_j$.

The multiprocessor scheduling problem over *MLS* model can be shown to be NP-complete. By [31], it is clear that list scheduling is NP-complete. One can reduce in polynomial time an instance of the list scheduling to an instance of scheduling over k sub-lists by considering $k = 1$, i.e., assuming that the scheduler receives jobs from a single source only.

3 Our Proposed Online Algorithm: LJLLM

We present the pseudocode of algorithm *LJLLM* for $m(\geq 2)$ identical machines, n jobs and $k = 2$ in *Algorithm 1.*

Algorithm 1 LJLLM

```
Initially, i = 1, l_1 = l_2 = ............l_m = 0
MIN()
  BEGIN
    l_min = min{l_j | j = 1, 2, ....m − 1, m}
    Select any M_j for which l_j = l_min
    Index M_j as M_1.
    RETURN M_1
  END
WHILE new jobs (J_i^1 AND J_i^2) arrives DO
  BEGIN
    CALL MIN()
    IF (p_i^1 ≥ p_i^2) DO
      Assign J_i^1 to M_1
      UPDATE l_1 = l_1 + p_i^1
      CALL MIN()
      Assign J_i^2 to M_1.
      UPDATE l_1 = l_1 + p_i^2
      i = i + 1
    END IF
    ELSE DO
      Assign J_i^2 to M_1
      UPDATE l_1 = l_1 + p_i^2
      CALL MIN()
      Assign J_i^1 to M_1.
      UPDATE l_1 = l_1 + p_i^1
      i = i + 1
    END ELSE
  END WHILE
WHILE new job (J_i^1 OR J_i^2) arrives DO
  BEGIN
    CALL MIN()
    Assign J_i^1 to M_1 OR J_i^2 to M_1.
    UPDATE l_1 = l_1 + p_i^1 OR l_1 = l_1 + p_i^2
    i = i + 1
  END WHILE
Return   l_j = max{l_j | j = 1, 2, ......m}
```

Algorithm *LJLLM* can further be extended for $k > 2$. Now, algorithm *LJLLM* assigns the largest J_i^r, where $1 \leq r \leq k$ among the available jobs to the least loaded machine M_1. Then the load of each machine is updated so that the second largest J_i^r can be assigned to M_1 of newly indexed machines and this continues iteratively until

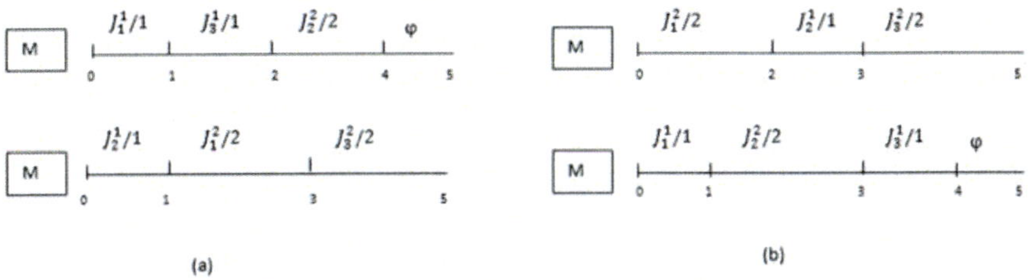

Fig. 2 **a** Online list scheduling. **b** Online multi lists scheduling

all r jobs get scheduled in one slot. Similarly, algorithm *LJLLM* takes at least 1 slot and at most $n - k + 1$ slots to schedule n jobs.

Significance of the MLS model and Algorithm LJLLM. To address the limitation of fair sharing of jobs with respect to distinct sources in the *LS* model with algorithm *LSA* as shown in Fig. 2a, we present our proposed algorithm *LJLLM* using *MLS* model through an illustration as shown in Fig. 2b. In Fig. 2, a special case is illustrated with both *LS* and *MLS* models. In particular, the case is considered, where a large number of jobs from a single source may be available upfront in the list in the *LS* model. However, the *MLS* model avoids this case as it facilitates simultaneous submissions from distinct sources. Let us consider $m = 2$, $n = 6$, $n_1 = 3$, $n_2 = 3$, the arrival sequence of jobs in the *LS* model is denoted by $L : \langle J_1^1/1, J_2^1/1, J_3^1/1, J_1^2/2, J_2^2/2, J_3^2/2\rangle$, the list L is partitioned into two sub-lists in the *MLS* model as follows: $L_1 : \langle J_1^1/1, J_2^1/1, J_3^1/1\rangle$ and $L_2 : \langle J_1^2/2, J_2^2/2, J_3^2/2\rangle$. Let us consider, two jobs are available simultaneously, one from each sub-list, in the order $\sigma : \langle (J_1^1/1, J_1^2/2), (J_2^1/1, J_2^2/2), (J_3^1/1, J_3^2/2)\rangle$.

4 Our Results on Competitive Analysis of Algorithm LJLLM for Online Scheduling over MLS Model

This section presents lower bound result for online scheduling of parallel submissions based on the *MLS* model followed by an upper bound proof of algorithm *LJLLM* for larger $m(\to \infty)$.

Let A be a deterministic online algorithm. Let $C_A(I)$ and $C_{OPT}(I)$ be the makespans obtained by algorithm A and OPT, respectively, after processing any input sequence I in the *MLS* model. Now, algorithm A can achieve a *lower bound* of c on the competitive ratio, if for smallest $c(\geq 1)$, there exists an input sequence I in the *MLS* model such that $\frac{C_A(I)}{C_{OPT}(I)} \geq c$.

Theorem 4.1 *Any deterministic online algorithm A for online scheduling over MLS model achieves a lower bound of* $2 - \frac{1}{m}$ *on the competitive ratio for* $m \geq 2$.

Proof Let C_A and C_{OPT} be the makespans achieved by algorithm A and OPT, respectively. Let us consider the following instance in the *MLS* model: algorithm

A is given m (≥ 2) identical machines and k sub-lists, where $k = m$. Each sub-list L_r has $n_r = m - 1$, where $1 \leq r \leq k - 1$. The list L_k has $n_k = m$. Each J_i^r has $p_i^r = 1$ for $1 \leq r \leq k$ and $1 \leq i \leq m - 1$. Job J_m^k has $p_m^k = m$. At each time step t_i, algorithm A receives m unit size jobs, where $1 \leq i \leq m - 1$. At the time step t_m, algorithm A incurs $l_j \geq m - 1, \forall M_j$ and receives the job J_m^k. Now, scheduling of J_m^k on any of the m machines would incur $C_A \geq l_j + m \geq 2m - 1$.

We now show the computation of *optimal offline algorithm (OPT)* as follows:

$$C_{OPT} = \frac{m(m-1)+m}{m} = m$$

Therefore, $$\frac{C_A}{C_{opt}} \geq \frac{2m-1}{m} \geq 2 - \frac{1}{m}.$$ □

Any deterministic online algorithm A achieves a upper bound $b(\geq 1)$ on the competitive ratio if for all input sequences in the *MLS* model, the following inequality holds true: $C_A \leq b \cdot C_{OPT}$.

Theorem 4.2 *Algorithm LJLLM for online scheduling over MLS model achieves a upper bound of* 2 *on the competitive ratio, where* $1 \leq k < n$ *and* $m \to \infty$.

Proof Let us consider a list $L : (J_1^1, J_2^1, \ldots, J_n^k)$ of n jobs that are partitioned into k sub-lists such that the number of jobs n_r in any k is $1 \leq n_r \leq (n - k + 1)$ and $\sum_{r=1}^{k} n_r = n$. Total load incurred by n jobs is: $\sum_{r=1}^{k} \sum_{i=1}(p_i^r)$. Without loss of generality, for simplicity in analysis, we consider J_i^r as J_i and p_i^r as p_i, where $1 \leq i \leq n$. Therefore, We have

$$\sum_{r=1}^{k} \sum_{i=1} p_i^r = \sum_{i=1}^{n} p_i \tag{1}$$

Let J_b be the largest job with *size* p_b, where $p_b = \max_{1 \leq i \leq n} p_i$. Let the optimal makespan obtained by *OPT* for L be $C_{OPT}(L)$ and makespan incurred by algorithm *LJLLM* be $C_{LJLLM}(L)$.
We now show the computation of OPT as follows:

$$C_{OPT}(L) \geq \frac{1}{m} \cdot \sum_{i=1}^{n} p_i \tag{2}$$

The following inequality is obvious:

$$C_{OPT}(L) \geq p_b. \tag{3}$$

Further, we prove Theorem 4.2 by considering different cases, which are presented in Lemma 4.2.1 and 4.2.2 as follows:

Lemma 4.2.1. Algorithm LJLLM is $(2 - \frac{1}{m})$-competitive, where each sub-list L_r has n_r number of jobs, where $1 \leq n_r \leq (n - k + 1)$.

Proof A machine M_j becomes idle, when it completes the execution of all jobs assigned to it and there is no unscheduled job left for processing. Therefore, the following inequality holds:

$$\sum_{j=1}^{m} (\varphi_j) \leq (m-1) \cdot p_b \tag{4}$$

Makespan obtained by algorithm *LJLLM* for k sequences of total n jobs is equivalent to the largest load incurred on any of the machines. Therefore, we have $C_{LJLLM}(L) \leq \max_{1\leq j\leq m} l_j \leq \frac{1}{m} \cdot (\sum_{i=1}^{n} p_i + \sum_{j=1}^{m} \varphi_j)$

By (2) and (4), we have $C_{LJLLM}(L) \leq C_{OPT}(L) + \frac{1}{m}(m-1) \cdot p_b$

By (3), we have $C_{LJLLM}(L) \leq C_{OPT}(L) + \frac{1}{m}(m-1) \cdot C_{OPT}(L)$

$\leq (2 - \frac{1}{m}) \cdot C_{OPT}(L)$ □

Lemma 4.2.2. Algorithm LJLLM is $(2 - \frac{1}{m})$-competitive for k sub-lists each with equal number of jobs.

Proof Let us consider that each list L_r has $n_r = \frac{n}{k}$. W.l.o.g, we assume that at the time of arrival of last k jobs, already $n - k$ jobs have been scheduled. At this time, let $l_1 = \min\{l_j | 1 \leq j \leq m\}$, then we have the following inequality:

$$l_1 \leq \frac{1}{m} \cdot \sum_{i=1}^{n-k} p_i \tag{5}$$

Case 1 $\{k \leq m\}$:

Let J_x, where $(n - k + 1) \leq x \leq n$ has the processing time p_x such that $p_x = \max\{p_i | (n - k + 1) \leq i \leq n\}$ and if $p_x = p_b$. Then, by (5), we have

$$C_{LJLLM}(L) \leq \left[\frac{1}{m} \cdot \sum_{i=1}^{n-k} p_i\right] + p_x \tag{6}$$

By the equal load sharing nature of algorithm *OPT* and if $p_x = p_b$, further (6) can be $C_{LJLLM}(L) \leq C_{OPT}(L) - \frac{p_b}{m} + p_b \leq C_{OPT}(L) - \frac{C_{OPT}(L)}{m} + C_{OPT}(L)$

$\leq (2 - \frac{1}{m}) \cdot C_{OPT}(L)$

Case 2 $\{k > m\}$:

Subcase 2.1. $\{p_x \geq \frac{1}{m-1} \cdot \sum_{i=n-k+2}^{n} p_i\}$:

By (5), we have $C_{LJLLM}(L) \leq \lfloor \frac{1}{m} \cdot \sum_{i=1}^{n\ k} p_i \rfloor + p_x$

Following the derivation of Case 1, we have $C_{LJLLM}(L) \leq (2 - \frac{1}{m}) \cdot C_{OPT}(L)$.

Subcase 2.2 $\{p_x < \frac{1}{m-1} \cdot \sum_{i=n-k+2}^{n} p_i\}$:

Suppose, job J_t with processing time of p_t completes at time T and let we have $C_{LJLL}(L) \leq T$. Let $T - p_t = t$ (Arrival time of J_t). Therefore, the following inequality holds: $t \leq \frac{1}{m} \cdot \sum_{i=1}^{n-k} p_i + \frac{1}{m} \cdot \sum_{i=n-k+1}^{n-1} p_i$, further, we have $t + p_t \leq \frac{1}{m} \cdot \sum_{i=1}^{n-1} p_i + p_t$. Now, if $p_t = p_b$, then we have $t + p_b \leq \frac{1}{m} \cdot \sum_{i+1}^{n-1} p_i + p_b$

$C_{LJLLM}(L) \le \frac{1}{m} \cdot \sum_{i=1}^{n-1} p_i + p_b \le \frac{1}{m} \cdot [\sum_{i=1}^{n-1} p_i + p_b] + (\frac{m-1}{m}) \cdot p_b$
By (2) and (3), we have $C_{LJLLM}(L) \le C_{OPT}(L) + (\frac{m-1}{m}) \cdot C_{OPT}(L) \le (2 - \frac{1}{m}) \cdot C_{OPT}(L)$.
Lemmas 4.2.1 and 4.2.2 obtain the following inequality for $m \to \infty$

$$C_{LJLLM}(L) \le 2 \cdot C_{OPT}(L)$$ □

Corollary 4.2.3. Algorithm LJLLM is at most $(2 - \frac{1}{m})$-competitive for online scheduling over MLS model, where all jobs are of unit size.

Lemma 4.3 *Algorithm LJLLM for online scheduling over MLS model is* 1.33-*competitive, where* $k = n$ *and* $m \to \infty$.

Proof Let us denote p_i^r, where $1 \le r \le k$ and $1 \le i \le n - k + 1$ as p_i, where $1 \le i \le n$. Let us update jobs' indices as $1, 2, \ldots, n$ and organize the jobs in a list L such that $p_1 \ge p_2 \ge \ldots \ge p_n$. Algorithm *LJLLM* schedules the jobs one by one in the following order $\sigma : \langle J_1, J_2, \ldots. J_{n-1}, J_n \rangle$ on the machine which has minimum load after the assignment of each job. Let the costs obtained by algorithms *LJLLM* and *OPT* be $C_{LJLLM}(L)$ and $C_{OPT}(L)$, respectively. By Theorem 2 of Graham [11], we can infer

$$C_{LJLLM}(L) \le \left(\frac{4}{3} - \frac{1}{3m}\right) \cdot C_{OPT}(L) \quad (7)$$

For $m \to \infty$, (7) can be $C_{LJLLM}(L) \le (1.33) \cdot C_{OPT}(L)$. □

Lemma 4.4 *Algorithm LJLLM is* 1-*competitive for online scheduling over MLS model, where* $m = 1$ *or* $n \le m$.

Proof It is quite obvious that for m=1, algorithms *OPT* and *LJLLM* obtain makespan

$$C_{OPT}(L) = C_{LJLLM}(L) \ge \sum_{r=1}^{k} \sum_{i} p_i^r \quad (8)$$

Let $p_b = \max\{p_i^r | 1 \le r \le k \, and \, i \ge 1\}$. If $n \le m$, then algorithms *OPT* and *LJLLM* obtain makespan

$$C_{OPT}(L) = C_{LJLLM}(L) \ge p_b \quad (9)$$

From Eqs. (8) and (9), we can deduce that algorithm *LJLLM* is 1-competitive. □

5 Conclusion and Future Scope

In this paper, to address the challenging issue of designing efficient deterministic online scheduling algorithms in the framework of parallel submissions of jobs from multiple sources, we have proposed a Multi List Scheduling Model. The novelty of our proposed model lies in the explicit consideration of multiple sources of job requests for ensuring fairness in scheduling. Furthermore, we achieve fairness in scheduling by designing the novel deterministic online algorithm *Largest Job on Least Loaded Machine (LJLLM)*, which intuitively follows the load balancing policy. By competitive analysis approach, we have shown that our proposed *LJLLM* algorithm based on the *MLS* model is 2-competitive for larger m.

Scope of Future Work. It is an interesting and non-trivial research challenge to determine the fairness of algorithm *LJLLM* based on a standard quantitative measure. Further work can be done by designing improved online algorithms with competitive ratio better than 2. It is interesting to explore, whether constant competitive online algorithms can be designed for scheduling framework in which machines are related or unrelated.

References

1. Albers S (2009) Online scheduling. In: Robert Y, Vivien F (eds) Introduction to scheduling. Chapman and Hall/CRC Press, pp 57–84
2. Borodin A, El-Yaniv R (1998) Online computation and competitive analysis, 2nd edn. Cambridge University Press, Cambridge
3. Albers S (2003) Online algorithms: a survey. Math Program 97(1):3–26
4. Albers S (1997) Competitive online algorithms. OPTIMA: Math Program Soc Newsl 54(1):1–8
5. Kay J, Lauder P (1988) A fair share scheduler. Commun ACM 31(1):44–55
6. Lenstra JK, Rinnooy Kan AHG, Brucker P (1977) Complexity of machine scheduling problems. Ann Discret Math 1:343–362
7. Sleator DD, Tarjan RE (1985) Amortized efficiency of list update and paging rules. Commun ACM 28(1):202–208
8. Brucker P (2006) Scheduling algorithms, 5th edn. Springer
9. Feitelson DG (1997) Job scheduling in multi-programmed parallel systems(Extended Version). IBM research report, RC 19790 (87657), Second Revision
10. Graham RL (1966) Bounds for certain multiprocessor anomalies. Bell Syst Tech J 45(1):1563–1581
11. Graham RL (1969) Bounds on multiprocessor timing anomalies. SIAM J Appl Math 17(2):416–429
12. Bartal Y, Fiat A, Karloff H, Vohra R (1992) New algorithms for an ancient scheduling problem. In: Proceedings of the 24th ACM symposium on the theory of computing (STOC), Victoria, Canada, pp 51–58
13. Galambos G, Woeginger GJ (1993) An online scheduling heuristic with better worst-case Ratio than Graham's list scheduling. SIAM J Comput 22(2):349–355
14. Karger DR, Phillips SJ, Torng E (1996) A better algorithm for an ancient scheduling problem. J Algorithm 20, article no:19:400–430
15. Bartal Y, Karloff H, Rabani Y (1994) A better lower bound for online scheduling. Inf Process Lett. 50(1):113–116
16. Chen B, Vliet AV, Woeginger GJ (1994) New lower and upper bound for online scheduling. Oper Res Lett 16(1):221–230
17. Albers S (1999) Better bounds for online scheduling. SIAM J Comput 29(1):459–473
18. Fleischer R, Wahl M (2000) Online scheduling revisited. J Sched 3(1):343–353
19. Rudin III JF (2001) Improved bounds for the online scheduling problem, PhD thesis. The University of Texas at Dallas, May

20. Englert M, Ozmen D, Westermann M (2008) The power of reordering for online minimum makespan scheduling. In: Proceedings of the 49th annual IEEE symposium on foundations of computer science (FOCS)
21. Seiden S, Sgall J, Woeginger G (2000) Semi-online scheduling with decreasing job sizes. Oper Res Lett 27(1):215–221
22. Cheng TCE, Kellerer H, Kotov V (2012) Algorithms better than LPT for semi-online scheduling with decreasing processing times. Oper Res Lett 40(1):349–352
23. Tang F, Nie J (2015) LS algorithm for semi-online scheduling jobs with non-decreasing processing times. International conference on computers, information system and industrial applications (CISIA)
24. Kellerer H, Kotov V, Speranza MG, Zs Tuza (1997) Semi-online algorithms for the partition problem. Oper Res Lett 21(1):235–242
25. Zhang G (1997) A simple semi-online algorithm for $P_2//C_{max}$ with a buffer. Inf Process Lett 61(1):145–148
26. Dosa G, He Y (2004) Semi-online algorithms for parallel machine scheduling problems. Computing 72(1):355–363
27. Lan Y, Chen X, Ding N, Dosa G, Han X (2012) Online Makespan scheduling with a buffer. Frontiers in algorithms and aspects in information and management, pp 161–171
28. Chen X, Xu Z, Dosa G, Han X, Jiang H (2013) Semi-online hierarchical scheduling problems with buffer or re-arrangements. Inf Process Lett 113(1):127–131
29. Albers S, Hellwig M (2016) Online Makespan minimization with parallel schedules. Algorithmica. https://doi.org/10.1007/s00453-016-0172-5
30. Pinheiro V, Rzadca K, Trystram D (2012) Campaign scheduling, 19th International conference on high performance computing (HiPC), pp 1–10
31. Karp RM (1972) Reducibility among Combinatorial Problems, TR3, Department of Computer Science. University of Califormia at Berkeley, April

Hardware Implementation of Multiplication over Quartic Extension Fields

Arthur Lavice, Nadia El Mrabet, Alexandre Berzati, and Jean-Baptiste Rigaud

Abstract New Number Field Sieves (NFS) attacks on the discrete logarithm problem have led to increasing the key size of pairing-based cryptography on most popular curves like BN. To ensure 128-bit security level, recent cost estimations between several pairing friendly elliptic curves recommend to switch for BLS24 curves. However, implementing a pairing on a BLS24 curve requires arithmetic operation over $\mathbb{F}_{p^4}$. In this paper, we transpose previous work on multiplication over extension fields using Newton's interpolation to construct a new formula for multiplication in $\mathbb{F}_{p^4}$ and propose efficient hardware implementation of this operation. Our Arithmetic Logic Unit (ALU) is implemented on Kintex-7 Xilinx® FPGA. The efficiency of our design in terms of $\boldsymbol{time} \times \boldsymbol{area}$ is almost 3 times better than previous specific architectures for multiplication in $\mathbb{F}_{p^4}$. We also used this new architecture to estimate full pairing implementations on BLS24 and KSS16 for 128 and 192-bit security level. These estimations highlight that the new $\mathbb{F}_{p^4}$ formula allows a performance increase between 5 and 10%. And especially for KSS16 which appears to provide the best performance at the 128-bit level.

Keywords Pairing-based cryptography · Multiplication over extension fields · Polynomial interpolation · Newton's formula · Exact divisions · Hardware implementation · FPGA

A. Lavice (✉) · N. El Mrabet · J.-B. Rigaud
Mines Saint-Etienne, CEA-Tech, Centre CMP, Gardanne, France
e-mail: arthur.lavice@emse.fr

N. El Mrabet
e-mail: nadia.el-mrabet@emse.fr

J.-B. Rigaud
e-mail: jean-baptiste.rigaud@emse.fr

A. Lavice · A. Berzati
Thales DIS Design Services SAS, Meyreuil, France
e-mail: alexandre.berzati@thalesgroup.com

© The Author(s), under exclusive license to Springer Nature Singapore Pte Ltd. 2022

D. Giri et al. (eds.), *Proceedings of the Seventh International Conference on Mathematics and Computing*, Advances in Intelligent Systems and Computing 1412,
https://doi.org/10.1007/978-981-16-6890-6_43

1 Introduction

Pairings are mathematical objects known for long time but recently considered for supporting cryptographic functions such as identity-based encryption [6], short signature [7] or zero-knowledge Succinct Non-interactive ARgument of Knowledge (zk-SNARK) [5]. These applications legitimate the search for efficient implementations of cryptographic pairings in both software [12] and hardware [20]. These researches make pairing-based cryptography a key element of real-life use cases as in Zcash^{RM} cryptocurrency.

Pairings are bilinear function $e : \mathbb{G}_1 \times \mathbb{G}_2 \to \mathbb{G}_3$ which relied on elliptic curves and extension fields arithmetic. The efficiency of pairings computation highly depends on the choice of the pairings family. Therefore, most attention was given on pairings BN [4] or BLS12 [3] because it allows several algorithmic optimizations. However, the Number Field Sieve (NFS) attack against discrete logarithms problems presented in [15] reduced the security of many pairings families including BN and BLS12 curves. Barbulescu et al. [1, 2] calculated the new key sizes required, and estimated the pairing calculation times on many curves at different security levels. Their results were confirmed at the 128-bit security level in [11].

Authors in [2] have estimated that pairings using extension fields of degree multiple of 4 like BLS24 [3] are theoretically more efficient than other pairings family at 128-bit and 192-bit security levels. Furthermore, KSS16 [14] curves are one of the best candidates at the 128-bit security level. Based on the cost details in [2], computing pairings on the BLS24 curve required 2387 multiplication over extension fields of degree 4 and it represents more than 75% of the modular multiplication required by the entire pairing. Hence, multiplication over extension fields of degree 4 is a key operation of pairings at the actual 128-bit security level. In [2], the authors estimate the complexity of multiplication over extension fields with the number of modular multiplications required by formulas given in [9, 16]. These formulas are based on Newton's interpolation which minimizes the number of modular multiplications required to compute multiplication over extension fields. However, Newton's interpolation requires many other operations to be properly implemented like modular addition or division.

Our contributions: The first contribution of the paper is to update cost estimations of multiplications over degree 4 extension fields considering all required operations. Hence, we have proposed a hardware design that minimizes the impact of underestimated extra operations. It implements the multiplier proposed in [13] and a custom "adder" used to accelerate the additional modular operations required for Newton's interpolation: addition, subtraction, double, division by 2, and the exact division by a small constant prime. This ALU has been designed to compute modular multiplication in parallel with other operations and to support multiplications over different extension fields required for pairing-based cryptography.

We also provide a way to preserve the divisibility properties of numbers during Newton's interpolation while using modular arithmetic. The implementation of divisions during Newton's interpolation is based on the method proposed in [9] to compute them within the time of modular additions. We implement Karatsuba's [16] formulas to compare hardware implementation of both methods for extension fields of degree 4.

The efficiency of our ALU in terms of ***time* x *area*** outperforms previous works on hardware implementation presented in [20] of multiplication over these fields.

Organization of the paper: Sect. 2 provides some mathematical background for pairing-based cryptography. An efficient formula for multiplication over extension fields of degree 4 is presented in Sect. 3. Section 4 details our ALU for modular arithmetic and divisions with concurrent custom adder and multiplier. In Sect. 5, we present the performances of our ALU for multiplications over extension fields of various degrees and compare our performance with previous work. Then Sect. 6 gives estimations of pairings performances on BLS24 and KSS16 curves. Finally, we summarize our work and discuss future research directions in the last section.

2 Background

2.1 Notation

$\mathbb{F}_p$: A finite fields of characteristic p, for p a prime number and $n = \log_2(p)$.
$\mathbb{F}_{p^k}$: Extension field of degree k of $\mathbb{F}_p$.
M_p (resp. $I_p, S_p, cS_p, A_p, Dbl_p$, $Div2_p$, F_p): Multiplication (resp. Inversion, Squaring, compressed Squaring, Addition, Double, Division by 2, Frobenius map) in $\mathbb{F}_p$.
D_i: Exact division by $i = 3$ or 5.
$\mathbb{G}$ (resp. $\mathbb{G}[r]$) a group (resp. a subgroup of order r of $\mathbb{G}$).
ω (resp, e): Size of the word (resp, number of words) used to represent numbers in $\mathbb{F}_p$.

2.2 Arithmetic for Pairing-Based Cryptography

This section briefly recalls the basics of pairings. The reader is invited to refer to [17] for more information about pairings and their applications. Let E be an elliptic curve defined over a finite field $\mathbb{F}_p$ and r a prime factor of the cardinal of E (denoted $\#(E)$) such that r^2 does not divide $\#(E)$. Let k be the smallest integer such that r divide $p^k - 1$, k is called *the embedding degree of* E with respect to r. The definition of a pairing is given in Definition 1.

Definition 1 *(Pairing)* Let $\mathbb{G}_1$, $\mathbb{G}_2$ be subgroups of order r of E. A pairing is a bilinear and non degenerate function e

$$e : \mathbb{G}_1 \times \mathbb{G}_2 \rightarrow \mathbb{G}_3, \quad (R, S) \rightarrow e(R, S).$$

Generally, we have $\mathbb{G}_1$ in $E(\mathbb{F}_p)[r]$, $\mathbb{G}_2$ in $E(\mathbb{F}_{p^k})[r]$ and $\mathbb{G}_3$ in $\mathbb{F}_{p^k}[r]$.

Pairing-based cryptography relies on the security of the Discrete Logarithm Problem (DLP) over elliptic curves ($E(\mathbb{F}_p)$ and $E(\mathbb{F}_{p^k})$) and finite fields ($\mathbb{F}_{p^k}$). The DLP can be computed, for instance, with Pollard's rho algorithm [18]. Efficient computations of pairings depend on the construction of pairing friendly elliptic curves over $\mathbb{F}_p$. The most studied constructions are KSS [14], BLS [3], and BN [4]. There exist other pairings friendly elliptic curves presented in the taxonomy of Freeman and al. [10]. An elliptic curve is defined by several parameters: its construction method (ex: BN), its embedding degree: k, and the order r of the groups $\mathbb{G}_1$, $\mathbb{G}_2$ and $\mathbb{G}_3$.

These parameters are used by the Extended Tower Number Field Sieve attack presented in [15] to reduce the security of many pairings families. Barbulescu and al. [1, 2] computed the new required key size and estimated pairing computation time on various curves and for different security levels. Their results were confirmed at 128-bit security level in [11].

2.3 Key Size Update and Complexity Evaluation

Before the key size update, much attention was given to Optimal Ate pairings [19] on BN or BLS12 curves. But the extended tower NFS attack presented in [15] is more efficient on these families than on families like BLS24. Since the efficiency of the attack is not the same for all curves, the security impact is also different. In [2], the authors estimate pairing computation complexity with the number of modular multiplications and inversions. They estimate the cost of multiplication in $\mathbb{F}_p$ as $M_p = e^2$ with e the number of words used to represent p. Finally, they show that BLS24 and KSS16 curves could be as interesting as BLS12 curves for the 128-bit security level.

Arithmetic used to implement these operations depends on the pairing family. Modular ($\mathbb{F}_p$) and extension fields ($\mathbb{F}_{p^k}$) arithmetic are the basis of pairings arithmetic. Hence, searching for cost-efficient hardware ALU to compute these operations is a way to enhance the efficiency of pairings implementation.

2.4 Multiplications over Extension Fields

The two main methods to compute multiplications over extension fields are Newton's interpolation from [9] and Karatsuba's formulas from [16].
An element A of an extension field of degree k of $\mathbb{F}_p$ (denoted $\mathbb{F}_{p^k}$) is a polynomial of degree n, with $0 \leq n \leq k-1$ with coefficient in $\mathbb{F}_p$. Let P be an irreducible polynomial in $\mathbb{F}_p$ (with no root in $\mathbb{F}_p$). Let A and B be elements of $\mathbb{F}_{p^k}$. The result C of the multiplication of A by B is defined as the Euclidean remainder of the polynomial $A \times B$ by the polynomial P. In this paper, we consider the same extension fields as

in [20]:

$$\mathbb{F}_{p^2} = \mathbb{F}_p[u],\ with\ u^2 = 2$$
$$\mathbb{F}_{p^4} = \mathbb{F}_{p^2}[v],\ with\ v^2 = u$$

Explicit formulas are given below.

Karatsuba Formulas

Let $A = a_0 + a_1 X$, $B = b_0 + b_1 X$ and $C = c_0 + c_1 X$, $A, B \in \mathbb{F}_{p^2}$. Then the Karatsuba formula to compute $C = A \times B \mod X^2 - 2$ is

$$\begin{aligned} C &= (a_0 + a_1 X)(b_0 + b_1 X), \\ &= a_0 b_0 + (a_0 + a_1)(b_0 + b_1)X + a_1 b_1 X^2. \end{aligned} \tag{1}$$

The cost of these computations is $3M_p + 2A_p$. We re-organize Eq. 1 to give an expression of the coefficients of the polynomial C. Products of each line are denoted c_i with $0 \leqslant i \leqslant 2$.

Equation 2 gives the new expression of C

$$C = c_2 X^2 + (c_1 - c_2 - c_0)X + c_0. \tag{2}$$

The cost of multiplication of degree $n = 1$ polynomial is $3M_p + 4A_p$. Then, the total cost of a multiplication over extension fields of degree 2 is $3M_p + 5A_p + Dbl_p$.

To compute multiplication over extension fields of degree 4, we apply Karatsuba formula twice.Hence, the cost of a multiplication in $\mathbb{F}_{p^4}$ is $9M_p + 25A_p + 4Dbl_p$.

Newton's Interpolation

The implementation cost of M_p is commonly considered more expensive than the implementation's cost of A_p or Dbl_p. Hence, the straightforward way to improve multiplications over extension fields is to reduce the number of modular multiplications. Newton's interpolation is an alternative way to do so. This part briefly recalls the method to obtain formulas for computing multiplications over extension fields using Newton's interpolation.

Let us consider an element of $\mathbb{F}_{p^k}$ as a polynomial in unknown X. Then $A(X) = a_0 + a_1 X + \cdots + a_{k-1} X^{k-1}$ and $B(X) = b_0 + b_1 X + \cdots + b_{k-1} X^{k-1}$ are two polynomials of degree $k-1$. Hence, the polynomial $C(X) = A(X) \times B(X)$ is of degree $2k-2$. To compute the polynomial C, we compute the products of $2k-1$ distinct images of A and B as explained below.

Note 1. Required steps of Newton's interpolation

1. **Choose** $2k-1$ **distinct points in** $\mathbb{F}_p$: $(\alpha_0, \alpha_1, \ldots \alpha_{2k-2})$, $\forall i \neq j, \alpha_i \neq \alpha_j$.

2. **Polynomial interpolation:** Evaluate polynomials A and B in these points by computing $A(\alpha_0), \ldots, A(\alpha_{2k-2}), B(\alpha_0), \ldots, B(\alpha_{2k-2})$.
3. **Evaluation of** C: Compute the image of polynomial C in these points ($C(\alpha_i) = A(\alpha_i) \times B(\alpha_i)$)
4. **Newton's interpolation + Reconstruction of** C **with Horner Scheme:** Reconstruct the polynomial C using Newton's interpolation formulas.

The complexity of multiplications using Newton's interpolation depends on the choice of interpolation points (α_i) because we have to compute multiplications and exact divisions by the ($\alpha_i - \alpha_j$) during the reconstruction step. By choosing the interpolation points as explained in [9], we can use Newton's interpolation to compute multiplications over different extension fields.

Newton's interpolation reduces the number of modular multiplications but requires extra modular divisions and more additions than Karatsuba formulas.

3 Our Proposition for Extensions-Fields of Degree 4

We follow the steps described in Note 1 to build an efficient formula for multiplication over extension fields of degree 4. Let $A(X) = a_0 + a_1X + a_2X^2 + a_3X^3$, $B(X) = b_0 + b_1X + b_2X^2 + b_3X^3$ and $C(X) = A(X) \times B(X) = c_0 + c_1X + c_2X^2 + c_3X^3$ be three elements of $\mathbb{F}_{p^4}$.

We choose the α_i's to be $\{0, 1, -1, 2, -2, 4, +\infty\}$, with $A(+\infty) = a_3$.

Polynomial Interpolation

$$
\begin{aligned}
sp1 &= a_0 + a_2, & sp2 &= a_0 + 2^2a_2, & A(0) &= a_0,\\
si1 &= a_1 + a_3, & si2 &= 2a_1 + 2^3a_3, & A(+\infty) &= a_3,\\
A(1) &= sp1 + si1, & A(2) &= sp2 + si2, & A(4) &= a_0 + 4(a_1 + 4(a_2 + 4a_3)).\\
A(-1) &= sp1 - si1, & A(-2) &= sp2 - si2, & &
\end{aligned}
$$

It requires $11A_p + 10Dbl_p$ to compute the interpolation of A.

Evaluation of C:

$$C(\alpha) = A(\alpha) \times B(\alpha) \; with \; \alpha \in \{0, \pm1, \pm2, 4, +\infty\}.$$

It requires $7M_p$.

Newton's Interpolation

$$
\begin{aligned}
c'_0 &= C(0), & c'_4 &= (((C(-2) - c'_0)\tfrac{1}{2} + c'_1)\tfrac{1}{3} - c'_2 + c'_3)\tfrac{1}{4},\\
c'_1 &= C(1) - c'_0, & c'_5 &= (((((C(4) - c'_0)\tfrac{1}{4} - c'_1)\tfrac{1}{3} - c'_2)\tfrac{1}{5} - c'_3)\tfrac{1}{2} - c'_4)\tfrac{1}{6},\\
c'_2 &= (C(-1) - c'_0 + c'_1)\tfrac{1}{2}, & c'_6 &= a_3b_3.\\
c'_3 &= ((C(2) - c'_0)\tfrac{1}{2} - c'_1 - c'_2)\tfrac{1}{3}, & &
\end{aligned}
$$

It requires $15A_p + 9Div2_p + 4Div_3 + 1Div_5$.

Reconstruction of C with Horner Scheme:
$C(x) = c'_0 + X(c'_1 + (X-1)(c'_2 + (X+1)(c'_3 + (X-2)(c'_4 + (X+2)(c'_5 + (X-4)c'_6)))))$
It requires $15A_p + 5Dbl_p$.
Finally, the cost of the polynomial reduction is $3A_p + 3Dbl_p$.

The overall cost is: $7M_p + 55A_p + 28Dbl_p + 9Div2_p + 4Div_3 + 1Div_5$. The proposed formula is more efficient than Karatsuba formulas when $2M_p > 30A_p + 25Dbl_p + 9Div2_p + 4Div_3 + 1Div_5$.

In practice, the extra costs brought by modular additions and divisions make these two methods almost equivalent for software implementations at the 128-bit security level as explained in [9]. Besides, Newton's interpolation requires divisions by small constants which cannot be computed by classical adder or subtractor.

Hence, our objective is to reduce the cost of additional operations to obtain efficient hardware implementation of multiplication over extension fields of degree 4.

4 A Dedicated Arithmetic Logic Unit (ALU) for Extension Field Operations

We present a generic ALU designed to accelerate multiplication over extension fields by reducing the penalty brought by additional operations.

4.1 Overall Presentation

Our ALU (see Fig. 1) is composed of two Main Processing Units (MPU). First, the multiplier to compute Montgomery's multiplications with reduction. Then, the custom "Adder" is the component used to compute modular additions, subtractions, divisions by 2 and exact divisions by 3 or 5. The RAM stores intermediate results. The ROM drives the two MPUs and replaces a dedicated FSM. The multiplexer selects the result of the multiplier or the one of the "Adder" to save it into the RAM. The proposed solution is generic in terms of word size (ω) and moduli size $n = \log_2(p)$. We choose to implement a 64-bit architecture to limit the number of memory-calls. We give further explanations of this choice in Sect. 5. Within this article, we use the color code in Fig. 1 to represent the different parts of the design.

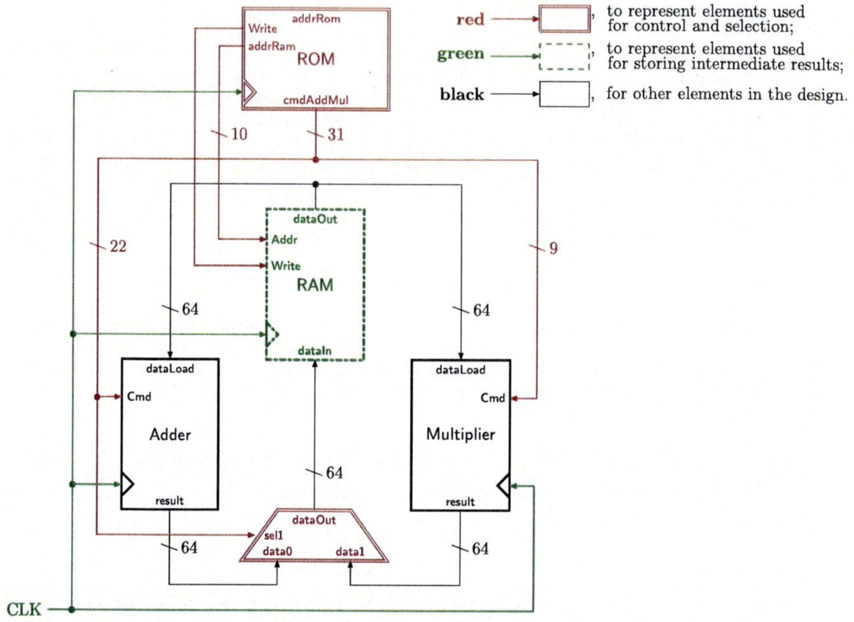

Fig. 1 Proposed ALU (for $\log_2(p) < 512$)

4.2 Multiplication

The modular multiplier is based on the optimized architecture of [13] (architecture 2). It is a systolic architecture composed of e Processing Unit (one for each word of the modulo). It used a variant of radix-2 Montgomery multiplication to compute modular multiplication. We add a Memory Unit (MU) to store the modulo and the operands and an Output Manager (OM) to format the output of a multiplier for the RAM.

4.3 Exact Division by a Small Constant

To efficiently use Newton's interpolation for multiplications over extension fields of degree $k = 4$, we need to compute exact divisions by 2, 3, or 5. For instance, if we need to divide δ in $\mathbb{N}$ by 3 during Newton's interpolation, δ will be divisible by 3. There is a number σ in $\mathbb{N}$ such as $\delta = 3\sigma \Leftrightarrow \delta - 2\sigma = \sigma$. Since the first bit of 2σ is 0, we can compute the division by 3 with a custom subtractor as explained in [9]. Our ALU implements this method to compute word divisions by 3 or 5. A dedicated component for each small constant has been implemented to enhance the maximum frequency of the design. These three components are instantiated in the component used to compute the division of a word in four clock cycles.

Divisions calculated during Newton's interpolation are supposed to be exact (the reminder is null) but using modular arithmetic, we lose the divisibility of numbers. A possible solution is to use classical arithmetic during Newton's interpolation and reduce modulo p after all exact division. In this case, the size of the operands will double during calculation due to successive multiplications and additions. We propose a solution to keep using modular arithmetic and save the divisibility property of numbers by working with a multiple of the prime p as demonstrated on Remark 1.

Remark 1 (*Divisibility of an integer and its residue*) Let $a, b \in \mathbb{N}$ and p be a prime number. If a is divisible by b (note $b|a$), then $a \mod (p)$ is not necessarily divisible by p but the number $a \mod (b.p)$ is divisible by b.

As a consequence of Remark 1, we have to increase the size of the modulus for each successive division to use modular multiplication and exact division.
Then, the modulus used with Newton's interpolation is

$$m = 3^4 \times 5 \times p, \text{ for multiplication over } \mathbb{F}_{p^4}. \tag{3}$$

A drawback is that increasing the size of the modulus will also increase the cost of multiplications using Newton's interpolation.

Division by 2 To compute Newton's interpolation, we need the modular division by 2. Since we work modulo p, with p odd, if A is odd then $A + p$ is even and a division by two corresponds to a shift. We choose to begin a division by 2 with the most significant word to reduce the component used for the division by 2 to a 1-bit register.

4.4 Design of Custom "Adder"

The modular adder-subtraction can compute additions (or subtractions) and reduction operations while limiting the number of memory accesses during modular additions. During Newton's interpolation, the majority of exact divisions follows a modular addition or a modular subtraction. We place our divider after components used for modular additions to save memory access by computing exact division during the saving step. In this way, our ALU can compute a modular addition/subtraction followed by an exact division.

We use two modular adder-subtractions. One to compute the addition (or subtraction) and the other to compute the modular reduction. The two possible results are stored in two registers RegAdd and RegRed and the component *selectRes* (on Fig. 2) is used to select the right result. If the component *selectRes* return 0, the right result is the operation without modular reduction otherwise, it is the result with modular reduction.

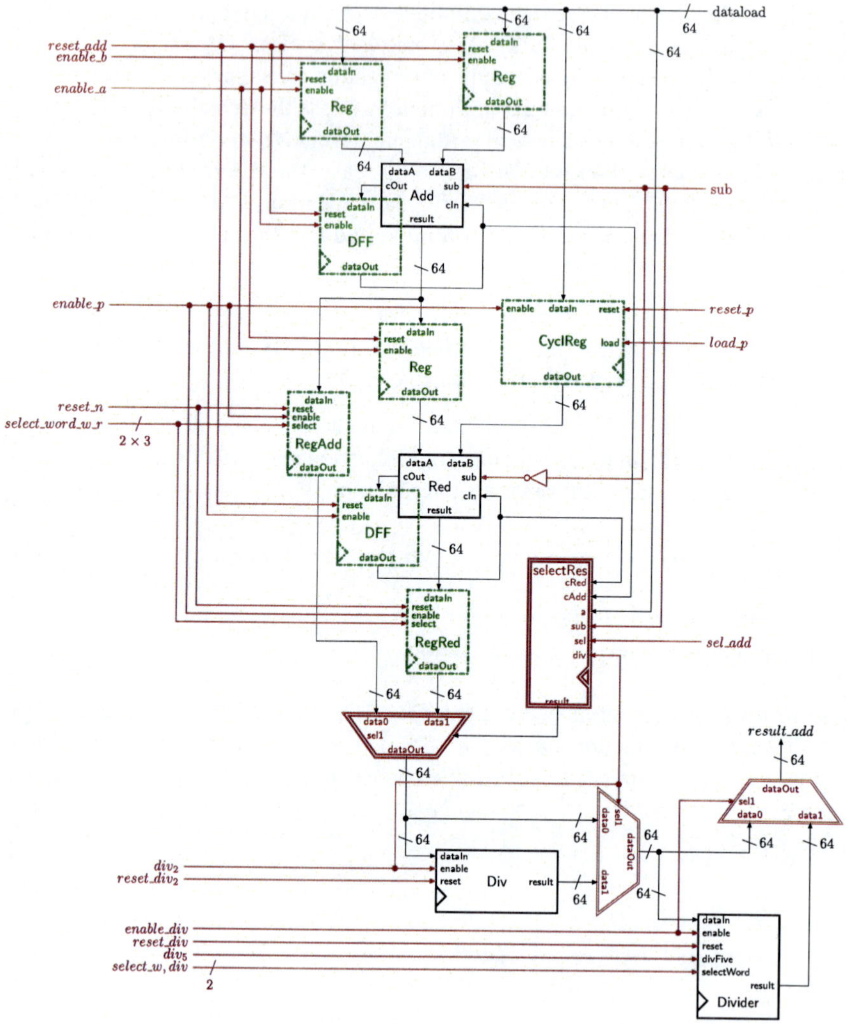

Fig. 2 Modular adder-subtractor and divider for $\log_2(p) < 512$

We also use several registers to store the carry for the next step of the computation and to store partial results during exact division. By doing this, we reduce the number of memory access to the minimum: load operands and store the final result.

Exact divisions by 3 or 5 during Newton's interpolation always follows a modular addition or subtraction. Hence, we place our divider to compute division during the saving step of an addition to reduce the number of memory accesses.

Table 1 Costs of base fields operations

Operation	Number of clock cycles
Montgomery multiplication	$n + 3e + 1$
Load Modulo	$e + 1$
Modular reduction	$2e + 4$
Modular addition/subtraction	$3e + 3$
Modular double	$2e + 3$
Modular division by 2	$2e + 3$
Modular add/sub + Exact division	$8e + 8$
Modular addition/subtraction and exact division	$6e + 5$

5 Hardware Implementation of Multiplications over Extension Fields

We implement and compare both Karatsuba and Newton's interpolation formulas for multiplications over extension fields of degree 4 with our ALU.

5.1 *Implementation of Base Fields Operations*

Multiplications over extension fields require different operations. To have a practical example, we use a single port RAM with a latency of one clock cycle for reading and writing.

Our design is generic in terms of words size (ω) and maximum modulo size ($\log_2(p)$). We denote $e = \lceil \frac{n}{\omega} \rceil$. The number of clock cycles required by each basis operation is express in Table 1.

Table 1 also shows that placing our exact divider after the adder allows us to save $2 \times e - 3$ clock cycles per division. Hence, we are able to compute a division during Newton's interpolation within the time of a modular addition.

5.2 *Multiplications over Extension Fields*

We design our ALU to parallelize multiplications and additions-like operations. For Karatsuba, we can also parallelize the polynomial reduction. It is because there is no division. For Newton's interpolation, we have to compute all divisions before applying Horner's scheme. Only the computation of $C(\infty)$ could be parallelized with the computation of the Horner scheme. We decided not to parallelize Horner's scheme nor polynomial reduction. It could be a way of further improvement.

Table 2 Implementation results of multiplication in $\mathbb{F}_{p^4}$

Security	Curves	$log_2(p)$	$time$ (μs)		area (slices)		$time \times area$ ($\mu s \times slices$)	
			K	N	K	N	K	N
128-bit	BLS24	318	32.84	29.06	825	883	27.09×10^3	**25.66**×10^3
	KSS16	339	35.44	30.46	883	31.29×10^3	**26.90**×10^3	
192-bit	BLS24	559	57.37	**48.57**	1306	74.93×10^3	**63.43**×10^3	
	KSS16	824	84.06	**71.45**	1643	1849	138.11×10^3	**132.11**×10^3

The ALU is designed to handle the different sizes of p depending on the chosen pairing family. Our ALU is implemented on Kintex-7 Xilinx® FPGA with Vivado design suite 18.1.

The critical path of our design is slightly impacted by the size of the modulo so we manage to run the different version of our ALU at 100 MHz. We use Vivado's IP to implement the RAM and the ROM. We use the FPGA only as a prototype for our ALU because our goal is to keep our design light without mapping Digital Signal Processing (DSP) units to make it more suitable for general-purpose integrated circuits.

Table 2 gives the post place and route area and timings of our design for multiplication over extension fields of degree 4 at 128-bit and 192-bit security levels. Timings are given in μs and areas are given in numbers of slices. We target a frequency of 100 MHz for prototyping but it will be possible to enhance the maximum frequency of our ALU when integrated into a System on Chip. Finally, performances are given in terms of $\boldsymbol{time \times area}$.

For multiplications over extension fields of degree 4, we manage to parallelize the majority of additions-like operations with modular multiplication that makes Newton's interpolation more interesting than Karatsuba formulas at both 128-bit and 192-bit security levels. At the 128-bit security level, we need five 64-bit words to represent a number modulo p and six 64-bit for a number modulo $m = 3^4 \times 5 \times p$. Therefore, we need to increase the size of the ALU to use Newton's interpolation at the 128-bit security level and it slows down every operation.

At the 192-bit security level, number modulo p and m required height 64-bit word. It explains why Newton's interpolation is more interesting at 192-bit than at the 128-bit security level.

5.3 Comparison with Previous Work

Multiplication over $\mathbb{F}_{p^4}$ is currently used for pairing over popular BN and BLS12 curves. We give the comparison for a modulus of size $\log_2(p) = 256$ which corresponds to the previously required size for 128-bit security level to give a fair comparison of our design with previous architecture.

Table 3 Comparison of our implementation with previous work for multiplication over $\mathbb{F}_{p^4}$

Platform	Design	*time* (μs)	F (MHz)	FF	LUT	DSP	Slices	*time* × *area* (μs × *slices*)
Virtex 7	Our	19.02	125	2564	2159	0	755	**14.36**×10^3
	[20]	**1.47**	115	55600	113700	128	28400	41.74×10^3

For the sake of fair comparison, we synthesized our ALU on a Virtex 7 to compare it with the ones proposed in [20]. Results of our comparison are shown in Table 3.

In [20], authors design a specific architecture to parallelize operations in $\mathbb{F}_p$ and $\mathbb{F}_{p^2}$ to compute pairings over BN curves. They integrated their design with a CPU in the FPGA. This leads to a time-efficient implementation of multiplication in $\mathbb{F}_{p^4}$. But in terms of ***time* x *area***, our ALU is 2.9 times more efficient for computing multiplication in $\mathbb{F}_{p^4}$. This shows that our generic ALU is as good as specific ALU. Multiplication in $\mathbb{F}_{p^4}$ is not a key operation for pairings on BN curves since the majority of operations are in $\mathbb{F}_{p^2}$. To our knowledge, Wang et al. [20] are the only ones which give the performances of their implementation of the multiplication in $\mathbb{F}_{p^4}$.

6 Refine Estimations of Pairings Implementations

This paper focuses on the computation of multiplication over degree 4 extension fields. To estimate performances of BLS24 and KSS16 implementations with our ALU, we use formulas described in [2]. We also estimate the cost of a modular inversion to $3/2 \log_2(p) M_p$ which corresponds to the number of modular multiplications required by inversion using Fermat's theorem. Then, we use the results presented in Sect. 5 to refine estimations of pairings complexity on KSS16 and BLS24 curves.

At the 128-bit security level, BLS24 pairings required $32893 M_p$. 21483 of these modular multiplications are computing during multiplication in $\mathbb{F}_{p^4}$. Hence, we estimate the cost of pairings on BLS24 curves as $(32893 - 21483) M_p + 21483/9 M_{p^4} = 11410 M_p + 2387 M_{p^4}$. We applied this method to estimate pairings computation times on BLS24 curves at the 192-bit security level and on KSS16 curves at both 128-bit and 192-bit security levels. Our estimations are given in Table 4.

Table 4 Refine estimation of pairings on BLS24 and KSS16 curves

Security	Curves	$log_2(p)$	*time (ms)*		area (slices)		*time* × *area* (*ms* × *slices*)	
			K	N	K	N	K	N
128-bit	BLS24	318	116.73	109.07	825	883	96.30×10^3	96.31×10^3
	KSS16	339	99.39	94.28	883		87.76×10^3	**83.25**×10^3
192-bit	BLS24	559	316.92	**284.02**	1306		413.90×10^3	**370.93**×10^3
	KSS16	824	536.57	516.90	1643	1849	905.81×10^3	983.14×10^3

In [2], authors estimate that pairing at the 128-bit security level is more efficient on BLS24 curves, whereas our estimation gives KSS16 pairing faster and more efficient in terms of ***time* x *area*** ratio than BLS24 pairing at the 128-bit security level.

The main reason is that the authors of [2] approximate the time complexity of a modular multiplication as $\log_{64}(p)^2$ or $\log_{32}(p)^2$. This estimation correspond to the time complexity of a software implementation, whereas the time complexity of a modular multiplication on our ALU is $\log_2(p) + 3\log_{64}(p) + 1$. Furthermore, authors in [2] do not take into account the complexity of modular inversions which require hundreds of modular multiplication when computing with Fermat's theorem.

Table 4 also shows that Newton's interpolation is a way to reduce pairings complexity of 5% at the 128-bit security level and 10% at the 192-bit security level.

7 Conclusion and Future Work

Multiplications over $\mathbb{F}_{p^4}$ became a key operation for pairings implementations since authors have estimated in [2] that BLS24 pairing is supposed to be the most efficient pairing at 128-bit and 192-bit security levels. Hence, the first contribution of this paper is an efficient formula to compute multiplications over these fields.

Optimization efforts have been focused on miss-study additional operations such as exact divisions required to efficiently implement multiplications over extension fields. Hence, we design an ALU to parallelize additional operations with modular multiplications and propose a way to keep the divisibility property during Newton's interpolation while using modular arithmetic. This allows to control the overhead generated by successive operations during Newton's interpolation. Our ALU is used to implement multiplications over extension fields of degree 4. With this ALU, Newton's interpolation is more efficient than Montgomery's formulas for multiplication over these fields. Our results are used to refine estimated pairings computation time at both 128-bit and 192-bit security level. These estimations show that KSS16 pairing should be considered as a good candidate for implementation at the 128-bit security level.

Future works. This paper proposes a hardware implementation of exact division for Newton's interpolation using the method proposed in [8]. This operation is miss-studied and it could be a source of information leakage. Future works will study the security of our ALU against side-channel attacks. We will use it to implement pairings on curves presented by [2] since there is no implementation of most of the new curves yet.

Finally, our ALU will be easily adaptable to use Newton's interpolation for multiplication over extension fields of degree 3, 5 which should bring to efficient implementations of pairings on promising BLS15 and BLS27 curves. Furthermore, multiplication over $\mathbb{F}_{p^{16}}$ can be computed by applying the formula proposed in this paper twice. It can be a way to optimize pairing on KSS16 curves which is a very promising candidate at the update 128-bit security level.

References

1. Barbulescu R, Duquesne S (2018) Updating key size estimations for pairings. J Cryptol. https://hal.archives-ouvertes.fr/hal-01534101
2. Barbulescu R, El Mrabet N, Ghammam L (2019) A taxonomy of pairings, their security, their complexity. IACR Cryptology ePrint Archive 2019:485
3. Barreto PSLM, Lynn B, Scott M (2002) Constructing elliptic curves with prescribed embedding degrees. In: SCN. Lecture notes in computer science, vol 2576. Springer, pp 257–267
4. Barreto PSLM, Naehrig M (2005) Pairing-friendly elliptic curves of prime order. In: Selected areas in cryptography. Lecture notes in computer science, vol 3897. Springer, pp 319–331
5. Ben-Sasson E, Chiesa A, Tromer E, Virza M (2014) Succinct non-interactive zero knowledge for a von neumann architecture. In: Proceedings of the 23rd USENIX conference on security symposium. SEC'14, USENIX Association, USA, pp 781–796
6. Boneh D, Franklin M (2001) Identity-based encryption from the weil pairing. In: Kilian J (ed) Advances in cryptology – CRYPTO 2001. Springer, Berlin, pp 213–229
7. Boneh D, Lynn B, Shacham H (2001) Short signatures from the weil pairing. In: Boyd C (ed) Advances in cryptology – ASIACRYPT 2001. Springer, Berlin, pp 514–532
8. El Mrabet N, Guillermin N, Ionica S (2009) A study of pairing computation for elliptic curves with embedding degree 15. IACR Cryptology ePrint Archive 2009:370
9. El Mrabet N, Guillevic A, Ionica S (2011) Efficient multiplication in finite field extensions of degree 5. In: Nitaj A, Pointcheval D (eds) Progress in cryptology - AFRICACRYPT 2011. Springer, Berlin, pp 188–205
10. Freeman D, Scott M, Teske E (2010) A taxonomy of pairing-friendly elliptic curves. J Cryptol 23(2):224–280. https://doi.org/10.1007/s00145-009-9048-z
11. Guillevic A (2020) A short-list of pairing-friendly curves resistant to special TNFS at the 128-bit security level. In: Public Key Cryptography (2). Lecture Notes in Computer Science, vol 12111. Springer, pp 535–564
12. Hankerson D, Menezes A, Scott M (2009) Software implementation of pairings. In: Identity-based cryptography, cryptology and information security series, vol 2. IOS Press, Amsterdam, pp 188–206
13. Huang M, Gaj K, El-Ghazawi T (2011) New hardware architectures for montgomery modular multiplication algorithm. IEEE Trans Comput 60(7):923–936. https://doi.org/10.1109/TC.2010.247
14. Kachisa EJ, Schaefer EF, Scott M (2008) Constructing brezing-weng pairing-friendly elliptic curves using elements in the cyclotomic field. In: Pairing. Lecture notes in computer science, vol 5209. Springer, pp 126–135
15. Kim T, Barbulescu R (2016) Extended tower number field sieve: a new complexity for the medium prime case. In: Robshaw M, Katz J (eds) Advances in cryptology - CRYPTO 2016. Springer, Berlin, pp 543–571
16. Knuth DE (1997) The art of computer programming, vol 1, 3rd edn. Fundamental algorithms. Addison Wesley Longman Publishing Co., Inc., USA
17. Menezes A (2009) An introduction to pairing-based cryptography. In: Recent trends in cryptography, vol 477, pp 47–65
18. Pollard JM (1978) Monte Carlo methods for index computation mod p. Math Comput 32:918–924
19. Vercauteren F (2010) Optimal pairings. IEEE Trans Inf Theory 56(1):455–461. https://doi.org/10.1109/TIT.2009.2034881
20. Wang AT, Guo BW, Wei CJ (2019) Highly-parallel hardware implementation of optimal ate pairing over barreto-naehrig curves. Integration 64:13–21. https://doi.org/10.1016/j.vlsi.2018.04.013, http://www.sciencedirect.com/science/article/pii/S0167926018300336

Deriving Bounds on $L(k_1, k_2)$ Labeling of Triangular Lattice by Exploring Underlined Graph Structures

Subhasis Koley and Sasthi C. Ghosh

Abstract The $L(k_1, k_2)$ labeling is a type of labeling of vertices of a graph $G(V, E)$ where absolute difference between labels of vertices at graph distances 1 and 2 are at least k_1 and k_2, respectively, where k_1 and k_2 are two non-negative real numbers. The objective of $L(k_1, k_2)$ labeling problem is to find a labeling f of the vertices of G such that the span is minimized, where span $\lambda(G; f)$ of a labeling f is defined as $\max_{u \in V} f(u) - \min_{v \in V} f(v)$. The motivation for $L(k_1, k_2)$ labeling problem for infinite triangular grid T_Δ comes from the fact that a class of frequency assignment problems in cellular networks can be modeled as $L(k_1, k_2)$ labeling problems of T_Δ. Existing bounds on $\lambda(T_\Delta; f)$ for $k_1 \leq k_2$ are based on enumerations through partial computer simulations. In this paper, we attempt to derive $\lambda(T_\Delta; f)$ theoretically by exploring the underlined graph structures without any computer simulations. Specifically, we first select a sub graph of T_Δ which holds certain structural properties and then using these properties, we find the lower bound of span of the concerned subgraph. This finally leads us to obtain the lower bound of $\lambda(T_\Delta; f)$. We establish that $\lambda(T_\Delta; f) \geq 3 + 2h$ when $h < 1/2$ and $\lambda(T_\Delta; f) \geq 4$ when $h \geq 1/2$, where $h = k_1/k_2$. Ours is the first attempt to derive $\lambda(T_\Delta; f)$ theoretically and our obtained results exactly coincide with that of the known bound obtained through computer simulations when $0 \leq h \leq 1/3$. For $h > 1/3$, such known bounds are actually finer than ours.

Keywords $L(k_1, k_2)$ labeling · Infinite triangular grid · Span

---

S. Koley (✉) · S. C. Ghosh
Advanced Computing and Microelectronics unit, Indian Statistical Institute, 203 B. T. Road, Kolkata 700108, India
e-mail: sasthi@isical.ac.in

© The Author(s), under exclusive license to Springer Nature Singapore Pte Ltd. 2022

D. Giri et al. (eds.), *Proceedings of the Seventh International Conference on Mathematics and Computing*, Advances in Intelligent Systems and Computing 1412,
https://doi.org/10.1007/978-981-16-6890-6_44

1 Introduction

Frequency channels are assigned to the transmitters of a cellular network in such a way that interference cannot occur during communication. The objective is to minimize the span where span represents the difference between the largest channel and the smallest channel used for communication. Hale [6] formulated the channel assignment problem as a graph coloring problem, where transmitters are represented as vertices of a graph and there is an edge between two vertices if their corresponding transmitters are interfering with each other. A class of channel assignment problems in radio networks [5, 11–13] can be formulated as an $L(k_1, k_2)$ labeling problem of a graph, where absolute difference between labels assigned to the vertices at graph distances 1 and 2 should be at least k_1 and k_2, respectively, [2, 9, 14–16]. Let us denote the span for $L(k_1, k_2)$ labeling of a graph G as $\lambda(G; k_1, k_2)$. Determining $\lambda(G; k_1, k_2)$ is a well-known NP-complete problem [1]. From the scaling lemma in [3], it follows that $\lambda(G; k_1, k_2) = k_2 * \lambda(G; k_1/k_2, 1)$. We denote $k_1/k_2 = h$ and hence $k_1 \geq k_2$ and $k_1 \leq k_2$ imply $h \geq 1$ and $h \leq 1$, respectively. To determine $\lambda(G; k_1, k_2)$, evaluating $\lambda(G; h, 1)$ is suffice as $\lambda(G; k_1, k_2)$ can be obtained by proper scaling of $\lambda(G; h, 1)$ using the scaling lemma stated in [3]. Thus, from now onward, our discussion will be based on $L(h, 1)$ labeling. We will use $\lambda(G; f)$ instead of $\lambda(G; h, 1)$ where f is an $L(h, 1)$ labeling of G.

In the context of cellular networks, $L(h, 1)$ labeling problem with $h \geq 1$ is widely studied [1, 2]. This is because practically at bigger distance, effect of interference is less than that of smaller distance, implying $h \geq 1$. However, in ad hoc network and device to device (D2D) communication, two proximity users can communicate with each other without the need of a base station. If two devices can sense each other (distance 1 apart in this context), proper parameter settings of multiple input multiple output (MIMO) antenna can be done so that the required frequency gap between those transmitting devices may be made shorten. If two transmitting devices cannot sense each other (distance 2 apart in this context), then due to the hidden terminal problem, the required frequency gap between them may be higher than that of the required gap when the devices can sense each other. Here, study of $L(h, 1)$ labeling with $h \leq 1$ is important. Therefore, calculating $\lambda(G; f)$ for $L(h, 1)$ labeling when $h \leq 1$ is also important from both theoretical and practical point of view. Infinite triangular lattice (T_Δ) is a form of regular graph which is widely used for modeling the channel assignment problem in cellular network [7, 8]. The problem of finding $\lambda(G; f)$ for $L(h, 1)$ labeling with $h \leq 1$ for T_Δ has been studied by several authors. Griggs and Jin [3] and Kral and Skoda [4] determined $\lambda(G; f)$ for $L(h, 1)$ labeling of T_Δ with

$0 \leq h \leq 1$ for different values of h. Their respective lower bounds λ_1 [3] and λ_2 [4] are as stated below.

$$\lambda_1 = \begin{cases} = 2h+3, & if\ 0 \leq h \leq 1/3, \\ \in [2h+3, 11h], & if\ 1/3 \leq h \leq 9/22, \\ \in [2h+3, 9/2], & if\ 9/22 \leq h \leq 3/7, \\ \in [9h, 9/2], & if\ 3/7 \leq h \leq 1/2, \\ \in [9/2, 16/3], & if\ 1/2 \leq h \leq 2/3, \\ \in [16/3, 23/4], & if\ 2/3 \leq h \leq 3/4, \\ \in [23/4, 6], & if\ 3/4 \leq h \leq 4/5, \\ = 6 & if\ 4/5 \leq h \leq 1. \end{cases} \qquad \lambda_2 = \begin{cases} 2h+3, & if\ 0 \leq h \leq 1/3, \\ 11h, & if\ 1/3 \leq h \leq 3/8, \\ 3h+3, & if\ 3/8 \leq h \leq 2/5, \\ 8h+1, & if\ 2/5 \leq h \leq 3/7, \\ h+4, & if\ 3/7 \leq h \leq 1/2, \\ 9h, & if\ 1/2 \leq h \leq 4/7, \\ 2h+4, & if\ 4/7 \leq h \leq 2/3, \\ 8h, & if\ 2/3 \leq h \leq 5/7, \\ h+5, & if\ 5/7 \leq h \leq 3/4, \\ 5h+2, & if\ 3/4 \leq h \leq 4/5, \\ 6 & if\ 4/5 \leq h \leq 1. \end{cases}$$

The bound λ_1 [3] is obtained by considering all possible $L(h, 1)$ labeling of three induced sub graphs of T_Δ having 7, 19, and 37 nodes using computer simulation. Later, the improved bound λ_2 [4] is also obtained through brute force computer simulations on the induced subgraphs of T_Δ having 81, 100, 169, and 225 nodes. In our work, we attempt to determine $\lambda(G; f)$ for $L(h, 1)$ labeling of T_Δ with $h \leq 1$ by examining the underlined graph structures. We have established that $\lambda(T_\Delta; f) \geq 3 + 2h$ when $h < 1/2$ and $\lambda(T_\Delta; f) \geq 4$ when $h \geq 1/2$. This result exactly coincides with the result obtained in [3, 4] through computer simulations when $h \leq 1/3$. When $h > 1/3$ results obtained in [3, 4] through computer simulations are finer than ours. In our approach, we first select a sub graph of T_Δ which holds certain structural properties and then using these properties, we find the lower bound of span of the concerned sub graph. It finally leads us to obtain the said lower bounds of $\lambda(T_\Delta; f)$. Thus, we made the first attempt to derive $\lambda(T_\Delta; f)$ theoretically without any computer simulations. We introduce the notion of *color classes* where all colors in $[k, k+1)$ are represented as a single color class $C_k, k \in \mathbb{N}$. Our approach is based on assignment of color classes instead of dealing with assignment of colors. This notion essentially eliminates the need for enumerations. Rather consideration of few cases will suffice.

2 Problem Statement and Key Ideas

Definition 1 For a graph $G(V, E)$ and a real number h $(0 \leq h \leq 1)$, an $L(h, 1)$ labeling of G is a labeling $f : V \to \mathbb{R}$ of the vertices of G with non-negative real numbers such that

- $|f(u) - f(v)| \geq h$, if $d(u, v) = 1$,
- $|f(u) - f(v)| \geq 1$, if $d(u, v) = 2$.

where $d(u, v)$ is the shortest distance (in terms of number of edges) between u and v in G.

Throughout the manuscript, by distance we mean to say shortest distance.

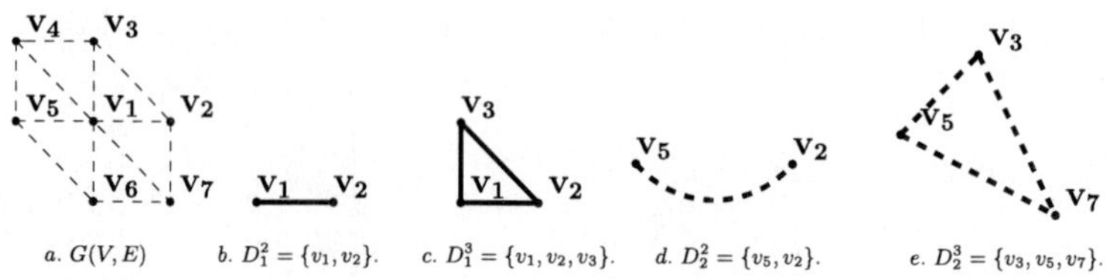

Fig. 1 $G(V, E)$ and its different D_k^ns

Definition 2 The span $\lambda(G; f)$ of a labeling $f : V \to \mathbb{R}$ of G is defined as $\max_{u \in V} f(u) - \min_{v \in V} f(v)$.

Throughout the manuscript, we use minimum label as 0. So the span of a labeling essentially indicates the maximum label used by the corresponding labeling. The objective of $L(h, 1)$ labeling is to find a labeling f' so that $\lambda(G; f') = \min_f \lambda(G; f)$. From now onwards, $f(v)$ will be alternatively termed as *a color* or *a label* of vertex v.

Definition 3 Suppose $G(V, E)$ is a graph. A subset V' of V is said to be a D_k^n, a distance k clique of size n, if $|V'| = n$ and every pair of vertices in V' is at distance k from each other in G, where $k \geq 1$ and $n \geq 2$ are two integers.

We have shown a graph G and its different D_k^n in Fig. 1. In this figure, a dotted edge between two vertices represents that they are at distance 2 apart from each other in G.

We introduce the notion of *color class* to partition the available colors into a set of disjoint color classes. Let us formally define the color class as follows:

Definition 4 A color class $C_k = \{i : k \leq i < k + 1, i \in \mathbb{R}, k \in \mathbb{N}\}$.

Note that the absolute difference between any two colors in a particular color class is strictly less than 1. Let f be an $L(h, 1)$ labeling of G, where $0 \leq h \leq 1$. Now the following facts are immediate for any pair of vertices u and v in G.

- When $d(u, v) = 2$, then $|f(u) - f(v)| \geq 1$, so $f(u)$ and $f(v)$ must belong to two different color classes.
- When $d(u, v) = 1$ then $|f(u) - f(v)| \geq h$, so $f(u)$ and $f(v)$ may or may not belong to the same color class.

Let u and v be two vertices in G and f be an $L(h, 1)$ labeling of G. We say that $\mathbf{u} \sim \mathbf{v}$ if $f(u)$ and $f(v)$ belong to the same color class and $\mathbf{u} \nsim \mathbf{v}$ if $f(u)$ and $f(v)$ do not belong to the same color class.

Definition 5 Let $G(V, E)$ be a graph and f be a $L(h, 1)$ labeling of it. Let $\{v_1, v_2, v_3\} \subseteq V$ be a D_1^3. We term D_1^3 as a monochromatic triangle if $f(v_1)$, $f(v_2)$, and $f(v_3)$ belong to the same color class.

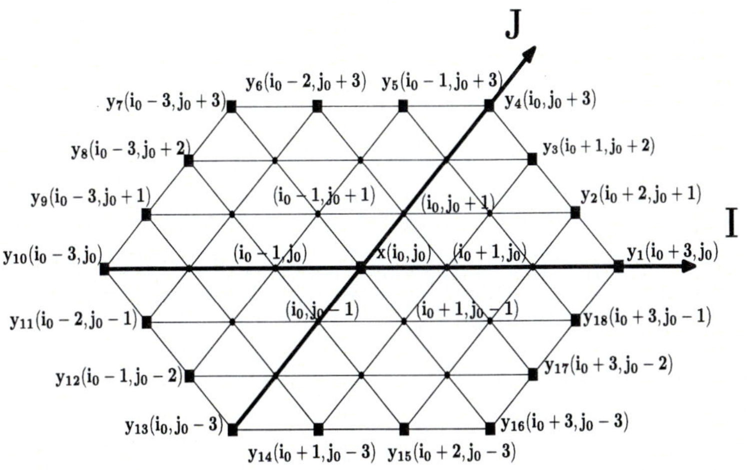

Fig. 2 Triangular lattice T_Δ

Suppose an $L(h, 1)$ labeling f of the D_1^2 shown in Fig. 1b produces $f(v_1) = 0$ and $f(v_2) = h$. If $h < 1$, then the edge (v_1, v_2) is said to be a *monochromatic edge* with color class C_0 as $0,\ h \in C_0$. Similarly, suppose an $L(h, 1)$ labeling f of the D_1^3 shown in Fig. 1c produces $f(v_1) = 0$, $f(v_2) = h$, and $f(v_3) = 2h$. If $h < 1/2$, then the triangle D_1^3 is said to be a *monochromatic triangle* with color class C_0 as $0,\ h,\ 2h \in C_0$.

As we are dealing with $L(h, 1)$ labeling of infinite triangular lattice T_Δ, we now introduce T_Δ and a coordinate system to represent its vertices. T_Δ is a lattice formed by tilling a $2D$ plane regularly with equilateral triangles [17] as shown in Fig. 2. For any vertex in T_Δ, it has 6 neighbors at distance 1, and hence 6 edges incident to v. Angle between any two consecutive edges in anticlockwise direction incident to any vertex v in T_Δ is 60°. Taking any vertex v as origin and two consecutive edges in anticlockwise direction incident to v as two coordinate axes, a coordinate system can be constructed in T_Δ. We follow one such coordinate system with axes (I, J) as shown in Fig. 2. In this coordinate system, for any vertex x with coordinate $(i_0,\ j_0)$, the coordinates of its six neighbors are $(i_0 + 1, j_0)$, $(i_0, j_0 + 1)$, $(i_0 - 1, j_0 + 1)$, $(i_0 - 1, j_0)$, $(i_0, j_0 - 1)$, and $(i_0 + 1, j_0 - 1)$ as shown in Fig. 2. The 18 vertices $y_i \in T_\Delta$, $i \in \{1, 2, \cdots, 18\}$ at distance 3 from x are also shown in Fig. 2. Among them, the 6 vertices y_1, y_4, y_7, y_{10}, y_{13}, and y_{16} are termed *corner* vertices of x. The remaining 12 vertices y_2, y_3, y_5, y_6, y_8, y_9, y_{11}, y_{12}, y_{14}, y_{15}, y_{17}, and y_{18} are termed *non-corner* vertices of x.

3 Results

Consider f as an $L(h, 1)$ labeling of T_Δ. For any two vertices $x, y \in T_\Delta$ with $d(x, y) = 3$, either $x \sim y$ or $x \nsim y$. In Sect. 3.1, we will determine $\lambda(T_\Delta; f)$ when $\mathbf{x} \nsim \mathbf{y}$ for all $x,\ y \in T_\Delta$ with $d(x, y) = 3$ and in Sect. 3.2 we will determine $\lambda(T_\Delta; f)$ when $\mathbf{x} \sim \mathbf{y}$ for at least a pair of vertices $x,\ y \in T_\Delta$ with $d(x, y) = 3$. Then we will state and prove our main result in Theorem 2.

3.1 *Determining* $\lambda(T_\Delta; f)$ *When* $\mathbf{x} \nsim \mathbf{y}$ *for All* $x,\ y \in T_\Delta$ *with* $d(x, y) = 3$.

Figure 3 shows a subgraph G_3 of T_Δ such that $\forall x, y \in G_3$, $d(x, y) \leq 3$. Such a subgraph can be constructed as follows: consider any D_1^3 in T_Δ and then G_3 is the subgraph induced by the vertices of the D_1^3 and all its neighbors. Figure 3 shows the G_3 constructed from $D_1^3 = \{f, g, j\}$.

Lemma 1 *In* G_3*, at most* 3 *vertices can be assigned colors from the same color class and these three vertices must form a* D_1^3 *when* $\mathbf{x} \nsim \mathbf{y}$ *for all* $x,\ y \in G_3$ *with* $d(x, y) = 3$.

Proof Two vertices at distance 2 in T_Δ cannot be assigned with the colors from the same color class. Therefore, for two vertices $v_1,\ v_2 \in G_3$, $v_1 \sim v_2$ only when $d(v_1, v_2) = 1$, as in the case under consideration, $\mathbf{x} \nsim \mathbf{y}$ for all $x,\ y \in T_\Delta$ with $d(x, y) = 3$. Let us consider $v_1 \sim v_2$ and color class of it is C_k. Any vertex in G_3 other than the vertices at distance 1 from both of v_1 and v_2 in G_3 are at distance 2 or 3 from either v_1 or v_2, hence these vertices cannot be assigned with the colors from the color class C_k. Therefore, C_k can be used in v_1, v_2 and the vertex v_3 which is at distance 1 from both of v_1 and v_2. Thus, v_1, v_2 and v_3 form a D_1^3. Any vertex other than v_1, v_2, and v_3 in G_3 is at distance 2 from any one of v_1, v_2, and v_3. Hence, except v_1, v_2, and v_3 any other vertex cannot be assigned with the colors from the color class C_k. Thereby Lemma 1 is proved.

Theorem 1 $\lambda(T_\Delta; f) \geq 4$ *if* $\mathbf{x} \nsim \mathbf{y}$ *for all* $x,\ y \in T_\Delta$ *with* $d(x, y) = 3$.

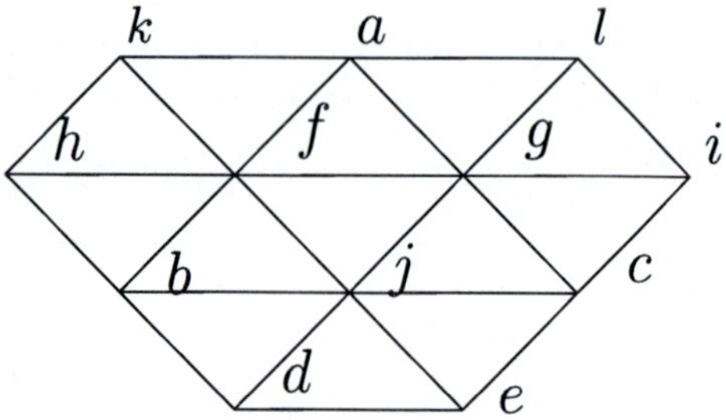

Fig. 3 A subgraph G_3 of T_Δ

Proof Consider the G_3 constructed from $\{f, g, j\}$ as shown in Fig. 3. It is evident that $\lambda(T_\Delta; f) \geq \lambda(G_3; f)$ as G_3 is a subgraph of T_Δ. From Lemma 1, at most three vertices can have colors from the same color class and they must form a D_1^3 when $x \nsim y$ for all $x, y \in T_\Delta$ with $d(x, y) = 3$. Note that any D_1^3 in G_3 must include at least one vertex from $\{f, g, j\}$. So at most three disjoint D_1^3s can be constructed in G_3. Let M_f be a D_1^3 which includes f. There are six such D_1^3s are possible. Consider M_f as any one of them. Any vertex in G_3 other than those belong to M_f is at distance 2 from at least one of the three vertices of M_f. Same thing holds for M_g and M_j as well. Therefore, M_f, M_g, and M_j must get colors from 3 different color classes. Let C_{k_1}, C_{k_2} and C_{k_3} be the color classes of M_f, M_g, and M_j, respectively, where $k_1 \neq k_2 \neq k_3$. So at most 9 vertices can be colored with three color classes. The rest three vertices other than the vertices of M_f, M_g, and M_j cannot form a D_1^3. Hence, at least two of them are at distance 2 or 3 from each other. Two vertices x and y with $d(x, y) = 2$ cannot be colored with the same color class. For the case under consideration, $\mathbf{x} \nsim \mathbf{y}$ for all $x, y \in T_\Delta$ with $d(x, y) = 3$. So these two vertices must have colors from two different color classes. Let C_{k_4} and C_{k_5} be the color classes of those two vertices, where $k_4 \neq k_5$. As any vertex other than M_f is at distance 2 from at least one of the three vertices of M_f, we get $k_4 \neq k_1$ and $k_5 \neq k_1$. The same argument holds for M_g and M_j as well, and hence $k_4 \neq k_2$, $k_4 \neq k_3$, $k_5 \neq k_2$, and $k_5 \neq k_3$. Therefore, at least 5 different color classes are required to color the vertices of G_3. Hence, $\lambda(T_\Delta; f) \geq 4$.

It is now evident that $\lambda(T_\Delta; f) \geq 4$ if $\mathbf{x} \nsim \mathbf{y}$ for all $x, y \in T_\Delta$ with $d(x, y) = 3$. So we will investigate whether $\lambda(T_\Delta; f)$ can be kept below 4 by making $\mathbf{x} \sim \mathbf{y}$ for at least a pair of vertices $x, y \in T_\Delta$ with $d(x, y) = 3$.

3.2 Determining $\lambda(T_\Delta; f)$ When $\mathbf{x} \sim \mathbf{y}$ for at Least a Pair of Vertices $x, y \in T_\Delta$ with $d(x, y) = 3$

Lemma 2 *For two vertices x and y in T_Δ where $d(x, y) = 3$, if $x \sim y$ then at least C_3 must be used to color the vertices of T_Δ.*

Proof Let $x(i, j)$ be any vertex in T_Δ and $f(x) \in C_k$. Now we consider the following two cases.

When y is a corner vertex of x: Let us consider the corner vertex $y(i+3, j)$ with axes (I, J) as shown in Fig. 4 and $f(y) \in C_k$. The three vertices $p_1(i-1, j+2)$, $p_2(i+1, j+2)$ and $p_3(i+1, j)$ form a D_2^3 as shown in Fig. 4. It can be verified that $d(p_1, x) = 2$, $d(p_2, y) = d(p_3, y) = 2$. Therefore, three distinct color classes other than C_k must be required to color the mentioned D_2^3. Hence, 4 distinct color classes must be required to color x, y and the D_2^3 implying at least C_3 must be used here. For other corner vertices of x, coordinates of corresponding D_2^3 can be found and same result can be obtained there also.

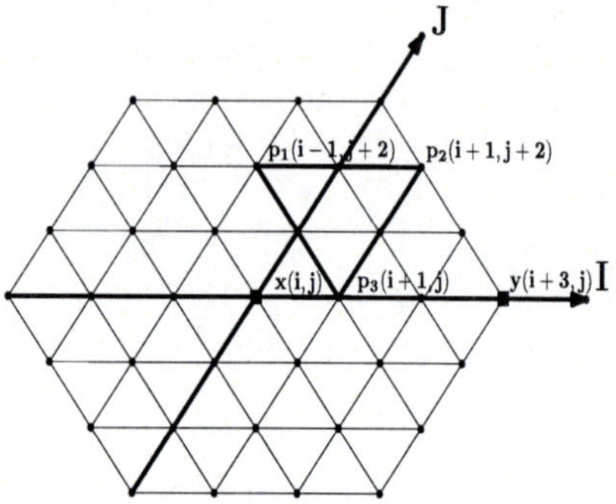

Fig. 4 Construction of a D_2^3 = $\{p_1, p_2, p_3\}$

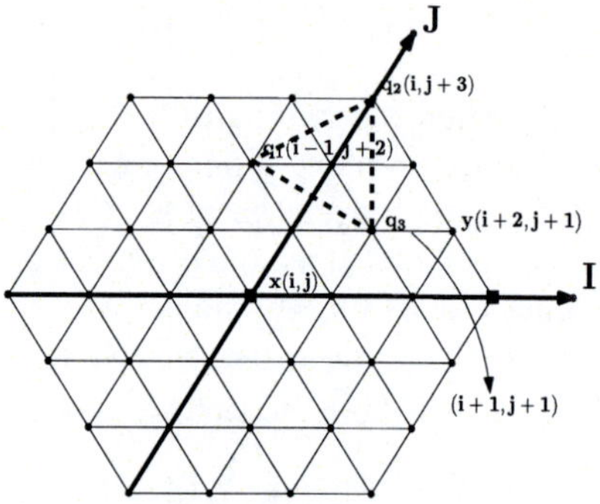

Fig. 5 Construction of a D_2^3 = $\{q_1, q_2, q_3\}$

When y is a non-corner vertex of x: Let us consider the non-corner vertex $y(i + 2, j + 1)$ with axes (I, J) as shown in Fig. 5 and $f(y) \in C_k$. The three vertices $q_1(i - 1, j + 2)$, $q_2(i, j + 3)$, and $q_3(i + 1, j + 1)$ form a D_2^3 as shown in Fig. 5. It can be verified that $d(q_1, x) = d(q_3, x) = 2$, $d(q_2, y) = 2$. Therefore, three distinct color classes other than C_k must be required to color the mentioned D_2^3. Hence, 4 distinct color classes must be required to color x, y and the D_2^3 implying at least C_3 must be used here. For other non-corner vertices of x, co-ordinates of corresponding D_2^3 can be found and same result can be obtained there also. Hence the proof.

From Lemma 2, C_3 must be used in a vertex if $x \sim y$ for a pair of vertices x and y in T_Δ with $d(x, y) = 3$. Therefore, let us consider a vertex $u \in T_\Delta$, where $f(u) \in C_3$.

Lemma 3 *Let us consider a vertex $u \in T_\Delta$ with $f(u) \in C_3$. If $v \nsim u$, $\forall v \in T_\Delta$ where $d(u, v) = 1$, then $\lambda(T_\Delta; f) \geq 4$.*

Proof Consider the sub graph G_6 of T_Δ as shown in Fig. 6. Assume $f(u) \in C_3$ and $f(v_i) \notin C_3$ for all vertices v_i, $i \in \{1, 2, \cdots, 6\}$ with $d(u, v_i) = 1$. Assume $f(v_1) \in C_{k_1}$, $f(v_3) \in C_{k_2}$ and $f(v_5) \in C_{k_3}$. As v_1, v_3, v_5 form a D_2^3, k_1, k_2, and k_3 are all distinct and none of them are 3. Note that $f(w_1) \notin C_3$ as $d(w_1, u) = 2$ and $f(u) \in C_3$. Consider $f(w_1) \in C_k$. Note that $k \notin \{k_1, k_2\}$ as w_1, v_1, v_3 form a D_2^3 and $f(v_1) \in C_{k_1}$, $f(v_3) \in C_{k_2}$. Therefore, $f(w_1) \in C_{k_3}$. So either $f(v_2) \in C_{k_1}$ or $f(v_2) \in C_{k_2}$,

Fig. 6 A subgraph G_6

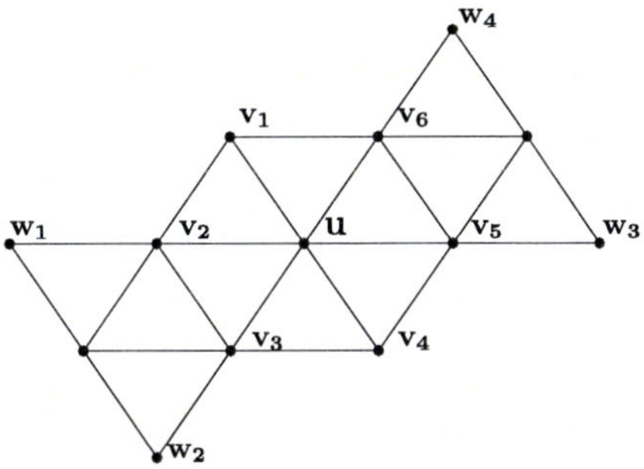

Fig. 7 A subgraph G_7

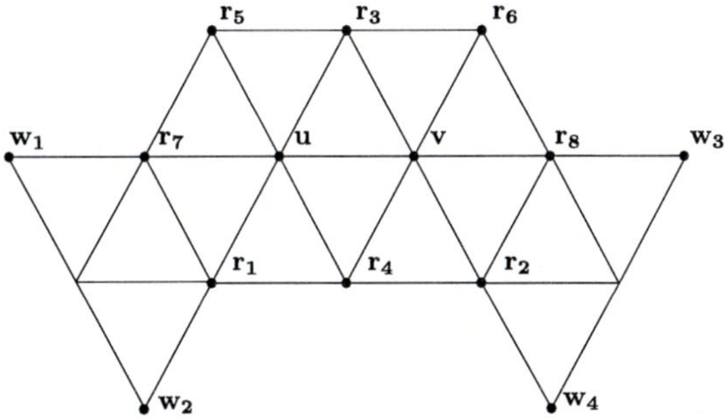

otherwise, $\lambda(G_6; f) \geq 4$. When $f(v_2) \in C_{k_1}$, we get $f(v_6) \in C_{k_3}$ and $f(v_4) \in C_{k_2}$. Note that $d(w_2, u) = d(w_2, v_2) = d(w_2, v_4) = d(w_2, w_1) = 2$. Therefore, $f(w_2) \in C_k$, where $k \notin \{3, k_1, k_2, k_3\}$. Hence, a fifth color class is required for w_2. Hence, $\lambda(G_6; f) \geq 4$ in this case. When $f(v_2) \in C_{k_2}$, arguing similarly, it can be shown that a fifth color class is required for w_4 resulting $\lambda(G_6; f) \geq 4$. Hence $\lambda(T_\Delta; f) \geq 4$.

From Lemma 3, we can conclude that there must exists a monochromatic edge with color class C_3 to keep $\lambda(T_\Delta; f) < 4$. So we now consider the case when for two vertices $u, v \in T_\Delta$ with $d(u, v) = 1$, $u \sim v$ and $f(u) \in C_3$.

Lemma 4 *Let us consider four vertices u,v, w_1, and w_2 in T_Δ where $d(u, v) = 1$, $u \sim v$, $f(u) \in C_3$, $f(v) \in C_3$, $d(w_1, u) = d(w_1, v) = 1$, and $d(w_2, u) = d(w_2, v) = 1$. If neither $w_1 \nsim u$, $w_1 \nsim v$ nor $w_2 \nsim u$, $w_2 \nsim v$ then $\lambda(T_\Delta; f) \geq 4$.*

Proof Consider the sub graph G_7 of T_Δ as shown in Fig. 7. Assume $f(u) \in C_3$ and $f(v) \in C_3$. As $d(r_3, u) = d(r_3, v) = 1$ and $d(r_4, u) = d(r_4, v) = 1$ we assume $f(r_3) \notin C_3$ and $f(r_4) \notin C_3$. Note that $f(r_1) \notin C_3$ and $f(r_2) \notin C_3$ as $d(r_1, v) = 2$ and $d(r_2, u) = 2$ respectively. Assume $f(r_1) \in C_{k_1}$, $f(r_2) \in C_{k_2}$ and $f(r_3) \in C_{k_3}$. As r_1, r_2, and r_3 form a D_2^3, k_1, k_2, and k_3 are all distinct and none of them are 3. Note that $f(r_4) \notin C_3$ by our assumption. Again, $f(r_4) \notin C_{k_3}$ as $d(r_4, r_3) = 2$ and $f(r_3) \in C_{k_3}$. So, either $f(r_4) \in C_{k_1}$ or $f(r_4) \in C_{k_2}$, otherwise, $\lambda(G_7; f) \geq 4$. First consider $f(r_4) \in C_{k_1}$. In that case, $f(r_5) \in C_{k_2}$ and $f(r_6) \in C_{k_3}$ as r_4, r_5 and r_6 form a D_2^3 and $f(r_2) \in C_{k_2}$. Now $f(w_1) \notin C_3$ as $d(w_1, u) = 2$ and $f(u) \in C_3$. Again, $f(w_1) \notin C_{k_1}$

and $f(w_1) \notin C_{k_2}$, as w_1, r_1, r_5 form a D_2^3 with $f(r_1) \in C_{k_1}$ and $f(r_5) \in C_{k_2}$. Therefore, $f(w_1) \in C_{k_3}$. Observe that $f(r_7) \in C_{k_2}$ because $f(r_7) \notin C_3$ as $d(r_7, v) = 2$, $f(v) \in C_3$; $f(r_7) \notin C_{k_1}$ as $d(r_7, r_4) = 2$, $f(r_4) \in C_{k_1}$; $f(r_7) \notin C_{k_3}$ as $d(r_7, r_3) = 2$, $f(r_3) \in C_{k_3}$. Note that $d(w_2, u) = d(w_2, r_4) = d(w_2, r_7) = d(w_2, w_1) = 2$. Therefore, $f(w_2) \in C_k$, where $k \notin \{3, k_1, k_2, k_3\}$. Hence, a fifth color class is required for w_2. Hence, $\lambda(G_7; f) \geq 4$ in this case. When $f(r_4) \in C_{k_2}$, arguing similarly, it can be shown that a fifth color class is required for w_4 resulting $\lambda(G_7; f) \geq 4$. Hence, $\lambda(T_\Delta; f) \geq 4$.

From Lemma 4, we can conclude that there must exists a monochromatic triangle with color class C_3 to keep $\lambda(T_\Delta; f) < 4$. Now we have the following theorem.

Theorem 2 $\lambda(T_\Delta; f) \geq 3 + 2h$ *when* $h < 1/2$ *and* $\lambda(T_\Delta; f) \geq 4$ *when* $h \geq 1/2$.

Proof If $x \nsim y$ for all pair of vertices $x,\ y \in T_\Delta$ with $d(x, y) = 3$, then from Theorem 1, we get $\lambda(T_\Delta; f) \geq 4$. If $x \sim y$ for at least one pair of vertices $x,\ y \in T_\Delta$ with $d(x, y) = 3$, from Lemma 2, it follows that C_3 must be used in T_Δ to keep $\lambda(T_\Delta; f) < 4$. Lemmas 3 and 4 state that there must exit a monochromatic edge and a monochromatic tringle with color class C_3 to keep $\lambda(T_\Delta; f) < 4$. Hence, $\lambda(T_\Delta; f) \geq 3 + 2h$ when $h < 1/2$ and $\lambda(T_\Delta; f) \geq 4$ when $h \geq 1/2$.

4 Conclusions

In this work, we derived $\lambda(T_\Delta; f)$ theoretically by exploring the underlined graph structures without any computer simulation. We introduced the notion of color class which essentially eliminates the need for enumerations and leads us to arrive at few cases. Our bound exactly coincides with that of the known bound obtained through computer simulations when $0 \leq h \leq 1/3$. For $h > 1/3$, existing bounds obtained through simulation are finer than ours. Our future research scope includes identifying appropriate subgraphs and accordingly obtaining tighter bounds by dividing the intervals into finer sub-intervals, when $h > 1/3$.

References

1. Calamoneri T (2006) Optimal L(h, k)-labeling of regular grids. Disc Math Theor Comput Sci 8:141–158
2. Calamoneri T (2011) The L(h, k)-labelling problem: an updated survey and annotated bibliography. Comput J 54(8):1344–1371
3. Griggs JR, Jin XT (2008) Real number channel assignment for lattices. Siam J Disc Math 22(3):996–1021
4. Kral D, Skoda P (2008) Bounds for the real number graph labellings and application to labellings of the triangular lattices. Siam J Disc Math 22(4):1559–1569
5. http://mathworld.wolfram.com/LabeledGraph.html

6. Hale WK (1980) Frequency assignment: theory and applications. Proc IEEE 68:1497–1514
7. Bertossi AA, Pinotti CM (2006) Channel assignment with separation in wireless networks based on regular plane tessellations. In: Proceedings of the international workshop on information security in wireless networks (ISWN)
8. Duan ZM, Lv P, Miao L, Miao Z (2010) Bounds of channel assignment problem for cellular networks modelled as triangular grids. In: Proceedings of 2nd IEEE international conferences on network security, wireless communications and trusted computing (NSWCTC), pp 20–23
9. Yeh RK (2006) A survey on labeling graphs with a condition at distance two. Disc Math 306(12):1217–1231
10. Griggs JR, Kral D (2009) Graph labellings with variable weights, a survey. Disc Appl Math 157(12):2646–2658
11. Audhya GK, Sinha K, Ghosh SC, Sinha BP (2011) A survey on the channel assignment problem in wireless networks. Wirel Commun Mob Comput 11(5):583–609
12. Aardal KI, van Hoesel SPM, Koster AMCA, Mannino C, Sassano A (2007) Models and solution techniques for frequency assignment problems. Ann Oper Res 153(1):79–129
13. Heuvel JVD, Leese RA, Shepherd MA (1998) Graph labeling and radio channel assignment. J Graph Theory 29(4):263–283
14. Griggs JR, Yeh RK (1992) Labelling graphs with a condition at distance 2. SIAM J Disc Math 5(4):586–595
15. Zhang X, Yang H, Li H (2017) L(2,1)-labeling of the brick product graphs. J Appl Math Phys 5(8):1529–1536
16. Paul S, Pal M, Pal A (2015) $L(2, 1)$-labeling of interval graph. J Appl Math Comput 49(1):419–432
17. http://mathworld.wolfram.com/TriangularGrid.html

The Spanning Maximal Planar Subgraph Problem for Some Classes of Complete 4-Partite Graphs

Tjaart Jan B. Estrada and Isagani B. Jos

Abstract Let T be a spanning subgraph of a finite undirected graph G. T is called a *spanning maximal planar subgraph* of G if it is also a maximal planar graph, that is, each of its regions is bounded by three edges. In this paper, we focus our attention on some classes of complete 4-partite graphs and determine when these graphs contain the spanning maximal planar subgraph T. In particular, we present some results on the spanning maximal planar subgraph problem for complete 4-partite graphs of the form $K_{1,1,1,z}$ and $K_{1,1,y,z}$ and utilize these results to attack the problem for general $K_{w,x,y,z}$ given some initial conditions for y and z.

Keywords Complete 4-partite graph · Maximal planar graph · Spanning maximal planar subgraph

1 Introduction

A graph is said to be a *planar graph* if there is a drawing of the graph on the plane such that there are no edge crossings. Otherwise, the graph is non-planar. Some examples of named graphs that are planar are paths, cycles, and trees. When a planar graph is drawn, the plane is divided into non-overlapping, open regions. This includes an infinite, exterior region. A well-known property of a planar graph was presented by Euler in 1754, and his proof was later on corrected by Legendre in 1794. It states the relationship between the order, size, and number of regions of any planar graph.

Supported by the Department of Science and Technology—Science Education Institute.

T. J. B. Estrada (✉)
Don Mariano Marcos Memorial State University, 2504 Agoo La Union, Philippines
e-mail: tjestrada@dmmmsu.edu.ph

I. B. Jos
De La Salle University, 2401 Taft Avenue, 0922 Manila, Philippines
e-mail: isagani.jos@dlsu.edu.ph

© The Author(s), under exclusive license to Springer Nature Singapore Pte Ltd. 2022

D. Giri et al. (eds.), *Proceedings of the Seventh International Conference on Mathematics and Computing*, Advances in Intelligent Systems and Computing 1412,
https://doi.org/10.1007/978-981-16-6890-6_45

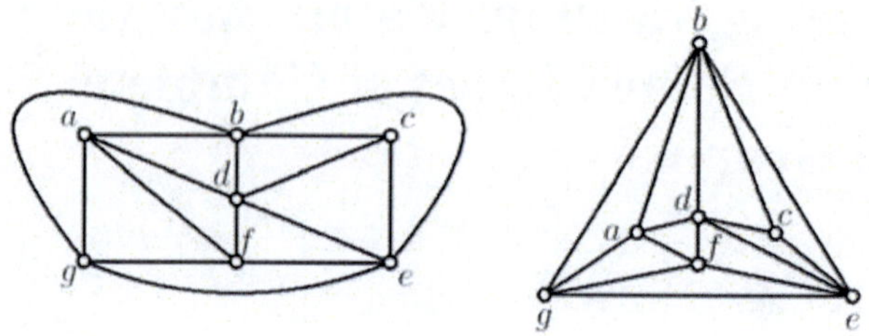

Fig. 1 Two drawings of a maximal planar graph

Theorem 1 (Euler) *If G is a connected planar graph with order v, size e, and has r regions, then*

$$v - e + r = 2$$

A special type of planar graph is a maximal planar graph. It has the property that any of its regions, including the exterior region, is triangular, that is, it is bounded by three edges. Maximal planar graphs can be drawn in such a way that the exterior region is bounded by a large triangle (as in Fig. 1), called the exterior boundary. Any region of a maximal planar graph, in fact, can be the exterior region by choosing its boundary to be this triangle. This becomes evident if the graph is drawn on a sphere, with the distances between the vertices being equal. Except for the complete graph K_3, any maximal planar graph contains only vertices with degree 3 or more. Also, adding an edge in a maximal planar graph causes the graph to be non-planar.

Two graphs known to be non-planar are $K_{3,3}$ and K_5. These two graphs play an important role in the study of graph planarity as they are used to invoke Kuratowski's Theorem [6]. Any graph containing a non-planar graph is also non-planar, and so the complete bipartite graph $K_{m,n}$ is non-planar for $m, n \geq 3$ since it contains the graph $K_{3,3}$.

Another important result on planar graphs was published by Euler, which can be proven using Theorem 1.

Theorem 2 (Euler) *If G is a planar graph with order v and size e, then*

$$e \leq 3v - 6$$

Theorem 2, when stated in the contrapositive, can be a useful tool to show that a graph is non-planar. Two maximal planar graphs of the same order will have same size as an implication of this theorem. However, it does not necessarily follow that the two are isomorphic as there can be more than one possible configuration. Moreover, a consequence of Theorem 2 is that any planar graph must have a vertex with degree at most 5. This is used to derive another important result in the study of maximal planar graphs.

Theorem 3 *If G is a maximal planar graph with four or more vertices, then either* $\delta(G) = 3$, $\delta(G) = 4$, *or* $\delta(G) = 5$.

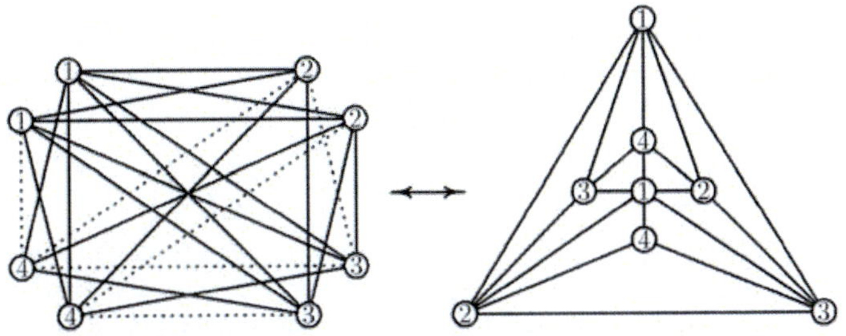

Fig. 2 Deleting edges of $K_{2,2,2,2}$ to obtain a spanning maximal planar subgraph

The minimum degree of a maximal planar graph must be considered, since each of the three possibilities induces different configurations of a maximal planar graph of a given order. The discussions of some results in this paper consider these possibilities.

Now, consider the complete 4-graph $K_{2,2,2,2}$. Is it possible to delete a number of edges from $K_{2,2,2,2}$ and obtain a spanning subgraph that is also a maximal planar graph? Figure 2 shows the existence of such a subgraph, which we refer to as a *spanning maximal planar subgraph*. A vertex labeled as 1, 2, 3, or 4 correspond to membership in V_1, V_2, V_3, or V_4, respectively. The problem of finding this subgraph, if it exists, belongs to the class of graph planarization problems. Resende and Ribeiro [9] discuss some problems in this class together with some applications. Some algorithms have also been developed to solve graph planarization problems, one of which is discussed by Djidjev [6].

In 2019, Almonte, et al. [1] tackled the spanning maximal planar subgraph problem in complete k-partite graphs, particularly for complete tripartite graphs, $K_{x,y,z}$. The main result found was a necessary and sufficient condition in order for a spanning maximal planar subgraph to exist in $K_{x,y,z}$. This also coincides with a result in the paper of Nakamoto, et al. [8].

The results presented in this paper take a step further on the spanning maximal planar subgraph problem for complete k-partite graphs. In particular, the class of complete 4-partite graphs of the form $K_{1,1,1,z}$ and $K_{1,1,y,z}$ is given much attention. It is shown that for $K_{1,1,1,z}$ and $K_{1,1,y,z}$, a spanning maximal planar subgraph will exist if and only if $z \in \{1, 2\}$ and $z - y \leq 1$, respectively. The spanning maximal planar subgraphs of these families of complete 4-partite graphs were utilized to construct spanning maximal planar subgraphs for more general complete 4-partite graphs given some initial conditions for y and z.

2 Results and Discussion

Throughout the discussion, we will abbreviate "spanning maximal planar subgraph" as "SMPS." Also, WLOG, assume that $w \leq x \leq y \leq z$, for all complete 4-partite graphs. We will first present the results for two families, $K_{1,1,1,z}$ and $K_{1,1,y,z}$. The vertex labels 1, 2, 3, and 4 denotes membership to partite sets V_1, V_2, V_3, and V_4, respectively. In cases where there are many vertices in a graph, the colors blue, red, green, and cyan will be used instead to denote memberships.

Theorem 4 *$K_{1,1,1,z}$ has an SMPS if and only if $z = 1$ or $z = 2$.*

Proof Suppose that $K_{1,1,1,z}$ has an SMPS T. Since T is a maximal planar graph of order $3 + z$, it must have size $e = 3(3 + z) - 6 = 3z + 3$ from Theorem 2. Note that $K_{1,1,1,z}$ also has size $3z + 3$, therefore no edge needs to be deleted from $K_{1,1,1,z}$ to obtain T. Hence $T = K_{1,1,1,z}$ since T is a spanning subgraph of $K_{1,1,1,z}$. Now if $z \geq 3$, $K_{3,3}$ is a subgraph, and so T will be non-planar. Therefore either $z = 1$ or $z = 2$.

If $z = 1$ or $z = 2$, then the given graphs are $K_{1,1,1,1}$ or $K_{1,1,1,2}$, respectively. Both of these graphs are SMPS themselves. □

The main result for the family $K_{1,1,y,z}$ states a relationship between y and z in order for an SMPS to exist. We require four lemmas to prove this result. Each lemma describes a structure of an SMPS of $K_{1,1,y,z}$

Lemma 1 *If T is an SMPS of $K_{1,1,y,z}$, then for any $x \in V_3 \cup V_4$, x is adjacent to both v_1 and v_2, where $V_1 = \{v_1\}$ and $V_2 = \{v_2\}$.*

Proof There are two cases that need to be considered.

Case 1: *x' is not adjacent to either v_1 or v_2.* Since T is a maximal planar graph, x' must belong to a set $\{x', y, y'\}$ that induce one triangular region in T, where $y, y' \in V_3 \cup V_4$. But this is impossible since $x' \in V_3 \cup V_4$ and is adjacent to both y and y', that is, there exists a pair of adjacent vertices that belong to the same partite set. In particular, the vertices in $\{x', y, y'\}$ are labeled 3 and 4 only and these vertices are in the boundary of a triangular region, a contradiction.

Case 2: *x' is adjacent to exactly one of v_1 and v_2.* WLOG, suppose $x' \in V_3$ and is not adjacent to v_1 and adjacent to v_2. Now from Theorem 3, the minimum degree in T can either be 3, 4, or 5. Hence x' is adjacent to at least 3, 4, or 5 vertices. This implies that x' is adjacent to v_2 and other vertices $u_1, \ldots, u_k$, where $k \geq 2$, $k \geq 3$, or $k \geq 4$ according to whether x' is adjacent to at least 3, 4 or 5 vertices, respectively. In any of these cases, we have that $u_1, \ldots, u_k \in V_4$ since $V_2 = \{v_2\}$. However the set of vertices $\{v_2, u_1, u_2, \ldots, u_k\}$ forms a $(k + 1)$-cycle in T, which implies that the set of vertices $\{u_1, u_2, \ldots, u_k\}$ induce a path in T whose vertices all belong to V_4. This is a contradiction since V_4 is a partite set. □

Lemma 2 *Let T be an SMPS of $K_{1,1,y,z}$. If there is a vertex $x \in V_3 \cup V_4$ in T such that $\deg(x) \neq 4$, then the edge v_1v_2, where $V_1 = \{v_1\}$ and $V_2 = \{v_2\}$, must be in $E(T)$.*

Proof Assume otherwise, that v_1 and v_2 are not adjacent in T. We consider the cases where the degree of v_1 is odd or even in T.

Case 1: *$deg(v_1) = k$ in T, where k is odd.* Since v_1 and v_2 are not adjacent, v_1 is adjacent with vertices in V_3 and V_4 only. Let $N(v_1) = \{c_1, c_2, \ldots, c_k\}$, k odd. Then there exists a planar drawing of T such that $(c_1, c_2, \ldots, c_k, c_1)$ is a k-cycle induced by $N(v_1)$. Further, $(c_1, c_2, \ldots, c_k, c_1)$ is a k-cycle containing vertices that alternate membership in V_3 and V_4. This is not possible since k is odd. In particular, there is pair of adjacent vertices in $\{c_1, c_2, \ldots, c_k\}$ that both belong to either V_3 or V_4.

Case 2: *$deg(v_1) = k$ in T, where k is even.* As in Case 1, there exists a k-cycle consisting of vertices in V_3 and V_4 induced by $N(v_1)$ given by $(c_1, c_2, \ldots, c_k, c_1)$, k even. Then the set $\{c_1, c_2, \ldots, c_k\}$ is a set of vertices in T that alternate membership in V_3 and V_4. Now since T is a maximal planar graph, every edge belongs to the boundary of exactly two triangular regions. Note that the edges of the form $c_i c_{i+1}$ already belong to the boundary of one triangular region R_i that is induced by the set of vertices $\{v_1, c_i, c_{i+1}\}$, where $1 \leq i \leq k$ and the subscript $i + 1$ is taken modulo k from the set $\{1, 2, \ldots, k\}$. Since c_i and c_{i+1} are in V_3 and V_4, they must both be adjacent to v_2 in view of Lemma 1, so that the edge $c_i c_{i+1}$ is now in the boundary of another triangular region R'_i induced by $\{v_2, c_i, c_{i+1}\}$. Observe that each region R_i and R'_i has both c_i and c_{i+1} as boundary vertices, and these vertices are in V_3 and V_4. Hence, there are no other vertices in V_3 and V_4 aside from the vertices in $\{c_1, c_2, \ldots, c_k\}$, that is, $V_3 \cup V_4 = \{c_1, c_2, \ldots, c_k\}$. Therefore for each $x \in V_3 \cup V_4$, we have $\deg(x) = 4$, since x is adjacent to v_1, v_2 and two other vertices in the cycle $(c_1, c_2, \ldots, c_k, c_1)$. However, this contradicts the assumption of the lemma. □

The proof of Lemma 2 Case 2, results to a configuration of an SMPS T of $K_{1,1,y,z}$, with the property that $\deg(x) = 4$ for all $x \in V_3 \cup V_4$, and $v_1 v_2 \notin E(T)$. This configuration will also be considered in proving our main result for the SMPS of $K_{1,1,y,z}$. Presented in Fig. 3 are some SMPS for $K_{1,1,2,2}$, $K_{1,1,3,3}$, and $K_{1,1,4,4}$, obtained by performing the steps elaborated in the proof of Lemma 2 Case 2, with $k = 4$, $k = 6$, and $k = 8$, respectively. A vertex colored in cyan, green, blue, or red corresponds to the vertex' membership to V_1, V_2, V_3 or V_4, respectively.

Lemma 3 *Let T be an SMPS of $K_{1,1,y,z}$. If there is a vertex $x \in V_3 \cup V_4$ in T such that* $\deg(x) \neq 4$, *then* $\delta(T) = 3$.

Proof From Theorem 3, any maximal planar graph G has minimum degree either $\delta(G) = 3$, $\delta(G) = 4$, or $\delta(G) = 5$. We will prove this lemma by showing that the cases where $\delta(T) = 4$ and $\delta(T) = 5$ are impossible.

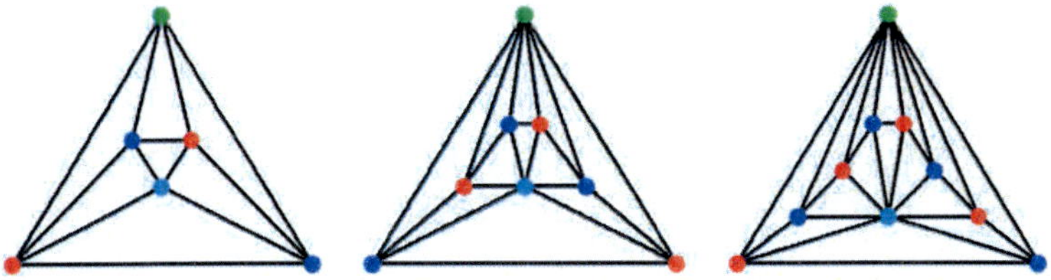

Fig. 3 Some SMPS of $K_{1,1,y,z}$, resulting from the proof of Lemma 2, Case 2

Case 1: $\delta(T) = 5$. Let a be the vertex with $deg(a) = 5$, and suppose that b, c, d, e, f are the neighbors of a. Then the set of vertices $\{b, c, d, e, f\}$ forms a 5-cycle in T. WLOG, let (b, c, d, e, f, b) be this 5-cycle. There are two subcases under Case 1: $a \notin V_1 \cup V_2$, and $a \in V_1 \cup V_2$.

Suppose that $a \notin V_1 \cup V_2$. Thus two vertices in $\{b, c, d, e, f\}$ are in $V_1 \cup V_2$. These two vertices are adjacent by Lemma 2. WLOG let $a \in V_3$. Then the remaining three vertices in $\{b, c, d, e, f\}$ must be in V_4. This is impossible since these three vertices form a path in V_4, which is a partite set. The other subcase can be shown to be impossible in a similar manner.

Case 2: $\delta(T) = 4$. Suppose that $deg(a) = 4$. Let b, c, d, and e be the vertices adjacent to a that form a 4-cycle. WLOG, let this 4-cycle be (b, c, d, e, b). Two subcases need to be considered for Case 2: $a \notin V_1 \cup V_2$, and $a \in V_1 \cup V_2$.

Assume that $a \notin V_1 \cup V_2$. Thus two vertices in $\{b, c, d, e\}$ must be in $V_1 \cup V_2$. By Lemma 2, these two vertices are adjacent. WLOG, assume that $a \in V_3$. Hence the two other adjacent vertices in $\{b, c, d, e\}$ must be in V_4, a contradiction since V_4 is a partite set.

Now suppose that $a \in V_1 \cup V_2$. Thus one vertex in $\{b, c, d, e\}$ is in $V_1 \cup V_2$, say b. Hence (c, d, e) is a path in T with vertices that alternate membership in V_3 and V_4. WLOG, suppose that $c \in V_3$ and $d \in V_4$. By assumption, we have $\delta(T) = 4$, and so there is at least one other vertex adjacent with c aside from a, b, and d. Let $c_1, c_2, \ldots, c_k$, be the other vertices adjacent with c. Now the set $\{a, d, c_1, \ldots, c_k, b\}$ forms a $(k+3)$-cycle. Thus the set $\{c_1, c_2, \ldots, c_k\}$ forms a path whose vertices belong to the same partite set, a contradiction.

Hence it follows that given an SMPS T of $K_{1,1,y,z}$, $\delta(T) = 3$. □

Lemma 4 *Let T be an SMPS of $K_{1,1,y,z}$. If there is a vertex $x \in V_3 \cup V_4$ in T such that* $\deg(x) \neq 4$*, then the set of all vertices in $V_3 \cup V_4$ forms a path in T. Further, each vertex in the path is adjacent to $v_1 \in V_1$ and $v_2 \in V_2$.*

Proof If the order of T is 4, then the conclusion follows clearly since T is $K_{1,1,1,1}$. Suppose that the order of T is greater than 4, hence there are at least three vertices in $V_3 \cup V_4$. By Lemma 1, all vertices in $V_3 \cup V_4$ are each adjacent to v_1 and v_2, and from Lemma 2, $v_1 v_2 \in E(T)$. Hence $\deg(v_1) \geq 4$ and $\deg(v_2) \geq 4$. Now by Lemma 3, there is a vertex in T with degree 3, say u_1, which must be in $V_3 \cup V_4$. Let u_2 be the third vertex adjacent with u_1. Therefore there exists a tetrahedron τ in T that is induced by the set of vertices $\{v_1, v_2, u_1, u_2\}$ where $v_1 \in V_1$, $v_2 \in V_2$, and $u_1, u_2 \in V_3 \cup V_4$.

Let $u_n \in V_3 \cup V_4$, $n > 2$, be a vertex outside τ. WLOG, assume that $\{v_1, v_2, u_n\}$ forms the exterior boundary of T. Suppose that $deg(u_n) > 3$. Then $N(u_n)$ forms a cycle C. However in C, v_1 and v_2 are adjacent, and they are the only vertices in V_1 and V_2, respectively. Therefore, the rest of the vertices in the cycle must belong to the same partite set, which is impossible. Hence $deg(u_n) = 3$. Let u_{n-1} be the third vertex adjacent to u_n which lies outside the tetrahedron τ.

Now we observe the rest of the vertices lying outside the tetrahedron τ and inside the exterior boundary of T. There exists a vertex $u_3 \in V_3 \cup V_4$ that is adjacent to u_2.

If there is no such vertex that is adjacent to u_2, a quadrangular region is formed since u_3 lies outside the tetrahedron τ and that u_3 is adjacent to v_1 and v_2.

Also, there exists u_4 outside τ that is adjacent to u_3. Similarly, if u_4 is not adjacent to u_3, a quadrangular region is formed since u_4 is adjacent to v_1 and v_2. Note that u_4 cannot be inside the region induced by $\{v_1, u_2, u_3\}$ or $\{v_2, u_2, u_3\}$ since an edge crossing will result. Hence it lies outside the separating triangle induced by $\{v_1, v_2, u_3\}$. In a similar manner, there exists u_5 that is adjacent to v_1,v_2, and u_4. u_5 cannot be in the regions induced by $\{v_1, u_3, u_4\}$ or $\{v_2, u_3, u_4\}$, and must lie outside the separating triangle induced by $\{v_1, v_2, u_4\}$. Following this pattern, there is a set of vertices in $\{u_4, u_5 \ldots, u_{n-1}\}$ such that for any j, $4 \leq j \leq n-1$, the vertex u_j is outside τ, lies outside the separating triangle induced by $\{v_1, v_2, u_{j-1}\}$, and is adjacent to v_1, v_2, and u_{j-1}.

The vertex u_{n-1} is adjacent to u_n, since if they were not adjacent, a quadrangular region in T is formed. This implies that the set $\{u_1, u_2, \ldots, u_n\}$ is a sequence of vertices that alternate membership in V_3 and V_4, that is, $V_3 \cup V_4$ forms a path in T with u_1 and u_n as its end vertices. □

Figure 4 is the resulting configuration after performing the steps elaborated in the proof of Lemma 4. Vertices in the set $\{u_1, u_2, \ldots, u_n\}$ alternate membership in the partite sets V_3 and V_4.

We are now ready to prove the next main result. The following theorem states the relationship between y and z so that an SMPS will exist for the complete 4-partite graph $K_{1,1,y,z}$.

Theorem 5 *$K_{1,1,y,z}$ has an SMPS if and only if $y = z$ or $z - y = 1$.*

Proof Let T be an SMPS of $K_{1,1,y,z}$. There are two cases that need to be considered.

Case 1: *There is a vertex $x \in V_3 \cup V_4$ in T such that* $\deg(x) \neq 4$. By Lemma 4, the set of vertices in V_3 and V_4 forms a path. Hence V_3 and V_4 either have equal cardinality or differ in cardinality by 1, that is, either $y = z$ or $z - y = 1$.

Case 2: *deg $(x) = 4$ for all $x \in V_3 \cup V_4$ in T*. Let $v \in V_3$. Since $\deg(v) = 4$, it must be adjacent to $v_1 \in V_1$, $v_2 \in V_2$, and two vertices in V_4, say u_1 and u_2. Now, the neighborhood $N(v) = \{v_1, v_2, u_1, u_2\}$ induces a 4-cycle. Further, u_1 and u_2 are non-adjacent in this 4-cycle since they both belong to V_4. WLOG, let the 4-cycle induced by $N(v)$ be given by $(u_1, v_1, u_2, v_2, u_1)$. Observe that v_1 and v_2 are not adjacent. If they were adjacent, then the join $P_3 + K_2$ is formed, where u_1 and u_2 are the end vertices of the path P_3. In particular, u_1 and u_2 will both be of degree 3, which is impossible from the assumption of this case. Now consider the k-cycle induced by $N(v_1)$, which is composed of vertices in V_3 and V_4 only. Note that k must be even, because if k was odd, there is a pair of adjacent vertices that belong to the same partite set. Let $t \in N(v_1) \setminus \{u_1, v, u_2\}$. WLOG, suppose that $t \in V_3$. Hence the two neighbors of t in the k-cycle induced by $N(v_1)$ are in V_4. By assumption, $\deg(t) = 4$. Since t is already adjacent to $v_1 \in V_1$ and two other vertices in V_4 in the k-cycle, it must be adjacent to exactly one more vertex in the partite set V_2. Thus, t is adjacent to $v_2 \in V_2$ because it is the only vertex in V_2. Therefore, the vertices in V_3 and V_4

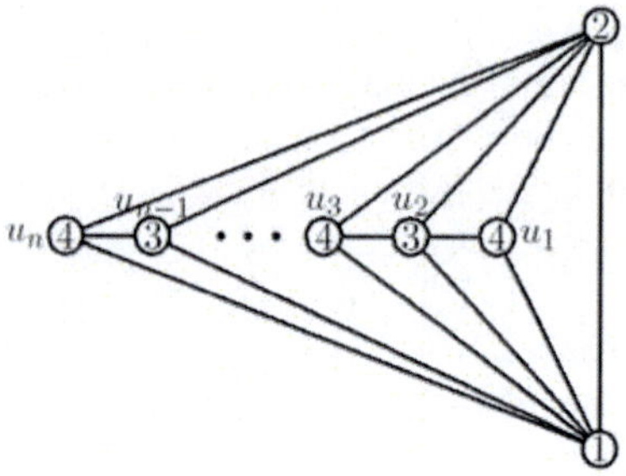

Fig. 4 An SMPS of $K_{1,1,y,z}$ containing vertices in $V_3 \cup V_4$ with degree 3

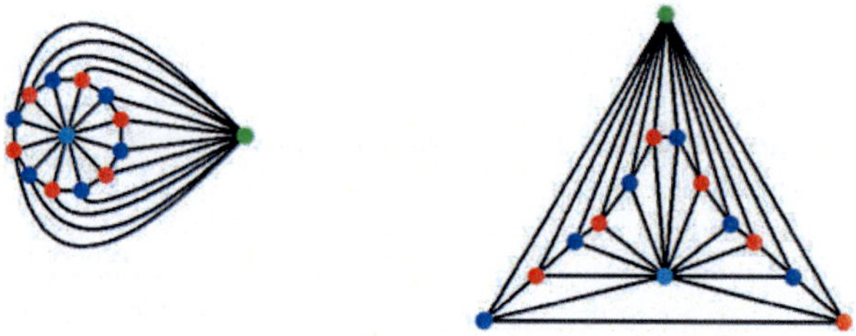

Fig. 5 Two drawings for an SMPS of $K_{1,1,6,6}$ with $\deg(x) = 4$, $x \in V_3 \cup V_4$

are also the members of the k-cycle induced by $N(v_1)$. Since k is even, then the cardinalities of V_3 and V_4 must be equal, that is, $y = z$.

Now assume that $y = z$ or $z - y = 1$. We will show that $K_{1,1,y,z}$ has an SMPS.
Case 1: $z - y = 1$. The graph in Fig. 4 shows that $K_{1,1,y,z}$ has an SMPS. Further, this graph is the join $P_{y+z} + K_2$, where vertices in P_{y+z} are in $V_3 \cup V_4$ and the two vertices in K_2 are in V_1 and V_2.
Case 2: $y = z$. There are two subcases that need to be considered. These subcases result from the proof of Lemma 2.

Suppose that there is a vertex $x \in V_3 \cup V_4$ such that $deg(x) \neq 4$. Hence the configuration considered in Case 1 is an SMPS. The path formed in the configuration is P_{2z}, since $y = z$.

Assume that all vertices in $V_3 \cup V_4$ are of degree 4. Consider the configuration resulting from Case 2 of the proof of Lemma 2. The configuration is obtained by forming a wheel W_k, $k = 2z$, with center $v_1 \in V_1$, and attaching each vertex of the k-cycle to the vertex $v_2 \in V_2$. The vertices in the k-cycle alternate membership in V_3 and V_4.

Therefore from these cases, if either $y = z$ or $z - y = 1$, the complete 4-partite graph $K_{1,1,y,z}$ contains an SMPS. □

Figure 5 is an example of an SMPS for $K_{1,1,y,z}$ in the second subcase of Case 2, in the necessity part of the proof.

The SMPS that can be obtained from the class $K_{1,1,y,z}$, can in fact be utilized to construct an SMPS for larger, more general classes of complete 4-partite graphs.

Bowen and Fisk [4] described some operations of adding or removing a vertex v from a maximal planar graph, and still preserve the maximal planarity. Other similar operations are discussed by Batagelj [3], Barnette [2], and Butler [5]. In complete 4-partite graphs, some of these operations are still admissible. Consider a region R of an SMPS of $K_{w,x,y,z}$, induced by vertices v_1, v_2, v_3, with $v_1 \in V_1$, $v_2 \in V_2$, $v_3 \in V_3$. A new vertex, $v \in V_4$, may be added in R, together with edges vv_1, vv_2, vv_3. This produces an SMPS of $K_{w,x,y,z+1}$. The other configurations of memberships to the partite sets could also be used to obtain other SMPS. In the next main results, we utilize this operation of "adding a vertex to a region" to attack the SMPS problem for larger classes of complete 4-partite graphs.

Theorem 6 *Suppose that* $z - y = 1$*. If* $w \leq y + z$ *and* $x \leq y + z$*, then* $K_{w,x,y,z}$ *contains an SMPS.*

Proof Suppose that $z - y = 1$. Consider the SMPS T in the proof of Theorem 5 where there is a vertex in $V_3 \cup V_4$ of degree 3. T contains a path P_k, $k = y + z$, whose vertices $u_1, u_2, \ldots, u_k$ alternate membership in $V_3 \cup V_4$, and the vertices in V_1 and V_2 are adjacent. Let $v_1 \in V_1$, $v_2 \in V_2$, and $u_i \in V_3 \cup V_4$, $1 \leq i \leq k$.

Let the SMPS T be drawn and labeled as in Fig. 4. We generate a larger SMPS by performing the following step: for some region induced by $\{v_1, u_{a-1}, u_a\}$, $a \in \{2, 3 \ldots, k-1, k\}$, add a new vertex u_j together with three edges. Then an SMPS of $K_{1,x,y,z}$ is obtained. Moreover, for each j, we have that $u_j \in V_2$ since it is adjacent to vertices in V_1, V_3, and V_4. Furthermore, the number of regions induced by $\{1, u_{a-1}, u_a\}$ is precisely the number of edges in the path P_k, which is $k - 1$. This is the maximum number of vertices u_j that can be added to the regions induced by $\{v_1, u_{a-1}, u_a\}$, that is, $j \leq k - 1$. The number of vertices u_j added in this process, together with the vertex v_2, gives the maximum number of vertices in V_2, given by $|V_2| = x \leq j + 1$. This implies that $x \leq k = y + z$. Therefore if $x \leq y + z$, then $K_{1,x,y,z}$ contains an SMPS that is generated from T.

In the same manner, vertices could be added to the other set of regions adjacent to the vertex v_2. If U is the SMPS of $K_{1,x,y,z}$ obtained from T by the previous process, U can be modified further into an SMPS T' of $K_{w,x,y,z}$, where $w \leq y + z$. Hence from the SMPS T, an SMPS T' of $K_{w,x,y,z}$ where $w \leq y + z$ and $x \leq y + z$ can be obtained. □

Figure 6 shows an SMPS of $K_{w,x,y,z}$ obtained from the SMPS shown in Fig. 4. $w - 1$ and $x - 1$ vertices were added to regions of the form $\{v_1, u_{a-1}, u_a\}$ and $\{v_2, u_{b-1}, u_b\}$, respectively.

Theorem 7 *Suppose that* $y = z$*. If* $w \leq 2z + 1$ *and* $x \leq 2z + 1$*, then* $K_{w,x,y,z}$ *contains an SMPS.*

Proof Suppose that $y = z$. Consider an SMPS T of $K_{1,1,y,z}$, where all vertices in $V_3 \cup V_4$ are of degree 4. It was shown in the proof of Theorem 5 that in this case, T has the configuration obtained from a wheel W_{2z}, $k = 2z$, with center $v_1 \in V_1$ and with vertices in the k-cycle attached to the vertex $v_2 \in V_2$ in the exterior of wheel W_k. Further, the vertices in the k-cycle alternate membership in $V_3 \cup V_4$.

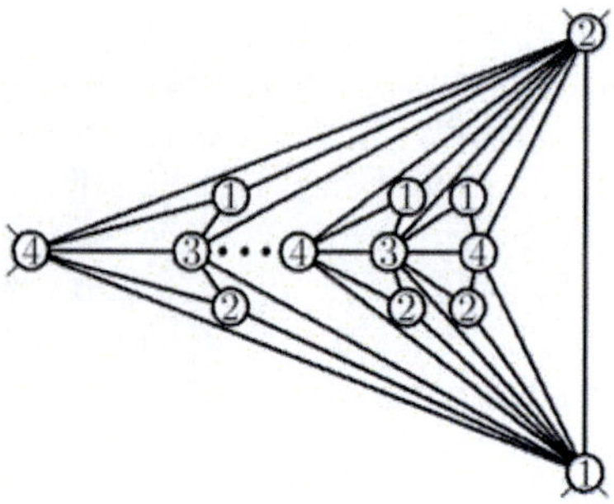

Fig. 6 An SMPS of $K_{w,x,y,z}$ with $w = x = y + z$, and $z - y = 1$

WLOG, let T be drawn and labeled, as in Fig. 5. Note that in this configuration, any region of T is induced by either $\{v_1, a, b\}$ or $\{v_2, a, b\}$ only, where $a \in V_3$ and $b \in V_4$. Hence, a vertex in V_1 may be added to a region induced by $\{v_2, a, b\}$. Similarly, we may add a vertex in V_2 to a region that is induced by $\{v_1, a, b\}$. We count the number of regions where the vertices may be added. Let $(u_1, u_2, \ldots, u_k, u_1)$ be the k-cycle of the wheel W_k. In the configuration, each pair of vertices u_i and u_{i+1} are adjacent to v_2, where $i + 1$ is taken modulo k from the set $\{1, 2, \ldots, k\}$. Hence, $u_i u_{i+1}$ is a boundary edge to each region induced by $\{v_1, a, b\}$. Therefore, the number of boundary edges is also the number of such regions. Since the number of edges of the form $u_i u_{i+1}$ is k, there are k regions induced by $\{v_1, a, b\}$. Hence addition of vertices may be applied at most $k = 2z$ times, that is, at most $2z$ vertices may be added to the partite set V_2. Hence if T is modified by adding vertices in V_2 in this manner, together with v_2, we must have $|V_2| = x \leq 2z + 1$. Hence we have shown that if $x \leq 2z + 1$, then $K_{1,x,y,z}$ contains an SMPS that can be generated from T, using this process.

Let U be the SMPS of $K_{1,x,y,z}$ obtained from T by adding vertices to the regions $\{v_1, a, b\}$, as elaborated in the previous process. In the same manner, regions of the form $\{v_2, a, b\}$ could be modified by the addition of vertices belonging to V_1. Using the same counting process done previously, at most $2z$ vertices in V_1 may be added. Thus together with v_1, it must be that $w \leq 2z + 1$. Therefore $K_{w,x,y,z}$ contains an SMPS that can be generated from U. □

To prove the last result in this paper, a method of "inserting a vertex to an edge" was utilized in the construction. This operation was also introduced by Bowen and Fisk [4], and it was modified to fit the structure of complete 4-partite graphs. The modified operation is described in Fig. 7.

Theorem 8 *Suppose that* $z - y = k$, *for some* $k \notin \{0, 1\}$. *If* $\frac{k+3}{2} \leq x + w$, *then* $K_{w,x,y,z}$ *contains an SMPS.*

Proof Let T be the SMPS of $K_{1,1,y,y+1}$, as in Fig. 4. Assume that t is the vertex in V_1, u is the vertex in V_2, and P_m is the path containing the vertices that alternate membership in V_3 and V_4.

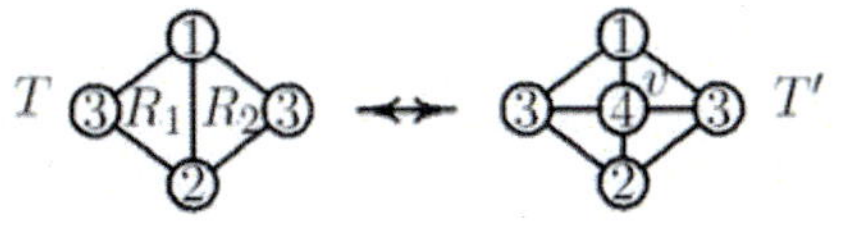

Fig. 7 Inserting $v \in V_4$ to the boundary edge of two regions that are induced by vertices in V_1, V_2 and V_3

Now we modify T by adding vertices in V_1 and V_2, to obtain an SMPS T', of the complete 4-partite graph $K_{w,x,y,y+1}$. There are $y-1$ edges of the form tv_i, where $t \in V_1$, $v_i \in V_4$, where v_i is not an end-vertex of P_m. Similarly, there are $y-1$ edges of the form uv_i, where $u \in V_2$, $v_i \in V_4$, and v_i is not an end-vertex of P_m. To each of the edges tv_i, we may insert a vertex $a_i \in V_2$ and attach this new vertex to the vertices in V_3 that are adjacent to v_i. At most $y-1$ vertices may be added in this manner. Therefore, counting the exterior boundary vertex $u \in V_2$, there is at most y vertices in V_2, that is, $x \leq y$. Similarly, to each of the edges uv_i, a vertex $b_i \in V_1$ may be inserted together with the necessary edges connecting b_i and two vertices in V_3 that are adjacent to v_i. This may also be done at most $y-1$ times. Hence including the exterior vertex $t \in V_1$, there are at most y vertices in V_1 also, that is, $w \leq y$. After the addition of vertices to the edges tv_i and uv_i, an SMPS T' of $K_{w,x,y,y+1}$ is obtained.

Since $z-y=k$, and $y+1$ vertices in V_4 are already in T', there are $k-1$ vertices in V_4 that will be added to T' in this last step. Note that for each vertex a_i added to the edge tv_i in the previous step, two new regions induced by vertices in V_1, V_2, and V_3 are formed. Similarly, each vertex b_i added to uv_i forms two new regions induced by vertices in V_1, V_2 and V_3. Hence, there are $2(x-1)+2(w-1)=2(w+x-2)$ new regions induced by vertices in V_1, V_2, and V_3. Adding a new vertex in V_4 inside each of these regions is still admissible, and attach each added vertex to the three boundary vertices in V_1, V_2 and V_3. Add $k-1$ vertices to V_4 in this manner, where $k-1 \leq 2(w+x-2)$, or equivalently, $\frac{k+3}{2} \leq w+x$. Therefore if $\frac{k+3}{2} \leq w+x$, the resulting graph, say T'', is now an SMPS of $K_{w,x,y,z}$. □

3 Conclusions

In this paper, a step further was taken in tackling the SMPS problem for complete k-partite graphs. The structure of any SMPS for some complete 4-partite graphs were studied, particularly in the class $K_{1,1,1,z}$ and $K_{1,1,y,z}$. The structures of the SMPS for these graphs were utilized to construct larger SMPS for more general complete 4-partite graphs, by adding a number of vertices and edges that will preserve maximal planarity. Since there were restrictions as to the number of vertices that can be added, some initial conditions were assumed, and relationships between the cardinalities w, x, y, and z of the four partite sets were found. It is interesting to find out if the converse of Theorems 6, 7, and 8 also hold true. This would require further study on various operations admissible for the SMPS of complete 4-partite graphs, to generate some SMPS with larger or smaller order.

References

1. Almonte V, Gervacio S, Natalio E (2019) Complete tripartite graphs with spanning maximal planar subgraphs. Matimyas Matematika 42(2):1–10
2. Barnette D (1974) On generating planare graphs. Discrete Math 7:199–208
3. Batagelj V (1989) An improved inductive definition of two restricted classes of triangulations of the plane. Comb Graph Theory 25:11–19
4. Bowen R, Fisk S (1967) Generation of triangulations of the sphere. Math Comput 21:250–252
5. Butler J (1974) A generation procedure for the simple 3-polytopes with cyclically 5-connected graphs. Canad J Math 26:686–708
6. Djidjev H (2006) A linear-time algorithm for finding a maximal planar subgraph. SIAM J Discret Math 20(2):444–462
7. Kuratowski K (1930) On the problem of skew curves in topology. Fund Math 15:271–283
8. Nakamoto A, Sakuma T, Suzuki Y (2006) N-Flips in even triangulations on the sphere. J Graph Theory 51(3):260–268
9. Resende M, Ribeiro C (2001) Graph planarization. Encyclopedia of optimization. Springer Link

On Column Competent Matrices and Linear Complementarity Problem

A. Dutta, R. Jana, and A. K. Das

Abstract We revisit the class of column competent matrices and study some matrix theoretic properties of this class. The local w-uniqueness of the solutions to the linear complementarity problem can be identified by the column competent matrices. We establish some new results on w-uniqueness properties in connection with column competent matrices. These results are significant in the context of matrix theory as well as algorithms in operations research. We prove some results in connection with locally w-uniqueness property of column competent matrices. Finally, we establish a connection between column competent matrices and column adequate matrices with the help of degree theory.

Keywords Linear complementarity problem · Column competent matrices · W-uniqueness · Column adequate matrices

1 Introduction

The w-uniqueness property is important in the context of dynamical systems under smooth unilateral constraints. Xu [18] introduced the column competent matrices. On uniqueness, quite a large number of results are available in the literature of operations research. The study of the uniqueness property of the solution is important in the context of the theory of the complementarity system as well as the method applied

Financially Supported by Department of Science and Technology, Govt. of India, INSPIRE Fellowship Scheme.
The author R. Jana presently working in an integrated steel plant.

A. Dutta (✉)
Jadavpur University, 188, Raja S.C. Mallick Rd, Kolkata 700032, India

R. Jana · A. K. Das
SQC & OR Unit, Indian Statistical Institute, 203 B.T Road, Kolkata 700108, India

© The Author(s), under exclusive license to Springer Nature Singapore Pte Ltd. 2022
D. Giri et al. (eds.), *Proceedings of the Seventh International Conference on Mathematics and Computing*, Advances in Intelligent Systems and Computing 1412,
https://doi.org/10.1007/978-981-16-6890-6_46

for finding the solution. For details see [4, 10, 11, 15]. Ingleton [6] studied the w-uniqueness solutions to linear complementarity problem in the context of adequate matrices.

The linear complementarity problem can be stated as follows: For $A \in R^{n\times n}$ and a vector $q \in R^n$, the *linear complementarity problem* denoted as LCP(q, A) finds the solution $w \in R^n$ and $z \in R^n$ to the following systems

$$w - Az = q \; ; \; z \geq 0 \; ; \; w \geq 0 \tag{1.1}$$

$$w^T z = 0 \tag{1.2}$$

or show that there does not exist any $z \in R^n$ and $w \in R^n$ satisfying the system of linear inequalities (1.1) and complementary condition (1.2).

Pang [14] studied local z-uniqueness of solutions of a linear complementarity problem. The LCP(q, A) has unique z-solution for all $q \in R^n$ iff A is a P-matrix [1]. The w-uniqueness property is identified by a condition on A related to the notion of sign-reversing. Motivated by the w-uniqueness results, we consider column competent matrices in the context of LCP(q, A). The sufficient matrices capture many properties of positive semidefinite matrices. The aim of this article is to study some matrix theoretic properties of this class and establish some new results which are useful to the solution of the LCP(q, A).

The paper is organized as follows. In Sect. 2, we include few related notations and results. Section 3 presents some new results related to column competent matrices. We develop several matrix theoretic results of column competent matrices which are related to solution of linear complementarity problem. Section 4 provides a conclusion about the article.

2 Preliminaries

Here, any vector $z \in R^n$ is a column vector and z^T is the row transpose of z. We write $z = z^+ - z^-$ where $z_i^+ = \max(0, z_i)$ and $z_i^- = \max(0, -z_i)$ for any index i. If A is a matrix of order n, $\alpha \subseteq \{1, 2, \ldots, n\}$ and $\beta \subseteq \{1, 2, \ldots, n\} \setminus \alpha$ then $A_{\alpha\beta}$ is the submatrix with the rows and columns of A whose indices are in α and β, respectively. A principal submatrix and a principal minor of A are denoted by $A_{\alpha\alpha}$ and det $A_{\alpha\alpha}$, respectively. For $A \in R^{n\times n}$ and $q \in R^n$, the feasible set of LCP(q, A) is defined by FEA(q, A) $= \{z \in R^n : z \geq 0, q + Az \geq 0\}$ and the solution set is also defined by SOL(q, A) $= \{z \in$ FEA(q, A) $: z^T(q + Az) = 0\}$. A z-solution, $\tilde{z}$ is called locally unique if $\exists$ a neighborhood of $\tilde{z}$ within which $\tilde{z}$ is the only z-solution. A w-solution, $\tilde{w} = A\tilde{z} + q$, is called locally unique if $\exists$ a neighborhood of $\tilde{w}$ in which $\tilde{w}$ is the only w-solution. Let $\psi : R^n \to R^n$ and the kernel of the function ψ is defined by ker $\psi = \{z \in R^n : \psi(z) = 0\}$. The kernel of a matrix $A \in R^{n\times n}$ is defined by ker $A = \{z \in R^n : Az = 0\}$. Now we define the column competent matrix.

Definition 2.1 ([18]) The matrix A is said to be column competent if $z_i(Az)_i = 0,\ i = 1, 2, \ldots, n \implies Az = 0$.

Column competent matrices can be singular or nonsingular matrices. Note that all singular matrices need not be column competent matrices. Consider $A = \begin{bmatrix} 1 & 0 \\ 1 & 0 \end{bmatrix}$ which is a singular matrix. For any $z = \begin{bmatrix} 0 \\ k \end{bmatrix}$, $k \in R,\ z_i(Az)_i = 0,\ i = 1, 2$ implies that $Az = 0$. Consider another $A = \begin{bmatrix} 1 & 1 \\ 0 & 0 \end{bmatrix}$. It is easy to show that $z_i(Az)_i = 0,\ i = 1, 2$ does not imply $Az = 0$. Hence A is not a column competent matrix. Let $A = \begin{bmatrix} 1 & 4 & 3 \\ 2 & 1 & 5 \\ 3 & 2 & 0 \end{bmatrix}$, for $z = \begin{bmatrix} 0 \\ 0 \\ 1 \end{bmatrix}$ $z_i(Az)_i = 0,\ i = 1, 2, 3$ but $Az = \begin{bmatrix} 3 \\ 5 \\ 0 \end{bmatrix} \neq 0$. Here, A is a nonsingular matrix but not a column competent matrix.

Now we define $\psi : R^n \to R^n$ where $\psi(z) = z * (Az)$ and $z * (Az)$ is the Hadamard product defined by $(z * (Az))_i = z_i * (Az)_i,\ \forall\, i$. Note that the product is associative, distributive and commutative.

Definition 2.2 ([18]) In view of Hadamard product, a matrix A is said to be column competent if $\ker \psi = \ker A$.

Column adequate matrices are related to column competent matrices. We start with definition of column adequate matrices.

Definition 2.3 ([11]) The matrix A is said to be column adequate if $z_i(Az)_i \leq 0,\ i = 1, 2, \ldots, n \implies Az = 0$.

We state the following lemma and theorems which are useful for the subsequent sections.

Lemma 2.1 ([18]) *The matrix A is said to be non-degenerate if and only if* $\ker \psi = \{0\}$.

Theorem 2.1 ([18]) *The following statements are equivalent.*

(i) A is column competent.
(ii) For all vector q, the LCP(q, A) has a finite number (possibly zero) of w-solutions.
(iii) For all vector q, any w-solution of the LCP(q, A), if it exists, must be locally w-unique.

Theorem 2.2 ([18]) *The following statements are equivalent.*

(i) (a) A is column competent.
(b) A is a P_0-matrix.
(ii) A is column adequate.

Theorem 2.3 ([1]) *Let $A \in R^{n\times n}$ be a E_0-matrix. Then, the following statements are equivalent.*

(i) $A \in R_0$.
(ii) $A \in R$.

We say that $A \in R^{n\times n}$ is a
– Q-matrix if for every $q \in R^n$, LCP(q, A) has a solution.
– Q_0-matrix if for any $q \in R^n$, feasibility implies solvability.
– P-matrix if for each vector $z \neq 0$ there exists an index i such that $\max_{(z_i \neq 0)} z_i(Az)_i > 0$.
– P_0-matrix if for each vector $z \neq 0$ there exists an index i such that $\max_{(z_i \neq 0)} z_i(Az)_i \geq 0$.
– R_0-matrix if LCP$(0, A)$ has unique solution.
– *principally non-degenerate* if it has no principal submatrix which has determinant zero.

For further details about the matrix classes in linear complementarity problem see [3, 7–9, 13].

The principal pivot transform (PPT) has an important role in the study of matrix classes and linear complementarity problem. The *principal pivot transform* of $A \in R^{n\times n}$ with real entries, with respect to $\alpha \subseteq \{1, 2, \ldots, n\}$ is defined as the matrix given by

$$A' = \begin{bmatrix} A'_{\alpha\alpha} & A'_{\alpha\bar{\alpha}} \\ A'_{\bar{\alpha}\alpha} & A'_{\bar{\alpha}\bar{\alpha}} \end{bmatrix}$$

where $A'_{\alpha\alpha} = (A_{\alpha\alpha})^{-1}$, $A'_{\alpha\bar{\alpha}}=-(A_{\alpha\alpha})^{-1}A_{\alpha\bar{\alpha}}$, $A'_{\bar{\alpha}\alpha} = A_{\bar{\alpha}\alpha}(A_{\alpha\alpha})^{-1}$ and $A'_{\bar{\alpha}\bar{\alpha}} = A_{\bar{\alpha}\bar{\alpha}} - A_{\bar{\alpha}\alpha}(A_{\alpha\alpha})^{-1}A_{\alpha\bar{\alpha}}$.

Here PPT is only identified with respect to those α for which $\det A_{\alpha\alpha} \neq 0$. When $\alpha = \emptyset$, by convention $\det A_{\alpha\alpha} = 1$ and $A' = A$. Here, $A'_{\bar{\alpha}\bar{\alpha}}$ is said to be Schur complement of A. We denote the PPT of A as $A' = \mathcal{P}_\alpha(A)$. The Schur complement of $A_{\alpha\alpha}$ in $A = \begin{bmatrix} A_{\alpha\alpha} & A_{\alpha\bar{\alpha}} \\ A_{\bar{\alpha}\alpha} & A_{\bar{\alpha}\bar{\alpha}} \end{bmatrix}$ is a principal submatrix of the principal pivot transform A'. For details of PPT see [2, 11, 12].

We establish a connection between competent matrices and adequate matrices using degree theoretic approach. We provide brief details about degree theory in the subsequent section.

2.1 Degree Theory

Let $f_A : R^n \to R^n$ be a piecewise linear mapping for a given matrix $A \in R^{n\times n}$ defined as $f_A(e_i) = e_i$ and $f_A(-e_i) = -Ae_i \ \forall\, i$. We write for any $z \in R^n$,

$$f_A(z) = z^+ - Az^-.$$

For details see [10]. It is clear that LCP(q, A) is equivalent to find a vector $z \in R^n$ such that $f_A(z) = q$. If z belongs to the interior of some orthants of R^n and det $A_{\alpha\alpha} \neq 0$ where $\alpha = \{i : z_i < 0\}$, then the index of $f_A(z)$ at z is well defined and can be written as

$$\text{ind} f_A(q, z) = sgn(\det A_{\alpha\alpha}).$$

Note that the cardinality of $f_A^{-1}(q)$ denotes the number of solutions of LCP(q, A). Particularly, if q is non-degenerate with respect to A, each index of f_A is well defined and we can define local degree of A at q. It can be denoted as $\deg_A(q)$. For details see ([1], Chap. 6). We state the following theorem from [10], which will be required to prove one of our result.

Theorem 2.4 *Let $A \in R^{n\times n}$. Let $K(A)$ denote the union of all the facets of the complementary cones of $(I, -A)$. Consider $q \in R^n \setminus k(A)$ where q is non-degenerate with respect to A. Let $\beta \subseteq \{1, 2, \ldots, n\}$ be such that* det $A_{\beta\beta} \neq 0$. *Suppose A' is a PPT of A with respect to β. Then $deg_{A'}(q') = sgn(\det A_{\beta\beta}) \cdot deg_A(q)$.*

3 Results on Column Competent Matrices

Hadamard product is important to characterize the complementary condition. Here we show that the property of column competent matrix is related to Hadamard product.

Theorem 3.1 *Suppose A is a column competent matrix and the function $\psi : R^n \to R^n$ defined by $\psi(z) = z * (Az)$ where $z * (Az)$ is the Hadamard product. Then $ker\ \psi = ker A$.*

Proof Let A be a column competent matrix. Then for a vector $z \in R^n$, $z_i(Az)_i = 0,\ i = 1, 2, \ldots, n \implies Az = 0$. Hence $z \in \ker \psi$ implies $z \in \ker A$. So we write $\ker \psi \subseteq \ker A$. Again by definition $\ker A \subseteq \ker \psi$. Therefore $\ker \psi = \ker A$.

The following result provides a characterization of non-degenerate column competent matrices.

Theorem 3.2 *Let $A \in R^{n\times n}$ be a non-degenerate column competent matrix. Then $A \in R_0$.*

Proof Let A be a non-degenerate column competent matrix. By Lemma 2.1, $\ker \psi = \{0\}$ where $\psi(z) = z * Az$. By Theorem 3.1, we can write $\ker \psi = \ker A = \{0\}$. Let z be the solution of LCP(0, A). Then $z_i(Az)_i = 0,\ i = 1, 2, \ldots, n$. This implies that $Az = 0$. Hence $z = 0$. Therefore, LCP(0, A) has only one solution zero. Hence A is a R_0- matrix.

Note that column competent matrix need not be a P_0- matrix in general. Consider the matrix $A = \begin{bmatrix} 2 & 1 \\ 1 & -1 \end{bmatrix}$. We show that A is a column competent matrix but not a P_0-matrix. Now we establish the following result.

Theorem 3.3 *Suppose A is a column competent matrix with* $A \in P_0$. *Then for* $0 \neq z \geq 0$, $(z, 0)$ *is the solution of LCP*$(0, A)$.

Proof Let $A \in R^{n\times n}$ be a column competent matrix with $A \in P_0$. Then for each $0 \neq z$, $\max_i z_i(Az)_i \geq 0$, $z_i \neq 0$. If $z_i(Az)_i = 0,\ i = 1, 2, \ldots, n$ implies that $Az = 0$. Then $(z, 0)$, $z \geq 0$ is the solution of LCP$(0, A)$.

Now we consider the matrix $A = \begin{bmatrix} 2 & -1 \\ -4 & 2 \end{bmatrix}$ is column competent as well as P_0 and $\left(\begin{bmatrix} 1 \\ 2 \end{bmatrix}, \begin{bmatrix} 0 \\ 0 \end{bmatrix}\right)$ is a solution of LCP$(0, A)$. Note that this can be explained using the Theorem 3.3.

Theorem 3.4 *Let A be a column competent matrix. Suppose* $z \geq 0$ *and* $z_i(Az)_i = 0,\ i = 1, 2, \ldots, n$. *Then LCP*$(0, A)$ *has the solution* $(z, 0)$.

Proof Since A is a column competent matrix, then for $z \geq 0$ and $z_i(Az)_i = 0,\ i = 1, 2, \ldots, n$. This implies that $Az = 0$. Therefore $(z, 0)$ is the solution of LCP$(0, A)$.

Xu [18] showed that if A is a column competent matrix then DAD^T is a column competent matrix where D is a diagonal matrix. In the next theorem, we prove that column competent matrices with some additional assumptions are invariant under principal rearrangement. For any principal submatrix $A_{\alpha\alpha}$ of A,it is possible to rearrange principally the rows and columns of A in such a way that $A_{\alpha\alpha}$ becomes a leading principal submatrix in the rearranged matrix PAP^T.

Theorem 3.5 *Suppose A is a column competent matrix. If for any* $z \in R^n$, *either* $z_i(Az)_i \geq 0$ *or* $z_i(Az)_i \leq 0$ *for all i, then* PAP^T *is also column competent where P is a permutation matrix.*

Proof Let for any $z \in R^n$, $y = Pz$. Consider $y_i(PAP^T y)_i = 0$ for all i. This implies that $(Pz)_i(PAP^T Pz)_i = 0$ for all i. We know that

$$z^T P^T PAP^T Pz = \textstyle\sum_{i=1}^n (Pz)_i(PAP^T Pz)_i = 0.$$

Hence $z^T Az = 0$ as $P^T P = I$. We write $\sum_{i=1}^n z_i(Az)_i = 0$. It means $z_i(Az)_i = 0,\ i = 1, 2, \ldots, n$. As A is a column competent matrix, $Az = 0$. Therefore $AP^T Pz = 0$. Hence $(PAP^T)(Pz) = 0$. Hence PAP^T is column competent.

Theorem 3.6 *Let A be a Z- matrix. Suppose* $z_i(Az)_i = 0$ *for all i,* $A|z| \geq 0$ *and* $Az \leq 0$. *Then A is a column competent matrix.*

Proof Suppose A is a Z- matrix. Consider $z_i(Az)_i = 0$ for all i, $A|z| \geq 0$ and $Az \leq 0$. As A is a Z- matrix, $Az \geq A|z| \geq 0$. Now this implies that $Az = 0$. Therefore A is a column competent matrix.

Consider $A = \begin{bmatrix} 1 & 1 & 4 \\ 2 & 2 & 5 \\ 3 & 4 & 1 \end{bmatrix}$. Note that A is a R_0-matrix. Now for $z = \begin{bmatrix} 1 \\ -1 \\ 0 \end{bmatrix}$, $z_i(Az)_i = 0,\ i = 1, 2, 3$ but $Az \neq 0$. Hence A is not a column competent matrix.

The class of non-degenerate matrices play an important role to characterize certain uniqueness properties of the solutions of LCP(q, A). We prove the following theorem to establish the relation between principally non-degenerate matrices and column competent matrices.

Theorem 3.7 *Let A be a principally non-degenerate matrix. Then A is column competent.*

Proof Let A be a principally non-degenerate matrix. Assume that A is not a column competent matrix. Hence $\exists$ a $0 \neq z \in R^n$ such that $z_i(Az)_i = 0,\ i = 1, 2, \ldots, n$ but $Az \neq 0$. Without loss of generality, consider $z = \begin{bmatrix} z_\alpha \\ z_\beta \end{bmatrix} \neq 0$ where $z_\alpha \neq 0$, $z_\beta = 0$ and $A = \begin{bmatrix} A_{\alpha\alpha} & A_{\alpha\beta} \\ A_{\beta\alpha} & A_{\beta\beta} \end{bmatrix}$. Then we consider the following cases.

Case 1 Let $\alpha = \{1, 2, \ldots, n\}$ and $\beta = \emptyset$. Then $z = z_\alpha$ and $z_i(Az)_i = 0,\ i \in \alpha$. It implies $Az = 0$, contradicts the fact that $Az \neq 0$.

Case 2 Let $\alpha \subset \{1, 2, \ldots, n\}$ and $\beta = \{1, 2, \ldots, n\} \setminus \alpha$. Consider $(z_\alpha)_i(A_{\alpha\alpha}z_\alpha)_i = 0,\ i \in \alpha$. This implies $A_{\alpha\alpha}z_\alpha = 0$. As $z_\alpha \neq 0$, $A_{\alpha\alpha}$ is a singular matrix. It contradicts that the matrix A is a principally non-degenerate matrix.

Therefore, A is a column competent matrix.

Here, we consider $A = \begin{bmatrix} 3 & -2 & 0 \\ -2 & 1 & 1 \\ 3 & 2 & 0 \end{bmatrix}$. For $z = \begin{bmatrix} 2k \\ 3k \\ k \end{bmatrix}$, $k \in R,\ z_i(Az)_i = 0,\ i = 1, 2, 3$ implies that $Az = 0$. Hence A is a column competent matrix. However, A is neither an adequate matrix nor a sufficient matrix. For details of sufficient matrices see [5, 16, 17].

Now we develop a necessary and sufficient condition for column competent matrices.

Theorem 3.8 *Let $A \in R^{n\times n}$. The following two statements are equivalent:*

(a) A is column competent.

(b) For $0 \neq z = \begin{bmatrix} z_\alpha \\ z_\beta \end{bmatrix} \geq 0$ with $z_\beta = 0$ and the submatrix $A_{\alpha\alpha}$ is singular with $A_{\alpha\alpha}z_\alpha = 0$ where $\alpha \cup \beta = \{1, 2, \ldots, n\}$ and $\alpha \cap \beta = \emptyset$, the system

$$\begin{bmatrix} A_{\alpha\alpha} & A_{\alpha\beta} \\ A_{\beta\alpha} & A_{\beta\beta} \end{bmatrix} \begin{bmatrix} z_\alpha \\ z_\beta \end{bmatrix} \not= 0 \tag{3.1}$$

has no solution.

Proof $(a) \implies (b)$. Suppose A is column competent and Eq. 3.1 is consistent. Let $\begin{bmatrix} z_\alpha \\ z_\beta \end{bmatrix}$ satisfies the Eq. 3.1 where $z_\beta = 0$ and the submatrix $A_{\alpha\alpha}$ is singular with

$A_{\alpha\alpha}z_\alpha = 0$ where $\alpha \cup \beta = \{1, 2, \ldots, n\}$ and $\alpha \cap \beta = \emptyset$. Here, $(z_\alpha)_i(A_{\alpha\alpha}z_\alpha)_i = 0$ and $(z_\beta)_i(A_{\beta\alpha}z_\alpha)_i = 0$. But $\begin{bmatrix} z_\alpha \\ z_\beta \end{bmatrix}$ satisfies Eq. 3.1 which contradicts that A is column competent.

$(b) \implies (a)$. Conversely, let $x \in R^n$ be a vector such that $x_i(Ax)_i = 0$ for all i. Consider A is not a column competent matrix. Suppose $x_\alpha = z_\alpha$ and $x_\beta = z_\beta$ with $0 \neq z = \begin{bmatrix} z_\alpha \\ z_\beta \end{bmatrix} \geq 0$, $z_\beta = 0$ and the submatrix $A_{\alpha\alpha}$ is singular with $A_{\alpha\alpha}z_\alpha = 0$ where $\alpha \cup \beta = \{1, 2, \ldots, n\}$ and $\alpha \cap \beta = \emptyset$. But the system Eq. 3.1 has no solution, i.e., x does not satisfy Eq. 3.1. Therefore A is column competent.

Now we prove the following sufficient condition related to the PPT of column competent matrices.

Theorem 3.9 *Let $A_{\alpha\alpha}$ and the Schur complement $A/A_{\alpha\alpha}$ be nonsingular of the square matrix $A = \begin{bmatrix} A_{\alpha\alpha} & A_{\alpha\beta} \\ A_{\beta\alpha} & A_{\beta\beta} \end{bmatrix}$ where $\alpha \cup \beta = \{1, 2, \ldots, n\}$ and $\alpha \cap \beta = \emptyset$. If A is column competent, then $A' = \mathcal{P}_\alpha(A)$ is column competent.*

Proof Let $w = A'z$ and $z * w = 0$ where $*$ is the Hadamard product. Thus we write

$$\begin{bmatrix} w_\alpha \\ w_\beta \end{bmatrix} = \begin{bmatrix} A'_{\alpha\alpha} & A'_{\alpha\beta} \\ A'_{\beta\alpha} & A'_{\beta\beta} \end{bmatrix} \begin{bmatrix} z_\alpha \\ z_\beta \end{bmatrix}. \tag{3.2}$$

The condition $z * w = 0$ means $\begin{bmatrix} z_\alpha \\ z_\beta \end{bmatrix} * \begin{bmatrix} w_\alpha \\ w_\beta \end{bmatrix} = \begin{bmatrix} w_\alpha * z_\alpha \\ w_\beta * z_\beta \end{bmatrix} = 0$. Since $A' = \mathcal{P}_\alpha(A)$, we have

$$\begin{bmatrix} z_\alpha \\ w_\beta \end{bmatrix} = \begin{bmatrix} A_{\alpha\alpha} & A_{\alpha\beta} \\ A_{\beta\alpha} & A_{\beta\beta} \end{bmatrix} \begin{bmatrix} w_\alpha \\ z_\beta \end{bmatrix}. \tag{3.3}$$

The matrix A is column competent implies that $\begin{bmatrix} A_{\alpha\alpha} & A_{\alpha\beta} \\ A_{\beta\alpha} & A_{\beta\beta} \end{bmatrix} \begin{bmatrix} w_\alpha \\ z_\beta \end{bmatrix} = 0$. It follows that $\begin{bmatrix} z_\alpha \\ w_\beta \end{bmatrix} = 0$. From Eq. 3.2, we get $A'_{\beta\alpha}z_\alpha + A'_{\beta\beta}z_\beta = 0$. Hence $A'_{\beta\beta}z_\beta = 0$ implies that $z_\beta = 0$ as $A'_{\beta\beta} = A/A_{\alpha\alpha}$ is nonsingular. Clearly, $w_\alpha = 0$. Hence $\begin{bmatrix} w_\alpha \\ w_\beta \end{bmatrix} = \begin{bmatrix} A'_{\alpha\alpha} & A'_{\alpha\beta} \\ A'_{\beta\alpha} & A'_{\beta\beta} \end{bmatrix} \begin{bmatrix} z_\alpha \\ z_\beta \end{bmatrix} = 0$. Therefore A' is column competent.

Theorem 3.10 *Let A be a column competent matrix where $A_{\alpha\alpha}$ and the Schur complement $A/A_{\alpha\alpha}$ be nonsingular of the square matrix $A = \begin{bmatrix} A_{\alpha\alpha} & A_{\alpha\beta} \\ A_{\beta\alpha} & A_{\beta\beta} \end{bmatrix}$. If $A \in E_0 \cap R_0$, then A is column adequate.*

Proof Suppose A is not a column adequate matrix but is column competent. By Theorem 2.2, A is not a P_0- matrix. Then there exists $\beta \subseteq \{1, 2, \ldots, n\}$ such that $\det A_{\beta\beta} < 0$. Let $A \in E_0 \cap R_0$. It follows from the Theorem 2.3 that $A \in R$. Then

$\deg_A(q) = 1$ for any q. Let A' be a principal pivot transform of A. Then $A' \in R$. Hence $\deg_{A'}(q') = 1$. By Theorem 2.4, $\deg_{A'}(q') = \deg_A(q).sgn(\det A_{\beta\beta})$. It implies that $\deg_{A'}(q') = -1$. This contradicts that A is not a P_0-matrix. Therefore, A is column adequate matrix.

3.1 Solution of Linear Complementarity Problem with Column Competent Matrices

We begin with some examples of w-uniqueness of the solution. Consider the column competent matrix $A = \begin{bmatrix} -1 & 3 \\ 2 & -6 \end{bmatrix}, q = \begin{bmatrix} 1 \\ -2 \end{bmatrix}$. This LCP$(q, A)$ has solution $z = \begin{bmatrix} 4 \\ 1 \end{bmatrix}$ and $w = \begin{bmatrix} 0 \\ 0 \end{bmatrix}$. In the neighborhood of z there is another solution $z' = \begin{bmatrix} 4.0100 \\ 1.0033 \end{bmatrix}$ and $w' = w = \begin{bmatrix} 0 \\ 0 \end{bmatrix}$.

We consider another matrix $A = \begin{bmatrix} -2 & 1 & 3 \\ 4 & -2 & -6 \\ 1 & -1 & -1 \end{bmatrix}, q = \begin{bmatrix} 1 \\ -2 \\ 1 \end{bmatrix}$. For $z = \begin{bmatrix} 2k \\ k \\ k \end{bmatrix}$, $k \in R$, $z_i(Az)_i = 0,\ i = 1, 2, \ldots, n$ implies that $Az = 0$. So A is a column competent matrix. This LCP(q, A) has solution $z = \begin{bmatrix} 4 \\ 4 \\ 1 \end{bmatrix}$ and $w = \begin{bmatrix} 0 \\ 0 \\ 0 \end{bmatrix}$. In the neighborhood of z there is another solution $z' = \begin{bmatrix} 4.02 \\ 4.01 \\ 1.01 \end{bmatrix}$ and $w' = w = \begin{bmatrix} 0 \\ 0 \\ 0 \end{bmatrix}$.

Now we prove the following two results in connection with locally w-uniqueness property of the column competent matrices. The following two results state the necessary and sufficient condition that A is a column competent matrix in the system of linear complementarity problem.

Theorem 3.11 *Suppose (w^*, z^*) is the solution of LCP(q, A) such that $w^* = q + Az^*$. Let $\alpha = \{i : w_i{}^* > 0\}$, $\beta = \{i : w_i{}^* = 0\}$ be the index set. Further consider that the submatrix $A_{\alpha\alpha}$ is nonsingular. If $A = \begin{bmatrix} A_{\alpha\alpha} & A_{\alpha\beta} \\ A_{\beta\alpha} & A_{\beta\beta} \end{bmatrix}$ is a column competent matrix, then $(w_\alpha, z_\beta) = (0, 0)$ is the only solution of the system:*

$$\begin{aligned} z_\alpha = A'_{\alpha\alpha} w_\alpha + A'_{\alpha\beta} z_\beta &= 0 \\ w_\beta = A'_{\beta\alpha} w_\alpha + A'_{\beta\beta} z_\beta &= 0 \\ w_\alpha &> 0 \\ z_\beta &> 0, \end{aligned} \tag{3.4}$$

where $A'_{\alpha\alpha} = (A_{\alpha\alpha})^{-1}$, $A'_{\alpha\beta} = -(A_{\alpha\alpha})^{-1}A_{\alpha\beta}$, $A'_{\beta\alpha} = A_{\beta\alpha}(A_{\alpha\alpha})^{-1}$ *and* $A'_{\beta\beta} = A_{\beta\beta} - A_{\beta\alpha}(A_{\alpha\alpha})^{-1}A_{\alpha\beta}$.

Proof Let A be a column competent matrix. Then by Theorem 2.1, it is locally w-unique. Suppose w^* is locally unique solution of LCP(q, A) such that $w^* = q + Az^*$ and the system (3.4) has a nonzero solution $(\bar{w}_\alpha, \bar{z}_\beta)$.
Now $\begin{bmatrix} \bar{z}_\alpha \\ \bar{w}_\beta \end{bmatrix} = \begin{bmatrix} A'_{\alpha\alpha} & A'_{\alpha\beta} \\ A'_{\beta\alpha} & A'_{\beta\beta} \end{bmatrix} \begin{bmatrix} \bar{w}_\alpha \\ \bar{z}_\beta \end{bmatrix} = 0$ implies that $\begin{bmatrix} \bar{w}_\alpha \\ \bar{w}_\beta \end{bmatrix} = \begin{bmatrix} A_{\alpha\alpha} & A_{\alpha\beta} \\ A_{\beta\alpha} & A_{\beta\beta} \end{bmatrix} \begin{bmatrix} \bar{z}_\alpha \\ \bar{z}_\beta \end{bmatrix}$.
Clearly, $\bar{w} = A\bar{z}$ and $(w^*)^T\bar{z} = 0$, $(\bar{w})^Tz^* = 0$. Hence $(w^* + \lambda\bar{w}, z^* + \lambda\bar{z})$ solves LCP(q, A) for all $\lambda \geq 0$. This contradicts the local uniqueness of w^*. Therefore, $(w_\alpha, z_\beta) = (0, 0)$ is the only solution of the system (3.4).

Theorem 3.12 *Suppose* (w^*, z^*) *is the solution of LCP*(q, A) *such that* $w^* = q + Az^*$ *where* $\alpha = \{i : w_i{}^* > 0\}$ *and* $\beta = \{i : w_i{}^* = 0\}$. *Further suppose* $\begin{bmatrix} z_\alpha \\ w_\beta \end{bmatrix} = \begin{bmatrix} A'_{\alpha\alpha} & A'_{\alpha\beta} \\ A'_{\beta\alpha} & A'_{\beta\beta} \end{bmatrix} \begin{bmatrix} w_\alpha \\ z_\beta \end{bmatrix} = 0$, $w_\alpha > 0$, $z_\beta > 0$. *If* $(z_\alpha, z_\beta) = (0, 0)$ *is the only solution of* $w_\beta = A_{\beta\alpha}z_\alpha + A_{\beta\beta}z_\beta = 0$ *then* $A = \begin{bmatrix} A_{\alpha\alpha} & A_{\alpha\beta} \\ A_{\beta\alpha} & A_{\beta\beta} \end{bmatrix}$ *is column competent.*

Proof Suppose the matrix A is not column competent. So w^* is not locally unique.
Now $\begin{bmatrix} z_\alpha \\ w_\beta \end{bmatrix} = \begin{bmatrix} A'_{\alpha\alpha} & A'_{\alpha\beta} \\ A'_{\beta\alpha} & A'_{\beta\beta} \end{bmatrix} \begin{bmatrix} w_\alpha \\ z_\beta \end{bmatrix} = 0$ implies that $\begin{bmatrix} w_\alpha \\ w_\beta \end{bmatrix} = \begin{bmatrix} A_{\alpha\alpha} & A_{\alpha\beta} \\ A_{\beta\alpha} & A_{\beta\beta} \end{bmatrix} \begin{bmatrix} z_\alpha \\ z_\beta \end{bmatrix}$ and $(w^*)^Tz = 0$, $(w)^Tz^* = 0$. Hence $(w^* + \lambda w, z^* + \lambda z)$ solves LCP(q, A) for all $\lambda \geq 0$. Hence $\exists$ a sequence of vectors $\{\bar{w}^k\}$ converging to w^* such that each $(\bar{w}^k, \bar{z}^k) = (w^* + \lambda^k w, z^* + \lambda^k z)$ is a solution of LCP(q, A) with $\bar{w}^k = q + A\bar{z}^k$. Since $\bar{w}^k \to w^*$ and $\bar{z}^k \to z^*$, it follows that $\bar{w}^k_\alpha > 0, \bar{z}^k_\beta > 0$. By complementarity $\bar{z}^k_\alpha = 0, \bar{w}^k_\beta = 0$. Consider $v^k = \bar{w}^k - w^*$ and $u^k = \bar{z}^k - z^*$. The normalized sequence $\{v^k/\|v^k\|\}$ is bounded and converges to $v^* \neq 0$ as $k \to \infty$. Similarly, the normalized sequence $\{u^k/\|u^k\|\}$ is bounded and converges to $u^* \neq 0$ as $k \to \infty$. Now for all large k, we have $\bar{w}^k_\beta - w^*_\beta = \lambda^k w_\beta = 0 = A_{\beta\alpha}u^k_\alpha + A_{\beta\beta}u^k_\beta$. Thus dividing by $\|u^k\|$ and $k \to \infty$, we have $A_{\beta\alpha}u_\alpha{}^* + A_{\beta\beta}u_\beta{}^* = 0$. Therefore, $u^* = \begin{bmatrix} u_\alpha{}^* \\ u_\beta{}^* \end{bmatrix} \neq 0$ is the nonzero solution of system $w_\beta = A_{\beta\alpha}z_\alpha + A_{\beta\beta}z_\beta = 0$. It contradicts that $(z_\alpha, z_\beta) = (0, 0)$ is the only solution of the system $w_\beta = A_{\beta\alpha}z_\alpha + A_{\beta\beta}z_\beta = 0$. Hence A is column competent.

4 Conclusion

The complementary condition is an important issue in operations research. The concept of matrix theoretic approach helps to develop many theories of linear complementary problems. In this study, we consider column competent matrix in the context of local w-uniqueness property which is important both for the theory as well as solution method of complementarity problem. The results based on w-uniqueness and column competent matrix class motivate future study and application in matrix theory.

References

1. Cottle RW, Pang JS, Stone RE (1992) The linear complementarity problem. SIAM
2. Das AK (2016) Properties of some matrix classes based on principal pivot transform. Ann Oper Res 243(1–2):375–382
3. Das AK, Jana R (2018) Deepmala: on generalized positive subdefinite matrices and interior point algorithm. In: Kar S, Maulik U, Li X (eds) Operations research and optimization. FOTA 2016. Springer proceedings in mathematics & statistics, vol 225. Springer, Singapore. https://doi.org/10.1007/978-981-10-7814-9_1
4. Das AK, Jana R (2017) Deepmala: finiteness of criss-cross method in complementarity problem. In: Giri D, Mohapatra R, Begehr H, Obaidat M (eds) Mathematics and computing. ICMC 2017. Communications in computer and information science, vol 655. Springer, Singapore. https://doi.org/10.1007/978-981-10-4642-1_15
5. Den HD, Roos C, Terlaky T (1993) The linear complimentarity problem, sufficient matrices, and the criss-cross method. Linear Algebra Appl 187:1–14
6. Ingleton AW (1966) A probelm in linear inequalities. Proc Lond Math Soc 3(1):519–536
7. Jana R, Das AK, Dutta A (2019) On hidden Z-matrix and interior point algorithm. Opsearch 56(4):1108–1116
8. Jana R, Dutta A, Das AK (2019) More on hidden Z-matrices and linear complementarity problem. Linear Multilinear Algebra 1–10
9. Mohan SR, Neogy SK, Das AK (2001) More on positive subdefinite matrices and the linear complementarity problem. Linear Algebra Appl 338(1–3):275–285
10. Mohan SR, Neogy SK, Das AK (2001) On the classes of fully copositive and fully semimonotone matrices. Linear Algebra Appl 323(1–3):87–97
11. Neogy SK, Das AK (2005) On almost type classes of matrices with Q-property. Linear Multilinear Algebra 53(4):243–257
12. Neogy SK, Das AK (2005) Principal pivot transforms of some classes of matrices. Linear Algebra Appl 400:243–252
13. Neogy SK, Das AK (2006) Some properties of generalized positive subdefinite matrices. SIAM J Matrix Anal Appl 27(4):988–995
14. Pang JS (1988) Two characterization theorems in complementarity theory. Oper Res Lett 7(1):27–31
15. Samelson H, Thrall RM, Wesler O (1958) A partition theorem for Euclidean n-space. Proc Amer Math Soc 9(5):805–807
16. Sun J, Huang ZH (2006) A smoothing Newton algorithm for the LCP with a sufficient matrix that terminates finitely at a maximally complementary solution. Optim Methods Softw 21(4):597–615
17. Väliaho H (1996) P*-matrices are just sufficient. Linear Algebra Appl 239:103–108
18. Xu S (1999) On local w-uniqueness of solutions to linear complementarity problem. Linear Algebra Appl 290(1–3):23–29

On Differential Uniformity and Nonlinearity of Permutations on $\mathbb{Z}_n$

Prachi Gupta, P. R. Mishra, and Atul Gaur

Abstract In this paper, we present an alternate expression for differential uniformity of permutations on $\mathbb{Z}_n$, the ring of integers modulo n, which may be useful for computational purposes. We also explore links between nonlinearity and differential uniformity of bijections on $\mathbb{Z}_n$. Further, we deduce some results on the graph of permutations on $\mathbb{Z}_n$. Based on our observation and experimental results, we present a couple of conjectures also.

Keywords Differential uniformity · Nonlinearity · Permutation

1 Introduction

Over the years cryptographers have relied on permutations to propose a robust cryptosystem. Stream ciphers like VPMC [11], RC4, and its variants [10], etc., depend upon permutations on n symbols. Nonlinearity [7] and differential uniformity [9] play an important role in investigating any permutation on $\mathbb{Z}_n$. Usually, permutations having low differential uniformity and high nonlinearity are desired in designing any cryptosystem to avert attacks like linear cryptanalysis [5], differential cryptanalysis [2] and other algebraic attacks. Some researchers took up the task of expanding the limited knowledge about the nonlinearity of permutations on $\mathbb{Z}_n$ and initial work focused on this can be seen in [4, 6, 7]. Some results on the nonlinearity of permutations on a finite commutative ring can be seen in [8].

In this paper, we focus on properties like nonlinearity and differential uniformity of permutations on $\mathbb{Z}_n$. There are many unanswered questions about the nonlinearity and differential uniformity of permutation on $\mathbb{Z}_n$. In Sect. 3, we derive a new expression for the differential uniformity of any permutation on $\mathbb{Z}_n$. In the next section, we

P. Gupta (✉) · A. Gaur
Department of Mathematics, University of Delhi, Delhi 110007, India

P. R. Mishra
Scientific Analysis Group, Defence Research and Development Organization,
Metcalfe House, Delhi 110054, India

© The Author(s), under exclusive license to Springer Nature Singapore Pte Ltd. 2022

D. Giri et al. (eds.), *Proceedings of the Seventh International Conference on Mathematics and Computing*, Advances in Intelligent Systems and Computing 1412,
https://doi.org/10.1007/978-981-16-6890-6_47

answer the question: *Is there any connection between the nonlinearity and differential uniformity of a permutation on* $\mathbb{Z}_n$?. In Sect. 5, we analyze the partitions of the graph of the permutations on $\mathbb{Z}_n$ and draw some fascinating conclusions on their sizes. In case of a permutation on $\mathbb{Z}_{2m}$ having zero nonlinearity, we give the exact value of the size of partitions.

2 Preliminaries

We now recall some definitions. For any natural number n, $\mathbb{Z}_n^*$ represents the set of all invertible elements in the ring $\mathbb{Z}_n = \{0, 1, \ldots, n-1\}$. We frequently use $|E|$ to denote the cardinality of the set E. For the convenience of the reader, we omit writing mod$n$ every time we deal with any bijection on $\mathbb{Z}_n$. However, it is kept in mind that the value always lies in the set $\{0, 1, \ldots, n-1\}$ to avoid any ambiguity arising while applying further mod function or any other operation. The number $d_H(f, g)$ gives the distance between any two permutations f and g on $\mathbb{Z}_n$ [7], i.e.,

$$d_H(f, g) := |\{x \in \mathbb{Z}_n : f(x) \neq g(x)\}|.$$

The symbol

$$\mathcal{A}_n := \{\alpha x + \beta : \alpha \in \mathbb{Z}_n^* \text{ and } \beta \in \mathbb{Z}_n\}$$

denotes the set of all affine functions on $\mathbb{Z}_n$. The nonlinearity of any bijective map f over $\mathbb{Z}_n$ [7] is denoted by $\mathcal{NL}(f)$ and is defined as

$$\mathcal{NL}(f) := \min\{d_H(f, g) : g \in \mathcal{A}_n\}.$$

The difference of n and the number of fixed points of f gives an upper bound on the $\mathcal{NL}(f)$ [6], i.e.,

$$\mathcal{NL}(f) \leq n - |\{\text{ fixed points of f }\}|.$$

We regularly use the fact that nonlinearity of a permutation is preserved under translation [6].

Let f be a bijection on $\mathbb{Z}_n$. Then for $a, b \in \mathbb{Z}_n \setminus \{0\}$ we define $\delta_f(a, b)$ [3] as

$$\delta_f(a, b) := |\{x \in \mathbb{Z}_n : f(x + a) - f(x) = b\}|.$$

The differential uniformity Δ_f [3, 9] of the permutation f is given by

$$\Delta_f := \max_{a,b \in \mathbb{Z}_n \setminus \{0\}} \{\delta_f(a, b)\}. \tag{1}$$

Note that $2 \leq \Delta_f \leq n$, for any bijection f on $\mathbb{Z}_n$. For a permutation f on $\mathbb{Z}_n$ if $\Delta_f = 2$, then we say f is an almost perfect nonlinear (APN) permutation [1].

By the abuse of the language we use δ or $\delta(a, b)$ to denote that the properties of symbol $\delta_f(a, b)$ are independent from the choice of the permutation f on $\mathbb{Z}_n$. The same applies for Δ_f. The symbol $\lceil . \rceil$ denotes the ceiling function.

3 An Alternate Expression for Δ

In this section, we derive a new expression for Δ when a permutation on n symbols is regarded as a bijection on the ring $\mathbb{Z}_n$. This alternate formulation can be advantageous for computational purposes as it requires lesser number of computations to obtain the value of Δ. The following result will play a crucial role in proving the main theorem of this section.

Proposition 1 *Let f be a permutation on $\mathbb{Z}_n$ and $0 < a, b < n$ then*

$$\delta_f(a, b) = \delta_f(n - a, n - b)$$

where $\delta_f(a, b) = |\{x \in \mathbb{Z}_n : f(x + a) - f(x) = b\}|$.

Proof As

$$\delta_f(a, b) = |\{x \in \mathbb{Z}_n : f(x + a) - f(x) = b\}|$$

and

$$\delta_f(n - a, n - b) = |\{x \in \mathbb{Z}_n : f(x + n - a) - f(x) = n - b\}|,$$

the map $\delta_f(a, b) \to \delta_f(n - a, n - b)$ defined as $x \mapsto x - a$ is a bijection. The result now follows immediately.

Given the above result we propose a new formula to calculate the differential uniformity of any bijection on $\mathbb{Z}_n$.

Theorem 1 *For any permutation f on $\mathbb{Z}_n$, the differential uniformity of f can be calculated as*

$$\Delta_f = \max_{\substack{0<a\leq\lceil \frac{n}{2}\rceil, \\ b\in\mathbb{Z}_n\setminus\{0\}}} \{\delta_f(a, b)\}.$$

Proof The differential uniformity Δ_f of f can be written as

$$\Delta_f = \max \left\{ \max_{\substack{0<a\leq\lceil \frac{n}{2}\rceil, \\ 0<b<n}} \{\delta_f(a, b)\}, \max_{\substack{\lceil \frac{n}{2}\rceil\leq a<n, \\ 0<b<n}} \{\delta_f(a, b)\} \right\}.$$

Using Proposition 1 in the second term of the right-hand side of the above expression and suitably adjusting the indices, the theorem follows.

In Theorem 1, as the choices of 'a' are reduced significantly when compared to the formula given by the Eq. (1), the new formula thus becomes more valuable.

4 Relation Between Differential Uniformity and Nonlinearity of Any Bijective Map on $\mathbb{Z}_n$

In the following theorem, we give a lower bound on the nonlinearity of any permutation on $\mathbb{Z}_n$ in terms of an expression involving its differential uniformity and the cardinality of the set $\mathbb{Z}_n$. Further, we explore that the differential uniformity of any permutation on $\mathbb{Z}_n$ is bounded below by a formula determined by its nonlinearity and n.

Theorem 2 *Let f be a permutation on $\mathbb{Z}_n$ for $n \geq 2$. Then we have the following:*

(i) If the nonlinearity of f is L, then the differential uniformity of f is at least $\left\lceil \frac{(n-L)(n-L-1)}{n-1} \right\rceil$.
(ii) If the differential uniformity of f is Δ, then the nonlinearity of f is greater than or equal to $\left\lceil n - \frac{1}{2} - \sqrt{(n-1)\Delta + \frac{1}{4}} \right\rceil$.

Proof Let f be a permutation on $\mathbb{Z}_n$ with nonlinearity L. It means that there exist $a \in \mathbb{Z}_n^*$ and $b \in \mathbb{Z}_n$ such that $|N| = t$ where $N = \{x \in \mathbb{Z}_n : f(x) = ax + b\}$ and $t = n - L$. Consider a subset T of $N \times N$ defined as

$$T = \{(x, y) \in N \times N : x \neq y\}$$

Then $|T| = t(t-1)$. It is known that $L < n$. If $L = n - 1$, then the inequality is trivially satisfied. Now, we may assume that $L < n - 1$. Define a map $\phi : T \to \mathbb{Z}_n \setminus \{0\}$ as $\phi(x, y) = x - y$. Observe that no ordered pair in T contains identical elements. Therefore, the map ϕ is well-defined. There exists an element $\alpha \in \mathbb{Z}_n \setminus \{0\}$ such that α is the image of at least u elements of T, where $u = \left\lceil \frac{t(t-1)}{n-1} \right\rceil$. It means

$$\begin{aligned} u &\leq |\{(x, y) \in T : \phi(x, y) = \alpha\}| \\ &= |\{(x, y) \in T : x - y = \alpha\}| \\ &= |\{y \in N : y + \alpha \in N\}| \\ &= |\{y \in \mathbb{Z}_n : f(y) = ay + b, f(y + \alpha) = a(y + \alpha) + b\}| \\ &\leq |\{y \in \mathbb{Z}_n : f(y + \alpha) - f(y) = a\alpha\}| \\ &= \delta(\alpha, a\alpha) \leq \Delta \end{aligned}$$

Therefore,

$$\Delta \geq \left\lceil \frac{(n-L)(n-L-1)}{n-1} \right\rceil$$

and (i) follows. As Δ is always a positive integer for any permutation we have

$$\Delta \geq \frac{(n-L)(n-L-1)}{n-1} \tag{2}$$

The inequality (2) can be reorganized as

$$(n-1)\Delta \geq (n-L)(n-L-1)$$
$$\text{i.e., } L^2-(2n-1)L+(n-1)(n-\Delta) \leq 0 \tag{3}$$

Let D be the discriminant of the above quadratic expression. Then

$$D=(2n-1)^2-4(n-1)(n-\Delta)=4(n-1)\Delta+1$$

Since $\Delta \geq 2$, we have

$$D \geq 8n-7>0$$

It means the equation $L^2-(2n-1)L+(n-1)(n-\Delta)=0$ has two distinct real roots (say r_1 and r_2 and $r_1 < r_2$). Then

$$L^2-(2n-1)L+(n-1)(n-\Delta)=(L-r_1)(L-r_2)$$

Clearly (3) holds when $r_1 \leq L \leq r_2$. We have $r_1=n-\frac{1}{2}-\sqrt{(n-1)\Delta+\frac{1}{4}}$ and $r_2=n-\frac{1}{2}+\sqrt{(n-1)\Delta+\frac{1}{4}}$. Observe that $r_2>n-1$. Therefore, $L \leq r_2$ is trivially satisfied and we are left with the inequality in (ii).

As an application of the above theorem, we deduce a lower bound on the nonlinearity of any almost perfect nonlinear permutation on $\mathbb{Z}_n$.

Corollary 1 *The nonlinearity of an APN permutation on* $\mathbb{Z}_n$ *is not less than*

$$\left\lceil n-\frac{1}{2}-\sqrt{2n-\frac{7}{4}}\right\rceil.$$

Proof For an APN permutation as $\Delta=2$. The result now follows from Theorem 2 (ii).

5 Some Results on the Graph of Permutations

The conventional way to calculate the nonlinearity of any permutation on $\mathbb{Z}_n$ is to measure its distance from all the affine functions on $\mathbb{Z}_n$ and declare the minimum value as its nonlinearity. Till now there is no simple method to segregate affine and non-affine permutations on $\mathbb{Z}_n$. When n is even, the last result of this section is beneficial in separating the permutations having nonlinearity zero from the permutations having nonlinearity non-zero solely by partitioning the given permutation on the basis of the point and its image. The aim of the results of this section is to broaden the limited knowledge about the graph of the permutations on $\mathbb{Z}_n$.

Theorem 3 *Let f be any bijection on $\mathbb{Z}_n$ and*

$$\begin{aligned} C_1 &= \{(x, f(x)) : x \in \mathbb{Z}_n, \textit{both} x \textit{ and } f(x) \textit{are odd}\}, \\ C_2 &= \{(x, f(x)) : x \in \mathbb{Z}_n, x \textit{ is odd and } f(x) \textit{is even}\}, \\ C_3 &= \{(x, f(x)) : x \in \mathbb{Z}_n, x \textit{ is even and } f(x) \textit{is odd}\}, \\ C_4 &= \{(x, f(x)) : x \in \mathbb{Z}_n, \textit{both } x \textit{ and } f(x) \textit{are even}\} \end{aligned}$$

be the partitions of the graph of f. Then $|C_2| = |C_3|$ and

$$|C_4| - |C_1| = \begin{cases} 0, & \textit{if } n \textit{ is even}, \\ 1, & \textit{otherwise}. \end{cases}$$

Proof Let n be even. Then there are exactly $n/2$ even numbers and $n/2$ odd numbers in the set $\{0, 1, \ldots, n-1\}$. Therefore, we have

$$\begin{aligned} |C_2| + |C_4| &= n/2, \\ |C_3| + |C_4| &= n/2, \\ |C_1| + |C_2| &= n/2, \\ |C_1| + |C_3| &= n/2. \end{aligned}$$

The above first two equations give $|C_2| = |C_3|$ and subtracting $|C_2| + |C_4| = n/2$ from $|C_1| + |C_2| = n/2$ gives $|C_1| = |C_4|$.

Now suppose that n is odd. As there are $(n+1)/2$ even and $(n-1)/2$ odd choices of x and $f(x)$, we have

$$\begin{aligned} |C_2| + |C_4| &= (n+1)/2, \\ |C_3| + |C_4| &= (n+1)/2, \\ |C_1| + |C_2| &= (n-1)/2, \\ |C_1| + |C_3| &= (n-1)/2. \end{aligned}$$

Arguing as above the result follows in this case also.

Remark Observe that the cardinality of one of $C_i's$ is greater than or equal to $n/4$ as there are total 4 classes and the sum of cardinalities of these classes is n.

Proposition 2 *Let f be a bijection on $\mathbb{Z}_{2m}$ of nonlinearity zero and*

$$\begin{aligned} C_1 &= \{(x, f(x)) : x \in \mathbb{Z}_{2m}, \text{both } x \text{ and } f(x) \text{are odd}\}, \\ C_2 &= \{(x, f(x)) : x \in \mathbb{Z}_{2m}, x \text{ is odd and } f(x) \text{is even}\}, \\ C_3 &= \{(x, f(x)) : x \in \mathbb{Z}_{2m}, x \text{ is even and } f(x) \text{is odd}\}, \\ C_4 &= \{(x, f(x)) : x \in \mathbb{Z}_{2m}, \text{both } x \text{ and } f(x) \text{are even}\} \end{aligned}$$

then either $|C_1| = m$ or $|C_2| = m$.

Proof As $2m \equiv 0 \pmod 2$, Theorem 3 gives $|C_1| = |C_4|$ and $|C_2| = |C_3|$. Note that the permutations of nonlinearity zero are essentially the affine permutations. Therefore, $f(x) = \alpha x + \beta$ where $\alpha \in \mathbb{Z}_{2m}^*$ and $\beta \in \mathbb{Z}_{2m}$. Observe that each $\alpha \equiv 1 \pmod 2$. Now if β is even we conclude that $x \equiv f(x) \pmod 2$. Hence $|C_1| = m$ and $|C_4| = m$. Otherwise, if β is odd then for $x \equiv 0 \pmod 2$ we see that $f(x) \equiv 1 \pmod 2$, and when $x \equiv 1 \pmod 2$ we have $f(x) \equiv 0 \pmod 2$. Thus $|C_2| = |C_3| = m$.

As an immediate consequence of the Proposition 2 we have the following result.

Corollary 2 *For any permutation f on $\mathbb{Z}_{2m}$, if $|C_i| \neq m$ for at least three distinct values of $i \in \{1, 2, 3, 4\}$, then $\mathcal{NL}(f) \geq 2$.*

It is worth noting that the converse of Proposition 2 is not true as is evident from the following example.

Example 1 For the permutation [0 3 6 9 8 5 4 7 2 1] on $\mathbb{Z}_{10}$. Note that

$$C_1 = \{(0, 0), (2, 6), (4, 8), (6, 4), (8, 2)\}$$

and

$$C_4 = \{(1, 3), (3, 9), (5, 5), (7, 7), (9, 1)\},$$

i.e., $|C_1| = |C_4| = 5$. Using Theorem 1, the differential uniformity can be easily calculated as 4. In view of Theorem 2 (ii), the nonlinearity is at least $\left\lceil 10 - \frac{1}{2} - \sqrt{(10-1)4 + \frac{1}{4}} \right\rceil = 4$. Clearly the given permutation has non-zero nonlinearity.

Let max $\mathcal{NL}_n$ denotes the maximum nonlinearity of any bijection on $\mathbb{Z}_n$. From now on p will denote the odd prime number. It has been proved that the max $\mathcal{NL}_p \leq p - 2$ [7]. We believe that upper bound on the nonlinearity of any permutation on $\mathbb{Z}_p$ can be improved. The nonlinearity of any permutation on $\mathbb{Z}_p$ can be atmost $p - 3$. Till now there is no such result for the permutations on $\mathbb{Z}_{2p}$. In case of permutations on $\mathbb{Z}_{2p}$, we feel that there does not exist any permutation of nonlinearity $2p$ and $2p - 1$. The above discussion can be summarized as

Conjecture 1 Let p be any odd prime then

(i) $\max \mathcal{NL}_p \leq p - 3$.
(ii) $\max \mathcal{NL}_{2p} \leq 2p - 2$.

We also tried to estimate the lower bound on $\max \mathcal{NL}_p$. For prime p between 7 and 1,500, we generated 1000 random permutations on $\mathbb{Z}_p$ for each p and studied their nonlinearity. If for a given p, there is an APN permutation, the bound given by Corollary 1 serves as a lower bound on $\max \mathcal{NL}_p$. We tried to arrive at an expression for a lower bound on $\max \mathcal{NL}_p$ with the following aim:

(i) It should provide an improvement over bound given by Corollary 1.
(ii) It should coincide with some of the $\max \mathcal{NL}_p$ values.
(iii) It should provide an approximate fit for experimental plot of n vs $\max \mathcal{NL}_p$ obtained from random permutations.

Based on the above criteria, we arrive at a lower bound. We state this lower bound in form of a conjecture stated below.

Conjecture 2 For any prime $p \geq 5$, there always exist a permutation of nonlinearity at least $\lceil p - 1.5 - \sqrt{p - 1.5} \rceil$.

This lower bound has been verified experimentally for primes between 5 and 1500. The experimental verification is shown in the following graph (Fig. 1).

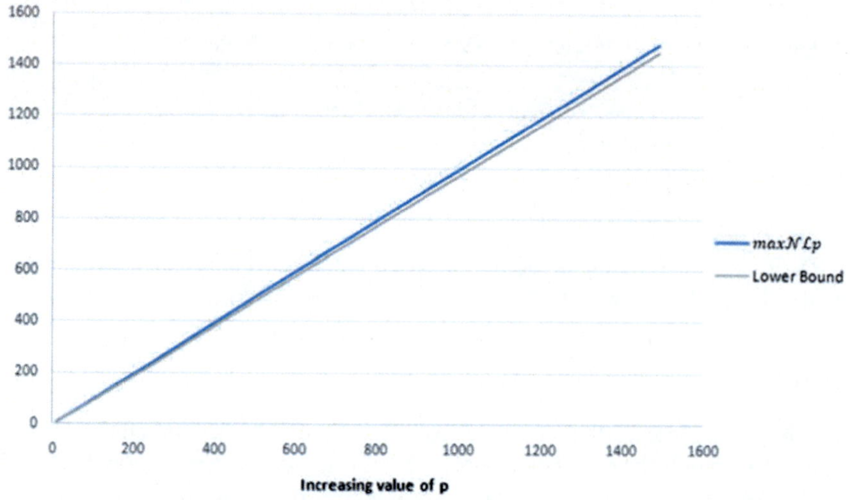

Fig. 1 Comparison of $\max \mathcal{NL}_p$ and the proposed lower bound

6 Conclusion

We obtain a new formula to calculate the differential uniformity of a bijection on the set of integer modulo n. The formula thus obtained takes lesser number of steps than the earlier one and hence can serve better in computational context. We estimate a lower bound on the nonlinearity of a permutation on $\mathbb{Z}_n$ in terms of its differential uniformity and n. This bound will be helpful in understanding the interdependence of differential uniformity and nonlinearity. Using this bound we conclude that an APN permutation on $\mathbb{Z}_n$ has a non-zero lower bound on its nonlinearity. At the end, we present some of our observations related to graph of a permutation on $\mathbb{Z}_n$ and state some open problems related to nonlinearity. We believe that these observations will play a crucial role in settling these open problems.

In future, we will try to prove the open problems mentioned in the paper and attempt to improve the bound on nonlinearity (if possible). We also would like to inspect similar results for differential uniformity of permutations on $\mathbb{Z}_n$.

Acknowledgements The authors are extremely thankful to the referees for their valuable comments and suggestions which led to the overall improvement in the quality and presentation of the paper. The first author would like to acknowledge the financial support by National Board for Higher Mathematics (NBHM), Department of Atomic Energy, India vide sanction no. 0203/6/2019/R&D-II/3659 and 0203/6/2019/R&D-II/10309.

References

1. Beth T, Ding C (1994) On almost perfect nonlinear permutations. Advances in cryptology - EUROCRYPT'93. In: Helleseth T (eds) LNCS, vol 765. Springer, Berlin, pp 65–76. https://doi.org/10.1007/3-540-48285-7_7
2. Biham E, Shamir A (1991) Differential cryptanalysis of DES-like cryptosystems. J Cryptology 4:3–72. https://doi.org/10.1007/BF00630563
3. Budaghyan L, Pott A (2009) On differential uniformity and nonlinearity of functions. Discret Math 309:371–384. https://doi.org/10.1016/j.disc.2007.12.025
4. Kumar Y, Mishra PR, Sharma RK (2018) Nonlinearity of k-cycle permutations on $\mathbb{Z}_n$. Asian-Europ J Math 11(02):1850020. https://doi.org/10.1142/S1793557118500201
5. Matsui M (1994) Linear cryptanalysis method for DES cipher. Advances in cryptology - EUROCRYPT'93. In: Helleseth T (eds) LNCS, vol 765. Springer, Berlin, pp 386–397. https://doi.org/10.1007/3-540-48285-7_33
6. Mishra PR, Gupta I, Pillai NR (2014) Non-linearity and affine equivalence of permutations. https://eprint.iacr.org/2014/974
7. Kumar Y, Mishra PR, Pillai NR, Sharma RK (2017) Affine equivalence and non-linearity of permutations over $\mathbb{Z}_n$. Appl Algebra Eng Commun Comput 28:257–279. https://doi.org/10.1007/s00200-016-0307-6
8. Mishra PR, Kumar Y, Pillai NR, Sharma RK (2018) On non-linearity and affine equivalence of permutations over an arbitrary finite commutative ring with unity. Cryptologia 42(1):81–94. https://doi.org/10.1080/01611194.2017.1325788

9. Nyberg K (1993) Differentially uniform mappings for cryptography. Advances in Cryptology - EUROCRYPT'93. In: Helleseth T (eds) LNCS, vol 765. Springer, Berlin, pp 55–64. https://doi.org/10.1007/3-540-48285-7_6
10. Paul G, Maitra S (2012) RC4 stream cipher and its variants. CRC Press, Boca Raton. https://doi.org/10.1201/b11310
11. Zoltak B (2004) VMPC one-way function and stream cipher. Fast software encryption 2004. LNCS, vol 3017. Springer, Berlin, pp 210–225. https://doi.org/10.1007/978-3-540-25937-4_14

Computing the 2-Adic Canonical Lift of Genus 2 Curves

Abdoulaye Maïga and Damien Robert

Abstract Let $\mathbb{k}$ be a field of even characteristic and $\mathcal{M}_2(\mathbb{k})$ the moduli space of the genus 2 curves defined over $\mathbb{k}$. We first compute modular polynomials with good reduction using Igusa's arithmetic invariants. These modular polynomials provide a canonical lifting of genus 2 curves in even characteristic. The lifted Frobenius is characterized by the reduction behaviors of the Weierstrass points over $\mathbb{k}$. When $\mathbb{k} = \mathbb{F}_{2^n}$, this allows us to compute the cardinality of the Jacobian variety in quasi-quadratic time $\tilde{O}(n^2)$. We give a detailed description with the necessary optimizations for an efficient implementation.

Keywords Arithmetic invariants of genus 2 curves · Modular polynomials · Canonical lift · Point counting

1 Introduction

Let $\mathcal{A}/\mathbb{F}_q$ (with $q = p^n$) be an ordinary abelian variety. A classical result due to Lubin, Serre, and Tate [24] states that there exists a unique abelian variety $\tilde{\mathcal{A}}$ over $\mathbb{Z}_q$ such that the modulo p reduction of $\tilde{\mathcal{A}}$ is $\mathcal{A}$ and $\mathrm{End}(\tilde{\mathcal{A}}) \cong \mathrm{End}(\mathcal{A})$ as a ring.

We thank the FAST team.

A. Maïga (✉)
Cheikh Anta Diop University, FST, B.P.5005 Dakar, Senegal
e-mail: abdoulaye.maiga@ucad.edu.sn

D. Robert
INRIA Bordeaux-Sud-Ouest, 33405 Talence Cedex, France
e-mail: damien.robert@inria.fr
URL: http://www.normalesup.org/~robert/

Institut de Mathématiques de Bordeaux, 33405 Talence Cedex, France

A. Maïga · D. Robert
Équipe FAST LIRIMA Laboratoire International de Recherche en Informatique et Mathématiques Appliquées, Bruz Cedex, France

© The Author(s), under exclusive license to Springer Nature Singapore Pte Ltd. 2022
D. Giri et al. (eds.), *Proceedings of the Seventh International Conference on Mathematics and Computing*, Advances in Intelligent Systems and Computing 1412,
https://doi.org/10.1007/978-981-16-6890-6_48

When $\Bbbk$ is a field, an absolutely irreducible principally polarized abelian surface over $\Bbbk$ is the Jacobian $\mathrm{Jac}(\mathcal{C})$ of a genus 2 hyperelliptic curve $\mathcal{C}$. If $\mathcal{A}$ is not absolutely simple, then it is isomorphic to the product of two elliptic curves, $\mathcal{A} = E_1 \times E_2$, and lifting $\mathcal{A}$ amount to lifting the two elliptic curves, which is well known. Hence, in this article, we focus on how to compute a canonical lift of $\mathrm{Jac}(\mathcal{C})$ (hence of $\mathcal{C}$), in the particular case of when the characteristic of $\Bbbk$ is equal to 2. We remark that lifting $\mathcal{C}$ may mean different things: lifting modular invariants, lifting modular forms, or lifting the curve itself. In this article, we explain how to lift a curve along with its canonical basis of differential form; from this data it is easy to compute lifts of modular functions or even lifts of vectorial modular forms.

A standard strategy to compute canonical lifts is to use modular polynomials for some specific modular invariants. Over $\mathbb{C}$, the isogeny class of a generic Jacobian variety of a genus 2 curves is given by its tuple of three Igusa's invariants (j_1, j_2, j_3). For p a prime number, the modular polynomials encode the p-isogenous abelian surfaces in term of these Igusa's invariants. More precisely, a birational model of the (coarse) moduli space $\mathfrak{A}_2(p) = \mathfrak{H}_2/\Gamma^0(p)$ is given by $\mathbb{C}(\mathfrak{A}_2(p)) = \mathbb{C}(j_1, j_2, j_3, j_{1,p}, j_{2,p}, j_{3,p})$, where $j_{i,p} = j_i(\tau/p)$, and $\mathfrak{H}_2$ is the Siegel moduli space. We can now define the pth-modular polynomials as follows: $\phi_{1,p}(X)$ is the minimal polynomial of $j_{1,p}$ over $K = \mathbb{C}(\mathfrak{A}_2) = \mathbb{C}(j_1, j_2, j_3)$, and $j_{2,p} = \phi_{2,p}(j_{1,p})$, $j_{3,p} = \phi_{3,p}(j_{1,p})$ for $\phi_{2,p}(X)$ and $\phi_{3,p}(X)$ monic polynomials in $\mathbb{C}(j_1, j_2, j_3)[X]$ of degree less than $\deg(\phi_{1,p}(X))$. If x goes through the roots over $\mathbb{C}$ of $\phi_{1,p}(X)$, then $(x, \phi_{2,p}(x), \phi_{3,p}(x))$ are the absolute j-invariants of a principally polarized abelian surfaces (p, p)-isogenous to an abelian variety with invariants (j_1, j_2, j_3). Hence, the schematic locus of the 2-dimensional p-modular polynomials:

$$\phi_{1,p}(X_1, X_2, X_3, Y_1) = 0, \quad Y_2 = \phi_{2,p}(X_1, X_2, X_3, Y_1) \text{ and } Y_3 = \phi_{3,p}(X_1, X_2, X_3, Y_1)$$

describe a birational model of $\mathfrak{A}_2(p)$.

These polynomials were first computed by R.Dupont in his thesis [11]. Some improvements have been proposed by others [3] and E.Milio in [29] computed modular polynomials using Streng's [38] and Gundlach [17] invariants in order to reduce their size. However, Streng's invariants have bad reduction modulo 2, hence these modular polynomials too.

An alternative in characteristic 2 is to use theta functions (of level 2 and 4) since they can be defined for ordinary abelian varieties [5, 7, 35]. Modular polynomials in theta functions are derived from the duplication formula [27, 28], a generalization of the AGM, and have also been used to compute canonical lifts for point counting [23]. However, these are not suitable for our purpose, since they require to take a field extension. Indeed, theta functions are modular forms of level $\Gamma(2, 4)$ or $\Gamma(4, 8)$, so they are not rational in general. While not necessarily a problem for applications to point counting (from the asymptotic point of view), computing curves from lifts of theta functions would give us a curve defined over an extension of $\mathbb{Z}_q$. One would then need to employ a delicate Galois descent algorithm, controlling the p-adic precision to descend back to $\mathbb{Z}_q$. Furthermore, the field extension is only really manageable when working over a finite field. Our whole algorithm can be adapted to functions fields of characteristic two, and in these cases, we really want to use modular invariants of level 1.

Our goal is as follows: firstly, use the specific arithmetic of genus 2 hyperelliptic curves, as described by Igusa, to compute modular polynomials with good reduction modulo 2. Secondly, use this arithmetic to lift the curves.

In characteristic 2, using Artin-Scheier theory, it is easy to see that a genus 2 curves is birationally equivalent to one of the three following curves :

$$Y^2 - Y = \begin{cases} \alpha X + \beta X^{-1} + \gamma(X-1)^{-1}, & (1,1,1) \\ X^3 + \alpha X + \beta X^{-1}, & (3,1) \\ X^5 + \alpha X^3, & (5) \end{cases}$$

according to the number and the degree of the ramified Weierstrass points [18, Sect. 2]. This classification can be seen on the Jacobian has the rank of the étale p-torsion. So type $(1, 1, 1)$ correspond to ordinary Jacobians (which we also call ordinary curves), type $(3, 1)$ correspond to a p-rank 1, and type (5) to supersingular Jacobians (or curves).

More generally, in all characteristics, a genus 2 curve has birationally equivalent equation

$$XY^2 + (1 + aX + bX^2)Y + X^2(c + dX + X^2) = 0$$

called its *normal form*; that is equivalent to the knowledge of the quintuple $(J_2, J_4, J_6, J_8, J_{10})$ with $J_2J_6 - J_4^2 - 4J_8 = 0$ called *Igusa arithmetic invariants* [18, Sect. 3]. Furthermore, J. Igusa [18, Sect. 7] has shown that the arithmetic variety of moduli $\mathcal{M}_2$ of genus 2 curves is generated over $\mathbb{Z}$ by ten absolute invariants in function of J_{2i}'s denoted by γ_i by Goren-Lauter [16].

These ten absolute invariants allow us to prove in Sect. 2 that in the case where $\mathrm{char}(\Bbbk) = 2$, the absolute arithmetic invariants $\mathfrak{a}_1 = J_4/J_2^2$, $\mathfrak{a}_2 = J_8/J_2^4$ and $\mathfrak{a}_3 = J_{10}/J_2^5$ generates the open set $\mathcal{M}_2[J_2^{-1}] \otimes \Bbbk$ of curves birationally equivalent to type $(1, 1, 1)$. The tuple $(0,\ J_8^3/J_6^4,\ J_{10}^3/J_6^5)$ generates the set (which lives inside $\mathcal{M}_2[J_6^{-1}] \otimes \Bbbk$) of curves birationally equivalent to type $(3, 1)$. And the set representing the type (5) (which, except for one point, live inside $\mathcal{M}_2[J_8^{-1}] \otimes \Bbbk$) can be defined using the tuple $(0, 0, J_{10}^4/J_8^5)$ of invariants.

These three subvarieties $\mathcal{M}_2[J_2^{-1}] \otimes \Bbbk$, $\mathcal{M}_2[J_6^{-1}] \otimes \Bbbk$ and $\mathcal{M}_2[J_8^{-1}] \otimes \Bbbk$, and the invariants above, admit a lifts in $\mathcal{M}_2$ over $\mathbb{Z}_q$. Our first goal is to compute, using these invariants with good reduction, the modular polynomials that parameterize $(2, 2)$-isogenous principally polarized abelian surfaces, with J_2 invertible. Thereafter, using those polynomials we want to compute the canonical lift of ordinary genus 2 curves defined over $\mathbb{F}_{2^n}$ with $n > 1$.

Like in Satoh's original approach [36] for elliptic curves, the computation of the canonical lift provides a point counting algorithm for the Jacobian variety of the genus 2 curves. As explained, we work directly with modular invariants of level 1, with have the advantage to stay on the field of definition in order to speed up the computation time.

Furthermore, we also explain how to directly lift the curve $\mathcal{C}$. Indeed, the conic equation used during Mestre's algorithm to reconstruct the curve from its invariant has bad reduction modulo 2, so we cannot use the point on the conic corresponding to $\mathcal{C}$ modulo 2 to find a rational point over $\mathbb{Z}_q$ via a Newton lift. Lifting the curve allows us to compute any covariant, hence modular form on it, which can be used to compute class polynomials [6].

We organize this paper as follows. In Sect. 2, we introduce the absolute Igusa's arithmetic invariants that describe each subvariety corresponding to the decomposition type of the curves. We propose a conversion of those lifted invariants in order to extend the known computation of modular polynomials [11, 31]. The computation of these Siegel modular polynomials is briefly detailed in Sect. 3, in particular, we explain how to adapt it to a given set of modular invariants. In Sect. 4, we propose an algorithm to lift ordinary Jacobian varieties over $\mathbb{Z}_q$ with $q = 2^n$ from the normal form of the associated genus 2 curves. We give the characteristic reduction properties of the Weierstrass points to compute the Frobenius using Richelot algorithm.

The last Sect. 5 exposes an extension to genus 2 of Satoh's point counting. We use Richelot [34] algorithm in order to evaluate the lifted Verschiebung and to reconstitute the characteristic polynomial using the product of Frobenius eigenvalues. Our lifting algorithm could also be used to compute class polynomials as in [14], without going through Rosenhain invariants (which can have negative valuations).

1.1 Notation and Convention

Let p be a prime, $\mathbb{Q}_p$ denotes the field of p-adic numbers and $\mathbb{Z}_p$ its ring of valuation. The field $\mathbb{Q}_q$ denotes the unramified extension of $\mathbb{Q}_p$ and $\mathbb{Z}_q$ its valuation ring which is also the ring $W(\mathbb{F}_q)$ of the Witt vectors over $\mathbb{F}_q$. And $\mathbb{Z}_q^{ur}$ be the maximal unramified extension of $\mathbb{Z}_q$.

Given $\mathbb{F}_q$ by $\mathbb{F}_p[X]/m(X)$, where $m(X)$ is a monic irreducible polynomial over $\mathbb{F}_p$, then $\mathbb{Q}_q$ is $\mathbb{Q}_p[X]/M(X)$ with M irreducible polynomial over $\mathbb{Z}_p[X]$ such that $M(X) = m(X) \mod p$. The complexity of an elementary operation over $\mathbb{Z}/p^k\mathbb{Z}[X]/M(X)$ requires $\tilde{O}(nk \log p)$ with Kronecker-Schönhage method.

The extension $\mathbb{Q}_q/\mathbb{Q}_p$ has a cyclic Galois group of order n, generated by an element Σ that reduces to the pth-power Frobenius automorphism σ on the residue field $\mathbb{F}_q$.

To obtain an efficient Frobenius substitution Σ, one takes m as sparse as possible and M its Teichmuller lift polynomial. We denote by : $\tilde{\mathcal{A}}$ the lift of any variety $\mathcal{A}$, and $\tilde{x}$ the lift of any element x of $\mathbb{F}_q$.

We let σ be the small Frobenius in $\mathbb{F}_q$, and σ_q the big Frobenius, and Σ, Σ_q be the lifts to $\mathbb{Q}_q$.

The coarce moduli space over $\mathbb{Z}$ of hyperelliptic curves is denoted by $\mathcal{M}_2$, it embeds via the Torelli morphism into the coarce moduli space $\mathfrak{A}_2$ of abelian schemes of relative dimension 2. We recall that a point of $\mathcal{M}_2 \otimes \mathbb{k}$ corresponds to the $\bar{\mathbb{k}}$-isomorphism class of an hyperelliptic curve C of genus 2 whose field of moduli is $\mathbb{k}$. It is of dimension 3, normal [18, Theorem 1] and birational to $\mathbb{A}^3_{\mathbb{k}}$ [18, Theorem 5].

2 Arithmetic Invariant Theory of Genus 2 Hyperelliptic Curves

We let $\mathbb{k}$ be a field and $\mathcal{C}$ a genus 2 curve over $\mathbb{k}$; then $\mathcal{C}$ admit an hyperelliptic model (which we call the standard model): $\mathcal{C} : y^2 + h(x)y = f(x)$ over $\mathbb{k}[x, y]$, where $\deg(h) \leqslant 2$, and f is a monic polynomial of degree 5 or 6.

In fact, $\mathcal{C}$ is completely determined by its function field $K = \mathbb{k}(\mathcal{C})$. Indeed, since $\mathcal{C}$ is smooth, it is the normalization of any birational model of K. The hyperelliptic involution shows that K is a degree 2 extension of $\mathbb{k}(\mathbb{P}^1_{\mathbb{k}}) = \mathbb{k}(x)$, and since $\mathcal{C}$ is smooth $K/\mathbb{k}(x)$ is separable. Hence, K can be described using Kummer theory in characteristic different from 2 or Artin-Schreier theory in characteristic 2.

2.1 *Invariants of Sextic Forms*

When the characteristic of $\mathbb{k}$ is $char(\mathbb{k}) \neq 2$, the equation of $\mathcal{C}$ reduces to $y^2 = f(x)$ and one defines the sextic :

$$F(x_1, x_2) = x_2^6 f(x_1/x_2)$$

and F has following form :

$$F(x_1, x_2) = a_0 x_1^6 + a_1 x_1^5 x_2 + \ldots + a_5 x_1 x_2^5 + a_6 x_2^6.$$

Following the *Algebraic Invariants Theory of Hilbert* [26], a *covariant* of order n, degree k and index j is a polynomial $C \in \mathbb{k}[a_0, \ldots, a_6, x_1, x_2]$ in function of the coefficients of the sextic form, of degree n in x_1, x_2 and k in a_i such that : every variable change

$$(x_1, x_2) \longrightarrow (\alpha x_1 + \beta x_2, \gamma x_1 + \delta x_2)$$

associated to the matrix $M = \begin{pmatrix} \alpha & \beta \\ \gamma & \delta \end{pmatrix} \in GL_2(\bar{\mathbb{k}})$, transforms C by a fixed power of $\det(M)$, i.e.,

$$C(a_0', \ldots, a_6', \alpha x_1 + \beta x_2, \gamma x_1 + \delta x_2) = \det(M)^j C(a_0, \ldots, a_6, x_1, x_2)$$

where the a_i''s with $i \in \{0, \cdots 6\}$ are the coefficients after the variable change. We then have $j = 3k - n/2$. Hence for a scalar covariant ($n = 0$), we also call k the weight.

The invariants I of a sextic form F are its covariants of order 0. We denote by $I(\mathcal{C})$ or $I(F)$ the value of I associated to a sextic F or to a genus 2 curves $\mathcal{C}$. Then the sextic forms F and F' are linearly equivalent if and only if there exists $r \in \overline{\mathbb{k}}^*$ such that for every invariant I, we have $I(F) = r^d I(F')$, where d is the degree of I.

In characteristic 0, let f and g be binary forms of degrees n and m, to compute covariants of sextic forms, Clebsh introduced [8] the *Ueberschiebung* operation defined by:

$$(fg)_k = \frac{(m-k)!(n-k)!}{m!n!}\left(\frac{\partial f}{\partial x}\frac{\partial g}{\partial y} - \frac{\partial f}{\partial y}\frac{\partial g}{\partial x}\right)^k$$

where in the binomial expression $\left(\frac{\partial f}{\partial x}\right)^r\left(\frac{\partial f}{\partial y}\right)^s$ means $\frac{\partial^{r+s} f}{\partial x^r \partial y^s}$. Using the Ueberschiebung, Clebsh showed that the algebra of invariants is generated by five invariants denoted as $A,\ B,\ C,\ D$, and R of respective degrees 2, 4, 6, 10, and 15 [8]. Since R^2 admits a polynomial expression in function of $A,\ B,\ C$, and D, then these four invariants suffice to characterize the linear equivalence classification of sextic forms. They are called *Clebsh invariants.*

Now let's consider a hyperelliptic model a of genus 2 curves $\mathcal{C} : y^2 = f(x) = u_0x^6 + u_1x^5 + \cdots + u_5x + u_6$ and $\alpha_1, \ldots, \alpha_6$ the distinct roots of f. If we denote by (ij) the difference $(\alpha_{\sigma(i)} - \alpha_{\sigma(j)})$ such that:

$$I_2 = u_0^2 \sum_{15} (12)^2(34)^2(56)^2\ ,$$

$$I_4 = u_0^4 \sum_{10} (12)^2(23)^2(31)^2(45)^2(56)^2(64)^2\ ,$$

$$I_6 = u_0^6 \sum_{60} (12)^2(23)^2(31)^2(45)^2(56)^2(64)^2(14)^2(25)^2(36)^2\ ,$$

$$I_{10} = u_0^{10} \prod_{i<j} (\alpha_i - \alpha_j)^2$$

where every α_i appears in each expression m times equal to the power of u_0. Then J. Igusa has shown in [18, p. 620] that I_2, I_4, I_6, and I_{10} are homogenous invariants of degree 2, 4, 6, and 10 with integers coefficients . They are called *Igusa-Clebsh invariants* because of the following relations with Clebsh invariants:

$$\begin{aligned}
I_2 &= -120\,A\,,\\
I_4 &= -720\,A^2 + 6750B\,,\\
I_6 &= 8640\,A^3 - 108000AB + 202500\,C\,,\\
I_{10} &= -62208\,A^5 + 972000\,A^3B + 1620000\,A^2C - 3037500AB^2 - 6075000BC - 4556250D.
\end{aligned}$$

2.2 Arithmetic Invariants

We are interested in Igusa's results [18] about the construction of the moduli of genus 2 curves. In characteristic 0, the knowledge of the quadruplet (A, B, C, D) (equivalently I_{2i}'s) in the weighted projective space corresponds exactly to a genus 2 curve in the coarse moduli space $\mathcal{M}_2$. However, those invariants have bad reduction in characteristic 2. In fact, the Clebsh invariants (A, B, C, D) have bad reduction in characteristic 2, 3, 5, and the Igusa-Clebsh invariants only have bad reduction in characteristic 2, 3 (their reduction is well defined, but they do not classify genus 2 curves anymore).

In general, every hyperelliptic curve of genus 2 admit a *universal normal form* [18, p. 617]

$$XY^2 + (1 + aX + bX^2)Y + X^2(c + dX + X^2) = 0$$

such that in characteristic different from 2, the Weierstrass points of the associated sextic form are the roots of the equation:

$$(1 + aX + bX^2)^2 - 4X^3(c + dX + X^2) = 0.$$

In characteristic 2, a genus 2 curve can also be defined via an Artin–Schreier equation: $y^2 - y = R(x)$, where R is a rational function in x with pole divisors. The isomorphism classes of these curves are in bijection with the orbits of $R(x)$ under the double actions by the Artin–Schreier group $\mathrm{AS}(\mathbb{k}(x))$ and the linear projective group $\mathrm{PGL}_2(\mathbb{k})$ [12, 18]. From the ramifications of the Weierstrass points, one deduces three types $(1, 1, 1)$, $(3, 1)$, (5) of birationnally equivalence classes defined by the following affine equations [18, p. 618]:

$$Y^2 - Y = \begin{cases} \alpha X + \beta X^{-1} + \gamma(X-1)^{-1}, \quad \alpha\beta\gamma \neq 0 & (1,1,1) \\ X^3 + \alpha X + \beta X^{-1}, \quad \beta \neq 0 & (3,1) \\ X^5 + \alpha X^3, & (5) \end{cases}$$

Recall that in Deuring's *Invariants Theory of Elliptic Curves* [10]: $\mathbb{Z}[j]$ is the ring of *characteristic preserving functions* depending on birationally equivalence classes of elliptic curves with values in an arithmetic universal domain. Igusa introduced an *integral invariant* as a continuous characteristic preserving function defined on the set of all sextic forms with values in an arithmetic universal domain and covariant under the orbit action [18, Sect. 3]. Those depending only on birational classes of hyperelliptic curves of genus 2 are called the *absolute invariants*. In modern terminology, an integral invariant is a global section of the stack quotient by this action, in other words, a section of the fine moduli space $\mathfrak{M}_2$, an algebraic stack, hence also of its coarse moduli space $\mathcal{M}_2$.

Igusa defines the following integral arithmetical invariants:

$$J_2 = 2^{-3} I_2, \quad J_4 = 2^{-5}3^{-1}(4J_2^2 - I_4),$$

$$J_6 = 2^{-6}3^{-2}(8J_2^3 - 160J_2J_4 - I_6),$$

$$J_8 = 2^{-2}(J_2J_6 - J_4^2), \quad J_{10} = 2^{-12} I_{10}.$$

are arithmetic invariants associated to the curves of equation :

$$XY^2 + (1 + aX + bX^2)Y + X^2(c + dX + X^2) = 0$$

[18, p. 621]. And every J_{2i} reduces well modulo 2, and $J_{10} \neq 0$. Indeed, J_{10} encode the fact that C is smooth (for instance in characteristic different from 2, I_{10} is the discriminant), so is invertible on $\mathcal{M}_2$, since a curve is smooth over its base. If we express J_{2i}, as polynomials in a, b, c, d, we get a quantity that depends only on the birational class of the genus 2 curves. In characteristic different from 2, it is a consequence of the invariance property of J_{2i}, and in the characteristic 2, it is a consequence of the explicit relations (5)–(7).

Reciprocally, Igusa shows that from the quintuplet $(J_2, J_4, J_6, J_8, J_{10})$, with $J_{10} \neq 0$, one can construct a normal form with arithmetic invariants those J_{2i}'s (possibly under a field extension) [18, Sects. 3, 5]. For the characteristic different from two, this can be done algorithmically using the Igusa-Clebsh construction [26] and in characteristic 2, it is a consequence of the three relations from Sect. 2 and Eq. (5) below. Also, by considering the relation $J_2J_6 - J_4^2 - 4J_8 = 0$, the invariant J_8 can be disregarded in characteristic different from 2 (but it is crucial in characteristic 2).

Let us consider the graded ring generated over $\mathbb{Z}$ by J_{2i}, $i = 1, 2, 3, 4, 5$, and localized at J_{10}. We denote by $\mathcal{R}$ the integral domain generated by its homogeneous elements of degree zero. Then Igusa shows [18, Sect. 7] that $\mathcal{R}$ is generated over $\mathbb{Z}$ by the elements of the form $J_2^{e_1} J_4^{e_2} J_6^{e_3} J_8^{e_4} J_{10}^{-e_5}$ with e_i a non negative integers verifying $e_1 + 2e_2 + 3e_3 + 4e_4 = 5e_5$.

In particular, if y_1, y_2, y_3, and y_4 are an independent variables with $4y_4 = y_1y_3 - y_2^2$, then the correspondence:

$$J_2^{e_1} J_4^{e_2} J_6^{e_3} J_8^{e_4} J_{10}^{-e_5} \longmapsto y_1^{e_1} y_2^{e_2} y_3^{e_3} y_4^{e_4}$$

from $\mathcal{R}$ to $\mathbb{Z}[y_1, y_2, y_3, y_4]^{\mu_5}$ defines an isomorphism between $\mathcal{R}$ and the elements of $\mathbb{Z}[y_1, y_2, y_3, y_4]$ that are invariant under the transformation $y^i \longmapsto \zeta_5{}^i y_i$ for $i = 1, 2, 3, 4$, where ζ_5 is a primitive fifth root of unity.

By considering the condition $J_2J_6 - J_4^2 - 4J_8 = 0$, Igusa shows that the monoid of the powers of e_i's appearing in the elements of $\mathbb{Z}[y_1, y_2, y_3, y_4]^{\mu_5}$ is generated by ten elements (only eight are needed in characteristics different form 2) [18, Sect. 7]. Therefore, $\mathcal{R}$ is generated by ten elements called γ_i in [15]:

$$\gamma_1 = J_2^5/J_{10}, \quad \gamma_2 = J_2^3J_4/J_{10}, \quad \gamma_3 = J_2^2J_6/J_{10}, \quad \gamma_4 = J_2J_8/J_{10},$$

$$\gamma_5 = J_2J_6/J_{10}, \quad \gamma_6 = J_4J_8^2/J_{10}^2, \quad \gamma_7 = J_6^2J_8/J_{10}^2, \quad \gamma_8 = J_6^5/J_{10}^3,$$

$$\gamma_9 = J_6 J_8^3 / J_{10}^3, \quad \gamma_{10} = J_8^5 / J_{10}^4.$$

In summary, we have

Theorem 1 *Let μ_5 be the group of the fifth root of unity, the moduli space $\mathcal{M}_2$ of the genus 2 curves is isomorphic to* Proj $(\mathbb{Z}[J_2, J_4, J_6, J_8, J_{10}])_{(J_{10})}$ =Spec $(\mathbb{Z}[y_1, y_2, y_3, y_4]^{\mu_5})$ = Spec $(\mathbb{Z}[\gamma_1, \ldots, \gamma_{10}])$ *[18, Theorem 2] (with a weighted grading on the* Proj*). And the variety $\mathcal{M}_2$ can be embedded as a subvariety of an affine space over $\mathbb{Z}$ of dimension ten and not less than ten [18, Theorem 6].*

Corollary 1 *Let $\gamma_i(\mathcal{C})$ be the evaluation of γ_i at a representative model of $\mathcal{C}$.*

- *If $\mathcal{C}$ is a curve defined over a number field $\mathbb{K}$, then $\mathcal{C}$ has good reduction modulo a prime $\mathfrak{p}$ of $\mathbb{K}$ if and only if :*

$$ord_{\mathfrak{p}}\,(\gamma_i(\mathcal{C})) \geqslant 0, \quad i = 1, \cdots, 10.$$

- *And if $\mathcal{C}_1$ and $\mathcal{C}_2$ are curves over $\overline{\mathbb{k}}$, then [18, Corollary p. 632] :*

$$\begin{aligned} \mathcal{C}_1 \simeq \mathcal{C}_2 &\Longleftrightarrow (\gamma_1(\mathcal{C}_1), \cdots, \gamma(\mathcal{C}_1)) = (\gamma_1(\mathcal{C}_2), \cdots, \gamma(\mathcal{C}_2)) \\ &\Longleftrightarrow (J_2(\mathcal{C}_1) : J_4(\mathcal{C}_1) : J_6(\mathcal{C}_1) : J_8(\mathcal{C}_1) : J_{10}(\mathcal{C}_1)) = (J_2(\mathcal{C}_2) : J_4(\mathcal{C}_2) : J_6(\mathcal{C}_2) : J_8(\mathcal{C}_2) : J_{10}(\mathcal{C}_2)) \end{aligned}$$

(with the weighted gradings).

2.3 Converting Between Models and Invariants

Converting between the standard hyperelliptic model $y^2 + h(x) = f(x)$ to the normal form is explained in Appendix 6. And conversely, the equations of the Weierstrass points of a normal form $(1 + aX + bX^2)^2 - 4X^3(c + dX + X^2) = 0$ (in characteristic different from 2) allows to recover the standard model from it.

Plugging the Igusa-Clebsch formula allows to express the I_2, I_4, I_6, I_{10}, hence the $J_2, J_4, J_6, J_8, J_{10}$ as rational functions in term of a, b, c, d. The rational functions expressing the J_i in terms of a, b, c, d then are still valid in characteristic 2, since both the covariants and the equation has good reduction modulo 2.

Also, in characteristic two, one can go from the standard hyperelliptic model to the normal form to the form defined by the three types $(1, 1, 1)$, $(3, 1)$, and (5) (and vice versa) using Artin–Schreier theory (see Appendix 6 and Sect. 2.4). Hence, we can also express the J_i in terms of the α, β, γ, and conversely using Eqs. (5)–(7).

From the J_i, which are covariants with weights, we can compute the 10 invariants γ_i. Conversely, from the knowledge of γ_i, we may rescale the covariants so that $J_{10} = 1$, hence recover J_6 (up to a 5-th root of unity) from γ_8. This allows to recover J_2, J_4, J_8. The correct 5-th root of unity may be found using γ_7 and γ_9. Then we scale back by J_{10} to get a rational weighted projective point.

However, using the γ_i is not convenient when defining modular polynomials. Indeed, we would like to describe the moduli space $\mathcal{M}_2$ using only three invariants (at least birationally). Indeed, by [18, Theorem 5], a generic point of the variety of moduli $\mathcal{M}_2 \otimes \Bbbk$ generates a purely transcendental extension of dimension three. As a corollary, we get that $\mathcal{M}_2 \otimes \Bbbk$ is birational to $\mathbb{A}^3_{\Bbbk}$.

In the next section, we will actually describe an affine cover of an open subscheme of the smooth locus of $\mathcal{M}_2$ by affines isomorphic to a standard open of $\mathbb{A}^3_{\mathbb{Z}}$ (so completely determined by only three invariants). This cover induce a full cover of the smooth locus of $\mathcal{M}_2 \otimes \Bbbk$ in all characteristics.

2.4 Absolute Invariants

We say that a set of three invariants (i_1, i_2, i_3) are absolute invariants for $\mathcal{M}_2 \otimes \Bbbk$ if they generate the function field $\Bbbk(\mathcal{M}_2 \otimes \Bbbk)$. Equivalently, an absolute invariants define a birational morphism $\mathcal{M}_2 \otimes \Bbbk \to \mathbb{A}^3_{\Bbbk}$, hence define coordinates on an open U of $\mathcal{M}_2 \otimes \Bbbk$. We say that (i_1, i_2, i_3) are well defined at C if C corresponds to a geometric point of U. It is stronger than just the i_j having no poles at C, we also require that the birationnal class of C can be recovered from the $(i_1(C), i_2(C), i_3(C))$, or equivalently by Theorem 1, that $(J_2(C) : J_4(C) : J_6(C) : J_8(C) : J_{10}(C)$ can be recovered.

A standard set of invariants used for computation of class or modular polynomials was I_2^5/I_{10}, $I_2^3 I4/I_{10}$, $I_2^2 I_6/I_{10}$, the corresponding U is $I_2 \neq 0$. To reduce the size of these polynomials, Streng introduced the absolute invariants $I_4 I_6'/I_{10}$, $I_2 I_4^2/I_{10}$, I_4^5/I_{10}^2, where $I_6' = 1/2(I_2 I_4 - 3I_6)$. The corresponding U is given by $I_4 \neq 0$. See Sect. 3.2 for the explanation of this choice: they correspond to nice modular forms defined in terms of theta constants.

These invariants have bad reduction modulo 2. (Their reductions are defined, but they are not absolute invariants on $\mathcal{M}_2 \otimes \Bbbk$ in characteristic 2).

To describe the whole of $\mathcal{M}_2 \otimes \Bbbk$, we need a system of invariants. We say that such a system of invariants $\{i, i', i" \ldots\}$ is fully covering if they give coordinates on a stratification of $\mathcal{M}_2 \otimes \Bbbk$. Typically, one use a system of absolute invariants i on an open U_0, then another set of invariants i' on an open U_1 of $\mathcal{M}_2 \otimes \Bbbk \setminus U_0$, and so on. Here, i_1 need not be an absolute invariant since we only require it to be defined on a locally closed subscheme rather than an open subscheme. We only require it to induce an isomorphism on U_1 to an open of $\mathbb{A}^{m_1}_{\Bbbk}$. In fact, in characteristic 2, we relax this last condition to a radicial morphism (that is an universally injective morphism) on U_1.

We say that a fully covering system of invariants is optimal if it induce a bijection $\mathcal{M}_2 \otimes \Bbbk \to \mathbb{A}^3_{\Bbbk}$ (be careful that since the system is defined on a stratification, the map is not a morphism).

In Characteristic Different From 2

One can define absolute invariants [4, Sect. 1], hence coordinates on the moduli space using the invariants J_{2i}'s. Knowing that the invariant J_{10} defines the discriminant of the curve, so is not null, we obtain that

Table 1 Number of isomorphism classes in char 2

Type	Number
(1,1,1)	$q^3 - q^2$
(3,1)	$q^2 - q$
(5)	q

- The class of genus 2 curves with a nonzero J_2, is an open set of $\mathcal{M}_2$ on which we have the tuple of absolute invariants : $\left(J_2^5/J_{10},\ J_2^3J_4/J_{10},\ J_2^2J_6/J_{10}\right)$.
- The genus 2 curves that annihilate in J_2 with a nonzero J_4, is a subspace of $\mathcal{M}_2$, where coordinates can be defined by $\left(0,\ J_4^5/J_{10}^2,\ J_4J_6/J_{10}\right)$.
- The other curves lie in set with coordinates defined by $\left(0,\ 0,\ J_6^5/J_{10}^3\right)$.

Then the set of points of $\mathcal{M}_2 \otimes \mathbb{k}$ is in bijection with the set of tuples defined previously, so in bijection with $\mathbb{A}^3_{\mathbb{k}}$. In other words, these invariants (k_1, k_2, k_3) on the above stratification are optimal.

Remark 1 Reciprocally, we can recover the projective J_{2i}'s from any tuple $(k_1, k_2, k_3) \in \mathbb{A}^3_{\mathbb{k}}$ (a point of $\mathcal{M}_2 \otimes \mathbb{k}$) using the following way from [4, Lemma 1] :

$$(J_2, J_4, J_6, J_{10}) = \begin{cases} (k_1, k_1k_2, k_1^2k_3, k_1^4), & \text{if } k_1 \neq 0, \\ (0, k_2, k_2k_3, k_2^2), & \text{else if } k_1 = 0,\ k_2 \neq 0, \\ (0, 0, k_3^2, k_3^3), & \text{otherwise.} \end{cases}$$

In Characteristic 2

We recall that every hyperelliptic curves of genus 2 is birationally equivalent to one of the three following types according to the number and the degree of the ramified Weierstrass points:

$$Y^2 - Y = \begin{cases} \alpha X + \beta X^{-1} + \gamma(X-1)^{-1}, & (1,1,1) \\ X^3 + \alpha X + \beta X^{-1}, & (3,1) \\ X^5 + \alpha X^3, & (5) \end{cases}$$

When $\mathbb{k}$ has q elements the number $\overline{\mathbb{k}}$-isomorphism classes of smooth projective curves of genus two defined over $\mathbb{k}$ is given in the following table according to the type [12, Theorem 20] (Table 1):

Let's consider the normal form equation :

$$XY^2 + (1 + aX + bX^2)Y + X^2(c + dX + X^2) = 0.$$

If ab is different from 0, we get three Weierstrass points; it corresponds to the type $(1, 1, 1)$. After a technical variable change in [18, Sect. 3], one can obtain

$$\alpha = ab^{-3}, \tag{1}$$
$$\beta = a^{-3}b\xi^{-2}\left(c + \xi^{-2} + a(c\xi + d + \xi^{-1})^{1/2}\right), \tag{2}$$
$$\gamma = a^{-3}b\eta^{-2}\left(c + \eta^{-2} + a(c\eta + d + \eta^{-1})^{1/2}\right) \tag{3}$$

in which $\xi + \eta = a$, $\eta\xi = b$;

But if a or b is nonzero and $ab = 0$, it corresponds the type (3, 1); and if $a \neq 0$ and $b = 0$, one can transform

$$XY^2 + (1 + aX + bX^2)Y + X^2(c + dX + X^2) = 0$$

into

$$Y^2 - Y = X^3 + \alpha X + \beta X^{-1}$$

with

$$\alpha = a^{5/3}\left(a^{-3}c + (a^{-5} + a^{-4}d)^{1/2}\right), \quad \beta = a^{-5/3}\left(a^{-5} + a^{-3}c + (a^{-5} + a^{-4}d + a^{-3}c)^{1/2}\right). \tag{4}$$

If $b \neq 0$ and $a = 0$, we obtain α, β in terms of b, c, d via a more complicated expressions for which we refer to Igusa.

The type (5) corresponds to $a = b = 0$ and the associated normal form $XY^2 + Y + X^2(c + dX + X^2) = 0$ can be transformed into $Y^2 - Y = X^5 + \alpha X^3$ with $\alpha = c$.

The open set $\mathcal{M}_2[J_2^{-1}] \otimes \Bbbk$ describes in $\mathcal{M}_2 \otimes \Bbbk$, the curves birationaly equivalent to a curve of type (1, 1, 1). It is characterized by the nonvanishing of J_2 modulo 2, and can be defined using the following three absolute arithmetic invariants: $\mathfrak{a}_1 = J_4/J_2^2$, $\mathfrak{a}_2 = J_8/J_2^4$ and $\mathfrak{a}_3 = J_{10}/J_2^5$. Indeed, one can recover these invariants from the coefficients of the normal form using the following relation [18, Sect. 3]:

$$\begin{cases} \alpha^2 + \beta^2 + \gamma^2 = J_4/J_2^2, \\ \alpha^2\beta^2\gamma^2 = J_{10}/J_2^5, \\ \alpha^2\beta^2 + \beta^2\gamma^2 + \gamma^2\alpha^2 = J_8/J_2^4 + (J_4/J_2^2)^3 + (J_4/J_2^2)^4 \end{cases} \tag{5}$$

And Igusa shows in [18, Sect. 2] that birational invariants are given by the three standard symmetric invariants: $\alpha + \beta + \gamma$, $\alpha\beta + \beta\gamma + \gamma\alpha$, $\alpha\beta\gamma$, they are also used by Cardona and al. in [12, Sect. 2]: Although they are ramified over the symmetric invariants, there is one important advantage to the three invariants $\mathfrak{a}_1, \mathfrak{a}_2, \mathfrak{a}_3$: they come from modular functions over $\mathbb{C}$ and are actually defined over $\mathbb{Z}$. By contrast, the symmetric invariants do not lift to modular forms (without characters) in characteristic 0. More precisely, it is proven in Theorem 2 that these invariants describe $\mathcal{M}_2[J_2^{-1}]$ over $\mathbb{Z}$. In particular, over $\mathbb{Z}_2$ these invariants describe the open set $\mathcal{M}_2[J_2^{-1}]$ of $\mathcal{M}_2$ that reduces to $\mathcal{M}_2[J_2^{-1}] \otimes \Bbbk$ modulo 2, in other words, of curves with good reduction modulo 2, and whose reduction is of type (1, 1, 1).

The type (3, 1) is characterized by $J_2 = 0$ and the nonvanishing of J_6 over $\Bbbk$, this corresponds to a closed subscheme of $\mathcal{M}_2[J_6^{-1}] \otimes \Bbbk$, hence a locally closed subscheme in $\mathcal{M}_2 \otimes \Bbbk$, of curves birationally equivalent to that type. Coordinates on this subscheme are defined using the following tuple of absolute invariants: $(0,\ J_8J_{10}/J_6^3,\ J_{10}^3/J_6^5)$. Indeed, from the equation (3, 1), one recovers this tuple using

$$\begin{cases} \alpha^6 &= J_8^{3/4}/J_6^1, \\ \beta^6 &= J_{10}^3/J_6^5, \\ \alpha^2\beta^2 &= J_8^{1/4}J_{10}/J_6^2. \end{cases} \tag{6}$$

And birational invariants are given by α^3, $\alpha\beta$ [18, Sect. 2].

The type (5) is characterized by $J_2 = J_6 = 0$. The corresponding closed subset of $\mathcal{M}_2 \otimes \Bbbk$, can be defined using the tuple $(0, 0, J_8^5/J_{10}^4)$ of invariants. And the following relation holds :

$$\left\{ \alpha^{10} = J_8^{5/4}/J_{10}. \right. \tag{7}$$

And α^5 is a birational invariant [18, Sect. 2].

From this discussion, we see that the invariants $(\mathfrak{a}_1, \mathfrak{a}_2, \mathfrak{a}_3) = (J_4/J_2^2,\ J_8/J_2^4,\ J_{10}/J_2^5)$ when $J_2 \neq 0$, $(0,\ J_8J_{10}/J_6^3,\ J_{10}^3/J_6^5)$ when $J_2 = 0$, $J_6 \neq 0$, and $(0, 0, J_8^5/J_{10}^4)$ induces a bijection between the set of points of $\mathcal{M}_2 \otimes \Bbbk$ and $\mathbb{A}_{\Bbbk}^3$, hence are optimal. Indeed, the above formula show how to recover α, β, γ from these invariants. A drawback compared to the optimal invariants defined above in characteristic different from 2 is that they do not share a common denominator. It would be interesting to combine the two versions to have optimal invariants working in all characteristic at once.

2.5 *Absolute Invariants in All Characteristic*

Igusa shows in [18, Theorem 4] that

- If $char(\Bbbk) \neq 2$, the variety $\mathcal{M}_2 \otimes \Bbbk$ has one and only one singular point, which corresponds to $J_2 = J_6 = J_8 = 0$.
- If $char(\Bbbk) = 2$, the singular locus of $\mathcal{M}_2 \otimes \Bbbk$ is a rational curve corresponding to $J_2 = J_6 = 0$, i.e., the curves of type (5) having only one Weierstrass point.

We give a set of three invariants over each of the four opens: $\mathcal{M}_2[J_2^{-1}]$, $\mathcal{M}_2[J_4^{-1}]$, $\mathcal{M}_2[J_6^{-1}]$, $\mathcal{M}_2[J_8^{-1}]$ in $\mathcal{M}_2$ over $\mathbb{Z}$. These opens reduce in all characteristic to a cover of $\mathcal{M}_2 \otimes \Bbbk$, except for the one curve defined by $J_2 = J_4 = J_6 = J_8 = 0$. This is the curve defined by the equation $y^2 + y = x^5$ over $\mathbb{Q}$, which has potential good reduction everywhere [33, Example 1], and reduces modulo 2 to a curve of type (5) with $\alpha = 0$.

Furthermore, these three invariants actually induce an isomorphism with a standard open of $\mathbb{A}^3$ for the first three opens (over $\mathbb{Z}[1/2]$ for $\mathcal{M}_2[J_4^{-1}]$, and over $\mathbb{Z}$ for the other two). In other words, they are absolute invariants, well defined on the above opens.

These three opens $\mathcal{M}_2[J_2^{-1}]$, $\mathcal{M}_2[J_4^{-1}]$, $\mathcal{M}_2[J_6^{-1}]$ are sufficient to cover $\mathcal{M}_2 \otimes \mathbb{k}$ in characteristic different from 2, minus the only non smooth point $y^2 + y = x^5$ from above. In characteristic 2, the first open corresponds to curves of type (1, 1, 1), and the third open includes the curves of type (3, 1). We cannot hope for the three invariants on the last open to yield an isomorphism to an open of $\mathbb{A}^3$, since $\mathcal{M}_2[J_8^{-1}]$ contains a non smooth locus.

From the preceding section, we see that our choice of cover and absolute invariants is particularly well adapted to the reduction of a curve in $\mathcal{M}_2 \otimes \mathbb{Z}_{(2)}$. Hence, in particular, in characteristic two and zero, and even more precisely to represent lifts over $\mathbb{Z}_q$ of the genus 2 curves defined over $\mathbb{F}_q$, with $q = 2^n$.

Theorem 2 *To the affine cover described above, we associate the set of three invariants:*

- $\mathfrak{a}_1 = J_4/J_2^2$, $\mathfrak{a}_2 = J_8/J_2^4$ *and* $\mathfrak{a}_3 = J_{10}/J_2^5$ *for* $\mathcal{M}_2[J_2^{-1}]$,
- $\mathfrak{d}_1 = J_2^2/J_4$, $\mathfrak{d}_2 = J_6^2/J_4^3$ *and* $\mathfrak{d}_3 = J_{10}^2/J_4^5$ *for* $\mathcal{M}_2[J_4^{-1}]$,
- $\mathfrak{u}_1 = J_4^3/J_6^2$, $\mathfrak{u}_2 = J_8 J_{10}/J_6^3$, *and* $\mathfrak{u}_3 = J_{10}^3/J_6^5$ *for* $\mathcal{M}_2[J_6^{-1}]$,
- $\mathfrak{w}_1 = J_2^4/J_8$, $\mathfrak{w}_2 = J_6^4/J_8^3$, *and* $\mathfrak{w}_3 = J_{10}^4/J_8^5$ *for* $\mathcal{M}_2[J_8^{-1}]$.

Then these invariants induce an isomorphism of $\mathcal{M}_2[J_2^{-1}]$, $\mathcal{M}_2[J_4^{-1}]$ *and* $\mathcal{M}_2[J_6^{-1}]$ *with the standard open of* $\mathbb{A}^3$ *defined by* $\mathfrak{a}_3^{-1}$, $\mathfrak{d}_3^{-1}$, $\mathfrak{u}_3^{-1}$, *respectively, over* $\mathbb{Z}$, $\mathbb{Z}[1/2]$ *and* $\mathbb{Z}$, *respectively.*

The last set of three invariants give a finite covering of $\mathcal{M}_2[J_8^{-1}]$ *to the standard open* $\mathbb{A}^3[\mathfrak{w}_3^{-1}]$ *over* $\mathbb{Z}$. *In characteristic two, they restrict on the non smooth locus* $J_2 = J_6 = 0$ *to a radicial morphism with image* $\mathbb{A}^1_{\mathbb{k}} \setminus \{0\}$.

Proof Since the J_{2i} have good reduction, these invariants too. We need to check that they define local coordinates, in other words, that they generate $\mathbb{Z}_{(2)}[\mathcal{M}_2]$ locally. By the results from Sect. 2.2, it suffices to prove that we can recover all the γ_i.

We can prove this result by calculation:

- For the set $\mathcal{M}_2[J_2^{-1}]$, we express the invariants γ_i's in function of $\mathfrak{a}_1$, $\mathfrak{a}_2$ and $\mathfrak{a}_3$.

$$\gamma_1 = \frac{1}{\mathfrak{a}_3}, \quad \gamma_2 = \frac{\mathfrak{a}_1}{\mathfrak{a}_3}, \quad \gamma_3 = \frac{\mathfrak{a}_1^2}{\mathfrak{a}_3} + 4\frac{\mathfrak{a}_2}{\mathfrak{a}_3}, \quad \gamma_4 = \frac{\mathfrak{a}_2}{\mathfrak{a}_3}, \quad \gamma_5 = \frac{\mathfrak{a}_1^3}{\mathfrak{c}} + 4\frac{\mathfrak{a}_1\mathfrak{a}_2}{\mathfrak{a}_3}, \quad \gamma_6 = \frac{\mathfrak{a}_1\mathfrak{a}_2^2}{\mathfrak{a}_3^2},$$

$$\gamma_7 = \frac{\mathfrak{a}_1^4\mathfrak{a}_2}{\mathfrak{a}_3^2} + 8\frac{\mathfrak{a}_1^2\mathfrak{a}_2^2}{\mathfrak{a}_3^2} + 16\frac{\mathfrak{a}_2^3}{\mathfrak{a}_3^2}, \quad \gamma_8 = \frac{1}{\mathfrak{a}_3^3}\left(\mathfrak{a}_1^{10} + 20\mathfrak{a}_1^8\mathfrak{a}_2 + 160\mathfrak{a}_1^6\mathfrak{a}_2^2 + 640\mathfrak{a}_1^4\mathfrak{a}_2^3 + 1280\mathfrak{a}_1^2\mathfrak{a}_2^4 + 1024\mathfrak{a}_2^5\right),$$

$$\gamma_9 = \frac{1}{\mathfrak{a}_3^3}\left(\mathfrak{a}_1^2\mathfrak{a}_2^3 + 4\mathfrak{a}_2^4\right), \quad \gamma_{10} = \frac{\mathfrak{a}_2^5}{\mathfrak{a}_3^4}.$$

This transformation makes sens over $\mathbb{Z}$. Therefore,

$$\mathcal{M}_2[J_2^{-1}] = \operatorname{Spec}\left(\mathbb{Z}[\mathfrak{a}_1, \mathfrak{a}_2, \mathfrak{a}_3][\mathfrak{a}_3^{-1}]\right)$$

and $\mathcal{M}_2[J_2^{-1}]$ is a subspace of $\mathcal{M}_2$ of dimension three over $\mathbb{Z}$.

• Over $\mathbb{Z}[1/2]$, the invariants γ_i's are expressed in function of $\mathfrak{d}_1$, $\mathfrak{d}_2$ and $\mathfrak{d}_3$ on $\mathcal{M}_2[J_4^{-1}]$ as follows:

$$\gamma_1 = \frac{\mathfrak{d}_1^5}{\mathfrak{d}_3}, \quad \gamma_2 = \frac{\mathfrak{d}_1^3}{\mathfrak{d}_3}, \quad \gamma_3 = \frac{\mathfrak{d}_1^2 s}{\mathfrak{d}_3}, \quad \gamma_4 = \frac{\mathfrak{d}_1^2\mathfrak{d}_2}{4\mathfrak{d}_3} - \frac{\mathfrak{d}_1}{4\mathfrak{d}_3}, \quad \gamma_5 = \frac{\mathfrak{d}_2}{\mathfrak{d}_3}, \quad \gamma_6 = \frac{1}{16\mathfrak{d}_3^2}\left(\mathfrak{d}_1^2\mathfrak{d}_2^2 - 2\mathfrak{d}_1\mathfrak{d}_2 + 1\right),$$
$$\gamma_7 = \frac{1}{4\mathfrak{d}_3^2}\left(\mathfrak{d}_1\mathfrak{d}_2^3 - \mathfrak{d}_2^2\right), \quad \gamma_8 = \frac{\mathfrak{d}_2^5}{\mathfrak{d}_3^3}, \quad \gamma_9 = \frac{1}{64\mathfrak{d}_3^3}\left(\mathfrak{d}_1^3\mathfrak{d}_2^4 - 3\mathfrak{d}_1^2\mathfrak{d}_2^3 + 3\mathfrak{d}_1\mathfrak{d}_2^2 - \mathfrak{d}_2\right),$$
$$\gamma_{10} = \frac{1}{1024\mathfrak{d}_3^4}\left(\mathfrak{d}_1^5\mathfrak{d}_2^5 - 5\mathfrak{d}_1^4\mathfrak{d}_2^4 + 10\mathfrak{d}_1^3\mathfrak{d}_2^3 - 10\mathfrak{d}_1^2\mathfrak{d}_2^2 + 5\mathfrak{d}_1\mathfrak{d}_2 - \mathfrak{d}_3^4\right).$$

• In the case of $\mathcal{M}_2[J_6^{-1}]$, we express the γ_i's in function of $\mathfrak{u}_1$, $\mathfrak{u}_2$ and $\mathfrak{u}_3$. This transformation also makes sens over $\mathbb{Z}$ and the subvariety $\mathcal{M}_2[J_6^{-1}]$ of $\mathcal{M}_2$ has dimension three over $\mathbb{Z}$.

$$\gamma_1 = \frac{1}{\mathfrak{u}_3^6}(\mathfrak{u}_3^5\mathfrak{u}_1^{10} + 20\mathfrak{u}_3^4\mathfrak{u}_2\mathfrak{u}_1^8 + 160\mathfrak{u}_3^3\mathfrak{u}_2^2\mathfrak{u}_1^6 + 640\mathfrak{u}_3^2\mathfrak{u}_2^3\mathfrak{u}_1^4 + 1280\mathfrak{u}_3\mathfrak{u}_2^4\mathfrak{u}_1^2 + 1024\mathfrak{u}_2^5),$$
$$\gamma_2 = \frac{1}{\mathfrak{u}_3^4}(\mathfrak{u}_3^3\mathfrak{u}_1^7 + 12\mathfrak{u}_3^2\mathfrak{u}_2\mathfrak{u}_1^5 + 48\mathfrak{u}_3\mathfrak{u}_2^2\mathfrak{u}_1^3 + 64\mathfrak{u}_2^3\mathfrak{u}_1), \quad \gamma_3 = \frac{1}{\mathfrak{u}_3^3}(\mathfrak{u}_3^2\mathfrak{u}_1^4 + 8\mathfrak{u}_3\mathfrak{u}_2\mathfrak{u}_1^2 + 16\mathfrak{u}_2^2),$$
$$\gamma_4 = \frac{1}{\mathfrak{u}_3^3}(\mathfrak{u}_3\mathfrak{u}_2\mathfrak{u}_1^2 + 4\mathfrak{u}_2^2), \quad \gamma_5 = \frac{\mathfrak{u}_1}{\mathfrak{u}_3}, \quad \gamma_6 = \frac{\mathfrak{u}_1\mathfrak{u}_2^2}{\mathfrak{u}_3^4}, \quad \gamma_7 = \frac{\mathfrak{u}_2}{\mathfrak{u}_3^3}, \quad \gamma_8 = \frac{1}{\mathfrak{u}_3^3}, \quad \gamma_9 = \frac{\mathfrak{u}_2^3}{\mathfrak{u}_3^6}, \quad \gamma_{10} = \frac{\mathfrak{u}_2^5}{\mathfrak{u}_3^9}.$$

• In the case of $\mathcal{M}_2[J_8^{-1}]$, we express the squares of γ_i's in function of $\mathfrak{w}_1$, $\mathfrak{w}_2$ and $\mathfrak{w}_3$.

$$\gamma_1^2 = \frac{\mathfrak{w}_1^{10}}{\mathfrak{w}_3^2}, \quad \gamma_2^2 = \frac{1}{\mathfrak{w}_3^2}\left(\mathfrak{w}_1^7\mathfrak{w}_2 - 4\mathfrak{w}_1^6\right), \quad \gamma_3^2 = \frac{\mathfrak{w}_1^4\mathfrak{w}_2^2}{\mathfrak{w}_3^2}, \quad \gamma_4^2 = \frac{\mathfrak{w}_1^2}{\mathfrak{w}_3^2}, \quad \gamma_5^2 = \frac{1}{\mathfrak{w}_3^2}\left(\mathfrak{w}_1\mathfrak{w}_2^3 - 4\mathfrak{w}_2^2\right),$$
$$\gamma_6^2 = \frac{1}{\mathfrak{w}_3^4}\left(\mathfrak{w}_1\mathfrak{w}_2 - 4\right), \quad \gamma_7^2 = \frac{\mathfrak{w}_2^4}{\mathfrak{w}_3^4}, \quad \gamma_8^2 = \frac{\mathfrak{w}_2^{10}}{\mathfrak{w}_3^6}, \quad \gamma_9^2 = \frac{\mathfrak{w}_2^2}{\mathfrak{w}_3^6}, \quad \gamma_{10}^2 = \frac{1}{\mathfrak{w}_3^8}.$$

As we have seen, we can't do better in characteristic 2, since $\mathcal{M}_2[J_8^{-1}] \otimes \Bbbk$ contain the singular locus $J_2 = J_6 = 0$ of curves of type (5). If we restrict to this locus in characteristic 2, we get that $\mathfrak{w}_3 = \alpha^{-40}$. Since α^5 is a birational invariant on this locus, we see that the coordinates $\mathfrak{w}_3$ induce a totally ramified cover of degree 8 on the curves of type (5), minus, as usual, the curve $y^2 - y = x^5$.

3 Modular Polynomials in Dimension 2

3.1 Theta Functions

In this section, we consider $\mathbb{C}$ as the base field. For an integer $g > 0$, we denote by $\mathfrak{H}_g$, the *Siegel upper half space*, and by Γ_g the *symplectic group* $\mathrm{Sp}_{2g}(\mathbb{Z})$ such that :

$$\mathfrak{H}_g = \left\{\Omega \in M_g(\mathbb{C}),\ {}^t\Omega = \Omega,\ Im(\Omega) > 0\right\}$$

$$\text{and}\ \ \mathrm{Sp}_{2g}(\mathbb{Z}) = \left\{\gamma \in M_{2g}(\mathbb{Z}) : \gamma J_{2g}\,{}^t\gamma = J_{2g}\right\},\ J_{2g} = \left(\begin{smallmatrix} 0 & 1 \\ -1 & 0 \end{smallmatrix}\right).$$

A complex abelian variety $\mathcal{A}$ of dimension g is analytically isomorphic to a torus $\mathcal{A}_\Omega = \mathbb{C}^g/(\Omega\mathbb{Z}^g + \mathbb{Z}^g)$, where $\Omega \in \mathfrak{H}_g$ is called its *period matrix*.

In general, the group Γ_g acts on $\mathfrak{H}_g$ by

$$\gamma\Omega = (A\Omega + B)(C\Omega + D)^{-1} \quad for \quad \gamma = \left(\begin{smallmatrix} A & B \\ C & D \end{smallmatrix}\right) \in \Gamma_g$$

And *two principally polarized abelian varieties Ω_1 and Ω_2 are isomorphic if and only if $\Omega_2 = \gamma\Omega_1$ where $\gamma \in \Gamma_{2g}$.*

We denote by $\mathcal{F}_g$ the fundamental domain of this action; therefore, for all $\Omega \in \mathfrak{H}_g$, there exist $\gamma\Gamma_g$ such that $\gamma\Omega \in \mathcal{F}_g$ and γ is unique if $\gamma\Omega$ is an inner point of $\mathcal{F}_g$.

We recall that

- The Riemann classical theta function is defined by

$$\theta : \mathbb{C} \times \mathfrak{H}_g \to \mathbb{C}, \quad (z, \Omega) \mapsto \sum_{n \in \mathbb{Z}^g} exp(i\pi\,{}^t(n)\Omega n + 2i\pi\,{}^t n z)$$

- For $a, b \in \mathbb{Q}$ the classical theta function with characteristic (a, b) is defined by

$$\theta\left[\begin{smallmatrix} a \\ b \end{smallmatrix}\right](z, \Omega) = \sum_{n \in \mathbb{Z}^g} exp(i\pi\,{}^t(n + a)\Omega(n + a) + 2i\pi\,{}^t(n + a)(z + b))$$

These functions satisfy the following fundamental property.

Proposition 1 (Functional Equation). *Let $\gamma = \left(\begin{smallmatrix} A & B \\ C & D \end{smallmatrix}\right) \in \Gamma_g$, and e' (reps. e'') the vector defined by the coefficients of the diagonal of $\frac{1}{2}\,{}^tAC$ (resp. $\frac{1}{2}\,{}^tDB$). Then for all $a, b \in \mathbb{Q}^g$, $z \in \mathbb{C}^g$ and $\Omega \in \mathfrak{H}_g$, we have*

$$\begin{aligned}\theta\left[\begin{smallmatrix} a \\ b \end{smallmatrix}\right](\gamma z, \gamma\Omega) =& \zeta_\gamma\sqrt{\det(C\Omega + D)}\exp\left(i\pi\,{}^t z(C\Omega + D)^{-1}Cz\right) \\ &\cdot \theta\left[{}^t\gamma\left(\begin{smallmatrix} a \\ b \end{smallmatrix}\right) + \left(\begin{smallmatrix} e' \\ e'' \end{smallmatrix}\right)\right](z, \Omega)\exp\left(-2i\pi\,{}^t({}^tAa + {}^tCb + e')e''\right) \\ &\cdot \exp(-i\pi\,{}^t aA\,{}^tBa)\exp(-i\pi\,{}^t bC\,{}^tDb)\exp(-2i\pi\,{}^t aB\,{}^tCb) \quad ,\end{aligned}$$

Proof See [9, Propriété 3.1.24.].

These theta functions are sections of powers of the canonical principal polarization associated to the choice of $\Omega \in \mathfrak{H}_2$ and a well known result from Lefschetz says

- For $n \geqslant 3$, any linearly independent set of $k = n^g$ theta functions of level n provides an embedding of $\mathbb{C}^g/(\Omega\mathbb{Z}^g + \mathbb{Z}^g)$ into $\mathbb{P}^{k-1}(\mathbb{C})$.
- For $n = 2$, if the polarization is irreducible, the theta functions of level 2 embed $(\mathbb{C}^g/(\Omega\mathbb{Z}^g + \mathbb{Z}^g))/\sim$, where $\sim$ is the equivalence relation $z \sim -z$.

When $\mathcal{C}$ is an hyperelliptic curves of genus g, the Abel–Jacobi theorem establishes that: the Jacobian $\mathrm{Jac}(\mathcal{C})$ of $\mathcal{C}$ is canonically isomorphic to the analytic Jacobian $\mathbb{C}^g/(\Omega\mathbb{Z}^g + \mathbb{Z}^g)$, where $\Omega \in \mathfrak{H}_g$ is called the period matrix of $\mathrm{Jac}(\mathcal{C})$ [1, Theorem 11.1.3]. Conversely, *a simple principally polarized abelian variety of dimension $g \leqslant 3$ is: the Jacobian of a smooth curve C of genus g, and C is an hyperelliptic curve if $g \leqslant 2$* [1, cor 11.8.2].

3.2 Modular Functions for Γ_2

We have seen that theta functions give an algebraic structure to the analytic abelian surfaces of Jacobian variety of genus 2 curves. They also provide an analytic definition of covariants associated to the corresponding curves. We define the theta constants of level n as the theta functions of level n evaluated at $z = 0$. We will use Dupont notation in [11] : for all $a = \binom{a_0}{a_1}$ and $b = \binom{b_0}{b_1}$ in $\{0, 1\}^2$,

$$\theta_{b_0+2b_1+4a_0+8a_1}(\Omega) := \theta_{a,b}(\Omega) := \theta\begin{bmatrix} a/2 \\ b/2 \end{bmatrix}(0, \Omega).$$

Let Γ be a subgroup of Γ_g of finite index

- A *modular form of weight k* for Γ is a holomorphic function f defined over $\mathfrak{H}_g$ such that $f(\gamma\Omega) = \det(C\Omega + D)^k f(\Omega)$ for all $\gamma = \left(\begin{smallmatrix} A & B \\ C & D \end{smallmatrix}\right) \in \Gamma$ and $\Omega \in \mathfrak{H}_g$. (By the Koecher principle holomorphy at the cusps is automatic when $g > 1$.)
- And we call *modular function* for Γ over $\mathfrak{H}_g$ a function defined by a quotient of two Γ-modular forms of the same weight.

Example 1 Let's denote by : $\mathcal{P} = \{0, 1, 2, 3, 4, 6, 8, 9, 12, 15\}$.

$$h_4 = \sum_{i \in \mathcal{P}} \theta_i^8, \quad h_6 = \sum_{60\, tuples\, (i,j,k) \in \mathcal{P}^3} (\theta_i \theta_j \theta_k)^4$$

$$h_{10} = \prod_{i \in \mathcal{P}} \theta_i^2, \quad h_{12} = \sum_{15\, tuples\, (i,j,k,l,m,n) \in \mathcal{P}^6} (\theta_i \theta_j \theta_k \theta_l \theta_m \theta_n)^4$$

and $h_{16} = \dfrac{1}{3}(h_{12}h_4 - 2h_6h_{10})$.

The h_i's with $i \in \{4, 6, 10, 12, 16\}$ are Γ_2-modular forms of weight i.

Furthermore, we have

$$h_4 = 2^2\psi_4, \quad h_6 = 2^2\psi_6, \quad h_{10} = -2^{14}\chi_{10} \quad \text{and} \quad h_{12} = 2^{17}3\chi_{12},$$

where ψ_k are the Eisenstein series of weight $k \geqslant 4$ defined by:

$$\psi_k(\Omega) = \sum_{\gamma\in\Gamma_2} \det(C\Omega + D)^{-k}$$

and χ_{10}, χ_{12} are cusp forms, expressed in function of ψ_k with $k \in \{4, 6, 10, 12\}$. Over $\mathbb{C}$, the graded ring of even-weight Siegel modular forms in genus 2 is generated by the four modular forms ψ_4 ,ψ_6 , χ_{10} , χ_{12} [19].

Therefore, we can define the well known Γ_2-modular functions called *j-invariants* [38, Chap. 2, Sect. 2.1] defined by

$$\mathrm{j}_1 = -2^{-10}\frac{\chi_4\chi_6}{\chi_{10}}, \quad \mathrm{j}_2 = 2^{-7}\cdot 3\frac{\chi_4^2\chi_{12}}{\chi_{10}^2}, \quad \mathrm{j}_3 = 2^{-18}\frac{\chi_4}{\chi_{10}^2}.$$

These invariants are birationally equivalent to Igusa's j-invariant [20] which are less convenient algorithmically since they have bigger denominators. We will see below that these are the same invariants as the absolute invariants (j_1, j_2, j_3) introduced in Sect. 2.4.

Theorem 3 *The field* $\mathbb{C}(\mathfrak{A}_2)$ *of Siegel modular functions in dimension 2, denoted by* $\mathbb{C}_{\Gamma_2}$*, is* $\mathbb{C}(\mathrm{j}_1, \mathrm{j}_2, \mathrm{j}_3)$.

Proof Proof see [19].

More generally given a finite-dimensional representation $\rho : \mathrm{GL}_2(\mathbb{C}) \to \mathrm{GL}(V)$, a (vectorial) Siegel modular form of weight ρ is a holomorphic function $f : \mathfrak{H}_2 \to V$ satisfying: $f(\gamma\Omega) = \rho(C\Omega + D)f(\Omega)$ for every $\Omega \in \mathfrak{H}_2$, $\gamma \in \Gamma_2$. Therefore, we say that f is *scalar-valued* if V has dimension 1, and otherwise f is said to be *vector-valued*. Furthermore, if f has no pole at $\Omega \in \mathfrak{H}_2$, for every complex principally polarized abelian surface $\mathcal{A}$ isomorphic to $\mathcal{A}_\Omega$ with a basis $\omega \in \Omega^1(\mathcal{A})$, we have the following well defined quantity

$$f(\mathcal{A}, \omega) := \rho({}^t(m^{-1}))f(\Omega)$$

where $m \in \mathrm{GL}_2(\mathbb{C})$ depends only on $\mathcal{A}$, ω and Ω, and is the base change matrix between ω and the canonical basis of differentials $(2\pi i dz_1, 2\pi i dz_2)$ on $\mathcal{A}_\Omega$. See [22, Sect. 2.2] for more detail about this construction. This is Katz's algebraic interpretation of modular forms as sections of representations of the Hodge vector bundle. Since $(\mathcal{A}, \omega)$ can be defined over any field $\Bbbk$, if f is defined over some $\mathbb{Z}[1/N]$ (as seen by its Fourier coefficients) with $\mathrm{char}(k) \nmid N$, then the quantity $f(\mathcal{A}, \omega)$ will make sense over $\Bbbk$.

In particular, given the equation of an hyperelliptic curve $\mathcal{C}$, the basis $\omega(\mathcal{C}) = (dx/y, xdx/y)$ provide a basis of $\Omega^1(\mathrm{Jac}(\mathcal{C}))$. In particular, a modular form induces a covariant on $\mathcal{C}$ via the following map:

$$\mathrm{Cot}(f) : \mathcal{C} \mapsto f(\mathrm{Jac}(\mathcal{C}), \omega(\mathcal{C})),$$

which allow to express $\mathrm{Cot}(f)$ in terms of the coefficients of the curve. In other words, seeing a modular form is a section of a representation of weight ρ of the Hodge vector bundle on $\mathfrak{A}_2$, the pullback of the Torelli morphism $\mathcal{M}_2 \to \mathfrak{A}_2$ induces a section of the representation of weight ρ of the Hodge vector bundle on $\mathcal{M}_2$.

Therefore, given f a Siegel modular function of weight ρ, $\mathrm{Cot}(f)$ is a fractional covariant (rational in term of the coefficients of the curve) of weight ρ. Note that $\mathrm{Cot}(f)$ is a polynomial covariant if and only if f has no poles at the divisor at infinity in $\overline{\mathfrak{A}_2}$ (a toroidal compactification of $\mathfrak{A}_2$). In particular, if f is a modular form, $\mathrm{Cot}(f)$ is a polynomial covariant by the Koecher principle. Reciprocally, if F is a fractional covariant, there exists a well defined Siegel modular meromorphic form f of the same weight associated to F [22, Sect. 3.2]. If F is polynomial, then f is a modular form if it has no pole at Ω such that $\chi_{10}(\Omega) = 0$.

The irreducible representations ρ are of the form $\det^k \mathrm{Sym}^n$ with $k \in \mathbb{Z}$ and $n \in \mathbb{N}$ (see [22, Sect. 2.2] for more details). A scalar modular form of weight k is thus a modular form of weight $\det^k$. So the covariants of order n and degree k in Hilbert's Invariant Theory of Sect. 2.2 will correspond to a polynomial covariant of weight $\det^{(k-n/2)} \mathrm{Sym}^n$. This is because the action of covariants on the sextic f is given by Sym^6, while the action of the curve $y^2 = f(x)$ preserving the differentials is given by $\det^{-2} \mathrm{Sym}^6$

In particular, the polynomial covariants J_{2i}'s have the following interpretation in term of modular forms (here we use the standard cusp forms χ_{10}, χ_{12} rather than the normalized ones from [22])

$$\begin{aligned} -3\mathrm{Cot}(\chi_{12}/\chi_{10}) &= J_2, \\ 4\mathrm{Cot}(\psi_4) &= I_4, \\ 4\mathrm{Cot}(\psi_6) &= \frac{1}{2}(I_2 I_4 - 3I_6) = I_6', \\ -2^2\mathrm{Cot}(\chi_{10}) &= J_{10}. \end{aligned}$$

One can then express Igusa's invariants j_i as the following quotients of covariants of the same weight:

$$\begin{aligned} \mathrm{Cot}(\mathrm{j}_1) &= \frac{1}{256 J_{10}} \left(J2^5 - 60 J_4 J_2^3 + 216 J_6 J_2^2 + 864 J_4^2 J_2 - 5184 J_6 J_4 \right), \\ \mathrm{Cot}(\mathrm{j}_2) &= \frac{1}{32 J_{10}} \left(J_2^5 - 48 J_4 J_2^3 + 576 J_4^2 J_2 \right), \\ \mathrm{Cot}(\mathrm{j}_3) &= \frac{1}{16384 J_{10}^2} \left(J_2^{10} - 120 J_4 J_2^8 + 5760 J_4^2 J_2^6 - 138240 J_4^3 J_2^4 + 1658880 J_4^4 J_2^2 - 7962624 J_4^5 \right). \end{aligned}$$

3.3 Siegel Modular Polynomials

The modular polynomials in dimension 2 were first computed by Regis Dupont in his thesis [11] for $p = 2$. The polynomials computed were in function of Igusa's invariants, and recently Enea Milio [29] proposed an extension of this construction to others invariants (not necessarily of level 1) and succeeded to reduce considerably their size. In this section, we briefly explain Milio's algorithm on how to compute modular polynomials in dimension 2, to show that it easily adapts to invariants with good reduction modulo 2.

For a subgroup $\Gamma \subset \Gamma_2$, $\mathbb{C}_\Gamma$ denotes the field of modular functions for Γ, $\mathbb{C}_\Gamma/\mathbb{C}_{\Gamma_2}$ is a finite algebraic extension of degree $[\Gamma_2 : \Gamma]$. Let p be a prime, $\Gamma_0(p) := \left\{\left(\begin{smallmatrix} A & B \\ C & D \end{smallmatrix}\right) \in \Gamma_2;\ C \equiv 0 \mod p\right\}$ and $\mathcal{C}_p$ the set of representative classes of $\Gamma_2/\Gamma_0(p)$ then $[\Gamma_2 : \Gamma_0(p)] = p^3 + p^2 + p + 1$.

The period matrices of the (p, p)-isogenous varieties of a variety Ω are the $p\gamma\Omega$ for $\gamma \in \mathcal{C}_p$. We consider the following definition : for $f \in \mathbb{C}_{\Gamma_2}$ and $\gamma \in \Gamma_2$, the functions f_p and f_p^γ from $\mathfrak{H}_2$ to $\mathbb{C}$ are defined by $f_p(\Omega) = f(p\Omega)$, $f_p^\gamma(\Omega) = f(p\gamma\Omega)$, where

$$\gamma_p := \begin{pmatrix} A & pB \\ C/p & D \end{pmatrix}.$$

Then $p\Omega$ is equivalent to $p\gamma\Omega$ only when $\gamma \in \Gamma_0(p)$, i.e. $\mathrm{j}_l(p\gamma\Omega) = \mathrm{j}_l(p\Omega)$ for $l = 1, 2, 3$;

Theorem 4 *For a prime p, the function* $\mathrm{j}_{l,p} := (\mathrm{j}_l)_p$ *is modular for* $\Gamma_0(p)$ *and* $\mathbb{C}_{\Gamma_0(p)}$ *equals* $\mathbb{C}(\mathrm{j}_1, \mathrm{j}_2, \mathrm{j}_3, \mathrm{j}_{l,p})$ *for any* $l = 1, 2, 3$.

Proof See [2, Lemma 4.2].

The pth modular polynomial for j_1 denoted by $\phi_{1,p}(X)$ is the minimal polynomial of $\mathrm{j}_{1,p}$ over $\mathbb{C}_{\Gamma_2} = \mathbb{C}(\mathrm{j}_1, \mathrm{j}_2, \mathrm{j}_3)$: $\phi_{1,p}(X) = \prod_{\gamma \in \mathcal{C}_p}(X - \mathrm{j}_{1,p}^\gamma)$. By Theorem 4 we know that $\mathrm{j}_{2,p}$ and $\mathrm{j}_{3,p}$ are in $\mathbb{C}(\mathrm{j}_1, \mathrm{j}_2, \mathrm{j}_3, \mathrm{j}_{1,p}) = \mathbb{C}(\mathrm{j}_1, \mathrm{j}_2, \mathrm{j}_3)[\mathrm{j}_{1,p}]$, so $\mathrm{j}_{2,p} = \phi_{2,p}(\mathrm{j}_{1,p})$ and $\mathrm{j}_{3,p} = \phi_{3,p}(\mathrm{j}_{1,p})$ for $\phi_{2,p}(X)$ and $\phi_{2,p}(X)$ monic polynomials in $\mathbb{C}(\mathrm{j}_1, \mathrm{j}_2, \mathrm{j}_3)[X]$ of degree less than $\deg(\phi_{1,p}(X))$. The modular polynomials $\phi_{1,p}(X)$, $\phi_{2,p}(X)$, and $\phi_{3,p}(X)$ lie in $\mathbb{Q}(\mathrm{j}_1, \mathrm{j}_2, \mathrm{j}_3)[X]$ [11, Theorem 5.1]. If x is a root over $\mathbb{C}$ of $\phi_{1,p}(X)$, then $(x, \phi_{2,p}(x), \phi_{3,p}(x))$ are the j-invariants of a principally polarized abelian surface (p, p)-isogenous to the variety with invariants $(\mathrm{j}_1(\Omega), \mathrm{j}_2(\Omega), \mathrm{j}_3(\Omega))$.

More generally, let f_1, f_2, and f_3 be three modular functions for Γ such that $\mathbb{C}_\Gamma$ is generated by f_1, f_2, and f_3 (we say that they are Γ-birational invariants), and the level of Γ is prime to $2p$. Then the pth-modular polynomials for f_1, f_2, and f_3 are

$$\phi_{1,p}(X) = \prod_{\gamma \in \mathcal{C}_p}(X - f_{1,p}^\gamma), \quad \text{and} \quad \phi_{l,p}(X) = \psi_{l,p}(X)/\phi_{1,p}'(X)$$

$$\text{where} \quad \psi_{l,p}(X) = \sum_{\gamma \in \mathcal{C}_p} f_{l,p}^\gamma \prod_{\gamma' \in \mathcal{C}_p \setminus \{\gamma\}} (X - f_{1,p}^{\gamma'}), \quad for \quad l = 2, 3,$$

with $\mathcal{C}_p$ a set of representative classes of $\Gamma/(\Gamma \cap \Gamma_0(p))$.

Let us consider the modular functions for $\Gamma(2,4)$ defined by

$$\mathfrak{b}_i(\Omega) := \frac{\theta_i(\Omega/2)}{\theta_0(\Omega/2)} \quad \text{for } i = 1, 2, 3 \quad \text{and} \quad \mathfrak{c}_i = \frac{\theta_i^2(\Omega)}{\theta_0^2(\Omega)} \quad \text{for } i \in \{1, \cdots, 15\} \quad (i \text{ even})$$

and linked by the following relations:

$$\mathfrak{b}_1 = (\mathfrak{c}_1 + \mathfrak{c}_9)(1 + \mathfrak{c}_4 + \mathfrak{c}_8 + \mathfrak{c}_{12})^{-1},$$
$$\mathfrak{b}_2 = (\mathfrak{c}_2 + \mathfrak{c}_6)(1 + \mathfrak{c}_4 + \mathfrak{c}_8 + \mathfrak{c}_{12})^{-1},$$
$$\mathfrak{b}_3 = (\mathfrak{c}_3 + \mathfrak{c}_{15})(1 + \mathfrak{c}_4 + \mathfrak{c}_8 + \mathfrak{c}_{12})^{-1}.$$

Then we have: $\mathbb{C}_{\Gamma(2,4)} = \mathbb{C}(\mathfrak{c}_1, ..., \mathfrak{c}_{15}) = \mathbb{C}(\mathfrak{b}_1, \mathfrak{b}_2, \mathfrak{b}_3)$.

One can compute the polynomials $\phi_{1,p}(X)$, $\phi_{2,p}(X)$, and $\phi_{3,p}(X)$ (or ψ_2, ψ_3) using Evaluation/Interpolation using fast evaluation (and computation of the period matrix) of the $\mathfrak{b}_i$ [11] and the following algorithm is a variant detailed in [29, Chap. 4]. We give a brief summary of the algorithm in Algorithm 3.1.

We assume that the modular relations $F(f_i, \mathfrak{j}_i)$ are known (alternatively that we have modular relations $F(f_i, \mathfrak{b}_i)$. They allow us to go from the $f_i(\Omega)$ to the j-invariants (or Rosenhain invariants) in the first step of the algorithm. Likewise, they allow us to recover the values of $f_{i,p}^{\gamma}(\Omega)$ from the values of $\mathfrak{j}_{i,p}^{\gamma}(\Omega)$ by using Newton's algorithm combined with a low precision computation for its initialization.

Let $n_p = p^3 + p^2 + p + 1$ be the degree of $\phi_{1,p}(X)$. The Algorithm 3.1 evaluates $\phi_{1,p}(X)$ in $O(\mathcal{M}(n_p) \log n_p)$.

In the applications of Algorithm 3.1, E. Milio has computed 2th, 3th, 5th modular polynomials for several system of invariants. However, his 2th modular polynomials using the $\mathfrak{j}_i$ modular invariants have bad reduction in characteristic 2. We instead used Algorithm 3.1 to compute the modular polynomials for the invariants $(\mathfrak{a}_1, \mathfrak{a}_2, \mathfrak{a}_3)$ that reduce well modulo 2. Such polynomials parameterize the 2-isogenous Jacobians in $\mathcal{M}_2[J_2^{-1}]$. This open locus is well adapted to curves of type $(1, 1, 1)$ in characteristic two.

Remark 2 • Since the invariants $(\mathfrak{a}_1, \mathfrak{a}_2, \mathfrak{a}_3)$ and the invariants $(\mathfrak{j}_1, \mathfrak{j}_2, \mathfrak{j}_3)$ are birationally equivalents, we could have compute the modular polynomials for $(\mathfrak{a}_1, \mathfrak{a}_2, \mathfrak{a}_3)$ using a change of variable in the modular polynomials of $(\mathfrak{j}_1, \mathfrak{j}_2, \mathfrak{j}_3)$. In practice, this change of variable involve rational function, this was too expensive to do it directly. We might have computed it using evaluation/interpolation, but it was easier to just reuse Algorithm 3.1.

- The common denominator of $(\mathfrak{j}_1, \mathfrak{j}_2, \mathfrak{j}_3)$ is χ_{10}, which is zero on the locus of product of elliptic curves. Denote by $\mathcal{L}_p$ the locus off the principal polarized abelian surfaces that are (p, p)-isogenous to a product of elliptic curves. It is a 2-dimensional algebraic subvariety of the moduli space $\Gamma_2\backslash\mathfrak{H}_2$ and can be parametrized by an equation $L_p = 0$. The denominator of the coefficients of $\phi_{1,p}(X, f_1(\Omega), f_2(\Omega), f_3(\Omega))$, $\psi_{1,p}(X, f_1(\Omega), f_2(\Omega), f_3(\Omega))$, and $\psi_{2,p}(X, f_1(\Omega), f_2(\Omega), f_3(\Omega))$ are thus all divisible by the polynomial L_p.

Algorithm 3.1 Evaluation of Modular Polynomials

Input $(f_1(\Omega), f_2(\Omega), f_3(\Omega))$ such that $\mathbb{C}_\Gamma = \mathbb{C}(f_1(\Omega), f_2(\Omega), f_3(\Omega))$, a prime p coprime with the level of Γ, $\mathcal{C}_p$ of the representative classes in $\Gamma/(\Gamma \cap \Gamma_0(p))$ and a pre-compute of the action of the representative classes in Γ_2/Γ, a precision $N \in \mathbb{N}$.

Output $\phi_{1,p}(X, f_1(\Omega), f_2(\Omega), f_3(\Omega))$ and $\psi_{l,p}(X, f_1(\Omega), f_2(\Omega), f_3(\Omega))$ at precision N for $l = 2, 3$.

- ➔ Deduce from $f_i(\Omega)$ the j-invariant $\mathfrak{j}_i(\Omega)$ or the Rosenhain invariants ;
- ➔ Using these invariants construct a hyperelliptic curve $Y^2 = f(X)$ at precision N ;
- ➔ Deduce the ten $\mathfrak{c}_i(\Omega)$ at precision N using Thomae's formulae and Numerical integration for the choice of roots;
- ➔ Reverse the functions to recover Ω' at precision N using [11];
- ➔ Determine Ω and $\gamma \in \Gamma_2/\Gamma$ such that $\Omega = \gamma\Omega'$ (comparing $f_i(\Omega)$'s and $f_i(\Omega')$'s);
- ➔ Compute at precision N the $\mathfrak{b}_{i,p}^{\gamma}(\Omega)$ for $\gamma \in \mathcal{C}_p$ using [11] and the corresponding $\mathfrak{j}_{i,p}^{\gamma}(\Omega)$;
- ➔ Compute at precision N the $f_{i,p}^{\gamma}(\Omega)$ for $\gamma \in \mathcal{C}_p$;
- ➔ Compute $\phi_{1,p}(X, f_1(\Omega), f_2(\Omega), f_3(\Omega))$ and $\psi_{l,p}(X, f_1(\Omega), f_2(\Omega), f_3(\Omega))$ at precision N using a tree of sub-product ;
- ➔ Compute $\phi_{1,p}, \psi_{l,p}$ at precision N for $l = 2, 3$ using fast interpolation of rational functions [31];

- Using $(\mathfrak{a}_1, \mathfrak{a}_2, \mathfrak{a}_3)$, the denominator of the modular polynomials contain the locus of Jacobians which are (p, p)-isogenous to an abelian surface A with $J_2(A) = 0$. It would be interesting to have a modular interpretation of this locus, similar to the description of $\mathcal{L}_p$ above. We denote by $\mathcal{D}_p$ the locus of the denominators of the modular polynomials $\Phi_{1,p}, \Psi_{1,p}, \Psi_{2,p}$.
- We only applied Algorithm 3.1 to birational invariants of level 1. It would be interesting to find good invariants how higher levels that give smaller polynomials.

4 2-Adic Canonical Lift of Genus 2 Curves

In this section, we use the modular polynomials to lift the absolute invariants, and then we explain how to lift the curves. The Frobenius induces an isogeny, and we use the modular polynomials to lift this isogeny.

The reduction modulo 2 of curves over $\mathbb{Z}_q$ follows the birational classification given by the type (1, 1, 1), (3, 1), and (5). Note that curves of different type are not isogenous by the Frobenius, so we only need to lift two isogenous curves of the same type.

In this section, we focus on computing a canonical lift, which amount to lifting the Frobenius as above by Lubin-Serre-Tate's theorem. So we assume that $A = \mathrm{Jac}C$

is ordinary over $\mathbb{F}_q$, or equivalently, that C is of type $(1, 1, 1)$. Hence we can use the modular polynomials with invariants $J = (\mathfrak{a}_1, \mathfrak{a}_2, \mathfrak{a}_3)$ as defined in Sect. 3.3. We give some indications on how to lift the Frobenius for the other types (be careful that in this case the lift is no longer unique).

4.1 Kronecker Conditions

Let J be absolute invariants such that $J = (f_1, f_2, f_3)$ where f_i are birational invariants with good reduction modulo 2. To lift the Frobenius, we need to lift the modular relation between J and J^σ.

Let $U = (x, y, z)$ be the variable representing the absolute invariants of the curve $\mathcal{C}$ and $V = (u, v, w)$ the absolute invariants of a (p, p)-isogenous curve to $\mathcal{C}$. We let

$$\begin{aligned}\Phi_{1,p} &= \mathrm{Num}(\phi_{1,p})(u, x, y, z)\\ \Psi_{2,p} &= \mathrm{Num}(\psi_{2,p})(u, x, y, z) - v \cdot \mathrm{Den}(\psi_{2,p})(u, x, y, z)\\ \Psi_{3,p} &= \mathrm{Num}(\psi_{3,p})(u, x, y, z) - w \cdot \mathrm{Den}(\psi_{3,p})(u, x, y, z)\end{aligned}$$

where Num(...) and Den(...) are the numerator and the denominator of the computed modular polynomials. We also denote by Φ_p the column matrix with components: $\Phi_{1,p}$, $\Psi_{2,p}$ and $\Psi_{3,p}$.

We now specialize to $J = (\mathfrak{a}_1, \mathfrak{a}_2, \mathfrak{a}_3)$.

Proposition 2 (Kronecker condition) *Let $p = 2$ and $\mathcal{A}$ an abelian surface over $\mathbb{F}_q$ such that $J_2(\mathcal{A})J_{10}(\mathcal{A}) \neq 0$. The $\mathcal{A}$ is ordinary, and is the Jacobian of a curve $\mathcal{C}$ of type* $(1, 1, 1)$. *Let $J = (\mathfrak{a}_1, \mathfrak{a}_2, \mathfrak{a}_3)$ be its absolute invariants (they are well defined by Theorem 2). Assume that $q > p^2$. Then (i) $\frac{\partial \Phi_p}{\partial V}(J, J^\sigma)$ is invertible;*

(ii) $\frac{\partial \Phi_p}{\partial U}(J, J^\sigma) \equiv 0 \mod p.$

Proof When $J = (\mathfrak{a}_1, \mathfrak{a}_2, \mathfrak{a}_3)$, this can be checked from the coefficients of the modular polynomials computed. Since a birational change of variable induce an invertible differentiable map (at the open locus of definition), Proposition 2 is thus also valid for any $J = (f_1, f_2, f_3)$ well defined on the locus $J_2 J_{10} \neq 0$. This Kronecker condition generalizes the well known reduction of elliptic modular polynomials modulo p, and one can prove that it is also valid for $p > 2$ (See [25]).

Remark 3 J is well defined at $\mathcal{A}$ exactly when $J_2(\mathcal{A})J_{10}(\mathcal{A}) \neq 0$, so in this case J is automatically well defined at $\mathcal{A}^\sigma$, since $J(\mathcal{A}^\sigma) = J(\mathcal{A})^\sigma$. It may seem that the modular equation is not well defined if $\mathcal{A}$ is in $\mathcal{D}_p$, the denominator locus. But we can always clear the denominator, and get a modular equation on $\mathcal{A}$ and $\mathcal{A}^\sigma$ since J is well defined at both. More precisely, we can interpret the denominator as a modular form [21] in order to remove the parasite factors [30] introduced by the rewriting procedure [21].

Algorithm 4.1 HarleyDim2

Input J, integer N the precision.
Output $\tilde{J}$ the lift of J at precision N.

➔ If $N = 1$ then return J at precision p;

➔ Else $N' = N/2$;

➔ J =HarleyDim2(J, N');

a. $G = \frac{\partial \Phi}{\partial U}(J, J^\sigma)$; $H = \frac{\partial \Phi}{\partial V}(J, J^\sigma)$; $Q = \Phi(J, J^\sigma)$;
b. $a = G.H^{-1}, b = Q.H^{-1}$;
c. e =ArtinSchreier(a, b, N');
d. $J = J + p^k e$;

➔ Return J ;

4.2 *Computing the Lift of Invariants*

Let J be the vector of birational invariants as above. It satisfy the Kronecker condition and we want to lift it from $\mathbb{F}_q$ to $\mathbb{Z}_q$. Suppose that we can compute efficiently the Frobenius automorphism σ of $\mathbb{Q}_q$ and $X \in \mathbb{Z}_q^3$ is an approximation of $\tilde{J}$ at precision p^k i.e. $\tilde{J} - X = p^k e$ for some error e a vector over $\mathbb{Z}_q$ that we want to find.

Since $\frac{\partial \Phi_p}{\partial V}(X, X^\sigma)$ is invertible to have the error e we must solve over $\mathbb{Z}_q$ an equation of the form: $e^\sigma + Ae + B = 0$ (with $A \equiv 0 \mod p$) called "*Artin-Schreier equation*" in [13, 25]. An extended Harley algorithm solves the equation $e^\sigma + Ae + B = 0$ over $\mathbb{Z}_q$ and we obtain J at a given precision.

Theorem 5 *Let $J = (\mathfrak{a}_1, \mathfrak{a}_2, \mathfrak{a}_3)$ be the absolute invariants of a polarized abelian surface $\mathcal{A}$ over $\mathbb{F}_q$. If not all $\mathfrak{a}_1$, $\mathfrak{a}_2$ and $\mathfrak{a}_3$ in $\mathbb{F}_p$ then the Algorithm 4.1 computes (by doubling precision) the absolute invariants of the lifting curve $\tilde{\mathcal{A}}$ over $\mathbb{Z}_q$, where $\Phi_{1,p}$ is minimal polynomial of the one element of the triple invariants not in $\mathbb{F}_p$.*

Proof This is a standard application of the Newton lifting described above. See also [25] for a generalization to $p > 2$.

4.3 *2-Adic Lift of the Curves*

To reconstruct the lifted curve using the lifted absolute invariants over $\mathbb{Z}_q$, we can avoid Mestre's algorithm by directly lifting a normal form equation of the curves from $\mathbb{F}_q$ to $\mathbb{Z}_q$. Indeed, the conic in Mestre's algorithm has bad reduction modulo 2, so it is not obvious how to find a rational point on it.

Let $q = 2^n$ and C be a genus 2 curve given over $\mathbb{F}_q$ by its normal form equation :

$$C : XY^2 + (1 + aX + bX^2)Y + X^2(c + dX + X^2) = 0.$$

Since the J_{2i}'s are expressed in term of the coefficients a, b, c, and d of the equation of C, we can also express the birational invariants (f_1, f_2, f_3) as rational functions in (a, b, c, d). Let $(\tilde{f}_1, \tilde{f}_2, \tilde{f}_3)$ be the lift of (f_1, f_2, f_3) computed in Sect. 4.2, and $(\tilde{a}, \tilde{b}, \tilde{c}, \tilde{d})$ be the lifts of the normal form equation of C. The relation above provide a system of equations for $(\tilde{a}, \tilde{b}, \tilde{c}, \tilde{d})$, to which we can apply Newton lifting from the solution (a, b, c, d) modulo 2 (since everything has good reduction).

In particular, since the model $\tilde{C}$ has good reduction modulo 2, the reduction w of a basis of differentials $\tilde{w}$ is well defined. Comparing w with a fixed basis w_C of C, we get that $w = Mw_C$ for a matrix in $\mathrm{GL}_2(\mathbb{F}_q)$. Lifting M to $\tilde{M} \in \mathrm{GL}_2(\mathbb{Z}_q)$, and setting $\tilde{w}_{\tilde{C}} = \tilde{M}^{-1}\tilde{w}$, we can lift w_C. As an application, let g be a vectorial modular form (with good reduction modulo 2), then $g(\tilde{C}, \tilde{w}_{\tilde{C}})$ is a lift of $g(C, w_C)$.

As an example we detail the lift of the curves of type (1, 1, 1) which concern all ordinary cases.

Lift of Curves of Type (1, 1, 1)

For the curve of type (1, 1, 1) over $\mathbb{F}_{2^n}$ we can compute the absolute invariants $(\mathfrak{a}_1, \mathfrak{a}_2, \mathfrak{a}_3)$ using the relation in Sect. 2.4 in term of the coefficients a, b, c, and d. The coefficients $\tilde{a}$, $\tilde{b}$, $\tilde{c}$, and $\tilde{d}$ of the normal form of the curve $\tilde{C}$ that reduce to C satisfy the following system of equations in the variables $\tilde{a}$, $\tilde{b}$, $\tilde{c}$, and $\tilde{d}$:

$$\begin{cases} \tilde{J}_4 - \tilde{J}_2^2\tilde{\mathfrak{a}}_1 = 0, \\ \tilde{J}_8 - \tilde{J}_2^4\tilde{\mathfrak{a}}_2 = 0, \\ \tilde{J}_{10} - \tilde{J}_2^5\tilde{\mathfrak{a}}_3 = 0 \end{cases} \tag{8}$$

where $(\tilde{\mathfrak{a}}_1, \tilde{\mathfrak{a}}_2, \tilde{\mathfrak{a}}_3)$ are the lift of $(\mathfrak{a}_1, \mathfrak{a}_2, \mathfrak{a}_3)$ given by the algorithm 4.1, and the expression of the J_{2i} in term of the coefficients a, b, c, and d is given by [18, Sect. 3]. On other hand one compute J_{2i}'s from the hyperelliptic model using the coefficients of the normal form:

$$y^2 + h(x)y = f(x) \quad \text{where} \quad h(x) = 1 + ax + bx^2 \quad \text{and} \quad f(x) = -x^3(c + dx + x^2).$$

Lemma 1 *Over $\mathbb{Z}_q$ the intersection of the three surfaces of Eq. (8) is nonsingular at (a, b, c, d) satisfying the conditions of Theorem 5 is nonsingular at precision > 1. These equations ramify modulo 2.*

Proof Given a genus 2 curve C over $\mathbb{Z}_q$, two characterizations of C reduce well modulo 2: its normal form (a, b, c, d) and the values of J_{2i}'s at C. Furthermore the J_{2i}'s can be expressed in function of (a, b, c, d) using [18, pp. 10–12]. We have modulo 2

$$J_2 \equiv a^2 b^2.$$

Over $\mathbb{Z}_q$ and if J_2 is nonzero modulo 2 (i.e. $ab \not\equiv 0 \mod 2$), the system (8) defines a nonsingular point in $\mathcal{M}_2[J_2^{-1}]$ [18, Theorem 4]. (representing the birational class of $\tilde{C}$). Furthermore at precision 1, we have the following factorization:

$$\begin{cases} J_4 = & J_2^2(\alpha^2+\beta^2+\gamma^2), \\ J_8 = & J_2^4(\alpha^2\beta^2+\alpha^2\gamma^2+\beta^2\gamma^2-\mathfrak{g}^3-\mathfrak{g}^4), \quad \text{with } \mathfrak{g}=\alpha^2+\beta^2+\gamma^2, \\ J_{10} = & J_2^5(\alpha^2\beta^2\gamma^2). \end{cases}$$

where α, β and γ are expressed in function of (a, b, c, d) using the relations Sects. 2 and 2.4. We can normalize the new equations of Eq. (8) with even powers of ab. Therefore the partial derivatives at a and b of the system Eq. (8) vanish. Then the determinant of the Jacobian of Eq. (8) (fixing a parameter to get three variables) vanishes too. We remark that the factorization above is only available modulo 2. On other hand at precision 2 when one parameter (for instance a) is fixed, the determinant of the Jacobian matrix of Eq. (8) is given by

$$2a^{12}\left[a^{10}+(b^3+db)a^9+(cb^2+b)a^8+(d^2b^5+c^2b)a^7+\right.$$
$$(cb^3+b^2)a^6+(\mathfrak{a}_3b^{15}+c^2b^7+b^5)a^5+\mathfrak{a}_3b^{18}a^4+$$
$$(\mathfrak{a}_3db^{19}+c^4b^5)a^3+(\mathfrak{a}_3cb^{20}+\mathfrak{a}_3b^{19})a^2+$$
$$\left.(\mathfrak{a}_3c^2b^{19}+c^2b^9+b^7)a+(\mathfrak{a}_3cb^{21}+\mathfrak{a}_3b^{20})\right].$$

The vanishing modulo 4 of the value above implies the vanishing modulo 2 of the value between the brackets. Plugging the value of $\mathfrak{a}_3$, we get that this polynomial is not identically zero, hence up to changing our (a, b, c, d) by an isomorphism it does not vanishes. Hence, the determinant of the system has only 2-valuation 1.

Remark 4 From [25, Sect. 4], we have a Newton algorithm to lift at precision N, (a, b, c, d) knowing that the valuation of the minor of the Jacobian matrix of Eq. (8) is 1.

So we know how to compute the canonical lift of every ordinary genus 2 curve defined over $\mathbb{F}_q$, its normal form equation (hence its canonical equation by Appendix 6), and lift Siegel modular forms. We now explain how to compute the Zeta function of C by giving equations of the isogeny induced by the lifted Frobenius. For this, we first need to explain how to lift its kernel. We first explain the reduction properties of the Weierstrass points.

4.4 Reduction of the Weierstrass Points

Let C be genus 2 curve given by $XY^2 + (1 + aX + bX^2)Y + X^2(c + dX + X^2) = 0$ over $\mathbb{F}_{2^n}$. Then the lifted coefficients $(\tilde{a}, \tilde{b}, \tilde{c}, \tilde{d})$ determine the curve $\tilde{C}$ defined over $\mathbb{Z}_q$ by :

$$Y^2 + (1 + \tilde{a}X + \tilde{b}X^2)Y = -X^3(\tilde{c} + \tilde{d}X + X^2).$$

And such a curve $\tilde{C}$ is equivalent over $\mathbb{Z}_q$ to the curve defined by :

$$Y^2 = (1 + \tilde{a}X + \tilde{b}X^2)^2 - 4X^3(\tilde{c} + \tilde{d}X + X^2).$$

A point P on $\tilde{C}$ of affine coordinates (x, y) is a Weierstrass point if and only if for $H(x) = 1 + \tilde{a}x + \tilde{b}x^2$ and $F(x) = x^3(\tilde{c} + \tilde{d}x + x^2)$, we have

$$2y + H(x) = 0, \quad and \quad y^2 + H(x)y + F(x) = 0.$$

The five Weierstrass points of $\tilde{C}$ reduce as follows:
The case of type (1, 1, 1) Two points reduce to P, two others reduce Q such that x_P and x_Q are the two roots modulo 2 of the equation : $H(x) = 0$. And the fifth point reduces to the point at infinity.
The case of type (3, 1) This case is characterized by ($a = 0$ and $b \neq 0$) or ($a \neq 0$ and $b = 0$), where one can remark that two points reduce to the unique point which abscissa a^{-1} or $1/\sqrt{b}$ depending on the case ($a = 0$ and $b \neq 0$) or ($a \neq 0$ and $b = 0$). And the three others points reduce all to the point at infinity.
The case of type (5) All the Weierstrass points reduce modulo 2 to the point at infinity.

4.5 Computing the Lifted Frobenius

We explain some details about the (2, 2)-isogeny algorithm introduced by Richelot in [34] and we give a simple way to compute the lifted Frobenius and the lifted Verschiebung. To express Richelot's construction in terms of quadratic splittings, we use B. Smith notation in [37].

For a polynomials G_1, G_2 and G_3 in $\mathbb{k}[X]$ we denote by : $[G_2, G_3] = G_2'G_3 - G_2G_3'$ and $\det(G1, G2, G3)$ the determinant of the matrix with as coefficient (ij) the jth-coefficient of G_i.

Richelot Algorithm
Let $\Pi(G_1, G_2, G_3)$ be the squarefree polynomial of degree 5 or 6 of a hyperelliptic model of a genus 2 curve $\mathcal{X}$. The Richelot operator $\mathcal{R}$ is defined by

$$\mathcal{R}(G_1, G_2, G_3) := (\delta[G_2, G_3], \delta[G_3, G_1], \delta[G_1, G_2]),$$

where $\det(G_1, G_2, G_3) \neq 0$ and $\delta = (\det(G_1, G_2, G_3))^{-1}$.

Let $(H_1, H_2, H_3) = \mathcal{R}(G_1, G_2, G_3)$ then

- $\Pi(H_1, H_2, H_3)$ is a squarefree polynomial of degree 5 or 6 .
- $\mathcal{R}(k_1G_1, k_2G_2, k_3G_3) = (k_1^{-1}H_1, k_2^{-1}H_2, k_3^{-1}H_3)$ for a nonzero k_i's in the base field ;
- $\mathcal{R}(G_2, G_3, G_1) = (H_2, H_3, H_1)$ and $\mathcal{R}(G_3, G_1, G_2) = (H_3, H_1, H_2)$;
- $\mathcal{R}(H_1, H_2, H_3) = (G_1, G_2, G_3)$.

Lemma 2 *Let $\mathcal{X} : v^2 = f_{\mathcal{X}}(u)$ be a curve of genus 2. For each nonsingular quadratic splitting $G = (G_1, G_2, G_3)$ of $f_{\mathcal{X}}$, we define $\mathcal{X}_G$ to be the curve given by $\mathcal{X}_G : v^2 = f_{\mathcal{X}_G} := \Pi(\mathcal{R}(G))$. Then $\mathcal{X}_G$ is a curve of genus 2 such that* Jac$\mathcal{X}$ *and* Jac$\mathcal{X}_G$ *are (2,2)-isogenous by the isogeny ϕ of kernel generated by divisors classes $[(G_1(u), 0)]$, $[(G_2(u), 0)]$ and $[(G_3(u), 0)]$. Further, $\mathcal{R}(G)$ defines the dual isogeny $\widehat{\phi}$ of ϕ on* Jac$\mathcal{X}_G$.

Proof By using the Theorem 8.4.11 and Proposition 8.4.12 in [37].

It remains to identify the nonsingular quadratic splitting $G = (G_1, G_2, G_3)$, such that $\Pi(\mathcal{R}(G))$ corresponds to the action of the lift Σ of the pth-Frobenius morphism.

Reduction of the Kernel

Let's consider a symplectic basis P, Q, R, and S of Jac$(\mathcal{C})[2]$, where P and Q étale, R and S moderately ramified, and R is the dual of P, and S the dual of Q for the Weil-Pairing. An isotropic kernel $\tilde{\mathcal{K}}$ of the lifted (2, 2)-isogeny from Jac$(\tilde{\mathcal{C}})$ has three possibilities of reduction on Jac$(\mathcal{C})$:

1. $\tilde{\mathcal{K}} = \langle \tilde{R}, \tilde{S} \rangle$, the only kernel which reduces to $\langle 0 \rangle$. It corresponds to the kernel of the canonical lift of the Frobenius.
2. $\tilde{\mathcal{K}}$ reduces to a cyclic group of cardinality 2 modulo p. There are six possibilities: $\langle \tilde{P}, \tilde{S} \rangle$, $\langle P \tilde{+} R, \tilde{S} \rangle$, $\langle \tilde{Q}, \tilde{R} \rangle$, $\langle Q \tilde{+} S, \tilde{R} \rangle$, $\langle P \tilde{+} Q, R \tilde{+} S \rangle$, $\langle P + \tilde{Q} + R, R \tilde{+} S \rangle$.
3. The remaining 8 kernels $\tilde{\mathcal{K}}$ reduce to $\langle P, Q \rangle$ corresponding to the Verschiebung. The canonical lift of the Verschiebung, defined as the dual of the canonical lift of the Frobenius, is the only kernels among these whose points are defined in an étale extension of $\mathbb{Q}_q$. The points of the other kernels live in some moderately ramified extension.

And we get the 15 isogenies (counting with multiplicity) corresponding to the roots of $\phi_{1,p}$.

Note that type (3,1) has p-rank 1 and the type (5) has p-rank 0. And in both the cases, several isogenies reduce to the Frobenius.

Proposition 3 *In the case of type (1,1,1): The conjugate genus two curve $\tilde{\mathcal{C}}^{\Sigma}$ over $\mathbb{Z}_q$ of $\tilde{\mathcal{C}}$ is defined by the equation : $Y^2 = \Pi\mathcal{R}(G_1, G_2, G_3)$, where : $G_1(u) = (u - x_1)(u - x_2)$ and $G_2(u) = (u - x_3)(u - x_4)$ and $x_1 = x_2 \mod 2$ and $x_3 = x_4 \mod 2$ are the solutions of the equation : $1 + ax + bx^2 = 0$.*

Proof Using Lemma 2 and the kernel's reduction properties, we deduce that the nonsingular quadratic splittings G of the equation $\tilde{C}$ corresponding to the Frobenius $\Sigma : \tilde{C} \to \tilde{C}^{\Sigma}$ is defined by the divisors that reduce modulo 2 to a principal divisor.

Set $G_1(u) = (u - x_1)(u - x_2)$ and $G_2(u) = (u - x_3)(u - x_4)$, where $x_1 = x_2 \mod 2$ and $x_3 = x_4 \mod 2$ are the solutions of the equation : $1 + ax + bx^2 = 0$. The corresponding divisors are $\tilde{D}_1 = P_1 + P_2 - 2\infty$ and $\tilde{D}_1 = P_3 + P_4 - 2\infty$ that reduce modulo 2 to $\mathrm{div}(u - x_1)$ and $\mathrm{div}(u - x_3)$. And the divisor defined by G_3 reduce to 0. Then $\mathcal{R}(G_1, G_2, G_3)$ corresponds to the Frobenius computation using Richelot.

In the case of type (3,1): Set $G_k(u) = (u - x_0)^2 \mod 2$ for a $k \in \{1, 2, 3\}$, where x_0 is the unique solution of the equation $1 + ax = 0$ or $1 + bx^2 = 0$ depending on the case. The divisor defined by $G_k(u)$ reduces to $\mathrm{div}(u - x_0)$ and the others divisors to 0. Thus, the nontrivial kernels modulo 2 are those defined by (G_1, G_2, G_3), where one $G_k(u) = (u - x_0) \mod 2$ defines the unique nontrivial divisor class. The other isogenies reduce to the Frobenius.

In the case of type (5): for every the nonsingular quadratic splittings G of the equation $\tilde{C}$ the divisors defined by $G_i(u)$ reduce to 0.

Remark 5 Let's consider,

$$\Sigma : \mathrm{Jac}\tilde{C} \longrightarrow \mathrm{Jac}\tilde{C}^{\Sigma} \quad \textit{and} \quad \widehat{\Sigma} : \mathrm{Jac}\tilde{C}^{\Sigma} \longrightarrow \mathrm{Jac}\tilde{C}$$

If G defines the kernel of the Frobenius Σ, then $\mathcal{R}(G)$ defines the kernel of the Verschiebung $\widehat{\Sigma}$ on $\mathrm{Jac}\tilde{C}^{\Sigma}$. Applying this to $\mathrm{Jac}C^{\Sigma^{-1}}$ this allows us to identify the Verschiebung $\widehat{\Sigma} : \mathrm{Jac}C \to \mathrm{Jac}C^{\Sigma^{-1}}$ on $\mathrm{Jac}C$.

5 Application to Point Counting

In addition to the computation of lift of modular forms using the lifted curves equations, the lifted Verschiebung also gives essential information about the characteristic polynomials of the Frobenius of the ordinary genus 2 curves. For the following, we detail an extension of Satoh's method to ordinary genus 2 curves.

5.1 Description and Complexity of the Algorithm

Let C be an ordinary genus 2 curve, $\mathcal{A}$ the Jacobian variety of C over $\mathbb{Z}_q$. In this section, we give a details of the evaluation of the Verschiebung the action on the 2-torsion group. The qth-power Frobenius morphism decomposes as follows:

$$\mathcal{A} \longrightarrow \mathcal{A}^{\sigma} \longrightarrow \cdots \longrightarrow \mathcal{A}^{\sigma^{n-1}}$$

Let Σ from $\tilde{\mathcal{A}}$ to $\tilde{\mathcal{A}}^{\Sigma}$ be the lift of the Frobenius morphism σ from $\mathcal{A}$ to $\mathcal{A}^{\sigma}$. Then $\hat{\Sigma}$ decomposes as follows:

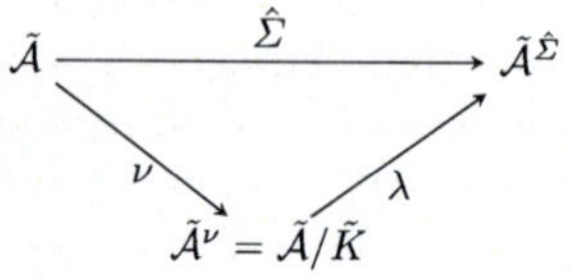

Where ν is normalized Verschiebung computed using the Richelot Formulea and λ is an isomorphism between $\tilde{\mathcal{A}}/\tilde{K}$ and $\tilde{\mathcal{A}}^{\hat{\Sigma}}$. Since ν is normalized its action is trivial on the canonical basis of differential forms, and we only need the action of the isomorphism λ on the differential forms to get the one of the Verschiebung $\hat{\Sigma}$.

More concretely, the abelian surfaces are described explicitly by curves in the algorithm: $\tilde{\mathcal{A}}$, $\tilde{\mathcal{A}}^{\hat{\Sigma}} = \mathrm{Jac}(\tilde{\mathcal{C}}^{\hat{\Sigma}})$, $\tilde{\mathcal{A}}^{\nu} = \mathrm{Jac}(\tilde{\mathcal{C}}^{\nu})$ where $\tilde{\mathcal{C}}^{\nu}$ is described by Richelot's algorithm. We recall that if $X : y^2 + h(x) = f(x)$ is the equation of an hyperelliptic curve of genus 2, its canonical basis of differentials is $w_X = (dx/(2y+h), xdx/(2y+h))$. We let M be the change of base matrix between $\hat{\Sigma}^* w_{\tilde{\mathcal{C}}^{\hat{\Sigma}}}$ and $w_{\tilde{\mathcal{C}}}$, by the above discussion, since ν is normalized, this is the same as the change of base matrix between $\lambda^* w_{\tilde{\mathcal{C}}^{\hat{\Sigma}}}$ and $w_{\tilde{\mathcal{C}}^{\nu}}$.

Then by the Galois action of Σ, it is easy to see that the change of base matrix between the canonical differential basis of $\tilde{\mathcal{C}}^{\hat{\Sigma}^n}$ and $\tilde{\mathcal{C}}^{\hat{\Sigma}^{n+1}}$ is given by $\Sigma^{-n} M$. So if we let $\hat{\Sigma}_q$ be the lift of the big Verschiebung $\hat{\sigma_q}$, the dual of the Frobenius $\pi = \pi_q$, we get that the matrix of the big Verschiebung acting on the differentials of $\tilde{\mathcal{C}}$ is $M_q = N_{\mathbb{Q}_q/\mathbb{Q}_p}(M)$. This matrix is diagonal, with eigenvalues π_1, π_2, the inversible roots of the characteristic polynomial χ_π. The other two roots are $q/\pi_1, q/\pi_2$.

Alternatively, rather than computing M directly, we may compute $I_k(\tilde{\mathcal{C}}^{\hat{\Sigma}})/I_k(\tilde{\mathcal{C}}^{\nu})$ to get $\det M^k$ where I_k is any covariant of degree k. This yields the following algorithm.

Initialization Let C be an hyperelliptic curve of genus 2 over a finite field $\mathbb{F}_q$ (with $q = 2^n$) given by its equation : $y^2 + h(x)y = f(x)$ where $h(x) = 1 + ax + bx^2$ and $f(x) = x^3(c + dx + x^2)$. Using relations (5), one can compute the corresponding vector of invariants $J = (\mathfrak{a}_1, \mathfrak{a}_2, \mathfrak{a}_3)$ modulo 2.

Lift Invariants Since the ordinary case corresponds to the type (1,1,1), Harley's algorithm in Sect. 4 and Algorithm 4.1 computes $\tilde{J}$ using the modular polynomials in function of $\mathfrak{a}_1$, $\mathfrak{a}_2$, and $\mathfrak{a}_3$. This computation can be done in $\tilde{O}(Nn)$, where N is the p-adic precision.

Lift of the Curves and the Verschiebung We can use Sect. 4.3 to lift the curve C by computing $(\tilde{a}, \tilde{b}, \tilde{c}, \tilde{d})$ via a Newton iteration. This costs $\tilde{O}(N)$.

Using the properties in Sects. 4 and 4.5 and Proposition 3 and Richelot's algorithm 2, one computes the normalized curve $\tilde{\mathcal{C}}^{\nu}$ whose Jacobian variety is isomorphic to the one of $\tilde{\mathcal{C}}^{\Sigma}$ in time $\tilde{O}(N)$.

Indeed, the isogenous curve $\tilde{\mathcal{C}}^{\nu}$ (whose Jacobian is $\tilde{\mathcal{A}}/\tilde{K}$) can be computed by applying Richelot algorithm 2 to the nonsingular quadratic splitting (G_1, G_2, G_3)

determined using Remarks 5. Concretely, if (H_1, H_2, H_3) is the kernel of the Frobenius on $\tilde{\mathcal{C}}^{\hat{\Sigma}}$, the kernel of the Verschiebung on $\tilde{\mathcal{C}}$ is $\mathcal{R}(H_1, H_2, H_3)$, since it is the image of the two torsion under

$$\Sigma : \mathcal{C}^{\widehat{\Sigma}} \longrightarrow \mathcal{C}_0 \simeq \mathcal{C}.$$

Alternatively, we can test directly at weak precision the eight combinations formed by $G_1 = P_1 P_3$, $G_2 = P_2 P_4$, and P_5 such that $P_1 \equiv P_2 \mod 2$ and $P_3 \equiv P_4 \mod 2$.

Computing the Characteristic Polynomial of the Frobenius Since the Frobenius decomposes as $\sigma \cdot \sigma^2 \cdots \sigma^{n-1}$, the product $\pi_1\pi_2$ of its eigenvalues is given by

$$(\pi_1\pi_2)^k = N_{\mathbb{Q}_q/\mathbb{Q}_p}\left(\frac{I_k(\tilde{\mathcal{C}}^{\hat{\Sigma}})}{I_k(\tilde{\mathcal{C}}^{\nu})}\right)$$

where I_k is any covariant of degree k. In practice we take the Igusa-Clebsch invariants.

Let χ be characteristic polynomial of the Frobenius of $\mathcal{C}$, then $\#\mathrm{Jac}\mathcal{C}(\mathbb{F}_q) = \chi_\pi(1)$ and the Hasse-Weil bound gives the following inequality:

$$\lceil(\sqrt{q} - 1)^2\rceil \leqslant \chi(1) \leqslant \lfloor(\sqrt{q} + 1)^2\rfloor.$$

Following [7, 35], one can use the LLL algorithm to recover P_{sym} knowing the $\pi_1\pi_2$. Using a quick algorithm in [35, Sect. 4.2], $\chi_\pi(\pm X)$ is deduced from the knowledge of P_{sym} when this one is irreducible. From the Hasse-Weil bound, we deduce that the needed precision is $N = 3n/2$ [28].

Theorem 6 *Let $\mathcal{C}$ be an ordinary hyperelliptic curve of genus 2 over a finite field $\mathbb{F}_{2^n}$ such that its invariants $(\mathfrak{a}_1, \mathfrak{a}_2, \mathfrak{a}_3)$ satisfy the Kronecker conditions. One computes $\#Jac\mathcal{C}(\mathbb{F}_{2^n})$ (and also the characteristic polynomial of the Frobenius) using our algorithm in $\tilde{O}(n^2)$ operations, using modular polynomials in function of $(\mathfrak{a}_1, \mathfrak{a}_2, \mathfrak{a}_3)$.*

5.2 *Example*

Let $\mathcal{C}$ be defined over $\mathbb{F}_2[T]/(m)$ with $m = T^{15} + T^5 + T^4 + T^2 + 1$ by : $y^2 + h(x)y = f(x)$ where $h(x) = 1 + ax + bx^2$ and $f(x) = -x^3(c + dx + x^2)$, and the coefficients are given by

$$\begin{aligned}
a &= T^{14} + T^{13} + T^{12} + T^{11} + T^9 + T^7 + T^5 + T^2 + T + 1,\\
b &= T^{14} + T^{13} + T^{12} + T^{10} + T^8 + T^7 + T^5 + T^4 + T^3,\\
c &= T^{14} + T^{13} + T^{10} + T^9 + T^4,\\
d &= T^{14} + T^{13} + T^{12} + T^{11} + T^9 + T^7 + T^5 + T^2 + T;
\end{aligned}$$

Then the birational invariants are given by

$$
\begin{aligned}
J =[&T^{13} + T^{11} + T^{10} + T^9 + T^8 + T^7 + T,\\
&T^{13} + T^{11} + T^{10} + T^9 + T^8 + T^7 + T,\\
&T^{14} + T^{13} + T^{11} + T^8 + T^3 + T^2 + T + 1];
\end{aligned}
$$

and their lifts by

$$
\begin{aligned}
\tilde{J} =[&13474940032272004850T^{14} + 16754559589254918889T^{13} + 8152007378073597672T^{12} + 14572772277029696995T^{11} +\\
&9119476433797133587T^{10} + 4240523924640418307T^9 + 6042690066170973759T^8 + 2446801185468953757T^7 +\\
&10879075386207885102T^6 + 16543022202561708534T^5 + 10457394664501256190T^4 + 10084398973515633314T^3 +\\
&12770643410383775408T^2 + 12442302921457674557T + 7907936886510960452,\ \ 9391731769952962454T^{14} +\\
&5272471816093258695T^{13} + 4694061109440793682T^{12} + 11604440828776285509T^{11} + 11540449592893237497T^{10} +\\
&365295094271749277T^9 + 4466186733562397311T^8 + 7829903155290047333T^7 + 11267196542271673434T^6 +\\
&17253646350720823144T^5 + 6033296500969149134T^4 + 17471757159738016832T^3 + 10216651861362684 16T^2 +\\
&12749389994979729051T + 17738635541368473904,\ \ 15031255427447906939T^{14} + 18038781498479259395T^{13} +\\
&6632965468075882578T^{12} + 10958722626511106823T^{11} + 14970482097439825396T^{10} + 16140758820844564488T^9 +\\
&5066524261002336475T^8 + 15608122093822831302T^7 + 6346841391962884778T^6 + 2569207608253317670T^5 +\\
&17560295176929269876T^4 + 7729919906268531181T^3 + 18211641834701810925T^2 + 3154110753804246243T +\\
&10813538131737859899];
\end{aligned}
$$

The Weierstrass points on the lift are given by:

$$
\begin{aligned}
P_1 =(x + (&-274993973066669T^{14} - 104108864093772T^{13} - 72190469780601T^{12} - 115434333597685T^{11} -\\
&130682791175686T^{10} + 71485344737819T^9 + 93810423080862T^8 - 79173306769954T^7 - 127418781818674T^6 +\\
&168210495993381T^5 - 65589614246955T^4 + 70251572308663T^3 - 80835683163668T^2 - 108825539834 11T -\\
&140522863371668));
\end{aligned}
$$

$$
\begin{aligned}
P_2 =(x + (&319313741718 63T^{14} - 86513099225224T^{13} - 93457672426045T^{12} + 102150671708127T^{11} -\\
&65305500898574T^{10} + 111042151260415T^9 - 111164785066626T^8 - 114569669306774T^7 - 50517044414922T^6 +\\
&35343102544249T^5 + 6828453986989T^4 + 178435263333367T^3 - 96792882371272T^2 + 134680560484 65T +\\
&16161264291160));
\end{aligned}
$$

$$
\begin{aligned}
P_3 =(x + (&37058400079907T^{14} - 90894798270813T^{13} + 136899047104414T^{12} + 123980821883020T^{11} +\\
&63786983405963T^{10} - 55522175792215T^9 - 47167405597672T^8 - 66232728309206T^7 - 13922670795072T^6 +\\
&114898854962428T^5 + 23815982188633T^4 + 19845818312694T^3 - 88612939847082T^2 - 93598522843036T +\\
&54854521291173));
\end{aligned}
$$

$$
\begin{aligned}
P_4 =(x + (&95876749096463T^{14} - 46282769547553T^{13} + 96099901407366T^{12} - 108440868522716T^{11} -\\
&108884878509769T^{10} - 28105265738995T^9 - 134819990078416T^8 + 36493467184354T^7 + 34610520131656T^6 +\\
&5866413448212T^5 - 40588267667491T^4 - 68511278224546T^3 + 44626765726166T^2 + 26926642085156T -\\
&19235354424143));
\end{aligned}
$$

$$P_5 = -(4x + (-5493864730017407T^{14} + 185416724591973T^{13} - 2687545533424494T^{12} - 9010553886254T^{11} - 1607448577143336T^{10} - 3955968403727444T^{9} + 3950068869701127T^{8} - 2331019477801557T^{7} - 4953920362475227T^{6} + 4360260438452967T^{5} + 3010356449841077T^{4} - 1565839243294297T^{3} - 2415808123007517T^{2} + 2552541706294387T + 354634709659596));$$

we have $P_1 = P_2 \mod 2$ and $P_3 = P_4 \mod 2$ with $\tilde{C} : y^2 = P_1 P_2 P_3 P_4 P_5$.

Using Sects. 5 and 5.1, one determines the nonsingular quadratic splitting corresponding to the Verschiebung isogeny And we get

$$G1 = P_1 P_3, \quad G_2 = P_4 P_5, \quad G_3 = P_2 \quad \text{and we get} \quad \tilde{C}^{\nu} : y^2 = \Pi\mathcal{R}(G_1, G_2, G_3).$$

And we can compute the characteristic polynomial

$$\chi(X) = \left(X^2 + 2482103 \cdot 2^{16} X + 157 \cdot 2^{30}\right)\left(X^2 - 81333551261 \cdot 2X + 129494369717\right)$$

Acknowledgements We thank Enea Milio for his comments and suggestion on the part concerning the computation of modular polynomials. We are supported by the FAST project and ANR CIAO. Experiments presented in this paper were carried out using PARI/GP [32] and the modular polynomials were computed using https://members.loria.fr/EMilio/modular-polynomials/ .

6 Computing the Normal Form from an Hyperelliptique Model

Let C be genus 2 curve over a field $\mathbb{k}$. We want to compute from its hyperelliptic model the normal equation of C i.e. the coefficients (a, b, c, d) such that C is defined over $\mathbb{k}$ by:

$$XY^2 + (1 + aX + bX^2)Y + X^2(c + dX + X^2) = 0$$

6.1 *If the Characteristic of* $\mathbb{k}$ *Is 2*

We recall that the relations between the hyperelliptic model and the coefficients (a, b, c, d) come from the invariants J_{2i}'s through the symmetric functions on α, β, γ (depending on the Artin-Shreier type).

Let $y^2 + h(x)y = f(x)$ be an hyperelliptic model of C over $\mathbb{k}$. First we compute the J_{2i}'s:

- If $J_2 \neq 0$ then C is ordinary which corresponds to type $(1, 1, 1)$;
- If $J_2 = 0$ and $J_6 \neq 0$ then C has p-rank 1 which corresponds to type $(3, 1)$;
- For the other cases C has p-rank 2 and the hyperelliptic model reduces $y^2 + y =$

$x^5 + a_3x^3$ which corresponds to the type (5).
For this class of curves, we have $(a, b, c, d) = (0, 0, a_3, 0)$ for the coefficients of the normal form of $\mathcal{C}$ over $\Bbbk$.
For the two others classes, the strategy is the same. We derive an isomorphism by considering the Weierstrass points for the both models of $\mathcal{C}$. In the hyperelliptic model, they correspond to the solutions of the equation $h(x) = 0$, and up to a variables change on x we may take them nonzero.
• For instance, when $J_2 \neq 0$, we have two nontrivial Weierstrass points with abscissa ξ and η, then we take $b = 1/(\eta\xi)$ and $a = b(\xi + \eta)$. To compute c and d, one can use together the relations Sect. 2.4 and Eq. (5) in Sect. 2.
• When $J_2 = 0$ with $J_6 \neq 0$ the curve $\mathcal{C}$ admits only one nontrivial Weierstrass point of abscissa x_0. Then one can take $a = 1/x_0$ and $b = 0$. Therefore, by solving the relations Sect. 2.4 and Eq. (6) in Sect. 2, we obtain c and d.

We refer to [18, pp. 6–8] for the conversion between the normal model and the Artin–Shreier models.

6.2 If the Characteristic of $\Bbbk$ Is Different from 2

Let $y^2 = u \cdot (x - \alpha_1) \cdots (x - \alpha_5)$ be the hyperelliptic model of $\mathcal{C}$ over $\overline{\Bbbk}$, and assume that all α_i's are different from 0 and 1. From [18, Pages 6-8], the α_i's are solutions of the equation

$$(1 + aX + bX^2)^2 - 4X^3(c + dX + X^2) = 0.$$

Pluggint the α_i, this induces a system of equation in (a, b, c, d) which is then easy to solve.

We can also recover the normal model from the hyperelliptic model directly, without computing the Weierstrass points. We have on one hand

$$\begin{aligned} y^2 &= (1 + ax + bx^2)^2 - 4x^3(c + dx + x^2) \\ &= -4x^5 + (b^2 - 4d)x^4 + (2ab - 4c)x^3 + (a^2 + 2b)x^2 + 2ax + 1 \cdot \end{aligned}$$

such that a has weight 1, b has weight 2, c has weight 3 and d has weight 4. And on the other hand, we have the starting equation:

$$y^2 = a_5x^5 + \cdots + a_1x + a_0 \quad \text{with} \quad a_0, a_5 \quad \text{non null.}$$

By applying successively the transformations: $y \longmapsto \sqrt{a_0}y$ and $x \longmapsto -\sqrt[5]{\dfrac{4a_0}{a_5}}$.
x to :

$$y^2 = a_5x^5 + a_4x^4 + a_3x^3 + a_2x^2 + a_1x + a_0$$

we get

$$y^2 = -4x^5 + \frac{a_4}{a_0}\left(\frac{4a_0}{a_5}\right)^{4/5} x^4 - \frac{a_3}{a_0}\left(\frac{4a_0}{a_5}\right)^{3/5} x^3 + \frac{a_2}{a_0}\left(\frac{4a_0}{a_5}\right)^{2/5} x^2 - \frac{a_4}{a_0}\left(\frac{4a_0}{a_5}\right)^{1/5} x + 1$$

Set $\zeta = \sqrt[5]{\frac{4a_0}{a_5}}$, then the quadruplet (a, b, c, d) is defined by

$$a = -\frac{a_4}{2a_0}\zeta, \quad b = \left(\frac{a_2}{2a_0} - \frac{a_4^2}{8a_0^2}\right)\zeta^2,$$

$$c = \left(\frac{a_3}{4a_0} + \frac{a_4^3}{32a_0^3} - \frac{a_2a_4}{8a_0^2}\right)\zeta^3, \quad d = \left[\frac{1}{4}\left(\frac{a_2}{2a_0} - \frac{a_4^2}{8a_0^2}\right)^2 - \frac{a_4}{4a_0}\right]\zeta^4.$$

Like the hyperelliptic model which is not smooth at infinity, the normal model is not smooth at the point $(0 : 1 : 0)$. By [18, pp. 6–8], the normal model is obtained from collapsing a Weierstrass point P and a non Weierstrass point Q together. Once P and Q are fixed, the normal model is unique up to an action of the form $(a', b', c', d') \longmapsto (a'\xi, b'\xi^2, c'\xi^3, d'\xi^4)$ for a fifth of unity ξ.

References

1. Birkenhake C, Lange H (2004) Complex abelian varieties, 2nd edn. vol. 302. Grundlehren der Mathematischen Wissenschaften [Fundamental Principles of Mathematical Sciences]. Springer, Berlin, pp xii+635. isbn: 3-540-20488-1
2. Bröker R, Lauter K (2009) Modular polynomials for genus 2. LMS J Comput Math 1(12):326–339
3. Bröker R, Lauter K (2009) Modular polynomials for genus 2. LMS J Comput Math 12:326–339. issn: 1461-1570. arXiv: 0804.1565
4. Cardona G, Quer J (2005) Field of moduli and field of definition for curves of genus 2. Computational aspects of algebraic curves. World Scientific, pp 71–83
5. Carls R (2003) Generalized AGM sequences and approximation of canonical lifts. PhD thesis. http://www.math.leidenuniv.nl/carls
6. Carls R, Kohel D, Lubicz D (2008) Higher-dimensional 3-adic CM construction. J Algebra 319(3):971–1006. issn: 0021-8693. https://doi.org/10.1016/j.jalgebra.2007.11.016
7. Carls R, Lubicz D (2008) A p-adic quasi-quadratic time and quadratic space point counting algorithm. Int Math Res Not
8. Clebsh A (1872) Theorie der Binärien Algebraischen Formen. Verlag von B. G
9. Cosset R (2011) Application des fonctions thêta à la cryptographie sur courbes hyperel- liptiques. PhD thesis
10. Deuring M (1958) Die Klassenkörper der komplexen Multiplikation, vol 2. Teubner Stuttgart
11. Dupont R (2006) Moyenne arithmetico-geometrique, suites de Borchardt et applications. PhD thesis
12. Gabriel C, Nart E, Pujolàs J () Curves of genus two over fields of even characteristic
13. Gaudry P (2004) Algorithmes de comptage de points d'une courbe définie sur un corps fini. http://www.loria.fr/~gaudry/publis/pano.pdf (cit. on p. 19)

14. Gaudry P, Houtmann T, Kohel D, Ritzenthaler C, Weng A (2006) The 2-adic CM method for genus 2 curves with application to cryptography. In:International conference on the theory and application of cryptology and information security. Springer, pp 114–129
15. Goren E, Lauter K (2010) Genus 2 curves with Complex Multiplication. https://arxiv.org/abs/1003.4759v1
16. Goren EZ (2012) Lauter KE (2012) Genus 2 curves with complex multiplication. Int Math Res Not 5:1068–1142
17. Gundlach K (1963) Die Bestimmung der Funktionen zur Hilbertschen Modulgruppe des Zahlkörpers $\mathbb{Q}\sqrt{5}$. Math Ann 152:226–256
18. Igusa J-I (1960) Arithmetic variety of moduli for genus two. Ann Math 72(3):612–649
19. Igusa J (1962) On Siegel modular forms of genus 2, 84(1). Johns Hopkins University Press
20. Igusa J-I (1972) Theta functions. Die Grundlehren der mathematischen Wissenschaften, Band 194. Springer, New York, pp x+232
21. Kieffer J (2020) Degree and height estimates for modular equations on PEL Shimura varieties. arXiv: 2001.04138 [math.AG] (cit. on p. 19)
22. Kieffer J, Page A, Robert D (2020) Computing isogenies from modular equations in genus two. arXiv: 2001.04137 [math.AG]
23. Lercier R, Lubicz D (2006) A quasi-quadratic time algorithm for hyperelliptic curve point counting. Ramanujan J 12(3):399–423
24. Lubin J, Serre J, Tate J (1964) El liptic curves and formal groups. http://ma.utexas.edu/users/voloch/lst.html
25. Maiga A, Robert D (2020) Computing the canonical lift of genus 2 curves in odd characteristic. In preparation
26. Mestre J (1991) Construction de Courbes de Genre 2 à partir de leurs Modules. In: Effective methods in algebraic geometry (Castiglioncel lo, 1990), pp 313–334
27. Mestre J-F (2001) Lettre à Gaudry et Harley. http://www.math.jussieu.fr/mestre
28. Mestre J-F (2002) Notes of a talk given at the cryptography seminar rennes. http://www.math.univ-rennes1.fr/crypto/2001-02/mestre.ps
29. Milio E (2015) Calcul de polynômes modulaires en dimension 2. PhD thesis. https://members.loria.fr/EMilio (cit. on pp. 2, 16, 17)
30. Milio E, Robert D (2020) Modular polynomials on Hilbert surfaces. J Num Theory
31. Milio E (2014) A quasi-linear algorithm for computing modular polynomials in dimension 2. arXiv:1411.0409
32. PARI Developers (2019) PARI/GP, version2.12.1. available from http://pari.math.u-bordeaux.fr/. The PARI Group
33. Qing L (1993) Courbes Stables de genre 2 et leur schéma de modules. Mthematische Annalen Springer-Verlag 295:201–222
34. Richelot F (1836) Essai sur une méthode générale pour déterminer la valeur des integrals ultra-elliptiques, fondée sur des transformations remarquables de ces transcendantes. C. R. Acad. Sci. Paris 2:622–627
35. Ritzenthaler C (2003) Problémes arithmétiques relatifs à certaines familles de courbes sur les corps finis. PhD thesis. Université Denis Diderot Paris VII
36. Satoh T (2000) The canonical lift of an ordinary elliptic curve over a finite field and its point counting. J Ramanujan Math Soc 15(4):247–270
37. Smith B (2005) Explicit endomorphisms and correspondences. PhD thesis
38. Streng M (2010) Complex multiplication of abelian surfaces. PhD thesis. Leiden, marco.streng@gmail.com. isbn: -13/EAN: 978-90-5335-291-5

Application of $\mathcal{T}$-Direct Codes in Multiple-Rate Codes

Ashwini Kumar and R. S. Raja Durai

Abstract A straightforward extension of linear codes with complementary dual is the class of $\mathcal{T}$-Direct codes. A systematic construction of $\mathcal{T}$-Direct codes along with the coding scheme, capable of handling variable-length q-ary user messages, is proposed. By inducing constituent codes, constructed using a *self-dual* basis of extension field $\mathbb{F}_q$, construction of $\mathcal{T}_{\mathbb{K}}$-Direct codes over $\mathbb{F}_q$ is defined. It is shown that the constituent codes of a $\mathcal{T}_{\mathbb{K}}$-Direct code are equipped with the necessary properties required for handling the variable-rate feature in a multiple-rate code. In the proposed coding scheme, q-ary user messages of lengths $1, 2, \ldots, \mathcal{T}_{\mathbb{K}}$ can be encoded and decoded efficiently, providing variable information rates $\mathcal{R} = \frac{k}{n^2}, \frac{k+1}{n^2}, \ldots, \frac{k\mathcal{T}_{\mathbb{K}}}{n^2}$. Furthermore, the encoding and decoding scheme for distinct as well as non-distinct constituent codes is illustrated with the help of an example.

Keywords $\mathcal{T}$-Direct codes · Multiple-rate codes · LCD codes · MRD codes

1 Introduction

Denote the finite field with $q\ (= p^n)$ elements by $\mathbb{F}_q$, where p is a prime. The existence of isomorphism between an extension field $\mathbb{F}_{q^n}$ of degree n and a vector space $\mathbb{F}_q^n$ of dimension n over the field $\mathbb{F}_q$ enables the alternative use of the notations $\mathbb{F}_{q^n}$ and $\mathbb{F}_q^n$. A linear code $C(n, k, d)$ is defined to be a k-dimensional subspace of the n-dimensional vector space $\mathbb{F}_q^n$ over the field $\mathbb{F}_q$. The *linear codes with complementary dual* (LCD codes) defined over $\mathbb{F}_q$ and meeting trivially with their dual, $C \cap C^{\perp} = \{\mathbf{0}\}$, were firstly defined by Massey in 1992 [10]. The class of LCD codes is generalized as $\mathcal{T}$-Direct codes [26], which is primarily composed of $\mathcal{T}$ constituent LCD codes C_i, $i = 1, 2, \ldots, \mathcal{T}$, defined over the field $\mathbb{F}_q$. The constituent codes of a $\mathcal{T}$-Direct code $(C_1, C_2, \ldots, C_{\mathcal{T}})$ have LCD property with respect to the subspace $\Omega = C_1 \oplus C_2 \oplus \cdots \oplus C_{\mathcal{T}} \subseteq \mathbb{F}_q^n$: $C_i \cap C_i^{\perp} = \{\mathbf{0}\}$ with

A. Kumar (✉) · R. S. Raja Durai
Jaypee University of Information Technology, Waknaghat-173234, Waknaghat, India
e-mail: rsraja.durai@juit.ac.in

© The Author(s), under exclusive license to Springer Nature Singapore Pte Ltd. 2022

D. Giri et al. (eds.), *Proceedings of the Seventh International Conference on Mathematics and Computing*, Advances in Intelligent Systems and Computing 1412,
https://doi.org/10.1007/978-981-16-6890-6_49

$C_i^{\perp} = C_1 \oplus C_2 \oplus \cdots \oplus C_{i-1} \oplus C_{i+1} \oplus \cdots \oplus C_{\mathcal{T}}$ being the dual of C_i with respect to Ω. An LCD code C and its dual $C^{\perp}$ together constitute a 2-Direct code $(C, C^{\perp})$ such that $\mathbb{F}_q^n = C \oplus C^{\perp}$. Furthermore, each *constituent* code in a $\mathcal{T}$-Direct code $(C_1, C_2, \ldots, C_{\mathcal{T}})$ possesses the *complementary-dual* property; but not conversely [22]. It must be noted that the necessary and sufficient condition for a linear code C generated by matrix $\mathbf{G}$ to be LCD is $|\mathbf{G}\mathbf{G}^{\mathbf{T}}| \neq 0$. An interesting application of $\mathcal{T}$-Direct codes can be found in multi-user *adder channel*, therein the efficacy of $\mathcal{T}$-Direct codes in coding schemes and retrieval of transmitted codewords from the received sequence over noiseless as well as noisy $\mathcal{T}$-user $\mathbb{F}_{q^n}$-Adder Channel (FAC) could be seen [26]. The class of LCD codes along with their algebraic properties [1, 15, 27] also have a wide range of applicability, especially in the field of cryptography which grabbed the attention of researchers [5, 8, 25]. In particular, LCD codes are applicable in the developments of prevention techniques against *side-channel attack* (SCA) and *fault injection attacks* (FIAs) [5]. For more information regarding LCD codes and their applications, the readers are encouraged to refer [16, 25, 28].

In recent years, studies concerning the error-correcting codes have also induced the deviation from the usual Hamming metric [24] to 'rank metric', introduced by Gabidulin in 1985 [7]. In contrast to the Hamming metric, rank metric recognizes linear dependence among symbols of a codeword [19]; *rank distance* codes and their properties are discussed in [11, 12, 18]. An important class of *rank distance* codes, analogous to the *maximum distance separable* (MDS) codes in the Hamming metric, is the class of *maximum rank distance* (MRD) codes. Due to their rich algebraic properties and applicability, linear codes equipped with both LCD as well as MRD properties are of great importance.

Conventionally, an error-correcting code $C(n, k, d)$ can encode a user message of length k into an n-length codeword, resulting in fixed-rate coding scheme. However, in many practical error-control systems, channel codes capable of handling flexible code rates are required. In achieving the robust and efficient communication in modern wireless channels, where fading is common, *adaptive coded modulation* technique is employed [21]. Moreover, in *automatic repeat-request* protocol, the coder is required to retransmit an adequate amount of redundancy to receiver which in terms require codes with flexible information rates. Constructions of multiple-rate codes along with coding scheme are presented in [30]. In order to address this issue, the present paper attempts to construct a class of q^n-ary block codes equipped with encoding and decoding scheme to handle q-ary user messages of variable lengths ranging between 1 and $\mathcal{T}_{\mathbb{K}}$ with $\mathcal{T}_{\mathbb{K}} = \max\{x | \frac{x(x+1)}{2} \leq n\}$. A systematically constructed subclass namely $\mathcal{T}_{\mathbb{K}}$-Direct codes of the existing $\mathcal{T}$-Direct codes is defined. The coding scheme employs an existing 'multiple-rate code' $C_{\mathrm{MR}}(n, k, d)$ over the composite field $\mathbb{F}_{q^n}$ and k LCD MRD 'constituent codes' $C_{i_1}(n, i_1), C_{i_2}(n, i_2), \ldots, C_{i_k}(n, i_k)$, $1 \leq i_j \leq \mathcal{T}_{\mathbb{K}}$, constructed using a *self-dual* basis over the base field $\mathbb{F}_q$. A necessary condition for $C_{i_j}(n, i_j)$ in order to provide efficient variable-rate feature is obtained such that $C_{i_1}, C_{i_2}, \ldots, C_{i_k}$ constitute a $\mathcal{T}_{\mathbb{K}}$-Direct code. The proposed coding scheme considers k user messages of length less than or equal to $\mathcal{T}_{\mathbb{K}}$ over $\mathbb{F}_q$; i_j-length user message is encoded through i_j-dimensional 'constituent code' $C_{i_j}(n, i_j)$ resulting in a k-length block of n-length vectors in $\mathbb{F}_q^n$. Further, considering the encoded n-

length vector in $\mathbb{F}_q^n$ as an element in $\mathbb{F}_{q^n}$, the k-length block is then encoded through $\mathcal{C}_{\mathrm{MR}}(n, k, d)$ for transmission over a q^n-ary noisy channel. The proposed coding scheme produces rates $\mathcal{R} = \frac{k}{n^2}, \frac{k+1}{n^2}, \ldots, \frac{k\mathcal{T}_{\mathbb{K}}}{n^2}$.

The contents in the paper are organized as follows: the following section recalls the required preliminary concepts. In Sect. 3, we introduce the systematized construction of a $\mathcal{T}_{\mathbb{K}}$-Direct code along with the definition of the generator matrix of the LCD MRD 'constituent codes'. Section 4 presents the construction and application of $\mathcal{T}_{\mathbb{K}}$-Direct codes in multiple-rate codes with the help of an illustration. Lastly, we conclude the results in Sect. 5.

2 Preliminaries

This section briefly summarizes the necessary background required for the current work. Basic concepts regarding *self-dual* basis, MRD codes and $\mathcal{T}$-Direct codes are recalled.

2.1 *Self-dual Bases*

Consider a basis $\boldsymbol{\alpha} = \{\alpha_1, \alpha_2, \ldots, \alpha_n\}$ of $\mathbb{F}_q^n$ over $\mathbb{F}_q$. Any basis $\boldsymbol{\beta} = \{\beta_1, \beta_2, \ldots, \beta_n\}$ of $\mathbb{F}_q^n$ satisfying the trace-orthogonal relation $\mathbf{tr}(\alpha_i\beta_j) = \delta_{ij}$ is called the dual basis of $\boldsymbol{\alpha}$, where the function $\mathbf{tr} : \mathbb{F}_{q^n} \to \mathbb{F}_q$, defined as $\mathbf{tr}(\gamma) = \gamma + \gamma^q + \cdots + \gamma^{q^{n-1}}$ for an element $\gamma \in \mathbb{F}_{q^n}$, is a symmetric bilinear form on $\mathbb{F}_{q^n}$. Consider the trace matrix of the bases $\boldsymbol{\alpha} and \boldsymbol{\beta}$:

$$\mathbf{T}_{[\boldsymbol{\alpha}:\boldsymbol{\beta}]} = \begin{pmatrix} \mathbf{tr}(\alpha_1\beta_1) & \mathbf{tr}(\alpha_1\beta_2) & \cdots & \mathbf{tr}(\alpha_1\beta_n) \\ \mathbf{tr}(\alpha_2\beta_1) & \mathbf{tr}(\alpha_2\beta_2) & \cdots & \mathbf{tr}(\alpha_2\beta_n) \\ \vdots & \vdots & \ddots & \vdots \\ \mathbf{tr}(\alpha_n\beta_1) & \mathbf{tr}(\alpha_n\beta_2) & \cdots & \mathbf{tr}(\alpha_n\beta_n) \end{pmatrix},$$

where $\mathbf{T}_{[\boldsymbol{\alpha}:\boldsymbol{\beta}]}$ is called the trace-orthogonal matrix of basis $\boldsymbol{\alpha}$ and $\boldsymbol{\beta}$. Interestingly, among the class of *dual-bases*, the present work is concerned mainly with *self-dual* bases. A basis $\boldsymbol{\alpha}$ of $\mathbb{F}_q^n$ is called a *self-dual basis*, if $\mathbf{tr}(\alpha_i\alpha_j) = 0$ for $i \neq j, i, j \in \{1, 2, \ldots, n\}$ and $\mathbf{tr}(\alpha_i^2) = 1$ for $i = 1, 2, \ldots, n$. For a *self-dual basis*, the trace-orthogonal matrix $\mathbf{T}_{[\boldsymbol{\alpha}:\boldsymbol{\alpha}]} = \mathbf{T}_{[\boldsymbol{\alpha}]}$ is indeed an identity matrix of order n [6].

A field extension $\mathbb{F}_{q^n}$ over $\mathbb{F}_q$ is said to possess a *self-dual basis* if either q is even or both q and n are odd [2]. Also there exists a *self-dual normal* basis in $\mathbb{F}_q^n$ (q even) if and only if n is odd or $n \equiv 2\ (mod 4)$ [3]. The significance of these bases is portrayed through their applicability in coding theory and cryptography.

2.2 Rank Distance Codes

The rank metric codes introduced by Gabidulin [7], due to their rich intrinsic algebraic properties and applicability in cryptography and network coding [13, 14], are of particular interest for researchers. In *rank distance* codes, the weight of a codeword in $\mathbb{F}^n_{q^N}$ is defined as the maximal number of linearly independent coordinates over $\mathbb{F}_q$ which makes them ideal for error-correction techniques [19]. Among the class of *rank distance* codes, *maximum rank distance* (MRD) codes being maximal in size, i.e., achieving the Singleton bond $d = n - k + 1$, are analogous to those of *maximum distance separable* (MDS) codes in the Hamming metric. An MRD code is defined to be $\mathbb{F}_{q^N}$-linear subspace of n-dimensional vector space $\mathbb{F}^n_{q^N}$. Hereafter, the present investigation is mainly concerned with the case when $n = N$ only. The intrinsic properties, bounds and decoding schemes of the MRD codes have been studied in [11, 12, 18].

Definition 2.1 A *maximum rank distance* (MRD) code $C(n, k, d)$ is a *rank distance* code which attains equality in the Singleton-like bound, i.e., code for which $d = n - k + 1$.

Definition 2.2 Let $\mathbf{v} = (v_1, v_2, \ldots, v_n) \in \mathbb{F}^n_{q^n}$ be linearly independent over $\mathbb{F}_q$. The MRD code $C(n, k, d)$ over $\mathbb{F}_{q^n}$ is the code generated by the matrix $\mathbf{G}$:

$$\mathbf{G} = \begin{pmatrix} v_1 & v_2 & \cdots & v_n \\ v_1^{[1]} & v_2^{[1]} & \cdots & v_n^{[1]} \\ \vdots & \vdots & \ddots & \vdots \\ v_1^{[k-1]} & v_2^{[k-1]} & \cdots & v_n^{[k-1]} \end{pmatrix},$$

where hereafter we use the notation $x^{[i]}$ to represent x^{q^i} for $x \in \mathbb{F}_{q^n}$, with i an integer.

2.3 $\mathcal{T}$-Direct Codes

Prior to the exclusive introduction of LCD codes by J. L. Massey in 1992, error-correcting codes with *complementary-dual* property were known as reversible codes [9]. For a code $C(n, k, d)$ described by the matrix $\mathbf{G}$ to be LCD requires the determinant of the $k \times k$ matrix $\mathbf{G}\mathbf{G}^{\mathrm{T}}$ to be non-zero. The studies associated with the constructions and properties of LCD codes came into the picture from 2004 onwards. Interestingly, in a $\mathcal{T}$-Direct code, each constituent code meets the *complementary-dual* property with the rest.

In a $\mathcal{T}$-Direct code $(C_1, C_2, \ldots, C_{\mathcal{T}})$, the constituent codes C_i have the LCD property with respect to $\Omega \subseteq \mathbb{F}^n_q$, i.e., they satisfy the condition $C_i \cap C_i^{\perp} = \{\mathbf{0}\}$ with respect to $\Omega \subseteq \mathbb{F}^n_q$. The *orthogonal projector* from $\Omega \to C_i$ for each $i = 1, 2, \ldots, \mathcal{T}$

is defined as $\Pi_{C_i} = \mathbf{G}_i^{\mathbf{T}}(\mathbf{G}_i\mathbf{G}_i^{\mathbf{T}})^{-1}\mathbf{G}_i$ [22]. The following theorems provide a better insight about the structure and properties of the $\mathcal{T}$-Direct codes.

Theorem 1 *[22] Let C_i be an (n, k_i) q^n-ary linear code with the generator matrix $\mathbf{G}_i$ such that $\mathbf{G}_i\mathbf{G}_j^{\mathbf{T}} = (\mathbf{0})$ for each $i = 1, 2, \ldots, \mathcal{T}$ with $i \neq j$. Then $(C_1, C_2, \ldots, C_{\mathcal{T}})$ is a $\mathcal{T}$-Direct code if and only if the $k_i \times k_i$ matrix $\mathbf{G}_i\mathbf{G}_i^{\mathbf{T}}$ is non-singular for each $i = 1, 2, \ldots, \mathcal{T}$.*

Theorem 2 *[22] Let C_i be an (n, k_i) q^n-ary linear code with generator matrix $\mathbf{G}_i$ for each $i = 1, 2, \ldots, \mathcal{T}$ and $\Omega = C_1 \oplus C_2 \oplus \cdots \oplus C_{\mathcal{T}}$. Then $(C_1, C_2, \ldots, C_{\mathcal{T}})$ is a $\mathcal{T}$-Direct code if the $k_1 + k_2 + \cdots + k_{\mathcal{T}}$ row vectors of $\mathbf{G}_1, \mathbf{G}_2, \ldots, \mathbf{G}_{\mathcal{T}}$ are orthonormal vectors in $\mathbb{F}_{q^n}^n$.*

The constituent codes constituting a $\mathcal{T}$-Direct code are such that $\mathbb{F}_{q^n}^n = C_1 \oplus C_2 \oplus \cdots \oplus C_{\mathcal{T}}$ and each C_i have the *complementary-dual* property for $i = 1, 2, \ldots, \mathcal{T}$ [22]. But the converse is not true in general, i.e., if $C_1, C_2, \ldots, C_{\mathcal{T}}$ are LCD codes, then $(C_1, C_2, \ldots, C_{\mathcal{T}})$ need not be a $\mathcal{T}$-Direct code. If the generator matrix $\mathbf{G}$ of a k-dimensional MRD code with row vectors consisting of the q-powers of a basis $\mathbf{v} = (v_1, v_2, \ldots, v_n)$ of $\mathbb{F}_{q^n}$ be abbreviated as $\mathbf{G}_k = \langle v_1, v_2, \ldots, v_n \rangle$, then the following theorem provides a criteria for $\mathcal{T}$ constituent codes to constitute a $\mathcal{T}$-Direct code and for the converse to hold true.

Theorem 3 *[31] Let C_i be an (n, k_i) MRD code with generator matrix $\mathbf{G}_i$ for each $i = 1, 2, \ldots, \mathcal{T}$. Suppose that $\alpha_1, \alpha_2, \ldots, \alpha_n$ be a self-dual basis in $\mathbb{F}_{q^n}$. Then $(C_1, C_2, \ldots, C_{\mathcal{T}})$ is a $\mathcal{T}$-Direct code if C_i is an LCD code with $\mathbf{G}_i = \left\langle \alpha_1^{[\sum_{r=0}^{i-1} k_r]}, \alpha_2^{[\sum_{r=0}^{i-1} k_r]}, \ldots, \alpha_n^{[\sum_{r=0}^{i-1} k_r]} \right\rangle$ for $i = 1, 2, \ldots, \mathcal{T}$ and $k_0 = 0$.*

3 A Systematized Construction of a $\mathcal{T}$-Direct Code

In order to employ the $\mathcal{T}$-Direct codes in a flexible-rate encoding and decoding scheme, a systematized construction of a $\mathcal{T}$-Direct code is required. In that, the constituent codes of distinct dimensions over $\mathbb{F}_q$ are considered to be LCD MRD codes.

Consider a primitive element β in $\mathbb{F}_q = \{0, \beta, \beta^2, \ldots, \beta^{q-1}\}$. Let $\mathcal{T}_{\mathbb{K}} = \max_{x \in \mathbb{N}} \left\{x \mid \frac{x(x+1)}{2} \leq n\right\}$, where $\mathbb{N}$ is the set of natural numbers. Consider $\mathcal{T}_{\mathbb{K}}$ constituent LCD MRD codes $C_1(n, 1), C_2(n, 2), \ldots, C_{\mathcal{T}_{\mathbb{K}}}(n, \mathcal{T}_{\mathbb{K}})$ of dimensions $i = 1, 2, \ldots, \mathcal{T}_{\mathbb{K}}$, respectively, constructed using a *self-dual* basis $(\beta_1, \beta_2, \ldots, \beta_n)$ of $\mathbb{F}_q$. The generator matrix $\mathbf{G}_i$ of the i-dimensional constituent code $C_i(n, i)$ is uniquely defined in such a manner that each constituent code satisfies the LCD condition with the rest: $C_i \cap C_j = \{\mathbf{0}\}$, for $i \neq j$, $i, j \in \{1, 2, \ldots, \mathcal{T}_{\mathbb{K}}\}$, with respect to the subspace $\Omega_{\mathbb{K}} = C_1 \oplus C_2 \oplus \cdots \oplus C_{\mathcal{T}_{\mathbb{K}}} \subseteq \mathbb{F}_q^n$.

Clearly, $(C_1, C_2, \ldots, C_{\mathcal{T}_{\mathbb{K}}})$ is a $\mathcal{T}_{\mathbb{K}}$-Direct code of dimension $\frac{\mathcal{T}_{\mathbb{K}}(\mathcal{T}_{\mathbb{K}}+1)}{2}$ over $\mathbb{F}_q$. Thus, the class of $\mathcal{T}_{\mathbb{K}}$-Direct codes is a subclass of the class of $\mathcal{T}$-Direct codes in the subspace $\Omega_{\mathbb{K}} = C_1 \oplus C_2 \oplus \cdots \oplus C_{\mathcal{T}_{\mathbb{K}}} \subseteq \mathbb{F}_q^n$. The following definition of a $\mathcal{T}_{\mathbb{K}}$-Direct code characterizes the systematic construction of a $\mathcal{T}$-Direct code.

Definition 3.1 Let $(\beta_1, \beta_2, \ldots, \beta_n)$ be a *self-dual* basis of $\mathbb{F}_q$ over $\mathbb{F}_p$ and $\mathcal{T}_{\mathbb{K}} = \max_{x \in \mathbb{N}} \left\{x \mid \frac{x(x+1)}{2} \leq n\right\}$, then the $\mathcal{T}$-Direct code $(C_1, C_2, \ldots, C_{\mathcal{T}_{\mathbb{K}}})$ is called a $\mathcal{T}_{\mathbb{K}}$-Direct code if the constituent codes $C_i(n, i), i = 1, 2, \ldots, \mathcal{T}_{\mathbb{K}}$ are LCD MRD codes and the generator matrix $\mathbf{G}_i$ of the i-dimensional constituent code is given by

$$\mathbf{G}_i = \begin{pmatrix} \beta_1^{\left[\frac{i(i-1)}{2}\right]} & \beta_2^{\left[\frac{i(i-1)}{2}\right]} & \cdots & \beta_n^{\left[\frac{i(i-1)}{2}\right]} \\ \beta_1^{\left[\frac{i(i-1)}{2}+1\right]} & \beta_2^{\left[\frac{i(i-1)}{2}+1\right]} & \cdots & \beta_n^{\left[\frac{i(i-1)}{2}+1\right]} \\ \vdots & \vdots & \ddots & \vdots \\ \beta_1^{\left[\frac{i(i+1)}{2}-1\right]} & \beta_2^{\left[\frac{i(i+1)}{2}-1\right]} & \cdots & \beta_n^{\left[\frac{i(i+1)}{2}-1\right]} \end{pmatrix}.$$

It is worth mentioning that the generator matrix of each constituent code C_i satisfies $\mathbf{G}_i \mathbf{G}_j^{\mathbf{T}} = (\mathbf{0})_{i \times j}$ for $i \neq j$ and $\mathbf{G}_k \mathbf{G}_k^{\mathbf{T}} = \mathbf{I}_k$ for the k-dimensional constituent code, where $\mathbf{I}_k$ is the identity matrix of order k. Clearly, $1 + 2 + \cdots + \mathcal{T}_{\mathbb{K}}$ rows of generator matrices $\mathbf{G}_1, \mathbf{G}_2, \ldots, \mathbf{G}_{\mathcal{T}_{\mathbb{K}}}$ constitute an orthonormal set of vectors in $\mathbb{F}_q^n$.

It should be emphasized that the above-defined systematic construction of the $\mathcal{T}_{\mathbb{K}}$-Direct codes by uniquely defining the generator matrices $\mathbf{G}_1, \mathbf{G}_2, \ldots, \mathbf{G}_{\mathcal{T}_{\mathbb{K}}}$ of the constituent codes $C_1, C_2, \ldots, C_{\mathcal{T}_{\mathbb{K}}}$ for a given *self-dual* basis $(\beta_1, \beta_2, \ldots, \beta_n)$ is equipped with the necessary properties required for handling the variable-length q-ary user messages.

Remark It must be noted that the $\mathcal{T}_{\mathbb{K}}$-Direct code $(C_1, C_2, \ldots, C_{\mathcal{T}_{\mathbb{K}}})$ defined above is in fact the largest systematically constructed $\mathcal{T}$-Direct code with dimension $\frac{\mathcal{T}_{\mathbb{K}}(\mathcal{T}_{\mathbb{K}}+1)}{2}$ and spans the subspace $\Omega_{\mathbb{K}} = C_1 \oplus C_2 \oplus \cdots \oplus C_{\mathcal{T}_{\mathbb{K}}}$ of $\mathbb{F}_q^n$. Further, if $1 + 2 + \cdots + \mathcal{T}_{\mathbb{K}} = n$, then $\Omega_{\mathbb{K}} = \mathbb{F}_q^n$.

In general, a systematically constructed $\mathcal{T}_{\sum i_j}$-Direct code $(C_{i_1}, C_{i_2}, \ldots, C_{i_j})$, $1 \leq i_1, i_2, \ldots, i_j \leq \mathcal{T}_{\mathbb{K}}$, composed of j distinct constituent LCD MRD codes and spanning a subspace $\Omega_{\sum i_j} = C_{i_1} \oplus C_{i_2} \oplus \cdots \oplus C_{i_j}$ of $\Omega_{\mathbb{K}}$ can be employed in variable-rate coding scheme. The necessary condition on the constituent codes of the $\mathcal{T}_{\sum i_j}$-Direct code $(C_{i_1}, C_{i_2}, \ldots, C_{i_j})$ can be stated as in the following theorem.

Theorem 4 *Let $(\beta_1, \beta_2, \ldots, \beta_n)$ be a self-dual basis of $\mathbb{F}_q$ and $C_{i_k}(n, i_k)$ be a q-ary MRD code with generator matrix $\mathbf{G}_{i_k}$ given by*

$$\mathbf{G}_{i_k} = \begin{pmatrix} \beta_1^{\left[\frac{i_k(i_k-1)}{2}\right]} & \beta_2^{\left[\frac{i_k(i_k-1)}{2}\right]} & \cdots & \beta_n^{\left[\frac{i_k(i_k-1)}{2}\right]} \\ \beta_1^{\left[\frac{i_k(i_k-1)}{2}+1\right]} & \beta_2^{\left[\frac{i_k(i_k-1)}{2}+1\right]} & \cdots & \beta_n^{\left[\frac{i_k(i_k-1)}{2}+1\right]} \\ \vdots & \vdots & \ddots & \vdots \\ \beta_1^{\left[\frac{i_k(i_k+1)}{2}-1\right]} & \beta_2^{\left[\frac{i_k(i_k+1)}{2}-1\right]} & \cdots & \beta_n^{\left[\frac{i_k(i_k+1)}{2}-1\right]} \end{pmatrix},$$

for distinct $1 \le i_k \le \mathcal{T}_{\mathbb{K}}$ *with* $\mathcal{T}_{\mathbb{K}} = \max_{x \in \mathbb{N}}\{x \mid \frac{x(x+1)}{2} \le n\}$ *and let* $\Omega_{\mathbb{K}} = C_1 \oplus C_2 \oplus \cdots \oplus C_{\mathcal{T}_{\mathbb{K}}}$. *Then* $(C_{i_1}, C_{i_2}, \ldots, C_{i_j})$ *is a* $\mathcal{T}_{\sum i_j}$*-Direct code. Further* $(C_{i_1}, C_{i_2}, \ldots, C_{i_j}) = \Omega_{\sum i_j}$, *where* $\Omega_{\sum i_j} = C_{i_1} \oplus C_{i_2} \oplus \cdots \oplus C_{i_j} \subseteq \Omega_{\mathbb{K}}$.

In view of the fact that $\mathcal{T}_{\sum i_j}$-Direct codes are particularly a subclass of the existing $\mathcal{T}$-Direct codes, the proof follows from Theorem 3. The constituent codes $C_{i_k}(n, i_k)$, $1 \le i_k \le \mathcal{T}_{\mathbb{K}}$ of distinct dimensions constituting a $\mathcal{T}_{\sum i_j}$-Direct code with *complementary-dual* property are equipped with the necessary properties for efficient handling of user messages of variable lengths, as discussed in the following section.

4 Construction of Multiple-Rate Codes Using $\mathcal{T}_{\mathbb{K}}$-Direct Codes

Consider the composite field $\mathbb{F}_{q^n} = \{0, \gamma, \gamma^2, \ldots, \gamma^{q^n-1}\}$, defined over the subfield $\mathbb{F}_q$, where $q = p^n$ and γ is a root of a primitive polynomial of degree n over $\mathbb{F}_q$. For conveniently performing the field operations over the base field $\mathbb{F}_q$, let $\mathbb{F}_q = \{0, \beta, \beta^2, \ldots, \beta^{q-1}\}$.

In order to manage variable-length encoding of q-ary user messages, consider k distinct codes $C_{J_1}(n, J_1)$, $C_{J_2}(n, J_2)$, $\ldots$, $C_{J_k}(n, J_k)$, for $1 \le J_1, J_2, \ldots, J_k \le \mathcal{T}_{\mathbb{K}}$ with $\mathcal{T}_{\mathbb{K}} = \max_{x \in \mathbb{N}}\left\{x \mid \frac{x(x+1)}{2} \le n\right\}$, over the base field $\mathbb{F}_q$ with the generator matrix $\mathbf{G}_{J_i}$ $(1 \le i \le k)$ of the J_i-dimensional constituent code:

$$\mathbf{G}_{J_i} = \begin{pmatrix} \beta_1^{\left[\frac{J_i(J_i-1)}{2}\right]} & \beta_2^{\left[\frac{J_i(J_i-1)}{2}\right]} & \cdots & \beta_n^{\left[\frac{J_i(J_i-1)}{2}\right]} \\ \beta_1^{\left[\frac{J_i(J_i-1)}{2}+1\right]} & \beta_2^{\left[\frac{J_i(J_i-1)}{2}+1\right]} & \cdots & \beta_n^{\left[\frac{J_i(J_i-1)}{2}+1\right]} \\ \vdots & \vdots & \ddots & \vdots \\ \beta_1^{\left[\frac{J_i(J_i+1)}{2}-1\right]} & \beta_2^{\left[\frac{J_i(J_i+1)}{2}-1\right]} & \cdots & \beta_n^{\left[\frac{J_i(J_i+1)}{2}-1\right]} \end{pmatrix},$$

where $(\beta_1, \beta_2, \ldots, \beta_n)$ is a *self-dual basis* of $\mathbb{F}_q$ over $\mathbb{F}_p$. Each constituent code $C_{J_i}(n, J_i)$ is a J_i-dimensional LCD MRD code of length n [20]. In order to provide variable-rate feature, the k distinct constituent codes $C_{J_1}(n, J_1), C_{J_2}(n, J_2), \ldots, C_{J_k}(n, J_k)$ with their respective generator matrices $\mathbf{G}_{J_1}, \mathbf{G}_{J_2}, \ldots, \mathbf{G}_{J_k}$ as defined above must form a $\mathcal{T}_{\sum J_k}$-Direct code with $(C_{J_1}, C_{J_2}, \ldots, C_{J_k}) \subseteq \Omega_{\mathbb{K}}$.

For a basis $\{\gamma_1, \gamma_2, \ldots, \gamma_n\}$ of $\mathbb{F}_{q^n}$, consider the 'multiple-rate code' $C_{\mathrm{MR}}(n, \{k, k+1, \ldots, k\mathcal{T}_{\mathbb{K}}\}_q, d\})_{q^n}$ defined by the k constituent codes of $\mathcal{T}_{\sum J_k}$-Direct code $(C_{J_1}, C_{J_2}, \ldots, C_{J_k})$ as 'children' codes and the code $C_{\mathrm{MR}}(n, k, d)$ generated by the following generator matrix as the 'mother' code:

$$\mathbf{G}_{\mathrm{MR}} = \begin{pmatrix} \gamma_1 & \gamma_2 & \cdots & \gamma_n \\ \gamma_1^q & \gamma_2^q & \cdots & \gamma_n^q \\ \vdots & \vdots & \ddots & \vdots \\ \gamma_1^{q^{k-1}} & \gamma_2^{q^{k-1}} & \cdots & \gamma_n^{q^{k-1}} \end{pmatrix}.$$

4.1 Encoding

Consider the q-ary message-block $\mathbf{u} = (\mathbf{u}_{1J_1}, \mathbf{u}_{2J_2}, \ldots, \mathbf{u}_{kJ_k}) \in \mathbb{F}_q^{J_1} \times \mathbb{F}_q^{J_2} \times \cdots \times \mathbb{F}_q^{J_k}$ of length $J = J_1 + J_2 + \cdots + J_k$ for distinct $1 \leq J_1, J_2, \ldots, J_k \leq \mathcal{T}_{\mathbb{K}}$, where $\mathbf{u}_{iJ_i} = (u_{i1}, u_{i2}, \ldots, u_{iJ_i}) \in \mathbb{F}_q^{J_i}$. Also consider the $\mathcal{T}_{\sum J_k}$-Direct code $(C_{J_1}, C_{J_2}, \ldots, C_{J_k})$ with J_i-dimensional LCD MRD constituent code $C_{J_i}(n, J_i)$ being described by the generator matrix $\mathbf{G}_{J_i}$.

The J_i-dimensional constituent code $C_{J_i}(n, J_i)$ of the $\mathcal{T}_{\sum J_k}$-Direct code is employed to obtain an n-length codeword $\mathbf{u}'_{in}$ (say) for encoding the ith message-segment $\mathbf{u}_{iJ_i}$ of length J_i in the message-block $\mathbf{u}$, for $1 \leq J_i \leq \mathcal{T}_{\mathbb{K}}$. Then by considering the constituent-encoded vector $(u'_{i1}, u'_{i2}, \ldots, u'_{in})$ in $\mathbb{F}_q^n$ as a symbol $\mathbf{u}'_{in}$ in $\mathbb{F}_{q^n}$ for each $i = 1, 2, \ldots, k$, the k-length constituent-encoded vector-block $\mathbf{u}' = (\mathbf{u}'_{1n}, \mathbf{u}'_{2n}, \ldots, \mathbf{u}'_{kn}) \in \mathbb{F}_{q^n}^k$ is obtained. Finally, the k-length q^n-ary vector-block $\mathbf{u}'$ is encoded through the 'multiple-rate code' $C_{\mathrm{MR}}\big(n, \{k, k+1, \ldots, k\mathcal{T}_{\mathbb{K}}\}_q, d\big)$ over $\mathbb{F}_{q^n}$, resulting in a codeword $\mathbf{c} \in \mathbb{F}_{q^n}^n$ for channel transmission over a noisy q^n-ary channel. The *complementary-dual* property of the constituent codes under consideration plays an important role in retrieving the original user messages of different lengths.

The decoding scheme requires the *orthogonal projectors* $\Pi_{C_{J_1}}, \Pi_{C_{J_2}}, \ldots, \Pi_{C_{J_k}}$ associated, respectively, with the constituent codes $C_{J_1}, C_{J_2}, \ldots, C_{J_k}$ of $\mathcal{T}_{\sum J_k}$-Direct code to recover $\mathbf{u}_{iJ_i}$ from $\mathbf{u}'_{in}$ at the decoder's end. The *orthogonal projectors* uniquely identify the correct constituent code $C_{J_i}(n, J_i)$ that is used for encoding. As the constituent codes $C_{J_1}, C_{J_2}, \ldots, C_{J_k}$ are chosen such that they constitute a $\mathcal{T}_{\sum J_k}$-Direct code $\big(C_{J_1}, C_{J_2}, \ldots, C_{J_k}\big) \subseteq \Omega_{\mathbb{K}}$ of dimension $\sum_i^k J_i$. Observe that $\mathbf{u}'_{in} \leftrightarrow (u'_{i1}, u'_{i2}, \ldots, u'_{in}) \in C_{J_i}(n, J_i)$ and $\mathbf{u}'_{in} \notin C_{J_j}(n, J_j)$ for $i \neq j$. Thus, the *orthogonal projectors* of the constituent codes satisfy the condition

$$\begin{cases} \mathbf{u}'_{in}\Pi_{C_{J_j}} = \mathbf{u}'_{in}, & \text{for } i = j \\ \mathbf{u}'_{in}\Pi_{C_{J_j}} = \{\mathbf{0}\}, & \text{for } i \neq j \end{cases}$$

The encoding procedure can be carried out in either of the following two ways: **(i)** By considering individual j_i-length segment $\mathbf{u}_{j_i}$ in the user message-block $\mathbf{u}$ and encoding conventionally with the corresponding j_i-dimensional constituent code C_{j_i} or **(ii)** By encoding the user message-block $\mathbf{u} = (\mathbf{u}_{1j_1}, \mathbf{u}_{2j_2}, \ldots, \mathbf{u}_{kj_k})$ with the bigger (block-diagonal) matrix $\mathbf{G}_{\mathcal{T}_{\sum j_k}}$ composed of generator matrices $\mathbf{G}_{j_1}, \mathbf{G}_{j_2}, \ldots, \mathbf{G}_{j_k}$ of the corresponding constituent codes. The generator matrix employed for encoding purpose and the associated *orthogonal projector* are of the form:

$$\mathbf{G}_{\mathcal{T}_{\sum j_k}} = \begin{pmatrix} \mathbf{G}_{j_1} & \mathbf{0} & \cdots & \mathbf{0} \\ \mathbf{0} & \mathbf{G}_{j_2} & \cdots & \mathbf{0} \\ \vdots & \vdots & \ddots & \vdots \\ \mathbf{0} & \mathbf{0} & \cdots & \mathbf{G}_{j_k} \end{pmatrix}; \; \Pi_{\mathbf{C}_{\mathcal{T}_{\sum j_k}}} = \begin{pmatrix} \Pi_{\mathbf{C}_{j_1}} & \mathbf{0} & \cdots & \mathbf{0} \\ \mathbf{0} & \Pi_{\mathbf{C}_{j_2}} & \cdots & \mathbf{0} \\ \vdots & \vdots & \ddots & \vdots \\ \mathbf{0} & \mathbf{0} & \cdots & \Pi_{\mathbf{C}_{j_k}} \end{pmatrix},$$

where $\mathbf{G}_{j_i}\mathbf{G}_{j_j}^{\mathbf{T}} = \begin{cases} \mathbf{I}_{j_i}, & \text{for } i = j \\ \mathbf{0}_{j_i \times j_j}, & \text{for} i \neq j \end{cases}$ and trivially $\mathbf{G}_{\mathcal{T}_{j_i}} = \mathbf{G}_{j_i} \forall\, j_i$ with respect to $\Omega_{\mathbb{K}}$.

Clearly, $\mathbf{u}' \in \Omega_{\mathbb{K}}^k$, and thus, the resultant k-length vector $\mathbf{u}'$ is channel encoded conventionally by the 'multiple-rate code' $C_{\text{MR}}(n,\ k,\ d)$ through $\mathbf{G}_{\text{MR}}$ to obtain an n-length q^n-ary codeword $\mathbf{c}$ for transmission.

Note: In case the message-segments $\mathbf{u}_{ij_i} \in \mathbf{u}$ are of non-distinct lengths, i.e., $j_1, j_2, \ldots, j_k$ are not distinct, then the constituent codes of the $\mathcal{T}_{\sum i_\ell}$-Direct code $(C_{i_1}, C_{i_2}, \ldots, C_{i_\ell})$ are employed for encoding, where $i_1, i_2, \ldots, i_\ell \in \{j_1, j_2, \ldots, j_k\}$ are the distinct lengths of user messages-segments.

4.2 Decoding

At the receiver's end, let the received n-length vector be $\mathbf{r} = \mathbf{c} + \mathbf{e}$ over the q^n-ary channel, where $\mathbf{e}$ is the error-vector. The receiver employs the conventional decoding technique associated with the 'multiple-rate code' $C_{\text{MR}}(n,\ k,\ d)$ to recover the transmitted codeword $\mathbf{c}$. As long as the rank of error-vector $\mathbf{e}$ is less than or equal to $\lfloor \frac{d-1}{2} \rfloor$, the decoder can correctly decode the received vector $\mathbf{r}$ to k-length vector $\mathbf{u}' = (\mathbf{u}'_{1n}, \mathbf{u}'_{2n}, \ldots, \mathbf{u}'_{kn}) \in \mathbb{F}_{q^n}^k$ through $C_{\text{MR}}\big(n, \{k, k+1, \ldots, k\mathcal{T}_{\mathbb{K}}\}_q, d\big)_{q^n}$. Now in the next phase, the receiver attempts to recover the actual j-length message $\mathbf{u} \in \mathbb{F}_q^{j_1} \times \mathbb{F}_q^{j_2} \times \cdots \times \mathbb{F}_q^{j_k}$ from the k-length vector $\mathbf{u}' \in \mathbb{F}_{q^n}^k$ with the help of *orthogonal projectors*, where $j = j_1 + j_2 + \cdots + j_k$.

Recall that in the encoding scheme $\mathbf{u}'_{ij_i}$ was encoded through the corresponding j_i-dimensional constituent code C_{j_i} as $\mathbf{u}'_{ij_i}\mathbf{G}_{j_i} = \mathbf{u}'_{in}$ to obtain $\mathbf{u}'_{in} \in \mathbb{F}_{q^n}$ and thus $\mathbf{u}' = (\mathbf{u}'_{1n}, \mathbf{u}'_{2n}, \ldots, \mathbf{u}'_{kn}) \in \mathbb{F}_{q^n}^k$. The *orthogonal projector* $\Pi_{C_{j_i}}$ associated with the constituent code C_{j_i} makes it possible to uniquely identify $\mathbf{u}'_{in}$ and thereby recover the actual message-segment $\mathbf{u}'_{ij_i}$ of length j_i. The decoder recovers message-segments $\mathbf{u}_{1j_1}, \mathbf{u}_{2j_2}, \ldots, \mathbf{u}_{kj_k}$ of variable lengths in two stages: in stage one, the receiver employs the *orthogonal projectors* associated with the constituent code C_{J_i} and its dual $C_{J_i}^{\perp}$ (with respect to $\Omega_{\mathbb{K}}$); secondly, the ith message-segment $\mathbf{u}_{iJ_i}$ is recovered conventionally from $\mathbf{u}'_{in}$. The decoding procedure can be understood as detailed below.

For $J_i = 1, 2, \ldots, \mathcal{T}_{\mathbb{K}}$: (for fixed J_i)

$$(i)\ \ \textbf{set}\ \ \mathbf{u}_{in}^{(1)} = \mathbf{u}'_{in}\Pi_{C_{J_i}};$$

$$(ii)\qquad \mathbf{u}_{in}^{(2)} = \mathbf{u}'_{in}\Pi_{C_{J_i}^{\perp}}.$$

In stage one, we have $\mathbf{u}_{jn}^{(1)} = \mathbf{u}'_{in}$ if and only if $i = j$ and secondly, $\mathbf{u}_{jn}^{(2)} = \mathbf{0}\,\forall i \neq j$, and thus, the only case for which the decoder outputs a non-zero vector is when the *orthogonal projector* corresponding to the correct constituent code is applied and none other. The advantage of employing the constituent codes of the $\mathcal{T}_{\mathbb{K}}$-Direct code $(C_{J_1}, C_{J_2}, \ldots, C_{J_k})$ for encoding is that at the decoder's end only one of the constituent code will satisfy the conditions, thus, uniquely determining the actual user message of length $J_i = 1$ or 2 or $\cdots$ or $\mathcal{T}_{\mathbb{K}}$. Starting with $J_i = 1$, the decoder checks for the fulfillment of two conditions at each step till $J_i = \mathcal{T}_{\mathbb{K}}$, if the conditions are met for $J_i = r$ (say), then the constituent code $C_r(n, r)$ is identified and hence, the user message $\mathbf{u}_{ir}$ is retrieved conventionally.

In a similar manner, for each $i = 1, 2, \ldots, k$, the ith message-segment $\mathbf{u}_{iJ_i}$ (of length J_i) is recovered from the constituent-encoded vector $\mathbf{u}'_{in}$ (of length n) to obtain the actual J_i-length segment $\mathbf{u}_{iJ_i} = (u_{i1}, u_{i2}, \ldots, u_{iJ_i}) \in \mathbb{F}_q^{J_i}$, and hence, the user message-block $\mathbf{u} = (\mathbf{u}_{1J_1}, \mathbf{u}_{2J_2}, \ldots, \mathbf{u}_{kJ_k}) \in \mathbb{F}_q^{J_1} \times \mathbb{F}_q^{J_2} \times \cdots \times \mathbb{F}_q^{J_k}$.

4.3 An Example

Consider $q = 2^3$ with $k = 2$ and $n = 3$. Let $\gamma \in \mathbb{F}_{8^3}$ be a root of the primitive polynomial $x^3 + x + \beta$ over $\mathbb{F}_8$, where β is a primitive element in $\mathbb{F}_{2^3}$ satisfying $\beta^3 + \beta + 1 = 0$. Here $\mathcal{T}_{\mathbb{K}} = 2$ $\left(\because\ \max\left\{x|\frac{x(x+1)}{2} \leq 3\right\} = 2\right)$. Consider the 'multiple-rate' MRD code $\mathcal{C}_{\text{MR}}(3, \{2, 3, 4\}_{2^3}, 2)$ with the following generator matrix:

$$\mathbf{G}_{\text{MR}} = \begin{pmatrix} 1 & \gamma & \gamma^2 \\ 1 & \gamma^8 & \gamma^{16} \end{pmatrix}.$$

Generator matrices of constituent LCD MRD codes $C_1(3,\ 1)_{2^3}, C_2(3,\ 2)_{2^3}$ satisfying $\mathbf{G}_i\mathbf{G}_j^{\mathbf{T}} = \begin{cases} \mathbf{I}_i, & \text{for } i = j, \\ (\mathbf{0})_{i\times j}, & \text{for } i \neq j \end{cases}$ and $\sum i = 1 + 2 = 3$, along with their associated *orthogonal projectors*:

$$\mathbf{G}_1 = \begin{pmatrix}\beta^3 & \beta^6 & \beta^5\end{pmatrix}; \qquad \Pi_{C_1} = \begin{pmatrix}\beta^6 & \beta^2 & \beta \\ \beta^2 & \beta^5 & \beta^4 \\ \beta & \beta^4 & \beta^3\end{pmatrix}$$

and

$$\mathbf{G}_2 = \begin{pmatrix}\beta^6 & \beta^5 & \beta^3 \\ \beta^5 & \beta^3 & \beta^6\end{pmatrix}; \qquad \Pi_{C_2} = \begin{pmatrix}\beta^2 & \beta^2 & \beta \\ \beta^2 & \beta^4 & \beta^4 \\ \beta & \beta^4 & \beta\end{pmatrix},$$

where $(\beta^3, \beta^6, \beta^5)$ is a *self-dual* basis in $\mathbb{F}_{2^3}$. Clearly, the code $\left(C_1(3, 1), C_2(3, 2)\right)$ is a $\mathcal{T}_{(1+2)}$-Direct code. The generator matrix for the encoding purpose and corresponding *orthogonal projector* are

$$\mathbf{G}_{\mathcal{T}_{(1+2)}} = \begin{pmatrix}\mathbf{G}_1 & \mathbf{0} \\ \mathbf{0} & \mathbf{G}_2\end{pmatrix}; \qquad \Pi_{C_{\mathcal{T}_{(1+2)}}} = \begin{pmatrix}\Pi_{C_1} & \mathbf{0} \\ \mathbf{0} & \Pi_{C_2} \\ . & \end{pmatrix}$$

Also, as $\mathbf{G}_{\mathcal{T}_{(2)}} = \mathbf{G}_2$ trivially,

$$\mathbf{G}_{\mathcal{T}_{(2+2)}} = \begin{pmatrix}\mathbf{G}_2 & \mathbf{0} \\ \mathbf{0} & \mathbf{G}_2\end{pmatrix}; \qquad \Pi_{C_{\mathcal{T}_{(2+2)}}} = \begin{pmatrix}\Pi_{C_2} & \mathbf{0} \\ \mathbf{0} & \Pi_{C_2}\end{pmatrix}.$$

The parity-check matrix $\mathbf{H}_i$ associated with the generator matrix $\mathbf{G}_i$ and the *orthogonal projector* $\Pi_{C_i^\perp}$ of the dual code $C_i^\perp$ $(i = 1, 2)$ are given as follows:

$$\mathbf{H}_1 = \begin{pmatrix}\beta^6 & \beta^5 & \beta^3 \\ \beta^5 & \beta^3 & \beta^6\end{pmatrix}; \qquad \Pi_{C_1^\perp} = \begin{pmatrix}\beta^2 & \beta^2 & \beta \\ \beta^2 & \beta^4 & \beta^4 \\ \beta & \beta^4 & \beta\end{pmatrix}$$

and

$$\mathbf{H}_2 = \begin{pmatrix}\beta^3 & \beta^6 & \beta^5\end{pmatrix}; \qquad \Pi_{C_2^\perp} = \begin{pmatrix}\beta^6 & \beta^2 & \beta \\ \beta^2 & \beta^5 & \beta^4 \\ \beta & \beta^4 & \beta^3\end{pmatrix}.$$

Encoding For $k = 2$, we have $1 \leq \jmath_1, \jmath_2 \leq 2$ so that message $\mathbf{u} = (\mathbf{u}_{1\jmath_1}, \mathbf{u}_{2\jmath_2}) \in \mathbb{F}_q^{\jmath_1} \times \mathbb{F}_q^{\jmath_2}$ of length $\jmath = \jmath_1 + \jmath_2 = 2$ or 3 can be encoded for transmission over the 8^3-ary channel. The following two cases demonstrate the encoding scheme for distinct as well as non-distinct lengths of segments in user message-block $\mathbf{u}$.

(i) For $\jmath_1 = \jmath_2 = 2$:

Consider the user message $\mathbf{u} = \left(\mathbf{u}_{12}, \mathbf{u}_{22}\right) = \left((\beta,\ \beta^3), (\beta^2,\ \beta^4)\right)$. The q-ary messages $\mathbf{u}_{12}$ and $\mathbf{u}_{22}$ are encoded by the constituent code of dimension 2, i.e., $C_2(3,\ 2)_{2^3}$, and the 3-length codewords $\mathbf{u}'_{13}$ and $\mathbf{u}'_{23}$ are obtained as follows:

$$\mathbf{u}_{12}\mathbf{G}_2 = (\beta,\ \beta^3)\mathbf{G}_2 \ = \left(\beta^2, \beta^3, \beta^5\right) \leftrightarrow \mathbf{u}'_{13},$$

which correspond to the element γ^{26} of $\mathbb{F}_{8^3}$. Further,

$$\mathbf{u}_{22}\mathbf{G}_2 = (\beta^2,\ \beta^4)\mathbf{G}_2 \ = \ \left(\beta^4,\ 0,\ \beta^2\right) \ \leftrightarrow \ \mathbf{u}'_{23},$$

corresponding to the element γ^{152} in $\mathbb{F}_{8^3}$. Or equivalently,

$$\mathbf{u}\mathbf{G}_{\mathcal{T}(2+2)} = ((\beta,\ \beta^3), (\beta^2,\ \beta^4))\mathbf{G}_{\mathcal{T}(2+2)} \ = \ \left(\left(\beta^2,\ \beta^3,\ \beta^5\right), \left(\beta^4,\ 0,\ \beta^2\right)\right) \ \leftrightarrow \ (\mathbf{u}'_{13}, \mathbf{u}'_{23}).$$

Thus $\mathbf{u}' = (\gamma^{26}, \gamma^{152})$ is obtained, further $\mathbf{u}'$ is encoded by the 'multiple-rate code' by $\mathbf{G}_{\mathrm{MR}}$ for transmission:

$$\mathbf{u}'\mathbf{G}_{\mathrm{MR}} = (\mathbf{u}'_{13}, \mathbf{u}'_{23})\mathbf{G}_{\mathrm{MR}} = \left(\gamma^{40},\ \gamma^{103},\ \gamma^{37}\right) = \mathbf{c}_1.$$

(ii) For $j_1 = 1$ and $j_2 = 2$:

Consider $\mathbf{u} = \left(\mathbf{u}_{11}, \mathbf{u}_{22}\right) = \left((\beta), (\beta^2,\ \beta^4)\right)$. This time the constituent code $C_1(3,\ 1)_{2^3}$ is used to encode $\mathbf{u}_{11}$ and $C_1(3,\ 2)_{2^3}$ to encode $\mathbf{u}_{22}$, respectively. Consequently, $\mathbf{u}'_{13}$ and $\mathbf{u}'_{23}$ are obtained as follows:

$$\mathbf{u}_{11}\mathbf{G}_1 = \ (\beta)\,\mathbf{G}_1 \ = \ \left(\beta^4,\ 1,\ \beta^6\right) \ \leftrightarrow \ \mathbf{u}'_{13},$$

which correspond to the element γ^{133} of $\mathbb{F}_{8^3}$. Also,

$$\mathbf{u}_{22}\mathbf{G}_2 = \left(\beta^2,\ \beta^4\right)\mathbf{G}_2 = \left(\beta^4,\ 0,\ \beta^2\right) \leftrightarrow \mathbf{u}'_{23},$$

which correspond to the element γ^{152} of $\mathbb{F}_{8^3}$. Or equivalently,

$$\mathbf{u}\mathbf{G}_{\mathcal{T}(1+2)} = ((\beta), (\beta^2,\ \beta^4))\mathbf{G}_{\mathcal{T}(1+2)} \ = \ \left(\left(\beta^4,\ 1,\ \beta^6\right), \left(\beta^4,\ 0,\ \beta^2\right)\right) \ \leftrightarrow \ (\mathbf{u}'_{13}, \mathbf{u}'_{23}).$$

Finally, $\mathbf{u}'$ is encoded with $\mathbf{G}_{\mathrm{MR}}$ for transmission:

$$\mathbf{u}'\mathbf{G}_{\mathrm{MR}} = ((\mathbf{u}'_{13}, \mathbf{u}'_{23}))\mathbf{G}_{\mathrm{MR}} = \left(\gamma^{37},\ \gamma^{379},\ \gamma^{196}\right) = \ \mathbf{c}_2.$$

Decoding By employing the 'multiple-rate code' $C_{\mathrm{MR}}(3, 2, 2)_{8^3}$, the transmitted codeword $\mathbf{c}_j$ (for fixed j) can be recovered from a received vector $\mathbf{r}$. After retrieving $\mathbf{c}_j$ and hence the messages $\mathbf{u}'_{13}$ and $\mathbf{u}'_{23}$, the receiver then employs the *orthogonal projectors* of the constituent codes to recover the actual user messages $\mathbf{u}_{1j_1}$ (of length j_1) from $\mathbf{u}'_{13}$ and $\mathbf{u}_{2j_2}$ (of length j_2) from $\mathbf{u}'_{23}$ $(1 \leq j_1, j_2 \leq 2)$ as discussed below.

(i) Recovering $\mathbf{u}_{12}$ and $\mathbf{u}_{22}$ from $\mathbf{c}_1$:
After decoding the message $\mathbf{u}'$ from $C_{\mathrm{MR}}(3, 2, 2)_{8^3}$ as $\mathbf{u}' = (\mathbf{u}'_{13}, \mathbf{u}'_{23}) = \left((\beta^2, \beta^3, \beta^5), (\beta^4,\ 0,\ \beta^2)\right)$, the decoder proceeds to identify the constitute codes as follows:

Set $j_1 = j_2 = 1:$

$$\begin{aligned} \mathbf{u}_{13}^{(1)} &= \mathbf{u}'_{13}\Pi_{C_1} = \left(0,\,0,\,0\right)\\ \mathbf{u}_{13}^{(2)} &= \mathbf{u}'_{13}\Pi_{C_1^{\perp}} = \left(\beta^2,\,\beta^3,\,\beta^5\right)\\ \text{and}\quad \mathbf{u}_{23}^{(1)} &= \mathbf{u}'_{23}\Pi_{C_1} = \left(0,\,0,\,0\right)\\ \mathbf{u}_{23}^{(2)} &= \mathbf{u}'_{23}\Pi_{C_1^{\perp}} = \left(\beta^4,\,0,\,\beta^2\right). \end{aligned}$$

The conditions are neither met for $j_1 = 1$ nor for $j_2 = 1$. Therefore, proceeding to next step, i.e., setting $j_1 = j_2 = 2$.

Set $j_1 = j_2 = 2:$

$$\begin{aligned} \mathbf{u}_{13}^{(1)} &= \mathbf{u}'_{13}\Pi_{C_2} = \left(\beta^2,\,\beta^3,\,\beta^5\right)\\ \mathbf{u}_{13}^{(2)} &= \mathbf{u}'_{13}\Pi_{C_2^{\perp}} = \left(0,\,0,\,0\right)\\ \text{and}\quad \mathbf{u}_{23}^{(1)} &= \mathbf{u}'_{23}\Pi_{C_2} = \left(\beta^4,\,0,\,\beta^2\right)\\ \mathbf{u}_{23}^{(2)} &= \mathbf{u}'_{23}\Pi_{C_2^{\perp}} = \left(0,\,0,\,0\right). \end{aligned}$$

Conditions are met for $j_1 = 2$ and $j_2 = 2$, i.e., $\mathbf{u}_{13}^{(1)} = \mathbf{u}'_{13}$, $\mathbf{u}_{23}^{(1)} = \mathbf{u}'_{23}$ and $\mathbf{u}_{13}^{(2)} = \mathbf{u}_{23}^{(2)} = \mathbf{0}$. In this case, the constituent code is identified as $C_2(3,2,2)_{2^3}$. Hence, $\mathbf{u}_{12}$ and $\mathbf{u}_{22}$ are recovered from $\left(\mathbf{u}'_{13}, \mathbf{u}'_{23}\right)$ as $\left((\beta,\ \beta^3), (\beta^2,\ \beta^4)\right)$.

(ii) Recovering $\mathbf{u}_{11}$ and $\mathbf{u}_{22}$ from $\mathbf{c}_2$:
After decoding the message $\mathbf{u}'$ from $C_{\text{MR}}(3,2,2)_{8^3}$ as $\mathbf{u}' = (\mathbf{u}'_{13}, \mathbf{u}'_{23}) = \left((\beta^4,\ 1,\ \beta^6), (\ \beta^4,\ 0,\ \beta^2)\right)$, the decoder proceeds to identify the constitute codes as follows:

Set $j_1 = j_2 = 1:$

$$\begin{aligned} \mathbf{u}_{13}^{(1)} &= \mathbf{u}'_{13}\Pi_{C_1} = \left(\beta^4,\,1,\,\beta^6\right)\\ \mathbf{u}_{13}^{(2)} &= \mathbf{u}'_{13}\Pi_{C_1^{\perp}} = \left(0,\,0,\,0\right)\\ \text{and}\quad \mathbf{u}_{23}^{(1)} &= \mathbf{u}'_{23}\Pi_{C_1} = \left(0,\,0,\,0\right)\\ \mathbf{u}_{23}^{(2)} &= \mathbf{u}'_{23}\Pi_{C_1^{\perp}} = \left(\beta^4,\,0,\,\beta^2\right). \end{aligned}$$

Conditions are met for $j_1 = 1$ but not for j_2. Therefore, moving to next step for j_2.

Set $j_2 = 2:$

$$\begin{aligned} \mathbf{u}_{23}^{(1)} &= \mathbf{u}'_{23}\Pi_{C_2} = \left(\beta^4,\,0,\,\beta^2\right)\\ \mathbf{u}_{23}^{(2)} &= \mathbf{u}'_{23}\Pi_{C_2^{\perp}} = \left(0,\,0,\,0\right). \end{aligned}$$

Finally, conditions are met for $\jmath_1 = 1$ and $\jmath_2 = 2$, i.e., $\mathbf{u}_{13}^{(1)} = \mathbf{u'}_{13}$, $\mathbf{u}_{23}^{(1)} = \mathbf{u'}_{23}$ and $\mathbf{u}_{13}^{(2)} = \mathbf{u}_{23}^{(2)} = \mathbf{0}$. In this case, the constituent codes are identified as $C_1(3, 1, 3)_{2^3}$ and $C_2(3, 2, 2)_{2^3}$. Hence, $\mathbf{u}_{11}$ and $\mathbf{u}_{22}$ are recovered from $\left(\mathbf{u}'_{13}, \mathbf{u}'_{23}\right)$ as $\left((\beta), (\beta^2, \beta^4)\right)$.

The above illustration shows the proposed encoding and decoding scheme for 2^3-ary user messages of lengths 1 or 2 and 'multiple-rate code' $C_{\mathrm{MR}}(3, 2, 2)_{8^3} \leftrightarrow C_{\mathrm{MR}}(3, \{2, 3, 4\}_{2^3}, 2)_{8^3}$ with 'constituent codes' $C_1(3, 1, 3)_{2^3}$ and $C_2(3, 2, 2)_{2^3}$ for transmission over 8^3-ary channel.

5 Conclusion

The need for channel codes that are capable of handling user messages of variable lengths also exists in communication systems. Addressing this issue, the paper proposed an encoding and decoding scheme that employs a class of $\mathcal{T}$-Direct codes in order to provide the variable-rate feature. For this, it considers k LCD MRD codes $C_{i_1}, C_{i_2}, \ldots, C_{i_k}$ of respective dimensions from 1 to $\mathcal{T}_{\mathbb{K}} = \max\left\{x \mid \frac{x(x+1)}{2} \leq n\right\}$, constructed using a *self-dual* basis in $\mathbb{F}_q$ constituting together a $\mathcal{T}_{\sum i_k}$-Direct code $(C_{i_1}, C_{i_2}, \ldots, C_{i_k})$ over $\mathbb{F}_q$. The q-ary $\jmath_i$-length segment of the user message-block is encoded with constituent code $C_{\jmath_i}$. Further, a 'multiple-rate code' $C_{\mathrm{MR}}(n, k, d)$ over the extension field $\mathbb{F}_{q^n}$ is considered for encoding the k-length constituent-encoded vector-block for transmission over the q^n-ary channel. The *orthogonal projectors* associated with the constituent codes are employed to uniquely determine and hence decode the q-ary messages of various lengths $1, 2, \ldots, \mathcal{T}_{\mathbb{K}}$, providing variable information rates $\frac{k}{n^2}, \frac{k+1}{n^2}, \ldots, \frac{k\mathcal{T}_{\mathbb{K}}}{n^2}$. Further construction of multiple-rate codes using $\mathcal{T}_{\mathbb{K}}$-Direct codes with enhanced information rates is under consideration by the authors.

References

1. Boripan A, Jitman S, Udomkavanich P (2018) Characterization and enumeration of complementary dual abelian codes. J Appl Math Comput 58(1–2):527–544
2. Lempel A, Seroussi G (1980) Factorization of symmetric matrices and trace-orthogonal bases in finite fields. SIAM J Comput 9:758–767
3. Lempel A, Weinberger M (1988) Self-complementary normal bases in finite fields. SIAM J Discret Math 1:193–198
4. Lempel A, Seroussi G (1991) Explicit formulas for self-complementary normal bases in certain finite fields. IEEE Trans Inf Theory 37:1220–1222
5. Carlet C, Guilley S (2016) Complementary dual codes for countermeasures to side-channel attacks. Adv Math Commun 10(1):131–150
6. Jungnickel D, Menezes AJ, Vanstone SA (1990) On the number of self-dual bases of $\mathbf{GF}(q^m)$ over $\mathbf{GF}(q)$. Proc Am Math Soc 109:23–29
7. Gabidulin EM (1985) Theory of codes with maximum rank distance. Probl Inf Transm 21:1–12

8. Bringer J, Carlet C, Chabanne H, Guilley S, Maghrebi H (2014) Orthogonal direct sum masking: a smartcard friendly computation paradigm in a code, with built-in protection against side-channel and fault attacks. In: Information security theory and practice international workshop 8501 (LNCS, 2014), vol 8501, pp 40–56
9. Massey JL (1964) Reversible codes. Inf Control 7(3):369–380
10. Massey JL (1992) Linear codes with complementary duals. Discret Math 106 and 107:337–342
11. Gadouleau M, Yan Z (2008) On the decoder error probability of bounded rank-distance decoders for maximum rank distance codes. IEEE Trans Inf Theory 54:3202–3206
12. Gadouleau M, Yan Z (2008) Packing and covering properties of rank metric codes. IEEE Trans Inf Theory 54:3873–3883
13. Silva D, Kschishang FR (2009) On metrics for error correction in network coding. IEEE Trans Inf Theory 55:5479–5490
14. Silva D, Kschischang FR (2011) Universal secure network coding via rank-metric codes. IEEE Trans Inf Theory 57:1124–1135
15. Sendrier M (2004) Linear codes with complementary duals meet the Gilbert-Varshamov bound. Discret Math 285:345–347
16. Shi M, Daitao H (2018) On LCD MRD codes. IEICE Trans Fundam E101-A(9):1599–1602
17. Devi M (2013) On the class of $\mathcal{T}$-Direct codes: constructions, properties and applications. Ph.D. dessertation, Department of Mathematics, Jaypee University of Information Technology, India. http://hdl.handle.net/10603/35221
18. Loidreau P (2008) Properties of codes in rank metric. In: Proceedings of the eleventh international workshop on algebraic and combinatorial coding theory. Bulgaria, pp 192–198
19. Roth RM (1991) Maximum-rank array codes and their application to crisscross error correction. IEEE Trans Inf Theory 37:328–336
20. Kandasamy WBV, Smarandache F, Sujatha R, Raja Durai RS (2012) Erasure techniques in MRD codes. ZIP Publishing, Columbus, OH
21. Tse D, Viswanath P (2005) Fundamentals of wireless communication. Cambridge University Press, Cambridge, U.K
22. Raja Durai RS (2004) On linear codes with rank metric: constructions, properties, and applications. Ph.D. dessertation, Department of Mathemtics, Indian Institute of Technology—Chennai, India
23. Raja Durai RS, Devi M (2013) On the class of $\mathcal{T}$-Direct codes over $\mathbf{GF}(2^N)$. Int J Comput Inf Syst Ind Manag Appl 5:589–596
24. Hamming RW (1950) Error detecting and error correcting codes. Bell Syst Tech J 29:147–160
25. Bhasin S, Danger, J-L, Guilley S, Najm Z, Ngo XT (2015) Linear complementary dual code improvement to strengthen encoded circuit against hardware Trojan horses. IEEE Int Symp Hardw Oriented Secur Trust 82–87
26. Vasantha WB, Raja Durai RS (2002) $\mathcal{T}$-Direct codes: an application to $\mathcal{T}$-user BAC. In: Proceedings of IEEE information theory workshop. Bangalore, p 214
27. Yang X, Massey JL (1994) The condition for a cyclic code to have a complementary dual. Discret Math 126(1–3):391–393
28. Liu X, Fan Y, Liu H (2018) Galois LCD codes over finite fields. Finite Fields Their Appl 49:227–242
29. Liu X, Liu H (2019) Rank-metric complementary dual codes. J Appl Math Comput 61(1–2):281–295
30. Raja Durai RS, Devi M, Kumar A (2020) Multiple-rate error-correcting coding scheme. Appl Algebr Eng, Commun Comput. https://doi.org/10.1007/s00200-020-00435-x
31. Devi M, Raja Durai RS, Kumar A, Hongjun Xu H Some important results on $\mathcal{T}$-Direct codes. Discret Math, Algorithms Appl https://doi.org/10.1142/S1793830921500671

Sufficient Optimality Criteria for Optimization Problem Involving Pseudoconvex Interval Objective Function

Priyanka Roy and Geetanjali Panda

Abstract In this article, the pseudoconvex property of a class of functions, whose parameters vary within some closed intervals, is studied under generalized Hukuhara differentiability concept. The necessary and sufficient condition for the existence of a critical point of interval-valued minimization problem is derived using descent property. Sufficient optimality conditions are established for a critical point to be a weak efficient point under pseudoconvexity assumption. Numerical illustrations are provided to justify the theoretical results.

Keywords Interval optimization · Generalized Hukuhara derivative · Pseudoconvex property · Optimality condition

1 Introduction

In most of the economic and engineering problems, the different parameters are not exact due to vagueness in data. Digital computers also have limited representation capacity in floating point systems, which causes significant accumulation errors. This type of impreciseness occurs due to different type of complications in environment and deep-rooted biased nature of human thought. These uncertain parameters can be expressed as intervals if bounds are known. The functions that are involved in this type of modelling process using uncertain data are known as interval functions.

For instance, the return and the variance risk of a financial portfolio can be expressed as an interval function. Suppose an investor wants to invest some capital amount P in n number of assets $A_1, A_2, \ldots A_n$, which are available in the financial market. Return of each asset is not fixed. From the historical data, it is

P. Roy (✉)
Indian Institute of Technology Kharagpur, Kharagpur, India
e-mail: proy180192@maths.iitkgp.ac.in

G. Panda
Department of Mathematics, Indian Institute of Technology Kharagpur, Kharagpur-721302, W.B, India

© The Author(s), under exclusive license to Springer Nature Singapore Pte Ltd. 2022
D. Giri et al. (eds.), *Proceedings of the Seventh International Conference on Mathematics and Computing*, Advances in Intelligent Systems and Computing 1412,
https://doi.org/10.1007/978-981-16-6890-6_50

observed that return of A_i varies between $\underline{r}_i$ and $\overline{r}_i$. Then the total return of the portfolio is a linear interval function, which is expressed in a mathematical form as $\sum_{i=1}^{n}[\underline{r}_i, \overline{r}_i]x_i = [\sum_{i=1}^{n}\underline{r}_i x_i, \sum_{i=1}^{n}\overline{r}_i x_i]$, where x_i is the amount of investment in the asset A_i, $\sum_{i=1}^{n} x_i = P$. Since the return parameter is expressed as interval, so the covariance of the returns of the assets A_i and A_j will lie in an interval $[\underline{\sigma}_{ij}, \overline{\sigma}_{ij}]$ say. In that case, the variance risk of the portfolio is $\sum_{i=1}^{n}\sum_{j=1}^{n}[\underline{\sigma}_{ij}, \overline{\sigma}_{ij}]x_i x_j$, which is a nonlinear interval-valued function.

In literature (Refs. [1, 4, 11]), one may observe that the algebraic operations in the set of intervals are defined as an extension over $\mathbb{R}$. The set of intervals is not closed with respect to usual arithmetic operations as additive inverse may not exist. Due to this difficulty, the concepts for real-valued function like limit, continuity, differentiability etc. cannot be directly extended to interval functions. Hukuhara [6] and Markov [9] defined H-difference and M-difference, respectively, to address this issue. These differences are dependent on the length of the intervals. Recently, Stefanini and Bede [16] introduced gH-difference between two intervals, which is the most generalized concept of interval differences and widely explored in literature (Refs. [2, 8, 12, 13] etc.)

Several interval orderings exist using the endpoint functions since interval space is not completely ordered. Therefore, the numerical techniques to find the minimum value of a real optimization model cannot be directly implemented for interval optimization problems. Convex and generalized convex property of objective function play important role to determine the minimizer. In recent years, several authors have tried to extend these properties in interval optimization models in different directions. Wu [17] has extended the concept of generalized convexity for endpoint functions to LU, UC convexity and pseudoconvexity, where LU and UC partial orderings are based on lower, upper bound functions and upper bound, midpoint functions, respectively. Chalco-Cano et al. [5] extended this concept for LS partial ordering, which is based on lower bound and width functions. Zhang et al. [18] extended the concept of invexity to interval objective functions, which are based on H-differentiability. Osuna Gomez et al. [10] derived the optimality conditions using pseudoinvexity property for interval single and multi-objective functions. Singh et al. [15] derived optimality conditions with the help of endpoint functions. These approaches are based on the algebraic operations of endpoint functions in different ways, which do not take care of the data at the interior points of the interval parameters. In this article, we have introduced gH-pseudoconvexity of interval function, which does not need the information of differentiability of endpoint functions. Hence, the proposed sufficient conditions totally deal with the interval objective function without transforming it into real-valued functions, which can be regarded as computational advantage over existing literature.

The article is prepared as follows: In Sect. 2, the mathematical background of interval calculus is provided. In Sect. 3, gH-pseudoconvex property of interval function is explained with an example. Sufficient condition for minimizing the interval objective functions is discussed in Sect. 4 using pseudoconvex property. Numerical illustrations are provided for the theoretical justification.

2 Prerequisites

Throughout this article, $\mathbf{K}(\mathbb{R})$ denotes the set of closed real intervals on $\mathbb{R}$. Any element $\breve{\alpha} \in \mathbf{K}(\mathbb{R})$ is of the form $[\underline{\alpha}, \overline{\alpha}]$ with $\underline{\alpha} \leq \overline{\alpha}$, where $\underline{\alpha}, \overline{\alpha}$ are bounds of $\breve{\alpha}$. $\breve{I} = [1, 1]$ and $\breve{0} = [0, 0]$ denote the unit and null interval, respectively.

In usual arithmetic operation of $\mathbf{K}(\mathbb{R})$, $\breve{\alpha} \ominus \breve{\alpha}$ is not necessarily $\breve{0}$. The generalized Hukuhara differece(gH difference) between two intervals $\breve{\alpha}$ and $\breve{\beta}$ is defined as follows. (See [16])

$$\breve{\alpha} \ominus_{gH} \breve{\beta} = \breve{\gamma} \Leftrightarrow \begin{cases} \breve{\alpha} = \breve{\beta} \oplus \breve{\gamma} \\ \text{ or} \\ \breve{\alpha} \oplus (-1)\breve{\gamma} = \breve{\beta}. \end{cases} \tag{1}$$

According to this difference, the following properties hold (See Ref. [16]).

$$(i) \ominus_{gH} \breve{\alpha} = \hat{0} \ominus_{gH} \breve{\alpha} = (-1)\breve{\alpha};\ (ii)\breve{\alpha} \ominus_{gH} \breve{\alpha} = \hat{0};\ (iii)(\breve{\alpha} \oplus \breve{\beta}) \ominus_{gH} \breve{\beta} = \breve{\alpha} \tag{2}$$

Following interval ordering from Ref. [7] is often used for computations.

For $\breve{\alpha} = [\underline{\alpha}, \overline{\alpha}]$, $\breve{\beta} = [\underline{\beta}, \overline{\beta}] \in \mathbf{K}(\mathbb{R})$, $\breve{\alpha} \preceq \breve{\beta} \Leftrightarrow \underline{\alpha} \leq \underline{\beta}$ and $\overline{\alpha} \leq \overline{\beta}$;

$\breve{\alpha} \preceq \breve{\beta} \Leftrightarrow \breve{\alpha} \preceq \breve{\beta}$ and $\breve{\alpha} \neq \breve{\beta}$;

$\breve{\alpha} \prec \breve{\beta} \Leftrightarrow \underline{\alpha} < \underline{\beta}$ and $\overline{\alpha} < \overline{\beta}$.

The product $d^T\breve{\alpha}$ where $d \in \mathbb{R}^n$ and $\breve{\alpha} = (\breve{\alpha}_1, \cdots, \breve{\alpha}_n)$, $\breve{\alpha}_i \in \mathbf{K}(\mathbb{R})$ is

$$d^T\breve{\alpha} = d_1\breve{\alpha}_1 \oplus d_2\breve{\alpha}_2 \oplus \cdots \oplus d_n\breve{\alpha}_n \triangleq \sum_{i=1}^{n} d_i\breve{\alpha}_i.$$

An interval function is of the form $\breve{F}(x) = [\underline{F}(x), \overline{F}(x)]$, where $\underline{F}, \overline{F} : \mathbb{R}^n \to \mathbb{R}$, $\underline{F}(x) \leq \overline{F}(x)\ \forall\ x$. The metric structure of gH difference is understood using Hausdorff distance between intervals, that is $D(\breve{\alpha}, \breve{\beta}) = \max\left\{|\underline{\alpha} - \underline{\beta}|, |\overline{\alpha} - \overline{\beta}|\right\}$ Limit and continuity of an interval function is defined using this metric (see Ref. [16]). Some existing results on calculus of interval function are provided as follows.

Definition 1 (*Stefanini and Bede* [16]) Generalized Hukuhara derivative of $\breve{F}$: $(t_1, t_2) \subseteq \mathbb{R} \to \mathbf{K}(\mathbb{R})$ at $x \in (t_1, t_2)$ is defined as $\breve{F}'(x) = \lim_{h \to 0} \frac{\breve{F}(x+h) \ominus_{gH} \breve{F}(x)}{h}$.

Definition 2 (*Osuna-Gómez et al.* [10]) Let $\breve{F} : X \subseteq \mathbb{R}^n \to \mathbf{K}(\mathbb{R})$ and $x = (x_1, x_2, \cdots, x_n) \in X$ and consider $\hat{\psi}_i(x_i) = \breve{F}(x_1, x_2, \cdots x_{i-1}, x_i, x_{i+1}, \cdots, x_n)$. If $\lim\limits_{h_i \to 0} \dfrac{\hat{\psi}_i(x_i + h_i) \ominus_{gH} \hat{\psi}_i(x_i)}{h_i}$ exists, then we say that the partial derivative of $\breve{F}$ with respect to x_i exists at x and the limiting value is denoted by $\frac{\partial \breve{F}(x)}{\partial x_i}$.

Gradient of an interval function $\breve{F}$ at a point $x \in \mathbb{R}^n$ is an interval vector. It is denoted by

$$\nabla \breve{F}(x) \triangleq (\frac{\partial \breve{F}(x)}{\partial x_1}, \frac{\partial \breve{F}(x)}{\partial x_2}, \dots, \frac{\partial \breve{F}(x)}{\partial x_n})^T$$

.

Definition 3 (*Roy and Panda* [13]) For $\breve{F} : X \to \mathbf{K}(\mathbb{R})$, $\breve{F}$ is called gH differentiable at $x_0 \in X$ if $\nabla \breve{F}(x_0)$ exists and $\lim_{\|h\| \to 0} \frac{\hat{w}(\breve{F}(x_0);h) \ominus_{gH} h^T \nabla \breve{F}(x_0)}{\|h\|} = \hat{0}$, where $\hat{w}(\breve{F}(x_0); h) \triangleq \breve{F}(x_0 + h) \ominus_{gH} \breve{F}(x_0)$.

3 gH-pseudoconvex Interval Function

Convexity has a significant part to ensure the existence of local minimum point in classical optimization theory. Convex property can be generalized in several directions to weaken the hypothesis. Wu extended this notion to define pseudoconvex property of interval function as follows, which is based on weak differentiability and partial ordering with respect to the endpoint functions.

Definition 4 (*Wu* [17]) An interval function $\breve{F} : X \subseteq \mathbb{R}^n \to \mathbf{K}(\mathbb{R})$ is called pseudoconvex at $x^* \in X$ if $\underline{F}$ and $\overline{F}$ are pseudoconvex at x^*.

In this section, we develop pseudoconvex concept for gH-differentiable interval function.

In Definition 4, $\underline{F}$ and $\overline{F}$ are assumed to be differentiable. If the endpoint functions are differentiable, $\breve{F}$ is differentiable. But the converse is not true.

Example 5 Consider the interval function as $\breve{F}(x_1, x_2) = [1, 3]x_1^2 \oplus [2, 4]x_2$.
$\breve{F}(x_1, x_2) = \begin{cases} [x_1^2 + 2x_2, 3x_1^2 + 4x_2]; \text{ if } x_2 \geq 0 \\ [x_1^2 + 4x_2, 3x_1^2 + 2x_2]; \text{ if } x_2 \leq 0 \end{cases}$

One may note that $\frac{\partial \underline{F}(0,0)}{\partial x_2} = \begin{cases} 2, \text{ if } x_2 \geq 0 \\ 4, \text{ if } x_2 \leq 0 \end{cases}$ and $\frac{\partial \overline{F}(0,0)}{\partial x_2} = \begin{cases} 4, \text{ if } x_2 \geq 0 \\ 2, \text{ if } x_2 \leq 0 \end{cases}$.

Therefore, $\frac{\partial \underline{F}(0,0)}{\partial x_2}$ and $\frac{\partial \overline{F}(0,0)}{\partial x_2}$ can not exist. So the endpoint functions are not differentiable at $(0, 0)$. Therefore, psudoconvex property of $\breve{F}$ cannot be defined using Defintion 4. But $\frac{\partial \breve{F}(0,0)}{\partial x_1} = [0, 0]$ and $\frac{\partial \breve{F}(0,0)}{\partial x_2} = [2, 4]$.

Hence, $\nabla \breve{F}(0, 0) = \begin{pmatrix} [0, 0] \\ [2, 4] \end{pmatrix}$.

Therefore, in this example, $\breve{F}$ is gH-differentiable and $\nabla \breve{F}(0, 0)$ exists, but $\breve{F}$ is not pseudoconvex from Defintion 4.

To overcome this difficulty, we used the notion of gH-pseudoconvexity for a class of gH-differentiable interval functions without bothering about the differentiability of endpoint functions (See Ref. [14]).

Definition 6 (*Roy and Panda* [14]) Suppose $\breve{F} : X \subseteq \mathbb{R}^n \to \mathbf{K}(\mathbb{R})$ be gH-differentiable. $\breve{F}$ is said to be gH-pseudoconvex at $\bar{x} \in X$ if $\breve{F}(x) \prec \breve{F}(\bar{x})$, then $(x - \bar{x})^T \nabla \breve{F}(\bar{x}) \prec \hat{0}$ hold for each $x \in X$.

$\breve{F}$ is said to be gH-pseudoconvex interval function on X if for each $x, y \in X$, $\breve{F}(y) \prec \breve{F}(x)$ implies $(y - x)^T \nabla \breve{F}(x) \prec \hat{0}$.

Example 7 The gH-pseudoconvex property of interval function does not depend on the endpoint functions. To justify this, consider $\breve{F}(x_1, x_2) = [1, 3]x_1^2 \oplus [2, 4]x_2$ from Example 5 in which $\breve{F}$ is gH differentiable at $x = (0, 0)^T$ and $\nabla \breve{F}(0, 0) = ([0, 0],\ [2, 4])^T$ though $\underline{F}$ and $\overline{F}$ are not differentiable at $x = (0, 0)^T$.

To verify gH-pseudoconvexity, consider $y \in \mathbb{R}^n$ such that $\breve{F}(y) \prec \breve{F}(x)$ holds. Then $[1, 3]y_1^2 \oplus [2, 4]y_2 \prec \hat{0}$. Using property (iii) of (2), this is same as $[2, 4]y_2 \prec \ominus_{gH}[1, 3]y_1^2 \preceq \hat{0}$.

Next, $(y - x)^T \nabla \breve{F}(x) = y_1[0, 0] \oplus y_2[2, 4] = [2, 4]y_2 \prec \ominus_{gH}[1, 3]y_1^2 \preceq \hat{0}$.

Therefore, $\breve{F}(y) \prec \breve{F}(x) \Rightarrow (y - x)^T \nabla \breve{F}(x) \prec \hat{0}$ holds. Hence, $\breve{F}$ is gH-pseudoconvex at $(0, 0)^T$.

If all the parameters of interval functions are degenerate, then gH-pseudoconvexity of an interval function coincides with the pseudoconvexity of real-valued functions.

Definition 8 Suppose $\breve{F} : X \subseteq \mathbb{R}^n \to \mathbf{K}(\mathbb{R})$ be gH-differentiable. $\breve{F}$ is said to be component-wise gH-pseudoconvex at $\bar{x} \in X$ if $\breve{F}(x) \prec \breve{F}(\bar{x})$ implies that for each $x \in X$, $(x_i - \bar{x}_i)\frac{\partial \breve{F}(\bar{x})}{\partial x_i} \preceq \hat{0}$ hold $\forall i \in \Lambda_n$ with $(x_i - \bar{x}_i)\frac{\partial \breve{F}(\bar{x})}{\partial x_i} \prec \hat{0}$ for at least one $i \in \Lambda_n$.

$\breve{F}$ is said to be component-wise gH-pseudoconvex interval function on X for each $x, y \in X$, if $\breve{F}(y) \prec \breve{F}(x)$ implies that $(y_i - x_i)\frac{\partial \breve{F}(x)}{\partial x_i} \prec \hat{0}$ hold $\forall i \in \Lambda_n$ with $(y_i - x_i)\frac{\partial \breve{F}(x)}{\partial x_i} \prec \hat{0}$ for at least one $i \in \Lambda_n$.

Note 1

1. Every component-wise gH-pseudoconvex interval function is also gH-pseudoconvex. From Definition 8, if $\breve{F}$ is component-wise gH-pseudoconvex interval function on X, for each $x, y \in X$, then

$$\breve{F}(y) \prec \breve{F}(x)$$
$$\Rightarrow (y_i - x_i)\frac{\partial \breve{F}(x)}{\partial x_i} \preceq \hat{0} \text{ hold } \forall i \in \Lambda_n \text{ with } (y_i - x_i)\frac{\partial \breve{F}(x)}{\partial x_i} \prec \hat{0}$$
$$\text{for at least one } i \in \Lambda_n$$
$$\Rightarrow \sum_{i \in \Lambda_n}(y_i - x_i)\frac{\partial \breve{F}(x)}{\partial x_i} \prec \hat{0}$$
$$\Rightarrow (y - x)^T \nabla \breve{F}(x) \prec \hat{0}.$$

Hence $\breve{F}$ is gH-pseudoconvex on X.

2. Converse of the above result is not necessarily true. Consider Example 7. Suppose $\breve{F}(y) \prec \breve{F}(x)$ holds at $x = (0, 0)^T$. One may note that $(y_1 - x_1)\frac{\partial \breve{F}(0,0)}{\partial x_1} = \hat{0} \nprec \hat{0}$. Therefore, from Definition 8, it is clear that $\breve{F}$ is not component-wise gH-pseudoconvex.

4 Solution of Interval-Valued Minimization Problem

Consider

$$(\breve{P}) : \min_{x \in X \subseteq \mathbb{R}^n} \breve{F}(x)$$

where $\breve{F}$ is considered to be gH differentiable interval function on an open subset X of $\mathbb{R}^n$ throughout the article. For this purpose, the descent property of $\breve{F}$ is posed with respect to interval ordering in $\mathbf{K}(\mathbb{R})$ as defined in Sect. 2. Every $x \in R^n$ provides a set of objective values lying in a closed interval. Hence, $(\breve{P})$ may be treated as particular type of set-valued optimization problem. As a result, minimum value of $\breve{F}(x)$ cannot be determined following classical optimization techniques. The notion of efficient point is stated as follows according to the dominance relations of intervals,

Definition 9 (*Bhurjee and Panda* [3]) A point $x^* \in \mathbb{R}^n$ is called a weak efficient point of $\breve{F}$ if $\nexists\, x \in B(x^*, \delta)$ such that $\breve{F}(x) \prec \breve{F}(x^*)$. $x^* \in \mathbb{R}^n$ is called a local weak efficient point of $\breve{F}$ if it is a weak efficient point restricted to some neighbourhood of x^*.

Before focusing on the necessary and sufficient conditions for the existence of efficient point of $\breve{F}$, we characterize the notion of descent direction of $\breve{F}$ at a point.

Definition 10 (*Roy and Panda* [14]) A nonzero vector $d \in \mathbb{R}^n$ is said to be a descent direction of $\breve{F}$ at x if there exists some $\delta > 0$ such that $\breve{F}(x + \alpha d) \prec \breve{F}(x)$ $\forall \alpha \in (0, \delta)$.

Theorem 11 *(Roy and Panda [14]) Suppose $\breve{F} : \mathbb{R}^n \to \boldsymbol{K}(\mathbb{R})$ is gH-differentiable at $x \in \mathbb{R}^n$. If $d^T \nabla \breve{F}(x) \prec \hat{0}$ for some nonzero $d \in \mathbb{R}^n$, then d is a descent direction of $\breve{F}$ at x.*

Corollary 1 *If $d_i \frac{\partial \breve{F}(x)}{\partial x_i} \prec \hat{0}$ for each $i \in \Lambda_n$ at x then d is a descent direction at x.*

Proof Since $d_i \frac{\partial \breve{F}(x)}{\partial x_i} \prec \hat{0}$ for each $i \in \Lambda_n$ at x, $\sum_{i=1}^{n} d_i \frac{\partial \breve{F}(x)}{\partial x_i} \prec \hat{0}$. That is, $d^T \nabla \breve{F}(x) \prec \hat{0}$ at x. Hence, d is a descent direction at x.

Next critical point of $(\breve{P})$ will be defined from this concept of descent direction.

Definition 12 A point x^* is said to be critical point of $(\breve{P})$ if there exist $\lambda_i,\ \mu_i \geq 0$ with $(\lambda_i,\ \mu_i) \neq (0, 0)$ such that $\lambda_i \left(\frac{\partial \breve{F}(x^*)}{\partial x_i}\right)_{\min} + \mu_i \left(\frac{\partial \breve{F}(x^*)}{\partial x_i}\right)_{\max} = 0$ hold for each $i \in \Lambda_n$.

Theorem 13 $0_n \in \nabla \breve{F}(x^*)$ *for some* x^* *if and only if* x^* *is a critical point of* $(\breve{P})$ *where* 0_n *denotes the n dimensional zero vector in* $\mathbb{R}^n$.

Proof Suppose $0_n \in \nabla \breve{F}(x^*)$ hold. If possible let x^* is not a critical point of $(\breve{P})$. Then from Definition 12, there exist $\lambda_i,\ \mu_i \geq 0$ with $(\lambda_i,\ \mu_i) \neq (0,0)$ such that

$$\lambda_i \left(\frac{\partial \breve{F}(x^*)}{\partial x_i}\right)_{\min} + \mu_i \left(\frac{\partial \breve{F}(x^*)}{\partial x_i}\right)_{\max} = 0$$

has no solution for each $i \in \Lambda_n$.

Using Gordan's theorem of alternative, $\exists\ \alpha_i \in \mathbb{R}$ such that $\alpha_i \left(\frac{\partial \breve{F}(x^*)}{\partial x_i}\right)_{\min} < 0$ and $\alpha_i \left(\frac{\partial \breve{F}(x^*)}{\partial x_i}\right)_{\max} < 0$. Therefore, $\left(\frac{\partial \breve{F}(x^*)}{\partial x_i}\right)_{\min}$ and $\left(\frac{\partial \breve{F}(x^*)}{\partial x_i}\right)_{\max}$ are of same sign for each $i \in \Lambda_n$. Hence, $0 \notin \frac{\partial \breve{F}(x^*)}{\partial x_i}$ for each $i \in \Lambda_n$. Hence $0_n \notin \nabla \breve{F}(x^*)$, which contradicts our hypothesis. Hence x^* is a critical point of$(\breve{P})$.

Conversely, suppose x^* is a critical point of $(\breve{P})$ and $0_n \notin \nabla \breve{F}(x^*)$. Then $0 \notin \frac{\partial \breve{F}(x^*)}{\partial x_i}$ for at least one $i \in \Lambda_n$. Hence, $\left(\frac{\partial \breve{F}(x^*)}{\partial x_i}\right)_{\min}$ and $\left(\frac{\partial \breve{F}(x^*)}{\partial x_i}\right)_{\max}$ are of same sign for at least one $i \in \Lambda_n$. Therefore, there exists $\beta_i \in \mathbb{R}$ such that $\beta_i \frac{\partial \breve{F}(x^*)}{\partial x_i} \prec \hat{0}$ for at least one $i \in \Lambda_n$. From Gordan's alternative theorem, there exist $\lambda_i,\ \mu_i \geq 0$ with $(\lambda_i,\ \mu_i) \neq (0,0)$ such that $\lambda_i \left(\frac{\partial \breve{F}(x^*)}{\partial x_i}\right)_{\min} + \mu_i \left(\frac{\partial \breve{F}(x^*)}{\partial x_i}\right)_{\max} = 0$ has no solution for at least $i \in \Lambda_n$. Hence, x^* is not a critical point of $(\breve{P})$, which is a contradiction. Therefore $0_n \in \nabla \breve{F}(x^*)$.

Theorem 14 (Sufficient optimality condition) *Suppose* $\breve{F} : X \subseteq \mathbb{R}^n \to \boldsymbol{K}(\mathbb{R})$ *be gH-pseudoconvex interval function at a point* $x^* \in X$. *If* $0_n \in \nabla \breve{F}(x^*)$ *at* x^*, *then* x^* *is a local weak efficient solution of* $(\breve{P})$.

Proof If possible let x^* be not a local weak efficient solution of $(\breve{P})$. Then there exists $y \neq x$ in the neighbourhood of x such that $\breve{F}(y) \prec \breve{F}(x)$ hold. Due to gH-pseudoconvexity of $\breve{F}$,

$$\begin{aligned}
\breve{F}(y) \prec \breve{F}(x) &\Rightarrow (y-x)^T \nabla \breve{F}(x) \prec \hat{0} \\
&\Rightarrow \sum_{i \in \Lambda_n} (y_i - x_i) \frac{\partial \breve{F}(x)}{\partial x_i} \prec \hat{0} \\
&\Rightarrow \sum_{i \in \Lambda_n} \left((y_i - x_i) \frac{\partial \breve{F}(x)}{\partial x_i} \right)_{\min} < 0; \\
&\sum_{i \in \Lambda_n} \left((y_i - x_i) \frac{\partial \breve{F}(x)}{\partial x_i} \right)_{\max} < 0.
\end{aligned}$$

Therefore

$$0 > \sum_{i\in\Lambda_n}\left((y_i - x_i)\frac{\partial \breve{F}(x)}{\partial x_i}\right)_{\max} \geq \sum_{i\in\Lambda_n}(y_i - x_i)\left(\frac{\partial \breve{F}(x)}{\partial x_i}\right)_{\max} \tag{3}$$

Since $0_n \in \nabla\breve{F}(x)$, from Theorem 13, x is a critical point of $(\breve{P})$.

Then from Definition 12, there exist $\lambda_i,\ \mu_i \geq 0$ with $(\lambda_i,\ \mu_i) \neq (0,0)$ such that $\lambda_i\left(\frac{\partial \breve{F}(x)}{\partial x_i}\right)_{\min} + \mu_i\left(\frac{\partial \breve{F}(x)}{\partial x_i}\right)_{\max} = 0$ has solution for each $i \in \Lambda_n$. Therefore,

$$\sum_{i\in\Lambda_n}(y_i - x_i)\left(\lambda_i\left(\frac{\partial \breve{F}(x)}{\partial x_i}\right)_{\min} + \mu_i\left(\frac{\partial \breve{F}(x)}{\partial x_i}\right)_{\max}\right) = 0$$

Hence

$$\begin{aligned}0 &= \sum_{i\in\Lambda_n}(y_i - x_i)\left(\lambda_i\left(\frac{\partial \breve{F}(x)}{\partial x_i}\right)_{\min} + \mu_i\left(\frac{\partial \breve{F}(x)}{\partial x_i}\right)_{\max}\right)\\ &\leq \sum_{i\in\Lambda_n}(y_i - x_i)(\lambda_i + \mu_i)\left(\frac{\partial \breve{F}(x)}{\partial x_i}\right)_{\max}.\end{aligned}$$

Since $(\lambda_i + \mu_i) > 0$ for each $i \in \Lambda_n$, $\sum_{i\in\Lambda_n}(y_i - x_i)\left(\frac{\partial \breve{F}(x)}{\partial x_i}\right)_{\max} \geq 0$, which gives a contradiction to (3). Therefore x is a local weak efficient point of $(\breve{P})$.

Theorem 15 (Sufficient optimality condition) *Suppose $\breve{F} : X \subseteq \mathbb{R}^n \to \boldsymbol{K}(\mathbb{R})$ be component-wise gH-pseudoconvex interval function at a point $x^* \in X$. If $0_n \in \nabla\breve{F}(x^*)$ at x^*, then x^* is a local weak efficient solution of $(\breve{P})$.*

Proof If possible let x^* be not a local weak efficient solution of $(\breve{P})$. Then $\exists y$ in neighbourhood of x such that $\breve{F}(y) \prec \breve{F}(x)$ hold. Again due to component-wise gH-pseudoconvexity of $\breve{F}$ on X,

$$\breve{F}(y) \prec \breve{F}(x) \Rightarrow (y_i - x_i)\frac{\partial \breve{F}(x)}{\partial x_i} \preceq \hat{0} \text{ hold for each } i \in \Lambda_n \text{ with}$$

$$(y_i - x_i)\frac{\partial \breve{F}(x)}{\partial x_i} \prec \hat{0} \text{ hold for at least one } i \in \Lambda_n$$

$$\text{Therefore } (y_i - x_i)\left(\frac{\partial \breve{F}(x)}{\partial x_i}\right)_{\min} < 0;\ (y_i - x_i)\left(\frac{\partial \breve{F}(x)}{\partial x_i}\right)_{\max} < 0 \tag{4}$$

hold for at least one $i \in \Lambda_n$.

Since $0_n \in \nabla \breve{F}(x)$, from Theorem 13, x is a critical point of $\breve{F}$. Then from Definition 12, there exist $\lambda_i,\ \mu_i \geq 0$ with $(\lambda_i,\ \mu_i) \neq (0,0)$ such that $\lambda_i \left(\frac{\partial \breve{F}(x)}{\partial x_i}\right)_{\min} + \mu_i \left(\frac{\partial \breve{F}(x)}{\partial x_i}\right)_{\max} = 0$ has solution for each $i \in \Lambda_n$.

Therefore, for each $i \in \Lambda_n$,

$$0 = (y_i - x_i)\left(\lambda_i \left(\frac{\partial \breve{F}(x)}{\partial x_i}\right)_{\min} + \mu_i \left(\frac{\partial \breve{F}(x)}{\partial x_i}\right)_{\max}\right)$$
$$\leq (y_i - x_i)(\lambda_i + \mu_i)\left(\frac{\partial \breve{F}(x)}{\partial x_i}\right)_{\max}.$$

Since $(\lambda_i + \mu_i) > 0$ for each $i \in \Lambda_n$, $(y_i - x_i)\left(\frac{\partial \breve{F}(x)}{\partial x_i}\right)_{\max} \geq 0$, which contradicts (4).

Therefore, x is a local weak efficient point of $(\breve{P})$.

Example 16 Consider $\min\limits_{x \in X} \quad \breve{F}(x) = [1,4]x_1^2 \oplus [-1,0]x_1 \oplus [0,2]x_2$ where $X = [0,\infty) \times [0,\infty)$. One may verify that $\breve{F}$ is gH-differentiable on X and $\nabla \breve{F}(x) = \begin{pmatrix} [2,8]x_1 \oplus [-1,0] \\ [0,2] \end{pmatrix}$.

At $x = (0,0)^T$, $\nabla \breve{F}(x) = ([-1,0],[0,2])^T$. Therefore $(0,0)^T \in \nabla \breve{F}(x)$. Using Theorem 13, one can conclude that $x^* = (0,0)^T$ is a critical point of $\breve{F}$.

In order to show the gH-pseudoconvexity of $\breve{F}$ at x^*, consider $y \in X$ such that $\breve{F}(y) \prec \breve{F}(x^*)$ hold. Then $[1,4]y_1^2 \oplus [-1,0]y_1 \oplus [0,2]y_2 \prec \hat{0}$. Now $(y - x^*)^T \nabla \breve{F}(x^*) = [-1,0]y_1 \oplus [0,2]y_2 \prec \ominus_{gH}[1,4]_1^2 \preceq \hat{0}$.

Thus, $\breve{F}(y) \prec \breve{F}(x^*) \Rightarrow (y - x^*)^T \nabla \breve{F}(x^*) \prec \hat{0}$ hold. Therefore, $\breve{F}$ is gH-pseudoconvex at x^*. Hence, from Theorem 14, one can conclude that x^* is weak efficient point of $\breve{F}$. Weak efficient point is also a critical point.

5 Conclusion

In this article, pseudoconvex property of interval function is explored using gH-differentiability, which is not necessarily dependent on the differentiability of end-point functions. Necessary and sufficient condition for the existence of critical point of interval optimization problem is derived in inclusion form using descent property. Sufficient conditions are established using gH-pseudoconvex property. This concept can be further extended to constrained optimization problems with interval parameters, which is the future scope of the present contribution.

References

1. Alefeld G, Mayer G (2000) Interval analysis: theory and applications. J Comput Appl Math 121(1):421–464
2. Bede B, Stefanini L (2013) Generalized differentiability of fuzzy-valued functions. Fuzzy Sets Syst 230:119–141
3. Bhurjee AK, Panda G (2012) Efficient solution of interval optimization problem. Math Methods Oper Res 76(3):273–288
4. Caprani O, Madsen K (1980) Mean value forms in interval analysis. Computing 25(2):147–154
5. Chalco-Cano Y, Lodwick WA, Rufian-Lizana A (2013) Optimality conditions of type KKT for optimization problem with interval-valued objective function via generalized derivative. Fuzzy Optim Decis Mak 12(3):305–322
6. Hukuhara M (1967) Integration des applications mesurables dont la valeur est un compact convexe. Funkcial Ekvac 10:205–223
7. Ishibuchi H, Tanaka H (1990) Multiobjective programming in optimization of the interval objective function. Eur J Oper Res 48(2):219–225
8. Lupulescu V (2013) Hukuhara differentiability of interval-valued functions and interval differential equations on time scales. Inf Sci 248:50–67
9. Markov S (1979) Calculus for interval functions of a real variable. Computing 22(4):325–337
10. Osuna-Gómez R, Chalco-Cano Y, Hernández-Jiménez B, Ruiz-Garzón G (2015) Optimality conditions for generalized differentiable interval-valued functions. Inf Sci 321:136–146
11. Rall LB (1983) Mean value and Taylor forms in interval analysis. SIAM J Math Anal 14(2):223–238
12. Roy P, Panda G (2019) Expansion of function with uncertain parameters in higher dimension. In Applied mathematics and scientific computing. Birkhuser, Cham, pp 29–36
13. Roy P, Panda G (2019) Expansion of generalized hukuhara differentiable interval valued function. New Math Nat Comput 15(3):553–570
14. Roy P, Panda G (2020) Existence of solution of constrained interval optimization problems with regularity concept. RAIRO—Oper Res https://doi.org/10.1051/ro/2020060
15. Singh D, Dar BA, Kim D (2016) KKT optimality conditions in interval valued multiobjective programming with generalized differentiable functions. Eur J Oper Res 254(1):29–39
16. Stefanini L, Bede B (2009) Generalized hukuhara differentiability of interval-valued functions and interval differential equations. Nonlinear Anal: Theory, Methods Appl 71(3):1311–1328
17. Wu HC (2007) The karush-kuhn-tucker optimality conditions in an optimization problem with interval-valued objective function. Eur J Oper Res 176(1):46–59
18. Zhang J, Liu S, Li L, Feng Q (2014) The KKT optimality conditions in a class of generalized convex optimization problems with an interval-valued objective function. Optim Lett 8(2):607–631

Sufficient Conditions for the Uniqueness of Solution of the Weighted Norm Minimization Problem

K. Z. Najiya, Munnu Sonkar, and C. S. Sastry

Abstract Prior support constrained compressed sensing, achieved via the weighted norm minimization, has recently become popular due to its potential for applications. The weighted norm minimization problem is given by

$$(P_{p,w}) \ \min \|x\|_{p,w} \text{ subject to } y = Ax, \ \ p = 0, 1, \ \text{ and } w \in [0, 1],$$

where $\|x\|_{p,w} = w\|x_T\|_p + \|x_{T^c}\|_p, \ p = 0, 1$ with T representing the partial support information. For $p = 0$, uniqueness results are known only for $w = 0, 1$. For $p = 1$, there are some uniqueness results for general $w \in [0, 1]$. The work reported in this paper presents a set of conditions ensuring the uniqueness of the solution of $P_{p,w}$ for $p = 0, \ \forall w \in [0, 1]$ and an alternative set of conditions for $p = 1, \ \forall w \in [0, 1]$. Further, we compare and contrast our uniqueness results with the existing ones.

Keywords Compressed sensing · Prior support · Weighted 1-norm minimization

1 Introduction

In Compressed Sensing (CS), a sparse signal $x \in \mathbb{R}^n$ can be recovered from a small set of measurements $y \in \mathbb{R}^m$ satisfying $y = Ax$ with $k \ll m$, where k is the number of nonzero elements in x. The results guaranteeing the uniqueness of the recovery process depend on the restricted isometry property (RIP) of the matrix A [3, 4, 12]. In recent years, CS with a priori support information caught the attention of

Supported by UGC, India

K. Z. Najiya (✉) · M. Sonkar · C. S. Sastry
Indian Institute of Technology, 502285 Hyderabad, India
e-mail: ma17resch11004@iith.ac.in

M. Sonkar
e-mail: ma17resch01001@iith.ac.in

C. S. Sastry
e-mail: csastry@iith.ac.in

© The Author(s), under exclusive license to Springer Nature Singapore Pte Ltd. 2022

D. Giri et al. (eds.), *Proceedings of the Seventh International Conference on Mathematics and Computing*, Advances in Intelligent Systems and Computing 1412,
https://doi.org/10.1007/978-981-16-6890-6_51

several researchers [2, 5–7, 9, 10, 13, 14], to name a few. The weighted norm (accommodating prior support information) minimization aims at providing signals, satisfying the data constraint, that are sparse inside and sparsest outside a given prior support. In [13], the authors modified the 1-norm by taking zero weights on the known partial support, minimizing thereby the terms in the complement of prior support set. The results in [13] presented the uniqueness of the solution of weighted norm minimization under the stated conditions. When all the weights are set to 1, the weighted 0-norm and the weighted 1-norm problems coincide, respectively, with their standard 0-norm and 1-norm counterparts, whose exact recovery conditions were established in [1]. To the best of our knowledge, however, the uniqueness of the solution of the general weighted 0-norm is not proposed to date. The authors of [5, 8] established the stability of recovery in noisy-setting for weighted 1-norm minimization problem in terms of RIP and mutual coherence, respectively. In the noise-free case and when the signal is sufficiently sparse, these two results give the unique recovery condition for the weighted 1-norm problem for general $w \in [0, 1]$. The authors of [11] established the unique recovery of weighted 1-norm problem for $w \in [0, 1]$ in terms of weighted null space property.

The present work proposes sufficient conditions for the uniqueness of the solution of the weighted 0, 1-norm minimization problems. To our knowledge, the proposed unique recovery condition for the weighted 0-norm minimization for general weights in (0, 1). Is first of its kind. We analyze and show that our unique recovery conditions mostly coincide with the known ones in such cases as $w = 0, 1$. Since the uniqueness conditions of proposed as well as existing results involve RIP-parameters of different orders, different coefficients and different parameters, a general comparison is not possible in other cases. Nevertheless, using the probabilistic approximation of the RIP-constant for Gaussian *i.i.d* matrices [1], we single out the cases wherein our results possess weaker conditions than their existing counterparts that involve RIP-based argument.

The paper is organized as six sections. In Sects. 2 and 3, we provide basic introduction to Compressed Sensing and existing uniqueness results, respectively. In Sects. 4 and 5, we discuss the uniqueness results with general weights for 0-norm and 1-norm problems, respectively. In Sects. 6 and 7, we provide an analysis and comparison of uniqueness conditions, and the concluding remarks, respectively.

2 Basics of Compressed Sensing

For simplicity in notation, we represent the set $\{1, 2, \ldots, n\}$ as $[n]$. For $A \in \mathbf{R}^{m\times n}$ with $m << n$, suppose $y = Ax$. One may recover the sparsest solution of this system from the following minimization problem:

$$(P_0)\quad \min \|x\|_0 \quad \textit{subject to} \quad y = Ax, \tag{1}$$

where $\|x\|_0 = |\{i : x_i \neq 0\}|$. Since P_0 becomes NP-hard as the dimension increases, the convex relaxation of P_0 problem was proposed [1] as

$$(P_1) \quad \min \|x\|_1 \quad \textit{subject to} \quad y = Ax. \tag{2}$$

The coherence $\mu(A)$ of a matrix A is the largest normalized absolute inner product between pairs columns of it. The k-th Restricted Isometry Constant (k-RIC), δ_k, of a matrix A is the smallest real number such that

$$(1-\delta_k)\|x\|_2^2 \leq \|Ax\|_2^2 \leq (1+\delta_k)\|x\|_2^2, \ \forall x \text{ with } \|x\|_0 \leq k.$$

The restricted orthogonality constant $\theta_{s,\tilde{s}}$ of a matrix A is the smallest real number such that

$$|\eta' A_T' A_{\tilde{T}} \tilde{\eta}| \leq \theta_{s,\tilde{s}} \|\eta\|_2 \|\tilde{\eta}\|_2,$$

for all disjoint sets T and $\tilde{T}$ with $|T| \leq s$ and $|\tilde{T}| \leq \tilde{s}$ such that $s + \tilde{s} \leq n$ and for all vectors $\eta \in \mathbb{R}^{|T|}$ and $\tilde{\eta} \in \mathbb{R}^{|\tilde{T}|}$. Here, A_T denotes the sub matrix of columns of A restricted to the indices in $T \subseteq [n]$. For simplicity, we denote $\theta_s := \theta_{s,s}$. In [1], E. Candes and T. Tao provided the following conditions for the exact recovery of x from the pair (A, y) in terms of RIC for (1) and (2):

Theorem 1 (Candes et al. *[1]*): *Suppose that* $s \geq 1$ *is such that*

$$\delta_{2s} < 1 \tag{3}$$

and let $N \subseteq [n]$ *be such that* $|N| \leq s$. *Let* $y := Ax$, *where* x *is an arbitrary vector supported on* N. *Then* x *is the unique minimizer to (1) so that* x *can be reconstructed from knowledge of the vector* y *(and* a_i *'s).* □

Theorem 2 (Candes et al. *[1]*): *Suppose that* $s \geq 1$ *is such that*

$$\delta_s + \theta_{s,s} + \theta_{2s,s} < 1 \tag{4}$$

and let x *be a real vector supported on a set* $N \subseteq [n]$ *obeying* $|N| \leq s$. *Put* $y := Ax$. *Then* x *is unique minimizer to (2).* □

The following result is relevant to the objective of the present work.

Lemma 1 (Candes et al. *[1]*): *Let* $s \geq 1$ *be such that* $\delta_s + \theta_{s,2s} < 1$, *and* c *be a real vector supported on* $N \subseteq [n]$ *obeying* $|N| \leq s$. *Then there exists a vector* $\gamma \in \mathbb{R}^n$ *such that* $\gamma' a_i = c_i$ *for all* $i \in N$ *where* a_i *is the* i^{th} *column of a matrix* $A \in \mathbb{R}^{m \times n}$. *Furthermore,* γ *obeys*

$$| < \gamma, a_i > | \leq \frac{\theta_s}{(1 - \delta_s - \theta_{s,2s})\sqrt{s}} . \|c\|, \ \forall i \notin N. \tag{5}$$

3 CS with Partial Support Constraint

It may be noted that the reconstruction method given by P_1 in (2) is nonadaptive as no information about x is used in P_1. It can, however, be made partially adaptive by imposing constraints on the support of the solution to be obtained. In [5, 8, 13] (and the references therein), the authors modified the cost function of P_1 by incorporating the partial support information into the reconstruction process as detailed below.

Consider that T is the known partial support information of signal x. Here T is considered in the general sense that it can have an error part which corresponds to the complement of support of x. In [13], the authors modified the P_0 problem by considering zero weights in T and posed it as follows:

$$(P_{0,0})\ \min \|x_{T^c}\|_0 \textit{ subject to } y = Ax. \tag{6}$$

This problem recovers the signal that satisfies the data constraint, whose support is sparsest outside T. The following result in [13] establishes the uniqueness of (6).

Theorem 3 (Vasawani et al. [13]): *Given a sparse vector x with support $N = T \cup \Delta/\Delta_e$ where Δ and T are unknown and known disjoint supports, respectively, and Δ_e is the error in known support such that $\Delta_e \subseteq T$. Consider reconstructing it from $y = Ax$ by solving (6). Then x is the unique minimizer of (6) if $\delta_{k+2u} < 1$,where $k := |T|$ and $u := |\Delta|$.* □

In [13], the convex relaxation of (6) was considered as

$$(P_{1,0})\ \min \|x_{T^c}\|_1 \textit{ subject to } y = Ax. \tag{7}$$

The unique recovery of (7) was established as follows:

Theorem 4 (Vasawani et al. [13]): *Given a sparse vector x whose support $N = T \cup \Delta/\Delta_e$ where Δ and T are unknown and known disjoint supports, respectively, and Δ_e is the error in known support such that $\Delta_e \subseteq T$. Consider reconstructing it from $y = Ax$ by solving (7). Then x is the unique minimizer of (7) if*

1. $\delta_{k+u} < 1$ *and* $\delta_{2u} + \delta_k + \theta^2_{k,2u} < 1$,
2. $\rho_k(2u, u) + \rho_k(u, u) < 1$, *with* $\rho_k(s, \tilde{s}) := \frac{\theta_{\tilde{s},s} + \frac{\theta_{\tilde{s},k}\theta_{s,k}}{1-\delta_k}}{1-\delta_s - \frac{\theta^2_{s,k}}{1-\delta_k}}$,

where $s := |N|$, $k := |T|$ and $u := |\Delta|$. □

Corollary 1 (Vasawani et al. [13]): *Given a sparse vector, x, whose support $N = T \cup \Delta/\Delta_e$ where Δ and T are unknown and known disjoint supports, respectively, and Δ_e is the error in known support such that $\Delta_e \subseteq T$. Consider reconstructing it from $y = Ax$ by solving (7). Then x is the unique minimizer of (7) if $u \le k$ and $\delta_{k+2u} < \frac{1}{5}$.* □

Since sparsity of a signal inside T is unconstrained in (6), the recovered signal may not be sparse in T. In order to recover a signal, satisfying the data constraint, which is in general sparse inside T and sparsest outside T, one may choose weights $w \in [0, 1]$ and propose the general weighted-zero-norm problem as

$$(P_{0,w}) \ \min \|x\|_{0,w} \ \textit{subject to} \ y = Ax, \tag{8}$$

where $\|x\|_{0,w} = w\|x_T\|_0 + \|x_{T^c}\|_0$. It may be noted that when $w = 0$, $P_{0,w}$ coincides with (6) and when $w = 1$, it coincides with the P_0 problem in (1). As stated in the previous section, the uniqueness results in these two cases were established in the literature. In [5], the authors convexified $P_{0,w}$ for a general weight $w \in [0, 1]$ and an arbitrary $T \subset [n]$ as

$$(P_{1,w}) \ \min \|x\|_{1,w} \ \textit{subject to} \ y = Ax, \tag{9}$$

where $\|x\|_{1,w} := \sum_i w_i|x_i|$ with $w_i = \begin{cases} w & \text{for } i \in T \\ 1 & \text{for } i \notin T \end{cases}$.

It was shown in [5] that a signal x can be stably and robustly recovered from $P_{1,w}$ problem in noisy case if at least 50% of the partial support information is accurate. The stability result proposed in [5] is as follows:

Theorem 5 (Friedlander et al. [5])*: Let $x \in \mathbf{R}^n$ and let x_k be its best k−term approximation supported on T_0. Let $T \subset [n]$ be an arbitrary set and define ρ, α such that $|T| = \rho k$ and $|T_0 \cap T| = \alpha\rho k$. Suppose that there exists an $a \in \frac{1}{k}\mathbb{Z}$, with $a \geq (1-\alpha)\rho$, $a > 1$, and the measurement matrix A has RIP with*

$$\delta_{ak} + \frac{a}{\gamma^2}\delta_{(a+1)k} < \frac{a}{\gamma^2} - 1, \tag{10}$$

where $\gamma = w + (1-w)\sqrt{1+\rho-2\alpha\rho}$ for some given $0 \leq w \leq 1$. Then the solution x^* to the (9) obeys

$$\|x^* - x\|_2 \leq C_0\epsilon + C_1\big(w\|x - x_k\|_1 + (1-w)\|x_{T^c \cap T_0^c}\big). \tag{11}$$

□

Theorem 5 implies the uniqueness of weighted 1-norm problem for general $w \in [0, 1]$ when $\epsilon = 0$ and x is k-sparse. The present work proposes an alternate set of conditions which ensures the unique recovery of $P_{1,w}$ problem. As such our conditions and the ones in Theorem 5 are not comparable. Nevertheless, we analyze and single out the cases where our uniqueness results involve less restrictive conditions. Further, we propose sufficient conditions for the uniqueness of the solution of $P_{0,w}$ problem as well.

4 Unique Recovery of Weighted 0-Norm Problem

Motivated by the results in [13], we present our uniqueness conditions for $P_{0,w}$ as follows:

Theorem 6 *Let x be a real sparse vector supported on $N \subseteq [n]$ with $|N| = s$ and $y = Ax$, where $A \in \mathbb{R}^{m\times n}$ with $m < n$. Let $T \subseteq [n]$ with $|T| = k$ and $\Delta_1 = T \cap N$ with $|\Delta_1| = t$ and $\Delta = T^c \cap N$ with $|\Delta| = u$. If*

$$\delta_{k+2u+\lceil wt \rceil} < 1, \tag{12}$$

then x is the unique minimizer to $P_{0,w}$ in (8) for $0 \leq w \leq 1$.

Proof Let $\tilde{x}$ be a minimizer of (8). Then $\|\tilde{x}\|_{0,w} \leq \|x\|_{0,w}$, which implies that $\|\tilde{x}_{T^c}\|_0 \leq w\|x_T\|_0 + \|x_{T^c}\|_0 - w\|\tilde{x}_T\|_0 \leq w\|x_T\|_0 + \|x_{T^c}\|_0 \leq wt + u$. Hence, $\tilde{x}_{T^c}$ has at most $wt + u$ number of nonzero elements. Therefore, $\tilde{x}$ remains supported on a subset of T of cardinality at most k and on a set $\tilde{\Delta} \subseteq T^c$ of cardinality at most $wt + u$. Similarly x is also supported on a subset $\Delta_1 \subseteq T$ of cardinality $t \leq k$ and on a set $\Delta \subseteq T^c$ of cardinality at most u. Then the support of $(\tilde{x} - x)$ remains contained in the union $T \cup \Delta \cup \tilde{\Delta}$, which is of cardinality at most $k + u + wt + u = k + 2u + wt$. Now $A(\tilde{x} - x) = 0$ reduces to $A_{T\cup\Delta\cup\tilde{\Delta}}(\tilde{x} - x) = 0$. As $0 < \delta_{k+2u+\lceil wt \rceil} < 1$, $A_{T\cup\Delta\cup\tilde{\Delta}}$ is a full rank matrix, which implies that $\tilde{x} = x$. □

Remark 1 Here, the ceiling operation $\lceil wt \rceil$ stands for the smallest integer greater than or equal to the real number wt.

Remark 2 When $w = 1$, $P_{0,w}$ in (8) coincides with P_0 in (1) and $k + 2u + wt = k - t + 2(t + u) = 2s + e$ with $e = |T \cap N^c|$. Further, if $T \subseteq N$, then $e = 0$. Hence, $\delta_{k+2u+\lceil wt \rceil} < 1$ coincides with the uniqueness condition $\delta_{2s} < 1$ of the P_0 problem. When $w = 0$, however, $P_{0,w}$ coincides with $P_{0,0}$ in (6) and the uniqueness condition in (12) of the $P_{0,w}$ coincides with $\delta_{k+2u} < 1$ of Theorem 3.

5 Unique Recovery of Weighted 1-Norm Problem

The following lemma helps us in realizing our objective.

Lemma 2 *Let $x \in \mathbb{R}^n$ be a real sparse vector supported on $N \subseteq [n]$ with $|N| = s$ and $A \in \mathbb{R}^{m\times n}$ with $m < n$. Let $c \in \mathbb{R}^n$ be such that*

$$c_i = \begin{cases} w.sgn(x_i) & \textit{for } i \in T \\ sgn(x_i) & \textit{for } i \in \Delta \\ 0 & \textit{otherwise}, \end{cases} \tag{13}$$

where $T \subseteq [n]$ *with* $|T| = k$, $\Delta = T^c \cap N$ *with* $|\Delta| = u$ *and* $w \in [0, 1]$. *If*

$$\left(\sqrt{\frac{kw^2 + u}{k + u}}\right)\theta_{k+u} + \delta_{k+u} + \theta_{k+u,2(k+u)} < 1, \tag{14}$$

then there exists a vector $\gamma \in \mathbb{R}^n$ *such that*

1. $\gamma' a_i = w.sgn(x_i)$ *for* $i \in T$
2. $\gamma' a_i = sgn(x_i)$ *for* $i \in \Delta$
3. $|\gamma' a_i| < 1$ *for* $i \in (T \cup \Delta)^c$.

Proof Since $\delta_{k+u} + \theta_{k+u,2(k+u)} < 1$ follows from (14), Lemma 1 implies that there exists a vector $\gamma \in \mathbb{R}^n$ such that $\gamma' a_i = c_i$ for $i \in T \cup \Delta$, that is, $\gamma' a_i = w.sgn(x_i)$ for $i \in T$ and $\gamma' a_i = sgn(x_i)$ for $i \in \Delta$. From (5) and (14), we have

$$\begin{aligned} |\gamma' a_i| &\le \frac{\theta_{k+u} \, \|c\|}{(1 - \delta_{k+u} - \theta_{k+u,2(k+u)})\sqrt{k+u})} \\ &= \frac{\theta_{k+u}(\sqrt{kw^2 + u})}{(1 - \delta_{k+u} - \theta_{k+u,2(k+u)})\sqrt{k+u}} < 1. \end{aligned} \tag{15}$$

□

The following result summarizes the uniqueness of solution of $P_{1,w}$ problem, whose proof is motivated by the results in [1].

Theorem 7 *Let* x *be a real sparse vector supported on* $N \subseteq [n]$ *with* $|N| = s$ *and* $y = Ax$, *where* $A \in \mathbb{R}^{m \times n}$ *with* $m < n$. *Let* $T \subseteq [n]$ *with* $|T| = k$ *and* $\Delta = T^c \cap N$ *with* $|\Delta| = u$. *If*

$$\left(\sqrt{\frac{kw^2 + u}{k + u}}\right)\theta_{k+u} + \delta_{k+u} + \theta_{k+u,2(k+u)} < 1, \tag{16}$$

then x *is the unique minimizer to* $P_{1,w}$ *in (9) for* $0 \le w \le 1$.

Proof By standard convex arguments, there exists one minimizer $\tilde{x}$ to the problem (9), which implies that $\|\tilde{x}\|_{1,w} \le \|x\|_{1,w}$. Note that $x_i = 0$ for $i \in (T \cup N)^c \subseteq N^c$. We have

$$
\begin{aligned}
\|\tilde{x}\|_{1,w} &= \sum_{i\in T} w|\tilde{x}_i| + \sum_{i\in T^c} |\tilde{x}_i| \\
&= \sum_{i\in T} w|\tilde{x}_i| + \sum_{i\in \Delta} |\tilde{x}_i| + \sum_{i\in (T\cup\Delta)^c} |\tilde{x}_i| \\
&= \sum_{i\in T} w|x_i + \tilde{x}_i - x_i| + \sum_{i\in \Delta} |x_i + \tilde{x}_i - x_i + \sum_{i\in (T\cup\Delta)^c} |\tilde{x}_i - x_i| \\
&\geq \sum_{i\in T} w.sgn(x_i)(x_i + \tilde{x}_i - x_i) + \sum_{i\in \Delta} sgn(x_i)(x_i + \tilde{x}_i - x_i) + \sum_{i\in (T\cup\Delta)^c} (\tilde{x}_i - x_i) \\
&= \sum_{i\in T} w|x_i| + \sum_{i\in \Delta} |x_i| + \sum_{i\in T} w.sgn(x_i)(\tilde{x}_i - x_i) + \sum_{i\in \Delta} sgn(x_i)(\tilde{x}_i - x_i) + \sum_{i\in (T\cup\Delta)^c} (\tilde{x}_i - x_i) \\
&\geq \|x\|_{1,w} + \sum_{i\in T} \gamma' a_i(\tilde{x}_i - x_i) + \sum_{i\in \Delta} \gamma' a_i(\tilde{x}_i - x_i) + \sum_{i\in (T\cup\Delta)^c} \gamma' a_i(\tilde{x}_i - x_i) \\
&= \|x\|_{1,w} + \gamma' A(\tilde{x} - x) = \|x\|_{1,w}.
\end{aligned} \tag{17}
$$

In the above chain of steps, the vector $\gamma \in \mathbb{R}^n$ is supposed to satisfy the following properties:

1. $\gamma' a_i = w.sgn(x_i)$ for $i \in T$
2. $\gamma' a_i = sgn(x_i)$ for $i \in \Delta$
3. $|\gamma' a_i| < 1$ for $i \in (T \cup \Delta)^c$.

In view of (16), the existence of such a vector γ is guaranteed by Lemma 2. From (17), it follows that $\|\tilde{x}\|_{1,w} = \|x\|_{1,w}$. Consequently, all the inequalities in (17) must be equalities. But then $\sum_{i\in(T\cup\Delta)^c} |\tilde{x}_i| = \sum_{i\in(T\cup\Delta)^c} (\gamma' a_i)\tilde{x}_i$ implies that $\tilde{x}_i = 0$ on $(T \cup \Delta)^c$ as $|\gamma' a_i| < 1$ on $(T \cup \Delta)^c$. Now $Ax = A\tilde{x}$ reduces to $A_{T\cup\Delta}(x - \tilde{x}) = 0$. By (16), we have $\delta_{k+u} < 1$ which implies that $\tilde{x}_i = x_i$ on $T \cup \Delta$. Thus, $\tilde{x} = x$ follows. □

6 Analysis and Comparison of Uniqueness Conditions

In this section, we compare the unique recovery conditions of the present work to the ones in [1, 5, 13]. Since our results are based on RIP-constants, their comparison to the results based on the mutual coherence [8] and weighted null space property [11] does not look plausible. In view of this, we make a comparison to those results that involve RIP-based arguments only. When $w = 1$, the weighted 1-norm problem coincides with the standard 1-norm problem in (2) and $k + u = t + u + k - t = s + e$, where $t = |T \cap N|$ and $e = |T \cap N^c|$. Further, if $T \subseteq N$, then $e = 0$. In this case, $k + u$ coincides with s and the uniqueness condition (16) of Theorem 7 coincides with the uniqueness condition $\theta_s + \delta_s + \theta_{s,2s} < 1$ of the standard 1-norm problem.

We now turn toward comparing Theorem 5 and our results. As stated already, in a special case where x is k-sparse and ϵ is zero, Theorem 5 also results in unique recovery conditions. In this case, T_0 in Theorem 5 coincides with the support N of the vector x. Incorporating the notation used in Theorem 5 (that is, $|N| = k$, $|T| = \rho k$, and $|T \cap N| = \alpha\rho k$), we see that our uniqueness condition reduces to the following:

$$\sqrt{\frac{\rho w^2 + 1 - \alpha\rho}{\rho + 1 - \alpha\rho}}\theta_{ck} + \delta_{ck} + \theta_{ck,2ck} < 1, \tag{18}$$

where $c = \rho + 1 - \alpha\rho$. As such (10) and (18) cannot be compared. In order to express (18) in terms of δ_k alone, we use $\theta_{s,\tilde{s}} \leq \delta_{s+\tilde{s}}$. Note that (18) is independent of a. Comparing these two conditions with a general a is difficult. The parameter a in (10) should be such that $a > 1$, $a \geq (1-\alpha)\rho$ and $ak \in \mathbb{Z}$. Consider the trivial case $a = 2$. In this case, (10) and a sufficient condition for (18), respectively, become

$$\frac{b}{2 - b^2}\delta_{2k} + \frac{2}{2 - b^2}\delta_{3k} < 1, \tag{19}$$

$$\delta_{ck} + \sqrt{\frac{\rho w^2 + 1 - \alpha\rho}{\rho + 1 - \alpha\rho}}\delta_{2ck} + \delta_{3ck} < 1, \tag{20}$$

where $b = (w + (1-w)\sqrt{\rho + 1 - 2\alpha\rho})^2$. A direct comparison between (19) and (20) does not look to make sense for a general matrix A. For the purpose of comparison, however, we consider a special case. In [1], it was proved that for an $m \times n$ matrix whose entries are i.i.d Gaussian with mean zero and variance $\frac{1}{m}$, for any $k < n$, the RIC-δ_k behaves like

$$\delta_k < -1 + (1 + f(r))^2 \tag{21}$$

with overwhelming probability where $f(r) = \sqrt{\frac{m}{n}}\big(\sqrt{r} + \sqrt{2H(r)}\,\big)$ *and* $H(r) = -rlog(r) - (1-r)log(1-r)$ with $r = \frac{k}{n}$. Consequently, (19) and (20) reduce, respectively, to the following:

$$\frac{b}{2 - b^2}(-1 + (1 + f(2r))^2) + \frac{2}{2 - b^2}(-1 + (1 + f(3r))^2) < 1, \tag{22}$$

and

$$(-1 + (1 + f(cr))^2) + \sqrt{\frac{\rho w^2 + 1 - \alpha\rho}{\rho + 1 - \alpha\rho}}.[-1 + (1 + f(2cr))^2] + (-1 + (1 + f(3cr))^2) < 1. \tag{23}$$

We refer to the bound in (22) as bound 1. We plot w against the maximum value of r (denoted $Max\ r$) such that the above conditions hold for various values of α, but for a fixed ρ. Note that the condition which satisfies the larger maximum value of r is less restrictive than the other. For our experiments, we fix $\frac{m}{n} = 0.4$ and $\rho = 1$ and 2. It can be verified that other values for $\frac{m}{n}$ lead to similar results.

Figures 1 and 2 show the cases where our sufficient conditions become weaker (in the sense of involving lighter conditions). It is clear from the plots that our conditions behave better than (22) for larger values of ρ. It may be seen that ρ takes larger values when the estimated prior support has a larger size than the actual support of the signal, which can be possible.

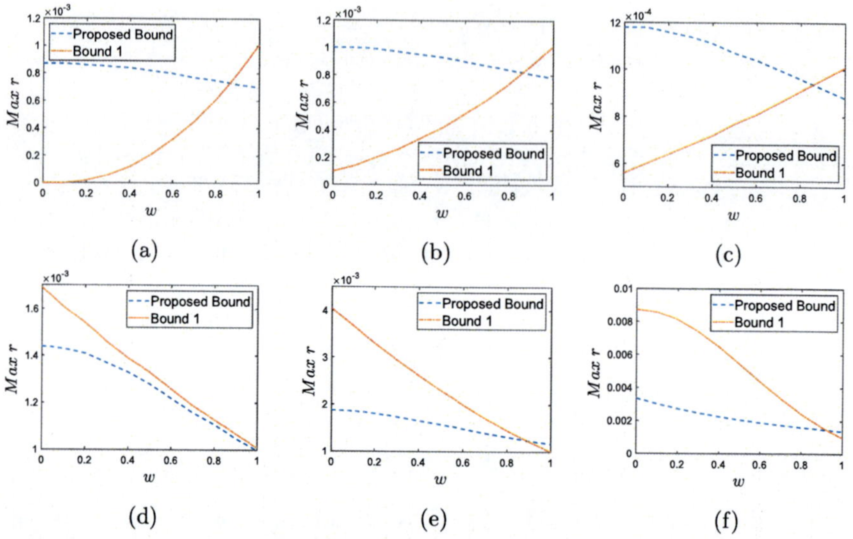

Fig. 1 w is plotted against $Max\ r$ for $\alpha = 0, 0.2, \cdots, 1$ for fixed $\rho = 1$. Note that for smaller values of α, our $Max\ r$ is larger for almost all values of w except for those being close to 1. For larger values of α, our $Max\ r$ is larger for w being close to 1

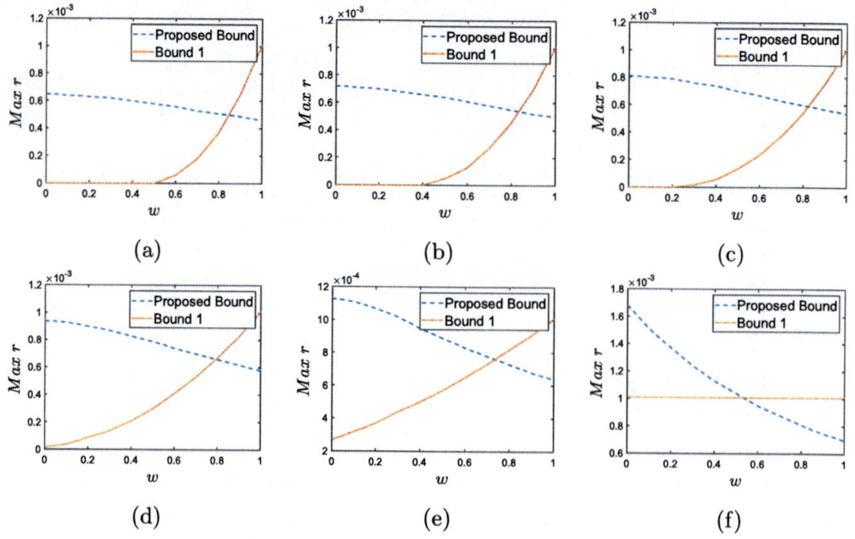

Fig. 2 w is plotted against $Max\ r$ for $\alpha = 0, 0.1, \cdots, 0.5$ for fixed $\rho = 2$. Note that for smaller values of α, our $Max\ r$ is larger for almost all values of w except for those being close to 1. For larger values of α, our $Max\ r$ is larger for smaller values of w

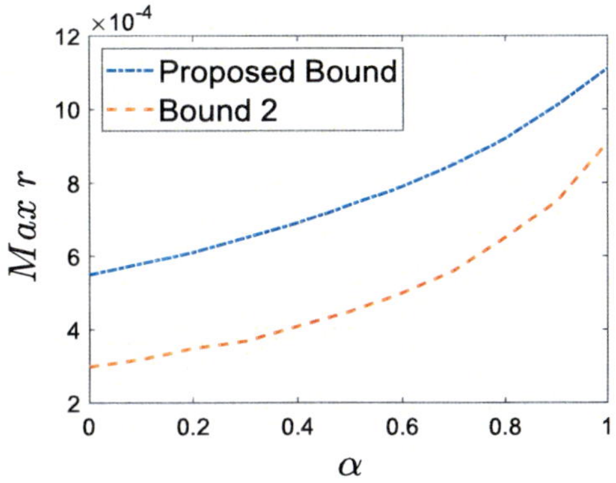

Fig. 3 Here $\rho = 1$, $\frac{m}{n} = 0.4$. It is clear that our $Max\ r$ is strictly greater than the bound 2 for all values of α. It can be verified that these plots behave similarly for other values of these parameters

We now compare our sufficient condition (by setting w to 0) to the one in Theorem 4, which essentially deals with (7). When $w = 0$, the uniqueness condition (16) reduces to

$$\left(\sqrt{\frac{u}{k+u}}\right)\theta_{k+u} + \delta_{k+u} + \theta_{k+u,2(k+u)} < 1. \tag{24}$$

As such, it is not possible to compare the above condition to the uniqueness condition of Theorem 4. This is because the proofs of both adopt different strategies. Nevertheless, in order to deduce a condition from (24) in terms of RIC (that is akin to the condition in Corollary 1), we use the inequality $\theta_{s,\tilde{s}} \leq \delta_{s+\tilde{s}}$. Again if $u \leq k$, then $\frac{u}{k+u} \leq \frac{1}{2}$. Hence, (24) holds if $(\frac{1}{\sqrt{2}} + 2)\delta_{3(k+u)} < 1$, that is, $\left(\frac{1+2\sqrt{2}}{\sqrt{2}}\right)\delta_{3(k+u)} < 1$. Since direct comparison between this and the sufficient condition given in Corollary 1 is not possible, as before for the purpose of comparison, we consider the Gaussian assumption on A along with a probabilistic approximation of the value of δ_k. Then the above condition reduces to $\left(\frac{1+2\sqrt{2}}{\sqrt{2}}\right)\delta_{3(\rho+1-\alpha\rho)k} < 1$ and the sufficient condition given by Corollary 1 reduces to $5\ \delta_{(\rho+2(1-\alpha\rho))k} < 1$. We refer the second bound as bound 2. We compare these bounds like in the previous case.

It is clear from Fig. 3 that our result involves weaker conditions than its counterpart in Theorem 4.

7 Conclusion

The current work proposed the conditions that guarantee the uniqueness of the solution of weighted 0-norm and weighted 1-norm minimization problems for $w \in [0, 1]$. Through an analysis of unique recovery conditions, we showed that our uniqueness conditions match with their known counterparts in the particular cases where (i). $w = 0, 1$ with 0-norm, (ii). $w = 1$ with 1-norm. As a direct comparison in other cases is not possible, we made a comparison by taking $i.i.d$ Gaussian matrices and using the approximate expression for δ_k and singled out the cases where our results involve weaker conditions.

Acknowledgements The first author is thankful to UGC, Govt. of India, (JRF/2016/409284), for its financial support. The second author is thankful for the support that he receives from MHRD, Government of India.

References

1. Candes EJ, Tao T (2005) Decoding by linear programming. IEEE Trans Inf Theory 51(12):4203–4215. https://doi.org/10.1109/TIT.2005.858979
2. Chen W, Ge H (2017) Recovery of signals by a weighted ℓ_2/ℓ_1 minimization under arbitrary prior support information. Signal Process 148(06). https://doi.org/10.1016/j.sigpro.2018.02.027
3. Elad M (2010) Sparse and redundant representations: from theory to applications in signal and image processing. Springer Science & Business Media
4. Foucart S, Rauhut H (2017) A mathematical introduction to compressive sensing. Bull Am Math 54:151–165
5. Friedlander MP, Mansour H, Saab R, Yilmaz Ö (2011) Recovering compressively sampled signals using partial support information. IEEE Trans Inf Theory 58(2):1122–1134
6. Ge H, Chen W (2019) An optimal recovery condition for sparse signals with partial support information via OMP. Circuits, Syst, Signal Process 38(07). https://doi.org/10.1007/s00034-018-01022-9
7. Jacques L (2010) A short note on compressed sensing with partially known signal support. Signal Process 90(12):3308–3312
8. Liu H, Song B, Tian F, Qin H (2014) Compressed sensing with partial support information: coherence-based performance guarantees and alternative direction method of multiplier reconstruction algorithm. IET Signal Process 8(7):749–758
9. Mahata K, Hyder MM (2018) Fast frequency estimation with prior information. IEEE Trans Signal Process 66(1):264–273. https://doi.org/10.1109/TSP.2017.2764865
10. Mangia M, Pareschi F, Rovatti R, Setti G (2020) Adapted compressed sensing: a game worth playing. IEEE Circuits Syst Mag 20(1):40–60. https://doi.org/10.1109/MCAS.2019.2961727
11. Mansour H, Saab R (2015) Weighted one-norm minimization with inaccurate support estimates: sharp analysis via the null-space property. In: 2015 IEEE international conference on acoustics, speech and signal processing (ICASSP). IEEE, pp 3317–3321
12. Naidu RR, Jampana P, Sastry CS (2016) Deterministic compressed sensing matrices: construction via Euler squares and applications. IEEE Trans Signal Process 64(14):3566–3575
13. Vaswani N, Lu W (2010) Modified-CS: modifying compressive sensing for problems with partially known support. IEEE Trans Signal Process 58(9):4595–4607
14. Von Borries R, Miosso CJ, Potes C (2007) Compressed sensing using prior information. In: 2007 2nd IEEE international workshop on computational advances in multi-sensor adaptive processing. IEEE, pp 121–124

On the Exponential Diophantine Equation $p^x + q^y = z^3$: Theorems and Conjectures

Ronald L. Aquino and Jerico B. Bacani

Abstract In this work, we solve the Diophantine equations $p^x + q^y = z^3$ in the set of nonnegative integers. We consider twin primes p and q having the following forms:

(a) twin primes of the form $8N + 1$ and $8N + 3$ for some $N \in \mathbb{N}_0$,
(b) twin primes of the form $8N + 3$ and $8N + 5$ for some $N \in \mathbb{N}_0$,
(c) twin primes of the form $8N + 5$ and $8N + 7$ for some $N \in \mathbb{N}_0$, and
(d) twin primes of the form $8N + 7$ and $8N + 9$ for some $N \in \mathbb{N}_0$.

Theorems are established. Conjectures are provided for further studies.

Keywords Exponential Diophantine equation · Twin primes · Mihailescu's theorem · Fermat's last theorem

1 Introduction

On the margin of his copy of the book *Arithmetica* by Diophantus, Pierre de Fermat of France once wrote the following text: *"Cubum autem in duos cubos, aut quadrato-quadratum in duos quadrato-quadratos, et generaliter nullam in infinitum ultra quadratum potestatem in duos eiusdem nominis fas est dividere cuius rei demonstrationem mirabilem sane detexi. Hanc marginis exiguitas non caperet"* [21]. This is translated as *"It is impossible for a cube to be the sum of two cubes, a fourth power to be the sum of two fourth powers, or in general for any number that is a power greater than the second to be the sum of two like powers. I have discovered a*

R. L. Aquino
Don Mariano Marcos Memorial State University, La Union 2504, Bacnotan, Philippines
e-mail: raquino@dmmmsu.edu.ph

R. L. Aquino · J. B. Bacani (✉)
University of the Philippines Baguio, 2600 Baguio City, Philippines
e-mail: jbbacani@up.edu.ph

© The Author(s), under exclusive license to Springer Nature Singapore Pte Ltd. 2022

D. Giri et al. (eds.), *Proceedings of the Seventh International Conference on Mathematics and Computing*, Advances in Intelligent Systems and Computing 1412,
https://doi.org/10.1007/978-981-16-6890-6_52

truly marvelous demonstration of this proposition that this margin is too narrow to contain." This proposition is about the nonsolvability of the *Diophantine equation*

$$x^n + y^n = z^n \tag{1}$$

for $n \geq 3$.

Equation (1) is just one of the many Diophantine equations that were studied by mathematicians across the globe. In the recent past, mathematicians got interested in solving exponential Diophantine equations (eDes) of the form $p^x + q^y = z^2$. In 2012, Sroysang [12] found out that there were no nonnegative integer solutions to the equation $31^x + 32^y = z^2$. In another paper, Sroysang [13] showed that the unique solution to the equation $3^x + 5^y = z^2$ is $(x, y, z) = (1, 0, 2)$. Peker and Cenberci [4] studied the equation $4^x + p^y = z^{2n}$. In a different paper, Peker and Cenberci [3] showed that the Diophantine equation $8^x + 19^y = z^2$ has no nonnegative integer solutions; however, Sroysang [14] disproved them by showing that (1, 0, 3) is a unique solution to the equation. In the same year, Rabago [9] proved that either of the Diophantine equations $3^x + 19^y = z^2$ or $3^x + 91^y = z^2$ has two nonnegative integer solutions. The solutions are (4, 1, 10) and (1, 0, 2), and (2, 1, 10) and (1, 0, 2), respectively.

In 2013, Sroysang published several results on eDes. He showed that the eDe $5^x + 7^y = z^2$ has no nonnegative integer solution [17]; $2^x + 3^y = z^2$ has only three solutions, namely (0, 1, 2), (3, 0, 3), and (4, 2, 5) [14]; $47^x + 49^y = z^2$ has no nonnegative integer solution [16]; and $23^x + 32^y = z^2$ and $89^x + 91^y = z^2$ have no nonnegative integer solutions [15, 20]. In the same year, Chotchaisthit [5] showed that $p^x + (p+1)^y = z^2$, where p is a Mersenne prime, has exactly two solutions (p, x, y, z); namely $(7, 0, 1, 3)$ and $(3, 2, 2, 5)$. He also proved that $(3, 0, 3)$ is the only nonnegative integer solution to the Diophantine equation $2^x + 11^y = z^2$ [6]. Rabago also published various works. He gave all solutions to the equation $8^x + 17^y = z^2$. The solutions are $(x, y, z) \in \{(1, 0, 3), (1, 1, 5), (2, 1, 9), (3, 1, 23)\}$ [10]. He also studied eDes of type $p^x + q^y = z^2$, where p and q are both odd primes. He proved that the eDes $17^x + 19^y = z^2$ and $71^x + 73^y = z^2$ have unique solutions (1, 1, 6) and (1, 1, 12), respectively, in the set of nonnegative integers [11].

In 2014, Sroysang [19] proved that the eDe $5^x + 43^y = z^2$ has no nonnegative integer solution. In the same year, he showed that (1, 0, 2) is a unique nonnegative integer solution to $3^x + 85^y = z^2$ [18].

In 2015, Bacani and Rabago [1] exhausted all solutions of the eDe $p^x + q^y = z^2$ for twin primes p and q.

In 2016, Khan [7] studied Diophantine equations of the type $p^x + 9^y = z^2$, where p is a prime number. Khan found out that $2^x + 9^y = z^2$ has exactly two solutions (3, 0, 3) and (4, 1, 5) and showed that $5^x + 9^y = z^2$ has no nonnegative integer solution.

In 2017, Burshtein [2] studied the eDe $p^x + q^y = z^2$, where $p = 2$ and $p = 3$. In his study, Burshtein found out that for each every integer $x \geq 1$ with $y = 1$ or $y = 2$, the equation has infinitely many solutions.

Motivated by vast researches mentioned above, this paper shall provide the non-negative integer solutions (x, y, z) of the eDe

$$p^x + q^y = z^3 \tag{2}$$

where p, q are twin primes of the following forms:

a. $p = 8N + 1$ and $q = 8N + 3$,
b. $p = 8N + 3$ and $q = 8N + 5$,
c. $p = 8N + 5$ and $q = 8N + 7$, and
d. $p = 8N + 7$ and $q = 8N + 9$,

where $N \in \mathbb{N}_0$.

2 Main Results

2.1 *Solving $p^x + q^y = z^3$, where $p = 8N + 1$ and $q = 8N + 3$*

Recall Catalan's conjecture, which dates back to 1844. It predicted that 8 and 9 were the only consecutive perfect powers among the positive integers. The claim was so simple but mathematicians found it too difficult to prove. In fact, it took one-and-a-half centuries to come up with a complete proof. It was only in 2002 that Swiss-Romanian mathematician Preda Mihailescu [8] was able to prove the theorem, and hence, the result was named after him.

Theorem 1 (Mihailescu's Theorem/Catalan's Conjecture) *Let $a, b, x,$ and y be positive integers satisfying the Diophantine equation $a^x - b^y = 1$. Then the 4-tuple $(a, b, x, y) = (3, 2, 2, 3)$ is the unique solution of this equation.*

The next theorem is also significant to our work. Its proof can be seen in any undergraduate texts in Number Theory.

Theorem 2 (Fermat's Last Theorem) *Let $a, b, c,$ and n be positive integers. Then, the equation $a^n + b^n = c^n$ admits no solution (a, b, c) if $n > 2$.*

By Fermat's Last Theorem, we can conclude the following:

1. $x^3 + y^3 = z^3$ has no positive integer solutions (x, y, z).
2. $s^3 + t^x = r^y$ has no positive integer solutions (s, t, r) if $x, y > 2$.
3. $27 + y^3 = w^3$ has no positive integer solutions (y, w).

The first preliminary result is useful in investigating the solution of (2).

Theorem 3 *The Diophantine equation*

$$1 + p^x = z^3, \tag{3}$$

where p is an odd prime, has a unique solution $(p, x, z) = (7, 1, 2)$ *in* $\mathbb{N}_0$.

Proof Let p be an odd prime number, and x and z be nonnegative integers.

Case 1 ($x = 0$): If $x = 0$, then eDe (3) becomes $z^3 = 2$, which implies that $z \notin \mathbb{N}_0$. Hence, there is no nonnegative integer solution.

Case 2 ($x = 1$): If $x = 1$, then eDe (3) becomes $p + 1 = z^3$. Subtracting 1 from both sides of (3), we get

$$p = z^3 - 1 = (z - 1)(z^2 + z + 1). \tag{4}$$

Since p is an odd prime whose factors are 1 and itself, then from eDe (4), it follows that $z - 1 = 1$ and $z^2 + z + 1 = p$. Solving these equations, we get $z = 2$ and $p = 7$.

Case 3 ($x > 1$): For $x > 1$, we will consider the equation $z^3 - p^x = 1$. However, this equation has no solution by Theorem 1.

Combining all cases, we see that $(7, 1, 2)$ is the only solution to (3). □

The succeeding discussion is about finding the solutions of (2) for different forms of twin primes p and q. We start with $p = 8N + 1$ and $q = 8N + 3$, $N \in \mathbb{N}$.

Theorem 4 *Consider twin primes p and q of the form* $8N + 1$ *and* $8N + 3$, *respectively, for some* $N \in \mathbb{N}$. *Then, eDe* (2) *has no nonnegative integer solution* (x, y, z).

Proof Let p and q be twin primes of the form $p = 8N + 1$ and $q = 8N + 3$, $N \in \mathbb{N}$. Also, assume that x, y, z are nonnegative integers. Applying the consequence of the binomial theorem, there should exist positive integers U, V such that $(8N + 1)^x = 8U + 1$ and $(8N + 3)^y = 8V + 3^y$. Thus, (2) becomes

$$8U + 8V + 1 + 3^y = z^3.$$

It is clear that z is even and so $z^3 \equiv 0 \pmod 8$. Hence, we have

$$1 + 3^y \equiv 0 (\text{mod } 8). \tag{5}$$

However, (5) has no solution because $3^y \equiv 3, 1 \pmod 8$ for any $y \in \mathbb{N}_0$.

Thus, there are no nonnegative integers (x, y, z) that would satisfy (2), whenever p and q are of the form $p = 8N + 1$ and $q = 8N + 3$, respectively. □

Using Theorem 4, the following eDes have no nonnegative integer solutions:

1. $17^x + 19^y = z^3$ $(N = 2)$,
2. $1049^x + 1051^y = z^3$ $(N = 131)$, and
3. $8009^x + 8011^y = z^3$ $(N = 1001)$.

2.2 *Solving* $p^x + q^y = z^3$*, where* $p = 8N + 3$ *and* $q = 8N + 5$

We have shown from Theorem 4 that eDe (2) does not admit nonnegative integer solutions (x, y, z) when twin primes are of the form $p = 8N + 1$ and $q = 8N + 3$. What about for other forms, such as twin primes of the form $p = 8N + 3$ and $q = 8N + 5$? To find out the answer, we first consider a special case. It focuses on the nonnegative integer solutions (x, y, z) of the particular Diophantine equation $3^x + 5^y = z^3$. We might think of this as the "simplest" case of (2).

Theorem 5 *The ordered triple* $(1, 1, 2)$ *is the unique nonnegative integer solution* (x, y, z) *of the Diophantine equation*

$$3^x + 5^y = z^3. \tag{6}$$

Proof Clearly, regardless of parity of x and y, eDe (6) is only possible for an even integer z. Since $1 + p^y = z^3$ only holds for $(p, x, z) = (7, 1, 2)$ over all odd primes p, and noting that $z^3 = 2$ is absurd, it suffices to consider only the case where $\min(x, y) \geq 1$. Moreover, it is easy to see that $3^1 + 5^1 = 2^3$. So, $(x, y, z) = (1, 1, 2)$ is a solution to (6).

Now the difficult part is to determine whether the given solution is unique or there are other triples (x, y, z) that satisfy (6). Nevertheless, we try to deal with the problem using the notion of modular arithmetic.

By taking modulo 8, we can reduce the problem to the case where x and y are both odd; i.e., we can focus on the following problem:

Find odd integers $u, v \geq 1$ and even integer z such that

$$3^u + 5^v = z^3. \tag{7}$$

Here, we note the following facts:

- If z is even, then $z^3 \equiv 1, 2, 3, 4, 5, 0 \pmod 6$;
- If u is odd, then $3^u \equiv 3 \pmod 6$; and
- If v is odd, then $5^v \equiv 5 \pmod 6$.

Hence, we deduce that $z = 2(3k + 1)$, for some $k \in \mathbb{N}$. Furthermore, we note that, for $i \in \mathbb{N}$, we have the following statements:

- $z^3 \equiv 1, 6, 0 \pmod 7$;
- $3^{2i-1} \equiv 3, 6, 5 \pmod 7$ when $i \equiv 1, 2, 0 \pmod 3$, respectively; and
- $5^{2i-1} \equiv 5, 6, 3 \pmod 7$ when $i \equiv 1, 2, 0 \pmod 3$, respectively.

From the above equivalences, we obtain the following observations:

- For $i \equiv 2 \pmod 3$, the Diophantine equations

$$3^{2i-1} + 5^v = z^3, \qquad 3^u + 5^{2i-1} = z^3,$$

where u and v are odd numbers, and z is an even number, have no solutions.
- For $i \equiv 0 \pmod 3$ and $j \equiv 1 \pmod 3$, the Diophantine equation

$$3^{2i-1} + 5^{2j-1} = z^3,$$

where z is even, has no solution.

We record the above results in the following lemmas.

Lemma 1 *For any odd integers k, x, y, the Diophantine equations $27^k + 5^y = z^3$ and $3^x + 125^k = z^3$ have no solution in $\mathbb{N}$.*

Lemma 2 *Let $s, t \in \mathbb{N}$ be such that $s \equiv 5 \pmod 6$ and $t \equiv 1 \pmod 6$. Then, the Diophantine equation $3^s + 5^t = z^3$ has no solution in $\mathbb{N}$.*

In addition to these lemmas, we also have the following valid statements:

- $(2(3k+1))^3 \equiv 8 \pmod 9$;
- Let u be an odd number greater than 1. Then, $3^u + 5^{2i-1} \equiv 5, 8, 2 \pmod 9$ if and only if $i \equiv 1, 2, 0 \pmod 3$, respectively.

These equivalences imply that the exponent of 5 must be of the form $3(2n+1)$, for $n \in \mathbb{N}$. This leads us to

$$3^{odd} + 125^{odd} = (6k+2)^3,$$

for some $k \in \mathbb{N}$. By virtue of Lemma 1, the above equations do not hold in $\mathbb{N}$. Thus, the only solution (x, y, z) to (6) is the triple $(1, 1, 2)$. □

The result of Theorem 5 is useful in determining the solutions (x, y, z) of the eDe (2), given twin primes p and q of the form $8N+3$ and $8N+5$, respectively, for $N \geq 0$. Results were not proven completely but verified numerically. So, instead of theorems, we would like to present some observations about the problem under consideration.

Conjecture 1 Let p, q be twin primes such that $p = 8N+3$ and $q = 8N+5$, for some $N \in \mathbb{N}$. Furthermore, suppose $2N+1$ is a perfect cube. Then the triple $(x, y, z) = (1, 1, 2\sqrt[3]{2N+1})$ is the solution to Eq. (2).

Table 1 shows the list of solutions (x, y, z) of (2), when twin primes are of the form $p = 8N+3$ and $q = 8N+5$. Here l is defined as $\sqrt[3]{2N+1}$. MATLAB software (version 2014) is used for the finite search of solutions. Since we are not using high-performance computing machine, we only bound x and y for up to 50 only. First, we provide values for p and q and compute for possible values of x, y, and z. Interestingly, results tell us that x and y are both 1 to have a positive integer z. To check that we are also having correct solutions, we also tried to fix x and y to 1, and we let the computer search for possible values of p and q of the desired form, as well as a positive integer z. Fortunately, we get the same primes p and q.

Table 1 List of Solutions (x, y, z) to eDe (2), where twin primes p and q are of the form $p = 8N + 3$ and $q = 8N + 5$ for some $N \in \mathbb{N}_0$

N	$2N + 1$	l	p	q	$x = y$	z
0	1	1	3	5	1	2
13	27	3	107	109	1	6
329251	658503	87	2634011	2634013	1	174
3707437	7414875	195	29659499	29659501	1	390
7174453	14348907	243	57395627	57395629	1	486
13099036	26198073	297	104792291	104792293	1	594

We have shown analytically and numerically that when twin primes are of the form $p = 8N + 3$ and $q = 8N + 5$, Eq. (2) may admit positive integer solutions (x, y, z). For numerical computation, due to computer memory issues, bounds for x and y are set only up to 50. With this constraint, we still found out that the solvable eDes under consideration have only one solution. For instance, $3^x + 5^y = z^3$ has only one solution $(1, 1, 2)$; $107^x + 109^y = z^3$ has the solution $(1, 1, 6)$; and $57395627^x + 57395629^y = z^3$ has the solution $(1, 1, 486)$. If the twin primes p and q are of the same forms but not given in Table 1, it means that their corresponding eDes do not have integer (positive) solutions. For example, to name a few, the following eDes do not have integer solutions: $11^x + 13^y = z^3$; $59^x + 61^y = z^3$; $1091^x + 1093^y = z^3$; and $7307^x + 7309^y = z^3$.

2.3 *Solving $p^x + q^y = z^3$, where $p = 8N + 5$ and $q = 8N + 7$*

We now study the solutions of (2), given twin primes of the form $p = 8N + 5$ and $q = 8N + 7$. We start with the case where $N = 0$.

Theorem 6 *The ordered triple* $(0, 1, 2)$ *is a unique nonnegative integer solution* (x, y, z) *of the Diophantine equation*

$$5^x + 7^y = z^3. \tag{8}$$

Proof Clearly, the triple $(0, 1, 2)$ is a solution to Eq. (8). Regardless of the parity of x and y, Eq. (8) is only possible for an even integer z. Since $1 + p^x = z^3$ only holds for $(p, x, z) = (7, 1, 2)$ over all odd primes p, and $z^3 = 2$ is absurd, it suffices to consider only the case where $\min(x, y) \geq 1$. By taking modulo 8, we can reduce the problem to the case where x is even and y is odd, reducing the problem to the case

$$5^{even} + 7^{odd} = z^3.$$

Now we consider modulo 14. The following facts are observed

$$z^3 \equiv 8, 6, 0 \pmod{14};$$

$$5^{2l} \equiv \begin{cases} 11 & \pmod{14} \text{ if } l \equiv 1 \pmod{3}, \\ 9 & \pmod{14} \text{ if } l \equiv 2 \pmod{3}, \\ 1 & \pmod{14} \text{ if } l \equiv 0 \pmod{3}. \end{cases}$$

$$7^{2j-1} \equiv 7 \pmod{14},$$

for $l, j \in \mathbb{N}$.

Hence, we deduce that $x = 6k$ for $k \in \mathbb{N}$. Now, we will consider the following manipulation:

$$5^{6k} + 7^y = z^3 \tag{9}$$
$$7^y = z^3 - 5^{6k} \tag{10}$$
$$7^y = (z - 5^{2k})(z^2 + (5^{2k})z + 5^{4k}). \tag{11}$$

From Eq. (11), we have

$$7^u = z - 5^{2k} \quad \text{and} \quad 7^{y-u} = z^2 + (5^{2k})z + 5^{4k}, \quad \text{for some} u \in \mathbb{N}_0.$$

We take the square of $7^u = z - 5^{2k}$ and subtract it from $7^{y-u} = z^2 + (5^{2k})z + 5^{4k}$ giving us

$$7^{y-u} - 7^{2u} = 3(5^{2k})z \tag{12}$$
$$7^{2u}(7^{y-3u} - 1) = 3(5^{2k})z. \tag{13}$$

Assuming $u > 0$, Eq. (13) implies $7|z$. But the original equation $5^x + 7^y = z^3$ would imply $7|5$, which is a contradiction. It remains to treat the case $u = 0$. If this is the case, then $z = 5^{2k} + 7^u = 5^{2k} + 1$. So $1 \equiv z^3 = 5^x + 7^y \equiv 7^y \pmod{5}$. This implies that $4|y$, a contradiction since y must be odd. Hence, the triple $(0, 1, 2)$ is a unique solution to Eq. (8). □

We now study another particular eDe—the eDe $3^x + 7^y = z^3$.

Theorem 7 *The Diophantine equation*

$$3^x + 7^y = z^3 \tag{14}$$

has a unique nonnegative integer solution $(x, y, z) = (0, 1, 2)$.

Proof First, show that there's a solution when $x = 0$, and that is $(0, 1, 2)$. Next, we assume x to be greater than 0. Note that $z^3 \equiv 1 \pmod{3}$. This implies that $z \equiv 1$

(mod 3). Also note that $3^x \equiv \pm 1 \pmod 4$. This implies that $3|x$. It means that there exists a $k \in \mathbb{N}$ such that $x = 3k$. It follows that $x > 1$.

Since we have established that $x > 1$, reducing now Eq. (14) modulo 9, we get

$$7^y \equiv z^3 \pmod 9.$$

However, $z \equiv 1 \pmod 3$ implies $z^3 \equiv 1 \pmod 9$. Thus, we obtain

$$7^y \equiv 1 \pmod 9.$$

This implies further that $3|y$ because 3 is the order of 7 mod 9.

With $3|x$ and $3|y$, by using Fermat's Last Theorem, we say that (14) has no solution (x, y, z) if $x > 0$. The conclusion follows. □

We have shown that the Diophantine equation $3^x + 7^y = z^3$ admits only a unique solution $(0, 1, 2)$. This result has something to do with the proof of the following observation. Since no complete proof will be given, we just consider it as a conjecture.

Conjecture 2 Let p, q be twin primes such that $p = 8N + 5$ and $q = 8N + 7$, for some $N \in \mathbb{N}_0$. Then, the only solution $(x, y, z) \in \mathbb{N}^3$ to Eq. (2) is the triple $(0, 1, 2)$; that is, $N = 0$.

We provide an initial justification about this observation.

Let N be fixed. If $N = 0$, it is clear that $(x, y, z) = (0, 1, 2)$ is a solution to (2). Now, we consider the case where $\min(N, x, y) \geq 1$. Suppose the equation holds for some integers N, x, y, z. Then, z is obviously even. Let $8N + 5 = 8M - 3$, where $M = N + 1$. By substitution and application of the consequence of binomial theorem, there should exist some positive integers U, V such that (2) is manipulated as follows:

$$\begin{aligned} (8M-3)^x + (8N+7)^y &= z^3 \\ 8U + 8V + (-3)^x + 7^y &= z^3. \end{aligned} \tag{15}$$

By taking modulo 8, we could deduce that Equation (2) only holds for x is even and y is odd. So,

$$3^{even} + 7^{odd} \equiv 0 \pmod 8.$$

In above equation, we established the fact that (2) only holds for even integers x and odd integers y. By virtue of Theorem 7, eDe (15) only holds for $(x, y, z) = (0, 1, 2)$ and will not hold in other modular classes. The authors are not sure if this justification would be sufficient to justify that the triple $(0, 1, 2)$ is indeed the only solution to (2), if $p = 8N + 5$ and $q = 8N + 7$ are given. If this is true, then this means that Eq. (2) is only solvable when $N = 0$.

Nevertheless, we determine the solutions of (2) by numerical computation. So far, no pairs give integer solutions except when $p = 5$ and $q = 7$. The solution obtained is (0, 1, 2). Based on Conjecture 2 and our numerical results, the following equations have no nonnegative integer solutions:

$$29^x + 31^y = z^3, \qquad 269^x + 271^y = z^3,$$

as well as all other equations $p^x + q^y = z^3$, whose p and q are of the form $p = 8N + 5$ and $q = 8N + 7$.

2.4 *Solving* $p^x + q^y = z^3$*, where* $p = 8N + 7$ *and* $q = 8N + 9$

We first consider the eDe (2), where $p = 7$ and $q = 9$. Note that $q = 9$ is a composite number. Thus, this is a case where p and q are not twin primes. We need this, however, to prove a result on twin primes of the form $p = 8N + 7$ and $q = 8N + 9$. We now state the result as follows:

Theorem 8 *The ordered triple* $(1, 0, 2)$ *is the unique nonnegative integer solution* (x, y, z) *of the Diophantine equation*

$$7^x + 9^y = z^3. \tag{16}$$

Proof Clearly, regardless of the parity of x and y in Eq. (16), z is always an even number. If $x = 0$, then we get no nonnegative integer solution (y, z) to $1 + 9^y = z^3$ by Theorem 1. If $y = 0$, then $7^x + 1 = z^3$ admits only one nonnegative integer solution $(x, z) = (1, 2)$ by Theorem 3. If x and y are both zero, then $z^3 = 2$, which has no integer solution. So, for $0 \leq \{x, y\} < 1$, we have $(1, 0, 2)$ as a solution to (16).

We next consider the case where $\min\{x, y\} \geq 1$. By taking modulo 4, we would lead to saying that x is odd.

Next, we consider the following facts:

$$even^3 \equiv 4, 8, 0 \pmod{12}; \qquad 7^{odd} \equiv 7 \pmod{12}; \qquad 9^y \equiv 9 \pmod{12}.$$

From here, we deduce that $z = 2(3k + 2)$, $k \in \mathbb{N}_0$. Moreover, we note that

$$(2(3k + 2))^3 \equiv 6, 8, 0 \pmod{14}; \quad 7^{odd} \equiv 7 \pmod{14};$$

$$9^y \equiv \begin{cases} 9 & \pmod{14} \text{ if } y \equiv 1 \pmod{3}, \\ 11 & \pmod{14} \text{ if } y \equiv 2 \pmod{3}, \\ 1 & \pmod{14} \text{ if } y \equiv 0 \pmod{3}. \end{cases}$$

Hence, we deduce that $y \equiv 0 \pmod{3}$. So, we let $y = 3l$, $l \in \mathbb{N}$.

Now, by considering the equation modulo 18, we obtain

$$(2(3k+2))^3 \equiv 10 \pmod{18}; \quad 9^{3l}9 \pmod{18};$$

$$7^x \equiv \begin{cases} 7 & \pmod{18} \text{ if } x \equiv 1 \pmod{3}, \\ 1 & \pmod{18} \text{ if } x \equiv 0 \pmod{3}, \\ 13 & \pmod{18} \text{ if } x \equiv 2 \pmod{3}. \end{cases}$$

Hence, the exponent of 7 is $x \equiv 0 \pmod{3}$. Let $x = 3(2m+1)$, $m \in \mathbb{N}_0$. Equation (16) becomes

$$7^{3(2m+1)} + 9^{3l} = z^3 \tag{17}$$

$$(7^{2m+1})^3 + (9^l)^3 = (2(3k+2))^3. \tag{18}$$

In summary, we get the structures of x, y, and z; that is, $x = 3(2m+1)$, $y = 3l$, *and* $z = 2(3k+2)$. By assumption $\min\{x, y\} \geq 1$. So, 7^{2m+1}, 9^l, and $2(3k+2)$ are all positive integers for all $m, l, k \in \mathbb{N}$. Applying Fermat's Last Theorem (Theorem 2), we conclude that there will be no nonnegative integers m, l, and k that would satisfy (18). Equivalently, we say that there are no positive integers (x, y, z) that satisfy (16), when $\min\{x, y\} \geq 1$.

Therefore, $(1, 0, 2)$ is the unique solution to (16). □

The following observation focuses on the solutions of (2), where $p = 8N + 7$ and $q = 8N + 9$, $N \in \mathbb{N}_0$.

Conjecture 3 Let p, q be twin primes such that $p = 8N + 7$ and $q = 8N + 9$, for some $N \in \mathbb{N}_0$. Then, the only solution (x, y, z) in $\mathbb{N}_0$ to Eq. (2) is the triple $(1, 1, 2\sqrt[3]{2N+2})$.

We may begin the justification of this observation by considering the following lines:

Let N be fixed. Clearly, $(x, y, z) = (1, 1, 2\sqrt[3]{2N+2})$ is a solution to (2). Hence, it suffices to consider the case $min(x, y) > 1$. Suppose the equation holds for some integers x, y, z. Then, z is obviously even. Then for some positive integers U and V, we have the following manipulations.

$$p^x + q^y = z^3,$$
$$(8N+7)^x + (8N+9)^y = z^3,$$
$$8U + 8V + 7^x + 9^y = z^3.$$

So,

$$7^x + 9^y \equiv 0 \pmod{8}.$$

Table 2 List of solutions (x, y, z) to eDe (2), where the given twin primes p and q are of the form $p = 8N + 7$ and $q = 8N + 9$ for some $N \in \mathbb{N}_0$

N	$2N + 2$	l	p	q	$x = y = 1$	z
71118323	142236648	522	568946591	568946593	1	1044
73598975	147197952	528	588791807	588791809	1	1056
107999999	216000000	600	863999999	864000001	1	1200
254084795	508169592	798	2032678367	2032678369	1	1596
559193435	1118386872	1038	4473547487	4473547489	1	2076
683815499	1367631000	1110	5470523999	5470524001	1	2220

Similar to what we have done previously, it can be shown that the above equation holds only for odd integer x. Thus, we have

$$7^{odd} + 9^y \equiv 0 \pmod 8). \tag{19}$$

The challenging task is to determine the general forms of x, y, and z, which might be expressed in terms of an integer and N. If the reader uses the modular technique, then he/she should determine an appropriate modulus to which the eDe (2) must be reduced. Afterwards, show by using Fermat's Last Theorem the non-existence of solutions when $\min\{x, y, z\} > 1$.

Due to the difficulty of the process, we just did numerical computations to somehow validate the claim.

Table 2 shows the list of some twin primes of the form $p = 8N + 7$ and $q = 8N + 9$, for which the eDe (2) has a solution (x, y, z).

By looking at the table, we see that when $p = 568946591$ and $q = 568946591$, eDe (2) has a solution $(1, 1, 1044)$. To check whether p and q are primes, you may use MATLAB or WolframAlpha. Note that the computer does not give any other values of x and y except for 1, as far as bound value is concerned (i.e., upper bound for x and y is 50).

Acknowledgements The authors would also like to convey their gratitude to Dr. Jerome Dimabayao and Dr. Julius Fergy Rabago for their assistance in polishing some proofs for this research study. The authors would also like to thank the University of the Philippines for its support in disseminating the results and in publishing the manuscript.

References

1. Bacani JB, Rabago JFT (2015) The complete set of solutions of the Diophantine equation $p^x + q^y = z^2$ for twin primes p and q. Int J Pure Appl Math 104(4):517–521. https://doi.org/10.12732/ijpam.v104i4.3

2. Burshtein N (2017) On the infinitude of solutions to the Diophantine equation $p^x + q^y = z^2$ when $p = 2$ and $p = 3$. Ann Pure Appl Math 13(2):207–2010. https://doi.org/10.22457/apam.v13n2a6
3. Peker B, Cenberci S (2012) On the Diophantine equations of $(2^n)^x + p^y = z^2$ type. Amer J Math Sci 1(1):195–199
4. Peker B, Cenberci S (2012) Solution of the Diophantine equation $4^x + p^y = z^{2n}$. Selcuk J Appl Math 13(2):31–34
5. Chotchaisthit S (2013) On the Diophantine equation $p^x + (p+1)^y = z^2$, where p is a Mersenne prime. Int J Pure Appl Math 88(2):169–172. https://doi.org/10.12732/ijpam.v88i2.2
6. Chotchaisthit S (2013) On the Diophantine equation $2^x + 11^y = z^2$. Maejo Int J Sci Technol 7(2):291–293. http://www.thaiscience.info/journals/Article/MJST/10896385.pdf
7. Khan MA, Rashid A, Uddin MS (2016) Non-negative integer solutions of two Diophantine equations $2^x + 9^y = z^2$ and $5^x + 9^y = z^2$. J Appl Math Phys 4(4):762–765. https://doi.org/10.4236/jamp.2016.44086
8. Metsänkylä T (2004) Catalan's conjecture: another old Diophantine problem solved. Bull Am Math Soc 41(1):43–57
9. Rabago JFT (2013) On two Diophantine equations $3^x + 19^y = z^2$ and $3^x + 91^y = z^2$. Int J Math Sci Comput 3(1):28–29
10. Rabago JFT (2013) On an open problem by B. Sroysang. Konuralp J Math 1(2):30–32. https://dergipark.org.tr/en/download/article-file/275620
11. Rabago JFT (2014) A note on two Diophantine equations $17^x + 19^y = z^2$ and $71^x + 73^y = z^2$. Math J Interdiscip Sci 2(1):19–24. https://doi.org/10.15415/mjis.2013.21002
12. Sroysang B (2012) On the Diophantine equation $31^x + 32^y = z^2$. Int J Pure Appl Math 81(4):609–612. https://ijpam.eu/contents/2012-81-4/8/8.pdf
13. Sroysang B (2012) On the Diophantine equation $3^x + 5^y = z^2$. Int J Pure Appl Math 81(4):605–608. https://ijpam.eu/contents/2012-81-4/7/7.pdf
14. Sroysang B (2013) More on the Diophantine equation $2^x + 3^y = z^2$. Int J Pure Appl Math 84(2):133–137. https://doi.org/10.12732/ijpam.v84i2.11
15. Sroysang B (2013) On the Diophantine equation $23^x + 32^y = z^2$. Int J Pure Appl Math 84(3):231–234. DOI:https://doi.org/10.12732/ijpam.v84i3.9
16. Sroysang B (2013) On the Diophantine equation $47^x + 49^y = z^2$. Int. J Pure Appl Math 89(2):279–282. https://doi.org/10.12732/ijpam.v89i2.11
17. Sroysang B (2013) On the Diophantine equation $5^x + 7^y = z^2$. Int J Pure Appl Math 89(1):115–118. DOI:https://doi.org/10.12732/ijpam.v89i1.14
18. Sroysang B (2014) More on the Diophantine equation $3^x + 85^y = z^2$. Int J Pure Appl Math 91(1):131–134. https://doi.org/10.12732/ijpam.v91i1.13
19. Sroysang B (2014) On the Diophantine equation $5^x + 43^y = z^2$. Int J. Pure Appl Math 91(4):537–540. https://doi.org/10.12732/ijpam.v91i4.10
20. Sroysang B (2013) On the Diophantine equation $89^x + 91^y = z^2$. Int J Pure Appl Math 89(2):283–286. https://doi.org/10.12732/ijpam.v89i2.12
21. Weisstein EW (2004) Fermat's last theorem. Wolfram Mathworld

On the Construction of 4 × 4 Lightweight Involutory MDS Matrices Over $\mathbb{F}_{2^8}$

Meltem Kurt Pehlivanoğlu, Fatma Büyüksaraçoğlu Sakalli, Sedat Akleylek, and Muharrem Tolga Sakalli

Abstract In this paper, we consider the problem of building lightweight maximum distance separable (MDS) linear layers for substitution permutation networks (SPNs) ciphers. We focus on the construction of 4×4 involutory MDS matrices which can be implemented by the lightest (to the best of our knowledge) exclusive OR (XOR) count over $\mathbb{F}_{2^8}$. We handle local optimization techniques as well as global optimization techniques, then we give 4×4 involutory MDS matrices which require the best implementation cost of 155 XOR gates, in terms of local optimization over $\mathbb{F}_{2^8}$. Moreover, we generate a new 4×4 involutory MDS matrix over $\mathbb{F}_{2^8}$ that can be implemented by only 90 XOR gates by using a global optimization technique.

Keywords MDS matrices · Involutory matrices · GHadamard matrix · Diffusion layer · Global optimization

1 Introduction

Confusion and diffusion are fundamental properties used in the design of secure block ciphers [1]. The SPN structure ensures to build secure block ciphers, and these networks have been well studied in the literature as in the standard Advanced Encryption Standard (AES) [2]. SPN ciphers include three main components which are confu-

M. Kurt Pehlivanoğlu (✉)
Kocaeli University, 41001 Kocaeli, Turkey
e-mail: meltem.kurt@kocaeli.edu.tr

F. Büyüksaraçoğlu Sakalli · M. Tolga Sakalli
Trakya University, 22030 Edirne, Turkey
e-mail: fbuyuksaracoglu@trakya.edu.tr

M. Tolga Sakalli
e-mail: tolga@trakya.edu.tr

S. Akleylek
Ondokuz Mayis University, 55200 Samsun, Turkey
e-mail: sedat.akleylek@omu.edu.tr

© The Author(s), under exclusive license to Springer Nature Singapore Pte Ltd. 2022

D. Giri et al. (eds.), *Proceedings of the Seventh International Conference on Mathematics and Computing*, Advances in Intelligent Systems and Computing 1412,
https://doi.org/10.1007/978-981-16-6890-6_53

sion layer (called substitution boxes (S-boxes)), diffusion layer, and key schedule algorithm. MDS matrices are the most preferable diffusion layers. When using an MDS matrix as a diffusion layer in a block cipher, it provides the maximum diffusion and achieves the desired number of active S-boxes. Moreover, it ensures secure design against differential and linear cryptanalysis [2]. However, it is a challenging task to construct lightweight MDS matrices used in ciphers targeting resource-constrained devices, especially involutory MDS ones which can be implemented by the same module in both encryption and decryption phases [3]. In particular, there are three construction methods to generate MDS matrices; direct construction methods (such as Cauchy matrices [5], Companion matrices [6], Vandermonde matrices [7], construction method given in [8]), search-based methods (such as recursive search [9], hybrid structures [4], some special matrix forms: Hadamard matrices, Circulant matrices, Circulant-like matrices, Toeplitz matrices, Toeplitz-like matrices [10]), and hybrid methods (such as Generalized Hadamard Matrix (GHadamard) form [11]). Moreover, additionally to these methods, there is also a different construction method [12] based on the ground field structure that generates isomorphic MDS matrices.

XOR count metric [13] is the number of XOR operations (gates) required for the hardware implementation of a diffusion matrix. The reduction of the number of gates used in the circuit reduces chip area and power consumption. The circuit depth which is the number of gates on the longest circuit path is another metric. The reduction of the depth ensures the design of faster circuits [14].

In the literature, there are two important hardware cost optimization techniques: local optimization and global optimization. The local optimization techniques are based on the selection of the matrix entries with low hardware cost (with minimum XOR count). In the literature, there are many local optimization techniques such as [4, 11, 15–18]. Before applying any global optimization, it is a good idea to find lighter involutory MDS matrices by using local optimization techniques. We choose the GHadamard matrix form for local optimization. It is a hybrid construction method that combines direct construction methods and search-based methods. If the GHadamard form is applied to a $k \times k$ involutory MDS matrix over $\mathbb{F}_{2^m}$, new $k \times k$ involutory MDS matrices can be generated. Namely, one $k \times k$ involutory MDS matrix over $\mathbb{F}_{2^m}$, defines totally $(2^m - 1)^{k-1}$ involutory MDS matrices, e.g., one 4×4 involutory MDS matrix over $\mathbb{F}_{2^8}$ generates $16,581,375$ ($\approx 2^{23.98}$) new 4×4 involutory MDS matrices over $\mathbb{F}_{2^8}$. It increases the chance of finding well-locally optimized matrices with minimum XOR count.

There are two important global optimization techniques which are cancellation-free programs and heuristic techniques. In 1997, Paar proposed PAAR1 and PAAR2 algorithms which eliminate iteratively common sub-expressions and these algorithms are also called cancellation-free programs [19]. In 2010, Boyar and Peralta [20] proposed a heuristic method that is slower than PAAR1 but it is good in practice in terms of finding implementations with lower XOR counts. In 2013, Boyar et al. [21] improved the results given in [20] by using a new heuristic method that allows cancellations. The authors used the SLP programs to optimize globally the implementation of the previously defined linear function, also showed that it is an

NP-hard problem. In 2019, Li et al. [22] modified the heuristic technique given in [21], it is simply called BP algorithm in this paper.

Our main research question is based on the fact given in [4], which is "hardware implementations suffer from the area requirement of the multiplication process over the Galois field (or finite field)". Therefore, we choose the finite field $\mathbb{F}_{2^8}$ to make it more challenging. We search new 4×4 involutory MDS matrices with lower XOR count and lower circuit depth which is superior to the AES diffusion matrix (called Mix Column matrix) over the same finite field $\mathbb{F}_{2^8}$.

1.1 Motivation and Contribution

The finding lightweight involutory MDS matrices with low XOR counts is a challenging problem [3]. Some special matrix forms are useful for finding involutory MDS matrices. Despite the fact that the search space can be reduced by using these matrix forms for small dimensions, the computation cost for checking the MDS condition is still too expensive for large dimensions [4]. Moreover, it is NP-hard to obtain the minimum number of XOR gates required for the hardware implementation of an (involutory) MDS matrix [3].

In this paper, we handle these two open problems, namely that we focus on the local optimization as well as global optimization of the 4×4 involutory MDS matrices over $\mathbb{F}_{2^8}$. We choose finite field $\mathbb{F}_{2^8}$, because it is really hard to generate involutory MDS matrices over this field while compared with $\mathbb{F}_{2^4}$ (in terms of computational time, cost and we have more huge search space).

- We first start with the construction of lighter 4×4 involutory MDS matrices requiring the minimum naive XOR count, i.e., local optimization. We generate new 4×4 involutory MDS matrices over $\mathbb{F}_{2^8}$ by using GHadamard matrix form.
- We give 4×4 involutory MDS matrices which require the best implementation cost of 155 XORs, in terms of local optimization over $\mathbb{F}_{2^8}$.
- We also apply global optimization techniques such as PAAR1, PAAR2, and BP to find the best XOR count result for 4×4 involutory MDS matrices over $\mathbb{F}_{2^8}$. We generate a new 4×4 involutory MDS matrix over $\mathbb{F}_{2^8}$ requires 90 XOR gates, which is the lightest one over this finite field.
- We give all of our experimental results (matrices in 32×32 binary representations with their optimization implementations) publicly available at https://github.com/mkurtpehlivanoglu/GHadamard_XORreduce.

1.2 Organization

In Sect. 2, some basic definitions are summarized. In Sect. 3, GHadamard matrix form is given. In Sect. 4, the experimental results by considering XOR count are presented for the new 4×4 involutory MDS matrices over $\mathbb{F}_{2^8}$ and these results are compared with the previous results given in the literature. The conclusion and future works are given in Sect. 5.

2 Preliminaries

In this paper, we focus on the construction of lightweight 4×4 involutory MDS matrices over the finite field $\mathbb{F}_{2^m}/p(x)$ constructed by using the irreducible polynomial $p(x)$. The degree of $p(x)$ is m and finite field $\mathbb{F}_{2^m}/p(x)$ has also 2^m elements. Any element over this finite field can be represented by $a_{m-1}\alpha^{m-1} + a_{m-2}\alpha^{m-2} + \cdots + a_1\alpha^1$ with $a_i \in \{0, 1\}$, where α is a primitive element of $\mathbb{F}_{2^m}$. In this paper, while denoting the finite field $\mathbb{F}_{2^m}/p(x)$, we use the hexadecimal notation of the irreducible polynomial $p(x)$, e.g., for $\mathbb{F}_{2^8}/\texttt{0x12d}$, 0x12d stands for the irreducible polynomial $p(x) = x^8 + x^5 + x^3 + x^2 + 1$.

MDS codes are used to construct MDS matrices. If an $[n, k, d]$ code C reaches the Singleton bound $d = n - k + 1$, C is called MDS code. MDS matrices have the maximum differential and linear branch number $k + 1$ for $k \times k$ dimension [11]. MDS matrices have some important properties given as follows:

- $\boldsymbol{M}$ is a $k \times k$ MDS matrix, if and only if every square sub-matrix of $\boldsymbol{M}$ is non-singular.
- If $\boldsymbol{M}$ is a $k \times k$ MDS matrix, the transpose matrix of $\boldsymbol{M}$ ($\boldsymbol{M}^T$) is also an MDS matrix.
- Let $c \in \mathbf{F}_{2^m}$ be a nonzero constant and let $\boldsymbol{M}$ be a $k \times k$ MDS matrix, then the multiplication of a row (or column) of $\boldsymbol{M}$ by c does not affect the MDS property of $\boldsymbol{M}$.

We evaluate the hardware cost of an MDS matrix as the number of XOR gates required in its implementation and it is known as the direct XOR count (or naive XOR count) metric. The XOR count of an element β over $\mathbb{F}_{2^m}/p(x)$ is the number of XOR gates required to implement the multiplication of β, more details can be found in [23].

Let a $k \times k$ MDS matrix over $\mathbb{F}_{2^m}/p(x)$, the XOR count of one row of this matrix can be computed by using the following formula (1):

$$\sum_{i=1}^{k} \gamma_i + (l - 1) \cdot m, \tag{1}$$

where γ_i is the XOR count of the i-th entry in the row, k is the dimension of the matrix, l is the number of nonzero elements in the row, and m is the dimension of the finite field, and also note that for all MDS matrices $l = k$ because all of the elements are nonzero.

Before the global optimization is applied, a $k \times k$ (involutory) MDS matrix over $\mathbb{F}_{2^m}/p(x)$ is transformed into its corresponding $mk \times mk$ binary matrix, the naive XOR count of this binary matrix is calculated by the number of 1s minus mk. After that, the $mk \times mk$ binary matrix is optimized by using global optimization techniques.

3 GHadamard Matrix Form

In this paper, we use the GHadamard matrix form to generate well-locally optimized 4×4 involutory MDS matrices with minimum XOR count over $\mathbb{F}_{2^8}$.

Definition 1 A $k \times k$ Hadamard matrix $\boldsymbol{H}$ over $\mathbb{F}_{2^m}$ with $k = 2^t$ for $t > 0$ can be expressed as follows (2):

$$\boldsymbol{H} = had(\boldsymbol{A}_0, \boldsymbol{A}_1) = \begin{bmatrix} \boldsymbol{A}_0 & \boldsymbol{A}_1 \\ \boldsymbol{A}_1 & \boldsymbol{A}_0 \end{bmatrix} \tag{2}$$

where sub-matrices $\boldsymbol{A}_0$ and $\boldsymbol{A}_1$ are also $2^{t-1} \times 2^{t-1}$ Hadamard matrices. While denoting a $k \times k$ Hadamard matrix, we use the notation had($a_0, a_1, \ldots, a_{k-1}$), where $a_i \in \mathbb{F}_{2^m}$ for $0 \leq i \leq k-1$. A 4×4 Hadamard matrix $\boldsymbol{H}$ can be given as follows (3):

$$\boldsymbol{H} = had(a_0, a_1, a_2, a_3) = \begin{bmatrix} a_0 & a_1 & a_2 & a_3 \\ a_1 & a_0 & a_3 & a_2 \\ a_2 & a_3 & a_0 & a_1 \\ a_3 & a_2 & a_1 & a_0 \end{bmatrix} \tag{3}$$

Definition 2 Let $\boldsymbol{H}$ = had(a_0, a_1) be a 2×2 Hadamard matrix over $\mathbb{F}_{2^m}$. Then, the 2×2 GHadamard matrix over $\mathbb{F}_{2^m}$ corresponding to $\boldsymbol{H}$ is as follows (4):

$$\boldsymbol{GH} = Ghad(a_0, a_1; b_1) = \begin{bmatrix} a_0 & a_1 b_1 \\ a_1 b_1^{-1} & a_0 \end{bmatrix} \tag{4}$$

where b_1 is any nonzero element of $\mathbb{F}_{2^m}$. A 4×4 GHadamard matrix $\boldsymbol{GH}$ = Ghad($a_0, a_1; b_1, a_2; b_2, a_3; b_3$) can be given as follows (5):

$$\boldsymbol{GH} = \begin{bmatrix} a_0 & a_1 b_1 & a_2 b_2 & a_3 b_3 \\ a_1 b_1^{-1} & a_0 & a_3 b_1^{-1} b_2 & a_2 b_1^{-1} b_3 \\ a_2 b_2^{-1} & a_3 b_2^{-1} b_1 & a_0 & a_1 b_2^{-1} b_3 \\ a_3 b_3^{-1} & a_2 b_3^{-1} b_1 & a_1 b_3^{-1} b_2 & a_0 \end{bmatrix} \tag{5}$$

In our previous paper [11], we proved that GHadamard matrix form ensures the property of Hadamard matrix, namely $\boldsymbol{H}^2 = c^2 \times \boldsymbol{I}$, where $c = \oplus_{i=0}^{k-1} a_i$ and $\boldsymbol{I}$ is the $k \times k$ identity matrix.

In the following section, we show that the matrices given in [24] can be generated by using GHadamard matrix form. We also generate new 4×4 involutory MDS matrices which require 155 XOR gates over $\mathbb{F}_{2^8}/\texttt{0x12d}$, in this way, we give the lightest result in terms of local optimization for this dimension. Moreover, we give globally optimized lightweight 4×4 involutory MDS matrix over $\mathbb{F}_{2^8}$ which requires only 90 XOR gates.

4 Experimental Results for 4 x 4 Involutory MDS Matrices Over $\mathbb{F}_{2^8}$

We give our experimental results under the following two subtitles.

4.1 Local Optimization Results

In Examples 1 and 2, we show that the 4×4 involutory MDS matrices over $\mathbb{F}_{2^8}/\texttt{0x12d}$ given in [24] can be generated by using 4×4 GHadamard matrix form.

Example 1 Let the element α over $\mathbb{F}_{2^8}/\texttt{0x12d}$ be a root of $\texttt{0x12d}$ (also primitive polynomial). Consider the 4×4 Hadamard involutory MDS matrix $\boldsymbol{H}_1$= had $(1, \alpha^{128}, \alpha^{109}, \alpha^{238})$ over $\mathbb{F}_{2^8}/\texttt{0x12d}$. GHadamard matrix $\boldsymbol{GH}_1$= Ghad $(1, \alpha^{128}; \alpha^{128}, \alpha^{109}; \alpha^{146}, \alpha^{238}; \alpha^{19})$ corresponding to $\boldsymbol{H}_1$ with parameters $b_1 = \alpha^{128}$, $b_2 = \alpha^{146}$, $b_3 = \alpha^{19}$ is given as follows (6):

$$\boldsymbol{GH}_1 = \begin{bmatrix} 1 & \alpha & 1 & \alpha^2 \\ 1 & 1 & \alpha & 1 \\ \alpha^{218} & \alpha^{220} & 1 & \alpha \\ \alpha^{219} & \alpha^{218} & 1 & 1 \end{bmatrix} \tag{6}$$

which is an involutory MDS matrix and can be implemented by $157(= 61 + 4 \times 3 \times 8)$ XOR gates. The same matrix can be considered as; let $\boldsymbol{A}_0 = \begin{bmatrix} 1 & \alpha \\ 1 & 1 \end{bmatrix}$ and $\boldsymbol{A}_1 = \begin{bmatrix} 1 & \alpha^2 \\ \alpha & 1 \end{bmatrix}$ then $\boldsymbol{GH}_1 = \begin{bmatrix} \boldsymbol{A}_0 & \boldsymbol{A}_1 \\ \alpha^{218}\boldsymbol{A}_1 & \boldsymbol{A}_0 \end{bmatrix}$.

Example 2 Let the element α over $\mathbb{F}_{2^8}/\texttt{0x12d}$ be a root of $\texttt{0x12d}$. Consider the 4×4 Hadamard involutory MDS matrix $\boldsymbol{H}_2$= had $(1, \alpha^{128}, \alpha^{238}, \alpha^{109})$ over $\mathbb{F}_{2^8}/\texttt{0x12d}$. GHadamard matrix $\boldsymbol{GH}_2$= Ghad $(1, \alpha^{128}; \alpha^{128}, \alpha^{238}; \alpha^{18}, \alpha^{109}; \alpha^{146})$ corresponding to $\boldsymbol{H}_2$ with parameters $b_1 = \alpha^{128}$, $b_2 = \alpha^{18}$, $b_3 = \alpha^{146}$ is given as follows (7):

$$GH_2 = \begin{bmatrix} 1 & \alpha & \alpha & 1 \\ 1 & 1 & \alpha^{254} & \alpha \\ \alpha^{220} & \alpha^{219} & 1 & \alpha \\ \alpha^{218} & \alpha^{220} & 1 & 1 \end{bmatrix} \tag{7}$$

which is an involutory MDS matrix and can be implemented by $156(= 60 + 4 \times 3 \times 8)$ XOR gates. The same matrix can be considered as; let $A_0 = \begin{bmatrix} 1 & \alpha \\ 1 & 1 \end{bmatrix}$ and $A_1 = \begin{bmatrix} \alpha & 1 \\ \alpha^{254} & \alpha \end{bmatrix}$ then $GH_2 = \begin{bmatrix} A_0 & A_1 \\ \alpha^{219} A_1 & A_0 \end{bmatrix}$.

In [11], we searched for 4 × 4 involutory MDS matrices over $\mathbb{F}_{2^8}$/0x1c3 and then we found a new involutory MDS matrix which can be implemented by 158 XOR gates, it was the lightest one. We emphasized that there might be lighter matrices for different finite field representations, so in this paper, we construct 4 × 4 involutory MDS matrices over $\mathbb{F}_{2^8}$/0x12d. Finally, we generate 16 different new 4 × 4 involutory MDS matrices over $\mathbb{F}_{2^8}$/0x12d with minimum XOR count 155 (best known-locally optimized) in GHadamard form (Ghad($a_0, a_1; b_1, a_2; b_2, a_3; b_3$)), which are presented in Table 1. We give one of them in Example 3.

Example 3 Let the element α over $\mathbb{F}_{2^8}$/0x12d be a root of 0x12d. Consider the 4 × 4 Hadamard involutory MDS matrix H_3= had $(1, \alpha^{128}, \alpha^{109}, \alpha^{238})$ over $\mathbb{F}_{2^8}$/0x12d. GHadamard matrix GH_3= Ghad $(1, \alpha^{128}; \alpha^{127}, \alpha^{109}; \alpha^{110}, \alpha^{238}; \alpha^{237})$ corresponding to H_3 with parameters $b_1 = \alpha^{127}$, $b_2 = \alpha^{110}$, $b_3 = \alpha^{237}$ is given as follows (8):

$$GH_3 = \begin{bmatrix} 1 & 1 & \alpha^{219} & \alpha^{220} \\ \alpha & 1 & \alpha^{221} & \alpha^{219} \\ \alpha^{254} & 1 & 1 & 1 \\ \alpha & \alpha^{254} & \alpha & 1 \end{bmatrix} \tag{8}$$

which is an involutory MDS matrix and can be implemented by only $155(= 59 + 4 \times 3 \times 8)$ XOR gates.

Table 1 4 × 4 involutory MDS Matrices in GHadamard form which can be implemented by 155 XORs over $\mathbb{F}_{2^8}$

a_0	a_1	b_1	a_2	b_2	a_3	b_3	a_0	a_1	b_1	a_2	b_2	a_3	b_3
1	α^{128}	α^{127}	α^{109}	α^{110}	α^{238}	α^{237}	1	α^{128}	α^{127}	α^{109}	α^{110}	α^{238}	α^{238}
1	α^{128}	α^{127}	α^{109}	α^{144}	α^{238}	α^{17}	1	α^{128}	α^{127}	α^{109}	α^{145}	α^{238}	α^{17}
1	α^{128}	α^{128}	α^{109}	α^{110}	α^{238}	α^{238}	1	α^{128}	α^{128}	α^{109}	α^{111}	α^{238}	α^{238}
1	α^{218}	α^{128}	α^{109}	α^{145}	α^{238}	α^{17}	1	α^{218}	α^{128}	α^{109}	α^{145}	α^{238}	α^{18}
1	α^{128}	α^{127}	α^{238}	α^{17}	α^{109}	α^{144}	1	α^{128}	α^{127}	α^{238}	α^{17}	α^{109}	α^{145}
1	α^{128}	α^{127}	α^{238}	α^{237}	α^{109}	α^{110}	1	α^{128}	α^{127}	α^{238}	α^{238}	α^{109}	α^{110}
1	α^{128}	α^{128}	α^{238}	α^{17}	α^{109}	α^{145}	1	α^{128}	α^{128}	α^{238}	α^{18}	α^{109}	α^{145}
1	α^{128}	α^{128}	α^{238}	α^{238}	α^{109}	α^{110}	1	α^{128}	α^{128}	α^{238}	α^{238}	α^{109}	α^{111}

4.2 Global Optimization Results

We also apply some global optimization techniques such as PAAR1, PAAR2, and BP to improve the best-known XOR count result for 4×4 involutory MDS matrices over $\mathbb{F}_{2^8}$. Firstly, we construct 4×4 involutory MDS matrices in GHadamard matrix form which require less then 161 XOR gates. After that, we eliminate matrices that can be implemented more than 110 XOR gates by using the PAAR1 algorithm. Then, we construct globally optimized matrices using PAAR2 and BP (for different circuit depths) algorithms. Finally, we generate a new lightest 4×4 involutory MDS matrix which requires 90 XOR gates for this finite field and present this matrix in the following Example 4.

Example 4 Let the element α over $\mathbb{F}_{2^8}/\texttt{0x12d}$ be a root of 0x12d. Consider the 4×4 Hadamard involutory MDS matrix $\boldsymbol{H}_4$= had $(1, \alpha^{128}, \alpha^{109}, \alpha^{238})$ over $\mathbb{F}_{2^8}/\texttt{0x12d}$. GHadamard matrix $\boldsymbol{GH}_4$= Ghad $(1, \alpha^{128}; \alpha^{127}, \alpha^{109}; \alpha^{109}, \alpha^{238}; \alpha^{236})$ corresponding to $\boldsymbol{H}_4$ with parameters $b_1 = \alpha^{127}$, $b_2 = \alpha^{109}$, $b_3 = \alpha^{236}$ is given as follows (9):

$$\boldsymbol{GH}_4 = \begin{bmatrix} 1 & 1 & \alpha^{218} & \alpha^{219} \\ \alpha & 1 & \alpha^{220} & \alpha^{218} \\ 1 & \alpha & 1 & 1 \\ \alpha^2 & 1 & \alpha & 1 \end{bmatrix} \tag{9}$$

which is an involutory MDS matrix and can be implemented by $157 (= 61 + 4 \times 3 \times 8)$ XOR gates. After using the BP algorithm, $\boldsymbol{GH}_4$ can be implemented by 98 XOR operations, 94 XOR operations, and 90 XOR operations for depth 4, depth 5, and depth 6, respectively. In Table 2, we compare our optimized results with matrices given in the literature which are defined in the same finite field. Table 2 shows that while $\boldsymbol{GH}_3$ can be implemented (locally optimized) with the lightest XOR count 155, $\boldsymbol{GH}_4$ also can be implemented (globally optimized) with the lightest XOR count 90 over $\mathbb{F}_{2^8}$.

In Table 3, we give the global optimization result of $\boldsymbol{GH}_4$ with 90 XORs and depth 6, where x_is $[(x_0, x_1, \ldots, x_{31})]$, y_is $[(y_0, y_1, \ldots, y_{31})]$ and t_js $[(t_1, t_2, \ldots, t_{90})]$ represent input bits, output bits and temporary variables, respectively. The values given in parentheses represent the related circuit depth.

5 Conclusion

In this paper, to the best our knowledge, we generate the lightest 4×4 involutory MDS matrices over $\mathbb{F}_{2^8}$. Finding lightweight 4×4 involutory MDS matrices over $\mathbb{F}_{2^8}$ finite field is really challenging problem because of the search space is too large. We firstly use the GHadamard matrix form, which is one of the local optimization

Table 2 Comparison of the 4 × 4 non-involutory MDS and involutory MDS matrices over $\mathbb{F}_{2^8}$

Ref.	Finite Field	Type	Naive XOR	PAAR1	PAAR2[a]	BP		
						depth 4	*depth 5*	*depth 6*
AES [2]	$\mathbb{F}_{2^8}$/0x11b	Circulant, non-inv.[b]	152	108	108	99	97	97
CLEFIA M_0 [25]	$\mathbb{F}_{2^8}$/0x11d	Hadamard, non-inv.	184	121	121	113	107	106
CLEFIA M_1 [25]	$\mathbb{F}_{2^8}$/0x11d	Hadamard, non-inv.	208	121	121	117	111	111
FOX MU4 [26]	$\mathbb{F}_{2^8}$/0x11b	non-inv.	219	144	143	140	137	137
Anubis [27]	$\mathbb{F}_{2^8}$/0x11d	Hadamard, inv.[c]	184	122	121	116	113	113
[4]	$\mathbb{F}_{2^8}$/0x165	Hadamard, inv.	160	104	104	99	99	99
Exam. 6 [11]	$\mathbb{F}_{2^8}$/0x1c3	GHadamard, inv.	158	121	119	113	111	111
Exam. 5.2 [24]	$\mathbb{F}_{2^8}$/0x12d	GHadamard, inv.	157	107	106	99	95	91
Exam. 5.2 [24]	$\mathbb{F}_{2^8}$/0x12d	GHadamard, inv.	156	106	106	102	97	94
GH_3	$\mathbb{F}_{2^8}$/0x12d	GHadamard, inv.	**155**	106	105	99	95	94
GH_4	$\mathbb{F}_{2^8}$/0x12d	GHadamard, inv.	157	107	107	**98**	**94**	**90**

[a] Stopped algorithm after eight hours runtime. [b] non-involutory. [c] involutory

techniques, to generate the well-locally optimized matrices which require less then XOR count 161. Then we select the initial threshold XOR count value as 110 and eliminate the matrices with higher ones than 110 by using the PAAR1 global optimization technique. Finally, we use PAAR2 and BP algorithms to find the lightest 4 × 4 involutory MDS matrices over $\mathbb{F}_{2^8}$.

As future work, we will focus on the construction of a new global optimization technique to generate lightweight MDS matrices over $\mathbb{F}_{2^8}$ with lower XOR count and minimum circuit depth. Moreover, the subfield constructions will be studied. These may give better results in terms of efficiency.

Table 3 The global optimization result of GH_4 with 90 XORs and depth 6, where x_is $[(x_0, x_1, \ldots, x_{31})]$, y_is $[(y_0, y_1, \ldots, y_{31})]$, and t_js $[(t_1, t_2, \ldots, t_{90})]$ represent input bits, output bits and temporary variables, respectively, and the values are given in parentheses represent the related circuit depth

No	Operation	No	Operation	No	Operation
1	$t_1 = x_{22} + x_{29}$ (1)	31	$t_{31} = t_{25} + t_{29}$(4)	61	$t_{61} = t_{14} + t_{60} * y_9$ (5)
2	$t_2 = x_{21} + x_{28}$ (1)	32	$t_{32} = t_{28} + t_{31} * y_1$ (5)	62	$t_{62} = x_{10} + t_{13}$ (5)
3	$t_3 = x_{20} + x_{23}$ (1)	33	$t_{33} = x_{11} + t_5$ (3)	63	$t_{63} = t_{31} + t_{62} * y_{10}$ (6)
4	$t_4 = x_{18} + x_{25}$ (1)	34	$t_{34} = x_3 + t_{33} * y_3$ (4)	64	$t_{64} = x_{11} + x_{20}$ (1)
5	$t_5 = x_{31} + t_2$ (2)	35	$t_{35} = x_6 + x_{13}$ (1)	65	$t_{65} = t_{23} + t_{64} * y_{20}$ (2)
6	$t_6 = x_{19} + t_1$ (2)	36	$t_{36} = x_{30} + t_{35}$(2)	66	$t_{66} = x_{13} + t_{23}$ (2)
7	$t_7 = x_{16} + x_{26}$ (1)	37	$t_{37} = x_{22} + t_{36} * y_{22}$ (3)	67	$t_{67} = t_{24} + t_{66} * y_{13}$ (3)
8	$t_8 = t_6 + t_7$ (3)	38	$t_{38} = x_3 + t_9$ (3)	68	$t_{68} = x_{14} + t_8$ (4)
9	$t_9 = x_{30} + t_3$ (2)	39	$t_{39} = x_{12} + t_{38} * y_{12}$ (4)	69	$t_{69} = t_{15} + t_{68} * y_{14}$ (5)
10	$t_{10} = x_6 + x_{23}$ (1)	40	$t_{40} = t_{10} + t_{13}$ (5)	70	$t_{70} = x_{15} + x_{16}$ (1)
11	$t_{11} = x_{27} + t_9$ (3)	41	$t_{41} = x_{15} + t_{40} * y_{15}$ (6)	71	$t_{71} = t_{14} + t_{70} * y_{16}$ (2)
12	$t_{12} = x_{17} + x_{24}$ (1)	42	$t_{42} = x_{10} + t_{11}$ (4)	72	$t_{72} = x_{21} + t_{16}$ (3)
13	$t_{13} = t_{11} + t_{12}$ (4)	43	$t_{43} = x_7 + x_{14}$ (1)	73	$t_{73} = t_{18} + t_{72} * y_{21}$ (4)
14	$t_{14} = x_0 + x_{24}$ (1)	44	$t_{44} = x_{31} + t_{43}$ (2)	74	$t_{74} = x_{28} + t_{16}$ (3)
15	$t_{15} = x_5 + x_{22}$ (1)	45	$t_{45} = x_{23} + t_{44} * y_{23}$ (3)	75	$t_{75} = t_{21} + t_{74} * y_{28}$ (4)
16	$t_{16} = x_{12} + t_1$ (2)	46	$t_{46} = x_0 + x_8$ (1)	76	$t_{76} = x_{29} + t_{36}$ (3)
17	$t_{17} = x_4 + t_{16} * y_4$ (3)	47	$t_{47} = t_{13} + t_{46} * y_0$ (5)	77	$t_{77} = t_{38} + t_{76} * y_{29}$ (4)
18	$t_{18} = x_{15} + t_{15}$ (2)	48	$t_{48} = x_2 + t_8$ (4)	78	$t_{78} = t_6 + t_{27}$ (3)
19	$t_{19} = x_{31} + t_{18} * y_{31}$ (3)	49	$t_{49} = t_{42} + t_{48} * y_2$ (5)	79	$t_{79} = t_{50} + t_{78} * y_{18}$ (6)
20	$t_{20} = x_2 + x_7$ (1)	50	$t_{50} = x_{15} + t_{48}$ (5)	80	$t_{80} = t_{36} + t_{67}$ (4)
21	$t_{21} = t_6 + t_{20}$ (3)	51	$t_{51} = t_{20} + t_{50} * y_7$ (6)	81	$t_{81} = t_{55} + t_{80} * y_{30}$ (6)
22	$t_{22} = x_{11} + t_{21} * y_{11}$ (4)	52	$t_{52} = x_5 + t_{36}$ (3)	82	$t_{82} = x_{15} + x_{19}$ (1)
23	$t_{23} = x_4 + x_{28}$ (1)	53	$t_{53} = t_{40} + t_{52} * y_5$ (6)	83	$t_{83} = t_{38} + t_{42}$ (5)
24	$t_{24} = x_7 + t_4$ (2)	54	$t_{54} = x_6 + t_{29}$ (4)	84	$t_{84} = t_{82} + t_{83} * y_{19}$ (6)
25	$t_{25} = x_1 + x_{25}$ (1)	55	$t_{55} = t_{44} + t_{54} * y_6$ (5)	85	$t_{85} = x_{26} + t_{14}$ (2)
26	$t_{26} = x_9 + x_{18}$ (1)	56	$t_{56} = x_8 + x_{17}$ (1)	86	$t_{86} = t_{42} + t_{85}$ (5)
27	$t_{27} = x_{16} + t_{26}$ (2)	57	$t_{57} = t_{25} + t_{56} * y_{17}$ (2)	87	$t_{87} = t_{40} + t_{86} * y_{26}$ (6)
28	$t_{28} = t_{24} + t_{27} * y_{25}$ (3)	58	$t_{58} = x_8 + x_{24}$ (1)	88	$t_{88} = x_{27} + t_{10}$ (2)
29	$t_{29} = t_5 + t_{24}$ (3)	59	$t_{59} = t_{10} + t_{58} * y_{24}$ (2)	89	$t_{89} = t_{31} + t_{33}$ (5)
30	$t_{30} = x_8 + t_{29} * y_8$ (4)	60	$t_{60} = x_9 + t_8$ (4)	90	$t_{90} = t_{88} + t_{89} * y_{27}$ (6)

Acknowledgements Meltem Kurt Pehlivanoğlu is partially supported by the Scientific Research Project Department of Kocaeli University under the Project No: 2017/078.

References

1. Shannon CE (1949) Communication theory of secrecy systems. Bell Syst Tech J 28:656–715
2. Daemen J, Rijmen V (2002) The design of Rijndael: AES- the advanced encryption standard. In: Information security and cryptography. 1st edn. Springer
3. Li S, Sun S, Li C, Wei Z, Hu L (2019) Constructing low-latency involutory MDS matrices with lightweight circuits. IACR Trans Symmetric Cryptol 1:84–117
4. Sim SM, Khoo K, Oggier F, Peyrin T (2015) Lightweight MDS involution matrices. In: Leander G (ed) Fast software encryption FSE 2015, LNCS, vol 9054. Springer, Heidelberg, pp 471–493. https://doi.org/10.1007/978-3-662-48116-5_23
5. Youssef AM, Mister S, Tavares SE (1997) On the design of linear transformation for substitution permutation encryption networks. In: Selected areas in cryptography (SAC). Ottawa Ontario, Canada, pp 40–48
6. Guo J, Peyrin T, Poschmann A (2011) The PHOTON family of lightweight hash functions. In: Rogaway P (ed) Advances in cryptology—CRYPTO 2011. CRYPTO 2011. LNCS, vol 6841. Springer, Heidelberg, pp 222–239. https://doi.org/10.1007/978-3-642-22792-9_13
7. Lacan J, Fimes J (2004) Systematic MDS erasure codes based on Vandermonde matrices. IEEE Commun Lett 8(9):570–572
8. Guzel GG, Sakalli MT, Akleylek S, Rijmen V, Cengellenmis Y (2019) A new matrix form to generate all 3 × 3 involutory MDS matrices over $\mathbb{F}_{2^m}$. Inf Process Lett 147:61–68
9. Sajadieh M, Dakhilalian M, Mala H, Sepehrdad P (2012) Recursive diffusion layers for block ciphers and hash functions. In: Canteaut A (ed) Fast software encryption. FSE 2012. LNCS, vol 7549. Springer, Heidelberg, pp 385–401. https://doi.org/10.1007/978-3-642-34047-5_22
10. Gupta KC, Pandey SK, Ray IG, Samanta S (2019) Cryptographically significant MDS matrices over finite fields: a brief survey and some generalized results. Adv Math Commun 13:779–843
11. Pehlivanoglu MK, Sakalli MT, Akleylek S, Duru N, Rijmen V (2018) Generalisation of Hadamard matrix to generate involutory MDS matrices for lightweight cryprography. IET Inf Secur 12(4):348–355
12. Sakalli MT, Akleylek S, Akkanat K, Rijmen V (2020) On the automorphisms and isomorphisms of MDS matrices and their efficient implementations. Turk J Electr Comput Sci 28(1):275–287
13. Khoo K, Peyrin T, Poschmann AY, Yap H (2014) FOAM: searching for hardware-optimal SPN structures and components with a fair comparison. In: Batina L, Robshaw M (eds) Cryptographic hardware and embedded systems—CHES 2014. CHES 2014. Lecture notes in computer science, vol 8731. Springer, Heidelberg, pp 433–450. https://doi.org/10.1007/978-3-662-44709-3_24
14. Boyar J, Find MG, Peralta R (2019) Small low-depth circuits for cryptographic application. Cryptogr Commun 11(109)
15. Beierle C, Kranz T, Leander G (2016) Lightweight multiplication in $GF(2^n)$ with applications to MDS matrices. In: Robshaw M, Katz J (eds) Advances in cryptology—CRYPTO 2016. CRYPTO 2016. Lecture notes in computer science, vol 9814. Springer, Heidelberg, pp 625–653. https://doi.org/10.1007/978-3-662-53018-4_23
16. Liu M, Sim SM (2016) Lightweight MDS generalized circulant matrices. In: Peyrin T (ed) Fast software encryption. FSE 2016. Lecture notes in computer science, vol 9783. Springer, Berlin, Heidelberg, pp 101–120. https://doi.org/10.1007/978-3-662-52993-5_6
17. Li C, Wang Q (2017) Design of lightweight linear diffusion layers from near-MDS matrices. IACR Trans Symmetric Cryptol 2017(1):129–155
18. Zhou L, Wang L, Sun Y (2018) On efficient constructions of lightweight MDS matrices. IACR Trans Symmetric Cryptol 2018(1):180–200

19. Paar C (1997) Optimized arithmetic for reed-solomon encoders. In: Proceedings of IEEE international symposium on information theory, p 250
20. Boyar J, Peralta R (2010) A new combinational logic minimization technique with applications to cryptology. In: Festa P (ed) Experimental algorithms. SEA 2010. LNCS, vol 6049. Springer, Heidelberg, pp 178–189. https://doi.org/10.1007/978-3-642-13193-6_16
21. Boyar J, Matthews P, Peralta R (2013) Logic minimization techniques with applications to cryptology. J Cryptol 26:280–312
22. Li S, Sun S, Li C, Wei Z, Hu L (2019) Constructing low-latency involutory MDS matrices with lightweight circuits. IACR Trans Symmetric Cryptol 2019(1):84–117
23. Sakalli FB, Aydin O, Tuncay G, Pehlivanoglu MK, Guzel GG, Sakalli MT (2020) On lightweight 4x4 MDS matrices over binary field extensions. Int J Inf Secur Sci 9(2):94–103
24. Bai J, Sun Y, Wang D (2020) On the construction of involutory MDS matrices over $\mathbb{F}_{2^m}$. J Syst Sci Complex 33:836–848
25. Shirai T, Shibutani K, Akishita T, Moriai S, Iwata T (2007) The 128-Bit blockcipher CLEFIA (extended abstract). In: Biryukov A (ed) Fast software encryption. FSE 2007. Lecture notes in computer science, vol 4593. Springer, Heidelberg, pp 181–195. https://doi.org/10.1007/978-3-540-74619-5_12
26. Junod P, Vaudenay S (2004) FOX : a new family of block ciphers. In: Handschuh H, Hasan MA (eds) Selected areas in cryptography. SAC 2004. LNCS, vol 3357. Springer, Heidelberg, pp 114–129. https://doi.org/10.1007/978-3-540-30564-4_8
27. Barreto P, Rijmen V (2000) The ANUBIS block cipher. In: First open NESSIE workshop

Factual Power Loss Reduction by Meadow Fritillary Butterfly Optimization Algorithm

Lenin Kanagasabai

Abstract In this work, Improved Meadow Fritillary Butterfly (IMB) optimization algorithm is designed for power loss reduction. In the Meadow Fritillary Butterfly optimization algorithm, the exploration method has two properties of Meadow Fritillary butterflies; in that butterfly adjusting operator in preliminary iterations remarkably directs the exploration procedure in the direction of the present most outstanding solution. To trounce this deficit, firefly's algorithm exploration method has been incorporated into the standard Meadow Fritillary optimization algorithm. Meadow Fritillary Butterfly (IMB) optimization algorithm validated in IEEE 30 and 57 bus test systems and Decline in loss has been attained.

Keywords Optimal Reactive Power · Transmission Loss · Improved Meadow Fritillary Butterfly optimization algorithm

1 Introduction

Real Power loss reduction is an important task in transmission and distribution. Copious conventional methods [1–6] are already utilized, but difficulties are found with constraints [7–9]. These decades Evolutionary techniques [10–19] are applied. In this work, the Improved Meadow Fritillary Butterfly (IMB) optimization algorithm has been applied. The Meadow Fritillary butterfly optimization algorithm exploration procedure possesses two properties of Meadow Fritillary butterflies: Levy flights and through female butterflies' offspring's are engendered. Standard Meadow Fritillary butterfly algorithm does not strike the exact part of the exploration space in preliminary iterations, so it leads to poor mean values with elevated scattering. In the projected algorithm, the exploration procedure has been enhanced by implementing the Firefly algorithm's search equation. In IEEE 30 and 57 bus test systems, Improved Meadow Fritillary Butterfly (IMB) optimization algorithm has been evaluated. Comparison of loss has been done with other standard reported algorithms.

L. Kanagasabai (✉)
Prasad V. Potluri Siddhartha Institute of Technology, Kanuru, Vijayawada, AP 520007, India

© The Author(s), under exclusive license to Springer Nature Singapore Pte Ltd. 2022
D. Giri et al. (eds.), *Proceedings of the Seventh International Conference on Mathematics and Computing*, Advances in Intelligent Systems and Computing 1412,
https://doi.org/10.1007/978-981-16-6890-6_54

The proposed Improved Meadow Fritillary Butterfly (IMB) optimization algorithm condensed the Power loss competently, and the Percentage of the power loss decline has been enhanced significantly.

2 Problem Formulation

Factual power loss decrease is mathematically defined as

$$F = P_L = \sum_{k \in Nbr} g_k \left(V_i^2 + V_j^2 - 2V_i V_j \cos\theta_{ij} \right) \tag{1}$$

$$F = P_L + \omega_v \times \text{Voltage Deviation} \tag{2}$$

$$\text{VoltageDeviation} = \sum_{i=1}^{Npq} |V_i - 1| \tag{3}$$

Equality and Inequality

$$P_G = P_D + P_L \tag{4}$$

$$P_{gslack}^{min} \leq P_{gslack} \leq P_{gslack}^{max} \tag{5}$$

$$Q_{gi}^{min} \leq Q_{gi} \leq Q_{gi}^{max}, i \in N_g \tag{6}$$

$$V_i^{min} \leq V_i \leq V_i^{max}, i \in N \tag{7}$$

$$T_i^{min} \leq T_i \leq T_i^{max}, i \in N_T \tag{8}$$

$$Q_c^{min} \leq Q_c \leq Q_C^{max}, i \in N_C \tag{9}$$

3 Improved Meadow Fritillary Butterfly Optimization Algorithm

The Meadow Fritillary butterfly optimization algorithm exploration procedure possesses two properties of Meadow Fritillary butterflies: Levy flights and through female butterflies' offspring's are engendered [20].

$$MF_{i,k}^{t+1} = MF1_{r1,k}^{t} \tag{10}$$

Computation of ratio “r” is done by

$$r = \text{Capricious}.p \tag{11}$$

Newly engendered Meadow Fritillary butterfly is computed by

$$MF_{i,k}^{t+1} = MF_{r2,k}^{t} \tag{12}$$

Meadow Fritillary butterfly exploration procedure is given by

$$MF_{j,k}^{t+1} = MF_{best,k}^{t} \tag{13}$$

Recently engendered Meadow Fritillary butterfly is computed by

$$MF_{i,k}^{t+1} = MF_{r3,k}^{t} \tag{14}$$

Adaptability rate is defined as

$$MF_{j,k}^{t+1} = MF_{j,k}^{t} + \gamma \times (AM_k - 0.464) \tag{15}$$

In the projected Improved Meadow Fritillary Butterfly (IMB) optimization algorithm, the exploration procedure is enhanced by implementing the Firefly algorithm’s search equation:

$$MF_i^{t+1} = MF_i^t + \beta_o r^{-\gamma r_{i,j}^2}\left(MF_j^t - MF_i^t\right) + \gamma(random - 0.464) \tag{16}$$

1. Initialization
2. Calculate the fitness value for each Meadow Fritillary butterfly
3. While t < Maximum Generations, do
4. According to the fitness value categorize the individuals in the population
5. Segregate the population
6. for i = 1 to NP1 (the entire butterflies in the subpopulation-1) do
7. By means of butterfly migration operator engender new-fangled individuals
8. End for
9. For j = 1 to NP2 (the entire butterflies in the subpopulation-2) do
10. If t < Maximum Generation 0:0.464 then engender subpopulation through customized butterfly adjusting operator or Else
11. By using butterfly adjusting operator engender new-fangled individuals in subpopulation 2
12. End if
13. End for
14. As one whole population combine the newly engendered subpopulations

Table 1 Comparison of values (IEEE 30 bus system)

Parameter	Base case [22]	MPSO [22]	PSO [23]	EP [24]	SARGA [24]	IMB
Percentage of reduction in power loss	0	8.4	7.4	6.6	8.3	9.28
Power loss (Mw)	17.55	16.07	16	16.0	16.09	15.921

Table 2 Comparison of loss (IEEE 57 bus system)

Parameter	Base case [22]	MPSO [22]	PSO [23]	CGA [24]	AGA [24]	IMB
Percentage of reduction in power loss	0	15.4	14.1	9.2	11.6	21.25
Power loss (Mw)	27.8	23.51	23	25	24.5	21.890

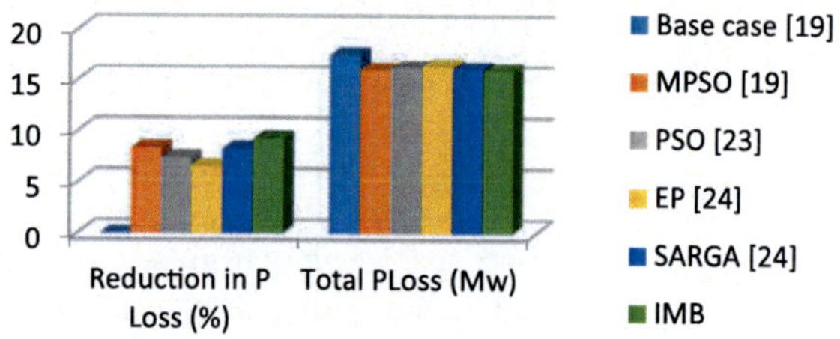

Fig. 1 Comparison of power loss (IEEE 30 bus system)

15. With reference to the recently modernized positions estimate the population
16. iteration counter t = 1
17. End while
18. Select the excellent entity in the entire population.

4 Simulation Results

The validity of the Improved Meadow Fritillary Butterfly (IMB) optimization algorithm is verified in IEEE 30 and 57 bus systems [21]. Tables 1 and 2 give the comparison results. Figures 1 and 2 show power loss comparison.

5 Conclusion

In this paper, the Improved Meadow Fritillary Butterfly (IMB) optimization algorithm proficiently reduced the power loss. Customary Meadow Fritillary butterfly algorithm does not strike the exact component of the exploration space in the preliminary phase,

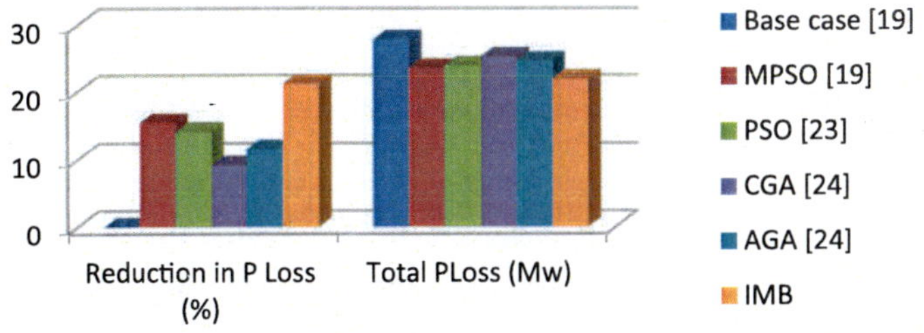

Fig. 2 Comparison of loss (IEEE 57 bus system)

so it leads to poor values with elevated scattering. In the projected algorithm, the exploration procedure has been enhanced by implementing the Firefly algorithm's search equation. In IEEE 30 and 57 bus test systems algorithm evaluated. Power loss reduction and percentage of reduction in power loss were enhanced.

References

1. Lee K (1984) Fuel-cost minimisation for both real and reactive-power dispatches. Proc Gen Trans Distrib Conf 131(3):85–93
2. Deeb N (1998) An efficient technique for reactive power dispatch using a revised linear programming approach. Electric Power Syst Res 15(2):121–134
3. Bjelogrlic M (1990) Application of Newton's optimal power flow in voltage/reactive power control. IEEE Trans Power Syst 5(4):1447–1454
4. Granville S (1994) Optimal reactive dispatch through interior point methods. IEEE Trans Power Syst 9(1):136–146
5. Grudinin N (1998) Reactive power optimization using successive quadratic programming method. IEEE Trans Power Syst 13(4):1219–1225
6. Khan I (2016) Distributed control algorithm for optimal reactive power control in power grids. Int J Electr Power Energy Syst 83(1):505–513
7. Reddy S (2017) Optimal reactive power scheduling using cuckoo search algorithm. Int J Electr Comput Eng 7(5):2349–2356
8. Reddy (2014) Faster evolutionary algorithm based optimal power flow using incremental variables. Electr Power Energy Syst 54(1):198–210
9. Dai C (2009) Seeker optimization algorithm for optimal reactive power dispatch. IEEE T Power Syst 24(3):1218–1231
10. Leke Z (2018) Concepts, methods, and performances of particle swarm optimization, backpropagation, and neural networks. Appl Comput Intell Soft Comput 7(1):1–7
11. Türkylmaz A (2015) A hybrid algorithm for total tardiness minimisation in flexible job shop: genetic algorithm with parallel VNS execution. Int J Prod Res 53(6):1832–1848
12. Rajput S (2016) ParasOptimization of benchmark functions and practical problems using Crow Search Algorithm. In: Proceedings of the 2016 fifth international conference on eco-friendly computing and communication systems (ICECCS), Bhopal, India, 8–9 Dec 2016; pp 73–78
13. Sahli Z (2014) Hybrid PSO-tabu search for the optimal reactive power dispatch problem. In: Proceedings of the IECON 2014–40th annual conference of the IEEE industrial electronics society, vol 40, pp 3536–3542. Dallas, TX, USA
14. Mouassa S (2017) Ant lion optimizer for solving optimal reactive power dispatch problem in power systems. Eng Sci Technol Int J 20(3):885–895

15. Mandal B (2013) Optimal reactive power dispatch using quasi-oppositional teaching learning based optimization. Int J Electr Power Energy Syst 53(1):123–134
16. Khazali H (2011) Optimal reactive power dispatch based on harmony search algorithm. Int J Electr Power Energy Syst 33(3):684–692
17. Tran H (2019) Finding optimal reactive power dispatch solutions by using a novel improved stochastic fractal search optimization algorithm. Telecommun Comput Electron Control 17(5):2517–2526
18. Polprasert J (2016) Optimal reactive power dispatch using improved pseudo-gradient search particle swarm optimization. Electric Power Compon Syst 44(5):518–532
19. Thanh L (2020) Optimal reactive power flow for large-scale power systems using an effective metaheuristic algorithm. Hindawi J Electr Comput Eng 20(1):1–11
20. Arora S, Singh S (2019) Butterfly optimization algorithm: a novel approach for global optimization. Soft Comput 23:715–734. https://doi.org/10.1007/s00500-018-3102-4
21. IEEE, The IEEE-test systems (1993). http://www.ee.washington.edu/trsearch/pstca/
22. Hussain AN (2018) Modified particle swarm optimization for solution of reactive power dispatch. Res J Appl Sci Eng Technol 15(8):316–327
23. Pandya S (2015) Particle swarm optimization based optimal reactive power dispatch. In Proceeding of the IEEE international conference on electrical, computer and communication technologies (ICECCT), pp 1–5
24. Subbaraj P (2009) Optimal reactive power dispatch using self-adaptive real coded Genetic algorithm. Electr Power Syst Res 79(2):374–38

A Two-Fold Multi-objective Multi-verse Optimization-Based Time Series Forecasting

Cherry Tandu, Mohit Kosuri, Sobhan Sarkar, and J. Maiti

Abstract In this study, to overcome error due to high-dimensional data and to get the best forecasting prediction for time series data, we employ a feature selection method to obtain the best exploitation and exploratory performance. Due to a large number of irrelevant factors within data, it is imperative to classify the tasks by using a feature selection method. Therefore, a two-fold multi-objective multi-verse optimization as a feature selection optimization method has been proposed to obtain a trade-off between minimization loss and minimization of the number of features selected. The Convolution Neural Network (CNN) has been used as a basic predictor. A dynamic error correction is also proposed to reduce the error further to the deep learning models to get the best time series forecasting. However, many Multi-Objective Optimization techniques have been used to deal with high-dimensional data, the proposed method showed the best trade-off for feature selection.

Keywords Time series forecasting · Feature selection · Multi-objective optimization · Deep learning

C. Tandu
Department of Ocean Engineering and Naval Architecture, IIT Kharagpur, Kharagpur 721302, India

M. Kosuri
Department of Mechanical Engineering, IIT Kharagpur, Kharagpur 721302, India

S. Sarkar (✉)
Business School, University of Edinburgh, 29 Buccleuch Pl, Edinburgh EH8 9JS, UK
e-mail: sobhan.sarkar@ed.ac.uk

J. Maiti
Department of Industrial and Systems Engineering, IIT Kharagpur, Kharagpur 721302, India

© The Author(s), under exclusive license to Springer Nature Singapore Pte Ltd. 2022

D. Giri et al. (eds.), *Proceedings of the Seventh International Conference on Mathematics and Computing*, Advances in Intelligent Systems and Computing 1412,
https://doi.org/10.1007/978-981-16-6890-6_55

1 Introduction

Time series is a method of analyzing historical data arranged sequentially to help predict future events and outcomes. A time series needs to give accurate predictions. Time series forecasting has various applications, including economic forecasting, sales forecasting, budgetary analysis, stock market analysis, yield projections, process and quality control, and air pollution forecasting. While stable forecasting has many advantages, instability and volatility in time series put tremendous challenges to achieving high precision forecasting.

Deep learning (DL) is one of the powerful machine learning (ML) approaches [23]. There are different ML models available, such as support vector machine [18, 28–30], artificial neural network [25, 26], decision tree [24, 27], clustering [21, 35], rule generation models [19, 22], and convolutional neural network [15, 16]. These models are comparatively more powerful than their traditional counterparts [7, 11, 31]. The DL models are used to obtain better features from the data by transforming original data to a more precise level by combining simple non-linear modules layer by layer. In the process of extracting various types of features, certain types of deep neural networks are effective. DL models have a great effect on the accuracy of time-series forecasting by improving its potential of extracting features from data. The DL model which provides an excellent contribution in developing the time series forecasting accuracy has a very different style of feature extraction ability. Feature selection (FS) is needed to improve performance with an increase in speed and simplicity of prediction. The difference and the partial opinions which are obtained through the decomposition errors are engaged as an objective function of the multi-objective optimizer. It also increases cost-effectiveness. Due to FS, we can remove irrelevant data and reduce noise and computational time. The reason for combining these two powerful models is to produce the best optimal ensemble model. The main objectives of GA-II are Bias and Variance [1].

The rest of the paper is organized as follows: Sect. 2 portrays foundation business related to strategies utilized in this investigation. In Sect. 3, the methodology proposed is explained. Section 4 discusses the experimentation and their results, and finally, the key findings and the scopes for future works are discussed briefly in Sect. 5.

2 Literature Review

Single-objective methods deal with only one objective and multi-objective techniques deal with two or more objectives, which are often in conflict [13, 33]. Thus, evolutionary computation (EC) can create multiple compromised arrangements in a single run, which makes them show great adequacy on multi-objective optimization [34]. As of late with development in man-made brainpower and AI, FS strategies have developed quickly and each factor's significance is not difficult to judge [8, 12]. Here, we clarify the most significant multi-target FS approaches. The multi-objective EC calculations (e.g., Genetic Algorithm (GA)) these days have been utilized to address the FS problem. Reference [32] proposed a multi-target micro-GA to shape a troupe enhancer that improves the FS issue and advances the neural organization classifier. A multi-objective GA-II (Non-dominated Sorting-based Multi-objective Genetic Algorithm II, i.e., NSGA-II) in light of non-ruled arranging was initially proposed by [4] to tackle multi-target optimization issues. Multi-Verse Optimization (MVO) proposed by Mirjalili [14] is a developmental methodology that has indicated its exploratory and manipulative presentation in managing a few useful applications in designing, science issues, and real life. Based on this the main use of the classification is to predict the class of an unidentified object.

NSGA-II was utilized in [8] as a multi-target FS approach. We compared the performance of TFMOMVO-pbest with these well-known methods in solving all datasets. Furthermore, all the settings and parameters that are used in the experiments for all popular multi-objective techniques are obtained from [8, 12, 32]. Numerous analysts have utilized this strategy for various DL and ML models to get an exact time series forecasting accuracy. An exceptional DL gathering model is built to utilize Multi-populace NSGA-II [3] for exactness. The fundamental points of interest of three different Multi-Objective Optimal Evolutionary Algorithms (MOOEA) [2, 5] for ascertaining and choosing the best exactness from given DL models are unmistakably clarified by Peimankar. For MOOEA [9], the most famous strategy is Ensemble Pruning (EP) as its key objective is to improve the forecasting and bring about terms of exactness and unwavering quality and to keep up the exhibition speed during running. Therefore, in this work, GA-II [10, 17] is utilized for EP to acquire accuracy and to keep up unwavering quality and strength during time series forecasting.

2.1 Contributions of the Work

Based on the above-mentioned discussion, the present study contributes to the body of literature as follows:

(i) The personal best two-fold multi-objective verse optimization (p-best multi-MVO) is proposed for the classification model to minimize the error by reducing the dimension of data (e.g., features).
(ii) Dynamic error correction is proposed to reduce the forecasting error. It corrects the forecasting error by observing the sequential error of the previous time points.

Algorithm 1 Pseudo-code of pbest TFMOMVO.

Input: number of chromosomes (universe), Termination condition (T), and partitioned datasets.
Output: The best feature sets.
Generate the random chromosomes (universe), W_{ep}, T_{dr}, and best universe.
Dominant = {}
Obtain the Moving rates of all universes
Update the Dominant using Dominance equation
Save the best solution
Save the best personal solution for each universe P_b
while $t \leq T$ **do**
  for (each Universei) **do**
    $Calculate\ (1 - accuracy)$
    $Calculate\ (ratio\ of\ feature\ selected\ and\ total features)$
    Update the Dominant using Dominance equation
    Update the best personal solution (P_b)
    Update W_{ep}, T_{dr} using Eqs. (1) and (2)
    for $each\ element_j$ **do**
      Calculate RWS
      Update the universe using Eq. (5)
    end for
  end for
end while
Return: The best feature set

3 Methodology

3.1 TFMOMVO Feature Selection with Personal Best (TFMOMVO-pbest)

In this subsection, we discuss the approach to manage FS using two-fold multi-objective MVO to examine the Pareto front of feature subsets. The standard MVO and MOMVO use the fittest universe/best universe to update all of the universe's positions. This action strengthens the behavior of the algorithm. To add more maltreatment limits concerning the universes, another term is added to the Wormhole Existence Probability as expressed in Eq. (1). This term is called personal fittest, which is the best position achieved by the genuine universe.

W_{ep} and T_{dr} represent versatile boundaries, in which W_{ep} is used in MVO to support the exploitation power and T_{dr} is utilized to improve misuse in the region of the best universe found so far. Wormhole existing probability is expressed in Eq. (1):

$$W_{ep} = min + t * \frac{max - min}{T} \tag{1}$$

T_{dr} denotes the voyaging distance rate. The advancement of refreshing the items in every universe from MVO to p-best MOMVO is expressed in Eq. (2):

$$T_{dr} = 1 - \frac{t^{1/p}}{T^{1/p}} \tag{2}$$

For the convenience of understanding the equations, we introduce some new terms which are expressed in Eqs. (5), (6), and (7), as follows:

$$X_j + T_{dr} = U_j^T \tag{3}$$

$$X_j - T_{dr} = L_j^T \tag{4}$$

where X_j is the element in the universe and T_{dr} is the voyaging distance rate.

At first, the MVO is a single-objective optimizer that considers the upper limit and the lower limit to update the object in the universe:

$$X_i^j(t+1) = \begin{cases} \begin{cases} (U_j^T \times ((U_l - L_l)r_3 + L_l), & r_2 < 0.5 \\ (L_j^T \times ((U_l - L_l)r_3 + L_l), & r_2 \geqslant 0.5 \end{cases}, & r_1 < W_{ep} \\ X_i^j(t), & r_2 \geqslant W_{ep} \end{cases} \tag{5}$$

where U_l denotes the upper limit, L_l indicates the lower limit, and r_1, r_2, r_3 are the arbitrary numbers set between (0, 1). Now, MVO is additionally modified into a multi-objective optimizer simply by eliminating the upper and lower limits with the condition for updating the p-best in every universe:

$$X_i^j(t+1) = \begin{cases} \begin{cases} U_j^T \times r_3, & r_2 < 0.5 \\ L_j^T \times r_3, & r_2 > 0.5 \end{cases} & r_1 < W_{ep} \\ X_i^j(t) & r_1 > W_{ep} \end{cases} \tag{6}$$

$$X_i^j(t+1) = \begin{cases} \begin{cases} U_j^T \times (P_{bj} - X_i^j(t)), & r_2 < 0.5 \\ L_j^T \times (P_{bj} - X_i^j(t)) & r_2 > 0.5 \end{cases} & r_1 < W_{ep} \\ X_i^j(t) & r_1 > W_{ep} \end{cases} \tag{7}$$

where P_b signifies the individual most important aspect of the ith universe, T denotes the end condition, and t demonstrates the current cycle. The P_b is updated based on dominance relationship, such as if the new P dominates the old one, then the new one is employed; otherwise, the old one continues to be used. In this approach, universes explore space and store the best position they have achieved so far. Each universe has momentum that allows the universe to explore more areas in the search space. Furthermore, the universe is also attracted to its personal best location in its memory. The main strength gained is that a universe can balance between its ability of exploration and exploitation, moreover, diversity can be maintained. The selection of universe from wormholes is conducted by Roulette Wheel Selection (RWS) [14].

3.2 Convolutional Neural Network (CNN)

CNN can adequately catch the matrix-like constructions to represent the fever calculations because of inadequately associated neurons and the pruning cycle. They are otherwise called move-invariant or space-invariant counterfeit neural organizations. CNN catches the loads as the 1-D cluster by y collecting all the local weights from the given input data to enhance and robustness of the model based on unique feature extraction. Over-fitting of the information is being evaded as a result of the complete connectedness of this organization. Progressive examples are distinguished and exploited to discover complex examples utilizing more easier examples. Numerous layers have been built in the engineering of CNN [6] which incorporate information layers, yield layers, and various arrangements of multiple layers. Actuation and pooling are essential for multiple layers. In this experiment, we have considered (1×3), (1×1) 1-D convolution kernels.

3.3 Dynamic Error Correction (DEC)

Each basic DL model can more effectively reduce the forecasting error by learning the new dynamic error correction (DEC) method. It is employed to reduce the forecasting error of the model as per the following Eq. (8). It reduces the forecasting error of a particular time point by calculating the errors of k time points before the particular point. Hence, it tracks the error sequence before the particular time point and it can dynamically reduce the forecasting error exponential of $-t$ that helps us to see the error dynamically which means time increases, the error decreases. It is important to analyze and correct error sequences of DL models before EDNN. DEC uses the previous few time points to achieve a new time point and automatically updates the present forecasting error at the present time point.

$$\overline{y}_{t+1}{}' = \overline{y}_{t+1}{}' + \sum_{t=1}^{k} (y_t - \overline{y}_t) * \frac{\left| y_t - \overline{y}_t \right|}{MaxAE} * exp(-t) \tag{8}$$

where $\overline{y'}_{t+1}$ = corrected forecasting value, $exp^{(-t)}$ = time delay factor, $k = 5$ (by trail and error method), and $MAXAE$ = Maximum absolute error. Here, $\overline{y}_{t+1}{}'$ refers to the forecast value.

4 Results and Discussions

In this study, we have studied the performance of our proposed algorithm on five different datasets of stock markets, namely DIJ, NASDAQ, NYSE, RUSSEL, and S&P. Each of them contains prices of various items which are the factor for predicting the closing value of the particular stock index. For each of them, the training, validation, and testing sizes are mentioned in Table 1. The total number of features of each dataset, number of instances or records, and number of selected features are mentioned in Table 2. In Table 3, the loss values obtained from each method on each dataset are given. From the table, we can observe that feature selection-based forecasting with dynamic error correction provides minimum loss for each dataset. Figures 1, 2, and 3 are the experimental results obtained using the proposed methodology applied on DIJ, NYSE, and Russell stock datasets, respectively. Each figure consists of 4 graphical representations. In Fig. 1a, a relationship map is introduced between the CNN and Actual yield. The light tone indicates the higher positive relationship between the real and anticipated worth, and the profound tone denotes the negative connection between them. In Fig. 1b, c, and d, the actual versus anticipated/estimated yields by CNN, FS-based CNN (i.e., CNN FS), and FS-based DEC CNN (i.e., CNN FS DEC) separately are illustrated. From these figures, it is evident that the proposed FS-based DEC CNN algorithm performs in a way that is better than others in estimating the DIJ stock price.

Table 1 Dataset description

Dataset	Train data	Validation data	Test data
DIJ	713	178	223
NASDAQ	713	178	223
NYSE	713	178	223
RUSSEL	713	178	223
S&P	713	178	223

Table 2 Number of feature selection for different datasets

Datasets	No. of features	No. of instances	No. of selected features
DIJ	81	1900	32
NASDAQ	81	1900	27
NYSE	81	1900	23
RUSSELL	81	1900	31
S&P	81	1900	42

Table 3 Loss values comparison

Method\dataset	DIJ	NASDAQ	NYSE	RUSSELL	S&P
CNN	22.63958	37.30023	41.00128	54.08834	39.17466
FS_CNN	10.92644	16.49902	15.73422	18.73449	24.84021
FS_CNN_DEC	9.72835	16.27777	11.91463	17.89253	21.74928

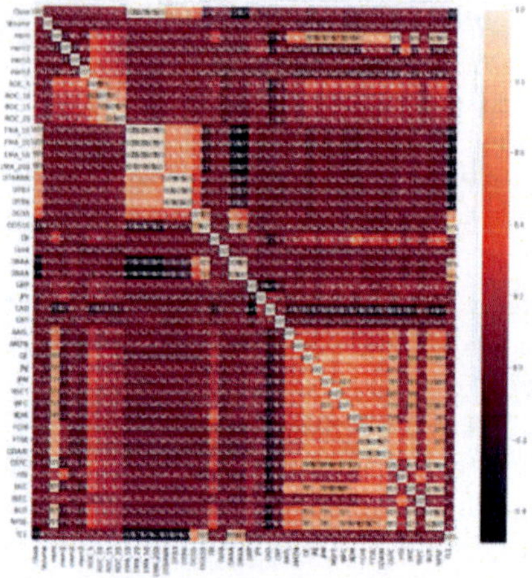

(a) CNN VS Actual

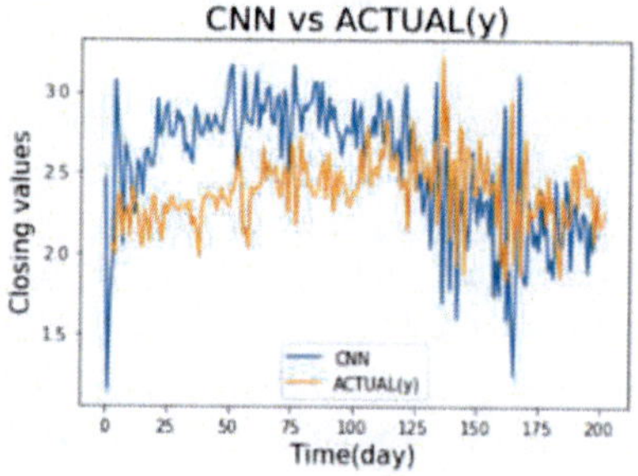

(b) CNN VS Actual

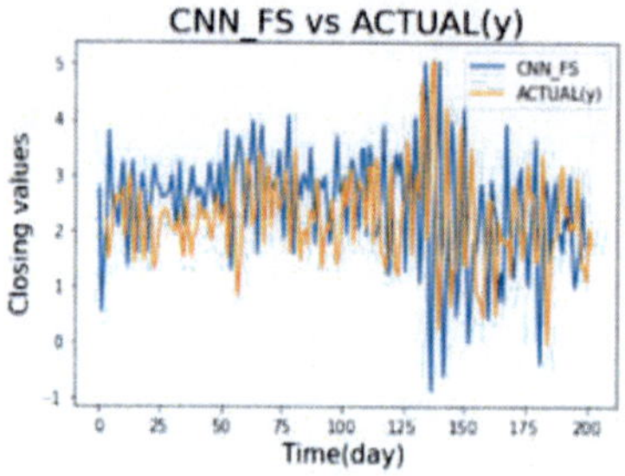

(c) CNN_FS VS Actual

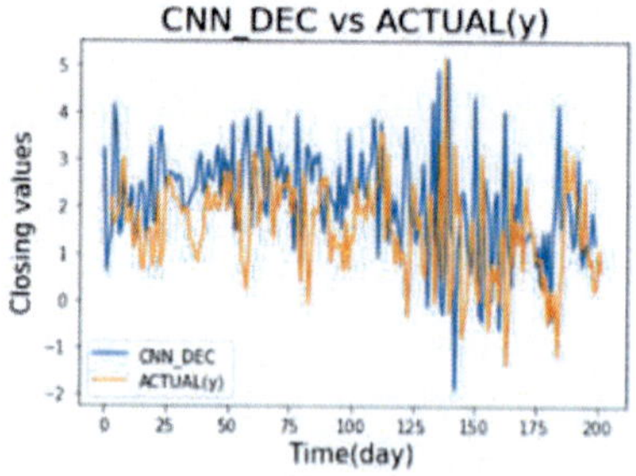

(d) CNN_FS_DEC VS Actual

Fig. 1 Experimental results on the DIJ stock dataset

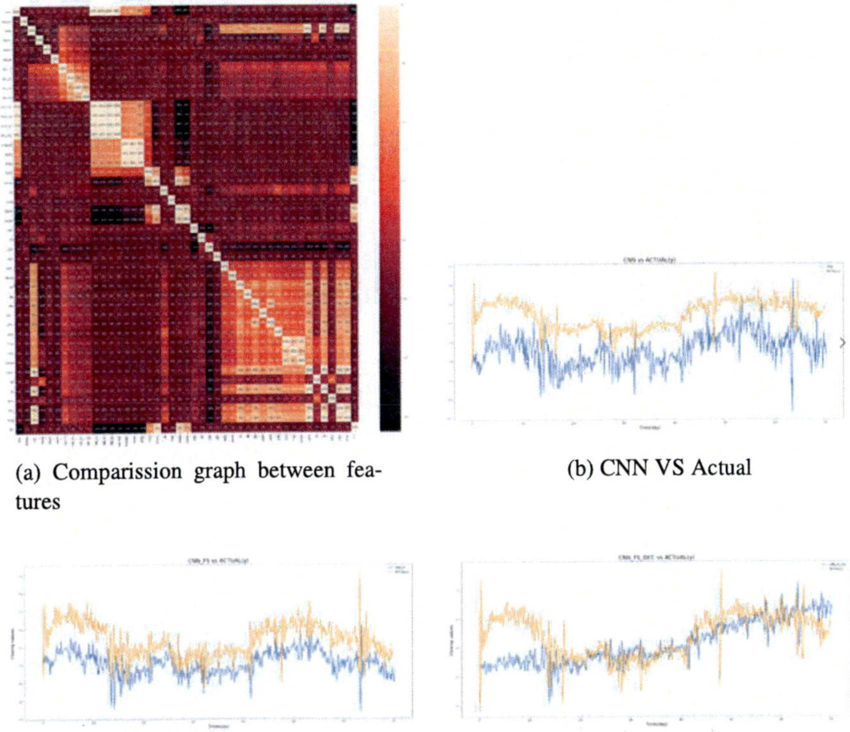

(a) Comparission graph between features

(b) CNN VS Actual

(c) CNN_FS VS Actual

(d) CNN_FS_DEC VS Actual

Fig. 2 Experimental results on the NYSE stock dataset

5 Conclusion

In this work, a p-best multi-objective multi-verse optimization (MOMVO) has been proposed for the feature selection task for forecasting time series. The MOMVO calculation has been planned dependent on using three ideas in cosmology which are white hole, black hole, and wormhole. The p-best variation of the MOMVO that joins the individual best situation has been proposed, also. The proposed MOMVO-based methodologies manage the precision of the model and the decrease in the dimensional as a multi-target advancement issue. A dynamic error correction (DEC) has been proposed to reduce the forecasting error. The results of the conducted experiments based on the five datasets reveal that TFMOMVO-pbest along with the DEC approach outperforms TFMOMVO-pbest. Moreover, TFMOMVO-pbest shows the best results when compared with the other multi-objective optimization algorithms for feature selection. In future, parameter tuning of the MOMVO algorithm can also be done for

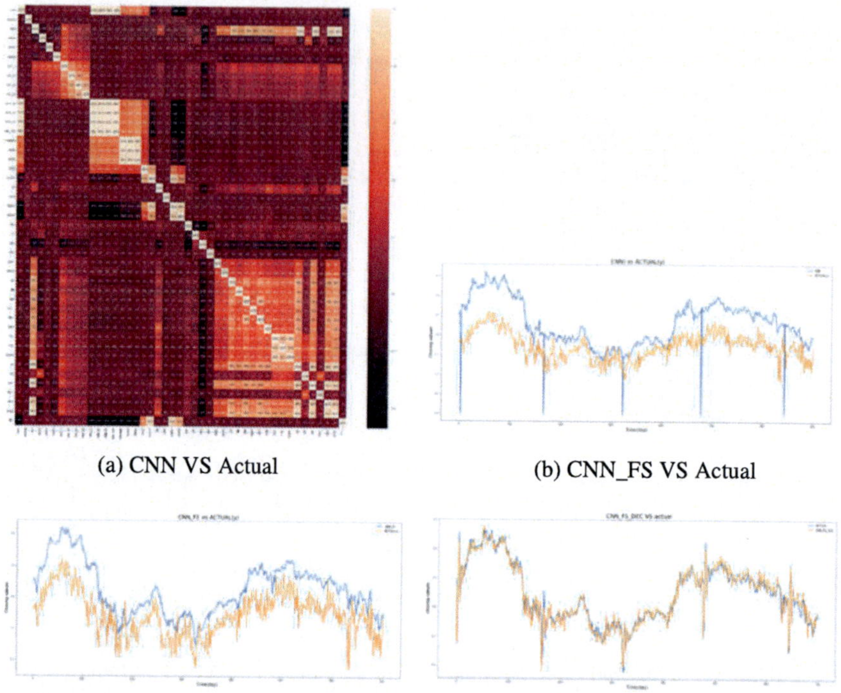

(a) CNN VS Actual

(b) CNN_FS VS Actual

(c) Comparission graph between features

(d) CNN_FS_DEC VS Actual

Fig. 3 Experimental results on the Russell stock dataset

better prediction results. A decision support system [20] can also be developed as a potential future scope for time series forecasting. Moreover, to validate the proposed model, it can be tested on a real-world datasets from different domains, such as air quality forecasting and weather forecasting.

References

1. Cheng KP, Mohan RE, Nhan NHK, Le AV (2020) Multi-objective genetic algorithm-based autonomous path planning for hinged-tetro reconfigurable tiling robot. IEEE Access 8:121267–121284
2. Corne DW, Lones MA (2018) Evolutionary algorithms. arXiv:1805.11014
3. Deb K, Pratap A, Agarwal S, Meyarivan T (2002) A fast and elitist multiobjective genetic algorithm: Nsga-ii. IEEE Trans Evol Comput 6(2):182–197
4. DEH K, Agrawal S, Pratap A, Meyarivan T (2000) A fast elitist non-dominated sorting genetic algorithm for multi-objective optimization: Nsga-ii. Lecture notes in computer science (2000), pp 849–858

5. Dutta S, Das KN (2019) A survey on pareto-based eas to solve multi-objective optimization problems. In: Soft computing for problem solving. Springer, pp 807–820
6. Ferreira A, Giraldi G (2017) Convolutional neural network approaches to granite tiles classification. Expert Syst Appl 84:1–11
7. Gautam S, Maiti J, Syamsundar A, Sarkar S (2017) Segmented point process models for work system safety analysis. Safety Sci 95:15–27
8. De la Hoz E, De La Hoz E, Ortiz A, Ortega J, Martínez-Álvarez A (2014) Feature selection by multi-objective optimisation: application to network anomaly detection by hierarchical self-organising maps. Knowl-Based Syst 71:322–338
9. Jiménez F, Sánchez G, García JM, Sciavicco G, Miralles L (2017) Multi-objective evolutionary feature selection for online sales forecasting. Neurocomputing 234:75–92
10. Knowles JD, Corne DW (2000) Approximating the nondominated front using the pareto archived evolution strategy. Evolut Comput 8(2):149–172
11. Krishna OB, Maiti J, Ray PK, Samanta B, Mandal S, Sarkar S (2015) Measurement and modeling of job stress of electric overhead traveling crane operators. Safety Health Work 6(4):279–288
12. Martín-Smith P, Ortega J, Asensio-Cubero J, Gan JQ, Ortiz A (2017) A supervised filter method for multi-objective feature selection in eeg classification based on multi-resolution analysis for bci. Neurocomputing 250:45–56
13. Miao R, Gong D, Yong Z (2018) A multi-direction prediction approach for dynamic multi-objective optimization. IEEE Trans Cybern 1–13
14. Mirjalili S, Mirjalili SM, Hatamlou A (2016) Multi-verse optimizer: a nature-inspired algorithm for global optimization. Neural Comput Appl 27(2):495–513
15. Pramanik A, Singh H, Djeddi C, Sarkar S, Maiti J (2020) Region proposal and object detection using hog-based cnn feature map. In: International conference on data analytics for business and industry 2020, Kingdom of Bahrain
16. Pramanik A, Pal SK, Maiti J, Mitra P (2021) Granulated rcnn and multi-class deep sort for multi-object detection and tracking. IEEE Trans Emerg Topics Comput Intell
17. Raymer ML, Punch WF, Goodman ED, Kuhn LA, Jain AK (2000) Dimensionality reduction using genetic algorithms. IEEE Trans Evol Comput 4(2):164–171
18. Sarkar S, Verma A, Maiti J (2018) Prediction of occupational incidents using proactive and reactive data: a data mining approach. In: Industrial safety management- 21st century perspective of Asia. Springer, Singapore, pp 65–79
19. Sarkar S, Baidya S, Maiti J (2017) Application of rough set theory in accident analysis at work: a case study. In: ICRCICN 2017, pp 245–250
20. Sarkar S, Chain M, Nayak S, Maiti J (2019) Decision support system for prediction of occupational accident: a case study from a steel plant. In: Emerging technologies in data mining and information security, vol 813. Springer, Singapore , pp 787–796
21. Sarkar S, Ejaz N, Maiti J (2018) Application of hybrid clustering technique for pattern extraction of accident at work: a case study of a steel industry. In: 2018 4th international conference on recent advances in information technology (RAIT), IIT Dhanbad. IEEE, pp 1–6
22. Sarkar S, Lohani A, Maiti J (2017) Genetic algorithm-based association rule mining approach towards rule generation of occupational accidents. In: Communications in computer and information science, vol 776. Springer, Singapore, pp 517–530
23. Sarkar S, Maiti J (2020) Machine learning in occupational accident analysis: a review using science mapping approach with citation network analysis. Safety Sci 131:104900
24. Sarkar S, Patel A, Madaan S, Maiti J (2017) Prediction of occupational accidents using decision tree approach. In: INDICON 2017. IEEE, pp 1–6
25. Sarkar S, Pateshwari V, Maiti J (2017) Predictive model for incident occurrences in steel plant in india. In: ICCCNT 2017, pp 1–5
26. Sarkar S, Pramanik A, Maiti J, Reniers G (2020) Predicting and analyzing injury severity: a machine learning-based approach using class-imbalanced proactive and reactive data. Safety Sci 125:104616

27. Sarkar S, Raj R, Sammangi V, Maiti J, Pratihar D (2019) An optimization-based decision tree approach for predicting slip-trip-fall accidents at work. Safety Sci 118:57–69
28. Sarkar S, Sammangi V, Raj R, Maiti J, Mitra P (2019) Application of optimized machine learning techniques for prediction of occupational accidents. Comput & Oper Res 106:210–224
29. Sarkar S, Vinay S, Maiti J (2016) Text mining based safety risk assessment and prediction of occupational accidents in a steel plant. In: ICCTICT 2017. IEEE, pp. 439–444
30. Sarkar S, Vinay S, Pateshwari V, Maiti J (2017) Study of optimized svm for incident prediction of a steel plant in india. In: INDICON 2017. IEEE, pp 1–6
31. Singh K, Raj N, Sahu S, Behera R, Sarkar S, Maiti J (2015) Modelling safety of gantry crane operations using petri nets. Int J Injury Control Safety Prom 1–12
32. Tan CJ, Lim CP, Cheah YN (2014) A multi-objective evolutionary algorithm-based ensemble optimizer for feature selection and classification with neural network models. Neurocomputing 125:217–228
33. Tian Y, Cheng R, Zhang X, Cheng F, Jin Y (2017) An indicator-based multiobjective evolutionary algorithm with reference point adaptation for better versatility. IEEE Trans Evol Comput 22(4):609–622
34. Tian Y, Zhang X, Cheng R, He C, Jin Y (2018) Guiding evolutionary multiobjective optimization with generic front modeling. IEEE Trans Cybern
35. Verma A, Chatterjee S, Sarkar S, Maiti J (2018) Data-driven mapping between proactive and reactive measures of occupational safety performance. In: Industrial safety management- 21st century perspective of Asia. Springer, Singapore, pp 53–63

A Novel Ranking-Based Non-linear Programming Approach to Solve Bi-matrix Games in Dense Fuzzy Environment

Shuvasree Karmakar and Mijanur Rahaman Seikh

Abstract In reality, occasionally, the players of a bi-matrix game problem are forced to bring little changes in their game strategies to manage the situation. As a result, this phenomenon could lead to a change in the pay-offs of a game problem. This paper aims to model such kind of bi-matrix games with pay-offs as dense fuzzy lock sets. A novel defuzzification function of this new set, viz., $Val_D(.)$ function is defined here. An auxiliary quadratic dense fuzzy programming problem (QDFPP) is established to solve the bi-matrix game. Later this QDFPP is transformed into an equivalent crisp programming problem by employing the induced defuzzification function and its linearity property. A notable facet of this approach is that players' profits increase with increased trials. The efficacy and applicability of the proposed methodology are explained by reflecting a tourism planning strategy problem.

Keywords Bi-matrix game · Triangular dense fuzzy lock sets · Triangular dense fuzzy sets · Defuzzification function · Tourism planning strategy problem

1 Introduction

In non-cooperative game theory, Nash [18] introduced bi-matrix game as a special tool to present a framework for competitive problems. Initially, the pay-offs of a bi-matrix game considered common numerical values. Later, considering the imprecise information, game theorists employed the notion of fuzzy set theory [26] and intuitionistic fuzzy set theory [1] to develop bi-matrix games. In the recent past, numerous endeavors have been made in the fuzzy bi-matrix game theory [16, 19, 27]. Li and Li [14] developed bi-matrix games with fuzzy entries in credibility space. Brikaa et al. [7] developed a tourism planning strategy model by considering it as a bi-matrix game through a rough interval approach. Based on mean-area-based ranking, An [2] executed a non-linear programming approach to solve intuitionistic fuzzy (I-fuzzy) bi-matrix games. An and Li [3] discussed a solution methodology for

S. Karmakar (✉) · M. Rahaman Seikh
Department of Mathematics, Kazi Nazrul University, Asansol 713 340, India

© The Author(s), under exclusive license to Springer Nature Singapore Pte Ltd. 2022

D. Giri et al. (eds.), *Proceedings of the Seventh International Conference on Mathematics and Computing*, Advances in Intelligent Systems and Computing 1412,
https://doi.org/10.1007/978-981-16-6890-6_56

constrained bi-matrix games with I-fuzzy pay-offs. Moreover, I-fuzzy sets appeared in the bi-matrix games by the researchers [13, 15, 17, 22, 23, 25] in several ways. Matrix games also experienced its extensions in different environmnents [4–6, 24].

Aiming to reduce the fuzziness of a situation with an increment of trials of an event, De and Beg [8] introduced the notion of the triangular dense fuzzy sets (TDFSs). A TDFS consists of three Cauchy sequences having converging point zero. Later, modifying the features of TDFSs, De [9] defined triangular dense fuzzy lock sets (TDFLSs). The concept of TDFS and TDFLS has been implemented in the economic production quantity model [11, 12] problems.

In reality, players are often forced to bring variations in their game strategy to cope-up with the situation. To portray such problems, here we have developed a bi-matrix game with TDFLS pay-offs. It is worth noting that, with the number of trials, players can reduce the fuzziness of the pay-offs in a bi-matrix game. Here we execute an auxiliary quadratic dense fuzzy programming problem (QDFPP) to find the nash equilibrium strategies of the players involved in the game. To make the solution a novel defuzzification function, viz., $Val_D(.)$ function is defined here. By applying the defuzzification function, the QDFPP is transformed into an equivalent crisp programming problem. Then it is solved by using LINGO 17.0 software. Here we show that the gain of the players rises with the increment of the trials. Finally, a tourism planning strategy problem validates and justified our proposed methodology.

The rest of the paper is sketched as follows. Section 2 revisits the notion of TDFS and TDFLS to develop the subsequent sections. Section 3 describes the defuzzification method to make a ranking order of TDFLSs. A bi-matrix game with TDFLS pay-offs is developed in Sect. 4. Also, the solution procedure of such a bi-matrix game is discussed here. Tourism planning strategy problem check the validity of the proposed methodology in Sect. 5. Finally, Sect. 6 concludes the observation and provides some outlook for future direction.

2 Preliminaries

To develop our discussion, here we revisit some definitions associated with dense fuzzy sets.

Definition 1 ([8]) Suppose $\tilde{\mathbb{D}}_1$ be a fuzzy set. The components of $\tilde{\mathbb{D}}_1$ are taken in the form of some sequences of functions (defined on the set of the natural number ($\aleph$), which maps on a crisp number z). If all the sequences of $\tilde{\mathbb{D}}_1$ converge to z as $r(\in \aleph) \to \infty$, then $\tilde{\mathbb{D}}_1$ is called dense fuzzy set (DFS) under consideration.

Definition 2 ([8]) Consider a triangular fuzzy number $\tilde{\mathbb{D}}_2 = \langle z_1, z_2, z_3 \rangle$, where z_1 and z_2 are expressed in terms of some sequences, as $z_1 = z_2 f^{(r)}$ and $z_3 = z_2 g^{(r)}$. Note that $f^{(r)}$ and $g^{(r)}$ are two cauchy sequences converge to 1 as $r \to \infty$. Then the set $\tilde{\mathbb{D}}_2$ is termed as triangular dense fuzzy set (TDFS).

Definition 3 ([9]) Suppose $\tilde{\mathbb{D}} = \langle z(1-\mu f^{(r)}), z, z(1+\nu g^{(r)})\rangle$ be a TDFS, where, μ and ν be the left and right fuzzy deviations, respectively with $0 \le \mu, \nu \in \Re$, and $f^{(r)}$, and $g^{(r)}$ be the cauchy sequences. If $f^{(r)}$ and $g^{(r)}$ converge to some other real numbers instead of 1, say, $\frac{1}{k^1}(k^1 \neq 0)$ and $\frac{1}{k^2}(k^2 \neq 0)$, respectively, as $r(\in \aleph) \to \infty$. Such special kind of TDFS $\tilde{\mathbb{D}}$ is called triangular dense fuzzy lock set (TDFLS), with double key vectors k^1 and k^2.

We should mention that the key vectors k^1, and k^2 depend on the fuzzy deviations μ, and ν. Assume that the set Ω be the collection of all the TDFLSs. Let us express the sequences $f^{(r)}$ and $g^{(r)}$ as, $f^{(r)} = \frac{1}{k^1} - \frac{1}{r+1}$, and $g^{(r)} = \frac{1}{k^2} - \frac{1}{r+1}$, respectively, so that the sequence $f^{(r)}$ converges to $\frac{1}{k^1}$, and the sequence $g^{(r)}$ converges to $\frac{1}{k^2}$ as $r \to \infty$. Thus the TDFLS $\tilde{\mathbb{D}} \in \Omega$ may be represented as $\tilde{\mathbb{D}} = \left\langle z\left\{1-\mu\left(\frac{1}{k^1}-\frac{1}{r+1}\right)\right\}, z, z\left\{1+\nu\left(\frac{1}{k^2}-\frac{1}{r+1}\right)\right\}\right\rangle$.

Definition 4 (*Mathematical Operations of TDFLSs*) Let $\tilde{\mathbb{D}}_1, \tilde{\mathbb{D}}_2, \tilde{\mathbb{D}} \in \Omega$ are expressed as $\tilde{\mathbb{D}}_1 = \left\langle z_1\left\{1-\mu_1\left(\frac{1}{k_1^1}-\frac{1}{r+1}\right)\right\}, z_1, z_1\left\{1+\nu_1\left(\frac{1}{k_1^2}-\frac{1}{r+1}\right)\right\}\right\rangle$, $\tilde{\mathbb{D}}_2 = \left\langle z_2\left\{1-\mu_2\left(\frac{1}{k_2^1}-\frac{1}{r+1}\right)\right\}, z_2, z_2\left\{1+\nu_2\left(\frac{1}{k_2^2}-\frac{1}{r+1}\right)\right\}\right\rangle$, and $\tilde{\mathbb{D}} = \left\langle z\left\{1-\mu\left(\frac{1}{k^1}-\frac{1}{r+1}\right)\right\}, z, z\left\{1+\nu\left(\frac{1}{k^2}-\frac{1}{r+1}\right)\right\}\right\rangle$ Then addition and scalar multiplication of TDFLSs are defined as follows.

Addition: $\tilde{\mathbb{D}}_1 + \tilde{\mathbb{D}}_2 = \left\langle z_1\left\{1-\mu_1\left(\frac{1}{k_1^1}-\frac{1}{r+1}\right)\right\} + z_2\left\{1-\mu_2\left(\frac{1}{k_2^1}-\frac{1}{r+1}\right)\right\}, z_1+z_2, z_1\left\{1+\nu_1\left(\frac{1}{k_1^2}-\frac{1}{r+1}\right)\right\} + z_2\left\{1+\nu_2\left(\frac{1}{k_2^2}-\frac{1}{r+1}\right)\right\}\right\rangle$

Scalar multiplication: For any real scalar ρ,

$$\rho\tilde{\mathbb{D}} = \begin{cases} \left\langle \rho z\left\{1-\mu\left(\frac{1}{k^1}-\frac{1}{r+1}\right)\right\}, \rho z, \rho z\left\{1+\nu\left(\frac{1}{k^2}-\frac{1}{r+1}\right)\right\}\right\rangle, & \text{if } \rho > 0 \\ \left\langle \rho z\left\{1+\mu\left(\frac{1}{k^2}-\frac{1}{r+1}\right)\right\}, \rho z, \rho z\left\{1-\nu\left(\frac{1}{k^1}-\frac{1}{r+1}\right)\right\}\right\rangle, & \text{if } \rho < 0 \end{cases}$$

As the sum of two convergent sequences and a sequence obtained by scalar multiplication of a convergent sequence are also convergent, $\tilde{\mathbb{D}}_1 + \tilde{\mathbb{D}}_2$, and $\rho\tilde{\mathbb{D}}$ are also TDFLSs.

3 Defuzzification of TDFLS

DFS has an exceptional character to disappear the uncertainty of imprecise information by increasing the trial over the phases of the decision-making problems. But a little conflict occurs in DFLS even after the overall ambiguities have been removed from the system. Hence to make a mathematical illustration of a real-world problem in an inadequate environment, DFLS works as a perfect tool. But to make an order of such kind of fuzzy sets, we have to defuzzify the TDFLSs. According to De [9], $\alpha-$cut set of the TDFLS $\tilde{\mathbb{D}}$, can be derived as an interval, $\left[\mathbb{D}_L, \mathbb{D}_R\right]$, where, $\mathbb{D}_L = z\left\{1-\mu\left(\frac{1}{k^1}-\frac{1}{r+1}\right)+\mu\alpha\left(\frac{1}{k^1}-\frac{1}{r+1}\right)\right\}$, and $\mathbb{D}_R = z\left\{1+\nu\left(\frac{1}{k^2}-\frac{1}{r+1}\right)-\nu\alpha\left(\frac{1}{k^2}-\frac{1}{r+1}\right)\right\}$.

To make the ranking order of TDFLSs, we propose a novel defuzzification function here. Motivated by the work of Qiu et al. [21], we can resolve the value of $\mathbb{D}$ as,

$$Val_D(\tilde{\mathbb{D}}) = \frac{1}{N}\sum_{r=1}^{N}\int_0^1 \{\mathbb{D}_L + \mathbb{D}_R\} f(\alpha) d\alpha \tag{1}$$

where $f(\alpha)$ is a continuous function and also monotonic increasing in the interval [0,1]. Further, it satisfies the boundary conditions $f(0) = 0$, and $f(1) = 1$, and also, $\int_0^1 f(\alpha)d\alpha = \frac{1}{2}$. For the sake of simplicity, we take $f(\alpha) = \alpha$. Clearly, $f(\alpha) = \alpha$ satisfies all the requirements assigned here. Substituting the values of $\mathbb{D}_L$ and $\mathbb{D}_R$ in Eq. (1), we obtain the value of the TDFLS $\tilde{\mathbb{D}}$ as,

$$\begin{aligned} Val_D(\tilde{\mathbb{D}}) &= \frac{1}{N}\sum_{r=1}^{N}\int_0^1 \{\mathbb{D}_L + \mathbb{D}_R\} f(\alpha) d\alpha \\ &= \frac{1}{N}\sum_{r=1}^{N}\int_0^1 \Big[z\Big\{1 - \mu\Big(\frac{1}{k^1} - \frac{1}{r+1}\Big) + \mu\alpha\Big(\frac{1}{k^1} - \frac{1}{r+1}\Big)\Big\} + \\ &\qquad z\Big\{1 + \nu\Big(\frac{1}{k^2} - \frac{1}{r+1}\Big) - \nu\alpha\Big(\frac{1}{k^2} - \frac{1}{r+1}\Big)\Big\}\Big] \\ &= z\Big[1 + \frac{1}{6}\Big(\frac{\nu}{k^2} - \frac{\mu}{k^1}\Big) - \frac{(\nu - \mu)}{6N}\Big(\frac{1}{2} + \frac{1}{3} + \ldots + \frac{1}{N+1}\Big)\Big] \end{aligned} \tag{2}$$

Based on the deffuzzification function $Val_D(.)$ of $\tilde{\mathbb{D}}$, the ranking order relation between two TDFLSs can be defined as follows.

Definition 5 Suppose $\tilde{\mathbb{D}}_1, \tilde{\mathbb{D}}_2 \in \Omega$ be two TDFLSs. Then the ranking order of these two TDFLSs are assigned as,

1. If $Val_D(\tilde{\mathbb{D}}_1) > Val_D(\tilde{\mathbb{D}}_2)$, then $\tilde{\mathbb{D}}_1 \succ \tilde{\mathbb{D}}_2$.
2. If $Val_D(\tilde{\mathbb{D}}_1) < Val_D(\tilde{\mathbb{D}}_2)$, then $\tilde{\mathbb{D}}_1 \prec \tilde{\mathbb{D}}_2$.
3. If $Val_D(\tilde{\mathbb{D}}_1) = Val_D(\tilde{\mathbb{D}}_2)$, then $\tilde{\mathbb{D}}_1 \sim \tilde{\mathbb{D}}_2$.

Where, "$\prec$","$\succ$","$\sim$" define usual "less than", "greater than", and, "equal to" in the sense of dense fuzzy environment.

On the basis of the notion of the $Val_D(.)$ function following theorem can be established easily.

Theorem 1 *$Val_D(.)$ is a linear function, i.e., if $\tilde{\mathbb{D}}_1, \tilde{\mathbb{D}}_2 \in \Omega$, then for any positive scalar quantity ρ, $Val_D(\rho\tilde{\mathbb{D}}_1 + \tilde{\mathbb{D}}_2) = \rho Val_D(\tilde{\mathbb{D}}_1) + Val_D(\tilde{\mathbb{D}}_2)$.*

The $Val_D(.)$ values of the TDFLSs are varied depending upon different values of the key vectors and the fuzzy deviation parameters, for which the following cases arose.

Case I: If $k^1 = k^2 = k$, then the TDFLS with double key vectors turns into a TDFLS with single key and the value of $Val_D(.)$ will be $z\Big[1 + \frac{\nu-\mu}{6K} - \frac{(\nu-\mu)}{6N}\Big(\frac{1}{2} + \frac{1}{3} + \ldots + \frac{1}{N+1}\Big)\Big]$.

Table 1 Difuzzified values of TDFLSs by two different methods

TDFLS	Defuzzifiers		Key vectors	Defuzzified value	
	μ	ν	(k^1, k^2)	De's approach [9]	Proposed approach
$\tilde{10}$	0.75	1.25	(0.95, 1.25)	10.5263	10.3509
$\tilde{15}$	0.80	1.05	(0.85, 1.02)	15.3309	15.2206
$\tilde{47}$	0.78	1.05	(1.05, 1.30)	47.7618	47.5079
$\tilde{84}$	0.72	1.15	(0.92, 1.38)	85.0652	84.7101

Case II: If $k^1 = k^2 = 1$, then the TDFLS $\tilde{\mathbb{B}}$ turns into a TDFS and $Val_D(.)$ value will be $z\left[1 + \frac{\nu-\mu}{6} - \frac{(\nu-\mu)}{6N}\left(\frac{1}{2} + \frac{1}{3} + \ldots + \frac{1}{N+1}\right)\right]$.

Case III: If $k^1 = k^2 = 1$ and $N \to \infty$, then the TDFLS $\tilde{\mathbb{B}}$ turns into a TFN whose defuzzified value will be $z\left[1 + \frac{\nu-\mu}{6}\right]$.

Case IV: If $\nu = \mu$, $k^1 = k^2$, and $N \to \infty$, then the TDFLS $\tilde{\mathbb{B}}$ turns into a crisp number and consequently, the $Val_D(.)$ of $\tilde{\mathbb{B}}$ gives the value z. From which we may infer that the $Val_D(.)$ value of a crisp number is nothing but itself.

De [9] schematized a way out for finding the key vectors of a TDFLS. According to De [9], if $\mathbb{D}_L$ be a TDFLS with double key vectors k^1 and k^2, then we must have $k^1 \geq \frac{\mu a}{2(z - 2z^L)}$ and $k^2 \geq \frac{\nu z}{2(2z^U - z)}$. (where z^L and z^U are considered as lower and upper bound of $\mathbb{D}_L$, respectively). Also, for a TDFLS $\mathbb{D}_L$ with single key vector (k), we have $k \geq \max\left\{\frac{z}{2}\left(1 - \frac{\mu}{2k}\right), \frac{z}{2}\left(1 + \frac{\nu}{2k}\right)\right\}$.

3.1 Comparison with the Existing Defuzzification Method of TDFLS

To defuzzify the TDFLS $\tilde{\mathbb{D}}$, De [9] defined the ranking index of TDFLSs as, $I(\tilde{\mathbb{D}}) = \frac{1}{2N}\sum_{r=0}^{N}\int_0^1\{\mathbb{D}_L + \mathbb{D}_R\}d\alpha$, the value of which tends to $z\left\{1 + \frac{1}{4}\left(\frac{\nu}{k^2} - \frac{\mu}{k^1}\right)\right\}$ as $N \to \infty$.

In the following table, we enlist some TDFLSs with their defuzzified values obtained by applying De's method [9], and proposed method. From Table 1, we observed that the defuzzified values obtained by the proposed method are closer enough to the central value of the TDFLSs than that of the value obtained by De's method. So we may say that our proposed defuzzification function is superior to the existing method.

4 Model Formulation of Bi-matrix Games in Dense Fuzzy Environment

Let us consider a bi-matrix game in dense fuzzy environment. Assume that two players Q_I, and Q_{II} are involved here. Suppose $\Gamma_m = \{\gamma_1, \gamma_2, \ldots, \gamma_m\}$, and $\Gamma_n = \{\delta_1, \delta_2, \ldots, \delta_n\}$, be the set of pure strategies of Player Q_I, and Player Q_{II}, respectively. Also assume two sets X_m, and Y_n, as the sets of mixed strategies of Player Q_I, and Player Q_{II}, respectively. Where, X_m, and Y_n are defined as, $X_m = \{\mathbf{x} = (x_1, x_2, \ldots, x_m)^T | \sum_{p=1}^{m} x_p = 1; x_p \geq 0 \text{ for, } p = 1, 2, \ldots, m\}$, and $Y_n = \{\mathbf{y} = (y_1, y_2, \ldots, y_n)^T | \sum_{q=1}^{n} y_q = 1; y_q \geq 0 \text{ for, } q = 1, 2, \ldots, n\}$. Actually, x_p, and y_q stand for the probabilities in which Player Q_I, and Player Q_{II} choose their pure strategies $\gamma_p (\in \Gamma_m)$, and $\delta_q (\in \Gamma_n)$, respectively. If Player Q_I chooses her/his pure strategy γ_p, and Player Q_{II} chooses her/his pure strategy δ_q, then at the situation $< \gamma_p, \delta_q >$, the gain of Player Q_I will be $\tilde{\Phi}_{pq}$, while the gain of Player Q_{II} will be $\tilde{\Psi}_{pq}$. Thus the pay-off matrix for Player Q_I can be expressed as, $\tilde{\Phi} = (\tilde{\Phi}_{pq})_{m\times n}$, and the pay-off matrix for Player Q_{II} be $\tilde{\Psi} = (\tilde{\Psi}_{pq})_{m\times n}$. Here $\tilde{\Phi}_{pq}$, and $\tilde{\Psi}_{pq}$ are estimated as some TDFLSs having expressions $\tilde{\Phi}_{pq} = \Big\langle \phi_{pq}\Big\{1 - \mu_{pq}^{\phi}\Big(\frac{1}{k_{pq}^{1\phi}} - \frac{1}{r+1}\Big)\Big\}, \phi_{pq}, \phi_{pq}\Big\{1 - \nu_{pq}^{\phi}\Big(\frac{1}{k_{pq}^{2\phi}} - \frac{1}{r+1}\Big)\Big\}\Big\rangle$, and $\tilde{\Psi}_{pq} = \Big\langle \psi_{pq}\Big\{1 - \mu_{pq}^{\psi}\Big(\frac{1}{k_{pq}^{1\psi}} - \frac{1}{r+1}\Big)\Big\}, \psi_{pq}, \psi_{pq}\Big\{1 - \nu_{pq}^{\psi}\Big(\frac{1}{k_{pq}^{2\psi}} - \frac{1}{r+1}\Big)\Big\}\Big\rangle$. Eventually, the bi-matrix game can be written as, $B_{\tilde{\Phi},\tilde{\Psi}} = \langle X_m, Y_n, \tilde{\Phi}, \tilde{\Psi} \rangle$.

If Player Q_I chooses the mixed strategy $\mathbf{x}(\in X_m)$ and Player Q_I chooses the mixed strategy $\mathbf{y}(\in Y_n)$, then the expected pay-off of Player Q_I is calculated as,

$$\tilde{E}_{Q_I}(\mathbf{x}, \mathbf{y}) = \sum_{p=1}^{m}\sum_{q=1}^{n} x_p \tilde{\Phi}_{pq} y_q = \Big\langle \sum_{p=1}^{m}\sum_{q=1}^{n} x_p \phi_{pq}\Big\{1 - \mu_{pq}^{\phi}\Big(\frac{1}{k_{pq}^{1\phi}} - \frac{1}{r+1}\Big)\Big\} y_q, \sum_{p=1}^{m}\sum_{q=1}^{n} x_p \phi_{pq} y_q, \sum_{p=1}^{m}\sum_{q=1}^{n} x_p \phi_{pq}\Big\{1 - \nu_{pq}^{\phi}\Big(\frac{1}{k_{pq}^{2\phi}} - \frac{1}{r+1}\Big)\Big\} y_q \Big\rangle. \quad (3)$$

Similarly, the expected pay-off for Player Q_{II} can be calculated as,

$$\tilde{E}_{Q_{II}}(\mathbf{x}, \mathbf{y}) = \sum_{p=1}^{m}\sum_{q=1}^{n} x_p \tilde{\Psi}_{pq} y_q = \Big\langle \sum_{p=1}^{m}\sum_{q=1}^{n} x_p \psi_{pq}\Big\{1 - \mu_{pq}^{\psi}\Big(\frac{1}{k_{pq}^{1\psi}} - \frac{1}{r+1}\Big)\Big\} y_q, \sum_{p=1}^{m}\sum_{q=1}^{n} x_p \psi_{pq} y_q, \sum_{p=1}^{m}\sum_{q=1}^{n} x_p \psi_{pq}\Big\{1 - \nu_{pq}^{\psi}\Big(\frac{1}{k_{pq}^{2\psi}} - \frac{1}{r+1}\Big)\Big\} y_q \Big\rangle. \quad (4)$$

Definition 6 (*Nash equilibrium solution*) A pair of mixed strategy $(\mathbf{x}^*, \mathbf{y}^*) \in X_m \times Y_n$ is said to be Nash equilibrium solution of the bi-matrix game $B_{\tilde{\Phi},\tilde{\Psi}}$ if the following inequalities are satisfied: $\mathbf{x}^T \tilde{\Phi} \mathbf{y}^* \preccurlyeq \mathbf{x}^{*T} \tilde{\Phi} \mathbf{y}^*, \forall \mathbf{x} \in X_m$ and $\mathbf{x}^{*T} \tilde{\Psi} \mathbf{y} \preccurlyeq \mathbf{x}^{*T} \tilde{\Psi} \mathbf{y}^*, \ \forall \mathbf{y} \in Y_n$.

The pair $(\mathbf{x}^*, \mathbf{y}^*)$ is also termed as the optimal solution of the bi-matrix game $B_{\tilde{\Phi},\tilde{\Psi}}$, and $\mathbf{x}^{*T} \tilde{\Phi} \mathbf{y}^*$, and $\mathbf{x}^{*T} \tilde{\Psi} \mathbf{y}^*$ are designated as the Nash equilibrium values of the bi-matrix game $B_{\tilde{\Phi},\tilde{\Psi}}$ for Player Q_I, and Player Q_{II}, respectively.

According to Nash existence theorem [20], every bi-matrix game has at least one equilibrium solution, and determine the Nash equilibrium solution of the bi-matrix game $B_{\tilde{\Phi},\tilde{\Psi}}$ is equivalent to resolve the following QDFPP:

$$\begin{aligned} &\max\{\mathbf{x}(\tilde{\Phi}+\tilde{\Psi})\mathbf{y}-\tilde{V}-\tilde{W}\} \\ &\text{subject to, } \tilde{\Phi}\mathbf{y} \preccurlyeq \tilde{V}e^m, \mathbf{x}^T\tilde{\Psi} \preccurlyeq \tilde{W}e^n \\ &\mathbf{x}\in X_m, \mathbf{y}\in Y_n \end{aligned} \tag{5}$$

Further, if the optimal solution of the QDFPP is determined in the form $(\mathbf{x}^*, \mathbf{y}^*, \tilde{V}^*, \tilde{W}^*)$, then $\tilde{V}^* = \mathbf{x}^{*T}\tilde{\Phi}\mathbf{y}^*$, $\tilde{W}^* = \mathbf{x}^{*T}\tilde{\Psi}\mathbf{y}^*$, and $\mathbf{x}^{*T}(\tilde{\Phi}+\tilde{\Psi})\mathbf{y}^* - \tilde{V} - \tilde{W} = 0$.

Now, the aforementioned QDFPP (5) can be rewritten as,

$$\begin{aligned} &\max\left\{\sum_{p=1}^{m}\sum_{q=1}^{n} x_p(\tilde{\Phi}_{pq}+\tilde{\Psi}_{pq})y_q - \tilde{V} - \tilde{W}\right\} \\ &\text{subject to, } \sum_{q=1}^{n}\tilde{\Phi}_{pq}y_q \preccurlyeq \tilde{V},\ \sum_{p=1}^{m}x_p\tilde{\Psi}_{pq} \preccurlyeq \tilde{W} \\ &\sum_{p=1}^{m}x_p = 1, \sum_{q=1}^{n}y_q = 1;\ x_p, y_q \geq 0,\ \text{for } p=1,2,\ldots,m; q=1,2,\ldots,n \end{aligned} \tag{6}$$

By exploiting the predefined ranking index and the linearity property of it, we can present Problem (6) as follows,

$$\begin{aligned} &\max\left\{\sum_{p=1}^{m}\sum_{q=1}^{n} x_p\Big(Val_D(\tilde{\Phi}_{pq}+\tilde{\Psi}_{pq})\Big)y_q - V - W\right\} \\ &\text{subject to, } \sum_{q=1}^{n} Val_D\Big(\tilde{\Phi}_{pq}\Big)y_q \leq V,\ \sum_{p=1}^{m}x_p Val_D\Big(\tilde{\Psi}_{pq}\Big)\mathbb{D} \leq W \\ &\sum_{p=1}^{m}x_p = 1, \sum_{q=1}^{n}y_q = 1;\ x_p, y_q \geq 0,\ \text{for } p=1,2,\ldots,m; q=1,2,\ldots,n \end{aligned} \tag{7}$$

Where, V, and W are the crisp equivalent of $\tilde{V}$, and $\tilde{W}$, respectively.

Problem (6) is equivalent to Problem (8).

$$\begin{aligned} &\max\Bigg\{\sum_{p=1}^{m}\sum_{q=1}^{n}x_p\Bigg(\phi_{pq}\Bigg\{1+\frac{1}{6}\Bigg(\frac{\nu_{pq}^{\phi}}{k_{pq}^{2\phi}}-\frac{\mu_{pq}^{\phi}}{k_{pq}^{1\phi}}\Bigg)-\frac{(\nu_{pq}^{\phi}-\mu_{pq}^{\phi})}{6N}\Bigg(\frac{1}{2}+\ldots+\frac{1}{N+1}\Bigg)\Bigg\}+ \\ &\psi_{pq}\Bigg\{1+\frac{1}{6}\Bigg(\frac{\nu_{pq}^{\psi}}{k_{pq}^{2\psi}}-\frac{\mu_{pq}^{\psi}}{k_{pq}^{1\psi}}\Bigg)-\frac{(\nu_{pq}^{\psi}-\mu_{pq}^{\psi})}{6N}\Bigg(\frac{1}{2}+\ldots+\frac{1}{N+1}\Bigg)\Bigg\}\Bigg)y_q - V - W\Bigg\} \\ &\text{subject to, } \sum_{q=1}^{n}\phi_{pq}\Bigg\{1+\frac{1}{6}\Bigg(\frac{\nu_{pq}^{\phi}}{k_{pq}^{2\phi}}-\frac{\mu_{pq}^{\phi}}{k_{pq}^{1\phi}}\Bigg)-\frac{(\nu_{pq}^{\phi}-\mu_{pq}^{\phi})}{6N}\Bigg(\frac{1}{2}+\ldots+\frac{1}{N+1}\Bigg)\Bigg\}y_q \leq V \\ &\sum_{p=1}^{m}x_p\psi_{pq}\Bigg\{1+\frac{1}{6}\Bigg(\frac{\nu_{pq}^{\psi}}{k_{pq}^{2\psi}}-\frac{\mu_{pq}^{\psi}}{k_{pq}^{1\psi}}\Bigg)-\frac{(\nu_{pq}^{\psi}-\mu_{pq}^{\psi})}{6N}\Bigg(\frac{1}{2}+\ldots+\frac{1}{N+1}\Bigg)\Bigg\} \leq W \end{aligned}$$

$$\sum_{p=1}^{m} x_p = 1, \sum_{q=1}^{n} y_q = 1;\ x_p,\ y_q \geq 0,\ \text{for } p = 1, 2, \ldots, m; q = 1, 2, \ldots, n \tag{8}$$

Now solving Problem (8), we obtain the Nash equilibrium solution $(\mathbf{x}^*, \mathbf{y}^*)$ and crisp-equivalent value of the game V, and W for Player Q_I, and Q_{II}, respectively.

4.1 Some Special Cases

Case I: If $k_{pq}^{2\phi} = k_{pq}^{1\phi} = k_{pq}^{\phi}$, $k_{pq}^{2\psi} = k_{pq}^{1\psi} = k_{pq}^{\psi}$, then the pay-offs of the bi-matrix game $B_{\tilde{\Phi},\tilde{\Psi}}$ are turned into TDFLSs with single keys and the crisp-equivalent pay-offs of such games will be $Val_D\left(\tilde{\Phi}_{pq}\right) = \phi_{pq}\left\{1 + \frac{1}{6}\left(\frac{\nu_{pq}^{\phi} - \mu_{pq}^{\phi}}{k_{pq}^{\phi}}\right) - \frac{(\nu_{pq}^{\phi} - \mu_{pq}^{\phi})}{6N}\left(\frac{1}{2} + \cdots + \frac{1}{N+1}\right)\right\}$, and $Val_D\left(\tilde{\Psi}_{pq}\right) = \psi_{pq}\left\{1 + \frac{1}{6}\left(\frac{\nu_{pq}^{\psi} - \mu_{pq}^{\psi}}{k_{pq}^{\psi}}\right) - \frac{(\nu_{pq}^{\psi} - \mu_{pq}^{\psi})}{6N}\left(\frac{1}{2} + \cdots + \frac{1}{N+1}\right)\right\}$, for Player Q_I, and Player Q_{II}, respectively. Solving which problem we can obtain the solution for a bi-matrix game whose pay-offs are considered as TDFLSs with single key.

Case II: If $k_{pq}^{2\phi} = k_{pq}^{1\phi} = k_{pq}^{\phi} = k_{pq}^{2\psi} = k_{pq}^{1\psi} = k_{pq}^{\psi} = 1$, then the pay-offs of the bi-matrix game $B_{\tilde{\Phi},\tilde{\Psi}}$ are turned into TDFSs and the crisp-equivalent pay-offs of such games will be $Val_D\left(\tilde{\Phi}_{pq}\right) = \phi_{pq}\left\{1 + \frac{\nu_{pq}^{\phi} - \mu_{pq}^{\phi}}{6} - \frac{(\nu_{pq}^{\phi} - \mu_{pq}^{\phi})}{6N}\left(\frac{1}{2} + \cdots + \frac{1}{N+1}\right)\right\}$, and $Val_D\left(\tilde{\Psi}_{pq}\right) = \psi_{pq}\left\{1 + \frac{\nu_{pq}^{\psi} - \mu_{pq}^{\psi}}{6} - \frac{(\nu_{pq}^{\psi} - \mu_{pq}^{\psi})}{6N}\left(\frac{1}{2} + \cdots + \frac{1}{N+1}\right)\right\}$, for Player Q_I, and Player Q_{II}, respectively. Solving which problem we can obtain the solution for a bi-matrix game whose pay-offs are considered in terms of TDFSs.

Case III: If $k_{pq}^{2\phi} = k_{pq}^{1\phi} = k_{pq}^{\phi} = k_{pq}^{2\psi} = k_{pq}^{1\psi} = k_{pq}^{\psi} = 1$, and also, $N \to \infty$, then the pay-offs of the bi-matrix game $B_{\tilde{\Phi},\tilde{\Psi}}$ are transformed into TFNs and the crisp-equivalent pay-offs of such games will be $Val_D\left(\tilde{\Phi}_{pq}\right) = \phi_{pq}\left\{1 + \frac{\nu_{pq}^{\phi} - \mu_{pq}^{\phi}}{6}\right\}$, and $Val_D\left(\tilde{\Psi}_{pq}\right) = \psi_{pq}\left\{1 + \frac{\nu_{pq}^{\psi} - \mu_{pq}^{\psi}}{6}\right\}$, for Player Q_I, and Player Q_{II}, respectively. Solving which problem we can obtain the solution for a bi-matrix game with TFN pay-offs.

Case IV: If $k_{pq}^{2\phi} = k_{pq}^{1\phi} = k_{pq}^{\phi} = k_{pq}^{2\psi} = k_{pq}^{1\psi} = k_{pq}^{\psi} = 1$, $N \to \infty$, and also, $\mu_{pq}^{\phi} = \nu_{pq}^{\phi}$, and $\mu_{pq}^{\psi} = \nu_{pq}^{\psi}$, then the pay-offs of the bi-matrix game $B_{\tilde{\Phi},\tilde{\Psi}}$ are turned into TDFSs and the crisp-equivalent pay-offs of such games will be $Val_D\left(\tilde{\Phi}_{pq}\right) = \phi_{pq}$, and $Val_D\left(\tilde{\Psi}_{pq}\right) = \psi_{pq}$, for Player Q_I, and Player Q_{II}, respectively. Solving which problem we can obtain the solution for a bi-matrix game whose pay-offs are considered in as crisp numbers.

5 Numerical Illustration

In this section, we have extended the idea of the tourism planning strategy problem from Fei and Li [10] in the dense fuzzy environment.

5.1 *Tourism Planning Strategy Problem*

Assume that two tourism companies TC_I and TC_{II} want to maximize their profit in a targeted market without cooperating. Suppose Company TC_I has two options: "expanding the landscape" and "covetable package for tourists" to maximize the profit. Company TC_{II} also takes similar two options. But in the present pandemic situation, they are forced to make some little changes in their strategies to keep up their annual profit. Such changes left a remarkable impact on the gain of both of the companies. Bi-matrix game is the perfect tool to analyze such kind of problem, involving two players Company TC_I and Company TC_{II}. The pay-offs of this game problem are considered in the forms of TDFLSs. Let $\tilde{\Phi}$, and $\tilde{\Psi}$ be the pay-off matrices for Company TC_I, and Company TC_{II}, respectively. Suppose the pay-off matrices are given as, $\tilde{\Phi} = \begin{pmatrix} \tilde{180} & \tilde{156} \\ \tilde{90} & \tilde{175} \end{pmatrix}$, and $\tilde{\Psi} = \begin{pmatrix} \tilde{175} & \tilde{160} \\ \tilde{100} & \tilde{180} \end{pmatrix}$. The entries of these matrices are enlisted in Table 2. Assume that the key vectors of the entries of the pay-off matrix are strong enough, so that we may take all the right fuzzy deviations $\nu_{180}, \nu_{175}, \nu_{160}, \nu_{156}, \nu_{100}, \nu_{90}$ have the same value ν and the left fuzzy deviations $\mu_{180}, \mu_{175}, \mu_{160}, \mu_{156}, \mu_{100}, \mu_{90}$ have the same value μ. Also, assume that the lower and upper bounds of the entries are $\tilde{180}^L = 175,\ \tilde{175}^L = 178,\ \tilde{160}^L = 157,\ \tilde{156}^L = 154,\ \tilde{100}^L = 97,\ \tilde{90}^L = 85$, and $\tilde{180}^U_L = 185,\ \tilde{175}^U_L = 178,\ \tilde{160}^U_L = 165,\ \tilde{156}^U_L = 158,\ \tilde{100}^U_L = 105,\ \tilde{90}^U_L = 95$. Here, we choose the left and right fuzzy deviations as $\mu = 0.8$, and $\nu = 1.5$, respectively, i.e., the lower bound and the upper bound of the entries may deviate up to 0.8 units and up to 1.5 units, respectively. Using the proposed defuzzification method, we defuzzify the entries of the pay-off matrices and enlisted in Table 2.

Utilizing the working formula for finding the key vectors explained in Sect. 3, let us set the key vectors for the entries as, $k^1_{180} = 0.45,\ k^1_{175} = 0.44,\ k^1_{160} = 0.47,\ k^1_{156} = 0.49,\ k^1_{100} = 0.44,\ k^1_{90} = 0.45$, and $k^2_{180} = 0.72,\ k^2_{175} = 0.70,\ k^2_{160} = 0.74,\ k^2_{156} = 0.74,\ k^2_{100} = 0.70,\ k^2_{90} = 0.68$. Substituting the defuzzified values depicted in Table 2 along with this key vectors in Problem (8), we find the Nash equilibrium strategies $(\mathbf{x}^*, \mathbf{y}^*)$ and the crisp-equivalent values of the value of the game V and W for the Company TC_I, and TC_{II}, respectively. The results obtained from Problem (8) for different trials are enlisted in Table 3.

Similarly, we can set the single key vectors for the entries as $k_{180} = 0.97,\ k_{175} = 0.95,\ k_{160} = 0.96,\ k_{156} = 0.95,\ k_{100} = 0.92,\ k_{90} = 0.94$, and solve the bi-matrix

Table 2 TDFLS expression of the entries of the pay-off matrices

Entries	Expressions	Defuzzified value
$\tilde{180}$	$\langle 180\{1-\mu_{180}(\frac{1}{k^1_{180}}-\frac{1}{r+1})\}, 180, 180\{1+\nu_{180}(\frac{1}{k^2_{180}}-\frac{1}{r+1})\}\rangle$	$180\left\{1+\frac{0.25}{k^2_{180}}-\frac{0.13}{k^1_{180}}-\frac{0.7}{N}\left(\frac{1}{2}+\cdots+\frac{1}{N+1}\right)\right\}$
$\tilde{175}$	$\langle 175\{1-\mu_{175}(\frac{1}{k^1_{175}}-\frac{1}{r+1})\}, 175, 175\{1+\nu_{175}(\frac{1}{k^2_{175}}-\frac{1}{r+1})\}\rangle$	$175\left\{1+\frac{0.25}{k^2_{175}}-\frac{0.13}{k^1_{175}}-\frac{0.7}{N}\left(\frac{1}{2}+\cdots+\frac{1}{N+1}\right)\right\}$
$\tilde{160}$	$\langle 160\{1-\mu_{160}(\frac{1}{k^1_{160}}-\frac{1}{r+1})\}, 160, 160\{1+\nu_{160}(\frac{1}{k^2_{160}}-\frac{1}{r+1})\}\rangle$	$160\left\{1+\frac{0.25}{k^2_{160}}-\frac{0.13}{k^1_{160}}-\frac{0.7}{N}\left(\frac{1}{2}+\cdots+\frac{1}{N+1}\right)\right\}$
$\tilde{156}$	$\langle 156\{1-\mu_{156}(\frac{1}{k^1_{156}}-\frac{1}{r+1})\}, 156, 156\{1+\nu_{156}(\frac{1}{k^2_{156}}-\frac{1}{r+1})\}\rangle$	$156\left\{1+\frac{0.25}{k^2_{156}}-\frac{0.13}{k^1_{156}}-\frac{0.7}{N}\left(\frac{1}{2}+\cdots+\frac{1}{N+1}\right)\right\}$
$\tilde{100}$	$\langle 100\{1-\mu_{100}(\frac{1}{k^1_{100}}-\frac{1}{r+1})\}, 100, 100\{1+\nu_{100}(\frac{1}{k^2_{100}}-\frac{1}{r+1})\}\rangle$	$100\left\{1+\frac{0.25}{k^2_{100}}-\frac{0.13}{k^1_{100}}-\frac{0.7}{N}\left(\frac{1}{2}+\cdots+\frac{1}{N+1}\right)\right\}$
$\tilde{90}$	$\langle 90\{1-\mu_{90}(\frac{1}{k^1_{90}}-\frac{1}{r+1})\}, 90, 90\{1+\nu_{90}(\frac{1}{k^2_{90}}-\frac{1}{r+1})\}\rangle$	$90\left\{1+\frac{0.25}{k^2_{90}}-\frac{0.13}{k^1_{90}}-\frac{0.7}{N}\left(\frac{1}{2}+\cdots+\frac{1}{N+1}\right)\right\}$

problem for the pay-offs of TDFLSs with single keys. The result is also written in Table 3 for different trials.

Aiming to compare the results obtained by the proposed methodology with different environment, we reduce the TDFLS entries into TDFS, TFN, and crisp entries in accordance to the cases explained in Sect. 4.1. The results obtained in these three cases are also inscribed in Table 3.

5.2 *Results and Discussions*

From Table 3, the following observations are made,

(i) Solving the bi-matrix game with TDLFS entries with double key vectors, we get the profit of Company TC_I, and Company TC_{II} as $V^* = 164.0358$, and $W^* = 165.8278$, respectively at 4th trial. Also, the Nash equilibrium strategies are $x^* = (0.8412, 0.1588)$ and $y^* = (0.1638, 0.8362)$ for the Company TC_I, and Company TC_{II}, respectively. These indicate that the profit for Company TC_I, and Company TC_{II} are "around 164.0358 units" and "around 165.8278 units", respectively, when Company TC_I uses strategy γ_1, with probability 84.12% and strategy γ_2, with probability 15.88% and Company TC_{II} uses strategy δ_1, with probability 16.38% and strategy δ_2, with probability 83.62%.

Table 3 Different game values by different approaches

Pay-offs	Trials	Value of the game		Optimal strategies	
		V	W	**x***	**y***
Crisp	…	162.5769	163.1579	(0.7895, 0.2105)	(0.2105, 0.7895)
TFNs	…	176.8716	178.1933	(0.7895, 0.2105)	(0.2105, 0.7895)
TDFS	1	169.5275	172.6751	(0.7895, 0.2105)	(0.2105, 0.7895)
	2	171.1161	174.2935		
	3	172.1439	175.3406		
	4	172.8913	176.1015		
TDFLS (Single key)	1	170.4534	172.6751	(0.8390, 0.1610)	(0.1752, 0.8248)
	2	172.2443	175.2749		
	3	173.0697	176.1619		
	4	173.8173	176.9236		
TDFLS (Double key)	1	160.6723	162.4015	(0.8412, 0.1588)	(0.1638, 0.8362)
	2	162.2607	164.0192		
	3	163.2885	164.8694		
	4	164.0358	165.8278		

(ii) Table 3 shows that the value of the game increases with the number of trials when the game is developed in the dense fuzzy environment either in dense fuzzy or in dense fuzzy lock environment with single or double key vectors.

(iii) Table 3 also shows that on account of maximizing the gain, the bi-matrix game with TDFLS pay-offs with a single key gives a better result than the other two environments. Therefore, it would be better to consider bi-matrix games with TDFLSs pay-offs with a single key vector.

(iv) Finally, we observe that we obtained almost the same Nash equilibrium strategies for both the players for bi-matrix games in different environments except TDFLS with single or double keys vectors. Henceforth, if decision-makers estimate pay-offs of a bi-matrix game in terms of TDFLSs, they have to adjust their strategies to obtain optimal outcomes. Furthermore, if one delineates the matrix game in any one of the dense fuzzy backgrounds, she/he achieves the same optimal strategies for different trials. Thus a decision-maker may improve his/her gain by increasing the trials and applying the same strategies. This result is one of the significant benefits of adopting dense fuzzy sets in game theory.

6 Conclusion

To mitigate the fuzziness in information in a decision-making problem and make the problem more realistic, TDFLS performs a vital role. In the present paper, we develop a bi-matrix game under the dense fuzzy environment. To solve such a problem, we have introduced a new ranking function of TDFLSs. We illustrate the rendered model through a tourism planning strategy problem to check the efficacy and applicability. Observing the result of this problem, it is has been seen that the game with pay-offs of TDFLSs with a single key gives the best outcome instead of others. Furthermore applying dense fuzzy model players can increase their profit by upgrading trials unaltering their strategies.

The proposed methodology is developed based on a ranking function. So, we need a more general methodology to obtain the fuzzy value of the game directly, which is left as the future study of our present work. Although the proposed model is illustrated through a tourism planning strategy problem, it can be performed in counter-terrorism issues, management problems, medical sciences, etc.

References

1. Atanassov KT (1986) Intuitionistic fuzzy sets. Fuzzy Sets Syst 20:87–96
2. An JJ, Li DF, Nan JX (2017) A mean-area ranking based non-linear programming approach to solve intuitionistic fuzzy bi-matrix games. J Intell Fuzzy Syst 33(1):563–573
3. An JJ, Li DF (2019) A linear programming approach to solve constrained bi-matrix games with intuitionistic fuzzy pay-offs. J Intell Fuzzy Syst 21:908–915
4. Bhaumik A, Roy SK (2019) Intuitionistic interval-valued hesitant fuzzy matrix games with a new aggregation operator for solving management problem. Granul Comput. https://doi.org/10.1007/s41066-019-00191-5
5. Bhaumik A, Roy SK, Weber GW (2020) Hesitant interval-valued intuitionistic fuzzy-linguistic term set approach in Prisoners' dilemma game theory using TOPSIS: a case study on Human-trafficking. Central Eur J Oper Res 28:797–816
6. Bhaumik A, Roy SK, Li DF (2020) $(\alpha, \beta, \gamma)-$ cut set based ranking approach to solving bi-matrix games in neutrosophic environment. Soft Comput. https://doi.org/10.1007/s00500-020-05332-6
7. Brikaa MG, Ammer ES, Zhang Z (2019) Solving bi-matrix games in tourism planning management under rough interval approach. Int J Math Sci Comput 4:44–62
8. De SK, Beg I (2016) Triangular dense fuzzy sets and new defuzzification methods. J Intell Fuzzy Syst 31(1):469–477
9. De SK (2018) Triangular dense fuzzy lock sets. Soft Comput 22(21):7243–7254
10. Fei W, Li DF (2016) Bilinear programming approach to solve interval bi-matrix games in tourism planning management. Int J Fuzzy Syst 18:504–510
11. Karmakar S, De SK, Goswami A (2017) A pollution sensitive dense fuzzy economic production quantity model with cycle time dependent production rate. J Clean Prod 154:139–150
12. Karmakar S, De SK, Goswami A (2018) A pollution sensitive remanufacturing model with waste items: triangular dense fuzzy lock set approach. J Clean Prod 187:789–803
13. Khan I, Mehra A (2020) A novel equilibrium solution concept for intuitionistic fuzzy bi-matrix games considering proportion mix of possibility and necessity expectations. Granular Comput 5:461–483

14. Li C, Li M (2019) A new bi-matrix game model with fuzzy pay-offs in credibility space. Int J Fuzzy Comput Model 10(5):556–563
15. Li DF, Yang J (2013) A difference-index based ranking bilinear programming approach to solving bi-matrix games with pay-offs of trapezoidal intuitionistic fuzzy numbers. J Appl Math 13:1–13
16. Liu K, Xing Y (2019) Solving bi-matrix games based on fuzzy payoffs via utilizing the interval value function method. Mathematics 7(5):469. https://doi.org/10.3390/math7050469
17. Nan JX, Li DF, An JJ (2017) Solving bi-matrix games with intuitionistic fuzzy pay-offs. J Intell Fuzzy Syst 33(6):3723–3732
18. Nash JE (1951) Non cooperative games. Ann Math 54(2):286–295
19. Nayak PK, Pal M (2011) Intuitionistic fuzzy optimization technique for nash equilibrium solution of multi-objective bi-matrix game. J Uncertain Syst 5(4):271–285
20. Owen G (1995) Game theory, 3rd edn. Academic, New York
21. Qiu D, Xing X, Chen S (2018) Solving fuzzy matrix games through a ranking value function method. J Math Comput Sci 18(2):175–183
22. Seikh MR, Nayak PK, Pal M (2015) Solving bi-matrix games with pay-offs of triangular intuitionistic fuzzy numbers. Eur J Pure Appl Math 8(2):153–171
23. Seikh MR, Nayak PK, Pal M (2016) An alternative approach to solve bi-matrix games with intuitionistic fuzzy goals. Int J Fuzzy Comput Model 1(4):362–381
24. Seikh MR, Karmakar S, Xia M (2016) Solving matrix games with hesitant fuzzy pay-offs. Iran J Fuzzy Syst 17(4):25–40
25. Yang J, Fei W, Li DF (2016) Non-linear programming approach to solve bi-matrix games with payoffs represented by I-fuzzy numbers. Int J Fuzzy Syst 18:492–503
26. Zadeh LA (1965) Fuzzy sets. Inf Control 8(33):338–352
27. Zhang W, Xing Y, Qiu D (2017) Multiobjective fuzzy bi-matrix game model: a multicriteria non-linear programming approach. Symmetry 9(8):159. https://doi.org/10.3390/sym9080159

Real-Time Error Correction Codes for Deletable Errors

Ghurumuruhan Ganesan

Abstract In this paper, we study codes for correcting deletable errors in binary words, where each bit is either retained, substituted, erased or deleted, and the total number of errors is much smaller compared to the length of the codeword. We construct codes capable of correcting errors in the received codeword in real time with a small delay and low probability of decoding error.

Keywords Deletable errors · redundancy of codes · real-time correction

1 Introduction

Consider the recovery of a binary word corrupted by deletable errors, where each error could either be an erasure, bit substitution (i.e., a one is converted to a zero and vice versa) or deletion. Regarding deletions, [5] used Varshamov-Tenengolts codes (VT codes) for correcting a single deletion. Later, [3] described codes for correcting an arbitrary number of deletions and recently, in [7], codes with near-optimal redundancy for combating a fixed number of deletions are constructed using hash functions. Error correction codes with time constraints have also been studied before. For example, [9] focused on developing bidirectional interactive algorithms with low information exchange for correcting deletions. An estimate of the number of rounds needed for synchronization was obtained using a divide and conquer strategy. Streaming codes with delay constraints for erasure and burst channels have been studied in [4] (see also references therein).

In this paper, we seek codes with low redundancy that are capable of correcting deletable errors (that could be an erasure, substitution or deletion) in real time with low delay, a notion made precise in the next section. We use modified Varshamov-Tenengolts codes to construct the desired codes and also obtain an estimate of the fraction of error patterns that are correctable by the proposed codes.

G. Ganesan (✉)
Institute of Mathematical Sciences, HBNI, Chennai, India

© The Author(s), under exclusive license to Springer Nature Singapore Pte Ltd. 2022
D. Giri et al. (eds.), *Proceedings of the Seventh International Conference on Mathematics and Computing*, Advances in Intelligent Systems and Computing 1412,
https://doi.org/10.1007/978-981-16-6890-6_57

The paper is organized as follows: In Sect. 2, we state our main result Theorem 1 regarding codes capable of real-time correction with small delay and also Theorem 2 describing codes for correcting $P-$far deletable error patterns. In Sect. 3, we prove Theorem 2 and also describe an error correction algorithm $\mathcal{A}$ for correcting $P-$far errors. In Sect. 4, we use the correction algorithm $\mathcal{A}$ to construct codes with real-time correction capability and prove Theorem 1.

2 Main Results

A word of length n is a vector $\mathbf{x} = (x_1, \ldots, x_n) \in \{0, 1\}^n$. A deletable error pattern of length n is an element $\mathbf{g} = (\mathbf{e}, \mathbf{v}) = ((e_1, \ldots, e_n), (s_1, \ldots, s_n)) \in \{0, 1\}^n \times \{D, E, F\}^n$, with D, E and F denoting deletion, erasure and substitution, respectively. The word $\mathbf{y} = F_{\mathbf{g}}(\mathbf{x})$ obtained after $\mathbf{x}$ is corrupted by the error pattern $\mathbf{g}$ is defined as follows: If $e_i = 1$, then the bit x_i is deleted, erased or substituted depending on whether $s_i = D$, E or F, respectively. If $e_i = 0$, no change occurs to the bit x_i. For integer $1 \leq t \leq n$, we define $F_{\mathbf{e}}(\mathbf{x}, t)$ to be the result of corrupting the first t bits of $\mathbf{x}$ according to the first t entries in $\mathbf{e}$.

For example, if $n = 5$ and $\mathbf{g} = ((1, 0, 1, 1, 1), (F, E, D, F, E))$, then $F_{\mathbf{g}}(\mathbf{x}) = (1 - x_1, x_2, 1 - x_4, \varepsilon)$, where ε is the erasure symbol and $F_{\mathbf{g}}(\mathbf{x}, 3) = (1 - x_1, x_2)$. For $r \geq 1$, we define $\mathcal{E}_n(r) := \left\{(\mathbf{e}, \mathbf{v}) : \sum_{i=1}^{n} e_i \leq r\right\}$ to be the set of all possible $n-$length patterns containing at most r deletable errors. We also let $\mathcal{E}_n := \cup_{1 \leq r \leq n} \mathcal{E}_n(r)$ be the set of all possible $n-$length error patterns.

An $n-$length code $\mathcal{C}$ of size $\#\mathcal{C} = q$ is a set $\{\mathbf{x}_1, \ldots, \mathbf{x}_q\} \subset \{0, 1\}^n$. We define the redundancy of $\mathcal{C}$ to be

$$R(\mathcal{C}) := n - \log(\#\mathcal{C}), \tag{2.1}$$

where all logarithms are to the base 2 throughout.

Let $\mathcal{F} \subseteq \mathcal{E}_n$ be any set of $n-$length error patterns. The code $\mathcal{C}$ is said to be capable of correcting all error patterns in $\mathcal{F}$ if for any $\mathbf{x}_1 \neq \mathbf{x}_2 \in \mathcal{C}$ and any $\mathbf{g}_1, \mathbf{g}_2 \in \mathcal{F}$, we must have that the corruptions $F_{\mathbf{g}_1}(\mathbf{x}_1) \neq F_{\mathbf{g}_2}(\mathbf{x}_2)$. We have the following definition regarding real-time error correction.

Definition 1 For integer $d \geq 1$, we say that $\mathcal{C}$ can correct all error patterns in $\mathcal{F} \subseteq \mathcal{E}_n$ in real time with delay at most d if for any $1 \leq z \leq n - d$, $\mathbf{g}_1, \mathbf{g}_2 \in \mathcal{F}$ and $\mathbf{x}_1 \neq \mathbf{x}_2 \in \mathcal{C}$, we have that

$$F_{\mathbf{g}_1}(\mathbf{x}_1, z + d) \neq F_{\mathbf{g}_2}(\mathbf{x}_2, z + d).$$

The consequence of Definition 1 is this: Suppose the word $\mathbf{x}$ is transmitted and is corrupted by the error pattern $\mathbf{g}$. Given the partial word $F_{\mathbf{g}}(\mathbf{x}, z + d)$ obtained by corrupting the first $z + d$ bits of $\mathbf{x}$ for any $1 \leq z \leq n - d$, we must be able to determine the first z bits of $\mathbf{x}$.

For integers $t, d \geq 1$, let $\mathcal{F}(t, d) \subset \mathcal{E}_n(t)$ be a set of maximum size containing at most t errors that are correctable by the code $\mathcal{C}$ in real time with delay d. We define the $(t, d)-$*real-time error probability* of $\mathcal{C}$ to be

$$\rho(t, d) := \frac{\#\left(\mathcal{E}_n(t) \setminus \mathcal{F}(t, d)\right)}{\#\mathcal{E}_n(t)}.$$

If all the error patterns in $\mathcal{E}_n(t)$ are equally likely, then $\rho(t, d)$ could be interpreted as the decoding error probability.

For a given t, we are interested in studying the trade-off involved in choosing the delay parameter d. If $d = \infty$, then we simply require a code that corrects t errors and we know (see, for example, [7]) there are codes with negligible redundancy for correcting any fraction of errors. Intuitively therefore if d is large, then our code should have small redundancy since the effect of the delay becomes lesser and lesser as d increases. What about the redundancy when d is small? In the extreme case of $d = 1$, we must simply have a code that is equivalent to a repetition code since the first bit must be deductable using the second bit and the first two bits must be inferred using the first three bits and so on. In this case, the redundancy is $1 - \frac{1}{n}$. In general, for small d therefore we expect the code to have a large redundancy.

The following result obtains bounds on the delay, redundancy and the error probability of codes capable of correcting deletable errors, as a function of the total number of errors.

Theorem 1 *Let $1 \leq t_n \leq \frac{n^{\frac{1}{3}}}{6}$ be any integer. For every integer d_n satisfying $4t_n \leq d_n \leq \frac{2n}{3t_n^2}$, there is an $n-$length code $\mathcal{C}_{frac} = \mathcal{C}_{frac}(n)$ with a real-time delay d_n, redundancy*

$$R(\mathcal{C}_{frac}) \leq n \cdot \frac{4\log(2d_n)}{d_n}, \tag{2.2}$$

and a $(t_n, d_n)-$real-time error probability

$$\rho(t_n, d_n) \leq 11\frac{t_n^2 d_n}{n}. \tag{2.3}$$

Thus, if the maximum number of deletable errors in an $n-$length word is t_n, then for every d_n satisfying the bounds in the theorem, we can obtain a code capable of real-time correction with delay at most d_n, redundancy as in (2.2) and total error probability as in (2.3). Ideally, we prefer that $d_n \longrightarrow \infty$ and $\frac{t_n^2 d_n}{n} \longrightarrow 0$ as $n \to \infty$ so that for all n large, the code $\mathcal{C}_{frac}$ has redundancy $o(n)$ and an error probability of $o(1)$, using the standard $o(.)$ notations. Theorem 1 also illustrates the redundancy-error probability trade-off in terms of the desired delay d_n. Choosing larger d_n reduces redundancy but increases the error probability. We illustrate our result with some examples.

Example 1 Suppose we have a communication channel that introduces at most n^α deletable errors in an $n-$length word for some constant $\alpha < \frac{1}{3}$. Setting $t_n = n^\alpha$, choosing $d_n = 4n^\alpha \cdot \log n$ and using $\alpha < \frac{1}{3}$, we see that the conditions in Theorem 1 are satisfied for all n large. From Theorem 1, we therefore see that there exists an $n-$length code with real-time delay d_n, redundancy $o(n^{1-\alpha})$ and a real-time error probability of at most $\frac{42\log n}{n^{1-3\alpha}}$.

Example 2 For a numerical example, suppose that an encoded file containing $n = 10^8$ bits is to be transmitted across a channel that introduces at most $t_n = 10$ deletable errors. Setting $d_n = 4000$, we see that the redundancy of the code $\mathcal{C}_{frac}$ is at most $10^6 \cdot \log(2000) \leq 12 \cdot 10^6$ and so the rate of $\mathcal{C}_{frac}$ is at least $1 - \frac{12 \cdot 10^6}{10^8} = 0.88$. If all error patterns are equally likely, then the real-time error probability is also the decoding error probability and is at most $\frac{44}{10000} = 4.4 \cdot 10^{-3}$.

The main ingredient in the construction of the real-time error correction code in Theorem 1 is $P-$far deletable error patterns which we describe next. Let $n \geq P \geq 2$ be any integers. If $\mathbf{g} = (\mathbf{e}, \mathbf{v}), \mathbf{e} = (e_1, \ldots, e_n) \in \{0, 1\}^n$ is an error pattern such that $|i - j| \geq P$ for any two nonzero e_i, e_j, we then say that $\mathbf{g}$ is an $n-$length *$P-$far deletable error pattern*. Letting

$$\delta(P) := \frac{2P + 1}{2^{P-1}}, \tag{2.4}$$

we have the following result regarding codes capable of correcting $3P-$far deletable error patterns.

Theorem 2 *For all integers $n \geq P \geq 2$, there exists an $n-$length code $\mathcal{C}_{far} = \mathcal{C}_{far}(n, P)$ with redundancy*

$$R(\mathcal{C}_{far}) \leq \left(\frac{n}{P} - 1\right) \log\left(\frac{2P + 1}{1 - \delta(P)}\right) + \log P + 2 \tag{2.5}$$

that is capable of correcting all $3P-$far deletable error patterns.

Conversely, if $\mathcal{A}$ is any $n-$length code capable of correcting all $3P-$far deletable error patterns, then

$$R(\mathcal{A}) \geq \frac{n}{64(3P + 6)} - 3 \tag{2.6}$$

for all large n.

Suppose $\{P_n\}$ is a sequence satisfying

$$\frac{P_n}{\log n} \longrightarrow \infty \text{ and } \frac{P_n}{\sqrt{n \log n}} \longrightarrow 0 \tag{2.7}$$

as $n \to \infty$. If $\mathcal{B}$ is any $n-$length code capable of correcting all $3P_n-$far deletable error patterns, then

$$R(\mathcal{B}) \geq \left(\frac{n}{6P_n} - 1\right) \log\left(\frac{3P_n}{64}\right) - 2 \tag{2.8}$$

for all large n.

For $P = P_n$ growing with n, we see from Theorem 2 that the redundancy grows at least of the order of $n \cdot \frac{\log P_n}{P_n} = o(n)$. For this range of P, we also see from the upper bound (2.5) that the code $\mathcal{C}_{far}$ has near-optimal redundancy.

Before we proceed with the proofs, we briefly contrast and compare our results with related existing schemes for deletion correction. In our proof of Theorem 1 below, we use Theorem 2 to first construct a block code that is capable of correcting all P_n-far deletable errors and then choose P_n appropriately so as to obtain the desired delay d_n. The advantage of using P_n-far deletable errors is that we can ensure that at most one deletable error occurs within any one block (defined formally in the next section). Such an approach is known as *segmentation* and has been studied before for edit channels (i.e., channels allowing only deletion or insertion of bits). The Liu-Mitzenmacher construction [6] employs a segment-by-segment code specified via sufficient conditions and finds a maximum size code by converting it into an independent set problem. Recently, [1] has studied coding for segmented edit channels by using subsets of Varshamov-Tenengolts (VT) codes and obtained bounds on the maximum possible size of such a code.

The approaches described above are not directly applicable for compound channels allowing deletable errors because, in the case of segmentation, we can no longer interpret a substitution as a deletion followed by insertion as this would cause two errors within a block. Below, we use a modified form of VT codes to obtain our desired codes and demonstrate how deletable error correction can be performed in a sequential manner.

In the next two sections, we prove Theorems 1 and 2 in that order.

3 Proof of Theorem 2

We begin with a description of codes capable of correcting $P-$far errors.

Codes for Correcting $P-$far Errors

For integers $m \geq 3$ and $0 \leq a \leq m$, we define the modified Varshamov-Tenengolts (VT) code $VT_a(m)$

$$VT_a(m) := \left\{ \mathbf{x} = (x_1, \ldots, x_m) \in \{0,1\}^m : \sum_{i=1}^{m} i \cdot x_i \equiv a \mod 2m+1 \right\}.$$

Arguing as in [8], the code $VT_a(m)$ is capable of correcting a single deletable error.

For integers $P \geq 2$ and $n \geq P$, write $n = tP + s$ where $0 \leq s < P - 1$. Let $\mathbf{0}$ and $\mathbf{1}$ denote the all zero and all ones words with length depending on the context. For integers $0 \le a_1 \le 2P$ and $0 \le a_2 \le 2(P+s)$, define the code $\mathcal{C}_{far} = \mathcal{C}_{far}(a_1, a_2, P, s, n)$ as

$$\mathcal{C}_{far} := \{\mathbf{x} = (\mathbf{x}(1), \ldots, \mathbf{x}(t)) : \mathbf{x}(i) \in VT_{a_1}(P) \setminus \{\mathbf{0}, \mathbf{1}\} \\ \text{for } 1 \le i \le t-1, \mathbf{x}(t) \in VT_{a_2}(P+s)\}, \tag{3.1}$$

so that $\mathcal{C}_{far}$ is obtained by appending together $t-1$ words from the code $VT_{a_1}(P) \setminus \{\mathbf{0}, \mathbf{1}\}$ and then appending a word containing $P+s$ bits from the code $VT_{a_2}(P+s)$. We denote the bits in the ith word as $\mathbf{x}(i) = (x_1(i), \ldots, x_P(i))$ for $1 \le i \le t-1$ and let $\mathbf{x}(t) = (x_1(t), \ldots, x_{P+s}(t))$. We choose optimal values of a_1 and a_2 later for minimizing the redundancy. The encoding and decoding described in this subsection hold for all values of a_1 and a_2.

We see below that the code $\mathcal{C}_{far}$ is capable of correcting all $3P$–far deletable patterns. The main idea is that if $\mathbf{x}$ is corrupted by a $3P$–far deletable error pattern, then at most one deletable error occurs in each word $\mathbf{x}(j)$ in (3.2). This allows us to correct the errors in a sequential manner. Let $\mathbf{y} = F_{\mathbf{g}}(\mathbf{x})$ where $\mathbf{g} = (\mathbf{e}, \mathbf{v})$ be the received word and suppose that $\mathbf{x}(j), j \le t-2$ is the first block of $\mathbf{x}$ to be corrupted by an error in $\mathbf{g}$;, i.e., $\mathbf{e} = (e_1, \ldots, e_n)$ is such that if $u = \min\{i : e_i = 1\}$, then $(j-1)P + 1 \le u \le jP$. We consider the cases $j = t-1$ or $j = t$ at the end.

Divide the corrupted word $\mathbf{y} = (\mathbf{y}(1), \ldots, \mathbf{y}(t_1))$ into blocks where the jth block $\mathbf{y}(j) = (y_1(j), \ldots, y_P(j)), 1 \le j \le t_1 - 1$ has P bits and $\mathbf{y}(t_1)$ has at least P and at most $2P-1$ bits. Since the first error position $(j-1)P + 1 \le u \le jP$ falls in block $\mathbf{x}(j)$, the blocks $\mathbf{x}(i), 1 \le i \le j-1$ are uncorrupted and $\mathbf{y}(i) = \mathbf{x}(i), 1 \le i \le j-1$. Consequently, the checksum

$$CS(l) := \sum_{i=1}^{P} i \cdot y_i(l) \equiv a_1 \mod (2P+1) \tag{3.2}$$

for $1 \le l \le j-1$ and so the checksum difference $|CS(l) - a_1| \mod (2P+1) \equiv 0$ for all blocks $1 \le l \le j-1$.

If the error corrupting the block $\mathbf{x}(j)$ is an erasure, then $y_u(j) = \varepsilon$ (the erasure symbol) and $y_k(j) = x_k(j)$ for all $1 \le k \neq u \le P$. Therefore the modified checksum

$$CS_{er}(j) := \sum_{i=1, i \neq u}^{P} i \cdot y_i(j) = \sum_{i=1, i \neq u}^{P} i \cdot x_i(j). \tag{3.3}$$

If $x_u(j) = 0$, then $CS_{er}(j) - a_1 \equiv 0 \mod (2P+1)$ and if $x_u(j) = 1$, then $CS_{er}(j) - a_1 \equiv -u \mod (2P+1) \neq 0$ and so the modified checksum difference $|CS_{er}(j) - a_1| \mod (2P+1) \equiv u\mathbb{1}(x_u(j) = 1)$ is nonzero if and only if the erased bit is a one. This allows us to correct the erasure.

If the error corrupting the block $\mathbf{x}(j)$ is a substitution or deletion, then it is not directly detectable as in the case of erasure above. We therefore use the checksums of both $\mathbf{y}(j)$ and $\mathbf{y}(j+1)$ to indirectly deduce the nature of the error. Indeed, if the error is a substitution then $y_u(j) = 1 - x_u(j)$ and $y_k(j) = x_k(j)$ for $1 \leq k \neq u \leq P$ and so the checksum of the block $\mathbf{y}(j)$ is

$$CS(j) := \sum_{i=1}^{P} i \cdot y_i(j) = \sum_{i=1, i\neq u} i \cdot x_i(j) + u \cdot (1 - x_u(j)). \quad (3.4)$$

Thus, $CS(j) - a_1 \equiv u(1 - 2x_u(j))$ and so the checksum difference for the block $\mathbf{y}(j)$,

$$|CS(j) - a_1| \equiv u \mod (2P + 1)$$

is nonzero. Moreover, since the errors are $3P$−far apart, no errors have corrupted the block $\mathbf{x}(j+1)$ of the original word (which also contains P bits since $j \leq t - 2$). Therefore, the block $\mathbf{y}(j+1) = \mathbf{x}(j+1)$ and the checksum difference $|CS(j+1) - a_1| \mod (2P+1)$ for the block $\mathbf{y}(j+1)$ is zero.

If, on the other hand, the error corrupting the block $\mathbf{x}(j)$ is a deletion, we may or may not get a nonzero checksum difference in block $\mathbf{y}(j)$. But we are guaranteed a nonzero checksum difference in block $\mathbf{y}(j+1)$. This is because in the block $\mathbf{y}(j+1)$, the corresponding bits of $\mathbf{x}(j+1)$ are shifted one position to the left; i.e., $y_i(j) = x_{i+1}(j)$ for $1 \leq i \leq P - 1$. Therefore, the checksum

$$\begin{aligned} CS(j+1) &= \sum_{i=1}^{P-1} i \cdot x_{i+1}(j+1) + P \cdot y_P(j+1) \\ &= \sum_{i=1}^{P-1} (i+1) \cdot x_{i+1}(j+1) + P \cdot y_P(j+1) - \sum_{i=2}^{P} x_i(j+1) \\ &= \sum_{i=1}^{P} i \cdot x_i(j+1) + P \cdot y_P(j+1) - \sum_{i=1}^{P} x_i(j+1) \end{aligned}$$

and the checksum difference

$$CS(j+1) - a_1 \equiv P \cdot y_P(j+1) - \sum_{i=1}^{P} x_i(j+1) \mod (2P+1).$$

If $y_P(j+1) = 1$, then $P \geq P - \sum_{i=1}^{P} x_i(j+1) \geq 1$, since not all bits in a block can be one. If $y_P(j+1) = 0$, then $-P \leq -\sum_{i=1}^{P} x_i(j+1) \leq -1$, since not all bits in a block can be zero. In other words, the checksum difference for the block $\mathbf{y}(j+1)$, $|CS(j+1) - a_1| \mod (2P+1) \neq 0$.

Summarizing, suppose we get the first nonzero checksum difference (mismatch) at some block $\mathbf{y}(j)$ containing P bits. The mismatch may be because of an error either in block $\mathbf{y}(j-1)$ or block $\mathbf{y}(j)$. If indeed the mismatch was because of an error in block $\mathbf{y}(j-1)$, then the error must be a deletion since a substitution would have already caused a mismatch for block $\mathbf{y}(j-1)$ (see discussion following (3.4)). We therefore remove the last bit of the block $j-1$ and perform single deletion correction to get a new word $\hat{\mathbf{x}}(j-1)$. If $\hat{\mathbf{x}}(j-1) \neq \mathbf{y}(j-1)$, we replace the first $P-1$ bits of the block $\mathbf{y}(j-1)$ with the block $\hat{\mathbf{x}}(j-1)$.

If $\hat{\mathbf{x}}(j-1) = \mathbf{y}(j-1)$, then we can assume that the error has occurred in block $\mathbf{y}(j)$ and use block $\mathbf{y}(j+1)$ to determine the nature of error: If $\mathbf{y}(j+1)$ is not the final block in $\mathbf{y}$, then we proceed as above and deduce that the error in $\mathbf{y}(j)$ is either a substitution or deletion depending on whether the checksum difference of the block $\mathbf{y}(j+1)$ is zero or not, respectively.

If, on the other hand, $\mathbf{y}(j+1)$ is the final block in $\mathbf{y}$, then it is necessarily true that $\mathbf{y}$ is obtained after corrupting $\mathbf{x}$ by a single error occurring in one of the last two blocks $\mathbf{x}(t-1)$ or $\mathbf{x}(t)$;, i.e., the first error position u defined in the paragraph prior to (3.2) falls in one of the last two blocks $\mathbf{x}(t-1)$ or $\mathbf{x}(t)$. Since errors are $3P-$far apart, the error at position u is the only error corrupting $\mathbf{x}$ and so the corrupted word $\mathbf{y}$ has either n or $n-1$ bits.

Split $\mathbf{y} = (\mathbf{y}(1), \ldots, \mathbf{y}(t-1), \mathbf{y}(t))$ where the first $t-1$ blocks contain P bits each and the final block $\mathbf{y}(t)$ has either $P+s$ or $P+s-1$ bits. One of the blocks in $\{\mathbf{y}(t-1), \mathbf{y}(t)\}$ has been corrupted by an error and we perform correction as follows. If the error is an erasure, then we correct the erased bit simply by computing the checksum difference as described before (see discussion prior to (3.3)). If there is no erasure but the received word $\mathbf{y}$ has n bits, then the error is a substitution and we obtain the location of the substitution by computing the checksum difference for each of the blocks $\mathbf{y}(t-1)$ and $\mathbf{y}(t)$, as described in the paragraph containing (3.4).

Finally, if $\mathbf{y}$ has $n-1$ bits, then the error is a deletion and we determine the block where the deletion has occurred as follows. We perform single deletion correction on the first $P-1$ bits $\{\mathbf{y}_i(t-1)\}_{1 \leq i \leq P-1}$ of $\mathbf{y}(t-1)$ and get a new P bit word $\hat{\mathbf{w}}$. If $\hat{\mathbf{w}} \neq \mathbf{y}(t-1)$, then the deletion necessarily has occurred in the block $\mathbf{x}(t-1)$ of the original word $\mathbf{x}$. We therefore finish the iteration by inserting $\hat{\mathbf{w}}$ into the word $\mathbf{y}$ giving the output $\hat{\mathbf{x}} = \left(\mathbf{y}(1), \ldots, \mathbf{y}(t-2), \hat{\mathbf{w}}, \ y_P(t-1), \mathbf{y}(t)\right)$.

If, on the other hand, $\hat{\mathbf{w}} = \mathbf{y}(t-1)$, then we assume that the deletion has occurred in the final block $\mathbf{x}(t)$ of the original word $\mathbf{x}$. We then perform single deletion correction on the $P+s-1$ bits of $\mathbf{y}(t)$ to get a new word $\hat{\mathbf{z}}$ containing $P+s$ bits and output $\hat{\mathbf{x}} = (\mathbf{y}(1), \ldots, \mathbf{y}(t-1), \hat{\mathbf{z}})$.

The procedure described above corrects a single error in the corrupted word $\mathbf{y}$ and after the iteration, the first error occurring at position u in the error pattern $\mathbf{g}$ (see discussion prior to (3.2)) is corrected. The received word after the first iteration $\mathbf{y}^{(1)} = F_{\mathbf{g}(1)}(\mathbf{x})$ is a lesser corrupted version of $\mathbf{x}$ than $\mathbf{y}$ and $\mathbf{g}(1) = (\mathbf{e}(1), \mathbf{v})$ is an error pattern whose first error occurs strictly after the first error in $\mathbf{g}$;, i.e., if $\mathbf{e}(1) = (e_1(1), \ldots, e_n(1))$ and $u_1 = \min\{i : e_i(1) = 1\}$, then $u_1 > u$.

We now repeat the above procedure to correct the error at position u_1 and continuing iteratively, we sequentially correct all the remaining errors. The procedure stops if there is no checksum mismatch and we output the final word as our estimate of $\mathbf{x}$. We summarize the above method as an Algorithm before the Appendices, where the function $BlockSplit(., P)$ splits the input into blocks of length P with the last block containing between P and $2P-1$ bits.

From the description, we also see that the above algorithm is capable of correcting any P−far error pattern in real time with a delay of at most $4P$. This is because knowing the corruption of the first $z + 4P$ bits of the codeword $\mathbf{x}$ for any integer $1 \leq z \leq n - 4P$, we can determine the value of the bits x_i, $1 \leq i \leq z$.

Redundancy of the Code $\mathcal{C}_{far}$

To compute an upper bound on the redundancy of the code $\mathcal{C}_{far}$, we choose a_1 and a_2 above such that

$$\#VT_{a_1}(P) \geq \frac{2^P}{2P+1} \text{ and } \#VT_{a_2}(s) \geq \frac{2^{P+s}}{2(P+s)+1}. \tag{3.5}$$

Since we append together $t - 1$ words each chosen from $VT_{a_1}(P) \setminus \{\mathbf{0}, \mathbf{1}\}$ and one word from $VT_{a_2}(P + s)$, we have that

$$\#\mathcal{C}_{far} \geq \left(\frac{2^P}{2P+1} - 2\right)^{t-1} \frac{2^{P+s}}{(2P+2s+1)} = 2^n \left(\frac{1-\delta(P)}{2P+1}\right)^{t-1} \frac{1}{2P+2s+1}, \tag{3.6}$$

where $\delta(P) = \frac{2P+1}{2^{P-1}}$ is as in (2.4) and relation (3.6) is true since $Pt + s = n$. Using $s + 1 \leq P$, we further get

$$\#\mathcal{C}_{far} \geq 2^n \left(\frac{1-\delta(P)}{2P+1}\right)^{t-1} \frac{1}{4P} \geq 2^n \left(\frac{1-\delta(P)}{2P+1}\right)^{\frac{n}{P}-1} \frac{1}{4P},$$

since $t \leq \frac{n}{P}$ and $\frac{1-\delta(P)}{2P+1} < 1$. From (2.1) we then get (2.5).

To obtain the redundancy lower bounds in Theorem 2, we use the fact that any code capable of correcting t deletable errors must also be capable of corrected t deletions. We also remark that redundancy lower bounds for segmented deletion channels have been studied in [9] and so for completeness, we give our proofs in Appendix 1.

4 Proof of Theorem 1

Let

$$\omega_n := \frac{4n}{t_n^2 \cdot d_n} \text{ and } P_n := \frac{n}{t_n^2 \omega_n} = \frac{d_n}{4} \tag{4.1}$$

so that $6 \leq \omega_n \leq \frac{n}{t_n^3}$ and assume that P_n is an integer greater than or equal to 6. Let $\mathcal{C}_{frac} := \mathcal{C}_{far}(n, P_n)$ be the code defined in Theorem 2 so that $\mathcal{C}_{frac}$ is capable of real-time correction of all $3P_n$−far deletable error patterns with a delay of at most $4P_n$ (see the final paragraph in the code description subsection of Sect. 3). Using (2.5), $P_n \geq 6$ and $\delta(P_n) \leq \frac{1}{2}$, we also get that the redundancy of $\mathcal{C}_{frac}$ is

$$R(\mathcal{C}_{frac}) \leq \left(\frac{n}{P_n} - 1\right) \log\left(\frac{2P_n + 1}{1 - \delta(P_n)}\right) + \log P_n + 2 \leq n \cdot \frac{\log 8P_n}{P_n} \tag{4.2}$$

where the second inequality in (4.2) above follows from $P_n \geq 6$. Setting $P_n = \frac{d_n}{4}$, we get the redundancy estimate (2.2) in Theorem 1.

We show below that if $\mathcal{F}_n$ is the set of all $3P_n$−far error patterns, then

$$\frac{\#\mathcal{F}_n}{\#\mathcal{E}_n(t_n)} \geq 1 - \frac{42}{\omega_n} \tag{4.3}$$

for all $n \geq 10$. Since $\mathcal{C}_{frac}$ is capable of correcting all $3P_n$−far deletable error patterns, this proves the estimate on the fraction of deletable errors corrected in Theorem 1. In what follows, we prove (4.3). First, the total number of error patterns in $\mathcal{E}_n(t_n)$ is $\#\mathcal{E}_n(t_n) = \sum_{k=0}^{t_n} \binom{n}{k} 3^k$. Suppose $\mathbf{g} = (\mathbf{e}, \mathbf{v}), \mathbf{e} = (e_1, \ldots, e_n)$ is not a $3P_n$−far deletable error pattern. Let $n = 3LP_n + s, 0 \leq s \leq 3P_n - 1$ and split $\mathbf{e} = (\mathbf{e}(1), \ldots, \mathbf{e}(L))$ into L blocks, where the first $L - 1$ blocks each have length $3P_n$ and the final block has length $3P_n + s$. Since $\mathbf{g}$ is not a $3P_n$−far error pattern, there are only two possibilities:
$(p1)$ One of the L blocks in $\mathbf{e}$ contains at least two nonzero entries.
$(p2)$ Two consecutive blocks in $\mathbf{e}$ contain exactly one nonzero entry each.
We bound the number of patterns for each possibility.

To bound the number of patterns $\#\mathcal{F}(p1)$ in possibility $(p1)$, we argue as follows: If $\mathcal{S}(l), 1 \leq l \leq L$ is the set of error patterns $\mathbf{g} = (\mathbf{e}, \mathbf{v})$ where the lth block $\mathbf{e}(l)$ of $\mathbf{e}$ has two or more nonzero entries, then

$$\#\mathcal{F}(p1) \leq \sum_{l=1}^{L} \#\mathcal{S}(l) = (L - 1)\#\mathcal{S}(1) + \#\mathcal{S}(L). \tag{4.4}$$

In Appendix 2 below, we use Binomial coefficient estimates to show that

$$\#\mathcal{S}(l) \leq \frac{84P_n^2 t_n^2 + 8t_n^4}{n^2}(\#\mathcal{E}_n(t_n)) \tag{4.5}$$

for all $1 \leq l \leq L$ and since $L \leq \frac{n}{3P_n}$, we get from (4.4) that

$$\#\mathcal{F}(p1) \leq \frac{n}{3P_n}\left(\frac{84P_n^2 t_n^2 + 8t_n^4}{n^2}\right)(\#\mathcal{E}_n(t_n)) = \left(\frac{28P_n t_n^2}{n} + \frac{8t_n^4}{3nP_n}\right)(\#\mathcal{E}_n(t_n)).$$

Since $P_n = \frac{n}{t_n^2 \omega_n}$, we get that $\frac{28P_n t_n^2}{n} + \frac{4t_n^4}{3nP_n} = \frac{28}{\omega_n} + \frac{4}{3\omega_n}\left(\frac{\omega_n t_n^3}{n}\right)^2 \leq \frac{28+\frac{4}{3}}{\omega_n}$ by (4.1) and so

$$\#\mathcal{F}(p1) \leq \frac{30}{\omega_n} \cdot (\#\mathcal{E}_n(t_n)). \tag{4.6}$$

To bound the number of patterns $\#\mathcal{F}(p2)$ in possibility $(p2)$ described prior to (4.4), we argue as follows: If $\mathcal{T}(l)$ is the set of error patterns $\mathbf{g} = (\mathbf{e}, \mathbf{v})$ where the blocks $\mathbf{e}(l)$ and $\mathbf{e}(l+1)$, $1 \leq l \leq L-1$ each has exactly one nonzero entry each, then

$$\#\mathcal{F}(p2) \leq \sum_{l=1}^{L-1} \#\mathcal{T}(l) = (L-2)\#\mathcal{T}(1) + \#\mathcal{T}(L-1). \tag{4.7}$$

We estimate $\#\mathcal{T}(1)$ as follows. Suppose $\mathbf{g} = (\mathbf{e}, \mathbf{v}) \in \mathcal{T}(1)$ and suppose that $\mathbf{e} = (e_1, \ldots, e_n)$ has k nonzero entries. We split $\mathbf{e} = (\mathbf{e}(1), \ldots, \mathbf{e}(L))$ into L blocks as before. Among the $3P_n$ bits in the first block $\mathbf{e}(1)$, we choose one bit; among the $3P_n$ bits of $\mathbf{e}(2)$ we choose one bit and choose $k-2$ bits from the remaining $n - 6P_n$ bits of $\mathbf{e}(j)$, $j \neq 1, 2$. This can be done in $\binom{3P_n}{1}\binom{3P_n}{1}\binom{n-6P_n}{k-2}$ ways and so

$$\#\mathcal{T}(1) = \sum_{k=2}^{t_n} 9P_n^2 \binom{n-6P_n}{k-2} 3^k \leq 9P_n^2 \sum_{k=2}^{t_n} \binom{n}{k-2} 3^k. \tag{4.8}$$

For $k \leq t_n$, the ratio

$$\frac{\binom{n}{k-2}}{\binom{n}{k}} = \frac{k(k-1)}{(n-k+2)(n-k+1)} \leq \frac{t_n^2}{(n-t_n)^2} = \left(\frac{t_n}{n}\right)^2 \left(1 - \frac{t_n}{n}\right)^{-2} \tag{4.9}$$

and using $\frac{t_n}{n} \leq \frac{1}{\sqrt{n}}$ (see statement of Theorem), we also have that $\left(1 - \frac{t_n}{n}\right)^{-2} \leq \frac{n}{(\sqrt{n}-1)^2} \leq 4$ for all $n \geq 4$. From (4.8), we therefore get that

$$\#\mathcal{T}(1) \leq \frac{36P_n^2 t_n^2}{n^2} \sum_{k=2}^{t_n} \binom{n}{k} 3^k \leq \frac{36P_n^2 t_n^2}{n^2} (\#\mathcal{E}_n(t_n)). \tag{4.10}$$

Performing an analogous analysis as above for the last two blocks with length $3P_n$ and $3P_n + s$, we get

$$\#\mathcal{T}(L-1) = \sum_{k=2}^{t_n} 3P_n(3P_n + s) \binom{n - 6P_n - s}{k-2} 3^k \leq 18P_n^2 \sum_{k=2}^{t_n} \binom{n}{k-2} 3^k, \tag{4.11}$$

since $s \leq 3P_n$. Again using (4.9), we get

$$\#\mathcal{T}(L-1) \leq \frac{36P_n^2 t_n^2}{n^2} \sum_{k=2}^{t_n} \binom{n}{k} 3^k \leq \frac{36P_n^2 t_n^2}{n^2} (\#\mathcal{E}_n(t_n)). \tag{4.12}$$

Using (4.10) and (4.12) in (4.7), we therefore get that

$$\#\mathcal{F}(p2) \leq L \cdot \frac{36P_n^2t_n^2}{n^2}(\#\mathcal{E}_n(t_n)) \leq \frac{12P_nt_n^2}{n}(\#\mathcal{E}_n(t_n)) = \frac{12}{\omega_n}(\#\mathcal{E}_n(t_n)), \quad (4.13)$$

where the second inequality in (4.13) is true since $L \leq \frac{n}{3P_n}$ and the final estimate in (4.13) is true since $P_n = \frac{n}{t_n^2\omega_n}$ (see (4.1)).

Recall from discussion prior to (4.4) that $\mathcal{F}_n$ denotes the set of all $3P_n$−far error patterns and using (4.6) and (4.13), we get that $\frac{\#\mathcal{F}_n}{\#\mathcal{E}_n(t_n)} \geq 1 - \frac{42}{\omega_n}$, proving (4.3). □

Conclusion

In this paper, we have studied the redundancy of codes capable of correcting deletable errors in real time with a low probability of error. We have also provided decoding algorithms that recover the errors in real time. It would be interesting to extend our results to other channel models, for example, magnetic channels with possible additional constraints.

Acknowledgements I thank Profs. Alberto Gandolfi, Federico Camia, C. R. Subramanian and the referees for crucial comments that led to an improvement of the paper. I also thank Profs. Federico Camia, C. R. Subramanian and IMSc for my fellowships.

Appendix 1: Redundancy Lower Bounds

In our proofs below, we use the following large deviation estimate. Let $\{X_j\}_{1\leq j\leq m}$ be independent Bernoulli random variables with

$$\mathbb{P}(X_j = 1) = p_j = 1 - \mathbb{P}(X_j = 0)$$

and fix $0 < \epsilon \leq \frac{1}{2}$. If $T_m = \sum_{j=1}^m X_j$ and $\mu_m = \mathbb{E}T_m$, then

$$\mathbb{P}\left(|T_m - \mu_m| \geq \mu_m\epsilon\right) \leq 2\exp\left(-\frac{\epsilon^2}{4}\mu_m\right) \quad (4.1)$$

for all $m \geq 1$. For a proof of (4.1), we refer to Corollary A.1.14, p. 312 of [2].

Proof of (2.6) in Theorem 2: Let $\mathcal{D}$ be any code capable of correcting all $3P$−far deletable error patterns. The code $\mathcal{D}$ is capable of correcting all $3P$−far deletable error patterns consisting only of deletions and so we henceforth consider only deletions. Let $\mathcal{F}$ be the set of all $3P$−far deletable error patters consisting only of deletions and for $\mathbf{x} \in \mathcal{D}$, set $\mathcal{N}(\mathbf{x}) = \cup_{\mathbf{g}\in\mathcal{F}} F_{\mathbf{g}}(\mathbf{x})$. Since the deletable errors are at least $3P$ apart, there are at most $\frac{n}{3P}$ deletable errors in $\mathbf{x}$ and so if T is the largest integer

Algorithm 1: Correcting a $3P$−far deletable error pattern

Input : Received word $\mathbf{y}$
Output : Estimated word $\hat{\mathbf{x}}$

SequentialDecode
  Initialization: $\hat{\mathbf{x}} \leftarrow \mathbf{y}, j \leftarrow 0, CS \leftarrow 0$.
  Preprocessing: $(\hat{\mathbf{x}}(1), \ldots, \hat{\mathbf{x}}(t-1), \hat{\mathbf{x}}(t)) \leftarrow BlockSplit(\hat{\mathbf{x}}, P)$,
  where $P \leq Length(\hat{\mathbf{x}}(t)) < 2P$.
  `/* Pass all blocks with checksum match */`
  while $CS = 0$ *and* $j \leq t$ **do**
    $j \leftarrow j+1, CS \leftarrow -1$;
    if *erasure in* j^{th} *block* **then**
      $\hat{\mathbf{x}}(j) \leftarrow CorrectErasure(\hat{\mathbf{x}}(j))$;
    if $j < t$ *or* $j = t$ *and* $Length(\hat{\mathbf{x}}(t)) = P + s$ **then**
      $CS \leftarrow CheckSumDifference(\hat{\mathbf{x}}(j))$;
  if $CS = 0$ *and* $j = t$ **then**
    output $(\hat{\mathbf{x}}(1), \ldots, \hat{\mathbf{x}}(t-1), \hat{\mathbf{x}}(t))$;
  `/* Checksum mismatch in` j^{th} `block; Check if deletion in` $(j-1)^{th}$ `block */`
  $l \leftarrow \min(j-1, 1), \mathbf{T} \leftarrow CorrectDeletion(RemoveLastBit(\hat{\mathbf{x}}(l)))$;
  if $\mathbf{T} \neq \hat{\mathbf{x}}(l)$ **then**
    $\mathbf{T} \leftarrow (\mathbf{T}, LastBit(\hat{\mathbf{x}}(l)))$;
  else
    `/* Error in` j^{th} `block; If final block then directly perform correction */`
    if $j = t$ **then**
      if $\hat{\mathbf{x}}(j)$ *has* $P + s$ *bits* **then**
        $\mathbf{T} \leftarrow CorrectFlip(\hat{\mathbf{x}}(j))$;
      else
        $\mathbf{T} \leftarrow CorrectDeletion(\hat{\mathbf{x}}(j))$;
      output $(\hat{\mathbf{x}}(1), \ldots, \hat{\mathbf{x}}(t-1), \mathbf{T})$;
    `/* In all other cases, determine if error is flipping or deletion using block` $j+1$ `*/`
    $l \leftarrow j, ErrorType \leftarrow 1$;
    if $l < t-1$ *or* $l = t-1$ *and* $Length(\hat{\mathbf{x}}(l+1)) = P + s$ **then**
      $ErrorType \leftarrow CheckSumDifference(\hat{\mathbf{x}}(l+1))$;
    if $ErrorType = 0$ **then**
      $\mathbf{T} \leftarrow CorrectFlip(\hat{\mathbf{x}}(l))$;
    else
      $\mathbf{T} \leftarrow CorrectDeletion(RemoveLastBit(\hat{\mathbf{x}}(l)))$;
      $\mathbf{T} \leftarrow (\mathbf{T}, LastBit(\hat{\mathbf{x}}(l)))$;
  `/* Update the estimate` $\hat{\mathbf{x}}$ `with the new block` $\mathbf{T}$ `*/`
  $\hat{\mathbf{x}} \leftarrow (\hat{\mathbf{x}}(1), \ldots, \hat{\mathbf{x}}(l-1), \mathbf{T}, \hat{\mathbf{x}}(l+1), \ldots, \hat{\mathbf{x}}(t))$;
  goto *Preprocessing*;

less than or equal to $\frac{n}{3P}$, then $\mathcal{N}(\mathbf{x}) \subseteq \cup_{0\le r\le T}\{0,1\}^{n-r}$. Moreover, if $\mathbf{x}_1 \neq \mathbf{x}_2 \in \mathcal{D}$, then $F_{\mathbf{e}}(\mathbf{x}_1) \neq F_{\mathbf{e}}(\mathbf{x}_2)$ by definition. So the sets $\{\mathcal{N}(\mathbf{x})\}$ are all disjoint and therefore

$$\sum_{\mathbf{x}\in\mathcal{D}} \#\mathcal{N}(\mathbf{x}) \le \sum_{r=0}^{T} 2^{n-r} = 2^{n+1}\left(1 - \frac{1}{2^{T+1}}\right) \le 2^{n+1}. \tag{4.2}$$

In what follows, we estimate the size of $\mathcal{N}(\mathbf{x})$ for each $\mathbf{x}$ belonging to a certain "nice" subset of $\mathcal{D}$. For $\mathbf{x} \in \{0,1\}^n$ and integer $r \ge 1$, let

$$\mathbf{x}_r := \mathbf{x}_r(\mathbf{x}) = \mathbf{x}((3P+5)(r-1)+1, (3P+5)(r-1)+5) \tag{4.3}$$

be a sequence of five bits of $\mathbf{x}$. Say that $\mathbf{x}_r$ is a good 5−block if the first and last bits are one and the remaining three bits are zero. Let $Q(\mathbf{x})$ be the number of good 5−blocks in $\{\mathbf{x}_r\}_{1\le r\le W}$ where W is the largest integer x satisfying $x \cdot (3P+5) \le n$ so that

$$\frac{n}{3P+6} \le \frac{n}{3P+5} - 1 \le W \le \frac{n}{3P+5}, \tag{4.4}$$

for all n large.

Defining

$$\mathcal{Y}_n := \left\{\mathbf{x} \in \{0,1\}^n : Q(\mathbf{x}) \ge \frac{n}{64(3P+6)}\right\}, \tag{4.5}$$

we show at the end of this proof that

$$\#\mathcal{Y}_n \ge 2^n\left(1 - \exp\left(-\frac{n}{16(3P+6)}\right)\right), \tag{4.6}$$

for all n large.

If $\mathbf{x} \in \mathcal{Y}_n \cap \mathcal{D}$, then there are at least $\frac{n}{64(3P+6)}$ good 5−blocks in $\mathbf{x}$. Let m_P be the largest integer less than or equal to $\frac{n}{64(3P+6)}$. For every one of the m_P good 5−blocks, we can either choose to remove a zero bit or not. There are 2^{m_P} ways of performing such a procedure and the resulting set of corrupted words are all distinct. Therefore, $\#\mathcal{N}(\mathbf{x}) \ge 2^{m_P}$ and using (4.2), we get

$$2^{m_P}\#\left(\mathcal{D}\bigcap\mathcal{Y}_n\right) \le \sum_{\mathbf{x}\in\mathcal{D}\bigcap\mathcal{Y}_n} \#\mathcal{N}(\mathbf{x}) \le 2^{n+1}.$$

From (4.6), we therefore get

$$\#\mathcal{D} \le \frac{2^{n+1}}{2^{m_P}} + \#\mathcal{Y}_n^c \le \frac{2^{n+1}}{2^{m_P}} + 2^n \exp\left(-\frac{n}{16(3P+6)}\right) \le \frac{2^{n+2}}{2^{m_P}}$$

for all n large, since $m_P \leq \frac{n}{64(3P+6)} \leq \frac{n \log e}{16(3P+6)}$. Therefore, the redundancy of $\mathcal{D}$ is at least $m_P - \log 4 \geq \frac{n}{64(3P+6)} - 3$ for all n large, since $m_P \geq \frac{n}{64(3P+6)} - 1$. □

Proof of (4.6): Let $\mathbf{X} = (X_1, \ldots, X_n)$ be a uniformly randomly chosen word in $\{0,1\}^n$ so that $X_i, 1 \leq i \leq n$ are independent and identically distributed (i.i.d.) with $\mathbb{P}(X_1 = 0) = \frac{1}{2} = \mathbb{P}(X_1 = 1)$.

To lower bound $Q(\mathbf{X})$, let $r \geq 1$ be an integer and define

$$I_r := \{X_{(3P+5)(r-1)+1} = 1\} \bigcap_{i=2}^{4} \{X_{(3P+5)(r-1)+i} = 0\} \bigcap \{X_{(3P+5)(r-1)+5} = 1\} \quad (4.7)$$

as the event that $\mathbf{X}_r$ defined in (4.3) is a good block. The events $\{I_r\}$ are i.i.d. with $\mathbf{P}(I_1) = \frac{1}{32}$ and so $Q(\mathbf{x}) = \sum_{r=1}^{W} I_r$ where W is as in (4.4) has mean $\frac{W}{32}$. Therefore, using the deviation estimate (4.1) with $\epsilon = \frac{1}{2}$, we get

$$\mathbb{P}\left(Q(\mathbf{x}) \geq \frac{W}{64}\right) \geq 1 - \exp\left(-\frac{W}{16}\right).$$

Using (4.4), we have that $\frac{W}{64} \geq \frac{n}{64(3P+6)}$ for all n large and so

$$\mathbb{P}\left(Q(\mathbf{x}) \geq \frac{n}{64(3P+6)}\right) \geq 1 - \exp\left(-\frac{n}{16(3P+6)}\right)$$

for all n large. □

Proof of (2.6) *in Theorem* 2: Let $\mathcal{D}$ be any code capable of correcting all $3P_n$−far deletable error patterns. The code $\mathcal{D}$ is capable of correcting all $3P_n$−far deletable error patterns consisting only of deletions and so we henceforth consider only deletions. Let $\mathcal{F}$ be the set of all $3P_n$−far deletable error patters consisting only of deletions and for $\mathbf{x} \in \mathcal{D}$, set $\mathcal{N}(\mathbf{x}) = \cup_{\mathbf{e} \in \mathcal{F}} F_{\mathbf{e}}(\mathbf{x})$. Since the deletions are at least $3P_n$ apart, there are at most $\frac{n}{3P_n}$ deletable errors in $\mathbf{x}$ and so if L is the largest integer less than or equal to $\frac{n}{3P_n}$, then $\mathcal{N}(\mathbf{x}) \subseteq \cup_{0 \leq r \leq L} \{0,1\}^{n-r}$. Moreover, if $\mathbf{x}_1 \neq \mathbf{x}_2 \in \mathcal{D}$, then $F_{\mathbf{e}}(\mathbf{x}_1) \neq F_{\mathbf{e}}(\mathbf{x}_2)$ by definition. So the sets $\{\mathcal{N}(\mathbf{x})\}$ are all disjoint and therefore

$$\sum_{\mathbf{x} \in \mathcal{D}} \#\mathcal{N}(\mathbf{x}) \leq \sum_{r=0}^{L} 2^{n-r} = 2^{n+1}\left(1 - \frac{1}{2^{L+1}}\right) \leq 2^{n+1}. \quad (4.8)$$

As before, we estimate the size of $\mathcal{N}(\mathbf{x})$ for a nice subset of $\mathcal{D}$. Let W be the largest integer x such that $x \cdot 3P_n \leq n$ so that

$$\frac{n}{3P_n} - 1 \leq W \leq \frac{n}{3P_n}. \quad (4.9)$$

For integer $1 \leq r \leq W$ and $\mathbf{x} \in \{0,1\}^n$, let

$$\mathbf{x}_r := \mathbf{x}_r(\mathbf{x}) = \mathbf{x}(3P_n(r-1)+1, 3rP_n) \tag{4.10}$$

be a sequence of $3P_n$ bits of $\mathbf{x}$. We have thus divided the first $3WP_n$ bits of $\mathbf{x}$ into W disjoint blocks $\{\mathbf{x}_r\}_{1\leq r\leq W}$, each containing $3P_n$ bits.

Let U be the largest integer y such that $5\cdot y \leq 3P_n$ so that $\frac{3P_n}{5} - 1 \leq U \leq \frac{3P_n}{5}$. Divide the first $5U$ bits of each block $\mathbf{x}_r$ into disjoint blocks $\{\mathbf{x}_{r,j}\}_{1\leq j\leq U}$ of five bits each. Say that $\mathbf{x}_{r,j}$ is a good 5−block if the first and last bits are one and the remaining three bits are zero. Let $T_r(\mathbf{x})$ be number of good 5−blocks in $\mathbf{x}_r$ and let

$$\mathcal{V}_n := \left\{\mathbf{x}\in\{0,1\}^n : T_r(\mathbf{x}) \geq \frac{3P_n}{64} \text{ for all } 1\leq r\leq W\right\}. \tag{4.11}$$

At the end of this proof, we prove that there are positive constants $\beta_i, i=1,2$ such that for all integers $n\geq\beta_1$,

$$\#\mathcal{V}_n \geq 2^n(1-e^{-\beta_2P_n}). \tag{4.12}$$

Let W be as in (4.9) and as before, split the first $3WP_n$ bits of $\mathbf{x}$ into W blocks $\{\mathbf{x}_r\}_{1\leq r\leq W}$ of length $3P_n$ each. If $\mathbf{x}\in\mathcal{V}_n\cap\mathcal{D}$, then there are at least $\frac{3P_n}{64}$ good 5−blocks in each $\mathbf{x}_r$. Choosing one good 5−block from each of $\mathbf{x}_1,\mathbf{x}_3,\ldots$ and deleting a middle zero from each such good block, we get a set of distinct words. Since there are at least $\frac{W-1}{2}$ blocks in $\mathbf{x}_1,\mathbf{x}_3,\ldots,$ the number of corrupted words obtained as above is at least $\left(\frac{3P_n}{64}\right)^{(W-1)/2}$. The deletable error patterns giving rise to these words are all $3P_n$−far apart and so we get

$$\#\mathcal{N}(\mathbf{x}) \geq \left(\frac{3P_n}{64}\right)^{(W-1)/2} \geq \left(\frac{3P_n}{64}\right)^{\frac{n}{6P_n}-1}, \tag{4.13}$$

since $W\geq\frac{n}{3P_n}-1$, (see (4.9)).

Letting

$$\Delta = \left(\frac{n}{6P_n}-1\right)\log\left(\frac{3P_n}{64}\right)$$

and using (4.13) in (4.8), we get

$$2^{\Delta}\#\left(\mathcal{D}\bigcap\mathcal{V}_n\right) \leq \sum_{\mathbf{x}\in\mathcal{C}\bigcap\mathcal{V}_n}\#\mathcal{N}(\mathbf{x}) \leq 2^{n+1}$$

and so from (4.12) we get

$$\#\mathcal{D} \leq \frac{2^{n+1}}{2^{\Delta}} + \#\mathcal{V}_n^c \leq \frac{2^{n+1}}{2^{\Delta}} + 2^ne^{-\beta_2P_n} \leq \frac{2^{n+2}}{2^{\Delta}}$$

for all n large, since using the second relation in (2.7) that $\frac{P_n^2}{n \log n} \longrightarrow 0$ as $n \to \infty$, we have

$$\Delta \le \frac{n}{P_n}\log(3P_n) \le \frac{n\log n}{P_n} \le \beta_2 P_n \log e$$

for all n large. Therefore, the redundancy of $\mathcal{D}$ is at least $\Delta - \log 4$ for all n large. This proves the redundancy lower bounds with the exception of (4.6) and (4.12), which we prove separately below. □

Proof of (4.12): Let $\mathbf{X} = (X_1, \ldots, X_n)$ be a uniformly randomly chosen word in $\{0, 1\}^n$ so that X_i, $1 \le i \le n$ are independent and identically distributed (i.i.d.) with $\mathbb{P}(X_1 = 0) = \frac{1}{2} = \mathbb{P}(X_1 = 1)$.

To lower bound $T_r(\mathbf{X})$ for $1 \le r \le W$, let $I_{r,j}$ be the event that $\mathbf{X}_{r,j}$ is a good 5−block. The events $\{I_{r,j}\}_{1\le j\le U}$ are i.i.d. with $\mathbf{P}(I_{r,j}) = \frac{1}{32}$ and so $T_r(\mathbf{X}) = \sum_{j=1}^{U} I_{r,j}$ has mean $\frac{U}{32}$. Therefore, from the deviation estimate (4.1), we get that

$$\mathbb{P}\left(T_r(\mathbf{X}) \ge \frac{U}{64}\right) \ge 1 - e^{-C_1 U} \tag{4.14}$$

for all n large and some constant $C_1 > 0$. Since $U \ge \frac{3P_n}{5} - 1 \ge \frac{3P_n}{8}$ for all large n, we get

$$\mathbb{P}\left(T_r(\mathbf{X}) \ge \frac{3P_n}{2^9}\right) \ge 1 - e^{-4C_2 P_n} \tag{4.15}$$

for some constant $C_2 > 0$ and all large n. The constant C_2 does not depend on r and so

$$\mathbb{P}\left(\bigcap_{1\le r\le W}\left\{T_r(\mathbf{X}) \ge \frac{3P_n}{2^9}\right\}\right) \ge 1 - W\cdot e^{-4C_2 P_n}. \tag{4.16}$$

Using (4.9) and the first relation in (2.7) that $\frac{P_n}{\log n} \longrightarrow \infty$, we get that

$$W\cdot e^{-4C_2 P_n} \le \frac{n}{2P_n}\cdot e^{-4C_2 P_n} \le e^{-2C_2 P_n}$$

for all n large. □

Appendix 2: Proof of (4.5)

We estimate $\#\mathcal{S}(1)$ as follows. Suppose $\mathbf{g} = (\mathbf{e}, \mathbf{v}) \in \mathcal{S}(1)$ and suppose that $\mathbf{e} = (e_1, \ldots, e_n)$ has $k \le t_n$ nonzero entries. We split $\mathbf{e} = (\mathbf{e}(1), \ldots, \mathbf{e}(L))$ into L blocks as before. Among the $3P_n$ bits in the first block $\mathbf{e}(1)$, we choose $2 \le r \le k$ of the bits in $\binom{3P_n}{r}$ ways and choose $k - r$ bits of the remaining $n - 3P_n$ bits in $\binom{n-3P_n}{k-r}$ ways. The binomial coefficient $\binom{3P_n}{r}$ is nonzero since $r \le k \le t_n$ and $P_n = \frac{n}{t_n^2\omega_n} \ge t_n$

(see (4.1)). Thus, $\#\mathcal{S}(1) = \sum_{k=2}^{t_n} \sum_{r=2}^{k} \binom{n-3P_n}{k-r}\binom{3P_n}{r} 3^k$ and using $\sum_{k=0}^{r} \binom{x}{k-r}\binom{y}{r} = \binom{x+y}{k}$ with $x = n - 3P_n$ and $y = 3P_n$, we get

$$\#\mathcal{S}(1) = \sum_{k=2}^{t_n} (1 - S(n,k)\Delta) \binom{n}{k} 3^k, \tag{4.1}$$

where $S(n,k) = \frac{\binom{n-3P_n}{k}}{\binom{n}{k}}$ and

$$\Delta = 1 + \frac{\binom{3P_n}{1} \cdot \binom{n-3P_n}{k-1}}{\binom{n-3P_n}{k}} = 1 + \frac{3P_n k}{n - 3P_n - k + 1} \geq 1 + \frac{3P_n k}{n}. \tag{4.2}$$

We now evaluate $S(n,k)$ by letting $x = n - 3P_n$ to get

$$S(n,k) = \frac{\binom{x}{k}}{\binom{n}{k}} = \frac{x(x-1)\dots(x-k+1)}{n(n-1)\dots(n-k+1)} = \left(\frac{x}{n}\right)^k \frac{L(x,k)}{L(n,k)}, \tag{4.3}$$

where $1 \geq L(x,k) := \prod_{i=0}^{k-1} \left(1 - \frac{i}{x}\right) \geq \left(1 - \frac{1}{x}\sum_{i=1}^{k-1} i\right) = 1 - \frac{k(k-1)}{2x}$. The term $x = n - 3P_n = n - \frac{3n}{t_n^2 \omega_n} \geq n - \frac{3n}{\omega_n} \geq \frac{n}{2}$ since $\omega_n \geq 6$ (see statement of Theorem) and so

$$1 - \frac{k(k-1)}{2x} = 1 - \frac{k(k-1)}{2n} - \frac{3P_n k(k-1)}{2nx} \geq 1 - \frac{k(k-1)}{2n} - \frac{3P_n k^2}{n^2}$$

and

$$L(x,k) \geq 1 - \frac{k(k-1)}{2n} - \frac{3P_n k^2}{n^2}. \tag{4.4}$$

To find an upper bound for $L(n,k)$, we use the fact that $1 - y < e^{-y}$ for $0 < y < 1$ to get $\log L(n,k) \leq -\sum_{i=1}^{k-1} \frac{i}{n} = -\frac{k(k-1)}{2n}$. But $\frac{k(k-1)}{2n} \leq \frac{t_n^2}{2n} \leq \frac{1}{2}$ since $t_n \leq \sqrt{n}$ (see statement of Theorem) and so using $e^{-z} \leq 1 - z + z^2$ for $z < \frac{1}{2}$, we get

$$L(n,k) \leq \exp\left(-\frac{k(k-1)}{2n}\right) \leq 1 - \frac{k(k-1)}{2n} + \left(\frac{k(k-1)}{2n}\right)^2 \leq 1 - \frac{k(k-1)}{2n} + \frac{k^4}{n^2}. \tag{4.5}$$

Using (4.5) and the lower bound (4.4), we have

$$\frac{L(x,k)}{L(n,k)} - 1 \geq \frac{1 - \frac{k(k-1)}{2n} - \frac{3P_n k^2}{n^2}}{1 - \frac{k(k-1)}{2n} + \frac{k^4}{n^2}} - 1 = -\frac{3P_n k^2 + k^4}{n^2}\left(1 - \frac{k(k-1)}{2n} + \frac{k^4}{n^2}\right)^{-1}. \tag{4.6}$$

Since $k \leq t_n \leq \sqrt{n}$ (see statement of Theorem), we have

$$1 - \frac{k(k-1)}{2n} + \frac{k^4}{n^2} \geq 1 - \frac{t_n(t_n-1)}{2n} \geq \frac{1}{2}$$

and so we get from (4.6) that $\frac{L(x,k)}{L(n,k)} - 1 \geq -2\frac{3P_n k^2 + k^4}{n^2}$. Substituting this estimate into (4.3) and using $\left(1 - \frac{3P_n}{n}\right)^k \geq 1 - \frac{3P_n k}{n}$, we get

$$S(n,k) \geq \left(1 - \frac{3P_n k}{n}\right)\left(1 - 2\frac{3P_n k^2 + k^4}{n^2}\right) \geq 1 - \frac{3P_n k}{n} - 2\frac{3P_n k^2 + k^4}{n^2}. \quad (4.7)$$

Using (4.7) and (4.2), we therefore get that $S(n,k) \cdot \Delta$ is bounded below by

$$1 - \left(\frac{3P_n k}{n}\right)^2 - 2\frac{3P_n k^2 + k^4}{n^2}\left(1 + \frac{3P_n k}{n}\right).$$

Since $k \leq t_n$, we have that $\frac{P_n k}{n} \leq \frac{P_n t_n}{n} = \frac{1}{t_n \omega_n} \leq 1$, because ω_n and t_n are both at least one. Thus, $1 + \frac{3P_n k}{n} \leq 4$ and so

$$S(n,k)\Delta \geq 1 - \frac{9P_n^2 k^2}{n^2} - 8\frac{3P_n k^2 + k^4}{n^2}.$$

Substituting this into (4.1) and using $k \leq t_n$, we get

$$\#\mathcal{S}(1) \leq \theta \cdot \sum_{k=2}^{t_n} \binom{n}{k} 3^k \leq \theta \cdot \#\mathcal{E}_n(t_n),$$

where $\theta := \frac{9P_n^2 t_n^2}{n^2} + 8\frac{3P_n t_n^2 + t_n^4}{n^2}$. The above estimate holds for all the sets $\mathcal{S}(l)$, $1 \leq l \leq L - 1$ (see discussion prior to (4.4)). Performing the same analysis with $3P_n + s$ instead of $3P_n$, we get

$$\#\mathcal{S}(L) \leq \left(\frac{(3P_n + s)^2 t_n^2}{n^2} + 8\frac{(3P_n + s)t_n^2 + t_n^4}{n^2}\right)(\#\mathcal{E}_n(t_n))$$

and using the fact that $s \leq 3P_n$, we therefore get

$$\#\mathcal{S}(l) \leq \left(\frac{(6P_n)^2 t_n^2}{n^2} + 8\frac{6P_n t_n^2 + t_n^4}{n^2}\right)(\#\mathcal{E}_n(t_n)) \leq \frac{84P_n^2 t_n^2 + 8t_n^4}{n^2}(\#\mathcal{E}_n(t_n))$$

for all $1 \leq l \leq L$

References

1. Abroshan M, Venkataramanan R, Guillén i Fàbregas A (2018) Coding for segmented edit channels. IEEE Trans Inf Theory 64:3086–3098
2. Alon N, Spencer J The probabilistic method. Wiley Interscience
3. Helberg ASJ, Ferreira HC (2002) On multiple insertion/deletable error correcting codes. IEEE Trans Inf Theory 48:305–308
4. Krishnan MN, Shukla D, Vijay Kumar P (2020) Rate-optimal streaming codes for channels with burst and random erasures. IEEE Trans Inf Theory 66:4869–4891
5. Levenshtein VI (1965) Binary codes capable of correcting deletable errors. Inser Reversals, Dokl Akad Nauk SSSR 163:845–848
6. Liu Z, Mitzenmacher M (2010) Codes for deletion and insertion channels with segmented errors. IEEE Trans Inf Theory 56:224–232
7. Sima J, Bruck J (2019) Optimal $k-$Deletion correcting codes. Int Symp Inf Theory 847–851
8. Tatwawadi K, Chandak S (2019) Tutorial on algebraic deletion correction codes. arXiv:1906.07887.pdf
9. Venkataramanan R, Swamy VN, Ramchandran K (2015) Low complexity interactive algorithms for synchronization from deletions. Inser, SubstitS, IEEE Trans Inf Theory 61:5670–5689

Lyapunov-Type Inequalities for Fractional Differential Operators with Non-singular Kernels

Debananda Basua and Jagan Mohan Jonnalagadda

Abstract In this paper, we deal with the newly introduced fractional differential operators, which involve exponential and Mittag-Leffler kernels. First, we find Green's functions and their properties for conjugate and anti-periodic boundary value problems (BVPs) involving Caputo–Fabrizio (CF) and Atangana–Baleanu–Caputo (ABC) fractional derivatives of order $1 < \varrho \leq 2$. Then, we establish Lyapunov-type inequalities (LTIs) for CF and ABC fractional boundary value problems (FBVPs) using the properties of their Green functions.

Keywords Fractional derivative · Boundary value problem · Lyapunov-type inequality · Exponential · Mittag-Leffler

1 Introduction

Riemann–Liouville (R-L) and Caputo derivatives are the most popular fractional derivatives which involve singular kernels [14, 17]. Recently, fractional derivatives involving exponential and Mittag-Leffler [3, 8] kernels were introduced. Fractional differential equations (FDEs) have been of great interest for many researchers due to their applications in distinct fields of engineering and science. Of these applications, derivatives involving non-singular kernels are more attractive [4, 18].

On the other hand, in 1907, Lyapunov [15] found a remarkable result, known as Lyapunov's inequality. In [15], the author proved the existence of a solution of 2nd-order ordinary differential equation (Hill's equation) with conjugate boundary conditions. Many extensions and generalizations of Lyapunov's inequality were developed due to its applicability in various problems of differential equations, including disconjugacy, eigenvalue problems, etc. In this direction, in 2013, Ferreira [9] generalized Lyapunov's inequality, in which the author replaced the 2nd-order ordinary derivative with the ϱth-order ($\varrho \in (1, 2]$) R-L fractional derivative. Further, in

D. Basua (✉) · J. M. Jonnalagadda
Department of Mathematics, BITS-Pilani, Hyderabad Campus, Hyderabad 500078, Telangana, India

© The Author(s), under exclusive license to Springer Nature Singapore Pte Ltd. 2022

D. Giri et al. (eds.), *Proceedings of the Seventh International Conference on Mathematics and Computing*, Advances in Intelligent Systems and Computing 1412,
https://doi.org/10.1007/978-981-16-6890-6_58

2014, Ferreira [10] developed LTIs for the Caputo fractional derivative. In 2018, Ntouyas et al. [16] surveyed several generalizations of Lyapunov's inequality for FDEs associated with a wide variety of boundary conditions. For more works related to this direction, we refer to [5–7, 11–13], and the references therein. In particular, recently, Abdeljawad [1, 2] studied LTIs for Caputo–Fabrizio–Riemann (CFR) and the Atangana–Baleanu–Riemann (ABR) FBVPs of order $2 \leq \varrho < 3$, associated with conjugate boundary conditions.

In this study, we prove LTIs for CF and ABC FBVPs of ϱth-order ($\varrho \in (1, 2]$), associated with conjugate boundary conditions and anti-periodic boundary conditions.

2 Preliminaries

In this section, we present some important definitions and results of FDEs which will be useful in the next sections.

Definition 1 ([14]) The Euler gamma function and the one parameter Mittag-Leffler function are given by

$$\Gamma(z) = \int_0^\infty t^{z-1} e^{-t} d\mu, \quad \Re(z) > 0$$

and

$$E_\varrho(z) = \sum_{k=0}^{\infty} \frac{z^k}{\Gamma(k\varrho + 1)}, \quad \Re(z) > 0.$$

Definition 2 ([19]) Let $\Omega \subset \mathbb{R}$ and $H^p(\Omega)$ be a Sobolev space defined by

$$H^p(\Omega) = \{f \in L^2(\Omega) : D^\varsigma f \in L^2(\Omega), \forall\, |\varsigma| \leq p, p \in [1, \infty)\}.$$

Definition 3 ([8]) Let $\vartheta \in H'(\imath, \jmath)$. The ϱth-order CF fractional derivative is defined by

$$({}^{CF}D^{\varrho}_{\imath}\vartheta)(\mu) = \frac{B(\varrho)}{1-\varrho} \int_\imath^\mu e^{\left[-\varrho \frac{(\mu-\nu)}{1-\varrho}\right]} \vartheta'(\nu) d\nu, \quad \imath \leq \mu \leq \jmath, \ 0 < \varrho < 1. \tag{1}$$

The CF fractional integral is given by

$$({}^{CF}I^{\varrho}_{\imath}\vartheta)(\mu) = \frac{1-\varrho}{B(\varrho)} \vartheta(\mu) + \frac{\varrho}{B(\varrho)} \int_\imath^\mu \vartheta(\nu) d\nu, \quad \imath \leq \mu \leq \jmath. \tag{2}$$

Definition 4 ([3]) Let $\vartheta \in H'(\imath, \jmath)$ and $0 < \varrho < 1$. The ϱth-order ABC fractional derivative is given by

$$({}^{ABC}D_{\imath}^{\varrho}\vartheta)(\mu) = \frac{B(\varrho)}{1-\varrho}\int_{\imath}^{\mu} E_{\varrho}\left[(\mu-\nu)^{\varrho}\frac{-\varrho}{1-\varrho}\right]\vartheta'(\nu)d\nu, \quad \imath \le \mu \le \jmath. \tag{3}$$

The AB fractional integral is given by

$$({}^{AB}I_{\imath}^{\varrho}\vartheta)(\mu) = \frac{1-\varrho}{B(\varrho)}\vartheta(\mu) + \frac{\varrho}{B(\varrho)}({}^{RL}I_{\imath}^{\varrho}\vartheta)(\mu), \quad \imath \le \mu \le \jmath; \tag{4}$$

the normalizing constant $B(\varrho) > 0$ has the property $B(0) = B(1) = 0$.

Definition 5 ([3, 8]) Let ϑ be such that $\vartheta^{(\upsilon)} \in H'(\imath, \jmath)$ and $\upsilon < \varrho \le \upsilon + 1$. Set $\varsigma = \varrho - \upsilon,\ \varsigma \in (0, 1]$. Then

$$({}^{CF}D_{\imath}^{\varrho}\vartheta)(\mu) = ({}^{CF}D_{\imath}^{\varsigma}\vartheta^{(\upsilon)})(\mu) \text{ and } ({}^{ABC}D_{\imath}^{\varrho}\vartheta)(\mu) = ({}^{ABC}D_{\imath}^{\varsigma}\vartheta^{(\upsilon)})(\mu).$$

Lemma 1 ([3, 8]) *If $\vartheta \in H'(\imath, \jmath)$ and $\upsilon < \varrho \le \upsilon + 1$, then we have*

$$({}^{CF}I_{\imath}^{\varrho}{}^{CF}D_{\imath}^{\varrho}\vartheta)(\mu) = \vartheta(\mu) - \sum_{k=0}^{\upsilon}\frac{\vartheta^{(k)}(\imath)}{k!}(\mu-\imath)^{k}$$

and

$$({}^{AB}I_{\imath}^{\varrho}{}^{ABC}D_{\imath}^{\varrho}\vartheta)(\mu) = \vartheta(\mu) - \sum_{k=0}^{\upsilon}\frac{\vartheta^{(k)}(\imath)}{k!}(\mu-\imath)^{k}.$$

3 CF Fractional BVPs

In this section, we derive Green's functions and their properties and then formulate the LTIs for the FVBPs (5) and (10). Thereafter, we develop the LTIs for the FBVPs (5) and (10).

Theorem 1 *Let $\vartheta \in H'(\imath, \jmath)$. The following FBVP*

$$\begin{cases} ({}^{CF}D_{\imath}^{\varrho}\vartheta)(\mu) + \xi(\mu)\vartheta(\mu) = 0, \ 1 < \varrho \le 2, \quad \imath < \mu < \jmath, \\ \vartheta(\imath) = \vartheta(\jmath) = 0 \end{cases} \tag{5}$$

is equivalent to

$$\vartheta(\mu) = \int_{\imath}^{\jmath} G(\mu, \nu)\xi(\nu)\vartheta(\nu)d\nu, \tag{6}$$

where $G(\mu, \nu)$ is given by

$$G(\mu,\nu)=\begin{cases}\frac{(2-\varrho)(\mu-\imath)+(\varrho-1)(\mu-\imath)(\jmath-\nu)-(\jmath-\imath)\{(2-\varrho)+(\varrho-1)(\mu-\nu)\}}{(\jmath-\imath)B(\varrho-1)}, & \imath\le\nu\le\mu\le\jmath,\\ \frac{(2-\varrho)(\mu-\imath)+(\varrho-1)(\mu-\imath)(\jmath-\nu)}{(\jmath-\imath)B(\varrho-1)}, & \imath\le\mu\le\nu\le\jmath.\end{cases} \tag{7}$$

Proof Set $\varsigma=\varrho-1$. Applying the ϱth-order Caputo–Fabrizio fractional integral operator on (5) and applying Lemma 1, we get

$$\vartheta(\mu)=-({}^{CF}I_\imath^\varrho\xi(.)\vartheta(.))(\mu)+C_1+C_2(\mu-\imath). \tag{8}$$

But $({}^{CF}I_\imath^\varrho\xi(.)\vartheta(.))(\mu)=\frac{1-\varsigma}{B(\varsigma)}\int_\imath^\mu\xi(\nu)\vartheta(\nu)d\nu+\frac{\varsigma}{B(\varsigma)}I_\imath^2\xi(\mu)\vartheta(\mu)$. Then, we have

$$\vartheta(\mu)=C_1+C_2(\mu-\imath)-\int_\imath^\mu\frac{(2-\varrho)+(\varrho-1)(\mu-\nu)}{B(\varrho-1)}\xi(\nu)\vartheta(\nu)d\nu. \tag{9}$$

Using the conjugate boundary conditions to (9), we get $C_1=0$ and

$$C_2=\frac{2-\varrho}{(\jmath-\imath)B(\varrho-1)}\int_\imath^\jmath\xi(\nu)\vartheta(\nu)d\nu+\frac{\varrho-1}{(\jmath-a)B(\varrho-1)}\int_\imath^\jmath(\jmath-\nu)\xi(\nu)\vartheta(\nu)d\nu.$$

Now, substituting the values of C_1 and C_2 in (9), we have

$$\begin{aligned}\vartheta(\mu)&=\int_\imath^\mu\Big[\frac{(2-\varrho)(\mu-\imath)+(\varrho-1)(\mu-\imath)(\jmath-\nu)}{(\jmath-\imath)B(\varrho-1)}\\&\quad-\frac{(2-\varrho)+(\varrho-1)(\mu-\nu)}{B(\varrho-1)}\Big]\xi(\nu)\vartheta(\nu)d\nu\\&\quad+\int_\imath^\jmath\frac{(2-\varrho)(\mu-\imath)+(\varrho-1)(\mu-\imath)(\jmath-\nu)}{(\jmath-\imath)B(\varrho-1)}\xi(\nu)\vartheta(\nu)d\nu\\&=\int_\imath^\jmath G(\mu,\nu)\xi(\nu)\vartheta(\nu)d\nu.\end{aligned}$$

The proof is complete.

Theorem 2 *Let $\vartheta\in H'(\imath,\jmath)$. The following FBVP*

$$\begin{cases}({}^{CF}D_\imath^\varrho\vartheta)(\mu)+\xi(\mu)\vartheta(\mu)=0,\ 1<\varrho\le 2, & \imath<\mu<\jmath,\\ \vartheta(\imath)+\vartheta(\jmath)=0,\ \vartheta'(\imath)+\vartheta'(\jmath)=0\end{cases} \tag{10}$$

is equivalent to

$$\vartheta(\mu)=\frac{(2-\varrho)\{\xi(\imath)\vartheta(\imath)+\xi(\jmath)\vartheta(\jmath)\}(2\mu-\imath-\jmath)}{4B(\varrho-1)}+\int_\imath^\jmath H(\mu,\nu)\xi(\nu)\vartheta(\nu)d\nu, \tag{11}$$

where $H(\mu,\nu)$ is given by

$$H(\mu,\nu)=\begin{cases}\frac{2(\varrho-1)\{(\mu-\imath)+(\jmath-\nu)\}-2(2-\varrho)-(\varrho-1)(\jmath-\imath)+(\varrho-1)(\mu-\nu)}{4B(\varrho-1)}, & \imath\le\nu\le\mu\le\jmath,\\ \frac{2(\varrho-1)\{(\mu-\imath)+(\jmath-\nu)\}+2(2-\varrho)-(\varrho-1)(\jmath-\imath)}{4B(\varrho-1)}, & \imath\le\mu\le\nu\le\jmath.\end{cases} \tag{12}$$

Proof The proof is the same as the previous theorem.

Theorem 3 *The property of $G(\mu,\nu)$ defined in (7) is*

$$\max_{\mu,\nu\in[\imath,\jmath]}\left|G(\mu,\nu)\right|=\frac{2-\varrho}{B(\varrho-1)}.$$

Proof Let us denote the functions

$$G_1(\mu,\nu)=G_2(\mu,\nu)-\frac{(2-\varrho)+(\varrho-1)(\mu-\nu)}{B(\varrho-1)}$$

and

$$G_2(\mu,\nu)=\frac{(2-\varrho)(\mu-\imath)+(\varrho-1)(\mu-\imath)(\jmath-\nu)}{(\jmath-a)B(\varrho-1)}.$$

For any fixed $\mu\in[\imath,\jmath]$, consider

$$\frac{\partial}{\partial\nu}G_1(\mu,\nu)=\frac{(\varrho-1)\{(\jmath-\imath)-(\mu-\imath)\}}{(\jmath-\imath)B(\varrho-1)}\ge 0.$$

Implying that $G_1(\mu,\nu)$ is an increasing function of ν, then we have

$$\max_{\nu\in[\imath,\mu]}\left|G_1(\mu,\nu)\right|=\max\left\{\left|G_1(\mu,\imath)\right|,\left|G_1(\mu,\mu)\right|\right\}.$$

Now,

$$\frac{d}{d\mu}G_1(\mu,\imath)=\frac{2-\varrho}{(\jmath-\imath)B(\varrho-1)}\ge 0,$$

implying that $G_1(\mu,\imath)$ is an increasing function of μ. Then, we have

$$\begin{aligned}\max_{\mu\in[\imath,\jmath]}\left|G_1(\mu,\imath)\right|&=\max\left\{\left|G_1(\imath,\imath)\right|,\left|G_1(\jmath,\imath)\right|\right\}\\&=\max\left\{\frac{2-\varrho}{B(\varrho-1)},0\right\}\\&=\frac{2-\varrho}{B(\varrho-1)}.\end{aligned}$$

Now, consider

$$\frac{d}{d\mu}G_1(\mu,\mu) = \frac{(2-\varrho)+(\varrho-1)(\jmath+\imath-2t)}{(\jmath-\imath)B(\varrho)},$$

$$\frac{d^2}{d\mu^2}G_1(\mu,\mu) = \frac{-2(\varrho-1)}{(\jmath-\imath)B(\varrho)} \le 0.$$

Therefore, we have $\frac{d}{d\mu}G_1(\mu,\mu)$ is a decreasing function of μ and $\frac{d}{d\mu}G_1(\imath,\imath) > 0$, implying that $G_1(\mu,\mu)$ is an increasing function of μ.

$$\begin{aligned}\max_{\mu\in[\imath,\jmath]}\left|G_1(\mu,\mu)\right| &= \max\left\{\left|G_1(\imath,\imath)\right|, \left|G_1(\jmath,\jmath)\right|\right\}\\ &= \max\left\{\frac{2-\varrho}{B(\varrho-1)}, 0\right\}\\ &= \frac{2-\varrho}{B(\varrho-1)}.\end{aligned}$$

Therefore, we get

$$\max_{\nu\in[\imath,\mu],\mu\in[\imath,\jmath]}\left|G_1(\mu,\nu)\right| = \frac{2-\varrho}{B(\varrho-1)}. \tag{13}$$

For any fixed $\nu \in [\imath, \jmath]$, consider

$$\frac{\partial}{\partial\mu}G_2(\mu,\nu) = \frac{(2-\varrho)+(\varrho-1)(\jmath-\nu)}{(\jmath-\imath)B(\varrho-1)} \ge 0,$$

implying that $G_2(\mu,\nu)$ is an increasing function of μ. Thus, we have

$$\max_{\mu\in[\imath,\nu]}\left|G_2(\mu,\nu)\right| = \max\left\{\left|G_2(\imath,\nu)\right|, \left|G_2(\nu,\nu)\right|\right\}.$$

Since $G_2(\imath,\nu) = 0$, then we check only for

$$G_2(\nu,\nu) = \frac{(2-\varrho)(\nu-\imath)+(\varrho-1)(\nu-\imath)(\jmath-\nu)}{(\jmath-\imath)B(\varrho-1)}.$$

Now, consider

$$\frac{d}{d\nu}G_2(\nu,\nu) = \frac{(2-\varrho)+(\varrho-1)(\jmath+\imath-2\nu)}{(\jmath-\imath)B(\varrho)},$$

$$\frac{d^2}{d\nu^2}G_2(\nu,\nu) = \frac{-2(\varrho-1)}{(\jmath-\imath)B(\varrho)} \le 0.$$

Therefore, we get $\frac{d}{d\nu}G_2(\nu,\nu)$ is a decreasing function of ν and $\frac{d}{d\nu}G_2(\imath,\imath) > 0$, implying that $G_2(\nu,\nu)$ is an increasing function of ν. Then, we have

$$\begin{aligned}\max_{\nu\in[\imath,\jmath]}|G_2(\nu,\nu)| &= \max\{|G_2(\imath,\imath)|,|G_2(\jmath,\jmath)|\}\\ &= \max\left\{0,\frac{2-\varrho}{B(\varrho-1)}\right\}\\ &= \frac{2-\varrho}{B(\varrho-1)}.\end{aligned}$$

Thus, we get

$$\max_{\mu\in[\imath,\nu],\nu\in[\imath,\jmath]}\left|G_2(\mu,\nu)\right| = \frac{2-\varrho}{B(\varrho-1)}. \tag{14}$$

Hence, the result follows from (13) and (14). This completes the proof of the theorem.

Theorem 4 *The property of $H(\mu,\nu)$ defined in (12) is*

$$\max_{\mu,\nu\in[\imath,\jmath]}\left|H(\mu,\nu)\right| = \frac{2(2-\varrho)+(\varrho-1)(\jmath-\imath)}{4B(\varrho-1)}.$$

Proof The proof is similar to the proof of the previous theorem.

Now, we develop the LTIs for the FBVPs (5) and (10).

Theorem 5 *If the following FBVP has a nontrivial solution*

$$\begin{cases}({}^{CF}D_\imath^\varrho\vartheta)(\mu)+\xi(\mu)\vartheta(\mu)=0, & 1<\varrho\le 2, \quad \imath<\mu<\jmath,\\ \vartheta(\imath)=\vartheta(\jmath)=0,\end{cases} \tag{15}$$

then

$$\int_\imath^\jmath\left|\xi(\nu)\right|d\nu > \frac{B(\varrho-1)}{2-\varrho}. \tag{16}$$

Proof Let $\vartheta\in\mathbb{U}=C[\imath,\jmath]$ be a nontrivial solution of the FBVP (5), where $\|\vartheta\| = \max_{\mu\in[\imath,\jmath]}|\vartheta(\mu)|$. By Theorem 1, u must satisfy

$$\vartheta(\mu)=\int_\imath^\jmath G(\mu,\nu)\xi(\nu)\vartheta(\nu)d\nu.$$

Then,

$$\begin{aligned}\|\vartheta\| = \max_{\mu,\nu\in[\imath,\jmath]}\left|\int_\imath^\jmath G(\mu,\nu)\xi(\nu)\vartheta(\nu)d\nu\right| &\le \max_{\mu,\nu\in[\imath,\jmath]}\left[\int_\imath^\jmath\left|G(\mu,\nu)\right|\left|\xi(\nu)\right|\left|\vartheta(\nu)\right|d\nu\right]\\ &\le \|\vartheta\|\left[\max_{\mu,\nu\in[\imath,\jmath]}\int_\imath^\jmath\left|G(\mu,\nu)\right|\left|\xi(\nu)\right|d\nu\right]\\ &\le \|\vartheta\|\left[\max_{\mu,\nu\in[\imath,\jmath]}\left|G(\mu,\nu)\right|\right]\int_\imath^\jmath\left|\xi(\nu)\right|d\nu,\end{aligned}$$

or, equivalently,

$$1 < \Big[\max_{\mu,\nu\in[\imath,\jmath]} \big|G(\mu,\nu)\big|\Big] \int_{\imath}^{\jmath} \big|\xi(\nu)\big| d\nu.$$

Hence, by using Theorem 3, we have the desired result.

Theorem 6 *Let $\xi(\imath)\vartheta(\imath) + \xi(\jmath)\vartheta(\jmath) = 0$. If the following FBVP has a nontrivial solution*

$$\begin{cases} ({}^{CFC}D_{\imath}^{\varrho}\vartheta)(\mu) + \xi(\mu)\vartheta(\mu) = 0, \quad 1 < \varrho \le 2, \quad \imath < \mu < \jmath, \\ \vartheta(\imath) + \vartheta(\jmath) = 0, \ \vartheta'(\imath) + \vartheta'(\jmath) = 0, \end{cases} \tag{17}$$

then

$$\int_{\imath}^{\jmath} \big|\xi(\nu)\big| d\nu > \frac{4B(\varrho - 1)}{2(2-\varrho) + (\varrho-1)(\jmath-\imath)}. \tag{18}$$

Proof The proof is similar to the previous theorem.

4 ABC Fractional BVPs

In this section, First, we derive Green's functions and their properties and then formulate the LTIs for the FVBPs (19) and (22). Thereafter, we develop the LTIs for the FBVPs (19) and (22).

Theorem 7 *Let $\vartheta \in H^1(\imath, \jmath)$. The following FBVP*

$$\begin{cases} ({}^{ABC}D_{\imath}^{\varrho}\vartheta)(\mu) + \xi(\mu)\vartheta(\mu) = 0, \quad 1 < \varrho \le 2, \quad \imath < \mu < \jmath, \\ \vartheta(\imath) = \vartheta(\jmath) = 0 \end{cases} \tag{19}$$

is equivalent to

$$\vartheta(\mu) = \int_{\imath}^{\jmath} \bar{G}(\mu,\nu)\xi(\nu)\vartheta(\nu) d\nu, \tag{20}$$

where $\bar{G}(\mu,\nu)$ is given by

$$\bar{G}(\mu,\nu) = \begin{cases} \bar{G}_1(\mu,\nu) = \bar{G}_2(\mu,\nu) - \frac{\Gamma(\varrho)(2-\varrho)+(\varrho-1)(\mu-\nu)^{\varrho-1}}{\Gamma(\varrho)B(\varrho-1)}, & \imath \le \nu \le \mu \le \jmath, \\ \bar{G}_2(\mu,\nu) = \frac{(2-\varrho)\Gamma(\varrho)(\mu-\imath)+(\varrho-1)(\mu-\imath)(\jmath-\nu)^{\varrho-1}}{\Gamma(\varrho)(\jmath-\imath)B(\varrho-1)}, & \imath \le \mu \le \nu \le \jmath. \end{cases} \tag{21}$$

Proof The proof is the same as Theorem 1.

Theorem 8 *The property of $\bar{G}(\mu,\nu)$ defined in (21) is*

$$\max_{\mu,\nu\in[\imath,\jmath]} \big|\bar{G}(\mu,\nu)\big| = \frac{2-\varrho}{B(\varrho-1)}.$$

Proof The proof is the same as Theorem 3.

Theorem 9 *Let $\vartheta \in H^1(\imath, \jmath)$. The following FBVP*

$$\begin{cases} ({}^{ABC}D_\imath^\varrho \vartheta)(\mu) + \xi(\mu)\vartheta(\mu) = 0, \quad 1 < \varrho \leq 2, \quad \imath < \mu < \jmath, \\ \vartheta(\imath) + \vartheta(\jmath) = 0, \ \vartheta'(\imath) + \vartheta'(\jmath) = 0 \end{cases} \tag{22}$$

is equivalent to

$$\vartheta(\mu) = \frac{(2-\varrho)\{\xi(\imath)\vartheta(\imath) + \xi(\jmath)\vartheta(\jmath)\}(2\mu - \imath - \jmath)}{4B(\varrho-1)} + \int_\imath^\jmath \bar{H}(\mu, \nu)\xi(\nu)\vartheta(\nu)d\nu, \tag{23}$$

where $\bar{H}(\mu, \nu)$ is given by

$$\bar{H}(\mu, \nu) = \begin{cases} \frac{2(\varrho-1)(\jmath-\nu)^{\varrho-1}+(\varrho-1)(2\mu-\jmath-\imath)(\jmath-\nu)^{\varrho-2}+(4-2\varrho)\Gamma(\varrho)}{4\Gamma(\varrho)B(\varrho-1)} \\ -\frac{\Gamma(\varrho)(2-\varrho)+(\varrho-1)(\mu-\nu)^{\varrho-1}}{\Gamma(\varrho)B(\varrho-1)}, & \imath \leq \nu \leq \mu < \jmath, \\ \frac{2(\varrho-1)(\jmath-\nu)^{\varrho-1}+(\varrho-1)(2\mu-\jmath-\imath)(\jmath-\nu)^{\varrho-2}+(4-2\varrho)\Gamma(\varrho)}{4\Gamma(\varrho)B(\varrho-1)}, & \imath \leq \mu \leq \nu < \jmath. \end{cases} \tag{24}$$

Proof The proof is the same as Theorem 1.

Theorem 10 *The property of $\bar{H}(\mu, \nu)$ defined in (24) is*

$$\max_{\mu,\nu \subset [\imath, \jmath)} \left|(\jmath - \nu)^{2-\varrho}\bar{H}(\mu, \nu)\right| = \frac{2(2-\varrho)\Gamma(\varrho)(\jmath - \imath)^{2-\varrho} + (\varrho^2 - 1)(\jmath - \imath)}{4\Gamma(\varrho)B(\varrho - 1)}.$$

Proof The proof is similar to the proof of Theorem 3.

Now, we develop the LTIs for the FBVPs (19) and (22).

Theorem 11 *If the following FBVP has a nontrivial solution*

$$\begin{cases} ({}^{ABC}D_\imath^\varrho \vartheta)(\mu) + \xi(\mu)\vartheta(\mu) = 0, \quad \imath < \mu < \jmath, \\ \vartheta(\imath) = \vartheta(\jmath) = 0, \ 1 < \varrho \leq 2, \end{cases} \tag{25}$$

then

$$\int_\imath^\jmath \left|\xi(\nu)\right|d\nu > \frac{B(\varrho - 1)}{2 - \varrho}. \tag{26}$$

Proof The proof is similar to Theorem 5.

Theorem 12 *If the following FBVP has a nontrivial solution*

$$\begin{cases} ({}^{ABC}D_\imath^\varrho \vartheta)(\mu) + \xi(\mu)\vartheta(\mu) = 0, \qquad \imath < \mu < \jmath, \\ \vartheta(\imath) + \vartheta(\jmath) = 0, \ \vartheta'(\imath) + \vartheta'(\jmath) = 0, \ 1 < \varrho \leq 2, \end{cases} \tag{27}$$

then

$$\int_{\imath}^{\jmath} (\jmath - \nu)^{\varrho-2} \big|\xi(\nu)\big| d\nu > \frac{4\Gamma(\varrho) B(\varrho - 1)}{2(2-\varrho)\Gamma(\varrho)(\jmath - \imath)^{2-\varrho} + (\varrho^2 - 1)(\jmath - \imath)}. \quad (28)$$

Proof The proof is similar to Theorem 5.

5 Application

In this section, we discuss applications of the results established in the previous sections. We estimate lower bounds on the smallest eigenvalues of the eigenvalue problems corresponding to (5), (10), (19) and (22).

Theorem 13 *Assume ϑ is a nontrivial solution to the CF fractional eigenvalue problem*

$$\begin{cases} ({}^{CF}D_{\imath}^{\varrho}\vartheta)(\mu) + \lambda\vartheta(\mu) = 0, \ 1 < \varrho \leq 2, \\ \vartheta(\imath) = \vartheta(\jmath) = 0, \qquad \imath < \mu < \jmath, \end{cases}$$

where $\vartheta(\mu) \neq 0$ for each $t \in (\imath, \jmath)$. Then,

$$|\lambda| > \frac{B(\varrho - 1)}{(\jmath - \imath)(2 - \varrho)}.$$

Theorem 14 *Assume ϑ is a nontrivial solution of the CF fractional eigenvalue problem*

$$\begin{cases} ({}^{CF}D_{\imath}^{\varrho}\vartheta)(\mu) + \lambda\vartheta(\mu) = 0, \ 1 < \varrho \leq 2, \quad \imath < \mu < \jmath, \\ \vartheta(\imath) + \vartheta(\jmath) = 0, \ \vartheta'(\imath) + \vartheta'(\jmath) = 0, \end{cases} \quad (29)$$

where $\vartheta(\mu) \neq 0$ for each $t \in (\imath, \jmath)$. Then,

$$|\lambda| > \frac{4B(\varrho - 1)}{2(2 - \varrho)(\jmath - \imath) + (\varrho - 1)(\jmath - \imath)^2}.$$

Theorem 15 *Assume ϑ is a nontrivial solution of the ABC fractional eigenvalue problem*

$$\begin{cases} ({}^{ABC}D_{\imath}^{\varrho}\vartheta)(\mu) + \lambda\vartheta(\mu) = 0, \ 1 < \varrho \leq 2, \\ \vartheta(\imath) = \vartheta(\jmath) = 0, \qquad \imath < \mu < \jmath, \end{cases}$$

where $\vartheta(\mu) \neq 0$ for each $t \in (\imath, \jmath)$. Then,

$$|\lambda| > \frac{B(\varrho - 1)}{(\jmath - \imath)(2 - \varrho)}.$$

Theorem 16 *Assume y is a nontrivial solution of the ABC fractional eigenvalue problem*

$$\begin{cases} ({}^{ABC}D_{\imath}^{\varrho}\vartheta)(\mu) + \lambda\vartheta(\mu) = 0, \ 1 < \varrho \leq 2, \quad \imath < \mu < \jmath, \\ \vartheta(\imath) + \vartheta(\jmath) = 0, \ \vartheta'(\imath) + \vartheta'(\jmath) = 0, \end{cases} \tag{30}$$

where $\vartheta(\mu) \neq 0$ *for each* $t \in (\imath, \jmath)$*. Then,*

$$|\lambda| > \frac{4(\varrho-1)\Gamma(\varrho)B(\varrho-1)}{2(2-\varrho)\Gamma(\varrho)(\jmath-\imath) + (\varrho^2-1)(\jmath-\imath)^2}.$$

6 Conclusion

In this paper, first, we found Green's functions for CF and ABC FBVPs associated with conjugate and anti-periodic boundary conditions. Then, we established LTIs for the CF and ABC FBVPs. Lastly, we estimated lower bound for the eigenvalues of the fractional eigenvalue problems corresponding to the CF and ABC FBVPs. As another application, we may discuss the disconjugacy property for CF and ABC FBVPs (5) and (19).

References

1. Abdeljawad T (2017) Fractional operators with non-singular exponential kernels and a Lyapunov-type inequality. Adv Differ Equ 2017:313
2. Abdeljawad T (2017) A Lyapunov-type inequality for fractional operators with non-singular Mittag–Leffler kernel. J Inequal Appl 2017:130
3. Atangana A, Baleanu D (2016) New fractional derivatives with non-local and non-singular kernel: theory and applications to heat transfer model. Therm Sci 20(2):763–769
4. Baleanu D, Jaiami A, Sajjadi S, Mozyrska D (2019) A new fractional model and optimal control of a tumor-immune surveillance with non-singular derivative operator. Chaos 28(8):83–127
5. Baleanu D, Basua D, Jonnalagadda JM, Green's function and an inequality of Lyapunov-type for conformable boundary value problem. Novi Sad J Math. https://doi.org/10.30755/NSJOM.10766
6. Basua D, Jonnalagadda JM (2019) Lyapunov-type inequality for Riemann-Liouville type fractional boundary value problems. Novi Sad J Math 49:137–152
7. Basua D, Jonnalagadda JM, Satpathi DK (2019) Lyapunov-type inequalities for Riemann-Liouville type fractional boundary value problems with fractional boundary conditions. Adv Th Non Anal Appl 3(2):53–63
8. Caputo M, Fabrizio M (2015) A new definition of fractional derivative without singular kernel. Prog Fract Differ Appl 1(2):73–85
9. Ferreira RAC (2013) A Lyapunov-type inequality for a fractional boundary value problem. Fract Calc Appl Anal 16:978–984
10. Ferreira RAC (2014) On a Lyapunov-type inequality and the zeros of a certain Mittag-Leffler function. J Math Anal Appl 412:1058–1063
11. Jonnalagadda JM, Debananda B (2020) Fractional analogue of the Lyapunov inequality in conformable sense. J Fract Calc Appl 11(2):23–31

12. Jonnalagadda JM, Debananda B (2020) Lyapunov-type inequalities for Hadamard type fractional boundary value problems. AIMS Math 5(2):1127–1146
13. Jonnalagadda JM, Debananda B, Satpathi DK (2021) Lyapunov-type inequality for an anti-periodic conformable boundary value problem. Kragujevac J Math 45:289–298
14. Kilbas AA, Srivastava HM, Trujillo JJ (2006) Theory and applications of fractional differential equations. Elsevier, Amsterdam
15. Liapounoff A (1907) Problème général de la stabilité du mouvement. Ann Fac Sci Toulouse Math 9:203–474
16. Ntouyas SK, Ahmad B, Horikis TP (2019) Recent developments of Lyapunov-type inequalities for fractional differential equations. In: Differential and Integral Inequalities. Springer, Cham, pp 619–686
17. Podlubny I (1999) Fractional differential equations. Academic Press, San Diego
18. Saad KM, Atangana A, Baleanu D (2018) New fractional derivatives with non-singular kernel applied to the Burger's equation. Chaos 28(6):63–109
19. Syam MI, Al-Refai M (2019) Fractional differential equations with Atangana-Baleanu fractional derivative: Analysis and applications. Chaos, Solitons Fractals: X 2:100013

Chebyshev Spectral Projection Methods for Fredholm Integral Equations of the Second Kind

Bijaya Laxmi Panigrahi and Jitendra Kumar Malik

Abstract In this paper, we will propose the Chebyshev spectral Galerkin and collocation methods for the Fredholm integral equations (FIEs) of the second kind with smooth kernel and its associated eigenvalue problem (EVPs). The convergence rates of approximated solutions, iterated solutions with exact solution in L^2_ω norm have been investigated. We will evaluate the errors between exact eigen-elements and approximated eigen-elements both in L^2_ω and L^∞_ω norms. We will show that eigenvalues and iterated eigenvectors have super-convergence rate in Chebyshev spectral Galerkin methods.

Keywords Fredholm integral equations · Eigenvalue problems · Compact integral operator · Chebyshev polynomials

1 Introduction

Let ω be a non negative, continuous and integrable real valued function on $[-1, 1]$. Then the weighted space $\mathbb{X} = L^2_\omega[-1, 1]$ is defined by

$$L^2_\omega[-1, 1] = \{u | u : [-1, 1] \to \mathbb{R} \text{ and } \|u\|_{L^2_\omega} < \infty\}$$

where

$$\|u\|_{L^2_\omega} = \left(\int_\Omega |u(t)|^2 \; w(t) \; \mathrm{d}t \right)^{\frac{1}{2}}.$$

Let $\mathcal{T}$ be an integral operator defined on $\mathbb{X}$ by

B. Laxmi Panigrahi (✉)
Department of Mathematics, Gangadhar Meher University, Sambalpur 768004, India

J. Kumar Malik
Department of Mathematics, Sambalpur University, Sambalpur 768019, Odisha, India

© The Author(s), under exclusive license to Springer Nature Singapore Pte Ltd. 2022

D. Giri et al. (eds.), *Proceedings of the Seventh International Conference on Mathematics and Computing*, Advances in Intelligent Systems and Computing 1412,
https://doi.org/10.1007/978-981-16-6890-6_59

$$\mathcal{T}u(s) = \int_{-1}^{1} k(s,t)u(t)\,\mathrm{d}t, \quad s \in [-1, 1], \tag{1}$$

where the kernel $k(s, t) \in \mathcal{C}([-1, 1] \times [-1, 1])$. Then $\mathcal{T}$ is a compact operator on $\mathbb{X}$.

The FIEs of the second kind is in the following operator form

$$(\mathcal{I} - \mathcal{T})u = f, \tag{2}$$

where f is known and $u \in \mathbb{X}$ are unknown functions to be determined and $\mathcal{I}$ be the identity operator defined on $\mathbb{X}$.

The associated eigenvalue problem is as follows: find $u \in \mathbb{X}$ and $\lambda \in \mathbb{C} \setminus \{0\}$ such that

$$\mathcal{T}u = \lambda u, \quad \|u\| = 1. \tag{3}$$

Assume λ be the eigenvalue of $\mathcal{K}$ with algebraic multiplicity q and ascent ℓ.

Integral equations arise naturally while trying to solve a large variety of IVPs and BVPs involving both ordinary as well as partial differential equations. Integral equations are formulated from potential problems, aerodynamics, fracture mechanics, Dirichlet problems, and electromagnetic scattering. A complete study of integral equations and their solutions require the development of both analytical and numerical methods of solution. In the literature, many different techniques are used to solve different types of integral equations. This paper concerns on the numerical solution of linear FIEs of second kind with smooth kernel and its associated EVPs.

The Chebyshev methods for different types of integral equations have been solved by many authors [3, 4, 8, 9] with piecewise polynomial based projection methods. However, the convergence analysis has not been discussed in many papers available in the literature using Chebyshev methods. The computation of sufficient accurate solutions of FIEs and eigenelements of EVPs may be needed the avail of much refine partition of the region employing piecewise polynomial as the basis function. As the norm of the partition turns into smaller, the size of the matrix becomes very large. We are concerned in the Projection based techniques for the FIEs and the associated EVPs employing global polynomial basis functions rather than piecewise polynomial basis functions. Recently spectral methods [5, 6, 14] have been applied to problem arises in different areas. Legendre spectral projection and multi projection methods have been applied to linear and nonlinear integral equations with smooth and weakly singular kernels, and also for EVPs of a compact integral operator [7, 11–13].

In this paper, Chebyshev spectral projection methods have been applied on FIEs of second kind with smooth kernel and its associated EVPs. Now, to solve numerically the FIEs of second kind with smooth kernel and its associated EVPs, we will discuss on the Chebyshev spectral Galerkin and Chebyshev spectral collocation methods in Sect. 2. The errors between approximated solutions as well as iterated solutions versus exact solution have been calculated in Sect. 3. The convergence rates between exact eigenelements with approximated eigenelements also have evaluated in both weighted L^2 and L^∞ norm in the same section.

2 Chebyshev Spectral Projection Methods

In this section, we will discuss on Chebyshev spectral projection methods first and then apply to both FIEs of second kind and its associated EVPs.

For $x = \cos\theta,\ 0 \le \theta \le \pi$, the function

$$P_m(x) = \cos(m\theta) = \cos(m\cos^{-1} x),$$

is a Chebyshev polynomial of x of degree m, where $m = 0, 1, 2, 3, \ldots$. When θ increases from 0 to π, x decreases from 1 to -1. Then the interval $[-1, 1]$ is domain of $P_m(x)$. $P_m(x)$ satisfies the orthogonality condition with respect to the weight $w(x) = (1-x^2)^{-\frac{1}{2}}$.

Let $P_0(x) = 1,\ \ P_1(x) = x$, and $P_i(x) = 2xP_{i-1}(x) - P_{i-2}(x),\ -1 \le x \le 1$, be the Chebyshev polynomial of degree i on $[-1, 1]$. By normalizing the polynomial $P_i(x)$, we get

$$\phi_0(x) = \frac{1}{\sqrt{\pi}} P_0(x),\ \ \phi_i(x) = \sqrt{\frac{2}{\pi}} P_i(x),\ i = 1, 2, 3, \ldots.$$

Let $\mathbb{X}_m = \text{span}\{\phi_i(x) : i = 0, 1, 2, \ldots, m\}$ where $\phi_i(x),\ x \in [-1, 1],\ i = 0, 1, 2, \ldots m$ denotes the sequence of orthonormal polynomials associated with the $L^2_\omega[-1, 1]$ inner product. Then dimension of $\mathbb{X}_m$ is $m+1$. The set $\{\phi_i : 0 \le i \le m\}$, forms an orthonormal basis for $\mathbb{X}_m$ and $\mathbb{X}_m \subset \mathbb{X}$ is the set of all polynomial of degree at most m on $[-1, 1]$.

For any function $u \in C^r[-1, 1]$ and $r \in \mathbb{N}$,

$$\|u\|_{r,\omega,\infty} = \sum_{i=0}^{r} \|D^{(i)}u\|_{L^\infty_\omega}, \qquad \|u\|_{r,\omega,2} = \sum_{i=0}^{r} \|D^{(i)}u\|_{L^2_\omega},$$

where $D^{(i)}u(t) = \frac{\partial^i}{\partial t^i} u(t)$.

For any non negative integer $r \in \mathbb{N}$, define

$$H^r_{2,\omega}[-1, 1] = \left\{ u(t) \middle| \frac{\partial^i}{\partial t^i} u(t) \in L^2_\omega[-1, 1], i = 0, 1, \ldots, r \right\}.$$

Let $C^{(i,j)}([-1, 1] \times [-1, 1])$ be the set of all continuously differentiable functions of order (i, j) on $[-1, 1] \times [-1, 1]$. For any $r \in \mathbb{N}$ and $k(s, t) \in C^{(r,r)}[-1, 1] \times [-1, 1]$, we set

$$\|k\|_{r,\infty} = \sum_{i=0}^{r} \sum_{j=0}^{r} \|D^{(i,j)} k(s, t)\|_\infty, \tag{4}$$

where $D^{(i,j)}k(s,t) = \dfrac{\partial^{i+j}k(s,t)}{\partial s^i \partial t^j}$.

Then by using Cauchy Schwarz inequality, we have

$$
\begin{aligned}
|D^{(i)}\mathcal{T}u(s)| &= \Big|\int_{-1}^{1} D^{(i,0)}k(s,t)u(t)\mathrm{d}t\Big| \\
&\le \max_{t\in[-1,1]} |D^{(i,0)}k(s,t)| \Big|\int_{-1}^{1} \frac{(1-t^2)^{\frac{1}{4}}}{(1-t^2)^{\frac{1}{4}}} u(t)\mathrm{d}t\Big| \\
&\le \max_{t\in[-1,1]} |D^{(i,0)}k(s,t)| \Big(\int_{-1}^{1} \frac{|u(t)|^2}{\sqrt{1-t^2}}\mathrm{d}t\Big)^{\frac{1}{2}} \Big(\int_{-1}^{1} |\sqrt{1-t^2}|\mathrm{d}t\Big)^{\frac{1}{2}} \\
&\le \sqrt{2} \max_{t\in[-1,1]} |D^{(i,0)}k(s,t)| \|u\|_{L^2_\omega}.
\end{aligned}
$$

Thus,

$$
\begin{aligned}
\|D^{(i)}\mathcal{T}u\|_\infty &= \sup_{s\in[-1,1]} |D^{(i)}\mathcal{T}u(s)| \\
&\le \sqrt{2} \sup_{s,t\in[-1,1]} |D^{(i,0)}k(s,t)| \|u\|_{L^2_\omega} \le \sqrt{2}\|k\|_{r,\infty}\|u\|_{L^2_\omega},
\end{aligned}
$$

and

$$
\|D^{(i)}\mathcal{T}u\|_{L^2_\omega} \le \|D^{(i)}\mathcal{T}u\|_\infty \|1\|_{L^2_\omega} \le \sqrt{2}\|1\|_{L^2_\omega}\|k\|_{r,\infty}\|u\|_{L^2_\omega}. \tag{5}
$$

For $k(s,t) \in \mathcal{C}^{(r,r)}([-1,1]\times[-1,1])$, we have

$$
\|\mathcal{T}u\|_{r,\omega,2} = \sum_{i=0}^{r} \|D^{(i)}\mathcal{T}u\|_{L^2_\omega} \le \sqrt{2}\|k\|_{r,\infty}\|u\|_{L^2_\omega}\|1\|_{L^2_\omega} \tag{6}
$$

and

$$
\|\mathcal{T}u\|_{r,\omega,\infty} = \sum_{i=0}^{r} \|D^{(i)}\mathcal{T}u\|_\infty \le \sqrt{2}\|k\|_{r,\infty}\|u\|_{L^2_\omega}. \tag{7}
$$

To discuss Chebyshev spectral Galerkin and Chebyshev spectral collocation methods for FIEs of the second kind and its associated EVPs, we need to define the Chebyshev orthogonal projection and Chebyshev interpolatory projection operator.

Let $\mathcal{C}_m^G : \mathbb{X} \to \mathbb{X}_m$ be the orthogonal projection operator defined by

$$
\mathcal{C}_m^G u(s) = \sum_{i=0}^{m} \langle u, \phi_i\rangle_\omega \phi_i(s), \quad u \in \mathbb{X}, \tag{8}
$$

where $\langle u, \phi_i\rangle_\omega = \int_{-1}^{1} w(x)u(x)\phi_i(x)\mathrm{d}x$, $i = 0, 1, 2, \ldots, m$.

Let $\mathbb{X} = \mathcal{C}[-1, 1]$. Let $\{\tau_0, \tau_1, \ldots, \tau_m\}$ be the zeros of Chebyshev polynomial of degree $m + 1$. Define interpolatory projection $\mathcal{C}_m^C : \mathbb{X} \to \mathbb{X}_m$ by

$$\mathcal{C}_m^C u \in \mathbb{X}_m, \ \ \mathcal{C}_m^C u(\tau_i) = u(\tau_i), \ \ i = 0, 1, 2, \ldots, m. \tag{9}$$

Now we quote the following properties of $\mathcal{C}_m^G$ and $\mathcal{C}_m^C$ from [6].

Lemma 1 ([6]) *For any* $u(s) \in H^r_{2,\omega}[-1, 1]$ *and* $r \geq 0$,

$$\|(\mathcal{I} - \mathcal{C}_m^G)u\|_{L^2_\omega} \leq c\, m^{-r} \|u\|_{r,\omega,2},$$
$$\|(\mathcal{I} - \mathcal{C}_m^G)u\|_{L^\infty_\omega} \leq c\,(1 + \log m)\, m^{-r} \|u\|_{r,\omega,\infty}.$$

For simplicity of notation, we define $\mathcal{C}_m$ be the projection operator either orthogonal $\mathcal{C}_m^G$ or interpolatory $\mathcal{C}_m^C$ defined on $\mathbb{X}$ into $\mathbb{X}_m$.

Using $\mathcal{C}_m$, the Chebyshev spectral projection methods for the FIEs of the second kind is of the form

$$u_m - \mathcal{C}_m \mathcal{T} u_m = \mathcal{C}_m f. \tag{10}$$

Let the iterated solution be $\widetilde{u}_m = f + \mathcal{T} u_m$, which gives $\mathcal{C}_m \widetilde{u}_m = u_m$. Then using this in iterated solution, we get

$$\widetilde{u}_m - \mathcal{T}\mathcal{C}_m \widetilde{u}_m = f. \tag{11}$$

This is iterated Chebyshev spectral projection method.

Using $\mathcal{C}_m$, the Chebyshev spectral projection method for the EVPs is to find $u'_m \in \mathbb{X}_m$ and $\lambda_m \in \mathbb{C} \setminus \{0\}$ such that

$$\mathcal{C}_m \mathcal{T} u'_m = \lambda_m u'_m, \ \ \|u'_m\| = 1, \tag{12}$$

The iterated eigenvector is defined by

$$\widetilde{u'}_m = \frac{1}{\lambda_m} \mathcal{T} u'_m. \tag{13}$$

If $\mathcal{C}_m = \mathcal{C}_m^G$, Eqs. (10) and (11) give rise to Chebyshev spectral Galerkin and iterated Chebyshev spectral Galerkin methods for FIEs of the second kind and Eq. (12) gives rise to Chebyshev spectral Galerkin method for EVPs of a compact integral operator. Similarly, if $\mathcal{C}_m = \mathcal{C}_m^C$, Eqs. (10) and (11) give rise to Chebyshev spectral collocation and iterated Chebyshev spectral collocation methods for FIEs of the second kind and Eq. (12) gives rise to Chebyshev spectral collocation methods for EVPs of a compact integral operator.

To discuss the convergence of Chebyshev spectral projection methods for FIEs of the second kind and its associated EVPs, we need the following results.

Theorem 1 *$\mathcal{C}_m \mathcal{T}$ is norm convergent to $\mathcal{T}$ and $\mathcal{T}\mathcal{C}_m$ is ν-convergent to $\mathcal{T}$ in L^2_ω norm.*

Proof Using Lemma 1 and estimate (6), we see that

$$\|(\mathcal{I}-\mathcal{C}_m)\mathcal{T}u\|_{L^2_\omega} \le c\, m^{-r}\|\mathcal{T}u\|_{r,\omega,2} \le c\sqrt{2}\, m^{-r}\|k\|_{r,\infty}\|u\|_{L^2_\omega}\|1\|_{L^2_\omega} \to 0,$$

as $m \to \infty$. This proves that $\mathcal{C}_m\mathcal{T}$ is norm convergent to $\mathcal{T}$ in L^2_ω norm.

Now, we need to prove that $\mathcal{T}\mathcal{C}_m$ is ν-convergent to $\mathcal{T}$ in L^2_ω norm. It is sufficient to show that (i) $\|\mathcal{T}\mathcal{C}_m\|_{L^2_\omega} \le M$ (ii) $\|(\mathcal{T}\mathcal{C}_m-\mathcal{T})\mathcal{T}\|_{L^2_\omega} \to 0$ as $m \to \infty$ (iii) $\|(\mathcal{T}\mathcal{C}_m-\mathcal{T})\mathcal{T}\mathcal{C}_m\|_{L^2_\omega} \to 0$ as $m \to \infty$.

Consider, $u(t) \in \mathcal{C}[-1,1]$. We have

$$\begin{aligned}
|\mathcal{T}(\mathcal{C}_m-\mathcal{I})u(s)| &= \left|\int_{-1}^{1} k(s,t)(\mathcal{C}_m-\mathcal{I})u(t)\mathrm{d}t\right| \\
&\le \max_{t\in[-1,1]}|\sqrt{1-t^2}|\left|\int_{-1}^{1}\frac{1}{\sqrt{1-t^2}}k(s,t)(\mathcal{C}_m-\mathcal{I})u(t)\mathrm{d}t\right| \\
&\le c\left|\int_{-1}^{1}\frac{1}{(1-t^2)^{\frac{1}{4}}}k(s,t)\frac{1}{(1-t^2)^{\frac{1}{4}}}(\mathcal{C}_m-\mathcal{I})u(t)\mathrm{d}t\right|.
\end{aligned}$$

By using Cauchy-Schwarz inequality, we have

$$\begin{aligned}
|\mathcal{T}(\mathcal{C}_m-\mathcal{I})u(s)| &\le c\left[\int_{-1}^{1}\frac{|k(s,t)|^2}{\sqrt{1-t^2}}\mathrm{d}t\right]^{\frac{1}{2}}\left[\int_{-1}^{1}\frac{|(\mathcal{C}_m-\mathcal{I})u(t)|^2}{\sqrt{1-t^2}}\mathrm{d}t\right]^{\frac{1}{2}} \\
&\le c\, M_1\|1\|_{L^2_\omega}\|(\mathcal{C}_m-\mathcal{I})u\|_{L^2_\omega},
\end{aligned}$$

where $M_1 = \sup_{s,t\in[-1,1]}|k(s,t)|$. It follows from the estimate (14) and Lemma 1 that

$$\begin{aligned}
\lim_{m\to\infty}\|\mathcal{T}(\mathcal{C}_m-\mathcal{I})u\|_{L^2_\omega} &\le \|1\|_{L^2_\omega}\lim_{m\to\infty}\|(\mathcal{T}\mathcal{C}_m-\mathcal{T})u\|_\infty \\
&\le c\, M_1\|1\|^2_{L^2_\omega}\|\mathcal{C}_m u-u\|_{L^2_\omega} \le c\, m^{-r}\|u\|_{r,\omega,2}. \qquad (14)
\end{aligned}$$

Since $\mathcal{T}\mathcal{C}_m$ is point-wise convergent to $\mathcal{T}$, for $u \in \mathcal{C}[-1,1]$ and from Banach-Steinhaus Theorem it follows that $\|\mathcal{T}\mathcal{C}_m\|_{L^2_\omega} \le M$, which proves (i).

Let B be a closed unit ball in $\mathcal{C}[-1,1]$. Since $\mathcal{T}$ is a compact operator and $\mathcal{C}_m$ is uniformly bounded in L^2_ω, it follows that $S = \{\mathcal{T}x : x \in B\}$ and $S' = \{\mathcal{T}\mathcal{C}_m u : m \ge 1, u \in B\}$ are relatively compact set in $\mathcal{C}[-1,1]$. Then by using [2], it follows that

$$\begin{aligned}
\|(\mathcal{T}\mathcal{C}_m-\mathcal{T})\mathcal{T}\|_{L^2_\omega} &= \sup\{\|(\mathcal{T}\mathcal{C}_m-\mathcal{T})u\|_{L^2_\omega} : u \in S\} \to 0, \\
\|(\mathcal{T}\mathcal{C}_m-\mathcal{T})\mathcal{T}\mathcal{C}_m\|_{L^2_\omega} &= \sup\{\|(\mathcal{T}\mathcal{C}_m-\mathcal{T})u\|_{L^2_\omega} : u \in S'\} \to 0.
\end{aligned}$$

as $m \to \infty$. This completes the proof.

By using the Theorem 1, the spectrum of $\mathcal{C}_m\mathcal{T}$ inside Γ consists $\lambda_{m,1}, \lambda_{m,2}, \ldots, \lambda_{m,q}$ as eigenvalues counted accordingly to their algebraic multiplicities inside Γ. Let $\widehat{\lambda}_m$ be the average of eigenvalues. Let $\mathcal{P}^S$, $\mathcal{P}^{S,G}_m$ and $\mathcal{P}^{S,C}_m$ be the spectral projections of

$\mathcal{T}$, $\mathcal{C}_m^G\mathcal{T}$ and $\mathcal{C}_m^C\mathcal{T}$, respectively, associated with their corresponding spectral inside Γ. Let $\mathcal{R}(\mathcal{P}^S)$, $\mathcal{R}(\mathcal{P}_m^{S,G})$ and $\mathcal{R}(\mathcal{P}_m^{S,C})$ be the ranges of the spectral projections $\mathcal{P}^S$, $\mathcal{P}_m^{S,G}$, and $\mathcal{P}_m^{S,C}$, respectively.

The closeness of eigenvectors will be evaluated by using the gap between the spectral subspaces. We quote the following definition from [1].

Gap between the spectral subspaces: For nonzero subspaces $\mathbb{Y}_1$ and $\mathbb{Y}_2$ of $\mathbb{X}$ and for L^p_ω where $p = 2$ or ∞, let

$$\delta_p(\mathbb{Y}_1, \mathbb{Y}_2) = \sup\{\text{dist}(y, \mathbb{Y}_2) : y \in \mathbb{Y}_1, \|y\|_{L^p_\omega} = 1\},$$
$$\widehat{\delta}_p(\mathbb{Y}_1, \mathbb{Y}_2) = \max\{\delta_p(\mathbb{Y}_1, \mathbb{Y}_2), \delta_p(\mathbb{Y}_2, \mathbb{Y}_1)\},$$

denotes the gap between $\mathbb{Y}_1$ and $\mathbb{Y}_2$ in p-norm.

3 Convergence Rates

In this section, the errors between exact eigenelements with the approximated eigenelements of EVPs (3) have been evaluated by using both Chebyshev spectral Galerkin and collocation methods in both weighted L^2 and weighted L^∞ norm. The convergence rates of approximated solutions and iterated solutions for FIEs of second kind in both the methods have been calculated. We will discuss this in next two subsections.

3.1 Chebyshev Spectral Galerkin Methods

We will first discuss the following results to calculate the errors between exact eigenelements and the approximated eigenelements of the EVP (3) using Chebyshev spectral Galerkin methods in L^p_ω norm where $p = 2, \infty$.

Theorem 2 *Let the kernel $k(.,.) \in \mathcal{C}^{(r,r)}[-1, 1] \times [-1, 1]$. Then the following results hold:*

$$\|\mathcal{T} - \mathcal{C}_m^G\mathcal{T}\|_{L^\infty_\omega} = \mathcal{O}((1 + \log m)m^{-r}), \qquad \|\mathcal{T}(\mathcal{I} - \mathcal{C}_m^G)\mathcal{T}\|_{L^2_\omega} = \mathcal{O}(m^{-2r}).$$

Proof By using Lemma-1 and estimate (7), we obtain

$$\begin{aligned}\|(\mathcal{I} - \mathcal{C}_m^G)\mathcal{T}u\|_{L^\infty_\omega} &\leq c\,(1 + \log m)\, m^{-r}\, \|\mathcal{T}u\|_{r,\omega,\infty} \\ &\leq c\,(1 + \log m)\, m^{-r}\, \|k\|_{r,\infty}\|u\|_\infty\|1\|_{L^2_\omega}.\end{aligned}$$

This completes the proof of the first estimate. Now using orthogonality of $(\mathcal{I} - \mathcal{C}_m^G)$, we obtain

$$
\begin{aligned}
|\mathcal{T}(\mathcal{I}-\mathcal{C}_m^G)\mathcal{T}u(s)| & \\
&= \left|\int_{-1}^{1}\frac{1}{\sqrt{1-t^2}}\sqrt{1-t^2}k(s,t)(\mathcal{I}-\mathcal{C}_m^G)\mathcal{T}u(t)\mathrm{d}t\right| \\
&\le c\left|\int_{-1}^{1}\frac{1}{\sqrt{1-t^2}}k(s,t)(\mathcal{I}-\mathcal{C}_m^G)\mathcal{T}u(t)\mathrm{d}t\right| \\
&= c\,|\langle k(s,.),(\mathcal{I}-\mathcal{C}_m^G)\mathcal{T}u(.)\rangle_\omega| \\
&= c|\langle(\mathcal{I}-\mathcal{C}_m^G)k(s,.),(\mathcal{I}-\mathcal{C}_m^G)\mathcal{T}u(.)\rangle_\omega|.
\end{aligned}
$$

Using estimate (6), Lemma 1 and Cauchy-Schwarz inequality, we obtain

$$
\begin{aligned}
&|\mathcal{T}(\mathcal{I}-\mathcal{C}_m^G)\mathcal{T}u(s)| \\
&\le c\left|\int_{-1}^{1}\frac{1}{(1-t^2)^{1/4}}(\mathcal{I}-\mathcal{C}_m^G)k(s,t)\frac{1}{(1-t^2)^{1/4}}(\mathcal{I}-\mathcal{C}_m^G)\mathcal{T}u(t)\mathrm{d}t\right| \\
&\le c\,\|(\mathcal{I}-\mathcal{C}_m^G)k(s,t)\|_{L^2_\omega}\|(\mathcal{I}-\mathcal{C}_m^G)\mathcal{T}u\|_{L^2_\omega} \\
&\le c\,m^{-r}\|k\|_{r,\omega,2}m^{-r}\,\|\mathcal{T}u\|_{r,\omega,2} \\
&\le c\,m^{-2r}\|k\|_{r,\omega,2}\|k\|_{r,\infty}\|u\|_{L^2_\omega}\|1\|_{L^2_\omega}.
\end{aligned}
$$

Thus, we obtain

$$
\|\mathcal{T}(\mathcal{I}-\mathcal{C}_m^G)\mathcal{T}u\|_{L^2_\omega}\le\|1\|_{L^2_\omega}\|\mathcal{T}(\mathcal{I}-\mathcal{C}_m^G)\mathcal{T}u\|_\infty\le c\,m^{-2r}\|k\|_{r,\omega,2}\|k\|_{r,\infty}\|u\|_{L^2_\omega}\|1\|^2_{L^2_\omega}.
$$

This completes the proof.

Theorem 3 *Let $\lambda^G_{m,j}$, be the eigenvalues of the operator $\mathcal{C}_m^G\mathcal{T}$ and $\widehat{\lambda}^G_m$ be the arithmetic mean of the eigenvalues $\lambda^G_{m,j}$, for $j=0,1,2,\ldots,q$. Then the following results hold:*

$$
|\lambda-\widehat{\lambda}^G_m|=\mathcal{O}(m^{-2r}),\qquad |\lambda-\lambda^G_{m,j}|^l=\mathcal{O}(m^{-2r}).
$$

Proof The proof follows directly by using Lemma 3.5 [11] and Theorem 2 for $j=1,2,\ldots,q$.

Theorem 4 *Then the following results hold for any $u'^G_m\in\mathcal{R}(\mathcal{P}^{S,G}_m)$ in L^2_ω norm:*

$$
\begin{aligned}
&\widehat{\delta}_2(\mathcal{R}(\mathcal{P}^{S,G}_m),\mathcal{R}(\mathcal{P}^S))=\mathcal{O}(m^{-r}), && \|u'^G_m-\mathcal{P}^S u'^G_m\|_{L^2_\omega}=\mathcal{O}(m^{-r}),\\
&\delta_2(\mathcal{T}\mathcal{R}(\mathcal{P}^{S,G}_m),\mathcal{R}(\mathcal{P}^S))=\mathcal{O}(m^{-2r}), && \|\mathcal{T}u'^G_m-\mathcal{P}^S\mathcal{T}u'^G_m\|_{L^2_\omega}=\mathcal{O}(m^{-2r}).
\end{aligned}
$$

Proof For any $u'^G_m\in\mathcal{R}(\mathcal{P}^{S,G}_m)$, it follows from Lemma 3.6 [11] and Theorem 1 that,

$$
\begin{aligned}
&\widehat{\delta}_2(\mathcal{R}(\mathcal{P}^{S,G}_m),\mathcal{R}(\mathcal{P}^S))\le c\,\|(\mathcal{C}_m^G\mathcal{T}-\mathcal{T})\mathcal{T}\|_{L^2_\omega}=\mathcal{O}(m^{-r}),\\
&\|u'^G_m-\mathcal{P}^S u'^G_m\|_{L^2_\omega}\le c\,\|(\mathcal{C}_m^G\mathcal{T}-\mathcal{T})\mathcal{T}\|_{L^2_\omega}=\mathcal{O}(m^{-r}).
\end{aligned}
$$

Now using Lemma 3.6 [11] and Theorem 2 for any $u'^{G}_{m} \in \mathcal{R}(\mathcal{P}^{S,G}_{m})$ and the iterated eigenvector $\widetilde{u'}^{G}_{m} = \mathcal{T}u'^{G}_{m}$, we obtain

$$\delta_2(\mathcal{T}\mathcal{R}(\mathcal{P}^{S,G}_{m}), \mathcal{R}(\mathcal{P}^{S})) \leq c\,\|(\mathcal{T}\mathcal{C}^{G}_{m} - \mathcal{T})\mathcal{T}\|_{L^2_\omega} = \mathcal{O}(m^{-2r}),$$
$$\|\widetilde{u}^{G}_{m} - \mathcal{P}^{S}\widetilde{u}^{G}_{m}\|_{L^2_\omega} = \|\mathcal{T}u'^{G}_{m} - \mathcal{P}^{S}\mathcal{T}u'^{G}_{m}\|_{L^2_\omega} = \mathcal{O}(m^{-2r}).$$

Thus, we obtain the desired result.

Theorem 5 *Then the following results hold for any $u'^{G}_{m} \in \mathcal{R}(\mathcal{P}^{S,G}_{m})$ in L^∞_ω norm.*

$$\widehat{\delta}_\infty(\mathcal{R}(\mathcal{P}^{S,G}_{m}), \mathcal{R}(\mathcal{P}^{S})) = \mathcal{O}((1+\log m)m^{-r}),$$
$$\|u'^{G}_{m} - \mathcal{P}^{S,G}_{m}u'^{G}_{m}\|_{L^\infty_\omega} = \mathcal{O}((1+\log m)m^{-r}),$$
$$\delta_\infty(\mathcal{T}\mathcal{R}(\mathcal{P}^{S,G}_{m}), \mathcal{R}(\mathcal{P}^{S})) = \mathcal{O}(m^{-2r}),$$
$$\|\mathcal{T}u'^{G}_{mn} - \mathcal{P}^{S,G}_{m}\mathcal{T}u'^{G}_{mn}\|_{L^\infty_\omega} = \mathcal{O}(m^{-2r}).$$

Proof For any $u'^{G}_{m} \in \mathcal{R}(\mathcal{P}^{S,G}_{m})$, we obtain the following errors by using Lemma 3.6 [11] and Theorem 2:

$$\widehat{\delta}_\infty(\mathcal{R}(\mathcal{P}^{S,G}_{m}), \mathcal{R}(\mathcal{P}^{S})) \leq c\,\|(\mathcal{C}^{G}_{m}\mathcal{T} - \mathcal{T})\mathcal{T}\|_{L^\infty_\omega} \leq c\,(1+\log m)m^{-r}\|k\|_{r,\infty}\|1\|^2_{L^2_\omega},$$
$$\|u'^{G}_{m} - \mathcal{P}^{S}u'^{G}_{m}\|_\infty \leq c\,\|(\mathcal{C}^{G}_{m}\mathcal{T} - \mathcal{T})\mathcal{T}\|_{L^\infty_\omega} \leq c\,(1+\log m)m^{-r}\|k\|_{r,\infty}\|1\|^2_{L^2_\omega}.$$

Now using Lemma 3.6 [11] and Theorem 2, we obtain the following for any $u'^{G}_{m} \in \mathcal{R}(\mathcal{P}^{S,G}_{m})$ and the iterated eigenvector $\widetilde{u}^{G}_{m} - \mathcal{T}u'^{G}_{m}$:

$$\delta_\infty(\mathcal{T}\mathcal{R}(\mathcal{P}^{S,G}_{m}), \mathcal{R}(\mathcal{P}^{S})) \leq c\,\|(\mathcal{T}\mathcal{C}^{G}_{m} - \mathcal{T})\mathcal{T}\|_{L^\infty_\omega} = \mathcal{O}(m^{-2r}),$$
$$\|\widetilde{u}^{G}_{m} - \mathcal{P}^{S}\widetilde{u'}^{G}_{m}\|_{L^\infty_\omega} = \|\mathcal{T}u'^{G}_{m} - \mathcal{P}^{S}\mathcal{T}u'^{G}_{m}\|_{L^\infty_\omega} = \mathcal{O}(m^{-2r}).$$

This completes the proof.

Now, we will obtain the convergence rates for the approximated and iterated solution in Chebyshev Galerkin method for FIEs of the second kind in L^2_ω norm.

Theorem 6 *Let u and $\widetilde{u}^{G}_{m}$ be the solutions of* (2) *and* (11)*, respectively. Assume that $(\mathcal{I}-\mathcal{T})^{-1}$ exists. Then for $m \geq N$ and $N \in \mathbb{N}$, such that*

$$\|u - \widetilde{u}^{G}_{m}\|_{L^2_\omega} = \mathcal{O}(m^{-2r}), \qquad \|u - u^{G}_{m}\|_{L^2_\omega} = \mathcal{O}(m^{-r}).$$

Proof From Eqs. (2) and (11), we obtain

$$\begin{aligned} u - \widetilde{u}^{G}_{m} &= (\mathcal{I}-\mathcal{T})^{-1}f - (\mathcal{I}-\mathcal{T}\mathcal{C}^{G}_{m})^{-1}\mathcal{C}^{G}_{m}f \\ &= (\mathcal{I}-\mathcal{T}\mathcal{C}^{G}_{m})^{-1}(\mathcal{I}-\mathcal{T}\mathcal{C}^{G}_{m}-\mathcal{I}+\mathcal{T})(\mathcal{I}-\mathcal{T})^{-1}f \\ &= (\mathcal{I}-\mathcal{T}\mathcal{C}^{G}_{m})^{-1}(\mathcal{T}-\mathcal{T}\mathcal{C}^{G}_{m})u. \end{aligned}$$

Since $\mathcal{T}\mathcal{C}_m^G$ is ν convergent to $\mathcal{T}$ in L^2_ω norm, the operator $(\mathcal{I}-\mathcal{T}\mathcal{C}_m^G)^{-1}$ exists and there exists a positive real number M and a positive integer N such that for all $m \geq N$, we have $\|(\mathcal{I}-\mathcal{T}\mathcal{C}_m^G)^{-1}\|_{L^2_\omega} \leq M$. Hence,

$$\|u-\widetilde{u}_m^G\|_{L^2_\omega} \leq M\|(\mathcal{T}-\mathcal{T}\mathcal{C}_m^G)u\|_{L^2_\omega}. \tag{15}$$

Using Theorem 2, we obtain

$$\|u-\widetilde{u}_m^G\|_{L^2_\omega} \leq c\|1\|_{L^2_\omega} m^{-2r}\|k\|_{r,\omega}\|u\|_{r,\omega}.$$

This proves the second estimate of the theorem. Using $u_m^G = \mathcal{C}_m^G\widetilde{u}_m^G$, we obtain

$$u-u_m^G = u-\mathcal{C}_m^G\widetilde{u}_m^G = u-\mathcal{C}_m^G u+\mathcal{C}_m^G u-\mathcal{C}_m^G\widetilde{u}_m^G$$

Using the fact that $\|\mathcal{C}_m^G\|_{L^2_\omega} \leq M$, second estimate and Lemma 1, we obtain

$$\begin{aligned}\|u-u_m^G\|_{L^2_\omega} &\leq \|u-\mathcal{C}_m^G u\|_{L^2_\omega} + \|\mathcal{C}_m^G\|_{L^2_\omega}\|u-\widetilde{u}_m^G\|_{L^2_\omega}\\ &\leq c\,m^{-r}\|u\|_{r,\omega} + c\,M\|1\|_{L^2_\omega} m^{-2r}\|k\|_{r,\omega}\|u\|_{r,\omega}.\end{aligned}$$

This proves the result.

3.2 Chebyshev Collocation Methods

The errors between exact eigenelements with the approximated eigenelements of the EVP (3) have been calculated by using Chebyshev spectral collocation methods in both weighted L^2 and weighted L^∞ norms. The convergence rates of approximated solutions and iterated solutions with exact solution for FIEs of second kind also have been evaluated in weighted L^2 norm.

Theorem 7 *Let $k(.,.) \in \mathcal{C}^{(r,r)}[-1,1]\times[-1,1]$. Then the following results hold:*

$$\|\mathcal{T}-\mathcal{C}_m^C\mathcal{T}\|_{L^\infty_\omega} = \mathcal{O}((1+\log m)m^{-r}), \qquad \|\mathcal{T}-\mathcal{C}_m^C\mathcal{T}\|_{L^2_\omega} = \mathcal{O}(m^{-r}),$$
$$\|\mathcal{T}(\mathcal{I}-\mathcal{C}_m^C)\mathcal{T}\|_{L^2_\omega} = \mathcal{O}(m^{-r}).$$

Proof By using Lemma 1 and estimate (7), we obtain

$$\|(\mathcal{I}-\mathcal{C}_m^C)\mathcal{T}u\|_{L^\infty_\omega} \leq c\,(1+\log m)\,m^{-r}\|\mathcal{T}u\|_{r,\omega,\infty} \leq c\,(1+\log m)m^{-r}\|k\|_{r,\infty}\|u\|_{L^2_\omega}.$$

This proves the first estimate of the theorem. By using Lemma 1 and estimate (6), we see that

$$\|(\mathcal{I}-\mathcal{C}_m^C)\mathcal{T}u\|_{L^2_\omega} \leq c\,m^{-r}\|\mathcal{T}u\|_{r,\omega,2} \leq c\,m^{-r}\|k\|_{r,\infty}\|1\|_{L^2_\omega}\|u\|_{L^2_\omega}. \tag{16}$$

This completes the proof of second estimate of the Theorem. Using estimate (16) and Cauchy-Schwarz's inequality, we obtain

$$
\begin{aligned}
|\mathcal{T}(\mathcal{I}-\mathcal{C}_m^C)\mathcal{T}u(s)| &= \Big|\int_{-1}^{1} k(s,t)(\mathcal{I}-\mathcal{C}_m^C)\mathcal{T}u(t)\,\mathrm{d}t\Big| \\
&= \Big|\int_{-1}^{1} \frac{1}{\sqrt{1-t^2}}\sqrt{1-t^2}k(s,t)(\mathcal{I}-\mathcal{C}_m^C)\mathcal{T}u(t)\,\mathrm{d}t\Big| \\
&\le \max_{t\in[-1,1]} |\sqrt{1-t^2}|\Big|\int_{-1}^{1} \frac{1}{\sqrt{1-t^2}}k(s,t)(\mathcal{I}-\mathcal{C}_m^C)\mathcal{T}u(t)dt\Big| \\
&\le c|\langle k(s,t),(\mathcal{I}-\mathcal{C}_m^C)\mathcal{T}u(t)\rangle_\omega| \\
&\le c\,\|k\|_{L^2_\omega}\|(\mathcal{I}-\mathcal{C}_m^C)\mathcal{T}u\|_{L^2_\omega} \le c\,\|k\|_{L^2_\omega}m^{-r}\|k\|_{r,\infty}\|1\|_{L^2_\omega}\|u\|_{L^2_\omega}.
\end{aligned}
$$

Thus, we obtain

$$
\|\mathcal{T}(\mathcal{I}-\mathcal{C}_m^C)\mathcal{T}u\|_{L^\infty_\omega} = \sup_{s\in[-1,1]} |\mathcal{T}(\mathcal{I}-\mathcal{C}_m^C)\mathcal{T}u(s)| \le c\,m^{-r}\|k\|_{L^2_\omega}\|k\|_{r,\infty}\|1\|_{L^2_\omega}\|u\|_{L^2_\omega}.
$$

Now using this estimate, we obtain

$$
\|\mathcal{T}(\mathcal{I}-\mathcal{C}_m^C)\mathcal{T}u\|_{L^2_\omega} \le \|1\|_{L^2_\omega}\|\mathcal{T}(\mathcal{I}-\mathcal{C}_m^C)\mathcal{T}u\|_{L^\infty_\omega} \le c\,m^{-r}\|k\|_{L^2_\omega}\|k\|_{r,\infty}\|1\|^2_{L^2_\omega}\|u\|_{L^2_\omega}.
$$

This completes the proof.

Now, we will evaluate the convergence rates for approximated eigenelements with exact eigenelements in Chebyshev spectral collocation methods.

Theorem 8 *Let $k(.,.) \in \mathcal{C}^{(r,r)}([-1,1]\times[-1,1])$ and $\widehat{\lambda}_m^C$ be the arithmetic mean of the eigenvalues $\lambda_{m,j}$, for $j = 0, 1, 2, \dots, q$, then*

$$
|\lambda - \widehat{\lambda}_m^C| = \mathcal{O}(m^{-r}), \qquad |\lambda - \lambda_{m,j}^C|^\ell = \mathcal{O}(m^{-r}).
$$

Proof By using Lemma 3.5 [11] and Theorem 7, for $j = 1, 2, \dots, q$, it follows that

$$
\begin{aligned}
|\lambda - \widehat{\lambda}_m^C| &\le \|\mathcal{T}(\mathcal{I}-\mathcal{C}_m^C)\mathcal{T}\|_{L^2_\omega} = \mathcal{O}(m^{-r}), \\
|\lambda - \lambda_{m,j}^C|^l &\le \|\mathcal{T}(\mathcal{I}-\mathcal{C}_m^C)\mathcal{T}\|_{L^2_\omega} = \mathcal{O}(m^{-r}).
\end{aligned}
$$

This completes the proof.

Theorem 9 *Then the following results hold in L^2_ω norm and in particular, for any $u_m^{'C} \in \mathcal{R}(\mathcal{P}_m^{S,C})$,*

$$
\begin{aligned}
\widehat{\delta}_2(\mathcal{R}(\mathcal{P}_m^{S,C}), \mathcal{R}(\mathcal{P}^S)) &= \mathcal{O}(m^{-r}), &\qquad \|u_{mn}^{'C} - \mathcal{P}^S u_{mn}^{'C}\|_{L^2_\omega} &= \mathcal{O}(m^{-r}), \\
\delta_2(\mathcal{T}\mathcal{R}(\mathcal{P}_m^S), \mathcal{R}(\mathcal{P}^S)) &= \mathcal{O}(m^{-r}), &\qquad \|\mathcal{T}u_{mn}^{'C} - \mathcal{P}^S\mathcal{T}u_{mn}^{'C}\|_{L^2_\omega} &= \mathcal{O}(m^{-r}).
\end{aligned}
$$

Proof The proof follows by similar steps as in Theorem 4 and using Theorem 7.

Theorem 10 *Then the following results hold in* L^{∞}_{ω} *norm and in particular, for any* $u^{'C}_m \in \mathcal{R}(\mathcal{P}^{S,C}_m)$:

$$\widehat{\delta}_{\infty}(\mathcal{R}(\mathcal{P}^{S,C}_m), \mathcal{R}(\mathcal{P}^S)) = \mathcal{O}((1+\log m)m^{-r}),$$
$$\|u^{'C}_m - \mathcal{P}^S u^{'C}_m\|_{L^{\infty}_{\omega}} = \mathcal{O}((1+\log m)m^{-r}),$$
$$\delta_{\infty}(\mathcal{T}\mathcal{R}(\mathcal{P}^{S,C}_m), \mathcal{R}(\mathcal{P}^S)) = \mathcal{O}(m^{-r}),$$
$$\|\mathcal{T}u^{'C}_m - \mathcal{P}^S\mathcal{T}u^{'C}_m\|_{L^{\infty}_{\omega}} = \mathcal{O}(m^{-r}).$$

Proof By using Theorem 7 and proceeding similar steps as in Theorem 5, we get the desired result.

Now, we will obtain the convergence rates for the approximated and iterated solution in Chebyshev spectral collocation method for FIEs of second kind.

Theorem 11 *Let* u *and* $\widetilde{u}^C_m$ *be the solutions of* (2) *and* (11)*, respectively. Assume that* $(\mathcal{I}-\mathcal{T})^{-1}$ *exists. Then for* $m, n \geq N$*, the following results hold:*

$$\|u - \widetilde{u}^C_m\|_{L^2_{\omega}} = \mathcal{O}(m^{-r}), \qquad \|u - u^C_m\|_{L^2_{\omega}} = \mathcal{O}(m^{-r}).$$

Proof By similar steps as in Theorem 11 and using Theorem 7, we get the required result.

4 Numerical Results

In this section we discuss the numerical results for FIEs of second with smooth kernel and its associated EVPs. We choose the Chebyshev polynomials $\{\psi_0, \psi_1, \ldots, \psi_n\}$ as an orthonormal basis subspace $\mathbb{X}_n$, where

$$\psi_0(x) = 1/\sqrt{\pi}, \quad \psi_1(x) = (\sqrt{2/\pi})x, \; x \in [-1, 1],$$

and for $i = 1, 2, \ldots n-1$,

$$\psi_{i+1}(x) = (\sqrt{2/\pi})(2x\psi_i(x) - \psi_{i-1}(x)), \; x \in [-1, 1].$$

FIEs: For different values of m, we compute u^G_m, $\widetilde{u}^G_m$, u^C_m and $\widetilde{u}^C_m$ and evaluate the errors between exact solution in L^2_{ω} norm is presented in Table 1.

Example 1 Consider the following FIEs

$$u(s) - \int_{-1}^{1} stu(t)dt = s, \; s \in [-1, 1].$$

Table 1 Chebyshev spectral Galerkin and collocation methods

n	$\|u - u_m^G\|_{L^2_\omega}$	$\|u - \widetilde{u}_m^G\|_{L^2_\omega}$	$\|u - u_m^C\|_{L^2_\omega}$	$\|u - \widetilde{u}_m^C\|_{L^2_\omega}$
2	5.25361e-03	3.47581e-04	5.26182e-03	2.35461e-03
3	3.45561e-05	6.35464e-07	4.26181e-05	3.47281e-05
4	5.67782e-07	5.36281e-09	9.35461e-07	5.27181e-07

Table 2 Chebyshev spectral Galerkin method

m	$\|\lambda - \widehat{\lambda}_m^G\|$	A_G	B_G	C_G	D_G
2	1.437756e-04	1.322080e-02	3.805042e-04	2.618893e-02	4.154016e-04
3	1.538258e-07	4.017100e-04	1.121864e-07	1.001950e-04	1.100000e-06
4	1792788e-12	1.184884e-06	1.867739e-10	4.502633e-06	2.003734e-07

Table 3 Chebyshev spectral collocation method

m	$\|\lambda - \widehat{\lambda}_m^C\|$	A_c	B_c	C_c	D_c
3	2.364791e-04	1.657598e-02	4.374821e-04	2.263829e-02	5.918271e-04
4	4.526811e-06	7.263738e-04	1.364718e-05	2.253691e-03	1.536471e-05
5	1.839732e-08	5.263712e-04	1.384901e-06	1.253618e-03	1.364791e-06

The exact solution is given by $u(s) = 3s$.

EVPs: For different values of m, we compute the approximate eigenfunctions $\widehat{\lambda}_m^G$, u'^G_m, $\widetilde{u'}^G_m$ in the Chebyshev spectral Galerkin methods and $\widehat{\lambda}_m^C$, u'^C_m, $\widetilde{u'}^C_m$ in Chebyshev spectral collocation methods and evaluate the error in both weighted two norm and weighted infinity norm are presented in Tables 2 and 3.

Example 2 Consider the following EVP

$$\int_{-1}^{1} e^{-|t-s|^2} u(t)dt = \lambda u(s), \ \ s \in [-1, 1].$$

For simplification of notations, we denote
$\|u'^G_m - \mathcal{P}^S u'^G_m\|_{L^2_\omega} = A_G$, $\|\widetilde{u'}^G_m - \mathcal{P}^S \widetilde{u'}^G_m\|_{L^2_\omega} = B_G$, $\|u'^G_m - \mathcal{P}^S u'^G_m\|_{L^\infty_\omega} = C_G$ and $\|\widetilde{u'}^G_m - \mathcal{P}^S \widetilde{u'}^G_m\|_{L^\infty_\omega} = D_G$ $\|u'^C_m - \mathcal{P}^S u'^C_m\|_{L^2_\omega} = A_c$, $\|\widetilde{u'}^C_m - \mathcal{P}^S \widetilde{u'}^C_m\|_{L^2_\omega} = B_c$, $\|u'^C_m - \mathcal{P}^S u'^C_m\|_{L^\infty_\omega} = C_c$ and $\|\widetilde{u'}^C_m - \mathcal{P}^S \widetilde{u'}^C_m\|_{L^\infty_\omega} - D_c$
From the above tables we observe that our theoretical results agree with the numerical results.

References

1. Ahues M, Largillier A, Limaye BV (2001) Spectral computations for bounded operators. Chapman and Hall/CRC, New York
2. Atkinson KE (1997) The numerical solution of integral equations of the second kind. Cambridge University Press, Cambridge
3. Avazzadeh Z, Heydari M (2012) Chebyshev polynomials for solving two dimensional linear and nonlinear integral equations of the second kind. Comput Appl Math 31(1):127–142
4. Baboliana E, Fattahzadeh F (2007) Numerical computation method in solving integral equations by using Chebyshev wavelet operational matrix of integration. Appl Math Comput 188(1):1016–1022
5. Ben-yu G (1998) Spectral methods and their applications. World Scientific
6. Canuto C, Hussaini MY, Quarteroni A, Zang TA (2006) Spectral methods. Springer, Berlin Heidelberg
7. Das P, Nelakanti G (2015) Convergence analysis of Legendre spectral projection methods for Hammerstein integral equations of mixed type. J Appl Math Comput 49:529–555
8. Elliott D (1963) A Chebyshev series method for the numerical solution of Fredholm integral equations. Comput J 6(1):102–112
9. Mason JC, Handscomb DC (2002) Chebyshev polynomials. Chapman and Hall/CRC, New York
10. Osborn JE (1975) Spectral approximation for compact operators. Math Comp 29:712–725
11. Panigrahi BL, Nelakanti G (2011) Superconvergence of Legendre projection methods for the eigenvalue problem of a compact integral operator. J Comput Appl Math 235:2380–2391
12. Panigrahi BL, Long G, Nelakanti G (2013) Legendre multi-projection methods for eigenvalue problem of a compact integral operator. J Comput Appl Math 239:135–151
13. Panigrahi BL, Mandal M, Nelakanti G (2019) Legendre multi-Galerkin methods for Fredholm integral equations with weakly singular kernel and the corresponding eigenvalue problem. J Comput Appl Math 346:224–236
14. Shen J, Tang T, Wang L (2011) Spectral methods: algorithms, analysis, applications, vol 41. Springer series in computational mathematics. Springer, New York, USA
15. Yousef A, Javad S, Babolian E, Morad E (2019) Convergence analysis of the Chebyshev-Legendre spectral method for a class of Fredholm fractional integro-differential equations. J Comput Appl Math 358:95–110

Some Sequence Spaces of Weight f

Ekrem Savaş

Abstract The aim of the paper is to present the sequence space that is strongly double lacunary p-summable respecting an Orlicz function. We also consider a relation between the family of lacunary double statistically convergent sequences of weight f and $\hat{w}^{f}_{\theta_{u,v}}[\varphi, p]$ −summable sequences of weight f.

Keywords Orlicz function · Double statistical convergence of weight f · Double lacunary statistical convergence

1 Introduction

Statistical convergence was given by Fast [2] (see also Schoenberg [23]) as an extension of the convergence as follows:

Let x be a real number sequence. We call that x is statistically convergent to the number ξ, if for every $\delta > 0$

$$\lim_{j} \frac{1}{j} |\{j < n : |x_j - \xi| \geq \delta\}| = 0.$$

In this action, we consider $S - \lim x = \xi$ or $x_k \to \xi(S)$.

In [1], Colak studied the idea of the statistical convergence of order β, $0 < \beta < 1$.

In this paper, we consider the lacunary double statistical convergence of weight f, where $f : [0, \infty) \to [0, \infty)$ such that where $f : [0, \infty) \times [0, \infty) \to [0, \infty)$ such that $f(x_{ij}) \to \infty$ if $x_{ij} \to \infty$. The family of all such functions shall be denoted as **F**.

Let s_2 be the set of all double sequences. A double sequence $x = (x_{ij})$ has a Pringsheim limit ξ denoted by $P - \lim x = \xi$ provided that given $\delta > 0$ and there exists $N \in \mathbb{N}$ such that $|x_{ij} - \xi| < \delta$ whenever $i, j \geq N$. We observe such a x briefly as "**P-convergent**", (see, [8]).

E. Savaş (✉)
Department of Mathematics, Uşak University, Uşak, Turkey

© The Author(s), under exclusive license to Springer Nature Singapore Pte Ltd. 2022

D. Giri et al. (eds.), *Proceedings of the Seventh International Conference on Mathematics and Computing*, Advances in Intelligent Systems and Computing 1412,
https://doi.org/10.1007/978-981-16-6890-6_60

We set by c_2, the space of P-convergent sequences.

Some recent papers concerning double sequences can be find in [6, 7, 12, 13, 16, 22].

If, for every $\delta > 0$,

$$P - \lim_{i,j} \frac{1}{ij} |\{(i, j) : i \leq k \text{ and } j \leq l : |x_{ij} - \xi| \geq \delta\}| = 0,$$

we call that a real double sequence $x = (x_{ij})$ is statistically convergent to ξ, (see, [4]).

We now define the double statistical convergence of weight f as follows: Let $f \in \mathbf{F}$. The sequence (x_{ij}) is said to be double statistically convergent of weight f, if there is a real number ξ such that

$$P - \lim_{i,j} \frac{1}{f(x_{ij})} \left|\{i \leq k \text{ and } j \leq l : |x_{ij} - \xi| \geq \delta\}\right| = 0,$$

for every $\delta > 0$, then we call that x is double statistically convergent of weight f to ξ. We write $S_2^f - \lim x_{ij} = \xi$. The family of all double statistically convergent sequences of weight f can be consider as S_2^f. If $f(ij) = (ij)$, we can have previous definition.

Definition 1 The double sequence $\theta_{u,v} = \{(k_u, l_v)\}$ is said **double lacunary** if there exist two increasing sequences of integers such that

$$k_0 = 0, h_u = k_u - k_{u-1} \to \infty \text{ as } u \to \infty$$

and

$$l_0 = 0, \bar{h}_v = l_v - l_{v-1} \to \infty \text{ as } v \to \infty.$$

Let's write $k_{u,v} = k_u l_v$, $h_{uv} = h_u \bar{h}_v$ and $\theta_{u,v}$ is determine by $I_{u,v} = \{(i, j) : k_{u-1} < k \leq k_u \,\&\, l_{v-1} < l \leq l_v\}$, $q_u = \frac{k_u}{k_{u-1}}$, $\bar{q}_v = \frac{l_v}{l_{v-1}}$, and $q_{u,v} = q_u \bar{q}_v$. We will represent the family of all double lacunary sequences by $\mathbf{N}_{\theta_{u,v}}$. Let $T \subseteq \mathbb{N} \times \mathbb{N}$ has double lacunary density $\delta_2^\theta(T)$ if

$$P - \lim_{u,v} \frac{1}{h_{uv}} |\{(i, j) \in I_{u,v} : (i, j) \in T\}|$$

exists.

The following definition was giving by Patterson and Savas, [6].

Definition 2 Let $\theta_{u,v}$ be a double lacunary sequence; the double sequence x is S_θ^2-convergent to ξ provided that for every $\delta > 0$,

$$P - \lim_{u,v} \frac{1}{h_{uv}} \left|\{(i, j) \in I_{u,v} : |x_{ij} - \xi| \geq \delta\}\right| = 0.$$

We can write $S_\theta^2 - \lim_{i,j} x_{i,j} = \xi$.

More papers concerning double lacunary and double sequences will be meet in [11, 13, 18, 19]. We now define

Definition 3 ([14]) Let $f \in \mathbf{F}$. If,

$$P - \lim_{u,v} \frac{1}{f(h_{uv})} \left|\left\{(i,j) \in I_{uv} : \left|x_{ij} - \xi\right| \geq \varepsilon\right\}\right| = 0,$$

we call that the sequence $x = \left(x_{ij}\right)$ is double lacunary statistical convergence of weight f to ξ. Then we set $S^f_{\theta_{u,v}} - \lim x_{ij} = \xi$. We demonstrate the family of all $S^f_{\theta_{u,v}}$ – statistically convergent sequences of weight f as $S^f_{\theta_{u,v}}$.

Remark 1 Taking $f(h_{uv}) = h^\beta_{uv}$, we observe the lacunary statistical convergence of order β, [22]. If $\theta = 2^{rs}$ and $f(kl) = (kl)^\beta$ then we write the double statistical convergence of order α. When $f(h_{uv}) = h_{uv}$, we get the lacunary double statistical convergence, [6].

2 A New Sequence Space

Recall in [3] that an Orlicz function $\varphi : [0,\infty) \to [0,\infty)$ is continuous, convex, non-decreasing function such that $\varphi(0) = 0$ and $\varphi(x) > 0$ for $x > 0$, and $\varphi(x) \to \infty$ as $x \to \infty$.

Later on the classes of Orlicz sequence spaces were interested by Parashar and Choudhary [5], Savas [9, 10, 15, 17, 19–21] and some others.

Definition 4 Let φ be an Orlicz function, $p = (p_{ij})$ be a sequence of strictly positive real numbers and take $f \in \mathbf{F}$. We define,

$$w^f_{\theta_{u,v}}[\varphi, p] = \{x = \left(x_{ij}\right) : P - \lim_{uv} \frac{1}{f(h_{uv})} \sum_{(i,j)\in I_{uv}} \left[\varphi\left(\frac{|x_{ij} - \xi|}{\rho}\right)\right]^{p_{ij}} = 0 \text{ for some } \xi \text{ and } \rho > 0\}.$$

When $x \in w^f_{\theta_{u,v}}[\varphi, p]$, then we say that x is strongly double lacunary summable of weight f respecting the Orlicz function φ.

Also;

1. If $\varphi(x) = x$, $\theta = 2^{uv}$, and $p_{ij} = 1$ for all (i,j) then $w^f_{\theta_{u,v}}[\varphi, p]$ reduces to $[w^f]$.
2. If $p_{i,j} = 1$ for all (i,j), then $w^f_{\theta_{u,v}}[\varphi, p]$ reduces to $w^f_{\theta_{u,v}}[\varphi]$.
3. If $p_{ij} = 1$ for all (i,j) and $\theta = 2^{uv}$, then $w^f_{\theta_{u,v}}[\varphi, p]$ reduces to $w^f[\varphi]$.
4. If $\theta = 2^{uv}$ then $w^f_{\theta_{u,v}}[\varphi, p]$ reduces to $w^f_p[\varphi]$.

For the next theorems we suppose that the double sequence is bounded and $0 < s = \inf_{i,j} p_{ij} \le p_{ij} \le \sup_{i,j} p_{ij} = S < \infty$.

Theorem 1 *Let $f_1, f_2 \in \boldsymbol{F}$ be such that there exist $R > 0$ and $(u_0, v_0) \in \mathbb{N} \times \mathbb{N}$ such that $f_1(h_{uv})/f_2(h_{uv}) \le R$ for all $u \ge u_0$ and $v \ge v_0$ and let φ be an Orlicz function then $w_{\theta_{u,v}}^{f_1}[\varphi, p] \subset S_{\theta_{u,v}}^{f_2}$.*

Proof Let $x \in w_{\theta}^{f_1}[\varphi, p]$, $\delta > 0$ be taken and $\sum_1$ and $\sum_2$ indicate the sums over $(i, j) \in I_{uv}$, $|x_{ij} - \xi| \ge \varepsilon$ and $(i, j) \in I_{uv}$, $|x_{ij} - \xi| < \delta$, respectively. We get

$$\begin{aligned}
&\frac{1}{f_1(h_{uv})} \sum_{(i,j)\in I_{uv}} \left[\varphi\left(\frac{|x_{ij} - \xi|}{\rho}\right)\right]^{p_{ij}} \\
&= \frac{1}{f_1(h_{uv})} \sum_1 \left[\varphi\left(\frac{|x_{ij} - \xi|}{\rho}\right)\right]^{p_{ij}} \\
&+ \frac{1}{f_1(h_{uv})} \sum_2 \left[\varphi\left(\frac{|x_{ij} - \xi|}{\rho}\right)\right]^{p_{ij}} \\
&\ge \frac{1}{R.f_1(h_{uv})} \left[\sum_1 \left[\varphi\left(\frac{|x_{ij} - \xi|}{\rho}\right)\right]^{p_{ij}} + \sum_2 \left[\varphi\left(\frac{|x_{ij} - \xi|}{\rho}\right)\right]^{p_{ij}}\right] \\
&\ge \frac{1}{R.f_1(h_{uv})} \sum_1 [\varphi(\delta)]^{p_{ij}} \\
&\ge \frac{1}{R.f_1(h_{uv})} \sum_1 \min\left([\varphi(\varepsilon_1)]^s, [\varphi(\delta_1)]^S\right), \epsilon_1 = \frac{\delta}{\rho} \\
&\ge \frac{1}{R.f_1(h_{uv})} \left|\left\{(i, j) \in I_{uv} : |x_{ij} - \xi| \ge \delta\right\}\right| \min\left([\varphi(\delta_1)]^s, [\varphi(\delta_1)]^S\right).
\end{aligned}$$

So it is to easy that $x \in S_{\theta_{u,v}}^{f_2}$.

We next consider

Theorem 2 *Let $f_1, f_2 \in \boldsymbol{F}$ and φ be an Orlicz function and $x = (x_{ij})$ be a bounded sequence. Then there is $R > 0$ and $(r_0, s_0) \in \mathbb{N} \times \mathbb{N}$ such that $f_1(h_{uv})/f_2(h_{uv}) \le R$ for all $u \ge u_0$, $v \ge v_0$. Then $S_{\theta_{u,v}}^{f_1} \subset w_{\theta_{u,v}}^{f_2}[\varphi, p]$.*

Proof Assume that $x \in \ell_2^{\infty}$ and $S_{\theta_{u,v}}^{f} - \lim x_{ij} = \xi$. Since $x \in \ell_2^{\infty}$, then there exists a constant $K > 0$ such that $|x_{ij}| \le K$. Writing $\delta > 0$, we have

$$\frac{1}{f_2(h_{uv})}\sum_{(i,j)\in I_{uv}}\left[\varphi\left(\frac{|x_{ij}-\xi|}{\rho}\right)\right]^{p_{ij}} = \frac{R}{f_1(h_{uv})}\sum_1\left[\varphi\left(\frac{|x_{ij}-\xi|}{\rho}\right)\right]^{p_{ij}}$$
$$+\frac{R}{f_1(h_{uv})}\sum_2\left[\varphi\left(\frac{|x_{ij}-\xi|}{\rho}\right)\right]^{p_{ij}}$$
$$\leq \frac{R}{f_1(h_{uv})}\sum_1 \max\left\{\left[\varphi\left(\frac{K}{\rho}\right)\right]^s, \left[\varphi\left(\frac{K}{\rho}\right)\right]^S\right\}$$

$$+\frac{1}{f_1(h_{uv})}\sum_2\left[\varphi\left(\frac{\delta}{\rho}\right)\right]^{p_{ij}}$$
$$\leq \max\left\{[\varphi(T)]^s, [\varphi(T)]^S\right\}$$
$$\times \frac{R}{f_1(h_{uv})}\left|\left\{(i,j)\in I_{uv} : |x_{ij}-L|\geq\delta\right\}\right|$$
$$+\max\left\{[\varphi(\delta_1)]^s, [\varphi(\epsilon_1)]^S\right\},\ \frac{K}{\rho}=T, \frac{\delta}{\rho}=\epsilon_1.$$

Hence $x\in w_{\theta_{u,v}}^{f_2}[\varphi, p]$.

If we consider $f(h_{uv}) = h_{uv}^{\alpha}$, the above theorems reduce to the following results, (see, [22]).

Corollary 1 *Let $\beta, \rho\in(0,1]$ be real numbers such that $\beta\leq\rho$ and φ be an Orlicz function, then $w_{\theta_{u,v}}^{\beta}[\varphi, p]\subset S_{\theta_{u,v}}^{\rho}$.*

Corollary 2 *Let $\beta\in(0,1]$ and φ be an Orlicz function, then $w_{\theta_{u,v}}^{\beta}[\varphi, p]\subset S_{\theta_{u,v}}^{\beta}$.*

Corollary 3 *Let φ be an Orlicz function and $x=(x_{ij})$ be a bounded sequence, then $S_{\theta_{u,v}}^{\beta}\subset w_{\theta_{u,v}}^{\beta}[\varphi, p]$.*

References

1. Colak R (2010) Statistical convergence of order α. In: Modern methods in analysis and its applications, New Delhi, India, Anamaya Pub, pp 121–129
2. Fast H (1951) Sur la convergence statistique. Colloq Math 2:241–244
3. Krasnoselskii MA, Rutisky YB (1961) Convex function and Orlicz spaces. Groningen, Netherlands
4. Mursaleen, Edely OH (2003) Statistical convergence of double sequences. J Math Anal Appl 288(1):223–231
5. Parashar SD, Choudhury B (1994) Sequence space defined by Orlicz function. Indian J Pure Appl Math 25(14):419–428
6. Patterson RF, Savaş E (2000) Lacunary statistical convergence of double sequences. Math Commun 10:55–61

7. Patterson RF, Savas R, Savas E (2020) Multidimensional linear functional connected with double strongly Cesàro summability. Indian J Pure Appl Math 51(1):143–153
8. Pringsheim A (1900) Zur theorie der zweifach unendlichen Zahlenfolgen. Mathematische Annalen 53:289–321
9. Savaş E (2008) (A, λ) - double sequence spaces defined by Orlicz function and double statistical convergence. Comput Math Appl 55(6):1293–1301
10. Savaş E (2009) On some new double lacunary sequences spaces via Orlicz function. J Comput Anal Appl 11(3):423–430
11. Savaş E (2009) Remark on double lacunary statistical convergence of fuzzy numbers. J Comput Anal Appl 11(1):64–69
12. Savaş E (2013) Double almost statistical convergence of order α. Adv Differ Equ 2013:62
13. Savaş E (2013) On lacunary double statistical convergence in locally solid Riesz spaces. J Ineq Appl 2013:99
14. Savaş E (Preprint) Double lacunary statistical convergence sequences of weight g. In: De Gruyter (ed) Proceedings in mathematics, operator theory (2021)
15. Savaş E, Patterson RF (2007) Double sequence spaces defined by Orlicz functions. Iran J Sci Technol Trans A Sci 31(2):183–188
16. Savaş E, Patterson RF (2007) Double sequence spaces defined by Orlicz functions. Iran J Sci Tech Trans A Sci 31(2):183–188
17. Savaş E, Patterson RF (2008) (σ, λ)- double sequence spaces via Orlicz function. J Comput Anal Appl 10(1):101–111
18. Savaş E, Patterson RF (2009) Double σ -convergence lacunary statistical sequences. J Comput Anal Appl 11(4):610–615
19. Savaş E, Patterson RF (2011) Some double lacunary sequence spaces defined by Orlicz functions. Southeast Asian Bull Math 35(1):103–110
20. Savaş E, Savaş R (2003) Some λ- sequence spaces defined by Orlicz functions. Indian J Pure Appl Math 34(12):1673–1680
21. Savaş E, Savaş R (2004) Some sequence spaces defined by Orlicz functions. Arch Math (Brno) 40(1):33–40
22. Savaş Eren R, Savaş E (2015) Double lacunary statistical convergence of order α. Indian J Math 57(1):11–15
23. Schoenberg IJ (1959) The integrability of certain functions and related summability methods. Amer Math Monthly 66:361–375

Generalized Libera Operator on Mixed-Norm Spaces

S. Naik and K. Rajbangshi

Abstract In this paper, we discuss the boundedness of generalized Libera operator Λ^{γ} on mixed-norm spaces $H^{p,q}_{\alpha,\nu}$. As a consequence, we find few results about the action of the operator Λ^{γ} on various function spaces such as Hardy, Zygmund, Lipschitz, Bloch type, and Besov spaces.

Keywords Mixed-norm spaces · Bloch spaces · Besov spaces · Zygmund spaces · Libera operator

1 Introduction

Let $\mathbf{D}$ denote the unit disc and $H(\mathbf{D})$ denote the space of all analytic functions on the unit disc $\mathbf{D}$ of the complex plane. In [4], Libera introduced the following transformation

$$f(z) \mapsto \frac{2}{z}\int_0^z f(t)dt$$

and call it Libera transformation. He also showed the importance of Libera transformation in the theory of univalent functions. He finds that Libera transformation is in the family of star-like functions. Since then many authors studied this aspect of the Libera transformation. Later, A. G. Siskakis defined the following: For $|a| \leq 1$

$$\Lambda_a f(z) = \frac{1}{a-z}\int_z^a f(t)dt$$

and studied it from the functional analytic point of view in [8, 9], and further in [3, 5, 7]. For $|a| < 1$, Λ_a is defined on $H(\mathbf{D})$ as well as on the classical spaces such

S. Naik (✉)
Department of Applied Sciences, Gauhati University, Guwahati, India

K. Rajbangshi
Department of Mathematics, Morigaon College, Morigaon, India

© The Author(s), under exclusive license to Springer Nature Singapore Pte Ltd. 2022

D. Giri et al. (eds.), *Proceedings of the Seventh International Conference on Mathematics and Computing*, Advances in Intelligent Systems and Computing 1412,
https://doi.org/10.1007/978-981-16-6890-6_61

as Hardy, Bergman and Besov. Λ_a has almost the same linear topological properties as the integration operator $f(z) \mapsto \int_0^z f(t)dt$. Therefore, from functional analytic point of view it is not so interesting [1]. So $|a| = 1$ is considered and defined the Libera operator as

$$\Lambda f(z) = \frac{1}{1-z} \int_z^1 f(t)dt.$$

The dual of $H(\mathbf{D})$ is $H(\overline{\mathbf{D}})$, is needed here where $g \in H(\overline{\mathbf{D}})$ means that g is holomorphic in a neighborhood of $\overline{\mathbf{D}}$ (depending on g).

The duality pairing is given by

$$\langle f, g \rangle = \sum_{n=0}^{\infty} a_n b_n,$$

where $f(z) = \sum_{n=o}^{\infty} a_n z^n \in H(\mathbf{D})$ and $g(z) = \sum_{n=o}^{\infty} b_n z^n \in H(\overline{\mathbf{D}})$.

Libera operator can be generalized as follows:

For every complex γ with $\mathrm{Re}(\gamma) > -1$ and for any nonnegative integer k, let A_k^γ be defined as the kth coefficient in the expression

$$(1-x)^{-(\gamma+1)} = \sum_{k=o}^{\infty} A_k^\gamma x^k,$$

where

$$A_k^\gamma = \frac{\Gamma(\gamma+k+1)}{\Gamma(\gamma+1)\Gamma(k+1)}.$$

For $f = \sum_{k=0}^{\infty} a_k z^k \in H(\mathbf{D})$, we define the linear operator $\Lambda^\gamma(f)$ by

$$\Lambda^\gamma(f)(z) = \sum_{n=o}^{\infty} \left(\sum_{k=n}^{\infty} \frac{A_{k-n}^\gamma}{A_k^{\gamma+1}} a_k \right) z^n.$$

The equivalent integral representation is given by

$$\Lambda^\gamma(f)(z) = \frac{1+\gamma}{(z-1)^{\gamma+1}} \int_1^z f(t)(z-t)^\gamma dt.$$

Taking a path the segment joining z and 1 we get

$$\Lambda^\gamma f(z) = (\gamma+1) \int_0^1 f(t+(1-t)z)t^\gamma dt. \tag{1}$$

We call operator (1) generalized Libera operator. We use the symbol $\overline{\Lambda^\gamma}$ to denote the operator $\Lambda^\gamma : H(\overline{\mathbf{D}}) \to H(\overline{\mathbf{D}})$.

Let $f \in H(\mathbf{D})$ and $r < 1$. For $0 < p < \infty$, the integral means of f are defined by

$$M_p(r, f) = \left(\frac{1}{2\pi}\int_0^{2\pi} |f(re^{i\theta})|^p d\theta\right)^{\frac{1}{p}}$$

and

$$M_\infty(r, f) = \max_{0\le\theta\le 2\pi} |f(re^{i\theta})|.$$

It is well known that $M_p(r, f)$ is an increasing function of r.

Suppose $0 < p \le \infty,\ \ 0 < \alpha, q < \infty$ and ν be a nonnegative integer. Let $H^{p,q}_{\alpha,\nu}$ denote the mixed-norm space consisting of $f \in H(\mathbf{D})$ such that

$$\|f\|_{H^{p,q}_{\alpha,\nu}} = \left(\int_0^1 (1-r)^{q\alpha-1} M_p^q(r, f^{(\nu)})dr\right)^{\frac{1}{q}} < \infty.$$

For $\nu = 0$, it is the space $H(p, q, \alpha)$.

The (quasi) norm in $H^{p,q}_{\alpha,\nu}$ is given by

$$\|g\|_{H^{p,q}_{\alpha,\nu}} = \sum_{j=0}^{\nu-1} |g^{(j)(0)}| + \left(\int_0^1 (1-r)^{q\alpha-1} M_p^q\left(r, g^{(\nu)}\right) dr\right)^{\frac{1}{q}},$$

where $\sum_{j=0}^{-1}$ should be interpreted as equal to zero.

The space $H^{p,\infty}_{\alpha,\nu} := H^p_{\alpha,\nu}$ is specific in that the set $\mathcal{P}$ of all analytic polynomials is not dense in it. The closure of $\mathcal{P}$ in $H^p_{\alpha,\nu}$ coincides with the "little oh" space

$$h^p_{\alpha,\nu} = \left\{g \in H(\mathbf{D}) : M_p(r, g^{(\nu)}) = o(1-r)^{-\alpha}(r \uparrow 1)\right\}.$$

If $q = p,\ \nu = 0$ and $\beta = \alpha p - 1$, then $A^p_\beta = H^{p,p}_{\frac{\beta+1}{p},0}$.

Let $\beta \in \mathbf{R}$, $0 < p, q < \infty$ and $\nu \ge 0$ be an integer such that $\nu - \beta > 0$. The Besov mixed-norm space $B^{p,q}_\beta$ is defined as

$$B^{p,q}_\beta = H^{p,q}_{\nu-\beta,\nu}.$$

If $q = \infty$, then we can assume that $\nu = \beta$.

If $p = q > 1$ and $\beta = \frac{1}{p}$, then the $B^{p,q}_\beta$ is equivalent to the classical Besov spaces. If $\beta < 0$, then $B^{p,q}_\beta = H^{p,q}_{-\beta}$ and if $\beta = 0$, then the space is called Hardy-Bloch space $B^{p,q}$. For $q = \infty$ we have the following spaces:

$$B^p_\beta = H^p_{\nu-\beta} = \{g \in H(\mathbf{D}) : M_p(r, g^{(\nu)}) = O((1-r)^{\beta-\nu})\} \qquad (0 < \beta < \nu).$$

and its subspace

$$b^p_\beta = \{g \in H(\mathbf{D}) : M_p(r, g^{(\nu)}) = o((1-r)^{\beta-\nu})\} \qquad (0 < \beta < \nu).$$

For $1 \leq p \leq \infty$, $0 < q \leq \infty$ and $\beta > 0$, the space $B^{p,q}_\beta = \Lambda^{p,q}_\beta$, called the Lipschitz space consisting of those $g \in H^p$ for which

$$\left(\int_0^1 \left(\frac{\|\Delta^\nu_t g\|_p}{t^\beta}\right)\frac{dt}{t}\right)^{\frac{1}{q}} < \infty,$$

where H^p is the Hardy space, $\|.\|$ is the H^p norm and $\Delta^\nu_t g$ denotes the symmetric νth difference with step t, defined as

$$\Delta^\nu_t h(\theta) = \sum_{k=0}^{\nu} \tbinom{\nu}{k}(-1)^{\nu-k} h(\theta + kt), \qquad h(\theta) = g(e^{i\theta}).$$

For $q = \infty$, we have

$$B^p_\beta = \Lambda^p_\beta := \Lambda^{p,\infty}_\beta = \{g \in H^p : \|\Delta^\nu_t g\|_p = O(t^\beta)\} \quad (0 < \beta < \nu),$$

$$b^p_\beta = \lambda^p_\beta = \{g \in H^p : \|\Delta^\nu_t g\|_p = o(t^\beta)(t \downarrow 0)\} \quad (0 < \beta < \nu).$$

For $p = q = \infty$ and $\beta = 1$, the space $\Lambda^{p,q}_\beta := \Lambda_*$ is called the Zygmund space and corresponding "little" space is denoted by λ_*.

For any positive integer ν, the Hardy-Sobolev space W^p_ν is defined as

$$W^p_\nu = H^{p,\infty}_{0,\nu} = \{g \in H^p : g^{(\nu)} \in H^p\}.$$

In particular, W^∞_1 coincides with the ordinary Lipschitz space.

For $p, \beta > 0$, Dirichlet space $\mathcal{D}^p_\beta$ is defined as

$$\mathcal{D}^p_\beta := H^{p,p}_{\frac{\beta+1}{p},1} = \{f \in H(\mathbf{D}) : \|f\|^p_{\mathcal{D}^p_\beta} = |f(0)|^p + \int_{\mathbf{D}} |f'(z)|^p (1-|z|)^\beta dA(z) < \infty\}.$$

The following result was proved in [6].

Theorem: Let $X = H^{p,q}_{\alpha,\nu}$ or $X = h^p_{\alpha,\nu}$. The operator Λ acts as a bounded operator from X into X if and only if $\nu - \alpha > \frac{1}{p} - 1$.

In view of this result, it is natural to consider the operator Λ^γ on mixed-norm spaces. In this paper, we give conditions on the parameters p, q, α, ν and γ which ensure the boundedness of generalized Libera operator on mixed-norm spaces. As a consequence, we find few results about the action of the operator Λ^γ on various function spaces such as Zygmund, Lipschitz, Bloch type, and Besov spaces.

Now onwards C denotes a positive constant, not necessarily the same on any two occurrences.

2 Basic Lemmas

We need the following lemmas to prove our main result.

Lemma 1 ([6]) *If a and b are positive real numbers such that $a+b \leq 1$ and if $g \in H^p$, $0 < p \leq \infty$, then*

$$\left(\int_0^{2\pi} |g(a+be^{i\theta})|^p \, d\theta\right)^{\frac{1}{p}} \leq \left(\frac{2a+b}{b}\right)^{\frac{1}{p}} \left(\int_0^{2\pi} \left|g\left((a+b)e^{i\theta}\right)\right|^p \, d\theta\right)^{\frac{1}{p}}.$$

Lemma 2 ([6]) *Let $0 < q < 1$, $\beta = 1+\delta > 0$, and let $u : (0,1) \mapsto [0,\infty)$ be a nondecreasing function. Then*

$$\left(\int_r^1 u(s)(1-s)^{\beta-1} ds\right)^q \leq C \int_r^1 u(s)^q (1-s)^{q\beta-1} ds,$$

where C is independent of u.

Lemma 3 ([6]) *Let $1 < q < \infty$, $\epsilon > 0$, and $u \geq 0$, a measurable function defined on $(r,1)$. Then*

$$\left(\int_r^1 u(s)(1-s)^{\delta} ds\right)^q \leq C(1-r)^{\epsilon q} \int_r^1 u(s)^q (1-s)^{(1+\delta-\epsilon)q-1} ds,$$

where C is independent of u.

Lemma 4 ([6]) *If $\alpha > 0$, then*

(a) $H^{p,q}_{\alpha,\nu} \subsetneqq h^p_{\alpha,\nu} \subsetneqq H^p_{\alpha,\nu}$ *for* $q < \infty$,
(b) $H^{p,q}_{\alpha,\nu} \subsetneqq H^{u,q}_{\beta,\nu}$, *where* $\beta = \frac{1}{p} - \frac{1}{u} + \alpha$, *where* $u > p$ *and* $0 < q \leq \infty$,
(c) $h^p_{\alpha,\nu} \subsetneqq h^u_{\beta,\nu}$ *where* $\beta = \frac{1}{u} - \frac{1}{p} + \alpha$, *where* $u > p$.

3 Generalized Libera Operator on Mixed-Norm Spaces

In this section we study the boundedness of generalized Libera operator on mixed-norm space.

Theorem 1 *Let $1 \leq p \leq \infty$, $0 < q \leq \infty$. For $Re(\gamma) \in [0,\infty)$, Λ^{γ} acts as a bounded operator from X to X where $X = H^{p,q}_{\alpha,\nu}$ or $X = h^p_{\alpha,\nu}$ if $\nu - \alpha > \frac{1}{p} - 1$.*

Proof Since

$$\Lambda^{\gamma} f(z) = (\gamma+1) \int_0^1 f(t+(1-t)z) t^{\gamma} dt$$

therefore,

$$(\Lambda^\gamma f)^{(\nu)}(z) = (\gamma+1)\int_0^1 f^{(\nu)}(t+(1-t)z)(1-t)^\nu t^\gamma dt.$$

Using Minkowski's inequality and Lemma 1 with $a = t$, $b = (1-t)r$ we obtain

$$\begin{aligned} M_p(r, (\Lambda^\gamma f)^{(\nu)}) &\le \int_0^1 (1-t)^\nu t^\gamma M_p\left(t+(1-t)r, f^{(\nu)}\right)\left(\frac{2t+(1-t)r}{(1-t)r}\right)^{\frac{1}{p}} dt \\ &\le Cr^{-\frac{1}{p}} \int_0^1 (1-t)^{\nu-\frac{1}{p}} t^\gamma M_p\left(t+(1-t)r, f^{(\nu)}\right) dt. \end{aligned}$$

For $\mathrm{Re}(\gamma) \ge 0$, we have

$$M_p(r, (\Lambda^\gamma f)^{(\nu)}) \le Cr^{-\frac{1}{p}} \int_0^1 (1-t)^{\nu-\frac{1}{p}} M_p\left(t+(1-t)r, f^{(\nu)}\right) dt. \tag{2}$$

Let $t + (1-t)r = s$ and $\delta = \nu - \frac{1}{p}$, then from (2) we get

$$r^{\frac{1}{p}} M_p(r, (\Lambda^\gamma f)^{(\nu)}) \le C(1-r)^{-\delta-1} \int_r^1 (1-s)^\delta M_p\left(s, f^{(\nu)}\right) ds.$$

$$r^{\frac{q}{p}} M_p^q(r, (\Lambda^\gamma f)^{(\nu)}) \le C(1-r)^{-(\delta+1)q} \left(\int_r^1 (1-s)^\delta M_p\left(s, f^{(\nu)}\right) ds\right)^q. \tag{3}$$

For $q > 1$ using Lemma 3 and for $q < 1$ using Lemma 2 to (3) we get

$$r^{\frac{q}{p}} M_p^q(r, (\Lambda^\gamma f)^{(\nu)}) \le C(1-r)^{-(\delta+1-\varepsilon)q} \left(\int_r^1 (1-s)^{(\delta+1-\varepsilon)q-1} M_p^q\left(s, f^{(\nu)}\right) ds\right), \tag{4}$$

where

$$\varepsilon = \begin{cases} = 0 & \text{if } q \le 1, \\ > 0 & \text{if } q > 1. \end{cases}$$

Let $q > 1$. Maximum modulus principle for analytic function gives

$$\int_0^1 M_p^q(r, (\Lambda^\gamma f)^{(\nu)})(1-r)^{q\alpha-1} dr \le C \int_0^1 r^{\frac{q}{p}} M_p^q(r, (\Lambda^\gamma f)^{(\nu)})(1-r)^{q\alpha-1} dr. \tag{5}$$

Combining (4) and (5) we have

$$\int_0^1 M_p^q(r,(\Lambda^\gamma f)^{(\nu)})(1-r)^{q\alpha-1}dr$$
$$\leq C\int_0^1 (1-r)^{(\alpha-\delta-1+\varepsilon)q-1}dr\left(\int_r^1 M_p^q(s,f^{(\nu)})(1-s)^{q(\delta+1-\varepsilon)-1}ds\right)$$
$$=C\int_0^1 M_p^q(s,f^{(\nu)})(1-s)^{(\delta+1-\varepsilon)q-1}ds\left(\int_0^s (1-r)^{(\alpha-\delta-1+\varepsilon)q-1}dr\right). \quad (6)$$

Since $\alpha-\delta-1=\alpha+\frac{1}{p}-\nu-1<0$, we can choose $\varepsilon>0$ so that $\alpha-\delta-1+\varepsilon<0$. Then

$$\int_0^s (1-r)^{(\alpha-\delta-1+\varepsilon)q-1}dr \leq C(1-s)^{(\alpha-\delta-1+\varepsilon)q}. \quad (7)$$

Combining (6) and (7) we get the required result.

For $q<1$, proceeding similarly we get the required result.

For $\gamma=0$ we get Theorem 11 ($(b)\Rightarrow(a)$) of [6].

Remark Theorem 1 fails for the cases $\nu-\alpha\leq\frac{1}{p}-1$. To show that we will prove the following two results:

Suppose $\kappa_{p,\alpha,\nu}=\nu-\alpha+1-\frac{1}{p}$.

(A) Let (i) $\kappa_{p,\alpha,\nu}\leq 0$, $\alpha>0$, and $X\in\{H^p_{\alpha,\nu},h^p_{\alpha,\nu}\}$
or
(ii) $\kappa_{p,\alpha,\nu}<0$ and $X=H^{p,q}_{\alpha,\nu}(q<\infty)$.
Then the operator $\overline{\Lambda^\gamma}$ cannot be extended to a bounded operator from X to $H(\mathbf{D})$.

(B) If $1<q<\infty$, $p(\nu+1)>1$ and $\kappa_{p,\alpha,\nu}=0$, then the following function

$$f_1(z)=(1-z)^{-1}\left(\ln\frac{e}{1-z}\right)^{-1}, \quad z\in\mathbf{D}$$

belongs to $H^{p,q}_{\alpha,\nu}$ but the function $\Lambda^\gamma f_1\notin H^{p,q}_{\alpha,\nu}$.

(A) **Case (i).** Let $\kappa_{p,\alpha,\nu}\leq 0$. By Lemma 4, we have $h^\infty_{\beta,\nu}\subset h^p_{\alpha,\nu}$ where $\beta=\alpha+\frac{1}{p}$. Since $\kappa_{\infty,\beta,\nu}=\kappa_{p,\alpha,\nu}$, it is enough to consider the case of $h^\infty_{\beta,\nu}$.
Let $f_\rho(z)=\frac{1}{1-\rho z}$, $0<\rho<1$. It is clear that $f_\rho\in H(\overline{\mathbf{D}})$.
A simple calculation gives that

$$f_\rho^\nu(z)=(1-\rho z)^{-\nu-1}\rho^\nu$$

and

$$M_\infty(r,f_\rho^{(\nu)})\leq(1-r)^{-\alpha},$$

therefore, the set $\{f_\rho:0<\rho<1\}$ is bounded in $h^\infty_{\alpha,\nu}$. Hence, if $\overline{\Lambda^\gamma}$ has an extension to a bounded operator from $h^\infty_{\alpha,\nu}$ to $H(\mathbf{D})$, then the set $\{\overline{\Lambda^\gamma}f_r(0)\}$ is

bounded because the functional $g \mapsto g(0)$ is bounded on $H(\mathbf{D})$.
However,

$$\overline{\Lambda^{\gamma}} f_r(0) = (\gamma+1)\sum_{n=0}^{\infty}\frac{r^n}{n+\gamma+1} \to \infty, \qquad r \to 1^-,$$

which is a contradiction. This gives the required result in the case $h^p_{\alpha,\nu}$. Since $h^p_{\alpha,\nu} \subset H^p_{\alpha,\nu}$, we see that the result holds for $H^p_{\alpha,\nu}$ space as well.
Case (ii). Let $\kappa_{p,\alpha,\nu} < 0$. We choose β such that $\kappa_{p,\beta,\nu} = 0$. This implies that $\beta < \alpha$. Since $H^{p,q}_{\alpha,\nu} \subset H^p_{\alpha,\nu}$, therefore case (i) gives the required result.

(B) Let $\kappa_{p,\alpha,\nu} = 0$. Let

$$f_1(z) = (1-z)^{-1}\left(\ln\frac{e}{1-z}\right)^{-1}, \quad z \in \mathbf{D}.$$

A simple calculation shows that

$$\left|f_1^{(\nu)}(z)\right| \le C\left|(1-z)^{-\nu-1}\left(\ln\frac{e}{1-z}\right)^{-1}\right|, \quad z \in \mathbf{D}.$$

Proceeding similarly as in Remark 2 of [2] we have that the integral $\|f_1\|_{\alpha,\nu}^{p,q}$ is equiconvergent to the integral

$$I := \int_{\frac{9}{10}}^{1}\left[\int_{-\frac{1}{2}}^{\frac{1}{2}}\frac{d\theta}{|1-re^{i\theta}|^{p(\nu+1)}|\ln\frac{e}{1-re^{i\theta}}|^p}\right]^{\frac{q}{p}}(1-r)^{\alpha q-1}dr.$$

Next we estimate the integral

$$J_p(r) := \int_{-\frac{1}{2}}^{\frac{1}{2}}\frac{d\theta}{|1-re^{i\theta}|^{p(\nu+1)}|\ln\frac{e}{1-re^{i\theta}}|^p}, \quad \frac{9}{10} < r < 1.$$

Let $p(\nu+1) > 1$. Proceeding similarly as in Remark 2 of [2] we have

$$
\begin{aligned}
J_p(r) \leq & \int_{-\frac{1}{2}}^{\frac{1}{2}} \frac{d\theta}{|1-re^{i\theta}|^{p(\nu+1)}\left(\ln\frac{e}{|1-re^{i\theta}|}\right)^p} \\
\leq & C\int_0^{\frac{1}{2}} \frac{d\theta}{(1-r+\theta)^{p(\nu+1)}\left(\ln\frac{1}{1-r+\theta}\right)^p} \\
= & C\int_{1-r}^{\frac{3}{2}-r} \frac{d\chi}{\chi^{p(\nu+1)}\left(\ln\frac{1}{\chi}\right)^p} \\
= & C\int_{\ln\frac{1}{\frac{3}{2}-r}}^{\ln\frac{1}{1-r}} \frac{e^{\{p(\nu+1)-1\}t}}{t^p}dt \leq C\int_1^{\ln\frac{1}{1-r}} \frac{e^{\{p(\nu+1)-1\}t}}{t^p}dt.
\end{aligned}
$$

Using the l'Hôspital rule

$$
\int_1^{\chi} \frac{e^{\{p(\nu+1)-1\}t}}{t^p}dt \sim \frac{e^{\{p(\nu+1)-1\}\chi}}{(p(\nu+1)-1)\chi^p}dt \quad \text{as} \quad \chi \to +\infty.
$$

Therefore, for all r sufficiently close to 1

$$
J_p(r) \leq C\frac{1}{(1-r)^{p(\nu+1)-1}\left(\ln\frac{1}{1-r}\right)^p}.
$$

Therefore,

$$
\begin{aligned}
I \leq & C\int_{\frac{9}{10}}^{1} \frac{dr}{(1-r)^{q(\nu+1-\frac{1}{p}-\alpha)+1}\left(\ln\frac{1}{1-r}\right)^q} \\
= & C\int_0^{\frac{1}{10}} \frac{dx}{x\left(\ln\frac{1}{x}\right)^q},
\end{aligned}
$$

which is convergent. But $(\Lambda^\gamma f_1(z))$ makes no sense at any point $z \in \mathbf{D}$ because

$$
(\Lambda^\gamma f_1(z)) = \frac{\gamma+1}{1-z}\int_0^1 \frac{(1-t)^\gamma dt}{t\ln\frac{e}{t(1-z)}} = \infty,
$$

and consequently $\Lambda^\gamma f_1 \notin H^{p,q}_{\alpha,\nu}$.

For some special choices of the parameters following results are obtained as corollaries to Theorem 1.

Corollary 1 *When* $\alpha = 0, q = \infty$, *we have following two results*

(a) *For* $\nu = 0$, Λ^γ *maps* H^p *to* H^p *if* $1 < p \leq \infty$.
(b) *For* $\nu \geq 1$, Λ^γ *maps* $W^p_\nu = H^p_{0,\nu}$ *to* W^p_ν *if* $\nu > \frac{1}{p} - 1$.

In particular Λ^γ *maps ordinary Lipschitz space into itself.*

For $\gamma = 0$, Λ maps Hardy and ordinary Lipschitz spaces into itself [6, 8].

Corollary 2 *If $\nu = 0$, then*

(a) *Λ^γ maps $H^{p,q}_\alpha$ to itself if $\alpha < 1$ and $p > \frac{1}{1-\alpha}$.*
(b) *If $p = q = \infty$, then Λ^γ maps H^∞_α to itself if $0 \le \alpha < 1$.*
(c) *If $q = \infty$, then Λ^γ maps $h^p_\alpha := h^p_{\alpha,0}$ to itself if $0 < \alpha < 1 - \frac{1}{p}$.*
For $\gamma = 0$, we get that Λ maps $H^{p,q}_\alpha$, H^∞_α and h^p_α spaces into itself [6].

Corollary 3 *If $\nu = 1$, then Λ^γ maps $\mathcal{D}^p_\beta$ to itself if $p > 1 + \frac{\beta}{2}$.*

In particular, Λ^γ maps $\mathcal{D}^1_\beta$ to itself if $-1 < \beta < 0$ and Λ^γ maps $\mathcal{D}^2_\beta$ to itself if and only if $\beta < 2$.

For $\gamma = 0$, we get that Λ maps $\mathcal{D}^p_\beta$ into itself [6].

Corollary 4 *For $1 \le p \le \infty, 0 < q \le \infty$ and $\beta > 0$, Λ^γ maps $\mathfrak{B}^{p,q}_\beta = \Lambda^{p,q}_\beta$ into itself. Same result is true for little space $\mathfrak{b}^p_\beta = \lambda^p_\beta$.*

In particular if $p = q = \infty$, then Λ^γ maps the Lipschitz space $\Lambda_\beta = \Lambda^\infty_\beta$ into itself.

For $\gamma = 0$, Λ maps $\mathfrak{B}^{p,q}_\beta$, $\mathfrak{b}^p_\beta$ and Λ_β spaces into itself [6].

Corollary 5 *For $1 \le p \le \infty$ and $\alpha > 0$, Λ^γ maps p-Bloch type spaces $\mathfrak{B}^p_\alpha = H^p_{\alpha,1}$ into itself if $\alpha + \frac{1}{p} < 2$.*

Same result is true for little space $\mathfrak{b}^p_\alpha = h^p_{\alpha,1}$.

In particular if $p = \infty$, then Λ^γ maps the α-Bloch space $\mathfrak{B}_\alpha$ into itself if $0 < \alpha < 2$.

For $\gamma = 0$, Λ maps $\mathfrak{B}^p_\alpha$, $\mathfrak{b}^p_\alpha$ and $\mathfrak{B}_\alpha$ spaces into itself [3, 6, 10].

Corollary 6 *Λ^γ maps Hardy-Bloch spaces $\mathfrak{B}^{p,q}$ into itself if $p > 1$.*

In particular, if $2 < p = q < \infty$, Λ^γ maps $\mathfrak{B}^{p,p}$ spaces into H^p.

For $\gamma = 0$, Λ maps Hardy-Bloch and $\mathfrak{B}^{p,p}$ spaces into itself [5, 6].

Corollary 7 *For $p = q = \infty$ and $\beta = 1$, Λ^γ maps Zygmund space spaces Λ_* into itself. Same is true for λ_*.*

For $\gamma = 0$, Λ maps Zygmund spaces into itself [6].

Corollary 8 *Λ^γ act as a bounded operator A^p_β into itself if $p > \beta + 2$.*

For $\gamma = 0$, A^p_β maps into itself [9].

Disclosure of Potential Conflicts of Interest

We confirm that we have no potential conflict of interest.

References

1. Aulaskari R, Zhao R (1999) Boundedness and compactness properties of the Libera transform. Complex analysis and differential equations (Uppsala, 1997), vol 64 of Acta Universitatis Upsaliensis. Skrifter rörande Uppsala Universitet. C. Organisation och Historia. Uppsala University, Uppsala, Sweden, pp 69–80
2. Avetisyan K, Stević S (2009) The generalized Libera transform is bounded on the Besov mixed-norm, BMOA and VMOA spaces on the unit disc. Appl Math Comput 213(2):304–311
3. Danikas N, Ruscheweyh S, Siskakis AG (1994) Metrical and topological properties of a generalized Libera transform. Arch Math 63(6):517–524
4. Libera RJ (1965) Some classes of regular univalent functions. Proc Amer Math Soc 16:755–758
5. Nowak M, Pavlović M (2010) On the Libera operator. J Math Anal Appl 370(2):588–599
6. Pavlović M, (2014) Definition and properties of the Libera operator on mixed norm spaces. Sci World J, Article ID 590656
7. Ruscheweyh S, Siskakis AG (2008) Corrigendum to: mertical and toplogical properties of a generalized Libera transform. Arch Math 91(3):254–255
8. Siskakis AG (1987) Composition semigroups and the Cesàro operator on H^p. J Lond Math Soc 36(2):153–164
9. Siskakis AG (1987) Semigroups of composition operators in Bergman spaces. Bull Aust Math Soc 35(2):397–406
10. Xiao J (1997) Cesàro-type operators on Hardy, BMOA and Bloch spaces. Archiv der Mathematik 68(5):398–406

Improving the Monitoring of Mortality Surveillance by Analyzing Regression Approach

Sarada Ghosh and G. P. Samanta

Abstract Nowadays, mortality due to corona virus (COVID-19) is uttermost important for surveillance and monitoring purpose. It is an illness caused by a virus which spreads from person to person throughout the World. The delay in reporting for such case purpose between the time of occurrence of the mortality and them being registered is unavoidable. This work proposes an approach for estimating the unknown surveillance in a timely manner, based on weekly reporting on mortality. It is also performed that the demographic characteristics such as gender, age, and various diseases responsible for death are significantly associated with the probability of the enlisting report of USA and also predict the influence of the variables. The current work can generate timely and accurate estimations on the weekly count of mortality surveillance only 1-week lag. The proposed technique has executed average measurements with threefold nested cross validation and among all splitting methods the best performance has high coefficients of determination, i.e., 0.99 with MAE (mean absolute error) 63.78 and RMSE (root mean squared error) 93.20. Simulation is based on R-software.

Keywords Logistic regression · Elastic net regression · Odds ratio · Coefficient of determination · Mean absolute error · Root mean squared error

1 Introduction

The epidemic of severe acute respiratory syndrome (SARS) which is a viral respiratory disease has showed how spontaneously new infectious diseases can spread out in worldwide. In 17 countries, an outbreak of SARS in southern China caused an

S. Ghosh
Department of Statistics, Gurudas college, 1/1, Suren Sarkar Rd, Jewish Graveyard, Phool Bagan, Narkeldanga, Kolkata 700054, India

G. P. Samanta (✉)
Department of Mathematics, Indian Institute of Engineering Science and Technology, Shibpur, Howrah 711103, India
e-mail: g_p_samanta@yahoo.co.uk; gpsamanta@math.iiests.ac.in

© The Author(s), under exclusive license to Springer Nature Singapore Pte Ltd. 2022

D. Giri et al. (eds.), *Proceedings of the Seventh International Conference on Mathematics and Computing*, Advances in Intelligent Systems and Computing 1412,
https://doi.org/10.1007/978-981-16-6890-6_62

incidence 8,098 cases which is relatively low but resulting in 774 deaths (i.e., the fatality rate 9.6%) which is relatively high with epidemics occurring in mainland China and Hong Kong between November 2002 and July 2003 [15]. Later, in 2017, Chinese scientists traced the virus through the intermediary of civets to cave-dwelling horseshoe bats in Yunnan Province. No cases of SARS have been reported worldwide since 2004. However, the related virus SARS corona virus (namely, SARS-CoV-2) is the cause of the ongoing 2019–2020 corona virus pandemic. Sometimes the resultant spreads the point of panic which is so strong as to dominate or prevent reason and logical thinking. Coupled with the economic costs resulting from restriction of movement placed on the affected countries, the epidemic highlights the necessity for a rapid international response to disease control. More recently, the outbreak of novel corona virus has again reinforced the potential for pandemic spread of newly emerging or evolving infectious agents. The primary symptoms are flu like fever, muscle pain, lethargy symptoms, cough, sore throat, and other non-specific symptoms. The only symptom common to all patients appears to be a fever above 100 °F or 38 °C. It may eventually lead to shortness of breath and pneumonia.

These viruses vary significantly in risk factor. Some of them can be able to kill more than 30% of those infected such as MERS-CoV and some of them are relatively harmless (such as the common cold). Corona viruses cause colds with major symptoms, such as fever, and sore throat from swollen adenoids, occurring primarily in the winter and early spring seasons. It can cause either direct viral pneumonia or a secondary bacterial pneumonia and same as for the bronchitis, i.e., either direct viral bronchitis or a secondary bacterial bronchitis. The much publicized human corona virus discovered in 2003, SARS-CoV, which caused severe acute respiratory syndrome (SARS) had a unique pathogenesis due to this virus caused both upper and lower respiratory tract infections.

In December 2019, a pneumonia outbreak was first pointed out at Wuhan in China [8]. On December 31, 2019, the outbreak has been traced to a novel strain of corona virus, giving the interim name 2019-nCoV by WHO but later it is renamed as SARS-CoV-2 by the International Committee on Taxonomy of Viruses. It is also known as COVID-19. Some researchers have suggested that the Huanan Seafood Wholesale Market may not be the original source of viral transmission to humans [17]. The Wuhan strain has been identified as a new strain of beta corona virus from group 2B with approximately 70% genetic similarity to the SARS-CoV. The virus has a 96% similarity to a bat corona virus. For this reason, it is widely suspected to originate from bats as well. The pandemic has resulted in serious travel restrictions.

The virus primarily spreads among people via exhaled respiratory droplets such as coughing or sneezing. The time between exposure and symptom onset is around 5 days, but may range from 2 to 14 days. Among those who died from the disease, the time from development of symptoms to death is between 6 and 41 days, with a median of 14 days. As of March 20, 2020 more than 10, 200 deaths have been attributed to COVID-19. Most of the people who died were elderly and about 80% of deaths were in those over 60 and 75% had pre-existing health conditions including cardiovascular diseases and diabetes. The study goes on that with a basic reproductive number of two, the majority of the population will be infected, except when public health interventions are set in force among other measures [17].

In Sect. 2, the materials and information generated dataset used in analytical framework of time-related mortality surveillance are presented. The analytical part is described in Sect. 3, including discussions on different views of the proposed models. In Sect. 4, proposed regression models in the literature are described based on underlying dataset and compared regarding model assumptions, applicability, and interpretation of obtained results. Various techniques of the basic models are mentioned and illustrated with real-life examples. Finally, the presented methods as well as the findings from the simulation study and the data analyses are discussed in last section and also provide general discussion of this work.

2 Materials

The dataset is collected from COVID-19 surveillance center named National Center for Health Statistics (NCHS) which provided centers for disease control and prevention. This center collects, analyzes, and disseminates information on the health of the nation. In response to the COVID-19 pandemic, NCHS is providing the most recent data available on deaths, access to health care and telemedicine from the vital statistics system, the NCHS Research and Development Survey, and through a partnership with the U.S. Census Bureau.

3 Analytical Parts

There are two main parts in this work: (i) the modeling of the consisting pattern and (ii) the prediction of mortality due to COVID-19. In the first part, the consisting pattern is obtained by using logistics regression (LR) and the second part contains an ordinary least squares (OLS) method with Elastic Net regression which provides weekly estimations of mortality due to COVID-19.

3.1 Consisting Pattern Using LR Technique

The consisting pattern is obtained using LR technique. This underlying dataset contains some missing values. As the number of missing observations is comparatively very low than the total number of observations, so, we have decided to exclude those missing observations. Thus, less imputation efforts are needed. Due to the sparsity of reporting on some combinations of underlying characters, the levels of ordered categorical variables such as age group and others are deaths due to COVID-19, pneumonia deaths (excluding pneumonia deaths involving influenza); pneumonia deaths involving COVID-19; influenza deaths; and deaths involving pneumonia, influenza, and COVID-19 by week ending date are aggregated in this work. The dataset includes the weekly provisional count of deaths in the United States due to COVID-19, deaths from all causes which are already mentioned. But we have considered only New York state excluding New York City and only New York City where it is being hit the hardest by the novel corona virus spreading across the U.S. More than 5,600 deaths in the city account for roughly one-third of all confirmed U.S. deaths from COVID-19, the illness caused by the corona virus reported by WHO (source: [21]). The binary LR technique is applied here to obtain the consisting pattern [22]. It is the uttermost important for ordered categorical data purpose [5, 16, 18]. Table 1 shows all variables, except the consisting status which is the responsible variable in this work used in LR technique. P-value (by Hosmer–Lemeshow test ≥ 0.05) points out the goodness-of-fit test of this model. The goodness of fit of a statistical model describes how well it fits a set of observations [10, 11, 13]. Multicollinearity among the explanatory variables is also obtained with help of variance inflation factors (VIFs).

Here we have considered Z as dependent variable and X as independent variable, let $\pi(x) = P(Z = 1|X = x)$. The LR model is given as follows [1–3]:

$$\pi(x) = \frac{\exp(a + bx)}{1 + \exp(a + bx)} \tag{1}$$

Similarly, the log odds (also known as logit) has a relationship which is linear such as

$$\text{logit}[\pi(x)] = \log \frac{\pi(x)}{1 - \pi(x)} = a + bx, \tag{2}$$

which equates the "logit link" function to the linear predictor [10–12]. Here, a is the intercept and we seek to investigate b in Eq. 2 since b's sign determines whether $\pi(x)$ is decreasing or increasing when x increases. If $b \to 0$, the curve flattens to a horizontal straight line. Z is independent of X, whenever $b = 0$. Logistic density is symmetric since $\pi(x)$ approaches 1 at the same rate that it approaches 0 [2, 4]. After taking exponential function in Eq. 2 (at both sides), the odds gives a basic interpretation of $|b|$ (where $|b|$ denotes the magnitude of b). For each 1-unit increase in x, the odds increase multiplicatively by e^b.

3.2 Weekly Estimation by Using the EN Model

The weekly mortality due to COVID-19 is estimated by obtaining the Elastic Net (EN) regression. In Table 1, the weekly number of mortality cases is shown with group of some variables. EN model is a penalized linear regression model which includes both the L_1 and L_2 penalties during training.

EN regression uses the penalties from both the Lasso and Ridge techniques to regularize regression models. The technique combines both the Lasso and Ridge

Table 1 List of variables with brief description

Name of variable	Description about variable
Consisted	Binary indicator: the person belongs to New York City, 1; or belongs to New York State other than New York, 0
Age group	Categorical pattern, it has been categorized into four groups: ≤24, 25–54, 55–84, ≥85
Gender	Male, 0 and Female, 1
Disease	Categorical pattern: COVID-19, Pneumonia, COVID with Pneumonia, Influenza, COVID with Pneumonia and Influenza

regression methods by learning from their shortcomings to improve on the regularization of statistical models [14]. It is as follows:

$$\hat{z}(\omega, x) = \omega_0 + \omega_1 x_1 + \cdots + \omega_\rho x_\rho \tag{3}$$

together with for minimization of the objective function as follows:

$$\min_{\omega} \frac{1}{2p} \|X\omega - y\|_2^2 + \theta\varrho\|\omega\|_1 + \frac{\theta(1-\varrho)}{2}\|\omega\|_2^2, \tag{4}$$

where $\hat{z}$ be the weekly estimated number of deaths due to COVID-19 and COVID-related purpose and x_1 be the inflated weekly number of consisted cases and $[x_2, x_3, \ldots, x_\rho]$ be the characteristics of all consisted case purpose on a week. In this work, $\|\omega\|_1$ and $\|\omega\|_2$ denote the L_1-norm and L_2-norm, respectively, of the parameter vector. Lastly, θ and L_1 ratio (ϱ) be the hyper-parameters for optimization.

The weekly number of mortality cases is used in this work as responsible variable. For obtaining the inflated weekly number of consisted cases (x_1), at first using LR technique we have estimated the probability of consisting mortality for each case. The inflation depends on the observed cases and the inflation number will still be underestimating the weekly mortality purpose. It cannot be inflated if on a certain week, any specific combination of observations is not perceived. So, other explanatory variables (besides the inflated numbers) are also included into the regression model at the week level for improving on the performance of the weekly estimation. For obtaining the explanatory variables at the week level, i.e., $(x_2, \ldots, x_\rho)$, firstly, transform the variables (shown in Table 1) by one-hot encoding which helps to convert a p-level factor into p variables and next for prevailing purpose, the weekly representations of the consisted mortality surveillance are investigated by summing up the observations on the same week.

Moreover, the linear dependency among the underlying variables (derived from the same factor) has been introduced by the method of one-hot encoding. For multicollinearity purpose, the least squares estimates (LSEs) become highly sensitive to

Table 2 Depiction of consisted and non-consisted of mortality dataset

Description	Non-consisted (N = 44367)	Consisted (N = 60111)
Age		
≤24	59	115
25–54	2844	6371
55–84	24192	37588
≥85	17272	16037
Gender		
Male	22334	32855
Female	22303	27256
Disease		
COVID-19	9166	15107
Pneumonia	11448	11506
COVID + Pneumonia	5684	7999
Influenza	194	943
COVID + Pneumonia + Influenza	17875	24556

random errors in case of observed response that leads to high variance. So, the EN model combines both Ridge and Lasso penalty functions [23]. Ridge model provides an alternative to this (about the impact of the variables that are excluded from the model on the response variable) hard selection of variables which can split them into "included in" and "excluded from" the model. Instead, it penalizes the coefficients to shrink them toward zero. Not exactly zero, as that would mean exclusion from the model, but in the direction of zero, which can be viewed as decreasing complexity of model in a continuous way, while keeping all variables in the model. In case of Ridge model, it imposes a penalty on the size of the coefficients to the objective function as follows:

$$\min_{\omega} \frac{1}{2p} \|X\omega - z\|_2^2 + \theta \|\omega\|_2^2, \tag{5}$$

where θ denotes the parameter which can control the shrinkage's amount. Next, L_2 norm of the parameter vector is denoted by $\|\omega\|_2$. The shrinkage's amount is increasing whenever the parameter (θ) is increasing. If shrinking coefficients from correlated variables proportionally tend to zero, the coefficients become more robust to collinearity. Lasso, a penalized least squares technique, imposes an L_1 penalty on the objective function to minimize such as [19]

$$\min_{\omega} \frac{1}{2p} \|X\omega - z\|_2^2 + \theta \|\omega\|_1, \tag{6}$$

where θ is a constant, the L_1-norm of the parameter vector is denoted by $\|\omega\|_1$. Lasso or least absolute shrinkage and selection operator is a shrinkage method of continuous process and by imposing the L_1-penalty on the absolute value of the coefficients it can select automatic variable simultaneously. For high values of θ, many coefficients are exactly zeroed under LASSO, which is never the case in Ridge regression. But it has some limitations after applying it to current situations which are as follows:

1. It has tendency to choose only one variable randomly from the group whenever the group of correlated variables are present.

2. Under simulations, the Ridge regression dominates the Lasso whenever $p > \rho$ and if there exist high correlations between the predictors.

So, Elastic Net model is a convex combination of both: (i) Ridge and (ii) Lasso penalty shown in Eq. 4.

For optimization purpose for a model fitting, distinct window sizes are explored. In this work, a threefold nested cross validation is executed. Window sizes are as follows: about 3 months, 6 months, and 9 months which are obtained with various starting points. For evaluation purpose of test set in the proposed model, 3 months following are considered as the training set. Such type of cross validation has an advantage, and it can provide an almost unbiased measurement of the true error [20]. Coefficient of determination (denoted as R^2), mean absolute error (MAE), and root mean squared error (RMSE) assess goodness-of-fit test [6] among underling models which are shown in Table 3. The model with the higher R^2 but lower RMSE and MAE is chosen in this work.

4 Result

4.1 Consisting Pattern

Totally 106045 deaths are observed according to NCHS in New York and New York City from February 1, 2020 to November 7, 2020. Nearly about 1.5% of cases in the dataset have unknown gender or ages of the surveillance of mortality. The profile of consisting and non-consisting purpose for mortality purpose related to COVID-19 is shown in Table 2. The VIF providing from dataset points out that multicollinearity is not present in this work and there is no sign of outlier by Bonferroni test. Table 3 provides the results of the LR techniques reported in NCHS. The results provide demographic characteristics of gender and age, the mortality surveillance for various reasons.

The NCHS provision is biased toward the aged $\geq$85 since OR = 3.01 (with 95% C.I. 0.90 – 9.97) and against those aged $\leq$24 which are frequently provided (where OR = 0.33 together with 95% C.I 0.1 – 1.10) after comparing to between 25 and 54 years. It also provides that for 55–84 years, the odds ratio be 1.02 (95% C.I. 0.33 – 3.01) which indicates a slightly biased for age purpose. Furthermore,

Table 3 Odds ratios by LR technique on mortality surveillance logistic

Description	OR	95% C.I.	P-Value
Age (ref = 25–54)			
≤24	0.33	(0.10, 1.10)	<0.05
55–84	1.02	(0.33, 3.01)	<0.05
≥85	3.01	(0.90, 9.97)	<0.05
Gender (ref = Female)			
Female	0.7	(0.68, 0.71)	<0.01
Disease (ref = Pneumonia)			
COVID-19	0.78	(0.74, 0.83)	<0.05
COVID + Pneumonia	1.37	(1.29, 1.45)	<0.05
Influenza	0.27	(0.25, 0.29)	<0.05
COVID + Pneumonia + Influenza	1.43	(1.35, 1.52)	<0.05

Table 4 Fit statistics with sliding windows

	1/2/2020–25/4/2020	26/4/2020–25/7/2020	26/7/2020–7/11/2020	R^2	MAE	RMSE
Split 1	Training	Test		0.998	135.16	193.11
Split 2	Training		Test	0.990	135.20	193.10
Split 3		Training	Test	0.999	63.78	93.20

COVID-19 together with pneumonia (OR = 1.43 with 95% C.I. 1.35 – 1.52) and influenza and COVID-19 with pneumonia (OR = 1.37 with 95% C.I. 1.29 – 1.45) are more consisted than other mortality surveillance. The estimation procedures are also made for each combination of age and mortality surveillance, where rest of the variables are considered as explanatory. From the current work, it is revealed that there is an under-representation of youth mortality and an over-representation of aged-person mortality.

4.2 *Weekly Estimation*

In Table 4, the splitting of test and training sets is depicted. The underlying splits (shown in Table 4) correspond to the threefold nested cross validation.

For such nested cross validation, in the underlying method let the average of RMSE be 159.80, MAE be 111.38, and R^2 be 0.995. Among all, the best estimate in most recent year is from May 2, 2020 to November 7, 2020 which rely on the 3-month training set, and the results provide the best performance where R^2 be 0.998, RMSE be 93.20, and MAE be 63.78.

5 Concluding Remarks

In this work, we have demonstrated that monitoring the emergence of COVID-19 and guiding public health response will also require accurate and timely death reporting. The purpose of this report is to provide guidance to death certifiers on proper cause-of-death certification for cases where confirmed or suspected COVID-19 infection resulted in death. It is investigated that the performance from the estimation is better whenever the current information is used that provides the continuous changing of enlisting pattern of mortality in NCHS. It also helps to assess accurately the effects of this pandemic and contributes positive direction about public health response. In public health emergencies, mortality surveillance provides crucial information about population-level disease progression, as well as guides the development of public health interventions and assessment of their impact. Monitoring and analysis of mortality data allow dissemination of critical information to the public and key stakeholder. One of the most important methods of mortality surveillance is through monitoring causes of death. Appropriate completion of death certificates yields accurate and reliable data for use in epidemiological analyses and public health reporting. When determining whether COVID-19 is playing an important role in the cause of death, follow the CDC (centers for disease control and prevention) [7] clinical criteria for evaluating a person under investigation for COVID-19.

This work has practical interpretation which can be used to places outside New York as well, where providing dataset is not so small or in case of delay reporting purpose in any state of USA. Besides, it can be also used in another country purpose where required variables are present. Since various countries have various reporting patterns of mortality surveillance, the LR method customized would be developed for such local context. If the proportion of mortality surveillance varies, the number of observations should not be very small. For this purpose, we have to extend the period of observed cases by considering several years. In this work, there are four types of age groups, two types of genders, and five types of mortality surveillance are considered which provide the deep impression about consisted related purpose. Not only that, but also suggest COVID-19-related and COVID-19-unrelated fact of mortality which depicts the critical conditions to ongoing public health surveillance and response. Furthermore, the works conclude the under-representation of youth mortality and an over-representation of aged-person mortality.

Nevertheless, several limitations are present in this work. Other mortality surveillance is not considered except COVID-19, COVID-19 related, pneumonia, and influenza-related flu and do not examine health condition of any person such as diabetes, heart disease [9], kidney disease, lung infections, etc. Therefore, a low R^2 might be the result of such a behavior of model. In future work, another important independent variable with theoretical bases such as patients having diabetes or not, having chronic heart disease or not, having high pressure or not can be included which allow precise estimation in statistical field.

One major task remains the transfer of available methods for the analysis of survival data to the medical community in order to avoid misinterpretation of study data, possibly leading to erroneous therapy decisions or risk stratifications, due to inadequate application of statistical methods in the presence of such pandemic situation. It is recommended that (i) interrupt human-to-human transmission including reducing secondary infections among close contacts; (ii) for healthcare workers, prevent transmission amplification events; and (iii) prevent further international spread. In this context, it is mentioned that WHO is providing guidance on early investigations, which are necessary in an outbreak of a new virus. The data collected from the protocols can be used to refine recommendations for surveillance and case definitions, to characterize the key epidemiological transmission features of COVID-19. It also helps to understand spread, severity, spectrum of disease, impact on the community, and to inform operational models for implementation of countermeasures such as case isolation, contact tracing, and isolation.

References

1. Agresti A (2002) Categorical data analysis, 2nd edn. Wiley, New York
2. Agresti A (2007) An introduction to categorical data analysis, 2nd edn. Wiley, New York
3. Agresti A (2010) The analysis of ordinal categorical data, 2nd edn. Wiley, New York
4. Agresti A (2010) Modeling ordinal categorical data. A link manual at www.stat.ufl.edu/~aa/ordinal/ord.html
5. Anderson JA (1984) Regression and ordered categorical variables. J R Stat Soc Ser B 46:1–30
6. Blizzard L, Hosmer DW (2006) Parameter estimation and goodness-of-fit in log binomial regression. Biom J 48:5–22
7. CDC: Centers for disease control and prevention. National Center for Health Statistics. https://www.cdc.gov
8. The Epidemiological Characteristics of an Outbreak of 2019 Novel Coronavirus Diseases (COVID-19) - China CDC (2020)
9. Chen Y, Gong X, Wang L, Guo J (2020) Effects of hypertension, diabetes and coronary heart disease on COVID-19 diseases severity: a systematic review and meta-analysis, medRxiv. BMJ. https://doi.org/10.1101/2020.03.25.20043133
10. Ghosh S, Samanta GP (2019) Fitting cumulative logit models for ordinal response variables in retail trends and predictions. Int J Ecol Econ Stat 20(1):32–49
11. Ghosh S, Samanta GP (2019) Statistical modeling for cancer mortality. Letters in Biomathematics. Taylor and Francis, vol 6, Issue 2. https://doi.org/10.1080/23737867.2019.1581104
12. Ghosh S, Samanta GP (2019) Model justification and stratification for confounding of chlamydia trachomatis disease. Letters in biomathematics. Taylor and Francis, vol 6, Issue 2. https://doi.org/10.1080/23737867.2019.1654418
13. Ghosh S, Samanta GP, Mubayi A (2021) Comparison of regression approaches for analyzing survival data in the presence of competing risks: an application to COVID-19. Let Biomath 8(1):29–47. https://lettersinbiomath.journals.publicknowledgeproject.org/index.php/lib/article/view/305
14. Hoerl AE, Kennard RW (2006) Ridge regression. In: Encyclopedia of statistical sciences. Wiley, Hoboken, NJ. https://onlinelibrary.wiley.com/doi/abs/10.1002/0471667196.ess2280
15. Lam WK, Zhong NS, Tan WC (2003) Overview on SARS in Asia and the World. Respirology 8:S2–S5
16. McCullagh P (1980) Regression models for ordinal data (with discussion). J R Stat Soc Ser B 42:109–142
17. Park SE (2020) Epidemiology, virology, and clinical features of severe acute respiratory syndrome -coronavirus-2 (SARS-CoV-2; Coronavirus Disease-19). Clin Exp Pediatr. https://doi.org/10.3345/cep.2020.00493

18. Powers DA, Xie Y (2000) Statistical models for categorical data analysis. Academic Press, SanDiego, California
19. Tibshirani R (1996) Regression shrinkage and selection via the lasso. J R Stat Soc Ser B Stat Methodol 58(1):267–288
20. Varma S, Simon R (2006) Bias in error estimation when using cross-validation for model selection. BMC Bioinform 7(1):91. https://doi.org/10.1186/1471-2105-7-91
21. The Washington post, Democracy Dies in Darkness (2020) Confirmed U.S. covid-19 death toll reaches 20,000, highest in the world. https://www.washingtonpost.com/world/2020/04/11/coronavirus-latest-news/
22. Williamson T, Eliasziw M, Fick GH (2013) Log-binomial models: exploring failed convergence Emerging Themes in Epidemiology
23. Zou H, Hastie T (2006) Regularization and variable selection via the elastic net. J R Stat Soc Ser B Stat Methodol 67(2):301–320

Motion Forecast of Dynamical Systems

Yuri Menshikov

Abstract In paper the problem of forecast dynamical systems motion is studied in a deterministic statement. The problem of synthesis of an adequate mathematical description of dynamical systems which is the best for the purpose of predicting the behavior of this system is considered. In general, this problem is reduced to solve number of integral equations of the first kind of Volterra (ill-posed problems) into which the change of the mathematical model parameters was considered. The special statement of problem is suggested.

Keywords Dynamical systems · Forecast motion · Deterministic statement · Adequate mathematical description

1 Introduction

Methods of forecasting of motion of real physical processses are expanding the field of applications constantly. Analysis of time series data is very common in many areas of practical life such as medical and health related applications, biometric or process industry, financial or economic analysis or weather prediction.

According to Lee and Lee [1] there are three different techniques for reliability prediction: statistical methods, similarity analyses using failure rate databases, and physic-of-failure methods. Goel and Graves classify the techniques into two groups: empirical-based models and physic-of-failure models. Many papers deal with extensions of analytic models that are based on Weibull exponential, Marshall-Olkin extended uniform and log-normal distributions. In some model new parameters are introduced that refer to the circumstances of usage production etc.

It could be said that properly modeled environmental stress is fundamental of varied environment oriented reliability prediction. During the past few years, besides adding new parameters or factors to existing models, new approaches that are founded on soft computing methods, such as fuzzy logic or artificial neural networks have been

Y. Menshikov (✉)
Dnepro University, Dnepro, Ukraine
e-mail: Menshikov2003@list.ru

© The Author(s), under exclusive license to Springer Nature Singapore Pte Ltd. 2022
D. Giri et al. (eds.), *Proceedings of the Seventh International Conference on Mathematics and Computing*, Advances in Intelligent Systems and Computing 1412,
https://doi.org/10.1007/978-981-16-6890-6_63

discovered. Nowadays, the mentioned techniques are widely used as mathematical tools that offer an alternative way to deal with complex systems. Recently, artificial neural networks have been widely used not only in the electronic industry, but also in several other industries to predict failure rates. In work [2, 3] presented a soft computing-based technique for acquiring a proper maintenance plan for individual parts in a complex system. They used a combination of neural network and evolutionary algorithm to discover the relationship between individual parts of a complex system to optimize its reliability. The developed forecasting method is founded on measuring fuzzy semi. In this sense this method is a hybrid one that combines times series, analytic curve fitting and soft computing techniques. The introduced method was tested on real life data and its goodness was compared to widely used forecasting techniques, such as the moving average, the exponential-smoothing, the linear regression, and the autoregressive integrated moving average (ARIMA) methods. Bivariate time series of counts arise in many application areas such as finance. epidemiology, criminology, sport studies and among others [4]. In these studies, the time series of counts are often subject to some time-varying explanatory variables that induce non-stationary moments. As a result, the bivariate discrete time series process becomes more complicated to model. Besides in literature most of the research in this field has laid focus on the modeling of stationary bivariate time series of counts due to its parsimonious nature. Box and Jenkins introduced new time series models derived from linear filter which usually are called ARIMA models. The fundamental aim of time series analysis is to understand the underlying mechanism that generates the observed data and, in tum, to forecast future values of the series. Given the unknowns that affect the observed values in time series, it is natural to suppose that the generating mechanism is probabilistic and to model time series as stochastic processes. ARIMA models com-bine autoregressive (AR), and moving average (MA) part. AR is a linear combination of previous values. Traditional forecasting methods are not able to handle forecasting problems in which initial data are given deterministic values [5].

In this work the problems of forecast in deterministic statement was considered.

2 Problem Statement

The practical significance of the results of mathematical modeling or simulation of the motion of dynamical systems depends on the degree of coincidence of the results of mathematical modeling of the selected mathematical description of the real process with experimental data [6]. If the coincidence of the results of mathematical modeling with experiment is bad, then further use of these results is problematic. An important property in this connection is the adequacy of the constructed mathematical description of the physical process being studied [6, 7].

2.1 Some Definitions

First of all, we introduce into consideration some useful definitions. We will call as *the mathematical model* of the physical process the set of equations or relations between the characteristics of the mathematical model that reflect the physical laws of the process under investigation. *External actions* (*loads*) will be called functions that are present in the mathematical model and which independent from motion of system. Initial and various boundary conditions for the process under study will be named as *additional conditions*. The totality of the mathematical model of the physical process, external actions and additional conditions will be called as *mathematical description* of the physical process [6, 7]. The process of solving the equations of a mathematical model of a physical process with allowance for external actions and additional conditions will be called *mathematical modeling or simulation*. Other definitions can be meeting in the scientific literature.

We will study problems associated with the synthesis of an adequate mathematical description and its further use on the example of a dynamical system whose motion is described by a system of ordinary differential equations.

2.2 Mathematical Descriptions Adequacy of Quantitative Type

Let the physical process be characterized in the general case by a certain number of variables (state variables) $\tilde{x}_1(t), \tilde{x}_2(t), ..., \tilde{x}_n(t)$ depending on an infinite number of initial parameters of the process. The choice of the characteristics of the physical process is determined by the ultimate goals of the research.

For simplicity of presentation, we will assume that the dynamical system is linear and described by a system of ordinary differential equations with constant coefficients

$$\dot{x} = Ax + Bz, \tag{1}$$

where $x_1(t), x_2(t), ..., x_n(t)$ is a vector function of state variables; $z(t) = (z_1(t), z_2(t), ..., z_l(t))^T$ is a vector function of external loads; A, B are matrices with constant coefficients of the corresponding dimension; $(.)^T$ is a sign of transposition.

Let us denote by $x_1^0, x_2^0, ..., x_n^0$ the initial conditions of the state variables $x(t) = (x_1(t), x_2(t), ..., x_n(t))^T$ at the initial moment of time.

If the original system is nonlinear, then one should go to the equations in variations. At the same time, the basic algorithms are preserved.

The conditions for constructing a linear mathematical description of the physical process imply limitations on the physical processes under study using such a description. The practical significance of the results of mathematical modeling depends on the degree of coincidence of the results of mathematical modeling for the chosen mathematical description of the real process with the experimental data [6, 7]. An important property in this connection is the adequacy of the constructed mathematical description of the physical process being studied [6, 7]. However, this definition of adequacy is fairly general and requires refinement and additions. In addition, the question of the possibility of using adequate mathematical descriptions to fulfill the forecast of the behavior of the physical process remains open.

An adequate mathematical description of a physical process of quantitative type, with respect to all variables $x_1(t), x_2(t), \ldots., x_n(t)$ of quantitative type, will be considered the mathematical description for which the results of the mathematical simulation of the variables $x_1(t), x_2(t), \ldots., x_n(t)$ conform with the results of experimental measurements $\tilde{x}_1(t), \tilde{x}_2(t), \ldots., \tilde{x}_n(t)$ of the characteristic $x_1(t), x_2(t), \ldots., x_n(t)$, with the accuracy of the experiment $\delta_1, \delta_2, \ldots., \delta_n$ in the selected functional metric on the given segment $t \in [a, b]$

$$\|x_i(t) - \tilde{x}_t(t)\|_{X[a,b]} \leq \delta_i, 1 \leq i \leq n, \quad (2)$$

where$\|\cdot\|_{X[a,b]}$ is the norm in functional space X. It is assumed that the error of the mathematical modelling is zero.

The refined definition of the adequacy of a mathematical description of quantitative type is based on experimental measurements, i.e., on objective data.

In practice, the measurement of the characteristics of state variables is limited by only one or two components. We formulate a refined definition of the adequacy of mathematical description for the case of a single variable $\tilde{x}_k(t)$. An *adequate mathematical description of a physical process of the type (1) with respect to the variable* $x_k(t)$ ($ALMD_{x_k}$) will be called the mathematical description for which the results of mathematical modeling of a variable $x_k(t)$ coincide with the results of experimental measurements $\tilde{x}_k(t)$ of the characteristic $x_k(t)$ with the accuracy of the experiment δ_k:

$$\Delta_{\delta k} = \|\tilde{x}_k(t) - x_k(t)\|_X \leq \delta k, \quad (3)$$

where$\|\cdot\|_X$ there is a priori the chosen norm in the functional space X, δ_k is const, $\delta_k > 0$.

If $\Delta_{x_k} > \delta_k$ then the property of adequacy of the mathematical description will be no objective.

It can be shown that there are infinitely many adequate mathematical descriptions of the variable $x_k(t)$ of the system (1). If the system (1) does not satisfy the condition of adequacy over the variable, $x_k(t)$ then the further use of the system (1) for mathematical modeling of the characteristic $\tilde{x}_k(t)$ will be problematic.

Two approach exist to· problem of construction of adequate mathematical description of physical process:

(1) mathematical model is given a priori with inexact parameters and then model of external actions is being determined for which the condition of mathematical description adequacy system (1) is valid [7, 8];

(2) some model of external actions is given a priori, and then mathematical model is being chosen for which the condition of adequacy for system (1) is valid [9, 10].

The solution of this problem (the synthesis of adequate mathematical description) in frame of first approach is reduced to solution of integral equation of first kind (ill-posed problem) [6–8] according to some unknown functions z_k^1:

$$\int_{t_0}^{t} N_k(t-\tau) z_k^1(\tau) d\tau = P_k(t), \tag{4}$$

where $N_k(t-\tau)$, $P_k(t)$ are known functions.

Rewrite the Eq. (4) in the form (the indices were removed) [6–8]:

$$A_p z = u_\delta, \tag{5}$$

where A_p is completely continuous operator (p is vector parameters of mathematical model), $A_p : Z \to U$, $z \in Z$, $u_\delta \in U$, $\rightleftarrows u_\delta$ is initial experimental data (graphic), z is unknown function, (Z, U are a functional spaces).

Further, we shall suppose that the element u_δ in the Eq. (5) is given with a known error:

$$\|u_\delta - u_{ex}\|_U \le \delta, \text{ where } \delta \text{ is const, } \delta > 0,$$

where u_{ex} is exact initial data.

Let's denote by $Q_{\delta,p}$ the set of possible solutions of Eq. (5): $Q_{\delta,p} = \{z : \|A_p z - u_\delta\| u \le \delta\}$.

The set $Q_{\delta,p}$ is not bounded at any. Thus, the initial mathematical model of physical process and any function from the set $Q_{\delta,p}$ provide an adequacy of results of mathematical simulation with accuracy. Besides the exact solution z_{ex} of the Eq. (5) may not belong to set of possible solutions $Q_{\delta,p}$ as the operator A_p is being described inexactly the real physical process [6–8].

By virtue of it the inverse problems of synthesis have a line of features:

- the approximate solution can considerably differ from the exact solution z_{ex};
- the size of an error of the approximate solution in relation to the exact solution z_{ex} has no importance for further use of the approximate solution as we will be use the inexact operator A_p for forecast of real process motion.
- the exact solution z_{ex} of an inverse problem in aggregate with initial mathematical model can give the worse results of mathematical modeling since the exact solution z_{ex} does not belong to set of possible solutions $Q_{\delta,p}$ with guarantee.
- it is possible to ignore the error of the operator A_p to the exact operator since the initial inexact mathematical model of physical process will be used at mathematical modeling in further.

- there is no sense to study behavior of the approximate solution of an inverse problem in dependence at reduction of an error of the initial data δ. In other words, the approximate solution cannot have properties of regularization.

For definition of the unique solution of an inverse problem (5) from set of the possible solutions it is possible to use approach which was offered by Phillips and Tikhonov [11].

Consider now the following extreme problem:

$$\Omega[z_{\delta,p}] = \underset{z \in Q_{\delta,p}}{inf} \; \Omega[z], \tag{6}$$

where functional $\Omega[z]$ has been defined on set Z [11].

It was shown that under certain conditions the solution of the extreme problem (6) exists, is unique and stable with respect to small change of initial data u_δ [11]. Note that there is no sense to investigate behavior of the obtained solution at $\delta \to 0$.

The examples of synthesis of steady adequate mathematical descriptions are given [6–8].

3 Guaranteed Estimation of Forecast of Dynamic System Motion

Here the problem of steady adequate mathematical model synthesis of special use for building forecast are investigated.

The adequate mathematical descriptions first of all are aimed at the forecast of motion of real processes. It is possible to predict of motion of real process in new conditions of operation with the help of adequate mathematical descriptions. For example, it is possible to test more intensive mode of operations of the real machine without risk of its destruction. Such tool (adequate mathematical descriptions) allows to simulate the characteristics of process in the unconventional modes of operations and to determine also optimum parameters of real process.

Suppose that the $\boldsymbol{ALMD}_{x_k}$ physical process is constructed using some algorithm. However, to use this description to study a dynamical system under new unknown conditions does not make sense, since the obtained $\boldsymbol{ALMD}_{x_k}$ in this new situation is not so one.

If the initial parameters of the physical process change little in the new conditions, then due to the inertia of the physical process, these small changes give small changes in the characteristics of the physical process under study. And, consequently, the results of the experiment under the new conditions will be close to the results of the previous experiment, with which $\boldsymbol{ALMD}_{x_k}$ was previously obtained. The practical usefulness of the simulation results will be insignificant in this case.

To construct a forecast of the motion of a physical process by mathematical modeling methods, it is necessary to have a correct mathematical model and all other components of the mathematical description in the future time. However, such knowledge cannot be obtained at the previous time in principle. Thus, the problem of forecasting using the methods of mathematical modeling cannot, in principle, be posed and solved in the usual sense. In other words, it is impossible obtain the result of mathematical modeling using some future (non-existent) mathematical description. It is necessary to change the formulation of the prediction problem in order to overcome these difficulties. First, the forecast of the motion of some physical process in the traditional sense is impossible to obtain since an adequate mathematical description does not have the property of adequacy in the future state of the physical process. Secondly, the use of a non-adequate mathematical description in the mathematical modeling of the forecast does not allow obtaining reliable information about the forecast.

3.1 Criterion of Qualitative Type Adequacy

Let the mathematical model of the physical process and the operator $\boldsymbol{A_p}$, in Eq. (5) depend continuously on the parameter vector $\boldsymbol{p}$ (for example, the mass of the elements, the rigidity of the elastic elements, etc.). In this case, there is a one-to-one correspondence between the components of the parameter vector p and real physical elements. This important correspondence will be called the *main correspondence* (MC). In addition, we will assume that the connections between the elements of the mathematical model correspond to the known physical laws (Hooke's law, Newton's laws, etc.). The fulfillment of such a correspondence will be considered the fulfillment of the *criterion of mathematical description adequacy of the qualitative type* [6, 7, 12].

Note that the two approaches (methods) considered earlier for constructing a mathematical description of a quantitative type have different possibilities for fulfilling the criterion of adequacy of a qualitative type. In the first approach, the fulfillment of the criterion of a qualitative type is guaranteed in the process of constructing the equations of motion and their parameters. In the second approach, the internal properties of the dynamic system change, that is, the criterion of the adequacy of the qualitative type may not be met in this case.

3.2 Possible Method of Forecast Estimation with Guaranty

Let us admission, something each vector component p can take values from some known segments $p_k^0 \leq p_i \leq \hat{p}_k, i = 1, n$; $p_k^0, \hat{p}_k$ are known. So, the vector parameters $p = (p_1, p_2, ..., p_n)^T \in D \subset R^n$, where D is closed parallelepiped in R^n (n is the dimension of the vector p). Scatter such parameters is determined by possible changes these parameters in the future. Each vector p corresponds to some operator A_p in Eq. (5). The set of completely continuous operators A_p is denoted as class

$K_A = \{A_p\}$. The exact operator in Eq. (5) does not belong to the class of operators K_A because their structure is unknown and therefore, they cannot be written in finite form (i.e., with by help a finite number of parameters).

We introduce the size of a possible deviation of operators A_p from the class $K_A = \{A_p\}$:

$$h_1 = \sup_{A_\alpha, A_\beta \in K_A} \left\| A_\alpha - A_\beta \right\|_{Z \to U} \tag{7}$$

The set of possible solutions of Eq. (5) $Q_{h_1,\delta}$ has the form in this case:

$$Q_{h_1,\delta} = \left\{ z : z \in Z, \left\| A_p z - u_\delta \right\|_U \leq \delta + h_1 \|z\|_z \right\}. \tag{8}$$

For problems of this type in [13] an algorithm for an approximate solution of an incorrect problem with an approximate operator based on Tikhonov's regularization method was proposed.

The problem of determining the regularized solution of Eq. (5) reduces to the $z_{p,h_1,\delta}$ which gives the exact lower bound of the smoothing functional

$$M^\alpha[z, u_\delta] = \left\| A_p z - u_\delta \right\|_U^2 + \alpha \Omega[z],$$

where $\Omega[z]$ is stabilizing functional.

The regularization parameter α is determined by the generalized discrepancy method [13]:

$$\left\| A_p z - u_\delta \right\|_U^2 = (\delta + h_1 \|z\|)^2$$

Necessary condition for the extreme of a given functional is the equality of zero of its first variation. In this case, finding a regularized solution $z_{p,h_1,\delta}$ reduces to finding the solution of the integrodifferential equation [11]. An approximate solution to this equation can being found by numerical methods on a computer. In practical use of this algorithm, it was established that the influence of the size of the error h_1 on the approximate solution is significant.

We now turn to a formulation of a possible algorithm for estimating the forecast of the motion of dynamical systems.

Let the prognosis of the motion of one variable $x_k(t)$ of the state of the dynamical system (1) be carried out.

The algorithm for the synthesis of an adequate mathematical description for the purposes of forecasting the motion of a dynamic system:

- the possible or desired spread of the parameters p of the mathematical model of the dynamic system is determined for future conditions, the external influence is not changed.

- for the given experimental measurements, a stable adequate mathematical description is synthesized for each variant of possible changes in the parameters of the mathematical model p of the dynamic system in the future.
- the value $\hat{x}_{p,k}$ of the deviation state variable $x_k(t)$ of interest for the time interval $[0, T]$ is being investigated for each variant of possible changes in the parameters p of the mathematical model of the dynamic system, for example, the maximum deviation of the variable $x_k(t)$ from zero on $[0, T]$

$$\hat{x}_{p,k} = F[x_k(t)] = \|x_k\|_{c[0,T]};$$

- then among the received $\hat{x}_{p,k}$ the greatest $\hat{x}_{p^{max}}$ is defined.
- further to obtain a guaranteed forecast of the behavior of the dynamic system with the indicated purpose, mathematical modeling using an adequate mathematical description with a vector of the parameter p^{max} is executed.

The results of mathematical modeling give the maximum deviation of the variable state $\hat{x}_{p,k}$ from zero for any vector $p \in D$ i.e., with guarantee.

We proceed to the justification of the prediction problem.

Suppose that the researcher knows that in the future some real parameters of the physical process will change, for example, mass number 1 of the real process will increase. By virtue of the MC in the mathematical model of this physical process, the corresponding parameter, for example, p_1 will increase. The union of physical parameters in some integral parameters of a mathematical model is inadmissible, since this violates the MC. For the same reason, the proposed forecasting algorithm cannot be used in cases where the physical parameter that is missing in the mathematical model will be change. In other words, in this case the forecast will not reflect real changes in the physical process. So let in the future, possible changes in the parameters of the vector p within the region D.

For each parameter of the mathematical model p and a given measurement of the variable state $\tilde{x}_k(t)$, a vector of external action function $z_p^1 = (z_{1,p}, z_{2,p}, \ldots, z_{l,p})^T$ is determined which together with the mathematical model (1), provides an adequate mathematical description of the physical process. The state variable $x_{p,k}(t)$ is determined by mathematical modeling and the desired characteristic $\hat{x}_{p,k}$.

Then we define a vector of parameters p^{sup} that satisfies the condition

$$\hat{x}_{p^{sup},k} = \sup_{p \in D} \hat{x}_{p,k}. \tag{9}$$

A mathematical description with a vector of parameters p^{sup} makes it possible to evaluate the forecast of a characteristic $\hat{x}_{p,k}$ with a guarantee if changes in the parameter vector within the region D are possible in the future. Physical parameters of external action are assumed to be unchanged.

It is possible to forecast the behavior of a dynamic system in order to estimate the minimum of the maximum deviation of a state variable $x_k(t)$. To this end it is necessary to change the condition (9):

$$\widehat{x}_p inf, k = \inf_{p \in D} \widehat{x}_{p,k} \tag{10}$$

To predict other characteristics of the physical process, it is necessary to change the definition of these characteristics and change the conditions of the type (9), (10).

To solve extreme problems of the type (9), (10), a method of special mathematical models was developed [14]. According to this method, problems of the type (9), (10) reduce in some cases to the examination of several points of the domain D.

If the parameters of the dynamic system do not change in the future, but significant changes in the external loads are possible, then forecasting algorithms are possible according to the following scheme:

- several problems of the synthesis of AMD_{x_1} are solved within the framework of the first approach for various experiments with parameter $p_0 \in D$ and we obtain models of external loads $z^1_{\delta, p_0}, z^2_{\delta, p_0}, \ldots, z^d_{\delta, p_0}$;
- by analyzing the results, general patterns in the behavior of these models can be found and a "general" model of external influence z^{av}_{δ, p_0} formed.
- we can predict the motion of variable $x_1(t)$ of system (4.36) with the model z^{av}_{δ, p_0} and parameter $p_0 \in D$.

An example of such an algorithm is the problem of assessing the coefficient of dynamism of mechanical vibrations in the main mechanical line of a rolling mill [6].

The analysis of forecasting methods showed that there is no basis for reliable forecasting without the use of mathematical methods. For example, well-known predictors such as Wangi, etc. exploit the features of human memory and have nothing to do with objective results. Typical "predictions" are statements that some fuzzy event will occur without specifying a specific place and time. Since a huge number of different events take place in the world, it is already a matter of time, technique, and trust to tie the "prediction" to a specific event.

4 Conclusion

The paper proposes a possible algorithm for the guaranteed estimation of the dynamic system behavior. For this purpose, an adequate mathematical description of the actual physical process is used. A mathematical description is understood as the totality of the mathematical model of the process, the functions of external action and additional conditions. One algorithm assumes that only certain parameters of the physical process can change in the future. Another algorithm assumes that the main parameters of the physical process will not change in the future. Several simplifying techniques have been developed.

References

1. Lee SW, Lee HK (2008) Reliability prediction system based on the failure rate models of electronics components. J Mech Sci Technol 22:957–964
2. Dreyfus G (2005) Neural networks: methodology and applications. Springer Sci Bus Media
3. Marcek D (2016) Statistical models and granular SOFI RBF neural network for Malaysia KLCI price index prediction. In: Proceedings of ITISE 2016 international work-conference on time series, Granada 27–29, June 2016, pp 530–540
4. Naushad MK, Yuvraj S, Vandna J (2016) GQL estimation in bivariate non-stationary poisson time series model based on copula approach. In: Proceedings of ITISE 2016 international work-conference on time series. Granada, 27–29, June 2016, pp 482–493
5. Chandola V, Banerjee A, Kumar V (2008) Anomaly detection: a survey. ACM Comput Surv (CSUR) 41 15:-15:58 (2008)
6. Menshikov Yu (2020) Synthesis of adequate mathematical descriptions of physical processes. Cambridge Scholars Publishing, Monograph
7. Menshikov Yu (2013) Synthesis of adequate mathematical description as solution of special inverse problems. Eur J Math Sci 2(3):256–271
8. Menshikov Y (2007) Identification of models of external loads. In: Proceedings of ICINCO 2007, Angers, France, pp 376–379
9. Gubarev V (2008) Method of iterative identification of many-dimensional systems with inexact data. Part 1. Theoretical basics. Problems of Control and Information, Kiev, 2, pp 8–26
10. Stepashko V (2008) Method of critical dispersions as analytical apparatus of theory of inductive modeling. Prob Control Inf 2:8–26
11. Tikhonov A, Arsenin V (1979) Methods of incorrect problems solution, Moscow, Science
12. Menshikov Yu (2019) Criteria for estimation the adequacy of mathematical descriptions of physical processes, chapter: modelling and simulation in engineering. Intech-Open, London
13. Goncharski A, Leonov A, Yagola A (1972) On a regularizing algorithm for ill-posed problems with an approximately given operator. J Comput Math Math Phys 12(6):1592-1594
14. Menshikov Y (2011) Inverse problems in non-classical statements. J Pure Appl Math 67(1):7–96

Modeling of Nonlinear Wave Phenomena in Interstellar Superthermal Plasma

Punam Kumari Prasad and Asit Saha

Abstract Modeling of dust-acoustic waves (DAWs) in a four-component plasma constituting negatively charged fluid dusts, superthermal electrons, positrons, and ions is analyzed. Employing the method of reductive perturbation, the Korteweg–de Vries (KdV) equation and modified KdV (mKdV) equations are formulated. Using a typical set of plasma parameters of interstellar medium, solutions of both equations are studied analytically and numerically. Each type of wave solution is confirmed by the plots of Hamiltonian energy function and potential energy function with specified set of physical parameters like superthermality index, number densities and temperature of plasma particles, and velocity of traveling wave in interstellar superthermal plasma.

Keywords Dynamical system · Hamiltonian function · Phase plane analysis · Small-amplitude waves · Superthermal plasma · Superperiodic waves

1 Introduction

The presence of electron–positron–ion–dust (e-p-i-d) plasma is widely observed in various space and astrophysical plasmas [1–3]. The nonlinearity due to the inertial adiabatic dust may lead to the varieties of nonlinear wave features which can be examined in the framework of many nonlinear evolution equations. Rahman et al. [4] analyzed the existence and genesis of dust-acoustic (DA) shock waves in a four-component plasma comprising of dust grains of opposite polarity. Recently, Demiraya and Abdikian [5] explored the modulational instability of DAWs in nonthermal plasma. Saini et al. [6] studied large amplitude nonlinear waves in an e-p-i-d plasma where plasma particles were assumed to follow the Maxwellian distribution.

P. Kumari Prasad (✉) · A. Saha
Department of Mathematics, Sikkim Manipal Institute of Technology, Sikkim Manipal University, Majitar, Rangpo 737136, East-Sikkim, India
e-mail: psikkim@ymail.com

© The Author(s), under exclusive license to Springer Nature Singapore Pte Ltd. 2022
D. Giri et al. (eds.), *Proceedings of the Seventh International Conference on Mathematics and Computing*, Advances in Intelligent Systems and Computing 1412,
https://doi.org/10.1007/978-981-16-6890-6_64

Various spacecraft in space and astrophysical plasma environments have observed the signature of highly energetic particles that deviate significantly from the Maxwellian distribution. These energized particles may arise either due to wave–particle interactions or the influence of an external force acting on the plasmas. Hence, the superthermal plasma particles of space and astrophysical plasmas are qualitatively modeled by kappa distribution. Low values of κ signify distributions with relatively large components of superthermal particles than thermal particles [7]. Moreover, Kappa distribution tends to Maxwellian distribution for higher values of spectral index (κ). Singh et al. [8] examined the behavior of collision between DA solitons that are defined by KdV equations but they did not discuss the higher order KdV equation which exists for the case when nonlinear term of the KdV equation dissolves. Thus, we aim to study both DA compressive and solitary waves (solitons) defined by KdV and mKdV equations.

Implementing the theory of planar dynamical systems, Samanta et al. [9] explored dynamical properties of dust-ion-acoustic waves in a magnetized dusty plasma. Selim et al. [10] reported bifurcations of nonlinear waves in a magnetized multi-component plasma with superthermal electrons. Recently, Monier and Atteya [11] explored the features of the KdVB equation in a multi-component dusty plasma. Rahim et al. [12] investigated the dynamical behaviors of DA solitary structures in Thomas Fermi dusty plasma. Supernonlinear waves like superperiodic waves are characterized by nontrivial topology of their phase portraits. Tamang et al. [13] explored the existence of nonlinear and supernonlinear waves under the mKdV and mG equations. Very recently, Saha et al. [14] showed the signature of supernonlinear waves in a two-component e-i Maxwellian plasma.

The manuscript is structured as follows: in Sect. 2, the guiding equations which describe the dynamics of DAWs in interstellar plasma system are considered. In Sect. 3.1, the derivation of nonlinear KdV equation is shown. For some critical set of plasma parameters, the coefficient of nonlinear term of KdV equation vanishes. So, for such a case higher order mKdv equation is formulated in Sect. 3.2. In Sect. 4, the planar Hamiltonian system analogous to KdV and mKdV equations is analyzed with their analytical wave solutions. The conclusion of the work is highlighted in Sect. 5.

2 Mathematical Modeling

Interstellar region stretches about 122 AU from the Sun and is composed of condensed clouds which contain material in the form of gas, cosmic rays, and dust particles of approximately 0.1 μm in radius. Although the dust particles contain only 1% of the matter, they scatter and absorb sufficient radiations emitted from stars [15]. Moreover, the region is also filled with superthermal electrons and positrons which are produced due to interactions of energetic gamma ray photons and atoms [16]. Thus, to study dust-acoustic waves (DAWs), we consider an interstellar plasma system that is composed of negatively charged fluid dust particles (n_d, T_d), inertialess superthermal electrons (n_e, T_e), ions (n_i, T_i), and positrons (n_p, T_p). Here n_j

and T_j are number density and temperature of jth species, where $j = d, e, p, i$ for dust particles, electrons, protons, and ions, respectively. The neutrality condition at equilibrium is $n_{e0} + Z_{d0}n_{d0} = n_{p0} + n_{i0}$, where the subscript "0" represents the unperturbed plasma quantities and Z_{d0} denotes elementary charge for dust particles. The dynamics of DAWs propagating with velocity of u_d can be studied by considering the following normalized fluid equations [8]:

$$\frac{\partial n_d}{\partial t} + \frac{\partial}{\partial x}(n_d u_d) = 0, \tag{1}$$

$$\frac{\partial u_d}{\partial t} + u_d\frac{\partial u_d}{\partial x} + 3\sigma n_d\frac{\partial n_d}{\partial x} - \frac{\partial \phi}{\partial x} = 0, \tag{2}$$

$$\frac{\partial^2 \phi}{\partial x^2} = n_e - n_p - n_i + n_d, \tag{3}$$

where $\sigma = \sigma_d/\alpha$, $\sigma_d = T_d/T_e$, $\alpha = T_{eff}/T_e$, and $\frac{1}{T_{eff}} = \frac{\mu_e}{T_e} + \frac{\mu_p}{T_p} + \frac{\mu_i}{T_i}$. Here, ϕ denotes the electrostatic wave potential. The normalized variables are as follows: $u_d \to u_d/C_d, n_j \to n_j/Z_{d0}n_{d0}$ (where $j = e, p, i, d$), $\phi \to k_B T_i/e, t \to \omega_{pd}t$, and $x \to x/\lambda_{Dd}$, where the DA wave speed $C_d = (Z_{d0}k_B T_{eff}/m_d)^{1/2}$, dust frequency $\omega_{pd}^{-1} = (m_d/4\pi Z_{d0}^2 e^2 n_{d0})^{1/2}$, and Debye length $\lambda_{Dd} = (k_B T_{eff}/4\pi Z_{d0}^2 n_{d0} e^2)^{1/2}$. Here k_B denotes the Boltzmann constant, e represents the electron charge, and m_d is the mass of dust particles. The pressure associated with dust particles is $p_d = cn_d^{\nu}$, where $\nu = (f+2)/f$ with f as the degree of freedom. For 1D case, $f = 1$ and hence $\nu = 3$.

The normalized number densities of electrons, positrons, and ions are given as follows:

$$n_e = \mu_e\left[1 - \frac{\alpha\phi}{(\kappa_e - 3/2)}\right]^{-\kappa_e+\frac{1}{2}}, \tag{4}$$

$$n_p = \mu_p\left[1 + \frac{\beta\phi}{(\kappa_p - 3/2)}\right]^{-\kappa_p+\frac{1}{2}}, \tag{5}$$

$$n_i = \mu_i\left[1 + \frac{\gamma\phi}{(\kappa_i - 3/2)}\right]^{-\kappa_i+\frac{1}{2}}, \tag{6}$$

where $\mu_e = n_{e0}/Z_{d0}n_{d0} = \delta_e/(1+\delta_p-\delta_e)$, $\mu_p = n_{p0}/Z_{d0}n_{d0} = \delta_p/(1+\delta_p-\delta_e)$, $\mu_i = n_{i0}/Z_{d0}n_{d0} = 1/(1+\delta_p-\delta_e)$, $\delta_{e,p} = n_{e0,p0}/n_{i0}$, $\alpha = T_{eff}/T_e = 1/(\mu_e + \mu_p\sigma_p + \mu_i\sigma_i)$, $\beta = \alpha\sigma_p$, and $\gamma = \alpha\sigma_i$ with $\sigma_p = T_e/T_p$ and $\sigma_i = T_e/T_i$. The spectral indices κ_e, κ_p, and κ_i represent the superthermality of electrons, protons, and ions, respectively. Evidently, one requires $\kappa_{e,p,i} > 3/2$ and if $\kappa_{e,p,i} \to \infty$ then distributions (4)–(6) reduce to Maxwell–Boltzmann distribution. For numerical simulation, we have considered $\kappa_e = \kappa_p = \kappa_i = \kappa$.

3 Formulation of Evolution Equations

3.1 KdV Equation

To study DAWs in the considered $e-p-i-d$ plasma system, we introduce the following stretching of independent variables:

$$\xi=\varepsilon^{1/2}(x-v_d t),\quad \tau=\varepsilon^{3/2}t, \tag{7}$$

where ε denotes extent of nonlinearity and v_d represents DAWs velocity.

We also consider the following expansions of dependent variables:

$$\begin{aligned} n_j &= 1+\varepsilon n_{j1}+\varepsilon^2 n_{j2}+\varepsilon^3 n_{j3}+\cdots\\ u_d &= 0+\varepsilon u_{d1}+\varepsilon^2 u_{d2}+\varepsilon^3 u_{d3}+\cdots\\ \phi &= 0+\varepsilon\phi_1+\varepsilon^2\phi_2+\varepsilon^3\phi_3+\cdots \end{aligned} \tag{8}$$

Substituting expressions (7) and (8) in system of normalized Eqs. (1)–(3) and equating the coefficients of various powers of ε to zero, we get

$$\begin{cases} u_{d1}=v_d n_{d1},\\ n_{d1}=\frac{1}{3\sigma-v_d^2}\phi_1,\\ u_{d1}=\frac{v_d}{3\sigma-v_d^2}\phi_1,\\ v_d^2=3\sigma+1/G, \end{cases} \tag{9}$$

$$\frac{\partial n_{d1}}{\partial\tau}+\frac{\partial u_{d2}}{\partial\xi}-v_d\frac{\partial n_{d2}}{\partial\xi}+\frac{\partial}{\partial\xi}(n_{d1}u_{d1})=0, \tag{10}$$

$$\frac{\partial u_{d1}}{\partial\tau}-v_d\frac{\partial u_{d2}}{\partial\xi}+u_{d1}\frac{\partial u_{d1}}{\partial\xi}-\frac{\partial\phi_2}{\partial\xi}+3\sigma n_{d1}\frac{\partial n_{d1}}{\partial\xi}+3\sigma\frac{\partial n_{d2}}{\partial\xi}=0, \tag{11}$$

$$\frac{\partial^2\phi_1}{\partial\xi^2}=G\phi_2+K\phi_1^2+n_{d2}, \tag{12}$$

where $G=(\alpha\mu_e+\beta\mu_p+\gamma\mu_i)\frac{2\kappa-1}{2\kappa-3}$ and $K=(\alpha^2\mu_e-\beta^2\mu_p-\gamma^2\mu_i)\frac{4\kappa^2-1}{2(2\kappa-3)^2}$. Using above set of relations (9)–(12), we formulate the following KdV equation:

$$\frac{\partial\phi_1}{\partial\tau}+A\phi_1\frac{\partial\phi_1}{\partial\xi}+B\frac{\partial^3\phi_1}{\partial\xi^3}=0, \tag{13}$$

where $A=-\frac{1}{2v_dG^2}(3(v_d^2+\sigma)G^3+2K)$, and $B=\frac{1}{2v_dG^2}$.

3.2 Modified KdV Equation

The study of DAWs described by KdV equation (13) becomes impossible at critical set of parameters where coefficient of nonlinear term (A) in Eq. (13) vanishes, i.e., for $3(v_d^2+\sigma)G^3 \approx -2K$. The wave solution of Eq. (13) produces compressive and rarefactive solitons for $A > 0$ and $A < 0$, respectively. However, the amplitude of DAW becomes infinitely large for $A \approx 0$ which makes the method of reductive perturbation invalid [17]. Thus, to study wave features at those critical points it becomes important to consider the modified KdV (mKdV) equation. Thus, to formulate mKdV equation, we take the same set of expansions of dependent variables (i.e., expansions (8)) with following stretching of independent variables:

$$\xi = \varepsilon(x - v_d t), \quad \tau = \varepsilon^3 t, \tag{14}$$

where v_d signifies velocity of DAWs.

Substituting relations (8) and (14) in Eqs. (1)–(3) and proceeding in the similar way as in previous section, we get same equations as in Eq. (9) and the following relations:

$$n_{d2} = \frac{1}{3\sigma - v_d^2}\phi_2 - \frac{3(v_d^2+\sigma)}{2(3\sigma - v_d^2)^3}\phi_1^2, \tag{15}$$

$$u_{d2} = \frac{v_d}{3\sigma - v_d^2}\phi_2 - \frac{v_d(v_d^2+9\sigma)}{2(3\sigma - v_d^2)^3}\phi_1^2, \tag{16}$$

$$\frac{\partial n_{d1}}{\partial \tau} - v_d\frac{\partial n_{d3}}{\partial \xi} + \frac{\partial u_{d3}}{\partial \xi} + \frac{\partial}{\partial \xi}\left(n_{d1}u_{d2} + n_{d2}u_{d1}\right) = 0, \tag{17}$$

$$\frac{\partial u_{d1}}{\partial \tau} - v_d\frac{\partial u_{d3}}{\partial \xi} + \frac{\partial}{\partial \xi}\left(u_{d1}u_{d2} + 3\sigma n_{d1}n_{d2}\right) - \frac{\partial \phi_3}{\partial \xi} + 3\sigma\frac{\partial n_{d3}}{\partial \xi} = 0, \tag{18}$$

$$\frac{\partial^2 \phi_1}{\partial \xi^2} = G\phi_3 + 2K\phi_1\phi_2 + L\phi_1^3 + n_{d3}, \tag{19}$$

where $L = (\alpha^3\mu_e + \beta^3\mu_p + \gamma^3\mu_i)\dfrac{(4\kappa^2-1)(2\kappa+3)}{6(2\kappa-3)^3}$.

Using Eqs. (9) and (15)–(19), we eliminate $n_{d1,d2,d3}$, $u_{d1,d2,d3}$, and $\phi_{2,3}$ to formulate the following mKdV equation:

$$\frac{\partial \phi_1}{\partial \tau} + A_1\phi_1^2\frac{\partial \phi_1}{\partial \xi} + B_1\frac{\partial^3 \phi_1}{\partial \xi^3} = 0, \tag{20}$$

where $A_1 = \frac{1}{4v_d G^2}\left(3(3\sigma G^5 + 1)(v_d^2 + \sigma) + 2G^5 v_d^2(v_d^2 + 9\sigma) - 6L\right)$, and $B_1 = \frac{1}{2v_d G^2}$.

4 Phase Plane Analysis

To analyze the phase plane of Hamiltonian function corresponding to Eqs. (13) and (20), we take a wave transformation $\chi = \xi - V\tau$, where V is DAW velocity. Thus, using transformation χ into Eqs. (13) and (20), we formulate the following dynamical systems:

$$\begin{cases} \frac{d\phi_1}{d\chi} = y, \\ \frac{dy}{d\chi} = \frac{V}{B}\phi_1 - \frac{A}{2B}\phi_1^2, \end{cases} \tag{21}$$

and

$$\begin{cases} \frac{d\phi_1}{d\chi} = y, \\ \frac{dy}{d\chi} = \frac{V}{B_1}\phi_1 - \frac{A_1}{3B_1}\phi_1^3. \end{cases} \tag{22}$$

Systems (21) and (22) are the Hamiltonian systems with Hamiltonian energy functions (total energy functions) given, respectively, as

$$H_1(\phi_1, y) = \frac{y^2}{2} - \frac{V}{2B}\phi_1^2 + \frac{A}{6B}\phi_1^3, \tag{23}$$

and

$$H_2(\phi_1, y) = \frac{y^2}{2} - \frac{V}{2B_1}\phi_1^2 + \frac{A_1}{12B_1}\phi_1^4. \tag{24}$$

The potential function ($\psi_{1,2}(\phi_1)$) associated with systems (21) and (22) are, respectively, defined as [13] follows:

$$\psi_1(\phi_1) = -\frac{V}{2B}\phi_1^2 + \frac{A}{6B}\phi_1^3 \tag{25}$$

and

$$\psi_2(\phi_1) = -\frac{V}{2B_1}\phi_1^2 + \frac{A_1}{12B_1}\phi_1^4. \tag{26}$$

Employing the bifurcation theory [18, 19], we investigate the phase plots of our dynamical system which may vary consequentially depending on numbers of fixed points and separatrix layers [20]. To distinguish various orbits in the phase portrait, we represent $NPT_{m,n}$, $NSPT_{m,n}$, and $NHT_{m,n}$ for nonlinear periodic, superperiodic, and homoclinic orbits, respectively, where m denotes number of center points

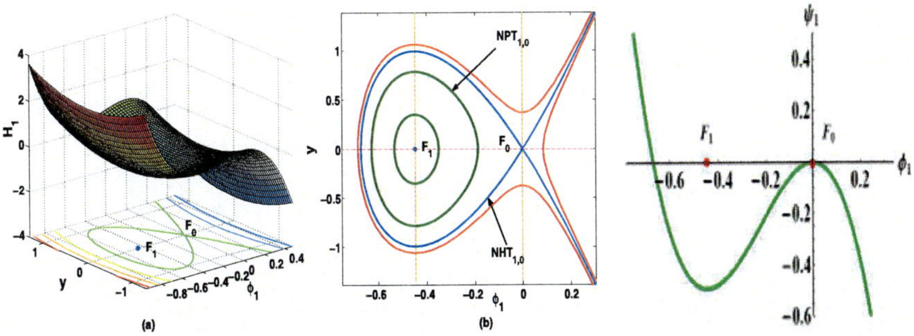

Fig. 1 **a** Total energy function, **b** Phase portrait of system (21), and **c** Potential energy function for $\kappa = 2$, $n_{i0} = 3 \times 10^{-3}$, $n_{e0} = n_{p0} = 10^{-3}$, $T_i = 1.2 \times 10^5$, $T_e = T_p = 2 \times 10^7$, $T_d = 2 \times 10^4$ and $V = 0.9$

and n indicates number of separatrix layers. Each orbit in phase portrait provides a traveling wave solution for the plasma system. For instance, superperiodic, periodic, and homoclinic orbits of the Hamiltonian system correspond to superperiodic, periodic, and solitonic solutions of the plasma system, respectively. Now, we analyze Eqs. (21)–(26) under the influence of superthermality index (κ), the number density ratio of positrons to ions ($\delta_p = n_{p0}/n_{i0}$), the number density ratio of electrons to ions ($\delta_e = n_{e0}/n_{i0}$), the temperature ratio of electrons to positrons ($\sigma_p = T_e/T_p$), the temperature ratio of electrons to ions ($\sigma_i = T_e/T_i$), the temperature ratio of fluid dust to electron ($\sigma_d = T_d/T_e$), and a traveling wave velocity (V). For numerical simulation, we take a specific range of plasma parameters observed in the interstellar medium [8, 21]: $n_i \sim 10^{-3} - 10^{-2}$ cm^{-3}, $n_{e,p} \sim 10^{-3}$ cm^{-3}, $n_d \sim 10^{-7}$ cm^{-3}, $T_i \sim 10^5 - 10^6$ K, $T_{e,p} \sim 10^7 - 10^8$ K.

Figure 1 shows the plots of Hamiltonian energy function, dynamical system, and potential function corresponding to Eq. (13) for $\kappa = 2$, $n_{i0} = 3 \times 10^{-3}$, $n_{e0} = n_{p0} = 10^{-3}$, $T_i = 1.2 \times 10^5$, $T_e = T_p = 2 \times 10^7$, $T_d = 2 \times 10^4$, and $V = 0.9$. The plot of Hamiltonian energy function (23) shown in Fig. 1a exhibits a local maxima at $F_0(0, 0)$ and a local minima at $F_1(\frac{2V}{A}, 0)$. The projection of this energy function in $\phi_1 - y$ plane gives the phase plane of system (21) where local minima corresponds to a stable center and maxima corresponds to an unstable saddle point. Figure 1b displays the phase portrait of system (21) for the specified set of parameters which shows the existence of two fixed points, namely, saddle point at $F_0(0, 0)$ and center at $F_1(\frac{2V}{A}, 0)$, where $A < 0$. There also exist two distinct kinds of phase orbits, i.e., homoclinic orbit ($NHT_{1,0}$) and a family of periodic orbits ($NPT_{1,0}$). Figure 1c illustrates the graph of potential function ψ_1 against ϕ_1. Here, local minima at F_1 in potential plots signifies the existence of solitary solutions and the potential dip at $\phi_1 > 0$ corresponds to a compressive solitonic solution whereas the potential dip at $\phi_1 < 0$ signifies a rarefactive solitonic solution. Therefore, for the condition $A < 0$ we obtain only rarefactive DAW solution of Eq. (13). The periodic wave solution analogous to a periodic orbit presented in Fig. 1b is displayed in Fig. 2.

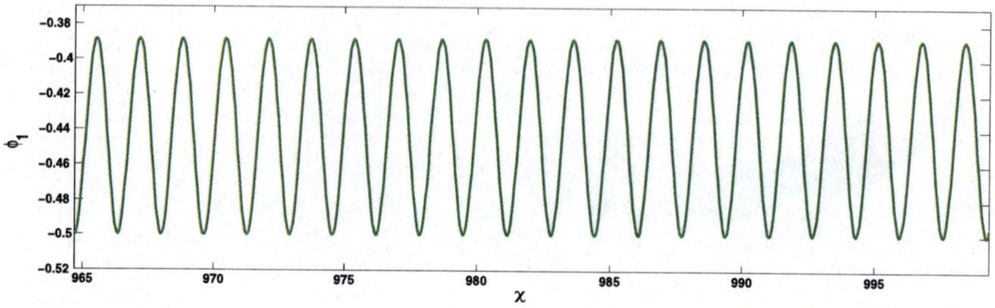

Fig. 2 Periodic wave solution for Eq. (13)

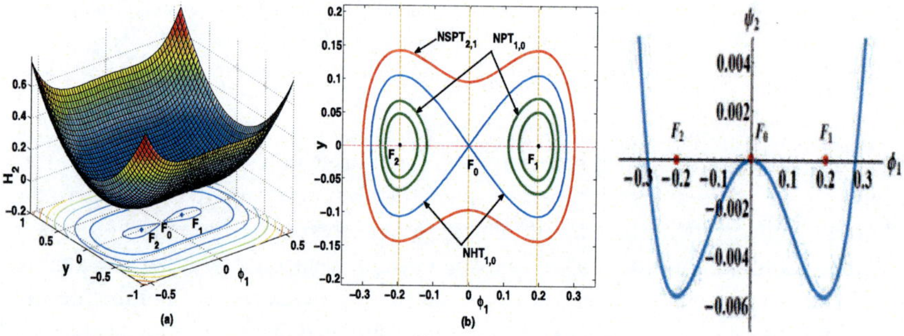

Fig. 3 **a** Total energy function, **b** Phase portrait of system (22), and **c** Potential energy function for $\kappa = 3$, $n_{i0} = 3 \times 10^{-3}$, $n_{e0} = n_{p0} = 10^{-3}$, $T_i = 1.2 \times 10^5$, $T_e = T_p = 2 \times 10^7$, $T_d = 2 \times 10^4$ and $V = 0.1$

Figure 3 displays the plots of Hamiltonian energy function, dynamical system, and potential function corresponding to Eq. (13) for $\kappa = 3$, $n_{i0} = 3 \times 10^{-3}$, $n_{e0} = n_{p0} = 10^{-3}$, $T_i = 1.2 \times 10^5$, $T_e = T_p = 2 \times 10^7$, $T_d = 2 \times 10^4$, and $V = 0.1$. The plot of Hamiltonian energy function (23) presented in Fig. 3a shows a local maxima at $F_0(0, 0)$ and a pair of local minima at $F_1(\sqrt{\frac{3V}{A_1}}, 0)$ and $F_2(-\sqrt{\frac{3V}{A_1}}, 0)$. The phase portrait of system (22) for the mentioned set of parameters in Fig. 3b shows the existence of three fixed points, namely, saddle point at $F_0(0, 0)$ and center points at $F_1(\sqrt{\frac{3V}{A_1}}, 0)$ and $F_2(-\sqrt{\frac{3V}{A_1}}, 0)$. There also exist three qualitatively different phase orbits, i.e., homoclinic orbit ($NHT_{1,0}$) and a family of periodic ($NPT_{1,0}$) and superperiodic orbits ($NSPT_{2,1}$). Importantly, the phase portrait for our considered plasma system shows the presence of periodic orbit containing separatrix layer and two center points. This type of periodic orbit with separatrix layer and more than two centers is called superperiodic orbit which corresponds to superperiodic wave solution. Hence, DAWs in considered plasma system exhibit superperiodic nature. The graph of potential function ψ_1 against ϕ_1 in Fig. 3c illustrates potential dips at F_1 and F_2 which signify the presence of both compressive and rarefactive soliton solutions of Eq. (20). Periodic and superperiodic wave solutions analogous to periodic and superperiodic orbits shown in Fig. 3b are presented in Fig. 4.

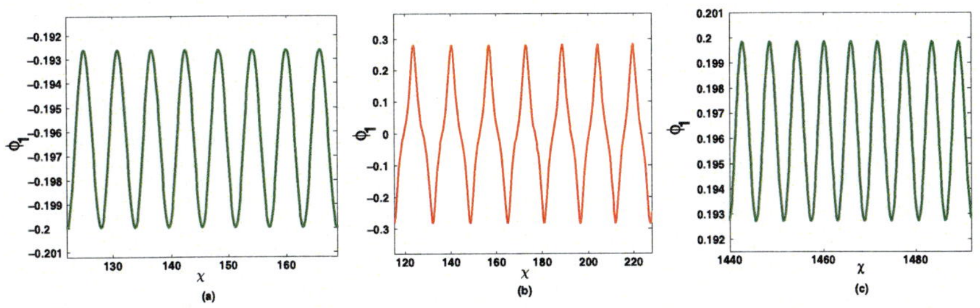

Fig. 4 Periodic and superperiodic wave solutions for Eq. (20)

4.1 Solitary Wave Solutions

Considering the transformation $\chi = \xi - v\tau$ and using boundary conditions $\phi_1 \to 0$ and $\frac{d\phi_1}{d\chi} \to 0$ as $\chi \to \pm\infty$, we derive rarefactive solitonic solution of Eq. (13) analogous to the homoclinic orbit at fixed point F_0 and enclosing point F_1 as

$$\phi_1 = \frac{3v}{A} sech^2\left(\frac{1}{2}\sqrt{\frac{v}{B}}\chi\right). \tag{27}$$

The compressive and rarefactive solitonic solutions of Eq. (20) for a pair of homoclinic orbit at fixed point F_0 and enclosing fixed point F_1 and F_2 on either sides of ϕ_1 are given as

$$\phi_1 = \pm\sqrt{\frac{6v}{A_1}} sech\left(\sqrt{\frac{v}{B_1}}\chi\right). \tag{28}$$

Figure 5 shows influence of plasma parameters κ, n_{e0}, n_{p0}, T_i, T_d, and V on the rarefactive solitonic solution of Eq. (13). It can be observed that the width and amplitude of DA solitary wave decrease as electrons, positrons, and ions become highly superthermal. Also, the same effect can be seen for number density of electrons and temperatures of dust. On the contrary, the width and amplitude of DA solitary wave increase with increasing values of number density of positrons and temperature of ion. The DA rarefactive solitary wave amplifies and becomes narrow as velocity of traveling wave increases.

Figure 6 depicts compressive and rarefactive solitonic solutions of Eq. (20). Here, it can be seen that if the magnitudes of parameters κ, n_{p0}, and T_i are raised then the amplitude and width of both DA compressive and rarefactive solitons increase whereas if parameters n_{e0} and T_d are raised then the width and amplitude of the wave are enhanced. Also, DA compressive and rarefactive solitons enlarge and become narrow as V increases.

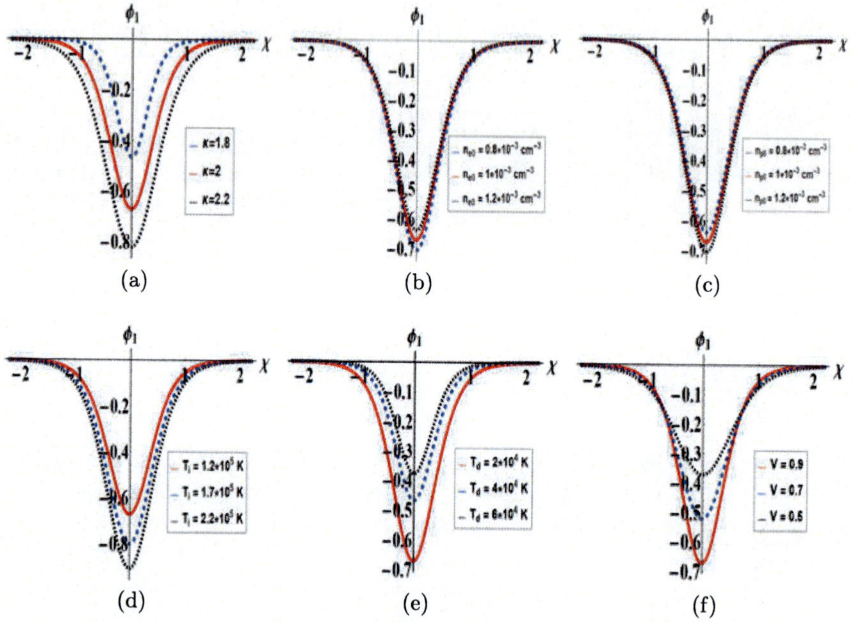

Fig. 5 Effects of **a** κ, **b** n_{e0}, **c** n_{p0}, **d** T_i, **e** T_d, and **f** V on solitonic solution of Eq. (13)

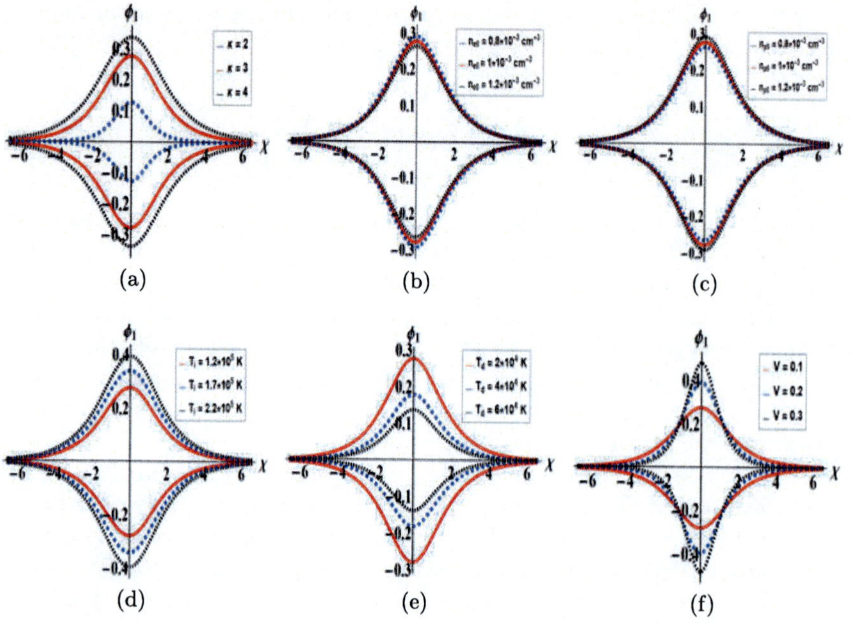

Fig. 6 Effects of **a** κ, **b** n_{e0}, **c** n_{p0}, **d** T_i, **e** T_d, and **f** V on solitonic solutions of Eq. (20)

5 Conclusion

We have studied DAWs in a four-component plasma of interstellar medium constituting mobile dust particles with superthermal electrons, positrons, and ions. Bifurcation analysis of DAWs in the framework of the KdV and mKdV equations is carried out. Typical set of plasma parameters of an astrophysical surroundings such as interstellar region [8, 21] is taken for numerical simulation. The main results of our work are mentioned below:

1. Phase plane corresponding to KdV equation (13) only defines a homoclinic orbit on the negative phase of wave potential, which signifies the existence of a rarefactive soliton in the interstellar region of astrophysical surroundings.
2. Phase plane corresponding to mKdV equation (20) explores three different phase orbits, namely, periodic, superperiodic, and homoclinic orbits. Thus, the presence of periodic, superperiodic, compressive, and rarefactive solitonic solutions of mKdV equation are shown.
3. On analysis of exact wave solutions of Eqs. (13) and (20), we observed that the plasma parameter has same influence on the wave solutions of KdV and mKdV equations.
4. The increasing values of parameters like spectral index (κ), number density of positrons (n_{p0}), and temperature of ion (T_i) are responsible in increasing width and amplitude of DA solitary wave, whereas number density of electrons (n_{e0}) and temperature of dust (T_d) shorten the wave by decreasing the width and amplitude of DA solitary waves. Also, the solitons amplify and become narrow as velocity of traveling wave increases.

From our analysis, we can interpret that the presence of nonlinear solitary, periodic, and superperiodic waves can be found in interstellar region where superthermal plasma particles exist in dusty plasmas. So, the observations of this work may be important for scientists working in the field of space and astronomy to understand the wave features in the interstellar region of astrophysical surroundings where plasma particles are superthermal in nature.

Acknowledgements The first author is grateful to the Sikkim Manipal Institute of Technology (SMIT) and Sikkim Manipal university (SMU) for sanctioning research fellowship under TMA Pai University Research Fund (Ref. No. 118/SMU/REG/ UOO/104/2019).

References

1. Jehan N, Masood W, Mirza AM (2009) Planar and nonplanar dust acoustic solitary waves in electron-positron-ion-dust plasmas. Physica Scripta 80(3):035506. https://doi.org/10.1088/0031-8949/80/03/035506
2. Zurek WH (1985) Annihilation radiation from the galactic center-Positrons in dust? Astrophys J 289:603–608. https://doi.org/10.1086/162921

3. Merlino RL (2006) Dusty plasmas and applications in space and industry. Plasma Phys Appl 81:73–110
4. Rahman A, Mamun AA, Khurshed Alam SM (2008) Shock waves in a dusty plasma with dust of opposite polarities. Astrophys Space Sci 315:243–247. https://doi.org/10.1007/s10509-008-9824-5
5. Demiray H, Abdikian A (2019) Modulational instability of acoustic waves in a dusty plasma with nonthermal electrons and trapped ions. Chaos Solitons Fractals 121:50–58. https://doi.org/10.1016/j.chaos.2019.02.005
6. Saini NS, Chahal BS, Bains AS (2013) Large amplitude dust ion-acoustic solitary waves in a plasma in the presence of positrons. Astrophys Space Sci 347:129–138. https://doi.org/10.1007/s10509-013-1502-6
7. Saini NS, Kourakis I, Hellberg MA (2009) Arbitrary amplitude ion-acoustic solitary excitations in the presence of excess superthermal electrons. Phys Plasmas 16(6):62903. https://doi.org/10.1063/1.3143036
8. Singh K, Kaur N, Saini NS (2017) Head-on collision between two dust acoustic solitary waves and study of rogue waves in multicomponent dusty plasma. Phys Plasmas 24:063703. https://doi.org/10.1063/1.4984996
9. Samanta UK, Saha A, Chatterjee P (2013) Bifurcations of dust ion acoustic travelling waves in a magnetized dusty plasma with a q-nonextensive electron velocity distribution. Phys Plasmas 20(2):022111. https://doi.org/10.1063/1.4791660
10. Selim MM, El-Depsy A, El-Shamy EF (2015) Bifurcations of nonlinear ion-acoustic travelling waves in a multicomponent magnetoplasma with superthermal electrons. Astrophys Space Sci 360:66. https://doi.org/10.1007/s10509-015-2574-2
11. El-Monier SY, Atteya A (2018) Bifurcation analysis for dust-acoustic waves in a four-component plasma including warm ions. IEEE Trans Plasma Sci 46:815–824. https://doi.org/10.1109/TPS.2017.2766097
12. Rahim Z, Adnan M, Qamar A, Saha A (2018) Nonplanar dust-acoustic waves and chaotic motions in Thomas Fermi dusty plasmas. Phys Plasmas 25:083706. https://doi.org/10.1063/1.5016893
13. Tamang J, Saha A (2020) Bifurcations of small-amplitude supernonlinear waves of the mKdV and modified Gardner equations in a three-component electron-ion plasma. Phys Plasmas 27(1):012105. https://doi.org/10.1063/1.5115821
14. Saha A, Chatterjee P, Banerjee S (2020) An open problem on supernonlinear waves in a two-component Maxwellian plasma. Eur Phys J Plus 135(10):801. https://doi.org/10.1140/epjp/s13360-020-00816-8
15. Herbst E (2001) The chemistry of interstellar space. Chem Soc Rev 30(3):168
16. Adriani O, Barbarino G, Bazilevskaya G, et al (2009) An anomalous positron abundance in cosmic rays with energies 1.5-100 GeV. Nature 458:607. https://doi.org/10.1038/nature07942
17. Das GC, Sarma J, Uberoi C (1997) Explosion of soliton in a multicomponent plasma. Phys Plasmas 4(6):2095. https://doi.org/10.1063/1.872545
18. Guckenheimer J, Holmes PJ (1983) Nonlinear oscillations dynamical systems and bifurcations of vector fields. Springer, New York
19. Chow SN, Hale JK (1981) Methods of bifurcation theory. Springer, New York
20. Dubinov AE, Kolotkov DY, Sazonkin MA (2012) Nonlinear theory of ion-sound waves in a dusty electron-positron-ion plasma. Tech Phys 57:585–593. https://doi.org/10.1134/S1063784212050088
21. Klein U, Kerp J (2008) Physics of the interstellar medium. Argelander, Institut fur Astronomie, Bonn

Application of Group Methods in Solving Wave Equations

Jervin Zen Lobo

Abstract We obtain the Lie invariance condition for partial differential equations of second order. This condition is used in obtaining the determining equations of the one-dimensional wave equation with constant speed. The determining equations are split to obtain an overdetermined system of partial differential equations, which are solved to obtain the symmetries of the wave equation. By making an appropriate transformation between the dependent and independent variables, the wave equation is reduced to an easily solvable ordinary differential equation. We solve this resulting differential equation to obtain the solutions of the wave equation. In particular, the one-dimensional wave equation with unit speed has been solved.

Keywords Determining equations · Lie group · Partial differential equations · Symmetries · Wave equation

1 Introduction

Over a hundred years ago, Sophus Lie introduced Lie groups that are invertible point transformations of the independent and dependent variables in differential equations. In such group analysis, these transformations lead the original differential equation of first order (which may be a complicated one) to become a separable one, which can be easily solved. In addition, for higher order differential equations, group analysis of the equation is used to reduce the order of the differential equation by one. Literature on symmetry analysis for differential equations can be found in [2, 9].

We call transformations leaving objects invariant or unchanged as *Symmetries*. In [21], it is pointed out that symmetries account for various laws in nature. Conservation law is an important implication of symmetry in physics and mathematics. This is because their reproduction at different times and places depends on invariance laws. In this connection, Emma Nöether [20] observed and proved a relation between conservation laws and continuous symmetries. Scientists like Kepler got interested

J. Z. Lobo (✉)
Department of Mathematics, St. Xavier's College, Mapusa, Goa, India

© The Author(s), under exclusive license to Springer Nature Singapore Pte Ltd. 2022
D. Giri et al. (eds.), *Proceedings of the Seventh International Conference on Mathematics and Computing*, Advances in Intelligent Systems and Computing 1412,
https://doi.org/10.1007/978-981-16-6890-6_65

in studying the motion of planets using symmetries while Newton studied the orbits of planets using the laws of mechanics as a symmetry principle.

In the past, when the solvability of a polynomial equation was a question, Galois associated an object with the polynomial (algebraic) equation. If this object called the Galois group was solvable, then the corresponding polynomial had a root or was said to be solvable. On similar lines, the solvability of differential equations reduced to the solvability of the associated Lie group. As Galois used Galois groups to model the discrete symmetries of algebraic equations, this motivated Lie to model the continuous symmetries of differential equations using Lie groups. Hence, Lie theory is the only unified understanding that explains the existence of solutions for differential equations.

Partial differential equations are of paramount importance in particle physics, nonlinear optics, fluid dynamics, in addition to their applications in general relativity to differential and algebraic geometry, topology, etc. Research dedicated to using symmetry analysis for partial differential equations arising in mathematical physics may be found in [19, 26]. In addition, symmetry analysis for time-fractional partial differential equations can be found in [6]. Lie symmetry analysis for the Kudryashov–Sinelshikov equation may be found in [17] while the Burgers' equation is analyzed using Lie symmetries in [14, 18]. Studies on the Korteweg-de Vries equation using Lie symmetry analysis are seen in [11, 16]. Interests in studying symmetries are so much that in [25], the definition of an admitted Lie group is developed for Stochastic differential equations. Softwares are often helpful in aiding the computation of symmetries particularly for partial differential equations as seen in [22]. Recently in [27], the Fokker–Planck equation, which is nonlinear is analyzed using Lie symmetries. Furthermore, the subalgebras of one dimension, which are constructed, are used to find similarity reductions and invariant solutions. In [3], the solution of the heat equation is found by developing an invariantized finite difference scheme.

We now state the definition of a Lie group which we will be using. We will also give some examples of Lie group to illustrate the definition.

Definition 1 ([5]) Lie groups are a set of transformations such that:

1. The composition of two transformations carried out successively yields another transformation, which belongs to the set.
2. There exists an identity transformation.
3. The inverse of every transformation exists.

Remark 1 Because of the *closure* property, the associativity law for groups holds.

Remark 2 The group is **abelian** if the order of carrying out the transformations does not matter.

Example 1 $\bar{t} = a^{\epsilon}t, t \in \mathbb{R} \setminus \{0\}$ is known as the *stretching group*.

Example 2 $\bar{t} = t + \epsilon$, is known as the *translational group*.

Example 3 $\bar{t}_1 = t_1 \cos\epsilon - t_2 \sin\epsilon,\ \ \bar{t}_2 = t_1 \sin\epsilon + t_2 \cos\epsilon$ is known as a *Rotational group*.

Wave equations find applications in physics, fluid mechanics, engineering sciences, etc.

Traditional methods to solve wave equations are inverse scattering method [1], Hirota bilinear method [8], Darboux transformation [15], Bӑckund method [7] and Painleve's extension method [4]. Other methods of solution using D'Alembert's formula and method of separation of variables can be found in [23, 24]. Group classification and exact solutions of the nonlinear wave equation are found in [12, 13]. More studies on the wave equation can be found in [10].

The invariance condition for second-order partial differential equations has been obtained by using the Taylor's theorem for a multivariable function. Using this, we obtain symmetries and hence the solutions of the wave equation, for which we have not found any literature. The advantage of using Taylor's theorem for a multivariable function as compared with the Lie-Bäcklund operator method is that the computation of symmetries using the former is way simpler as compared with the latter. It may be noted that the invariance condition can be obtained very quickly by using the Taylor series approach.

2 Invariance Condition for Partial Differential Equations of Second Order

Let us suppose that $u = u(t, x)$. We shall consider transformations with h_1, h_2, h_3 as smooth functions in t, x, u having convergent Taylor series in ϵ which are represented as,

$$\begin{cases} \bar{t} = h_1(t, x, u) = t + \epsilon\theta(t, x, u) + O(\epsilon^2), \\ \bar{x} = h_2(t, x, u) = x + \epsilon\phi(t, x, u) + O(\epsilon^2), \\ \bar{u} = h_3(t, x, u) = u + \epsilon\psi(t, x, u) + O(\epsilon^2), \end{cases} \tag{1}$$

where $\theta(t, x, u) = \left.\frac{\partial h_1}{\partial \epsilon}\right|_{\epsilon=0}$, $\phi(t, x, u) = \left.\frac{\partial h_2}{\partial \epsilon}\right|_{\epsilon=0}$, $\psi(t, x, u) = \left.\frac{\partial h_3}{\partial \epsilon}\right|_{\epsilon=0}$.

We differentiate (1) with respect to each of the parameters t and x, in order to calculate the prolongation of a given transformation. We define the following *total derivatives*:

$$D_t = \frac{\partial}{\partial t} + u_t\frac{\partial}{\partial u} + u_{xt}\frac{\partial}{\partial u_x} + u_{tt}\frac{\partial}{\partial u_t} + \cdots, \tag{2}$$

$$D_x = \frac{\partial}{\partial x} + u_x\frac{\partial}{\partial u} + u_{xx}\frac{\partial}{\partial u_x} + u_{xt}\frac{\partial}{\partial u_t} + \cdots. \tag{3}$$

The first two equations of (1) may be inverted (locally) to give t and x in terms of $\bar{t}$ and $\bar{x}$, provided that the Jacobian is non-zero, that is,

$$\mathcal{J} = \begin{vmatrix} D_t\bar{t} & D_t\bar{x} \\ D_x\bar{t} & D_x\bar{x} \end{vmatrix} \neq 0, \quad \text{where} \quad u = u(x, t). \tag{4}$$

Since the last equation of (1) contains $\bar{u}$ as a function of some variables, one of which is $u(t, x)$, we see that if Eq. (4) is satisfied, then by local inverse function theorem, the last equation of (1) can be rewritten as

$$\bar{u} = \bar{u}(\bar{t}, \bar{x}). \tag{5}$$

Applying the chain rule to Eq. (5), we obtain,

$$\begin{bmatrix} D_t\bar{u} \\ D_x\bar{u} \end{bmatrix} = \begin{bmatrix} D_t\bar{t} & D_t\bar{x} \\ D_x\bar{t} & D_x\bar{x} \end{bmatrix} \begin{bmatrix} \bar{u}_{\bar{t}} \\ \bar{u}_{\bar{x}} \end{bmatrix},$$

and therefore by (Cramer's rule)

$$\bar{u}_{\bar{t}} = \frac{1}{\mathcal{J}} \begin{vmatrix} D_t\bar{u} & D_t\bar{x} \\ D_x\bar{u} & D_x\bar{x} \end{vmatrix}, \quad \bar{u}_{\bar{x}} = \frac{1}{\mathcal{J}} \begin{vmatrix} D_t\bar{t} & D_t\bar{u} \\ D_x\bar{t} & D_x\bar{u} \end{vmatrix}. \tag{6}$$

Equation (6) can be simplified to get the extended infinitesimal representation,

$$\bar{u}_{\bar{t}} = u_t + \epsilon\psi_{[t]} + O(\epsilon^2), \quad \bar{u}_{\bar{x}} = u_x + \epsilon\psi_{[x]} + O(\epsilon^2), \tag{7}$$

where

$$\psi_{[t]} = D_t(\psi) - u_x D_t(\phi) - u_t D_t(\theta), \tag{8}$$

$$\psi_{[x]} = D_x(\psi) - u_x D_x(\phi) - u_t D_x(\theta). \tag{9}$$

The explicit expression for Eq. (9) is

$$\psi_{[t]} = \psi_t - \phi_t u_x + (\psi_u - \theta_t)u_t - \phi_u u_x u_t - \theta_u u_t^2,$$

$$\psi_{[x]} = \psi_x + (\psi_u - \phi_x)u_x - \theta_x u_t - \phi_u u_x^2 - \theta_u u_x u_t.$$

Continuing the procedure, we can obtain the second-order prolongations as follows:

$$\bar{u}_{\bar{t}\bar{t}} = \frac{1}{\mathcal{J}} \begin{vmatrix} D_t\bar{x} & D_t\bar{u}_{\bar{t}} \\ D_x\bar{x} & D_x\bar{u}_{\bar{t}} \end{vmatrix}, \quad \bar{u}_{\bar{x}\bar{x}} = \frac{1}{\mathcal{J}} \begin{vmatrix} D_t\bar{u}_{\bar{x}} & D_t\bar{t} \\ D_x\bar{u}_{\bar{x}} & D_x\bar{t} \end{vmatrix} \tag{10}$$

$$\bar{u}_{\bar{x}\bar{t}} = \frac{1}{\mathcal{J}} \begin{vmatrix} D_t\bar{x} & D_t\bar{u}_{\bar{x}} \\ D_x\bar{x} & D_x\bar{u}_{\bar{x}} \end{vmatrix} = \frac{1}{\mathcal{J}} \begin{vmatrix} D_t\bar{u}_{\bar{t}} & D_t\bar{t} \\ D_x\bar{u}_{\bar{t}} & D_x\bar{t} \end{vmatrix}. \tag{11}$$

On simplifying (10) and (11), we get the extended infinitesimal representations, namely

$$\bar{u}_{\bar{t}\bar{t}} = u_{tt} + \epsilon\psi_{[tt]} + O(\epsilon^2), \quad \bar{u}_{\bar{x}\bar{x}} = u_{xx} + \epsilon\psi_{[xx]} + O(\epsilon^2), \quad \bar{u}_{\bar{t}\bar{x}} = u_{tx} + \epsilon\psi_{[tx]} + O(\epsilon^2), \tag{12}$$

where

$$\psi_{[tt]} = D_t(\psi_{[t]}) - u_{tx}D_t(\phi) - u_{tt}D_t(\theta), \quad \psi_{[xx]} = D_x(\psi_{[x]}) - u_{xx}D_x(\phi) - u_{tx}D_x(\theta), \tag{13}$$

and

$$\begin{aligned} \psi_{[tx]} &= D_t(\psi_{[x]}) - u_{xx}D_t(\phi) - u_{tx}D_t(\theta) \\ &= D_x(\psi_{[t]}) - u_{tx}D_x(\phi) - u_{tt}D_x(\theta) \end{aligned}. \tag{14}$$

The explicit expressions for $\psi_{[tt]}$, $\psi_{[xx]}$, $\psi_{[tx]}$ given by Eqs. (13) and (14) are

$$\begin{aligned} \psi_{[tt]} &= \psi_{tt} - \phi_{tt}u_x + (2\psi_{tu} - \theta_{tt})u_t - 2\phi_{tu}u_xu_t + (\psi_{uu} - 2\theta_{tu})u_t^2 - \phi_{uu}u_xu_t^2 \\ &\quad - \theta_{uu}u_t^3 - 2\phi_t u_{xt} - 2\phi_u u_t u_{xt} + (\psi_u - 2\theta_t)u_{tt} - \phi_u u_x u_{tt} - 3\theta_u u_t u_{tt}, \end{aligned} \tag{15}$$

$$\begin{aligned} \psi_{[xx]} &= \psi_{xx} + (2\psi_{xu} - \phi_{xx})u_x - \theta_{xx}u_t + (\psi_{uu} - 2\phi_{xu})u_x^2 - 2\theta_{xu}u_xu_t - \phi_{uu}u_x^3 \\ &\quad - \theta_{uu}u_x^2u_t + (\psi_u - 2\phi_x)u_{xx} - 2\theta_x u_{xt} - 3\phi_u u_x u_{xx} - \theta_u u_t u_{xx} - 2\theta_u u_x u_{xt}, \end{aligned} \tag{16}$$

$$\begin{aligned} \psi_{[xt]} &= \psi_{xt} + (\psi_{tu} - \phi_{xt})u_x + (\psi_{xu} - \theta_{xt})u_t - \phi_{tu}u_x^2 + (\psi_{uu} - \phi_{xu} - \theta_{tu})u_xu_t \\ &\quad - \theta_{xu}u_t^2 - \phi_{uu}u_x^2u_t - \theta_{uu}u_xu_t^2 - \phi_t u_{xx} - \phi_u u_t u_{xx} + (\psi_u - \phi_x - \theta_t)u_{xt} \\ &\quad - 2\phi_u u_x u_{xt} - 2\theta_u u_t u_{xt} - \theta_x u_{tt} - \theta_u u_x u_{tt}. \end{aligned}$$

For any interval I in $\mathbb{R}$ and for any open set D, if $G : I \times D^7 \to \mathbb{R}$ is a differentiable function, then by the invariance condition we should have,

$$\begin{aligned} 0 &= H(\bar{t}, \bar{x}, \bar{u}, \bar{u}_{\bar{t}}, \bar{u}_{\bar{x}}, \bar{u}_{\bar{t}\bar{t}}, \bar{u}_{\bar{x}\bar{x}}, \bar{u}_{\bar{t}\bar{x}}) \\ &= H(t + \epsilon\theta + O(\epsilon^2), x + \epsilon\phi + O(\epsilon^2), u + \epsilon\psi + O(\epsilon^2), u_t + \epsilon\psi_{[t]} + O(\epsilon^2), \\ &\quad u_x + \epsilon\psi_{[x]} + O(\epsilon^2), u_{tt} + \epsilon\psi_{[tt]} + O(\epsilon^2), u_{xx} + \epsilon\psi_{[xx]} + O(\epsilon^2), \\ &\quad u_{tx} + \epsilon\psi_{[tx]} + O(\epsilon^2)) \\ &= H(t, x, u, u_t, u_x, u_{tt}, u_{xx}, u_{tx}) + \epsilon\Big(\theta H_t + \phi H_x + \psi H_u + \psi_{[t]}H_{u_t} + \psi_{[x]}H_{u_x} \\ &\quad + \psi_{[tt]}H_{u_{tt}} + \psi_{[xx]}H_{u_{xx}} + \psi_{[tx]}H_{u_{tx}}\Big) + O(\epsilon^2). \end{aligned}$$

On comparing the coefficient of ϵ, we see that

$$\theta H_t + \phi H_x + \psi H_u + \psi_{[t]}H_{u_t} + \psi_{[x]}H_{u_x} + \psi_{[tt]}H_{u_{tt}} + \psi_{[xx]}H_{u_{xx}} + \psi_{[tx]}H_{u_{tx}} = 0.$$

For Eq. (16), the admitted Lie group has the infinitesimal generator given by

$$\zeta^* = \theta\frac{\partial}{\partial t} + \phi\frac{\partial}{\partial x} + \psi\frac{\partial}{\partial u}.$$

We have the extension of the prolongation given by,

$$\zeta^{(1)} = \theta G_t + \phi G_x + \psi G_u + \psi_{[t]}G_{u_t} + \psi_{[x]}G_{u_x} + \psi_{[tt]}G_{u_{tt}} + \psi_{[xx]}G_{u_{xx}} + \psi_{[tx]}G_{u_{tx}}. \quad (17)$$

The condition $\zeta^{(1)}H\mid_{G=0} = 0$, where $H(t, x, u, u_t, u_x, u_{tt}, u_{xx}, u_{tx}) = 0$, represents the invariance condition for second-order partial differential equations.

3 Symmetry Analysis of the Wave Equation

In this section, we perform symmetry analysis and use group methods to obtain solutions of,

$$\frac{\partial^2 u}{\partial t^2}(t, x) - c^2\frac{\partial^2 u}{\partial x^2}(t, x) = 0, \quad (18)$$

where for any open internal I in $\mathbb{R}$ and any open set D in $\mathbb{R}$, the real valued function u is defined on $I \times D$. Here, c is a real constant.

We now obtain symmetries and hence the solutions of

$$u_{tt}(t, x) - c^2 u_{xx}(t, x) = 0. \quad (19)$$

The invariance condition for Eq. (19) gives

$$\psi_{[tt]} - c^2\psi_{[xx]} = 0. \quad (20)$$

Substituting the values of $\psi_{[tt]}$ and $\psi_{[xx]}$ from Eqs. (15) and (16), we get the determining equations given by

$$\begin{aligned}
&\psi_{tt} - c^2\psi_{xx} - (\phi_{tt} + c^2(2\psi_{xu} - \phi_{xx}))u_x + (2\psi_{tu} - \theta_{tt} + c^2\theta_{xx})u_t - c^2(\psi_{uu} - 2\phi_{xu})u_x^2 \\
&- 2(\phi_{tu} + \theta_{xu})u_x u_t + (\psi_{uu} - 2\theta_{ut})u_t^2 + c^2\phi_{uu}u_x^3 + c^2\theta_{uu}u_x^2 u_t - \phi_{uu}u_x u_t^2 - \theta_{uu}u_t^3 \\
&- c^2(\psi_u - 2\phi_x)u_{xx} + (2c^2\theta_x - 2\phi_t)u_{xt} + c^2(\psi_u - 2\theta_t)u_{xx} + 3c^2\phi_u u_x u_{xx} + c^2\theta_u u_t u_{xx} \\
&\qquad + 2c^2\theta_u u_x u_{tx} - 2\phi_u u_t u_{xt} - c^2\phi_u u_x u_{xx} - 3c^2\theta_u u_t u_{xx} = 0. \quad (21)
\end{aligned}$$

Splitting Eq. (21) with respect to u_x, u_t, u_x^2, $u_x u_t$, u_t^2, u_x^3, $u_x^2 u_t$, $u_x u_t^2$, u_t^3, u_{xx}, u_{xt}, $u_x u_{xx}$, $u_t u_{xx}$, $u_x u_{tx}$, $u_t u_{xt}$, and the constant term, respectively, we get

$$-(\phi_{tt} + c^2(2\psi_{xu} - \phi_{xx})) = 0, \quad (22)$$

$$2\psi_{tu} - \theta_{tt} + c^2\theta_{xx} = 0, \quad (23)$$

$$-c^2(\psi_{uu} - 2\phi_{xu}) = 0, \quad (24)$$

$$-2(\phi_{tu} + \theta_{xu}) = 0, \tag{25}$$

$$\psi_{uu} - 2\theta_{ut} = 0, \tag{26}$$

$$c^2\phi_{uu} = 0, \tag{27}$$

$$c^2\theta_{uu} = 0, \tag{28}$$

$$-\phi_{uu} = 0, \tag{29}$$

$$-\theta_{uu} = 0, \tag{30}$$

$$c^2(\phi_x - \theta_t) = 0, \tag{31}$$

$$c^2\theta_x - \phi_t = 0, \tag{32}$$

$$2c^2\phi_u = 0, \tag{33}$$

$$-2c^2\theta_u = 0 = 0, \tag{34}$$

$$2c^2\theta_u = 0 = 0, \tag{35}$$

$$-2\phi_u = 0 = 0, \tag{36}$$

$$\psi_{tt} - c^2\psi_{xx} = 0. \tag{37}$$

From Eqs. (33), (34) and hence (26), we get,

$$\theta(t, x, u) = A(t, x), \quad \phi(t, x, u) = B(t, x), \quad \psi(t, x, u) = P(t, x)u + Q(t, x),$$

where A, B, P and Q are arbitrary functions of t and x.

From Eqs. (31), (32), (22) and (23), we get, $P_t = 0, \quad P_x = 0 \Rightarrow P(t, x) = p$, a constant.

Hence, $\psi = pu + Q(t, x)$, where Q satisfies the wave equation.

The generators of the four-dimensional admitted Lie group of the wave equation are given by

$$\zeta_1^* = A(t, x)\frac{\partial}{\partial t}, \quad \zeta_2^* = B(t, x)\frac{\partial}{\partial x}, \quad \zeta_3^* = u\frac{\partial}{\partial u}, \quad \zeta_4^* = Q(t, x)\frac{\partial}{\partial u}.$$

We now seek a solution of the wave equation by making a special choice of the infinitesimals, namely, $A = x, \quad B = t, \quad Q = 0$.

Table 1 Solutions of Eq. (38) for some integer values of p

p	F_1	F_2
1	1	r
2	$\frac{c^2r^2 - 2cr + 1}{c^2}$	$\frac{c^2r^2 + 2cr + 1}{c^2}$
3	$\frac{c^3r^3 - 3c^2r^2 + 3cr - 1}{c^3}$	$\frac{c^3r^3 + 3c^2r^2 + 3cr + 1}{c^3}$
4	$\frac{c^4r^4 - 4c^3r^3 + 6c^2r^2 - 4cr + 1}{c^4}$	$\frac{c^4r^4 + 4c^3r^3 + 6c^2r^2 + 4cr + 1}{c^4}$
−1	$\frac{c}{cr - 1}$	$\frac{c}{cr + 1}$
−2	$\frac{c^2}{c^2r^2 - 2cr + 1}$	$\frac{c^2}{c^2r^2 + 2cr + 1}$
−3	$\frac{c^3}{c^3r^3 - 3c^2r^2 + 3cr - 1}$	$\frac{c^3}{c^3r^3 + 3c^2r^2 + 3cr + 1}$
−4	$\frac{c^4}{c^4r^4 - 4c^3r^3 + 6c^2r^2 - 4cr + 1}$	$\frac{c^4}{c^4r^4 + 4c^3r^3 + 6c^2r^2 + 4cr + 1}$

The associated invariant surface condition is $xu_x + tu_t = pu$, which is solved to get $u = x^p F\left(\frac{t}{x}\right)$, where F is an arbitrary function.

Substituting in the one-dimension wave equation, we get,

$$(c^2r^2 - 1)F''(r) - 2c^2r(p-1)F'(r) + c^2p(p-1)F(r) = 0 \quad \text{where} \quad r = \frac{t}{x}. \tag{38}$$

This can be integrated easily giving,

$$F(r) = \begin{cases} \frac{c_1}{c_2} log\left(\frac{rc-1}{rc+1}\right) & \text{if } p = 0 \\ c_1\left(r - \frac{1}{c}\right)^p + c_2\left(r + \frac{1}{c}\right)^p & \text{if } p \neq 0 \end{cases}.$$

If $p = 0$, exact solution is

$$u(x,t) = \frac{c_1}{c^2} log\left(\frac{ct - x}{ct + x}\right) + c_2. \tag{39}$$

If $p \neq 0$, exact solution is

$$u(x,t) = x^p\left[c_1F_1\left(\frac{t}{x}\right) + c_2F_2\left(\frac{t}{x}\right)\right] \tag{40}$$

where $F_i, i = 1, 2$ are solutions of Eq. (38).

The solutions of Eq. (38) for some integer values of p are presented in Table 1.

We shall conclude the analysis for the wave equation by giving as an example, the particular solution of the wave equation with unit speed.

Example 4 Consider the wave equation with unit speed given by

$$u_{tt}(t,x) - u_{xx}(t,x) = 0.$$

To obtain a particular solution, consider, $A = t, \quad B = x, \quad P = Q = 0.$

The associated invariant surface condition is $tu_x + xu_t = 0$.

The solution of this linear partial differential equation is $u(x,t) = F(t^2 - x^2)$, where F is arbitrary.

To find F, we use the wave equation $u_{tt} = u_{xx}$, and get $rF''(r) + F'(r) = 0$, where $r = t^2 - x^2$.

The solution of this Clairaut differential equation is
$F(r) = c_3 ln(r) + c_4$, where c_3, c_4 are arbitrary constants.

Resubstituting r, we get,

$$u(x,t) = c_3 ln(t^2 - x^2) + c_4, \tag{41}$$

which is the exact solution of the one-dimensional unit speed wave equation.

4 Conclusion

We have used group methods for the wave equation with constant speed and have obtained its equivalent symmetries and explicit solutions.

1. The invariance condition for partial differential equations of the second order has been found using Taylor's theorem for a multivariable function.
2. The wave Eq. (19) admits the four-dimensional Lie group with the most general infinitesimal generator given by

$$\zeta^* = A(t,x)\frac{\partial}{\partial t} + B(t,x)\frac{\partial}{\partial x} + (pu + Q(t,x))\frac{\partial}{\partial u}.$$

3. The invariant surface condition has been obtained using which the exact solutions of the wave equation have been found.
4. The wave equation with general constant speed admits solutions (found by symmetry analysis) and is given by Eqs. (39) and (40).
5. The wave equation with unit speed admits solutions (found by symmetry analysis) and is given by Eq. (41).

References

1. Ablowitz MJ, Clarkson PA (1991) Soliton. Nonlinear evolution equations and inverse scattering. Cambridge University Press, Cambridge
2. Bluman GW, Kumei S (1989) Symmetries and differential equations. Springer, New York
3. Bibi K, Feroze T (2020) Discrete symmetry group approach for numerical solution of the heat equation. Symmetry J 12(359). https://doi.org/10.3390/sym12030359
4. Cariello F, Tabor M (1989) Painleve expansions for nonintegrable evolution equations. Phys D 39:77–94
5. Dressner L (1999) Applications of lie's theory of ordinary and partial differential equations. Institute of Physics Publishing, Bristol and Philadelphia
6. El Kinani EH, Ouhadan A (2015) Lie symmetry analysis of some time fractional partial differential equations. Int J Mod Phys: Conf Ser 38:1–8
7. Gu CH (1995) Solition theory and its applications. Springer, Berlin
8. Hirota R (1971) Exact solution of the Korteweg-de Vries equation for multiple collisions of solitons. Phys Rev Lett 27:1192–1194
9. Ibragimov NH (1996) CRC handbook of lie group analysis of differential equations 2(1), 3(1). Boca Raton: CRC Press, Florida
10. Khurshudyan AZ (2018) Nonlinear green's function for wave equation with quadratic and hyperbolic potentials. Advaces Math Phys 7179160. https://doi.org/10.1155/2018/7179160
11. Kumar V, Koksal ME (2018) Lie symmetry based-analytical and numerical approach for modified burgers-KdV equation. Results Phys 8:1136–1142
12. Lahno L, Zhdanov R, Magda O (2006) Group classification and exact solutions of nonlinear wave equations. Acta Appl Math 21:253–313
13. Lahno L, Zhdanov R (2005) Group classification of nonlinear wave equations. J Math Phys 46. https://doi.org/10.1063/1.1884886
14. Liu H, Zhang Q (2009) Lie symmetry analysis and exact explicit solutions for general Burgers' equation. J Comput Appl Math 228(1):1–9
15. Matveev VB, Salle MA (1991) Darboux transformations and solitons. Springer, Berlin
16. Molati M, Khalique CM (2013) Lie group analysis of a forced KdV equation. Math Probl Eng 845843. https://doi.org/10.1155/2013/845843
17. Nadjafikhah M, Shirvani-Sh V (2011) Lie symmetry analysis of Kudryashov-Sinelshchikov equation. Math Probl Eng 457697. https://doi.org/10.1155/2011/457697
18. Nadjafikhah M (2009) Lie symmetries of inviscid Burgers' equation. Adv Appl Clifford Algebr 19(1):101–112
19. Nadjafikhah M, Mahdavi A (2013) Approximate symmetry analysis of a class of perturbed nonlinear Reaction-Diffusion equations. Abstr Appl Anal 395847:1–7
20. Nöether E (1918, 1971) Invariante variationsprobleme. Math Phy Kl, 235–257. English translation, Tramp Th Stat, Phys 1:189–207
21. Oliveri F (2010) Lie symmetries of differential equations: classical results and recent contributions. Symmetry J 2(2):658–706
22. Pulov V, Uzunov IM (2008) Finding lie symmetries of partial differential equations with MATHEMATICA. In: Proccedings of conference on geometry integrability and qunatization. Bulgaria, vol 9, pp 280–291
23. Sabri M, Rasheed M (2017) On the solution of wave equation in three dimensions using D'Alembert formula. Int J Math Trends Technol 49(5):311–315
24. Singh R, Chandra M, Singh BK (2015) Solution of 3-dimensional wave equation by method of separation of variables. Int J Curr Res Rev 7(14):54–56
25. Srihirun B, Meleshko SV, Schulz E (2006) On the definition of an admitted lie group for stochastic differential equations with multi-Brownian motion. J Phys A 39:13951–13966
26. Vorobyova A (2002) Symmetry analysis of equations of mathematical physics. Proc Inst Math NAS Ukr 43(1):252–255
27. Zheng J (2020) Lie symmetry analysis and invariant solutions of a nonlinear Fokker-Planck equation describing cell population growth. Adv Math Phys 4975943. https://doi.org/10.1155/2020/4975943

Traveling Wave Solutions of Some Coupled Nonlinear Systems by Sine–Cosine Approach

Sidheswar Behera and Jasvinder Singh Virdi

Abstract In this work, we make an analysis of some coupled nonlinear systems by sine–cosine approach and obtained distinct physical structures as traveling wave solutions. The mathematical formulation of traveling wave solutions and 3D graphical simulations for different solutions includes periodic solutions, soliton solutions, kink solutions, and some hybrid-type solutions are reported. It is appraised that the newly obtained trigonometric and hyperbolic function solutions reported in this paper may assist to describe some complex physical features in the nonlinear physical world.

Keywords Sine–cosine approach · Maccari system · Coupled Higgs equation · Coupled Konno–Oono equation

1 Introduction

It is commonplace to say that science has made astonishing progress during the past few decades, mathematics remains the language of science. Mathematical developments on nonlinear evolution equations (NLEEs) help to understand many nonlinear phenomena of nature such as non-differentiable problems arising in the areas of mechanics of solid-state physics ,transfer of heat and wave propagation, diffusion dynamics, hydrodynamics, vehicular traffic flow, and many more. Evidently, exact solutions to NLEEs always play a major role in nonlinear science, they bring more information and understanding about different aspects of the problem that can direct to fresh applications. Recently, a large number of research efforts have been speculated for the study of NLEEs by numerous methods. Some notable methods include variational iteration method [1–3], the $(\frac{G'}{G}, \frac{1}{G})$-expansion method [4, 5], truncated painlevae expansion method [6], the tanh-function method [7], Adomionos decomposition method [8, 9], the $(\frac{G'}{G})$ method [10–12], the Jacobi elliptic functions method [13], the sine–cosine method [14, 15], and many more. Among these methods, we are motivated to implement the sine–cosine method which is based on solving NLEEs

S. Behera (✉) · J. Singh Virdi
Department of Physics, Veer Surendra Sai University of Technology, Burla 768018, Odisha, India

© The Author(s), under exclusive license to Springer Nature Singapore Pte Ltd. 2022

D. Giri et al. (eds.), *Proceedings of the Seventh International Conference on Mathematics and Computing*, Advances in Intelligent Systems and Computing 1412,
https://doi.org/10.1007/978-981-16-6890-6_66

with less computational work and provides the nearly exact solution. The following nonlinear systems are taken to investigate with their graphical simulation, they are listed below.

1. *The nonlinear Maccari system (NMS)* [16]: this system is an important candidate in applied science, especially in mathematical physics, nonlinear optics, hydrodynamics, etc. Sometimes, it is used to model the motion of isolated waves trapped in a small isolated space. The mathematical structure of NMS

$$\begin{aligned} i g_t + g_{xx} + gh &= 0, \\ h_t + h_y + \left(|g|^2\right)_x &= 0. \end{aligned} \tag{1}$$

2. *The coupled nonlinear Higgs equation (CNHE)* [16]: with the help of this equation, we can study scalar nuclear interaction with neutral scalar mesons.

$$\begin{aligned} g_{tt} - g_{xx} + |g|^2 g - 2gh &= 0, \\ h_{tt} + h_{xx} - (|g|^2)_{xx} &= 0. \end{aligned} \tag{2}$$

3. *The coupled nonlinear Konno–Oono equation (CNKOE)* [17]; The CNKOE finds its application in mathematical physics such as in group theory and particle physics.

$$\begin{aligned} g_{xt} - 2gh &= 0, \\ g_t + 2g g_x &= 0. \end{aligned} \tag{3}$$

This article is structured as follows: Sect. 2 describes the method for obtaining exact traveling wave solutions of aforementioned NLEEs. Section 3 describes application of the proposed methods to three coupled equations: the nonlinear Maccari system, the coupled nonlinear Higgs equation and the coupled nonlinear Konno–Oono equation. Section 4 describes concluding remarks.

2 Mathematical Formulation of the Sine–Cosine Approach

We start our discussion with NLEE which can be expressed as a combination of different order-dependent and -independent terms and their partial derivatives.

$$S(g, g_t, g_x, g_{tt}, g_{xt}, g_{xx}, \ldots) = 0, \tag{4}$$

where $g = g(x, t)$ is a test function. The effective steps of the sine–cosine approach are illustrates as below:

Step 1:To find out the traveling wave solution of Eq. (4), the wave variable may have as below;

$$(\phi) = (x - ct), \tag{5}$$

therefore,

$$g(x, t) = G(\phi). \tag{6}$$

This may lead the following change:

$$\frac{\partial}{\partial t} = -c\frac{\partial}{\partial \phi}, \quad \frac{\partial^2}{\partial t^2} = c^2\frac{\partial^2}{\partial \phi^2}, \quad \frac{\partial}{\partial x} = \frac{\partial}{\partial \phi}, \quad \frac{\partial^2}{\partial x^2} = \frac{\partial^2}{\partial \phi^2} \tag{7}$$

and can be extended to higher orders. Eq. (4) can be written as an ordinary differential equation (ODE) by implementing Eq. (7),

$$S(G, G_\phi, G_{\phi\phi}, G_{\phi\phi\phi}, \ldots) = 0. \tag{8}$$

here G_ϕ denotes $\frac{dG}{d\phi}$.

We have to integrate the obtained ODE (8) as many times to get a comparatively simpler equation, and for simplicity assume the integration as zero.

Step 2: The solutions of nonlinear systems by sine–cosine approach can be expressed as below, the (cos) form of the solution

$$G(x, t) = \lambda \cos^\beta(\mu\phi), [\phi] < \frac{\pi}{2\mu} \tag{9}$$

and the (sin) form of the solution

$$G(x, t) = \lambda \sin^\beta(\mu\phi), [\phi] < \frac{\pi}{2\mu}, \tag{10}$$

where λ, β and μ are constants, can be determined later, μ stands for the wave number, and c stands for the wave speed. Equations (9) and (10) can be generalized.

$$\begin{aligned}
&G(\phi) = \lambda \cos^\beta(\mu\phi), \\
&G^n(\phi) = \lambda^n \cos^{n\beta}(\mu\phi), \\
&(G^n)_\phi = n\mu\beta\lambda^n \cos(\mu\phi)\sin^{n\beta-1}(\mu\phi) \\
&(G^n)_{\phi\phi} = -n^2\mu^2\beta^2\lambda^n \cos^{n\beta}(\mu\phi) + n\mu^2\lambda^n\beta(n\beta - 1)\cos^{n\beta-2}(\mu\phi),
\end{aligned} \tag{11}$$

$$
\begin{aligned}
&G(\phi) = \lambda \sin^{\beta}(\mu\phi), \\
&G^n(\phi) = \lambda^n \sin^{n\beta}(\mu\phi), \\
&(G^n)_{\phi} = n\mu\beta\lambda^n \sin(\mu\phi)\cos^{n\beta-1}(\mu\phi) \\
&(G^n)_{\phi\phi} = -n^2\mu^2\beta^2\lambda^n \sin^{n\beta}(\mu\phi) + n\mu^2\lambda^n\beta(n\beta-1)\sin^{n\beta-2}(\mu\phi),
\end{aligned} \tag{12}
$$

Step 3: After substituting Eq. (12) or Eq. (11) in Eq. (8), we will get trigonometric equations interns of $\cos^{\beta}(\mu\phi)$ or $\sin^{\beta}(\mu\phi)$. Successively, we will get other parameters such as α, β, and μ by solving algebraic system, they steps pointed as below:

1. First, we will determine β by balancing the exponents of trigonometric sine or cosine term,
2. Then the coefficients can be collected by same power rule of $\cos^{\beta}(\mu\phi)$ or $\sin^{\beta}(\mu\phi)$ terms after equating them zero separately, we will get algebraic systems,
3. Simplifying algebraic systems we will get the values of λ, μ, and β.

3 Sine–Cosine Approach and Exact Solutions

In this section, we used the sine–cosine approach, explained above is implemented to compute analytical traveling wave solutions of three aforementioned coupled system of equations.

3.1 *The Nonlinear Maccari System (NMS)*

We start our tour with the NMS, in order to get traveling wave solution of Eq. (1) by the sine–cosine approach, we are taking the wave transformation $g(x,t) = e^{i\theta}G(\phi), h = H(\phi)$ with the wave variable $\theta = px + qy + rt, \phi = (x + y + ct)$, Eq. (1) takes the form of an ODE.

$$
\begin{aligned}
&G'' - (r + p^2)G + GH = 0, \\
&(c+1)H' + 2GG'' = 0.
\end{aligned} \tag{13}
$$

We have to integrate the second equation of the system (13) to find out the value of $'H'$,

$$
-(c+1)H = G^2. \tag{14}
$$

Now substituting Eq. (14) into first equation of system (13) and by successive integration, we obtain (Fig. 1)

$$-(c+1)G'' - (c+1)(r+p^2)G - G^3 = 0. \tag{15}$$

Substituting Eq. (12) in Eq. (15), we will get

$$\begin{aligned}(c+1)\lambda\mu^2\beta(\beta-1)\cos^{\beta-2}(\mu\phi) - (c+1)\lambda\mu^2\beta^2\cos^{\beta}(\mu\phi)\\ -(c+1)(r+p^2)\lambda\cos^{\beta}\mu\phi - \lambda^3\cos^{3\beta}\mu\phi = 0.\end{aligned} \tag{16}$$

Now collecting the cosine terms pair wise by homogeneous balance rule and equating them to zero, we will get algebraic systems as below:

$$\begin{aligned}&(\beta-1) \neq 0,\\ &(\beta-2) = 3\beta,\\ &(r+p^2) = \mu^2\beta^2,\\ &\lambda^2 = -(c+1)\mu^2\beta(\beta-1).\end{aligned} \tag{17}$$

After solving these, we will get

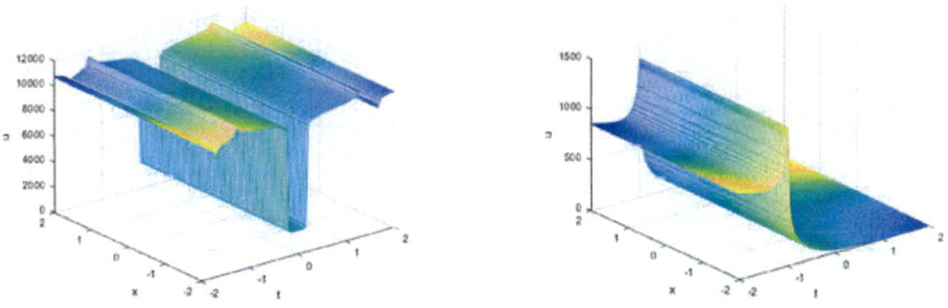

(a) Plot of $g_1(left)$ and $g_2(right)$ when $c = -0.5, r = 1, p = 2$.

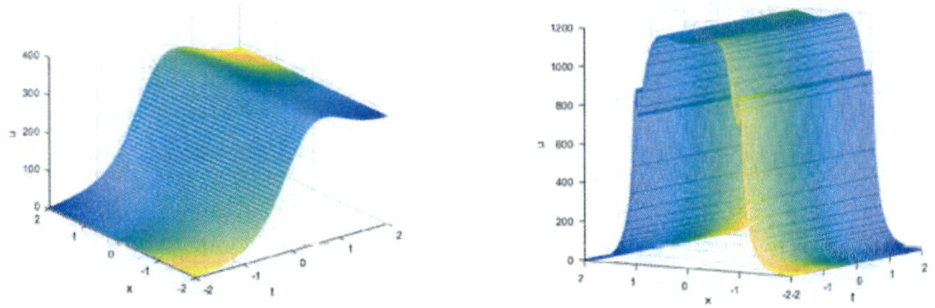

(b) Plot of $g_3(left)$ and $g_4(right)$ when $c = 2, r = 3, p = -2$.

Fig. 1 Traveling wave solution corresponding to the nonlinear Maccari system

$$\beta = -1, \mu = \sqrt{r + p^2}, \lambda = -2(c+1)(r + p^2). \tag{18}$$

The traveling wave solutions can be successfully written as, for $c < 0$

$$g_1(x,t) = e^{i(px+qy+rt)}\left[-2(c+1)(r+p^2)\sec\left[\sqrt{r+p^2}(x+y+ct)\right]\right], c < 0, \tag{19}$$

$$g_2(x,t) = e^{i(px+qy+rt)}\left[-2(c+1)(r+p^2)\csc\left[\sqrt{r+p^2}(x+y+ct)\right]\right]. c < 0, \tag{20}$$

Similarly, for $c > 0$,

$$g_3(x,t) = e^{i(px+qy+rt)}\left[-2(c+1)(r+p^2)\,\mathrm{sech}\left[\sqrt{r+p^2}(x+y+ct)\right]\right], c > 0, \tag{21}$$

$$g_4(x,t) = e^{i(px+qy+rt)}\left[-2(c+1)(r+p^2)\,\mathrm{csch}\left[\sqrt{r+p^2}(x+y+ct)\right]\right], c > 0. \tag{22}$$

3.2 The Coupled Nonlinear Higgs Equation (CNHE)

Now we consider the CNHE, in order to construct traveling wave solutions of Eq. (2) by the sine–cosine approach, we are taking the wave transformation $g(x,t) = e^{i\theta}G(\phi), h = H(\phi)$ with the wave variable $\theta = px + rt, \phi = (x + ct)$ Eq. (2) takes the form of an ODE.

$$\begin{aligned} &(c^2-1)G'' + (p^2 - r^2)G - 2GH + G^3 = 0, \\ &(c^2+1)H'' - 2(G')^2 - 2GG'' = 0. \end{aligned} \tag{23}$$

We have to integrate the second equation of the system (23) to find out the value of $'H'$,

$$(c^2+1)H = G^2. \tag{24}$$

Now substitute Eq. (24) into first part of the system (23) and assume integration constant as zero. We obtain

$$(c^4-1)G'' + (c^2+1)(p^2-r^2)G + (c^2-1)G^3 = 0. \tag{25}$$

substituting Eq. (12) in Eq. (25), we will get

$$(c^2+1)(p^2-r^2)\lambda\cos^{\beta}\mu\phi+(c^4-1)\lambda^3\cos^{3\beta}\mu\phi+(c^4-1)\lambda\mu^2\beta(\beta-1)\cos^{\beta-2}(\mu\phi)$$
$$-(c^4-1)\lambda\mu^2\beta^2\cos^{\beta}(\mu\phi)=0. \quad (26)$$

Now collecting the cosine terms pair wise by homogeneous balance rule and equating them to zero, we will get algebraic systems as below:

$$\begin{aligned}&(\beta-1)\neq 0,\\&(\beta-2)=3\beta,\\&(p^2-r^2)=(1-c^2)\mu^2\beta^2,\\&\lambda^2=(c^4-1)\mu^2\beta(\beta-1).\end{aligned} \quad (27)$$

After solving these, we will get

$$\beta=-1,\ \mu=\frac{p^2-r^2}{1-c^2}\sqrt{a-c},\ \lambda=\sqrt{3(c^2+1)(r^2-p^2)}. \quad (28)$$

The following traveling wave solutions can be constructed, for $c^2>0$ (Fig. 2)

$$g_5(x,t)=e^{i(px+rt)}\left[\sqrt{3(c^2+1)(r^2-p^2)}\sec\left[\sqrt{\frac{r^2-p^2}{c^2-1}}(x+ct)\right]\right],c^2>0 \quad (29)$$

$$g_6(x,t)=e^{i(px+rt)}\left[\sqrt{3(c^2+1)(r^2-p^2)}\csc\left[\sqrt{\frac{r^2-p^2}{c^2-1}}(x+ct)\right]\right].c^2>0 \quad (30)$$

Similarly, for $c^2<0$

$$g_7(x,t)=e^{i(px+rt)}\left[\sqrt{3(c^2+1)(r^2-p^2)}\,\mathrm{sech}\left[\sqrt{\frac{r^2-p^2}{c^2-1}}(x+ct)\right]\right],c^2<0, \quad (31)$$

$$g_8(x,t)=ie^{i(px+rt)}\left[\sqrt{3(c^2+1)(r^2-p^2)}\,\mathrm{csch}\left[\sqrt{\frac{r^2-p^2}{c^2-1}}(x+ct)\right]\right],c^2<0. \quad (32)$$

(a) Plot of $g_5(left)$ and $g_6(right)$ when $c = 2, r = 3, p = 1$.

(b) Plot of $g_7(left)$ and $g_8(right)$, when $c = 3i, r = 2, p = 1$.

Fig. 2 Traveling wave solution corresponding to nonlinear Higgs equation

3.3 The Coupled Nonlinear Konno–Oono Equation (CNKOE)

Finally, we consider the CNKOE, in order to construct traveling wave solutions of Eq. (3) by the sine–cosine approach, we are considering the wave transformation $g(x,t) = G(\phi), h(x,t) = H(\phi)$ with the wave variable $\phi = (x - ct)$ Eq. (3) takes the form of an ODE.

$$\begin{aligned} -cG'' - 2GH &= 0, \\ -cH' + 2GG' &= 0. \end{aligned} \tag{33}$$

After integrating second equation of the system (33) with respect to (ϕ), we obtain

$$H = \frac{1}{c}(G^2 + d), \tag{34}$$

where d is an integral constant. We will get the following equation when we substitute Eq. (34) into the first equation of the system (33):

$$c^2G'' + Gd + 2G^3 = 0. \tag{35}$$

Substituting Eq. (12) in Eq. (35), we will get

$$2d\lambda\cos^{\beta}\mu\phi + 2\lambda^3\cos^{3\beta}\mu\phi + c^2\lambda\mu^2\beta(\beta-1)\cos^{\beta-2}(\mu\phi) - c^2\lambda\mu^2\beta^2\cos^{\beta}(\mu\phi) = 0. \quad (36)$$

Now collecting the cosine terms pair wise by homogeneous balance rule and equating them to zero, we will get algebraic systems as below:

$$\begin{aligned} &(\beta-1) \neq 0, \\ &3\beta = (\beta-2), \\ &2d = -c^2\mu^2\beta^2, \\ &2\lambda^2 = c^2\mu^2\beta(\beta-1). \end{aligned} \quad (37)$$

After solving the above Eq. (37), we will get

$$\beta = -1, \mu = \sqrt{\frac{-2d}{c^2}}, \lambda = \mp 2\sqrt{d}. \quad (38)$$

The following traveling wave solutions can be constructed, for $\lambda > 0$ (Fig. 3)

$$g_9(x,t) = 2\sqrt{d}\sec\left[\sqrt{\frac{-2d}{c^2}}(x-ct)\right], \lambda > 0, \quad (39)$$

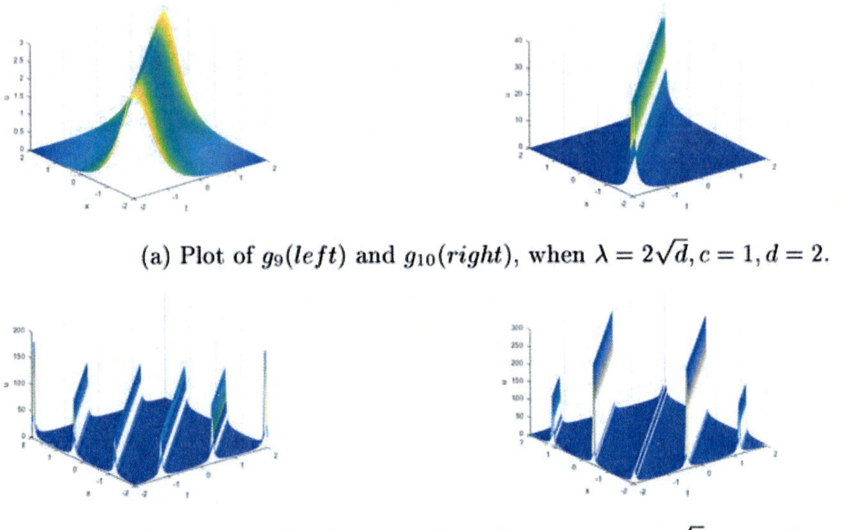

(a) Plot of $g_9(left)$ and $g_{10}(right)$, when $\lambda = 2\sqrt{d}, c = 1, d = 2$.

(b) Plot of $g_{11}(left)$ and $g_{12}(right)$, when $\lambda = -2\sqrt{d}, c = 1, d = 2$.

Fig. 3 Traveling wave solution corresponding to the Konno–Oono equation

$$g_{10}(x, t) = 2\sqrt{d} \csc\left[\sqrt{\frac{-2d}{c^2}}(x - ct)\right], \lambda > 0. \quad (40)$$

Similarly, for $\lambda < 0$

$$g_{11}(x, t) = 2\sqrt{d} \operatorname{sech}\left[\sqrt{\frac{-2d}{c^2}}(x - ct)\right], \lambda < 0, \quad (41)$$

$$g_{12}(x, t) = 2\sqrt{d} \operatorname{csch}\left[\sqrt{\frac{-2d}{c^2}}(x - ct)\right], \lambda < 0. \quad (42)$$

4 Conclusion

It is fascinating to notice that the general traveling wave solutions give some soliton, periodic soliton, kink solution, and hybrid solutions under some special conditions. This method can be used for seeding solutions for other methods. This method is more effective, reliable, compact, concise, and involves less computational work compare to other such approaches.

However, the trigonometric and hyperbolic functions discussed in this work having different physical meanings as well. It is observed that the solutions presented in this work have certain physical meanings, for example, the secant hyperbolic basically appears in the laminar jet profile. It is believed that the proposed hyperbolic solutions in this work may also be used to describe certain complex physical phenomena in the nonlinear physical world and some of their properties. To the best of our knowledge, the application of the sine–cosine method to the proposed coupled system of equations has not been submitted elsewhere.

References

1. Safari M, Ganji DD, Moslemi M (2009) Computers mathematics with applications 58:2091–2097. https://doi.org/10.1016/j.camwa.2009.03.043
2. He JH (2004) Variational principles for some non-linear partial differential equations with variable coefficients. Chaos, Solitons Fractals 19:847–851. https://doi.org/10.1016/S0960-0779(03)00265-0
3. He JH (2000) Variational iteration method for autonomous ordinary differential systems. Appl Math Comput 114:115–123. https://doi.org/10.1016/S0096-3003(99)00104-6
4. Miah MM, Ali HMS, Akbar MA, Wazwaz AM (2017) Some applications of the $(\frac{G'}{G}, \frac{1}{G})$ - expansion method to find new exact solutions of NLEEs. Eur Phys J Plus 132(252). https://doi.org/10.1140/epjp/i2017-11571-0

5. Zayed EME, Abdelaziz MAM (2012) Two-variable the, $(\frac{G'}{G}, \frac{1}{G})$-expansion method for solving the nonlinear KdV-mKdV equation Hindawi publishing corporation, vol 725061, 14 p. https://doi.org/10.1155/2012/725061
6. Kudryashov NA (2018) Exact solutions and integrability of the duffing—van der pol equation. Eur Phys J Plus 23:471–479. https://doi.org/10.1134/S156035471804007X
7. Zahrana EM, Khater MMA (2016) Modified extended tanh-function method and its applications to the Bogoyavlenskii equation. Appl Math Modeling 40:1769–1775. https://doi.org/10.1016/j.apm.2015.08.018
8. El-Wakil SA, Abdou MA (2007) New applications of Adomian decomposition method. Chaos, Solitons Fractals 33:513–522. https://doi.org/10.1016/j.chaos.2005.12.037
9. Bildik N, Konuralp A (2006) The use of variational iteration method, differential transform method and adomian decomposition method for solving different types of nonlinear partial differential equations. Int J Non Sci Num Simltn 7:65–70. https://doi.org/10.1515/IJNSNS.2006.7.1.65
10. Jabbari A, Kheiri H, Bekir A (2011) Exact solutions of the coupled Higgs equation and the Maccari system using He's semi-inverse method and $(\frac{G'}{G})$-expansion method. Comp Math Appl 62:2177–2186. https://doi.org/10.1016/j.camwa.2011.07.003
11. Bian C, Pang J, Jin L, Ying X (2010) Solving two fifth order strong nonlinear evolution equations by using the $(\frac{G'}{G})$-expansion method. Comm Non Sci Num Smln 15:2337–2343. https://doi.org/10.1016/j.cnsns.2009.10.006
12. Zayed EME, Khaled A (2009) Gepreel, Some applications of the $(\frac{G'}{G})$ -expansion method to non-linear partial differential equations. App Math Compun 212:1–13. https://doi.org/10.1016/j.amc.2009.02.009
13. Cveticanin L (2011) The approximate solving methods for the cubic duffing equation based on the Jacobi elliptic functions. Int J Non Sci Numl Simltn 10:1491–1516. https://doi.org/10.1515/IJNSNS.2009.10.11-12.1491
14. Yusufoglu E, Bekir A (2008) The tanh and the sine-cosine methods for exact solutions of the MBBM and the Vakhnenko equations. Chaos, Solitons Fractals 38:1126–1133. https://doi.org/10.1016/j.chaos.2007.02.004
15. Wazaz AM (2004) A sine-cosine method for handling nonlinear wave equations. Math Comput Model 40:499–508. https://doi.org/10.1016/j.mcm.2003.12.010
16. Hafez MG, Alam MN, Akbar MA (2015) Traveling wave solutions for some important coupled nonlinear physical models via the coupled Higgs equation and the Maccari system. J King Saud Univ Sci 27(2):105–112. https://doi.org/10.1016/j.jksus.2014.09.001
17. Konno K, Oono H (1994) New coupled integrable dispersionless equations. J Phys Soc Jpn 63:377–378. https://doi.org/10.1143/jpsj.63.377

Several Families of New Exact Solutions for Wave-Like Equations with Variable Coefficients

Subin P. Joseph

Abstract Several families of new exact solutions for second-order wave-like partial differential equations with variable coefficients are derived in this paper. To derive these solutions, certain recurrence relations for powers of the variables and coefficients are formed using the method of balancing powers of the variables. Solving these recurrence relations, several exact solutions are derived. We derive all polynomial solutions of these partial differential equations. Different families of new exact solutions of these equations which can be expressed in terms of hypergeometric functions are also constructed. Euler–Tricomi equation, Keldysh equation, and generalized Euler–Tricomi equations are special cases of this general equations.

Keywords Exact solutions · Polynomial solutions · Hypergeometric functions · Wave equation · Euler–Tricomi equation · Keldysh equation

1 Introduction

Several real-world physical problems can be formulated in terms of partial differential equations. Exact solutions of these partial differential equations are inevitable because they give more insight in to the problem compared to numerical or approximate solutions. We derive new families of exact polynomial solutions and non-polynomial solutions of certain second-order partial differential equations with variable coefficients. Some of the polynomial solutions of constant coefficient partial differential equations have been obtained in [5–8, 10–12, 16, 20, 23]. The polynomial solutions of variable coefficient ordinary differential equations of second degree have been discussed in [1, 21]. But the polynomial solutions of variable coefficient partial differential equations are little in the literature. In this paper, we consider the generalizations of two important partial differential equations in the field of applied

S. P. Joseph (✉)
Government Engineering College, Wayanad, Thalapuzha P.O., Wayanad, KL, India
e-mail: subinpj@gecwyd.ac.in

© The Author(s), under exclusive license to Springer Nature Singapore Pte Ltd. 2022

D. Giri et al. (eds.), *Proceedings of the Seventh International Conference on Mathematics and Computing*, Advances in Intelligent Systems and Computing 1412,
https://doi.org/10.1007/978-981-16-6890-6_67

mathematics and mechanics. The partial differential equation that we consider is the following wave-like equation with variable coefficients of the form

$$\alpha y^m \frac{\partial^2 f}{\partial x^2} + \beta x^n \frac{\partial^2 f}{\partial y^2} = 0, \tag{1}$$

which generalizes the usual one-dimensional wave equation. Here, α, β, m, and n are real parameters and f is a function of x and y. This equation can also be considered as a generalization of two-dimensional Laplace equation. Several other partial differential equations in applied mathematics and mechanics are particular cases of these generalized equations. Euler–Tricomi Equation

$$\frac{\partial^2 f}{\partial x^2} + x \frac{\partial^2 f}{\partial y^2} = 0 \tag{2}$$

is a particular case of the wave-like Eq. (1). This equation is an example of a mixed-type equation. It is clearly elliptic in the half plane $x > 0$, parabolic at $x = 0$ and hyperbolic in the half plane $x < 0$. There are several applications for this equation in fluid mechanics and differential geometry. Several problems like transonic flow problems and isometric embedding problems can be reduced to corresponding problems for the Euler–Tricomi equation [2, 9, 13, 19, 23].

Another important equation is the Keldysh equation which is given by

$$x \frac{\partial^2 f}{\partial x^2} + \frac{\partial^2 f}{\partial y^2} = 0, \tag{3}$$

where f is a function of the variables x and y. This equation is also having many applications in the fields such as non-geometrical optics and models of zero temperature plasma to which a magnetic field has been applied [2, 3, 13, 18]. In addition to these equations, we can also consider the generalized version of the Euler–Tricomi equation given by

$$\frac{\partial^2 f}{\partial x^2} + \beta x^p \frac{\partial^2 f}{\partial y^2} = 0, \tag{4}$$

where p can be any real number. These equations have been studied by several authors with applications to the Theory of Plane Transonic Flow [4, 9, 24].

These equations and their generalizations have a large number of applications in different fields of mathematical physics such as fluid mechanics, mechanics, elasticity, propagation of sounds, relativity theory, etc. So finding new exact solutions of these equations will lead us to a better understanding of theories related to these equations.

In this paper, we are deriving new exact solution of the wave-like Eq. (1). First of all, we derive all polynomial solutions of these equations and then we derive the other new exact solutions in terms of hypergeometric functions. A new method of balancing powers of the variables is proposed to obtain the polynomial solutions of these equations in the variables x and y. Assuming a polynomial solution for these equations we derive recurrence formula for the powers of the variables as well as recurrence formula for the coefficients of the variables. We solve these recurrence relations to find the required polynomial solutions. The obtained solutions are not always polynomial solutions. Depending upon the value of the parameters m and n, the derived solutions can have different forms. Some of them become polynomials multiplied by some power functions of the variables x and y. In some other cases, we get solutions in which the powers of the variables are non-positive integers. We can call all these solutions as polynomial-like solutions. After obtaining these polynomial or polynomial-like solutions, they are used to generate other families of exact solutions which generalize the polynomial solutions. We derive different set of families of exact solutions of these equations in the following sections. In the second section, we derive all polynomial and polynomial-like solutions of wave-like Eq. (1). Also we derive other exact solutions which can be expressed in terms of hypergeometric functions. From these solutions, we can find all polynomial solutions and other exact solutions of the Euler–Tricomi equation, Keldysh Equation, generalized Euler–Tricomi equation, and wave equation, which are having important applications in the field of applied mathematics and mechanics. Finally, we conclude our paper with a discussion on the results obtained.

2 Exact Solutions of Wave-Like Equation

We are going to derive several families of exact solutions of the wave-like equation with variable coefficients given by Eq. (1) in this section. All possible polynomial solutions in the variables x and y are derived for this equation. We derive two sets of families of such polynomial or polynomial-like solutions of the Eq. (1) in the following two subsections. We generalize all these solutions to new families of exact solutions of Eq. (1) expressed in terms of hypergeometric functions. All the polynomial or polynomial-like solutions are particular cases of these family of general solutions.

2.1 First Set of Exact Solutions

Here, we derive polynomial solutions of the wave-like Eq. (1). After finding the polynomial solutions, we derive the exact solutions which generalizes the polynomial solutions. We characterize the polynomial solutions of this equation in the following theorem.

Theorem 1 *The different families of exact polynomial solutions of the wave-like Eq. (1) are given by*

$$f_1(x, y) = \sum_{k=1}^{N_1} c_k\, x^{(k-1)(n+2)}\, {}_2F_1\left(1-k, -k+\frac{1}{n+2}+1; \frac{m+1}{m+2}; u\right), \tag{5}$$

$$f_2(x, y) = \sum_{k=1}^{N_2} c_k\, yx^{(k-1)(n+2)}\, {}_2F_1\left(1-k, -k+\frac{1}{n+2}+1; 1+\frac{1}{m+2}; u\right), \tag{6}$$

$$f_3(x, y) = \sum_{k=1}^{N_3} c_k\, x^{k(n+2)-n-1}\, {}_2F_1\left(1-k, \frac{n+1}{n+2}-k; \frac{m+1}{m+2}; u\right), \tag{7}$$

and

$$f_4(x, y) = \sum_{k=1}^{N_4} c_k\, yx^{k(n+2)-n-1}\, {}_2F_1\left(1-k, \frac{n+1}{n+2}-k; 1+\frac{1}{m+2}; u\right), \tag{8}$$

where

$$u = -\frac{\alpha\ (n+2)^2 y^{m+2}}{\beta\ (m+2)^2 x^{n+2}}$$

and for each summation, c_k are any constants and N_1, N_2, N_3, and N_4 are arbitrary positive integers.

Here the hypergeometric function [15] ${}_2F_1(a, b; c; z)$ is given by

$${}_2F_1(a, b; c; z) = \sum_{r=0}^{\infty} \frac{(a)_r (b)_r}{r!(c)_r} z^r$$

with $(a)_r = a(a+1)(a+2)\dots(a+r-1)$ is the Pochhammer symbol. The function ${}_2F_1(a, b; c; z)$ will be a polynomial of degree k in z when a or b is a non-positive integer, $-k$, and $c \neq 0, -1, -2, -3, \dots$ [15]. So it follows from the above equations that the given solutions are always polynomials in x and y, provided that the parameters m and n are positive integers. Otherwise, we get polynomial-like solutions, that is, solutions in which the powers of the variables need not be positive integers.

Proof We consider a general polynomial in the variables x and y of the form

$$a_1 x^{m_1} y^{n_1} + a_2 x^{m_2} y^{n_2} + \cdots + a_k x^{m_k} y^{n_k}. \tag{9}$$

This is a solution to the wave-like equation if this polynomial satisfies the partial differential equation (1). We have to obtain the conditions under which this polynomial is a solution to this partial differential equation. Substituting this in the equation and rearranging the terms, we get

$$\sum_{i=2}^{k}\left(\alpha a_{i-1}\left(m_{i-1}-1\right)m_{i-1}x^{m_{i-1}-2}y^{n_{i-1}+m}+a_i\beta\left(n_i-1\right)n_iy^{n_i-2}x^{m_i+n}\right)$$
$$+a_1\beta\left(n_1-1\right)n_1y^{n_1-2}x^{m_1+n}+\alpha a_k\left(m_k-1\right)m_kx^{m_k-2}y^{n_k+m}=0. \quad (10)$$

The above grouping of terms have done in such a way that the recurrence relation of powers of the variables and the recurrence relation of coefficients can be easily derived. We can also get another possible grouping of terms different from above so that we are able to obtain polynomial solutions. This possible grouping will be discussed in the next subsection. The above equation is satisfied if each term in the summation vanish and also the last two terms vanish. First, we consider the terms given in the summation. We have to determine the recurrence relations so that balancing of the terms in the parentheses are possible. When this is made possible, we get the following relations among the powers and coefficients. To balance the powers of x and y in the braces, the powers should be such that $m_i=m_{i-1}-n-2$ and $n_i=n_{i-1}+m+2$. Then the terms in the braces will vanish if we choose $a_i=-\frac{\alpha a_{i-1}(m_{i-1}-1)m_{i-1}}{\beta(n_i-1)n_i}$. The second last term will vanish if we choose $n_1=0$ or $n_1=1$. Now the last term vanishes only when $m_k=0$ or $m_k=1$. So, we need to consider four different cases according to the values of m_k and n_1.

Case 1: When $\mathbf{m_k=0}$ and $\mathbf{n_1=0}$ In this case, the recurrence relation for the power of x is $m_i=m_{i-1}-n-2$, for $2\le i\le k$, with $m_k=0$. Solving this recurrence relation, we get $m_i=(n+2)(k-i)$, for $1\le i\le k$. So $m_1=(k-1)(n+2)$. Next we consider the recurrence relation for the power of y. We have the recurrence relation $n_i=n_{i-1}+m+2$, $1\le i\le k-1$, with $n_1=0$. Solving this recurrence relation, we get $n_i=(i-1)(m+2)$, for $1\le i\le k$.

From the recurrence relation of the coefficients, we get

$$a_i=\frac{(-1)^{i+1}\prod_{q=1}^{i-1}\alpha m_q(m_q-1)}{\prod_{q=2}^{i}\beta n_q(n_q-1)}a_1, \quad (11)$$

where $2\le i\le k$. Hence, the corresponding solution can be written as

$$\sum_{i=1}^{k}a_ix^{m_i}y^{n_i}=a_1\left(x^{m_1}y^{n_1}+\sum_{i=2}^{k}\frac{(-1)^{i+1}\prod_{q=1}^{i-1}\alpha m_q(m_q-1)}{\prod_{q=2}^{i}\beta n_q(n_q-1)}x^{m_i}y^{n_i}\right), \quad (12)$$

where $m_i=(n+2)(k-i)$ and $n_i=(i-1)(m+2)$ for $1\le i\le k$. Once again, we will expand the products in the expression given by Eq. (11) and simplify to obtain

$$a_i=\frac{(-1)^{i+1}(1-k)_{i-1}\left(\alpha(n+2)^2\right)^{i-1}\left(-k+\frac{1}{n+2}+1\right)_{i-1}}{\Gamma(i)\left(\beta(m+2)^2\right)^{i-1}\left(\frac{m+1}{m+2}\right)_{i-1}}a_1. \quad (13)$$

This equation is valid for $1\le i\le k$. Then the Eq. (12) can be written as

$$a_1 \sum_{i=1}^{k} \frac{(-1)^{i+1}(1-k)_{i-1}\left(\alpha(n+2)^2\right)^{i-1}\left(-k+\frac{1}{n+2}+1\right)_{i-1}}{\Gamma(i)\left(\beta(m+2)^2\right)^{i-1}\left(\frac{m+1}{m+2}\right)_{i-1}} x^{(n+2)(k-i)} y^{(i-1)(m+2)}. \tag{14}$$

This can be written as

$$a_1\, x^{(k-1)(n+2)} \sum_{i=1}^{k} \frac{(1-k)_{i-1}\left(-k+\frac{1}{n+2}+1\right)_{i-1}}{\Gamma(i)\left(\frac{m+1}{m+2}\right)_{i-1}} \left(\frac{-\alpha(n+2)^2 y^{m+2}}{\beta(m+2)^2 x^{n+2}}\right)^{i-1}. \tag{15}$$

Since k and i are positive integers, this equation becomes an infinite summation of the form

$$a_1\, x^{(k-1)(n+2)} \sum_{i=1}^{\infty} \frac{(1-k)_{i-1}\left(-k+\frac{1}{n+2}+1\right)_{i-1}}{(i-1)!\left(\frac{m+1}{m+2}\right)_{i-1}} \left(\frac{-\alpha(n+2)^2 y^{m+2}}{\beta(m+2)^2 x^{n+2}}\right)^{i-1}. \tag{16}$$

Putting $j = i - 1$, this becomes

$$a_1\, x^{(k-1)(n+2)} \sum_{j=0}^{\infty} \frac{(1-k)_{j}\left(-k+\frac{1}{n+2}+1\right)_{j}}{j!\left(\frac{m+1}{m+2}\right)_{j}} \left(\frac{-\alpha(n+2)^2 y^{m+2}}{\beta(m+2)^2 x^{n+2}}\right)^{j}. \tag{17}$$

This infinite sum is nothing but the hypergeometric function given by

$$a_1 x^{(k-1)(n+2)}\, {}_2F_1\left(1-k, -k+\frac{1}{n+2}+1; \frac{m+1}{m+2}; -\frac{\alpha(n+2)^2 y^{m+2}}{\beta(m+2)^2 x^{n+2}}\right). \tag{18}$$

So, in this case, the polynomial or polynomial-like solution to the wave-like equation becomes (18). Hence, it follows that (5) is a solution to Eq. (1), since this equation is a linear partial differential equation. Here, it is to be noted that the solution is not defined for the values of m for which $\frac{m+1}{m+2} = 0, -1, -2, \ldots$, etc. [15]. So the solution is not defined for $m = \frac{-2i-1}{i+1}$, where i is a non-negative integer. Also it is not defined for $m = -2$ and $n = -2$.

Case 2: When $m_k = 0$ and $n_1 = 1$ In this case, the recurrence relation for the power of x is same as in the previous case. So we have $m_i = (n+2)(k-i)$, for $1 \le i \le k$. So $m_1 = (k-1)(n+2)$. Next, we consider the recurrence relation for the power of y. We have the recurrence relation $n_i = n_{i-1} + m + 2$, $1 \le i \le k-1$, with $n_1 = 1$. Solving this recurrence relation, we get $n_i = i(m+2) - m - 1$, for $1 \le i \le k$. Now, proceeding as in the previous case, we can derive the polynomial or polynomial-like solution

$$a_1 y x^{(k-1)(n+2)}\, {}_2F_1\left(1-k, -k+\frac{1}{n+2}+1; 1+\frac{1}{m+2}; -\frac{\alpha(n+2)^2 y^{m+2}}{\beta(m+2)^2 x^{n+2}}\right). \tag{19}$$

So, in this case, the polynomial or polynomial-like solution to the wave-like equation becomes (19). Hence, it follows that (6) is a solution to Eq. (1), since this equation is a linear partial differential equation. Here, it is to be noted that the solution is not defined for the values of m for which $1+\frac{1}{m+2}=0,-1,-2,\ldots$, etc. [15]. So the solution is not defined for $m=\frac{-2i-3}{i+1}$, where i is a non-negative integer. Also it is not defined for $m=-2$ and $n=-2$.

Case 3: When $m_k=1$ and $n_1=0$ Here, the recurrence relation for the power of y is as in the first case. So we have $n_i=(i-1)(m+2)$, for $1\leq i\leq k$. Now we consider the power of x. The recurrence relation for the power of x is $m_i=m_{i-1}-n-2$, for $2\leq i\leq k$, with $m_k=1$. Solving this recurrence relation, we get $m_i=-i(n+2)+k(n+2)+1$, for $1\leq i\leq k$. Now, proceeding as in the first case we can derive the polynomial or polynomial-like solution

$$a_1x^{k(n+2)-n-1}\,{}_2F_1\left(1-k,\frac{n+1}{n+2}-k;\frac{m+1}{m+2};-\frac{\alpha(n+2)^2y^{m+2}}{\beta(m+2)^2x^{n+2}}\right). \tag{20}$$

So, in this case, the polynomial or polynomial-like solution to the wave-like equation becomes (20). Hence, it follows that (7) is a solution to Eq. (1), since this equation is a linear partial differential equation. Here, it is to be noted that the solution is not defined for the values of m for which $\frac{m+1}{m+2}=0,-1,-2,\ldots$, etc. [15]. So, the solution is not defined for $m=\frac{-2i-1}{i+1}$, where i is a non-negative integer. Also it is not defined for $m=-2$ and $n=-2$.

Case 4: When $m_k=1$ and $n_1=1$ In this case, the recurrence relation for the power of x is same as in the previous case. So we get $m_i=-i(n+2)+k(n+2)+1$, for $1\leq i\leq k$. Also, the recurrence relation for the power of y is as in the second case. So, we have $n_i=i(m+2)-m-1$, for $1\leq i\leq k$.

Now, proceeding as in the case 1, we can derive the polynomial or polynomial-like solution

$$a_1yx^{k(n+2)-n-1}\,{}_2F_1\left(1-k,\frac{n+1}{n+2}-k;1+\frac{1}{m+2};-\frac{\alpha(n+2)^2y^{m+2}}{\beta(m+2)^2x^{n+2}}\right). \tag{21}$$

So, in this case, the polynomial or polynomial-like solution to the wave-like equation becomes (21). Hence, it follows that (8) is a solution to Eq. (1), since this equation is a linear partial differential equation. Here, also it is to be noted that the solution is not defined for the values of m for which $1+\frac{1}{m+2}=0,-1,-2,\ldots$, etc. [15]. So the solution is not defined for $m=\frac{-2i-3}{i+1}$, where i is a non-negative integer. Also it is not defined for $m=-2$ and $n=-2$.

In Theorem 1, we have derived the polynomial solutions or polynomial-like solutions of the wave-like Eq. (1). All these solutions are expressed in terms of hypergeometric functions, where k is a positive integer. Now the striking feature is that all these functions are solutions of wave-like Eq. (1) even when k is not a positive integer. That is, k can be any real numbers or a complex number and the summation is taken over arbitrary set of real number or complex numbers. Such exact solutions need not be polynomial solutions or polynomial-like solutions when k is not a positive integer. So, we can extend the above family of polynomial solutions or polynomial-like solutions to include non-polynomial solutions expressible in terms of hypergeometric functions, which are given in the following theorem.

Theorem 2 *Four different families of exact solutions to the wave-like Eq. (1) are given by*

$$f_1(x, y) = \sum_k c_k\, x^{(k-1)(n+2)}\, {}_2F_1\left(1-k, -k+\frac{1}{n+2}+1; \frac{m+1}{m+2}; u\right), \tag{22}$$

$$f_2(x, y) = \sum_k c_k\, yx^{(k-1)(n+2)}\, {}_2F_1\left(1-k, -k+\frac{1}{n+2}+1; 1+\frac{1}{m+2}; u\right), \tag{23}$$

$$f_3(x, y) = \sum_k c_k\, x^{k(n+2)-n-1}\, {}_2F_1\left(1-k, \frac{n+1}{n+2}-k; \frac{m+1}{m+2}; u\right), \tag{24}$$

and

$$f_4(x, y) = \sum_k c_k\, yx^{k(n+2)-n-1}\, {}_2F_1\left(1-k, \frac{n+1}{n+2}-k; 1+\frac{1}{m+2}; u\right), \tag{25}$$

where

$$u = -\frac{\alpha\,(n+2)^2 y^{m+2}}{\beta\,(m+2)^2 x^{n+2}}$$

and for each summation, c_k are constants and the summation is taken over any finite set of real numbers or complex numbers k.

2.2 Second Set of Exact Solutions

We have derived first set of exact solutions of wave-like Eq. (1) in the previous section. Now we are going to explore another set of exact solutions of these equations. Here, the required exact solutions of these equations are derived from the second possible grouping of the terms in Eq. (10), which facilitate the balancing procedure. The grouping is given below.

$$\sum_{i=2}^{k}\left(\alpha a_i m_i\left(m_i-1\right)x^{m_i-2}y^{n_i+m}+\beta a_{i-1}n_{i-1}\left(n_{i-1}-1\right)y^{n_{i-1}-2}x^{m_{i-1}+n}\right)$$
$$+\left(\alpha a_1 m_1\left(m_1-1\right)x^{m_1-2}y^{m+n_1}\right)+\left(\beta a_k n_k\left(n_k-1\right)y^{n_k-2}x^{m_k+n}\right)=0. \quad (26)$$

The above grouping of terms also have done in such a way that the recurrence relation of powers of the variables and the recurrence relation of coefficients can be easily derived. The above equation is satisfied if each term in the summation vanish and also the last two terms vanish. First, we consider the terms given in the summation. We have to determine the recurrence relations so that balancing of the terms in the parentheses are possible. When this is made possible, we get the following relations among the powers and coefficients. To balance the powers of x and y in the braces, the powers should be such that, $m_i = m_{i-1} + n + 2$ and $n_i = n_{i-1} - m - 2$. Then the terms in the braces will vanish if we choose $a_i = -\frac{\beta a_{i-1}n_{i-1}(n_{i-1}-1)}{\alpha m_i(m_i-1)}$. The second last term will vanish if we choose $m_1 = 0$ or $m_1 = 1$. Now the last term vanishes only when $n_k = 0$ or $n_k = 1$. So, we need to consider four different cases according to the values of m_1 and n_k. Since the remaining procedure is similar to the case described in the previous section, we summarize the result in the following theorem.

Theorem 3 *Four different families of exact solutions to the wave-like Eq. (1) are given by*

$$f_1(x,y)=\sum_k c_k\, y^{(k-1)(m+2)}\,{}_2F_1\left(1-k,-k+\frac{1}{m+2}+1;\frac{n+1}{n+2};u\right), \quad (27)$$

$$f_2(x,y)=\sum_k c_k\, y^{k(m+2)-m-1}\,{}_2F_1\left(1-k,\frac{m+1}{m+2}-k;\frac{n+1}{n+2};u\right), \quad (28)$$

$$f_3(x,y)=\sum_k c_k\, xy^{(k-1)(m+2)}\,{}_2F_1\left(1-k,-k+\frac{1}{m+2}+1;1+\frac{1}{n+2};u\right), \quad (29)$$

and

$$f_4(x,y)=\sum_k c_k\, xy^{k(m+2)-m-1}\,{}_2F_1\left(1-k,\frac{m+1}{m+2}-k;1+\frac{1}{n+2};u\right), \quad (30)$$

where

$$u=-\frac{(\beta(m+2))^2x^{n+2}}{\alpha(n+2)^2y^{m+2}}$$

and for each summation, c_k are constants and the summation is taken over any finite set of real numbers or complex numbers k.

So we have derived all polynomial-like solutions of wave-like equation in the Sects. 2.1 and 2.2. Also we have constructed eight families exact solutions of these partial differential equations in terms of hypergeometric functions. Polynomial or polynomial-like solutions of the wave-like Eq. (1) are obtained except in the case of the parameters $m = -2$ and $n = -2$.

3 Conclusion

There are several problems in physics and mechanics that can be represented and analyzed using partial differential equations which are special cases of the Eq. (1). Usually, these equations are called wave-like or Laplace-like equations. The important special cases of these equations include Euler–Tricomi equation, generalized Euler–Tricomi equations, Keldysh equation, Laplace equation, and wave equation. There are several other special cases of these equations discussed in the literature [22, 25]. So, obtaining several classes of new exact solutions to these equation will in turn give several solutions to the well-known partial differential equations mentioned above and other similar equations. Availability of exact solutions are fundamental to

Table 1 Exact solution of the Euler–Tricomi equation (2) given by the solutions in Theorem 2, for certain values of k (putting $m = 0, n = 1, \alpha = 1$ and $\beta = 1$ in Theorem 2). For positive integer values of k, we get polynomial solutions. For some other values, we get solutions expressible in terms of elementary functions

k	f_1	f_2	f_3	f_4
1	1	y	x	xy
2	x^3-3y^2	$y(x^3-y^2)$	x^4-6xy^2	$xy(x^3-2y^2)$
3	$x^6-15x^3y^2+\frac{15y^4}{2}$	$x^6y-5x^3y^3+\frac{3y^5}{2}$	$x^7-21x^4y^2+21xy^4$	$x^7y-7x^4y^3+\frac{21xy^5}{5}$
-1	$\frac{{}_2F_1\left(2,\frac{7}{3};\frac{1}{2};-\frac{9y^2}{4x^3}\right)}{x^6}$	$\frac{y\,{}_2F_1\left(2,\frac{7}{3};\frac{3}{2};-\frac{9y^2}{4x^3}\right)}{x^6}$	$\frac{{}_2F_1\left(\frac{5}{3},2;\frac{1}{2};-\frac{9y^2}{4x^3}\right)}{x^5}$	$\frac{y\,{}_2F_1\left(\frac{5}{3},2;\frac{3}{2};-\frac{9y^2}{4x^3}\right)}{x^5}$
-2	$\frac{{}_2F_1\left(3,\frac{10}{3};\frac{1}{2};-\frac{9y^2}{4x^3}\right)}{x^9}$	$\frac{y\,{}_2F_1\left(3,\frac{10}{3};\frac{3}{2};-\frac{9y^2}{4x^3}\right)}{x^9}$	$\frac{{}_2F_1\left(\frac{8}{3},3;\frac{1}{2};-\frac{9y^2}{4x^3}\right)}{x^8}$	$\frac{y\,{}_2F_1\left(\frac{8}{3},3;\frac{3}{2};-\frac{9y^2}{4x^3}\right)}{x^8}$
-3	$\frac{{}_2F_1\left(4,\frac{13}{3};\frac{1}{2};-\frac{9y^2}{4x^3}\right)}{x^{12}}$	$\frac{y\,{}_2F_1\left(4,\frac{13}{3};\frac{3}{2};-\frac{9y^2}{4x^3}\right)}{x^{12}}$	$\frac{{}_2F_1\left(\frac{11}{3},4;\frac{1}{2};-\frac{9y^2}{4x^3}\right)}{x^{11}}$	$\frac{y\,{}_2F_1\left(\frac{11}{3},4;\frac{3}{2};-\frac{9y^2}{4x^3}\right)}{x^{11}}$
$\frac{1}{2}$	$\frac{1}{x^{3/2}\left(\frac{9y^2}{4x^3}+1\right)^{5/6}}$	$\frac{y\,{}_2F_1\left(\frac{1}{2},\frac{5}{6};\frac{3}{2};-\frac{9y^2}{4x^3}\right)}{x^{3/2}}$	$\frac{1}{\sqrt{x}\sqrt[6]{\frac{9y^2}{4x^3}+1}}$	$\frac{y\,{}_2F_1\left(\frac{1}{6},\frac{1}{2};\frac{3}{2};-\frac{9y^2}{4x^3}\right)}{\sqrt{x}}$
$\frac{2}{3}$	$\frac{2\cosh\left(\frac{1}{3}\sinh^{-1}\left(\frac{3y}{2x^{3/2}}\right)\right)}{x\sqrt{\frac{9y^2}{x^3}+4}}$	$2\sqrt{x}\sinh\left(\frac{\sinh^{-1}\left(\frac{3y}{2x^{3/2}}\right)}{3}\right)$	1	y
$-\frac{1}{2}$	$\frac{x^3-6y^2}{x^{15/2}\left(\frac{9y^2}{4x^3}+1\right)^{17/6}}$	$\frac{y}{x^{9/2}\left(\frac{9y^2}{4x^3}+1\right)^{11/6}}$	$\frac{x^3-3y^2}{x^{13/2}\left(\frac{9y^2}{4x^3}+1\right)^{13/6}}$	$\frac{y}{x^{7/2}\left(\frac{9y^2}{4x^3}+1\right)^{7/6}}$
$-\frac{3}{2}$	$\frac{x^6-21x^3y^2+21y^4}{x^{27/2}\left(\frac{9y^2}{4x^3}+1\right)^{29/6}}$	$\frac{y\left(x^3-2y^2\right)}{x^{21/2}\left(\frac{9y^2}{4x^3}+1\right)^{23/6}}$	$\frac{2x^6-30x^3y^2+15y^4}{2x^{25/2}\left(\frac{9y^2}{4x^3}+1\right)^{25/6}}$	$\frac{y\left(x^3-y^2\right)}{x^{19/2}\left(\frac{9y^2}{4x^3}+1\right)^{19/6}}$
$-\frac{2}{3}$	$\frac{{}_2F_1\left(\frac{5}{3},2;\frac{1}{2};-\frac{9y^2}{4x^3}\right)}{x^5}$	$\frac{y\,{}_2F_1\left(\frac{5}{3},2;\frac{3}{2};-\frac{9y^2}{4x^3}\right)}{x^5}$	$\frac{{}_2F_1\left(\frac{4}{3},\frac{5}{3};\frac{1}{2};-\frac{9y^2}{4x^3}\right)}{x^4}$	$\frac{y\,{}_2F_1\left(\frac{4}{3},\frac{5}{3};\frac{3}{2};-\frac{9y^2}{4x^3}\right)}{x^4}$

the conceptual understanding of behavior of the physical system or phenomena and also can be utilized to check the correctness of approximate methods.

We have obtained several closed form exact solutions of these equations in this paper. In Sect. 2, first of all, we have derived all the polynomial solutions of wave-like Eq. (1). Any polynomial solution or polynomial-like solution of this equation will be in any one of the family of polynomial solutions given in Theorems 1 and 3 or linear combinations of these solutions, provided k is positive integer. These polynomial solutions are having compact form expressed in terms of hypergeometric functions. Moreover, the families of solutions given in Theorems 2 and 3 are exact solutions to the given wave-like equation, which also include all the families of polynomial or polynomial-like solutions.

The polynomial and non-polynomial exact solutions that we have derived can be used to investigate the problems in mathematical physics which can be formulated as particular cases of these wave-like equations. Euler–Tricomi equation (2) is a special case of wave-like equation. Some of the polynomial and other solutions for this equation are illustrated in Table 1. The several applications of this equation in fluid mechanics, differential geometry, transonic flow problems, and isometric embedding problems can be found in [2, 9, 13, 19, 23]. Using these exact solutions, it may be possible to solve certain boundary value problems and initial value problems in these fields. We have used the method of balancing powers of the variables to find the recurrence relation among the powers of the variables and it is used to obtain exact solutions of all the above equation. The large classes of new exact non-polynomial and polynomial or polynomial-like solutions that we have derived are expressed in terms of hypergeometric functions. This method can also be applied to find exact solutions for higher order partial differential equations which appear in the field of mechanics.

References

1. Azad H, Laradji A, Mustafa MT (2011) Polynomial solutions of differential equations. Adv. Differ. Equ. 58:58
2. Chen GQ, Slemrod M, Wang D (2008) Vanishing viscosity method for transonic flow. D Arch Ration Mech Anal 189:159–188
3. Chen S (2009) The fundamental solution of the Keldysh type operator. Sci China Ser A-Math 52:1829–1843
4. Chen S (2012) A mixed equation of tricomi-keldysh type. J Hyper Differ Equ 09(3):545–553
5. Hayman WK, Shanidze ZG (1999) Polynomial solutions of partial differential equations. Methods Appl Anal 6:97–108
6. Joseph SP (2018) Polynomial solutions and other exact solutions of axisymmetric generalized Beltrami flows. Acta Mech **229**, 2737–2750
7. Keşan C (20) Taylor polynomial solutions of second order linear partial differential equations. Appl Math Comput 152:29–41
8. Lashkarian E, Hejazi SR (2016) Polynomial and non-polynomial solutions set for wave equation using Lie point symmetries. Comput Methods Differ Equ 4(4):298–308
9. Manwell AR (1979) Tricomi equation: with applications to the theory of plane transonic flow. Chapman and Hall/CRC Research Notes in Mathematics Series

10. Miles EP, Williams E (1955) A basic set of homogeneous harmonic polynomials in k variables. Proc Am Math Soc 6(2):191–194
11. Miles EP, Williams E (1956) A basic set of polynomial solutions for the Euler-Poisson-Darboux and Beltrami equations. Am Math Mon 63:401–404
12. Miles EP, Young EC (1967) Basic sets of polynomials for generalized Beltrami and Euler-Poisson-Darboux equations and their iterates. Proc Am Math Soc 18:981–986
13. Morawetz CS (2004) Mixed equations and transonic flow. J Hyper Differ Equ 01(1):1–26
14. Olde Daalhuis AB (2010) Confluent Hypergeometric Functions. In: Olver FWJ (ed) NIST handbook of mathematical functions. Springer, pp 322–349
15. Olde Daalhuis AB (2010) Hypergeometric function. In: Olver FWJ (ed) NIST handbook of mathematical functions. Springer, pp 383–401
16. Pedersen P (1996) A basis for polynomial solutions to systems of linear constant coefficient PDE's. Adv Math 117(1):157–163
17. Polyanin AD (2002) Handbook of linear partial differential equations for engineers and scientists. Chapman and Hall
18. Otway TH (2012) The dirichlet problem for elliptic-hyperbolic equations of Keldysh Type. Springer
19. Otway TH (2015) Elliptic-Hyperbolic partial differential equations: a mini-course in geometric and quasilinear methods. Spinger
20. Reznick B (1996) Homogeneous polynomial solutions to constant coefficient PDE's. Adv Math 117:179–192
21. Saad N, Hall RL, Trentona VA (2014) Polynomial solutions for a class of second-order linear differential equations. Appl Math Comput 226:615–634
22. Shou D, He J (2008) Beyond Adomian method: the variational iteration method for solving heat-like and wave-like equations with variable coefficients. Phys Lett A 372:233–237
23. Smith SP (1992) Polynomial solutions to constant coefficient differential equations. Trans Am Math Soc 329(2):551–569
24. Yagdjian K (2004) A note on the fundamental solution for the Tricomi-type equation in the hyperbolic domain. J Differ Equ 206(1):227–252
25. Wazwaz AM, Gorguis A (2004) Exact solutions for heat-like and wave-like equations with variable coefficients. Appl Math Comput 149:15–29

Coexistence of Chaotic, Quasiperiodic and Multiperiodic Features in Quantum Plasma

Barsha Pradhan and Asit Saha

Abstract Coexistence of chaotic, quasiperiodic and multiperiodic features of arbitrary amplitude quantum ion-acoustic waves are rigorously discussed in a quantum plasma consisting of electrons and ions. A conservative dynamical system of four dimensions is proposed to investigate the dynamics of quantum ion-acoustic waves. The system shows qualitatively different dynamical features at different initial conditions with specific values of physical parameters for quantum ion-acoustic waves. Tools, such as, coexisting phase spaces and plots of time series and Lyapunov exponents are employed to validate the results.

Keywords Modelling of quantum plasma · Coexistence of trajectories · Dense plasma

1 Introduction

Investigations revealed that some dynamical systems have more than one simultaneous solution for fixed set of parameters and different initial conditions. This phenomenon was termed as multistability [1–4] or coexistence of trajectories and was found to exist in many areas of discipline, such as, optics, electronics, biology and mechanics. Multistable chaotic system was first experimentally reported by Arecchi et al. [5] in a laser model of Q-switched gas. In the field of plasmas, coexisting features or simply multistability was investigated in plasma diodes [6], novel lunar wake plasma [7], solar wind plasma [8], electron–ion (e–i) plasma [9], Thomas–Fermi plasma [10] and experimentally observed in discharge plasmas [11]. Particularly, in the field of quantum plasmas, multistability was investigated in e–i plasma [12] and pair-ion plasma [13]. Some of the methods of showing existence

Barsha Pradhan is grateful to Sikkim Manipal University for TMA Pai University research grant (Ref. No. 118/SMU/REG/UOO/104/2019) dated 15-07-2019.

B. Pradhan (✉) · A. Saha
Department of Mathematics, Sikkim Manipal Institute of Technology, Sikkim Manipal University, Majitar, Rangpo East-Sikkim 737136, India
e-mail: asit_saha123@rediffmail.com

© The Author(s), under exclusive license to Springer Nature Singapore Pte Ltd. 2022
D. Giri et al. (eds.), *Proceedings of the Seventh International Conference on Mathematics and Computing*, Advances in Intelligent Systems and Computing 1412,
https://doi.org/10.1007/978-981-16-6890-6_68

of multistability in a system are by drawing coexisting phase spaces, time series plots or bifurcation diagram for a fixed set of parameter values and different initial conditions. Another such method is to draw cross section of basins of attraction to choose initial values and to validate multistability.

American meteorologist, Edward Lorenz [14], first described chaotic phenomenon while trying to predict weather using computers for a simplified model. Chaos is a long-term aperiodic phenomenon shown by deterministic system that is highly sensitive on initial conditions [15]. Lyapunov exponent (LE) quantifies the rate of repulsion or attraction of infinitesimally close orbits [15]. If a system is highly sensitive on initial conditions, then orbits that are initially close move apart exponentially after sufficiently long time. A positive value of largest Lyapunov exponent (LLE) implies that neighbouring orbits diverge exponentially and is therefore taken as a signature of chaos [16]. Wolf's algorithm [17] is popularly used to determine Lyapunov exponents of a dynamical system.

Quantum plasmas are naturally found in dense astrophysical objects, such as, in magnetars, interior of white dwarfs and neutron stars and cores of planets like Jupiter and Saturn. While, some applications of laboratory quantum plasmas are semiconductor devices, quantum diodes, carbon nanotubes and quantum dots. e–i quantum plasma may exist in compact astrophysical objects such as interior of white dwarfs [18–20]. Hossen et al. [21] investigated ion-acoustic shock waves in an e–i quantum plasma and discussed the relevance of their findings in astrophysical objects such as neutron stars and white dwarfs. In low-frequency ion-acoustic waves, ions supplies the driving force needed to maintain the wave, while pressure of inertialess electrons supplies the restoring forces. Linear and nonlinear properties of quantum ion-acoustic waves (QIAWs) have been rigorously studied using quantum hydrodynamical models [12, 22–24] in the recent years.

The use of concepts of dynamical system in quantum plasmas has gained popularity in the recent years. The bifurcation theory of planar dynamical system in quantum plasma was the first used by Samanta et al. [25] to examine dynamical features of nonlinear waves. Quasiperiodicity of ion-acoustic solitary waves was investigated by Sahu et al. [26] in an e–i quantum plasma. But, they did not investigate chaotic phenomenon or coexisting features in their work. Multistability phenomenon for small-amplitude QIAWs was investigated in e–i quantum plasma under the framework of Schrödinger equation in the presence of an external force [12]. But coexistence of chaotic, quasiperiodic and multiperiodic features of arbitrary amplitude quantum ion-acoustic waves has not been addressed in e–i quantum plasma. Therefore, in our manuscript, we wish to investigate coexisting phenomena for arbitrary amplitude QIAWs in an e–i quantum plasma.

We have arranged the manuscript as follows: In Sect. 2, we have presented the normalized basic equations that describe the dynamics of ion-acoustic wave. The dynamical system is formed in Sect. 3. Multistability phenomena are comprehensively addressed in Sect. 4. Finally, the conclusion is given in Sect. 5.

2 Normalized Governing Equations

In order to study the propagation of arbitrary-amplitude QIAWs, we consider an unmagnetized quantum plasma consisting of ions and electrons. The following normalized governing equations [27] describe the dynamics of low-frequency QIAWs:

$$\frac{\partial n_i}{\partial t} + \frac{\partial}{\partial x}(n_i u_i) = 0, \tag{1}$$

$$\frac{\partial u_i}{\partial t} + u_i \frac{\partial u_i}{\partial x} = -\frac{\partial \phi}{\partial x}, \tag{2}$$

$$\frac{\partial \phi}{\partial x} - \frac{1}{2}\frac{\partial n_e^2}{\partial x} + \frac{\alpha H_e^2}{2}\frac{\partial}{\partial x}\left(\frac{1}{\sqrt{n_e}}\frac{\partial^2}{\partial x^2}\sqrt{n_e}\right) = 0, \tag{3}$$

$$\frac{\partial^2 \phi}{\partial x^2} = n_e - n_i. \tag{4}$$

Here, $n_i(n_e)$ represents ion (electron) number density, u_i is the ion fluid velocity and ϕ represents electrostatic potential. n_i and n_e have been normalized by n_0, which is the unperturbed number density of electrons and ions at equilibrium. The quantities, viz. space variables x, time variable t, ion fluid velocity u_i and electrostatic potential ϕ have been, respectively, normalized by $\frac{c_s}{\omega_{pi}}$, $\frac{1}{\omega_{pi}}$, c_s and $\frac{2K_B T_{Fe}}{e}$, where $\omega_{pi} = \left(\frac{n_0 e^2}{\epsilon_0 m_i}\right)^{1/2}$ is the ion-plasma frequency and $c_s = \left(\frac{2k_B T_{Fe}}{m_i}\right)$ is the ion-acoustic speed. Here, $H_e = \frac{\hbar \omega_{pe}}{2K_B T_{F_e}}$ denotes the non-dimensional quantum parameter for electrons. In reference [27], the prefactor of the Bohm potential is taken as 1, but this prefactor was strongly criticized for giving unphysical prediction[28]. Here in our manuscript, α represents prefactor of the Bohm potential and is taken as $-1/3$ [28–30] for low-frequency ion-acoustic wave in an one-dimensional quantum plasma.

3 Dynamical System

We implement the following Galilean transformation to study the propagation of arbitrary amplitude QIAWs in an e–i quantum plasma:

$$\xi = x - Mt, \tag{5}$$

where M represents Mach number.

Using this transformation in Eqs. (1)–(4) and defining $n_e = A^2$, we get system of second-order differential equations:

$$\frac{d^2A}{d\xi^2} = \frac{2A}{\alpha H_e^2}\left(\phi - \frac{1}{2} + \frac{A^4}{2}\right) \tag{6}$$

$$\frac{d^2\phi}{d\xi^2} = A^2 - \left(1 + \frac{\phi}{M^2} + \frac{3}{2}\frac{\phi^2}{M^4} + \frac{5}{2}\frac{\phi^3}{M^3}\right). \tag{7}$$

This system of equations can be equivalently written as a four dimension dynamical system given by

$$\begin{cases} \dfrac{dx_1}{d\xi} = x_2, \\ \dfrac{dx_2}{d\xi} = \dfrac{2x_1}{\alpha H_e^2}\left(x_3 - \dfrac{1}{2} + \dfrac{x_1^4}{2}\right), \\ \dfrac{dx_3}{d\xi} = x_4, \\ \dfrac{dx_4}{d\xi} = x_1^2 - \left(1 + \dfrac{x_3}{M^2} + \dfrac{3}{2}\dfrac{x_3^2}{M^4} + \dfrac{5}{2}\dfrac{x_3^3}{M^6}\right), \end{cases} \tag{8}$$

where $x_1 = A,\ x_2 = \frac{dA}{d\xi},\ x_3 = \phi$ and $x_4 = \frac{d\phi}{d\xi}$.

The rate of volume contraction of the four-dimensional dynamical system (8) is

$$\vec{\nabla}.\vec{F} = \frac{\partial}{\partial x_1}\left(\frac{dx_1}{d\xi}\right) + \frac{\partial}{\partial x_2}\left(\frac{dx_2}{d\xi}\right) + \frac{\partial}{\partial x_3}\left(\frac{dx_3}{d\xi}\right) + \frac{\partial}{\partial x_4}\left(\frac{dx_4}{d\xi}\right) = 0. \tag{9}$$

Since $\vec{\nabla}.\vec{F} = 0$, the dynamical system represented by Eqs. (8) is conservative which can be further supported by zero sum of Lyapunov exponents.

One can obtain critical points of the dynamical system (8) by solving the following system of equations:

$$\begin{cases} x_2 = 0, \\ x_1\left(x_3 - \frac{1}{2} + \frac{x_1^4}{2}\right) = 0, \\ x_4 = 0, \\ x_1^2 - \left(1 + \frac{x_3}{M^2} + \frac{3}{2}\frac{x_3^2}{M^4} + \frac{5}{2}\frac{x_3^3}{M^6}\right) = 0, \end{cases} \tag{10}$$

If one assumes $x_3 = 0$, then one can obtain $x_1 = \pm 1$. Therefore, the critical points are

$$\begin{cases} C_1(-1, 0, 0, 0) \\ C_2(1, 0, 0, 0). \end{cases} \tag{11}$$

To examine the stability of an arbitrary critical point $C_i(x_1^*, x_2^*, x_3^*, x_4^*)$, where $i = 1, 2$, we linearise dynamical system (8) at C_i. The Jacobian matrix is as follows:

$$J_{C_i} = \begin{pmatrix} 0 & 1 & 0 & 0 \\ \frac{5x_1^{*4}}{\alpha H_e^2} + \frac{(2x_3^*-1)}{\alpha H_e^2} & 0 & \frac{2x_1^*}{\alpha H_e^2} & 0 \\ 0 & 0 & 0 & 1 \\ 2x_1^* & 0 & 0 & 0 \end{pmatrix}.$$

The eigenvalues of the Jacobian matrix at C_i are obtained by solving $|\lambda I - J_{C_i}| = 0$. For the critical point C_1, the eigenvalues are of the form $\pm\sqrt{\frac{-6\pm\sqrt{36-48H_e^2}}{2H_e^2}}$, while for the critical point C_2, the eigenvalues are of the form $\pm\sqrt{\frac{18\pm\sqrt{324+48H_e^2}}{2H_e^2}}$. It needs to be mentioned that the critical point C_i is stable only when all the eigenvalues obtained have negative real parts. Otherwise, the critical point C_i is considered to be unstable. Hence, we can say that the critical points C_1 and C_2 are unstable.

4 Coexisting Features

Multistability is a phenomenon in which for a particular set of parameters, a dynamical system has more than one simultaneous solution for different initial states. Such multistability phenomenon can be vividly depicted with the help of phase spaces showing two or more qualitatively different trajectories each corresponding to separate solution arising from an initial state. The trajectories may be chaotic, quasiperiodic, periodic or multiperiodic. Chaos is basically a long-term aperiodic behaviour shown by a deterministic system which shows sensitive dependence on initial condition [15]. Aperiodic and erratic trend in a time series plot can be observed for such chaotic phenomenon. Further, chaos is confirmed by a positive Lyapunov exponent (LE) in a LE spectrum. On the other hand, quasiperiodic trajectory wind around endlessly on a torus without intersecting with itself and never quite closing [15]. In fact, quasiperiodic motion is shown by a dynamical system having finite number of incommensurable frequencies. Finally, periodic trajectory is a closed orbit and is a solution to a dynamical system that repeats itself after a finite interval.

To investigate multistability phenomenon, we fix parameters at $M = 7$ and $H_e = 0.55$ and vary only the initial states. One can observe a chaotic trajectory at $z0(0.77,\ 0.4,\ 0.2,\ 0)$ (blue), a quasiperiodic trajectory at $z0(0.77,\ 0.01,\ 0.2,\ 0)$ (red) and a multiperiodic trajectory at $z0(0.85,\ 0.001,\ 0.2,\ 0)$ (black) for QIAWs coexisting together. This coexistence is shown by phase spaces (Fig. 1). Time series plot for QIAWs corresponding to these initial states is depicted in Fig. 2a. Clearly, one can observe random aperiodic trend shown by blue curve, irregular periodic behaviour shown by red curve and periodic behaviour shown by black curve. Chaotic phenomenon is further assured by a positive LE at $H_e = 0.55$ given in Fig. 2b. Here, Lyapunov exponents are plotted against quantum diffraction parameter H_e with $M = 7$. We can also clearly observe that the sum of LEs is equal to zero, which confirms the conservative nature of the system (8).

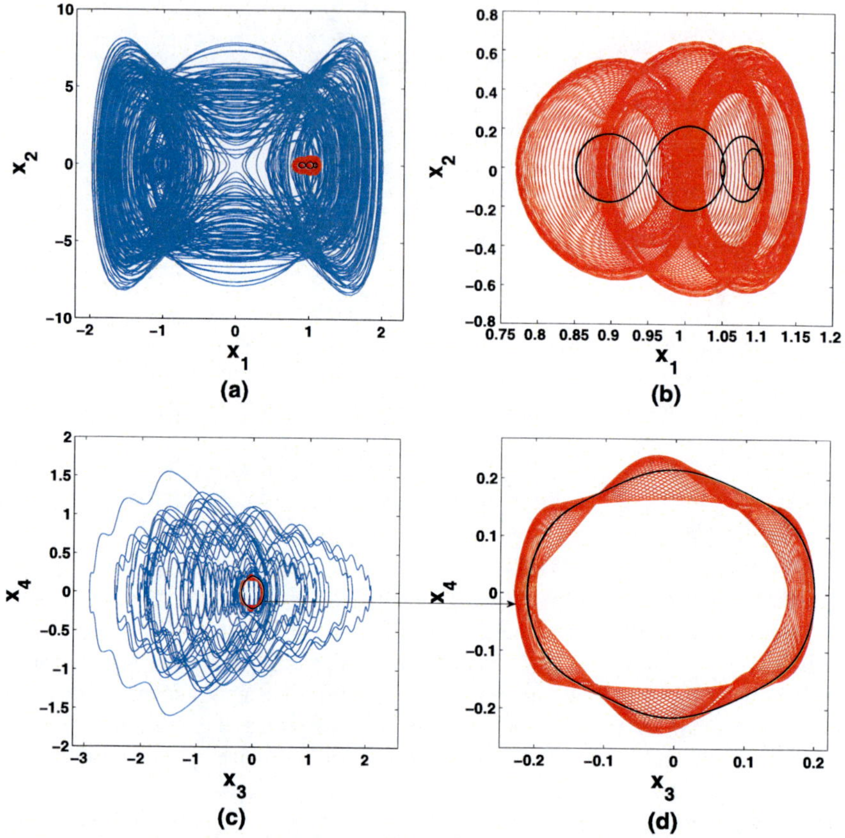

Fig. 1 Phase spaces presenting coexistence of chaotic trajectory at $z0(0.77,\ 0.4,\ 0.2,\ 0)$ (blue), quasiperiodic trajectory at $z0(0.77,\ 0.01,\ 0.2,\ 0)$ (red) and multiperiodic trajectory at $z0(0.85,\ 0.001,\ 0.2,\ 0)$ (black) for $M = 7$ and $H_e = 0.55$

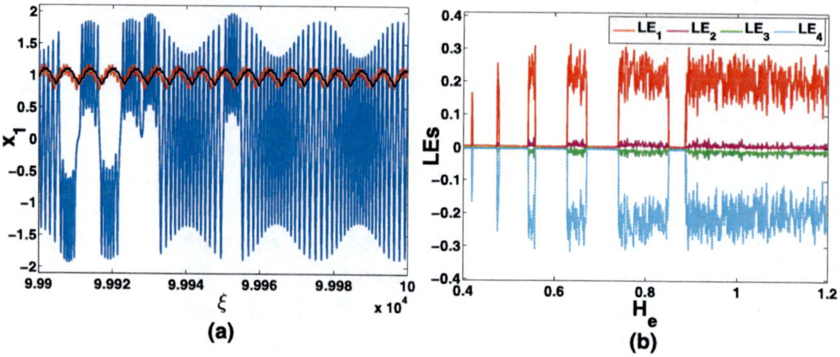

Fig. 2 **a** Time series plot for QIAWs initiated by $z0(0.77,\ 0.4,\ 0.2,\ 0)$ (blue), $z0(0.77,\ 0.01,\ 0.2,\ 0)$ (red), $z0(0.85,\ 0.001,\ 0.2,\ 0)$ (black) for $M = 7$ and $H_e = 0.55$. **b** LE spectrum w.r.t. H_e for $M = 7$ with initial state $z0(0.77,\ 0.4,\ 0.2,\ 0)$

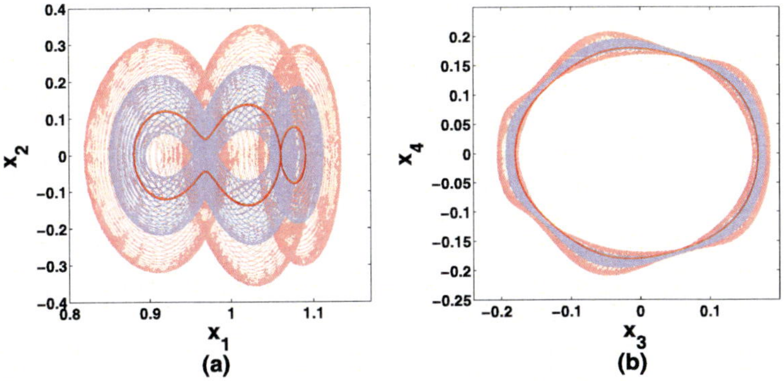

Fig. 3 Phase projections showing coexistence of quasiperiodic trajectories at $z0(0.88, 0.1, 0.17, 0)$ (peach) and $z0(0.894, 0.002, 0.17, 0)$ (lavender) and multiperiodic trajectory at $z0(0.881, 0.002, 0.17, 0)$ (red) for $M = 7$ and $H_e = 0.65$

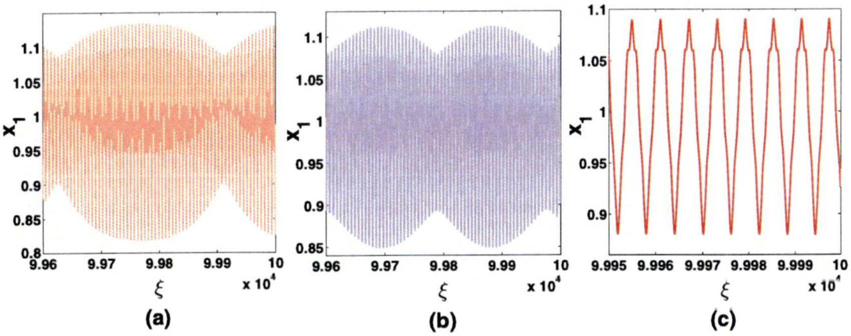

Fig. 4 Time series plots of the system (8) initiated by **a** $z0(0.88, 0.1, 0.17, 0)$ (peach), **b** $z0(0.894, 0.002, 0.17, 0)$ (lavender) and **c** $z0(0.881, 0.002, 0.17, 0)$ (red) for $M = 7$ and $H_e = 0.65$

Coexistence of quasiperiodic and multiperiodic trajectories in $x_1 - x_2$ and $x_3 - x_4$ planes are presented in Figs. 3. Parameters are fixed at $M = 7$ and $H_e = 0.65$. Two types of quasiperiodic trajectories for QIAWs initiated by $z0(.88, 0.1, 0.17, 0)$ and $z0(.894, 0.002, 0.17, 0)$ are, respectively, shown by peach and lavender orbits. A multiperiodic trajectory for QIAWs initiated by $z0(0.881, 0.002, 0.17, 0)$ is represented by red orbit. The corresponding time series plots of the two quasiperiodic trajectories and multiperiodic trajectory is given in Fig. 4a–c with initial states same as Fig. 3.

By fixing parameter values at $M = 7$ and $H_e = 1.1$, we alter only the initial states. Chaotic trajectory at $z0(-0.8, 0.002, 0.17, 0)$ (purple), quasiperiodic trajectories at $z0(-0.95, 0.001, 0.17, 0)$ (teal) and $z0(0.94, 0.001, 0.17, 0)$ (red) and multiperiodic trajectories at $z0(-0.8966, 0.002, 0.17, 0)$ (black) and $z0(0.8966, 0.002, 0.17, 0)$ (blue) for QIAWs are seen to coexist together. This

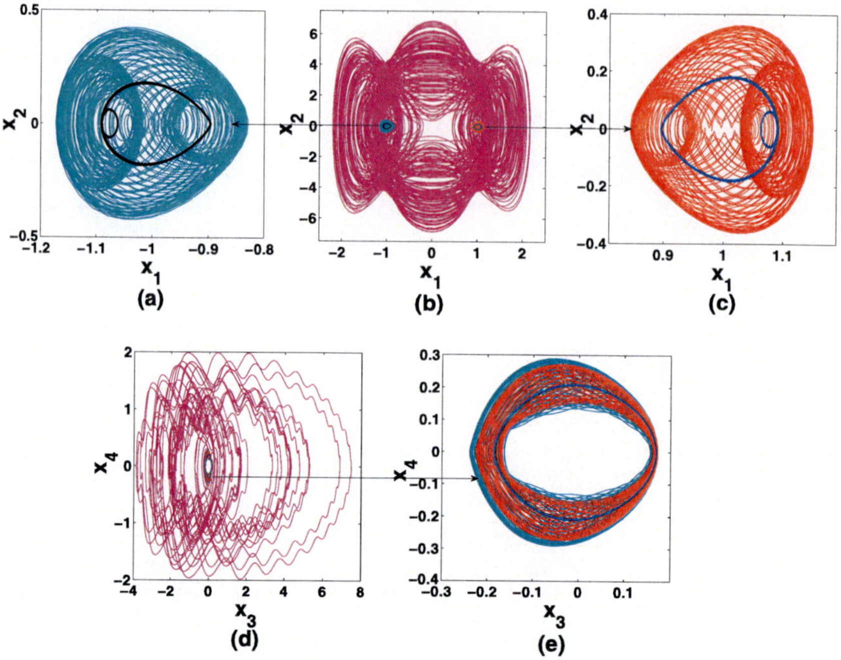

Fig. 5 Phase projections in $x_1 - x_2$ and $x_3 - x_4$ planes depicting coexistence of chaotic trajectory at $z0(-0.8,\ 0.002,\ 0.17,\ 0)$ (purple), quasiperiodic trajectories at $z0(-0.95,\ 0.001,\ 0.17,\ 0)$ (teal) and $z0(0.94,\ 0.001,\ 0.17,\ 0)$ (red) and multiperiodic trajectories at $z0(-0.8966,\ 0.002,\ 0.17,\ 0)$ (black) and $z0(0.8966,\ 0.002,\ 0.17,\ 0)$ (blue) for $M = 7$ and $H_e = 1.1$

multistability phenomenon is clearly displayed in Fig. 5. Corresponding time series plots are presented in Fig. 6a–e. A LE spectrum w.r.t. H_e is given in Fig. 6d. Value of the largest LE at $H_e = 1.1$ is positive which clearly confirms chaotic behaviour of the purple trajectory shown in Fig. 5.

In Fig. 7, three types of quasiperiodic trajectories for QIAWs are seen to coexist with a multiperiodic trajectory for QIAWs when the parameter values are fixed at M=7 and $H_e = 0.6$. Quasiperiodic trajectories are initiated by $z0(0.89,\ 0.002,\ 0.1, 0)$ (icy blue), $z_0(0.84,\ 0.002,\ 0.179,\ 0)$ (red) and $z_0(0.89,\ 0.002,\ 0.1,\ 0)$ (pink) while multiperiodic trajectory is initiated by $z_0(0.8760.0020.1790)$ (blue). Time series plot corresponding to these trajectories for QIAWs are given in Fig. 8a–d.

5 Conclusion

Investigation of coexisting phenomena has been meticulously carried out for arbitrary amplitude ion-acoustic waves in an e–i quantum plasma. The divergence of vector field of the system formed and the sum of Lyapunov exponents of the system

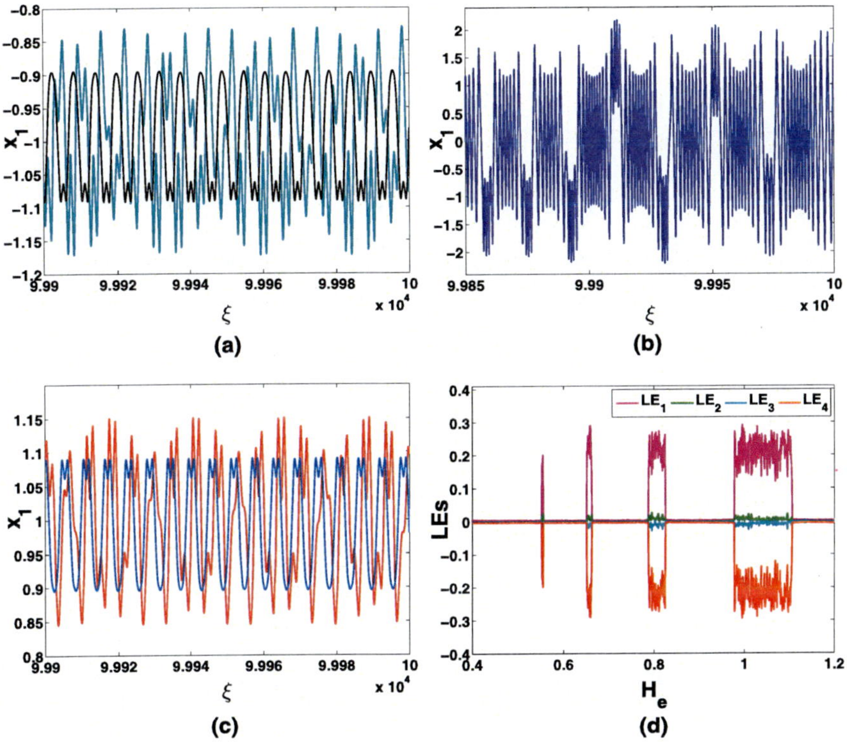

Fig. 6 Time series plots of the system (8) for initial states **a** $z0(-0.95,\ 0.001,\ 0.17,\ 0)$ (teal) and $z0(-0.8966,\ 0.002,\ 0.17,\ 0)$ (black), **b** $z0(-0.8,\ 0.002,\ 0.17,\ 0)$ (purple), **c** $z0(0.94,\ 0.001,\ 0.17,\ 0)$ (red) and $z0(0.8966,\ 0.002,\ 0.17,\ 0)$ (blue) with $M = 7$ and $H_e = 1.1$. (d) LE spectrum w.r.t. H_e for $M = 7$ initiated by $z0(-0.8,\ 0.002,\ 0.17,\ 0)$

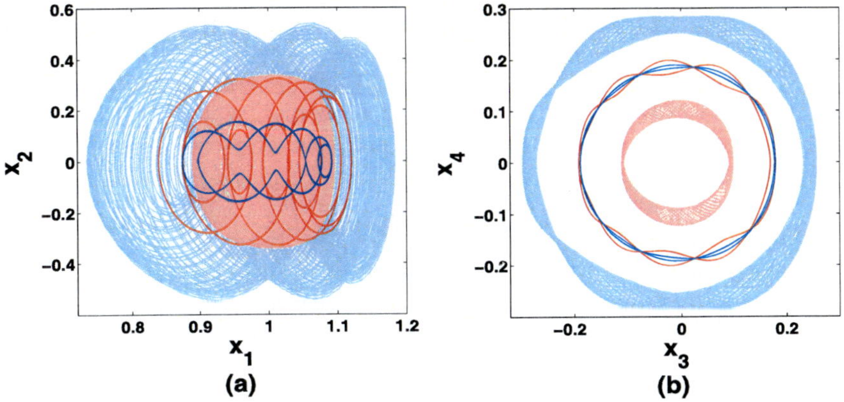

Fig. 7 Phase spaces displaying coexistence of quasiperiodic trajectories at $z0(0.89,\ 0.002,\ 0.1, 0)$ (icy blue), $z_0(0.84,\ 0.002,\ 0.179,\ 0)$ (red) and $z_0(0.89,\ 0.002,\ 0.1,\ 0)$ (pink) and multiperiodic trajectory at $z_0(0.8760.0020.1790)$ (blue) for $M = 7$ and $H_e = 0.6$

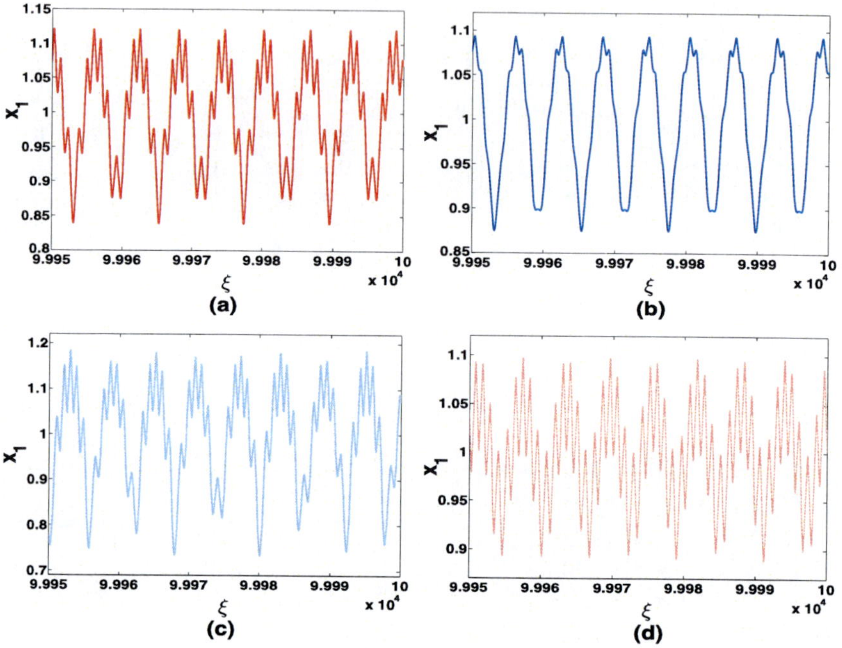

Fig. 8 Time series of the system (8) for initial states **a** $z_0(0.84, \ 0.002, \ 0.179, \ 0)$, **b** $z_0(0.8760.0020.1790)$, (c) $z0(0.89, \ 0.002, \ 0.1, 0)$, and (d) $z_0(0.89, \ 0.002, \ 0.1, \ 0)$ with $M = 7$ and $H_e = 0.6$

have been found to be zero. These indicate that the four dimensional dynamical system formed is conservative. The system has two unstable equilibrium points. Four different types of coexisting features have been observed in the plasma system: Coexistence of chaotic, multiperiodic and quasiperiodic trajectories, coexistence of two types of quasiperiodic and a multiperiodic trajectories, coexistence of a chaotic, two kinds of quasiperiodic and two kind of multiperiodic trajectories and coexistence of three types of quasiperiodic and a multiperiodic trajectories for quantum ion-acoustic waves. To check these phenomena of coexistence, phase spaces of the system for different initial values and fixed parameter sets have been portrayed. Dynamical features like chaos, multiperiodicity and quasiperiodicity depicted in coexisting phase spaces have been further supported by time series plots. Finally, chaotic phenomena has been verified by a positive Lyaunov exponent. From Sect. 4, it has been observed that electrostatic potential of this electron–ion quantum plasma system is sensitive to initial conditions and qualitatively different features like multiperiodic, chaotic and quasiperiodic trajectories were seen at different initial conditions with fixed values of physical parameters. The results are useful to investigate coexisting features in dense astrophysical objects and laboratory quantum plasmas consisting of electrons and ions.

References

1. He S, Banerjee S, Sun K (2019) Complex dynamics and multiple coexisting attractors in a fractional-order microscopic chemical system. Eur Phys J Spec Top 228:195
2. Natiq H, Banerjee S, Misra AP, Said MRM (2019) Degenerating the butterfly attractor in a plasma perturbation model using nonlinear controllers. Chaos, Solitons Fractals 122:58
3. Natiq H, Said MRM, Ariffin MRK, He S, Rondoni L, Banerjee S (2018) Self-excited and hidden attractors in a novel chaotic system with complicated multistability. Eur Phys J Plus 133:557
4. Rahim MFA, Natiq H, Fataf NAA, Banerjee S (2019) Dynamics of a new hyperchaotic system and multistability. Eur Phys J Plus 134:499
5. Arecchi FT, Meucci R, Puccioni G, Tredicce J (1982) Experimental evidence of subharmonic bifurcations, multistability and turbulence in a Q-switched gas laser. Phys Rev Lett 49(17):1217
6. Hahn SJ, Pae KH (2003) Competing multistability in a plasma diode. Phys Plasmas 10(1):314
7. Yan B, Prasad PK, Mukherjee S, Saha A, Banerjee S (2020) Dynamical complexity and multistability in a novel lunar wake plasma system. Complexity 2020:5428548
8. Prasad PK, Gowrishankar A, Saha A, Banerjee S (2020) Dynamical properties and fractal patterns of nonlinear waves in solar wind plasma. Phys Scr 95:6
9. Abdikian A, Tamang J, Saha A (2020) Electron-acoustic supernonlinear waves and their multistability in the framework of the nonlinear Schrödinger equation. Commun Theor Phys 72:075502
10. Saha A, Sarkar S, Banerjee S, Mondal KK (2020) Signature of chaos and multistability in a Thomas-Fermi plasma. Eur Phys J Spec Top 229:979
11. Yong J, Haida W, Changxuan Y (1988) Multistability phenomena in a discharge plasma. Chin Phys Lett 5(5):200
12. Saha A, Pradhan B, Banerjee S (2020) Multistability and dynamical properties of ion-acoustic wave for the nonlinear Schrödinger equation in an electron-ion quantum plasma. Phys Scr 95:055602
13. Pradhan B, Mukherjee S, Saha A, Natiq H, Banerjee S (2021) Multistability and chaotic scenario in a quantum pair-ion plasma. Zeitschrift für Naturforschung A 76:109
14. Lorenz EN (1963) Deterministic nonperiodic flow. J Atmos Sci 20:130
15. Strogatz SH (2016) Nonlinear dynamics and Chaos. 2nd edn, Avalon Publishing
16. Fell J, Roschke J, Beckmann P (1993) Deterministic chaos and the first positive Lyapunov exponent: a nonlinear analysis of the human electroencephalogram during sleep. Biol Cybern 69:139
17. Wolf A, Swift JB, Swinney HL, Vastono JA (1985) Deterministic Lyapunov exponents from a time series. Phys D 16:285
18. Shapiro SL, Teukolsky SA (1983) Black holes, white dwarfs and neutron stars. Wiley, New York
19. Koester D, Chanmugam G (1990) Physics of white dwarf stars. Rep Prog Phys 53:837
20. Akbari-Moghanjoughi M (2011) Propagation of arbitrary-amplitude ion waves in relativistically degenerate electron-ion plasmas. Astrophys Space Sci 332:187
21. Hossen MR, Nahar L, Sultana S, Mamun AA (2014) Relativistic ion-acoustic solitary waves in a magnetized pair ion dense plasma with nuclei of heavy elements. High Energy Density Phys 13:13
22. Haas F, Garcia LG, Goedert J, Manfredi G (2003) Quantum ion-acoustic waves. Phys Plasmas 10:3858
23. El-Labany SK, El-Taibany WF, Atteya A (2018) Bifurcation analysis for ion acoustic waves in a strongly coupled plasma including trapped electrons. Phys Lett A 382:412
24. Saha A, Pradhan B, Banerjee S (2020) Bifurcation analysis of quantum ion-acoustic kink, anti-kink and periodic waves of the Burgers equation in a dense quantum plasma. Eur J Phys 135:216
25. Samanta UK, Saha A, Chatterjee P (2013) Bifurcations of dust acoustic solitary waves and periodic waves in an unmagnetized plasma with nonextensive ions. Astrophys Space Sci 347:293

26. Sahu B, Poria S, Ghosh UN, Roychoudhury R (2012) Quasi-periodic behavior of ion acoustic solitary waves in electron-ion quantum plasma. Phys Plasmas 19:052306
27. Mahmood S, Haas F (2014) Ion-acoustic cnoidal waves in a quantum plasma. Phys Plasmas 21:102308
28. Michta D, Graziani F, Bonitz M (2015) Quantum hydrodynamics for plasmas -a thomas-fermi theory perspective. Contrib Plasma Phys 55(6):437
29. Moldabekov ZhA, Bonitz M, Ramazanov TS (2017) Gradient correction and Bohm potential for two- and one-dimensional electron gases at a finite temperature. Contrib Plasma Phys 57:499
30. Moldabekov ZhA, Bonitz M, Ramazanov TS (2018) Theoretical foundations of quantum hydrodynamics for plasmas. Phys Plasmas 25:031903

New Classes of Exact Solutions to Three-dimensional Schrodinger Equation

Subin P. Joseph

Abstract New classes of exact solutions for the steady state Schrodinger equation are derived in this paper. The solutions are derived for both non-zero energy and zero energy problems. The sum of the inverse square law potential and a power law potential are considered in deriving the exact solutions. The solutions are derived in cartesian coordinates and without separating the variables.

Keywords Schrodinger equation · Exact solution · Inverse square and power law potentials

1 Introduction

Exact solutions to Schrodinger equations are having many applications in several fields such as quantum information and econophysics as well as in modern physics [7, 9, 11, 15, 36]. The well known potentials that are discussed in the literature and for which certain exact solutions are available are given by Coulomb, inverse square, Morse, Mie-type, Eckart, Rosen–Morse, Poschl–Teller, Woods–Saxen, Manning–Rosen, Hulthén Potential, Scarf and Gendenshtein potentials [1–6, 13, 24, 25, 30–32, 35]. These equations can be solved by converting them in to ordinary differential equations by means of a suitable transformation. The required solutions are derived in terms of well known special functions such as Bessel, Hermite, Legendre, Heun, Whittaker, and confluent hypergeometric functions. When exact solutions are not available, several approximation methods are there for obtaining approximate solutions [14, 17]. Some of the recent works in finding exact solutions of Schrodinger equation are [4–6, 18–20, 27, 28, 33]. The usual method in finding the exact solutions of Schrodinger equation in three-dimensional case is by assuming the solution in the separation of variables form and by solving the corresponding radial part of the equation.

S. P. Joseph (✉)
Government Engineering College, Wayanad, Thalapuzha P.O., Wayanad, Kerala, India
e-mail: subinpj@gecwyd.ac.in

© The Author(s), under exclusive license to Springer Nature Singapore Pte Ltd. 2022

D. Giri et al. (eds.), *Proceedings of the Seventh International Conference on Mathematics and Computing*, Advances in Intelligent Systems and Computing 1412,
https://doi.org/10.1007/978-981-16-6890-6_69

The Schrodinger equation with zero energy have been discussed by various authors [8, 10, 12, 20–22, 26, 29]. In statistical mechanics, the zero energy eigenstates in the parabolic potential barrier play an important role in Gelfand triplet [21]. The other problems where zero energy exact solutions can be applied includes the studies in super symmetric quantum mechanics, scattering of ultracold particles, and coupling parameter calculations [8].

Several new exact solutions for Schrodinger equation with zero energy are derived in the second section, where the potential is taken to be the sum of an inverse square potential and a power law potential. We derive the required three-dimensional exact solutions using certain ansatz forms and without performing separation of variables. Several exact solutions for Schrodinger equation with non-zero energy are also derived in the case of inverse square potential in the third section. Only the bound state exact solutions are discussed in the paper. The paper is concluded with a discussion on the results obtained.

2 Exact Solutions

Usually, the method of separation of variables in the spherical coordinates is applied to obtain exact solutions to Schrodinger equation, which will lead to an ordinary differential equation corresponding to radial part. This differential equation is then solved to find various exact solutions in the case of potentials which are usually functions of the radial vector $r = \sqrt{x^2 + y^2 + z^2}$. Instead of applying the method of separation of variables, we will find out certain exact solutions to the three-dimensional Schrodinger equations directly in cartesian coordinates. In this section, zero energy level solutions are derived for inverse square potential and a power law potential.

The Schrodinger equation in cartesian coordinates is given by

$$\frac{\partial^2 f}{\partial x^2} + \frac{\partial^2 f}{\partial y^2} + \frac{\partial^2 f}{\partial z^2} = \left(\frac{a}{x^2 + y^2 + z^2} - b\left(x^2 + y^2 + z^2\right)^{k(\neq -1)}\right) f(x, y, z) \quad (1)$$

where a, b are arbitrary parameters. Assuming particular ansatz forms for the required exact solutions, we will solve the Schrodinger equation (1). We are interested to derive only the bound state exact solutions $f(x, y, z)$ to Schrodinger equation which are zero at origin and also tends to zero as x or y or z tends to $\pm\infty$.

2.1 *Case I*

First of all, we seek the solution for the Eq. (1) in the ansatz form given by

$$f(x, y, z) = (cx + dy + ez)g(u) \quad (2)$$

where $u = x^2 + y^2 + z^2$ and c, d, e are arbitrary constants. Substituting this in the Eq. (1), we get an ordinary differential equation in terms of $g(u)$ given by

$$(cx + dy + ez)\left(g(u)\left(bu^k - \frac{a}{u}\right) + 4ug''(u) + 10g'(u)\right) = 0 \tag{3}$$

So, $g(u)$ is a solution to the Schrodinger equation if it satisfies the ordinary differential equation

$$g(u)\left(bu^k - \frac{a}{u}\right) + 4ug''(u) + 10g'(u) = 0 \tag{4}$$

Now we use the following change of variables to the independent and dependent variables

$$t = \frac{\sqrt{bu^{k+1}}}{k+1} \tag{5}$$

and $g(u) = u^{-\frac{3}{4}}w(t)$. On substitution and simplification, the differential Eq. (4) becomes

$$t^2w''(t) + tw'(t) + w(t)\left(t^2 - \frac{4a+9}{4(k+1)^2}\right) = 0 \tag{6}$$

This is a Bessel's equation whose solutions in terms of Bessel functions of first and second kinds are given by

$$w(t) = C_1 J_{\frac{\sqrt{4a+9}}{2(k+1)}}(t) + C_2 Y_{\frac{\sqrt{4a+9}}{2(k+1)}}(t) \tag{7}$$

where C_1 and C_2 are arbitrary constants. Hence, particular solutions to Schrodinger equation (1) are given by

$$\begin{aligned} f = &(cx + dy + ez)\left(x^2 + y^2 + z^2\right)^{-\frac{3}{4}} \left(C_1 J_{\frac{\sqrt{4a+9}}{2(k+1)}}\left(\frac{\sqrt{b}\left(x^2 + y^2 + z^2\right)^{\frac{k+1}{2}}}{k+1}\right)\right. \\ &\left. + C_2 Y_{\frac{\sqrt{4a+9}}{2(k+1)}}\left(\frac{\sqrt{b}\left(x^2 + y^2 + z^2\right)^{\frac{k+1}{2}}}{k+1}\right)\right) \end{aligned} \tag{8}$$

The solution given by Bessel function of first kind satisfies the conditions for a bound state wave function for real numbers $a \geq -9/4$ and $b > 0$, that is, they are going to zero as x, y or z tends to $\pm\infty$. Also, it becomes zero at origin. Graphical representation of a particular wave function is given in Fig. 1 for fixed values for the coordinate variable z.

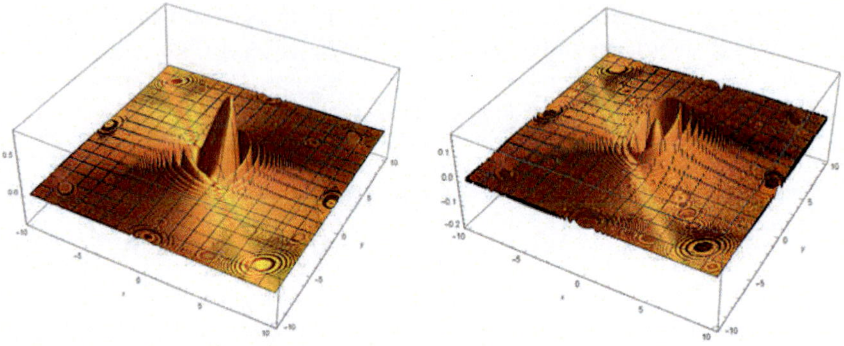

Fig. 1 Graphical representation of wave function given by Eq. (8) with $c_1 = 1, c_2 = 0, a = 3, b = 1, k = 2, c = 1, d = 1, e = 1$, and for $z = 1$ and $z = 3$, respectively (Case I)

2.2 *Case II*

Second solution is assumed in the form

$$f(x, y, z) = (cx + dy + ez)(c_1x + d_1y + e_1z)g\,(u) \tag{9}$$

where $u = x^2 + y^2 + z^2$ and c, d, e, c_1, d_1, e_1 are constants to be determined. Substituting this in the Eq. (1), we get an equation in terms of $g(u)$ given by

$$\begin{aligned} 2\left(\left(2ug''(u) + 7g'(u)\right)(cx + dy + ez)(c_1x + d_1y + e_1z) + g(u)(cc1 + dd1 + ee1)\right) \\ - g(u)\left(\frac{a}{u} - bu^k\right)(cx + dy + ez)(c1x + d1y + e1z) = 0 \end{aligned} \tag{10}$$

This equation is not easily solvable in this form for the function $g(u)$. To make this solvable, we assume that $(cc_1 + dd_1 + ee_1) = 0$. Solving this equation we get

$$e_1 = \frac{-cc_1 - dd_1}{e}.$$

Substituting this in the Eq. (10), we get an ordinary differential equation

$$g(u)\left(bu^{k+1} - a\right) + 2u\left(2ug''(u) + 7g'(u)\right) = 0 \tag{11}$$

Now we use the following change of variables to the independent and dependent variables

$$t = \frac{\sqrt{bu^{k+1}}}{k+1} \tag{12}$$

and $g(u) = u^{-\frac{5}{4}}w(t)$. On substitution and simplification, the differential Eq. (11) becomes

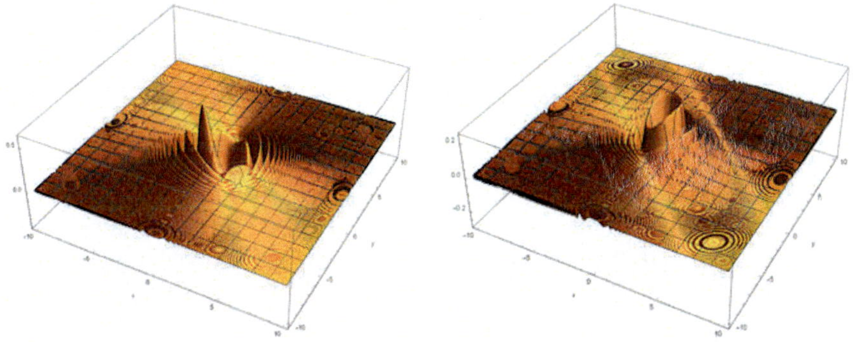

Fig. 2 Graphical representation of wave function given by Eq. (15) with $C_1 = 1, C_2 = 0, a = 3, b = 1, k = 2, c = 1, d = 1, e = 1, c_1 = 1, d_1 = 1$, and for $z = 1$ and $z = 3$, respectively (Case II)

$$t^2 w''(t) + t w'(t) + w(t)\left(t^2 - \frac{4a+25}{4(k+1)^2}\right) = 0 \tag{13}$$

This is a Bessel's equation whose solutions in terms of Bessel functions of first and second kinds are given by

$$w(t) = C_1 J_{\frac{\sqrt{4a+25}}{2(k+1)}}(t) + C_2 Y_{\frac{\sqrt{4a+25}}{2(k+1)}}(t) \tag{14}$$

where C_1 and C_2 are arbitrary constants. Hence, particular solutions to Schrodinger equation (1) are given by

$$\begin{aligned} f = &(cx + dy + ez)\left(c_1 x + d_1 y - \frac{(cc_1 + dd_1)z}{e}\right)\left(x^2 + y^2 + z^2\right)^{-\frac{5}{4}} \\ &\left(C_1 J_{\frac{\sqrt{4a+25}}{2(k+1)}}\left(\frac{\sqrt{b}\left(x^2+y^2+z^2\right)^{\frac{k+1}{2}}}{k+1}\right)\right. \\ &\left.+ C_2 Y_{\frac{\sqrt{4a+25}}{2(k+1)}}\left(\frac{\sqrt{b}\left(x^2+y^2+z^2\right)^{\frac{k+1}{2}}}{k+1}\right)\right) \end{aligned} \tag{15}$$

The solution given by Bessel function of first kind satisfies the conditions for a bound state wave function for real numbers $a \geq -25/4$ and $b > 0$, that is, they are going to zero as x, y or z tends to $\pm\infty$. Also, it becomes zero at origin. Graphical representation of a particular wave function is given in Fig. 2 for fixed values for the coordinate variable z.

2.3 Case III

Here, we assume the solution in the form

$$f(x, y, z) = (cx + dy + ez)(c_1x + d_1y + e_1z)(c_2x + d_2y + e_2z)g(u) \quad (16)$$

where $u = x^2 + y^2 + z^2$ and $c, d, e, c_1, d_1, e_1, c_2, d_2, e_2$ are constants to be determined. Substituting this in the Eq. (1), we get an equation in terms of $g(u)$ given by

$$h_1(x, y, z)g(u) + h_2(x, y, z)\left(\left(bu^k - \frac{a}{u}\right) + 2\left(2ug''(u) + 9g'(u)\right)\right) = 0 \quad (17)$$

where

$$\begin{aligned} h_1 = {} & (c_2dd_1 + c_1dd_2 + cd_1d_2 + c_2ee_1 + c_1ee_2 + ce_1e_2 + 3cc_1c_2)\,x \\ & + (c_1c_2d + cc_2d_1 + cc_1d_2 + d_2ee_1 + d_1ee_2 + de_1e_2 + 3dd_1d_2)\,y \\ & + (c_1c_2e + cc_2e_1 + cc_1e_2 + d_1d_2e + dd_2e_1 + dd_1e_2 + 3ee_1e_2)\,z \end{aligned} \quad (18)$$

and

$$h_2 = 2(cx + dy + ez)(c_1x + d_1y + e_1z)(c_2x + d_2y + e_2z)$$

This equation is not easily solvable in this form for the function $g(u)$. To make this solvable, we assume that $h_1(x, y, z) = 0$. This equation is identically satisfied if the following set of equations are satisfied.

$$\begin{aligned} &(c_2dd_1 + c_1dd_2 + cd_1d_2 + c_2ee_1 + c_1ee_2 + ce_1e_2 + 3cc_1c_2) = 0, \\ &(c_1c_2d + cc_2d_1 + cc_1d_2 + d_2ee_1 + d_1ee_2 + de_1e_2 + 3dd_1d_2) = 0, \\ &(c_1c_2e + cc_2e_1 + cc_1e_2 + d_1d_2e + dd_2e_1 + dd_1e_2 + 3ee_1e_2) = 0. \end{aligned} \quad (19)$$

Solving this system of nonlinear equation simultaneously, using any computational algebra system, we get three set of solutions

SET1:

$$c = -\frac{\Delta e}{\Lambda}, \quad d = -\frac{\Gamma e}{\Lambda}, \quad e_1 = \frac{4c_1c_2e_2 + 4d_1d_2e_2 + \phi}{{c_2}^2 + {d_2}^2 - 3{e_2}^2} \quad (20)$$

SET2:

$$c = \frac{\hat{\Delta} e}{\Lambda}, \quad d = -\frac{\hat{\Gamma} e}{\Lambda}, \quad e_1 = \frac{4c_1c_2e_2 + 4d_1d_2e_2 - \phi}{c_2^2 + d_2^2 - 3e_2^2} \quad (21)$$

SET3:

$$c = -\frac{e\left(c_1c_2d_2 + d_1\left(d_2^2 + e_2^2\right)\right)}{e_2\left(c_2d_1 - c_1d_2\right)}, \quad d = \frac{e\left(c_2d_1d_2 + c_1\left(c_2^2 + e_2^2\right)\right)}{e_2\left(c_2d_1 - c_1d_2\right)}, \quad e_1 = -\frac{c_1c_2 + d_1d_2}{e_2} \quad (22)$$

where

$$\begin{aligned}
\Gamma &= 2c_2c_1\left(d_1e_2\left(c_2^2+5d_2^2\right)+d_1e_2^3+d_2\phi\right)\\
&\quad+d_1\left(\left(d_2^2+e_2^2\right)\left(4d_1d_2e_2+\phi\right)-c_2^2\phi\right)+2c_1^2d_2e_2\left(c_2^2-d_2^2+e_2^2\right),\\
\hat{\Gamma} &= 2c_2c_1\left(d_1e_2\left(c_2^2+5d_2^2\right)+d_1e_2^3-d_2\phi\right)\\
&\quad+d_1\left(c_2^2\phi-\left(d_2^2+e_2^2\right)\left(\phi-4d_1d_2e_2\right)\right)+2c_1^2d_2e_2\left(c_2^2-d_2^2+e_2^2\right),\\
\Delta &= c_1\left(c_2^2\left(10d_1d_2e_2+\phi\right)+\phi\left(e_2^2-d_2^2\right)+2d_1d_2e_2\left(d_2^2+e_2^2\right)\right)\\
&\quad+2c_2d_1\left(d_1e_2\left(d_2^2-c_2^2\right)+d_1e_2^3+d_2\phi\right)+4c_2c_1^2e_2\left(c_2^2+e_2^2\right),\\
\hat{\Delta} &= c_1\left(c_2^2\left(\phi-10d_1d_2e_2\right)+\phi\left(e_2^2-d_2^2\right)-2d_1d_2e_2\left(d_2^2+e_2^2\right)\right)\\
&\quad+2c_2d_1\left(d_1e_2\left(c_2^2-d_2^2\right)-d_1e_2^3+d_2\phi\right)-4c_2c_1^2e_2\left(c_2^2+e_2^2\right),\\
\Lambda &= e_2^4\left(-\left(c_1^2+d_1^2\right)\right)+2e_2^2\left(c_1^2+d_1^2\right)\left(c_2^2+d_2^2\right)\\
&\quad+\left(c_2^2+d_2^2\right)\left(c_1^2\left(3c_2^2-d_2^2\right)+8c_2c_1d_1d_2-d_1^2\left(c_2^2-3d_2^2\right)\right),\\
\phi &= \sqrt{\left(c_2^2+d_2^2+e_2^2\right)\left(c_1^2\left(3c_2^2-d_2^2+3e_2^2\right)+d_1^2\left(3\left(d_2^2+e_2^2\right)-c_2^2\right)+8c_2c_1d_1d_2\right)}.
\end{aligned} \tag{23}$$

Corresponding to each set of the above solutions, we can find bounded solutions for Schrodinger equation. We will explicitly discuss the exact solutions to Schrodinger equation only in the case of first set of above solutions. In that case, substituting the solution (20) in the Eq. (17), we get an ordinary differential equation

$$g(u)\left(bu^{k+1}-a\right)+2u\left(2ug''(u)+9g'(u)\right)=0 \tag{24}$$

Now we use the following change of variables to the independent and dependent variables

$$t=\frac{\sqrt{bu^{k+1}}}{k+1} \tag{25}$$

and $g(u)=u^{-\frac{7}{4}}w(t)$. On substitution and simplification, the differential Eq. (24) becomes

$$t^2w''(t)+tw'(t)+w(t)\left(t^2-\frac{4a+49}{4(k+1)^2}\right)=0 \tag{26}$$

This is a Bessel's equation whose solutions in terms of Bessel functions of first and second kinds are given by

$$w(t)=C_1J_{\frac{\sqrt{4a+49}}{2(k+1)}}(t)+C_2Y_{\frac{\sqrt{4a+49}}{2(k+1)}}(t) \tag{27}$$

where C_1 and C_2 are arbitrary constants. Hence, particular solutions to Schrodinger equation (1) are given by

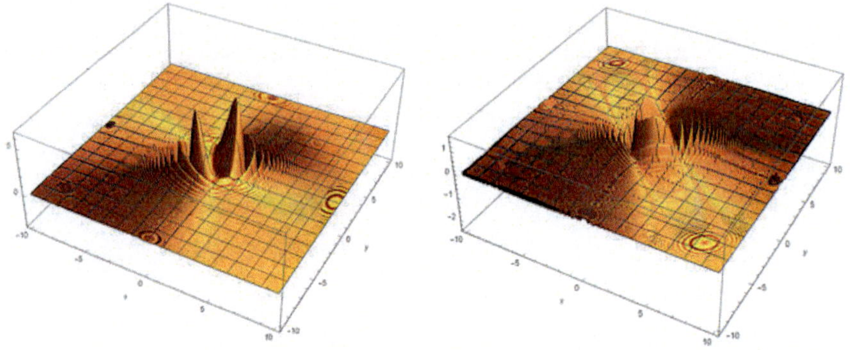

Fig. 3 Graphical representation of wave function given by Eq. (15) with $C_1 = 1, C_2 = 0, a = 1, b = 1, k = 2, e = 1, c_1 = 1, d_1 = 1, c_2 = 1, d_2 = 1, e_2 = 1$, and for $z = 1$ and $z = 3$, respectively (Case III)

$$\begin{aligned} f =& \frac{1}{\Lambda} \left(e \left(c_2 x + d_2 y + e_2 z\right) \left(-\Delta x - \Gamma y + \Lambda z\right) \right. \\ & \times \left(\frac{z \left(4 c_1 c_2 e_2 + 4 d_1 d_2 e_2 + \phi\right)}{c_2^2 + d_2^2 - 3 e_2^2} + c_1 x + d_1 y \right)\Bigg) \\ & \times \left(x^2 + y^2 + z^2\right)^{-\frac{7}{4}} \left(C_1 J_{\frac{\sqrt{4a+49}}{2(k+1)}} \left(\frac{\sqrt{b} \left(x^2 + y^2 + z^2\right)^{\frac{k+1}{2}}}{k+1} \right) \right. \\ & \left. + C_2 Y_{\frac{\sqrt{4a+49}}{2(k+1)}} \left(\frac{\sqrt{b} \left(x^2 + y^2 + z^2\right)^{\frac{k+1}{2}}}{k+1} \right) \right). \end{aligned} \tag{28}$$

The solution given by Bessel function of first kind satisfies the conditions for a bound state wave function for real numbers $a \geq -25/4$ and $b > 0$. Graphical representation of a particular wave function is given in Fig. 3 for fixed values for the coordinate variable z.

3 Non-zero Energy Solutions

Now we will consider three-dimensional Schrodinger equation with non-zero energy and derive several exact solutions for it. We consider the potential to be an inverse square potential. Then the corresponding Schrodinger equation is given by

$$\frac{\partial^2 f}{\partial x^2} + \frac{\partial^2 f}{\partial y^2} + \frac{\partial^2 f}{\partial z^2} = \left(\frac{a}{x^2 + y^2 + z^2} - E \right) f(x, y, z) \tag{29}$$

Assuming particular forms for the solution, as was done in the previous section, we will find out certain bound state exact solution of the Eq. (29). For each case in the previous section, the corresponding solutions are obtained here by simply putting $b = E$ and $k = 0$ in the exact solutions derived in the previous section. These exact solutions are summarized below.

CASE I:

$$f = (cx + dy + ez)\left(x^2 + y^2 + z^2\right)^{-\frac{3}{4}} \left(C_1 J_{\frac{\sqrt{4a+9}}{2}}\left(\sqrt{E}\left(x^2 + y^2 + z^2\right)^{\frac{1}{2}}\right) + C_2 Y_{\frac{\sqrt{4a+9}}{2}}\left(\sqrt{E}\left(x^2 + y^2 + z^2\right)^{\frac{1}{2}}\right)\right) \tag{30}$$

CASE II:

$$f = (cx + dy + ez)\left(c_1 x + d_1 y - \frac{(cc_1 + dd_1)z}{e}\right)\left(x^2 + y^2 + z^2\right)^{-\frac{5}{4}} \left(C_1 J_{\frac{\sqrt{4a+25}}{2}}\left(\sqrt{E}\left(x^2 + y^2 + z^2\right)^{\frac{1}{2}}\right) + C_2 Y_{\frac{\sqrt{4a+25}}{2}}\left(\sqrt{E}\left(x^2 + y^2 + z^2\right)^{\frac{1}{2}}\right)\right) \tag{31}$$

CASE III:

$$f = \frac{1}{\Lambda}\left(e\,(c_2 x + d_2 y + e_2 z)(-\Delta x - \Gamma y + \Lambda z) \times \left(\frac{z\,(4c_1c_2e_2 + 4d_1d_2e_2 + \phi)}{c_2^2 + d_2^2 - 3e_2^2} + c_1 x + d_1 y\right)\right) \times \left(x^2 + y^2 + z^2\right)^{-\frac{7}{4}} \left(C_1 J_{\frac{\sqrt{4a+49}}{2}}\left(\sqrt{E}\left(x^2 + y^2 + z^2\right)^{\frac{1}{2}}\right) + C_2 Y_{\frac{\sqrt{4a+49}}{2}}\left(\sqrt{E}\left(x^2 + y^2 + z^2\right)^{\frac{1}{2}}\right)\right) \tag{32}$$

4 Conclusion

Several exact solutions for three-dimensional Schrodinger equation with zero energy and non-zero energy are derived in this paper. The solutions of three-dimensional Schrodinger equation are usually derived in spherical coordinates, assuming the solution in the variable separable form. Then, for different potentials expressed as function of only the radial variable r, the resulting radial equation is solved. Here, we have derived the solutions in cartesian coordinates and without using separation of variables. There are several papers which discuss the application of zero energy potentials [8, 10, 12, 21, 22, 26, 29].

First of all, we derived zero energy solutions of Schrodinger equation for the potentials which are sum of an inverse square potential and a power law potential. All these solutions are bounded solutions which are vanishing at origin and at infinity. We have also derived certain non-zero energy solutions for the inverse square potentials in the third section. The applications of the potentials that are used here are recently discussed in [16, 23, 34]. In the case of non-zero energy equation, the obtained exact solutions are continuum wave functions for all positive values of energy E. The new exact solutions that are derived here can be used to analyze the physical situations where the inverse square potentials and power law potentials play a major role. Since Schrodinger equation is not only applied to the areas of modern physics and chemistry, but also in the emerging fields such as quantum information and econophysics [7, 9, 11, 15, 36], these exact solutions can be used in such fields and also can be used as benchmarking solutions.

References

1. Abdalla MS, Eleuch H (2014) Exact analytic solutions of the Schrödinger equations for some modified q-deformed potentials. J Appl Phys 115:234906. https://doi.org/10.1063/1.4883296
2. Ahmadov AI, Aydin C, Huseynova NSh, Uzun O (2013) Analytical Solutions of the Schrödinger equation with Manning-Rosen potentialplus a Ring-Shaped-like potential. Int J Modern Phys E 22(10):1350072
3. Ahmadov AI, Aydin C, Uzun O (2014) Analytical solutions of the Klein-Fock- Gordon equation with the Manning-Rosen potential plus a Ring-Shaped like potential. Int J Modern Phys A 29:1450002
4. Ahmadov AI, Jafarzade ShI, Qocayeva MV (2015) Analytical solutions of the Schrödinger equation for the Hulthén potential within SUSY quantum mechanics. Int J Mod Phys A 30(32):1550193
5. Ahmadov AI, Naeem M, Qocayeva MV, Tarverdiyeva VA (2018) Analytical bound state solutions of the Schrödinger equation for the manning-rosen plus Hulthén potential within SUSY quantum mechanics. Int J Mod Phys A 33(3):1850021
6. Ahmadov AI, Qocayev MV, Huseynova NSh (2017) The bound state solutions of the D-dimensional Schrödinger equation for the Hulthén potential within SUSY quantum mechanics. Int J Modern Phys E 2(5):1750028
7. Ahn K,Choi MY, Dai B, Sohn S, Yang B (2018) Modeling stock return distributions with a quantum harmonic oscillator. Europhys Lett 120(3). https://doi.org/10.1209/0295-5075/120/38003
8. Alhaidari AD (2002) Exact solutions of Dirac and Schrödinger equations for a large class of power-law potentials at zero energy. Int J Modern Phys A (Particles and Fields; Gravitation; Cosmology) 17(30):4551–4566
9. Baaquie BE (2004) Quantum finance. Cambridge University Press, Cambridge
10. Bagchi B, Quesne C (1997) Zero-energy states for a class of quasi-exactly solvable rational potentials. Phys Lett A 230(1–2):1–6
11. Chen G, Church DA, Englert BG, Henkel C, Rohwedder B, Scully MO, Zubairy MS (2007) Quantum computing devices: principles, designs, and analysis. Chapman and Hall/CRC, New York
12. Daboul J, Nieto MM (1995) Exact, E=0, classical solutions for general power-law potentials. Phys Rev E 52. https://doi.org/10.1103/PhysRevE.52.4430
13. Eckart C (1930) The penetration of a potential barrier by electrons. Phys Rev 35(11):1303–1309

14. Farizky MN, Suparmi A, Cari C, Yunianto M (2016) Solution of three dimensional Schrodinger equation for Eckart and Manning-Rosen non-central potential using asymptotic iteration method. J Phys: Conf Ser 776:012085
15. Gao T, Chen Y (2017) A quantum anharmonic oscillator model for the stock market. Phys A 468:307–314
16. Guillaumín-España E, Núez-Yñpez HN, Salas-Brito AL (2014) Classical and quantum dynamics in an inverse square potential. J Math Phys 55:103509. https://doi.org/10.1063/1.4899083
17. Hamzavi M, Ikhdair SM (2012) Approximate l-state solution of the trigonometric Pöschl–Teller potential. Mol Phys 110(24). https://doi.org/10.1080/00268976.2012.695029
18. Iacob F, Lute M (2015) Exact solution to the Schrödinger's equation with pseudo-Gaussian potential. J Math Phys 56:121501. https://doi.org/10.1063/1.4936309
19. Ishkhanyan AM (2015) Exact solution of the Schrödinger equation for the inverse square root potential $V_0/\sqrt{x}$. Europhys Lett 112(1). https://doi.org/10.1209/0295-5075/112/10006
20. Joseph SP (2020) Several exact solutions for three dimensional Schrodinger equation involving inverse square and power law potentials. Malaya J Mathematik 8(2):650–656
21. Kobayashi T, Shimbori T (2002) Zero-energy solutions and vortices in Schrödinger equations. Phys Rev A 65. https://doi.org/10.1103/PhysRevA.65.042108
22. Makowski AJ (2009) Exact, zero-energy, square-integrable solutions of a model related to the Maxwell's fish-eye problem. Ann Phys 324(12):2465–2472
23. Martínez-y-Romero RP, Núez-Yñpez HN, Salas-Brito AL (2013) The two dimensional motion of a particle in an inverse square potential: classical and quantum aspects. J Math Phys 54:053509. https://doi.org/10.1063/1.4804356
24. Millard F (1933) Manning and nathan rosen a potential function for the vibrations of diatomic molecules. Phys Rev 44(10):953
25. Morse PM (1929) Diatomic molecules according to the wave mechanics. II. Vibrational levels. Phys Rev 34:5764. https://doi.org/10.1103/PhysRev.34.57
26. Mustafa O (2013) Auxiliary quantization constraints on the von roos ordering-ambiguity at zero binding energies; azimuthally symmetrized cylindrical coordinates. Modern Phys Lett A: Particles Fields Grav Cosmol Nucl Phys 28(19). https://doi.org/10.1142/S021773231350082X
27. Naeimi IH, Batle J, Abdalla S (2017) Solving the two-dimensional Schrödinger equation using basis truncation: a hands-on review and a controversial case. Pramana J Phys 89:70. https://doi.org/10.1007/s12043-017-1467-z
28. Olsen T, Latini S, Rasmussen F, Thygesen KS (2016) Simple screened hydrogen model of excitons in two-dimensional materials. Phys Rev Lett 116. https://doi.org/10.1103/PhysRevLett.116.056401
29. Pade J (2009) Exact solutions of the Schrödinger equation for zero energy. Eur Phys J D 53. https://doi.org/10.1140/epjd/e2009-00074-0
30. P'oschl G, Teller E (1933) Bemerkungen zur Quantenmechanik des anharmonischen Oszillators. Zeitschrift f'ur Physik 83(34):143–151
31. Rosen N, Morse PM (1932) On the vibrations of polyatomic molecules. Phys Rev 42:210–217
32. Scarf FL (1958) New soluble energy band problem. Phys Rev 112. https://doi.org/10.1103/PhysRev.112.1137
33. Tayari V, Hemsworth N, Fakih I, Favron A, Gaufrès E, Gervais G, Martel R, Szkopek T (2015) Two-dimensional magnetotransport in a black phosphorus naked quantum well. Nat Commun 6:7702. https://doi.org/10.1038/ncomms8702
34. Vasyuta VM, Tkachuk VM (2016) Falling of a quantum particle in an inverse square attractive potential. Eur Phys J D 70:267. https://doi.org/10.1140/epjd/e2016-70463-3
35. Woods RD, Saxon DS (1954) Diffuse surface optical model for nucleon-nuclei scattering. Phys Rev 95(2). https://doi.org/10.1103/PhysRev.95.577
36. Zhang C, Huang L (2010) A quantum model for the stock market. Phys A 389(24):5769–5775

Dynamical Properties of Shock and Snoidal Waves in a Superthermal Multi-ion Dusty Plasma

Satyajit Sarkar, Ruchi Thapa, Asit Saha, and Kajal Kumar Mondal

Abstract The phase plane analysis of the dust-ion-acoustic (DIA) shock and snoidal waves in the framework of Burgers equation is investigated in a superthermal multi-ion dusty plasma. The reductive perturbation method (RPT) is utilized to deduce the nonlinear Burgers equation. Analytical solutions of the DIA kink, anti-kink, and snoidal waves are derived based on heteroclinic and periodic trajectories, respectively, in the phase plots. The fundamental characteristics of the DIA kink, anti-kink, and snoidal waves, like width, amplitude, etc. are investigated based on the physical parameters. It is seen that the intrinsic parameters, like, kinematic viscosity, electron-to-light ion temperature ratio, heavy-to-light ion number density ratio, etc. play a significant role in the modification of DIA kink, anti-kink, and snoidal waves of the plasma system. Three types of quasi-periodic behavior of DIA are presented via phase plots and time series plots. The results of this investigation may be used to further understand the propagation of waves in space and laboratory plasma environments.

Keywords Dynamical system · Burgers equation · Modeling · Solitary wave · Snoidal wave · Quasi-periodicity

1 Introduction

Since the previous few years, the investigation of dynamical properties of dust-ion-acoustic (DIA) waves [1–5] has been the most discussed topic and has been contributing a lot in the growth of dusty plasmas. Shukla [6] made a conclusion on

Supported by organization x

S. Sarkar (✉) · R. Thapa · A. Saha
Department of Mathematics, Sikkim Manipal Institute of Technology, Sikkim Manipal University, Majitar, Rangpo, East-Sikkim 737136, India

S. Sarkar · K. K. Mondal
Department of Mathematics, Cooch Behar Panchanan Barma University, Cooch Behar, West bengal, India

© The Author(s), under exclusive license to Springer Nature Singapore Pte Ltd. 2022

D. Giri et al. (eds.), *Proceedings of the Seventh International Conference on Mathematics and Computing*, Advances in Intelligent Systems and Computing 1412,
https://doi.org/10.1007/978-981-16-6890-6_70

the addition of stationary bigger and weighty charged dust grains in electron-ion plasma can bring about massive changes in the wave dynamics. The inertia is given by the mass of the ions while the thermal pressure due to the presence electrons gives the restoring force. The phase velocity of the nonlinear DIA waves will be in between the ion and electron velocity [7, 8]. However, it was observed [9] that the effect of the charged immobile dust grains modifies the quasi-neutrality at equilibrium. It is to be noted that some works were done experimentally [10] and theoretically [11] in the field of the DIA shock waves in a dusty plasma consisting of electrons, negative dusts, and positive ions.

The research done in the field of dusty multi-ion plasmas and its wave propagation has been focused mainly by the plasma physics researchers since the time it was confirmed that the multi-ion plasma existed in both space [12] and laboratory environments [7, 13]. Earth's ionosphere [14], cometary comae comes under the plasma environments and contain both positive and negative ion species including the electrons. Low-temperature laboratory experiments, plasma processing reactors [15], and neutral beam sources contain the positive-negative ion plasmas. On investigation during plasma etching, the negative ions outshined the positive ions. Hence, in the arena of plasma physics, the negative ion plasmas are getting more significance. Multi-ion plasmas were first studied by Hines [16] after which it has been examined by many more researchers both experimentally and theoretically.

The velocity distribution takes place in a thermal equilibrium which is known as Maxwell velocity distribution. It was noticed frequently that the velocity distribution of plasma components in laboratory and space plasmas is not precisely Maxwellian and that the particles obey distribution that differ from those of Maxwellian distribution, like kappa, nonthermal, superthermal, etc. In nature, various plasma particles especially undergo long-range interaction (extensivity property is often broken) whereas in space and laboratory plasmas, particles interact through short-range interactions. A new statistics called "non-extensive statistics" was first introduced by Tsallis [17] to deal with those kind of dynamical problems.

Using the theory of nonlinear dynamics, nonlinear wave phenomena in plasmas were reported by some researchers [18–21]. Applying bifurcation theory, Monier [22] investigated the DAWs in a multicomponent plasma, J. Tamang and A. Saha investigated dust-acoustic waves for the Burgers equation in a strongly coupled dusty plasma [23]. Very recently Saha et al. explored ion-acoustic kink, anti-kink, and periodic waves of the Burgers equation in a dense quantum plasma [24]. But, there is no study on the phase plane analysis of the DIA kink, anti-kink, and snoidal waves in the framework of the Burgers equation in a superthermal multi-ion dusty plasma. So, in the present work, our concern is to investigate the phase plane analysis of the DIA kink, anti-kink, and snoidal waves using the theory of nonlinear dynamics in a collisionless dusty plasma composing of kappa distributed electrons, mobile heavy ions with negative charge, and positively charged Maxwell distributed light ions with stationary dusts having negative charge. Furthermore, in existence of a periodic forcing, three-dimensional perturbed dynamical structure has been formed. Considering numerical imitations quasi-periodic behavior of DIA waves is measured and theoretical explanations have also been given.

The manuscript is planned as: in Sect. 2, model equations of the plasma system are presented. In Sect. 3, the model equations are deduced into a nonlinear Burgers equation. In Sect. 4, the Burgers equation is further transformed into planar dynamical system. In Sect. 5, the DIA kink and anti-kink waves are presented. The DIA snoidal wave is undertaken in Sect. 6. Perturbed dynamical structure and quasi-periodicity are depicted in Sect. 7. The conclusion is given in Sect. 8.

2 Model Equations

Let us contemplate a collisionless dusty plasma composing of kappa distributed electrons, mobile heavy ions with negative charge, and positively charged Maxwell distributed light ions with stationary dusts having negative charge. Hence, at equilibrium $n_{i0} - Z_h n_{h0} - n_{e0} - Z_d n_{d0} = 0$, where n_{i0} denotes unperturbed number density for positive light ions, n_{h0} denotes number density for positive heavy ions at equilibrium, n_{e0} is number density for electrons at equilibrium, n_{d0} denotes number density for immobile dust at equilibrium, and Z_d denotes number of electrons present on the surface of the dust grains. The number densities of Kappa distributed electrons [25, 26] and Maxwellian light ions are as follows:

$$n_e = n_{e0}\left[1 - \frac{\frac{e\phi}{k_B T_e}}{\kappa - \frac{3}{2}}\right]^{-\kappa+\frac{1}{2}}. \tag{1}$$

$$n_i = n_{i0}\, e^{-\frac{e\phi}{k_B T_i}}. \tag{2}$$

The normalized equations of the considered plasma system whose electron (ion) thermal speed is much larger (smaller) than the phase speed are given by

$$\frac{\partial n_h}{\partial t} + \frac{\partial}{\partial x}(n_h u_h) = 0, \tag{3}$$

$$\frac{\partial u_h}{\partial t} + u_h \frac{\partial u_h}{\partial x} = \frac{\partial \phi}{\partial x} + \eta \frac{\partial^2 u_h}{\partial x^2}, \tag{4}$$

$$\frac{\partial^2 \phi}{\partial x^2} = \mu_e\left[1 - \frac{\phi}{\kappa - \frac{3}{2}}\right]^{-\kappa+\frac{1}{2}} + \mu_h n_h - e^{-\sigma\phi} + \mu_d, \tag{5}$$

where κ is the parameter describing the kappa distributed electrons. The negative heavy-ion number density n_h is normalized to n_{h0}, the heavy-ion fluid u_h is normalized to $C_h = (k_B T_e/m_h)^{1/2}$, the electrostatic wave potential ϕ is normalized to $k_B T_e/e$, viscosity coefficient η is normalized to $m_h n_{h0} \omega_{ph} \lambda_{Dh}^2$, where k_B denotes the Boltzmann constant, T_e denotes temperature of electrons, and e denotes electron charge. Time t is normalized to $\lambda_{Dh} = (k_B T_e/4\pi e^2 n_{h0})^{1/2}$. We define the parameters that have occurred in (5) as $\mu_e = n_{e0}/n_{i0}$ (ratio of electron-to-light ion number densities), $\mu_h = Z_h n_{h0}/n_{i0}$ (ratio of heavy-to-light ion number densities), and $\sigma = T_e/T_i$ (temperature ratio of electron-to-light ion).

3 Burgers Equation

In order to explore DIA snoidal, kink, and anti-kink waves, we may use Eqs. (3)–(5) and a stretching $\xi = \epsilon(x - vt)$ and $\tau = \epsilon^2 t$, where v gives phase velocity of the DIA wave and ϵ measures weakness of the nonlinearity. We take following expansions of n_h, u_h, and ϕ as

$$\begin{cases} n_h = 1 + \epsilon n_1 + \epsilon^2 n_2 + \cdots, \\ u_h = 0 + \epsilon u_1 + \epsilon^2 u_2 + \cdots, \\ \phi = 0 + \epsilon \phi_1 + \epsilon^2 \phi_2 + \cdots, \end{cases} \tag{6}$$

and using the boundary conditions $n_h \to 1, u_h \to 0$, and $\phi \to 0$ as $\xi \to \pm\infty$, we can generate relations in different powers of ϵ. To the smallest power of ϵ, Eqs. (3)–(5) provide

$$n_1 = -\frac{\phi_1}{v^2}, \tag{7}$$

$$u_1 = -\frac{\phi_1}{v}, \tag{8}$$

$$\{\sigma + \mu_e \frac{(\kappa - \frac{1}{2})}{(\kappa - \frac{3}{2})}\}\phi_1 + \mu_h n_1 = 0. \tag{9}$$

From Eqs. (7)–(9), we can obtain

$$v = \sqrt{\frac{\mu_h}{\sigma + \mu_e \frac{(\kappa - \frac{1}{2})}{(\kappa - \frac{3}{2})}}}. \tag{10}$$

Equation (10) provides the dispersion relation for the DIA waves. Corresponding to the higher power of ϵ, Eqs. (3)–(5) give

$$-v\frac{\partial n_2}{\partial \xi} + \frac{\partial n_1}{\partial \tau} + \frac{\partial}{\partial \xi}(u_2 + n_1 u_1) = 0, \tag{11}$$

$$-v\frac{\partial u_2}{\partial \xi} + \frac{\partial u_1}{\partial \tau} + u_1\frac{\partial u_1}{\partial \xi} = \frac{\partial \phi_2}{\partial \xi} + \eta\frac{\partial^2 u_1}{\partial \xi^2}, \tag{12}$$

$$\{\sigma + \mu_e \frac{(\kappa - \frac{1}{2})}{(\kappa - \frac{3}{2})}\}\phi_2 + \{\frac{\mu_e(\kappa - \frac{1}{2})(\kappa + \frac{1}{2})}{2(\kappa - \frac{3}{2})^2} - \frac{\sigma^2}{2}\}\phi_1^2 + \mu_h n_2 = 0. \tag{13}$$

Now, following some mathematical calculations, we procure the nonlinear Burgers equation from Eqs. (3)–(13) as

$$\frac{\partial \phi_1}{\partial \tau} + A\phi_1 \frac{\partial \phi_1}{\partial \xi} = B\frac{\partial^2 \phi_1}{\partial \xi^2}, \tag{14}$$

where $A = \frac{v^3}{2\mu_h}[\sigma^2 - \frac{\mu_e(\kappa-\frac{1}{2})(\kappa+\frac{1}{2})}{(\kappa-\frac{3}{2})^2} - \frac{3\mu_h}{v^4}]$ and $B = \frac{\eta}{2}$.

4 Planar Dynamical System (PDS)

Taking traveling wave transformation $\chi = \xi - V\tau, \phi_1(\xi, \tau) = \psi(\chi)$, with traveling wave velocity V and $\psi \to 0$, $\frac{d\psi}{d\chi} \to 0$ as $\chi \to \infty$, the Burgers equation (14) may be expressed as

$$\begin{cases} \frac{d\psi}{d\chi} = z, \\ \frac{dz}{d\chi} = \frac{1}{B^2}\psi(\frac{A\psi}{2} - V)(A\psi - V). \end{cases} \tag{15}$$

The Hamiltonian function for the system (15) is

$$H(\psi, z) = \frac{z^2}{2} - \frac{V^2}{2B^2}\psi^2 + \frac{AV}{2B^2}\psi^3 - \frac{A^2}{8B^2}\psi^4 = h. \ (say) \tag{16}$$

System (15) is a planar dynamical system (PDS). In order to examine bifurcation of PDS (15) for DIA waves, the concept of phase plot may be used depending on variation of physical parameters κ, μ_e, μ_h, η, σ, and V. One can note that the phase plot of PDS (15) may differ depending on nature of equilibrium points. In a phase plot of PDS (15), any trajectory refers to a traveling wave solution of Eq. (14). One can find the different equilibrium points by the method of phase plane analysis. There exist three equilibrium points of the PDS (15) at $P_0(\psi_0, 0)$, $P_1(\psi_1, 0)$, and $P_2(\psi_2, 0)$, where $\psi_0 = 0$, $\psi_1 = \frac{V}{A}$, and $\psi_2 = \frac{2V}{A}$. The determinant of the Jacobian matrix of the PDS (15) is given by

$$J(\psi_i, 0) = -\left(\frac{3A^2}{2B^2}\psi_i^2 - \frac{3AV}{B^2}\psi_i + \frac{V^2}{B^2}\right). \tag{17}$$

According to phase plane analysis [27–30], if $J(\psi_i, 0) < 0$, then $P_i(\psi_i, 0)$ is a saddle point, and for $J(\psi_i, 0) > 0$, $P_i(\psi_i, 0)$ is a center. Phase plot of the PDS (15) is represented for specific values of parameters κ, μ_e, μ_h, η, σ, and V shown in Fig. 1.

In Fig. 1, we represent a phase plot of the PDS (15) with $\kappa = 1.8$, $\mu_e = 0.05$, $\mu_h = 0.5$, $\sigma = 0.01$, $\eta = 0.1$, and $V = 1.49$. There exist three equilibrium points P_0, P_1, and P_2 of the PDS (15) with one center and two saddles. In this figure, we can notice that P_0 (origin) and P_2 are saddles, while P_1 is a center. There is a pair of heteroclinic trajectories from P_0 to P_2 and P_2 to P_0 with a family of nonlinear periodic trajectories surrounding equilibrium point P_1. We also observe that the two

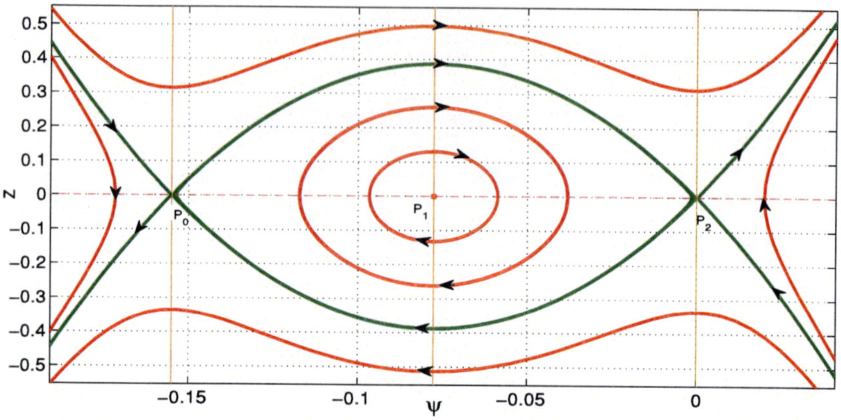

Fig. 1 Phase plot of the PDS (15) for $\kappa = 1.8$, $\mu_h = 0.5$, $\mu_e = 0.05$, $\eta = 0.1$, $V = 1.49$, and $\sigma = 0.01$

equilibrium points, namely, P_1 (center) and P_2 (saddle) lie on the left-hand side from the origin.

5 Kink and Anti-kink Waves

We take the traveling wave transformation $\zeta = c(\xi - V\tau)$, so as to deduce the shock wave solution. For simplicity, we have represented ϕ_1 by ϕ. Then, Eq. (14) becomes

$$-V\frac{d\phi}{d\zeta} + A\phi\frac{d\phi}{d\zeta} - cB\frac{d^2\phi}{d\zeta^2} = 0. \tag{18}$$

Integrating (18), we get

$$-V\phi + A\frac{\phi^2}{2} - cB\frac{d\phi}{d\zeta} = c_1. \tag{19}$$

Initially, $\phi \to 0$, therefore $\frac{d\phi}{d\zeta} \to 0$. Then, $c_1 = 0$. Therefore, we can express Eq. (19) as

$$-V\phi + A\frac{\phi^2}{2} - cB\frac{d\phi}{d\zeta} = 0. \tag{20}$$

Let $G = tanh(\gamma)$ and considering $\phi(\gamma) = L(G) = \sum_{m=0}^{R} a_m G^m$. Then, Eq. (20) becomes

$$-VL(G) + A\frac{L^2(G)}{2} - Bc(1 - G^2)\frac{dL}{dG} = 0. \tag{21}$$

Balancing the linear term of highest order derivative with the nonlinear term of highest degree (power), we get

$$2R = R + 1 \Rightarrow R = 1. \tag{22}$$

Hence,

$$L = a_0 + a_1 G. \tag{23}$$

Then, $\frac{dL}{dG} = a_1$. Therefore, using the value of L from Eq. (22) in Eq. (21), we get

$$-Va_0 + \frac{Aa_0^2}{2} - cBa_1 + (Aa_0a_1 - Va_1)Y + \left(\frac{Aa_1^2}{2} + cBa_1\right)G^2 = 0. \tag{24}$$

Equating to zero the coefficients of G^0, G, and G^2, we get

$$-Va_0 + \frac{Aa_0^2}{2} - cBa_1 = 0, \tag{25}$$

$$Aa_0a_1 - Va_1 = 0, \tag{26}$$

$$\frac{Aa_1^2}{2} + cBa_1 = 0. \tag{27}$$

From Eqs. (23)–(24), we get $a_0 = \frac{V}{A}$, $a_1 = -\frac{V}{A}$ and $c = \frac{V}{2B}$. Using these values in (23), we get

$$L = \frac{V}{A}(1 - tanh\gamma), \tag{28}$$

$$\Rightarrow \phi(\xi, \tau) = \frac{V}{A}\left[1 - tanh\left\{\frac{V}{2B}(\xi - V\tau)\right\}\right]. \tag{29}$$

Equation (29) is the required kink wave solution of Eq. (14) corresponding to the heteroclinic trajectory from P_0 to P_2. Similarly, we can obtain the required anti-kink wave solution of Eq. (14) corresponding to the trajectory from P_2 to P_0 as

$$\phi(\xi, \tau) = -\frac{V}{A}\left[1 - tanh\left\{\frac{V}{2B}(\xi - V\tau)\right\}\right]. \tag{30}$$

Therefore, shock wave solution is

$$\phi(\xi, \tau) = \pm\frac{V}{A}\left[1 - tanh\left\{\frac{V}{2B}(\xi - V\tau)\right\}\right]. \tag{31}$$

We present variations of DIA kink and anti-kink waves of Burgers equation (14) for physical parameters κ, μ_h, μ_e, η, σ, and V in Figs. 2, 3, 4, 5, 6, and 7 for $\tau = 1$.

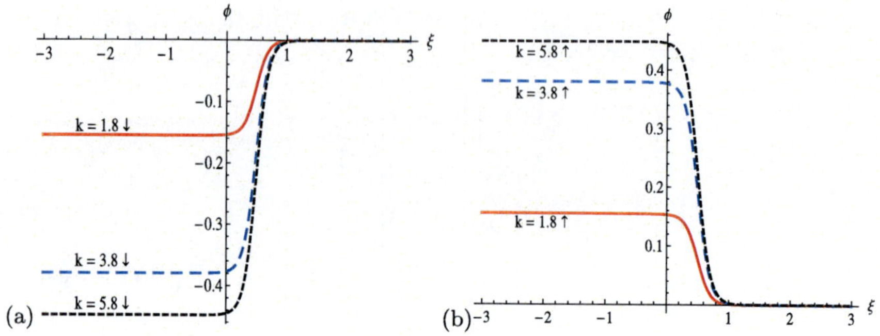

Fig. 2 **a** DIA kink wave **b** DIA anti-kink wave of Eq. (14) for distinct κ with $\mu_h = 0.5$, $\mu_e = 0.05$, $\eta = 0.1$, $V = 0.5$, and $\sigma = 0.01$

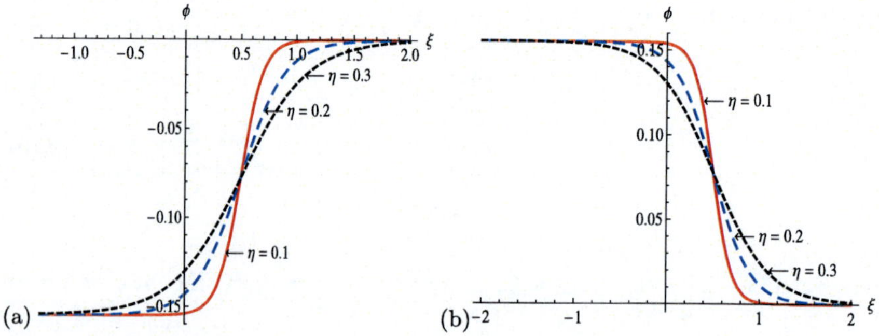

Fig. 3 **a** DIA kink wave **b** DIA anti-kink wave of Eq. (14) for distinct η with $\mu_h = 0.5$, $\mu_e = 0.05$, $\kappa = 1.8$, $V = 0.5$, and $\sigma = 0.01$

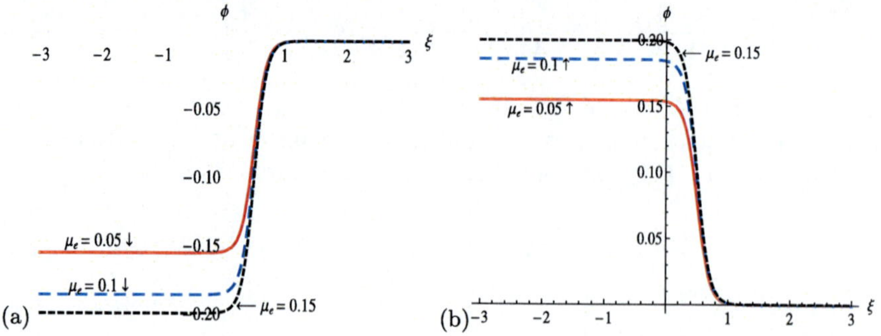

Fig. 4 **a** DIA kink wave **b** DIA anti-kink wave of Eq. (14) for distinct μ_e with $\mu_h = 0.5$, $\eta = 0.1$, $\kappa = 1.8$, $V = 0.5$, and $\sigma = 0.01$

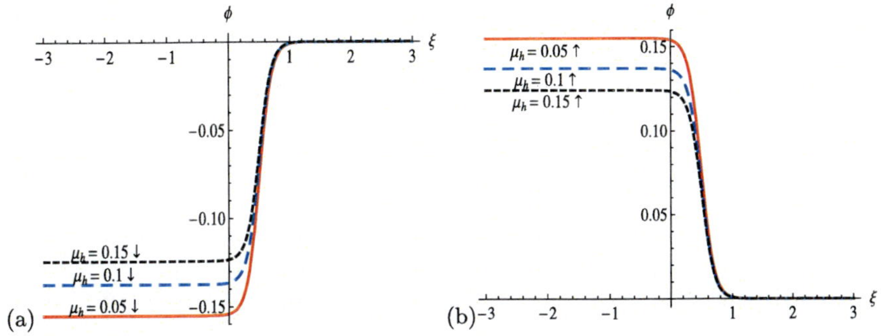

Fig. 5 **a** DIA kink wave **b** DIA anti-kink wave of Eq. (14) for distinct μ_h with $\mu_e = 0.05,\ \eta = 0.1,\ \kappa = 1.8,\ V = 0.5$, and $\sigma = 0.01$

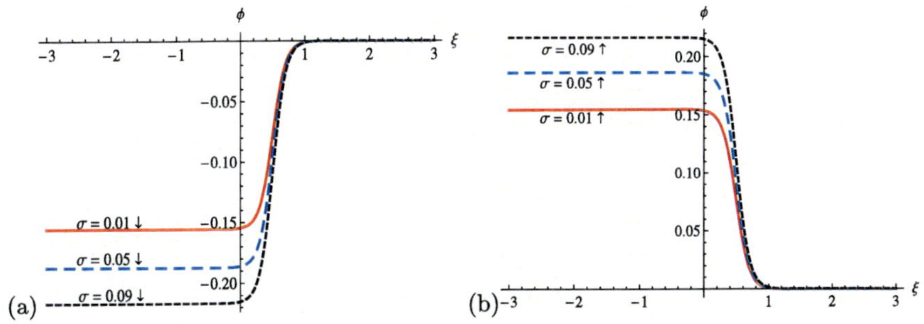

Fig. 6 **a** DIA kink wave **b** DIA anti-kink wave of Eq. (14) for distinct σ with $\mu_e = 0.05,\ \eta = 0.1,\ \kappa = 1.8,\ V = 0.5$, and $\mu_h = 0.5$

In Fig. 2a and b, we present DIA shock waves of Eq. (14) for distinct spectral index κ with $\mu_h = 0.5,\ \mu_e = 0.05,\ \eta = 0.1,\ V = 0.5$, and $\sigma = 0.01$. It is seen that amplitude of DIA shock wave grows as κ increases.

In Fig. 3a and b, we depict DIA shock waves of Eq. (14) for distinct kinematic viscosity η with $\mu_h = 0.5,\ \mu_e = 0.05,\ \kappa = 1.8,\ V = 0.5$, and $\sigma = 0.01$. We notice that smoothness of DIA kink wave of Burgers equation (14) increases as η increases, while amplitude remains the same.

In Fig. 4a and b, we show DIA shock waves of Eq. (14) for distinct μ_e (ratio of electron-to-light ion number densities) with $\mu_h = 0.5,\ \eta = 0.1,\ \kappa = 1.8,\ V = 0.5$, and $\sigma = 0.01$. It is seen that amplitude of DIA kink wave grows as μ_e increases.

In Fig. 5a and b, we convey DIA shock waves of Burgers equation (14) for distinct μ_h (ratio of heavy-to-light ion number densities) with $\mu_e = 0.05,\ \eta = 0.1,\ \kappa = 1.8,\ V = 0.5$, and $\sigma = 0.01$. We discover that amplitude of the DIA kink wave diminishes as μ_h grows. Thus, if concentration of heavy ions with negative charge grows compared to as that of lighter positive ions, the DIA shock waves of Eq. (14) diminish in amplitude.

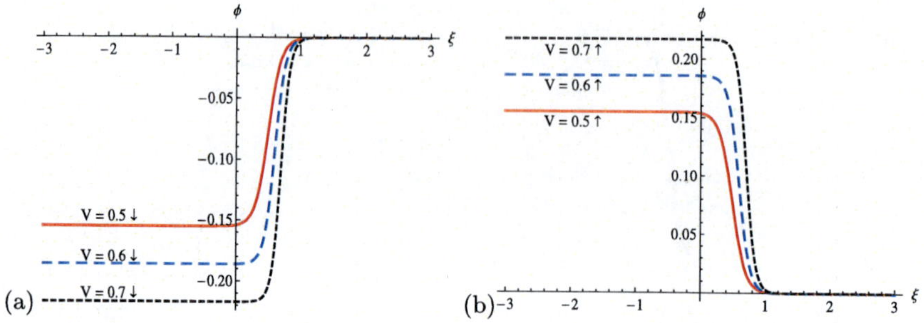

Fig. 7 **a** DIA kink wave **b** DIA anti-kink wave of Eq. (14) for distinct V with $\mu_e = 0.05,\ \eta = 0.1,$ $\kappa = 1.8,\ \sigma = 0.01$, and $\mu_h = 0.5$

In Fig. 6a and b, we reveal DIA shock waves of Eq. (14) for distinct σ (temperature ratio of electron-to-light ion) with $\mu_e = 0.05,\ \eta = 0.1,\ \kappa = 1.8,\ V = 0.5$, and $\mu_h = 0.5$. We detect that amplitude of DIA shock waves increases as σ increases.

In Fig. 7a and b, we present DIA shock waves of Eq. (14) for distinct velocities V with $\mu_e = 0.05,\ \eta = 0.1,\ \kappa = 1.8,\ \sigma = 0.01$, and $\mu_h = 0.5$. We perceive that amplitude of DIA shock wave significantly increases and smoothness decreases as velocity V increases.

6 Snoidal Wave

The DIA snoidal wave for the periodic orbit depicted in phase portrait (Fig. 1) of the dynamical structure (15) is given by

$$\psi = \frac{a + d\left[\left(\frac{a-b}{b-d}\right) sn^2\left(\frac{A}{2Bg}\chi, k\right)\right]}{1 + \left[\left(\frac{a-b}{b-d}\right) sn^2\left(\frac{A}{2Bg}\chi, k\right)\right]}, \tag{32}$$

where $a, b, c, andd$ are solutions of equation $H(\psi, z) = h_*$ with $h_* = H(-0.1, 0)$, $g = \frac{2}{\sqrt{(a-c)(b-d)}}$, and $k^2 = \frac{(a-b)(c-d)}{(a-c)(b-d)}$. Here sn denotes the Jacobi elliptic function [31].

In Fig. 8a, we present influence of spectral index (κ) on DIA snoidal wave solution with $\mu_h = 0.5$, $\mu_e = 0.05,\ \eta = 0.1,\ V = 1.49$, and $\sigma = 0.01$. In this figure, one can clearly note that for increment in κ, width of the DIA snoidal wave decreases but amplitude increases. Because of this fact DIA snoidal wave becomes spiky. Again, in Fig. 8b, we depict the effect of μ_e on DIA snoidal wave solution with $\kappa = 1.8$ whereas values of other parameters remain unchanged as given in Fig. 8a. It is found that as μ_e increases amplitude of snoidal wave decreases and width increases. Hence, the DIA snoidal wave becomes smooth as μ_e grows. On the other hand, if heavy-

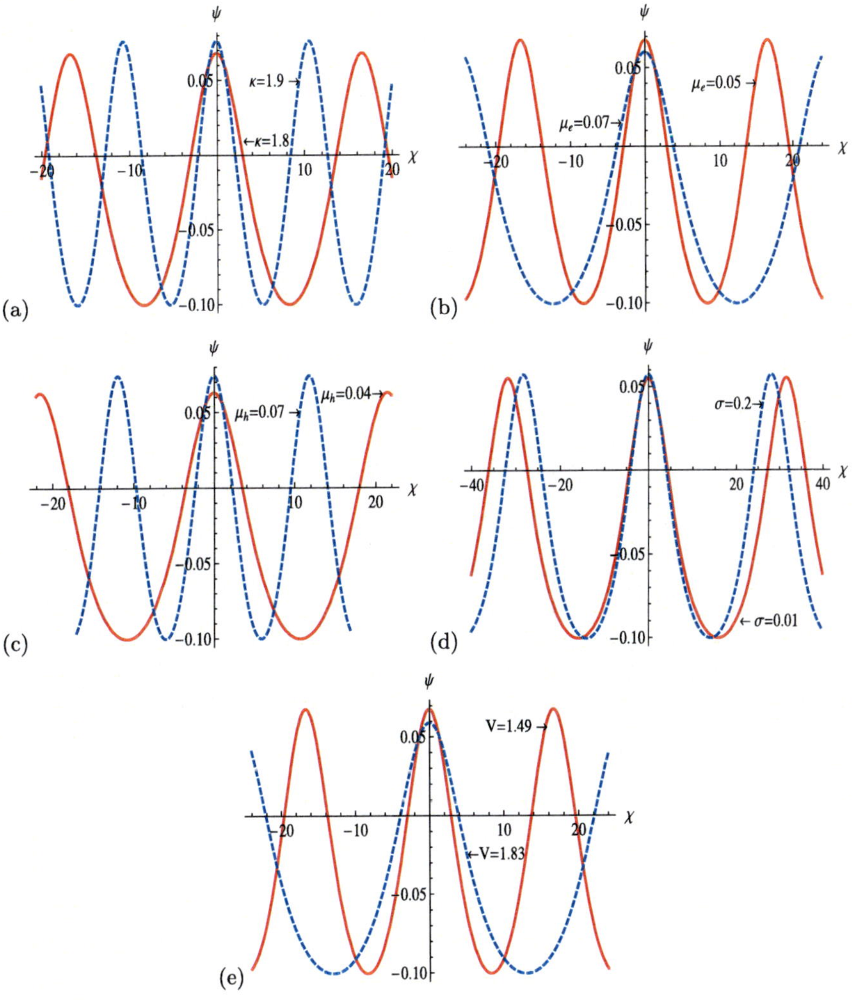

Fig. 8 DIA snoidal wave of Eq. (14) for distinct values of **a** κ with $\mu_h = 0.5$, $\mu_e = 0.05$, $\eta = 0.1$, $V = 1.49$, and $\sigma = 0.01$, **b** μ_e with $\mu_h = 0.5$, $\kappa = 1.8$, $\eta = 0.1$, $V = 1.49$, and $\sigma = 0.01$, **c** μ_h with $\mu_e = 0.05$, $\kappa = 1.8$, $\eta = 0.1$, $V = 1.49$, and $\sigma = 0.01$, **d** σ with $\mu_h = 0.5$, $\mu_e = 0.05$, $\kappa = 2.1$, $\eta = 0.1$, and $V = 3.4$, **e** V with $\mu_h = 0.5$, $\mu_e = 0.05$, $\kappa = 1.8$, $\eta = 0.1$,, and $\sigma = 0.01$

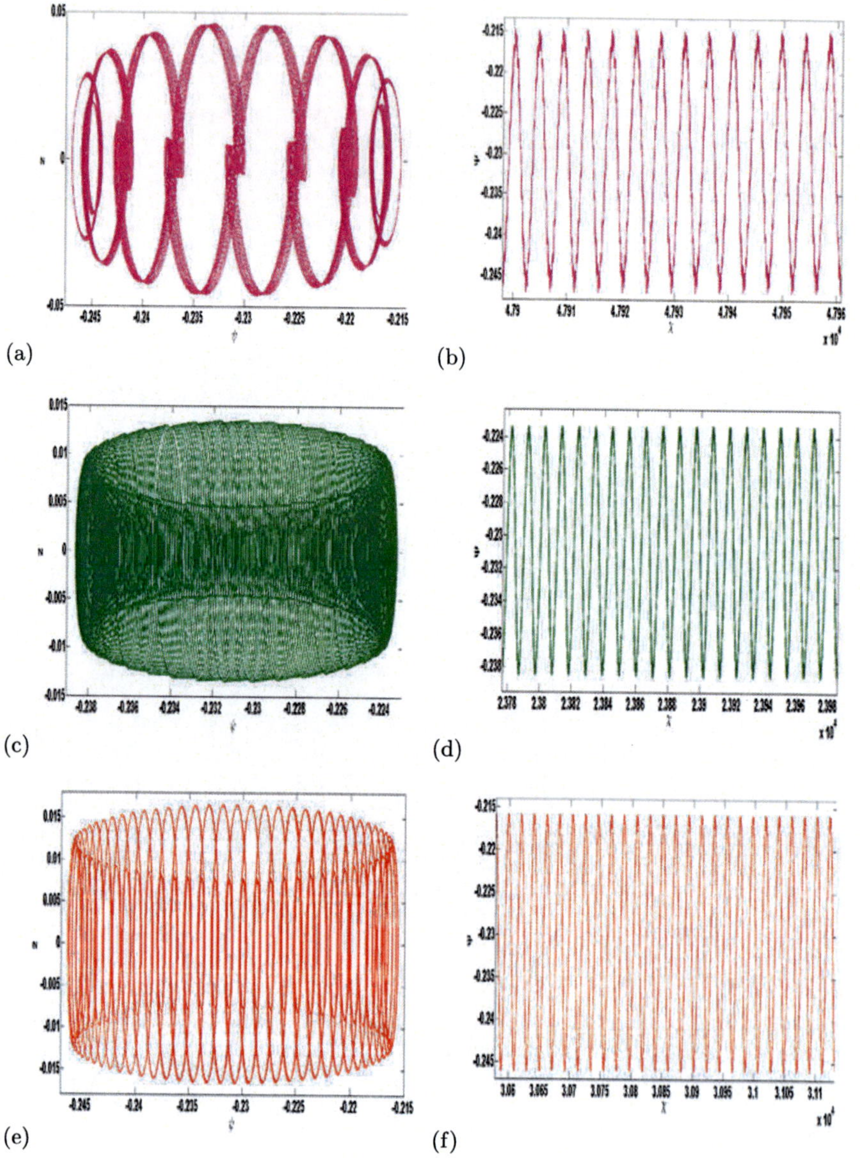

Fig. 9 Distinct quasi-periodic attractors and corresponding time series plots for PRDS (34)

to-light ion number density ratio (μ_h) grows with $\kappa = 1.8$ and same values of other parameters as Fig. 8a, amplitude of DIA snoidal wave increases and width decreases (see Fig. 8c). Again it seen from Fig. 8d that as electron-to-light ion temperature ratio (σ) increases amplitude of the DIA snoidal wave increases and width decreases slowly (see Fig. 8d). Furthermore, it is found that if we increase velocity (V) of the DIA snoidal wave with $\kappa = 1.8$ and same values of other parameters as Fig. 8a, amplitude of the DIA snoidal wave decreases and width increases (see Fig. 8e). Thus, the DIA snoidal wave becomes smooth.

7 Perturbed Dynamical Structure (PRDS) and Quasi-periodicity

In this part, by theoretical explanations and numerical imitations, dynamical behavior of DIAs in existence of a periodic force is reported by means of phase portraits and time series plots. Presence of space debris in plasmas can cause this type of periodic force with sine and cosine functions [32, 33].

Let us consider the non-autonomous PRDS of the DIAs, with an periodic force $f_0\, cos(\omega\chi)$ as

$$\begin{cases} \frac{d\psi}{d\chi} = z, \\ \frac{dz}{d\chi} = P\psi^3 + Q\psi^2 + R\psi + f_0\, cos(\omega\chi), \end{cases} \tag{33}$$

where $P = \frac{A^2}{2B^2}$, $Q = -\frac{3AV}{2B^2}, and R = \frac{V^2}{B^2}$. Here f_0 is strength and ω is frequency of the periodic force. Autonomous structure corresponding to the non-autonomous PRDS (33) can be formed as

$$\begin{cases} \frac{d\psi}{d\chi} = z, \\ \frac{dz}{d\chi} = P\psi^3 + Q\psi^2 + R\psi + f_0\, cos(\xi), \\ \frac{d\xi}{d\chi} = \omega \end{cases} \tag{34}$$

where ψ, χ, $and \xi$ are state variables.

To study the dynamics of perturbed non-autonomous model (34) of the DIAs, we depict its trajectory in the phase space, as shown in Fig. 9. We can observe that the system (34) under the periodic forcing shows three different types of quasi-periodic trajectories, obtained by varying the parameters f_0, ω and keeping other parameters fixed. Time series plots corresponding to each quasi-periodic trajectories have also been presented. Here quasi-periodic trajectories and corresponding time series are initiated with following parameter values as fixed $V = 1.49,\ A = -6.45179,\ B = 0.05,\ \sigma = 0.01,\ \mu_e = 0.05,\ \mu_h = 0.5,\ \kappa = 1.8,\ \eta = 0.1$ and varying parameter values $f_0 = 6.53,\ \omega = 1.4$ resulted in Fig. 9a, b; $f_0 = 3.21,\ \omega = 0.6$ resulted in Fig. 9c, d, whereas $f_0 = 6.53,\ \omega = 0.3$ resulted in Fig. 9e, f.

8 Conclusions

In this study, we have examined the bifurcation of DIA kink, anti-kink, and snoidal waves in a superthermal multi-ion dusty plasma with kappa distributed electrons, heavy ions with negative charge, Maxwellian light ions with positive charge, and stationary dusts with negative charge. Using RPT, we have obtained the Burgers equation for our plasma system. Using traveling wave transformation, we have formed a planar dynamical system from the Burgers equation. Phase plots of the dynamical system have been presented which exhibit the heteroclinic trajectories and nonlinear snoidal trajectories varying the values of the physical parameters, namely, spectral index κ, μ_h (ratio of heavy-to-light ion number densities), σ (temperature ratio of electron-to-light ion), and μ_e (ratio of electron-to-light ion number densities) while keeping values of the parameter η (viscosity coefficient) unchanged. It has been noticed that with increase in values of κ, μ_e, and σ, amplitude of the DIA shock waves has been increased, while in case of increasing μ_h, amplitude of the DIA shock waves has been decreased. In case of increasing V, amplitude of the DIA shock waves has been increased and smoothness has been decreased. Furthermore, in case of increasing η, there is a significant growing change on the width of the shock waves and as a result smoothness has been increased. The results of this contribution may be relevant in nonlinear DIA waves in space and laboratory plasmas.

Acknowledgements This research work is supported by TMA Pai University Research Grant: (6100/SMIT/R & D/Project/05/2018).

References

1. Verheest F, Hellberg MA, Kourakis I (2013) Dust-ion-acoustic supersolitons in dusty plasmas with nonthermal electrons. Phys Rev E 87:043107
2. Chatterjee P, Roy K (2008) Large amplitude solitary waves in a four-component dusty plasma with nonthermal ions. Zeitschrift für Naturforschung A 63:393–399
3. Bacha M, Tribeche M, Shukla PK (2012) Dust ion-acoustic solitary waves in a dusty plasma with nonextensive electrons. Phys Rev E 85:056413
4. El-Shewy EK, Zaghbeer SK, El-Rahman AA (2019) Three-dimensional instability of opposite polarity nonthermal dusty plasma. Zeitschrift für Naturforschung A 74:131–138
5. Baluku TK, Hellberg MA, Kourakis I, Saini NS (2010) Dust ion acoustic solitons in a plasma with kappa-distributed electrons. Phys Plasmas 17:053702
6. Shukla PK, Silin VP (1992) Dust-ion acoustic wave. Phys Scr 45(5):508
7. D'Angelo N, Merlino RL (1996) Current-driven dust-acoustic instability in a collisional plasma. Planet Space Sci 44(12):1593–1598
8. Mamun AA, Shukla PK, Eliasson B (2009) Arbitrary amplitude dust ion-acoustic shock waves in a dusty plasma with positive and negative ions. Phys Plasmas 16(11):114503
9. Shukla PK, Mamun AA (2002) Introduction to dusty plasma physics. Institute of Physics Publishing, Bristol
10. Luo QZ, D'Angelo N, Merlino RL (1999) Experimental study of shock formation in a dusty plasma. Phys Plasmas 6(9):3455
11. Eliasson B, Shukla PK (2005) Formation of large-amplitude dust ion-acoustic shocks in dusty plasmas. Phys Plasmas 12(2):024502

12. Shukla PK (2001) A survey of dusty plasma physics. Phys Plasmas 8(5):1791
13. Merlino RL, Barkan A, Thompson C, D'Angelo N (1998) Laboratory studies of waves and instabilities in dusty plasmas. Phys Plasmas 5(5):1607
14. Massey H (1976) Negative ions. Cambridge University Press, Cambridge
15. Gottscho RA, Gaebe CE (1986) Negative ions kinetics in RF glow discharges. IEEE Trans Plasma Sci 14(2):92–102
16. Hines CO (1957) 1957 Heavy-ion effects in audio-frequency radio propagation. J Atmos Terr Phys 11(1):36–42
17. Tsallis C (1988) Possible generalization of Boltzmann-Gibbs statistics. J Stat Phys 52:479–487
18. Saha A, Tamang J (2019) Effect of q-nonextensive hot electrons on bifurcations of nonlinear and supernonlinear ion-acoustic periodic waves. Adv Space Res 63(5):1596–1606
19. Saha A, Chatterjee P (2015) Solitonic, periodic and quasiperiodic behaviors of dust ion acoustic waves in superthermal plasmas. Braz J Phys 45:419–426
20. Selim MM, El-Depsy A, El-Shamy EF (2015) Bifurcations of nonlinear ion-acoustic travelling waves in a multicomponent magnetoplasma with superthermal electrons. Astrophys Space Sci 360:66
21. Saha A, Chatterjee P (2014) Bifurcations of dust acoustic solitary waves and periodic waves in an unmagnetized plasma with nonextensive ions. Astrophys Space Sci 351:533–537
22. El-Monier SY, Atteya A (2018) Bifurcation analysis for dust-acoustic waves in a four-component plasma including warm ions. IEEE Trans Plasma Sci 46(4):815–824
23. Tamang J, Saha A, Phase plane analysis of the dust-acoustic waves for the Burgers equation in a strongly coupled dusty plasma. Indian J Phys https://doi.org/10.1007/s12648-020-01733-3
24. Saha A, Pradhan B, Banerjee S, Bifurcation analysis of quantum ion-acoustic kink, anti-kink and periodic waves of the Burgers equation in a dense quantum plasma. Eur Phys J Plus https://doi.org/10.1140/epjp/s13360-020-00235-9
25. Vasyliunas VM (1968) A survey of low-energy electrons in the evening sector of the magnetosphere with OGO 1 and OGO 3. J Geophys Res 73:2839–2884
26. Verheest F, Hellberg MA, Kourakis I (2013) Ion-acoustic supersolitons in plasmas with two-temperature electrons: Boltzmann and kappa distributions. Phys Plasmas 20(8):082309
27. Chow SN, Hale JK (1981) Methods of bifurcation theory. Springer, New York
28. Tamang J, Saha A (2019) Dynamical behavior of supernonlinearpositron-acoustic periodic waves and chaosin nonextensive electron-positron-ion plasmas. Zeitschrift für Naturforschung A. https://doi.org/10.1515/zna-2018-0476
29. Strogatz SH (2007) Nonlinear dynamics and chaos. Westview Press (A member of the Pressus Books Group, USA)
30. Guckenheimer J, Holmes PJ (1983) Nonlinear oscillations dynamical systems and bifurcations of vector fields. Springer, New York
31. Byrd PF, Friedman MD (1971) Handbook of elliptic integrals for engineers and scientists. Springer, Berlin
32. Aslanov VS, Yudintsev VV (2015) Dynamics, analytical solutions and choice of parameters for towed space debris with flexible appendages. Adv Space Res 55(2):660–667
33. Mandi L, Saha A, Chatterjee P (2019) Dynamics of ion-acoustic waves in Thomas-Fermi plasmas with source term. Adv Space Res 64(2):427–435

A Simple Model on Streamflow Management with a Dynamic Risk Measure

Hidekazu Yoshioka and Yumi Yoshioka

Abstract We present an exactly solvable risk-minimizing stochastic differential game for flood management in rivers for sustainable and adaptive water management. The streamflow dynamics follow stochastic differential equations driven by a Lévy process. An entropic dynamic risk measure is employed to evaluate a flood risk under model uncertainty. The problem is solved via a Hamilton–Jacobi–Bellman–Isaacs equation. We explicitly derive an optimal flood mitigation policy along with its existence criteria and the worst-case probability measure. A backward stochastic differential representation as an alternative formulation is also presented. Our contribution provides a new mathematical approach for better understanding water management.

Keywords River management under uncertainty · Lévy process · Zero-sum differential game · Hamilton-Jacobi-Bellman-Isaacs equation

1 Introduction

River is a dynamic environment driven by stochastic streamflow. Stochastic differential equations (SDEs) of the Ornstein–Uhlenbeck types driven by Gaussian [1] and Poisson noises [2, 3] have been efficient models representing streamflow time series. Recently, a tempered stable Ornstein–Uhlenbeck model [4] has been proposed as a more realistic alternative to represent both continuous and jump noises in the streamflow time series [5].

Modeling streamflow itself is not a goal but rather a pathway toward streamflow management as dynamic optimization problems like dam and hydropower management [6, 7]. Stochastic control based on the dynamic programming with SDEs [8]

H. Yoshioka (✉) · Y. Yoshioka
Shimane University, Nishikawatsu-cho 1060, Matsue 690-8504, Japan
e-mail: yoshih@life.shimane-u.ac.jp

Y. Yoshioka
e-mail: yyoshioka@matsu.shimane-u.ac.jp

© The Author(s), under exclusive license to Springer Nature Singapore Pte Ltd. 2022

D. Giri et al. (eds.), *Proceedings of the Seventh International Conference on Mathematics and Computing*, Advances in Intelligent Systems and Computing 1412,
https://doi.org/10.1007/978-981-16-6890-6_71

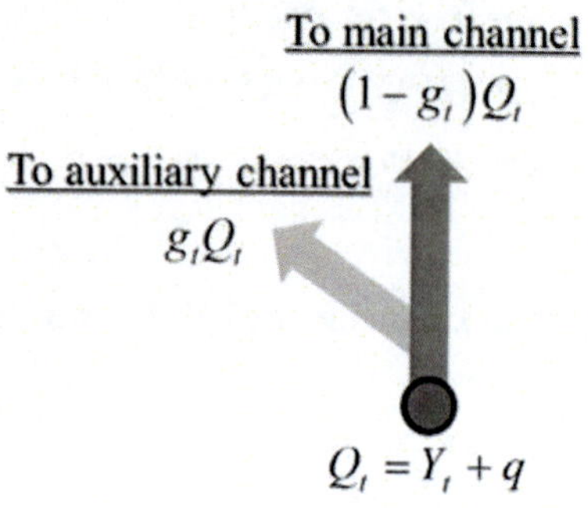

Fig. 1 Conceptual diagram of the model

is a versatile tool for dynamic optimization under uncertainty. Especially, a risk-minimizing problem where the decision-maker is not fully confident about his/her reference model is a game-type control problem with a dynamic risk measure [9]. In river environmental management, it is often difficult to accurately identify the target model due to limitations such as technical constraints and lack of knowledge [10, 11]. The dynamic risk measure is a mathematically rigorous tool harmonizing with this situation; however, such an approach to streamflow management does not exist to the best of our knowledge.

Solving a stochastic differential game reduces to finding an appropriate solution to a corresponding Hamilton–Jacobi–Bellman–Isaacs (HJBI) equation [8]. This kind of non-linear partial integro-differential equation can be solved explicitly only in limited cases [12–15]. The use of such explicit solutions is of course limited; however, they provide clear and useful insights into more complicated problems.

We present an exactly solvable streamflow management problem subject to stochastic streamflow driven by a pure jump Lévy process [16]. This is a streamflow diversion problem to minimize both flooding and diversion effort (Fig. 1, see the next section for more detail). The performance functional to be minimized by optimizing the diversion effort is set as a dynamic risk measure of an entropic type so that the deviation between the true and believed model is penalized [8]. The resulting HJBI equation is non-linear, non-local, and high dimensional. Nevertheless, we show that it is possible to derive its explicit solution along with its existence criteria.

2 Problem Formulation

2.1 System Dynamics

We consider a streamflow management under uncertainty as a simplified flood mitigation problem. There are the two channels, main and auxiliary channels (Fig. 1). The goal here is to simultaneously find the optimal diversion policy and the risk measure under model uncertainty.

We assume that the streamflow $Q = (Q_t)_{t \geq 0}$ at a point in a main channel follows a Lévy-driven Ornstein–Uhlenbeck model:

$$\mathrm{d}Y_t = -\lambda Y_t \mathrm{d}t + \int_0^{\infty} zN(\mathrm{d}z, \mathrm{d}t), \tag{1}$$

with $Q_t = Y_t + q$, where $q \geq 0$ is the baseflow and $Y = (Y_t)_{t \geq 0}$ is the fluctuation, $\lambda > 0$ is the recession rate, N is the Poisson measure of a Lévy process having positive jumps (subordinator), and the Lévy measure v. We assume $\int_0^{\infty} \min\{1, z\} v(\mathrm{d}z) < +\infty$ as in the usual setting [16]. Set the compensated Poisson measure $\tilde{N}(\mathrm{d}z, \mathrm{d}t) = N(\mathrm{d}z, \mathrm{d}t) - v(\mathrm{d}z)\mathrm{d}t$. A natural filtration generated by N is $\mathbb{F} = (\mathbb{F}_t)_{t \geq 0}$.

At each time t, the ratio $g_t \in \omega \subseteq [0, 1]$ of Q_t (ω: a compact set) flowing into the main channel is diverted into the auxiliary channel with the instantaneous effort $n^{-1} g_t^n (Y_t + q)$, $n > 1$. This means a gradual increase of the effort with respect to g_t. The cumulative water volume flowing into the main channel $X = (X_t)_{t \geq 0}$ and cumulative diversion effort $U = (U_t)_{t \geq 0}$ satisfy

$$\mathrm{d}X_t = (1 - g_t)(Y_t + q)\mathrm{d}t, \ t > 0, \ X_0 = 0 \tag{2}$$

and

$$\mathrm{d}U_t = n^{-1} g_t^n (Y_t + q)\mathrm{d}t, \ t > 0, \ U_0 = 0. \tag{3}$$

The diversion effort has a specific functional shape to formulate a tractable model. Problems with general parameterizations would require some numerical methods.

Other than the streamflow dynamics, a martingale transformation between equivalent probability measures representing model uncertainty is formulated [8]. We consider a problem where the measure v is the source of uncertainty. Set the reference and distorted measures as $\mathbb{P}$ and $\mathbb{Q} \in P$, respectively, where P is the collection of probability measures absolutely continuous with respect to $\mathbb{P}$. Set the Radon–Nikodým derivative $\frac{\mathrm{d}\mathbb{Q}}{\mathrm{d}\mathbb{P}}$. Consider the measure transform $M = (M_t)_{t \geq 0}$ with $\frac{\mathrm{d}\mathbb{Q}}{\mathrm{d}\mathbb{P}} = M_T$ for a terminal time $T > 0$. Its governing SDE is

$$\mathrm{d}M_t = M_{t-} \int_0^{\infty} \theta(t, z) \tilde{N}(\mathrm{d}z, \mathrm{d}t), \ t > 0, \ M_0 = 1, \tag{4}$$

with $\theta : [0, T] \times \mathbb{R}_+ \to (-1, +\infty)$. An element $\mathbb{Q} \in P$ is specified by θ. There is no misspecification if $\theta \equiv 0$.

We have a high-dimensional (4-D) stochastic system to be optimized. The variables g and θ are the control variables belonging to

$$G = \left\{ (g_t)_{t \geq 0} \middle| g_t \text{ is - predictable and valued in } \omega \text{ for each } t \geq 0 \right\} \tag{5}$$

and

$$T = \left\{ (\theta(t, \cdot))_{t \geq 0} \middle| \theta(t, z) \text{ is - predictable and valued in } (-1, +\infty) \text{ for each } t \geq 0 \right\}. \tag{6}$$

Assume $\int_0^\infty \theta^2(t, z)v(\mathrm{d}z) < +\infty$ a.s. to well-pose (4) and also the path-wise unique existence of X, Y, U, M for each $(g, \theta) \in G \times T$.

2.2 Performance Functional

The performance functional $J : [0, T] \times \mathbb{R}_+^4 \times G \times T \to \mathbb{R}$ is a dynamic risk measure considering the minimization of downstream flood damage (first term) and diversion effort (second term) subject to an entropic penalization of the misspecification (third term):

$$\begin{aligned} &J(t, x, y, u, m; g, \theta) \\ &= \mathbb{E}_{\mathbb{Q}}\left[wX_T + w'U_T - \ln M_T \middle| (X_t, Y_t, U_t, M_t) = (x, y, u, m) \right], \\ &= \mathbb{E}_{\mathbb{P}}\left[M_T\left(wX_T + w'U_T - \ln M_T\right) \middle| (X_t, Y_t, U_t, M_t) = (x, y, u, m) \right] \end{aligned} \tag{7}$$

where $\mathbb{E}_{\mathbb{P}}$ (resp., $\mathbb{E}_{\mathbb{Q}}$) is the expectation under $\mathbb{P}$ (resp., $\mathbb{Q}$) and $w,\ w' > 0$ are weights as inverses of risk-aversion parameters; larger $w,\ w'$ represent more risk aversion. The value function is the saddle point

$$\Phi(t, x, y, u, m) = \inf_{g \in G} \sup_{\theta \in T} J(t, x, y, u, m; g, \theta). \tag{8}$$

Our goal is to find the optimizer $(g, \theta) = (g^*, \theta^*)$ realizing (8). A practical interest would be $\Phi(0, 0, y, 0, 1)$: the value starting from $(X_0, Y_0, U_0, M_0) = (0, y, 0, 1)$.

2.3 HJBI Equation

By a dynamic programming principle, we formally get the HJBI equation

$$\inf_{g \in \omega} \sup_{\theta : \mathbb{R}_+ \to (-1, +\infty)} L^{g,\theta}\Phi = 0,\ t \in [0, T),\ x, y, u, m > 0 \tag{9}$$

with $\Phi(T, x, y, u, m) = m\left(wx + w'u - \ln m\right)$ for $x, y, u, m > 0$,

$$L^{g,\theta}\Phi = \frac{\partial \Phi}{\partial t} + (1-g)(y+q)\frac{\partial \Phi}{\partial x} + \frac{g^n}{n}(y+q)\frac{\partial \Phi}{\partial u} - \lambda y \frac{\partial \Phi}{\partial y}$$
$$+ \int_0^\infty \left(\Phi(t,x,y+z,u,m(1+\theta(z))) - \Phi(t,x,y,u,m) - \theta(z) m \frac{\partial \Phi}{\partial m} \right) v(\mathrm{d}z). \tag{10}$$

We can exchange the order of inf and sup in (9) because they are decoupled. With an abuse of notations, the optimizers are expressed as

$$g_t^* = \arg\min_{g \in \omega} \left\{ (1-g)(y+q)\frac{\partial \Phi}{\partial x} + \frac{g^n}{n}(y+q)\frac{\partial \Phi}{\partial u} \right\}, \tag{11}$$

$$\theta^*(t,z) = \arg\max_{\theta:\mathbb{R}_+ \to (-1,+\infty)} \int_0^\infty \left(\Phi(t,x,y+z,u,m(1+\theta(z))) - \theta(z) m \frac{\partial \Phi}{\partial m} \right) v(\mathrm{d}z), \tag{12}$$

where the right-hand sides are evaluated at $(t,x,y,u,m) = (t, X_t, Y_t, U_t, M_t)$.

3 Mathematical Analysis

3.1 Explicit Solution

Proposition 1 briefly summarizes the main result of this paper.

Proposition 1 *System* (9)–(10) *admits a smooth solution.*

$$\Phi(t,x,y,u,m) = wxm + w'um - m\ln m + A_t my + B_t m \in C^1\left([0,T] \times \mathbb{R}_+^4\right) \tag{13}$$

with

$$A_t = \frac{\hat{w}}{\lambda}\left(1 - e^{\lambda(t-T)}\right) \text{ and } B_t = \int_t^T \left(q\hat{w} + \int_0^\infty \left(e^{A_t z} - 1\right) v(\mathrm{d}z) \right) \mathrm{d}t, \tag{14}$$

$\hat{w} = (1-\overline{g})w + n^{-1}\overline{g}^n w'$ *and* $\overline{g} = \arg\min_{g \in \omega}\left\{(1-g)w + n^{-1}g^n w'\right\}$, *if* $\left(e^{A_t z} - 1\right)v(\mathrm{d}z)$ *is integrable in* $(0,\infty)$.

Proof We use a guessed solution technique. Substituting (13) into (9) yields

$$\inf_{g\in\omega}\left\{\frac{\partial\Phi}{\partial t}+(1-g)(y+q)\frac{\partial\Phi}{\partial x}+\frac{g^n}{n}(y+q)\frac{\partial\Phi}{\partial u}-\lambda y\frac{\partial\Phi}{\partial y}\right\}$$
$$=\left(\frac{\mathrm{d}A_t}{\mathrm{d}t}ym+\frac{\mathrm{d}B_t}{\mathrm{d}t}m+(1-\overline{g})(y+q)wm+n^{-1}\overline{g}^n(y+q)w'm-\lambda yA_tm\right)$$
$$=m\left(\frac{\mathrm{d}A_t}{\mathrm{d}t}y+\frac{\mathrm{d}B_t}{\mathrm{d}t}+\hat{w}(y+q)-\lambda yA_t\right) \tag{15}$$

and

$$\int_0^\infty\left(\Phi(t,x,y+z,u,m(1+\theta(z)))-\Phi(t,x,y,u,m)-m\theta(z)\frac{\partial\Phi}{\partial m}\right)v(\mathrm{d}z)$$
$$=-m\int_0^\infty((1+\theta(z))\ln(1+\theta(z))-z\theta(z)A_t-\theta(z))v(\mathrm{d}z)+m\int_0^\infty zv(\mathrm{d}z)A_t. \tag{16}$$

The optimal $\theta=\theta^*(\cdot)$ should maximize

$$I(\theta)=-\int_0^\infty((1+\theta(z))\ln(1+\theta(z))-A_tz\theta(z)-\theta(z))v(\mathrm{d}z). \tag{17}$$

Taking the variation of (17) yields

$$\delta I(\theta)=-m\int_0^\infty(\ln(1+\theta(z))-A_tz)\delta\theta v(\mathrm{d}z). \tag{18}$$

Therefore, we get

$$\ln\big(1+\theta^*(z)\big)-A_tz=0\text{ or equivalently }\theta^*(z)=e^{A_tz}-1(>-1) \tag{19}$$

Then, (9) becomes

$$0=m\left(\begin{array}{c}\frac{\mathrm{d}A_t}{\mathrm{d}t}y+\frac{\mathrm{d}B_t}{\mathrm{d}t}+(y+q)\hat{w}-\lambda yA_t\\ -\int_0^\infty\big((1+\theta^*(z))\ln\big(1+\theta^*(z)\big)-z\theta^*(z)A_t-\theta^*(z)\big)v(\mathrm{d}z)+\int_0^\infty zv(\mathrm{d}z)A_t\end{array}\right). \tag{20}$$

Considering (20), we get

$$\frac{\mathrm{d}A_t}{\mathrm{d}t}=\lambda A_t-\hat{w},\ \frac{\mathrm{d}B_t}{\mathrm{d}t}=-q\hat{w}-\int_0^\infty\big(e^{A_tz}-1\big)v(\mathrm{d}z,)\,A_T,\ B_T=0 \tag{21}$$

and thus (14). This completes the proof. □

The risk-neutral case follows under $\min\{w, w'\} \to +\infty$. **Corollary** 1 is a consequence of **Proposition** 1 that can be proven in the conventional way Theorem 6.1 in Øksendal and Sulem [8] owing to the smoothness of Φ.

Corollary 1 *If* $\omega = [0, 1]$ *and* $\int_0^{\infty}\left(e^{A_t z} - 1\right)v(\mathrm{d}z) < +\infty$ *for* $0 \le t \le T$, *then the optimizers are continuously weight-dependent:*

$$g^* = \begin{cases} 1 & (w \ge w') \\ \left(\frac{w}{w'}\right)^{\frac{1}{n-1}} & (w < w') \end{cases} \text{ and } \theta^* = e^{A_t z} - 1. \tag{22}$$

The optimal diversion ratio g^* is increasing (resp., decreasing) with respect to the weight w (resp., w'), and is simple to implement. By Laeven and Stadjie [17], the worst-case Lévy measure is $e^{A_t z}v(\mathrm{d}z)$, meaning that the reference model should be distorted to overestimate both the jump (flood) frequency and size. The coefficients, A_t, B_t, both are concave, are increasing and decreasing in t, respectively. In the long run limit $T \gg 1$, we get $A_0 \approx \frac{\hat{w}}{\lambda}$, $\frac{B_0}{T} \approx q\hat{w} + \int_0^{\infty}\left(e^{A_0 z} - 1\right)v(\mathrm{d}z)$, and $\Phi(0, 0, y, 0, 1) = A_0 y + B_0$.

Remark 1 A maximization problem under uncertainty like a draught mitigation problem can also be analyzed similarly.

3.2 *Brief Application*

A critical issue is the well-definiteness of the integral in (21). Recently, we have identified the tempered stable model for the streamflow in a river [5]

$$v(\mathrm{d}z) = \frac{\lambda a}{z^{1+\alpha}} e^{-bz}\mathrm{d}z \text{ with } a, b > 0,\ 0 < \alpha < 1. \tag{23}$$

By the asymptotic estimates

$$\left(e^{A_t z} - 1\right)v(\mathrm{d}z) \approx O\left(z^{-\alpha}\right)\mathrm{d}z,\ 0 < z \ll 1 \tag{24}$$

and

$$\left(e^{A_t z} - 1\right)v(\mathrm{d}z) \approx O\left(e^{-(b-A_t)z} z^{-1-\alpha}\right)\mathrm{d}z,\ z \gg 1 \tag{25}$$

the condition $A_t \le b$ guarantees integrability of $\left(e^{A_t z} - 1\right)v(\mathrm{d}z)$. **Proposition** 1 then means that $\hat{w}$ (or $\min\{w, w'\}$) or T should be sufficiently small; the decision-maker should be sufficiently close to risk-neutral or the time horizon be sufficiently short. Notice that we may require the stronger one $A_t \le 0.5b$ by the requirement

$\int_0^\infty \theta^2(t,z)v(\mathrm{d}z) < +\infty$. Interestingly, these are hard thresholds, for example, we get the locally bounded Φ if $A_t \leq b$, while we get $\Phi \equiv +\infty$ if $A_t > b$. The problem structure thus suddenly changes across the threshold. Therefore, the presented model is robust for decision-makers close to risk-neutral.

We conclude this section with a backward SDE (BSDE)-based alternative of the HJBI equation. It is well known that HJBI equations and BSDEs are closely linked with each other [18]. In our case, setting $V_t = \Phi(t, X_t, Y_t, U_t, M_t)$ with $g = g^*$ yields

$$\mathrm{d}V_t = -\int_0^{+\infty} \left(e^{Z(t,z)} - Z(t,z) - 1\right) v(\mathrm{d}z)\mathrm{d}t + \int_0^{+\infty} Z(t,z)\tilde{N}(\mathrm{d}t,\mathrm{d}z),\ 0 \leq t < T \tag{26}$$

with a $\mathbb{F}$-predictable $Z(t,z) = A_t z$. This kind of alternative representation provides a dual view of both the presented and further extended problems in future.

4 Conclusion

We presented an exactly solvable model of river water management using a dynamic risk measure based on the entropic penalization. The explicit nature of the solution to the HJBI equation clearly demonstrated its parameter dependence and the threshold nature of the risk-aversion behavior.

We used an entropic penalty, but non-entropic ones can also be used with slight modifications [19]. The BSDE representation opens a door to numerically handle cases with a larger number of variables. Accurate approximations of Z and $\tilde{N}$ under infinite-activities Lévy processes would be an important task.

Acknowledgements Kurita Water and Environment Foundation 19B018 and 20K004 support this research.

References

1. Lundström NL, Olofsson M, Önskog T (2020) Management strategies for run-of-river hydropower plants-an optimal switching approach (2020). arXiv:2009.10554
2. Botter G, Zanardo S, Porporato A, Rodriguez-Iturbe I, Rinaldo A (2008) Ecohydrological model of flow duration curves and annual minima. Water Resour Res 44(8):W08418
3. Ramirez JM, Constantinescu C (2020) Dynamics of drainage under stochastic rainfall in river networks. Stochastics Dyn 20(03):2050042
4. Bianchi ML, Rachev ST, Fabozzi FJ (2017) Tempered stable Ornstein-Uhlenbeck processes: a practical view. Commun Stat-Simul Comput 46(1):423–445

5. Yoshioka H, Yoshioka Y (2020) Tempered stable Ornstein–Uhlenbeck model for river discharge time series with its application to dissolved silicon load analysis. In: International conference on water security and management, Dec 15–18, Tokyo, Japan. Accepted 15 Oct 2020. Full paper 10pp, in press
6. Yoshioka H, Yoshioka Y (2020) Regime switching constrained viscosity solutions approach for controlling dam-reservoir systems. Comput Math Appl 80(9):2057–2072
7. Zakaria A, Ismail FB, Lipu MH, Hannan MA (2020) Uncertainty models for stochastic optimization in renewable energy applications. Renew Energy 145:1543–1571
8. Øksendal B, Sulem A (2019) Applied stochastic control of jump diffusions. Springer, Cham
9. Föllmer H, Schied A (2002) Convex measures of risk and trading constraints. Financ Stochast 6(4):429–447
10. Herman JD, Quinn JD, Steinschneider S, Giuliani M, Fletcher S (2020) Climate adaptation as a control problem: review and perspectives on dynamic water resources planning under uncertainty. Water Res Res 56(2):e24389
11. Wellen C, Van Cappellen P, Gospodyn L, Thomas JL, Mohamed MN (2020) An analysis of the sample size requirements for acceptable statistical power in water quality monitoring for improvement detection. Ecol Ind 118, 106684
12. Wei C, Luo C (2020) A differential game design of watershed pollution management under ecological compensation criterion. J Clean Prod 274:122320
13. Jiang K, You D, Li Z, Shi S (2019) A differential game approach to dynamic optimal control strategies for watershed pollution across regional boundaries under eco-compensation criterion. Ecol Ind 105:229–241
14. Komaee A (2020) An inverse optimal approach to design of feedback control: exploring analytical solutions for the Hamilton-Jacobi-Bellman equation. Optimal Control Applications and Methods, in press
15. Yoshioka H, Yoshioka Y (2019) Modeling stochastic operation of reservoir under ambiguity with an emphasis on river management. Optimal Control Applications and Methods 40(4):764–790
16. Cont R, Tankov P (2003) Financial modelling with jump processes. CRC Press, Boca Raton, London, New York, Washington, D.C.
17. Laeven RJ, Stadje M (2014) Robust portfolio choice and indifference valuation. Math Oper Res 39(4):1109–1141
18. Delong Ł (2013) Backward stochastic differential equations with jumps and their actuarial and financial applications. Springer, London
19. Faidi W, Matoussi A, Mnif M (2017) Optimal stochastic control problem under model uncertainty with nonentropy penalty. Int J Theore Appl Financ 20(03):1750015

Implementation Report of the Kohel–Lauter–Petit–Tignol Algorithm for the Constructive Deuring Correspondence

Yuta Kambe, Yusuke Aikawa, Momonari Kudo, Masaya Yasuda, Katsuyuki Takashima, and Kazuhiro Yokoyama

Abstract For an odd prime p, let E_0 be a supersingular elliptic curve over $\mathbb{F}_{p^2}$ with $\mathcal{O}_0 = \mathrm{End}(E_0)$. The Deuring correspondence gives a one-to-one correspondence between isogenies $E_0 \longrightarrow E$ and left $\mathcal{O}_0$-ideals. In 2014, Kohel–Lauter–Petit–Tignol provided a probabilistic algorithm, called the KLPT algorithm, that finds an equivalent ideal of a given left $\mathcal{O}_0$-ideal with some powersmooth norm. It is useful for both cryptanalyses and constructions in supersingular isogeny-based cryptography. In this paper, we modify the original KLPT algorithm to improve its output quality so that an output ideal has smaller norm. This would give an efficiency for the constructive Deuring correspondence, in which we compute the supersingular elliptic curve corresponding to a given left $\mathcal{O}_0$-ideal via the Deuring correspondence. We also report implementation results of our modified KLPT algorithm for primes p up to around 45 bits. This is the largest scale of implementation reports for the original KLPT algorithm in the literature.

Keywords The KLPT algorithm · The Deuring correspondence · Supersingular elliptic curves · Endomorphism rings · Quaternion algebras

Y. Kambe · M. Yasuda (✉) · K. Yokoyama
Department of Mathematics, Rikkyo University, Tokyo, Japan
e-mail: myasuda@rikkyo.ac

Y. Kambe
e-mail: y.kambe.021@ms.saitama-u.ac.jp

K. Yokoyama
e-mail: kazuhiro@rikkyo.ac.jp

Y. Aikawa · K. Takashima
Information Technology R&D Center, Mitsubishi Electric Corporation, Kanagawa, Japan
e-mail: Aikawa.Yusuke@bcjpMitsubishiElectric.co.jp

K. Takashima
e-mail: Takashima.Katsuyuki@ajMitsubishiElectric.co.jp

M. Kudo
Department of Mathematical Informatics, The University of Tokyo, Tokyo, Japan
e-mail: kudo@mist.i.u-tokyo.ac.jp

© The Author(s), under exclusive license to Springer Nature Singapore Pte Ltd. 2022
D. Giri et al. (eds.), *Proceedings of the Seventh International Conference on Mathematics and Computing*, Advances in Intelligent Systems and Computing 1412,
https://doi.org/10.1007/978-981-16-6890-6_72

1 Introduction

Supersingular isogeny-based cryptography has received much attention as post-quantum cryptography (PQC) since proposals of the hash function of [2] and the key exchange of [7]. Since 2015, the National Institute of Standards and Technology (NIST) has proceeded PQC standardization [17]. For the NIST PQC standardization process, Jao et al. [8] submitted algorithms of supersingular isogeny key encapsulation, called SIKE, which is based on [7]. In July 2020, NIST selected 15 proposals for the third round of the standardization process, of which SIKE was selected as an alternate candidate [13].

The security of supersingular isogeny-based cryptography is based on the hardness of finding an isogeny connecting two given supersingular elliptic curves. In [10], Kohel–Lauter–Petit–Tignol considered the quaternion analogue of the isogeny problem via the Deuring correspondence [4], and they provided a probabilistic algorithm solving the quaternion isogeny problem. For an odd prime p, let $B_{p,\infty}$ denote a quaternion algebra over $\mathbb{Q}$ ramified at p and infinity. A maximal order $\mathcal{O}_0$ of $B_{p,\infty}$ is given by the endomorphism ring of some supersingular elliptic curve E_0 over $\mathbb{F}_{p^2}$. Via the Deuring correspondence, there is a one-to-one correspondence between isogenies $E_0 \longrightarrow E$ and left $\mathcal{O}_0$-ideals. Given a left $\mathcal{O}_0$-ideal I, the KLPT algorithm finds an equivalent left $\mathcal{O}_0$-ideal J of I with some powersmooth norm. (This means that two elliptic curves corresponding to I and J are isomorphic.) The KLPT algorithm is useful for both cryptanalyses and cryptographic constructions. In [6], Eisenträger et al. used the KLPT algorithm as a main tool to provide several reductions among computational problems related to supersingular elliptic curves, their isogeny graphs, and endomorphism rings. On the other hand, Galbraith et al. [12] made use of the KLPT algorithm to construct signature schemes. Recently, De Feo et al. [5] gave a generalization of the KLPT algorithm and proposed a signature scheme from supersingular isogeny graphs.

Despite usefulness of the KLPT algorithm, there are only a few works of implementation and improvements of the KLPT algorithm. In his thesis [15], Ray implemented the original KLPT algorithm in SageMath [18], but his implementation report was only for small characteristics p around 10 bits. Petit and Smith in [14] proposed an improvement of the original KLPT algorithm, but there is no implementation report for the improvement. In this paper, in addition to Petit–Smith's improvement, we modify the original KLPT algorithm [10] to improve its output quality. (We do not handle the generalized KLPT algorithm by [5] in this paper.) Given an input left $\mathcal{O}_0$-ideal I, our main idea is an exhaustive search for primes $N = \mathrm{Nrd}(\delta)$ with some $\delta \in I$, where $\mathrm{Nrd}(\delta)$ denotes the reduced norm of an element δ. We then take the minimum prime N, whose size affects both the output quality and processing performance. In particular, our modified KLPT algorithm outputs a left $\mathcal{O}_0$-ideal with smaller reduced norm. This would give an efficiency in computing the constructive Deuring correspondence, in which we compute the supersingular elliptic curve corresponding to a given left $\mathcal{O}_0$-ideal via the Deuring correspondence. We also show implementation results of our modified KLPT algorithm in SageMath for characteristics p up to around 45 bits.

2 Mathematical Background

In this section, we summarize definitions and properties on the Deuring correspondence over supersingular elliptic curves.

2.1 *Quaternion Algebras*

Throughout this paper, we let p denote a prime number satisfying $p \equiv 3 \pmod 4$, and we only treat quaternion algebras over $\mathbb{Q}$ ramified at both p and infinity. Such an algebra can be represented as $B_{p,\infty} := \mathbb{Q}\langle \mathbf{i}, \mathbf{j} \rangle$ with $\mathbf{i}^2 = -1$, $\mathbf{j}^2 = -p$ and $\mathbf{ij} = -\mathbf{ji}$. Letting $\mathbf{k} := \mathbf{ij}$, every element of $B_{p,\infty}$ can be expressed as $\alpha = a + b\mathbf{i} + c\mathbf{j} + d\mathbf{k}$ with $a, b, c, d \in \mathbb{Q}$. The *canonical involution* or *conjugation* of $\alpha \in B_{p,\infty}$ is given by $\bar{\alpha} := a - b\mathbf{i} - c\mathbf{j} - d\mathbf{k}$. The *reduced trace* and the *reduced norm* of α are, respectively, defined as

$$\mathrm{Trd}(\alpha) := \alpha + \bar{\alpha} = 2a \quad \text{and} \quad \mathrm{Nrd}(\alpha) := \alpha \cdot \bar{\alpha} = a^2 + b^2 + p(c^2 + d^2). \tag{1}$$

The reduced trace and norm are additive and multiplicative, respectively.

An *order* of $B_{p,\infty}$ is a $\mathbb{Z}$-lattice of rank 4 that forms a subring of $B_{p,\infty}$. In particular, an order of $B_{p,\infty}$ is said *maximal* if it is not properly contained in any other order. Every order is represented by a $\mathbb{Z}$-basis with 4 elements in $B_{p,\infty}$ like $\langle \omega_0, \omega_1, \omega_2, \omega_3 \rangle_{\mathbb{Z}}$. (We can always take $\omega_0 = 1$.) The quaternion algebra $B_{p,\infty}$ has several maximal orders such as $\left\langle 1, \mathbf{i}, \frac{1+\mathbf{k}}{2}, \frac{\mathbf{i}+\mathbf{j}}{2} \right\rangle_{\mathbb{Z}}$ and $\left\langle 1, \mathbf{i}, \frac{1+\mathbf{j}}{2}, \frac{\mathbf{i}+\mathbf{k}}{2} \right\rangle_{\mathbb{Z}}$. (cf., The order $\langle 1, \mathbf{i}, \mathbf{j}, \mathbf{k} \rangle_{\mathbb{Z}}$ is never maximal.)

Fix a maximal order $\mathcal{O}$ of $B_{p,\infty}$. An (integral) *left $\mathcal{O}$-ideal* is a $\mathbb{Z}$-lattice $I \subseteq \mathcal{O}$ that satisfies $\alpha I \subseteq I$ for every $\alpha \in \mathcal{O}$ (see [19, Chap. 16] for quantum ideals). The reduced norm of a left $\mathcal{O}$-ideal I is defined as $\mathrm{Nrd}(I) := \gcd(\{\mathrm{Nrd}(\alpha) : \alpha \in I\})$. Every left $\mathcal{O}$-ideal can be written as $I = \mathcal{O}N + \mathcal{O}\alpha$ with $N = \mathrm{Nrd}(I)$ for some element α of I. Here we define the equivalent class of ideals. Two non-zero left $\mathcal{O}$-ideals I and J are equivalent if and only if there exists an element q of $B_{p,\infty}$ such that $J = Iq$. The following is a key lemma [10, Lemma 5] for the construction of KLPT algorithm.

Lemma 1 ([10]) *Let I be a left $\mathcal{O}$-ideal and α an element of I. Then the equivalent ideal $J = I\gamma$ with $\gamma = \bar{\alpha}/\mathrm{Nrd}(I)$ is a left $\mathcal{O}$-ideal of reduced norm* $\mathrm{Nrd}(\alpha)/\mathrm{Nrd}(I)$.

2.2 Elliptic Curves, Isogenies, and Endomorphism Rings

An elliptic curve over a finite field $\mathbb{F}_q$ of characteristic $p \geq 5$ is given by a (short) Weierstrass equation $E : y^2 = x^3 + ax + b$ with $a, b \in \mathbb{F}_q$. The discriminant and j-invariant of E are defined as $\Delta(E) = -16(4a^3 + 27b^2) \neq 0$ and $j(E) = -1728\frac{(4a)^3}{\Delta(E)}$, respectively. Two curves are isomorphic over the algebraic closure $\overline{\mathbb{F}}_q$ of $\mathbb{F}_q$ if and only if they have the same j-invariant. Moreover, there exists an elliptic curve E over $\mathbb{F}_q$ with j-invariant $j(E)$ equal to a given element $j \in \mathbb{F}_q$. The set of $\mathbb{F}_q$-rational points

$$E(\mathbb{F}_q) = \{(x, y) \in \mathbb{F}_q^2 : y^2 = x^3 + ax + b\} \cup \{O_E\}$$

forms an abelian group, where O_E denotes the infinity point on E (see [16, Chap. III].) The order of the group $E(\mathbb{F}_q)$ is represented as $\#E(\mathbb{F}_q) = q + 1 - t$, where t denotes the trace of the qth-power Frobenius map. It satisfies $|t| \leq 2\sqrt{q}$ by Hasse's theorem. An elliptic curve E over $\mathbb{F}_q$ is said *supersingular* if the characteristic p divides the trace t. Otherwise E is said *ordinary*. Every supersingular curve has its j-invariant defined over $\mathbb{F}_{p^2}$, and it is isomorphic (over $\overline{\mathbb{F}}_q$) to one defined over $\mathbb{F}_{p^2}$. For every $n \geq 2$, let $E[n] = \{P \in E(\overline{\mathbb{F}}_q) \mid nP = O_E\}$ denote the subgroup of $E(\overline{\mathbb{F}}_q)$ of n-torsion points.

Let E and E' be two elliptic curves over $\mathbb{F}_q$. A morphism $\phi : E \longrightarrow E'$ satisfying $\phi(O_E) = O_{E'}$ is called an *isogeny*. Two curves E and E' are called *isogenous* if there is a non-zero isogeny between them. Tate's theorem states that E and E' are isogenous over $\mathbb{F}_q$ if and only if $\#E(\mathbb{F}_q) = \#E'(\mathbb{F}_q)$. Every non-zero isogeny $\phi : E \longrightarrow E'$ induces an injection of function fields $\phi^* : \overline{\mathbb{F}}_q(E') \longrightarrow \overline{\mathbb{F}}_q(E)$. The degree of a non-zero isogeny is defined as $\deg \phi = [\overline{\mathbb{F}}_q(E) : \phi^*\overline{\mathbb{F}}_q(E')]$. We say that ϕ is separable if the extension $\overline{\mathbb{F}}_q(E)/\phi^*\overline{\mathbb{F}}_q(E')$ is separable. A non-zero isogeny $\phi : E \longrightarrow E'$ induces a surjective group homomorphism from $E(\overline{\mathbb{F}}_q)$ to $E'(\overline{\mathbb{F}}_q)$ [16, Thm. 4.8, Chap. III], and its kernel is a finite subgroup of $E(\overline{\mathbb{F}}_q)$, denoted by $E[\phi]$. It satisfies $\deg \phi = \#E[\phi]$ if ϕ is separable. Conversely, given any finite subgroup C of $E(\overline{\mathbb{F}}_q)$, there are a unique elliptic curve E' and a separable isogeny $\phi : E \longrightarrow E'$ with $E[\phi] = C$ [16, Prop. 4.12, Chap. III]. The curve E' and the isogeny ϕ are denoted by the quotient E/C and ϕ_C, respectively. Fix an elliptic curve E over $\mathbb{F}_q$. An isogeny $\phi : E \longrightarrow E$ is called an *endomorphism*, and the set of endomorphisms, denoted by $\mathrm{End}(E)$, has a ring structure (see [16]). In particular, if E is supersingular, the endomorphism ring of E is a maximal order O of the quaternion algebra $B_{p,\infty}$ ramified at p and infinity (i.e., $O \otimes_{\mathbb{Z}} \mathbb{Q} \simeq B_{p,\infty}$).

2.3 *The Deuring Correspondence over Supersingular Elliptic Curves*

Deuring [4] showed that for each prime p, there is a bijection between the j-invariants of supersingular elliptic curves over $\mathbb{F}_{p^2}$ (up to Galois conjugacy in $\mathbb{F}_{p^2}$), and the maximal orders in the quaternion algebra $B_{p,\infty}$ (up to the equivalence relation given by $\mathcal{O} \sim \mathcal{O}'$ if and only if $\mathcal{O} = \alpha^{-1}\mathcal{O}'\alpha$ for some $\alpha \in B_{p,\infty}$). Fix a supersingular elliptic curve E_0 over $\mathbb{F}_{p^2}$ with $\mathcal{O}_0 = \mathrm{End}(E_0)$. The Deuring correspondence is in fact an equivalence of categories between supersingular elliptic curves over $\mathbb{F}_{p^2}$ and left $\mathcal{O}_0$-ideals (see [9, Chapter 5]). In particular, there is a one-to-one correspondence between isogenies $\varphi : E_0 \longrightarrow E$ and left $\mathcal{O}_0$-ideals I. A key concept is *kernel ideals* for isogenies, and the explicit correspondence between isogenies and ideals is as follows [19, Chapter 42]. For a left $\mathcal{O}_0$-ideal I, we define the corresponding kernel $E_0[I] \subseteq E_0(\overline{\mathbb{F}}_p)$ to be the set[1]

$$E_0[I] := \left\{ P \in E_0(\overline{\mathbb{F}}_p) : \alpha(P) = O_{E_0}, \forall \alpha \in I \right\}.$$

Then the isogeny corresponding to I is given by $\varphi_I : E_0 \longrightarrow E_I := E_0/E_0[I]$. In particular, it holds $\deg \varphi_I = \mathrm{Nrd}(I)$ [19, Proposition 42.2.16]. Moreover, for two left $\mathcal{O}_0$-ideals, two curves E_I and E_J are isomorphic if I and J are equivalent [19, Lemma 42.2.13]. Conversely, for an isogeny $\varphi : E_0 \longrightarrow E$, the corresponding kernel ideal is given by $I_\varphi = \left\{ \alpha \in \mathcal{O}_0 : \alpha(P) = O_{E_0}, \forall P \in \ker \varphi \right\}$. The *constructive Deuring correspondence* asks to compute the supersingular elliptic curve E_I for a given left $\mathcal{O}_0$-ideal I.

3 Improvements and Implementation of the KLPT Algorithm

In this section, we briefly review the original KLPT algorithm [10], and improve the algorithm. We also implement the algorithm with improvements in SageMath [18] and show experimental results on the output quality and processing performance.

3.1 *The Original KLPT Algorithm*

It is the quaternion isogeny algorithm proposed by Kohel–Lauter–Petit–Tignol in [10]. Given a bound $B > 0$, a positive number $n = \prod_{i=1}^{r} \ell_i^{e_i}$ with distinct primes ℓ_i is said to be *B-powersmooth* if $\ell_i^{e_i} \leq B$ for all $1 \leq i \leq r$. The KLPT algorithm

[1] Alternatively, we can define $E_0[I]$ to be the scheme-theoretic intersection $\bigcap_{\alpha \in I} \ker \alpha$ as a group scheme over $\overline{\mathbb{F}}_p$.

takes a maximal order $\mathcal{O}_0$ of $B_{p,\infty}$ and a left $\mathcal{O}_0$-ideal I as an input, and finds an equivalent left $\mathcal{O}_0$-ideal J of I with B-powersmooth reduced norm $\mathrm{Nrd}(J)$ for a given bound $B \approx \frac{7}{2}\log p$ (see [10] for their experimental results on a bound B). The output depends on the ideal classes of I [12, Lemma 3]. The basic procedure is given below (see also [12, Algorithm 1]):

- **Input**: A maximal order $\mathcal{O}_0$ of $B_{p,\infty}$ and a left $\mathcal{O}_0$-ideal I.
- **Output**: An equivalent left $\mathcal{O}_0$-ideal J of the input ideal I such that the reduced norm $\mathrm{Nrd}(J)$ is B-powersmooth for $B \approx \frac{7}{2}\log p$.

Step 1. Find an element $\delta \in I$ such that the equivalent left $\mathcal{O}_0$-ideal $I' := I\bar{\delta}/\mathrm{Nrd}(I)$ of I has prime reduced norm N. (See Sect. 3.2 for details.)

(a) We first compute a reduced basis $\{\alpha_1, \alpha_2, \alpha_3, \alpha_4\}$ of I as a $\mathbb{Z}$-lattice.
(b) *Prime norm algorithm*: We then generate random elements $\delta = \sum_{i=1}^{4} x_i\alpha_i$ with small integers x_1, x_2, x_3, x_4, until the reduced norm of δ is equal to $\mathrm{Nrd}(I)$ times a prime N. Then $\mathrm{Nrd}(I') = \mathrm{Nrd}(\delta)/\mathrm{Nrd}(I) = N$ by Lemma 1.

Step 2. Find $\beta \in I'$ with reduced norm NS for some odd B-powersmooth S.

(a) We find $\alpha \in I'$ such that $I' = \mathcal{O}_0 N + \mathcal{O}_0\alpha$, by taking α as a small linear combination of a (reduced) basis of I' until the condition $\gcd(\mathrm{Nrd}(\alpha), N^2) = N$ is met. If $\gcd(\mathrm{Nrd}(\alpha), N^2) = N$, then $J := \mathcal{O}_0 N + \mathcal{O}\alpha \subseteq I'$ satisfies $J \neq \mathcal{O}_0 N$. Since the reduced norm of $\mathcal{O}_0 N$ is N^2 and N is prime, the reduced norm of J is N and hence $J = I'$.
(b) We next find $\beta_1 = a + b\mathbf{i} + c\mathbf{j} + d\mathbf{k} \in \mathcal{O}_0$ with odd reduced norm NS_1 for some B-powersmooth number S_1. If a large enough B-powersmooth number is given a priori for S_1, we generate small random integers c and d until the norm equation $a^2 + b^2 = NS_1 - p(c^2 + d^2)$ can be efficiently solved by Cornacchia's algorithm [3] to find integral solutions of a and b.
(c) Find $\beta_2 \in \mathbb{Z}\mathbf{j} + \mathbb{Z}\mathbf{k}$ such that $\alpha \equiv \beta_1\beta_2 \pmod{N\mathcal{O}_0}$ by linear algebra. It has a negligible probability of failure. (The equation is solvable if the prime N is inert in $B_{p,\infty}$.) In failure case, go back to Step 2-(b).
(d) We find $\beta_2' \in \mathcal{O}_0$ with odd powersmooth reduced norm S_2 and $\lambda \in (\mathbb{Z}/N\mathbb{Z})^*$ such that $\beta_2' \equiv \lambda\beta_2 \pmod{N\mathcal{O}_0}$. Write $\beta_2' = \lambda\beta_2 + N(a + b\mathbf{i} + c\mathbf{j} + d\mathbf{k})$. For a large enough powersmooth number S_2, we search five integers $a.b, c, d, \lambda$ with $\lambda \notin N\mathbb{Z}$ satisfying

$$N^2(a^2 + b^2) + p\left\{(\lambda C + cN)^2 + (\lambda D + dN)^2\right\} = S_2, \tag{2}$$

where $\beta_2 = C\mathbf{j} + D\mathbf{k}$. Specifically, we perform the below procedures:

i We first consider Equation (2) modulo N as $p\lambda^2(C^2 + D^2) \equiv S_2 \pmod{N}$, from which a solution of λ is obtained. (If the equation cannot be solved modulo N, we multiply S_2 by small primes.)

ii Once λ is obtained, we consider Equation (2) modulo N^2 as

$$p\lambda^2(C^2 + D^2) + 2p\lambda N(cC + dD) \equiv S_2 \pmod{N^2}. \tag{3}$$

We randomly pick $c \in \mathbb{Z}$ and solve the equation for d (or vice versa).

iii Given a triple of integers (λ, c, d), we finally solve the equation

$$a^2 + b^2 = \frac{S_2 - p\left\{(\lambda C + cN)^2 + (\lambda D + dN)^2\right\}}{N^2}$$

by Cornacchia's algorithm for integers a and b. (If we cannot solve the equation, we pick a different pair of integers (c, d).)

(e) We set $\beta = \beta_1 \beta_2'$, whose reduced norm is $N S_1 S_2$. (Note that by construction $\beta \equiv \lambda\alpha \pmod{N\mathcal{O}_0}$, and hence $\beta \in I' = \mathcal{O}_0 N + \mathcal{O}_0 \alpha$.)

Step 3. We output the ideal $J := I'\bar{\beta}/N$, whose reduced norm is $S := S_1 S_2$ by Lemma 1.

3.2 *Improvements of the KLPT Algorithm*

Here we present some improvements for the KLPT algorithm to reduce the size of the powersmooth number S_2, which can reduce the reduced norm $\mathrm{Nrd}(J) = S$ of an output ideal J. For an ideal J with smaller powersmooth reduced norm, we can compute the isogeny $\varphi_J : E_0 \longrightarrow E_J \simeq E_I$ more efficiently in the constructive Deuring correspondence since it holds $\deg \varphi_J = \mathrm{Nrd}(J)$.

3.2.1 (A) Petit–Smith's Improvement

In [14], Petit and Smith proposed an improvement for Step 2-(d) to obtain a small integral solution (c, d) of Eq. (3), rather than taking it randomly. This can reduce the size of S_2. (Unfortunately, [14] was just a conference talk, and there is no detailed description in the reference.) For the prime N in Step 1 and the element $\beta_2 = C\mathbf{j} + D\mathbf{k}$ in Step 2-(c), consider the $\mathbb{Z}$-lattice of rank 2

$$L := \mathbb{Z}(N^2, 0) + \mathbb{Z}(-DN, N). \tag{4}$$

Their idea is to regard the set of integral solutions (c, d) of Eq. (3) as a parallel transported lattice of L. Then small solutions (c, d) can be recovered by performing a typical algorithm for the closest vector problem (CVP), such as Babai's algorithms [1].

Proposition 1 ([14]) *Assume that $C = 1$ for the element β_2, namely, $\beta_2 = \mathbf{j} + D\mathbf{k}$. Then there exists a parallel transported lattice L' of L such that (c, d) is an integral solution of Eq. (3) if and only if the corresponding vector $(\lambda + cN, \lambda D + dN)$ belongs to L'.*

Proof Since a proof is not given in [14], we give a proof. Write $S_2 - p\lambda^2(1 + D^2) = eN$ for some $e \in \mathbb{Z}$. Then Eq. (3) is equivalent to the equation

$$2p\lambda(c + dD) \equiv e \pmod{N}.$$

Take $v \in \mathbb{Z}$ satisfying $2p\lambda v \equiv 1 \pmod{N}$. Any integral pair (c, d) satisfies Equation (3) if and only if there exists $y \in \mathbb{Z}$ such that $c = -dD + ve + yN$. Consider a parallel transported lattice of L as

$$L' := (\lambda + veN, \lambda D) + L. \tag{5}$$

Then (c, d) is an integral solution of Eq. (3) if and only if $(\lambda + cN, \lambda D + dN) \in L'$. Indeed, if the integral pair (c, d) holds Eq. (3), then we have

$$\begin{aligned}(\lambda + cN, \lambda D + dN) &= (\lambda - dDN + veN + yN^2, \lambda D + dN)\\ &= (\lambda + veN, \lambda D) + y(N^2, 0) + d(-DN, N) \in L'.\end{aligned}$$

Conversely, if $\mathbf{w} = (\lambda + cN, \lambda D + dN) \in L'$, then there exists a unique $y \in \mathbb{Z}$ satisfying $\mathbf{w} = (\lambda + veN, \lambda D) + y(N^2, 0) + d(-DN, N)$, which clearly gives $c = -dD + ve + yN$. This implies that the integral pair (c, d) gives a solution of Equation (3). □

As in Proposition 1, we may always assume that $C = 1$ for $\beta_2 = C\mathbf{j} + D\mathbf{k}$. Indeed, since the element $\mathbf{i}\beta_2 = -D\mathbf{j} + C\mathbf{k}$ is also valid for the element β_2 in Step 2-(c), we can assume $C \not\equiv 0 \pmod{N}$ and replace β_2 by $C^{-1}\beta_2$ in modulo $N\mathcal{O}_0$. To obtain a small integral solution (c, d) of Equation (3), we find a shortest vector $\mathbf{w} = (\lambda + cN, \lambda D + dN)$ in the parallel transported lattice L'. In other words, $\mathbf{w} - \mathbf{t}$ is the closest vector in the lattice L to the target $-\mathbf{t}$, where $\mathbf{t} = (\lambda + veN, \lambda D)$. Since the volume of L is equal to N^3, the squared Euclidean norm $\|\mathbf{w}\|^2 = (\lambda + cN)^2 + (\lambda D + dN)^2$ has size $O(N^3)$. This implies that the powersmooth number S_2 in Step 2-(d) must be greater than $p\|\mathbf{w}\|^2$ since $p\|\mathbf{w}\|^2 \equiv S_2 \pmod{N^2}$.

3.2.2 (B) Our Additional Modification: Exhaustive Search in the Prime Norm Algorithm

Since the size of S_2 is related with $O(pN^3)$, we hope that a reduction of the prime N in Step 1 would lead to a direct reduction of the size of S_2. For an input left $\mathcal{O}_0$-ideal I, let

$$q_I : I \longrightarrow \mathbb{Z}, \quad q_I(\alpha) = \frac{\mathrm{Nrd}(\alpha)}{\mathrm{Nrd}(I)}$$

denote the normalized norm map. Assuming that integers represented by q_I behave like random numbers, several primes must be contained in the set of integers $q_I(\delta)$ with $\delta = \sum_{i=1}^{4} x_i\alpha_i \in I$ and $x_1, x_2, x_3, x_4 \in [-m, m]$ for a sufficiently large bound m, where $\{\alpha_1, \alpha_2, \alpha_3, \alpha_4\}$ is a $\mathbb{Z}$-basis of I. For a Minkowski-reduced basis [2] $\{\alpha_1, \alpha_2, \alpha_3, \alpha_4\}$ of I, the original method in [10] for Step 1 is to randomly choose integers $x_1, x_2, x_3, x_4 \in [-m, m]$ for $m = \lceil \log p \rceil$, until a prime of form $q_I(\delta)$ is met. In contrast, we perform an exhaustive search for the primes of form $q_I(\delta)$ in the same search space for elements $\delta \in I$, and take the minimum prime N in Step 1. In our implementation, we only consider primes N *inert* in the quaternion algebra $B_{p,\infty}$, in which case Step 2-(c) is always solvable [15, Proposition 3.2]. (cf., if N is split in $B_{p,\infty}$, Step 2-(c) is solvable with probability $\frac{N^2-2N+3}{(N+1)^2}$.) See Algorithm 1 for our prime norm algorithm.

Example 1 Let us take the maximal order of $B_{p,\infty}$

$$\mathcal{O}_0 = \left\langle 1, \mathbf{i}, \frac{1+\mathbf{k}}{2}, \frac{\mathbf{i}+\mathbf{j}}{2} \right\rangle_{\mathbb{Z}} \tag{6}$$

and put $\alpha_1 = 1, \alpha_2 = \mathbf{i}, \alpha_3 = \frac{1+\mathbf{k}}{2}$ and $\alpha_4 = \frac{\mathbf{i}+\mathbf{j}}{2}$. Consider $p = 20627$ and

$$I = \langle -2\alpha_1 + 2\alpha_2, -2\alpha_1 - 2\alpha_2, -2\alpha_3 + 2\alpha_4, -2\alpha_4 - 2\alpha_4 \rangle_{\mathbb{Z}}.$$

Then I is a left $\mathcal{O}_0$-ideal with $\mathrm{Nrd}(I) = 8$. This basis of I is Minkowski-reduced. In this case, we get the minimum odd prime norm $N = 5$, not inert in $B_{p,\infty}$. However, there is no solution of the KLPT algorithm for $N = 5$. On the other hand, the minimum *inert* prime norm is $N = 5167$ and there exists a valid solution in Step 2-(c) with this N.

3.3 Implementation and Experiments

Here we report our implementation details and results of the KLPT algorithm.

3.3.1 Input Data

As in [12, 15], we take the maximal order (6) of $B_{p,\infty}$ as an input for the KLPT algorithm. It is isomorphic to the endomorphism ring of the supersingular elliptic curve $E_0 : y^2 = x^3 + x$ over $\mathbb{F}_{p^2}$ with $j(E_0) = 1728$ and $\#E_0(\mathbb{F}_{p^2}) = (p+1)^2$. (Recall that the prime p satisfies $p \equiv 3 \pmod 4$ throughout this paper.) An explicit isomorphism of quaternion algebras is given by

$$B_{p,\infty} \longrightarrow \mathrm{End}(E_0) \otimes_{\mathbb{Z}} \mathbb{Q}, \quad (1, \mathbf{i}, \mathbf{j}, \mathbf{k}) \longmapsto (1, \phi, \pi, \pi\phi),$$

---

[2] For a positive integer n, a basis $\{\alpha_1, \ldots, \alpha_n\}$ of a lattice L of rank n is said to be *Minkowski-reduced* if the first basis element α_1 is a shortest non-zero vector in L and the basis element α_i is a shortest lattice vector which is linearly independent of $\alpha_1, \ldots, \alpha_{i-1}$ for all $i = 2, \ldots, n$.

Algorithm 1 Minimum inert prime norm algorithm (in Step 1)

Input: A Minkowski-reduced $\mathbb{Z}$-basis $\{\alpha_1, \alpha_2, \alpha_3, \alpha_4\}$ of a left $\mathcal{O}_0$-ideal I
Output: An equivalent left $\mathcal{O}_0$-ideal I' of I such that its reduced norm $N = \mathrm{Nrd}(I')$ is a prime inert in $B_{p,\infty}$ and the prime norm N is the minimum in a search space

Set $N = \infty$ and $m = \lceil \log p \rceil$
for $x_1, x_2, x_3, x_4 \in [-m, m]$ **do**
  Set $\alpha = x_1\alpha_1 + x_2\alpha_2 + x_3\alpha_3 + x_4\alpha_4 \in I$ and compute $M = q_I(\alpha) \in \mathbb{Z}$
  if M is a prime satisfying $M \equiv 3 \pmod 4$ and $M < N$ **then**
    $N \leftarrow M$ and $\delta \leftarrow \alpha$
  end if
end for
return $I' = I\bar{\delta}/\mathrm{Nrd}(I)$ {An equivalent ideal of I with reduced norm N by Lemma 1}

where $\pi : (x, y) \mapsto (x^p, y^p)$ is the pth Frobenius endomorphism, and $\phi : (x, y) \mapsto (-x, \iota y)$ with $\iota^2 = -1$. For another input of the KLPT algorithm, we follow the process in [15, Chap. 4] to generate a random left $\mathcal{O}_0$-ideal. Specifically, we first construct such an ideal by generating a random upper triangular integral square matrix $\mathbf{U}$ of size 4. Putting the generators of $\mathcal{O}_0$ in a vector as $\mathbf{b} = \left(1, \mathbf{i}. \frac{1+\mathbf{k}}{2}, \frac{\mathbf{i}+\mathbf{j}}{2}\right)^\top$, we then check whether the $\mathbb{Z}$-lattice spanned by the row vectors of $\mathbf{Ub}$ forms a left $\mathcal{O}_0$-ideal. The determinant of $\mathbf{U}$ is required to be at least a square, and it holds $\mathrm{Nrd}(I) = \sqrt{|\det(\mathbf{U})|}$ if $\mathbf{Ub}$ generates an ideal I. (cf., test ideals of ℓ-power norms were generated in [10] for a small prime ℓ via a random walk from a given maximal order $\mathcal{O}_0$.)

3.3.2 Implementation Details

Below we present modifications in our implementation.

Exhaustive search for inert prime norms Algorithm 1 is our prime norm algorithm adopted in Step 1, in which we search the minimum inert prime norm among all coefficients in $[-m, m]^4$ with $m = \lceil \log p \rceil$. Note that since $p \equiv 3 \pmod 4$, any prime number N is inert in $B_{p,\infty}$ if and only if $N \equiv 3 \pmod 4$.

Getting random powersmooth numbers In our implementation, given an input powersmooth bound $B > 0$, we randomly choose two B-powersmooth numbers S_1 and S_2 by using Algorithm 2. Note that if we set $B = \frac{7}{2} \log p$ as an initial bound, there is not any B-powersmooth number S_2 satisfying $S_2 > p^3 \log p$ in most cases. In such case, we replace B by $1.5B$ to increase the powersmooth bound.

Computation of β_2 We adopt the method in [15, Sect. 4.1.4] to find $\beta_2 \in \mathbb{Z}\mathbf{j} + \mathbb{Z}\mathbf{k}$ in Step 2-(c). Specifically, given $\beta_1 \in \mathcal{O}_0$ and $\alpha \in I'$, finding such β_2 is equivalent to solving the homogeneous equation $\beta_1\beta_2 - u\alpha \equiv 0 \pmod{N\mathcal{O}_0}$ for some unit u in the quotient ring $\mathcal{O}_0/N\mathcal{O}_0$. We reduce this equation to a homogeneous system of 4 equations modulo N by expressing $u = u_1 + u_2\mathbf{i} + u_3\mathbf{j} + u_4\mathbf{k}$, $\beta_1 = b_1 + b_2\mathbf{i} + b_3\mathbf{j} + b_4\mathbf{k}$, $\beta_2 = C\mathbf{j} + D\mathbf{k}$ and $\alpha = a_1 + a_2\mathbf{i} + a_3\mathbf{j} + a_4\mathbf{k}$. We then solve the system

Algorithm 2 Getting random powersmooth numbers (in Step 2)

Input: A powersmooth bound $B > 0$
Output: Two B-powersmooth numbers S_1 and S_2 such that $S_1 > p \log p$, $S_2 > p^3 \log p$ and $S = S_1 S_2$ is also B-powersmooth (the input bound B is increased as needed)
Take the set of prime numbers less than B as $P = \{p_1, p_2, \dots, p_\ell\}$
while $B^\ell/2 \le p \log p$ or $B^\ell/2 \le p^3 \log p$ **do**
  $B \leftarrow 1.5B$
end while
while $S_1 > p \log p$ and $S_2 > p^3 \log p$ **do**
  Take random integers $e_i \in [0, \log_{p_i}(B)]$ and $f_i \in [0, \log_{p_i}(B) - e_i]$ for every $i = 1, \dots, \ell$
  $S_1 = \prod_{i=1}^{\ell} p_i^{e_i}$, $S_2 = \prod_{i=1}^{\ell} p_i^{f_i}$
end while
return S_1 and S_2

Algorithm 3 Computation of β_2' (in Step 2-(d))

Input: An element $\beta_2 = \mathbf{j} + D\mathbf{k} \in \mathcal{O}_0$, an odd prime N, a powersmooth bound $B > 0$, and two B-powersmooth numbers S_1 and S_2
Output: An element $\beta_2' = \lambda\beta_2 + N(a + b\mathbf{i} + c\mathbf{j} + d\mathbf{k})$ with $\lambda \in (\mathbb{Z}/N\mathbb{Z})^*$ and $a, b, c, d \in \mathbb{Z}$ such that $\mathrm{Nrd}(\beta_2') = S_2$ (it satisfies $\beta_2' \equiv \lambda\beta_2 \pmod{N\mathcal{O}_0}$)
Make a list of numbers $[n_1, n_2, \dots, n_\ell]$ like $[3, 3^2, 3^0 \cdot 5, 3 \cdot 5, 3^2 \cdot 5, 3^0 \cdot 5^0 \cdot 7, \dots]$ such that $n_i S_1 S_2$ is B-powersmooth for all $i = 1, \dots \ell$. Put $n_0 = 1$.
for $i = 0, 1, 2, \dots, \ell$ **do**
  $S_2' = n_i S_2$
  if $S_2' p(1 + D^2)$ is not square in $\mathbb{Z}/N\mathbb{Z}$ **then**
    continue
  end if
  Find a solution λ of the equation $p\lambda^2(1 + D^2) \equiv S_2' \pmod{N}$ {See Lemma 2}
  Find a shortest vector $(\lambda + c_0 N, \lambda D + d_0 N)$ in the parallel transported lattice L' defined by (5) (In other words, find a vector in the lattice L defined by (4), closest to the vector $(\lambda + veN, \lambda D)$ by Babai's algorithms, where v, e are the same as in the proof of Proposition 1)
  if $S_2' < p\{(\lambda + c_0 N)^2 + (\lambda D + d_0 N)^2\}$ **then**
    continue
  end if
  for integral pairs (c, d) satisfying $p\{(\lambda + cN)^2 + (\lambda D + dN)^2\} < S_2'$ **do**
    Solve $a^2 + b^2 = \dfrac{S_2' - p\{(\lambda + cN)^2 + (\lambda D + dN)^2\}}{N}$ for (a, b) by Cornacchia's algorithm
    if there exists an integral solution of (a, b) **then**
      return $\beta_2' = \lambda\beta_2 + N(a + b\mathbf{i} + c\mathbf{j} + d\mathbf{k})$
    end if
  end for
end for
Replace the powersmooth bound B by $1.5B$, and retry from the beginning

by using Gaussian elimination. Moreover, we can modify $\beta_2 = C\mathbf{j} + D\mathbf{k}$ so that $C = 1$ for Petit–Smith's improvement.

Computation of β_2' Given an element $\beta_2 = \mathbf{j} + D\mathbf{k}$ and a large enough powersmooth number S_2, we compute $\beta_2' \in \mathcal{O}_0$ such that $\mathrm{Nrd}(\beta_2') = S_2$ and $\beta_2' \equiv \lambda\beta_2 \pmod{N\mathcal{O}_0}$ for some $\lambda \in (\mathbb{Z}/N\mathbb{Z})^*$ in Step 2-(d). From the discussion in Sect. 3.2,

Table 1 Implementation results of the KLPT algorithm by using (A) Petit–Smith's improvement or (B) Petit–Smith's improvement and our additional modification

Bit size of p	Improvements (A) or (B)	Sizes of intermediate data N	S_1	S_2	Maximum prime factor of $S_1 S_2$	Time (s)
15	(A)	25 bit	23 bit	114 bit	79	1.89
	(B)	**14 bit**	31 bit	**66 bit**	**53**	16.82
25	(A)	38 bit	38 bit	173 bit	127	244
	(B)	**24 bit**	41 bit	**119 bit**	**83**	384
35	(A)	50 bit	59 bit	231 bit	167	596
	(B)	**36 bit**	63 bit	**175 bit**	**127**	613
45	(A)	64 bit	74 bit	291 bit	199	6823
	(B)	**43 bit**	91 bit	**216 bit**	**151**	**2530**

the size of S_2 must be grater than p times the squared Euclidean norm of a shortest vector in the parallel transported lattice L' defined by (5). Since the parallel transported lattice L' is determined by the element β_2, the size of S_2 must be increased if needed, after computation of β_2 in Step 2-(c). For example, when the powersmooth bound B is set to 10, we multiply S_2 with small smooth numbers in order as

$$3 \to 3^2 \to 3^0 \cdot 5 \to 3 \cdot 5 \to 3^2 \cdot 5 \to 3^0 \cdot 5^0 \cdot 7 \to \cdots,$$

until we find a solution of β_2'. If we do not find any solution of β_2', we reset the powersmooth bound B by $1.5B$. See Algorithm 3 for our detailed procedure. Below we also give a lemma on computation of Step-(d-i).

Lemma 2 *Given three integers C, D, and S_2 with $C^2 + D^2 \not\equiv 0 \pmod N$, the equation $p\lambda^2(C^2 + D^2) \equiv S_2 \pmod N$ has a solution for $\lambda \in (\mathbb{Z}/N\mathbb{Z})^*$ if and only if the element $S_2 p(C^2 + D^2)$ is square in $\mathbb{Z}/N\mathbb{Z}$.*

Proof A solution $\lambda \in (\mathbb{Z}/N\mathbb{Z})^*$ of the equation $p\lambda^2(C^2 + D^2) \equiv S_2 \pmod N$ satisfies

$$p^2\lambda^2(C^2 + D^2)^2 \equiv S_2 p(C^2 + D^2) \pmod N.$$

Therefore $S_2 p(C^2 + D^2)$ is square in $\mathbb{Z}/N\mathbb{Z}$. Conversely, if $S_2 p(C^2 + D^2) \equiv \mu^2 \pmod N$ for some $\mu \in \mathbb{Z}$, then we have $S_2 \equiv p(C^2 + D^2)(p^{-1}(C^2 + D^2)^{-1}\mu)^2 \pmod N$. Therefore, $\lambda = p^{-1}(C^2 + D^2)^{-1}\mu$ satisfies the equation $p\lambda^2(C^2 + D^2) \equiv S_2 \pmod N$. □

3.3.3 Implementation Results

We implemented the KLPT algorithm in SageMath [18], and our experiments ran on Intel Core i7-8750H@2.20GHz with 32GByte RAM. In Table 1, we show our implementation results on the output quality and processing performance of the KLPT algorithm. We adopted Petit–Smith's improvement for the item (A), and both Petit–Smith's improvement and our additional modification for (B). We ran our program five times for every characteristic p, and show the average running time and the median value for sizes of N, S_1, S_2 in Table 1.

We see from Table 1 that sizes of both N and S_2 are reduced due to our additional modification. This results to reduce the size of maximum prime factor of $S = S_1S_2$, which would give an efficiency in computing the constructive Deuring correspondence. In contrast, our additional modification makes the running time slower for small characteristics p, because of an exhaustive search in Step 1. However, for large characteristics such as 45 bit, our additional modification makes it faster in total since Step 2 can be much faster by smaller sizes of N and S_2.

Remark 1 For generic ideals, the reduced norms of Minkowski-reduced basis elements are in $O(\sqrt{p})$, which results in taking a prime N of size $\widetilde{O}(\sqrt{p})$ (see [10, Sect. 3.1]). Different from [10], we only consider primes N inert in $B_{p,\infty}$, whose minimum size roughly equals to p from Table 1. The complexity of the KLPT algorithm is heuristically given as $\widetilde{O}(\log^3 p)$ in [12, Lemma 4]. We see from Table 1 that the running time of our program follows the complexity for small characteristics p up to 35 bit. However, it does not follow for large characteristics, since a number of re-computations of Step 2-(d) are required in practice. (The complexity analysis in [12] is only for an ideal case. It seems very difficult in practice to take suitable intermediate data N, S_1, and S_2 used in the KLPT algorithm.)

4 Conclusion and Future Work

In addition to Petit–Smith's improvement [14], we modified the original KLPT algorithm [10] to improve the output quality. Our implementation results (Table 1) demonstrated that our modification enables us to find an equivalent left $\mathcal{O}_0$-ideal J of an input ideal I with smaller reduced norm $\mathrm{Nrd}(J)$. Our modification also makes the KLPT algorithm faster for large characteristics p of base fields, due to taking smaller sizes of intermediate data N and S_2 used in the algorithm. As a future work, we would like to apply our modified KLPT algorithm to the constructive Deuring correspondence that asks us to compute the supersingular elliptic curve E_I corresponding to an input left $\mathcal{O}_0$-ideal I. Since our modified KLPT algorithm finds an equivalent ideal J of I with small reduced norm $\mathrm{Nrd}(J)$, it would help us to efficiently compute the isogeny $\varphi_J : E_0 \longrightarrow E_J$ of $\deg \varphi_J = \mathrm{Nrd}(J)$ and the target curve $E_I \simeq E_J$.

Acknowledgements The second author was supported by JST, ACT-X Grant Number JPM-JAX2001, Japan. This work was also supported by JSPS KAKENHI Grant Numbers 19K22847 and 20K14301, Japan.

References

1. Babai L (1986) On Lovász' lattice reduction and the nearest lattice point problem. Combinatorica 6(1):1–13
2. Charles D, Lauter K, Goren E (2009) Cryptographic hash functions from expander graphs. J Cryptol 22(1):93–113
3. Cornacchia G (1903) Su di un metodo per la risoluzione in numeri interi dell' equazione $\sum_{h=0}^{n} c_h x^{n-h} y^h = p$. Giornale di Mathematiche di Battaglini 46:33–90
4. Deuring M (1941) Die Typen der Multiplikatorenringe elliptischer Funktionenkörper. Abhandlungen aus dem Mathematischen Seminar der Universität Hamburg 14:197–272
5. De Feo L, Kohel D, Leroux A, Petit C, Wesolowski B (2020) SQISign: Compact post-quantum signatures from quaternions and isogenies. In: Advances in Cryptology–ASIACRYPT 2020 (To appear). https://eprint.iacr.org/2020/1240.pdf
6. Eisenträger K, Hallgren S, Lauter K, Morrison T, Petit C (2018) Supersingular isogeny graphs and endomorphism rings: Reductions and solutions. In: Advances in Cryptology–EUROCRYPT 2018, LNCS, vol 10822. Springer, pp 329–368
7. Jao D, De Feo L (2011) Towards quantum-resistant cryptosystems from supersingular elliptic curve isogenies. In: International Workshop on Post-Quantum Cryptography (PQCrypto 2011), LNCS, vol 7071. Springer, pp 19–34
8. Jao D et al (2021) SIKE: Supersingular Isogeny Key Encapsulation, Submission to the NIST standardization process on post-quantum cryptography. https://sike.org/
9. Kohel D (1996) Endomorphism rings of elliptic curves over finite fields, PhD thesis, University of California, Berkeley
10. Kohel D, Lauter K, Petit C, Tignol J-P (2014) On the quaternion ℓ-isogeny path problem. LMS J Comput Math 17A:418–432
11. Galbraith SD, Petit C, Shani B, Ti YB (2016) On the security of super singular isogeny cryptosystems. In: Advances in cryptology–ASIACRYPT 2016, LNCS, vol 10031. Springer, pp 63–91
12. Galbraith SD, Petit C, Silva J (2017) Identification protocols and signature schemes based on super singular isogeny problems. In: Advances in cryptology–ASIACRYPT 2017, LNCS, vol 10624. Springer, pp 3–33
13. Moody D et al (2021) NISTIR 8309: status report on the second round of the NIST post-quantum cryptography standardization process. https://nvlpubs.nist.gov/nistpubs/ir/2020/NIST.IR.8309.pdf
14. Petit C, Smith S (2018) An improvement to the quaternion analogue of the ℓ-isogeny problem (slides) presented at MathCrypt
15. Ray D (2018) Constructing the Deuring correspondence with applications to super singular isogeny-based cryptography, Master Thesis, Eindhoven University of Technology
16. Silverman JH (2009) The arithmetic of elliptic curves, 2nd edn. Springer GTM 106,
17. The National Institute of Standards and Technology (NIST) (2021) Post-quantum cryptography. https://csrc.nist.gov/projects/post-quantum-cryptography
18. The Sage Developers, SageMath, the Sage Mathematics Software System (Version 9.0). https://www.sagemath.org/
19. Voight J (2021) Quaternion algebras (v.0.9.21). https://math.dartmouth.edu/~jvoight/quat-book.pdf

New Results on Competitive Analysis of Online SRPT Scheduling

Sheetal Swain, Rakesh Mohanty, and Debasis Dwibedy

Abstract Shortest Remaining Processing Time (SRPT) is a well-known preemptive scheduling algorithm for uni-processor and multiprocessor systems. Besides processor scheduling, algorithm SRPT finds applications in the emerging areas such as scheduling of client's requests that are submitted to a web server for accessing static web pages, managing the access requests to files in multi-user database systems and routing of packets across several links as per bandwidth availability in data communications. It has been proved in the literature that algorithm SRPT is optimal for the settings, where the objective is to minimize the mean response time of a list of jobs. However, algorithm SRPT has not been well studied with respect to the minimization of makespan, i.e., completion time of a list of jobs. In this paper, we study algorithm SRPT for online scheduling in multiprocessor systems with makespan minimization objective. The performance of SRPT is analyzed by competitive analysis method. We obtain improved constant competitive results for algorithm SRPT by judiciously chosen special classes of online job sequences.

1 Introduction

The inherent nature of input availability in current practical applications has paved the way for study of online algorithms [1]. In real-world applications, inputs are given one at a time and an online algorithm is asked to process the available input irrevocably prior to the availability of the next input. Usually, the online algorithm has no knowledge of future inputs and makes decision based on the current and past inputs. A sequence of inputs are processed on the fly to produce the final output. Hence, the performance of an online algorithm depends on the order of the input sequence rather than the size of the input sequence. Therefore, design and analysis of efficient online algorithms have been prompted as a non-trivial research challenge.

Competitive Analysis [2] has been considered as the standard and most widely used performance measure for online algorithms. In competitive analysis, the cost

S. Swain · R. Mohanty (✉) · D. Dwibedy
Veer Surendra Sai University of Technology, Burla 768018, Odisha, India

© The Author(s), under exclusive license to Springer Nature Singapore Pte Ltd. 2022
D. Giri et al. (eds.), *Proceedings of the Seventh International Conference on Mathematics and Computing*, Advances in Intelligent Systems and Computing 1412,
https://doi.org/10.1007/978-981-16-6890-6_73

obtained by an online algorithm is compared with the cost incurred by the optimum offline algorithm [2]. An optimum offline algorithm always incurs minimum cost among all the offline algorithms. The maximum ratio between the cost of an online algorithm and the cost of the optimum offline algorithm is defined as the *competitive ratio*. The competitive ratio is valid for all permutations of online job sequences. For minimization problems, the *lower value* (≥ 1) of the competitive ratio implies improved online algorithm. Efficient online algorithms have competitive ratios bounded by a constant. Our objective is to obtain optimal competitive ratio, which is bounded by a constant.

The classical problem of scheduling is of great practical and research interest. In case of uni-processor scheduling, there is only one processor and multiple jobs. Here, the scheduler has to decide which job to schedule next on the processor, depending upon certain parameters and constraints. In multiprocessor scheduling, we have multiple processors and multiple jobs. The scheduler has to decide which job to schedule next on which processor [1]. In case of offline scheduling, the arrival time of all jobs is assumed to be zero, i.e., all jobs are available before the execution of the algorithm. Here, the scheduler has prior information regarding the whole input sequence of jobs. In contrast, in case of online scheduling, the jobs arrive over time, and at any instance of time, the scheduler has the knowledge of current and past jobs only [3, 4], the future jobs being unknown. Online Scheduling finds applications in a wide range of real-life scenarios such as university time-tabling, client–server systems, airline and train scheduling, etc.

The problem of multiprocessor scheduling has been proved to be NP-Hard [5]. Many researchers have attempted to find near-optimal solutions for online scheduling [6, 7, 9–12]. To understand the variants of online scheduling and overview of important results, one can go through the recent survey article in [8].

In this paper, we study the Shortest Remaining Processing Time (SRPT) [13] online scheduling algorithm in the multiprocessor setting and perform the competitive analysis based on some non-trivial online job sequences.

2 Background and SRPT Algorithm

2.1 Basic Terminologies and Notations

- $J_i, 1 \leq i \leq n$, denote ith job, where n is the total number of jobs. Here jobs are independent to each other.
- $P_j, 1 \leq j \leq m$, denote the jth processor, where m is the number of processors. All processors are identical (i.e., processing speed of all processors is equal).
- $T_i, 1 \leq i \leq n$, denote the processing time of J_i.
- Makespan (w) denotes the completion time of the job that finishes last in the schedule.
- w_{SRPT} is the makespan incurred by algorithm SRPT.

- w_{OPT} is the makespan obtained by optimal offline algorithm OPT.
- $CR_{SRPT} = \frac{w_{SRPT}}{w_{OPT}}$ denotes the competitive ratio of SRPT.

2.2 SRPT Algorithm

Shortest Remaining Processing Time (SRPT) is a preemptive scheduling algorithm, in which the job having the least remaining processing time is executed first. Here, we are given m identical processors and a set of n independent jobs with equal processing time (each with t units), where each job is given at an interval of 1 unit each. The job which arrives first is executed on the processor P_1. If multiple jobs arrive simultaneously, then the job having smallest processing time is scheduled on the processor having smaller index among all the idle processors. Once a processor P_i starts executing a job, it is interrupted only when a new job arrives. The processor P_i then scans the list of all available jobs in search of the job with the shortest remaining processing time and starts its execution (or resumes execution of the current job). The job with shorter remaining processing time is always assigned to the idle processor having smaller index. All the processors start scanning for jobs at time $t = 0$, and proceed in the manner mentioned above, until the execution of all jobs is completed. Furthermore, we consider the following constraints for execution of SRPT algorithm:

- Number of jobs is greater than or equal to the number of processors ($n \geq m$).
- Processing time of each job is greater than or equal to the number of processors ($t \geq m$).

3 Our Competitive Analysis Results for Algorithm SRPT Based on Special Classes of Input Job Sequences

We analyze the performance of algorithm SRPT based on some special classes of online job sequences. We consider that a sequence of jobs arrive one by one at an unit time interval. The special classes of input job sequences are judiciously chosen and presented as follows:

Class S_1: The online job sequence consists of n jobs, where each job has its processing time of n unit.

Class S_2: The online job sequence consists of n jobs, where the processing time of each job is $n + 1$ unit.

Class S_3: The online job sequence consists of n jobs, where the processing time of each job is $2n$ unit.

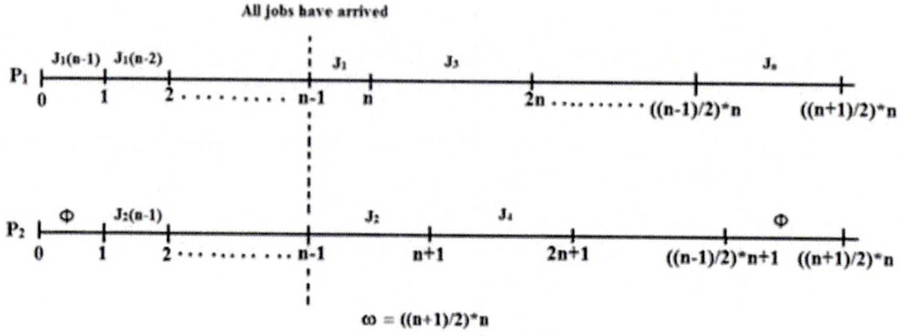

Fig. 1 Timing Diagram of SRPT for Class S_1 with even number of Jobs

Class S_4: The online job sequence consists of $2n$ jobs, where each job has its processing time of n unit.

Theorem 1 *Algorithm SRPT is* 1.5*-competitive for Class* S_1 *with* $m = 2$*, where* n *is an even integer.*

Proof Let w_{SRPT} and w_{OPT} be the makespans obtained by algorithms SRPT and OPT, respectively.

Computation of w_{SRPT}: Suppose the job J_1 arrives at time $t = 0$, and is scheduled on the processor P_1. Subsequently, job J_2 arrives at time $t = 1$. In this interval of time, processor P_2 is idle. When J_2 arrives, both the processors start scanning the list of available jobs for the shortest remaining processing time. Since J_1 has the shortest remaining processing time, it is scheduled on the smaller index processor (i.e., P_1) and J_2 is scheduled on processor P_2. Similarly, when job J_3 arrives at time $t = 2$, the processors again scan the list of available jobs, and this continues till time $t = n - 1$ (i.e., till availability of all jobs). After time $t > n - 1$, each scheduled job is executed on the processor, only after the execution of the previously scheduled job. Since the number of jobs is even, the makespan incurred by algorithm SRPT is $w_{SRPT} = \frac{n}{2}(n + 1)$. We present the execution schedule for S_1, constructed by algorithm SRPT through a timing diagram in Fig. 1.

Computation of w_{OPT}: In the offline scenario, the arrival time of all the jobs is assumed to be zero, and the jobs are scheduled on the processors according to their processing times. In this case, since the processing time of all jobs is same, so they are scheduled on the processors according to their index numbers. J_1 having the smallest index number is scheduled on the smaller index processor (i.e., P_1) and so on. Each scheduled job is executed, only after the execution of the previously scheduled job. Since the number of jobs is even, the makespan incurred by optimal offline algorithm is $w_{OPT} = \frac{n}{2}(n)$. The optimum execution schedule for S_1 is shown by a timing diagram in Fig. 2.

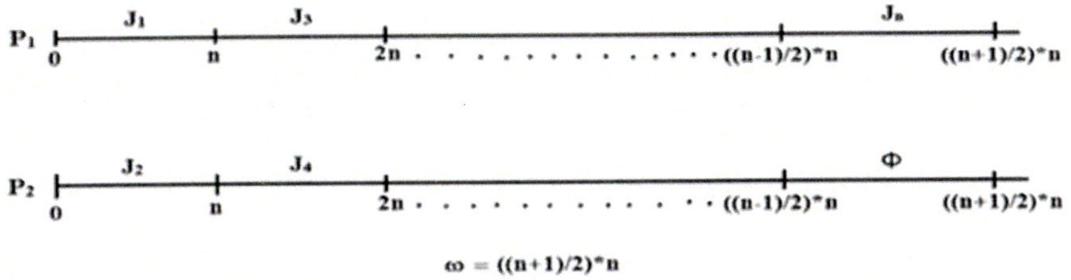

Fig. 2 Timing Diagram of OPT for Class S_1 with Even number of Jobs

$$CR_{SRPT} = \frac{w_{SRPT}}{w_{OPT}} = \frac{\frac{n}{2}(n+1)}{\frac{n}{2}(n)} = \frac{n^2+2}{n^2} = \frac{3}{2} - \frac{n^2-4}{2n^2} \leq 1.5. \qquad \square$$

Theorem 2 *Algorithm SRPT is $(2 - \frac{1}{n})$-competitive for Class S_1, where $m = n$.*

Proof. *Computation of w_{SRPT}.* Job J_1 is given at time $t = 0$ and is scheduled on processor P_1. Job J_2 is given at time $t = 1$. In the time interval from 0 to 1, processor P_1 executes job J_1 and at this time all other processors from P_2 to P_n are in idle state, since there is no other job to be executed. When job J_2 is given, P_1 halts the execution of J_1 and scans the list of available jobs for the shortest remaining time. All other processors also scan the list of available jobs. Since J_1 has the shortest remaining processing time, i.e., $n - 1$ unit, it is scheduled on the smaller index processor, i.e., on P_1. Similarly, job J_2 is scheduled on processor P_2, and all other processors are idle until any other job arrives. We continue scheduling in the same fashion till all jobs arrive, i.e., up to time $t = n - 1$. After time $t = n - 1$, each job is scheduled on the processor, only after the completion of the previously scheduled job. From time $t = n$ till the final completion time, P_1 remains idle, since there is no other job to be scheduled. Similar scenario happens in case of other processors also. Therefore, SRPT obtains a makespan of $w_{SRPT} = 2n - 1$. We present the schedule construction of SRPT in Fig. 3.

Computation of w_{OPT}: All jobs are given at time $t = 0$. Thus, according to the principle of SRPT algorithm, the job with the shortest remaining processing time must be scheduled on the lower index processor. But in this case, all jobs have equal processing time. Therefore, jobs are scheduled on the processors according to their index numbers. J_1 is scheduled on P_1, J_2 is scheduled on P_2 and following this J_n is scheduled on P_n. Therefore, the makespan incurred by OPT is $w_{OPT} = n$. We present the optimum schedule construction in Fig. 4.

We now have $CR_{SRPT} = \frac{w_{SRPT}}{w_{OPT}} = \frac{2n-1}{n} = 2 - \frac{1}{n}$. $\square$

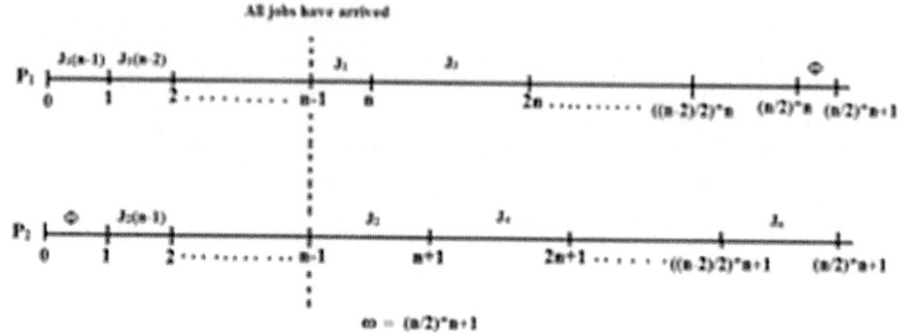

Fig. 3 Timing diagram of SRPT for class S_1 with $m = n$

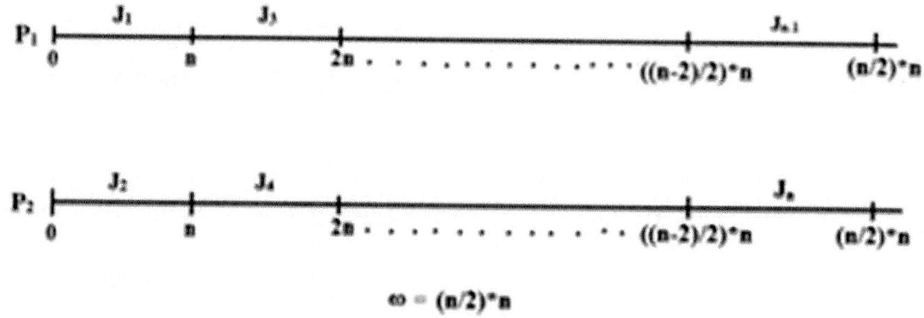

Fig. 4 Timing diagram of OPT for class S_1 with $m = n$

Theorem 3 *Algorithm SRPT is* $(2 - \frac{2}{n+1})$***-competitive for Class*** S_2

Proof. *Computation of* w_{SRPT}: Job J_1 is given at time $t = 0$ and is scheduled on processor P_1. Job J_2 arrives at time $t = 1$. In the time interval from 0 to 1, processor P_1 executes J_1 and all other processors from P_2 to P_n are remains idle, since there is no other job to be scheduled. When J_2 arrives, P_1 halts the execution of J_1 and scans the list of available jobs for the shortest remaining time. All other processors also scan the list of available jobs. Since J_1 has the shortest remaining processing time, that is, $n - 1$ units, it is scheduled on the smaller index processor, that is, P_1. Similarly, J_2 is scheduled on processor P_2, and all other processors are idle till any other process arrives. This process is carried out till all the jobs have arrived, that is, till time $t = n - 1$. After time $t = n - 1$, each job is scheduled on the processor, only after the execution of the previously scheduled job is completed. From time $t = n$ till the final completion time, P_1 remains idle, since there is no other job to be scheduled. Similar scenario happens in case of other processors also. Therefore, the makespan incurred by SRPT is $w_{SRPT} = 2n$. We now sketch the schedule obtained by algorithm SRPT in Fig. 5.

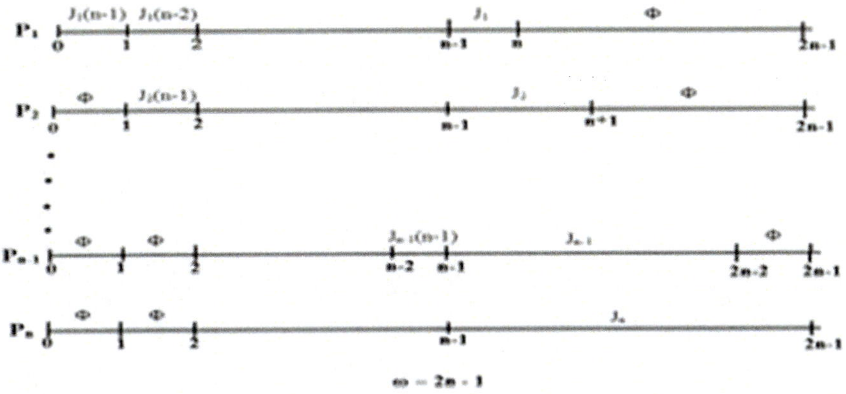

Fig. 5 Timing diagram of SRPT for class S_2

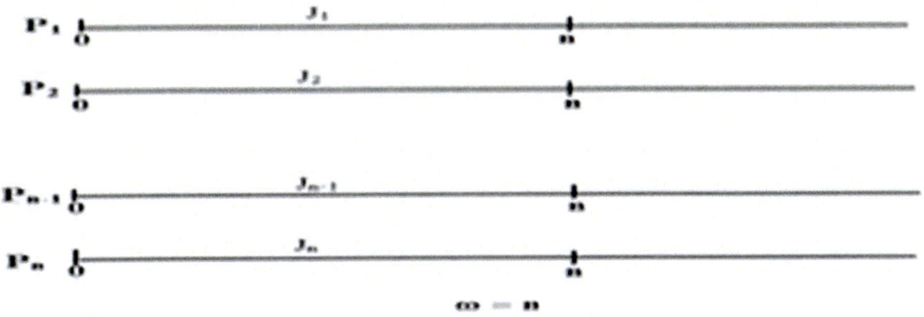

Fig. 6 Timing Diagram of OPT for Class S_2

Computation of w_{OPT}: All the jobs have arrived at time $t = 0$. Thus, according to the principle of SRPT algorithm, the job with the shortest remaining processing time must be scheduled on the lower index processor. But, in this case, all the jobs have equal processing time, and thus the jobs are scheduled on the processors according to their index numbers. J_1 is scheduled on P_1, J_2 is scheduled on P_2, and so on. Therefore, the makespan obtained by OPT is $w_{OPT} = n + 1$. The schedule obtained by OPT for S_2 is presented in Fig. 6

We now have $CR_{SRPT} = \frac{w_{SRPT}}{w_{OPT}} = \frac{2n}{n+1} = 2 - \frac{2}{n+1}$. □

Theorem 4 *SRPT is* $(2 - \frac{3}{n+2})$*-competitive for Class* S_3.

Proof Sketch. Following the same pattern described in Theorem 1 to Theorem 3, we sketch the schedule for algorithm SRPT in Fig. 7. The timing diagram at Fig. 7 shows that algorithm SRPT achieves $w_{SRPT} = 2n + 1$ for the input sequence S_3.

Computation of Optimum Offline: We construct the schedule produced by OPT in Fig. 8. The makespan obtained by OPT, i.e., $w_{OPT} = n + 2$.

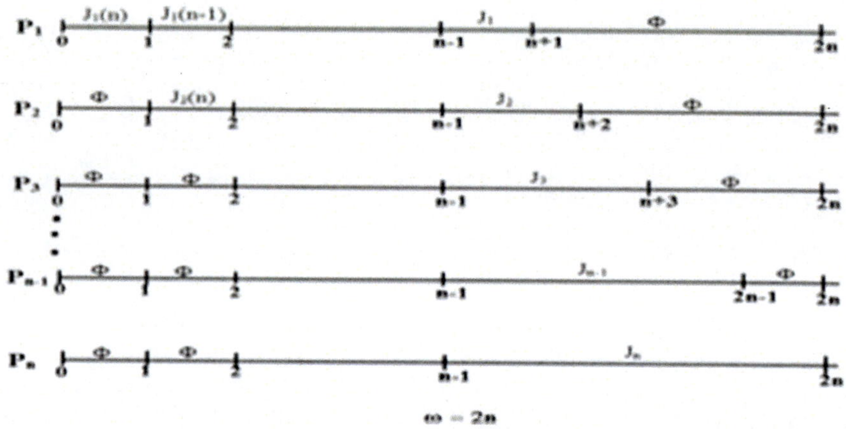

Fig. 7 Timing diagram of SRPT for class S_3

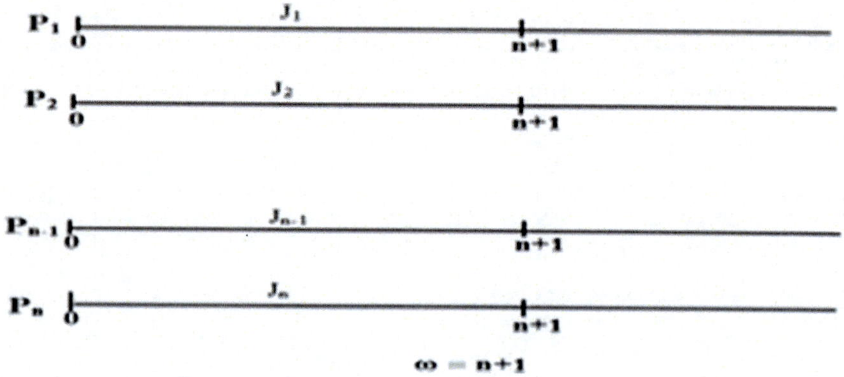

Fig. 8 Timing Diagram of OPT for Class S_3

We now have $CR_{SRPT} = \frac{w_{SRPT}}{w_{OPT}} = \frac{2n+1}{n+2} = 2 - \frac{3}{n+2}$

Theorem 5 *Algorithm SRPT is* $(\frac{3}{2} - \frac{1}{2n})$*-competitive for Class* S_4.

Proof Sketch. Computation of w_{SRPT}. The schedule achieved by algorithm SRPT for class S_4 is shown in Fig. 9. In this setting, algorithm SRPT obtains the makespan $w_{SRPT} = 3n - 1$.

Computation of w_{OPT}: The schedule construction by OPT for class S_4 is shown in Fig. 10. Algorithm OPT incurs a makespan $w_{OPT} = 2n$.

We now have $CR_{SRPT} = \frac{w_{SRPT}}{w_{OPT}} = \frac{3n-1}{2n} = \frac{3}{2} - \frac{1}{2n}$. □

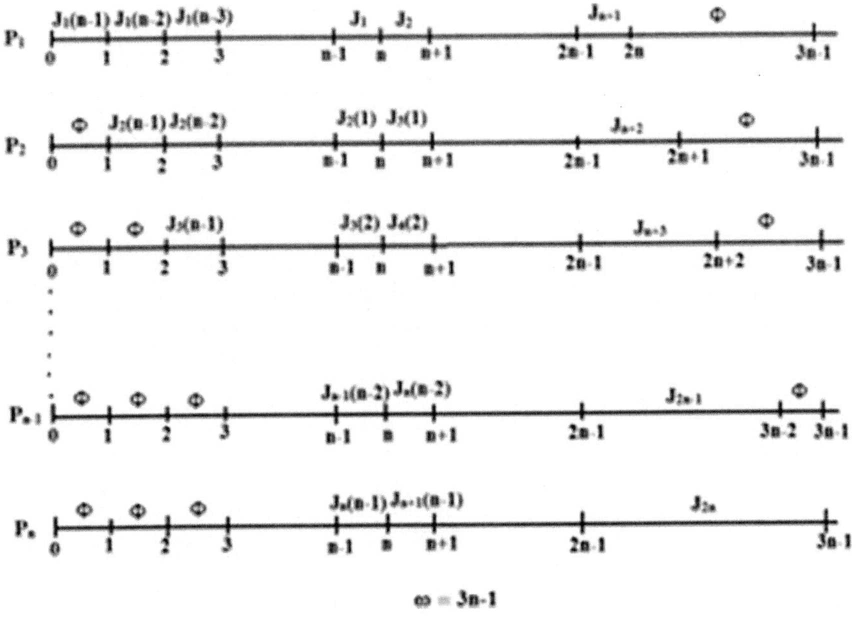

Fig. 9 Timing diagram of SRPT for class S_4

4 Conclusion

We have studied algorithm SRPT for preemptive online scheduling in multiprocessor systems by considering four special classes of online job sequences based on the number of processors, number of jobs, and processing times of the jobs. The performance of SRPT has been analyzed by competitive analysis method. We have obtained improved competitive results for algorithm SRPT based on our considered online job sequences.

Future Scope. Improved lower and upper bounds can be explored for algorithm SRPT, which will be valid for all online job sequences in preemptive online scheduling on multiprocessor systems.

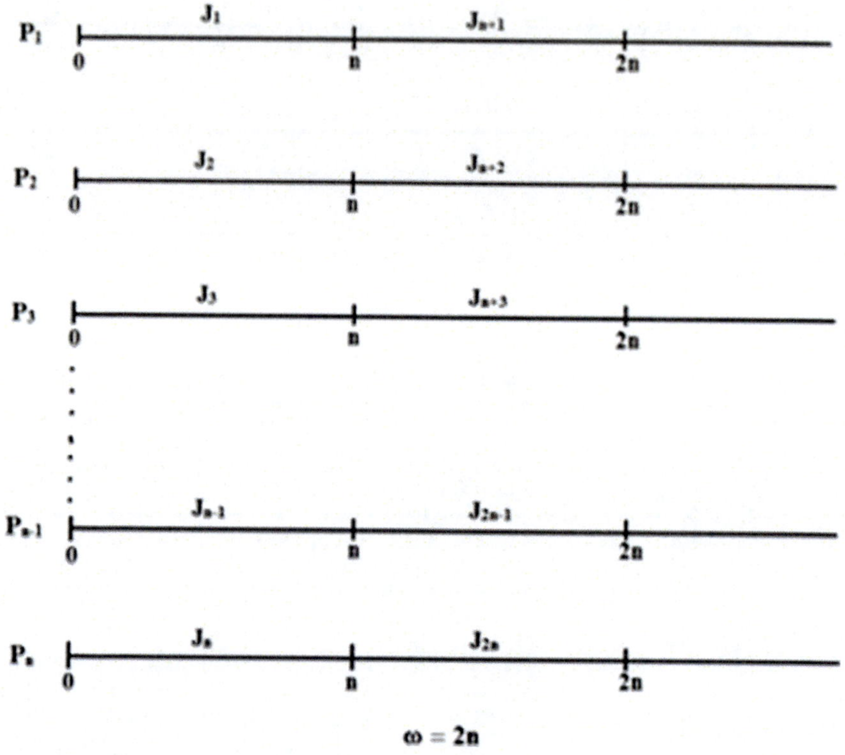

Fig. 10 Timing diagram of OPT for class S_4

References

1. Borodin A, El-Yaniv R (1998) Online computation and competitive analysis. Cambridge University Press, Cambridge
2. Tarjan RE, Sleator SS (1985) Amortized computational complexity. SIAM J Algebric Discret Methods 6(2):306–318
3. Graham RL (1966) Bounds for certain multiprocessor anomalies. Bell Syst Tech J 45:1563–1581
4. Graham RL (1969) Bounds on multiprocessing timing anomalies. SIAM J Appl Math 17:416–429
5. Garey MR, Johnson DS (1979) Computers and intractability: a guide to the theory of np-completeness, 1st edn. Freeman
6. Fleischer R, Wahl M (2000) Online scheduling revisited. J Sched 3:343–353
7. Bartal Y, Fiat A, Karloff H, Vohra R (1992) New algorithms for an ancient scheduling problem. In: Proceedings of the 24th ACM symposium on the theory of computing (STOC), Victoria, Canada, pp 51–58
8. Dwibedy D, Mohanty R (2020) Online scheduling with makespan minimization: state of the art results, research challenges and open problems. arXiv. 2001.04698 [cs.OS]
9. Albers S (2003) Online algorithms: a survey. Math Prog 97:3–26
10. Galambos G, Woeginger GJ (1993) An online scheduling heuristic with better worst-case ratio than Graham's list scheduling. SIAM J Comput 22(2):349–355

11. Albers S (1999) Better bounds for online scheduling. SIAM J Comput 29:459–473
12. Phillips C, Stein C, Wein J (1995) Scheduling jobs that arrive over time. In: Proceedings of the 4thworkshop on algorithms and data structures (WADS 95), Lecture Notes in Computer Science (LNCS), vol 955. Springer, pp 86 – 97
13. Chung C, Nonner T, Souza A (2010) SRPT is 1.86-competitive for completion time scheduling. In: Proceedings of SODA, pp 1373–1388

Perturbation Technique on MHD Slip Flow of an Accelerated Plate with Hall Current

T. Linga Raju and P. Satish

Abstract The current work focused on an unsteady incompressible viscous electrically conducting magnetohydrodynamic fluid flow of a periodically moving plate having partial slippage with Hall currents in a rotating system using perturbation technique. This work is applicable to biomedical science and biological systems. The solutions for the governing equation of momentum are obtained for the velocity profiles. The numerical outcomes for various estimations of the governing parameters are acquired. The profiles of both the velocity distributions have been drawn and their behavior is talked about. Variation of the skin friction components is presented in the form of graphs. A comparison of these numerical results with the exact solution derived by Laplace transform approach is also performed.

Keywords MHD fluid flow · Hall Effect · Slippage · Accelerated plate · Rotating fluids

1 Introduction

In medical and biological sciences, the application of magnetic field is one of the major important topics. Some of the examples of these applications are cancer tumors treatment, adjusting the blood flow in the surgery, magneto endoscopy, gastro-intestinal disorders controlled by magnetic field, transporting complex bio-waste fluids and hyperthermia created in cells by an alternating magnetic field.

T. Linga Raju
Department of Engineering Mathematics, AU College of Engineering (A), Visakhapatnam, AP, India

P. Satish (✉)
Department of Mathematics, Aditya Engineering College (A), Surampalem, AP, India

© The Author(s), under exclusive license to Springer Nature Singapore Pte Ltd. 2022
D. Giri et al. (eds.), *Proceedings of the Seventh International Conference on Mathematics and Computing*, Advances in Intelligent Systems and Computing 1412,
https://doi.org/10.1007/978-981-16-6890-6_74

Magnetohydrodynamics is concerned with the study of the intercommunication between motion of the fluid conductors by electricity and magnetic fields. Bigelow [1] investigated that the existence of magnetic field around the sun and later Larmor [2] advised that the sun having the magnetic fields and other supernatural bodies are controlled due to action of dynamics where the conducting quality material of the star behaved as the stator and armature of the self-exciting dynamo. An electric current is induced in the fluid flow due to motion of an electrically conducting fluid near the magnetic field. The interaction between the impressed magnetic field and these induced currents generates a body force (Lorentz force) on the fluid. These MHD phenomena have significant applications in boundary layer control in aerodynamics, MHD generators, Hall accelerators, geophysics and petroleum industries.

The theory of rotating fluids is important to the motion of fluids on the rotating earth on the other planets and stars. The atmospheric cyclones illustrate most striking the effect of rotation on fluid motions. The study of an electrically conducting incompressible viscous fluid flow of accelerated plates in a rotating frame of a magnetic field has attracted many researchers' interest in view of its applications in several engineering and scientific problems such as oceanography, lubrication, viscometers, centrifugal instruments, etc.

Von-Karman [3] first discussed the flow through a single infinite flat plate. Katagiri [4] investigated the effect of Hall parameter in the MHD fluid flow near a semi-infinite plate. Debnath [5] studied the solution of an unsteady hydro-magnetic and hydrodynamic fluid flow equation in a rotating system. Muhim and Deka [6] investigated the numerical solution of an unsteady MHD flow with Hall current. Pop [7] analyzed the influence of Hall parameter on hydro-magnetic flow past a moving plate.

Recently, many researchers focused on the simultaneous effect of rotation and Hall parameter on magnetohydrodynamic fluid flow. Also the concept of MHD fluid flow in rotating plates includes the effect of Coriolis forces and Hall current. Deka [8] has investigated the combined effect of rotational parameter and Hall parameter on magnetohydrodynamic fluid flow near a moving plate relative to the rotating fluid. Many researchers analyzed the MHD flow problems without considering the slippage. Takhar et al. [9] studied MHD flow in a rotating frame without slip condition. Recently, researchers are being concentrated to study the flow distribution problems under the consideration of slip parameter at the plates. Farhed et al. [10] obtained the exact solution for magnetohydrodynamic fluid flow with slippage by using transformation methods.

The main treatment in cancer tumors is the delivery of drug to the affected area of the patient's body through vessels that are involving slippage. In this task, ferrofluids are utilized for the perfect replacement of drug inside the human body. In view of these implementations, an important study on slip flow has been analyzed by many investigators like Lance and Rogers [11], Bhatt and Sacheti [12], Matthews and Hill [13], Ahmad et al. [14], Turkyilmazoglu [15] and many more.

In the current work, the problem of an unsteady viscous incompressible electrically conducting MHD flow of a periodically moving plate having partial slippage with Hall currents in a rotating system using perturbation technique is considered. The solutions for the governing equation of momentum are obtained for both the velocity profiles. The numerical outcomes for various estimations of the governing variables are acquired. The profiles of the velocity distributions have been drawn and their behavior is talked about. Variation of the skin friction components is presented in the form of diagrams. A comparison of these numerical results with the exact solution derived by Laplace transform technique is also performed.

2 Formulation and Mathematical Analysis

Consider a viscous incompressible unsteady electrically conducting fluid flow through a periodically accelerated infinite plate with partial slippage (see Fig. 1). Initially, the plate and the fluid flow are at rest. At $t > 0$, the infinite flat plate starts accelerated motion in its own plane with the prescribed velocity field $AC_1 sinK_0 t$, where C_1 is the amplitude of the plate, K_0 is the frequency of the oscillated plate and A is a constant. The fluid flow direction is assumed along x-axis and y-axis lies on the plate perpendicular to the x-axis and z-axis. The plate and the fluid are rotating with an angular velocity Ω_1 about the z- axis.

From the continuity equation $\nabla.\overline{q} = 0$, where $\overline{q} =$ (u, v, w), we obtain that w = 0 everywhere in the fluid field such that boundary condition w = 0 is satisfied. As the electrically non-conducting infinite plate is extent in the x, y directions, all physical parameters are functions of t and z only. Here, u is the primary velocity distribution and v is the secondary velocity distribution. A strong strength of continous magnetic field B_0 is applied in a direction parallel to z-axis, where B = (0, 0, B_0). An electric current is induced in the flow normally to both the magnetic and electric field, which generates a transverse motion of the fluid. Thus Hall effect induces a secondary fluid flow and so there will be only two components of the velocity flow field.

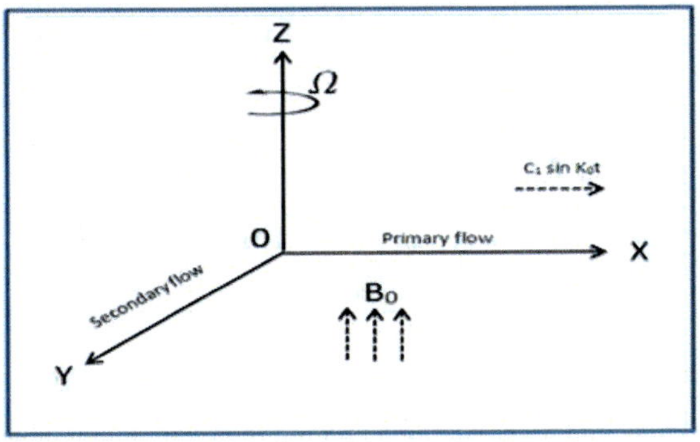

Fig. 1 Geometry of the flow problem

In comparison with the applied magnetic field B_0, the induced magnetic field is negligible, which tends to very small magnetic Reynolds number.

The modified Ohm's law with the inclusion of Hall parameter is given by

$$\overline{J} + \frac{m}{B_0}\left(\overline{J} \times \overline{B}\right) = \sigma\left(\overline{E} + \overline{q} \times \overline{B}\right) \tag{1}$$

where $m(= \omega_e \tau_e)$ is the Hall parameter with ω_e is the frequency of the electron, τ_e is the collision time of electrons, $\overline{J} = (J_x, J_y, J_z)$ is the current density vector, $\overline{q}$ is the velocity field vector and σ is the electrical conductivity. Under the above considerations, we obtain the following equations as

$$j_x + m j_y = \sigma(E_x + B_0 v) \tag{2}$$

$$j_y - m j_x = \sigma(E_y + B_0 u) \tag{3}$$

Since the magnetic field is neglected, from the Maxwell's equation $\nabla \times \overline{E} = -\frac{\partial \overline{B}}{\partial t}$. So that we take $\nabla \times \overline{E} = 0$, which gives $\frac{\partial E_x}{\partial z} = 0$ and $\frac{\partial E_y}{\partial z} = 0$. This indicates that E_x= constant, E_y= constant, everywhere in the fluid flow field. Again, choosing that this constant is equal to zero (i.e.,$E_x = E_y = 0$). We obtain,

$$j_x = \frac{\sigma B_0}{1 + m^2}(v + mu) \tag{4}$$

$$j_y = \frac{-\sigma B_0}{1 + m^2}(u - mv) \tag{5}$$

With these assumptions, the momentum equation for an unsteady MHD flow in a rotational system can take the following form

$$\frac{\partial u}{\partial t} = \vartheta \frac{\partial^2 u}{\partial z^2} + 2\Omega_1 v - \frac{\sigma {\mu_e}^2 {B_0}^2 (u + mv)}{\rho(1 + m^2)} \tag{6}$$

$$\frac{\partial v}{\partial t} = \vartheta \frac{\partial^2 v}{\partial z^2} - 2\Omega_1 u + \frac{\sigma {\mu_e}^2 {B_0}^2 (mu - v)}{\rho(1 + m^2)} \tag{7}$$

where B_0 is applied magnetic field, ϑ is the kinematic viscosity, m is Hall parameter, μ_e is magnetic permeability and ρ is the fluid density. In the presence of partial slippage, the velocity of the fluid is directly proportional to the shearing stress, the corresponding boundary conditions under the consideration of the periodic acceleration with partial slippage are:

$$t \leq 0;\ u = 0,\ v = 0 \text{ For } z \geq 0$$

$$t > 0; u = AC_1 sinK_0 t + \lambda_0 \frac{\partial u}{\partial z}, v = 0 \text{ At } z = 0$$

$$t > 0; u \to 0, v \to 0 \quad At\ z \to \infty \tag{8}$$

where λ_0 is the slip length (parameter) of the plate. The non-dimensional form of the Eq. (8), it is introduced the following dimensionless quantities:

$$Z = \frac{zc_1}{\vartheta}, T = \frac{tc_1^2}{\vartheta}, U = \frac{u}{c_1}, V = \frac{v}{c_1},$$

$$M^2 = \frac{\sigma\mu_e^2 B_0^2 \vartheta}{\rho c_1^2}, \Omega^2 = \frac{\Omega_1 \vartheta}{c_1^2}, K = \frac{\vartheta k_0}{c_1^2}, \lambda = \frac{c_1 \lambda_0}{\vartheta}, \tag{9}$$

where M^2 is magnetic parameter which represents the ratio of electro-magnetic force to the viscous force, Ω^2 is the rotational parameter which is the reciprocal of Ekman number, λ is the dimensionless slip parameter, K is the dimensionless frequency parameter. Then the Eqs. (6) and (7) in dimensionless form become:

$$\frac{\partial U}{\partial T} = \frac{\partial^2 U}{\partial Z^2} + 2V\left(\Omega - \frac{mM^2}{1+M^2}\right) - 2U\left(\frac{M^2}{1+m^2}\right) \tag{10}$$

$$\frac{\partial V}{\partial T} = \frac{\partial^2 V}{\partial Z^2} - 2U\left(\Omega - \frac{mM^2}{1+M^2}\right) - 2V\left(\frac{M^2}{1+m^2}\right) \tag{11}$$

The corresponding boundary conditions become:

$$T \le 0 : U = 0,\ V = 0 \quad \text{At}\, z \ge 0$$

$$T > 0 : U = AsinKT + \lambda \frac{\partial U}{\partial Z}, \quad V = 0 \quad \text{At } Z = 0$$

$$T > 0 : U \to 0,\ V \to 0\ At\ Z \to \infty \tag{12}$$

Further introducing F = U + iV as the fluid velocity in complex form, the Eqs. (10) and (11) are converted into the following single equation:

$$\frac{\partial F}{\partial T} = \frac{\partial^2 F}{\partial Z^2} - aF \tag{13}$$

where $a = 2i\left(\Omega - \frac{mM^2}{1+m^2}\right) + 2\left(\frac{M^2}{1+m^2}\right)$

The boundary conditions in complex notation are

$$T \leq 0 : F = 0 \quad \text{For} \;\; Z \geq 0$$

$$T > 0 : F = AsinKT + \lambda \frac{\partial F}{\partial Z} \quad \text{For} \;\; \text{Z} = 0$$

$$T > 0 : F \to 0 \quad \text{For} \;\; Z \to \infty \tag{14}$$

3 Mathematical Solutions

The mathematical solution of the fluid problem is carried out in the following two methods, namely, perturbation technique and Laplace transform method.

3.1 Solution Using Perturbation Analysis

Equation (13) represents a non-linear second-order partial differential equation. By using the perturbation technique, it can be separeted into a set of ordinary non-dimensional differential equations also solved analytically. The fluid velocity is assumed as

$$F = f_0(Z) + \in \exp(iKT) f_1(Z) + O(\in^2) \tag{15}$$

With the help of the above Eq. (15), Eq. (13) gives the following equations containing both the non-harmonic and harmonic terms, omitting the higher order terms of $O(\in^2)$, as

$$f_1{}^{11}(Z) - (a + ik) f_1(z) = 0 \tag{16}$$

$$f_0{}^{11}(z) - a f_0(z) = 0 \tag{17}$$

Applying the following expanded boundary conditions derived from the assumed perturbation expansion to the Eqs. (16) and (17), we obtain

$$T \leq 0 : f_0 = 0; \; f_1 = 0, \; For \; Z \geq 0$$

$$T > 0 : f_0 = AsinKT + \lambda f_0{}^1, \; f_1 = AsinKT + \lambda f_1{}^1 For \; Z = 0$$

$$T > 0 : f_0 \to 0;\ f_1 \to 0\ At\ Z \to \infty \tag{18}$$

By solving the above Eqs. (16) and (17) using the boundary conditions (18), we obtain

$$f_0(Z, T) = \frac{AsinKT}{1 + \lambda\sqrt{a}}\exp(-Z\sqrt{a}) \tag{19}$$

$$f_1(Z, T) = \frac{AsinKT}{1 + \lambda\sqrt{a + iK}}\exp(-Z\sqrt{a + iK}) \tag{20}$$

The final expression for the velocity distributions given in complex form for simplicity is as follows

$$\boldsymbol{F(Z, T) = \frac{AsinKT}{1 + \lambda\sqrt{a}}\exp\left(-Z\sqrt{a}\right) + \in \frac{AsinKT}{1 + \lambda\sqrt{a + iK}}\exp\left(-Z\sqrt{a + iK}\right)} \tag{21}$$

where $a = 2i\left[\Omega - \frac{mM^2}{1+m^2}\right] + 2\frac{M^2}{1+m^2}$

3.2 Solution Using Laplace Transformation Technique

By employing the Laplace transformation method to the Eq. (13), we obtain the following ordinary linear differential with constant coefficients as

$$\frac{d^2\overline{F}(Z, s)}{dZ^2} - (s + a)\overline{F}(Z, s) = 0 \tag{22}$$

where s is the TRansformation parameter and

$$\overline{F}(Z, s) = L[F(Z, T)] = \int_0^\infty F(Z, T)e^{-sT}dT \tag{23}$$

The corresponding boundary conditions for $\overline{F}(Z, s)$ are

$$\overline{F}(Z, s) = \frac{A}{2i}\left[\frac{1}{s - iK} - \frac{1}{s + iK}\right] + \lambda\frac{d\overline{F(Z, T)}}{dZ} \quad \mathrm{At\,Z} = 0$$

$$\overline{F}(Z, s) \to 0 \ \ \mathrm{At}\ \ Z \to \infty \tag{24}$$

The solution to Eq. (22) using the boundary condition (24) is given by

$$\overline{F(Z,s)} = \frac{iA}{2\lambda^2}\left\{\begin{array}{l} e^{-Z\sqrt{s+a}}\left[\frac{2iK}{(r^2+K^2)(s-r)} + \frac{1}{(r+iK)\,(s+iK)} - \frac{1}{(r-iK)\,(s-iK)}\right] \\ +\frac{\lambda e^{-Z\sqrt{s+a}}}{\sqrt{s+a}}\left[\frac{a+iK}{(r-iK)(s-Ik)} - \frac{a-iK}{(r+ik)\,(s+iK)} + \frac{(a+r)2iK}{(r^2+K^2)(s-r)}\right] \end{array}\right\} \tag{25}$$

where $r = \frac{1}{\lambda} - a$. Subsequently, by applying the inverse Laplace transformation of $F(\bar{Z}, s)$, we get the following solution for the velocity field on par with the results of Hetnarski [16], Narahari and Debnath [17].

$$F(Z,T) = \frac{iA}{2\lambda^2}\left\{\begin{array}{l} \left[\frac{2ik}{(r^2+K^2)} f_1(Z,a,r,T) - \frac{1}{(r-iK)} f_1(Z,a,iK,T) + \frac{1}{(r+iK)} f_1(Z,a,-iK,T)\right] \\ +\lambda\left[\frac{-2iK(a+r)}{(r^2+K^2)} f_2(Z,a,r,T) - \frac{(a-iK)}{(r+iK)} f_2(Z,a,-iK,T) + \frac{a+iK}{r-iK} f_2(Z,a,iK,T)\right] \end{array}\right\} \tag{26}$$

where

$$f_1(\alpha,\beta,\gamma,T) = \frac{e^{\gamma T}}{2}\left\{\begin{array}{l} e^{\alpha\sqrt{\beta+\gamma}} erfc\left[\frac{\alpha}{2\sqrt{T}} + \sqrt{(\beta+\gamma)T}\right] \\ +e^{-\alpha\sqrt{\beta+\gamma}} erfc\left[\frac{\alpha}{2\sqrt{T}} - \sqrt{(\beta+\gamma)T}\right] \end{array}\right\}$$

$$f_2(\alpha,\beta,\gamma,T) = \frac{e^{\gamma T}}{2\sqrt{\beta+\gamma}}\left\{\begin{array}{l} e^{-\alpha\sqrt{\beta+\gamma}} erfc\left[\frac{\alpha}{2\sqrt{T}} - \sqrt{(\beta+\gamma)T}\right] \\ +e^{\alpha\sqrt{\beta+\gamma}} erfc\left[\frac{\alpha}{2\sqrt{T}} + \sqrt{(\beta+\gamma)T}\right] \end{array}\right\}$$

Here α, β, γ are notation variables and f_1,f_2 are the notation functions, erfc(x) is complementary error function. Since the domains of integration for both the methods of approaches are different. The analytical solution using Laplace transformation and the approximate solution using perturbation analysis are obtained in different forms, but we observe that their numerical results are very close to each other, which are shown in Table 2.1. The velocity profiles U and V are obtained individually by divided F into real part and imaginary part. The non-dimensional form of the skin-friction components τ at the plate Z = 0 is obtained from Eq. (21) as

$$\frac{dF(Z,T)}{dZ} = -\left\{\begin{array}{l} \frac{A\sqrt{a}\, sinKT}{1+\lambda\sqrt{a}} \exp(-Z\sqrt{a}) \\ +\in \frac{A\sqrt{a+iK}\, sinKT}{1+\lambda\sqrt{a+iK}} \exp\left(-Z\sqrt{a+iK}\right) \end{array}\right\} \tag{27}$$

4 Results and Discussions

The perturbation technique has been performed to obtain the problem an unsteady MHD fluid flow of a periodically oscillated plate with Hall current and partial slippage in a rotating system. The velocity distributions, the primary skin friction coefficient and secondary skin friction coefficient have been represented graphically for varies values of the Hall parameter (m), Hartmann number (M), Time (T), rotation parameter (Ω) and slip parameter (λ).

From Fig. 2a, b, it is clear that both the primary and secondary velocity distribution increase due to an increase in the Hall current parameter (m). It is evident that the Hall current is directly proportional to both velocities.

Figure 3a, b shows the variation of the Hartmann parameter M on the velocity components in which both the velocity profiles decrease more rapidly with an increase of Hartmann number M. It is demonstrated that both the velocity profiles are significantly influenced by magnetic field.

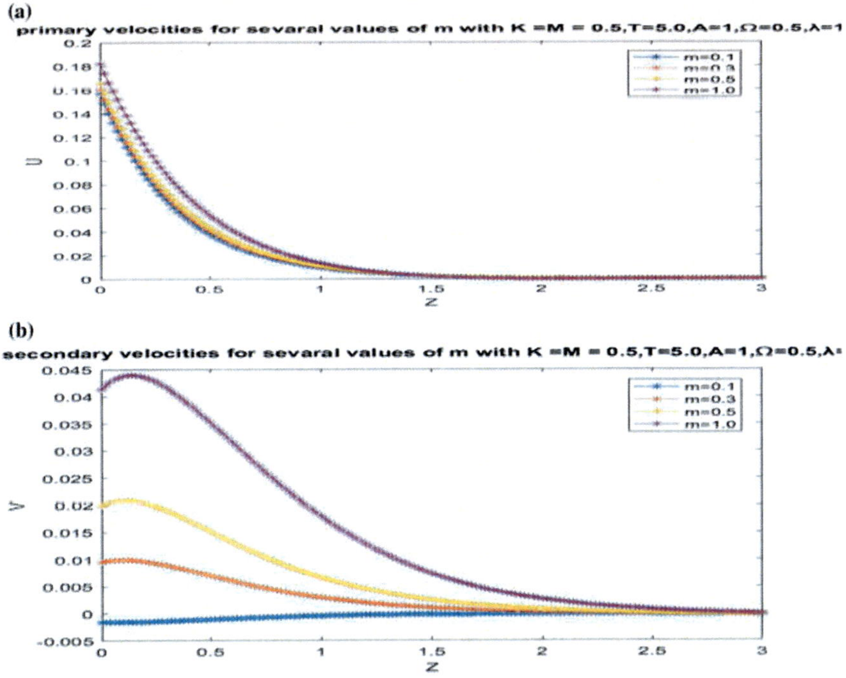

Fig. 2 (**a**), (**b**) Velocities for different values of Hall parameter 'm'

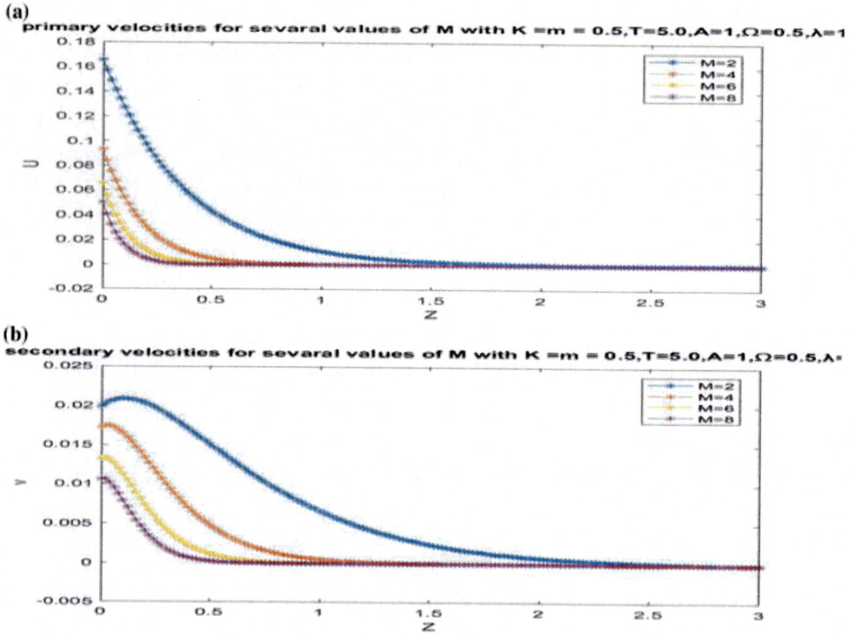

Fig. 3 (**a**), (**b**) Velocities for different values of Hartmann parameter ‘M’

Figure 4a, b shows the effect of the varying slippage value λ on the velocity distributions. It is observed that the large values of slippage value λ reveal less roughness of the surface. But, the behavior of the velocity components is different for smaller values of slip parameter λ.

The effect of the varying rotation parameter Ω is shown in Fig. 5a, b. These figures indicate that both the velocity distributions oscillate for large values of rotational parameter Ω. Since the Coriolis forces are enhanced due to increased rotational parameter Ω and hence it gives oscillatory behavior in the flow field.

Figure 6a, b represents the effect of the time T on the velocity profiles. It is observed that with an increase in the time T, the secondary velocity profile decreases whereas the primary velocity profile increases.

(a)

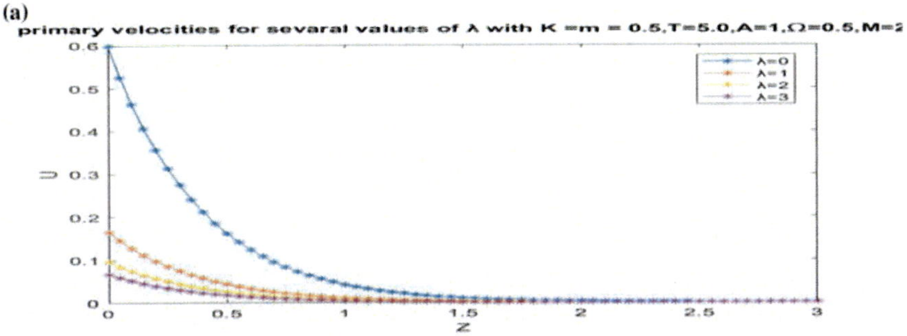

(b)

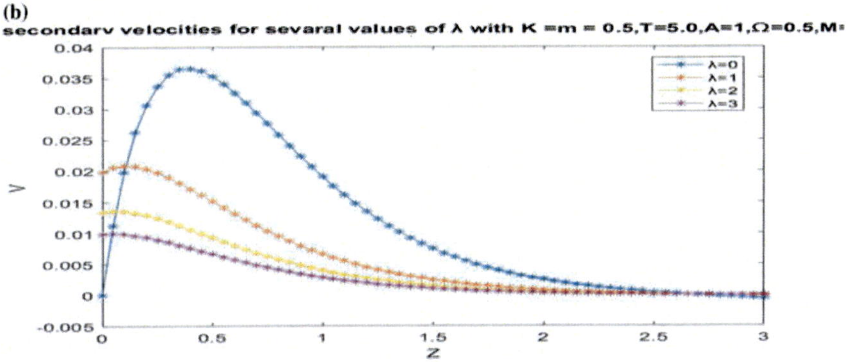

Fig. 4 (**a**), (**b**) Velocities for different values of slip parameter λ

Figure 7a, b shows the effect of the slippage λ on the non-dimensional skin friction τ. It is seen that the magnitude of skin friction for both the velocity profiles decreases due to an increase in the slip parameter λ.

The behavior of velocity fields for various values of Z and T for a selected set of the governing parameters is shown in Fig. 8a, b. It is noticed that the nature of the velocity distribution is in oscillatory manner.

Finally, it is performed a comparison of the numerical results to the velocity distributions by the Laplace transform method with that of the approximate solution using perturbation analysis as shown in Table 1. It is clear that both the velocity profiles are numerically almost close to each other by these methods.

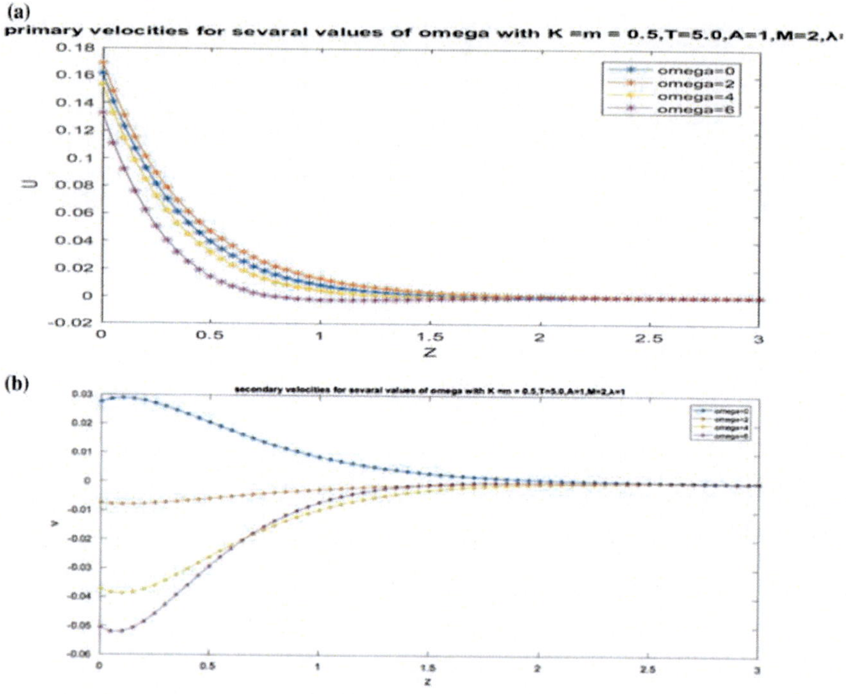

Fig. 5 (**a**), (**b**) Velocities for different values of Rotation parameter Ω

5 Conclusions

An unsteady viscous incompressible magnetohydrodynamic (MHD) fluid flow through a periodically oscillated plate under the effect of a strong magnetic field with Hall currents in a slip flow regime has been investigated. The methods of Laplace transforms and perturbations are being utilized to solve the mathematical equations of flow. The velocity components and the skin friction components are represented graphically for different sets of values of the flow parameters. Hence, it is summarized the following observations:

- Due to Hall current, the velocity profiles are increased in a region nearer to the plate.
- The velocity distributions are decreased in the presence of magnetic field.
- The velocity distributions act in an oscillatory way due to an increase in the rotational parameter.
- The velocity profiles are enhanced for increasing values of slip parameter.
- The magnitude of the skin friction components is increased due to the Hall effect.

(a)

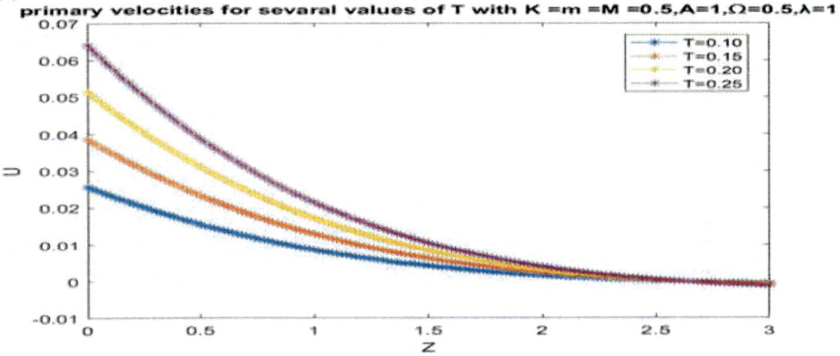

(b)

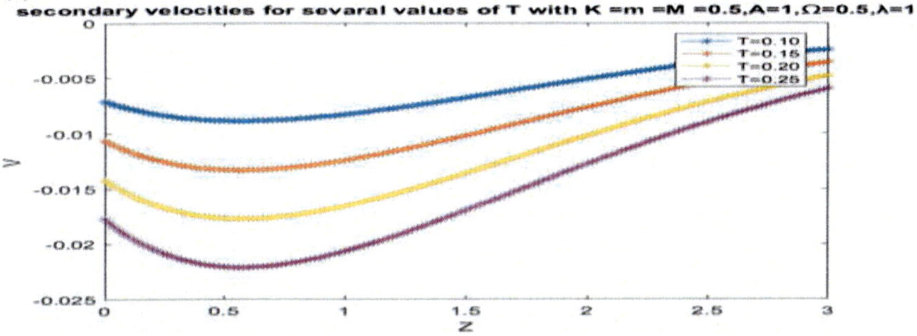

Fig. 6 (**a**), (**b**) Velocities for different values of Time T

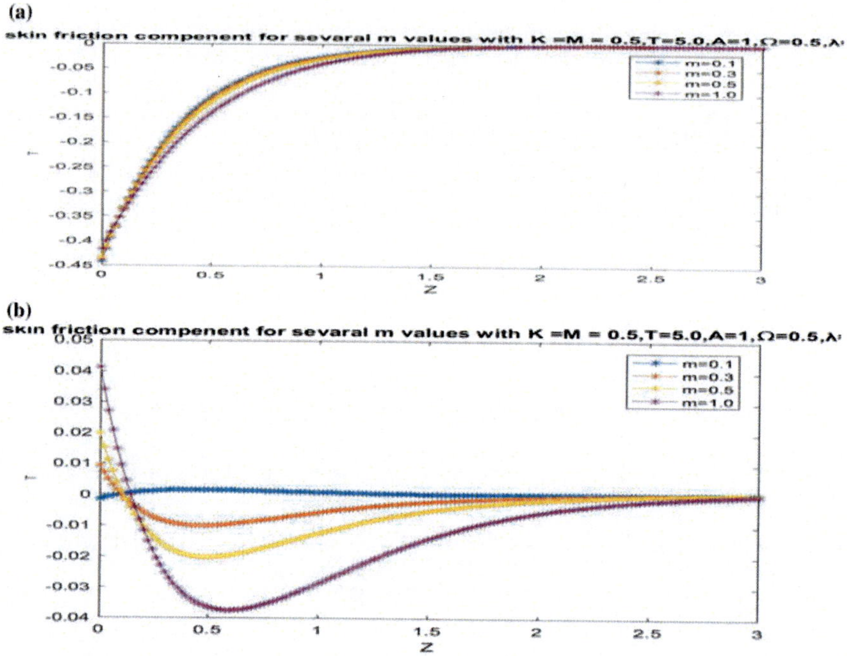

Fig. 7 (**a**), (**b**) Skin friction compenents of different values of m

(a)

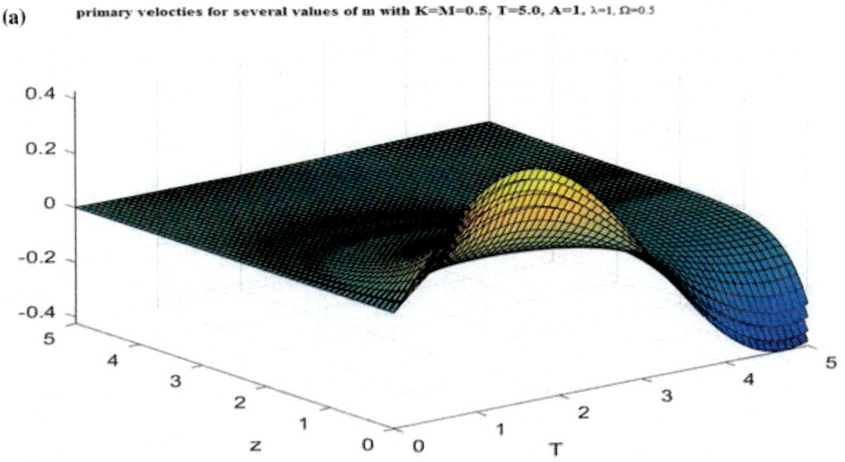

(b)

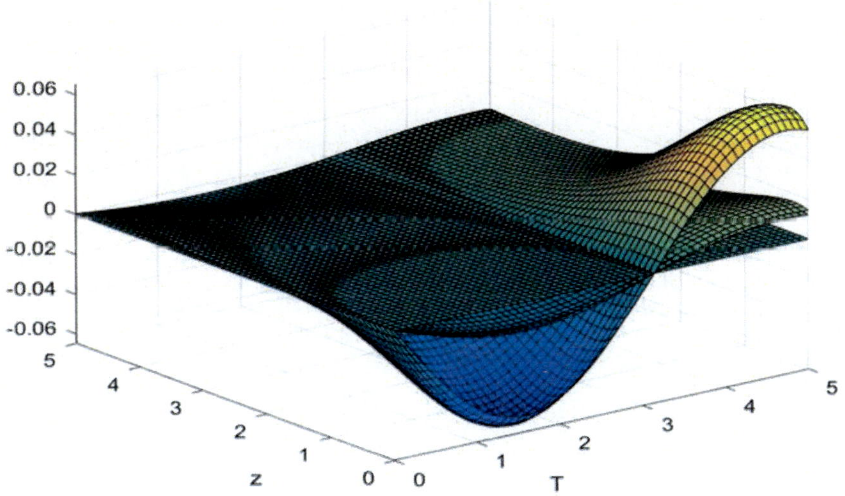

Fig. 8 (**a**), (**b**) Velocities for different values of Hall parameter ‘m’ and Time ‘T’

Table 1 Comparison of solutions for different values of Z at m = K = Ω = 0.5, T = 5.0, M = 2, A = 1 and λ = 1

Z	U(Z, T) by Laplace	U(Z, T) by Perturbation	V(Z, T) by Laplace	V(Z, T) by Perturbation
0.5	0.0474	0.0447	0.0173	0.0151
1.0	0.0124	0.0110	0.0079	0.0067
1.5	0.0030	0.0.0026	0.0030	0.0024
2.0	0.0006	0.0007	0.0010	0.0008

Acknowledgements Authors would like to thank the anonymous reviewers for their encouraging comments and constructive suggestions for the improvement of our paper.

References

1. Bigelow, F.H.: The Solar Crana. Smithsonian Institution, Washington D.C. (1889)
2. Larmor, Joseph,: Geophysical and Strophysical Dynamo Theory. Gunther Rudiger, Rainer Hollerbach (edtrs), Wiley-VCH (1919)
3. Von Karman, Biot,M.A.: Mathematical methods in Engineering. Mc Graw-Hill, New York (1940)
4. Katagiri M (1969) The effect of Hallcurrent on the magneto hydrodynamic boundary layerflow past a semi-infinite flatplate. J. Phys. Soc. Japan. 27:1051
5. Debnath L, Ray SC, Chatterjee AK (1979) Effect of Hall current and rotation on unsteady hydromagnetic flow past a porous plate in a rotating system. ZAMM. 59:469–471
6. Pop, I.: The effect of Hall current on hydromagnetic flow near accelerated plate. J. Math. Phys. Sci. (1971) 375
7. Muhim Chutia., Deka, P.N.: Numerical solution of unsteady MHD couette flow in the presence of uniform sucton and injection with Hall effect. Iranian. J. Sci. Tech., Transactions. Mech. Eng. (2020) 1- 12
8. Deka, R.K: Hall Effect on MHD Flow Passed an Accelerated plate. The. App. Mec. Vol. 35(4). (2008) 333–346
9. Takhar HS, Chamkha AJ, Nath G (2003) Flow and heattransfer on a stretching surface in a rotating fluid with a magnetic field. Int. J. Ther. Sci. 423(1):23–31
10. Farhad A, Narzieha M, Sharidan S (2012) Hydro-magnetic rotating flow in a porous medium with slip condition and Hall current. Int. J. Phys. Sci. 7:1540–1548
11. Lance, G.N., Rogurs, M.H.: Unsteady slip flow over a flat plate. Proc. R. Soc. A. (1962) 109
12. Bhatt BS, Sacheti NS (1979) On the analogy in slip flows. Int. J. Pure. Appl. Math. 10:303–306
13. Matthews, Miccal, Hill, James, M.: Effect of slip on the linear stability of flow through a tube. Z. Angew. Math. Phys. Vol. 59 (2008) 360–379
14. S. Ahmad, S., Chishtie, F., Mahmood, A.: Analytical technique for magnetohydrodynamic (MHD) fluid flow of a periodically accelerated plate with slippage. Eur. J. Mech. B/Fluids. Vol. 65. (2017) 192–198.
15. Turkyilmazoglu M (2018) Unsteady flow over a accelerating rotating sphere. Phys Fluids 30:1–9
16. Hetnarski, R.B.: An algorithm for generating some Inverse Laplace Transforms of exponential form. J. Appl. Math. Phys. (ZAMP). (1975) 249–253
17. Narahari M, Debnath M (2014) Some new convolution properties and inversion formulas of Laplace Transform. Integral Transforms Spec. Funct. 25:412–422

Simulation of Two-Dimensional Nonlinear Model for Washing of Pulp Fiber Bed Using Quintic Hermite Interpolation Polynomials

Satinder pal Kaur, Ajay Kumar Mittal, and Vijay Kumar Kukreja

Abstract Mathematical modeling is an area via which researchers can develop strategies for the industries. A lot of work is going on to develop and study the behavior of models related to the pulp and paper industry. The process of removal of impurities present on the particle surface through the introduction of water is covered under displacement washing. A two-dimensional nonlinear model of displacement washing with particle and bulk fluid phase is exhibited in this study. The model equations are solved using the orthogonal collocation finite element method with quintic Hermite interpolation polynomials as a basis function. Further, the numerical results derived are compared with experimental results available in the literature. The comparison with published results proves the efficacy of the method. The behavior of the exit solute concentration curve for parameters such as diffusion, distribution ratio, and bed porosity are also presented.

Keywords Quintic Hermite · Diffusion · Porosity · Displacement washing

1 Introduction

Displacement washing is a process that deals with the removal of impurities present on particle surfaces by the introduction of water or weak wash liquor. Solute displacement from fiber pores happens due to the occurrence of diffusion and dispersion phenomena. In this process, diffusion takes place in particle pores, while dispersion takes place when the liquor present inside the bed is displaced by wash liquor. It

S. pal Kaur (✉)
Research Scholar, Department of Mathematics, Maharaja Ranjit Singh Punjab Technical University, Bathinda 151001, PB, India

A. K. Mittal
Department of Mathematics, Aryabhatta Group of Institutes, Barnala 148101, PB, India

V. K. Kukreja
Department of Mathematics, Sant Longowal Institute of Engineering and Technology, Longowal 148106, PB, India

© The Author(s), under exclusive license to Springer Nature Singapore Pte Ltd. 2022

D. Giri et al. (eds.), *Proceedings of the Seventh International Conference on Mathematics and Computing*, Advances in Intelligent Systems and Computing 1412,
https://doi.org/10.1007/978-981-16-6890-6_75

reduces the concentration of solute and cleans the impurities (mainly Na salts and lignin derivatives) of the pulp fiber bed. Arora and Potůček [1] described that the convective dispersion and molecular diffusion of the bulk fluid in the axial direction take place due to the porous space present in the particles.

The researchers, engineers, and mathematicians have put their effort to develop mathematical models for such phenomena. The set of mathematical equations are based on physical conditions of the process, porous structure, particle geometry, and the nature of particles [5]. The investigators have proposed different models to study the behavior of pulp washing of porous structured packed bed. References [16, 19] have described the washing behavior of pulp fiber bed by considering the intrafiber diffusion and axial dispersion with mathematical models. Okhovat et al. [17] developed and validated the model for the prediction of erosion–corrosion phenomenon in two-phase fluid transport. de Pablo et al. [18] described in detail the two-dimensional (2D) porous medium structure. Nadukandi et al. [15] employed Petrov Galerkin method for a diffusion–reaction problem and discussed that the suitable element size and diffusion coefficient plays a leading role in this process. References [4, 12, 13, 20] solved the 2-D model for linear isotherm, analytically. Sridhar et al. [23] solved the model using the moving finite element method by dividing the axial domain. Jannesari and Tatari [9] solved the models using the adaptive element-free Galerkin technique. Kar et al. [10] derived the effect of important parameters for diffusion models. Mohebbi and Dehghan [14] applied fourth-order compact finite difference scheme for the solution of diffusion models. Guraslan and Sari [7] derived the numerical solution of a nonlinear diffusion model using the differential quadrature method. Roininen and Alopaeus [21] presented the moment method based on weighted residual methods to derive the results for the axial dispersion model. Saritha and Madras [22] explained a simple model using linear adsorption isotherm while ignoring the axial dispersion coefficient effect. Arora and Potůček [1] described a 2-D nonlinear model of displacement washing with particle and bulk fluid phase in their study. Further, they mentioned in detail the physical properties of the washing cell. Arora et al. [3] derived the numerical solution of the model using the orthogonal collocation finite element method (OCFE). Gupta et al. [6] solved the 2-D model equations by applying the cubic spline collocation method (CSCM).

In the current study, a 2-D nonlinear model considering the axial and radial direction is considered. The model is solved using the quintic Hermite collocation method (QHCM) in which quintic Hermite interpolating polynomials are used to discretize the model equations. These polynomials are chosen to estimate the trial function and the residual is equated to zero at collocation points. These polynomials are of a special type that interpolates function at each grid point as well as its consecutive derivative [11]. The study aims to solve the model equation describing the diffusion process and to demonstrate the efficiency of the numerical technique. Besides this, the effect of some important parameters such as distribution ratio, Peclet number, bed porosity, and Biot number on the concentration of solute at exit level are also discussed.

2 Mathematical Model

Over the years, various researchers have established mathematical models to describe the displacement washing behavior of packed bed with porous particles. Arora and Potůček [1] exhibited the 2-D model involving axial dispersion coefficient, particle diffusion, film resistance mass transfer coefficient, bed porosity, and pore radius of particles to study the flow of fluid for two phases through a packed bed of porous particles. The authors used Langmuir adsorption isotherm to relate the solute concentration of inter-particle and particle surface. In this work, the 2-D nonlinear model of displacement washing with particle and bulk fluid phase is solved.

The concentration of external fluid is considered as a function of time and axial distance whereas the concentration of inter-particle solute and concentration of solute adsorbed surface of the particle is considered as a function of time, axial distance, and pore radial distance. The description of certain assumptions of related models is discussed by [1] in detail. The model equations describing different phases are presented below:

2.1 *Model Equation for the Particle Phase*

The solute present on the fiber surface is represented by the particle diffusion model. Arora and Potůček [1] stated that equation for particle phase in terms of radial direction can be written as:

$$D_F\left(\frac{\partial^2 q}{\partial r^2} + \frac{1}{r}\frac{\partial q}{\partial r}\right) - \frac{\partial q}{\partial t} - C_F\frac{(1-\beta)}{\beta}\frac{\partial n}{\partial t} = 0 \tag{1}$$

The boundary conditions depending on the mass transfer rate explain the exchange of solute between the bulk fluid and the fiber surface. References [16, 20] assumed the boundary condition describing the relation of bulk fluid phase and particle phase as:

$$\frac{\partial q}{\partial r} = 0 \text{ at } r = 0, \tag{2}$$

$$k_f\beta(q|_{r=R} - c) = -K D_F\left(\frac{\partial q}{\partial r}\right) \text{ at } r = R \tag{3}$$

They considered the solute deposition rate (in the forward direction) as 'k_1' of second order and detachment rate (in the reverse direction) as 'k_2' of the first order. The concentrations of solute 'n' and 'q' are related as:

$$\frac{\partial n}{\partial t} = k_1\frac{q}{C_F}(N_i - n) - k_2 n \tag{4}$$

At equilibrium, the relation reduced to Langmuir isotherm as under:

$$n = \frac{qkN_i}{C_F + qk} \tag{5}$$

2.2 *Model Equation for Fluid Phase*

The model equation for fluid phase is represented by the one-dimensional axial dispersion model associated with the axial distance (z) and the concentration of intrapore solute with the fiber surface is described by [1] as follows:

$$u\varepsilon\frac{\partial c}{\partial z} + \varepsilon\frac{\partial c}{\partial t} + \frac{2(1-\varepsilon)}{R}D_F\left.\frac{\partial q}{\partial r}\right|_{r=R} = D_L\varepsilon\frac{\partial^2 c}{\partial z^2} \tag{6}$$

The rate of change of inside pores solute concentration w.r.t. inside the radial position of the particle is supposed to zero, i.e.,

$$\frac{\partial q}{\partial r} = 0 \text{ at } r = 0 \tag{7}$$

and the rate of change w.r.t particle boundary radial position is supposed to be regulated by film resistance mass transfer coefficient, i.e.,

$$-D_F\frac{\partial q}{\partial r} = \frac{k_f w}{k}(q|r = R - c) \text{ at } r = R \tag{8}$$

Also at the entry of the bed, it is considered that there is no loss of solute from the bed in the axial direction ($z = 0$) at the point of introduction of the displacing fluid:

$$uc - D_L\frac{\partial c}{\partial z} = 0 \text{ at } z = 0 \tag{9}$$

Similarly, the rate of change of concentration is taken as zero at the exit point of the bed in the axial direction ($z = L$) and given as:

$$\frac{\partial c}{\partial z} = 0 \text{ at } z = L \tag{10}$$

It is presumed that initially, the solute concentration of bulk fluid and solute concentration of intrapore are identical to the solute concentration inside the vat. Also, the solute concentration that is adsorbed on fibers is assumed to equal to the solute concentration that is adsorbed on the inside vat of the fibers

$$c = C_i = q \text{ and } n = N_i \text{ at } t = 0 \tag{11}$$

The models, boundary conditions, and initial conditions from (1) to (11) are converted into dimensionless form. The dimensionless parameters are explained in nomenclature. Following equations are obtained:

$$\left(\frac{\partial^2 Q}{\partial \eta^2} + \frac{1}{\eta}\frac{\partial Q}{\partial \eta}\right) = \frac{\partial Q}{\partial \tau} + N_1 \frac{(1-\beta)}{\beta}\frac{\partial N}{\partial \tau} \tag{12}$$

$$\frac{\partial Q}{\partial \eta} = 0 \text{ at } \eta = 0 \tag{13}$$

$$\frac{\partial Q}{\partial \eta} = Bi(C - Q)\ at\ \eta = 1 \tag{14}$$

$$\frac{\partial N}{\partial \tau} = \frac{R^2 k_1}{D_F}\left(C_1 Q(1-N) - \frac{N}{K}\right) \tag{15}$$

$$\frac{\partial C}{\partial \tau} = \frac{\psi}{Pe}\frac{\partial^2 C}{\partial \xi^2} - \psi\frac{\partial C}{\partial \xi} - \theta Bi\left(C - Q|_{\eta=1}\right) \tag{16}$$

$$C - \frac{1}{Pe}\frac{\partial C}{\partial \xi} = 0 \text{ at } \xi = 0 \tag{17}$$

$$\frac{\partial C}{\partial \xi} = 0 \text{ at } \xi = 1 \tag{18}$$

$$C = Q = N = 1 \text{ at } \tau = 0 \tag{19}$$

3 Numerical Procedure

The radial domain is divided into ω subdomains with $\eta_1 = 0$ and $\eta_{\omega+1} = 1$. Each part of the subdomain is mapped onto [0, 1] using the transformation $u = (\eta - \eta_l)/h_l$, where $l = 1, 2, \ldots, \omega$ and $h_l = \eta_{l+1} - \eta_l$. Similarly, the axial domain is divided into υ subdomains with $\xi_1 = 0$ and $\xi_{\upsilon+1} = 1$. Each subdomain is mapped onto [0, 1] by transformation $v = (\xi - \xi_m)/h_m$, where $m = 1, 2, \ldots, \upsilon$ and $h_m = \xi_{m+1} - \xi_m$. The interior collocation points are considered as roots of fourth-order shifted Legendre polynomials because these polynomial give better results at the center [6].

In the present technique, an unknown solution function is approximated using quintic Hermite interpolation as the trial function. These have the property of continuity of first and second derivatives at nodal points that help to derive the solution at a steep gradient. In the current technique, the domain is partitioned into finite elements and then the collocation method is applied within each element for the assumed trial function. In this case, the residuals derived for the approximate solution of model equations are satisfied at interior collocation points and the boundary conditions are satisfied at extreme node points. Then, the derived system of the differential equation along with the boundary conditions are solved using MATLAB ode15s.

Let $Q_l^m(u, v, \tau)$ denotes the trial solution of $Q(u, v, \tau)$, then

$$Q_l^m(u, v, \tau) = \sum_{i,j=1}^{6} q_{i+3(l-1)}^{j+3(m-1)}(\tau) H_i^j(u, v) \tag{20}$$

The six quintic Hermite basis functions are given as follows:

$$\begin{gathered} H_1^k(\eta) = (1 + 3\eta + 6\eta^2)(1-\eta)^3, \quad H_2^k(\eta) = \eta(1 + 3\eta)(1-\eta)^3 \\ H_3^k(\eta) = \frac{1}{2}\eta^2(1-\eta)^3, \quad H_4^k(\eta) = (10 - 15\eta + 6\eta^2)\eta^3 \\ H_5^k(\eta) = (1-\eta)\eta^3(3\eta - 4), \quad H_6^k(\eta) = \frac{1}{2}\eta^3(1-\eta)^2 \end{gathered}$$

where $H_4^k(1) = 1$, $H_1^k(0) = 1$, $H_2'^k(0) = 1$, $H_5'^k(1) = 1$, while others are zero.

$$\left.\begin{aligned} Q_l^m(1, v, \tau) &= Q_{l+1}^m(0, v, \tau), & Q_l^m(u, 1, \tau) &= Q_l^{m+1}(u, 0, \tau), \\ \frac{\partial Q_l^m}{\partial \eta}(1, v, \tau) &= \frac{\partial Q_{l+1}^m}{\partial \eta}(0, v, \tau), & \frac{\partial Q_l^m}{\partial \xi}(u, 1, \tau) &= \frac{\partial Q_{l+1}^m}{\partial \xi}(u, 0, \tau). \end{aligned}\right\} \tag{21}$$

Similarly

$$N_l^m(u, v, \tau) = \sum_{i,j=1}^{6} n_{i+3(l-1)}^{j+3(m-1)}(\tau) H_i^j(u, v) \tag{22}$$

And

$$C^m(v, \tau) = \sum_{j=1}^{6} c^{j+3(m-1)}(\tau) H_j(v). \tag{23}$$

Substituting the approximate solutions (20), (22), and (23) into the differential Eqs. (12), (15), and (16), following residual equations are obtained:

$$R_Q(u, v, \tau) = \sum_{j=1}^{6} \frac{dq_{i+3(l-1)}^{j+3(m-1)}(\tau)}{d\tau} H_i^j(u, v) - \frac{1}{h_l^2} \sum_{i,j=1}^{6} q_{i+3(l-1)}^{j+3(m-1)}(\tau) \frac{\partial^2 H_i^j(u, v)}{\partial u^2}$$

$$
- \frac{1}{(uh_l + \eta_l)h_l} \sum_{i,j=1}^{6} q_{i+3(l-1)}^{j+3(m-1)}(\tau) \frac{\partial H_i^j(u, v)}{\partial u}
$$

$$
+ N_1\left(\frac{1-\beta}{\beta}\right) \sum_{i,j=1}^{6} \frac{dn_{i+3(l-1)}^{j+3(m-1)}(\tau)}{d\tau} H_i^j(u, v), \tag{24}
$$

$$
R_N(u, v, \tau) = \sum_{j=1}^{6} \frac{dn_{i+3(l-1)}^{j+3(m-1)}(\tau)}{d\tau} H_i^j(u, v) - \frac{R^2 k_1}{D_F}\left[C_1 \sum_{i,j=1}^{6} q_{i+3(l-1)}^{j+3(m-1)}(\tau) H_i^j(u, v) \times \left(1 - \sum_{j=1}^{6} n_{i+3(l-1)}^{j+3(m-1)}(\tau) H_i^j(u, v)\right) - \frac{1}{k^*} \sum_{i,j=1}^{6} n_{i+3(l-1)}^{j+3(m-1)}(\tau) H_i^j(u, v)\right], \tag{25}
$$

$$
R_C(v, \tau) = \sum_{j=1}^{6} \frac{dc^{j+3(m-1)}(\tau)}{d\tau} H_j(v) - \frac{\psi}{Peh_m^2} \sum_{j=1}^{6} c^{j+3(m-1)}(\tau) H_j''(v)
$$

$$
+ \frac{\psi}{h_m} \sum_{j=1}^{6} c^{j+3(m-1)}(\tau) H_j'(v)
$$

$$
+ \theta Bi \left(\sum_{j=1}^{6} c^{j+3(m-1)}(\tau) H_j(v) - \sum_{i,j=1}^{6} q_{3w+1}^{j+3(m-1)}(\tau) H_j(v) \right) \tag{26}
$$

The residuals R_Q, R_N, and R_C are equated to zero at collocation points. The evaluation of $Q(u, v, \tau)$ and $C(v, \tau)$ at boundary collocation points yield:

$$
\sum_{j=1}^{6} q_{3l-1}^{j+3(m-1)}(\tau) H_j(v_s) = 0 \tag{27}
$$

$$
\sum_{i,j=1}^{6} q_{3w+2}^{j+3(m-1)}(\tau) H_j(v_s)
$$

$$
= -Bi(h_w)\left[\sum_{j=1}^{6} q_{3w+1}^{j+3(m-1)}(\tau) H_j(v_s) - \sum_{j=1}^{6} c^{j+3(m-1)}(\tau) H_j(v_s) \right] \tag{28}
$$

where for $m = 1,\ s = 1, 2, 3, 4, 5$; $m = 2, 3, \ldots, \upsilon - 1,\ s = 2, 3, 4, 5$; and $m = \upsilon,\ s = 2, 3, 4, 5$.

$$
c^1(\tau) - \frac{1}{Pe} c^2(\tau) = 0, \tag{29}
$$

$$c^{3ne+2}(\tau) = 0 \tag{30}$$

Let the index set $A = H_i(u)\{(i, j)|i, j = 1, 2, 3, 4, 5, 6\}$. The condition $Q(\eta, \xi, 0) = 1$ implies:

$$\sum_{i,j=1}^{6} q_{i+3(l-1)}^{j+3(m-1)}(0) H_i(u) H_j(v) = 1. \tag{31}$$

Using the points $(u, v) \in D = \{(0, 0), (1, 0), (0, 1), (1, 1)\}$, Eq. (20) implies: $q_{i+3(l-1)}^{j+3(m-1)}(0) = 1$.

For $(i, j) \in B = \{(1, 1), (1, 4), (4, 1), (4, 4)\}$, Eq. (20) implies:

$$\sum_{i,j=1}^{6} q_{i+3(l-1)}^{j+3(m-1)}(0) H_i'(u) H_j(v) = 0 \tag{32}$$

$$\sum_{i,j=1}^{6} q_{i+3(l-1)}^{j+3(m-1)}(0) H_i(u) H_j'(v) = 0 \tag{33}$$

$$\sum_{i,j=1}^{6} q_{i+3(l-1)}^{j+3(m-1)}(0) H_i'(u) H_j'(v) = 0 \tag{34}$$

Using the points $(u, v) \in D$ in the above equations, one can easily show that $q_{i+3(l-1)}^{j+3(m-1)}(0) = 0$. Similarly,

$$n_{i+3(l-1)}^{j+3(m-1)}(0) = \begin{cases} 1, & (i, j) \in B \\ 0, & (i, j) \in A - B \end{cases} \tag{35}$$

The initial condition $C(\xi, 0) = 1$ implies:

$$c^{3m+1}(0) = 1,\ c^{3m+2}(0) = 0,\ c^{3m+3}(0) = 0 \tag{36}$$

for $m = 0, 1, ..., \upsilon$. The initial conditions:

$$c^1(0) = 1,\ c^2(0) = 0,\ c^{3m+3}(0) = 0 \tag{37}$$

$$q_{3w+1}^1(0) = c^1(0) = 0 \tag{38}$$

4 Results and Discussions

The numerical values are calculated by QHCM and the results for the solution profile are presented for distinct values of parameters. The validity is checked by applying the present technique to the above model equations for experimental data of [2]. The comparisons with previously published results in the radial and axial domain are also performed in this work.

4.1 Comparison Between Experimental and Numerical Values

The given model is solved using the present technique, i.e., QHCM. The numerical values for exit solute concentration profiles using the present technique are determined and compared with experimental values given by [2]. These results are derived for Peclet number (Pe) = 20.81, Biot number (Bi) = 10, and bed porosity (ε) = 0.6711. The results obtained using QHCM are compared with the previously published results of [3, 6]. It is mentioned that [3] used OCFE and [6] used CSCM to solve the above-discussed model. Both studies evidenced that numerical results obtained are nearer to experimental results given by [2]. However, the present method displays that there are still chances of improvement in the numerical results. The improvements are proved by making comparisons in terms of absolute and relative errors calculated using the below-mentioned relations:

Absolute Error = $C_E - C_N$,

Relative Error = absolute error/ C_E,

where C_E and C_N represent the experimental and the numerical values, respectively.

Notably, the relative error presented in Table 1 is found to be very less using QHCM in comparison to OCFE and CSCM in all cases of experimental results. This substantiates that QHCM is superior to these techniques. Besides this, the results obtained using the present method for Pe = 4.91, ε = 0.61, Bi = 4.5 are compared in the form of absolute error with the numerical results derived with OCFE by [3]. Figure 1 displays that the numerical results for exit solute concentration using QHCM get less absolute error as compared to results of [3].

4.2 Effect of Some Important Parameters

Practically, it is not possible that the washing process depends on a single parameter. However, it depends on various parameters such as a change in bed porosity, pore radius, permeability, axial dispersion coefficient, interstitial velocity, Biot number, and Peclet number. The effect of some of the essential parameters using experimental values of [3] is shown graphically in this part.

Table 1 Comparison between experimental and numerical values with published results

Time	Experimental [2]	OCFE [3]	Relative error [3]	CSCM [6]	Relative error [6]	QHCM	Relative error (QHCM)
0.0	1.000E+00	1.000E+00	0.000E+00	1.000E+00	1.000E+00	1.000E+00	0.000E+00
0.25	9.900E-01	9.997E-01	9.818E-03	9.907E-01	7.070E-04	9.901E-01	1.010E-04
0.50	9.385E-01	9.395E-01	1.097E-03	9.370E-01	6.393E-04	9.380E-01	5.327E-04
0.75	8.169E-01	8.165E-01	4.651E-04	8.158E-01	1.346E-03	8.171E-01	2.448E-04
1.00	6.743E-01	6.736E-01	9.491E-04	6.736E-01	1.038E-03	6.750E-01	1.038E-03
1.25	5.335E-01	5.357E-01	4.179E-03	5.343E-01	1.499E-03	5.338E-01	5.623E-04
1.50	4.560E-01	4.524E-01	7.807E-03	4.571E-01	2.412E-03	4.553E-01	1.535E-03
1.75	3.556E-01	3.577E-01	5.989E-03	3.546E-01	2.812E-03	3.561E-01	1.406E-03
2.00	2.685E-01	2.664E-01	7.523E-03	2.674E-01	4.096E-03	2.691E-01	2.234E-03
2.25	1.849E-01	1.859E-01	5.516E-03	1.891E-01	2.271E-02	1.853E-01	2.163E-03
2.50	1.095E-01	1.086E-01	7.671E-03	1.056E-01	3.561E-02	1.090E-01	4.566E-03
2.75	4.294E-02	4.289E-02	1.094E-03	–	–	4.298E-02	9.315E-04
3.00	5.423E-03	5.441E-03	3.374E-03	–	–	5.430E-03	1.290E-03
3.25	1.875E-03	1.861E-03	7.360E-03	–	–	1.868E-03	3.733E-03
3.50	9.580E-04	9.638E-04	6.137E-03	–	–	9.640E-04	6.263E-03
3.75	5.850E-04	5.850E-04	1.709E-05	–	–	5.840E-04	1.709E-03
4.00	5.070E-04	5.088E-04	3.609E-03	–	–	5.060E-04	1.972E-03

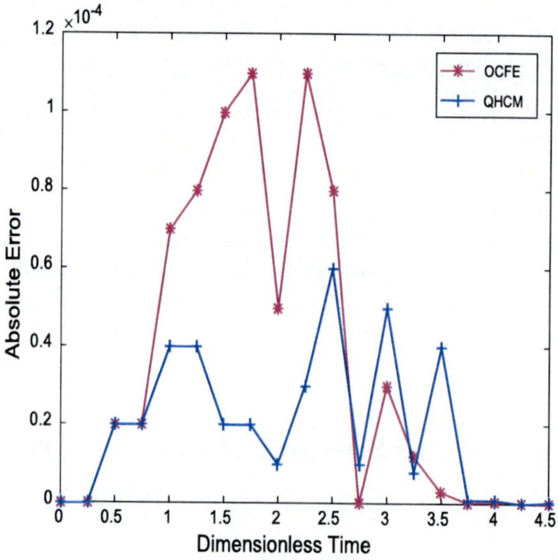

Fig. 1 Comparison of OCFE and QHCM in terms of absolute error

4.2.1 Influence of the Distribution Ratio

The influence of the distribution ratio on concentration profiles is represented in Fig. 2a. The influence of distribution ratio on solution profiles at exit level is presented for parameter Pe = 20.81, Bi = 10, and $\varepsilon = 0.67$. A small distribution ratio causes more bulging in pores and more diffusion is caused due to particle pores which results in improved washing.

4.2.2 Influence of Peclet Number (Pe)

Peclet number is proportional to the ratio of advection and diffusion. The Peclet number is inversely proportional to dispersion and it is lower when the dispersion is higher. Hasson et al. [8] also proved that the dispersion coefficient has a significant effect on concentration profiles. When the Peclet number is increased, the dispersion coefficient decreases and it causes a reduction in back mixing. Owing to this, the adsorbed impurities of particle surface are leached in a better manner and improved washing is attained. Figure 2b demonstrates the effect of the Peclet number in the form of a breakthrough curve. The results are attained for the range of Peclet number from 2 to 20 and Biot number (Bi = 10), bed porosity 0.6711. It is observed that the concentration decreases with an increase in time. The figure also declares that the tail of concentration profiles gets longer for a small value of Pe as compared to a large value of Pe that causes an increase in back-mixing and time for washing.

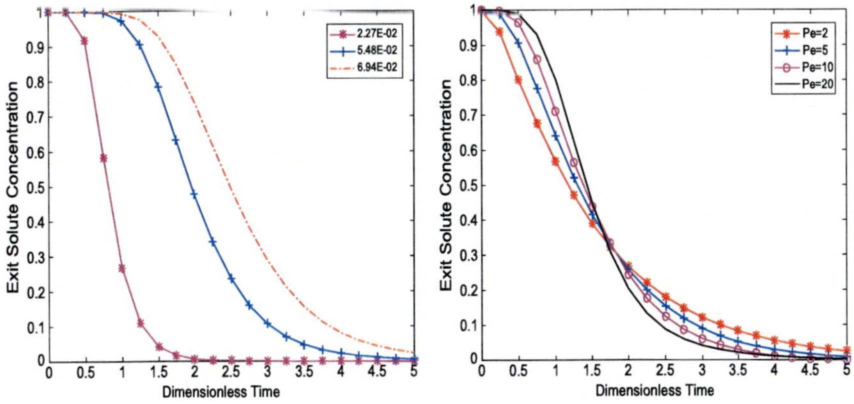

Fig. 2 **a**. Influence of distribution ratio on exit solute concentration. **b.** Influence of Peclet number on exit solute concentration

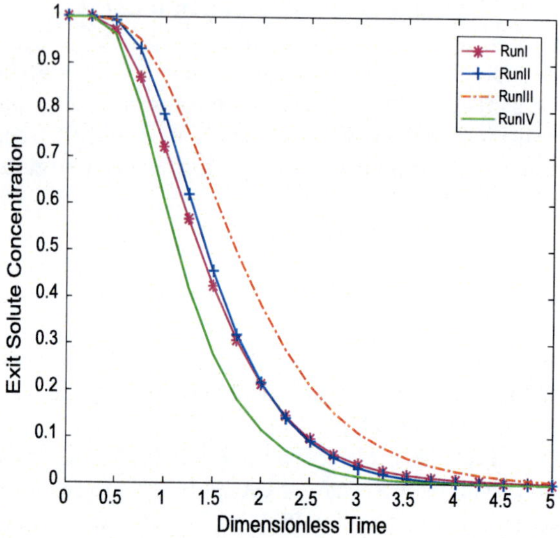

Fig. 3 Effect of different washing run of Table 2 on solution profiles

4.2.3 Influence of Biot Number (Bi)

Besides the Peclet number, there also exist other parameters that affect the whole process. The solution profile using QHCM are obtained for experimental conditions of washing runs by [1] given in Table 2. The results are presented as breakthrough curves in Fig. 3. It is seen from this figure that the concentration of solute presented as solution profiles for washing run 4 and Pe = 12.96 needs more time for converges to a steady-state condition in comparison with other experimental values for Pe = 12.25, 16.92, 20.8. The figure shows that the behavior of solution profiles for washing run 1 and 2 are nearly similar but the Pe = 12.25 and Pe = 20.81 are not the same. The reason is Bi which is also a significant parameter used to represent the mass transfer resistances on and inside the particle surface in the washing process. For washing run 2, the Bi is more than the run 1 and this causes an increase in the removal rate of adsorbed impurities on the particle surface. Also, it is noticed that large values of Bi cause an increase in mass transfer rate and indicate the fast convergence of solution profiles as compared to small Bi.

Table 2 Experimental data for different washing runs by [1]

Washing runs	Pe	Bi	C_F	ε
Run I	12.25	7.4	12.09	0.6898
Run II	20.81	10	11.91	0.6711
Run III	16.92	7.5	13.14	0.5561
Run IV	12.96	6.3	7.96	0.8120

4.2.4 Influence of Bed Porosity (ε)

Although from Fig. 3 it is clear that the Pe and Bi are nearly the same in washing runs 1 and 4 but concentration profiles at exit level are not the same. This happens due to bed porosity which is lesser in the former case in comparison to the later ones. The solution profile needs extra time to converge for the steady-state condition. Bed porosity (ε) indicates the ratio of volume present for flow wash liquid to the total volume. Also, the greater bed porosity shows the availability of more volume for displacement, and hence the absorbed impurities on the particle surface are leached away in a better manner.

5 Conclusions

A comprehensive study of 2-D nonlinear model with nonlinear adsorption isotherm for the porous structure of particles is done in this research. QHCM is followed in the discretization process for the model equations of the displacement washing process. Numerical results attained using QHCM are compared to the experimental results of [2]. The relative error is noticed least in the present method when results are compared with [2, 3, 6]. Also, the absolute error is less when compared with [3]. This fact verifies the applicability of the present method. After validating the results using the QHCM method, the effect of different parameters on concentration at exit level for solution profiles are analyzed and presented graphically in the form of breakthrough curves. It is observed that the Pe, Bi, ε, and distribution ratio are the significant parameters that collectively influence the displacement washing process. The small distribution ratio helps to improve the washing process. It is further seen that better washing can be achieved with the increase in Pe. From the present work, it is also determined that to achieve better results for pulp washing, large Pe, Bi, ε, and small distribution ratios can be preferred.

Nomenclature

c = Concentration of solute in the liquor (kg/m^3).
c_0 = Inlet solute concentration in the vat (kg/m^3).
c_s = Solute concentration in wash liquor (kg/m^3).
q = Inside Pore solute concentration (kg/m^3).
n = Concentration of solute in fibers (kg/m^3).
C = Dimensionless concentration $(c - c_s)/(c_0 - c_s)$.
N = Dimensionless concentration $(n - c_s)/(c_0 - c_s)$.
Q = Dimensionless concentration $(q - c_s)/(c_0 - c_s)$.
D_L = Longitudinal dispersion coefficient (m^2/s).
D_F = Interafiber diffusion coefficient (m^2/s).
C_F = Fiber consistency (kg/m^3).
z = Axial distance from point of introduction of solvent (m).
u = Interstitial velocity through bed (m/s).

B = Permeability coefficient (m^2).
k_1, k_2 = Mass transfer coefficient $(1/s)$.
k_f = Film mass transfer coefficient (m/s).
L = Thickness of the bed (m).
t = Time (s).
Pe = Peclet number, dimensionless (uL/D_L).
τ = Dimensionless time (ut/L).
K = Equilibrium constant (Dimensionless),
ξ = Dimensionless axial distance (z/L).
r = Particle solute radial position (m).
R = Fiber radius (m).
Bi = Biot number, dimensionless $(k_f \omega R/D_F K)$.
$\psi = R^2 u/LD_F$.

Greek Symbols.

ε = Porosity of cake, dimensionless.
μ = Ratio of Porosity $((1-\varepsilon)/\varepsilon)$.
β = Porosity of particle.

References

1. Arora S, Potůček F (2009) Modelling of displacement of washing of pulp: comparison between model and experimental data. Cell Chem Technol 43(7–8):307–315
2. Arora S, Potůček F (2012) Verification of mathematical model for displacement washing of kraft pulp fibres. Indian J Chem Tech 19:140–148
3. Arora S, Alemar DA, Potůček F (2020) Study of Two-Phase Nonlinear Advection Dispersion Model for Displacement Washing of Porous Particles Using OCFE with Lagrangian Basis. Arab J Sci Eng 45(2):531–542
4. Cocero MJ, Garcia J (2001) Mathematical model of supercritical extraction applied to oil seed extraction by CO2+ saturated alcohol—I. Desorption Model J Supercrit Fluids 20:229–243
5. Dehghan M, Shirzadi M (2015) Meshless simulation of stochastic advection–diffusion equations based on radial basis functions. Eng Analy Bound Elem 53:18–26
6. Gupta B, Kukreja VK, Parumasur N, Singh P (2015) Numerical Study of a Nonlinear Diffusion Model for Washing of Packed Bed of Cylindrical Fiber Particles. Arab J Sci Eng 40(5):1279–1287
7. Guraslan G, Sari M (2011) Numerical solutions of linear and nonlinear diffusion equations by a differential quadrature method. Int J Numer Methods Biomed Eng 27:69–77
8. Hasson D, Drak A, Yang Q, Semiat R (2006) Effect of axial dispersion on the concentration polarization level in spiral wound modules. Desalination 199:451–453
9. Jannesari Z, Tatari M (2020) An adaptive strategy for solving convection dominated diffusion equation. Comput Appl Math 39(2):1–15
10. Kar M, Sahoo SN, Rath PK, Dash GC (2014) Heat and mass transfer effects on a dissipative and radiative visco-elastic MHD flow over a stretching porous sheet. Arab J Sci Eng 39(05):3393–3401
11. Kaur SP, Mittal AK, Kukreja VK, Parumasur N, Singh P (2018) An efficient technique for solution of linear and nonlinear diffusion-dispersion models. AIP Conf Proc 1975 (1):030031, 1–10

12. Kill S, Bhatia SK (1998) Solution of adsorption problems involving steep moving profiles. Comput Chem Eng 22(7–8):893–896
13. Liu F, Bhatia SK (1999) Computationally efficient solution techniques for adsorption problems involving steep gradients bidisperse particles. Comput Chem Eng 23:933–943
14. Mohebbi A, Dehghan M (2010) High-order compact solution of the one-dimensional heat and advection–diffusion equations. Appl Math Model 34:3071–3084
15. Nadukandi P, Onate E, Garcia J (2012) A high resolution petrovgalerkin method for the convection–diffusion–reaction problem. Part II—a multidimensional extension. Comput Methods Appl Mech Eng 213–216, 327–352
16. Neretnieks I (1974) A mathematical model for continuous counter current adsorption. Svensk Papper Stid 11:407–411
17. Okhovat A, Heris SZ, Asgarkhani MAH, Fard KM (2014) Modeling and simulation of erosion–corrosion in disturbed two-phase flow through fluid transport pipelines. AJSE 39(03):1497–1505
18. de Pablo A, Quirós F, Rodríguez A, Vázquez JL (2011) A fractional porous medium equation. Adv Math 226(2):1378–1409
19. Pellett GL (1966) Longitudinal dispersion, intraparticle diffusion, and liquid-phase mass transfer during flow through multiparticle systems. Tappi 49(2):75–82
20. Raghvan NS, Ruthven DM (1983) Numerical simulation of a fixed bed adsorption column by the method of orthogonal collocation. AIChE 29(6):922–925
21. Roininen J, Alopaeus V (2011) The moment method for one dimensional dynamic reactor models with axial dispersion. Comp Chem Eng 35:423–433
22. Saritha NV, Madras G (2001) Modeling the chromatographic response of inverse size-exclusion chromatography. Chem Eng Sci 56:6511–6524
23. Sridhar P, Sastri NVS, Modak JM, Mukherjee AK (1994) Mathematical simulation of bioseparation in an affinity packed column. Chem Eng Technol 17:422–429

Semi-Markov Decision Processes with Vector Pay-Offs

Kushal Guha Bakshi

Abstract Semi-Markov decision processes with vector rewards are studied here under discounted pay-off criterion. The notion of optimality is replaced by Pareto optimality here. We show that a pure and stationary Pareto-optimal strategy exists along with the existence of an algorithm to construct an approximate version of Pareto curves, i.e., ϵ-approximate Pareto curve in polynomial time as in a multi-objective linear programming problem. Semi-Markov decision processes with multiple objectives find applications in situations like dynamic goal programming where the decision-maker has more than one objective to be optimized simultaneously. Further, we also investigate the Pareto-realizability problem as well as the NP-completeness of pure stationary Pareto-realizability problem.

Keywords Semi-Markov decision process · Pareto optimality · Pareto realizability

1 Introduction

Semi-Markov decision processes (SMDPs) were introduced by Jewell [1] and Howard [2] and later generalized by Ross [3], Lippman [4], Federgruen, and Tijms et al. [5]. It is a generalization of Markov decision processes (MDPs) in a larger class of dynamical models. SMDPs are used in modeling stochastic control problems arising in MDPs where the sojourn (transition) time is a random variable depending on the present state, the next state, and the action chosen in the present state. A finite SMDP is similar to an MDP, where the state changing occurs according to Markov property. The study of multiple objective SMDP model is an extension of the SMDP model, where several reward functions are used. The notion of Pareto curve is studied in this paper to analyze the discounted SMDP model with vector pay-offs. Pareto

Supported by Department of Science and Technology, Govt. of India, INSPIRE Fellowship Scheme

K. Guha Bakshi (✉)
Jadavpur University, 188, Raja S.C. Mallick Rd, Kolkata 700032, India

© The Author(s), under exclusive license to Springer Nature Singapore Pte Ltd. 2022
D. Giri et al. (eds.), *Proceedings of the Seventh International Conference on Mathematics and Computing*, Advances in Intelligent Systems and Computing 1412,
https://doi.org/10.1007/978-981-16-6890-6_76

optimality is generally studied in multi-criterion optimization theory. For details, see [6]. Owen [7] showed the usefulness of Pareto optimality in the field of co-operative game theory. We also study about the ϵ-approximate Pareto curve of a discounted SMDP model. Chatterjee et al. [8] showed that in a Markov decision process (discounted) with vector pay-offs stationary strategies are sufficient for Pareto optimality and constructed an algorithm for an ϵ-approximate Pareto curve in size of a multi-objective linear programming problem. This paper is a generalization of this work to a discounted SMDP model.

The results of this paper are organized as follows. Section 2 contains preliminaries and basic definitions about SMDPs and Pareto optimality. In Sect. 3, we show that stationary strategies are sufficient strategies for Pareto optimality. In Sect. 4, the basic definitions and properties of Pareto curves are studied. Section 5 investigates about the multi-objective linear programming problem formulation of a discounted SMDP with vector pay-offs and a conclusion is drawn that if there exists an algorithm to construct an ϵ-approximate Pareto curve in polynomial time to answer all queries about pay-off (value) profile of the discounted SMDP with vector pay-offs. Finally, in Sect. 6, we discuss about Pareto-realizability problem (pure stationary Pareto-realizability problem).

2 Preliminaries

We denote $\mathbb{P}(B)$ to be the set of probability distributions on the set B and $\mathbb{R}^n$ as the set of n tuples of real numbers, where $n \in \mathbb{N}$. For $x = (x_1, x_2, \ldots, x_n)$, $y = (y_1, y_2, \ldots, y_n) \in \mathbb{R}^n$, we assume that comparison (equality) between these two vectors is comparison (equality) between their respective co-ordinates, i.e., for any $x, y \in \mathbb{R}^n$, $x \geq y$ holds if $x_i \geq y_i$, $i = 1, 2, \ldots, n$. In this section, we discuss about some basic definitions about finite semi-Markov decision process under discounted pay-off criterion and Pareto optimality which will be used later in this article.

Definition 2.1 [9] A finite semi-Markov decision process is defined by the pentuple $< S, A = \{A(s) : s \in S\}, p, P, r >$ where the state space $S = \{1, 2, \ldots, N\} (\neq \phi)$ is a non-empty finite set and A is the finite action set for the decision-maker. A subset $A(s) \subseteq A$ is defined as the set of admissible actions for the state s of the decision-maker. The set of admissible state–action pairs is defined by $\kappa_A = \{(s, a) : a \in A(s), s \in S\} \subseteq S \times A$. For each $x \in \kappa_A$, $p(. \mid x)$ is the transition law. Given $(s, a) \in \kappa_A$ and $s^{'} \in S$, let $\tau^a_{ss^{'}}$ be the sojourn (transition) time random variable (discrete) denoting the time from state s to state $s^{'}$ by an action $a \in A(s)$. For each $(s, a) \in k_A$, $s^{'} \in S$, $P^a_{ss^{'}}(t) = Prob(\tau^a_{ss^{'}} \leq t)$ is a probability distribution function on $(0, \infty)$ given $\kappa_A \times S$ and it is called the conditional transition time distribution function. Without loss of generality we write P instead of $P^a_{ss^{'}}(t)$ in some cases. r is a real-valued function on κ_A (i.e., r: $\kappa_A \to \mathbb{R}$ is the reward function) which is called the immediate reward function.

At the first decision epoch, the process starts at a state $s_1 \in S$. The decision-maker looks at s_1 and chooses an action $a_1 \in A(s_1)$. Then he/she gets an immediate reward $r(s_1, a_1)$. At next decision epoch the system moves to a new state $s_2 \in S$ with probability $p(s_2 \mid s_1, a_1)$. $P^{a_1}_{s_1 s_2}(.)$, which denotes the transition time distribution function, determines the sojourn time from the state s_1 to state s_2, when the decision-maker chooses action a_1. After the transition to the state s_2 on the next decision epoch, the above-described process with s_1 replaced by s_2 is repeated all over again, and thus SMDP proceeds over infinite time period. Clearly, for identical transition times, an SMDP reduces to an MDP.

Definition 2.2 ([10]) By a strategy (behavioral) b_s of the decision-maker, we mean a sequence $\{(b_s)_n(. \mid hist_n)\}_{n=1}^{\infty}$, where $(b_s)_n$ specifies which action is to be chosen on the n-th decision epoch by associating with each history $hist_n$ of the system up to nth decision epoch (where $hist_n = (s_1, a_1, s_2, a_2, \ldots, s_{n-1}, a_{n-1}, s_n)$ for $n \geq 2$, $hist_1 = (s_1)$, and s_k, a_k are, respectively, the state and action of the decision-maker at the k-th decision epoch) a probability distribution $(b_s)_n(. \mid hist_n)$ on $A(s_n)$. Generally by any unspecified strategy, we mean behavioral strategy here.

A strategy $f = \{f_n\}_{n=1}^{\infty}$ is called stationary if $\exists$ a map $\xi : S \to \mathbb{P}(A)$, such that (i) $f_n = \xi \; \forall n \in \mathbb{N}$, i.e., $f_n(. \mid s_1, \ldots, s_n = s) = \xi(s)$ and (ii) $\xi(s)$ has a finite "support" in $A(s)$, i.e., $\xi(s) \in \mathbb{P}(A(s))$ for each $s \in S$.

A strategy is called a policy, when for any $b_s = \{(b_s)_n\}_{n=1}^{\infty}$, each $(b_s)_n$ is degenerate probability distribution.

Let us denote B_s and $\Im$ as the respective classes of all behavioral and stationary strategies of the decision maker. We consider a sequence of co-ordinates $(X_1, A_1, X_2, A_2, \ldots)$ in $S \times (A \times S)^{\infty}$. For a specified probability distribution η on $(S, 2^S)$ and a strategy $b_s \in B_s$, we can conclude from Kolmogorov's extension theorem that there exists an unique probability measure on the Borel-σ field of $S \times (A \times S)^{\infty}$, such that the marginal distribution of X_1 is η and for each $n \in \mathbb{N}$ and almost surely $A_n \in A(X_n)$. For a degenerate probability distribution η at $s \in S$, we denote $P_{b_s}(. \mid X_1 = s)$ to be the probability measure determined by η and b_s and the corresponding expectation operator is $E_{b_s}(. \mid X_1 = s)$. For every strategy $b_s \in B_s$, we have a sequence of random rewards $r(X_m, A_m)$ which denotes the immediate reward on the mth decision epoch ($m = 1, 2, 3, \ldots$), where X_m and A_m are, respectively, the state and action chosen on that decision epoch. The expectation of $r(X_m, A_m)$ is well defined and will be denoted by

$$E_{b_s}[r(X_m, A_m) \mid X_1 = s]. \tag{2.1}$$

In this paper, we consider SMDP model where rewards are discounted by a discount factor $d \in [0, 1)$. It means that a reward r at time t is equivalent to a reward rd^t at $t = 0$, i.e., the initial time.

Definition 2.3 For each $d \in [0, 1)$ and a fixed initial state $s \in S$, the total discounted pay-off for a strategy $b_s \in B_s$ is defined by

$$V_d^{b_s}(r,s) = E_{b_s}\left[\sum_{m=1}^{\infty} d^{(\tau_1+\tau_2+\cdots+\tau_m)} r(X_m, A_m) \mid X_1 = s\right], \tag{2.2}$$

where $\tau_1 = 0$ and $\tau_i (i \geq 2)$ is the time between $(i-1)$th and i-th decision epochs. τ_i is a random variable depending on X_{i-1}, A_{i-1} and X_i, where $X_i \in S$ and $A_i \in A(X_i)$ are, respectively, the state and action at the ith decision epoch (each X_i and A_i is a random variable $\forall i \geq 1$). For an SMDP with a fixed initial state s, strategy b_s and l different reward functions $r_1, r_2, \ldots, r_l$, the discounted value profile (or pay-off profile) w.r.t. $\bar{r} = (r_1, r_2, \ldots, r_l)$ is defined by

$$V_d^{b_s}(\bar{r}, s) = (V_d^{b_s}(r_1, s), \ldots, V_d^{b_s}(r_l, s)). \tag{2.3}$$

Definition 2.4 ([8]) For a discounted SMDP $< S, A, p, P, \bar{r} >$ with reward vector $\bar{r} = (r_1, r_2, \ldots, r_l)$ a strategy $b_{s^*} \in B_s$ is called Pareto optimal if there is no strategy $b_{s'} \in B_s$ such that both $V_d^{b_{s^*}}(\bar{r}, s) \leq V_d^{b_{s'}}(\bar{r}, s)$ and $V_d^{b_{s^*}}(\bar{r}, s) \neq V_d^{b_{s'}}(\bar{r}, s)$ hold, i.e., $\nexists$ any strategy $b_{s'}$ such that $\forall\, 1 \leq j \leq l$, we have $V_d^{b_{s^*}}(r_j, s) \leq V_d^{b_{s'}}(r_j, s)$ and $\exists 1 \leq j \leq l$, we have $V_d^{b_{s^*}}(r_j, s) < V_d^{b_{s'}}(r_j, s)$.

For a Pareto-optimal strategy b_{s^*}, the corresponding value profile or discounted value profile $V_d^{b_{s^*}}(\bar{r}, s)$ is referred to as a Pareto-optimal point.

Definition 2.5 ([8]) Given a discounted SMDP $S_m =< S, A, p, P, \bar{r} >$ with vector reward function $\bar{r} = (r_1, r_2, \ldots, r_l)$ and a fixed initial state s, a strategy $b_s \in B_s$ is called sufficient for Pareto optimality for discounted reward objectives if for every discount factor $d \in [0, 1)$ and a Pareto-optimal strategy b_{s^*} such that

$$V_d^{b_{s^*}}(\bar{r}, s) \leq V_d^{b_s}(\bar{r}, s) \tag{2.4}$$

where, $\bar{r} = (r_1, r_2, \ldots, r_l)$.

Definition 2.6 A strategy b_{s^*} is called an optimal strategy for a discounted SMDP with single reward function r, if

$$V_d^{b_{s^*}}(r, s) \geq V_d^{b_s}(r, s) \tag{2.5}$$

$\forall\, b_s \in B_s$ and $s \in S$.

Let us denote $V_d(r, s) = sup_{b_s \in B_s} V_d^{b_s}(r, s) \forall s \in S$. Now for a stationary strategy $f \in \Im$, initial state $s \in S$ and reward function r, we have from (2.2)

$$
\begin{aligned}
V_d^f(r,s) &= E_f\left[\sum_{m=1}^{\infty} d^{(\tau_1+\tau_2+\cdots+\tau_m)} r(X_m, A_m) \mid X_1 = s\right] \\
&= E_f[r(X_1, A_1) \mid X_1 = s] + E[E_f\{\sum_{m=2}^{\infty} d^{(\tau_2+\cdots+\tau_m)} r(X_m, Am) \mid hist_2, \tau_2\}] \\
&= r(s,f) + E[d^{\tau_2} E_f\{\sum_{m=3}^{\infty} d^{(\tau_3+\cdots+\tau_m)} r(X_m, A_m) \mid hist_2, \tau_2\}] \\
&= r(s,f) + E[d^{\tau_2} V_d^f(r, X_2)]\ \forall s \in S, \textit{being an initial state.}
\end{aligned}
\tag{2.6}
$$

This implies

$$
V_d^f(r,s) = r(s,f) + \sum_{a\in A(s)}\left[\sum_{s'\in S} p(s' \mid s,a) V_d^f(r,s') \sum_{y=0}^{T} d^y P(\tau_{ss'}^a = y)\right] f(s,a), \tag{2.7}
$$

where $f(s,a)$ is the probability of choosing action a when system is in state s.

Now putting Eq. (2.7) in matrix form, we get

$$
V_d^f = r(f) + P_d(f) V_d^f, \tag{2.8}
$$

where $V_d^f = \left[V_d^f(r,s)\right]_{N\times 1}$ is the value vector, $r(f) = [r(s,f)]_{N\times 1}$ and $P_d(f) = \left[\sum_{a\in A(s)} p(s' \mid s,a) \sum_{y=0}^{T} d^y P(\tau_{ss'}^u = y) f(s,a)\right]_{s,s'=1}^{N}$. $r(s,f)$ can be calculated as $\sum_{a\in A(s)} r(s,a) f(s,a)$. Here $P_d(f)$ is a sub-stochastic matrix whose all rows have a sum strictly less than 1. Using this fact and by induction, we can further write (2.8) as

$$
V_d^f = (I - P_d(f))^{-1} r(f). \tag{2.9}
$$

Theorem 2.1 ([2, 3]) *The value vector of a discounted SMDP with discount factor d is the unique solution of the optimality equation*

$$
V_d(r,s) = max_{a\in A(s)}\left\{r(s,a) + \sum_{s'\in S} p(s' \mid s,a) V_d(r,s') \sum_{y=0}^{T} d^y P(\tau_{ss'}^a = y)\right\},\ s \in S. \tag{2.10}
$$

Further a strategy $f^* \in \Im$ is an optimal stationary strategy iff $f^*(s)$ is an optimal action of (2.10).

Definition 2.7 A strategy b_s is called pure or deterministic if $b_s(s,a) \in \{0,1\}\ \forall s \in S, a \in A(s)$.

3 Main Results in Pareto-Optimal Strategy Spaces

Theorem 3.1 ([1]) *In an SMDP with single reward function r, there exist pure stationary strategies that are optimal for discounted reward objectives.*

Corollary 1 *There exist pure stationary Pareto-optimal strategies for discounted SMDPs, with vector pay-offs.*

Proof Suppose $r_1, r_2, \ldots, r_l$ are the given reward functions of the discounted SMDP with reward vector $\bar{r}$ and discount factor d. Now we construct a reward function r_{sum} as follows:

$$r_{sum} = r_1 + r_2 + \cdots + r_l.$$

Let f^* be a pure stationary strategy, which is optimal for the reward function r_{sum}. Our objective is to show that f^* is Pareto optimal, which proves the corollary. Let us assume by contradiction that f^* is not Pareto optimal. Then there exists a strategy b_s such that $V_d^{f^*}(\bar{r}, s) \leq V_d^{b_s}(\bar{r}, s)$ and for some $1 \leq j \leq l$, $V_d^{f^*}(r_j, s) < V_d^{b_s}(r_j, s)$, where $\bar{r} = (r_1, r_2, \ldots, r_l)$. Thus, we have

$$\begin{aligned} V_d^{f^*}(r_{sum}, s) &= \sum_{j=1}^{l} V_d^{f^*}(r_j, s) \\ &< \sum_{j=1}^{l} V_d^{b_s}(r_j, s) \\ &= V_d^{b_s}(r_{sum}, s). \end{aligned} \tag{3.1}$$

This leads to the contradiction that f^* is optimal for r_{sum}. Thus, f^* is a pure stationary Pareto-optimal strategy.

Note that pure strategies are not sufficient strategies for Pareto optimality. This holds for pure stationary strategies also. The following example illustrates this fact.

Example 3.1 Consider an SMDP with reward vector $\bar{r} = (r_1, r_2)$, discount factor $d \in [0, 1)$ with state space $S = \{s_1, s_2\}$. $A(s_1) = \{a_1^1, a_1^2\}$, $A(s_2) = \{a_2^1\}$ are the action sets chosen on state s_1 and s_2, respectively. The reward vectors, sojourn times, and transition probabilities are given as

	state s_1
a_1^1	$(1, 0)$ $(1\{2(1)\}, 0)$
a_1^2	$(0, 1)$ $(0, 1\{3(1)\})$

	state s_2
a_2^1	$(0, 0)$ $(0, 1\{1(1/2), 2(1/2)\})$

where the rectangle

(r_1, r_2)
$(p_1\{c_1(t_1), c_2(t_2)\}, p_2\{g_1(t_1^{'}), g_2(t_2^{'})\})$

represents that the reward vector is $\bar{r} = (r_1, r_2)$, p_1, p_2 are transition probabilities which denote that next states are s_1, s_2, respectively. $c_1, c_2; g_1, g_2$ are the discrete transition times and $t_1, t_2; t_1^{'}, t_2^{'}$ are the transition time probability distributions for the states s_1, s_2, respectively, if this cell is chosen at present.

Let us fix the initial state s_1 once and for all and consider the stationary strategy $f = \{(\frac{1}{2}, \frac{1}{2}), (1)\}$. This means the decision-maker chooses action a_1^1 and a_1^2 each with probability $\frac{1}{2}$ in the state s_1. In state s_2, the decision-maker chooses action a_2^1 with probability 1.

Now from (2.8) we get that $P_d(f) = \begin{bmatrix} \frac{d^2}{2} & \frac{d^3}{2} \\ 0 & \frac{d(1+d)}{2} \end{bmatrix}$. Here $\bar{r}(f) = [\bar{r}(s_1, f), \bar{r}(s_2, f)]$ where $\bar{r}(s_i, f) = [r_1(s_i, f), r_2(s_i, f)]$, $(i = 1, 2)$. The entries of $\bar{r}(s_1, f)$ can be calculated as $r_1(s_1, f) = \frac{1}{2}.1 + \frac{1}{2}.0$ and $r_2(s_1, f) = \frac{1}{2}.0 + \frac{1}{2}.1$. Similarly, we can calculate $\bar{r}(s_2, f)$. Thus, from (2.9), we get the value profile of the discounted SMDP $V_d^f(s_1, \bar{r}) = ((1 - \frac{d(1+d)}{2})(\frac{1}{2}, \frac{1}{2}))$. Now clearly f is a Pareto-optimal strategy and we cannot get a better value profile than $V_d^f(s_1, \bar{r})$ by any other pure stationary strategy. This example shows that pure stationary strategies are not sufficient for Pareto optimality.

Now we introduce some special kinds of reward functions in context of the above SMDP model to clarify the counter of state–action pair later defined in this section. We define reward function r_t by

$$r_t(s) = 1 \; if \; s = t; \; where \; s, t \in S$$
$$= 0 \; otherwise$$

and

$$r_{t,b}(s, a) = 1 \; if \; s = t \; and \; b = a; \; where \; a, b \in A(s); \; s, t \in S$$
$$= 0 \; otherwise.$$

Let us consider $1_{(X_m = s^{'}, A_m = a^{'})}$ to be the random variable, which denotes that the reward is 1 when the system is in the state $s^{'}$ and action $a^{'}$ is chosen on the m th decision epoch (i.e., when the system is on state $s^{'}$) and 0 otherwise. For a strategy $b_s \in B_s$, a fixed initial state $s \in S$ once and for all, and a specified discount factor d, we define the counter of state–action pair $(s^{'}, a^{'})$ (where $s^{'} \in S, a^{'} \in A(s^{'})$) as

$$Count_d^{s,b_s}(s^{'}, a^{'}) = V_d^{b_s}(r_{s^{'},a^{'}}, s)$$
$$= \sum_{m=1}^{\infty} d^{(\tau_1+\tau_2+\cdots+\tau_m)} E_{b_s}\left[1_{(X_m = s^{'}, A_m = a^{'})} \mid X_1 = s\right] \quad (3.2)$$

and for state s', we define the counter of the state s' as

$$Count_d^{s,b_s}(s') = V_d^{b_s}(r_{s'}, s)$$
$$= \sum_{m=1}^{\infty} d^{(\tau_1+\tau_2+\cdots+\tau_m)} E_{b_s}\left[1_{(X_m=s')} \mid X_1 = s\right]. \tag{3.3}$$

Obviously, $\sum_{a' \in A(s')} Count_d^{s,b_s}(s', a') = Count_d^{s,b_s}(s') \; \forall s, s' \in S$ and $a' \in A(s')$.

Lemma 3.1 *Suppose a stationary strategy $f \in \Im$ is given, the initial state s and discount factor $d \in [0, 1)$ is fixed in the above-discussed SMDP model. Consider the vector $\bar{x} = (x_1, x_2, \ldots, x_N)$ of N variables, where $S = \{1, 2, \ldots, N\}$ is the finite state space, then the set of N equations*

$$x_{s'} = r_s(s') + \sum_{s' \in S} p(s' \mid s, f) x_{s'} \sum_{y=0}^{T} d^y P(\tau_{ss'}^a = y) \tag{3.4}$$

has a unique solution $Count_d^{s,f}(s')$, $\forall s, s' \in S$, where $p(s' \mid s, f) = \sum_{a \in A(s)} p(s' \mid s, a).f(s, a)$.

Proof We claim that, $Count_d^{s,f}(s') = r_s(s') + \sum_{s' \in S} p(s' \mid s, f) Count_d^{s,f}(s') \sum_{y=0}^{T} d^y P(\tau_{ss'}^a = y)$. The proof of the above claim proves the theorem. Now, we have

$$Count_d^{s,f}(s') = \sum_{m=1}^{\infty} d^{(\tau_1+\tau_2+\cdots+\tau_m)} E_f[1_{(X_m=s')} \mid X_1 = s]$$
$$= r_s(s') + E\left[E_f\left\{\sum_{m=2}^{\infty} d^{(\tau_2+\cdots+\tau_m)} 1_{(X_m=s')} \mid hist_2, \tau_2\right\}\right]$$
$$= r_s(s') + E[d^{\tau_2} E_f\{\sum_{m=3}^{\infty} d^{(\tau_3+\cdots+\tau_m)} 1_{(X_m=s')} \mid hist_2, \tau_2\}]$$
$$= r_s(s') + E[d^{\tau_2} Count_d^{s,f}(X_2)]$$
$$= r_s(s') + \sum_{s' \in S} p(s' \mid s, f).Count_d^{s,f}(s') \sum_{y=0}^{T} d^y P(\tau_{ss'}^a = y). \tag{3.5}$$

So, our claim is proved and the uniqueness follows from Theorem 2.1.

For a strategy $b_s \in B_s$ and a fixed initial state $s \in S$ in the above model of SMDP, we can define a stationary strategy $f_{b_s,s}$ as follows:

$$f_{b_s,s} = \frac{Count_d^{s,b_s}(s', a')}{Count_d^{s,b_s}(s')} \tag{3.6}$$

$\forall s' \in S$ and $a' \in A(s')$. As we know $\sum_{a'} Count_d^{s,b_s}(s', a') = Count_d^{s,b_s}(s')$ and $Count_d^{s,b_s}(s', a') \geq 0$, it follows that $f_{b_s,s}$ is a probability distribution and it is surely stationary by construction.

Lemma 3.2 *For all strategies $b_s \in B_s$ and $s' \in S$, with a fixed initial state $s \in S$ in the above-discussed SMDP model, we have $Count_d^{s,b_s}(s') = Count_d^{s,f_{b_s,s}}(s')$.*

Proof We claim that, $Count_d^{s,b_s}(s') = r_s(s') + \sum_{s'' \in S} p(s'' \mid s', f_{b_s,s}).Count_d^{s,b_s}(s')$ $\sum_{y=0}^{T} d^y P(\tau^{a'}_{s's''} = y)$, then the proof follows from Lemma 3.1.

$$\begin{aligned}
Count_d^{s,b_s}(s') &= \sum_{m=1}^{\infty} d^{(\tau_1+\tau_2+\cdots+\tau_m)} E_{b_s}[1_{(X_m=s')} \mid X_1 = s] \\
&= r_s(s') + \sum_{s' \in S} \sum_{a \in A(s)} Count_d^{s,b_s}(s', a').p(s' \mid s, a) \sum_{y=0}^{T} d^y P(\tau^a_{ss'} = y) \\
&= r_s(s') + \sum_{s' \in S} \sum_{a \in A(s)} Count_d^{s,b_s}(s') \frac{Count_d^{s,b_s}(s', a')}{Count_d^{s,b_s}(s')} .p(s' \mid s, a) \sum_{y=0}^{T} d^y P(\tau^a_{ss'} = y) \\
&= r_s(s') + \sum_{s' \in S} p(s' \mid s, f_{b_s,s}).Count_d^{s,b_s}(s') \sum_{y=0}^{T} d^y P(\tau^a_{ss'} = y).
\end{aligned} \tag{3.7}$$

This proves our claim.

Lemma 3.3 *For all strategies $b_s \in B_s$, $s' \in S$ and actions $a' \in A(s')$, with a fixed initial state $s \in S$, in the above-discussed SMDP model, we have $Count_d^{s,b_s}(s', a') = Count_d^{s,f_{b_s,s}}(s', a')$.*

Proof We prove this using (3.6) as follows:

$$\begin{aligned}
Count_d^{s,f_{b_s,s}}(s', a') &= Count_d^{s,f_{b_s,s}}(s').f_{b_s,s}(s', a') \\
&= Count_d^{s,b_s}(s').\frac{Count_d^{s,b_s}(s', a')}{Count_d^{s,b_s}(s')} \\
&= Count_d^{s,b_s}(s', a').
\end{aligned} \tag{3.8}$$

Theorem 3.2 *For all single reward functions r, strategies $b_s \in B_s$ and a fixed initial state $s \in S$ in the above-discussed SMDP model, we have $V_d^{b_s}(r, s) = V_d^{f_{b_s,s}}(r, s)$.*

Proof We have

$$\begin{aligned}
V_d^{b_s}(r, s) &= \sum_{m=1}^{\infty} d^{(\tau_1+\tau_2+\cdots+\tau_m)} E_{b_s}[r(X_m, A_m) \mid X_1 = s] \\
&= \sum_{m=1}^{\infty} d^{(\tau_1+\tau_2+\cdots+\tau_m)} E_{b_s}[\sum_{s' \in S} \sum_{a' \in A(s')} r(s', a').1_{(X_m=s', A_m=a')} \mid X_1 = s]
\end{aligned}$$

$$= \sum_{s' \in S} \sum_{a' \in A(s')} \sum_{m=1}^{\infty} d^{(\tau_1+\tau_2+\cdots+\tau_m)} E_{b_s}[1_{(X_m=s',A_m=a')} \mid X_1 = s].r(s', a')$$

$$= \sum_{s' \in S} \sum_{a' \in A(s')} Count_d^{s,b_s}(s', a').r(s', a')$$

$$= \sum_{s' \in S} \sum_{a' \in A(s')} Count_d^{s,f_{b_s,s}}(s', a').r(s', a'). \tag{3.9}$$

Similarly, $V_d^{f_{b_s,s}}(r, s) = \sum_{s' \in S} \sum_{a' \in A(s')} Count_d^{s,f_{b_s,s}}(s', a').r(s', a')$. Thus, $V_d^{b_s}(r, s) = V_d^{f_{b_s,s}}(r, s)$.

Theorem 3.3 *Suppose $S_m =< S, A, p, P, \bar{r} >$ is a discounted SMDP with discount factor d, fixed initial state $s \in S$, and reward vector $\bar{r} = (r_1, r_2, \ldots, r_l)$. Then for all strategies $b_s \in B_s$ and the stationary strategy $f_{b_s,s} \in \Im$, we have $V_d^{b_s}(\bar{r}, s) = V_d^{f_{b_s,s}}(\bar{r}, s)$.*

Proof As Theorem 3.2 is true for all single reward functions r, we can extend the result to the reward vector $\bar{r} = (r_1, r_2, \ldots, r_l)$, which claims our proof.

Consequently, we conclude that stationary strategies are sufficient strategies for Pareto optimality for a discounted SMDP with vector pay-offs.

4 Pareto Curves and Multi-objective Linear Programming Problem

This section deals with some basic definitions about Pareto curve and its approximate version, i.e., ϵ-approximate Pareto curves. We also discuss about the standard form of a multi-objective linear programming problem as well as the Pareto curve of a multi-objective linear programming problem. These concepts will be used subsequently to prove some results in the later section.

Definition 4.8 ([8]) Let $S_m =< S, A, p, P, \bar{r} >$ (where $\bar{r} = (r_1, r_2, \ldots, r_l)$ is the vector reward function) be a discounted SMDP with discount factor d $(0 \leq d < 1)$ with vector reward. Fix the initial state s once and for all in the SMDP model S_m. The Pareto curve $Par^d(S_m, s, \bar{r})$ of the SMDP $S_m =< S, A, p, P, \bar{r} >$ can be defined as the set of all of value profiles with l co-ordinates such that for each $p \in Par^d(S_m, s, \bar{r})$ there is a Pareto-optimal strategy b_{s^*} such that $V_d^{b_{s^*}}(\bar{r}, s) = p$.

We mainly consider Pareto curve $Par^d(S_m, s, \bar{r})$ as a set of strategies that obtains the Pareto-optimal points.

Definition 4.9 Considering the SMDP $S_m =< S, A, p, P, \bar{r} >$ we define the ϵ-approximate Pareto curve $Par^{d,\epsilon}(S_m, s, \bar{r})$ as a set of strategies, such that there is no other strategy $b_{s'} \in B_s$ such that

$$V_d^{b_{s'}}(r_i, s) \geq (1+\epsilon)V_d^{b_{s^\epsilon}}(r_i, s);\ \forall\, b_{s^\epsilon} \in Par^{d,\epsilon}(S_m, s, \bar{r}), i = 1, 2, \ldots, l.$$

The existence of an algorithm to construct an ϵ-approximate Pareto-optimal curve in polynomial time bound is well known [11].

Definition 4.10 ([12]) Consider following linear optimization problem:

$$\begin{aligned}
&\text{Maximize } z_1(\bar{x}) = \bar{c_1}\bar{x}\\
&\text{Maximize } z_2(\bar{x}) = \bar{c_2}\bar{x}\\
&\quad\vdots\\
&\text{Maximize } z_l(\bar{x}) = \bar{c_l}\bar{x}
\end{aligned}$$

subject to $A.\bar{x} \geq \bar{b}, \bar{x} \geq 0$. A is a matrix of order $m \times l$ and $\bar{b}$ is a vector of order $m \times 1$. $\bar{x} \geq 0$ is a vector of l variables. Here $\bar{c_i}$ $(i = 1, 2, \ldots, l)$s are vectors of coefficients.

The set of equations above is the basic set up of multi-objective linear programming problem. We often use the abbreviation MOLP instead of multi-objective linear programming problem to avoid complexity. In the above problem, l objective functions are to be optimized (here, maximized). In a multi-objective optimization, unlike a single-objective optimization, there is no typical optimal solution, i.e., which is optimal for all objective functions. Here the notion of efficient solution is used. We can define them as follows.

Definition 4.11 ([12]) A feasible point $\bar{x}^* \in \mathbb{R}^l$ is known as efficient solution of the multi-objective linear programming problem defined above, if $\nexists$ any other feasible solution $\bar{x} \in \mathbb{R}^l$ such that $z_i(\bar{x}) \geq z_i(\bar{x}^*)$ for $i = 1, 2, \ldots, l$ and $z_i(\bar{x}) \neq z_i(\bar{x}^*)$ for at least one i.

Definition 4.12 ([13]) Let Q be any non-empty set in $\mathbb{R}^l$. A point $y \in Q$ is said to be a Pareto-maximal point of the set Q if $\nexists$ any point $y' \in Q$ such that $y' \geq y$ and $y' \neq y$. The set of Pareto-maximal points of Q is traditionally called the efficient or admissible set of solutions of Q.

Definition 4.13 ([13]) A point $\bar{x}^* \in \mathbb{R}^l$ is known as Pareto-optimal solution of the multi-objective linear programming problem defined above, if $\nexists$ any other feasible solution $\bar{x}' \in \mathbb{R}^l$ such that both $(z_1(\bar{x}^*), \ldots, z_k(\bar{x}^*)) \leq (z_1(\bar{x}'), \ldots, z_k(\bar{x}'))$ and $(z_1(\bar{x}^*), \ldots, z_k(\bar{x}^*)) \neq (z_1(\bar{x}'), \ldots, z_k(\bar{x}'))$ hold.

Pareto-maximal points defined above are directly related to efficient solutions. Given a multi-objective linear programming problem L we discussed earlier, the Pareto curve for L is denoted by $Par(L)$ is the set of vector V of values with l co-ordinates such that $\exists$ a Pareto-optimal solution $\bar{x}^* \in \mathbb{R}^l$ of L such that $V = ((z_1(\bar{x}^*), \ldots, z_l(\bar{x}^*))$. ϵ-approximate Pareto curves for MOLP can be defined similarly to the definition of ϵ-approximate Pareto curves for discounted SMDP with vector pay-offs and it will be denoted by $Par^\epsilon(L)$.

Definition 4.14 Consider the linear programming problem

$$\max \bar{c}^T \bar{x}$$

with respect to the constraints

$$A\bar{x} = \bar{b}$$
$$\bar{x} \geq 0.$$

Assume $A, \bar{b}, \bar{c}$ all have integer entries. Encoding an integer n in binary mode takes $log_2(n+1)+1$ bites which we denote as $size(n)$. Encoding a vector $\bar{v}$ in binary mode takes $size(\bar{v})$ bits. Similar approaches are applicable for a matrix A and vector $\bar{b}$ and $\bar{c}$. So we can say that the size of linear programming problem L is $size(A) + size(\bar{b}) + size(\bar{c})$. This is the input size of the linear programming problem.

5 Multi-objective Linear Programming Formulation of a Discounted SMDP with Vector Pay-Offs

Given an SMDP $< S, A, p, P, r >$ with a single reward function r and a discount factor $0 \leq d < 1$, the discounted SMDP model can be converted into a linear programming problem where an optimal solution can also be achieved [14, 15]. For a discounted SMDP with vector pay-offs, we can extend the standard linear programming problem as follows.

Let $S_m =< S, A, p, P, \bar{r} >$ be an SMDP with discount factor $0 \leq d < 1$, reward vector $\bar{r} = (r_1, r_2, \ldots, r_l)$, and fixed initial state $s \in S$, the multi-objective linear programming problem has the set $\{x(s', a') \mid s' \in S, a' \in A(s')\}$ of decision variables. We can consider the set of variables as the counter of state–action pair (s', a') defined earlier. The l number of objective functions of the MOLP over the variables $x(s', a')$ is given by

$\max \sum_{s' \in S} \sum_{a' \in A(s')} r_i(s', a').x(s', a')$ for $i = 1, 2, \ldots, l$,

with respect to the constraints

$$(i) \sum_{a' \in A(s')} x(s', a') = r_s(s') + \sum_{a'' \in A(s'')} \{\sum_{s''' \in S} p(s''' \mid s'', a'').x(s'', a'')$$

$$\sum_{y=0}^{T} d^y P(\tau^{a''}_{s'' s'''} = y)\} f(s'', a'')$$

$(ii)\ x(s', a') \geq 0$ for all $s' \in S,\ a' \in A(s')$.

For a stationary strategy f, $f(s'', a'')$ denotes the probability of choosing action a'' in state s''. The above constraints are satisfied by "counter" of state–action pair (s', a') discussed in Sect. 3. It can be easily proved that the above MOLP has finite solutions. For any solution $x(s', a')(s' \in S, a' \in A(s'))$ of the above multi-objective linear programming problem, let us denote $x(s') = \sum_{a' \in A(s')} x(s', a')$. Then as discussed earlier, it derives a stationary strategy which chooses action a' at state s' with probability $\frac{x(s', a')}{x(s')}$. The ith objective function of the multi-objective linear programming problem maximizes the discounted pay-off for ith reward function r_i over the set of stationary strategies. The optimal solution of the ith objective function also gives us a stationary optimal strategy for reward function r_i, where $i = 1, 2, \ldots, l$. The following theorem proves the existence of an ϵ-approximate Pareto curve of a given multi-objective linear programming problem L, which is bounded in size of the MOLP L.

Theorem 5.1 ([8]) *Given a multi-objective linear programming problem L with l number of objectives and for all $\epsilon \in \mathbb{R}^+$, (where $\mathbb{R}^+$ denotes the set of all positive real numbers) $\exists$ an ϵ-approximate Pareto curve $Par^{\epsilon}(L)$, whose size is polynomial in $\mid L \mid$ and $\frac{1}{\epsilon}$, and exponential in l, i.e., the number of objective functions in the multi-objective linear programming problem.*

Let us denote by $MOLP^d(S_m, s, \bar{r})$ to be the multi-objective linear programming problem for a discounted SMDP S_m, with discount factor d, fixed initial state s and having reward vector $\bar{r} = (r_1, r_2, \ldots, r_l)$. Denote $Par(MOLP^d(S_m, s, \bar{r}))$ to be the Pareto curve for the multi-objective linear programming problem.

Lemma 5.1 *For an SMDP $S_m =< S, A, p, P, \bar{r} >$ with reward vector $\bar{r} = (r_1, r_2, \ldots, r_l)$ and a discount factor $0 \le d < 1$, the results of the corresponding MOLP are true*

(i) $Par(MOLP^d(S_m, s, \bar{r})) = Par^d(S_m, s, \bar{r})$.

(ii) For all $\epsilon \in \mathbb{R}^+$ and all ϵ-approximate Pareto curves $Par^{\epsilon}(MOLP^d(S_m, s, \bar{r}))$ of the multi-objective linear programming problem $MOLP^d(S_m, s, \bar{r})$, $\exists$ an ϵ-approximate Pareto curve
$Par^{d,\epsilon}(S_m, s, \bar{r})$ such that $Par^{\epsilon}(MOLP^d(S_m, s, \bar{r})) = Par^{d,\epsilon}(S_m, s, \bar{r})$.

Proof From previous discussions, we may denote $Par(MOLP^d(S_m, s, \bar{r}))$ as the set of stationary strategies which are derived from Pareto-optimal solutions of the $MOLP^d(S_m, s, \bar{r})$. So, the set of Pareto-optimal points of the SMDP model S_m can be characterized by $Par(MOLP^d(S_m, s, \bar{r}))$, as $MOLP^d(S_m, s, \bar{r})$ is the the multi-objective linear programming problem corresponding to the SMDP S_m. Now following Theorem 3.3 we know that stationary strategies are sufficient strategies for Pareto optimality, which claims the proof.

Theorem 5.2 *(i) Let us consider an SMDP $S_m =< S, A, p, P, \bar{r} >$ with reward vector $\bar{r} = (r_1, r_2, \ldots, r_l)$ and a discount factor $0 \le d < 1$. For all $\epsilon \in \mathbb{R}^+$, $\exists$ an ϵ-approximate Pareto curve $Par^{d,\epsilon}(S_m, s, \bar{r})$ whose size is polynomial in $\mid S_m \mid$, $\frac{1}{\epsilon}$, and exponential in l.*

(ii) There exists an algorithm to construct the above ϵ-approximate Pareto curve $Par^{d,\epsilon}(S_m, s, \bar{r})$ polynomially in $|S_m|$, $\frac{1}{\epsilon}$, and exponentially in l.

Proof Proof follows from Lemma 5.1 and Theorem 5.1.

6 Pareto-Realizability Problem for Discounted SMDP with Vector Pay-Offs

Pareto realizability problem for a discounted SMDP with vector pay-off is defined as follows. Given an SMDP $S_m =< S, A, p, P, \bar{r} >$ with reward vector $\bar{r} = (r_1, r_2, \ldots, r_l)$ and discount factor $0 \le d < 1$ and a fixed initial state $s \in S$, if $\exists$ a strategy $b_s \in B_s$, such that $V_d^{b_s}(\bar{r}, s) \ge \bar{z}$, where $\bar{z} \in \mathbb{R}^l$ is given by $\bar{z} = (z_1, z_2, \ldots, z_l)$.

Theorem 6.1 *A value profile $\bar{z} = (z_1, z_2, \ldots, z_l)$ is Pareto realizable iff $\exists$ a Pareto-optimal strategy b_{s^*} such that $V_d^{b_{s^*}}(\bar{r}, s) \ge \bar{z}$, where $\bar{r} = (r_1, r_2, \ldots, r_l)$ is the reward vector.*

Proof Firstly, consider $\bar{z} = (z_1, z_2, \ldots, z_l)$ to be a Pareto-realizable value profile. Then by the definition, $\exists$ a strategy $b_s \in B_s$ such that $V_d^{b_s}(\bar{r}, s) \ge \bar{z}$. Now for a Pareto-optimal strategy b_{s^*} we have $V_d^{b_{s^*}}(\bar{r}, s) \ge V_d^{b_s}(\bar{r}, s) \ge \bar{z}$. Conversely, let b_{s^*} be a Pareto-optimal strategy. Let $V_d^{b_{s^*}}(\bar{r}, s) < \bar{z}$ and $\exists\ b_s \in B_s$ such that $\bar{z} = V_d^{b_s}(\bar{r}, s)$. Then, from the above discussion we arrive at a contradiction that b_{s^*} is Pareto optimal. So, our previous assumption was false, and b_{s^*} serves as the strategy which makes $\bar{z}$ to be Pareto realizable.

Corollary 2 *$\bar{z} \in \mathbb{R}^l$ is Pareto realizable iff it is realizable by a stationary strategy.*

Proof Suppose that $\bar{z}$ is Pareto realizable. Now by Theorem 6.1, $\exists$ a Pareto-optimal strategy b_{s^*} such that $V_d^{b_{s^*}}(\bar{r}, s) \ge \bar{z}$. By contradiction, let $\bar{z}$ not to be realizable by a stationary strategy. This implies $V_d^f(\bar{r}, s) < \bar{z}$ for a stationary strategy f. Thus, we have the inequality that $V_d^{b_{s^*}}(\bar{r}, s) > V_d^f(\bar{r}, s)$. Now, as stationary strategies suffice for Pareto optimality, we arrive at a contradiction from the above inequality. So, our assumption was wrong. The converse part is obvious, as the stationary strategy is our desired strategy by which $\bar{z}$ is realizable.

The pure stationary Pareto-realizability problem needs the above strategy to be both pure and stationary.

Theorem 6.2 *Given a discounted SMDP $S_m =< S, A, p, P, \bar{r} >$ with discount factor $0 \le d < 1$ and reward vector $\bar{r} = (r_1, r_2, \ldots, r_l)$, the Pareto-realizability problem can be solved in polynomial time and the pure stationary Pareto-realizability problem of such SMDP model is NP-complete.*

Proof We reduce the Pareto-realizability problem to a multi-objective linear programming problem by introducing constraints to the existing multi-objective linear programming problem we described earlier. The reduced MOLP looks like the following:

$\max \sum_{s' \in S} \sum_{a' \in A(s')} r_i(s', a').x(s', a')$ for $i = 1, 2, \ldots, l$,

with respect to the new set of constraints

$$(i) \sum_{a' \in A(s')} x(s', a') = r_s(s') + \sum_{a'' \in A(s'')} \{ \sum_{s''' \in S} p(s''' \mid s'', a).x(s'', a'') \sum_{y=0}^{T} d^y P(\tau^{a''}_{s'' s'''} = y)\} f(s'', a'')$$

$$(ii)\ x(s', a') \geq 0 \text{ for all } s' \in S,\ a' \in A(s').$$

$$(iii)\ z_i \leq \sum_{s' \in S} \sum_{a' \in A(s')} r_i(s', a').x(s', a') \text{ for } i = 1, 2, \ldots, l. \quad (6.1)$$

The last constraint suggests that for each reward function r_i $(i = 1, 2, \ldots, l)$, the pay-off is greater than or equal to the corresponding co-ordinates of the value profile $\bar{z} = (z_1, z_2, \ldots, z_l)$. Now, if the above inequality becomes true, then the set of strategies that realizes the value profile is consistent, otherwise the set is inconsistent. We can check consistency of the set in polynomial time [16], which proves the first part. From the theory of Pareto-optimal strategy and Theorem 5.2, we conclude that the pure stationary Pareto-realizability problem is in NP class. For the above Pareto-realizability problem, we introduce the subset-sum problem as follows. Let $\{m_1, m_2, \ldots, m_n\}$ be a set of natural numbers, where $n \in \mathbb{N}$. The subset-sum problem asks if $\exists$ numbers $y_1, y_2, \ldots, y_n \in \{0, 1\}$ such that $m_1 y_1 + m_2 y_2 + \cdots + m_n y_n = q$, where $q \in \mathbb{N}$ is given. The above problem is NP-complete [17]. We consider the following example to illustrate the concept.

Example 6.2 Consider the discounted SMDP model having state space $S = \{s_1, s_2\}$, where s_1 is the fixed initial state. $A(s_1) = \{a_1^1, a_1^2\}$, $A(s_2) = \{a_2^1, a_2^2\}$ are the action sets chosen on state s_1 and s_2, respectively. The transition probability and the sojourn time (mean) are given below:

state s_1	
a_1^1	$(m_1, 0)$ $(0, 1\{2(1)\})$
a_1^2	$(0, m_2)$ $(0, 1\{3(1)\})$

state s_2	
a_2^1	$(m_1, 0)$ $(0, 1\{1(1)\})$
a_2^2	$(0, m_2)$ $(0, 1\{2(1)\})$

where the rectangle

(r_1, r_2)
$(p_1\{c_1(t_1), c_2(t_2)\}, p_2\{g_1(t_1^{'}), g_2(t_2^{'})\})$

represents that the reward vector is $\bar{r} = (r_1, r_2)$, p_1, p_2 are transition probabilities denoting that next states are s_1, s_2, respectively. $c_1, c_2; g_1, g_2$ are the discrete transition times and $t_1, t_2; t_1^{'}, t_2^{'}$ are the transition time probability distributions for the states s_1 and s_2, respectively, if this cell is chosen at present. Let us fix discount factor $d = \frac{1}{2}$. Suppose that the decision-maker uses the pure stationary strategy $f = \{(1, 0), (0, 1)\}$. Then by the same method as we calculated in Example 3.1, we get the value profile $[m_1(1 - d(1 + d)) + m_2(d^2(1 + d)), 0] = V_d^f(\bar{r}, s_1)$. We enquire if $\bar{z} = (q, (y_1 m_1 + y_2 m_2) - q)$ is Pareto realizable or not. The answer is yes, if we implement the rule that if action a_i^1 is chosen in state s_i, then each $y_i = 1$, otherwise $y_i = 0$ $\forall i = 1, 2$. Thus, $\bar{z} = (q, (y_1 m_1 + y_2 m_2) - q)$ becomes $(m_1, 0)$. Now suitably scaling m_1 and m_2, we get that $\bar{z}$ is realizable by the pure stationary strategy f.

Thus, we conclude that if the subset-sum problem is satisfied, then $\bar{z}$ is pure Pareto-realizable value profile, which gives us the NP-completeness of the problem.

7 Conclusion

In this article, we deal with the existence of an algorithm to construct an ϵ-approximate Pareto curve in polynomial time. Given an SMDP $S_m =< S, A, p, P, \bar{r} >$ with vector reward $\bar{r} = (r_1, r_2, \ldots, r_l)$, discount factor $d \in [0, 1)$, and a pay-off profile $\bar{v} = (v_1, v_2, \ldots, v_l)$, we can check whether $\bar{v}$ is ϵ-close to a Pareto-optimal point at state $s \in S$ by constructing $Par^{d,\epsilon}(S_m, s, \bar{r})$ in polynomial time. Also, from the construction of the stationary strategy f in Example 3.1, we can conclude that set of Pareto-optimal points may be uncountable, so this approximation is quite useful. Finally, Pareto-realizability problem investigates about the existence of a strategy, corresponding to whose value profile is greater than or equal to a given l-tuple $\bar{z} \in \mathbb{R}^l$. Pareto-realizability problem finds applications in various optimization problems, such as circuit implementations [18]. Although for the undiscounted SMDP model, the existence of pure stationary optimal strategy is still not proved. So, pure stationary Pareto-optimal strategy might not exist in undiscounted SMDP with vector rewards. Further investigations can be done for constructing an ϵ-approximate Pareto curve in the undiscounted case, which is an open problem yet.

References

1. Jewell WS (1963) Markov-renewal programming. I: formulation, finite return models. Oper Res
2. Howard RA (1963) Semi-Markovian decision-processes. Bull Int Stat Inst
3. Ross SM (2013) Applied probability models with optimization applications. Courier Corporation
4. Lippman SA, et al (1971) Maximal average-reward policies for semi-Markov decision processes with arbitrary state and action space. Ann Math Stat
5. Federgruen A, Hordijk A, Tijms HC (1978) A note on simultaneous recurrence conditions on a set of denumerable stochastic matrices. J Appl Probab
6. Yang P, Catthoor F (2003) Pareto-optimization-based run-time task scheduling for embedded systems. In: Proceedings of the 1st IEEE/ACM/IFIP international conference on Hardware/software codesign and system synthesis
7. Owen G (1995) Game theory academic press. San Diego
8. Chatterjee K, Majumdar R, Henzinger TA (2006) Markov decision processes with multiple objectives. In: Annual symposium on theoretical aspects of computer science
9. Mondal P (2016) On undiscounted semi-Markov decision processes with absorbing states. Math Methods Oper Res
10. Sinha S, Mondal P (2017) Semi-Markov decision processes with limiting ratio average rewards. J Math Anal Appl
11. Papadimitriou CH, Yannakakis M (2000) On the approximability of trade-offs and optimal access of web sources. In: Proceedings 41st annual symposium on foundations of computer science
12. Belenson SM, Kapur KC (2017) An algorithm for solving multicriterion linear programming problems with examples. J Math Anal Appl
13. Luc DT, Luc DT (2016) Pareto optimality. Multiobjective linear programming: an introduction
14. Puterman ML (2014) Markov decision processes: discrete stochastic dynamic programming. Wiley
15. Wessels J, Nunen JAEE van (1975) Discounted semi-Markov decision processes: linear programming and policy iteration. Wiley Online Library
16. Goldstein AA, Cheney W, et al (1958) A finite algorithm for the solution of consistent linear equations and inequalities and for the Tchebycheff approximation of inconsistent linear equations. Pacific J Math
17. Garey MR, Johnson DS (1979) Computers and intractability. Freeman San Francisco
18. Brayton RK, Hachtel GD, Sangiovanni-Vincentelli AL (1981) A survey of optimization techniques for integrated-circuit design. Proc IEEE

Sonar Sensing-Based Human Leg Localization Using Gaussian Process Regression

Pritam Paral, Amitava Chatterjee, and Anjan Rakshit

Abstract Human detection and following problems are considered as one of the crucial problems in a human-robot coexisting environment. The present work focuses on a significant subproblem within this genre, where the measurements obtained with the onboard range sensing system of a mobile robot are used for localizing the positions of the leg pair of a walking person at a time instance, during human tracking. This paper demonstrates how such a human leg localization problem can be formulated as a nonlinear regression problem where two distinct prediction models are learned by using two different training datasets, and how the learned models can be used to localize the positions of the human legs. Although sonar sensors produce readings fraught with uncertainty and imprecision, they are still considered as an attractive range sensing solution for robotic applications, especially in developing countries, where cost, size, and power consumption associated with the sensors are the major constraints. Keeping the sonar behavior in mind, in this paper, we propose to employ *Gaussian Process Regression* (GPR), a popular Bayesian nonparametric method, to solve our formulated nonlinear regression problem. Here, we have aimed to suitably adapt the basic GPR framework as per our formulated problem to perform a statistical model calibration and then localize the human leg positions based on the calibrated model. Performance evaluations carried out for localizing the positions of human legs based on real-life data show the superiority of the proposed GPR-based approach over the conventional position estimation method.

Keywords Human leg localization · Sonar sensing · Gaussian process regression (GPR)

Supported by the "VISVESVARAYA PHD SCHEME FOR ELECTRONICS & IT", Ministry of Electronics & IT, Government of India.

P. Paral (✉) · A. Chatterjee · A. Rakshit
Department of Electrical Engineering, Jadavpur University, Kolkata 700032, India

© The Author(s), under exclusive license to Springer Nature Singapore Pte Ltd. 2022

D. Giri et al. (eds.), *Proceedings of the Seventh International Conference on Mathematics and Computing*, Advances in Intelligent Systems and Computing 1412,
https://doi.org/10.1007/978-981-16-6890-6_77

1 Introduction

In recent years, detecting people in motion and subsequently tracking them has emerged as one of the major problems in human-robot coexisting environments [1–7]. Such people detection and tracking schemes primarily involve estimating the kinematics of a moving human being in a surveillance area based on a set of measurements acquired by a single or multiple sensor(s). The present work proposes a new approach for addressing a significant subproblem within this genre, where the readings gathered by the onboard range sensing system of a mobile robot is employed for localizing the positions of the leg pair of a walking person at a specific time instance [1–3], in the course of human tracking.

Such a human leg localization problem can be formulated as a nonlinear regression problem where two separate prediction models are learned by using two different training datasets [8], both of which contain same input data, namely, various angular orientations with respect to a certain robot pose along which the human being is positioned and the range readings acquired at those angles by the onboard range sensor, whereas output data of one are the corresponding x coordinates and the other are the y-coordinates of the leg postures, measured with respect to the *robot pose frame* [9] located at that certain robot pose. Then the learned models can be used to localize the (x, y) coordinates of a given query leg posture that corresponds to a certain input combination of angular orientation and range measurement which does not usually coincide with the pre-specified sets of input data belonging to the training datasets.

The most popular and commonly used range sensors in different human detection and tracking approaches are the laser scanners. Also, ultrasonic range sensors, predominantly in developing countries, are considered as appealing and effective options, especially from the perspectives of weight, price, and power consumption. However, localizing human leg position with sonar sensor is not as reliable as it is with laser scanner, primarily on account of the low angular resolution of the sonar sensors and the sparseness of the sonar readings [9]. Besides, phenomena such as *multiple reflections* or *cross-talking* [10] regularly take place in sonar sensing, which produce significant amount of spurious readings.

Keeping the sonar behavior in mind, in this paper, we propose to employ *Gaussian Process Regression* (GPR) [8, 11] to solve our formulated nonlinear regression problem. On account of their tractable posterior computation and appealing analytical properties, Gaussian processes (GPs) are extensively used by the statistics, machine learning, and signal processing communities for solving various identification, multi-class classification, and nonlinear regression problems, subjected to real-life noisy data. The GP, demonstrated to be a robust Bayesian nonparametric method for finding solutions of such problems, enables to realize a probabilistic prediction within a sophisticated and elegant inference framework while holding magnificent resilience to over-fitting that frequently takes place in machine learning. We apply GPR to our problem where the objective is to predict the x and y coordinates for a certain query input when two sets of training data are provided.

However, contrary to the standard GP, where the input data are generally assumed to be fixed and certain but output data are considered to be noisy or often fraught with uncertainty, the input data in our approach, i.e., angular orientations and ranges suffer from a certain degree of imprecision and uncertainty whereas the output data, i.e., the x/y coordinates are presumed to be fixed and stable. Unlike the majority of existing sonar sensor-based visual target detection and tracking approaches, where a set of ultrasonic sensors at fixed poses with respect to the robot is employed, in our approach, we build a cost-effective sensing system, where a single ultrasonic sensor driven by a servo motor is used, which is rotated within a certain angular range to acquire range readings. The servo motor performs the actuation mechanism to drive the sonar sensor to rotate based on a set of user-given commands. However, in reality, due to different nonlinearities (e.g., friction, dead zone, saturation, etc.) commonly occurring in a servo motor-based system, the sonar sensor may not rotate precisely to a specific angle through the set of commands. Hence, there is every possibility of uncertainty being associated with the angles. Also, the range measurements obtained at the corresponding angles undergo a certain degree of imprecision and uncertainty because of the constraints of the sonar sensing discussed earlier. On the other hand, the x/y coordinates are measured on a two-dimensional real plane using a conventional distance measurement methodology, and thus can be assumed to be fixed and reliable. So, in this paper, we have aimed to suitably adapt the fundamental input-output characteristics of the GPR to perform a statistical model calibration and then localize human leg positions based on the calibrated model.

The rest of this paper is organized as follows. Section 2 discusses an overview of the Gaussian process regression (GPR). Section 3 describes our proposed human position estimation approach. Section 4 presents the experimental results and performance evaluations of our proposed research work. Then, Sect. 5 concludes the paper.

2 Gaussian Process Regression

The objective of nonlinear regression is to retrieve an unknown function $\mathsf{g} : \mathbb{R}^n \rightarrow \mathbb{R}$ from an arbitrary dataset $\mathscr{D} = \{\mathbf{z}_j, w_j\}_{j=1}^{u}$, where for the n-dimensional input vector $\mathbf{z}_j \in \mathbb{R}^n$, the observation $w_j \in \mathbb{R}$ (a scalar) can be considered as related to the underlying function value $\mathsf{g}(\mathbf{z}_j)$ through a Gaussian noise model [12]:

$$w_j = \mathsf{g}(\mathbf{z}_j) + \epsilon_j \text{ with } \epsilon_j \sim \mathcal{N}(0, \mathsf{h}^2(\mathbf{z}_j)) \tag{1}$$

where ϵ_j, termed as the observation error (or measurement noise), is commonly assumed to be independently and normally distributed with a mean of 0 and a variance of $\mathsf{h}^2(\mathbf{z}_j)$ [12]. For ease of notation, the true function value and the noise standard deviation are indicated here by $\mathsf{g}_j = \mathsf{g}(\mathbf{z}_j)$ and $\mathsf{h}_j = \mathsf{h}(\mathbf{z}_j)$, respectively. Also, the sets of inputs, outputs, true function values and noise standard deviations are populated into $\mathsf{Z} = [\mathbf{z}_1, \ldots, \mathbf{z}_u]^\mathsf{T}$, $\mathbf{w} = [w_1, \ldots, w_u]^\mathsf{T}$, $\mathbf{g} = [\mathsf{g}_1, \ldots, \mathsf{g}_u]^\mathsf{T}$, and $\mathbf{h} = [\mathsf{h}_1, \ldots, \mathsf{h}_u]^\mathsf{T}$, respectively.

The Gaussian process (GP) with which a nonparametric Bayesian modeling for an unknown function is performed, delineates a distribution of the function values $\mathsf{g}(\mathbf{z})$ on the given input space $\mathbf{z}$ by its mean function $\mu(\mathbf{z})$ and covariance function $\mathsf{c}(\mathbf{z}, \mathbf{z}')$. For the finite set of inputs, $\mathbf{z}_1, \ldots, \mathbf{z}_u$, the corresponding function values, $\mathsf{g}(\mathbf{z}_1), \ldots, \mathsf{g}(\mathbf{z}_u)$, are then fully specified by a multivariate Gaussian distribution as follows [8]:

$$\begin{bmatrix} \mathsf{g}(\mathbf{z}_1) \\ \vdots \\ \mathsf{g}(\mathbf{z}_u) \end{bmatrix} \sim \mathcal{N}(\boldsymbol{\mu}, \mathbf{C}), \tag{2}$$

where

$$\boldsymbol{\mu} = \begin{bmatrix} \mu(\mathbf{z}_1) \\ \vdots \\ \mu(\mathbf{z}_u) \end{bmatrix} \text{ and } \mathbf{C} = \begin{bmatrix} \mathsf{c}(\mathbf{z}_1, \mathbf{z}_1) & \cdots & \mathsf{c}(\mathbf{z}_1, \mathbf{z}_u) \\ \vdots & \ddots & \vdots \\ \mathsf{c}(\mathbf{z}_u, \mathbf{z}_1) & \cdots & \mathsf{c}(\mathbf{z}_u, \mathbf{z}_u) \end{bmatrix} \tag{3}$$

A usual assumption is to place a GP with $\boldsymbol{\mu} = \mathbf{0}$ prior over the function value, especially when no preceding information about the form of $\mathsf{g}(\mathbf{z})$ is available. This is given as [12]

$$p(\mathbf{g}|\mathbf{Z}) \sim \mathcal{N}(\mathbf{0}, \mathbf{C}) \tag{4}$$

The covariance (kernel) function $\mathsf{c}(\mathbf{z}_k, \mathbf{z}_l)$ is considered as a design parameter which delineates correlations between the function values g_k and g_l corresponding to any two given inputs $\mathbf{z}_k$ and $\mathbf{z}_l$. A popular choice among the various available covariance functions is the *squared exponential* (SE) function. The infinitely differentiable SE kernel is expressed as [12]:

$$\mathsf{c}(\mathbf{z}_k, \mathbf{z}_l) = \alpha_\mathsf{g}^2 \exp\left[-\frac{\|\mathbf{z}_k - \mathbf{z}_l\|^2}{2\gamma^2}\right], \tag{5}$$

where $\|\cdot\|$ indicates the Euclidean distance between $\mathbf{z}_k$ and $\mathbf{z}_l$, α_g^2 is the prior variance of the signal amplitude, and γ is the characteristic length-scale. The SE kernel is parameterized by $\boldsymbol{\phi}_\mathsf{g} = \{\alpha_\mathsf{g}, \gamma\}$.

Now, the joint distribution for the observed output $\mathbf{w}$ and the function value g_{test} at a test input $\mathbf{z}_{\text{test}}$ is given as [12]

$$\begin{bmatrix} \mathbf{w} \\ \hline \mathsf{g}_{\text{test}} \end{bmatrix} \sim \mathcal{N}\left(\begin{bmatrix} \mathbf{0} \\ \hline 0 \end{bmatrix}, \begin{bmatrix} \mathbf{C} + \mathbf{V} & \mathbf{C}_{\text{test}} \\ \hline \mathbf{C}_{\text{test}}^\mathsf{T} & \mathsf{c}(\mathbf{z}_{\text{test}}, \mathbf{z}_{\text{test}}) \end{bmatrix}\right), \tag{6}$$

where $\mathbf{V}$ denotes the diagonal matrix of noise variances whose entries are $[\mathbf{V}]_{jj} = \mathsf{h}^2(\mathbf{z}_j)$ and $\mathbf{C}_{\text{test}} = [\mathsf{c}(\mathbf{z}_{\text{test}}, \mathbf{z}_1) \;\; \mathsf{c}(\mathbf{z}_{\text{test}}, \mathbf{z}_2) \;\; \cdots \;\; \mathsf{c}(\mathbf{z}_{\text{test}}, \mathbf{z}_u)]^\mathsf{T}$. Based on the conditional identity of a multivariate Gaussian distribution, the posterior distribution of the function value g_{test} at the test input $\mathbf{z}_{\text{test}}$ is computed as [12]

$$p\left(\mathsf{g}_{\text{test}}|\mathbf{z}_{\text{test}},\ \boldsymbol{\phi}_{\mathsf{g}},\ \mathbf{h},\ \mathscr{D}\right) \sim \mathcal{N}\left(\mathbf{C}_{\text{test}}^{\mathsf{T}}(\mathbf{C}+\mathbf{V})^{-1}\mathbf{w},\ \mathsf{c}(\mathbf{z}_{\text{test}},\ \mathbf{z}_{\text{test}}) - \mathbf{C}_{\text{test}}^{\mathsf{T}}(\mathbf{C}+\mathbf{V})^{-1}\mathbf{C}_{\text{test}}\right) \quad (7)$$

Then, we can obtain the posterior distribution over the test output w_{test} simply by summing up the noise variance $\mathsf{h}_{\text{test}}^2$ at test input $\mathbf{z}_{\text{test}}$ with the posterior variance of g_{test} as [12]

$$p\left(w_{\text{test}}|\mathbf{z}_{\text{test}},\ \boldsymbol{\phi}_{\mathsf{g}},\ \mathbf{h},\ \mathsf{h}_{\text{test}},\ \mathscr{D}\right) \sim \mathcal{N}\left(\mathbf{C}_{\text{test}}^{\mathsf{T}}(\mathbf{C}+\mathbf{V})^{-1}\mathbf{w},\ \mathsf{c}(\mathbf{z}_{\text{test}},\ \mathbf{z}_{\text{test}}) - \mathbf{C}_{\text{test}}^{\mathsf{T}}(\mathbf{C}+\mathbf{V})^{-1}\mathbf{C}_{\text{test}} + \mathsf{h}_{\text{test}}^2\right). \quad (8)$$

This conditional distribution presents the statistical property of the predicted test output g_{test}, where the best estimate is generally selected as its conditional mean $\mathbf{C}_{\text{test}}^{\mathsf{T}}(\mathbf{C}+\mathbf{V})^{-1}\mathbf{w}$.

Note that in practical modeling scenarios, neither the noise level $\mathsf{h}(\mathbf{z})$ nor kernel parameters $\boldsymbol{\phi}_{\mathsf{g}}$ are generally accessible, and they have to be learned from the given data. In this paper, similar to the standard GP, the noise level is supposed to be constant across the input space, and hence we define the noise variance $\mathsf{h}^2(\mathbf{z}) \equiv \alpha_n^2$ and accordingly the noise matrix $\mathbf{V} \equiv \alpha_n^2\mathbf{I}$. The unknown parameters of the Gaussian process regression (GPR), including $\boldsymbol{\phi}_{\mathsf{g}}$ and α_n, are then together referred to as *GP hyperparameters*, represented as $\boldsymbol{\phi}_w = \{\boldsymbol{\phi}_{\mathsf{g}},\ \alpha_n\} \equiv \{\alpha_g,\ \gamma,\ \alpha_n\}$.

3 Proposed Human Position Estimation Approach

3.1 Framework for Data Acquirement

During human following, we need to estimate the positions of the leg pair of the target person at successive time instances. A wheeled mobile robot endowed with a servo motor-driven sonar sensor is used for this purpose. In our proposed approach, the onboard sonar sensor, within an angular range of 180°, gathers a group of range readings corresponding to the leg posture of the target person at a certain time instant, and using that group of measurements, we aim to estimate the two-dimensional human leg position at that time instant.

With this objective, we first generate two comprehensive datasets, both of which share same input information, i.e., the angular orientations of various human leg postures with respect to a definite robot pose and the range measurements acquired at those angles, but contain different output information, i.e., one comprising the corresponding x coordinates and the other the y coordinates of the leg postures, measured with respect to the *robot pose frame* located at that specific robot pose, where the position of the robot in a two-dimensional plane is considered as the *reference point* with coordinate (0, 0) and the heading of the robot is considered along the y-axis. The robot, by using its onboard servo-driven ultrasonic range finder, acquires a set of range readings $\tilde{r}_{d_i} \approx r_d$, $i = 1, 2, \ldots, M$ corresponding to the leg poses of a target person, when the person is positioned around a radial distance r_d along a pre-specified set of equally spaced angles $\psi_i \in [0°, 180°]$, $i = 1, 2, \ldots, M$, and, at the same time, we measure the positions of the human legs in that two-dimensional real plane using a conventional distance measurement technique. This mechanism is repeated for a set of P number of different radial distances $r_d^{(j)}$, $j = 1, 2, \ldots, P$. Therefore, we can obtain a total of $N = P * M$ number of different $\{\tilde{r}_d, \psi\}$ combinations, and correspondingly same number of x and y coordinates, which are aggregated into two datasets $\mathscr{D}_x$ and $\mathscr{D}_y$ as

$$\mathscr{D}_x = \{(r_1, \theta_1), (r_2, \theta_2), \ldots, (r_N, \theta_N), x_1, x_2, \ldots, x_N\}$$
$$\mathscr{D}_y = \{(r_1, \theta_1), (r_2, \theta_2), \ldots, (r_N, \theta_N), y_1, y_2, \ldots, y_N\}$$

where $r_1 = \tilde{r}_{d_1}^{(1)}$, $\theta_1 = \psi_1, \ldots$, $r_M = \tilde{r}_{d_M}^{(1)}$, $\theta_M = \psi_M$; $r_{M+1} = \tilde{r}_{d_1}^{(2)}$, $\theta_{M+1} = \psi_1, \ldots$, $r_{2M} = \tilde{r}_{d_M}^{(2)}$, $\theta_{2M} = \psi_M$; ......; $r_{(P-1)M+1} = \tilde{r}_{d_1}^{(P)}$, $\theta_{(P-1)M+1} = \psi_1, \ldots$, $r_N = \tilde{r}_{d_M}^{(P)}$, $\theta_N = \psi_M$ (as $N = PM$).

Note that when we acquire range data for the leg pair positioned around $r_d^{(q)}$ along a particular angle ψ_k, we gather a group of range readings around the angle instead of acquiring a single reading at the angle. This is primarily done to obtain a reliable range estimation for the above configuration and to marginalize the sonar uncertainty, which may not be achieved with a single reading. However, for generating dataset, we take into account a single representative reading $\tilde{r}_{d_k}^{(q)}$ corresponding to that specific configuration, which is defined as

$$\tilde{r}_{d_k}^{(q)} = \frac{1}{L}\left(\texttt{range}^{(q)}(\psi_{\mathsf{L}_k}) + \texttt{range}^{(q)}(\psi_{\mathsf{R}_k}) + \sum_{\delta \in \mathsf{V}} \texttt{range}^{(q)}(\delta)\right)$$
$$\text{with } \psi_{\mathsf{L}_k} < \psi_k < \psi_{\mathsf{R}_k}, \tag{9}$$

where

$$\mathsf{V} = \{\delta : \psi_{\mathsf{L}_k} < \delta < \psi_{\mathsf{R}_k} \text{ and } \texttt{range}^{(q)}(\delta) \neq \text{NaN}\}$$
$$L = \texttt{cardinality}(\mathsf{V}) + 2.$$

The variables ψ_{L_k} and ψ_{R_k} denote the angular orientations at which the left most and right most sonar readings are detected, respectively, corresponding to the leg pair positioned around $r_d^{(q)}$ along ψ_k, and $\texttt{range}^{(q)}(\psi_{\mathsf{L}_k})$ and $\texttt{range}^{(q)}(\psi_{\mathsf{R}_k})$ are the corresponding sonar readings. Also, for this specific arrangement, we have measured the (x, y) coordinates of the center of two feet of the target person as the real location of a human body usually lies at that position. Thus, each of such N arrangements can be characterized by a single set of values for r, θ, x, and y. Figure 1 illustrates

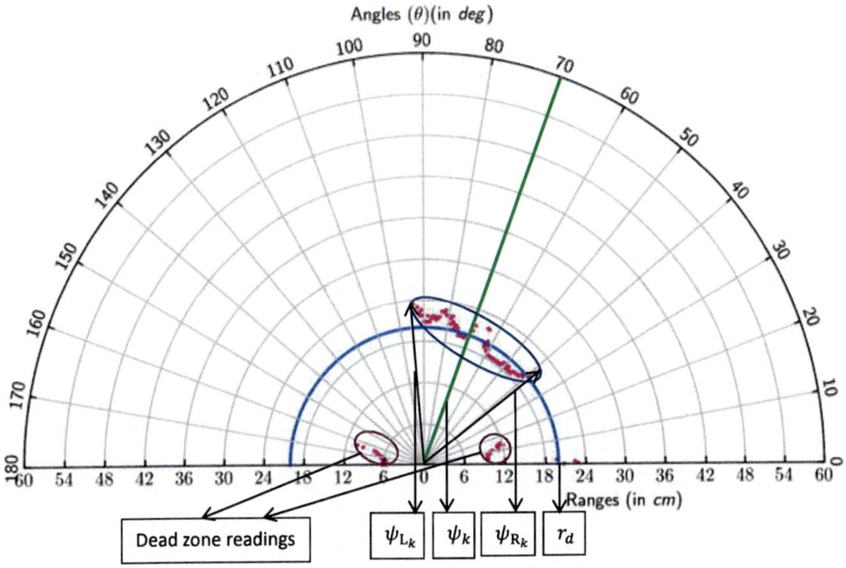

Fig. 1 Acquiring range readings corresponding to the leg pair positioned around r_d along ψ_k. Here, $r_d = 20$ cm and $\psi_k = 70°$. The *magenta* points indicate the range readings

the scheme of acquiring range readings corresponding to a specific leg posture along a certain angle with a representative example.

3.2 *Proposed Measurement Model*

As mentioned earlier, because of various nonlinearities usually occurring in a servo motor-based system, the sonar sensor may not rotate precisely to a specific angle through a user-given set of commands. Hence, there is every possibility of uncertainty being associated with the angle measurements. Also, it is well known that the performance of the sonar sensors is significantly affected by the environmental phenomena, sensor design and target characteristics, and sonar readings usually undergo some degree of uncertainty and imprecision. However, errors involved in measuring the *ground truth* (x, y) coordinates have been considered negligible, taking into account the reliability of the method adopted for measurement. Therefore, the inputs (r, θ) can be considered fraught with imprecision and uncertainty, whereas the outputs (x, y) can be considered stable and reliable in our framework.

Now, our human leg localization problem can be posed as a nonlinear regression problem, where two separate prediction models, best represented by the unknown functions $\mathsf{fx} : \mathbb{R}^2 \rightarrow \mathbb{R}$ and $\mathsf{fy} : \mathbb{R}^2 \rightarrow \mathbb{R}$, are learned by using two distinct training datasets, $\mathscr{D}_x = \{\mathbf{s}_j, x_j\}_{j=1}^N$ and $\mathscr{D}_y = \{\mathbf{s}_j, y_j\}_{j=1}^N$, respectively, where

$\mathbf{s}_j = [r_j, \theta_j] \in \mathbb{R}^2$ indicates a *two*-dimensional input vector, and $x_j \in \mathbb{R}$ and $y_j \in \mathbb{R}$ are the scalar measured outputs such that

$$\begin{aligned} x_j &= \mathsf{fx}(\mathbf{s}_j) + \epsilon_{xj} \text{ with } \epsilon_{xj} \sim \mathcal{N}(0,\, \mathsf{h}_x^2(\mathbf{s}_j)) \\ y_j &= \mathsf{fy}(\mathbf{s}_j) + \epsilon_{yj} \text{ with } \epsilon_{yj} \sim \mathcal{N}(0,\, \mathsf{h}_y^2(\mathbf{s}_j)), \end{aligned} \tag{10}$$

where both ϵ_{xj} and ϵ_{yj} signify measurement uncertainties which can be represented as the *equivalent modeling errors*. These equivalent modeling errors which are caused because of the uncertainties in input variables are random variables and are assumed to be independently and normally distributed, as mentioned in (10). The noise variances $\mathsf{h}_x^2(\mathbf{s}_j)$ and $\mathsf{h}_y^2(\mathbf{s}_j)$ are supposed to be constant throughout the input space, and defined as $\mathsf{h}_x^2(\mathbf{s}) \equiv \alpha_{nx}^2$ and $\mathsf{h}_y^2(\mathbf{s}) \equiv \alpha_{ny}^2$, respectively. Also, we define $\mathsf{h}_{xj} = \mathsf{h}_x(\mathbf{s}_j)$ and $\mathbf{h}_x = [\mathsf{h}_{x1}, \ldots, \mathsf{h}_{xN}]^\mathsf{T}$. Then these two learned models are utilized to localize the $(x_t,\ y_t)$ coordinates of a given query leg posture which corresponds to a certain input $\mathbf{s}_t = [r_t, \theta_t]$. It should be noted that $\mathbf{s}_t = [r_t, \theta_t]$ does not in general coincide with the pre-specified $\mathbf{s}_j = [r_j, \theta_j]$, $j = 1, 2, \ldots, N$, and hence x_t and y_t are usually not known but need to be predicted from the sets of other available variables x_j and y_j, $j = 1, 2, \ldots, N$.

We propose to apply GPR to our discussed problem in which the objective is to predict $x_t = \mathsf{fx}(\mathbf{s}_t) + \epsilon_{xt}$ and $y_t = \mathsf{fy}(\mathbf{s}_t) + \epsilon_{yt}$ for a query input $\mathbf{s}_t = [r_t, \theta_t]$, given two sets of (training) data $[x_1, x_2, \ldots, x_N]^\mathsf{T}$ and $[y_1, y_2, \ldots, y_N]^\mathsf{T}$ available for the corresponding inputs $\mathbf{s}_1, \mathbf{s}_2, \ldots, \mathbf{s}_N$. However, we will discuss here in the context of estimating x_t, which can be followed in the same way to estimate y_t also, except the fact that the training dataset $\mathcal{D}_y$ will be involved in the later case.

In this work, we have selected SE kernel as the covariance function. It is worth mentioning that, we have standardized the input data so that each dimension is weighted on the same scale. The SE kernel, for any two given inputs $\mathbf{s}_k$ and $\mathbf{s}_l$, is here defined as

$$\mathsf{c}(\mathbf{s}_k,\ \mathbf{s}_l) = \alpha_{\mathsf{fx}}^2 \exp\left[-\frac{\|\mathbf{s}_k - \mathbf{s}_l\|^2}{2\gamma_{\mathsf{fx}}^2}\right] = \alpha_{\mathsf{fx}}^2 \exp\left[-\frac{\left(\hat{r}_k - \hat{r}_l\right)^2 + \left(\hat{\theta}_k - \hat{\theta}_l\right)^2}{2\gamma_{\mathsf{fx}}^2}\right],$$

where $\hat{r}_j$ and $\hat{\theta}_j$ are the updated versions of r_j and θ_j, respectively, after standardization. Following the discussions of the last section, the SE kernel is here parameterized by $\boldsymbol{\phi}_{\mathsf{fx}} = \{\alpha_{\mathsf{fx}},\ \gamma_{\mathsf{fx}}\}$, and the GP hyperparameters are given by $\boldsymbol{\phi}_x = \{\alpha_{\mathsf{fx}},\ \gamma_{\mathsf{fx}},\ \alpha_{nx}\}$. If we define $\mathbf{X} = [x_1,\ x_2, \ldots, x_N]^\mathsf{T}$, a multivariate Gaussian distribution can be formed similar to (6) as

$$\begin{bmatrix} \mathbf{X} \\ \mathsf{fx}_t \end{bmatrix} \sim \mathcal{N}\left(\begin{bmatrix} \mathbf{0} \\ 0 \end{bmatrix}, \begin{bmatrix} \mathbf{C} + \mathbf{V} & \mathbf{C}_{\mathbf{s}_t} \\ \mathbf{C}_{\mathbf{s}_t}^\mathsf{T} & \mathsf{c}(\mathbf{s}_t,\ \mathbf{s}_t), \end{bmatrix}\right) \tag{11}$$

where fx_t denotes the value of the function fx at the query input $\mathbf{s}_t = [r_t, \theta_t]$,

$$\mathbf{C} = \begin{bmatrix} \mathsf{c}(\mathbf{s}_1, \mathbf{s}_1) & \cdots & \mathsf{c}(\mathbf{s}_1, \mathbf{s}_N) \\ \vdots & \ddots & \vdots \\ \mathsf{c}(\mathbf{s}_N, \mathbf{s}_1) & \cdots & \mathsf{c}(\mathbf{s}_N, \mathbf{s}_N) \end{bmatrix}$$

$$\mathbf{C}_{\mathbf{s}_t} = [\mathsf{c}(\mathbf{s}_t, \mathbf{s}_1) \;\; \mathsf{c}(\mathbf{s}_t, \mathbf{s}_2) \;\; \cdots \;\; \mathsf{c}(\mathbf{s}_t, \mathbf{s}_N)]^{\mathsf{T}} \tag{12}$$

and $\mathbf{V}$ is the diagonal matrix of noise variances whose entries are $[\mathbf{V}]_{jj} = \mathsf{h}_x^2(\mathsf{s}_t) = \alpha_{nx}^2$.

Similar to (8), the conditional distribution $p\left(x_t|\mathsf{s}_t, \boldsymbol{\phi}_x, \mathbf{h}_x, \mathsf{h}_{xt}, \mathscr{D}_x\right)$ is here obtained as

$$p\left(x_t|\mathsf{s}_t, \boldsymbol{\phi}_x, \mathbf{h}_x, \mathsf{h}_{xt}, \mathscr{D}_x\right) \sim \mathcal{N}\left(\mathbf{C}_{\mathbf{s}_t}^{\mathsf{T}}(\mathbf{C}+\mathbf{V})^{-1}\mathbf{X},\, \mathsf{c}(\mathbf{s}_t, \mathbf{s}_t) - \mathbf{C}_{\mathbf{s}_t}^{\mathsf{T}}(\mathbf{C}+\mathbf{V})^{-1}\mathbf{C}_{\mathbf{s}_t} + \alpha_{nx}^2\right). \tag{13}$$

As mentioned in the previous section, our best estimate for x_t is the mean of this distribution:

$$\overline{x_t} = \mathbf{C}_{\mathbf{s}_t}^{\mathsf{T}}(\mathbf{C}+\mathbf{V})^{-1}\mathbf{X} \tag{14}$$

and the uncertainty associated with our estimate is captured in its variance:

$$\mathsf{var}(x_t) = \mathsf{c}(\mathbf{s}_t, \mathbf{s}_t) - \mathbf{C}_{\mathbf{s}_t}^{\mathsf{T}}(\mathbf{C}+\mathbf{V})^{-1}\mathbf{C}_{\mathbf{s}_t} + \alpha_{nx}^2 \tag{15}$$

4 Experimental Results

4.1 Experimental Setup

The proposed human leg localization approach is tested using real-world range data acquired by the onboard HC-SR04 ultrasonic range sensor of an inexpensive Raspberry Pi 3 model B+ based *differentially driven* wheeled mobile robot [7]. The robot, depicted in Fig. 2, has been indigenously developed in our Electrical Measurement and Instrumentation Laboratory of Electrical Engineering Department, Jadavpur University, India [7]. In this scheme, `Python` 2.7 (along with `Numpy`, `Scipy` and `Matplotlib`) has been used as the experimental platform.

4.2 Dataset Generation

Based on the framework discussed in section, we generate two real-world datasets $\mathscr{D}_x = \{(r_j\, \theta_j),\, x_j\}_{j=1}^N$ and $\mathscr{D}_y = \{(r_j\, \theta_j),\, y_j\}_{j=1}^N$ corresponding to various human leg positions. Considering an effective sonar range of $5 - 55$ cm in this work, the

Fig. 2 The wheeled mobile robot used in this paper for testing the proposed GPR-based approach [7]

set of radial distances $r_d^{(j)}$, $j = 1, 2, \ldots, P$ is adopted as {5 cm, 7 cm, 9 cm, ..., 51 cm, 53 cm, 55 cm}, where the distances are separated by an interval of 2 cm. On the other hand, because of the presence of dead zone nonlinearity in the sensing system, we consider an effective angular range of $[10°, 170°]$, and taking into account the shape of the object and the lower angular resolution of sonar sensor, the set of angular orientations ψ_i, $i = 1, 2, \ldots, M$ are selected as $\{10°, 25°, \ldots, 160°, 170°\}$, with the consecutive angles separated by a span of 15° (except the last angle which is separated from the preceding one by 10°). Thus, the parameters P and M are set to 26 and 12 respectively. Accordingly, we obtain the parameter $N = 26 * 12 = 312$. In this paper, we have used approximately 80% of the observations for the purpose of training and the remaining 20% for the purpose of testing.

4.3 Performance Evaluations

The performance of our proposed human leg localization approach has been evaluated quantitatively by means of two metrics, namely *mean absolute percentage error* (MAPE) and *root mean square error* (RMSE), and compared with traditional method of computing (x, y) for a given (r, θ) combination.

1. **Mean Absolute Percentage Error (MAPE):** The MAPE is a measure of the prediction accuracy achieved by a forecasting method, such as regression. In this paper, the MAPE is defined as

$$\lambda = \frac{1}{S}\sum_{i=1}^{S}\left|\frac{\mathcal{G}_i - \mathcal{E}_i}{\mathcal{G}_i},\right| \times 100\% \tag{16}$$

where S is the testing dataset size, $\mathcal{G}_i$ and $\mathcal{E}_i$ are the ground truth observation and estimated output corresponding to the ith test input, respectively. Our pro-

Table 1 Comparison of performance metrics between proposed and traditional approach in estimating x and y

Method	Estimation of x		Estimation of y	
	λ (in %)	ξ (in m)	λ (in %)	ξ (in m)
Traditional approach	20.14	0.0357	20.61	0.0396
Proposed approach	**06.22**	**0.0155**	**06.51**	**0.0137**

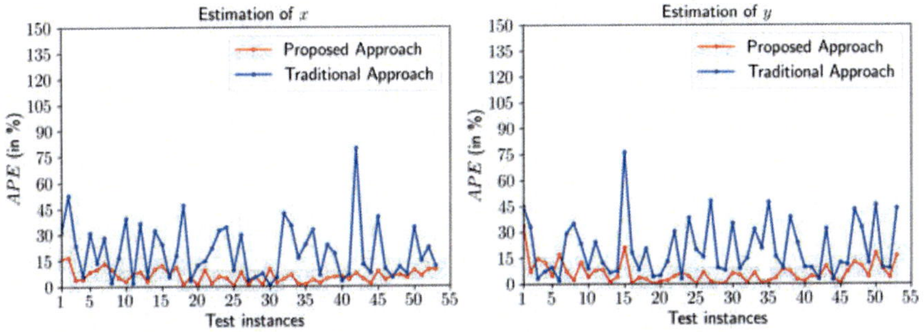

(a) (b)

Fig. 3 APEs for the estimated outputs corresponding to the testing inputs: **a** in the case of estimating x-coordinates, **b** in the case of estimating y-coordinates

posed approach outperforms the traditional approach in localizing both x and y coordinates, in terms of this quantitative performance metric, as presented in Table 1. Also, Fig. 3a, b show the absolute percentage errors (APEs) for the estimated outputs corresponding to the testing inputs, in the cases of estimating x and y coordinates respectively. As can be observed from Fig. 3, our GPR-based proposed approach and traditional method show better results for approximately 85% and 15% of the test instances, respectively, in the case of estimating x, and approximately 88% and 12% of the test instances, respectively, in the case of estimating y, which reaffirms the overall usefulness of our proposed GPR-based approach. In our experimentation, the proposed approach is found to improve the MAPE performance over the traditional method by 69.11% and 68.41% in the case of estimating $x-$ and y-coordinates, respectively.

2. **Root Mean Square Error (RMSE):** The RMSE is a standard statistical parameter to measure the error of a model in predicting quantitative data. In this paper, the RMSE is defined as follows:

$$\xi = \sqrt{\frac{\sum_{i=1}^{S} (\mathscr{E}_i - \mathscr{G}_i)^2}{S}} \tag{17}$$

As revealed by the experimental results in Table 1, our proposed approach is superior to the traditional approach, also in terms of this metric. In our experimentation, the proposed approach is found to improve the RMSE performance over the traditional method by 56.58% and 65.40% in the case of estimating $x-$ and y-coordinates, respectively.

We have evaluated the performance of our proposed GPR-based approach with the parameter values $\alpha_{\mathrm{fx}} = \alpha_{\mathrm{fy}} = 1$, and $\gamma_{\mathrm{fx}} = \gamma_{\mathrm{fy}} = 0.7$.

5 Conclusions

This paper demonstrates how a human leg localization problem can be formulated as a nonlinear regression problem where two separate prediction models are learned by using two different training datasets, and how the models can be used to localize the $x-$ and y-coordinates of a leg posture with respect to a specific robot pose frame. Also, in this paper, we have adopted sonar sensing-based system for acquiring range data and considering the uncertainty and imprecision associated with the sonar readings, we have proposed to employ a popular Bayesian nonparametric method, called *Gaussian Process Regression* (GPR), to solve our formulated nonlinear regression problem. We have suitably adapted the basic I/O characteristics of the GPR according to our formulated problem, performed a statistical model calibration, and then predict human positions based on the calibrated model. Experimental results and performance evaluations carried out for estimating the positions of human legs based on real-life data aptly demonstrate the superiority of the proposed approach over the traditional method.

References

1. Yuan J, Chen H, Sun F, Huang Y (2015) Multisensor information fusion for people tracking with a mobile robot: a particle filtering approach. IEEE Trans Instrum Meas 64(9):2427–2442
2. Bellotto, N, Hu H (2009) Multisensor-based human detection and tracking for mobile service robots. IEEE Trans Syst Man Cybern Part B (Cybern) 39(1):167–181
3. Li D, Li L, Li Y, Yang F, Zuo X (2018) A multi-type features method for leg detection in 2-D laser range data. IEEE Sens J 18(4):1675–1684
4. Wang M, Liu Y, Su D, Liao Y, Shi L, Xu J, Miro JV (2018) Accurate and real-time 3-D tracking for the following robots by fusing vision and ultrasonar information. IEEE/ASME Trans Mechatron 23(3):997–1006
5. Paral P, Chatterjee A, Rakshit A (2019) Vision sensor-based shoe detection for human tracking in a human-robot coexisting environment: a photometric invariant approach using DBSCAN algorithm. IEEE Sens J 19(12):4549–4559
6. Paral P, Chatterjee A, Rakshit A (2019) OPTICS-based template matching for vision sensor-based shoe detection in human-robot coexisting environments. IEEE Trans Instrum Meas 68(11):4276–4284

7. Paral P, Chatterjee A, Rakshit A (2020) Chaos-based random sampling for photometric invariant shoe detection with vision sensor in human-robot coexisting environments. IEEE Sens J 20(10):5424–5434
8. Lee S, McBride J (2019) Extended object tracking via positive and negative information fusion. IEEE Trans Signal Process 67(7):1812–1823
9. Burguera A, Gonzàlez Y, Oliver G (2008) A probabilistic framework for sonar scan matching localization. Adv Robot 22(11):1223–1241
10. Burguera A, Gonzàlez Y, Oliver G (2009) On the use of likelihood fields to perform sonar scan matching localization. Auton Robot 26(4):203–222
11. Bishop CM (2006) Pattern recognition and machine learning, 1st edn. Springer, Berlin
12. Zhang Q-H, Ni Y-Q (2020) Improved most likely heteroscedastic Gaussian process regression via Bayesian residual moment estimator. IEEE Trans Signal Process 68:3450–3460

Steady-State Analysis of a Vacation Queue with Random Serving Capacity and Batch-Size-dependent Service

Nilanjan Nandy and Sourav Pradhan

Abstract This article analyzes a queue in which the service time varies with the batch size, with infinite waiting space, random service rule, single and multiple vacation policy. Both the service and vacation times are generally distributed. The major concern is to acquire the bivariate probability generating function of queue and server content distribution. We extract the joint probabilities in an efficient and elegant way using the partial fraction technique. We also acquire the distribution at arbitrary epochs using which performance measures have been deduced. By evoking significant distributions, the feasibility and applicability of those results are illustrated numerically.

Keywords Queue · Roots · Distributions · Vacation · Batch-size-dependent

1 Introduction

In recent times, queueing systems with batch service and different vacation policies have become one of the major concerns of the researchers. Moreover, batch-size-dependent service mechanism also has been gained notable attention as it reduces congestion. Such queues have widespread applicability in transportation systems, group testing (blood screening for detecting HIV/influenza etc.), food processing technology, inventory systems, wireless telecommunication, mobile crowd-sourcing app etc.

Banerjee et al. [1–3] have focused on queues with group service where service time varies with the batch size and the waiting space is finite. They have acquired the joint probabilities of queue size as well as server size using the embedded Markov chain technique (EMCT) and supplementary variable technique (SVT). For discrete-time queues, Claeys et al. [4, 5] attained probability generating function (pgf) or vector generating function (vgf) of queue and server content distribution at random slot. They discussed the impact of correlation to study the nature of the system.

N. Nandy (✉) · S. Pradhan
Visvesvaraya National Institute of Technology,Nagpur, India

© The Author(s), under exclusive license to Springer Nature Singapore Pte Ltd. 2022
D. Giri et al. (eds.), *Proceedings of the Seventh International Conference on Mathematics and Computing*, Advances in Intelligent Systems and Computing 1412,
https://doi.org/10.1007/978-981-16-6890-6_78

Claeys et al. [6] also discussed the customer delay in terms of the tail probabilities. However, the procedure of extracting the joint probabilities was not provided by them. Pradhan and Gupta [7–9] have discussed batch-size-dependent service queues where they obtained the bivariate pgf/vgf along with detailed extraction procedure. It is worthwhile to mention here that none of these analyses includes any type of vacation policies, which may arise in several real-life circumstances.

There exists numerous circumstances, where after completing the service of a batch, if the size of the queue is less than the seeded minimum value, the server may leave the system for an arbitrary amount of time to do some other work. These situations are framed by vacation models and possess widespread applications in scheming of local area networks (LAN), computer and switching system, manufacturing system with server breakdown, shared resource maintenance etc. Doshi [10] provided a survey on various queueing models with vacation for both the discrete- and continuous- time set-up. Reddy et al. [11] derived the pgf of queue length distribution for batch-service queues with different vacation policies. Gupta et al. [12] and Sikdar et al. [13] analyzed finite- and infinite- buffer bulk queues with single/multiple vacation rules, where they employed SVT for computing the distribution of queue size at different epochs. Chang and Choi [14] analyzed $Geo^X/G^Y/1$ vacation queue, where they focused on the distribution of queue size at the steady-state. Samanta et al. [15] discussed $Geo^X/G^{(a,b)}/1$ vacation queue with finite-buffer where they employed both EMCT and SVT to acquire queue length distributions. The server content distribution, a key ingredient from designing perspective, was lacking in their analysis. Also, batch-size-dependent service mechanism was not adopted by them.

From the above literature works, it is clear that no one focuses on queue length along with server content for a group-service queue with random service rule with varying service time along with single as well as multiple vacation policies, i.e., $M/G_n^Y/1$ queue with single and multiple vacation. Keeping this in mind, we analyze an infinite-buffer $M/G_n^Y/1$ queue with single and multiple vacation. Firstly, we originate the steady-state difference-differential equations of the concerned queue. After taking the Laplace-Stiljies transform of those equations, we move toward the derivation of the bivariate pgf. We apply Rouche's theorem to determine the unknowns arising in that pgf. The joint probabilities are then extracted and expressed in a presentable form. We also acquire arbitrary epoch probabilities using which pertinent performance measures have been deduced. At the endmost, assorted numerical examples with noteworthy distributions have been presented which brings out an inner feeling of the readers.

The random service rule along with the dependence of batch size on service time and two types of vacation policies makes the analysis quite interesting as well as complicated. The server content distribution is a key ingredient in view of mean server content by which one can think of enlarging the serving capability. It is inescapable to work out on the system content distribution. More importantly, both the information about queue length and server content is required to speed up the process. This model has application in a vegetable oil refinery, where one major process consists of the bleaching of oil. Precisely, the neutralized oil (generally free from fat) is poured into the bleacher where firstly bleaching process leads to the removal of the color of the oil with the perfect use of chemicals. Using thermal fluid coils, oil is heated up to 120°C along with continued stirring. At the filter press, the bleached oil is collected where the separation of bleaching earth and chemicals are made in the form of oil cakes. For the filter process, suppose a protocol has been designed in such a way that it will serve minimum (the entire queue, i) customers, with $i \in [1, B]$ is the chosen capacity and B is the highest capacity of customers that can avail service at a time. Whenever at least one customer is not available to start the service, the process enters into the well-seeded vacation period and that particular duration is used to accomplish some auxiliary works such as collecting the unpurified oil for refinery, checking the quality of the refined oil etc., which is exactly the use of vacation policy. After returning to filter press, if the volume of aggregated oil is still less than the desired batch quantity, it continues to do those works until it finds enough proportion of oil for the processing which may be thought of as multiple vacation.

The structure of the paper is thus and so. Section 2 gives detailed model description. The steady-state equation of the model is described in Sect. 3. The mathematical derivation of the bivariate pgf and procedure of extracting the joint probabilities at departure epoch is given in Sect. 4. The relation between joint probabilities at random and departure epoch is given in Sect. 5, followed by performance measures and the numerical examples.

2 Model Description

- Arrival process: It is assumed that the customers arrive in accordance with the classical Poisson process with rate λ.
- Batch-service rule: The accumulated customers are served in groups in accordance with random capacity Y rule, i.e., with probability mass function (pmf) $Pr(Y = i) = y_i, i = 1, 2, ..., B$, where $y_B > 0$ and B designates the highest capacity of server, pgf $Y(z) = \sum_{i=1}^{B} y_i z^i$, and mean $\bar{y}$. When the service is about to initiate if the $queue\ length > 0$ but less than the selected capacity i, it does not stand by until queue length reaches i. Instead, it serves all the customers present in the queue, i.e., it serves $min(entire\ queue > 0, i)$. Whenever the queue is empty, the server enters to the ideal period.
- Service process: It is assumed that the service time follows any arbitrary distribution and is dependent on the size of the batch. This special service mechanism reduces the congestion as well as improves the system productivity. For a batch of i customers, the service time has probability density function (pdf) $s_i(t)$, $t \geq 0$, distribution function (DF) $S_i(t)$ and the Laplace-Stieltjes transform (L.S.T) $S_i^*(\theta) = \int_0^\infty e^{-\theta u} dS_r(u) = \int_0^\infty e^{-\theta u} s_r(u) du$. Also, the mean service time of batch of size i is $s_i = \frac{1}{\mu_i} = -S^{*(1)}(0)$, where $S^{*(1)}(0)$ is the derivative of order 1 evaluated at $\theta = 0$.
- Vacation process: We analyze the system with two types of vacation policies, viz., single and multiple vacation, with an indicator function δ_p defined as

$$\delta_p = \begin{cases} 0, & \text{for single vacation,} \\ 1, & \text{for multiple vacation.} \end{cases}$$

The server opts for the pre-assumed vacation policy prior to commencement of service. After substituting the appropriate value of the indicator function one can obtain the corresponding results with single and multiple vacation policies.

- Single vacation policy: After the accomplishment of the service, if the server observes at least one customer standing, it starts the service, else it departs for a vacation. After finishing off the vacation time, if the queue length is 0, then the server waits until a customer arrives.
- Multiple vacation policy: After finishing off the service, when the server observes at least one customer standing, it starts the service, else it departs for a vacation. After completing the vacation time, if still the queue is empty, it departs for another vacation and the process continues unless the queue length is > 0. For both the vacation policies, it has the pdf $v(t), t \geq 0$, DF $V(t)$, L.S.T $V^*(\theta) = \int_0^\infty e^{-\theta u} dV(u) = \int_0^\infty e^{-\theta u} v(u) du$ and mean $E(V) = \bar{v} = -V^{*(1)}(0)$.

- The traffic intensity of this queue is $\rho = \frac{\lambda \sum_{i=1}^{B} \frac{y_i}{\mu_i}}{\bar{y}} < 1$, which sets the seal on the system's stability.

3 Steady-State System Equations

In order to present the governing equations of the model, we first define the state of the system at time t:

- $N_q(t) \equiv$ Number of customers standing in the queue,
- $N_s(t) \equiv$ Number of customer with the server,
- $U(t) \equiv$ Left over service time of a batch in service (if any),
- $V(t) \equiv$ Left over vacation time of the server,
- $\zeta(t) \equiv$ State of the server:

$$\zeta(t) = \begin{cases} 0, & \text{the server belongs to dormant state,} \\ 1, & \text{the server is affiliated to vacation,} \\ 2, & \text{the server belongs to busy state.} \end{cases}$$

Let us define the following probabilities:

$$\begin{aligned} p_{0,0}(t) &= Pr\{N_q(t) = 0, N_s(t) = 0, \zeta(t) = 0\}, \\ p_{n,r}(u,t)du &= Pr\{N_q(t) = n, N_s(t) = r, u < U(t) < u + du, \zeta(t) = 2\}, \\ & \quad u \geq 0,\ n \geq 0,\ 1 \leq r \leq B, \\ q_n(u,t)du &= Pr\{N_q(t) = n, u < V(t) < u + du, \zeta(t) = 1\}, \quad u \geq 0,\ n \geq 0. \end{aligned}$$

We also define the following limiting probabilities:

$$p_{0,0} = \lim_{t\to\infty} p_{0,0}(t),$$
$$p_{n,r}(u) = \lim_{t\to\infty} p_{n,r}(u,t), \quad u \ge 0,\ n \ge 0,\ 1 \le r \le B,$$
$$q_n(u) = \lim_{t\to\infty} q_n(u,t), \quad u \ge 0,\ n \ge 0.$$

We are led to the following steady-state equations after looking into the states at two ensuing epochs t and $t + dt$, and using the SVT:

$$0 = \left\{ -\lambda p_{0,0} + q_0(0) \right\} \left(1 - \delta_p\right) \tag{1}$$

$$-\frac{d}{du} p_{0,1}(u) = -\lambda p_{0,1}(u) + \sum_{i=1}^{B} p_{1,i}(0) \sum_{j=1}^{B} y_j s_1(u) + q_1(0) \sum_{j=1}^{B} y_j s_1(u) \tag{2}$$
$$+\lambda \left(1 - \delta_p\right) p_{0,0} \sum_{j=1}^{B} y_j s_1(u)$$

$$-\frac{d}{du} p_{0,r}(u) = -\lambda p_{0,r}(u) + \sum_{i=1}^{B} p_{r,i}(0) \sum_{j=r}^{B} y_j s_r(u) \tag{3}$$
$$+q_r(0) \sum_{j=r}^{B} y_j s_r(u),\ 2 \le r \le B$$

$$-\frac{d}{du} p_{n,r}(u) = -\lambda p_{n,r}(u) + \lambda p_{n-1,r}(u) + \sum_{i=1}^{B} p_{n+r,i}(0) y_r s_r(u) \tag{4}$$
$$+q_{n+r}(0) y_r s_r(u),\ n \ge 1,\ 1 \le r \le B$$

$$-\frac{d}{du} q_0(u) = -\lambda q_0(u) + \left(\sum_{i=1}^{B} p_{0,i}(0) + \delta_p q_0(0) \right) v(u) \tag{5}$$

$$-\frac{d}{du} q_n(u) = -\lambda q_n(u) + \lambda q_{n-1}(u),\ n \ge 1. \tag{6}$$

The L.S.Ts of the probabilities are defined as

$$p^*_{n,r}(\theta) = \int_0^\infty e^{-\theta u} p_{n,r}(u)du,\ n \geq 0,\ 1 \leq r \leq B;\quad q^*_n(\theta) = \int_0^\infty e^{-\theta u} q_n(u)du,\ n \geq 0.$$

which implies $p_{n,r} \equiv p^*_{n,r}(0) = \int_0^\infty p_{n,r}(u)du,\ n \geq 0,\ 1 \leq r \leq B;\quad q_n \equiv q^*_n(0) = \int_0^\infty q_n(u)du,\ n \geq 0.$

Taking L.S.Ts of Eqs. (3)–(6), we obtain

$$(\lambda - \theta)\, p^*_{0,1}(\theta) = \left\{\sum_{i=1}^{B} p_{1,i}(0) + q_1(0)\right\} \sum_{j=1}^{B} y_j S^*_1(\theta) + \lambda\left(1 - \delta_p\right) p_{0,0} \sum_{j=1}^{B} y_j S^*_1(\theta) - p_{0,1}(0) \tag{7}$$

$$(\lambda - \theta)\, p^*_{0,r}(\theta) = \left\{\sum_{i=1}^{B} p_{r,i}(0) + q_r(0)\right\} \sum_{j=r}^{B} y_j S^*_r(\theta) - p_{0,r}(0),\ 2 \leq r \leq B \tag{8}$$

$$(\lambda - \theta)\, p^*_{n,r}(\theta) = \lambda p^*_{n-1,r}(\theta) + \left\{\sum_{i=1}^{B} p_{n+r,i}(0) + q_{n+r}(0)\right\} y_r S^*_r(\theta) - p_{n,r}(0),\ n \geq 1,\ 1 \leq r \leq B \tag{9}$$

$$(\lambda - \theta)\, q^*_0(\theta) = \left(\sum_{i=1}^{B} p_{0,i}(0) + \delta_p q_0(0)\right) V^*(\theta) - q_0(0) \tag{10}$$

$$(\lambda - \theta)\, q^*_n(\theta) = \lambda q^*_{n-1}(\theta) - q_n(0),\ n \geq 1. \tag{11}$$

Now, we come up with two results that will be used to deduce further results.

Lemma 1 *The relation between the probabilities* $\left(p^+_{n,r}, p_{n,r}(0)\right)$ *and* $\left(q^+_n, q_n(0)\right)$ *is given by*

$$p^+_{n,r} = \tau^{-1} p_{n,r}(0) \tag{12}$$

$$q^+_n = \tau^{-1} q_n(0) \tag{13}$$

where $\tau = \sum_{m=0}^{\infty} \sum_{r=1}^{B} p_{m,r}(0) + \sum_{m=0}^{\infty} q_m(0)$.

Proof Let $p_{n,r}(0)$ be the probability of having n customers in the queue and r with the server when there is no service time left. To find the relation between $p^+_{n,r}$ and

$p_{n,r}(0)$ we apply Bayes theorem

$$\begin{aligned}p^+_{n,r} &= Pr\{\text{queue contains } n \text{ customers and } r \text{ with the server just precedent to the service completion} \mid \text{total number waiting in the queue just precedent to the service/vacation completion epoch}\}, \quad n \geq 0, \quad 1 \leq r \leq B,\\ &= \frac{1}{\tau} p_{n,r}(0)\end{aligned}$$

where $\tau = \sum_{m=0}^{\infty} \sum_{r=1}^{B} p_{m,r}(0) + \sum_{m=0}^{\infty} q_m(0)$. Similarly, Eq. (13) may be proved.

Lemma 2 *The value of* τ *is*

$$\tau = \sum_{\ell=0}^{\infty} \sum_{j=1}^{B} p_{\ell,j}(0) + \sum_{\ell=0}^{\infty} q_\ell(0) = \frac{1 - \left(1 - \delta_p\right) p_{0,0}}{\omega} \tag{14}$$

where

$$\begin{aligned}\omega = p_0^+ \bar{\upsilon} + q_0^+ \left(\delta_p \bar{\upsilon} + \left(1 - \delta_p\right) s_1\right) + \sum_{n=1}^{B} \Bigg\{ \left(p_n^+ + q_n^+\right) \sum_{j=n}^{B} y_j s_n \\ + \sum_{r=1}^{B} \left(p_{n+r}^+ + q_{n+r}^+\right) y_r s_r \Bigg\} \\ + \sum_{n=B+1}^{\infty} \sum_{r=1}^{B} \left(p_{n+r}^+ + q_{n+r}^+\right) y_r s_r\end{aligned} \tag{15}$$

Proof Putting $\delta_p = 0$ in (1), using it in (8) and then taking sum over r from 1 to B and n from 0 to ∞ in (8)–(11), we get

$$\begin{aligned}&\sum_{n=0}^{\infty} \left\{ \sum_{r=1}^{B} p^*_{n,r}(\theta) + q^*_n(\theta) \right\} \\ &= \left(\frac{1 - V^*(\theta)}{\theta}\right) \sum_{i=1}^{B} p_{0,i}(0) + \left(\frac{1 - \delta_p V^*(\theta) - \left(1 - \delta_p\right) \sum_{j=1}^{B} y_j S_1^*(\theta)}{\theta}\right) q_0(0) \\ &+ \sum_{n=1}^{B} \Bigg\{ \left(\sum_{i=1}^{B} p_{n,i}(0) + q_n(0) \right) \left(\frac{1 - \sum_{j=n}^{B} y_j S_n^*(\theta)}{\theta}\right) + \sum_{r=1}^{B} \left(\sum_{i=1}^{B} p_{n+r,i}(0) + q_{n+r}(0) \right) \\ &\left(\frac{1 - y_r S_r^*(\theta)}{\theta}\right) \Bigg\} \\ &+ \sum_{n=B+1}^{\infty} \sum_{r=1}^{B} \left(\sum_{i=1}^{B} p_{n+r,i}(0) + q_{n+r}(0) \right) \left(\frac{1 - y_r S_r^*(\theta)}{\theta}\right)\end{aligned}$$

We take the limit when $\theta \to 0$, and use L'Hospitals rule. In addition, the use of the normalizing condition $\left(1 - \delta_p\right) p_{0,0} + \sum_{n=0}^{\infty} \sum_{r=1}^{B} p_{n,r} + \sum_{n=0}^{\infty} q_n = 1$ leads to the desired result.

4 Joint Probabilities of Queue Length and Server Content

Our salient aim is on the mathematical derivation of the bivariate pgf for which we define the pgfs as

$$P(z, x, \theta) = \sum_{n=0}^{\infty} \sum_{r=1}^{B} p_{n,r}^{*}(\theta) z^n x^r, \quad |z| \leq 1, \quad |x| \leq 1 \tag{16}$$

$$P^{+}(z, x) = \sum_{n=0}^{\infty} \sum_{r=1}^{B} p_{n,r}^{+} z^n x^r, \quad |z| \leq 1, \quad |x| \leq 1 \tag{17}$$

$$P^{+}(z, 1) = \sum_{n=0}^{\infty} \sum_{r=1}^{B} p_{n,r}^{+} z^n = \sum_{n=0}^{\infty} p_n^{+} z^n = P^{+}(z), \quad |z| \leq 1 \tag{18}$$

$$Q^{+}(z) = \sum_{n=0}^{\infty} q_n^{+} z^n, \quad |z| \leq 1. \tag{19}$$

We multiply (8)–(10) with suitable powers of z and x, then taking the sum over n from 0 to ∞ and r from 1 to B, and using (16) and (17), we get

$$\begin{aligned}(\lambda - \lambda z - \theta)\, P(z, x, \theta) = & \sum_{n=1}^{B} \left\{ \sum_{r=1}^{B} p_{n,r}(0) + q_n(0) \right\} x^n \sum_{j=n}^{B} y_j S_n^{*}(\theta) \\ & + \sum_{n=1}^{\infty} \sum_{r=1}^{B} \left\{ \sum_{i=1}^{B} p_{n+r,i}(0) + q_{n+r}(0) \right\} z^n x^r y_r S_r^{*}(\theta) \\ & + (1 - \delta_p)\, q_0(0) \sum_{j=1}^{B} y_j S_1^{*}(\theta) - \sum_{n=0}^{\infty} \sum_{r=1}^{B} p_{n,r}(0) z^n x^r \end{aligned} \tag{20}$$

Now substitution of $\theta = \lambda - \lambda z$ in (20) and use of (12), (13) and (17) leads to

$$\begin{aligned} P^{+}(z, x) = & \sum_{n=1}^{B} \left\{ p_n^{+} + q_n^{+} \right\} x^n \sum_{j=n}^{B} y_j K^{(n)}(z) + \sum_{n=1}^{\infty} \sum_{r=1}^{B} \left\{ p_{n+r}^{+} + q_{n+r}^{+} \right\} z^n x^r y_r K^{(r)}(z) \\ & + (1 - \delta_p)\, q_0^{+} \sum_{j=1}^{B} y_j K^{(1)}(z) \end{aligned} \tag{21}$$

where $K^{(r)}(z) = S_r^{*}(\lambda - \lambda z)$ is the pgf of $k_j^{(r)}$, $(1 \leq r \leq B,\ j \geq 0)$ and $k_j^{(r)} = Pr\{j$ customers arrive during service of a r sized batch$\}$.

Now multiplying (10)–(11) with suitable powers of z and taking sum over n from 0 to ∞, we get

$$(\lambda - \lambda z - \theta)\sum_{n=0}^{\infty} q_n^*(\theta)z^n = \left\{\sum_{r=1}^{B} p_{0,r}(0) + \delta_p q_0(0)\right\}V^*(\theta) - \sum_{n=0}^{\infty} q_n(0)z^n \tag{22}$$

Putting $\theta = \lambda - \lambda z$ in (22) and using (12), (13), and (19), we get

$$Q^+(z) = \left(p_0^+ + \delta_p q_0^+\right) H(z) \tag{23}$$

where $H(z) = \sum_{j=0}^{\infty} h_j z^j = V^*(\lambda - \lambda z)$ is the pgf of h_j together with $h_j = Pr\{j$ customers arrive during the random vacation taken by the server$\}$. Substitution of $x = 1$ in (21), use of (18), (23) and some algebraic manipulation leads to

$$P^+(z) = \frac{\begin{array}{c} p_0^+\left\{H(z) - 1\right\}\left(\sum_{i=1}^{B} y_i z^{B-i}K^{(i)}(z)\right) + q_0^+\left\{\delta_p H(z) - 1\right\}\left(\sum_{i=1}^{B} y_i z^{B-i}K^{(i)}(z)\right) \\ + z^B\left(1 - \delta_p\right) q_0^+ K^{(1)}(z) + \sum_{n=1}^{B-1}\left(p_n^+ + q_n^+\right)\sum_{j=n+1}^{B} y_j\left\{z^B K^{(n)}(z) - z^{B-j+n}K^{(j)}(z)\right\} \end{array}}{z^B - \sum_{i=1}^{B} y_i z^{B-i}K^{(i)}(z)} \tag{24}$$

The expression stands for the pgf of only queue content at service culmination epoch, and to the best of authors' knowledge this is a new contribution in the queueing literature.

Now looking back into (21) and using (23), (24), we obtain

$$P^+(z,x) = \frac{\begin{array}{c} p_0^+\left\{H(z) - 1\right\}\left(\sum_{i=1}^{D} y_i x^i z^{B-i}K^{(i)}(z)\right) + q_0^+\left\{\delta_p H(z) - 1\right\}\left(\sum_{i=1}^{B} y_i x^i z^{B-i}K^{(i)}(z)\right) \\ + \left(1 - \delta_p\right) q_0^+\left[x\left(z^B - \sum_{i=1}^{B} y_i z^{B-i}K^{(i)}(z)\right) + \sum_{i=1}^{B} y_i x^i z^{B-i}K^{(i)}(z)\right]K^{(1)}(z) \\ + \sum_{n=1}^{B-1}\left(p_n^+ + q_n^+\right)\left[\left\{z^B - \sum_{i=1}^{B} y_i z^{B-i}K^{(i)}(z)\right\}\sum_{j=n+1}^{B} y_j\left\{x^n K^{(n)}(z) - x^j z^{n-j}K^{(j)}(z)\right\}\right. \\ \left. + \sum_{j=n+1}^{B} y_j\left\{K^{(n)}(z) - z^{n-j}K^{(j)}(z)\right\}\sum_{i=1}^{B} y_i x^i z^{B-i}K^{(i)}(z)\right] \end{array}}{z^B - \sum_{i=1}^{B} y_i z^{B-i}K^{(i)}(z)} \tag{25}$$

The expression signifies the bivariate pgf of queue length along with server content distribution at departure epoch, which was the main focus of this analysis. The literature exhibits the unavailability of such result so far, to the best of the authors' knowledge. We feel that the use of SVT leads this in a simple and elegant way rather than using a transition probability matrix.

4.1 Determination of Unknowns in the Numerator of (25)

The first requirement is to acquire the unknown quantities p_n^+, q_n^+, $(0 \leq n \leq B-1)$, arising in the numerator of (25). Since both (24) and (25) contains the same unknowns, we use (24) without any loss of generality. We discuss this routine work for the sake of completeness. From Rouché's theorem it is known that $D(z) = z^B - \sum_{i=1}^{B} y_i z^{B-i} K^{(i)}(z)$ (say) possesses B zeroes in the unit circle $|z| \leq 1$. We have total $(B+1)$ unknown quantities as: p_0^+, q_0^+, $c_n = p_n^+ + q_n^+$, $1 \leq n \leq B-1$, which will be reduced to B unknowns using $q_0^+ = \frac{h_0 p_0^+}{1-\delta_p h_0}$. Use of these zeroes of $D(z)$ in the unit circle and the normalizing condition leads to B linear simultaneous equations. The requirement of the desired unknowns can be accomplished by solving them.

4.2 Extraction Procedure of Joint Probabilities

Now we possess the known bivariate pgf in our grasp. We now extract the joint probabilities for which we assemble the coefficients of x^l, $(1 \leq l \leq B)$ from the bivariate pgf.

Coefficient of x:

$$\sum_{n=0}^{\infty} p_{n,1}^+ z^n = \frac{\begin{aligned} & y_1 z^{B-1} K^{(1)}(z) \left\{ p_0^+ \left(H(z) - 1 \right) + q_0^+ \left(\delta_p H(z) - 1 \right) \right\} + (1-\delta_p) q_0^+ D(z) K^{(1)}(z) \\ & + (1-\delta_p) q_0^+ y_1 z^{B-1} \left(K^{(1)}(z) \right)^2 + K^{(1)}(z) \left\{ D(z) (p_1^+ + q_1^+) \sum_{j=2}^{B} y_j \right\} \\ & + y_1 z^{B-1} \sum_{n=1}^{B-1} \left(p_n^+ + q_n^+ \right) \sum_{i=n+1}^{B} y_i \left(K^{(n)}(z) - z^{n-i} K^{(i)}(z) \right) \end{aligned}}{z^B - \sum_{i=1}^{B} y_i z^{B-i} K^{(i)}(z)} \tag{26}$$

Coefficient of x^l, $2 \leq l \leq B-1$:

$$\sum_{n=0}^{\infty} p_{n,l}^+ z^n = \frac{\begin{aligned} & y_l z^{B-l} K^{(l)}(z) \left\{ p_0^+ \left(H(z) - 1 \right) + q_0^+ \left(\delta_p H(z) - 1 \right) \right\} + (1-\delta_p) q_0^+ y_l z^{B-l} K^{(l)}(z)\, K^{(1)}(z) \\ & + K^{(l)}(z) \Big[D(z) \left(\left(p_l^+ + q_l^+ \right) \sum_{i=l+1}^{B} y_i - \sum_{n=1}^{B-1} \left(p_n^+ + q_n^+ \right) z^{n-l} \right) \\ & + y_l z^{B-l} \sum_{n=1}^{B-1} \left(p_n^+ + q_n^+ \right) \sum_{i=n+1}^{B} y_i \left(K^{(n)}(z) - z^{n-i} K^{(i)}(z) \right) \Big] \end{aligned}}{z^B - \sum_{i=a}^{B} y_i z^{B-i} K^{(i)}(z)} \tag{27}$$

Coefficient of x^B:

$$\sum_{n=0}^{\infty} p_{n,B}^{+} z^n = \frac{y_B K^{(B)}(z)\left\{p_0^{+}\left(H(z)-1\right)+q_0^{+}\left(\delta_p H(z)-1\right)\right\}+(1-\delta_p)\, q_0^{+} y_B K^{(B)}(z)K^{(1)}(z) \sum_{n=1}^{B-1}\left(p_n^{+}+q_n^{+}\right)\left(y_B \sum_{i=n+1}^{B} y_i\left(K^{(n)}(z)-z^{n-i}K^{(i)}(z)\right)-D(z)K^{(B)}(z)z^{n-B}\right)}{z^B-\sum_{i=a}^{B} y_i z^{B-i} K^{(i)}(z)} \tag{28}$$

For the known service and vacation time distributions, we substitute $K^{(r)}(z) = S_r^{*}(\lambda - \lambda z),\ 1 \le r \le B$, and $H(z) = V^{*}(\lambda - \lambda z)$ in the above expressions. The numerator and denominator of $\sum_{n=0}^{\infty} p_{n,j}^{+} z^n,\ 1 \le j \le B$ are denoted by $\Lambda_j(z)$ and $D_j(z)$, possessing the degrees as L_j and M_j, respectively. Let $\alpha_{1,j}, \alpha_{2,j}, \ldots, \alpha_{M_j,j}$ be the total roots of $D_j(z) = 0$. After canceling out B insides roots, we are left with $M_j - B$ distinct roots in $|z| > 1$.

For $L_j \ge M_j$, the function $\sum_{n=0}^{\infty} p_{n,j}^{+} z^n$ has been expressed as

$$\sum_{n=0}^{\infty} p_{n,j}^{+} z^n = \sum_{i=0}^{L_j-M_j} \tau_{i,j} z^i + \sum_{k=1}^{M_j-B} \frac{\pi_{k,j}}{\alpha_{k,j}-z}, \tag{29}$$

for several constants $\tau_{i,j}$ and $\pi_{k,j}$'s, with $\tau_{i,j}$ as the coefficients of the quotient after the division of $\Lambda_j(z)$ by $D_j(z)$, and $\pi_{k,j} = -\frac{\Lambda_j(\alpha_{k,j})}{D_j'(\alpha_{k,j})},\quad k = 1, 2, \ldots, M_j - B$. Collection of coefficient of z^n from (29), we have

$$p_{n,j}^{+} = \begin{cases} \tau_{n,j} + \sum_{k=1}^{M_j-B} \frac{\pi_{k,j}}{\alpha_{k,j}^{n+1}}, & 0 \le n \le L_j - M_j, \\ \sum_{k=1}^{M_j-B} \frac{\pi_{k,j}}{\alpha_{k,j}^{n+1}}, & n > L_j - M_j\,. \end{cases} \tag{30}$$

Whenever $L_j < M_j$, the first summation term of (29) will not appear and another remaining term will contribute to the joint distribution.

Remark 1 When $D_j(z) = 0$ has multiple roots, we need to modify the procedure slightly, which is elaborately discussed by Pradhan [9]. Evoking similar approach, one can find the probabilities $q_n^{+},\ n \ge 0$ from $Q^{+}(z)$.

5 Probabilities of Queue Length Along with Server Content at Random Epoch

From the joint distributions at random/arbitrary epoch, one can compute several performance measures. We now generate a correspondence between the joint distribution at departure and random epochs.

Theorem 1 *The probabilities at departure and arbitrary epochs are related by*

$$p_{0,0} = \frac{(1-\delta_p)q_0^+}{E^*} \tag{31}$$

$$p_{0,1} = \frac{1}{E^*}\left\{(p_1^+ + q_1^+) - p_{0,1}^+\right\} + (1-\delta_p)\, p_{0,0} \tag{32}$$

$$p_{0,r} = \frac{1}{E^*}\left\{(p_r^+ + q_r^+)\sum_{j=r}^{B} y_j - p_{0,r}^+\right\},\ 2 \le r \le B \tag{33}$$

$$p_{n,r} = \frac{1}{E^*}\left\{(p_{n+r}^+ + q_{n+r}^+)\, y_r - p_{n,r}^+\right\} + p_{n-1,r},\ n \ge 1,\ 1 \le r \le B \tag{34}$$

$$q_0 = \frac{1}{E^*}\left\{p_0^+ - (1-\delta_p)\, q_0^+\right\} \tag{35}$$

$$q_n = q_{n-1} - \frac{q_n^+}{E^*},\ n \ge 1 \tag{36}$$

where $E^* = \lambda\omega + (1-\delta_p)\, q_0^+$ *and* ω *is given in lemma 2.*

Proof Using Eq. (1) and setting $\theta = 0$ in (8)–(11), after simplification, we arrive at our desired results.

6 Performance Measures

In order to upgrade the system productivity, performance measures play a supreme role. Below we exhibit some pertinent performance measures which are key ingredients from designing perspective.

- Mean system length $(L) = \sum_{n=0}^{\infty} n p_n^{system}$, where p_n^{system}, $n \ge 0$ designates the probability of aggregated number in the system (includes the number in the queue and server), and given as

$$p_n^{system} = \begin{cases} (1-\delta_p)p_{0,0} + q_0\ , & \\ \sum_{r=1}^{min(B,n)} p_{n-r,r} + q_n, & 1 \le n \le B, \\ \sum_{r=1}^{B} p_{n-r,r} & n \ge B+1. \end{cases}$$

- Mean length of the queue at random time $(L_q) = \sum_{n=0}^{\infty} n p_n^{queue}$, where p_n^{queue}, $n \ge 0$ designates the probability distribution of the number in the queue, and are given as

$$p_n^{queue} = \begin{cases} (1-\delta_p)p_{0,0} + \sum_{r=1}^{B} p_{0,r} + q_0 \,, & \\ \sum_{r=1}^{B} p_{n,r} + q_n & n \geq 1. \end{cases}$$

- Use of Little's law leads to the average waiting time of a customer in the queue $(W_q) = \frac{L_q}{\lambda}$ as well as in the system $(W) = \frac{L}{\lambda}$.
- Mean length of the server $(L_s) = \sum_{r=1}^{B} r p_r^{server}$, where $p_r^{server} = \sum_{n=0}^{\infty} p_{n,r}/P_{busy}$ $(1 \leq r \leq B)$ with the probability that the server is busy, i.e., $P_{busy} = \sum_{n=0}^{\infty} \sum_{r=1}^{B} p_{n,r}$.

7 Numerical Illustrations

In order to find the unknown probabilities, we compute the zeroes of the characteristic equation and solve simultaneous equations using Maple 15. Here we provide one numerical example to make apparent the qualitative feature of the concerned queue. We consider $B = 9$ and service batch-size distribution as uniform with $y_i = \frac{1}{9}$, $i = 1, 2, \ldots, 9$, arrival rate $\lambda = 1.5$. Both the service and vacation time follow exponential distribution with pdf $s_i(t) = \mu_i \ e^{-\mu_i t}$, $t \geq 0$ with $\mu_1 = 3.5000, \mu_2 = 1.7500, \mu_3 = 1.1667, \mu_4 = 0.8750, \mu_5 = 0.7000, \mu_6 = 0.5833, \mu_7 = 0.5000$, $\mu_8 = 0.4375$, $\mu_9 = 0.3889$, which are decreasing with the increase of batch size, and $v(t) = \eta \ e^{-\eta t}$, $t \geq 0$ with $\eta = 0.4$, mean $\bar{v} = 2.5$ so that $\rho = 0.428571$. In Tables 1, 2, 3, 4, the joint distributions at departure together with random epochs are displayed.

8 Conclusion

This article successfully analyzed a vacation queueing model with random service rule where service time varies with the batch size. We have derived the joint probabilities of queue length together with server content at two epochs. Theoretical results of the corresponding queue are validated with numerical examples. It is expected that the joint probabilities and performance measures will be worthwhile in the designing of chemical processing industries, shared resources maintenance etc.

Table 1 Departure epoch probabilities in case of single vacation

n	$p^+_{n,1}$	$p^+_{n,2}$	$p^+_{n,3}$	$p^+_{n,4}$	$p^+_{n,5}$	$p^+_{n,6}$	$p^+_{n,7}$	$p^+_{n,8}$	$p^+_{n,9}$	p^+_n	q^+_n
0	0.166650	0.053870	0.026929	0.014590	0.008157	0.004541	0.002427	0.001172	0.000429	0.278769	0.058688
1	0.058749	0.029598	0.018035	0.011103	0.006851	0.004175	0.002469	0.001379	0.000687	0.133051	0.046332
2	0.023780	0.017214	0.012388	0.008506	0.005701	0.003733	0.002374	0.001447	0.000825	0.075972	0.036578
3	0.011754	0.010706	0.008742	0.006565	0.004713	0.003272	0.002201	0.001427	0.000880	0.050263	0.028877
4	0.007115	0.007124	0.006333	0.005102	0.003878	0.002827	0.001990	0.001352	0.000881	0.036605	0.022798
5	0.004972	0.005030	0.004698	0.003992	0.003179	0.002415	0.001766	0.001246	0.000846	0.028148	0.017998
15	0.000516	0.000470	0.000438	0.000416	0.000396	0.000369	0.000333	0.000290	0.000243	0.003475	0.001692
30	0.000019	0.000018	0.000016	0.000015	0.000015	0.000015	0.000016	0.000016	0.000016	0.000150	0.000048
≥60	0.000000	0.000000	0.000000	0.000000	0.000000	0.000000	0.000000	0.000000	0.000000	0.000000	0.000000

Table 2 Random epoch probabilities in case of single vacation

n	$p_{n,1}$	$p_{n,2}$	$p_{n,3}$	$p_{n,4}$	$p_{n,5}$	$p_{n,6}$	$p_{n,7}$	$p_{n,8}$	$p_{n,9}$	p_n^{queue}	q_n
0	0.036965	0.023898	0.017920	0.012945	0.009046	0.006043	0.003768	0.002080	0.000858	0.257808	0.113906
1	0.013031	0.013130	0.012001	0.009852	0.007599	0.005557	0.003834	0.002447	0.001373	0.158752	0.089925
2	0.005274	0.007636	0.008243	0.007547	0.006323	0.004968	0.003686	0.002569	0.001648	0.118892	0.070994
3	0.002607	0.004749	0.005817	0.005824	0.005227	0.004355	0.003417	0.002532	0.001758	0.092339	0.056047
4	0.001578	0.003160	0.004214	0.004527	0.004301	0.003763	0.003090	0.002399	0.001759	0.073041	0.044248
5	0.001103	0.002231	0.003126	0.003542	0.003526	0.003214	0.002742	0.002211	0.001689	0.058320	0.034932
15	0.000114	0.000208	0.000291	0.000369	0.000439	0.000492	0.000518	0.000515	0.000485	0.006721	0.003285
30	0.000004	0.000008	0.000011	0.000014	0.000017	0.000020	0.000024	0.000029	0.000032	0.000257	0.000094
≥60	0.000000	0.000000	0.000000	0.000000	0.000000	0.000000	0.000000	0.000000	0.000000	0.000000	0.000000

L=5.500075, L_q=3.700748, L_s=4.198430,

$p_{0,0} = 0.030374$, W=2.467165, W_q=3.666716

Table 3 Departure epoch probabilities in case of multiple vacation

n	$p^+_{n,1}$	$p^+_{n,2}$	$p^+_{n,3}$	$p^+_{n,4}$	$p^+_{n,5}$	$p^+_{n,6}$	$p^+_{n,7}$	$p^+_{n,8}$	$p^+_{n,9}$	p^+_n	q^+_n
0	0.123038	0.055690	0.028675	0.015762	0.008873	0.004956	0.002653	0.001283	0.000470	0.241403	0.064374
1	0.045961	0.030745	0.019249	0.012010	0.007458	0.004559	0.002700	0.001509	0.000753	0.124946	0.050821
2	0.020342	0.018029	0.013268	0.009215	0.006210	0.004078	0.002597	0.001584	0.000904	0.076231	0.040122
3	0.011094	0.011324	0.009399	0.007124	0.005138	0.003576	0.002408	0.001562	0.000964	0.052593	0.031675
4	0.007232	0.007609	0.006835	0.005546	0.004231	0.003091	0.002177	0.001480	0.000965	0.039169	0.025007
5	0.005267	0.005417	0.005087	0.004345	0.003471	0.002641	0.001933	0.001364	0.000927	0.030456	0.019742
15	0.000565	0.000515	0.000480	0.000456	0.000434	0.000405	0.000365	0.000318	0.000266	0.003808	0.001856
30	0.000021	0.000019	0.000018	0.000017	0.000017	0.000017	0.000017	0.000018	0.000018	0.000165	0.000053
≥60	0.000000	0.000000	0.000000	0.000000	0.000000	0.000000	0.000000	0.000000	0.000000	0.000000	0.000000

Table 4 Probabilities at random epoch in case of multiple vacation

n	$p_{n,1}$	$p_{n,2}$	$p_{n,3}$	$p_{n,4}$	$p_{n\,5}$	$p_{n,6}$	$p_{n,7}$	$p_{n,8}$	$p_{n,9}$	p_n^{queue}	q_n
0	0.026277	0.023788	0.018372	0.013466	0.009475	0.006351	0.003967	0.002192	0.000904	0.225096	0.120300
1	0.009816	0.013132	0.012333	0.010260	0.007364	0.005842	0.004037	0.002579	0.001447	0.162388	0.094974
2	0.004344	0.007701	0.008501	0.007873	0.006532	0.005225	0.003882	0.002708	0.001738	0.123586	0.074979
3	0.002369	0.004837	0.006022	0.006086	0.005487	0.004583	0.003600	0.002669	0.001854	0.096705	0.059194
4	0.001544	0.003250	0.004379	0.004737	0.004518	0.003961	0.003256	0.002529	0.001855	0.076766	0.046732
5	0.001124	0.002313	0.003259	0.003712	0.003706	0.003385	0.002890	0.002331	0.001782	0.061401	0.036894
15	0.000120	0.000220	0.000307	0.000390	0.000464	0.000519	0.000546	0.000543	0.000512	0.007095	0.003470
30	0.000004	0.000008	0.000011	0.000014	0.000018	0.000021	0.000026	0.000030	0.000034	0.000271	0.000100
≥60	0.000000	0.000000	0.000000	0.000000	0.000000	0.000000	0.000000	0.000000	0.000000	0.000000	0.000000

L=5.757474, L_q=3.891789, L_s=4.353265,
W=3.838316, W_q=2.594526

References

1. Banerjee A, Gupta UC (2012) Reducing congestion in bulk-service finite-buffer queueing system using batch-size-dependent service. Perform Eval 69(1):53–70
2. Banerjee A, Gupta UC, Sikdar K (2013) Analysis of finite-buffer bulk-arrival bulk-service queue with variable service capacity and batch-size-dependent service. Int J Math Oper Res 5(3):358–386
3. Banerjee A, Gupta UC, Goswami V (2014) Analysis of finite-buffer discrete-time batch-service queue with batch-size-dependent service. Comput Ind Eng 75, 121–128
4. Claeys D, Walraevens J, Laevens K, Bruneel H (2011) Analysis of threshold-based batch-service queueing systems with batch arrivals and general service times. Perform Eval 68(6):528–549
5. Claeys D, Steyaert B, Walraevens J, Laevens K, Bruneel H (2013a) Analysis of a versatile batch-service queueing model with correlation in the arrival process. Perform Eval 70(4):300–316
6. Claeys D, Steyaert B, Walraevens J, Laevens K, Bruneel H (2013b) Tail probabilities of the delay in a batch-service queueing model with batch-size dependent service times and a timer mechanism. Comput Oper Res 40(5):1497–1505
7. Pradhan S, Gupta UC (2019) Analysis of an infinite-buffer batch-size-dependent service queue with Markovian arrival process. Ann Oper Res 277:161–196
8. Pradhan S, Gupta UC (2020) Stationary queue and server content distribution of a batch-size-dependent service queue with batch Markovian arrival process: $BMAP/G_n^{(a,b)}/1$. Commun Stat-Theory Methods https://doi.org/10.1080/03610926.2020.1813304
9. Pradhan S (2020) A discrete-time batch transmission channel with random serving capacity under batch-size-dependent service: $Geo^X/G_n^Y/1$. Int J Comput Math: Comput Syst Theory 5:175–197
10. Doshi BT (1986) Queueing systems with vacationsa survey. Queueing Syst 1(1):29–66
11. Reddy GVK, Anitha R (1998) Markovian bulk service queue with delayed vacations. Comput Oper Res 25(12):1159–1166
12. Gupta UC, Sikdar K (2004) The finite-buffer $M/G/1$ queue with general bulk-service rule and single vacation. Perform Eval 57(2):199–219
13. Sikdar K, Gupta UC (2005) Analytic and numerical aspects of batch service queues with single vacation. Comput Oper Res 32(4):943–966
14. Chang SH, Choi DW (2005) Performance analysis of a finite-buffer discrete-time queue with bulk arrival, bulk service and vacations. Comput Oper Res 32(9):2213–2234
15. Samanta SK, Chaudhry ML, Gupta UC (2007) Discrete-time $Geo^X/G^{(a,b)}/1/N$ queues with single and multiple vacations. Math Comput Model 45(1):93–108

Computing Prices for Target Profits in Contracts

Ghurumuruhan Ganesan

Abstract Price discrimination for maximizing expected profit is a well-studied concept in economics and there are various methods that achieve the maximum given the user type distribution and the budget constraints. In many applications, particularly with regard to engineering and computing, it is often the case than the user type distribution is unknown or not accurately known. In this paper, we therefore propose and study a mathematical framework for price discrimination with *target* profits under the contract-theoretic model. We first consider service providers with a given user type profile and determine sufficient conditions for achieving a target profit. Our proof is constructive in that it also provides a method to compute the quality-price tag menu. Next we consider a dual scenario where the offered service qualities are predetermined and describe an iterative method to obtain nominal demand values that best match the qualities offered by the service provider while achieving a target profit-user satisfaction margin. We also illustrate our methods with design examples in both cases.

Keywords Price discrimination · Target profits · Computational methods

1 Introduction

Pricing of a product with different qualities is an important topic that has been studied extensively in economics under various scenarios. To determine the price tags for the various qualities, the company making the product usually performs extensive market research and creates a segmentation of the market into distinct groups, depending on the number of qualities available. Using a contract-theoretic model [1], it is then possible to provide enough incentive for the customers to buy the best quality product within their budget constraints.

The three main issues in any segmentation problem are service quality, user demand, and the corresponding price tag. In this paper, we are interested in deter-

G. Ganesan (✉)
Institute of Mathematical Sciences HBNI, Chennai, India

© The Author(s), under exclusive license to Springer Nature Singapore Pte Ltd. 2022

D. Giri et al. (eds.), *Proceedings of the Seventh International Conference on Mathematics and Computing*, Advances in Intelligent Systems and Computing 1412,
https://doi.org/10.1007/978-981-16-6890-6_79

mining segmentations that achieve a target profit. There are many reasons for such a consideration. For example, suppose the user type distribution is not exactly known *a priori* and therefore it might not be feasible to compute the true expected profit. This happens in practice when a new technology comes into practice and companies providing the technology do not have sufficient background data regarding user preferences in order to estimate the user type distribution and thereby maximize the overall expected profit.

Another scenario where target profit may be useful, is when the user types providing the largest profit to the service provider, occur with very low frequency. Thus while the expected profit may be large for the service provider, the actual (random) profit obtained from a randomly chosen user type might itself be low, with large probability. Ensuring a (minimum) target profit helps offset such vagaries.

Related Work

In economics, [2] studied the multiproduct monopolist problem where a seller wants to provide a certain service at various qualities to cater to users of different types. The user type distribution has a density and the monopolist seeks to maximize the overall expected profit. Later in [3] price discrimination for a many-product firm for customers with unobservable tastes (actions) is investigated. The goal there is to determine two-part tariffs that maximize the profits for the seller by extracting as much consumer surplus as possible. The customer actions are modeled as random variables and the tariff chosen maximizes the expected profit when the user action distribution is known. A related continuous parameter version of the problem is considered in [4] who introduce a sweeping operator in the context of multi-dimensional monopolist problem.

Similar economic issues arise in technological products as well. For example, [5] uses contract-theory to study the problem of pricing the available qualities of transmission channels for the secondary (cognitive) users satisfying predetermined satisfaction levels. The type of a user is determined by its distance from the base station and the goal is to determine the prices for a given set of qualities and types of users (see also the book [6] for more details). Another example is the problem of spectrum trading in cognitive radio networks where primary users *lease* spectrum to the secondary cognitive users with strict constraints regarding the time of usage and transmission power level. An adaptive learning algorithm is used in [7] to estimate the instantaneous price of the spectrum and adjust individual demands accordingly.

Cloud and utility computing services are yet another area where pricing pays an important role with regard to usage of computing resources. In [8], the pros and cons of charging fixed prices for metered usage of the services is analyzed and it is demonstrated via implementations that a variable pricing system which adjusts rates according to demand and quality of service requirements (QoS) is more beneficial in terms of revenue for the service. Finally, we remark that recently [9] has studied resource allocation for network slices in 5G systems via network pricing and [10] investigate a multi-dimensional contract approach for data rewarding in mobile networks.

Our contributions

To achieve target profits, we consider two possible cases: In the first case, the user type profile is fixed and we create a quality-price segmentation that achieves the target profit for the provider. In Theorem 1 stated in Sect. 2, we determine conditions on the cost and the user budget functions that ensure that a target profit margin B is achieved. Our proof also provides a computational method to determine the service qualities offered and the corresponding price tags that best match the customer budget.

In the second case (Theorem 2, Sect. 3), the service quality profile is fixed and we describe an iterative computational procedure to obtain the corresponding nominal demand values that achieve the desired profit. We then set mutually beneficial prices for each of these nominal values so that any user can afford the nominal service quality closest to its own demand. We also study tradeoffs between the achievable profit for the service provider and the customer satisfaction margin.

The paper is organized as follows: In Sect. 2 we study achievable target profits for a given user demand profile and in Sect. 3 we describe conditions for achieving a target profit for a given service quality profile while also ensuring customer satisfaction.

2 Target Profits with Given Demand Profile

A certain service provider wishes to offer services at L different qualities to cater to L different types of buyers with the aim of achieving a target profit. It costs $C(s)$ to offer service at quality $s > 0$ and users of type i, $1 \leq i \leq L$ are willing to pay at most $P_i(s)$ for this service quality. The service provider would like to make a profit of $B(s)$ for offering service at quality s. Typically $B(s)$ is a fraction of $C(s)$ and we say that the profit margin $B = B(.)$ is *achievable* if there are quality values $0 < s_1 < s_2 < \ldots < s_L$ and corresponding price tags $0 < p_1 < \ldots < p_L$ satisfying the following for each $1 \leq k \leq L$:

$$\begin{aligned} &(x) \quad P_k(s_k) \geq p_k \geq C(s_k) + B(s_k) \text{ and} \\ &(y) \quad P_k(s_k) - p_k \geq P_k(s_l) - p_l \text{ for all } 1 \leq l \neq k \leq L. \end{aligned} \tag{1}$$

We refer to condition (x) as the modified *individual rationality* (IR) constraint that ensures that the price tag for service quality s_k lies within the budget constraint of the buyer type k, while also making a target profit $B(s_k)$ for the service provider. If $B(s_k) = 0$, then condition (x) is simply called the individual rationality constraint [1] and ensures that the service provider does not make a loss while providing the service at the given quality. Condition (y) is called the *incentive compatibility* (IC) constraint that provides the necessary incentive (in terms of savings) for the buyer of type k to actually buy service at quality s_k (among all the qualities offered).

The following result determines conditions under the target profit B is achievable. Throughout $f'(.)$ denotes the derivative of a function $f(.)$.

Theorem 1 *Suppose the functions $\{P_i(.)\}$ and $C(.)$ satisfy the following conditions:*
$(a1)$ The cost and profit functions $C(s)$ and $B(s)$ are differentiable, strictly increasing and convex in s with $C(0) = B(0) = 0$.
$(a2)$ For $1 \leq i \leq L$, the price function $P_i(s)$ is differentiable, strictly increasing and strictly concave in s with $P_i(0) = 0$. Moreover for $i < j$, the derivatives $P_i'(s) < P_j'(s)$ for all $s > 0$.
$(a3)$ There exists $x_1, y_L > 0$ such that $P_1(x_1) \geq C(x_1) + B(x_1)$ and $P_L(y_L) < C(y_L) + B(y_L)$.

If conditions $(a1) - (a3)$ hold, then the target profit $B = B(.)$ is achievable.

Conditions $(a1) - (a2)$ in Theorem 1 are usually true in practice since we expect that both the cost and the allowed budget to increase with the service quality. Condition $(a2)$ is also called the *single crossing condition* [11]. Condition $(a3)$ in Theorem 1 ensures that the buyer with the smallest type finds a service quality within its budget and even the buyer with the largest type cannot afford very high-quality services.

Proof of Theorem 1

For $1 \leq i \leq L$ the function

$$f_i(s) := P_i(s) - C(s) - B(s)$$

is differentiable and strictly concave with $f_i(0) = 0$. Moreover, the set

$$A_i := \{f_i \geq 0\}$$

is a closed and bounded (i.e., compact) interval containing at least two points: Indeed, using condition $(a2)$, we have for $i > j$ that

$$P_i(s) = \int_0^s P_i'(y)dy > \int_0^s P_j'(y)dy = P_j(s)$$

and so

$$A_1 \subset A_2 \subset \ldots \subset A_L.$$

Using $P_1(0) = 0 = C(0) + B(0)$ and $P_1(x_1) \geq C(x_1) + B(x_1)$ (condition $(a3)$) we then get that $\{0, x_1\} \subseteq A_1$.

If $x, y \in A_i$, then using concavity of f_i, we have for any $0 < \lambda < 1$ that

$$f_i(\lambda x + (1-\lambda)y) \geq \lambda f_i(x) + (1-\lambda) f_i(y) \geq 0.$$

Thus A_i is an interval and using continuity of f_i, we have that A_i is closed. The boundedness of A_L follows from the fact that $f_L(y_L) < 0$ (condition $(a3)$).

We now set $s_i > 0$ to be such that

$$f_i(s_i) = \max_{s \in A_i} f_i(s) = P_i(s_i) - C(s_i) - B(s_i) \geq 0 \tag{2}$$

and let $p_i := C(s_i) + B(s_i)$. The derivative f_i' is strictly decreasing and so s_i is uniquely defined, since it must satisfy $f_i'(s_i) = 0$. Also, by condition $(a2)$ we have that

$$f_2'(s_2) = 0 = f_1'(s_1) < f_2'(s_1)$$

and so $s_1 < s_2$. This also implies that

$$p_1 = C(s_1) + B(s_1) < C(s_2) + B(s_2) = p_2$$

since both $C(.)$ and $B(.)$ are strictly increasing (condition $(a1)$). Continuing iteratively we obtain that

$$0 < s_1 < s_2 < \ldots < s_L \text{ and } 0 < p_1 < p_2 < \ldots < p_L.$$

Moreover, from (2) we get that $P_i(s_i) \geq p_i = C(s_i) + B(s_i)$ so that condition (x) in (1) is true and again using (2), we get

$$P_i(s_i) - p_i = f_i(s_i) > f_i(s_j) = P_i(s_j) - p_j$$

for all $1 \leq i \neq j \leq L$. This implies that condition (y) in (1) is also true. □

We now illustrate the above proof with an example.

Example

Suppose the cost for providing service at quality s increases linearly with s and the service provider sets a target profit of 10%. In our notation we set

$$C(s) = D_c \cdot s \text{ and } B(s) = \frac{C(s)}{10} = D_c \cdot \frac{s}{10}$$

be the cost and profit functions, respectively, for some constant $D_c > 0$. Also suppose that there are L types of users and the user budget increases *logarithmically* in s and is proportional to the type number. Formally, we set

$$P_i(s) = D_b \cdot i \cdot \log(1+s)$$

for some positive constant D_b.

With the above definitions, conditions $(a1)-(a2)$ in Theorem 1 are satisfied. Moreover, if $D_b > \frac{11D_c}{10}$ strictly, then the condition $(a3)$ in Theorem 1 is true as well, since

$$P_1(s) = D_b \cdot \log(1+s) \geq \frac{11 \cdot D_c}{10} \cdot s = C(s) + B(s)$$

for all small $s > 0$ and

$$P_L(s) = D_b \cdot L \cdot \log(1+s) < \frac{11D_c s}{10} = C(s) + B(s)$$

for all large s. From (2), we therefore get that the quality s_i that achieves the target profit for the service must satisfy $f_i(s_i) = \max_s f_i(s)$ where

$$f_i(s) := D_b \cdot i \cdot \log(1+s) - \frac{11D_c s}{10}.$$

Differentiating $f_i(s)$ and equating to zero, we get that $s_i = \frac{10D_b}{11D_c} \cdot i - 1$. Thus the service quality offered to a particular user varies linearly with its type.

3 Target Profits with Given Quality Profile

Consider a service provider who already offers services at predetermined qualities and would like to assign price tags to these qualities based on the user demands. There are L different qualities $s_{low} \leq s_1 < s_2 < \ldots s_{L-1} < s_L \leq s_{up}$ and users with different demands request corresponding qualities from the service provider, taking into account the budget constraint. We assume that all user demands lie within an interval $[\theta_{low}, \theta_{up}]$ in the positive real line. A user with demand $\theta \in [\theta_{low}, \theta_{up}]$ is willing to pay up to $F(\theta, s)$ for service quality s, while it costs $C(s)$ for the service provider to deliver at such quality.

It is of interest to determine nominal demand values

$$\theta_{low} \leq \theta_1 < \theta_2 < \ldots \theta_L \leq \theta_{up}$$

tagged with corresponding prices $p_1 < p_2 < \ldots < p_L$ so that any user with demand $\theta \in [\theta_k, \theta_{k+1})$ is able to pay price p_k and utilize service at quality s_k. If the demand θ exactly equals θ_k, then the user also is able to save much more by choosing quality s_k than any of the other qualities. Formally, we prefer that conditions $(x) - (y)$ stated in (1) hold with $P_k(s_k) = F(\theta_k, s_k)$.

One bottleneck in the above formulation is that users with demands not equal to one of the nominal values are not guaranteed savings though they can still afford the services. It might possibly result in few number of users (those with demands exactly equal to one of the nominal values) being satisfied in terms of savings. In addition, the service provider itself may have a target profit in mind. Taking these into account, we have the following modified definition.

Let $\mathbf{b} = (b_1, \ldots, b_L)$, $\mathbf{m} = (m_1, \ldots, m_L)$ be any two $L-$tuples satisfying

$$0 < b_i < b_{i+1}, 0 < m_i < m_{i+1}, 1 \le i \le L - 1.$$

The service provider is said to achieve a profit-satisfaction margin $(\mathbf{b}, \mathbf{m})$ if there exists a demand-price profile $\{(\theta_k, p_k)\}_{1 \le k \le L}$ satisfying $\theta_{low} \le \theta_1 < \theta_2 < \ldots < \theta_L \le \theta_{up}$ and $0 < p_1 < p_2 < \ldots < p_L$ such that the following holds: For any $(\hat{\theta}_1, \ldots, \hat{\theta}_L)$ satisfying $|\hat{\theta}_k - \theta_k| \le m_k$, $1 \le k \le L$ we have

$$\begin{aligned} &(i)\ F(\hat{\theta}_k, s_k) \ge p_k \ge C(s_k) + b_k \text{ (IR) and} \\ &(ii)\ F(\hat{\theta}_k, s_k) - p_k \ge F(\hat{\theta}_k, s_l) - p_l \ \ \forall\ l \ne k \text{ (IC)}. \end{aligned} \tag{3}$$

Essentially conditions $(i) - (ii)$ ensure that users with demand in the range $[\theta_k - m_k, \theta_k + m_k]$ make savings if they choose service at quality s_k. At the same time, the service provider also secures a profit of b_k for providing the service at quality s_k. The quantities $\{\theta_k\}$ are the nominal demand values determined by the service provider that results in its net profit.

Throughout we assume that the function $F(\theta, s)$ has support in the compact interval $[\theta_{low}, \theta_{up}] \times [s_{low}, s_{up}]$, and that both its partial derivatives are positive, increasing and jointly continuous in both θ and s.. We also assume that

$$\frac{\partial^2 F(\theta, s)}{\partial \theta \partial s} = \frac{\partial^2 F(\theta, s)}{\partial s \partial \theta} > 0$$

is jointly continuous in both θ and s and for simplicity denote

$$F_s(\theta, s) := \frac{\partial F(\theta, s)}{\partial s}, \ F_\theta(\theta, s) := \frac{\partial F(\theta, s)}{\partial \theta} \text{ and } F_2(\theta, s) := \frac{\partial^2 F(\theta, s)}{\partial s \partial \theta}.$$

For $2 \le j \le L$, we define

$$\epsilon_j := \sup_\theta F_\theta(\theta, s_{j-1}) \text{ and } \delta_j := \inf_\theta \left(F_\theta(\theta, s_j) - F_\theta(\theta, s_{j-1})\right), \tag{4}$$

where the infimum and supremum are taken over all $\theta \in [\theta_{low}, \theta_{up}]$ and have the following result regarding the achievability of a target profit-satisfaction margin.

Theorem 2 *Suppose that*

$$\inf_{\theta} F_s(\theta, s) \geq \frac{dC(s)}{ds} \quad \forall s, \tag{5}$$

where the infimum and supremum are taken over all $\theta \in [\theta_{low}, \theta_{up}]$. *If the profit-satisfaction margin* $(\mathbf{b}, \mathbf{m})$ *satisfies*

$$\sum_{j=1}^{L} \Delta_j + m_L < \theta_{up} - \theta_{low}$$

and $P(\theta_{low}, s_1) \geq C(s_1) + b_1$,, *where* $\Delta_1 = m_1$ *and*

$$\Delta_j := (m_j + m_{j-1}) \left(1 + \frac{2\epsilon_j}{\delta_j}\right) + \frac{b_j - b_{j-1}}{\delta_j} \tag{6}$$

for $2 \leq j \leq L$, *then* $(\mathbf{b}, \mathbf{m})$ *is achievable.*

The condition in (5) (which we call the marginal budget increase condition) essentially says that the marginal budget increase for any user type is at least as large as the marginal cost increase. We obtain the desired profit-satisfaction margin by introducing an additional *profit constraint* that provides minimum incentives for users with demand θ_k to select service quality s_k than at a slightly better quality s_{k+1}. We then use an induction procedure analogous to [5] to demonstrate the achievability.

Example

For $\delta > 0$ and $0 < b, m < 1$ suppose

$$s_i = i \cdot \delta, b_i = b \cdot s_i \text{ and } m_i = m \cdot s_i$$

for $1 \leq i \leq L$ so that the service would like to achieve a flat profit of $(100b)\%$.. The cost of providing service $C(s) = D_c \cdot s$ for some constant $D_c > 0$ and we assume that the budget $F(\theta, s) = D_p \cdot \theta \cdot s$ where $D_p \geq \theta_{low}^{-1} D_c$ so that condition (5) is true.

From (4), we have for $2 \leq j \leq L$ that $\delta_j = D_p \cdot \delta$ and $\epsilon_j = D_p(j-1)\delta$ so and substituting into (4) we get that $\Delta_1 = ms_1 = m\delta$ and $\Delta_j = m\delta(2j-1)^2 + \frac{b}{D_p}$ for $2 \leq j \leq L$. Letting $\Delta(S) := s_{up} - s_{low}$ be the service quality range and $\Delta(\theta) := \theta_{up} - \theta_{low}$ denote the demand range, we then get

$$\sum_{j=1}^{L} \Delta_j + m_L < 4m\delta \sum_{j=1}^{L} j^2 + mL\delta < 4m\delta L^3 = 4m\Delta(S)L^2.$$

Therefore from Theorem 2, we get that the set of achievable profit-customer satisfaction margins is given by

$$\mathcal{A} := \{(b, m) : b > 0, 0 < m < 1 \text{ and}$$
$$m \cdot \left(4\Delta(S)L^2\right) + b \cdot \left(\frac{L}{D_p}\right) \leq \Delta(\theta)\Big\}. \quad (7)$$

The expression in (7) also determines the tradeoff between profit and customer satisfaction: For a given range of user demands, higher customer satisfaction can be obtained at the cost of reduced profit. For zero profit $b = 0$, the maximum achievable customer satisfaction margin is $m_0 := \min\left(1, \frac{\Delta(\theta)}{4\Delta(S)L^2}\right)$ and grows as the square of the inverse of the number of types L. Similarly for zero customer satisfaction margin $m = 0$, the maximum achievable profit is $b_0 := \frac{\Delta(\theta)}{D_p L}$ which grows inversely as the number of types.

To illustrate the tradeoff in (7), we consider the case of $L = 3$ service qualities with $D_c = 1$ and $D_p = \frac{1}{\theta_{low}} = 3$.. From (7) we then have that the tradeoff curve is

$$36m \cdot \Delta(S) + b = \Delta(\theta)$$

and in Fig. 1, we plot the tradeoff curves as a function of the normalized customer satisfaction margin 36m and the normalized profit margin b for various values of $\Delta(S)$ and $\Delta(\theta)$. From the solid and dashed lines in Fig. 1, we see that if the quality range $\Delta(S)$ is fixed, then any increase in the demand range $\Delta(\theta)$ results in increased profit for the same customer satisfaction margin. On the other hand, from the solid line and the marked line, we infer that increasing the quality range while the demand range remains fixed, might not be profitable.

Proof of Theorem 2

We prove a slightly stronger result. For $1 \leq k \leq L - 1$, let $f_k(\mathbf{b}) \geq b_{k+1} - b_k \geq 0$ be functions such that

$$F(\theta_{low}, s_1) \geq C(s_1) + b_1 \text{ and } \sum_{j=1}^{L} \Delta_j < \theta_{low} - \theta_{up}, \quad (8)$$

where $\Delta_1 = m_1$ and for $2 \leq j \leq L$,

$$\Delta_j := (m_j + m_{j-1})\left(1 + \frac{2\epsilon_j}{\delta_j}\right) + \frac{f_{j-1}(\mathbf{b})}{\delta_j} \quad (9)$$

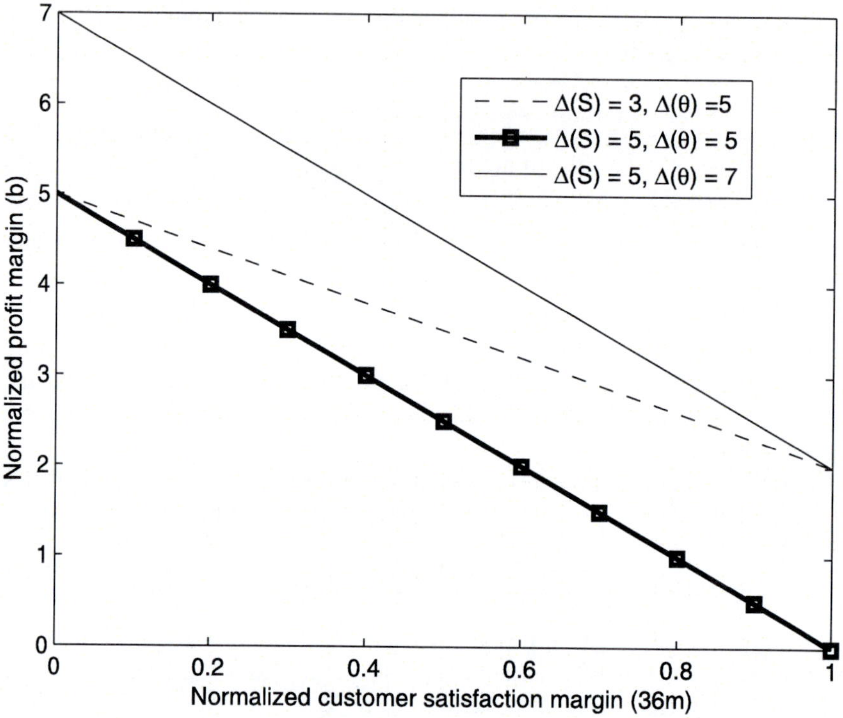

Fig. 1 Tradeoff curves for various values of $\Delta(S)$ and $\Delta(\theta)$

and ϵ_j and δ_j are as in (4). Letting $\omega_k := F(\theta_k - m_k, s_k) - p_k$, we prove that there are constants $\{(\theta_k, p_k)\}_{1\leq k\leq L}$ such that for each $1 \leq k, l \leq L$:

$$\begin{aligned} &(a)\ \omega_k \geq 0 \text{ and } p_k \geq C(q_k) + b_k.\\ &(b)\ \omega_k \geq f_k(\mathbf{b}) + F(\theta_k + m_k, s_{k+1}) - p_{k+1}.\\ &(c)\ \omega_k \geq F(\theta_k + m_k, s_l) - p_l, l \neq k. \end{aligned} \tag{10}$$

Condition (b) is an extra condition which we impose and call as the *profit constraint*. The profit constraint ensures that extra incentive is provided for user with demand θ_k to use service quality s_k than a slightly better quality s_{k+1}. Conditions (a) and (c) are modifications of the individual rationality constraint and the incentive compatibility constraints, respectively, discussed in Sect. 2. Because $F(\theta, s)$ is increasing in θ, Conditions (a) and (c), respectively, ensure that Conditions (i) and (ii) in (3) are satisfied.

Analogous to [5], we perform the desired pricing iteratively. Let $\theta_1 := \theta_{low} + m_1$ and let the price p_1 be such that $\omega_1 \geq p_1 \geq C(s_1) + b_1$.. This is possible by the statement of Theorem 2. Suppose now Conditions $(a) - (c)$ in (10) are true with L replaced by $j-1$ for some $2 \leq j \leq L-1$.. We determine (θ_j, p_j) as follows.

If $(a) - (c)$ were to be satisfied with L replaced by j, then from the profit constraint (b) we must have

$$
\begin{aligned}
p_j &\geq p_{j-1} + F(\theta_{j-1} + m_{j-1}, s_j) - \omega_{j-1} + f_{j-1}(\mathbf{b}) \\
&= p_{j-1} + \int_{s_{j-1}}^{s_j} F_s(\theta_{j-1} + m_{j-1}, s) \\
&\qquad + \int_{\theta_{j-1}-m_{j-1}}^{\theta_{j-1}+m_{j-1}} F_\theta(\theta, s_{j-1}) + f_{j-1}(\mathbf{b}) \\
&=: A_j.
\end{aligned} \tag{11}
$$

Similarly from IC constraint (c) we must have

$$
\begin{aligned}
p_j &\leq p_{j-1} + \omega_j - F(\theta_j + m_j, s_{j-1}) \\
&= p_{j-1} + \int_{s_{j-1}}^{s_j} F_s(\theta_j - m_j, s) - \int_{\theta_j - m_j}^{\theta_j + m_j} F_\theta(\theta, s_{j-1}) \\
&=: B_j.
\end{aligned} \tag{12}
$$

Thus we need to choose θ_j such that $J_1 - J_2 \geq f_{j-1}(\mathbf{b})$, where

$$
\begin{aligned}
J_1 &:= \int_{s_{j-1}}^{s_j} F_s(\theta_j - m_j, s) - \int_{s_{j-1}}^{s_j} F_s(\theta_{j-1} + m_{j-1}, s) \\
&= \int_{s_{j-1}}^{s_j} \int_{\theta_{j-1}+m_{j-1}}^{\theta_j - m_j} F_2(\theta, s) \\
&= \int_{\theta_{j-1}+m_{j-1}}^{\theta_j - m_j} F_\theta(\theta, s_j) - F_\theta(\theta, s_{j-1})
\end{aligned}
$$

and

$$
J_2 := \int_{\theta_j - m_j}^{\theta_j + m_j} F_\theta(\theta, s_{j-1}) + \int_{\theta_{j-1}-m_{j-1}}^{\theta_{j-1}+m_{j-1}} F_\theta(\theta, s_{j-1}).
$$

Recalling the definition of δ_j and ϵ_j from (4), we know that

$$
J_1 - J_2 \geq \delta_j(\theta_j - \theta_{j-1} - (m_j + m_{j-1})) - \epsilon_j(2m_j + 2m_{j-1})
$$

and so setting

$$
\theta_j := \theta_{j-1} + \Delta_j, \tag{13}
$$

where Δ_j is as in (6), we get the desired inequality $J_1 - J_2 \geq f_{j-1}(\mathbf{b})$. From (11) and (12) we also get that $A_j \leq B_j$ and so we choose p_j such that $A_j \leq p_j \leq B_j$.

With this choice of (θ_j, p_j) we now check if conditions $(a) - (c)$ in (10) are satisfied with L replaced by j and this would prove the induction step. Using (11) and the fact that $F_\theta > 0$ we get

$$p_j \geq p_{j-1} + \int_{s_{j-1}}^{s_j} F_s(\theta_{j-1} + m_{j-1}, s) + f_{j-1}(\mathbf{b}). \tag{14}$$

By the marginal budget increase condition (5), we know that

$$\int_{s_{j-1}}^{s_j} F_s(\theta_{j-1} + m_{j-1}, s) \geq C(s_j) - C(s_{j-1})$$

and so we get $p_j \geq C(s_j) + f_{j-1}(\mathbf{b}) + b_{j-1}$ since condition (a) in (10) holds with $k = j - 1$. Using $f_{j-1}(\mathbf{b}) \geq b_j - b_{j-1}$, we then get that $p_j \geq C(s_j) + b_j$, verifying the second condition of (a) with k replaced by j. To verify the first condition of (a), we have from (12) that

$$p_j \leq p_{j-1} + \omega_j - F(\theta_j + m_j, s_{j-1}) \leq p_{j-1} + \omega_j - \omega_{j-1}$$

since $F(\theta, s)$ is increasing in θ and $\theta_{j-1} - m_{j-1} < \theta_j + m_j$.. Using condition (a) with $L = j - 1$, we have $p_{j-1} \leq \omega_{j-1}$ and so $p_j \leq \omega_j$.. This verifies the first condition in (a) with L replaced by j.

By construction $A_j \geq B_j$ (see (11) and (12)) and so the profit constraint (b) in (10) is true with $k = j - 1$.. From (12) we have that Condition (c) in (10) is true with k replaced by j and l replaced by $j - 1$.. Further using $\theta_{j-1} - m_{j-1} < \theta_j + m_j$ and the fact that $F(\theta, s)$ is increasing in θ, we have for $1 \leq l \leq j - 1$ that $\omega_j - p_j$ is bounded below by

$$F(\theta_j + m_j, s_{j-1}) - p_{j-1} \geq \omega_{j-1} - p_{j-1} \geq F(\theta_l + m_l, s_l) - p_l,$$

since by induction assumption, Condition (c) in (10) is true when L and k are replaced by $j - 1$ and $l \leq j - 1$,, respectively.

For the reverse direction, we again use induction to get for $1 \leq l \leq j - 1$ that $\omega_l \geq F(\theta_l + m_l, s_{j-1}) - p_{j-1}$. Using (14) with $f_{j-1}(\mathbf{b}) \geq 0$ to estimate p_{j-1}, we get that $F(\theta_l + m_l, s_l) - p_l$ is at least

$$\begin{aligned} & F(\theta_l + m_l, s_{j-1}) - p_j + \int_{s_{j-1}}^{s_j} F_s(\theta_{j-1} + m_{j-1}, s) \\ \geq\ & F(\theta_l + m_l, s_{j-1}) - p_j + \int_{s_{j-1}}^{s_j} F_s(\theta_l + m_l, s) \\ =\ & F(\theta_l + m_l, s_j) - p_j, \end{aligned}$$

since $F_s(\theta, s)$ is increasing in θ and $\theta_l + m_l < \theta_{j-1} + m_{j-1}$.. This verifies condition (c) in (10) with L replaced by j, completing the induction step with L replaced by $j \leq L - 1$.

Finally, we get from the discussion following (13) that

$$\sum_{j=1}^{L-1} \Delta_j \leq \theta_{up} - \theta_{low} - \Delta_L - m_L$$

by (9). Thus $\theta_{up} - \theta_{L-1} \geq \Delta_L + m_L$ and setting $\theta_L := \theta_{L-1} + \Delta_L$ as in (13), we get that $\theta_L + m_L \leq \theta_{up}$.. Proceeding as in the discussion following (13), we then complete the induction step for $j = L$. □

4 Conclusion and Future Work

In this paper, we have described a computational framework for securing target profits for service providers under the contract-theoretic model for a given user type profile and a given service quality profile. We also illustrated our results with design examples. In the future, we plan to analyse further challenges that are unique to applications. As a typical example, consider a wireless network where some users are far from the base station and therefore may not have as good service quality as those close to the station. But if these far away users are willing to pay extra money, then how could the provider ensure high-quality service for such users? It would be interesting to study the profit-satisfaction tradeoff under such situations.

Acknowledgements I thank Professors Rahul Roy, Arunava Sen, C. R. Subramanian, and the referees for crucial comments that led to an improvement of the paper. I also thank IMSc for my fellowships.

References

1. Martimort D (2006) Contract Theory. In: Arrow KJ, Blume LE, Durlauf SN (eds) The New Palmgrave
2. Mussa M, Rosen S (1978) Monopoly and product quality. J Econ Theory 18:301–317
3. Armstrong M (1999) Price discrimination by a many-product firm. Rev Econ Stud 66:151–168
4. Rochet J-C, Choné P (1998) Ironing, sweeping and multidimensional screening. Econometrica 66:782–826
5. Gao L, Wang X, Xu Y, Zhang Q (2011) Spectrum trading in cognitive radio networks: a contract-theoretic modelling approach. IEEE J Sel Areas Commun 29:843–855
6. Huang J, Gao L (2013) Wireless network pricing, Morgan and Claypool
7. Niyato D, Hossain E (2008) Spectum trading in cognitive radio networks: a market-equilibrium-based approach. IEEE Trans Wireless Commun 15:71–80
8. Yeo CS, Venugopal S, Chu X, Buyya R (2010) Autonomic metered pricing for a utility computing service. Futur Gener Comput Syst 26:1368–1380
9. Wang G, Feng G, Tan W, Qin S, Wen R, Sun S (2017) In: Resource allocation for network slices in 5g with network resource pricing IEEE global communications conference (GLOBECOM). Singapore, pp 1–6

10. Xiong Z, Kang J, Niyato D, Wang P, Poor HV, Xie S (2020) A multi-dimensional contract approach for data rewarding in mobile networks. IEEE Trans Wireless Commun 19:5779–5793
11. Varian H (2009) Microeconomic analysis. Viva Books

On the Solution of Bush and Wilson's Stochastic Model for Two-Choice Behavior of the Paradise Fish Approached by the Fixed-Point Method

Ali Turab and Wutiphol Sintunavarat

Abstract The model of choice action is the study that defines the theological thought phase in relation to the system of assessing the significance of the various choices and dealing with the consequences to evaluate one for practice. This paper attempts to examine a particular form of choice preference study suggested by Bush and Wilson (Two-choice behavior of paradise fish. J Exp Psych 51, 315-322 (1956)) for paradise fish training in specific circumstances. The existence and uniqueness of the suggested choice preference model's solution are explored by using the fixed point theory tools.

Keywords Paradise fish · Probability · Choice preference · Banach fixed point theorem

1 Introduction

Mathematical psychology is an approach to psychological study focused on mathematical modeling of perceptual, thinking, cognitive, and motor processes and developing laws-like rules that link quantifiable stimuli characteristics to measurable behavior. The mathematical methods are used to derive more reliable theories and thus produce more rigorous empirical validations. The biggest issue with today's application of mathematics to psychological problems, and most likely for some time to come, is modeling these problems. Further secondary endeavors are the solution of complicated yet well-formulated mathematical issues and the study of complex problems implemented in terms of reliable and well-confirmed theories.

This work was supported by Thammasat University Research Unit in Fixed Points and Optimization.

A. Turab (✉) · W. Sintunavarat
Department of Mathematics and Statistics, Faculty of Science and Technology, Thammasat University Rangsit Center, Pathum Thani 12120, Thailand
e-mail: taurusnoor@yahoo.com

W. Sintunavarat
e-mail: wutiphol@mathstat.sci.tu.ac.th

© The Author(s), under exclusive license to Springer Nature Singapore Pte Ltd. 2022
D. Giri et al. (eds.), *Proceedings of the Seventh International Conference on Mathematics and Computing*, Advances in Intelligent Systems and Computing 1412,
https://doi.org/10.1007/978-981-16-6890-6_80

In an animal or human being, the learning phase may often be viewed as a series of choices between multiple possible reactions. Even in basic repetitive experiments under strictly regulated conditions, preference sequences are typically volatile, suggesting that the probability governs the choice of responses. It is also useful to identify structural adjustments in the series of alternatives that reflect changes in trial-to-trial outcomes probabilities. From that perspective, most of the learning analysis describes the probability of a trial-to-test occurrence that describes a stochastic mechanism.

In modern mathematical learning experiments, the researchers concluded that a basic learning experiment was compatible with any stochastic process. This is not only a new idea. After 1950, two critical features emerged mainly in the research initiated by Bush, Estes, and Mosteller. In the first instance, the learning method's egalitarian essence was a core feature of the developed model. Second, these frameworks were studied and applied in areas that did not conceal their quantitative aspects. Different models were suggested, and approximations to the stochastic learning processes studied when possible.

Recently, Turab and Sintunavarat discussed such types of learning models (see, e.g., [7, 10, 11]) and observed the behavior of a subject in a two-choice situation. Our aims are to discuss the experimental work of Bush and Wilson [4] and propose a suitable mathematical model for it which can cover all the aspects of the proposed learning model. The Banach fixed point theorem is being used to find the existence and uniqueness of a solution to the proposed model. In the end, we raise an open problem for those researchers who are interested in this type of research.

2 The Experimental Work of Bush and Wilson on Paradise Fish

Bush and Wilson [4] interacted with a specific fish, offering two swimming options. The two swimming alternatives were to swim on the favorable side as the starting gate was lifted or was floating on the unfavorable side of the tank's far end (see in Fig. 1).

Two options occurring in this paradise-fish experiment are as follows:

(1) opaque splitter among two containers;
(2) a transparent splitter.

The opaque splitter stopped the fish from seeing the target container it does not select, but the other visible box enabled the fish to see the food in another target container. It was a question of action because the fish could see the other target box's food but could not get it.

Data is collected on the number of trials per fish in the last 49 of its 140 attempts (see Table 1). Almost every fish goes to one side. In their latest experiments, 10 goes to the same advantageous hand, while 4 has gone unfavorable. This outcome is in greater alignment with the concept of habit-forming than the model of extinction. Its estimate that the fish will gather approximately 37 trials to the advantageous hand

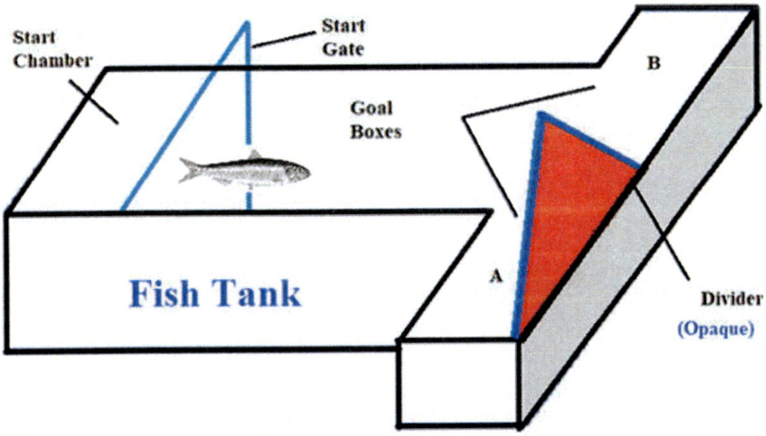

Fig. 1 Behavior of a fish in a opaque divider situation

Table 1 A behavior comparison of the paradise fish in the situation of transparent and opaque divider [4]

Favorable side trials	Real fish: transparent divider	Stat-fish	Real fish opaque divider
0–4	4	4	1
5–9	1	2	0
10–14	2	0	0
15–19	0	0	0
20–24	0	0	2
25–29	0	1	0
30–34	1	0	2
35–39	2	2	3
40–44	2	3	7
45–49	10	10	11

goes against the data collected at the end of either side of the distribution. Monte Carlo used the habit-forming model, culminating in Table 1 stat-fish estimates. It is relevant for certain fish operating an explicit divider parameter (ζ, η). Exploring this experiment he explained stochastic models' usage to relate uncommon theoretical positions and distinguish qualitatively amongst them.

On a given side, if the fish were awarded, there will be a common expectation that the chance of selecting that side will be an increase in numbers on the next trial. Except in the case of unrewarded experiments, there is uncertainty. According to a knowledge or extinction hypothesis, there will be a recommendation on the subsequent trial, and the risk of being to the unrewarded compartment would decrease. By habit-forming theory or secondary reinforcement, picking the same side chosen previously will be possible.

Table 2 Operators for choice experiments

Event	p_{n+1}	Probability of occurrence
Favorable side, food	$\zeta p_n + (1 - \zeta)$	$0.75 p_n$
Favorable side, no food	$\eta p_n + (1 - \eta)\vartheta$	$0.25 p_n$
Unfavorable side, food	ζp_n	$0.25(1 - p_n)$
Unfavorable side, no food	$\eta p_n + (1 - \eta)(1 - \vartheta)$	$0.75(1 - p_n)$

The four events of theoretical operators in the above experiment will vary based on the individual adjustment to these compositions. Let us presume at some stage, p is the probability of choosing favorable and unfavorable sides, while ω is the probability of being rewarding and non-rewarding in terms of food. Furthermore, if fish is choosing the favorable side then the probability of choosing that side has the form $\zeta p + 1 - \zeta$ and $\eta p + (1 - \eta)\vartheta$ depending on the placement of food, where $\zeta, \eta \in (0, 1)$ are the learning parameters appropriate to the particular outcomes and $\vartheta \in [0, 1]$ is the fixed point of that event. Moreover, if the fish chooses the unfavorable side then its probability of choosing that side would be ζp and $\eta p + (1 - \eta)(1 - \vartheta)$ which depend on the placement of food. These events are listed in Table 2.

The habit-forming model suggests that animals may be stabilized on all sides, but individual animals settle on the auspicious compartment, and few on the undesirable corner. Anyhow, it's a comforting thought. The number of incentive changes would improve the likelihood of heading to one hand. If a strong probability for one compartment is attained, the animal will select the specific side consistently.

3 Mathematical Modeling of the Proposed Model

Let us suppose that if the habit-forming model is considered suitable for the experiment, then which fish range would be reinforced on the advantageous side? This is a common question among several mathematical problems that emerged in studying mathematical analysis. We take a simple query to represent a mathematical problem. Suppose there are A and B as two results, and the consequence of each is positive but not the same.

For instance, if the probability of A is x and A is attained, then the subsequent probability of A is $\zeta x + 1 - \zeta$ and $\eta x + (1 - \eta)\vartheta$ depending on getting food or

Table 3 Event probabilities

Event	Probability
Favorable side, food	ωx
Favorable side, no food	$(1-\omega x)$
Unfavorable side, food	$\omega(1-x)$
Unfavorable side, no food	1-$\omega(1 - x)$

not. But if fish gets the food, then ζx is the new probability of A and vice versa. Depending on the sides, the four events' probability is listed in Table 3.

It can be noticed that after this characterization, an organism is practically on top, providing one of the results and only reacting with the other. Now let us name this option as $\mathcal{P}(x, \zeta, \eta)$. Based on the events ($A$ and B), placement of food and the corresponding probabilities, we get the following functional equation:

$$\begin{aligned}\mathcal{P}(x, \zeta, \eta, \vartheta) &= \omega x \mathcal{P}(\zeta x + (1-\zeta), \zeta, \eta, \vartheta) + (1-\omega) x \mathcal{P}(\eta x + (1-\eta)\vartheta, \zeta, \eta, \vartheta) \\ &\quad + \omega(1-x)\mathcal{P}(\zeta x, \zeta, \eta, \vartheta) \\ &\quad + (1-\omega)(1-x)\mathcal{P}(\eta x + (1-\eta)(1-\vartheta), \zeta, \eta, \vartheta). \qquad (1)\end{aligned}$$

Now, we present a notable result that will be needed in the next section.

Theorem 1 (Banach Fixed Point Theorem in [3]) *Let (M, d) be a complete metric space. A Banach contraction mapping $\mathcal{P} : M \to M$ has precisely one fixed point. Moreover, the Picard iteration $\{x_n\}$ in M which is defined by $x_n = \mathcal{P}x_{n-1}$ for all $n \in \mathbb{N}$, where $x_0 \in M$ converges to the unique fixed point of $\mathcal{P}$.*

4 Main Results

Throughout this paper, the class of all continuous real-valued functions $\mathcal{P} : [0, 1] \to \mathbb{R}$ such that $\mathcal{P}(0) = 0$ and

$$\sup_{m \neq n} \frac{|\mathcal{P}(m) - \mathcal{P}(n)|}{|m-n|} < \infty$$

is denoted by $\mathbb{B}$. It is straightforward that $(\mathbb{B}, \|\cdot\|_{\mathbb{B}})$ is a Banach space (see [2]), where $\|\cdot\|_{\mathbb{B}}$ is classified by

$$\|\mathcal{P}\|_{\mathbb{B}} = \sup_{m \neq n} \frac{|\mathcal{P}(m) - \mathcal{P}(n)|}{|m-n|}$$

for all $\mathcal{P} \in \mathbb{B}$.

First, we interpret the functional equation (1) as

$$\begin{aligned}\mathcal{P}(x) &= \omega x \mathcal{P}(\zeta x + (1-\zeta)) + (1-\omega) x \mathcal{P}(\eta x + (1-\eta)\vartheta) \\ &\quad + \omega(1-x)\mathcal{P}(\zeta x) + (1-\omega)(1-x)\mathcal{P}(\eta x + (1-\eta)(1-\vartheta)), \qquad (2)\end{aligned}$$

for all $x \in [0, 1]$, where $0 < \zeta, \eta < 1$, $\vartheta, \omega \in [0, 1]$ and $\mathcal{P} : [0, 1] \to \mathbb{R}$ is an unknown function such that $\mathcal{P}(0) = 0$. Our objective is to discuss the existence and uniqueness results of the proposed functional equation (2)'s solution by utilizing the Banach fixed-point theorem. Here, we present the subsequent main result.

Theorem 2 *Suppose that* $0 < \zeta, \eta < 1$ *and* $\vartheta, \omega \in [0, 1]$ *with*

$$|\omega(\zeta - \eta + 2) + (1 - \eta)(2\vartheta - 1) + 2\omega\vartheta(1 + \eta)| < 1.$$

If there exists a closed subset Υ *of* $\mathbb{B}$ *such that* Υ *is* $\mathcal{Z}$ *-invariant, that is,* $\mathcal{Z}(\Upsilon) \subseteq \Upsilon$, *where* $\mathcal{Z}$ *is the operator from* Υ *defined for each* $\mathcal{P} \in \Upsilon$ *by*

$$\begin{aligned}(\mathcal{Z}\mathcal{P})(x) &= \omega x \mathcal{P}(\zeta x + (1 - \zeta)) + (1 - \omega) x \mathcal{P}(\eta x + (1 - \eta)\vartheta) \\ &\quad + \omega(1 - x)\mathcal{P}(\zeta x) + (1 - \omega)(1 - x)\mathcal{P}(\eta x + (1 - \eta)(1 - \vartheta)), \quad (3)\end{aligned}$$

for all $x \in [0, 1]$, *then* $\mathcal{Z}$ *is a Banach contraction mapping with the metric d induced by* $\|\cdot\|_{\mathbb{B}}$.

Proof Let $\mathcal{P}_1, \mathcal{P}_2 \in \Upsilon$. For each distinct points $x, y \in [0, 1]$, we obtain

$$
\begin{aligned}
&\frac{|(\mathcal{Z}\mathcal{P}_1-\mathcal{Z}\mathcal{P}_2)(x)-(\mathcal{Z}\mathcal{P}_1-\mathcal{Z}\mathcal{P}_2)(y)|}{|x-y|}\\
&=\Bigg|\frac{1}{x-y}[\omega x\,(\mathcal{P}_1-\mathcal{P}_2)\,(\zeta x+(1-\zeta))+(1-\omega)x\,(\mathcal{P}_1-\mathcal{P}_2)\,(\eta x+(1-\eta)\vartheta)\\
&\quad+\omega(1-x)\,(\mathcal{P}_1-\mathcal{P}_2)\,(\zeta x)+(1-\omega)(1-x)\,(\mathcal{P}_1-\mathcal{P}_2)\,(\eta x+(1-\eta)(1-\vartheta))\\
&\quad-\omega y\,(\mathcal{P}_1-\mathcal{P}_2)\,(\zeta y+(1-\zeta))-(1-\omega)y\,(\mathcal{P}_1-\mathcal{P}_2)\,(\eta y+(1-\eta)\vartheta)\\
&\quad-\omega(1-y)\,(\mathcal{P}_1-\mathcal{P}_2)\,(\zeta y)-(1-\omega)(1-y)\,(\mathcal{P}_1-\mathcal{P}_2)\,(\eta y+(1-\eta)(1-\vartheta))]\Bigg|\\
&=\Bigg|\frac{1}{x-y}[\omega x\,(\mathcal{P}_1-\mathcal{P}_2)\,(\zeta x+(1-\zeta))-\omega x\,(\mathcal{P}_1-\mathcal{P}_2)\,(\zeta y+(1-\zeta))]\\
&\quad+\frac{1}{x-y}[(1-\omega)x\,(\mathcal{P}_1-\mathcal{P}_2)\,(\eta x+(1-\eta)\vartheta)-(1-\omega)x\,(\mathcal{P}_1-\mathcal{P}_2)\,(\eta y+(1-\eta)\vartheta)]\\
&\quad+\frac{1}{x-y}[\omega(1-x)\,(\mathcal{P}_1-\mathcal{P}_2)\,(\zeta x)-\omega(1-x)\,(\mathcal{P}_1-\mathcal{P}_2)\,(\zeta y)]\\
&\quad+\frac{1}{x-y}[(1-\omega)(1-x)\,(\mathcal{P}_1-\mathcal{P}_2)\,(\eta x+(1-\eta)(1-\vartheta))\\
&\quad-(1-\omega)(1-x)\,(\mathcal{P}_1-\mathcal{P}_2)\,(\eta y+(1-\eta)(1-\vartheta))]\\
&\quad+\frac{1}{x-y}[\omega x\,(\mathcal{P}_1-\mathcal{P}_2)\,(\zeta y+(1-\zeta))-\omega y\,(\mathcal{P}_1-\mathcal{P}_2)\,(\zeta y+(1-\zeta))]\\
&\quad+\frac{1}{x-y}[(1-\omega)x\,(\mathcal{P}_1-\mathcal{P}_2)\,(\eta y+(1-\eta)\vartheta)-(1-\omega)y\,(\mathcal{P}_1-\mathcal{P}_2)\,(\eta y+(1-\eta)\vartheta)]\\
&\quad+\frac{1}{x-y}[\omega(1-x)\,(\mathcal{P}_1-\mathcal{P}_2)\,(\zeta y)-\omega(1-y)\,(\mathcal{P}_1-\mathcal{P}_2)\,(\zeta y)]\\
&\quad+\frac{1}{x-y}[(1-\omega)(1-x)\,(\mathcal{P}_1-\mathcal{P}_2)\,(\eta y+(1-\eta)(1-\vartheta))\\
&\quad-(1-\omega)(1-y)\,(\mathcal{P}_1-\mathcal{P}_2)\,(\eta y+(1-\eta)(1-\vartheta))]\Bigg|\\
&\le \omega\zeta x\|\mathcal{P}_1-\mathcal{P}_2\|_{\mathbb{B}}+(1-\omega)\eta x\|\mathcal{P}_1-\mathcal{P}_2\|_{\mathbb{B}}+\omega\zeta x\|\mathcal{P}_1-\mathcal{P}_2\|_{\mathbb{B}}\\
&\quad+(1-\omega)(1-x)\eta\|\mathcal{P}_1-\mathcal{P}_2\|_{\mathbb{B}}+\omega\,|(\mathcal{P}_1-\mathcal{P}_2)\,(\zeta y+(1-\zeta))-(\mathcal{P}_1-\mathcal{P}_2)\,(\zeta y)|\\
&\quad+(1-\omega)\,|(\mathcal{P}_1-\mathcal{P}_2)\,(\eta y+(1-\eta)\vartheta)-(\mathcal{P}_1-\mathcal{P}_2)\,(\eta y+(1-\eta)(1-\vartheta))|\\
&\le |\omega(\zeta-\eta+2)+(1-\eta)(2\vartheta-1)+2\omega\vartheta(1+\eta)|\,\|\mathcal{P}_1-\mathcal{P}_2\|_{\mathbb{B}}.
\end{aligned}
$$

This gives that

$$
\begin{aligned}
d(\mathcal{Z}\mathcal{P}_1,\mathcal{Z}\mathcal{P}_2)&=\|\mathcal{Z}\mathcal{P}_1-\mathcal{Z}\mathcal{P}_2\|_{\mathbb{B}}\\
&\le |\omega(\zeta-\eta+2)+(1-\eta)(2\vartheta-1)+2\omega\vartheta(1+\eta)|\,\|\mathcal{P}_1-\mathcal{P}_2\|_{\mathbb{B}}\\
&=|\omega(\zeta-\eta+2)+(1-\eta)(2\vartheta-1)+2\omega\vartheta(1+\eta)|\,d(\mathcal{P}_1,\mathcal{P}_2).
\end{aligned}
$$

It comes through $0<|\omega(\zeta-\eta+2)+(1-\eta)(2\vartheta-1)+2\omega\vartheta(1+\eta)|<1$ that $\mathcal{Z}$ is a Banach contraction mapping with the metric d induced by $\|\cdot\|_{\mathbb{B}}$.

From the above Theorem 2, we obtain the subsequent outcome.

Theorem 3 *The functional equation (2) has a unique solution provided that*

$$|\omega(\zeta - \eta + 2) + (1 - \eta)(2\vartheta - 1) + 2\omega\vartheta(1 + \eta)| < 1,$$

and there exists a closed subset Υ of $\mathbb{B}$ such that Υ is $\mathcal{Z}$-invariant, that is, $\mathcal{Z}(\Upsilon) \subseteq \Upsilon$, where $\mathcal{Z}$ is the operator from Υ defined for each $\mathcal{P} \in \Upsilon$ by

$$\begin{aligned}(\mathcal{Z}\mathcal{P})(x) = {} & \omega x\mathcal{P}(\zeta x + (1 - \zeta)) + (1 - \omega)x\mathcal{P}(\eta x + (1 - \eta)\vartheta) \\ & + \omega(1 - x)\mathcal{P}(\zeta x) + (1 - \omega)(1 - x)\mathcal{P}(\eta x + (1 - \eta)(1 - \vartheta)), \quad (4)\end{aligned}$$

for all $x \in [0, 1]$. Furthermore, the iteration $\{\mathcal{P}_n\}$ in $\mathbb{B}$ which is defined by

$$\begin{aligned}(\mathcal{P}_n)(x) = {} & \omega x\mathcal{P}_{n-1}(\zeta x + (1 - \zeta)) + (1 - \omega)x\mathcal{P}_{n-1}(\eta x + (1 - \eta)\vartheta) \\ & + \omega(1 - x)\mathcal{P}_{n-1}(\zeta x) + (1 - \omega)(1 - x)\mathcal{P}_{n-1}(\eta x + (1 - \eta)(1 - \vartheta)) \quad (5)\end{aligned}$$

for all $n \in \mathbb{N}$, where $\mathcal{P}_0 \in \mathbb{B}$, converges to the unique solution of the functional equation (2) in the sense of the metric d induced by $\|\cdot\|_{\mathbb{B}}$.

Proof As Υ is a closed subset of $\mathbb{B}$, we get Υ is complete. Thus, by utilizing the Banach contraction principle with Theorem 2, we obtain the outcome of this theorem.

In several experimental models, a conclusive complementarity between two events is being introduced by psychologists. For instance, a subject can be requested to assume which of two light bulbs will be flashed on each trial. The outcomes will be based on the two desirable assumptions, and the alternatives will be flashing on/off the light bulbs. Our expectations will be such that the flashing off the light bulb will result in the possibility of envisioning "left," just in the same way as the flashing off the right bulb would change the possibility of predicting "right." In different words, if we interchange the acts of events and the response labels move backward, the real representation of the experiment would not get any change. Hence, in many experiments, the symmetric events strongly proposed the "same response condition ($\zeta = \eta$)," but for the investigation of this condition, there is another motivation. We often confront severe mathematical problems which can be untangled by imposing the same response condition. So, we can rewrite Eq. (10) as

$$\begin{aligned}\mathcal{P}(x) = {} & \omega x\mathcal{P}(\zeta x + (1 - \zeta)) + (1 - \omega)x\mathcal{P}(\zeta x + (1 - \zeta)\vartheta) \\ & + \omega(1 - x)\mathcal{P}(\zeta x) + (1 - \omega)(1 - x)\mathcal{P}(\zeta x + (1 - \zeta)(1 - \vartheta)), \quad (6)\end{aligned}$$

for all $x \in [0, 1]$, where $0 < \zeta < 1$, $\vartheta, \omega \in [0, 1]$ and $\mathcal{P} : [0, 1] \to \mathbb{R}$ is an unknown function such that $\mathcal{P}(0) = 0$.

From Theorems 2 and 3, we obtain the subsequent corollaries.

Corollary 1 *Suppose that $0 < \zeta, \eta < 1$ and $\vartheta, \omega \in [0, 1]$ with*

$$|2\omega + (1-\zeta)(2\vartheta - 1) + 2\omega\vartheta(1+\zeta)| < 1.$$

If there exists a closed subset Υ *of* $\mathbb{B}$ *such that* Υ *is* $\mathcal{Z}$ *-invariant, that is,* $\mathcal{Z}(\Upsilon) \subseteq \Upsilon$*, where* $\mathcal{Z}$ *is the operator from* Υ *defined for each* $\mathcal{P} \in \Upsilon$ *by*

$$(\mathcal{Z}\mathcal{P})(x) = \omega x \mathcal{P}(\zeta x + (1-\zeta)) + (1-\omega)x\mathcal{P}(\zeta x + (1-\zeta)\vartheta) \\ +\omega(1-x)\mathcal{P}(\zeta x) + (1-\omega)(1-x)\mathcal{P}(\zeta x + (1-\zeta)(1-\vartheta)), \quad (7)$$

for all $x \in [0, 1]$*, then* $\mathcal{Z}$ *is a Banach contraction mapping with the metric d induced by* $\|\cdot\|_{\mathbb{B}}$*.*

Corollary 2 *The functional equation (6) has a unique solution provided that*

$$|2\omega + (1-\zeta)(2\vartheta - 1) + 2\omega\vartheta(1+\zeta)| < 1,$$

and there exists a closed subset Υ *of* $\mathbb{B}$ *such that* Υ *is* $\mathcal{Z}$*-invariant, that is,* $\mathcal{Z}(\Upsilon) \subseteq \Upsilon$*, where* $\mathcal{Z}$ *is the operator from* Υ *defined for each* $\mathcal{P} \in \Upsilon$ *by*

$$(\mathcal{Z}\mathcal{P})(x) = \omega x \mathcal{P}(\zeta x + (1-\zeta)) + (1-\omega)x\mathcal{P}(\zeta x + (1-\zeta)\vartheta) \\ +\omega(1-x)\mathcal{P}(\zeta x) + (1-\omega)(1-x)\mathcal{P}(\zeta x + (1-\zeta)(1-\vartheta)), \quad (8)$$

for all $x \in [0, 1]$*. Furthermore, the iteration* $\{\mathcal{P}_n\}$ *in* $\mathbb{B}$ *which is defined by*

$$(\mathcal{P}_n)(x) = \omega x \mathcal{P}_{n-1}(\zeta x + (1-\zeta)) + (1-\omega)x\mathcal{P}_{n-1}(\zeta x + (1-\zeta)\vartheta) \\ +\omega(1-x)\mathcal{P}_{n-1}(\zeta x) + (1-\omega)(1-x)\mathcal{P}_{n-1}(\zeta x + (1-\zeta)(1-\vartheta))(9)$$

for all $n \in \mathbb{N}$*, where* $\mathcal{P}_0 \in \mathbb{B}$*, converges to the unique solution of the functional equation (6) in the sense of the metric d induced by* $\|\cdot\|_{\mathbb{B}}$*.*

5 Question and Open Problem

Here, we give the following interesting question to readers.

Question: What if a fish doesn't go around and stays in its position?
Also, we pose the stability (for the detailed study on stability, we refer [1, 5, 6, 8, 9]) of the following functional equation as an open question:

$$\mathcal{P}(x) = \omega x \mathcal{P}(\zeta x + (1-\zeta)) + (1-\omega)x\mathcal{P}(\eta x + (1-\eta)\vartheta) \\ +\omega(1-x)\mathcal{P}(\zeta x) + (1-\omega)(1-x)\mathcal{P}(\eta x + (1-\eta)(1-\vartheta)), \quad (10)$$

for all $x \in [0, 1]$, where $0 < \zeta, \eta < 1$, $\vartheta, \omega \in [0, 1]$ and $\mathcal{P} : [0, 1] \to \mathbb{R}$ is an unknown function such that $\mathcal{P}(0) = 0$.

References

1. Aoki T (1950) On the stability of the linear transformation in Banach spaces. J Math Soc Jpn 2:64–66
2. Turab A, Sintunavarat W (2020) On the solution of the traumatic avoidance learning model approached by the Banach fixed point theorem. J Fixed Point Theory Appl 22:50
3. Banach S (1922) Sur les opérations dans les ensembles abstraits et leur applications aux équations intégrales. Fund Math 3:133–181
4. Bush AA, Wilson TR (1956) Two-choice behavior of paradise fish. J Exp Psych 51:315–322
5. Găvruta P (1994) A generalization of the Hyers-Ulam-Rassias stability of approximately additive mappings. J Math Anal Appl 184:431–436
6. Hyers DH, Isac G, Rassias ThM (1998) Stability of functional equations in several variables. Birkhäuser, Boston
7. Turab A, Sintunavarat W (2020) On the existence and uniqueness of the solution of a probabilistic functional equation approached by the Banach fixed point theorem. Thai J Math 18(4):1983–1994
8. Hyers DH (1941) On the stability of the linear functional equation. Proc Nat Acad Sci USA 27:222–224
9. Ulam SM (1960) a collection of the mathematical problems. Interscience Publishers, New York
10. Turab A, Sintunavarat W (2020) On a solution of the probabilistic predator–prey model approached by the fixed point methods. J Fixed Point Theory Appl 22:64
11. Turab A, Sintunavarat W (2019) On analytic model for two-choice behavior of the paradise fish based on the fixed point method. J Fixed Point Theory Appl 21:56

A Fine-tuning Retrieval System for Mathematical Information

Pankaj Dadure, Partha Pakray, and Sivaji Bandyopadhyay

Abstract The web is a rich repository of mathematical information, the task of finding relevant documents in such collection is a laborious one. Although multiple approaches have been proposed to retrieve relevant documents for a queried formula, the poor values of evaluation measures depict existing limitations of such systems. To improve the performance of this systems, this paper proposes a novel approach of formula indexing by employing formula embedding and generalization techniques. The formula embedding and the generalization modules of the proposed system transform the formulas into the fixed-size vectors by counting the occurrence of different entities in formulas. Subsequently, the formula vectors are indexed by an indexer. The documents retrieved by both the modules have higher priorities in comparison to those retrieved by individual ones. The obtained results have been compared with the state-of-the-art existing approaches, and the comparison study reveals that the proposed approach gives better retrieval accuracy in terms of P_5=0.522, P_10=0.478, P_15=0.352, and P_20=0.289 measures.

Keywords Mathematical information retrieval · Formula embedding · Formula generalization · Bit position information table · Canonicalization tokenization · Structural unification.

1 Introduction

Mathematical Information Retrieval (MIR) plays a significant role in the area of Natural Language Processing (NLP) and is coming under the stardom from the past decade. The remarkable growth in the area of scientific research increases the mathematical data on the web [1], and such data is significantly different from the conventional textual data. The data generated from the scientific research comprises mathematical notations, scientific terms, and symbolic structures which are written in

P. Dadure (✉) · P. Pakray · S. Bandyopadhyay
Department of Computer Science and Engineering, National Institute of Technology Silchar, Assam, India

© The Author(s), under exclusive license to Springer Nature Singapore Pte Ltd. 2022

D. Giri et al. (eds.), *Proceedings of the Seventh International Conference on Mathematics and Computing*, Advances in Intelligent Systems and Computing 1412,
https://doi.org/10.1007/978-981-16-6890-6_81

L^AT_EX, XML and/or MathML format [7]. The previous studies [5] state that the text information retrieval system is not enough to retrieve the scientific information, thus it needs a noteworthy improvement to deal with such kind of data. The followings are the key points which depict that the retrieval of mathematical information is different from textual information retrieval: Mathematical expressions are written in L^AT_EX and/or MathML format which has some predefined structure instead of a string of character. The text-based search engine is inefficient in dealing with such formats. The meaning of some mathematical notations varies with context. For example, f(x) may either denote the function of x or multiplication of variable "f" with another variable "x". It shows the polysemic nature of mathematical notation. A large number of mathematical symbols present in a scientific document have different scripting and typefaces which give rise to the symbol identification problem. These challenges show the varied range of problems faced by the mathematical information retrieval system.

To ease the retrieval of mathematical information, there are some of the preprocessing operations which make search results more relevant to needed information [11]. The documents on the web can emerge from different sources, which contains similar mathematical expressions with trivial differences. Canonicalization accesses the single canonical form for such syntactically similar formulas or sub-formulas and indexes them in the same position. However, the canonicalization operation is unable to perform the syntactic manipulation in the mathematical notation. For example, although the notations "x+y" and "y+x" are semantically same, canonicalization fails to consider equivalence of such notations. Hence, ordering of operand is recorded and such notations are converted into single canonical form. Moreover, the MIR system does not restrict for exact match formula, tokenization is performed to obtain the partial similar formula. Sometimes same mathematical equations can be represented with the different variable names, for example, $x^2 + y^2 = z^2$ which can be represented as $a^2 + b^2 = c^2$. The meaning of both equations is the same but different in their representation. The unification operation replaced the variable name with an identifier and numeric value with constant to find the unified formula. The unified form for both the mathematical equation $x^2 + y^2 = z^2$ and $a^2 + b^2 = c^2$ is $id_1^{const} + id_2^{const} = id_3^{const}$ which makes the bridge for the retrieval of semantically similar formula.

In this paper, we have proposed a well established combined approach of formula embedding and generalization to represent and index the formula effectively. The proposed system generates the vectors of math formulas using the formula embedding and the generalizations modules contained in the scientific documents. Each bit of a vector represents the number of times a particular entity occurred in the formula. The similarity score is measured by comparing each bit of the indexed formulas and the user query. The performance of the proposed approach is tested using a Wikipedia corpus of NTCIR-12 [15] and obtained results compared with the state-of-the-art MIR approaches.

The rest of the paper is organized as follows: Sect. 2 describes related works. Section 3 provides a detailed description of the system architecture. Section 4 describes the experimental results of the proposed approach and their comparative analysis. Section 5 concludes the paper and points direction for future research.

2 Related Works

Over the past few years, there has been a surge of interest in mathematical information retrieval and many researchers around the globe have been working in this domain. Mathematical information retrieval has several irresistible properties that rekindle the interest of the AI community. In order to achieve robust results, the researcher has to brush-up the traditional linguistics and deep learning techniques which depicts the research scope in the field of document retrieval for math queries. The keyword based approach of team BRKLY [5] has shown that the retrieval of mathematical information is qualitatively and quantitatively different from textual information retrieval. It converts the XML data into the conventional textual form and used full text indexing mechanism. The performance evaluation of this approach has proved that the text searching system needs significant improvement to deal with mathematical information. To ease the retrieval of a mathematical expression, Math Indexer and Searcher (MIaS) system [11] performed the canonicalization and unification operations on formulas. For evaluating the relevance of the documents and the users query, a heuristic weighting approach has been adopted. At NTCIR-11, team MIRMU [10] introduced the second generation of scalable full text search engine Math Indexer and Searcher (MIaS) system which indexed the canonicalized form of math formulae, where the formulae are represented in both the Content MathML and the Presentation MathML format. MIaS has used the query expansion technique for obtaining the relevant results related to sub-queries. The approach of combining the textual information and the math expression [6] contained in the documents and the queries has shown the cutting edge performance of the MIR system. In addition to this, the query expansion technique has achieved a remarkable performance to retrieve the exact match, similar match and sub-formulas.

The MCAT search system [4] performed three different operations at the preprocessing stage, i.e., encode math expression, extract textual information, and extract dependency graph of math expression. The extracted textual information has been indexed into three different indexes and used unification as post-processing operation which considered each symbol present in a formula as a constant except free variable (qvar). To explore the potential of neural network approaches in MIR, the symbol2vec method [3] converts the formulas into the vector representation. Formula to vector transformation, Distributed Memory Model of Paragraph Vector (PV-DM) has been used. Afterward, the symbol2vec approach combined with the language model and the obtained results has shown much promising results as compared to an individual one. The type embedding with the query expansion approach [13] has shown that the semantic relationship between the types enriches the retrieval efficiency of the

MIR system. In which, detection and extraction of technical terms present in the documents have been done using the C-Value algorithm.

The ability to handle the text and the mathematical information into a single searching portal [12], MIaS introduced one approach where the text and the mathematical information is preprocessed separately. MIaS handles both the textual and mathematical information by means of basic apache lucene system and a set of preprocessing operations respectively. MIaS has performed canonicalization, ordering, tokenization, and unification operations on mathematical information to achieve effective retrieval results. To retrieve the math information effectively, the formula embedding approach [8] has used the bit position information table where binary vectors of formulas are generated. In which, 1 represents the presence and 0 represents the absence of a particular entity in a formula. For refinement in the mechanism of the formula embedding approach, Binary Vector Transformation of Math formula (BVTMF) [9] has used mathematical as well as textual information to achieve the more promising results. The rank-safe dynamic pruning strategy [16] introduced an upperbound formulation of structural similarity to speedup formula search using subtree matching and shown that the dynamic pruning strategies with a specialized inverted index different from traditional linear text search pruning methods and they further associate query structure representation with posting lists. To understand mathematical texts, the system needs to analyze the semantics of their mathematical constructs and to link their internal elements to the respective concepts or premises. For instance, the approach of natural language premise selection [2] used to retrieve supporting definitions and supporting propositions that are useful for generating an informal mathematical proof for particular statements.

3 Proposed Architecture

3.1 Corpus Description

The proposed approach has been tested on the MathTagArticles of Wikipedia corpus of NTCIR-12 [15] of size 2.1 GB which contained 31,840 scientific articles and around 579,610 mathematical formula. In comparison to the other corpus such as ArXiv which are found to be relatively more domain-specific or academic, the Wikipedia data set is used in different backgrounds. The scientific documents of MathTagArticles contained textual as well as mathematical information and each of which holds at least one mathematical notation.

3.2 *System Description*

The system architecture of the proposed approach is shown in Fig. 1 which constitute the following components and each component work collaboratively for leading-edge math information retrieval system.

3.2.1 Preprocessing

Dealing with math notation is the most strenuous part in mathematical information retrieval approaches. To develop an effective retrieval system, the preprocessing of documents play an important role which refined and structures the data into a unified form. The preprocessing phase of proposed system architecture constitutes two

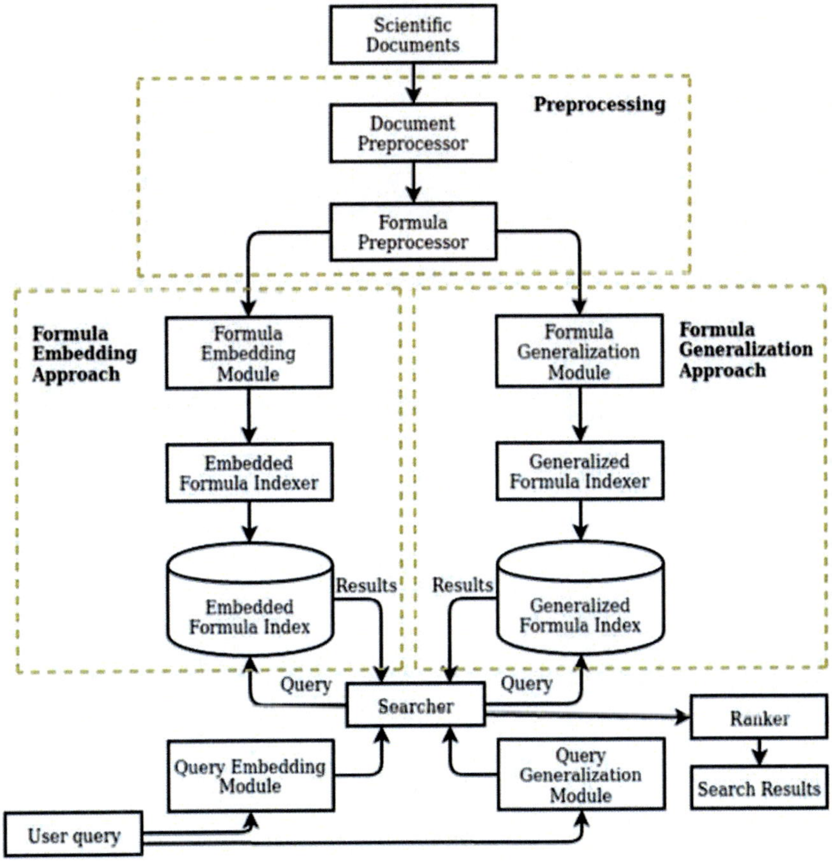

Fig. 1 Proposed system architecture

modules, namely: document preprocessor and formula preprocessor. In scientific documents, the mathematical information are tagged by "$" and "$" and are written in three distinct forms: Presentation MathML, Content MathML, and T_EX. The document preprocessor module of preprocessing phase extracts the math information from the scientific documents which are tagged by "$" and "$". Afterward, the outcome of the document preprocessor passed to the formula preprocessor module which extracts the Presentation MathML format of math notation from the preprocessed document and removes the irrelevant MathML notations like <mspace>, <mstyle>, which are considered as noise in vector formation.

3.2.2 Formula Embedding Module

Formula embedding module extracts the entities from the preprocessed Presentation MathML formulae and uses a bit position information table (Refer Table 1) to generate the formulae vector of size 202. In this, the formula embedding module estimate the count of a particular entity occurs in a formula and assigns the count of that entity to their bit position. The process of vector generation is shown in Fig. 2a.

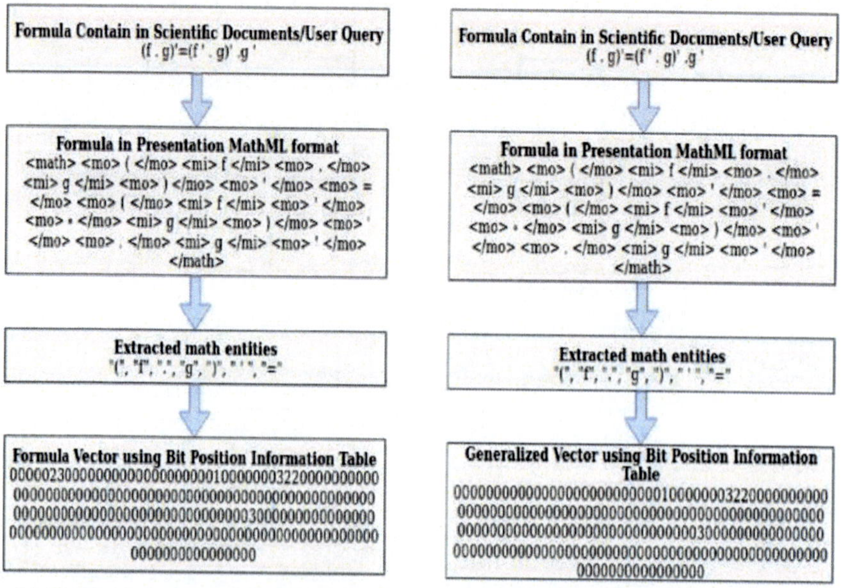

(a) Embedded Vector Representation (b) Generalized Vector Representation

Fig. 2 Vector generation using formula embedding and generalization module

3.2.3 Formula Generalization Module

For the retrieval of semantically similar formula, the proposed system architecture deploys the formula generalization module. The formula generalization module extracts all the entities present in formula and estimates their count. The estimated count of each entity is then placed to their bit position. For generalized vector form, bit position 0–25 and lexical entities present at bit positions 150–201 are set to “0”. The primary task of formula generalization approach is to represent the semantically similar formula into a unified form. For example, $a^2 + b^2$ and $x^2 + y^2$ are represented as $0^2 + 0^2$, respectively. The process of formula generalization is shown in Fig. 2b.

3.2.4 Query Embedding Module

In a proposed system, the user enters the query in a Presentation MathML format. The prime task of the query embedding module is to convert the user query into a query vector of size 202. Similar to the formula embedding module, it counts the number of times a particular entity present in a query and placed their count to their bit position. Subsequently, the searcher module takes the output of the query embedding module as its input.

3.2.5 Query Generalization Module

Similar to the formula generalization, the query generalization module converts the user query into a generalized query vector of size 202. Query generalization module replaces all the entities present between the bit position 0–25 and lexical entities present at bit positions 150–201 to “0”. The key approach of query generalization is to retrieve the semantically similar formula.

3.2.6 Embedded Formula Indexer

In any retrieval system, indexing is an imperative process and backbone of the effective search results. In the proposed approach to execute the search more effectively, an index for embedded formula vectors is deployed where each formula is indexed. The index contained 579,610 vectors extracted from 31,840 documents of size 787.6 MB, which is ~37% of the corpus size. Each index instance comprises three attributes, namely: vector representation, the document from which the formula extracted, and the math formula in Presentation MathML format. The structure of the embedded formula index is shown in Fig. 3.

Vector	Document	Formula
000000000000000100000000000000000000000000000 000000000000000000000000000000000000000000000 000000000000000000000000000000000000000000000 000000000000000000000000000000000000000000000 000000000000000000000000000000000000	Local_search_(const raint_satisfaction).ht ml	<math><semantics>< mi>P</mi></math></ semantics>
000000000001000000000000000000000000000000000 000000100000000000000000000000000000000000000 000000001000000000000000000000000000000000000 000000000000000000000000000000000000000000000 000000000000000000000000000000000000	Equivalent_potential _temperature.html	<math><semantics>< msub><mi>θ</ mi><mi>L</mi> </msub> </semantics></math>

Fig. 3 Structure of embedded formula index

3.2.7 Generalized Formula Indexer

Similar to the embedded formula indexer, a generalized formula indexer generates the index for preprocessed generalized formula present in scientific documents. The generalized formula index contained 579,610 generalized vectors from 31,840 documents of size 787.6 MB which is ~37% of the corpus size. Same as the embedded formula index, generalized formula index constitutes three attributes: generalized vector representation, the document from which the formula extracted, and the math formula in Presentation MathML format.

3.2.8 Searcher Module

The prime task of searcher module is to search the relevant documents which hold the syntactic and/or semantic meaning with respect to user need. This module compared the user query with all the indexed formulas and measured the maximum similarity for each of them. The user query matched with more than one formula in the same document, then that document gets the higher priority as compared to those documents where only one formula is matched.

3.2.9 Ranker Module

The ranker module ranks the retrieve documents based on the maximum similarity between the document and the user query. Those documents that are retrieved by both the module, i.e., formula embedding and the formula generalization module, get the higher rank as compared to the documents that are retrieved from an individual module. After that, priority goes to those documents that are retrieved through the formula embedding module. The documents retrieved from the formula generalization module have less priority as compared to the combined and the formula embedding approach. The ranker module retrieves the top 25 documents present in the corpus as a final search result of the user query.

Bit Position Information Table (BPIT) (Refer Table 1) shows the entities which are normally occurred in the mathematical formulas with their corresponding position in a vector used in the proposed approach. In BPIT, 150 distinct symbols, scientific terms, alphanumeric values, and Presentation MathML tags are defined. Bit positions 0–25, 57–65, and 71–100 represents the lexical entities or variables occurred in $< mi >$ tag, bit positions 26–45, 66–70, and 101–149 represents the mathematical operators occurred in $< mo >$ tag and bit positions 46–56 represent the some Presentation MathML tags which contribute to the semantic of formula. Bit position 65 defined the multipurpose variable whose use is not defined like log, det, xor, and so on. Bit positions 150–175 and 176–201 represents the lexical entities which are present in the superscript and subscript. Moreover, the upper case and the lower case of lexical entities are considered as equivalent.

4 Experimental Design and Results

4.1 Queryset Description

The queryset containing 30 scientific queries of the NTCIR-12 MathIR Wikipedia Task for evaluating the performance of the proposed system. The queryset contained simple as well as complex mathematical queries and each query is coupled with query ID. To retrieve the relevant documents, queries are converted into a vector of entities using query embedding module and query generalization module as described in Sects. 3.2.4 and 3.2.5.

4.2 Gold Dataset Description

The performance validation of the proposed approach is done against the gold dataset released by NTCIR-12 [15] and the structure of the gold dataset adheres to the Text REtrieval Conference (TREC) qrel format [14]. The gold dataset contained a set of documents that have been judged as relevant (1) or irrelevant (0) by humans. Gold dataset has four attribute: (i) QueryID, (ii) Iteration, (iii) Document#, and (iv) Relevance. QueryId represents the specific query in queryset. Iteration is an immaterial attribute which is disregard by TREC tool. Document attribute specifies the name of the document and relevance attribute specify the humans judgment in 0 (relevant) and 1 (irrelevant). The format of the gold dataset is shown in Table 2.

Table 1 Bit position information Table

Entity	Position	Entity	Position	Entity	Position	Entity	Position
a/A to z/Z	0-25	$< msqrt >$	56	log, ln	88	±	120
exp	4	ℤ	57	!	89	∫	121
=	26	ℕ	58	Trigo. Ratios	90	∘	122
Product	27	ℚ	59	ℝ	91	′	123
−	28	¬	60	θ	92	⋀	124
,	29	α	61	gcd	93	∃	125
+	30	γ	62	xor	94	¬	126
∇	31	ω	63	τ	95	lim	127
∂	32	ϑ	64	η	96	<, ⟨	128
→	33	Var. Name	65	σ	97	>, ⟩	129
·	34	Null	66	Ω	98	⊗	130
(	35	†	67	#	99	⊢	131
)	36	:	68	⌜	100	⊓	132
≡	37	dist	69	≠	101	⊔	133
≫	38	∓	70	{	102	‖	134
∝	39	ϕ φ, Φ	71	}	103	⋃	135
≈	40	ℏ	72	⊙	104	⋂	136
/	41	π	73	≤	105	⟼	137
⊆	42	Δ	74	∈	106	⊂	138
⊕	43	μ	75	[	107	det	139
∼	44	Σ	76	]	108	Π	140
|	45	ϵ	77	∗, ×	109	mod	141
<mfrac>	46	. . .	78	∉	110	sup	142
<mn>	47	δ	79	^	111	≥, ⩾, ≳	143
<msub>	48	ψ, Ψ	80	∑	112	dim	144
<msup>	49	Γ	81	;	113	:=	145
<msubsup>	50	∞	82	‾	114	≅	146
<mover>	51	ρ	83	⟺, ⇔	115	max	147
<munderover>	52	β	84	⇒	116	inf	148
<munder>	53	λ	85	⌈	117	min	149
<mtable>	54	ξ	86	⌉	118		
<mmultiscripts>	55	□	87	∀	119		

Table 2 Format of gold dataset

QueryID	Iteration	Documents#	Relevance
NTCIR12-MathWiki-3	xxx	Semi-local_ring	1
NTCIR12-MathWiki-3	xxx	Monoidal_functor	0
NTCIR12-MathWiki-3	xxx	MoorePenrose_pseudoinverse	1
NTCIR12-MathWiki-3	xxx	MV-algebra	1
NTCIR12-MathWiki-3	xxx	Persymmetric_matrix	0
NTCIR12-MathWiki-5	xxx	Abstract_index_notation	0
NTCIR12-MathWiki-5	xxx	Abuse_of_notation	0
NTCIR12-MathWiki-5	xxx	Actuarial_notation	0

4.3 *Format of Result Set*

The result set enclosed the search results, i.e., documents retrieved by the proposed approach. The result set follows the TREC format [14] and it constitutes 6 distinct fields namely QueryID, Iteration, Document#, Rank, Similarity, and RunID to denote the documents for each of the query term that has been retrieved. Out of these 6 fields, 3 of them, namely: Query ID, Document, and Similarity have been examined through an assessment tool, whereas the rest of the fields have been discarded. The similarity score is a floating-point value that ranges from 0.0 to 1.0 and attains distinguished values for each document which contains exact query term (and/or) sub-query term. The rank field is implicitly defined and is being ruled out. The format of the result set is shown in Table 3.

4.4 *Evaluation Parameters*

The efficiency of the proposed approach is measured in terms of P_5, P_10, P_15, and P_20. All these parameters are calculated for each query and then the average is taken for all queries present in queryset. The parameter P_K, where K=5, 10, 15, and 20 depict the precision among top-K retrieved documents.

Table 3 Format of result set

QueryID	Iteration	Documents#	Rank	Similarity	RunID
NTCIR12-MathWiki-3	Q0	Direct_sum_of_modules	4	0.014925373	demo
NTCIR12-MathWiki-3	Q0	Linear_logic	4	0.014925373	demo
NTCIR12-MathWiki-3	Q0	Exclusive_or	4	0.014925373	demo
NTCIR12-MathWiki-3	Q0	Additive_category	4	0.014925373	demo
NTCIR12-MathWiki-3	Q0	DempsterShafer_theory	4	0.014925373	demo
NTCIR12-MathWiki-5	Q0	Riemann's_differential_equation	4	0.03880597	demo
NTCIR12-MathWiki-5	Q0	Confluent_hypergeometric_function	4	0.03880597	demo
NTCIR12-MathWiki-5	Q0	Barnes_integral	4	0.03880597	demo
NTCIR12-MathWiki-5	Q0	Generalized_hypergeometric_function	4	0.03880597	demo
NTCIR12-MathWiki-5	Q0	Gauss's_continued_fraction	4	0.03880597	demo

Table 4 Values of evaluation parameter

Parameters	Text-Based	MIaS	Formula embedding approach	Proposed approach
P_5	0.260	0.392	0.407	**0.522**
P_10	0.198	0.290	0.353	**0.478**
P_15	0.121	0.203	0.246	**0.352**
P_20	0.091	0.173	0.188	**0.289**

4.5 Results Analysis

For evaluating the performance of the proposed approach, trec_tool is used, which compares the gold dataset with the result set obtained from the proposed system. The proposed approach delivers the best precision result compared to text-based search engine implemented using apache nutch platform, MIaS [11] and formula embedding [8] as shown in Table 4 and Fig. 4. The proposed model is able to outperform with three other state-of-the-art formula search engines to retrieve the exact match formula, sub-formula, and parent formula.

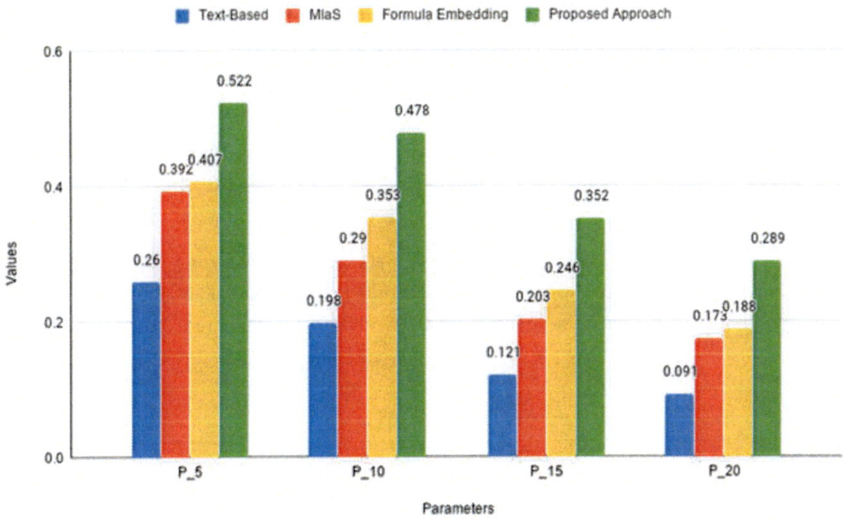

Fig. 4 Comparative analysis of evaluation parameters

1. The retrieval efficiency of the proposed approach for few queries is shown in Table 5 where the first attributes represent the user query, the second one is the retrieved search results using the proposed approach, and the third one is the document ID which constitutes the formula.
2. The outcomes of $x^2 + y^2 = z^2$ retrieved the syntactically and semantically similar formula. This observation shows the positive impact of the formula embedding and generalization approach in the retrieval of mathematical information.
3. The proposed approach efficiently retrieves all those documents which contain exact match formula, sub-formula, and/or parent formula and rank them based on the maximum similarity.
4. The number of times a particular entity present in a formula is a significant point instead of measuring the presence or absence of a particular entity in a formula.
5. The proposed approach effectively retrieves the relevant documents as well but in some cases, it retrieves irrelevant documents also. The search result for the query $x^2 + y^2 = z^2$ is shown in Table 5, where the first four results are relevant to user query but the last one is not.

Table 5 Search results from the proposed approach

Sr. No	User query	Search results	Document
1	O(mn log m)	O(mn log m)	Hunt_McIlroy_algorithm.html
		O(n log m)	kinetic_sorted_list.html
		O(m log n)	LCP_array.html
		O(mn log (mn))	frechet_distance.html
2	$x^2 + y^2 = z^2$	$x^4 + y^4 = z^4$	Number_theory.html
		$a^2 + b^2 = c^2$	Pythagoras.html
		$a^2 + b^2 = c^2$	Hilbert's_Theorem_90.html
		$x^2 + y^2 + z^2 = 0$	rational_singularity.html
		$\frac{\partial^2 Hz}{\partial x^2} + \frac{\partial^2 Hz}{\partial y^2} = \mu\varepsilon \frac{\partial^2 Hz}{\partial t^2}$	Transmission-line_matrix_method.html
3	$F = G\frac{m_1 m_2}{r^2}$	$F = G\frac{m_1 m_2}{r^2}$	kepler_orbit.html
		$F(r) = -G\frac{m_1 m_2}{(r_1 - r_2)^2}$	Test_particle.html
		$F = \frac{G_m M_m}{d^2}$	Schiehallion_experiment.html
4	$(f.g)' = (f'.g)'.g'$	$(f.g).h = f.(g.h)$	Composition_ring.html
		$D(f.g) = Df.Dg$	Chain_rule.html
		$(g.f)(x) = g(f(x))$	Function_(mathematics).html
		$(f.g).h = f(g.h)$	Associative_property.html
5	$\nabla \times E = -\frac{\delta B}{\delta t}$	$\nabla \times E = -\frac{\delta B}{\delta t}$	Lorentz_Heaviside_units.html
		$\frac{\delta B}{\delta t} = - \nabla \times E$	Magnetohydrodynamics.html
		$\nabla \times Eg = -\frac{\delta Bg}{\delta t}$	Gravitoelectromagneticsm.html
		$\nabla \times E = -\frac{\delta B}{\delta t} = 0$	Poissons_equation.html
6	$F = ma$	$F = ma$	Braking_distance.html
		$F = m.a$	Newtons_cradle.html
		$F \rightarrow= ma \rightarrow$	Force.html
		$a \rightarrow= F \rightarrow m$	Net_force.html
7	$\frac{dP}{dt} = r(1 - \frac{P}{k})P$	$\frac{dP}{dt} = rP(1 - \frac{P}{k})$	Logistic_function.html
		$\frac{dP}{dt} = kP(1 - \frac{P}{k})$	Separation_of_variables.html
		$\frac{dN}{dt} = rN(1 - \frac{N}{k})$	Allee_effect.html
		$\frac{dx}{dt} = rx(1 - \frac{x}{k})$	Competitive_Lotka_volterra_equation.html

5 Conclusions and Future Scope

This paper reports a novel approach for the retrieval of mathematical information by employing the formula embedding and generalization technique. Formula embedding and generalization modules transformed the formula into the fixed-size vector, where each bit represents the occurrence count of a particular entity in a formula. Overall, the combined model produced a better rankings mechanism by assigning the priorities to the individual module. The feasibility of this approach has been tested on the MathTagArticles of Wikipedia corpus of NTCIR-12 and the obtained results revealed the robustness and frailness of the same. The performance measures P_5=0.522, P_10=0.478, P_15=0.352, and P_20=0.289 claim the efficiency of the proposed approach for retrieving the relevant documents from the large scientific corpus.

Our future target is to improve the ranking mechanism by assigning weights to entities while calculating the similarities. Also, incorporate the machine learning techniques for formulae disambiguation and ranking mechanism.

Acknowledgements The authors would like to express gratitude to the Department of Computer Science and Engineering and Center for Natural Language Processing, National Institute of Technology, Silchar, India, for providing infrastructural facilities and support.

References

1. Asperti A, Padovani L, Coen CS, Guidi F, Schena I (2003) Mathematical knowledge management in helm. Ann Math Artif Intell 38(1–3):27–46
2. Ferreira D, Freitas A (2020) Natural language premise selection: finding supporting statements for mathematical text. arXiv:2004.14959
3. Gao L, Jiang Z, Yin Y, Yuan K, Yan Z, Tang Z (2017) Preliminary exploration of formula embedding for mathematical information retrieval: can mathematical formulae be embedded like a natural language? arXiv:1707.05154
4. Kristianto GY, Topic G, Aizawa A (2016) Mcat math retrieval system for ntcir-12 mathir task. In: Proceedings of the 12th NTCIR conference on evaluation of information access technologies, June 7–10, Tokyo, Japan, pp 323–330
5. Larson RR, Reynolds C, Gey FC (2013) The abject failure of keyword IR for mathematics search: Berkeley at NTCIR-10 math. In: Proceedings of the 10th NTCIR conference on evaluation of information access technologies, June 18–21, Tokyo, Japan, pp 662–666
6. Liska M, Sojka P, Ruzicka M (2015) Combining text and formula queries in math information retrieval: evaluation of query results merging strategies. In: Proceedings of the first international workshop on novel web search interfaces and systems, Oct 23. ACM, Melbourne, Australia, pp 7–9
7. Melis E, Büdenbender J, Goguadze G, Libbrecht P, Ullrich C et al (2003) Knowledge representation and management in activemath. Ann Math Artif Intell 38(1–3):47–64
8. Pathak A, Pakray P, Gelbukh A (2018) A formula embedding approach to math information retrieval. Computación y Sistemas 22(3)
9. Pathak A, Pakray P, Gelbukh A (2019) Binary vector transformation of math formula for mathematical information retrieval. J Intell Fuzzy Syst 36(5):4685–4695
10. Ruzicka M, Sojka P, Liska M (2014) Math indexer and searcher under the hood: history and development of a winning strategy. In: Proceedings of the 11th NTCIR Conference on Evaluation of Information Access Technologies, Dec 9–12, Tokyo, Japan, pp 127–134. Citeseer
11. Sojka P, Liska M (2011) The art of mathematics retrieval. In: Proceedings of the 11th ACM symposium on document engineering, pp 57–60
12. Sojka P, Ruzicka M (2018) Novotny: Mias: Math-aware retrieval in digital mathematical libraries. In: Proceedings of the 27th ACM international conference on information and knowledge management, pp 1923–1926
13. Stathopoulos Y, Teufel S (2016) Mathematical information retrieval based on type embedding and query expansion. In: Proceedings of COLING 2016, the 26th international conference on computational linguistics: technical papers, pp 2344–2355

14. Voorhees EM, Harman DK et al (2005) TREC: experiment and evaluation in information retrieval, vol 563–567. Computational Linguistics, MIT Press Cambridge
15. Zanibbi R, Aizawa A, Kohlhase M, Ounis I, Topic G, Davila K (2016) Ntcir-12 mathir task overview. In: Proceedings of the 12th NTCIR conference on evaluation of information access technologies, Tokyo, Japan, pp 299–308
16. Zhong W, Rohatgi S, Wu J, Giles CL, Zanibbi R (2020) Accelerating substructure similarity search for formula retrieval. In: European conference on information retrieval, pp 714–727. Springer (2020)

Wind Response Evaluation of Limb-Shifted U-Shaped Building Using CFD

Shanku Mandal, Sujit Kumar Dalui, and Soumya Bhattacharjya

Abstract The present study implemented the computational fluid dynamics (CFD) module of Ansys CFX to understand the differences in wind effect on U-shaped building with different limb positions at 0°, 30°, 60° and 90° wind angle. The remarkable variation in flow pattern, pressure and force coefficient has been observed due to change in limb position by 25, 50 and 75 mm from both ends. The mean pressure and horizontal pressure profile of the building faces have been presented for each case to represent the pressure variation. The drag force has been decreased in most of the cases, but the lift force increases with the repositioning of the limbs to the centre. Maximum drag force has been observed at a typical U shape for 0° angle, and the maximum lift force is noticed at 60° angle due to 25 mm limb shifting. Remarkable wind pressure differences have been noticed on the building faces due to the shifting as well as the change in wind angle. To get accurate results from the simulation, the mesh sensitivity study has been carried out. A validation study is also done to indicate the precision of the extracted data. This research reveals the variation in wind behaviour due to repositioning and also points out the critical factors for the structural design.

Keywords Computational fluid dynamics · Limb position · Wind angle · Flow pattern · Force coefficient · Pressure coefficient · Mesh sensitivity study · Wind behaviour

1 Introduction

Expanding the building height to accommodate the growth in population is the latest trend. It also ensures that land scarcity will not affect urban development. Nowadays,

---

S. Mandal (✉) · S. K. Dalui · S. Bhattacharjya
Department of Civil Engineering, Indian Institute of Engineering Science and Technology, Shibpur, Howrah, India

S. Bhattacharjya
e-mail: soumya@civil.iiests.ac.in

© The Author(s), under exclusive license to Springer Nature Singapore Pte Ltd. 2022
D. Giri et al. (eds.), *Proceedings of the Seventh International Conference on Mathematics and Computing*, Advances in Intelligent Systems and Computing 1412,
https://doi.org/10.1007/978-981-16-6890-6_82

the structures are being built considering both regular and irregular shapes. The wind interaction in irregular shape is quite complex as compared with regular shape. The guidelines of irregular shape design are not widely available. The wind effect on those shapes can be explored either by a numerical approach or by an experimental approach. A few researchers have researched wind behaviour in an irregular shape, which mainly follows the alphabetical form. Deng et al. [1] investigated the pattern of viscous flow on several arbitrary shapes and also developed an elliptic numerical method along with a two-layer turbulence model. Jain and Mandal [2] proposed an analytical model to explore the dynamic behaviour of V-shaped building with flexible floor diaphragms. Stathopoulos and Zhou [3] developed a procedure to calculate the pressure of wind on any irregular shape, which can be composed of two rectangular shapes. Swaddiwudhipong and Khan [4] studied the interaction between buildings and moving air numerically by computational fluid dynamics (CFD) of FLUENT 5 software. Gu and Quan [5] experimented with the dynamic across wind response by wind tunnel testing on the bevel and concave corner square shape, basic square and basic rectangular shape. Gomes et al. [6] applied CFD on PHOENICS package to study the wind pressure on U and L shape. Lam et al. [7] investigated the effect of wind due to the interference of the closely placed square buildings. Amin and Ahuja [8] calculated the pressure differences on the different plan shape of the same plan area and mentioned that the presented data could be integrated into the Indian wind design code. Lam et al. [9] revealed that the cavities in H shape are the primary factor for the wind behaviour variation at most of the wind angle. P et al. [10] calculated the pressure coefficients on the T, square and rectangular shape and discussed the effect of spacing between individual shapes. A different turbulence model has been adopted in CFD by Mukherjee et al. [11] to calculate the wind pressure on Y shape. Li and Li [12] evaluated the wind effect on L shape due to the dynamic load across wind direction. Cheng et al. [13] adopted proper orthogonal decomposition (POD) analysis to measure the wind forces on H and square shape. Elshaer et al. [14] predicted the optimal shape of a twisted corner alternated building with the help of artificial neural network (ANN), CFD and optimization technique. Paul and Dalui [15] examined the pressure and force coefficient of Z shape in the different wind flow. Zheng et al. [16] explained the benefits of combined aerodynamic technique to minimize the wind load on buildings. Sanyal and Dalui [17] compared the wind pressure, moment and forces between different types of the Y shape. Bhattacharyya and Dalui [18] explained the aerodynamics of asymmetric E-shaped building.

The modern commercial and residential buildings are mostly U types shaped in view of both architectural and functional requirements. The land scarcity is the major issue of current urbanization. So, most often the proper placement of U shape is not possible. This study mainly focuses on this aspect. In this study, the wind aerodynamics on the different U shape of various limb positions has been discussed. The CFD is adopted to calculate the building responses on 0°, 30°, 60° and 90° wind angles and also illustrates the difference in those responses.

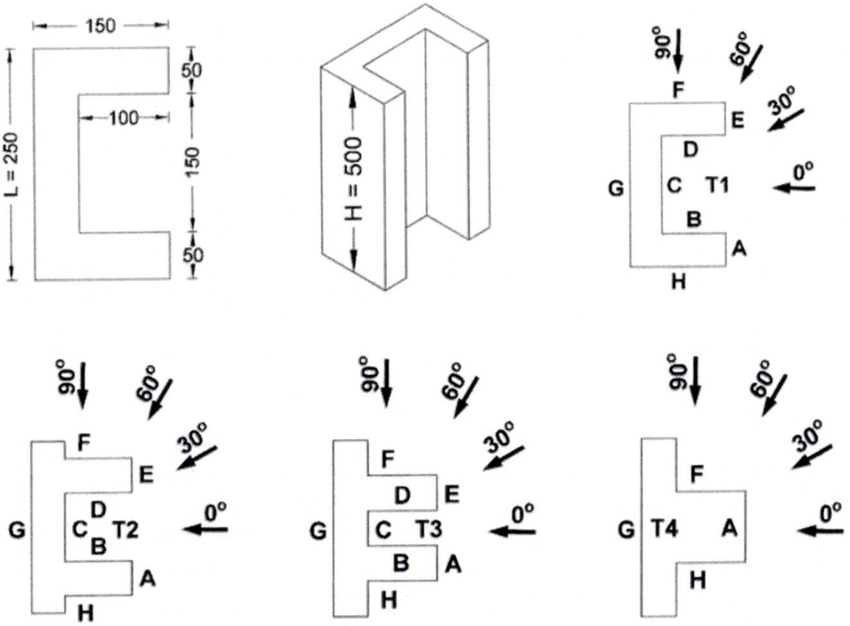

Fig. 1 The building models adopted in the current study (all dimensions are in mm)

2 Details of Building Models

The responses of typical U shape type (T1), limb-shifted U shape type (T2, T3 and T4) have been calculated in the above-mentioned wind angle using CFD. The 1:300 scale factor is adopted to model those buildings. The pictorial representation of the models is shown in Fig. 1. The two limbs of the Type 1 model are placed at the corners. The two limbs are shifted by 0.1, 0.2 and 0.3 L for T2, T3 and T4 models, respectively. L is the length of the building, and the value is adopted as 250 mm. The height (H = 500 mm) and the plan area are the same for each case. The building limb length and building width are taken 50 mm and 150 mm, respectively. The two limbs' gap in a typical U shape type (T1) is considered as 150 mm.

3 Solution Methodology

The virtual wind environment has been generated by the computational fluid dynamics (CFD) module of Ansys CFX. The implementation of CFD is one of the most efficient techniques to understand the interaction between fluid and structure (Löhner et al. [19]).

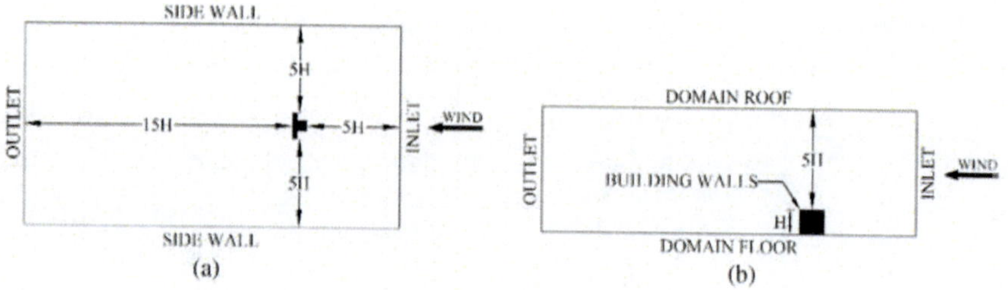

Fig. 2. **a** Plan **b** Elevation view of the domain used for the study

3.1 Computational Domain Setup

The different buildings (T1, T2, T3 and T4) are placed in a suitable domain (Revuz et al. [20]) for the analysis, which is free from any correction due to blockage. The turbulence between the domains is introduced by *k*-ε turbulence model. The domain (Fig. 2) maintained the 5H distance from the roof, sidewalls and inlet from the extreme corner edges building. For the outlet, the distance is fixed at 15H. The free slip boundary is adopted for the roof and the sidewalls, and no slip boundary has been placed on the floor and the building faces. The wind flow at 10 m/s is produced at the entry point of the domain. The formula (Eq. 1) for boundary layer flow follows the power-law equation. The atmospheric boundary layer (ABL) depth for this study is fixed to 1 m.

$$\frac{V}{V_0} = \left(\frac{h}{h_0}\right)^{\alpha} \tag{1}$$

Here, 'h_0' and 'h' represent the depth and any reference height above the floor, respectively. 'V_0' and 'V' signify the velocity at 'h_0' reference height and velocity at any reference height (h) respectively. The value of 'α' is considered as 0.133, which depends on the exposed region.

3.2 Generation of Mesh

The meshing for the building and domain has been executed using tetrahedral sections (Sanyal and Dalui [17]). The nearby regions of the building model are covered with small mesh elements (Fig. 3) to generate accurate wind characteristics around the models. So, the reliable results from the analysis can be calculated.

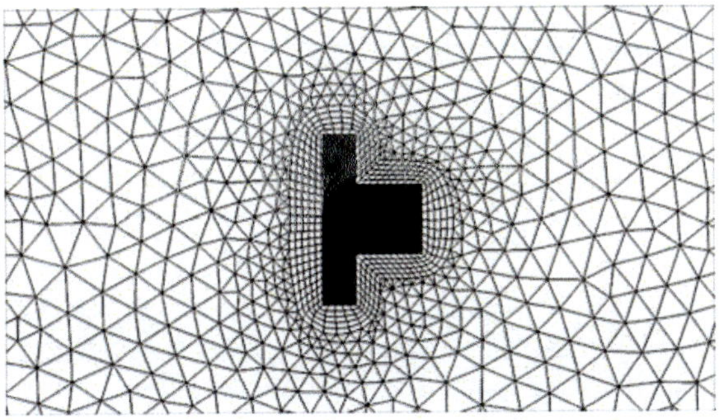

Fig. 3 The typical mesh pattern for T4 type building

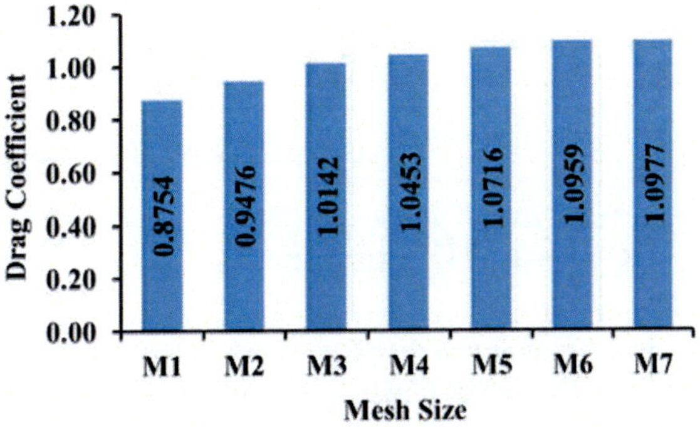

Fig. 4 The mesh sensitivity study for T1 type model

3.3 *Mesh Sensitivity Study*

The preferable mesh element size has been selected using the mesh sensitivity study on T1 type model at 0° wind angle. Seven types of mesh size (M1, M2, M3, M4, M5, M6 and M7) have been considered. The number of total elements for the seven mesh size is 2884984 (M1), 5,165,448 (M2), 12,546,748 (M3), 19,853,472 (M4), 26,709,214 (M5), 33,935,947 (M6) and 41,093,375 (M7). The drag coefficient for each mesh size has been calculated and compared in Fig. 4. The comparison shows that the drag coefficient converges at M6 (0.182% error) mesh size. So, the M6 mesh size is considered for all other models.

3.4 *Validation*

The analysis results have been validated with a published study [6]. The similar model with other boundary conditions has been adopted in the present study and analyzed in Ansys CFX. The pressure coefficient profile of face E at an angle of 45° has been compared with the past article. The profile almost matches with the published study. Hence, the analysis done in Ansys CFX is reliable (Fig. 5).

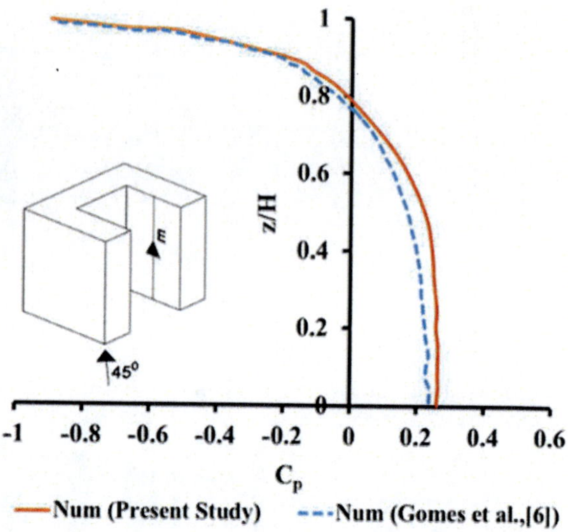

Fig. 5 The pressure coefficient profile of Face E at 45° angle

4 Results and Discussion

The placement of two limbs at new locations and the shift in flow direction induced remarkable variation in wind flow behaviour. The various wind responses of those shapes (T1, T2, T3 and T4) have been illustrated rigorously in this section. The comparison of some important factors is also presented to highlight the effect of shifting.

4.1 Variation in Wind Flow Streamline

The patterns of the flow streamlines (Fig. 6) depend on the outer geometry, which changes with the shifting and this produced irregular wind flow. The uniform vortices are visible on both sides at 0° angle because of axisymmetry in each model.

The streamlines varied at 30°, 60° and 90° flow due to the existence of limbs. The long vortices are formed at the wake region in T1 model as compared with other types. The rise in velocity has been noted near the flow separator building corners.

4.2 Variation in Force Coefficients

The differences in wind force on the T1, T2, T3 and T4 models at 0°, 30°, 60° and 90° angles have been demonstrated in Fig. 7. The drag coefficient decreases with the shifting. At 0° angle, the drag coefficient decreases, but in the case of the T4 model,

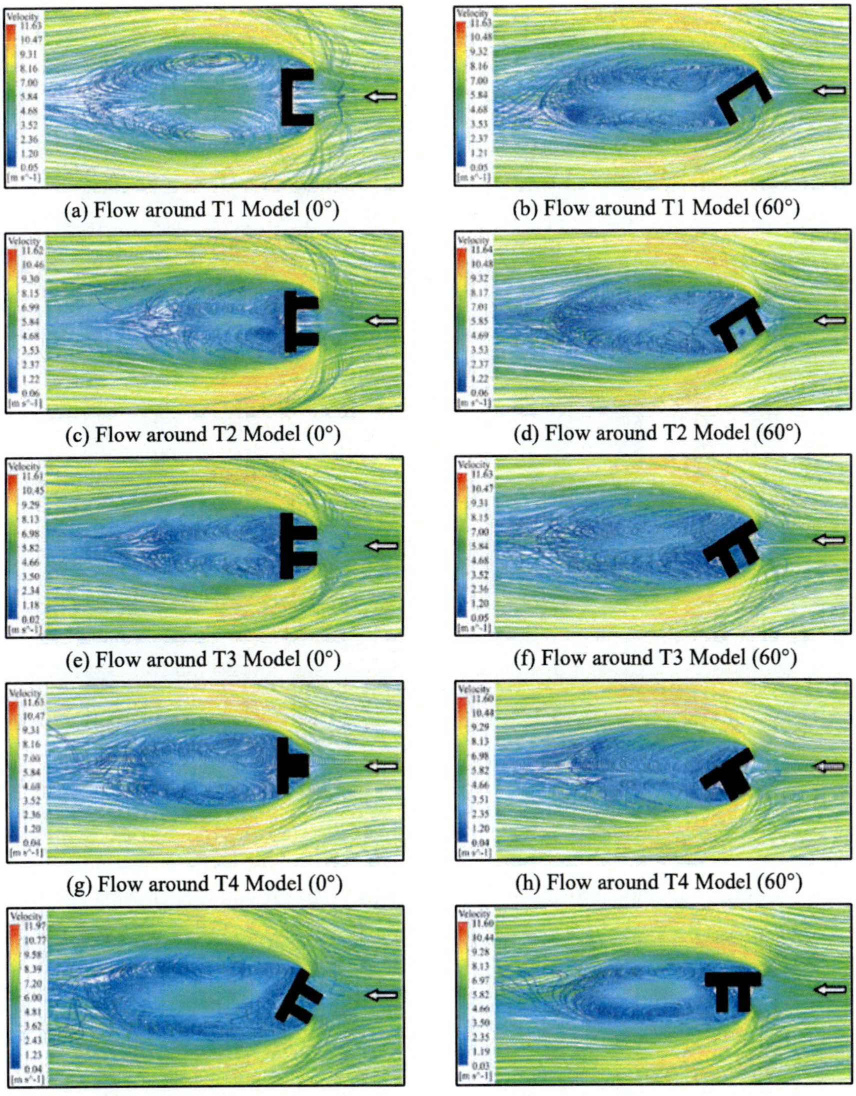

(a) Flow around T1 Model (0°)

(b) Flow around T1 Model (60°)

(c) Flow around T2 Model (0°)

(d) Flow around T2 Model (60°)

(e) Flow around T3 Model (0°)

(f) Flow around T3 Model (60°)

(g) Flow around T4 Model (0°)

(h) Flow around T4 Model (60°)

(i) Flow around T3 Model (30°)

(j) Flow around T3 Model (90°)

Fig. 6 The patterns of the flow streamline around the models at various angle

it is slightly increased as compared with the T3 model. At 30° and 60° wind angles, this shifting shows a remarkable dip in drag coefficient. The slight increase in this coefficient value is noted for T2, T3 and T4 models at 90° angle. In the case of the lift coefficient, the negligible wind response is observed in each model. The gradual rise in lift coefficient is noted at 0°, 30° and 90° angles for T2, T3 and T4 models. At 60° angle, the T2 model shows the maximum lift coefficient. At 60° angle, all

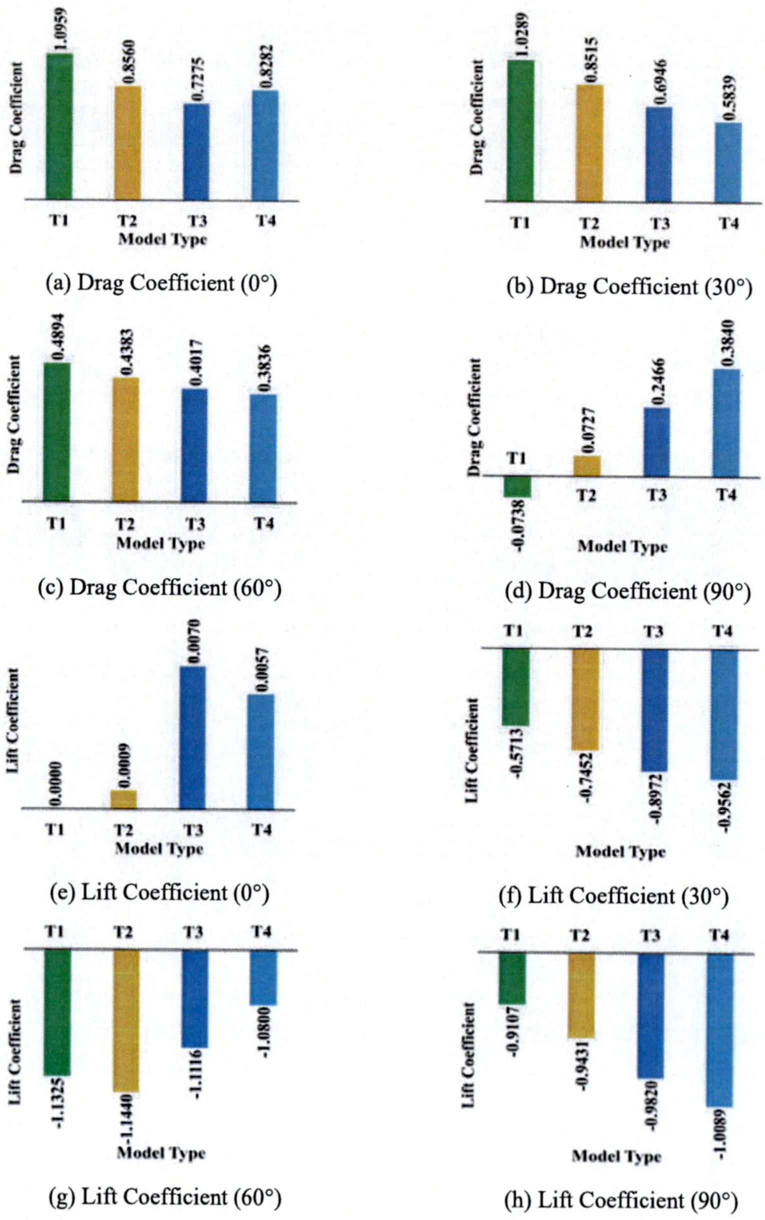

(a) Drag Coefficient (0°) (b) Drag Coefficient (30°)

(c) Drag Coefficient (60°) (d) Drag Coefficient (90°)

(e) Lift Coefficient (0°) (f) Lift Coefficient (30°)

(g) Lift Coefficient (60°) (h) Lift Coefficient (90°)

Fig. 7 The Drag and Lift coefficient for the different model the models are experiencing massive lift force. It is also visible that when the angle changes the lift force progressively increases in T4 model, but the opposite trend is noticed for T1, T2 and T3 models.

4.3 Variation in Pressure Coefficients

The mean pressure coefficient on the model faces (Fig. 8) has been measured for each angle and tabulated in Tables 1, 2, 3, 4, 5 and 6.

The variation in pressure values indicates the direct influence of the flow distribution pattern around the models. For 0° angle, faces F, G and H (T1, T2 and T3) experience enormous suction whereas positive pressure occurs at the other faces. Because of the shifting, the rise in pressure has been noticed at face A, but at the B, C and D faces, the positive pressure decreases. The negative values on face H1

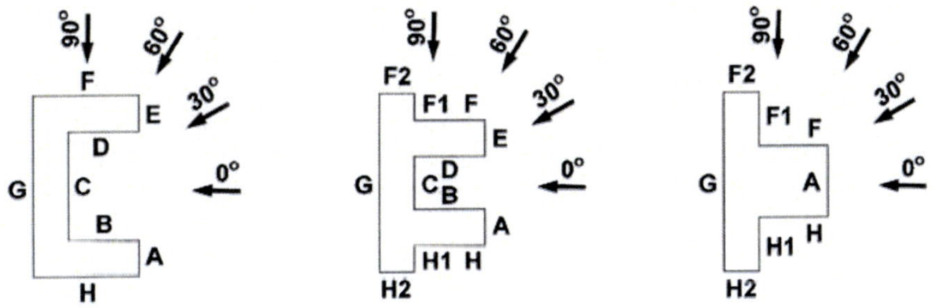

Fig. 8 The face names of the typical models

Table 1 The mean pressure coefficient of different faces at 0° angle

Model type	Face A	Face B	Face C	Face D	Face E	Face F	Face G	Face H
T1	0.410	0.828	0.870	0.828	0.410	−0.561	−0.381	−0.561
T2	0.444	0.822	0.862	0.822	0.444	−0.662	−0.409	−0.662
T3	0.553	0.791	0.826	0.791	0.553	−0.400	−0.396	−0.400
T4	0.700	–	–	–	–	0.079	−0.377	0.079

Table 2 The mean pressure coefficient of different faces at a 30° angle

Model type	Face A	Face B	Face C	Face D	Face E	Face F	Face G	Face H
T1	0.000	0.790	0.737	0.620	0.721	−0.073	−0.424	−0.511
T2	0.089	0.692	0.664	0.559	0.789	0.491	−0.402	−0.562
T3	0.102	0.540	0.601	0.456	0.735	0.749	−0.382	−0.587
T4	0.341	–	–	–	–	0.803	−0.376	−0.698

Table 3 The mean pressure coefficient of different faces at a 60° angle

Model type	Face A	Face B	Face C	Face D	Face E	Face F	Face G	Face H
T1	−0.235	0.397	0.145	0.147	0.211	0.535	−0.428	−0.387
T2	−0.237	0.160	0.090	0.012	0.012	0.813	−0.433	−0.473
T3	−0.239	−0.017	0.007	−0.127	−0.281	0.786	−0.431	−0.477
T4	−0.480	–	–	–	–	0.728	−0.476	−0.538

Table 4 The mean pressure coefficient of different faces at a 90° angle

Model type	Face A	Face B	Face C	Face D	Face E	Face F	Face G	Face H
T1	−0.437	−0.573	−0.545	−0.549	−0.675	0.567	−0.476	−0.280
T2	−0.481	−0.579	−0.557	−0.563	−0.716	0.629	−0.488	−0.307
T3	−0.495	−0.550	−0.536	−0.539	−0.675	0.620	−0.504	−0.340
T4	−0.512	–	–	–	–	0.606	−0.489	−0.362

Table 5 The mean pressure coefficient of the shifted faces (F1 and F2) at a different angle

Model type	Face F1	Face F2	Face F1	Face F2	Face F1	Face F2	Face F1	Face F2
	0°		30°		60°		90°	
T1	–	–	–	–	–	–	–	–
T2	−0.428	−0.489	0.428	−0.601	0.792	0.095	0.848	0.532
T3	−0.233	−0.545	0.614	−0.542	0.829	0.149	0.787	0.548
T4	0.207	−0.550	0.679	−0.542	0.830	0.180	0.705	0.557

Table 6 The mean pressure coefficient of the shifted faces (H1 and H2) at a different angle

Model type	Face H1	Face H2	Face H1	Face H2	Face H1	Face H2	Face H1	Face H2
	0°		30°		60°		90°	
T1	–	–	–	–	–	–	–	–
T2	−0.428	−0.489	−0.510	−0.479	−0.474	−0.393	−0.292	−0.296
T3	−0.233	−0.545	−0.513	−0.457	−0.484	−0.376	−0.343	−0.327
T4	0.207	−0.550	−0.569	−0.444	−0.515	−0.411	−0.383	−0.356

and F1 reduced with the shifting but it promotes the increment in suction at face H2 and F2. Moreover, a similar trend has been noticed at face A–E for 30° angle. At 30° angle, the increment in positive and negative pressure has been recorded for face F, F1 and H, H1, respectively. The dip in negative pressure has been listed for face G, F2 and H2 for the said angle. At 60°, positive pressure has been noticed at face F, F1 and F2. The suction on face A, G, H, H1 and H2 shows a gradual rise trend with shifting. However, the rise trend in pressure has been measured at face F1 and F2. At 90° angle face A–E, G–H2 experiences suction, and the pressure is noted in remaining faces. The magnitude of suction increases with shifting in face H, H1, H2 and A. The gradual pressure rise and pressure fall have been noticed at face F2 and F1, respectively.

The horizontal pressure profile has been illustrated in Fig. 10 for the close observation on pressure variation. The diagram follows the arrow direction (Fig. 9).

The profile has been generated at mid-height (250 mm). The horizontal pressure coefficient profile for T1, T2 and T3 model at 0° angle follows an almost similar pattern but remarkable variation is noticed in the T4 model. At 30° and 60° angles, the

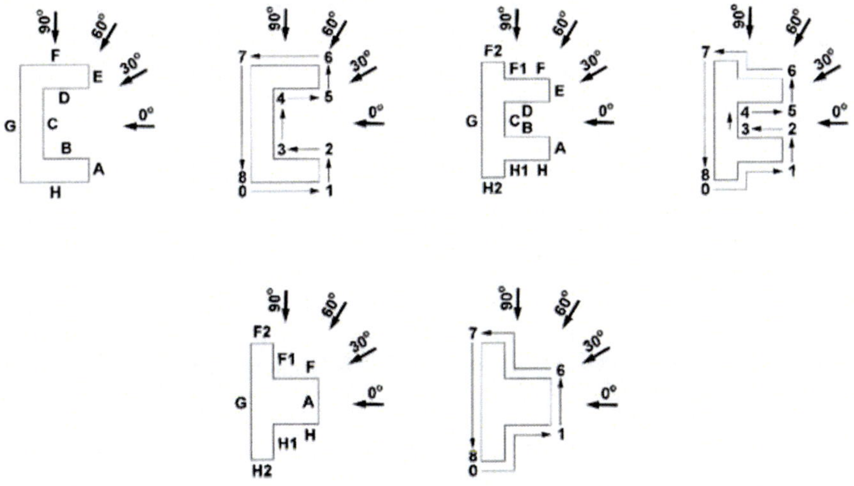

Fig. 9. The faces of typical models and the horizontal pressure profile direction

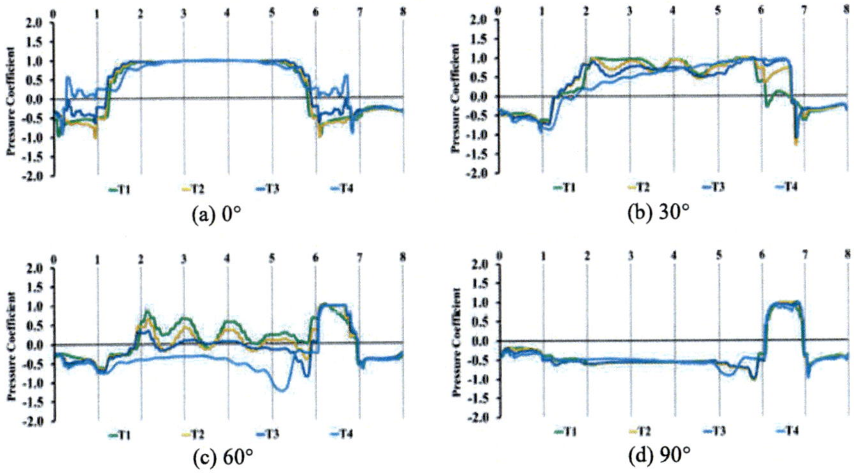

Fig. 10 The variation in horizontal pressure coefficient profile at various angles

T4 model is showing notable dissimilarity. This indicates that the shifting introduces massive wind pressure or suction on the building faces.

5 Conclusion

The wind responses have been evaluated on the building having limb-shifted U shape at 0° to 90° angle at 30° gap. The axisymmetric body remains constant though two limbs are moved together. The responses have been recorded by calculating the mean pressure, drag and lift coefficient. The horizontal pressure profile is also extracted, and a comparison has been presented for each model. The major observations from the study have been pointed out as follows.

- The gradual dip in drag coefficient has been observed up to the shifting distance of 50 mm at 0°, but the 75 mm shifting produces a little bit more drag force than previous shifts. However, at 30° and 60°, the gradual dip pattern is noted for each shifting. At 90°, the gradual rise pattern is recorded. In the case of the Lift coefficient, the basic U shape shows the minimum response at each flow angle.
- The rise in both suction and pressure has been noted at the outer faces, whereas the variation is insignificant on the inner faces. The tremendous variation has been identified at the T4 model (75 mm shift). The careful cladding design is necessary to overcome the issue.

This study pointed out that the shifting is effective in drag force reduction but in case of lift force reduction, it is ineffective. So, whenever the limb-shifted U-shaped buildings have to be constructed, those factors should be considered in the design for safe and economical construction.

References

1. Deng GB, Piquet J, Queutey P, Visonneau M (1991) Three-dimensional full Navier-Stokes solvers for incompressible flows past arbitrary geometries. Int J Numer Methods Eng 31:1427–1451. https://doi.org/10.1002/nme.1620310713
2. Jain BSK, Mandal UK (1992) Dynamics of buildings with v-shaped plan. J Eng Mech 118:1093–1112. https://doi.org/10.1061/(ASCE)0733-9399(1992)118:6(1093)
3. Stathopoulos T, Zhou YS (1993) Numerical simulation of wind-induced pressures on buildings of various geometries. J Wind Eng Ind Aerodyn 46–47:419–430. https://doi.org/10.1016/0167-6105(93)90308-B
4. Swaddiwudhipong S, Khan M (2002) Dynamic Response of Wind-Excited Building Using Cfd. J Sound Vib 253:735–754. https://doi.org/10.1006/jsvi.2000.3508
5. Gu M, Quan Y (2004) Across-wind loads of typical tall buildings. J Wind Eng Ind Aerodyn 92:1147–1165. https://doi.org/10.1016/j.jweia.2004.06.004
6. Gomes MG, Moret Rodrigues A, Mendes P (2005) Experimental and numerical study of wind pressures on irregular-plan shapes. J Wind Eng Ind Aerodyn 93:741–756. https://doi.org/10.1016/j.jweia.2005.08.008

7. Lam KMÃ, Leung MYH, Zhao JG (2008) Interference effects on wind loading of a row of closely spaced tall buildings. J Wind Eng Ind Aerodyn 96:562–583. https://doi.org/10.1016/j.jweia.2008.01.010
8. Amin JA, Ahuja AK (2008) Experimental study of wind pressures on irregular-plan shape buildings. In: BBAA VI international colloquium on: bluff bodies aerodynamics and applications, Milano, Italy, pp 1–9
9. Lam KM, Wong SY, To AP (2009) Dynamic wind loading of H-shaped tall buildings. In: 7th Asia-Pacific conference on wind engineering APCWE-VII, Taipei, Taiwan (2009)
10. P, KT, Ahuja AK, Chakrabarti A (2013) An experimental investigation of wind pressure developed in tall buildings for different plan shape. Int J Innov Res Stud 2:605–614
11. Mukherjee S, Chakraborty S, Dalui SK, Ahuja AK (2014) Wind induced pressure on 'Y' plan shape tall building. Wind Struct An Int J 19:523–540. http://dx.doi.org/https://doi.org/10.12989/was.2014.19.5.523
12. Li Y, Li QS (2016) Across-wind dynamic loads on L-shaped tall buildings. Wind Struct An Int J 23:385–403. https://doi.org/10.12989/was.2016.23.5.385
13. Cheng L, Lam KM, Wong SY (2015) POD analysis of crosswind forces on a tall building with square and H-shaped cross sections. Wind Struct An Int J 21:63–84. http://dx.doi.org/https://doi.org/10.12989/was.2015.21.1.063
14. Elshaer A, Bitsuamlak G, Damatty AE (2016) Aerodynamic shape optimization of tall buildings using twisting and corner modifications. In: 8th international colloquium on bluff body aerodynamics and applications, MA, USA
15. Paul R, Dalui SK (2016) Wind effects on 'Z' plan-shaped tall building: a case study. Int J Adv Struct Eng 8:319–335. https://doi.org/10.1007/s40091-016-0134-9
16. Zheng C, Xie Y, Khan M, Wu Y, Liu J (2018) Wind-induced responses of tall buildings under combined aerodynamic control. Eng Struct 175:86–100. https://doi.org/10.1016/j.engstruct.2018.08.031
17. Sanyal P, Dalui SK (2020) Comparison of aerodynamic coefficients of various types of Y-plan-shaped tall buildings. Asian J Civ Eng 21:1109–1127. https://doi.org/10.1007/-s42107-020-00265 9
18. Bhattacharyya B, Dalui SK (2020) Experimental and numerical study of wind pressure distribution on irregular-plan-shaped building. J Struct Eng (US) 146:1–14. https://doi.org/10.1061/(ASCE)ST.1943-541X.0002686
19. Löhner R, Haug E, Michalski A, Muhammad B, Drego A, Nanjundaiah R, Zarfam R (2015) Recent advances in computational wind engineering and fluid-structure interaction. Jnl Wind Eng Ind Aerodyn 144:14–23. https://doi.org/10.1016/j.jweia.2015.04.014
20. Revuz J, Hargreaves DM, Owen JS (2012) On the domain size for the steady-state CFD modelling of a tall building. Wind Struct An Int J **15**, 313–329. https://doi.org/10.12989/was.2012.15.4.313

Author Index

© The Editor(s) (if applicable) and The Author(s), under exclusive license to Springer Nature Singapore Pte Ltd. 2022
D. Giri et al. (eds.), *Proceedings of the Seventh International Conference on Mathematics and Computing*, Advances in Intelligent Systems and Computing 1412,
https://doi.org/10.1007/978-981-16-6890-6